CONCEPTS OF
GENETICS

CONCEPTS OF
GENETICS

TENTH EDITION

William S. Klug
THE COLLEGE OF NEW JERSEY

Michael R. Cummings
ILLINOIS INSTITUTE OF TECHNOLOGY

Charlotte A. Spencer
UNIVERSITY OF ALBERTA

Michael A. Palladino
MONMOUTH UNIVERSITY

PEARSON

Boston   Columbus   Indianapolis   New York   San Francisco   Upper Saddle River
Amsterdam   Cape Town   Dubai   London   Madrid   Milan   Munich   Paris   Montréal   Toronto
Delhi   Mexico City   São Paulo   Sydney   Hong Kong   Seoul   Singapore   Taipei   Tokyo

Editor-in-Chief: Beth Wilbur
Executive Director of Development: Deborah Gale
Senior Acquisitions Editor: Michael Gillespie
Project Editor: Dusty Friedman
Assistant Editor: Leslie Allen
Managing Editor: Michael Early
Production Project Manager: Camille Herrera
Production Management and Compositor:
    Cenveo Publisher Services/Nesbitt Graphics, Inc.
Production Editor: Rose Kernan, RPK Editorial Services, Inc.
Copyeditor: Betty Pessagno
Proofreader: Michael Rossa and Debra Gates
Interior and Cover Designer: Seventeenth Street Studios
Illustrators: Imagineering Media Services
Image Lead: Donna Kalal
Photo Researcher: Maureen Spuhler
Director of Editorial Content: Tania Mlawer
Senior Media Producer: Laura Tommasi
Media Development Editor: Matt Lee
Media Project Editor: Juliana Tringali
Manufacturing Buyer: Michael Penne
Cover Printer: Lehigh Phoenix
Printer and Binder: Courier Kendallville
Director of Marketing: Christy Lesko
Executive Marketing Manager: Lauren Harp
Cover Photo Credit: Evelin Schrock, Stan du Manoir and Tom Re
National Institutes of Health

PEARSON

About the Authors

William S. Klug is Professor of Biology at The College of New Jersey (formerly Trenton State College) in Ewing, New Jersey, where he served as Chair of the Biology Department for 17 years. He received his B.A. degree in Biology from Wabash College in Crawfordsville, Indiana, and his Ph.D. from Northwestern University in Evanston, Illinois. Prior to coming to The College of New Jersey, he was on the faculty of Wabash College as an Assistant Professor, where he first taught genetics, as well as general biology and electron microscopy. His research interests have involved ultrastructural and molecular genetic studies of development, utilizing oogenesis in *Drosophila* as a model system. He has taught the genetics course as well as the senior capstone seminar course in Human and Molecular Genetics to undergraduate biology majors for over four decades. He was the recipient in 2001 of the first annual teaching award given at The College of New Jersey, granted to the faculty member who "most challenges students to achieve high standards." He also received the 2004 Outstanding Professor Award from Sigma Pi International, and in the same year, he was nominated as the Educator of the Year, an award given by the Research and Development Council of New Jersey.

Michael R. Cummings is Research Professor in the Department of Biological, Chemical, and Physical Sciences at Illinois Institute of Technology, Chicago, Illinois. For more than 25 years, he was a faculty member in the Department of Biological Sciences and in the Department of Molecular Genetics at the University of Illinois at Chicago. He has also served on the faculties of Northwestern University and Florida State University. He received his B.A. from St. Mary's College in Winona, Minnesota, and his M.S. and Ph.D. from Northwestern University in Evanston, Illinois. In addition to this text and its companion volumes, he has also written textbooks in human genetics and general biology for nonmajors. His research interests center on the molecular organization and physical mapping of the heterochromatic regions of human acrocentric chromosomes. At the undergraduate level, he teaches courses in Mendelian and molecular genetics, human genetics, and general biology, and has received numerous awards for teaching excellence given by university faculty, student organizations, and graduating seniors.

Charlotte A. Spencer is a retired Associate Professor from the Department of Oncology at the University of Alberta in Edmonton, Alberta, Canada. She has also served as a faculty member in the Department of Biochemistry at the University of Alberta. She received her B.Sc. in Microbiology from the University of British Columbia and her Ph.D. in Genetics from the University of Alberta, followed by postdoctoral training at the Fred Hutchinson Cancer Research Center in Seattle, Washington. Her research interests involve the regulation of RNA polymerase II transcription in cancer cells, cells infected with DNA viruses, and cells traversing the mitotic phase of the cell cycle. She has taught courses in biochemistry, genetics, molecular biology, and oncology, at both undergraduate and graduate levels. In addition, she has written booklets in the Prentice Hall Exploring Biology series, which are aimed at the undergraduate nonmajor level.

Michael A. Palladino is Dean of the School of Science and Associate Professor in the Department of Biology at Monmouth University in West Long Branch, New Jersey. He received his B.S. degree in Biology from Trenton State College (now known as The College of New Jersey) and his Ph.D. in Anatomy and Cell Biology from the University of Virginia. He directs an active laboratory of undergraduate student researchers studying molecular mechanisms involved in innate immunity of mammalian male reproductive organs and genes involved in oxygen homeostasis and ischemic injury of the testis. He has taught a wide range of courses for both majors and nonmajors and currently teaches genetics, biotechnology, endocrinology, and laboratory in cell and molecular biology. He has received several awards for research and teaching, including the New Investigator Award of the American Society of Andrology, the 2005 Distinguished Teacher Award from Monmouth University, and the 2005 Caring Heart Award from the New Jersey Association for Biomedical Research. He is co-author of the undergraduate textbook *Introduction to Biotechnology*, Series Editor for the Benjamin Cummings *Special Topics in Biology* booklet series, and author of the first booklet in the series, *Understanding the Human Genome Project*.

Brief Contents

Conceptual Understanding

The book's conceptual focus emphasizes the fundamental ideas of genetics, helping students comprehend and remember the key ideas.

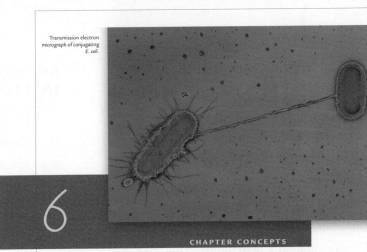

Transmission electron micrograph of conjugating *E. coli.*

6

Genetic Analysis and Mapping in Bacteria and Bacteriophages

CHAPTER CONCEPTS

- Bacterial genomes are most often contained in a single circular chromosome.
- Bacteria have developed numerous ways of exchanging and recombining genetic information between individual cells, including conjugation, transformation, and transduction.
- The ability to undergo conjugation and to transfer the bacterial chromosome from one cell to another is governed by genetic information contained in the DNA of a "fertility," or F, factor.
- The F factor can exist autonomously in the bacterial cytoplasm as a plasmid, or it can integrate into the bacterial chromosome, where it facilitates the transfer of the host chromosome to the recipient cell, leading to genetic recombination.
- Bacteriophages are viruses that have bacteria as their hosts. Viral DNA is injected into the host cell, where it replicates and directs the reproduction of the bacteriophage and the lysis of the bacterium.
- During bacteriophage infection, replication of the phage DNA may be followed by its recombination, which may serve as the basis for intergenic and intragenic mapping.
- Rarely, following infection, bacteriophage DNA integrates into the host chromosome, becoming a prophage, where it is replicated along with the bacterial DNA.

Key Concepts are listed within chapter ▶ openers to help students focus on the core ideas of each chapter.

The key concepts are ▶ revisited in greater detail in the end of chapter **Summary Points**.

Summary Points

 For activities, animations, and review quizzes, go to the study area at www.masteringgenetics.com

1. Genetic recombination in bacteria takes place in three ways: conjugation, transformation, and transduction.
2. Conjugation may be initiated by a bacterium housing a plasmid called the F factor in its cytoplasm, making it a donor cell. Following conjugation, the recipient cell receives a copy of the F factor and is converted to the F^+ status.
3. When the F factor is integrated from the cytoplasm into the chromosome, the cell remains as a donor and is referred to as an Hfr cell. Upon mating, the donor chromosome moves unidirectionally into the recipient, initiating recombination and providing the basis for time mapping of the bacterial chromosome.
4. Plasmids, such as the F factor, are autonomously replicating DNA molecules found in the bacterial cytoplasm, sometimes containing unique genes conferring antibiotic resistance as well as the genes necessary for plasmid transfer during conjugation.
5. Transformation in bacteria, which does not require cell-to-cell contact, involves exogenous DNA that enters a recipient bacterium and recombines with the host's chromosome. Linkage mapping of closely aligned genes is possible during the analysis of transformation.
6. Bacteriophages, viruses that infect bacteria, demonstrate a well-defined life cycle where they reproduce within the host cell and can be studied using the plaque assay.

7. Bacteriophages can be lytic, meaning they infect the host cell, reproduce, and then lyse it, or in contrast, they can lysogenize the host cell, where they infect it and integrate their DNA into the host chromosome, but do not reproduce.
8. Transduction is virus-mediated bacterial DNA recombination. When a lysogenized bacterium subsequently reenters the lytic cycle, the new bacteriophages serve as vehicles for the transfer of host (bacterial) DNA.
9. Various mutant phenotypes, including mutations in plaque morphology and host range, have been studied in bacteriophages and have served as the basis for mapping in these viruses.
10. Transduction is also used for bacterial linkage and mapping studies.
11. Various mutant phenotypes, including mutations in plaque morphology and host range, have been studied in bacteriophages. These have served as the basis for investigating genetic exchange and mapping in these viruses.
12. Genetic analysis of the *rII* locus in bacteriophage T4 allowed Seymour Benzer to study intragenic recombination. By isolating *rII* mutants and performing complementation analysis, recombinational studies, and deletion mapping, Benzer was able to locate and map more than 300 distinct sites within the two cistrons of the *rII* locus.

New and Updated Content

This edition has been thoroughly updated to include the latest discoveries that students need to know about.

NEW!

Special Topics in Modern Genetics are four unique mini-chapters (located between Parts 3 and 4) that explore cutting-edge topics, including Epigenetics, Genomics and Personalized Medicine, Stem Cells, and DNA Forensics. ▼

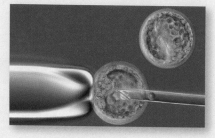

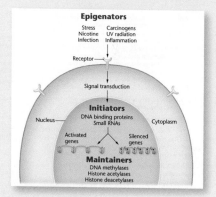

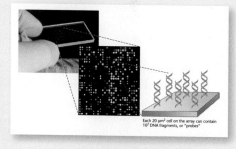

Each 20 μm² cell on the array can contain 10⁷ DNA fragments, or "probes"

SPECIAL TOPICS IN MODERN GENETICS

DNA Forensics

Genetics is arguably the most influential science today—dramatically affecting technologies in fields as diverse as agriculture, archaeology, medical diagnosis, and disease treatment.

One of the areas that has been the most profoundly altered by modern genetics is forensic science. **Forensic science** (or *forensics*) uses technological and scientific approaches to answer questions about the facts of criminal or civil cases. Prior to 1986, forensic scientists had a limited array of tools with which to link evidence to specific individuals or suspects. These included some reliable methods such as blood typing and fingerprint analysis, but also many unreliable methods such as bite mark comparisons and hair microscopy.

Since the first forensic use of **DNA profiling** in 1986 (Box 1), DNA forensics (also called **forensic DNA fingerprinting** or **DNA typing**) has become an important method for police to identify sources of biological materials. DNA profiles can now be obtained from saliva left on cigarette butts or postage stamps, pet hairs found at crime scenes, or bloodspots the size of pinheads. Even biological samples that are degraded by fire or time are yielding DNA profiles that help the legal system determine identity, innocence, or guilt. Investigators now scan large databases of stored DNA profiles in order to match profiles generated from crime scene evidence. DNA profiling has proven the innocence of hundreds of people who were convicted of serious crimes and even sentenced to death. Forensic scientists have used DNA profiling to identify victims of mass disasters such as the Asian Tsunami of 2004 and the September 11, 2001 terrorist attacks in New York. They have also used forensic DNA analysis to identify endangered species and animals trafficked in the illegal wildlife trade. The power of DNA forensic analysis has captured the public imagination, and DNA forensics is featured in several popular television series.

The applications of DNA profiling extend beyond forensic investigations. These include paternity and family relationship testing, identification of plant materials, verification of military casualties, and evolutionary studies.

> Even biological samples degraded by fire or time are yielding DNA profiles that help determine identity, innocence, or guilt.

NEW!

Updated coverage throughout the text of advancing fields, including:

- Role of cohesin and shugoshin during mitosis and meiosis
- Role of TERRA during replication of telomeric DNA
- Eukaryotic ribosomes and translation
- Mutational effect of natural and man-made radiation
- Mismatch DNA repair and cancer
- Riboswitches and gene regulation

- Chromatin remodeling
- *C. elegans* development
- Clonal selection and the origin of cancer
- Next generation DNA sequencing technology
- Systems biology
- Genome-wide association studies (GWAS)
- Synthetic genomes
- Molecular mechanisms underlying behavior

For a complete list of updated coverage, see Preface pp. xxxi-xxxiii.

Problem Solving

The book's problem-solving emphasis encourages students to apply their acquired knowledge and fine-tune their analytical skills.

EXPANDED! Now Solve This problems, expanded to include ▶ the complete problem statement, are integrated throughout each chapter to help students test their knowledge while learning chapter content.

NOW SOLVE THIS

8–3 What is the effect of a rare double crossover (a) within a chromosome segment that is heterozygous for a pericentric inversion; and (b) within a segment that is heterozygous for a paracentric inversion?

■ HINT: *This problem involves an understanding of how homologs synapse in the presence of a heterozygous inversion, as well as the distinction between pericentric and paracentric inversions. The key to its solution is to draw out the tetrad and follow the chromatids undergoing a double crossover.*

▲ Exercises include a **hint** to guide students, and a brief answer is provided in the appendix.

INSIGHTS AND SOLUTIONS

As a student, you will be asked to demonstrate your knowledge of transmission genetics by solving various problems. Success at this task requires not only comprehension of theory but also its application to more practical genetic situations. Most students find problem solving in genetics to be both challenging and rewarding. This section is designed to provide basic insights into the reasoning essential to this process.

Genetics problems are in many ways similar to word problems in algebra. The approach to solving them is identical: (1) analyze the problem carefully; (2) translate words into symbols and define each symbol precisely; and (3) choose and apply a specific technique to solve the problem. The first two steps are the most critical. The third step is largely mechanical.

The simplest problems state all necessary information about a P_1 generation and ask you to find the expected ratios of the F_1 and F_2 genotypes and/or phenotypes. Always follow these steps when you encounter this type of problem:

(a) Determine insofar as possible the genotypes of the individuals in the P_1 generation.

(b) Determine what gametes may be formed by the P_1 parents.

(c) Recombine the gametes by the Punnett square or the forked-line method, or if the situation is very simple, by inspection. From the genotypes of the F_1 generation, determine the phenotypes. Read the F_1 phenotypes.

(d) Repeat the process to obtain information about the F_2 generation.

Determining the genotypes from the given information requires that you understand the basic theory of transmission

genetics. Consider this problem: *A recessive mutant allele, black, causes a very dark body in Drosophila when homozygous. The normal wild-type color is described as gray. What F_1 phenotypic ratio is predicted when a black female is crossed to a gray male whose father was black?*

To work out this problem, you must understand dominance and recessiveness, as well as the principle of segregation. Furthermore, you must use the information about the male parent's father. Here is one logical approach to solving this problem:

The female parent is black, so she must be homozygous for the mutant allele (*bb*). The male parent is gray and must therefore have at least one dominant allele (*B*). His father was black (*bb*), and he received one of the chromosomes bearing these alleles, so the male parent must be heterozygous (*Bb*).

From this point, solving the problem is simple:

Homozygous black female   Heterozygous gray male

1/2 Heterozygous gray males and females, *Bb*
1/2 Homozygous black males and females, *bb*

Apply the approach we just studied to the following problems.

1. Mendel found that full pea pods are dominant over constricted pods, while round seeds are dominant over wrinkled seeds. One

◀ **Insights and Solutions** sections strengthen students' problem solving skills by showing step-by-step solutions and rationales for select problems.

Problems and Discussion Questions ▶ at the end of every chapter include several levels of difficulty, and most are assignable in MasteringGenetics. "How Do We Know?" questions ask students to identify and examine the experimental basis underlying important concepts. New problems have been added to the Tenth Edition.

Problems and Discussion Questions

HOW DO WE KNOW?

1. In this chapter, we first focused on the information that showed DNA to be the genetic material and then discussed the structure of DNA as proposed by Watson and Crick. We concluded the chapter by describing various techniques developed to study DNA. Along the way, we found many opportunities to consider the methods and reasoning by which much of this information was acquired. From the explanations given in the chapter, what answers would you propose to the following fundamental questions:

(a) How were scientists able to determine that DNA, and not some other molecule, serves as the genetic material in bacteria and bacteriophages?

(b) How do we know that DNA also serves as the genetic material in eukaryotes such as humans?

(c) How was it determined that the structure of DNA is a double helix with the two strands held together by hydrogen bonds formed between complementary nitrogenous bases?

(d) How do we know that G pairs with C and that A pairs with T as complementary base pairs are formed?

(e) How do we know that repetitive DNA sequences exist in eukaryotes?

Extra-Spicy Problems

34. *Newsdate: March 1, 2030.* A unique creature has been discovered during exploration of outer space. Recently, its genetic material has been isolated and analyzed. This material is similar in some ways to DNA in its chemical makeup. It contains in abundance the 4-carbon sugar erythrose and a molar equivalent of phosphate groups. In addition, it contains six nitrogenous bases: adenine (A), guanine (G), thymine (T), cytosine (C), hypoxanthine (H), and xanthine (X). These bases exist in the following relative proportions:

$$A = T = H \text{ and } C = G = X$$

X-ray diffraction studies have established a regularity in the molecule and a constant diameter of about 30 Å. Together, these data have suggested a model for the structure of this molecule.

(a) Propose a general model of this molecule. Describe it briefly.

(b) What base-pairing properties must exist for H and for X in the model?

(c) Given the constant diameter of 30 Å, do you think that *either* (i) both H and X are purines or both pyrimidines, *or* (ii) one is a purine and one is a pyrimidine?

Extra-Spicy Problems challenge students to solve complex problems, many based on data derived from primary genetics literature. ▲

NEW! Practice Problem-Solving with MasteringGenetics™

This book is now available with MasteringGenetics, a powerful online learning and assessment system proven to help students learn problem-solving skills.

In-depth tutorials, focused on key genetics concepts, reinforce problem solving skills with **hints and feedback** specific to students' misconceptions. Tutorial topics include pedigree analysis, sex linkage, gene interactions, DNA replication, and more.

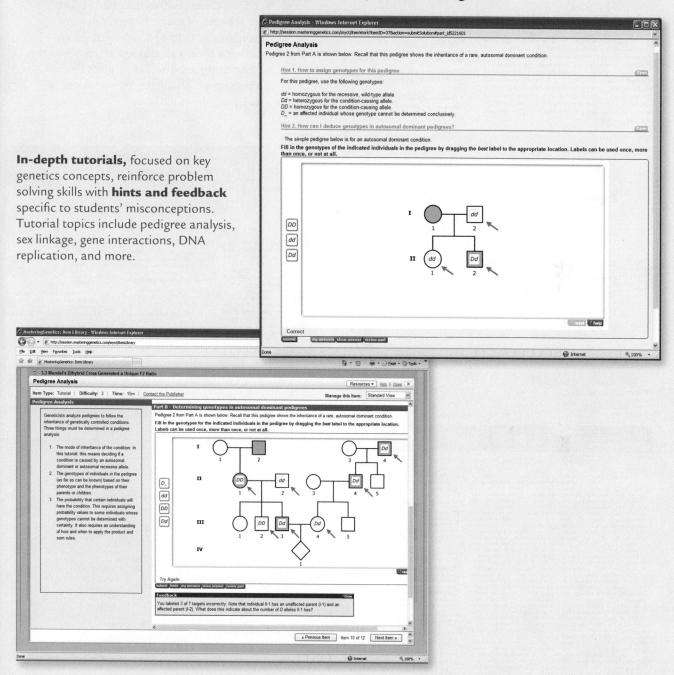

Selected questions with randomized values automatically provide individual students with different values for a given question, thereby ensuring students do their own work. These questions are identified with an icon in the MasteringGenetics item library.

Active and Cooperative Learning

This edition includes more opportunities for instructors and students to engage in active and cooperative learning.

◀ NEW! **Case studies** have been added to the end of each chapter, allowing students to read and answer questions about a short scenario related to one of the chapter topics. The Case Studies link the coverage of formal genetic knowledge to everyday societal issues.

CASE STUDY | To test or not to test

Thomas first discovered a potentially devastating piece of family history when he learned the medical diagnosis for his brother's increasing dementia, muscular rigidity, and frequency of seizures. His brother, at age 49, was diagnosed with Huntington disease (HD), a dominantly inherited condition that typically begins with such symptoms around the age of 45 and leads to death in one's early 60s. As depressing as the news was to Thomas, it helped explain his father's suicide. Thomas, 38, now wonders what his chances are of carrying the gene for HD, leading he and his wife to discuss the pros and cons of him undergoing genetic testing. Thomas and his wife have two teen-age children, a boy and a girl.

1. What role might a genetic counselor play in this real-life scenario?

2. How might the preparation and analysis of a pedigree help explain the dilemma facing Thomas and his family?

3. If Thomas decides to go ahead with the genetic test, what should be the role of the health insurance industry in such cases?

4. If Thomas tests positive for HD, and you were one of his children, would you want to be tested?

Exploring Genomics boxes help students ▶ apply genetics to modern techniques such as genomics, bioinformatics, and proteomics. These boxes illustrate how genomic studies have an impact on every aspect of genetics. Exercises provide thoughtful questions and direct students to related on line resources, allowing them to increase their awareness of genomics.

EXPLORING GENOMICS

Online Mendelian Inheritance in Man
(MG) Study Area: Exploring Genomics

The Online Mendelian Inheritance in Man (OMIM) database is a catalog of human genes and human genetic disorders that are inherited in a Mendelian manner. Genetic disorders that arise from major chromosomal aberrations, such as monosomy or trisomy (the loss of a chromosome or the presence of a superfluous chromosome, respectively), are not included. The OMIM database is a daily-updated version of the book *Mendelian Inheritance in Man*, edited by Dr. Victor McKusick of Johns Hopkins University. Scientists use OMIM as an important information source to accompany the sequence data generated by the **Human Genome Project**.

The OMIM entries will give you links to a wealth of information, including DNA and protein sequences, chromosomal maps, disease descriptions, and relevant scientific publications. In this exercise, you will explore OMIM to answer questions about the recessive human disease sickle-cell anemia and other Mendelian inherited disorders.

■ Exercise I – Sickle-cell Anemia

In this chapter, you were introduced to recessive and dominant human traits. You will now discover more about sickle-cell anemia as an autosomal recessive disease by exploring the OMIM database.

1. To begin the search, access the OMIM site at: www.ncbi.nlm.nih.gov/entrez/query.fcgi?db=OMIM&itool=toolbar.

2. In the "SEARCH" box, type "sickle-cell anemia" and click on the "Go" button to perform the search.

3. Select the first entry (#603903).

4. Examine the list of subject headings in the left-hand column and read some of the information about sickle-cell anemia.

5. Select one or two references at the bottom of the page and follow them to their abstracts in PubMed.

6. Using the information in this entry, answer the following questions:

a. Which gene is mutated in individuals with sickle-cell anemia?

b. What are the major symptoms of this disorder?

c. What was the first published scientific description of sickle-cell anemia?

d. Describe two other features of this disorder that you learned from the OMIM database and state where in the database you found this information.

■ Exercise II – Other Recessive or Dominant Disorders

Select another human disorder that is inherited as either a dominant or recessive trait and investigate its features, following the general procedure presented above. Follow links from OMIM to other databases if you choose.

Describe several interesting pieces of information you acquired during your exploration and cite the information sources you encountered during the search.

◀ **Genetics, Technology, and Society Essays** reflect recent findings in genetics and their impact on society. It includes a section called *Your Turn*, which directs students to related resources of short readings and websites to support deeper investigation and discussion.

GENETICS, TECHNOLOGY, AND SOCIETY

A Question of Gender: Sex Selection in Humans

Throughout history, people have attempted to influence the gender of their unborn offspring by following varied and sometimes bizarre procedures. In medieval Europe, prospective parents would place a hammer under the bed to help them conceive a boy, or a pair of scissors to conceive a girl. Other practices were based on the ancient belief that semen from the right testicle created male offspring and that from the left testicle created females. As late as the eighteenth century, European men might tie off or remove their left testicle to increase the chances of getting a male heir. In some cultures, efforts to control the sex of offspring has had a darker outcome—female infanticide. In ancient Greece, the murder of female infants was so common that the male:female ratio in some areas approached 4:1. In some parts of rural India, hundreds of families admitted to female infanticide as late as the 1990s. In 1997, the World Health Organization reported population data showing that about 50 million women were "missing" in China, likely because of selective abortion of female fetuses and institutionalized neglect of female children.

In recent times, sex-specific abortion has replaced much of the traditional female infanticide. For a fee, some companies offer amniocentesis and ultrasound tests for prenatal sex determination. Studies in India estimate that hundreds of thousands of fetuses are aborted each year because they are female. As a result of sex-selective abortion, the female:male ratio in India was 927:1000 in 1991. In some northern states, the ratio was as low as 600:1000.

In Western industrial countries, new genetics and reproductive technologies offer parents ways to select their children's gender prior to implantation of the embryo in the uterus—called *preimplantation gender selection (PGS)*. Following *in vitro* fertilization, embryos are biopsied and assessed for gender. Only sex-selected embryos are then implanted. The

simplest method involves separating X and Y chromosome-bearing spermatozoa based on their DNA content. Because of the difference in size of the X and Y chromosomes, X-bearing sperm contain 2.8 to 3.0 percent more DNA than Y-bearing sperm. Sperm samples are treated with a fluorescent DNA stain, then passed through a laser beam in a Fluorescence-Activated Cell Sorter machine that separates the sperm into two fractions based on the intensity of their DNA-fluorescence. The sorted sperm are then used for standard intrauterine insemination.

The emerging PGS methods raise a number of legal and ethical issues. Some people feel that prospective parents have the legal right to use sex-selection techniques as part of their fundamental procreative liberty. Proponents state that PGS will reduce the suffering of many families. For example, people at risk for transmitting X-linked diseases such as hemophilia or Duchenne muscular dystrophy can now enhance their chance of conceiving a female child, who will not express the disease.

The majority of people who undertake PGS, however, do so for nonmedical reasons—to "balance" their families. A possible argument in favor of this use is that the ability to intentionally select the sex of an offspring may reduce overpopulation and economic burdens for families who would repeatedly reproduce to get the desired gender. By the same token, PGS may reduce the number of abortions. It is also possible that PGS may increase the happiness of both parents and children, as the children would be more "wanted."

On the other hand, some argue that PGS serves neither the individual nor the common good. They argue that PGS is inherently sexist, having its basis in the idea that one sex is superior to the other, and leads to an increase in linking a child's worth to gender. Other critics fear that social approval of PGS will open the door to other genetic manipulations of children's characteristics. It is difficult to

predict the full effects that PGS will bring to the world. But the gender-selection genie is now out of the bottle and is unwilling to return.

Your Turn

Take time, individually or in groups, to answer the following questions. Investigate the references and links to help you understand some of the issues that surround the topic of gender selection.

1. What do you think are valid arguments for and against the use of PGS?

2. A generally accepted moral and legal concept is that of reproductive autonomy—the freedom to make individual reproductive decisions without external interference. Are there circumstances under which reproductive autonomy should be restricted?

The above questions, and others, are explored in a series of articles in the American Journal of Bioethics, Volume 1 (2001). See the article by J. A. Robertson on pages 2–9, for a summary of the moral and legal issues surrounding PGS.

3. What do you think are the reasons that some societies practice female infanticide and prefer the birth of male children?

For a discussion of this topic, visit the "Gendercide Watch" Web site http://www.gendercide.org.

4. If safe and efficient methods of PGS were available to you, do you think that you would use them to help you with family planning? Under what circumstances might you use them?

The Genetics and IVF Institute (Fairfax, Virginia) is presently using PGS techniques based on sperm sorting, in an FDA-approved clinical trial. As of 2008, over 1000 human pregnancies have resulted, with an approximately 80 percent success rate. Read about these methods on their Web site: http://www.microsort.net.

Contents

Mendelian Genetics   42

Extensions of Mendelian Genetics   71

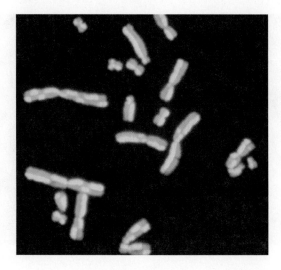

5
Chromosome Mapping in Eukaryotes   105

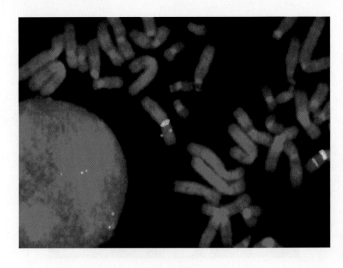

6

Genetic Analysis and Mapping in Bacteria and Bacteriophages   143

7

Sex Determination and Sex Chromosomes   174

8

Chromosome Mutations: Variation in Number and Arrangement   *197*

9

Extranuclear Inheritance   *221*

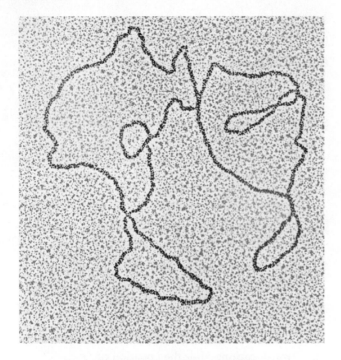

PART TWO
DNA: STRUCTURE, REPLICATION, AND VARIATION

10
DNA Structure and Analysis   238

11
DNA Replication and Recombination 269

12
DNA Organization in Chromosomes 294

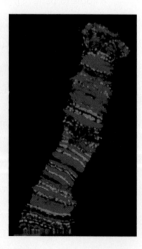

15

Gene Mutation, DNA Repair, and Transposition *374*

18

Developmental Genetics   *451*

19

Cancer and Regulation
of the Cell Cycle   *473*

21

Genomics, Bioinformatics, and Proteomics 574

24

Genetics of Behavior 680

25

Population and Evolutionary Genetics 697

26

Conservation Genetics   *725*

Preface

It is essential that textbook authors step back and look with fresh eyes as each edition of their work is planned. In doing so, two main questions must be posed: (1) How has the body of information in their field—in this case Genetics—grown and shifted since the last edition; and (2) What pedagogic innovations might be devised and incorporated into the text that will unquestionably enhance students' learning? The preparation of the 10th edition of *Concepts of Genetics*, a text now well into its third decade of providing support for students studying in this field, has occasioned still another fresh look. And what we focused on in this new edition, in addition to the normal updating that is inevitably required, were two things: (1) the need to increase the opportunities for instructors and students to engage in **active and cooperative learning approaches**, either within or outside of the classroom; and (2) the need to provide more **comprehensive coverage of important, emerging topics** that do not yet warrant their own traditional chapters.

Regarding the first point, and as discussed in further detail below, we have added a new feature called **Case Study**, which appears at the end of every chapter. In addition, we have converted the **Genetics, Technology, and Society** essays that appear at the end of many chapters to an active learning format by adding a *Your Turn* portion to each essay. These features join our unique **Exploring Genomics** entries, and together, these all may serve as the basis for interactions between small groups of students, either in or out of the classroom. Regarding the second point of covering emerging topics, we have devised a unique approach in genetics textbooks that offers readers a set of four abbreviated, highly focused chapters that we label **Special Topics in Modern Genetics**. As described below, these provide uniquely cohesive coverage of four important topics: *DNA Forensics, Genomics and Personalized Medicine, Epigenetics,* and *Stem Cells.*

Clearly, the field of genetics has grown tremendously since our book was first published, both in what we know and what we want beginning students to comprehend. In creating the current edition, we sought not only to continue to familiarize students with the most important discoveries of the past 150 years, but also to help them relate this information to the underlying genetic mechanisms that explain cellular processes, biological diversity, and evolution. We have also emphasized connections that link transmission genetics, molecular genetics, genomics and proteomics, and population-evolutionary genetics.

As we enter the second decade of this new millennium, discoveries in genetics continue to be numerous and profound. For students of genetics, the thrill of being part of this era must be balanced by a strong sense of responsibility and careful attention to the many scientific, social, and ethical issues that have already arisen, and others that will undoubtedly arise in the future. Policy makers, legislators, and an informed public will increasingly depend on detailed knowledge of genetics in order to address these issues. As a result, there has never been a greater need for a genetics textbook that clearly explains the principles of genetics.

Goals

In the 10th edition of *Concepts of Genetics*, as in all past editions, we have five major goals. Specifically, we have sought to:

- Emphasize the basic concepts of genetics.

- Write clearly and directly to students in order to provide understandable explanations of complex, analytical topics.

- Maintain our strong emphasis on and provide multiple approaches to problem solving.

- Propagate the rich history of genetics, which so beautifully illustrates how information is acquired during scientific investigation.

- Create inviting, engaging, and pedagogically useful full-color figures enhanced by equally helpful photographs to support concept development.

These goals collectively serve as the cornerstone of *Concepts of Genetics*. This pedagogic foundation allows the book to be used in courses with many different approaches and lecture formats. Although the chapters are presented in a coherent order that represents one approach to offering a course in genetics, they are nevertheless written to be as independent of one another as possible, allowing instructors to utilize them in various sequences. We believe that the varied approaches embodied in these goals together provide students with optimal support for their study of genetics.

Writing a textbook that achieves these goals and having the opportunity to continually improve on each new edition has been a labor of love for us. The creation of each of the ten editions is a reflection not only of our passion for teaching genetics, but also of the constructive feedback and encouragement provided by adopters, reviewers, and our students over the past three decades.

New to This Edition

- **Special Topics in Modern Genetics**—As new research topics in genetics gain stature and evolve, they gradually find their way into textbooks as either a short section in one chapter (when they are very specific), or they are mentioned briefly in many chapters (when they are more general). Some of these topics may be of great interest, having a genetic foundation; others represent major applications of genetic knowledge; and still others are ancillary to the coverage normally possible in an introductory textbook. In all cases, the topics are difficult for students and adopters of the text to find among all of the other coverage, and sometimes they are barely covered at all. This scenario is a source of major frustration to us as authors.

  New to this edition is a feature that we hope overcomes these limitations—the creation of a series of shorter, more specialized chapters that we call **Special Topics in Modern Genetics**. We have procured space for them in the text by providing abbreviated, cohesive coverage that focuses on core content and by eliminating features that appear in traditional chapters. In short, our goal is to provide concise support for both the construction and delivery of a lecture on each topic, as well as support for students who have heard such a lecture on any of these topics. And should the topics not be assigned in class, we are confident that they are of sufficient general interest that students will wish to read them on their own.

  For this edition, we have selected four important topics that are valuable, unique additions to the text, providing modern in-depth coverage that would otherwise not be present:

  1. DNA Forensics
  2. Genomics and Personalized Medicine
  3. Epigenetics
  4. Stem Cells

  The strong supporting figures that accompany each Special Topics chapter are available in PowerPoint to facilitate their use in classroom presentations. Special Topics are written to stand alone, so they also may be easily assigned without an accompanying lecture. We feel that these Special Topics chapters present subject matter that is important for an understanding of modern genetics and that is deserving of featured treatment in short, discrete chapters. Although we have inserted the Special Topics coverage following Chapter 19, where they can be identified by the colored margin tabs, they can be utilized at any time during the use of the book whenever the instructor feels it appropri-

ate. We hope that all users of this text find these to be highly practical additions to their Genetics course.

- **MasteringGenetics™**—This new supplement provides a robust online homework and assessment program, guiding students through complex topics in genetics, using in-depth tutorials that coach students to correct answers with hints and feedback specific to their misconceptions.

- **Case Study**—As elaborated upon below, each chapter now presents a Case Study, including several discussion questions. This feature is part of our larger effort to provide ample opportunities for active and cooperative learning.

Updated Topics

While we have updated each chapter in the text to present the most current findings in genetics, below is a list of some of the most significant, specific topics that have received attention.

Ch. 1: Introduction to Genetics
- New section on early history of genetics
- Added section on Charles Darwin and Evolution
- Streamlined discussion of genetics concepts

Ch. 2: Mitosis and Meiosis
- New coverage and a new figure involving the role of cohesin and shugoshin during mitosis and meiosis

Ch. 3: Mendelian Genetics
- New coverage on the molecular basis of Tay–Sachs disease

Ch. 4: Extensions of Mendelian Genetics
- Revised figure depicting the biochemical basis of the ABO blood groups
- New coverage of the molecular basis of dominance and recessiveness, illustrated in the agouti gene in mice

Ch. 5: Chromosome Mapping in Eukaryotes
- Streamlined coverage of haploid mapping

Ch. 6: Genetic Analysis and Mapping in Bacteria and Bacteriophages
- Revised discussion and terminology involving transduction
- Reworked figures involving conjugation

Ch. 7: Sex Determination and Sex Chromosomes
- Updated coverage of mammalian sex determination
- Updated coverage of the human Y chromosome
- Updated coverage of the mechanism of X chromosome inactivation
- New information regarding sex determination in chickens included as two new Extra-Spicy problems

Ch. 8: Chromosome Mutations: Variation in Number and Arrangement

- Additional coverage of the Down Syndrome Critical Region (DSCR)
- Noninvasive prenatal genetic diagnosis (NIPGD) introduced
- New coverage of the TERT (telomerase reverse transcriptase) gene and its link to cri du chat syndrome
- New discussion of copy number variants (CNVs) at the molecular level

Ch. 9: Extranuclear Inheritance

- New section added on mitochondria, human health, and aging

Ch. 10: DNA Structure and Analysis

- Continued classical coverage of DNA structure and analysis

Ch. 11: DNA Replication and Recombination

- Updated coverage of telomerase
- Updated coverage of eukaryotic DNA replication and chromatin assembly factors (CAFs)

Ch. 12: DNA Organization in Chromosomes

- Introduction to the role of chromatin remodeling in epigenetic modifications
- New section on telomeric DNA sequences and TERRA (telomere repeat-containing RNA)
- New photographs of polytene chromosomes and of nucleosomes

Ch. 13: The Genetic Code and Transcription

- New figure and updated coverage depicting the action of RNA polymerase during prokaryotic transcription
- Updated coverage of transcription in eukaryotes

Ch. 14: Translation and Proteins

- New coverage on the dynamic role of the eukaryotic ribosome during translation
- An introduction to amino acids selenocysteine (Sec) and pyrrolysine (Pyl) found in archaea, bacteria, and eukaryotes

Ch. 15: Gene Mutation, DNA Repair, and Transposition

- New section—Single-Gene Mutations Cause a Wide Range of Human Disease—describes the types of human disorders caused by the various types of single-gene mutations described in the chapter
- Beta-thalassemia is presented as an example of a prevalent human disease that can be caused by many different types of mutations in a single gene
- New material describes the link between defective mismatch repair and cancers such as leukemias, lymphomas, and tumors of the ovary, prostate, and endometrium
- New section on transposable elements, including a new subsection on transposons, mutations, and evolution

Ch. 16: Regulation of Gene Expression in Prokaryotes

- New figure and revised discussion of attenuation in the tryptophan operon
- New coverage, including a new figure, of riboswitches as metabolite-sensing RNAs

Ch. 17: Regulation of Gene Expression in Eukaryotes

- Updates on chromatin remodeling, posttranscriptional regulation, regulation of mRNA stability, translational control, and RNA silencing
- New section—Programmed DNA Rearrangements Regulate Expression of a Small Number of Genes—introduces immunoglobulin gene rearrangements and mating type switching in yeast
- New and updated material on core promoters, focused and dispersed promoters, and promoter elements

Ch. 18: Developmental Genetics

- Combined sections on evolutionary conservation of developmental mechanisms and model organisms
- Consolidated sections on determination and differentiation in *Drosophila*
- New section on programmed cell death

Ch. 19: Cancer and Regulation of the Cell Cycle

- New information on chromatin modifications and cancer epigenetics
- Expansion of metastasis section and clarification of the process of invasion
- New subsection on the cancer stem cell hypothesis

Ch. 20: Recombinant DNA Technology

- Major revision and reorganization of recombinant DNA techniques
- Eliminated focus on host cells and provided basic summary of different vectors and their applications
- Increased emphasis on gene libraries
- Changed emphasis on radioactive labeling techniques to indicate more widespread current usage of nonradioactive detection and labeling methods (e.g., probe-labeling, sequencing)
- Addition of RT-PCR and quantitative real-time PCR (qPCR) techniques, including new figure
- New material on FISH and spectral karyotyping
- Major revision of DNA sequencing technologies to include capillary electrophoresis-based computer automated sequencing and next generation sequencing technologies including new figure
- Major revision of PDQ content and additional new questions

Ch. 21: Genomics, Bioinformatics, and Proteomics

- New section on "10 years after HGP" including a new section on the Human Microbiome Project
- Updated content on the human genome including new information about copy number variations (CNVs)
- Expanded content on "stone age" genomics and new data on the Neanderthal genome
- Expanded content on comparative genomics to include comparisons of model organism genomes and the human genome
- New section on personal genomes including new figure on genome sequence costs, progress, and sequencing of individual diploid genomes
- New section on Genome 10K
- Updated content on systems biology, including new figure comparing human disease gene interaction network

Ch. 22: Applications and Ethics of Genetic Engineering and Biotechnology

- Updated discussions on synthetic genomes, direct to consumer genetic testing (DTC), and patenting genetic information

Ch. 24: Genetics of Behavior

- Reorganized to emphasize the behavior-first and gene-first approaches to the study of behavior
- Increased focus on the nervous system in behavior genetics
- Revised and updated coverage of human behavior genetics
- Discussion of the link between autism and schizophrenia
- New coverage of genomic techniques in behavior genetics

Ch. 25: Population and Evolutionary Genetics

- New integrated coverage of population genetics and evolution combined into a single chapter
- New coverage of linked factors affecting allele frequency to evolution
- New section on phylogeny and its application to the study of evolution
- Added material on comparative genomics of Neanderthals and modern humans

Chapter 26: Conservation Genetics

- Updated coverage of threatened species

This list reflects the rapid growth of information in genetics.

Emphasis on Concepts

The title of our textbook—*Concepts of Genetics*—was purposefully chosen, reflecting our fundamental pedagogic approach to teaching and writing about genetics. However, the word "concept" is not as easy to define as one might think. Most simply put, we consider a concept to be a cognitive unit of meaning—an abstract representation that encompasses a related set of scientifically-derived findings and ideas. Thus, a concept provides a broad mental image which, for example, might reflect a straightforward snapshot in your mind's eye of what constitutes a chromosome, a dynamic vision of the detailed processes of replication, transcription, and translation of genetic information, or just an abstract perception of varying modes of inheritance.

We think that creating such mental imagery is the very best way to teach science, in this case, genetics. Details that might be memorized, but soon forgotten, are instead subsumed within a conceptual framework that is easily retained. Such a framework may be expanded in content as new information is acquired and may interface with other concepts, providing a useful mechanism to integrate and better understand related processes and ideas. An extensive set of concepts may be devised and conveyed to eventually encompass and represent an entire discipline—and this is our goal in this genetics textbook.

To aid students in identifying the conceptual aspects of a major topic, each chapter begins with a section called *Chapter Concepts*, which identifies the most important ideas about to be presented. Each chapter ends with a new section called *Summary Points*, which enumerates the five to ten key points that have been discussed. And in the *How Do We Know?* question that starts each chapter's problem set, students are asked to connect concepts to experimental findings. Collectively, these features help to ensure that students easily become aware of and understand the major conceptual issues as they confront the extensive vocabulary and the many important details of genetics. Carefully designed figures also support this approach throughout the book.

Strengths of This Edition

- **Organization**—We have continued to attend to the organization of material by arranging chapters within major sections to reflect changing trends in genetics. A few important organizational changes have been made in this edition. First, the chapter on Recombinant DNA Technology, previously linked to DNA Structure and Analysis in Part II, has been moved to Part IV (Chapter 20) where it now directly precedes our coverage of genomics and biotechnology, for which it serves as an experimental foundation. We have also more carefully integrated all of our coverage of genomics into the two chapters that follow this introduction in Part IV. The second major change that we have made is designed to facilitate an instructor's coverage of Population and Evolutionary genetics. Previously, these topics have been presented in separate chapters. In the revised edition, we have integrated our discussions of them into a single chapter (Chapter 25), thus enhancing the close link between the two.

- **Model Organisms**—We have continued to emphasize the use by geneticists of model organisms, where coverage is woven throughout many chapters, but especially in Chapter 1—Introduction to Genetics, and in Chapter 18—Developmental Genetics. Clearly, the use of model organisms in genetic studies provides the underpinning of all current genetic research.

- **Pedagogy**—As discussed above, one of the major pedagogic goals of this edition is to provide features within each chapter that small groups of students can use either in the classroom or as assignments outside of class. Pedagogic research continues to support the value and effectiveness of such *active and cooperative learning experiences*. To this end, there are three features that greatly strengthen this edition.
  - **Case Study** This new feature, at the end of each chapter, introduces a short vignette (a "Case") of an everyday encounter related to genetics, followed by a series of discussion questions. Use of the Case Study should prompt students to relate their newly acquired information in genetics to issues that they may encounter away from the course.
  - **Genetics, Technology, and Society** This feature provides a synopsis of a topic related to a current finding in genetics that impacts directly on our current society. It now includes a new section called *Your Turn*, which directs students to related resources of short readings and Web sites to support deeper investigation and discussion of the main topic of each essay.
  - **Exploring Genomics** This feature extends the discussion of selected topics present in the chapter by exploring new findings resulting from genomic studies. Students are directed to Web sites that provide the "tools" that research scientists around the world rely on for current genomic information.

Whether instructors use these activities as active learning in the classroom or as assigned interactions outside of class, the above features will stimulate the use of current pedagogic approaches to students' learning. The activities help engage students, and the content of each feature ensures that they will become knowledgeable about cutting-edge topics in genetics.

Emphasis on Problem Solving

As authors and teachers, we have always recognized the importance of teaching students how to become effective problem solvers. Students need guidance and practice if they are to develop strong analytical thinking skills. To that end, we present a suite of features in every chapter to optimize opportunities for student growth in the important areas of problem solving and analytical thinking.

- **Now Solve This** Several times within the text of each chapter, each entry provides a problem similar to those found at the end of the chapter that is closely related to the current text discussion. In each case, a pedagogic hint is provided to offer insight and to aid in solving the problem. This feature closely links the text discussion to the problem.

- **Insights and Solutions** As an aid to the student in learning to solve problems, the *Problems and Discussion Questions* section of each chapter is preceded by what has become an extremely popular and successful section. *Insights and Solutions* poses problems or questions and provides detailed solutions or analytical insights as answers are provided. The questions and their solutions are designed to stress problem solving, quantitative analysis, analytical thinking, and experimental rationale. Collectively, these constitute the cornerstone of scientific inquiry and discovery.

- **Problems and Discussion Questions** Each chapter ends with an extensive collection of *Problems and Discussion Questions*. These include several levels of difficulty, with the most challenging (*Extra-Spicy Problems*) located at the end of each section. Often, Extra-Spicy problems are derived from the current literature of genetic research, with citations. Brief answers to all even-numbered problems are presented in Appendix B. The *Student Handbook and Solutions Manual* answers every problem and is available to students whenever faculty decide that it is appropriate. As the reader familiar with previous editions will see, about 50 new problems appear throughout the text.

- **How Do We Know?** Appearing as the first entry in the *Problems and Discussion Questions* section, this question asks the student to identify and examine the experimental basis underlying important concepts and conclusions that have been presented in the chapter. Addressing these questions will aid the student in more fully understanding, rather than memorizing, the end-point of each body of research. This feature is an extension of the learning approach in biology first formally descibed by John A. Moore in his 1999 book *Science as a Way of Knowing—The Foundation of Modern Biology*.

- **MasteringGenetics** Tutorials in MasteringGenetics help students strengthen their problem-solving skills while exploring challenging activities about key genetics content. In addition, end-of-chapter problems are also available for instructors to assign as online homework. Students will also be able to access materials in the Study Area that help them assess their understanding and prepare for exams.

For the Instructor

MasteringGenetics—
http://www.masteringgenetics.com

MasteringGenetics engages and motivates students to learn and allows you to easily assign automatically graded activities. Tutorials provide students with personalized coaching and feedback. Using the gradebook, you can quickly monitor and display student results. MasteringGenetics easily captures data to demonstrate assessment outcomes. Resources include:

- In-depth tutorials that coach students with hints and feedback specific to their misconceptions

- An item library of thousands of assignable questions including reading quizzes and end-of-chapter problems. You can use publisher-created prebuilt assignments to get started quickly. Each question can be easily edited to match the precise language you use.

- A gradebook that provides you with quick results and easy-to-interpret insights into student performance

Instructor Resource DVD
(032175400X)

The Instructor Resource DVD for the 10th edition offers adopters of the text convenient access to the most comprehensive and innovative set of lecture presentation and teaching tools offered by any genetics textbook. Developed to meet the needs of veteran and newer instructors alike, these resources include:

- The JPEG files of all text line drawings with labels individually enhanced for optimal projection results (as well as unlabeled versions) and all text tables.

- Most of the text photos, including all photos with pedagogical significance, as JPEG files.

- The JPEG files of line drawings, photos, and tables preloaded into comprehensive PowerPoint® presentations for each chapter.

- A second set of PowerPoint® presentations consisting of a thorough lecture outline for each chapter augmented by key text illustrations.

- An impressive series of concise instructor animations adding depth and visual clarity to the most important topics and dynamic processes described in the text.

- The instructor animations preloaded into PowerPoint® presentation files for each chapter.

- PowerPoint® presentations containing a comprehensive set of in-class Classroom Response System (CRS) questions for each chapter.

- In Word files, a complete set of the assessment materials and study questions and answers from the testbank, the text's in-chapter text questions, and the student media practice questions.

- Finally, to help instructors keep track of all that is available in this media package, a printable Media Integration Guide in PDF format that lists each chapter's media offerings.

TestGen EQ Computerized Testing Software
(0321754344)

Test questions are available as part of the TestGen EQ Testing Software, a text-specific testing program that is networkable for administering tests. It also allows instructors to view and edit questions, export the questions as tests, and print them out in a variety of formats.

For the Student

Student Handbook and Solutions Manual
Authored by Harry Nickla, Creighton University (Emeritus)
(0321754425)

This valuable handbook provides a detailed step-by-step solution or lengthy discussion for every problem in the text. The handbook also features additional study aids, including extra study problems, chapter outlines, vocabulary exercises, and an overview of how to study genetics.

MasteringGenetics—
http://www.masteringgenetics.com

Used by over one million science students, the Mastering platform is the most effective and widely used online tutorial, homework, and assessment system for the sciences; it helps students perform better on homework and exams. As an instructor-assigned homework system, MasteringGenetics is designed to provide students with a variety of assessment to help them understand key topics and concepts and to build problem solving skills. MasteringGenetics tutorials guide students through the toughest topics in genetics with self-paced tutorials that provide individualized coaching with hints and feedback specific to a student's individual misconceptions. Students can also explore the MasteringGenetics Study Area, which includes animations, the eText, *Exploring Genomics* exercises, and other study aids. The interactive eText allows students to highlight text, add study notes, review instructor's notes, and search throughout the text.

Acknowledgments

Contributors

We begin with special acknowledgments to those who have made direct contributions to this text. We particularly thank Sarah Ward at Colorado State University for initially creating Chapter 26 on Conservation Genetics several editions ago. We thank Joan Redd of Walla Walla University and Jutta Heller of the University of Washington—Tacoma for their work on the media program. We also thank David Kass of Eastern Michigan University, Chaoyang Zeng of the University of Wisconsin at Milwaukee, and Virginia McDonough of Hope College for their useful input into both text-related topics and the media program. In addition, Amanda Norvell, Janet Morrison, and Katherine Uyhazi provided input leading to revisions of earlier editions. Amanda and Janet are colleagues from The College of New Jersey, while Katherine has moved on to Yale University. Tamara Mans, currently teaching at North Hennepin Community College, has also provided help during earlier revisions. As with previous editions, Elliott Goldstein from Arizona State University was always readily available on the current edition to consult with us concerning the most modern findings in molecular genetics. We also express special thanks to Harry Nickla, retired from Creighton University. In his role as author of the *Student Handbook and Solutions Manual* and the *Instructor's Resource Manual with Tests*, he has reviewed and edited the problems at the end of each chapter, and has written many of the new entries as well. He also provided the brief answers to selected problems that appear in Appendix B.

We are grateful to all of these contributors not only for sharing their genetic expertise, but for their dedication to this project as well as the pleasant interactions they provided.

Proofreaders and Accuracy Checking

Reading the manuscript of an 800+ page textbook deserves more thanks than words can offer. Our utmost appreciation is extended to Darrell Killian at The College of New Jersey, who provided accuracy checking, and to Michael Rossa and Debra Gates who proofread the entire manuscript. They confronted this task with patience and diligence, contributing greatly to the quality of this text.

Reviewers

All comprehensive texts are dependent on the valuable input provided by many reviewers. While we take full responsibility for any errors in this book, we gratefully acknowledge the help provided by those individuals who reviewed the content and pedagogy of this and the previous edition:

Robert A. Angus, *University of Alabama, Birmingham*
Bruce Bejcek, *Western Michigan University*
Peta Bonham-Smith, *University of Saskatchewan*
Michael A. Buratovich, *Spring Arbor University*
Aaron Cassill, *University of Texas, San Antonio*
Alan H. Christensen, *George Mason University*
Bert Ely, *University of South Carolina*
Elliott S. Goldstein, *Arizona State University*
Edward M. Golenberg, *Wayne State University*
Ashley Hagler, *University of North Carolina, Charlotte*
Jocelyn Krebs, *University of Alaska, Fairbanks*
Paul F. Lurquin, *Washington State University*
Virginia McDonough, *Hope College*
Kim McKim, *Rutgers University*
Clint Magill, *Texas A&M University*
Harry Nickla, *Creighton University*
Mohamed Noor, *Duke University*
Margaret Olney, *Saint Martin's University*
John C. Osterman, *University of Nebraska–Lincoln*
Gloria Regisford, *Prairie View A&M University*
Rodney Scott, *Wheaton College*
Barkur Shastry, *Oakland University*
Linda Sigismondi, *University of Rio Grande*
Tara Turley Stoulig, *Southeastern Louisiana University*
Kenneth Wilson, *University of Saskatchewan*
Fang-sheng Wu, *Virginia Commonwealth University*
Chaoyang Zeng, *University of Wisconsin, Milwaukee*

Special thanks go to Mike Guidry of LightCone Interactive and Karen Hughes of the University of Tennessee for their original contributions to the media program.

As these acknowledgments make clear, a text such as this is a collective enterprise. All of the above individuals deserve to share in any success this text enjoys. We want them to know that our gratitude is equaled only by the extreme dedication evident in their efforts. Many, many thanks to them all.

Editorial and Production Input

At Pearson, we express appreciation and high praise for the editorial guidance and seminal input of Gary Carlson, and more recently of Michael Gillespie, whose ideas and efforts have helped to shape and refine the features of this edition of the text. Dusty Friedman, our Project Editor, has worked tirelessly to keep the project on schedule and to maintain our standards of high quality. In addition, our editorial team—Deborah Gale, Executive Director of Development, Laura Tommasi, Senior Media Producer, and Tania Mlawer, Director of Editorial Content for Mastering Genetics—has provided valuable input into the current edition. They have worked creatively to ensure that the pedagogy and design of the book and media package are at the cutting edge of a rapidly changing discipline. Sudhir Nayak of The College of New Jersey provided outstanding work for the new MasteringGenetics program and his input regarding genomics was appreciated. Camille Herrera supervised all of the production intricacies with great attention to detail and perseverance. Outstanding copyediting was performed by Betty Pessagno, for which we are most grateful. Lauren Harp has professionally and enthusiastically managed the marketing of the text. Finally, the beauty and consistent presentation of the art work is the product of Imagineering of Toronto. Without the work ethic and dedication of the above individuals, the text would never have come to fruition.

Newer model organisms in genetics include the roundworm *Caenorhabditis elegans*, the plant *Arabidopsis thaliana*, and the zebrafish, *Danio rerio*.

1

Introduction to Genetics

CHAPTER CONCEPTS

- Transmission genetics encompasses the general process by which traits controlled by factors (genes) are transmitted through gametes from generation to generation. Its fundamental principles were first put forward by Gregor Mendel in the mid-nineteenth century. Later work by others showed that genes are on chromosomes and that mutant strains can be used to map genes on chromosomes.

- The recognition that DNA encodes genetic information, the discovery of DNA's structure, and elucidation of the mechanism of gene expression form the foundation of molecular genetics.

- Recombinant DNA technology, which allows scientists to prepare large quantities of specific DNA sequences, has revolutionized genetics, laying the foundation for new fields—and for endeavors such as the Human Genome Project—that combine genetics with information technology.

- Biotechnology includes the use of genetically modified organisms and their products in a wide range of activities involving agriculture, medicine, and industry.

- Some of the model organisms employed in genetic research since the early part of the twentieth century are now used in combination with recombinant DNA technology and genomics to study human diseases.

- Genetic technology is developing faster than the policies, laws, and conventions that govern its use.

Following months of heated debate in 1998, the Icelandic Parliament passed a law granting deCODE Genetics, a biotechnology company with headquarters in Iceland, a license to create and operate a database containing detailed information drawn from medical histories of all of Iceland's 270,000 residents. The records in this Icelandic Health Sector Database (HSD) were encoded to ensure anonymity. The new law also allowed deCODE Genetics to cross-reference medical information from the HSD with a comprehensive genealogical database from the National Archives. In addition, deCODE Genetics would be able to correlate information in these two databases with results of deoxyribonucleic acid (DNA) profiles collected from Icelandic donors. This combination of medical, genealogical, and genetic information constitutes a powerful resource available exclusively to deCODE Genetics for marketing to researchers and companies for a period of 12 years, beginning in 2000.

This scenario is a typical example of the increasingly complex interaction of genetics and society we are witnessing in the early part of twenty-first century. The development and use of these databases in Iceland has generated similar projects in other countries as well. The largest is the "UK Biobank" effort launched in Great Britain in 2003. There, a huge database containing the genetic information of 500,000 Britons is being compiled from an initial group of 1.2 million residents. The database will be used to search for susceptibility genes that control complex traits. Other projects have since been announced in Estonia, Latvia, Sweden, Singapore, and the Kingdom of Tonga, while in the United States, smaller-scale programs, involving tens of thousands of individuals, are underway at the Marshfield Clinic in Marshfield, Wisconsin; Northwestern University in Chicago, Illinois; and Howard University in Washington, D.C.

deCODE Genetics selected Iceland for this unprecedented project because the people of Iceland have a level of genetic uniformity seldom seen or accessible to scientific investigation. This high degree of genetic relatedness derives from the founding of Iceland about 1000 years ago by a small population drawn mainly from Scandinavian and Celtic sources. Subsequent periodic population reductions by disease and natural disasters further reduced genetic diversity there, and until the last few decades, few immigrants arrived to bring new genes into the population. Moreover, because Iceland's health-care system is state-supported, medical records for all residents go back as far as the early 1900s. Genealogical information is available in the National Archives and church records for almost every resident and for more than 500,000 of the estimated 750,000 individuals who have ever lived in Iceland. For all these reasons, the Icelandic data are a tremendous asset for geneticists in search of genes that control complex disorders. The project already has a number of successes to its credit. Scientists at deCODE

Genetics have isolated genes associated with over a dozen common diseases including asthma, heart disease, stroke, and osteoporosis.

On the flip side of these successes are questions of privacy, consent, and commercialization—issues at the heart of many controversies arising from the applications of genetic technology. Scientists and nonscientists alike are debating the fate and control of genetic information and the role of law, the individual, and society in decisions about how and when genetic technology is used. For example, how will knowledge of the complete nucleotide sequence of the human genome be used? Should genetic technology such as prenatal diagnosis or gene therapy be available to all, regardless of ability to pay? More than at any other time in the history of science, addressing the ethical questions surrounding an emerging technology is as important as the information gained from that technology.

This introductory chapter provides an overview of genetics in which we survey some of the high points of its history and give preliminary descriptions of its central principles and emerging developments. All the topics discussed in this chapter will be explored in far greater detail elsewhere in the book. Later chapters will also revisit the controversies alluded to above and discuss many other issues that are current sources of debate. There has never been a more exciting time to be part of the science of inherited traits, but never has the need for caution and awareness of social consequences been more apparent. This text will enable you to achieve a thorough understanding of modern-day genetics and its underlying principles. Along the way, enjoy your studies, but take your responsibilities as a novice geneticist very seriously.

1.1
Genetics Has a Rich and Interesting History

We don't know when people first recognized the existence of heredity, but archaeological evidence (e.g., primitive art, preserved bones and skulls, and dried seeds) documents the successful domestication of animals and cultivation of plants thousands of years ago by artificial selection of genetic variants within populations. Between 8000 and 1000 B.C. horses, camels, oxen, and various breeds of dogs (derived from the wolf family) had been domesticated, and selective breeding soon followed. Cultivation of many plants, including maize, wheat, rice, and the date palm, began around 5000 B.C. Remains of maize dating to this period have been recovered in caves in the Tehuacan Valley of Mexico. Such evidence documents our ancestors' successful attempts to manipulate the genetic composition of species.

While few, if any, significant ideas were put forward to explain heredity during prehistoric times, during the Golden

Age of Greek culture, philosophers wrote about this subject as it relates to humans. This is evident in the writings of the Hippocratic School of Medicine (500–400 B.C.), and of the philosopher and naturalist Aristotle (384–322 B.C.). The Hippocratic treatise *On the Seed* argued that active "humors" in various parts of the male body served as the bearers of hereditary traits. Drawn from various parts of the male body to the semen and passed on to offspring, these humors could be healthy or diseased, the diseased condition accounting for the appearance of newborns with congenital disorders or deformities. It was also believed that these humors could be altered in individuals before they were passed on to off-spring, explaining how newborns could "inherit" traits that their parents had "acquired" because of their environment.

Aristotle, who studied under Plato for some 20 years, extended Hippocrates' thinking and proposed that the gen-erative power of male semen resided in a "vital heat" con-tained within it that had the capacity to produce offspring of the same "form" (i.e., basic structure and capacities) as the parent. Aristotle believed that this heat cooked and shaped the menstrual blood produced by the female, which was the "physical substance" that gave rise to an offspring. The em-bryo developed not because it already contained the parts in miniature (as some Hippocratics had thought) but because of the shaping power of the vital heat. Although the ideas of Hippocrates and Aristotle sound primitive and naïve today, we should recall that prior to the 1800s neither sperm nor eggs had been observed in mammals.

1600–1850: The Dawn of Modern Biology

During the ensuing 1900 years (from 300 B.C. to A.D. 1600), our understanding of genetics was not extended by any new or significant ideas. However, between 1600 and 1850, major strides provided insight into the biological basis of life, setting the scene for the revolutionary work and principles presented by Gregor Mendel and Charles Darwin. In the 1600s, the English anatomist William Harvey (1578–1657) wrote a treatise on reproduction and development patterned after Aristotle's work. He is credited with the earliest statement of the theory of **epigenesis**, which posits that an organism is derived from substances present in the egg that differentiate into adult structures during embryonic development. Epigenesis holds that structures such as body organs are not initially present in the early embryo but instead are formed *de novo* (anew).

This theory directly conflicted with that of **preforma-tion**, first proposed in the seventeenth century, which stated that sex cells contain a complete, miniature adult, perfect in every form, called a **homunculus** (Figure 1–1). Although this theory was later discounted, other significant chemical and biological discoveries made during this same period affected future scientific thinking. Around 1830, Matthias Schleiden

FIGURE 1–1   Depiction of the "homunculus," a sperm containing a miniature adult, perfect in proportion and fully formed. (Hartsoeker, N. Essay de dioptrique Paris, 1694, p. 230. National Library of Medicine)

and Theodor Schwann proposed the cell theory, stating that all organisms are composed of basic units called cells, which are derived from similar preexisting structures. The idea of **spontaneous generation**, the creation of living organisms from nonliving components, was disproved by Louis Pasteur later in the century, and living organisms were considered to be derived from preexisting organisms and to consist of cells.

Another influential notion prevalent in the nineteenth century was the **fixity of species**. According to this doctrine, animal and plant groups have remained unchanged in form since the moment of their appearance on Earth. This doctrine was particularly embraced by those who believed in special creation, including the Swedish physician and plant taxono-mist, Carolus Linnaeus (1707–1778), who is better known for devising the binomial system of species classification.

Charles Darwin and Evolution

With this background, we turn to a brief discussion of the work of Charles Darwin, who published the book-length statement of his evolutionary theory, *The Origin of Spe-cies*, in 1859. Darwin's geological, geographical, and bio-logical observations convinced him that existing species arose by descent with modification from other ancestral species. Greatly influenced by his voyage on the HMS *Beagle*

(1831–1836), Darwin's thinking culminated in his formulation of the theory of **natural selection**, which presented an explanation of the causes of evolutionary change. Formulated and proposed independently by Alfred Russel Wallace, natural selection was based on the observation that populations tend to consist of more offspring than the environment can support, leading to a struggle for survival among them. Those organisms with heritable traits that allow them to adapt to their environment are better able to survive and reproduce than those with less adaptive traits. Over a long period of time, slight but advantageous variations will accumulate. If a population bearing these inherited variations becomes reproductively isolated, a new species may result.

Darwin, however, lacked an understanding of the genetic basis of variation and inheritance, a gap that left his theory open to reasonable criticism. However, as we will see next, the work of Gregor Mendel in the 1850s, and its rediscovery in the early twentieth century, would soon provide the foundation for interpreting Darwin's proposal. It gradually became clear that inherited variation is dependent on genetic information residing in genes contained in chromosomes.

1.2

Genetics Progressed from Mendel to DNA in Less Than a Century

The true starting point of our understanding of genetics began in a monastery garden in central Europe in the 1860s, where Gregor Mendel, an Augustinian monk, conducted a decade-long series of experiments using pea plants. He applied quantitative data analysis to his results and showed that traits are passed from parents to offspring in predictable ways. He further concluded that each trait in the plant is controlled by a pair of genes and that during gamete formation (the formation of egg cells and sperm) members of a gene pair separate from each other. His work was published in 1866 but was largely unknown until it was partially duplicated and cited in papers by Carl Correns and others around 1900. Having been confirmed by others, Mendel's findings became recognized as explaining the transmission of traits in pea plants and all other higher organisms. His work forms the foundation for **genetics**, which is defined as the branch of biology concerned with the study of heredity and variation. Mendelian genetics will be discussed extensively in Chapter 3.

The Chromosome Theory of Inheritance

Mendel conducted his experiments before the structure and role of chromosomes were known. About 20 years after his work was published, advances in microscopy allowed researchers to identify chromosomes and establish that, in most eukaryotes, members of each species have a characteristic number of chromosomes called the **diploid number (2n)**

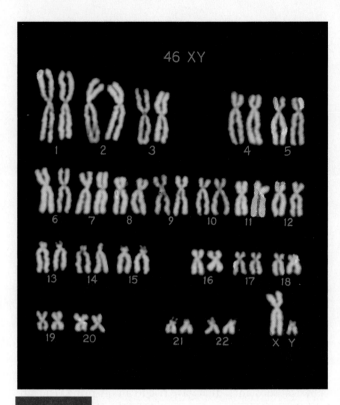

FIGURE 1–2   A colorized image of the human male chromosome set. Arranged in this way, the set is called a karyotype.

in most of its cells. For example, humans have a diploid number of 46 (Figure 1–2). Chromosomes in diploid cells exist in pairs, called **homologous chromosomes**. Members of a pair are identical in size and location of the centromere, a structure to which spindle fibers attach during cell division.

Researchers in the last decades of the nineteenth century also described the behavior of chromosomes during two forms of cell division, **mitosis** and **meiosis**. In mitosis, chromosomes are copied and distributed so that each daughter cell receives a diploid set of chromosomes. Meiosis is associated with gamete formation. Cells produced by meiosis receive only one chromosome from each chromosome pair, in which case the resulting number of chromosomes is called the **haploid (n) number**. This reduction in chromosome number is essential if the offspring arising from the union of two parental gametes are to maintain, over the generations, a constant number of chromosomes characteristic of their parents and other members of their species.

Early in the twentieth century, Walter Sutton and Theodor Boveri independently noted that genes, as hypothesized by Mendel, and chromosomes, as observed under the microscope, have several properties in common and that the behavior of chromosomes during meiosis is identical to the presumed behavior of genes during gamete formation described by Mendel. For example, genes and chromosomes exist in pairs, and members of a gene pair and members of a chromosome pair separate from each other during gamete

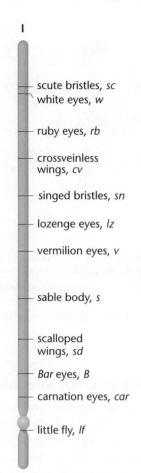

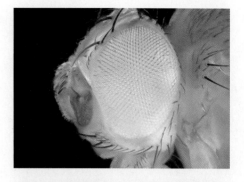

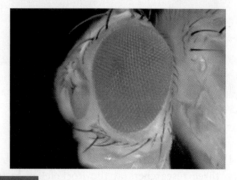

FIGURE 1–4 The normal red eye color in *D. melanogaster* (bottom) and the white-eyed mutant (top).

FIGURE 1–3 A drawing of chromosome I (the X chromosome, meaning one of the sex-determining chromosomes) of *D. melanogaster*, showing the locations of various genes. Chromosomes can contain hundreds of genes.

formation. Based on these parallels, Sutton and Boveri independently proposed that genes are carried on chromosomes (Figure 1–3). This proposal is the basis of the **chromosome theory of inheritance**, which states that inherited traits are controlled by genes residing on chromosomes faithfully transmitted through gametes, maintaining genetic continuity from generation to generation.

Geneticists encountered many different examples of inherited traits between 1910 and about 1940, allowing them to test the theory over and over. Patterns of inheritance sometimes varied from the simple examples described by Mendel, but the chromosome theory of inheritance could always be applied. It continues to explain how traits are passed from generation to generation in a variety of organisms, including humans.

Genetic Variation

At about the same time as the chromosome theory of inheritance was proposed, scientists began studying the inheritance of traits in the fruit fly, *Drosophila melanogaster*.

A white-eyed fly (Figure 1–4) was discovered in a bottle containing normal (wild-type) red-eyed flies. This variation was produced by a **mutation** in one of the genes controlling eye color. Mutations are defined as any heritable change and are the source of all genetic variation.

The variant eye color gene discovered in *Drosophila* is an **allele** of a gene controlling eye color. Alleles are defined as alternative forms of a gene. Different alleles may produce differences in the observable features, or **phenotype**, of an organism. The set of alleles for a given trait carried by an organism is called the **genotype**. Using mutant genes as markers, geneticists were able to map the location of genes on chromosomes (Figure 1–3).

The Search for the Chemical Nature of Genes: DNA or Protein?

Work on white-eyed *Drosophila* showed that the mutant trait could be traced to a single chromosome, confirming the idea that genes are carried on chromosomes. Once this relationship was established, investigators turned their attention to identifying which chemical component of chromosomes carried genetic information. By the 1920s, scientists were aware that proteins and DNA were the major chemical components of chromosomes. Of the two, proteins are the most abundant in cells. There are a large number of different proteins, and because of their universal distribution in the nucleus and cytoplasm, many researchers thought proteins would be shown to be the carriers of genetic information.

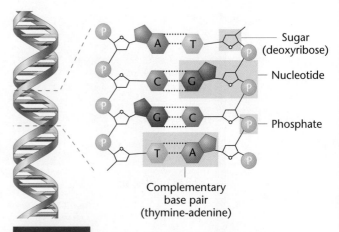

FIGURE 1–5   An electron micrograph showing T phage infecting a cell of the bacterium *E. coli.*

FIGURE 1–6   Summary of the structure of DNA, illustrating the arrangement of the double helix (on the left) and the chemical components making up each strand (on the right). The dotted lines between the bases represent weak chemical bonds, called hydrogen bonds, that hold together the two strands of the DNA helix.

In 1944, Oswald Avery, Colin MacLeod, and Maclyn McCarty, three researchers at the Rockefeller Institute in New York, published experiments showing that DNA was the carrier of genetic information in bacteria. This evidence, though clear-cut, failed to convince many influential scientists. Additional evidence for the role of DNA as a carrier of genetic information came from other researchers who worked with viruses that infect and kill cells of the bacterium *Escherichia coli* (Figure 1–5). Viruses that attack bacteria are called **bacteriophages**, or **phages** for short, and like all viruses, consist of a protein coat surrounding a DNA core. Experiments showed that during infection the protein coat of the virus remains outside the bacterial cell, while the viral DNA enters the cell and directs the synthesis and assembly of more phage. This evidence that DNA carries genetic information, along with other research over the next few years, provided solid proof that DNA, not protein, is the genetic material, setting the stage for work to establish the structure of DNA.

1.3

Discovery of the Double Helix Launched the Era of Molecular Genetics

Once it was accepted that DNA carries genetic information, efforts were focused on deciphering the structure of the DNA molecule and the mechanism by which information stored in it is expressed to produce an observable trait, called the **phenotype**. In the years after this was accomplished, researchers learned how to isolate and make copies of specific regions of DNA molecules, opening the way for the era of recombinant DNA technology.

The Structure of DNA and RNA

DNA is a long, ladder-like macromolecule that twists to form a double helix (Figure 1–6). Each strand of the helix is a linear polymer made up of subunits called **nucleotides**. In DNA, there are four different nucleotides. Each DNA nucleotide contains one of four nitrogenous bases, abbreviated A (adenine), G (guanine), T (thymine), or C (cytosine). These four bases, in various sequence combinations, ultimately specify the amino acid sequences of proteins. One of the great discoveries of the twentieth century was made in 1953 by James Watson and Francis Crick, who established that the two strands of DNA are exact complements of one another, so that the rungs of the ladder in the double helix always consist of A $=$ T and G $\equiv$ C base pairs. Along with Maurice Wilkins, Watson and Crick were awarded a Nobel Prize in 1962 for their work on the structure of DNA. A first-hand account of the race to discover the structure of DNA is told in the book *The Double Helix*, by James Watson. We will discuss the structure of DNA in Chapter 10.

RNA, another nucleic acid, is chemically similar to DNA but contains a different sugar (ribose rather than deoxyribose) in its nucleotides and contains the nitrogenous base uracil in place of thymine. In addition, in contrast to the double helix structure of DNA, RNA is generally single stranded. Importantly, an RNA strand can form complementary structures with strands of either DNA or RNA.

Gene Expression: From DNA to Phenotype

As noted earlier, nucleotide complementarity is the basis for gene expression, the chain of events that causes a gene to produce a phenotype. This process begins in the nucleus

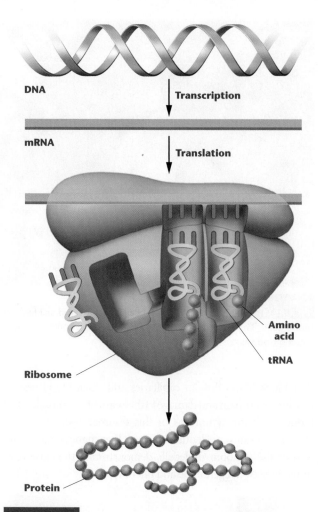

FIGURE 1–7   Gene expression consists of transcription of DNA into mRNA (top) and the translation (center) of mRNA (with the help of a ribosome) into a protein (bottom).

with **transcription**, in which the nucleotide sequence in one strand of DNA is used to construct a complementary RNA sequence (top part of Figure 1–7). Once an RNA molecule is produced, it moves to the cytoplasm. In protein synthesis, the RNA—called **messenger RNA**, or **mRNA** for short—binds to a **ribosome**. The synthesis of proteins under the direction of mRNA is called **translation** (middle part of Figure 1–7). Proteins, the end product of many genes, are polymers made up of amino acid monomers. There are 20 different amino acids commonly found in proteins.

How can information contained in mRNA direct the addition of specific amino acids into protein chains as they are synthesized? The information encoded in mRNA and called the **genetic code** consists of a linear series of nucleotide triplets. Each triplet, called a **codon**, is complementary to the information stored in DNA and specifies the insertion of a specific amino acid into a protein. Protein assembly is accomplished with the aid of adapter molecules called **transfer RNA (tRNA)**. Within the ribosome, tRNAs recognize the information encoded in the mRNA codons and carry the proper amino acids for construction of the protein during translation.

As the preceding discussion shows, *DNA makes RNA, which most often makes protein*. This sequence of events, known as the **central dogma** of genetics, occurs with great specificity. Using an alphabet of only four letters (A, T, C, and G), genes direct the synthesis of highly specific proteins that collectively serve as the basis for all biological function.

Proteins and Biological Function

As we have mentioned, proteins are the end products of gene expression. These molecules are responsible for imparting the properties of living systems. The diversity of proteins and of the biological functions they can perform—the diversity of life itself—arises from the fact that proteins are made from combinations of 20 different amino acids. Consider that a protein chain containing 100 amino acids can have at each position any one of 20 amino acids; the number of unique protein sequences consisting of 100 amino acids is therefore equal to

$$20^{100}$$

Because 20^{10} exceeds 5×10^{12}, or 5 trillion, imagine how large a number 20^{100} is! The tremendous number of possible amino acid sequences in proteins leads to enormous variation in their possible three-dimensional conformations. Obviously, proteins are molecules with the potential for enormous structural diversity and serve as the mainstay of biological systems.

The **enzymes** form the largest category of proteins. These molecules serve as biological catalysts, essentially causing biochemical reactions to proceed at the rates that are necessary for sustaining life. By lowering the energy of activation in reactions, enzymes enable cellular metabolism to proceed at body temperatures, when otherwise those reactions would require intense heat or pressure in order to occur.

Proteins other than enzymes are also critical components of cells and organisms. These include hemoglobin, the oxygen-binding pigment in red blood cells; insulin, the pancreatic hormone; collagen, the connective tissue molecule; keratin, the structural molecule in hair; histones, proteins integral to chromosome structure in eukaryotes (that is, organisms whose cells have nuclei); actin and myosin, the contractile muscle proteins; and immunoglobulins, the antibody molecules of the immune system. A protein's shape and chemical behavior are determined by its linear sequence of amino acids, which is dictated by the stored information in the DNA of a gene that is transferred to RNA, which then directs the protein's synthesis. To repeat, *DNA makes RNA, which then makes protein*.

Linking Genotype to Phenotype: Sickle-Cell Anemia

Once a protein is constructed, its biochemical or structural behavior in a cell plays a role in producing a phenotype. When mutation alters a gene, it may modify or even eliminate the encoded protein's usual function and cause an altered phenotype. To trace the chain of events leading from the synthesis of a given protein to the presence of a certain phenotype, we will examine sickle-cell anemia, a human genetic disorder.

Sickle-cell anemia is caused by a mutant form of hemoglobin, the protein that transports oxygen from the lungs to cells in the body. Hemoglobin is a composite molecule made up of two different polypeptides, α-globin and β-globin, each encoded by a different gene. Each functional hemoglobin molecule contains two α-globin and two β-globin chains. In sickle-cell anemia, a mutation in the gene encoding β-globin causes an amino acid substitution in 1 of the 146 amino acids in the protein. Figure 1–8 shows part of the DNA sequence, and the corresponding mRNA codons and amino acid sequence, for the normal and mutant forms of β-globin. Notice that the mutation in sickle-cell anemia consists of a change in one DNA nucleotide, which leads to a change in codon 6 in mRNA from GAG to GUG, which in turn changes amino acid number 6 in β-globin from glutamic acid to valine. The other 145 amino acids in the protein are not changed by this mutation.

Individuals with two mutant copies of the β-globin gene have sickle-cell anemia. Their mutant β-globin proteins cause hemoglobin molecules in red blood cells to polymerize when the blood's oxygen concentration is low, forming long chains of hemoglobin that distort the shape of red blood cells (Figure 1–9). The deformed cells are fragile and break easily, so that the number of red blood cells in circulation is reduced, causing anemia. Moreover, when blood cells are sickle shaped,

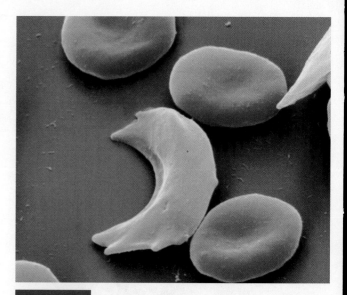

FIGURE 1–9   Normal red blood cells (round) and sickled red blood cells. The sickled cells aggregate, blocking capillaries and small blood vessels.

they block blood flow in capillaries and small blood vessels, causing severe pain and damage to the heart, brain, muscles, and kidneys. All the symptoms of this disorder result from the change in a single nucleotide in a gene that changes one amino acid of the β-globin molecule, demonstrating the close relationship between genotype and phenotype.

1.4

Development of Recombinant DNA Technology Began the Era of Cloning

The era of recombinant DNA began in the early 1970s, when researchers discovered that bacteria protect themselves from viral infection by producing enzymes that cut viral DNA at specific sites. When cut by these enzymes, the viral DNA cannot direct the synthesis of phage particles. Scientists quickly realized that such enzymes, called **restriction enzymes**, could be used to cut any organism's DNA at specific nucleotide sequences, producing a reproducible set of fragments. This set the stage for the development of DNA cloning, a way of making large numbers of copies of DNA sequences.

Soon after researchers discovered that restriction enzymes produce specific DNA fragments, methods were developed to insert these fragments into carrier DNA molecules called **vectors** to make **recombinant DNA** molecules and transfer them into bacterial cells. As the bacterial cells reproduce, thousands of copies, or **clones**, of the combined vector and DNA fragments are produced (Figure 1–10). These cloned copies can be recovered from the bacterial cells, and large amounts of the cloned DNA fragment can be isolated.

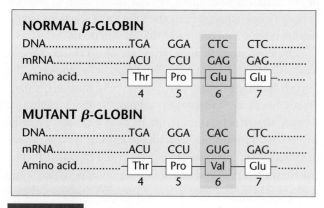

NORMAL β-GLOBIN				
DNA...........................TGA	GGA	CTC	CTC............
mRNA.......................ACU	CCU	GAG	GAG............
Amino acid..............– Thr	Pro	Glu	Glu –.........
4	5	6	7

MUTANT β-GLOBIN				
DNA...........................TGA	GGA	CAC	CTC............
mRNA.......................ACU	CCU	GUG	GAG............
Amino acid..............– Thr	Pro	Val	Glu –.........
4	5	6	7

FIGURE 1–8   A single-nucleotide change in the DNA encoding β-globin (CTC → CAC) leads to an altered mRNA codon (GAG → GUG) and the insertion of a different amino acid (glu → val), producing the altered version of the β-globin protein that is responsible for sickle-cell anemia.

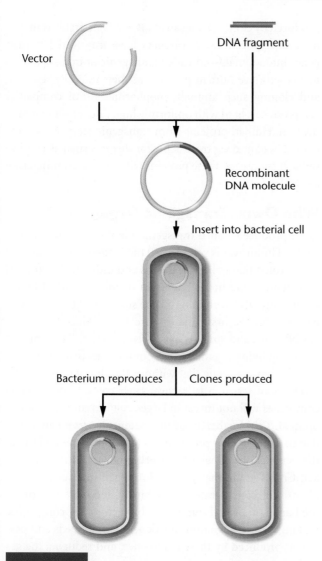

FIGURE 1–10   In cloning, a vector and a DNA fragment produced by cutting with a restriction enzyme are joined to produce a recombinant DNA molecule. The recombinant DNA is transferred into a bacterial cell, where it is cloned into many copies by replication of the recombinant molecule and by division of the bacterial cell.

Once large quantities of specific DNA fragments became available by cloning, they were used in many different ways: to isolate genes, to study their organization and expression, and to examine their nucleotide sequence and evolution.

As techniques became more refined, it became possible to clone larger and larger DNA fragments, paving the way to establish collections of clones that represented an organism's **genome**, which is the complete haploid content of DNA specific to that organism. Collections of clones that contain an entire genome are called genomic libraries. Genomic libraries are now available for hundreds of organisms.

Recombinant DNA technology has not only greatly accelerated the pace of research but has also given rise to the biotechnology industry, which has grown over the last 30 years to become a major contributor to the U.S. economy.

1.5
The Impact of Biotechnology Is Continually Expanding

Without arousing much notice in the United States, biotechnology has revolutionized many aspects of everyday life. Humans have used microorganisms, plants, and animals for thousands of years, but the development of recombinant DNA technology and associated techniques allows us to genetically modify organisms in new ways and use them or their products to enhance our lives. **Biotechnology** is the use of these modified organisms or their products. It is now in evidence at the supermarket; in doctors' offices; at drug stores, department stores, hospitals, and clinics; on farms and in orchards; in law enforcement and court-ordered child support; and even in industrial chemicals. There is a detailed discussion of the applications of biotechnology in Chapter 22, but for now, let's look at biotechnology's impact on just a small sampling of everyday examples.

Plants, Animals, and the Food Supply

The genetic modification of crop plants is one of the most rapidly expanding areas of biotechnology. Efforts have been focused on traits such as resistance to herbicides, insects, and viruses; enhancement of oil content; and delay of ripening (Table 1.1). Currently, over a dozen genetically modified crop plants have been approved for commercial use in the United States, with over 85 more being tested in field trials. Herbicide-resistant corn and soybeans were first planted in the mid-1990s, and now about 45 percent of the U.S. corn crop and 95 percent of the U.S. soybean crop is genetically modified. In addition, more than 60 percent of the canola crop and 85 percent of the cotton crop are grown from genetically modified strains. It is estimated that more than 75 percent of the processed food in the United States contains ingredients from genetically modified crop plants.

TABLE 1.1

Some Genetically Altered Traits in Crop Plants

Herbicide Resistance
Corn, soybeans, rice, cotton, sugarbeets, canola
Insect Resistance
Corn, cotton, potato
Virus Resistance
Potato, yellow squash, papaya
Nutritional Enhancement
Golden rice
Altered Oil Content
Soybeans, canola
Delayed Ripening
Tomato

This agricultural transformation has now become a source of controversy. Critics are concerned that the use of herbicide-resistant crop plants will lead to dependence on chemical weed management and may eventually result in the emergence of herbicide-resistant weeds. They also worry that traits in genetically engineered crops could be transferred to wild plants in a way that leads to irreversible changes in the ecosystem.

Biotechnology is also being used to enhance the nutritional value of crop plants. More than one-third of the world's population uses rice as a dietary staple, but most varieties of rice contain little or no vitamin A. Vitamin A deficiency causes more than 500,000 cases of blindness in children each year. A genetically engineered strain, called **golden rice**, has high levels of two compounds that the body converts to vitamin A. Golden rice should reduce the burden of this disease. Other crops, including wheat, corn, beans, and cassava, are also being modified to enhance nutritional value by increasing their vitamin and mineral content.

Livestock such as sheep and cattle have been commercially cloned for more than 30 years, mainly by a method called embryo splitting. In 1996, Dolly the sheep (Figure 1–11) was cloned by nuclear transfer, a method in which the nucleus of an adult cell is transferred into an egg that has had its nucleus removed. This nuclear transfer method makes it possible to produce large numbers of offspring with desirable traits. Cloning by nuclear transfer has many applications in agriculture, sports, and medicine. Some desirable traits, such as high milk production in cows or speed in race horses, do not appear until adulthood; rather than mating two adults and waiting to see if their offspring inherit the desired characteristics, animals known to have these traits can now be produced by cloning using an adult with a desirable trait. For medical applications, researchers have transferred human genes into animals—so-called **transgenic animals**—that as adults, produce human proteins in their milk. By selecting and cloning such animals, biopharmaceutical companies can produce a herd with uniformly high rates of protein production. Human proteins from transgenic animals are now being tested as drug treatments for diseases such as emphysema. If successful, these proteins will soon be commercially available.

Who Owns Transgenic Organisms?

Once produced, can a transgenic plant or animal be patented? The answer is yes. In 1980 the United States Supreme Court ruled that living organisms and individual genes can be patented, and in 1988 a strain of mice modified by recombinant DNA technology to be susceptible to cancer was patented for the first time (Figure 1–12). Since then, dozens of plants and animals have been patented. The ethics of patenting living organisms is a contentious issue. Supporters of patenting argue that without the ability to patent the products of research to recover their costs, biotechnology companies will not invest in large-scale research and development. They further argue that patents represent an incentive to develop new products because companies will reap the benefits of taking risks to bring new products to market. Critics argue that patents for organisms such as crop plants will concentrate ownership of food production in the hands of a small number of biotechnology companies, making farmers economically dependent on seeds and pesticides produced by these companies, and reducing the genetic diversity of crop plants as farmers discard local crops that might harbor important genes for resistance to pests and disease. Resolution of these and other issues raised

FIGURE 1–11   Dolly, a Finn Dorset sheep cloned from the genetic material of an adult mammary cell, shown next to her first-born lamb, Bonnie.

FIGURE 1–12   The first genetically altered organism to be patented, the *onc* strain of mouse, genetically engineered to be susceptible to many forms of cancer. These mice were created for studying cancer development and the design of new anticancer drugs.

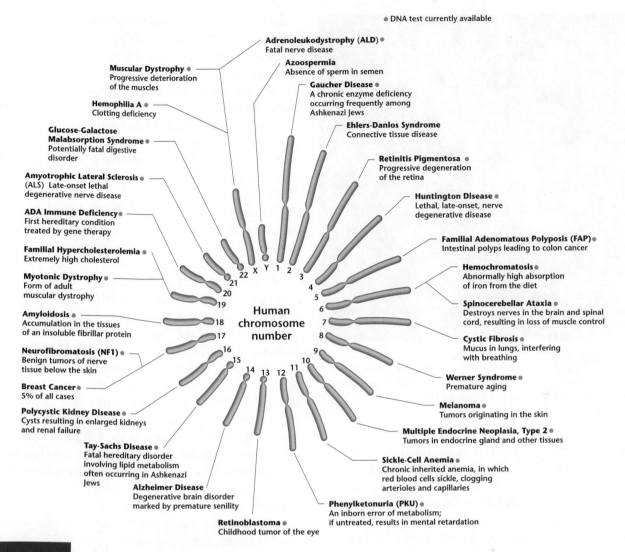

● DNA test currently available

Adrenoleukodystrophy (ALD) ●
Fatal nerve disease

Azoospermia
Absence of sperm in semen

Gaucher Disease ●
A chronic enzyme deficiency
occurring frequently among
Ashkenazi Jews

Ehlers-Danlos Syndrome
Connective tissue disease

Retinitis Pigmentosa ●
Progressive degeneration
of the retina

Huntington Disease ●
Lethal, late-onset, nerve
degenerative disease

Familial Adenomatous Polyposis (FAP) ●
Intestinal polyps leading to colon cancer

Hemochromatosis ●
Abnormally high absorption
of iron from the diet

Spinocerebellar Ataxia ●
Destroys nerves in the brain and spinal
cord, resulting in loss of muscle control

Cystic Fibrosis ●
Mucus in lungs, interfering
with breathing

Werner Syndrome ●
Premature aging

Melanoma ●
Tumors originating in the skin

Multiple Endocrine Neoplasia, Type 2 ●
Tumors in endocrine gland and other tissues

Sickle-Cell Anemia ●
Chronic inherited anemia, in which
red blood cells sickle, clogging
arterioles and capillaries

Phenylketonuria (PKU) ●
An inborn error of metabolism;
if untreated, results in mental retardation

Retinoblastoma ●
Childhood tumor of the eye

Alzheimer Disease
Degenerative brain disorder
marked by premature senility

Tay-Sachs Disease ●
Fatal hereditary disorder
involving lipid metabolism
often occurring in Ashkenazi
Jews

Polycystic Kidney Disease ●
Cysts resulting in enlarged kidneys
and renal failure

Breast Cancer ●
5% of all cases

Neurofibromatosis (NF1) ●
Benign tumors of nerve
tissue below the skin

Amyloidosis ●
Accumulation in the tissues
of an insoluble fibrillar protein

Myotonic Dystrophy ●
Form of adult
muscular dystrophy

Familial Hypercholesterolemia ●
Extremely high cholesterol

ADA Immune Deficiency ●
First hereditary condition
treated by gene therapy

Amyotrophic Lateral Sclerosis ●
(ALS) Late-onset lethal
degenerative nerve disease

**Glucose-Galactose
Malabsorption Syndrome** ●
Potentially fatal digestive
disorder

Hemophilia A ●
Clotting deficiency

Muscular Dystrophy ●
Progressive deterioration
of the muscles

Human
chromosome
number

FIGURE 1–13   Diagram of the human chromosome set, showing the location of some genes whose mutant forms cause hereditary diseases. Conditions that can be diagnosed using DNA analysis are indicated by a red dot.

by biotechnology and its uses will require public awareness and education, enlightened social policy, and carefully written legislation.

Biotechnology in Genetics and Medicine

Biotechnology in the form of genetic testing and gene therapy, already an important part of medicine, will be a leading force deciding the nature of medical practice in the twenty-first century. More than 10 million children or adults in the United States suffer from some form of genetic disorder, and every childbearing couple stands an approximately 3 percent risk of having a child with some form of genetic anomaly. The molecular basis for hundreds of genetic disorders is now known (Figure 1–13). Genes for sickle-cell anemia, cystic fibrosis, hemophilia, muscular dystrophy, phenylketonuria, and many other metabolic disorders have been cloned and are used for the prenatal detection of affected fetuses. In addition, tests are now available to inform parents of their status as "carriers" of a large number of inherited disorders. The combination

of genetic testing and genetic counseling gives couples objective information on which they can base decisions about childbearing. At present, genetic testing is available for several hundred inherited disorders, and this number will grow as more genes are identified, isolated, and cloned. The use of genetic testing and other technologies, including gene therapy, raises ethical concerns that have yet to be resolved.

Instead of testing one gene at a time to discover whether someone carries a mutation that can produce a disorder in his or her offspring, a new technology is now being used to screen whole genomes to determine an individual's risk of developing a genetic disorder or of having a child with a genetic disorder. This technology uses devices called **DNA microarrays**, or **DNA chips** (Figure 1–14). Each microarray can carry thousands of genes. In fact, microarrays carrying all human genes are now commercially available and are being used to test for gene expression in cancer cells as a step in developing therapies tailored to specific forms of malignancy. As the technology develops further, it will be

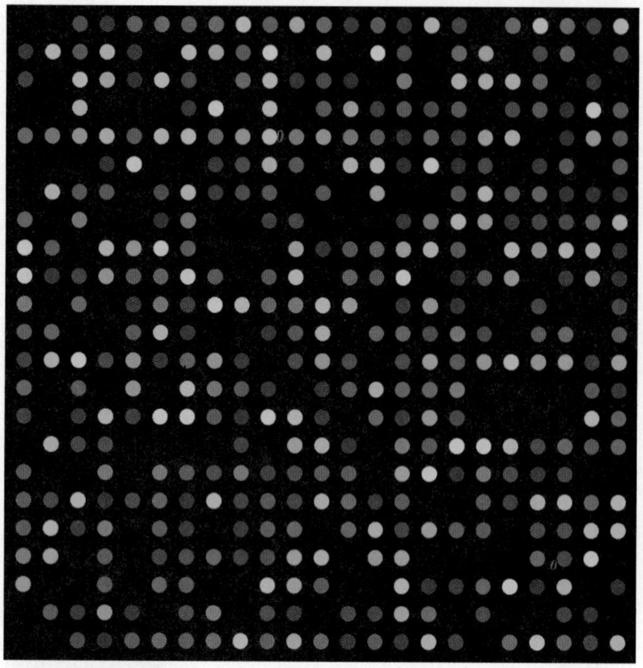

FIGURE 1–14   A portion of a DNA microarray. These arrays contain thousands of fields (the circles) to which DNA molecules are attached. Mounted on a microarray, DNA from an individual can be tested to detect mutant copies of genes.

possible to utilize microarrays to identify risks for genetic and environmental factors that may trigger disease.

In **gene therapy**, clinicians transfer normal genes into individuals affected with genetic disorders. Although many attempts at gene therapy initially appeared to be successful, therapeutic failures and patient deaths have slowed the development of this technology. New methods of gene transfer are expected to reduce these risks, and recently, gene therapy has enjoyed a number of successes.

1.6
Genomics, Proteomics, and Bioinformatics Are New and Expanding Fields

Once genomic libraries became available, scientists began to consider ways to spell out the nucleotide sequence of an organism's genome. Laboratories around the world initiated projects to sequence and analyze the genomes of different organisms, including those that cause human diseases. To date, the genomes of over 1000 organisms have been sequenced, and over 5000 additional genome projects are underway.

The Human Genome Project began in 1990 as an international, government-sponsored effort to sequence the human genome and the genomes of five of the model organisms used in genetics research (the importance of model organisms is discussed in the next section). Shortly thereafter,

a number of industry-sponsored genome projects also got underway. The first sequenced genome from a free-living organism, a bacterium, was published in 1995 by scientists at a biotechnology company.

In 2001, the publicly funded Human Genome Project and a private genome project sponsored by Celera Corporation reported the first draft of the human genome sequence, covering about 96 percent of the gene-containing portion of the genome. In 2003, the remaining portion of the gene-coding sequence was completed and published. The five model organisms whose genomes were also sequenced by the Human Genome Project are *Escherichia coli* (a bacterium), *Saccharomyces cerevisiae* (a yeast), *Caenorhabditis elegans* (a roundworm), *Drosophila melanogaster* (the fruit fly), and *Mus musculus* (the mouse).

As genome projects multiplied and more and more genome sequences were acquired, several new biological disciplines arose. One, called **genomics** (the study of genomes), sequences genomes and studies the structure, function, and evolution of genes and genomes. A second field, **proteomics**, is an outgrowth of genomics. Proteomics identifies the set of proteins present in a cell under a given set of conditions and additionally studies the post-translational modification of these proteins, their location within cells, and the protein–protein interactions occurring in the cell. To store, retrieve, and analyze the massive amount of data generated by genomics and proteomics, a specialized subfield of information technology called **bioinformatics** was created to develop hardware and software for processing, storing, and retrieving nucleotide and protein data. Consider that the human genome contains over 3 billion nucleotides, representing some 20,000 genes encoding tens of thousands of proteins, and you can appreciate the need for databases to store this information.

These new fields are drastically changing biology from a laboratory-based science to one that combines lab experiments with information technology. Geneticists and other biologists now use information in databases containing nucleic acid sequences, protein sequences, and gene-interaction networks to answer experimental questions in a matter of minutes instead of months and years. A feature called "Exploring Genomics," located at the end of this chapter and many other chapters in this textbook gives you the opportunity to explore these databases for yourself while completing an interactive genetics exercise.

1.7
Genetic Studies Rely on the Use of Model Organisms

After the rediscovery of Mendel's work in 1900, genetic research on a wide range of organisms confirmed that the principles of inheritance he described were of universal

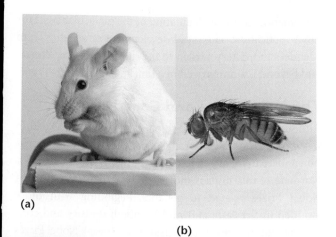

(a)

(b)

FIGURE 1–15   The first generation of model organisms in genetic research included (a) the mouse, *Mus musculus* and (b) the fruit fly, *Drosophila melanogaster*.

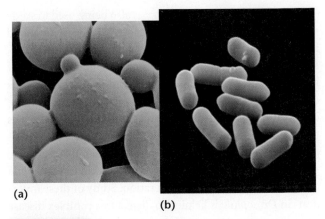

(a)

(b)

FIGURE 1–16   Microbes that have become model organisms for genetic studies include (a) the yeast *Saccharomyces cerevisiae* and (b) the bacterium *Escherichia coli*.

significance among plants and animals. Although work continued on the genetics of many different organisms, geneticists gradually came to focus particular attention on a small number of organisms, including the fruit fly (*Drosophila melanogaster*) and the mouse (*Mus musculus*) (Figure 1–15). This trend developed for two main reasons. First, it was clear that genetic mechanisms were the same in most organisms, and second, these species have several characteristics that make them especially suitable for genetic research. They are easy to breed, have relatively short life cycles, and their genetic analysis is fairly straightforward. Over time, researchers created a large catalog of mutant strains for these organisms, and the mutations were carefully studied, characterized, and mapped. Because of their well-characterized genetics and because of the ease with which they may be manipulated experimentally, these species are considered to be **model organisms**.

The Modern Set of Genetic Model Organisms

Gradually, geneticists added other species to their collection of model organisms: viruses (such as the T phage and lambda phage) and microorganisms (the bacterium *Escherichia coli* and the yeast *Saccharomyces cerevisiae*) (Figure 1–16).

More recently, additional species have been developed as model organisms. To study the nervous system and its role in behavior, the nematode *Caenorhabditis elegans* was chosen as a model system because it has a nervous system with only a few hundred cells. *Arabidopsis thaliana*, a small plant with a short life cycle, has become a model organism for the study of many other aspects of plant biology. The zebrafish, *Danio rerio*, is used to study vertebrate development; it is small, it reproduces rapidly, and its egg, embryo, and larvae are all transparent.

Model Organisms and Human Diseases

The development of recombinant DNA technology and the results of genome sequencing have confirmed that all life has a common origin. Because of this common origin, genes with similar functions in different organisms tend to be similar or identical in structure and nucleotide sequence. Much of what scientists learn by studying the genetics of other species can therefore be applied to humans and serve as the basis for understanding and treating human diseases. In addition, the ability to transfer genes between species has enabled scientists to develop models of human diseases in organisms ranging from bacteria to fungi, plants, and animals (Table 1.2).

The idea of studying a human disease such as colon cancer by using *E. coli* may strike you as strange, but the basic steps of DNA repair (a process that is defective in some forms of colon cancer) are the same in both organisms, and the gene involved (*mutL* in *E. coli* and *MLH1* in humans) is found in both organisms. More importantly, *E. coli* has the advantage of being easier to grow (the cells divide every 20 minutes), so that researchers can easily create and study new mutations in the bacterial *mutL* gene to figure out how it works. This knowledge may eventually lead to the development of drugs and other therapies to treat colon cancer in humans.

TABLE 1.2

Model Organisms Used to Study Human Diseases

Organism	Human Diseases
E. coli	Colon cancer and other cancers
S. cerevisiae	Cancer, Werner syndrome
D. melanogaster	Disorders of the nervous system, cancer
C. elegans	Diabetes
D. rerio	Cardiovascular disease
M. musculus	Lesch–Nyhan disease, cystic fibrosis, fragile-X syndrome, and many other diseases

The fruit fly, *D. melanogaster*, is also being used to study specific human diseases. Mutant genes have been identified in *D. melanogaster* that produce phenotypes with abnormalities of the nervous system, including abnormalities of brain structure, adult-onset degeneration of the nervous system, and visual defects such as retinal degeneration. The information from genome sequencing projects indicates that almost all these genes have human counterparts. As an example, genes involved in a complex human disease of the retina called retinitis pigmentosa are identical to *Drosophila* genes involved in retinal degeneration. Study of these mutations in *Drosophila* is helping to dissect this complex disease and identify the function of the genes involved.

As you read this textbook, you will encounter these model organisms again and again. Remember that each time you meet them they not only have a rich history in basic genetics research and technology, but are also at the forefront in the study of human genetic disorders and diseases. As discussed in the next section, however, we have yet to reach a consensus on how and when this technology is determined to be safe and ethically acceptable.

1.8

We Live in the Age of Genetics

Mendel described his decade-long project on inheritance in pea plants in an 1865 paper presented at a meeting of the Natural History Society of Brünn in Moravia. Just 100 years later, the 1965 Nobel Prize was awarded to François Jacob, André Lwoff, and Jacques Monod for their work on the molecular basis of gene regulation in bacteria. This time span encompassed the years leading up to the acceptance of Mendel's work, the discovery that genes are on chromosomes, the experiments that proved DNA encodes genetic

information, and the elucidation of the molecular basis for DNA replication. The rapid development of genetics from Mendel's monastery garden to the Human Genome Project and beyond is summarized in a timeline in Figure 1–17.

The Nobel Prize and Genetics

Although other scientific disciplines have also expanded in recent years, none has paralleled the explosion of information and excitement generated by the discoveries in genetics. Nowhere is this impact more apparent than in the list of Nobel Prizes related to genetics, beginning with those awarded in the early and mid-twentieth century and continuing into the present (see inside front cover). Nobel Prizes in the categories of Medicine or Physiology and Chemistry have been consistently awarded for work in genetics and associated fields. The first Nobel Prize awarded for such work was given to Thomas Morgan in 1933 for his research on the chromosome theory of inheritance. That award was followed by many others, including prizes for the discovery of genetic recombination, the relationship between genes and proteins, the structure of DNA, and the genetic code. In the current century, geneticists continue to be recognized for their impact on biology in the current millennium. The 2002 Prize for Medicine or Physiology was awarded for work on the genetic regulation of organ development and programmed cell death, while the 2006 prize was for the discovery that RNA molecules play an important role in regulating gene expression and for work on the molecular basis of eukaryotic transcription. The Nobel Prize was awarded in 2007 for the development of gene-targeting technology essential to the creation of knockout mice serving as animal models of human disease, and in 2009 for the discovery of how the DNA sequence making up telomeres (the structures that protect the ends of chromosomes) is replicated.

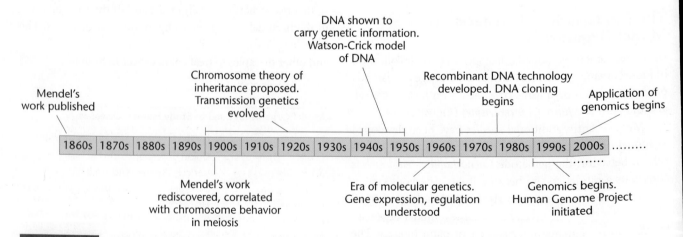

FIGURE 1–17 A timeline showing the development of genetics from Gregor Mendel's work on pea plants to the current era of genomics and its many applications in research, medicine, and society. Having a sense of the history of discovery in genetics should provide you with a useful framework as you proceed through this textbook.

Genetics and Society

Just as there has never been a more exciting time to study genetics, the impact of this discipline on society has never been more profound. Genetics and its applications in biotechnology are developing much faster than the social conventions, public policies, and laws required to regulate their use. As a society, we are grappling with a host of sensitive genetics-related issues, including concerns about prenatal testing, ownership of genes, access to and safety of gene therapy, and genetic privacy. By the time you finish this course, you will have seen more than enough evidence to convince you that the present is the Age of Genetics, and you will understand the need to think about and become a participant in the dialogue concerning genetic science and its uses.

GENETICS, TECHNOLOGY, AND SOCIETY

The Scientific and Ethical Implications of Modern Genetics

One of the special features of this text is the series of essays called *Genetics, Technology, and Society* that you will find at the conclusion of many chapters. These essays explore genetics-related topics that have an impact on the lives of each of us and on society in general.

Today, genetics touches all aspects of life, bringing rapid changes in medicine, agriculture, law, biotechnology, and the pharmaceutical industry. Physicians use hundreds of genetic tests to diagnose and predict the course of disease and to detect genetic defects *in utero*. Scientists employ DNA-based methods to trace the path of evolution taken by many species, including our own. Farmers grow disease-resistant and drought-resistant crops, and raise more productive farm animals, created by gene transfer techniques. Law enforcement agencies apply DNA profiling methods to paternity, rape, and murder investigations. The biotechnology industry itself generates over 700,000 jobs and $50 billion in revenue each year and doubles in size every decade.

Along with these rapidly changing gene-based technologies comes a challenging array of ethical dilemmas. Who owns and controls genetic information? Are gene-enhanced agricultural plants and animals safe for humans and the environment? How can we ensure that genomic technologies will be available to all and not just to the wealthy? What are the likely social consequences of the new reproductive technologies? It is a time when everyone needs to understand genetics in order to make complex personal and societal decisions.

Each *Genetics, Technology, and Society* essay is presented in a new interactive format. In the *Your Turn* section at the end of each essay, you will find several thought-provoking questions along with resources to help you explore each question. These questions are designed to act as entry points for individual investigations and as topics to explore in a classroom or group learning setting.

We hope that you will find the *Genetics, Technology, and Society* essays interesting, challenging, and an effective way to begin your lifelong studies in modern genetics. The first one appears at the end of Chapter 4. Below is a list of the titles of these essays and the chapters where they are found. Good reading!

Improving the Genetic Fate of Purebred Dogs	Chapter 4
From Cholera Genes to Edible Vaccines	Chapter 6
A Question of Gender: Sex Selection in Humans	Chapter 7
Down Syndrome, Prenatal Testing, and the New Eugenics	Chapter 8
Mitochondrial DNA and the Mystery of the Romanovs	Chapter 9
Telomeres: The Key to Immortality?	Chapter 11
Nucleic Acid–based Gene Silencing: Attacking the Messenger	Chapter 13
Quorum Sensing: Social Networking in the Bacterial World	Chapter 16
Stem Cell Wars	Chapter 18
Personal Genome Projects and the Race for the $1000 Genome	Chapter 22
The Green Revolution Revisited: Genetic Research with Rice	Chapter 23
Tracking Our Genetic Footprints Out of Africa	Chapter 25

EXPLORING GENOMICS

Internet Resources for Learning about Genomics, Bioinformatics, and Proteomics

MG *Study Area: Exploring Genomics*

Genomics is one of the most rapidly changing disciplines of genetics. As new information in this field accumulates at an astounding rate, keeping up with current developments in genomics, proteomics, bioinformatics, and other examples of the "omics" era of modern genetics is a challenging task indeed. As a result, geneticists, molecular biologists, and other scientists rely on online databases to share and compare new information.

The purpose of the "Exploring Genomics" feature, which appears at the end of many chapters, is to introduce you to a range of Internet databases that scientists around the world depend on for sharing, analyzing, organizing, comparing, and storing data from studies in genomics, proteomics, and related fields. We will explore this incredible pool of new information—comprising some of the best publicly available resources in the world—and show you how to use bioinformatics approaches to analyze the sequence and structural data that are found there. Each set of Exploring Genomics exercises will provide a basic introduction to one or more especially relevant or useful databases or programs and then guide you through exercises that use the databases to expand on or reinforce important concepts discussed in the chapter. The exercises are designed to help you learn to navigate the databases, but your explorations need not be limited to these experiences. Part of the fun of learning about genomics is exploring these outstanding databases on your own, so that you can get the latest information on any topic that interests you. Enjoy your explorations!

In this chapter, we discussed the importance of model organisms to both classic and modern experimental approaches in genetics. In our first set of Exploring Genomics exercises, we introduce you to a number of Internet sites that are excellent resources for finding up-to-date information on a wide range of completed and ongoing genomics projects involving model organisms.

Exercise I – Genome News Network

Since 1995, when scientists unveiled the genome for *Haemophilus influenzae*, making this bacterium the first organism to have its genome sequenced, the sequences for more than 500 organisms have been completed. Genome News Network is a site that provides access to basic information about recently completed genome sequences.

1. Visit the Genome News Network at www.genomenewsnetwork.org.

2. Click on the "Quick Guide to Sequenced Genomes" link. Scroll down the page; click on the appropriate links to find information about the genomes for *Anopheles gambiae*, *Lactococcus lactis*, and *Pan troglodytes*; and answer the following questions for each organism:

   a. Who sequenced this organism's genome, and in what year was it completed?

   b. What is the size of each organism's genome in base pairs?

   c. Approximately how many genes are in each genome?

   d. Briefly describe why geneticists are interested in studying this organism's genome.

Exercise II – Exploring the Genomes of Model Organisms

A tremendous amount of information is available about the genomes of the many model organisms that have played invaluable roles in advancing our understanding of genetics. Following are links to several sites that are excellent resources for you as you study genetics. Visit the site for links to information about your favorite model organism to learn more about its genome!

- Ensembl Genome Browser: www.ensembl.org/index.html. Outstanding site for genome information on many model organisms.

- FlyBase: flybase.org. Great database on *Drosophila* genes and genomes.

- Gold™ Genomes OnLine Database: www.genomesonline.org. Comprehensive access to completed and ongoing genome projects worldwide.

- Model Organisms for Biomedical Research: www.nih.gov/science/models. National Institutes of Health site with a wealth of resources on model organisms.

- Science Functional Genomics: www.sciencemag.org/feature/plus/sfg. Hosted by the journal *Science,* this is a good resource for information on model organism genomes and other current areas of genomics.

- The *Arabidopsis* Information Resource: www.arabidopsis.org. Genetic database for the model plant *Arabidopsis thaliana*.

- WormBase: www.wormbase.org. Genome database for the nematode roundworm *Caenorhabditis elegans*.

Summary Points

 For activities, animations, and review quizzes, go to the study area at www.masteringgenetics.com

1. Mendel's work on pea plants established the principles of gene transmission from parents to offspring that form the foundation for the science of genetics.
2. Genes and chromosomes are the fundamental units in the chromosomal theory of inheritance. This theory explains that inherited traits are controlled by genes located on chromosomes and shows how the transmission of genetic information maintains genetic continuity from generation to generation.
3. Molecular genetics—based on the central dogma that DNA is a template for making RNA, which encodes the order of amino acids in proteins—explains the phenomena described by Mendelian genetics, referred to as transmission genetics.
4. Recombinant DNA technology, a far-reaching methodology used in molecular genetics, allows genes from one organism to be spliced into vectors and cloned, producing many copies of specific DNA sequences.
5. Biotechnology has revolutionized agriculture, the pharmaceutical industry, and medicine. It has made possible the mass production of medically important gene products. Genetic testing allows detection of individuals with genetic disorders and those at risk of having affected children, and gene therapy offers hope for the treatment of serious genetic disorders.
6. Genomics, proteomics, and bioinformatics are new fields derived from recombinant DNA technology. These fields combine genetics with information technology and allow scientists to explore genome sequences, the structure and function of genes, the protein set within cells, and the evolution of genomes. The Human Genome Project is one example of genomics.
7. The use of model organisms has advanced the understanding of genetic mechanisms and, coupled with recombinant DNA technology, has produced models of human genetic diseases.
8. The effects of genetic technology on society are profound, and the development of policy and legislation to deal with issues derived from the use of this technology is lagging behind the resulting innovations.

Problems and Discussion Questions

 For instructor-assigned tutorials and problems, go to www.masteringgentics.com

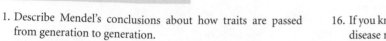

1. Describe Mendel's conclusions about how traits are passed from generation to generation.
2. What is the chromosome theory of inheritance, and how is it related to Mendel's findings?
3. Define genotype and phenotype, and describe how they are related.
4. What are alleles? Is it possible for more than two alleles of a gene to exist?
5. Contrast chromosomes and genes.
6. How is genetic information encoded in a DNA molecule?
7. Describe the central dogma of molecular genetics and how it serves as the basis of modern genetics.
8. How many different proteins, each composed of 5 amino acids, can be constructed using the 20 different amino acids found in proteins?
9. Outline the roles played by restriction enzymes and vectors in cloning DNA.
10. How has biotechnology changed agriculture in the United States?
11. DNA microarrays or DNA chips are commercially available diagnostic tools used to test for gene expression. How might such chips be applied in human medicine?
12. In what ways is the discipline called genomics similar to proteomics? How are they different? What is meant by the term *bioinformatics*?
13. Summarize the arguments for and against patenting genetically modified organisms.
14. We all carry 20,000 to 25,000 genes in our genome. So far, patents have been issued for more than 6000 of these genes. Do you think that companies or individuals should be able to patent human genes? Why or why not?
15. How have model organisms advanced our knowledge of genes that control human diseases?
16. If you knew that an untreatable devastating late-onset inherited disease runs in your family (in other words, a disease that does not appear until later in life) and you could be tested for it at the age of 20, would you want to know whether you will develop that disease? Would your answer be likely to change when you reach age 40?
17. The Age of Genetics was created by remarkable advances in the use of biotechnology to manipulate plant and animal genomes. Given that the world population has topped 6 billion and is expected to reach 9.2 billion by 2050, some scientists have proposed that only the worldwide introduction of genetically modified (GM) foods will increase crop yields enough to meet future nutritional demands. Pest resistance, herbicide, cold, drought, and salinity tolerance, along with increased nutrition, are seen as positive attributes of GM foods. However, others caution that unintended harm to other organisms, reduced effectiveness to pesticides, gene transfer to nontarget species, allergenicity, and as yet unknown effects on human health are potential concerns regarding GM foods. If you were in a position to control the introduction of a GM primary food product (rice, for example), what criteria would you establish before allowing such introduction?
18. The BIO (Biotechnology Industry Organization) meeting held in Philadelphia in June 2005 brought together worldwide leaders from the biotechnology and pharmaceutical industries. Concurrently, BioDemocracy 2005, a group composed of people seeking to highlight hazards from widespread applications of biotechnology, also met in Philadelphia. The benefits of biotechnology are outlined in your text. Predict some of the risks that were no doubt discussed at the BioDemocracy meeting.

Chromosomes in the prometaphase stage of mitosis, derived from a cell in the flower of *Haemanthus*.

2

Mitosis and Meiosis

CHAPTER CONCEPTS

- Genetic continuity between generations of cells and between generations of sexually reproducing organisms is maintained through the processes of mitosis and meiosis, respectively.

- Diploid eukaryotic cells contain their genetic information in pairs of homologous chromosomes, with one member of each pair being derived from the maternal parent and one from the paternal parent.

- Mitosis provides a mechanism by which chromosomes, having been duplicated, are distributed into progeny cells during cell reproduction.

- Mitosis converts a diploid cell into two diploid daughter cells.

- The process of meiosis distributes one member of each homologous pair of chromosomes into each gamete or spore, thus reducing the diploid chromosome number to the haploid chromosome number.

- Meiosis generates genetic variability by distributing various combinations of maternal and paternal members of each homologous pair of chromosomes into gametes or spores.

- During the stages of mitosis and meiosis, the genetic material is condensed into discrete structures called chromosomes.

E very living thing contains a substance described as the genetic material. Except in certain viruses, this material is composed of the nucleic acid, DNA. DNA has an underlying linear structure possessing segments called genes, the products of which direct the metabolic activities of cells. An organism's DNA, with its arrays of genes, is organized into structures called **chromosomes**, which serve as vehicles for transmitting genetic information. The manner in which chromosomes are transmitted from one generation of cells to the next and from organisms to their descendants must be exceedingly precise. In this chapter we consider exactly how genetic continuity is maintained between cells and organisms.

Two major processes are involved in the genetic continuity of nucleated cells: **mitosis** and **meiosis**. Although the mechanisms of the two processes are similar in many ways, the outcomes are quite different. Mitosis leads to the production of two cells, each with the same number of chromosomes as the parent cell. In contrast, meiosis reduces the genetic content and the number of chromosomes by precisely half. This reduction is essential if sexual reproduction is to occur without doubling the amount of genetic material in each new generation. Strictly speaking, mitosis is that portion of the cell cycle during which the hereditary components are equally partitioned into daughter cells. Meiosis is part of a special type of cell division that leads to the production of sex cells: **gametes** or **spores**. This process is an essential step in the transmission of genetic information from an organism to its offspring.

Normally, chromosomes are visible only during mitosis and meiosis. When cells are not undergoing division, the genetic material making up chromosomes unfolds and uncoils into a diffuse network within the nucleus, generally referred to as **chromatin**. Before describing mitosis and meiosis, we will briefly review the structure of cells, emphasizing components that are of particular significance to genetic function. We will also compare the structural differences between the prokaryotic (nonnucleated) cells of bacteria and the eukaryotic cells of higher organisms. We then devote the remainder of the chapter to the behavior of chromosomes during cell division.

2.1

Cell Structure Is Closely Tied to Genetic Function

Before 1940, our knowledge of cell structure was limited to what we could see with the light microscope. Around 1940, the transmission electron microscope was in its early stages of development, and by 1950, many details of cell ultrastructure had emerged. Under the electron microscope, cells were seen as highly varied, highly organized structures whose form and function are dependent on specific genetic expression by each cell type. A new world of whorled membranes, organelles, microtubules, granules, and filaments was revealed. These discoveries revolutionized thinking in the entire field of biology. Many cell components, such as the nucleolus, ribosome, and centriole, are involved directly or indirectly with genetic processes. Other components— the mitochondria and chloroplasts—contain their own unique genetic information. Here, we will focus primarily on those aspects of cell structure that relate to genetic study. The generalized animal cell shown in Figure 2–1 illustrates most of the structures we will discuss.

All cells are surrounded by a **plasma membrane**, an outer covering that defines the cell boundary and delimits the cell from its immediate external environment. This membrane is not passive but instead actively controls the movement of materials into and out of the cell. In addition to this membrane, plant cells have an outer covering called the **cell wall** whose major component is a polysaccharide called *cellulose*.

Many, if not most, animal cells have a covering over the plasma membrane, referred to as the **glycocalyx**, or **cell coat**. Consisting of glycoproteins and polysaccharides, this covering has a chemical composition that differs from comparable structures in either plants or bacteria. The glycocalyx, among other functions, provides biochemical identity at the surface of cells, and the components of the coat that establish cellular identity are under genetic control. For example, various cell-identity markers that you may have heard of—the AB, Rh, and MN antigens—are found on the surface of red blood cells, among other cell types. On the surface of other cells, histocompatibility antigens, which elicit an immune response during tissue and organ transplants, are present. Various **receptor molecules** are also found on the surfaces of cells. These molecules act as recognition sites that transfer specific chemical signals across the cell membrane into the cell.

Living organisms are categorized into two major groups depending on whether or not their cells contain a nucleus. The presence of a nucleus and other membranous organelles is the defining characteristic of **eukaryotic organisms**. The **nucleus** in eukaryotic cells is a membrane-bound structure that houses the genetic material, DNA, which is complexed with an array of acidic and basic proteins into thin fibers. During nondivisional phases of the cell cycle, the fibers are uncoiled and dispersed into chromatin (as mentioned above). During mitosis and meiosis, chromatin fibers coil and condense into chromosomes. Also present in the nucleus is the **nucleolus**, an amorphous component where ribosomal RNA (rRNA) is synthesized and where the initial stages of ribosomal assembly occur. The portions of DNA that encode rRNA are collectively referred to as the **nucleolus organizer region**, or the **NOR**.

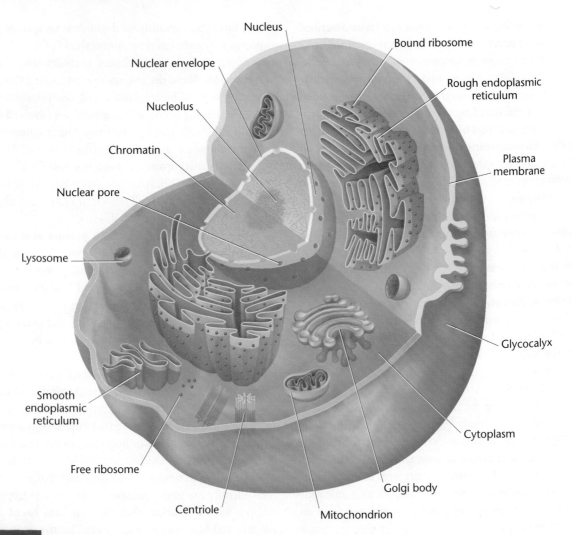

Nucleus

Nuclear envelope

Nucleolus

Chromatin

Nuclear pore

Lysosome

Smooth endoplasmic reticulum

Free ribosome

Centriole

Mitochondrion

Golgi body

Cytoplasm

Glycocalyx

Plasma membrane

Rough endoplasmic reticulum

Bound ribosome

FIGURE 2–1   A generalized animal cell. The cellular components discussed in the text are emphasized here.

Prokaryotic organisms, of which there are two major groups, lack a nuclear envelope and membranous organelles. For the purpose of our brief discussion here, we will consider the *eubacteria,* the other group being the more ancient bacteria referred to as *archaea.* In eubacteria, such as *Escherichia coli,* the genetic material is present as a long, circular DNA molecule that is compacted into an unenclosed region called the **nucleoid.** Part of the DNA may be attached to the cell membrane, but in general the nucleoid extends through a large part of the cell. Although the DNA is compacted, it does not undergo the extensive coiling characteristic of the stages of mitosis, during which the chromosomes of eukaryotes become visible. Nor is the DNA associated as extensively with proteins as is eukaryotic DNA. Figure 2–2, which shows two bacteria forming by cell division, illustrates the nucleoid regions containing the bacterial chromosomes. Prokaryotic cells do not have a distinct nucleolus but do contain genes that specify rRNA molecules.

The remainder of the eukaryotic cell within the plasma membrane, excluding the nucleus, is referred to as **cytoplasm** and includes a variety of extranuclear cellular organelles. In the

cytoplasm, a nonparticulate, colloidal material referred to as the **cytosol** surrounds and encompasses the cellular organelles. The cytoplasm also includes an extensive system of tubules and

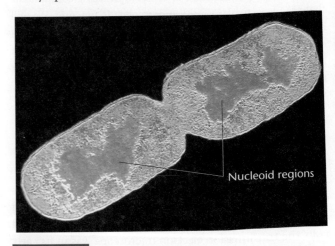

Nucleoid regions

FIGURE 2–2   Color-enhanced electron micrograph of *E. coli* undergoing cell division. Particularly prominent are the two chromosomal areas (shown in red), called nucleoids, that have been partitioned into the daughter cells.

filaments, comprising the cytoskeleton, which provides a lattice of support structures within the cell. Consisting primarily of **microtubules**, which are made of the protein **tubulin**, and **microfilaments**, which derive from the protein **actin**, this structural framework maintains cell shape, facilitates cell mobility, and anchors the various organelles.

One organelle, the membranous **endoplasmic reticulum (ER)**, compartmentalizes the cytoplasm, greatly increasing the surface area available for biochemical synthesis. The ER appears smooth in places where it serves as the site for synthesizing fatty acids and phospholipids; in other places, it appears rough because it is studded with ribosomes. **Ribosomes** serve as sites where genetic information contained in messenger RNA (mRNA) is translated into proteins.

Three other cytoplasmic structures are very important in the eukaryotic cell's activities: mitochondria, chloroplasts, and centrioles. **Mitochondria** are found in most eukaryotes, including both animal and plant cells, and are the sites of the oxidative phases of cell respiration. These chemical reactions generate large amounts of the energy-rich molecule adenosine triphosphate (ATP). **Chloroplasts**, which are found in plants, algae, and some protozoans, are associated with photosynthesis, the major energy-trapping process on Earth. Both mitochondria and chloroplasts contain DNA in a form distinct from that found in the nucleus. They are able to duplicate themselves and transcribe and translate their own genetic information. It is interesting to note that the genetic machinery of mitochondria and chloroplasts closely resembles that of prokaryotic cells. This and other observations have led to the proposal that these organelles were once primitive free-living organisms that established symbiotic relationships with primitive eukaryotic cells. This theory concerning the evolutionary origin of these organelles is called the **endosymbiont hypothesis**.

Animal cells and some plant cells also contain a pair of complex structures called **centrioles**. These cytoplasmic bodies, located in a specialized region called the **centrosome**, are associated with the organization of spindle fibers that function in mitosis and meiosis. In some organisms, the centriole is derived from another structure, the basal body, which is associated with the formation of cilia and flagella (hair-like and whip-like structures for propelling cells or moving materials). Over the years, many reports have suggested that centrioles and basal bodies contain DNA, which could be involved in the replication of these structures. This idea is still being investigated.

The organization of **spindle fibers** by the centrioles occurs during the early phases of mitosis and meiosis. These fibers play an important role in the movement of chromosomes as they separate during cell division. They are composed of arrays of microtubules consisting of polymers of

Chromosomes Exist in Homologous Pairs in Diploid Organisms

As we discuss the processes of mitosis and meiosis, it is important that you understand the concept of homologous chromosomes. Such an understanding will also be of critical importance in our future discussions of Mendelian genetics. Chromosomes are most easily visualized during mitosis. When they are examined carefully, distinctive lengths and shapes are apparent. Each chromosome contains a constricted region called the **centromere**, whose location establishes the general appearance of each chromosome. Figure 2–3 shows chromosomes with centromere placements at different distances along their length. Extending from either side of the centromere are the arms of the chromosome. Depending on the position of the centromere, different arm ratios are produced. As Figure 2–3 illustrates, chromosomes are classified as **metacentric**, **submetacentric**, **acrocentric**, or **telocentric** on the basis of the centromere location. The shorter arm, by convention, is shown above the centromere and is called the **p arm** (p, for "petite"). The longer arm is shown below the centromere and is called the **q arm** (q because it is the next letter in the alphabet).

In the study of mitosis, several other observations are of particular relevance. First, all somatic cells derived from members of the same species contain an identical number of chromosomes. In most cases, this represents the **diploid number (2n)**, whose meaning will become clearer below. When the lengths and centromere placements of all such chromosomes are examined, a second general feature is apparent. With the exception of sex chromosomes, they exist in pairs with regard to these two properties, and the members of each pair are called **homologous chromosomes**. So, for each chromosome exhibiting a specific length and centromere placement, another exists with identical features.

There are exceptions to this rule. Many bacteria and viruses have but one chromosome, and organisms such as yeasts and molds, and certain plants such as bryophytes (mosses), spend the predominant phase of their life cycle in the haploid stage. That is, they contain only one member of each homologous pair of chromosomes during most of their lives.

Figure 2–4 illustrates the physical appearance of different pairs of homologous chromosomes. There, the human mitotic chromosomes have been photographed, cut out of the print, and matched up, creating a display called a **karyotype**. As you can see, humans have a 2n number of 46 chromosomes, which on close examination exhibit a diversity of sizes and centromere placements. Note also that each of

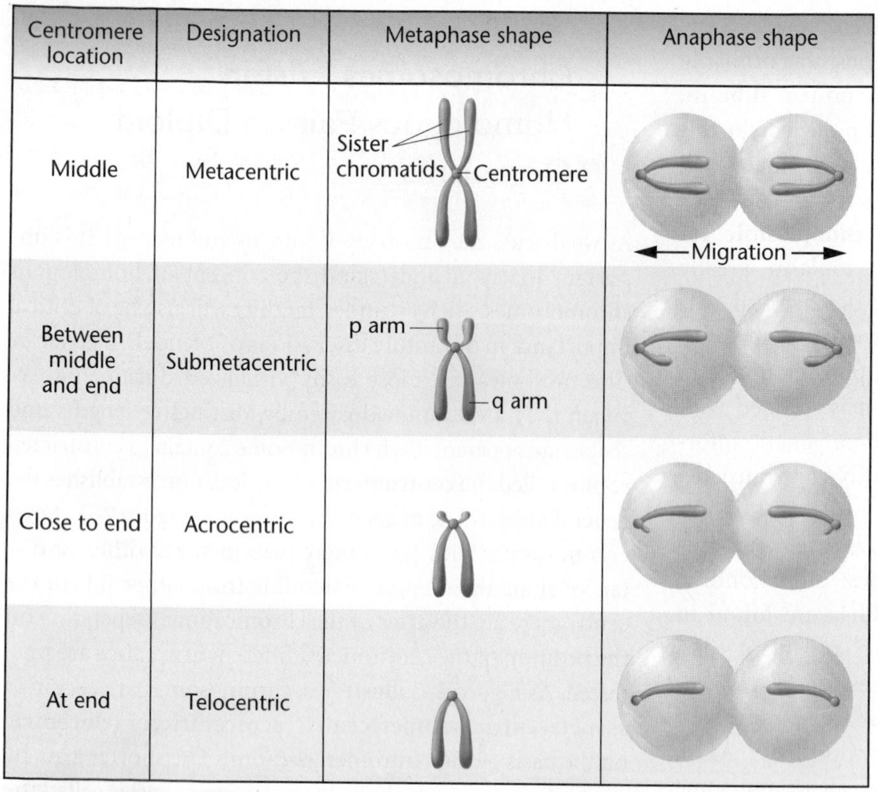

Centromere location	Designation	Metaphase shape	Anaphase shape
Middle	Metacentric	Sister chromatids — Centromere	◄— Migration —►
Between middle and end	Submetacentric	p arm — q arm	
Close to end	Acrocentric		
At end	Telocentric		

FIGURE 2–3   Centromere locations and the chromosome designations that are based on them. Note that the shape of the chromosome during anaphase is determined by the position of the centromere during metaphase.

The **haploid number** (*n*) of chromosomes is equal to one-half the diploid number. Collectively, the genetic information contained in a haploid set of chromosomes constitutes the **genome** of the species. This, of course, includes copies of all genes as well as a large amount of noncoding DNA. The examples listed in Table 2.1 demonstrate the wide range of *n* values found in plants and animals.

Homologous chromosomes have important genetic similarities. They contain identical gene sites along their lengths; each site is called a **locus** (pl. loci). Thus, they are identical in the traits that they influence and in their genetic potential. In sexually reproducing organisms, one member of each pair is derived from the maternal parent (through the ovum) and the other member is derived from the paternal parent (through the sperm). Therefore, each diploid organism contains two copies of each gene as a consequence of **biparental inheritance**, inheritance from two parents. As we shall see in the chapters on transmission genetics, the members of each pair of genes, while influencing the same characteristic or trait, need not be identical. In a population of members of the same species, many different alternative forms of the same gene, called **alleles**, can exist.

structure consisting of two parallel *sister chromatids* connected by a common centromere. Had these chromosomes been allowed to continue dividing, the sister chromatids, which are replicas of one another, would have separated into the two new cells as division continued.

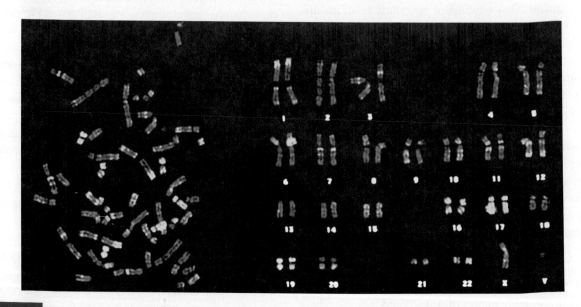

FIGURE 2–4   A metaphase preparation of chromosomes derived from a dividing cell of a human male (left), and the karyotype derived from the metaphase preparation (right). All but the X and Y chromosomes are present in homologous pairs. Each chromosome is clearly a double structure consisting of a pair of sister chromatids joined by a common centromere.

TABLE 2.1

The Haploid Number of Chromosomes for a Variety of Organisms

Common Name	Scientific Name	Haploid Number
Black bread mold	Aspergillus nidulans	8
Broad bean	Vicia faba	6
Cat	Felis domesticus	19
Cattle	Bos taurus	30
Chicken	Gallus domesticus	39
Chimpanzee	Pan troglodytes	24
Corn	Zea mays	10
Cotton	Gossypium hirsutum	26
Dog	Canis familiaris	39
Evening primrose	Oenothera biennis	7
Frog	Rana pipiens	13
Fruit fly	Drosophila melanogaster	4
Garden onion	Allium cepa	8
Garden pea	Pisum sativum	7
Grasshopper	Melanoplus differentialis	12
Green alga	Chlamydomonas reinhardtii	18
Horse	Equus caballus	32
House fly	Musca domestica	6
House mouse	Mus musculus	20
Human	Homo sapiens	23
Jimson weed	Datura stramonium	12
Mosquito	Culex pipiens	3
Mustard plant	Arabidopsis thaliana	5
Pink bread mold	Neurospora crassa	7
Potato	Solanum tuberosum	24
Rhesus monkey	Macaca mulatta	21
Roundworm	Caenorhabditis elegans	6
Silkworm	Bombyx mori	28
Slime mold	Dictyostelium discoideum	7
Snapdragon	Antirrhinum majus	8
Tobacco	Nicotiana tabacum	24
Tomato	Lycopersicon esculentum	12
Water fly	Nymphaea alba	80
Wheat	Triticum aestivum	21
Yeast	Saccharomyces cerevisiae	16
Zebrafish	Danio rerio	25

The concepts of haploid number, diploid number, and homologous chromosomes are important for understanding the process of meiosis. During the formation of gametes or spores, meiosis converts the diploid number of chromosomes to the haploid number. As a result, haploid gametes or spores contain precisely one member of each homologous pair of chromosomes—that is, one complete haploid set. Following fusion of two gametes at fertilization, the diploid number is reestablished; that is, the zygote contains two complete haploid sets of chromosomes. The constancy of genetic material is thus maintained from generation to generation.

There is one important exception to the concept of homologous pairs of chromosomes. In many species, one pair, consisting of the **sex-determining chromosomes**, is often not homologous in size, centromere placement, arm ratio, or genetic content. For example, in humans, while females carry two homologous X chromosomes, males carry one Y chromosome in addition to one X chromosome (Figure 2–4). These X and Y chromosomes are not strictly homologous. The Y is considerably smaller and lacks most of the gene loci contained on the X. Nevertheless, they contain homologous regions and behave as homologs in meiosis so that gametes produced by males receive either one X or one Y chromosome.

2.3

Mitosis Partitions Chromosomes into Dividing Cells

The process of mitosis is critical to all eukaryotic organisms. In some single-celled organisms, such as protozoans and some fungi and algae, mitosis (as a part of cell division) provides the basis for asexual reproduction. Multicellular diploid organisms begin life as single-celled fertilized eggs called **zygotes**. The mitotic activity of the zygote and the subsequent daughter cells is the foundation for the development and growth of the organism. In adult organisms, mitotic activity is the basis for wound healing and other forms of cell replacement in certain tissues. For example, the epidermal cells of the skin and the intestinal lining of humans are continuously sloughed off and replaced. Cell division also results in the continuous production of reticulocytes that eventually shed their nuclei and replenish the supply of red blood cells in vertebrates. In abnormal situations, somatic cells may lose control of cell division, and form a tumor.

The genetic material is partitioned into daughter cells during nuclear division, or **karyokinesis**. This process is quite complex and requires great precision. The chromosomes must first be exactly replicated and then accurately partitioned. The end result is the production of two daughter nuclei, each with a chromosome composition identical to that of the parent cell.

Karyokinesis is followed by cytoplasmic division, or **cytokinesis**. This less complex process requires a mechanism that partitions the volume into two parts, then encloses each new cell in a distinct plasma membrane. As the cytoplasm is reconstituted, organelles either replicate themselves, arise from existing membrane structures, or are synthesized *de novo* (anew) in each cell.

Following cell division, the initial size of each new daughter cell is approximately one-half the size of the parent cell. However, the nucleus of each new cell is not appreciably smaller than the nucleus of the original cell. Quantitative measurements of DNA confirm that there is an amount of genetic material in the daughter nuclei equivalent to that in the parent cell.

Interphase and the Cell Cycle

Many cells undergo a continuous alternation between division and nondivision. The events that occur from the completion of one division until the completion of the next division constitute the **cell cycle** (Figure 2–5). We will consider **interphase**, the initial stage of the cell cycle, as the interval between divisions. It was once thought that the biochemical activity during interphase was devoted solely to the cell's growth and its normal function. However, we now know that another biochemical step critical to the ensuing mitosis occurs during interphase: *the replication of the DNA of each chromosome.* This period, during which DNA is synthesized, occurs before the cell enters mitosis and is called the **S phase**. The initiation and completion of synthesis can be detected by monitoring the incorporation of radioactive precursors into DNA.

Investigations of this nature demonstrate two periods during interphase when no DNA synthesis occurs, one before and one after the S phase. These are designated **G1 (gap I)** and **G2 (gap II)**, respectively. During both of these intervals, as well as during S, intensive metabolic activity, cell growth, and cell differentiation are evident. By the end of G2, the volume of the cell has roughly doubled, DNA has been replicated, and mitosis (M) is initiated. Following mitosis, continuously dividing cells then repeat this cycle (G1, S, G2, M) over and over, as shown in Figure 2–5.

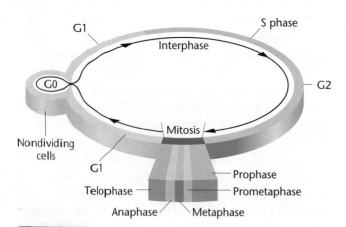

FIGURE 2-5   The stages comprising an arbitrary cell cycle. Following mitosis, cells enter the G1 stage of interphase, initiating a new cycle. Cells may become nondividing (G0) or continue through G1, where they become committed to begin DNA synthesis (S) and complete the cycle (G2 and mitosis). Following mitosis, two daughter cells are produced, and the cycle begins anew for both of them.

Interphase			Mitosis
G1	S	G2	M
5	7	3	1

Hours

Pro	Met	Ana	Tel
36	3	3	18

Minutes

FIGURE 2–6   The time spent in each interval of one complete cell cycle of a human cell in culture. Times vary according to cell types and conditions.

Much is known about the cell cycle based on *in vitro* (literally, "in glass") studies. When grown in culture, many cell types in different organisms traverse the complete cycle in about 16 hours. The actual process of mitosis occupies only a small part of the overall cycle, often less than an hour. The lengths of the S and G2 phases of interphase are fairly consistent in different cell types. Most variation is seen in the length of time spent in the G1 stage. Figure 2–6 shows the relative length of these intervals as well as the length of the stages of mitosis in a human cell in culture.

G1 is of great interest in the study of cell proliferation and its control. At a point during G1, all cells follow one of two paths. They either withdraw from the cycle, become quiescent, and enter the **G0 stage** (see Figure 2–5), or they become committed to proceed through G1, initiating DNA synthesis, and completing the cycle. Cells that enter G0 remain viable and metabolically active but are not proliferative. Cancer cells apparently avoid entering G0 or pass through it very quickly. Other cells enter G0 and never reenter the cell cycle. Still other cells in G0 can be stimulated to return to G1 and thereby reenter the cell cycle.

Cytologically, interphase is characterized by the absence of visible chromosomes. Instead, the nucleus is filled with chromatin fibers that are formed as the chromosomes uncoil and disperse after the previous mitosis [Figure 2–7(a)]. Once G1, S, and G2 are completed, mitosis is initiated. Mitosis is a dynamic period of vigorous and continual activity. For discussion purposes, the entire process is subdivided into discrete stages, and specific events are assigned to each one. These stages, in order of occurrence, are prophase, prometaphase, metaphase, anaphase, and telophase. They are diagrammed with corresponding photomicrographs in Figure 2–7.

Prophase

Often, over half of mitosis is spent in **prophase** [Figure 2–7(b)], a stage characterized by several significant occurrences. One of the early events in prophase of all animal cells is the

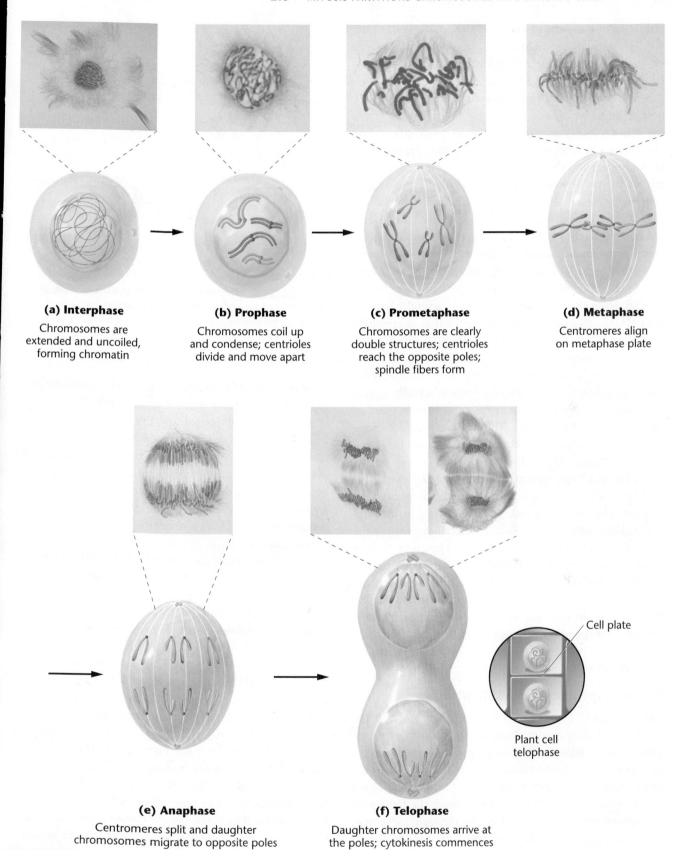

(a) Interphase

Chromosomes are extended and uncoiled, forming chromatin

(b) Prophase

Chromosomes coil up and condense; centrioles divide and move apart

(c) Prometaphase

Chromosomes are clearly double structures; centrioles reach the opposite poles; spindle fibers form

(d) Metaphase

Centromeres align on metaphase plate

Cell plate

Plant cell telophase

(e) Anaphase

Centromeres split and daughter chromosomes migrate to opposite poles

(f) Telophase

Daughter chromosomes arrive at the poles; cytokinesis commences

FIGURE 2–7 Drawings depicting mitosis in an animal cell with a diploid number of 4. The events occurring in each stage are described in the text. Of the two homologous pairs of chromosomes, one pair consists of longer, metacentric members and the other of shorter, submetacentric members. The maternal chromosome and the paternal chromosome of each pair are shown in different

migration of two pairs of centrioles to opposite ends of the cell. These structures are found just outside the nuclear envelope in an area of differentiated cytoplasm called the centrosome (introduced in Section 2.1). It is believed that each pair of centrioles consists of one mature unit and a smaller, newly formed centriole.

The centrioles migrate to establish poles at opposite ends of the cell. After migrating, the centrioles are responsible for organizing cytoplasmic microtubules into the spindle fibers that run between these poles, creating an axis along which chromosomal separation occurs. Interestingly, the cells of most plants (there are a few exceptions), fungi, and certain algae seem to lack centrioles. Spindle fibers are nevertheless apparent during mitosis. Therefore, centrioles are not universally responsible for the organization of spindle fibers.

As the centrioles migrate, the nuclear envelope begins to break down and gradually disappears. In a similar fashion, the nucleolus disintegrates within the nucleus. While these events are taking place, the diffuse chromatin fibers have begun to condense, until distinct threadlike structures, the chromosomes, become visible. It becomes apparent near the end of prophase that each chromosome is actually a double structure split longitudinally except at a single point of constriction, the centromere. The two parts of each chromosome are called **sister chromatids** because the DNA contained in each of them is genetically identical, having formed from a single replicative event. Sister chromatids are held together by a multi-subunit protein complex called **cohesin**. This molecular complex is originally formed between them during the S phase of the cell cycle when the DNA of each chromosome is replicated. Thus, even though we cannot see chromatids in interphase because the chromatin is uncoiled and dispersed in the nucleus, the chromosomes are already double structures, which becomes apparent in late prophase. In humans, with a diploid number of 46, a cytological preparation of late prophase reveals 46 chromosomes randomly distributed in the area formerly occupied by the nucleus.

Prometaphase and Metaphase

The distinguishing event of the two ensuing stages is the migration of every chromosome, led by its centromeric region, to the equatorial plane. The equatorial plane, also referred to as the *metaphase plate*, is the midline region of the cell, a plane that lies perpendicular to the axis established by the spindle fibers. In some descriptions, the term **prometaphase** refers to the period of chromosome movement [Figure 2–7(c)], and the term **metaphase** is applied strictly to the chromosome configuration following migration.

Migration is made possible by the binding of spindle fibers to the chromosome's **kinetochore**, an assembly of multilayered plates of proteins associated with the centromere. This structure forms on opposite sides of each paired

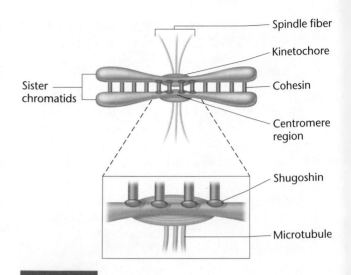

FIGURE 2–8   The depiction of the alignment, pairing, and disjunction of sister chromatids during mitosis, involving the molecular complexes cohesin and shugoshin and the enzyme separase.

centromere, in intimate association with the two sister chromatids. Once properly attached to the spindle fibers, cohesin is degraded by an enzyme, appropriately named *separase,* and the sister chromatid arms disjoin, except at the centromere region. A unique protein family called **shugoshin** (from the Japanese meaning guardian spirit) protects cohesin from being degraded by separase at the centromeric regions. The involvement of the cohesin and shugoshin complexes with a pair of sister chromatids during mitosis is depicted in Figure 2–8.

We know a great deal about spindle fibers. They consist of microtubules, which themselves consist of molecular subunits of the protein tubulin (we noted earlier that tubulin-derived microtubules also make up part of the cytoskeleton). Microtubules seem to originate and "grow" out of the two centrosome regions (which contain the centrioles) at opposite poles of the cell. They are dynamic structures that lengthen and shorten as a result of the addition or loss of polarized tubulin subunits. The microtubules most directly responsible for chromosome migration make contact with, and adhere to, kinetochores as they grow from the centrosome region. They are referred to as **kinetochore microtubules** and have one end near the centrosome region (at one of the poles of the cell) and the other end anchored to the kinetochore. The number of microtubules that bind to the kinetochore varies greatly between organisms. Yeast (*Saccharomyces*) has only a single microtubule bound to each plate-like structure of the kinetochore. Mitotic cells of mammals, at the other extreme, reveal 30 to 40 microtubules bound to each portion of the kinetochore.

At the completion of metaphase, each centromere is aligned at the metaphase plate with the chromosome arms extending outward in a random array. This configuration is shown in Figure 2–7(**d**).

Anaphase

Events critical to chromosome distribution during mitosis occur during **anaphase**, the shortest stage of mitosis. During this phase, sister chromatids of each chromosome, held together only at their centromere regions, *disjoin* (separate) from one another—an event described as **disjunction**—and are pulled to opposite ends of the cell. For complete disjunction to occur: (1), shugoshin must be degraded, reversing its protective role; (2), the cohesin complex holding the centromere region of each sister chromosome is then cleaved by separase; and (3) sister chromatids of each chromosome are pulled toward the opposite poles of the cell (Figure 2–8). As these events proceed, each migrating chromatid is now referred to as a **daughter chromosome.**

Movement of daughter chromosomes to the opposite poles of the cell is dependent on the centromere–spindle fiber attachment. Recent investigations reveal that chromosome migration results from the activity of a series of specific molecules called motor proteins found at several locations within the dividing cell. These proteins, described as **molecular motors**, use the energy generated by the hydrolysis of ATP. Their effect on the activity of microtubules serves ultimately to shorten the spindle fibers, drawing the chromosomes to opposite ends of the cell. The centromeres of each chromosome *appear* to lead the way during migration, with the chromosome arms trailing behind. Several models have been proposed to account for the shortening of spindle fibers. They share in common the selective removal of tubulin subunits at the ends of the spindle fibers. The removal process is accomplished by the molecular motor proteins described above.

The location of the centromere determines the shape of the chromosome during separation, as you saw in Figure 2–3. The steps that occur during anaphase are critical in providing each subsequent daughter cell with an identical set of chromosomes. In human cells, there would now be 46 chromosomes at each pole, one from each original sister pair. Figure 2–7(**e**) shows anaphase prior to its completion.

Telophase

Telophase is the final stage of mitosis and is depicted in Figure 2–7(**f**). At its beginning, two complete sets of chromosomes are present, one set at each pole. The most significant event of this stage is cytokinesis, the division or partitioning of the cytoplasm. Cytokinesis is essential if two new cells are to be produced from one cell. The mechanism of cytokinesis differs greatly in plant and animal cells, but the end result is the same: two new cells are produced. In plant cells, a **cell plate** is synthesized and laid down across the region of the metaphase plate. Animal cells, however,

undergo a constriction of the cytoplasm, much as a loop of string might be tightened around the middle of a balloon.

It is not surprising that the process of cytokinesis varies in different organisms. Plant cells, which are more regularly shaped and structurally rigid, require a mechanism for depositing new cell wall material around the plasma membrane. The cell plate laid down during telophase becomes a structure called the **middle lamella**. Subsequently, the primary and secondary layers of the cell wall are deposited between the cell membrane and middle lamella in each of the resulting daughter cells. In animals, complete constriction of the cell membrane produces the **cell furrow** characteristic of newly divided cells.

Other events necessary for the transition from mitosis to interphase are initiated during late telophase. They generally constitute a reversal of events that occurred during prophase. In each new cell, the chromosomes begin to uncoil and become diffuse chromatin once again, while the nuclear envelope reforms around them, the spindle fibers disappear, and the nucleolus gradually reforms and becomes visible in the nucleus during early interphase. At the completion of telophase, the cell enters interphase.

Cell-Cycle Regulation and Checkpoints

The cell cycle, culminating in mitosis, is fundamentally the same in all eukaryotic organisms. This similarity in many diverse organisms suggests that the cell cycle is governed by a genetically regulated program that has been conserved throughout evolution. Because disruption of this regulation may underlie the uncontrolled cell division characterizing malignancy, interest in how genes regulate the cell cycle is particularly strong.

A mammoth research effort over the past 15 years has paid high dividends, and we now have knowledge of many genes involved in the control of the cell cycle. This work was recognized by the awarding of the 2001 Nobel Prize in Medicine or Physiology to Lee Hartwell, Paul Nurse, and Tim Hunt. As with other studies of genetic control over essential biological processes, investigation has focused on the discovery of mutations that interrupt the cell cycle and on the effects of those mutations. As we shall return to this subject in Chapter 19 during our consideration of cancer, what follows is a very brief overview.

Many mutations are now known that exert an effect at one or another stage of the cell cycle. First discovered in yeast, but now evident in all organisms, including humans, such mutations were originally designated as *cell division cycle (cdc)* **mutations**. The normal products of many of the mutated genes are enzymes called **kinases** that can add phosphates to other proteins. They serve as "master control" molecules functioning in conjunction with proteins called **cyclins**. Cyclins bind to these kinases (creating *cyclin-dependent kinases*), activating them at appropriate times during the cell cycle. Activated kinases then phosphorylate other target proteins that regulate the progress of the cell cycle. The study of *cdc* mutations has established that the cell cycle contains at least three major *checkpoints*, where the processes culminating in normal mitosis are monitored, or "checked," by these master control molecules before the next stage of the cycle commences.

Checkpoints are named according to where in the cell cycle monitoring occurs (Figure 2–5). The first of three, the **G1/S checkpoint**, monitors the size the cell has achieved since its previous mitosis and also evaluates the condition of the DNA. If the cell has not reached an adequate size or if the DNA has been damaged, further progress through the cycle is arrested until these conditions are "corrected." If both conditions are "normal" at G1/S, then the cell is allowed to proceed from G1 to the S phase of the cycle. The second checkpoint is the **G2/M checkpoint**, where DNA is monitored prior to the start of mitosis. If DNA replication is incomplete or any DNA damage is detected and has not been repaired, the cell cycle is arrested. The final checkpoint occurs during mitosis and is called the **M checkpoint** (sometimes referred to as the *Spindle Assembly checkpoint*). Here, both the successful formation of the spindle fiber system and the attachment of spindle fibers to the kinetochores associated with the centromeres are monitored. If spindle fibers are not properly formed or if attachment is inadequate, mitosis is arrested.

The importance of cell-cycle control and these checkpoints can be demonstrated by considering what happens when this regulatory system is impaired. Let's assume, for example, that the DNA of a cell has incurred damage leading to one or more mutations impairing cell-cycle control. If allowed to proceed through the cell cycle as one of the population of dividing cells, this genetically altered cell would divide uncontrollably—precisely the definition of a cancerous cell. If instead the cell cycle is arrested at one of the checkpoints, the cell may effectively be removed from the population of dividing cells, preventing its potential malignancy.

2–1 With the initial appearance of the feature we call "Now Solve This," a short introduction is in order. The feature occurs several times in this and all ensuing chapters, each time providing a problem related to the discussion just presented. A "Hint" is then offered that may help you solve the problem. Here is the first problem:

(a) If an organism has a diploid number of 16, how many chromatids are visible at the end of mitotic prophase?

(b) How many chromosomes are moving to each pole during anaphase of mitosis?

■ HINT: *This problem involves an understanding of what happens to each pair of homologous chromosomes during mitosis. The key to its solution is to understand that throughout mitosis, the members of each homologous pair do not pair up, but instead behave*

2.4

Meiosis Reduces the Chromosome Number from Diploid to Haploid in Germ Cells and Spores

The process of meiosis, unlike mitosis, reduces the amount of genetic material by one-half. Whereas in diploids mitosis produces daughter cells with a full diploid complement, meiosis produces gametes or spores with only one haploid set of chromosomes. During sexual reproduction, gametes then combine through fertilization to reconstitute the diploid complement found in parental cells. Figure 2–9 compares the two processes by following two pairs of homologous chromosomes.

The events of meiosis must be highly specific since by definition, haploid gametes or spores contain precisely one member of each homologous pair of chromosomes. If successfully completed, meiosis ensures genetic continuity from generation to generation.

The process of sexual reproduction also ensures genetic variety among members of a species. As you study meiosis, you will see that this process results in gametes that each contain unique combinations of maternally and paternally derived chromosomes in their haploid complement. With such a tremendous genetic variation among the gametes, a huge number of maternal-paternal chromosome combinations are possible at fertilization. Furthermore, you will see that the meiotic event referred to as **crossing over** results in genetic exchange between members of each homologous pair of chromosomes. This process creates intact chromosomes that are mosaics of the maternal and paternal homologs from which they arise, further enhancing the potential genetic variation in gametes and the offspring derived from them. Sexual reproduction therefore reshuffles the genetic material, producing offspring that often differ greatly from either parent. Thus, meiosis is the major source of genetic recombination within species.

An Overview of Meiosis

In the preceding discussion, we established what might be considered the goal of meiosis: the reduction to the haploid complement of chromosomes. Before considering the phases of this process systematically, we will briefly summarize how diploid cells give rise to haploid gametes or spores. You should

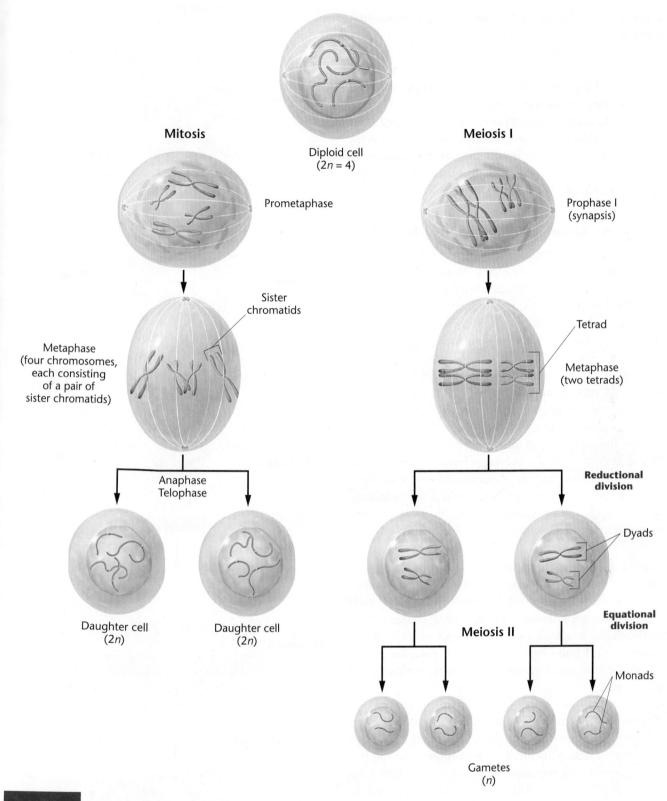

FIGURE 2–9 Overview of the major events and outcomes of mitosis and meiosis. As in Figure 2–7, two pairs of homologous chromosomes are followed.

We have established that in mitosis each paternally and maternally derived member of any given homologous pair of chromosomes behaves autonomously during division. By contrast, early in meiosis, homologous chromosomes synapsed structure, initially called a **bivalent**, eventually gives rise to a **tetrad** consisting of four chromatids. The presence of four chromatids demonstrates that both homologs (making up the bivalent) have in fact duplicated. Therefore, division occurs in meiosis I and is described as a **reductional division** (because the number of centromeres, each representing one chromosome, is *reduced* by one-half). Components of each tetrad—representing the two homologs—separate, yielding two **dyads**. Each dyad is composed of two sister chromatids joined at a common centromere. The second division occurs during meiosis II and is described as an **equational division** (because the number of centromeres remains *equal*). Here each dyad splits into two **monads** of one chromosome each. Thus, the two divisions potentially produce four haploid cells.

The First Meiotic Division: Prophase I

We turn now to a detailed account of meiosis. Like mitosis, meiosis is a continuous process. We assign names to its stages and substages only to facilitate discussion. From a genetic standpoint, three events characterize the initial stage, **prophase I** (Figure 2–10). First, as in mitosis, chromatin present in interphase thickens and coils into visible chromosomes. And, as in mitosis, each chromosome is a double structure, held together by the molecular complex called *cohesin*. Second, unlike mitosis, members of each homologous pair of chromosomes pair up, undergoing synapsis. Third, crossing over occurs between chromatids of synapsed homologs. Because of the complexity of these genetic events, this stage of meiosis is divided into five substages: leptonema, zygonema, pachynema, diplonema,* and diakinesis. As we discuss these substages, be aware that, even though it is not immediately apparent in the earliest phases of meiosis, the DNA of chromosomes has been replicated during the prior interphase.

Leptonema During the **leptotene stage**, the interphase chromatin material begins to condense, and the chromosomes, though still extended, become visible. Along each chromosome are **chromomeres**, localized condensations that resemble beads on a string. Evidence suggests that a process called **homology search**, which precedes and is essential to the initial pairing of homologs, begins during leptonema.

Zygonema The chromosomes continue to shorten and thicken during the **zygotene stage**. During the process of homology search, homologous chromosomes undergo initial alignment with one another. This so-called *rough pairing* is complete by the end of zygonema. In yeast, homologs are separated by about 300 nm, and near the end of zygonema, structures called *lateral elements* are visible between paired homologs. As meiosis proceeds, the overall length of the lateral elements along the chromosome increases, and a more extensive ultrastructural component called the

---

*These are the noun forms of these substages. The adjective forms (leptotene, zygotene, pachytene, and diplotene) are also used in the text.

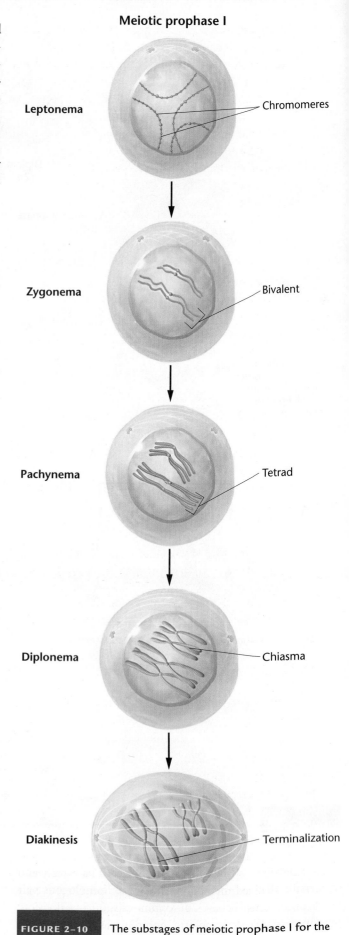

Meiotic prophase I

Leptonema — Chromomeres

Zygonema — Bivalent

Pachynema — Tetrad

Diplonema — Chiasma

Diakinesis — Terminalization

FIGURE 2–10 The substages of meiotic prophase I for the chromosomes depicted in Figure 2–9.

synaptonemal complex begins to form between the homologs. This complex is believed to be the vehicle responsible for the pairing of homologs. In some diploid organisms, this synapsis occurs in a zipper-like fashion, beginning at the ends of chromosomes attached to the nuclear envelope.

It is upon completion of zygonema that the paired homologs are referred to as bivalents. Although both members of each bivalent have already replicated their DNA, it is not yet visually apparent that each member is a double structure. The number of bivalents in each species is equal to the haploid (n) number.

Pachynema In the transition from the zygotene to the **pachytene stage**, the chromosomes continue to coil and shorten, and further development of the synaptonemal complex occurs between the two members of each bivalent. This leads to synapsis, a more intimate pairing. Compared to the rough-pairing characteristic of zygonema, homologs are now separated by only 100 nm.

During pachynema, each homolog is now evident as a double structure, providing visual evidence of the earlier replication of the DNA of each chromosome. Thus, each bivalent contains four member chromatids. As in mitosis, replicates are called *sister chromatids*, whereas chromatids from maternal and paternal members of a homologous pair are called *nonsister chromatids*. The four-membered structure, also referred to as a tetrad, contains two pairs of sister chromatids.

Diplonema During the ensuing **diplotene stage**, it is even more apparent that each tetrad consists of two pairs of sister chromatids. Within each tetrad, each pair of sister chromatids begins to separate. However, one or more areas remain in contact where chromatids are intertwined. Each such area, called a **chiasma** (pl. **chiasmata**), is thought to represent a point where nonsister chromatids have undergone genetic exchange through the process referred to above as *crossing over*. Although the physical exchange between chromosome areas occurred during the previous pachytene stage, the result of crossing over is visible only when the duplicated chromosomes begin to separate. Crossing over is an important source of genetic variability, and as indicated earlier, new combinations of genetic material are formed during this process.

Diakinesis The final stage of prophase I is **diakinesis**. The chromosomes pull farther apart, but nonsister chromatids remain loosely associated at the chiasmata. As separation proceeds, the chiasmata move toward the ends of the tetrad. This process of **terminalization** begins in late diplonema and is completed during diakinesis. During this final

and the two centromeres of each tetrad attach to the recently formed spindle fibers. By the completion of prophase I, the centromeres of each tetrad structure are present on the metaphase plate of the cell.

Metaphase, Anaphase, and Telophase I

The remainder of the meiotic process is depicted in Figure 2–11. After meiotic prophase I, stages similar to those of mitosis occur. In the first division, **metaphase I**, the chromosomes have maximally shortened and thickened. The terminal chiasmata of each tetrad are visible and appear to be the major factor holding the nonsister chromatids together. Each tetrad interacts with spindle fibers, facilitating its movement to the metaphase plate. The alignment of each tetrad prior to the first anaphase is random: Half of the tetrad (one of the dyads) will subsequently be pulled to one or the other pole, and the other half moves to the opposite pole.

During the stages of meiosis I, a single centromeric region holds each pair of sister chromatids together. It appears as a single unit, and a kinetechore forms around each one. As in our discussion of mitosis (see Figure 2–8), cohesin plays the major role in keeping sister chromatids together. At **anaphase I**, cohesin is degraded between sister chromatids, except at the centromere region, which, as in mitosis, is protected by a shugoshin complex. Then, one-half of each tetrad (a dyad) is pulled toward each pole of the dividing cell. This separation process is the physical basis of disjunction, the separation of homologous chromosomes from one another. Occasionally, errors in meiosis occur and separation is not achieved. The term **nondisjunction** describes such an error. At the completion of the normal anaphase I, a series of dyads equal to the haploid number is present at each pole.

If crossing over had not occurred in the first meiotic prophase, each dyad at each pole would consist solely of either paternal or maternal chromatids. However, the exchanges produced by crossing over create mosaic chromatids of paternal and maternal origin.

In many organisms, **telophase I** reveals a nuclear membrane forming around the dyads. In this case, the nucleus next enters into a short interphase period. If interphase occurs, the chromosomes do not replicate because they already consist of two chromatids. In other organisms, the cells go directly from anaphase I to meiosis II. In general, meiotic telophase is much shorter than the corresponding stage in mitosis.

The Second Meiotic Division

A second division, referred to as **meiosis II**, is essential if each gamete or spore is to receive only one chromatid from each original tetrad. The stages characterizing meiosis II are shown on the right side of Figure 2–11. During **prophase II**, each

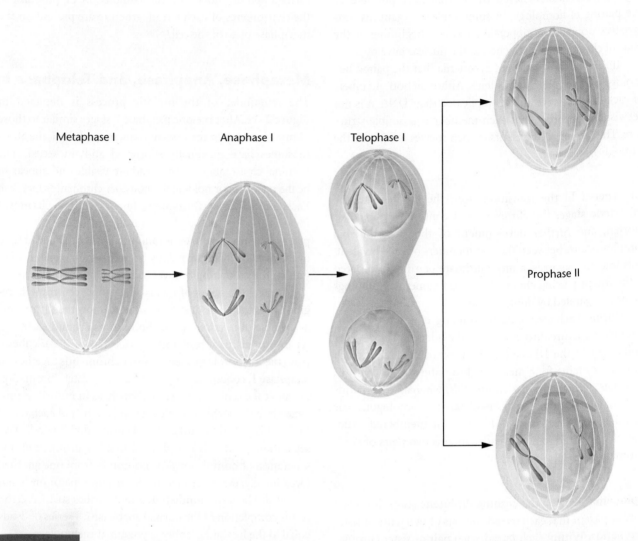

Metaphase I   Anaphase I   Telophase I

Prophase II

FIGURE 2–11   The major events in meiosis in an animal cell with a diploid number of 4, beginning with metaphase I. Note that the combination of chromosomes in the cells following telophase II is dependent on the random orientation of each tetrad and dyad when they align on the equatorial plate during metaphase I and metaphase II. Several other combinations, which are not shown, can also be produced. The events depicted here are described in the text.

by the common centromeric region. During **metaphase II**, the centromeres are positioned on the equatorial plate. When the shugoshin complex is degraded, the centromeres separate, **anaphase II** is initiated, and the sister chromatids of each dyad are pulled to opposite poles. Because the number of dyads is equal to the haploid number, **telophase II** reveals one member of each pair of homologous chromosomes present at each pole. Each chromosome is now a monad. Following cytokinesis in telophase II, four haploid gametes may result from a single meiotic event. At the conclusion of meiosis II, not only has the haploid state been achieved, but if crossing over has occurred, each monad contains a combination of maternal and paternal genetic information. As a result, the offspring produced by any gamete will receive a mixture of genetic information originally present in his or her grandparents.

NOW SOLVE THIS

2–2 An organism has a diploid number of 16 in a primary oocyte. (a) How many tetrads are present in the first meiotic prophase? (b) How many dyads are present in the second meiotic prophase? (c) How many monads migrate to each pole during the second meiotic anaphase?

■ HINT: *This problem involves an understanding of what happens to the maternal and paternal members of each pair of homologous chromosomes during meiosis. The key to its solution is to understand that maternal and paternal homologs synapse during meiosis. Once each chromatid has duplicated, creating a tetrad in the early phases of meiosis, each original pair behaves as a unit and leads to two dyads during anaphase I.*

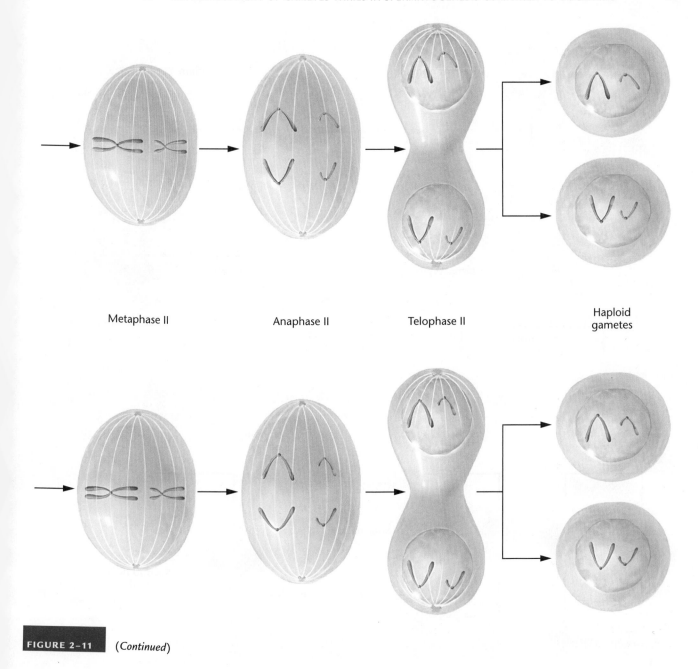

| Metaphase II | Anaphase II | Telophase II | Haploid gametes |

FIGURE 2–11   (*Continued*)

Meiosis thus significantly increases the level of genetic variation in each ensuing generation.

2.5

The Development of Gametes Varies in Spermatogenesis Compared to Oogenesis

Although events that occur during the meiotic divisions are similar in all cells participating in gametogenesis in most animal species, there are certain differences between the production of a male gamete (spermatogenesis) and a female gamete (oogenesis). Figure 2-12 summarizes these processes.

Spermatogenesis takes place in the testes, the male reproductive organs. The process begins with the enlargement of an undifferentiated diploid germ cell called a **spermatogonium**. This cell grows to become a **primary spermatocyte**, which undergoes the first meiotic division. The products of this division, called **secondary spermatocytes**, contain a haploid number of dyads. The secondary spermatocytes then undergo meiosis II, and each of these cells produces two haploid **spermatids**. Spermatids go through a series of developmental changes, **spermiogenesis**, to become highly specialized, motile **spermatozoa**, or **sperm**. All sperm cells produced during spermatogenesis contain the haploid number of chromosomes and equal amounts of cytoplasm.

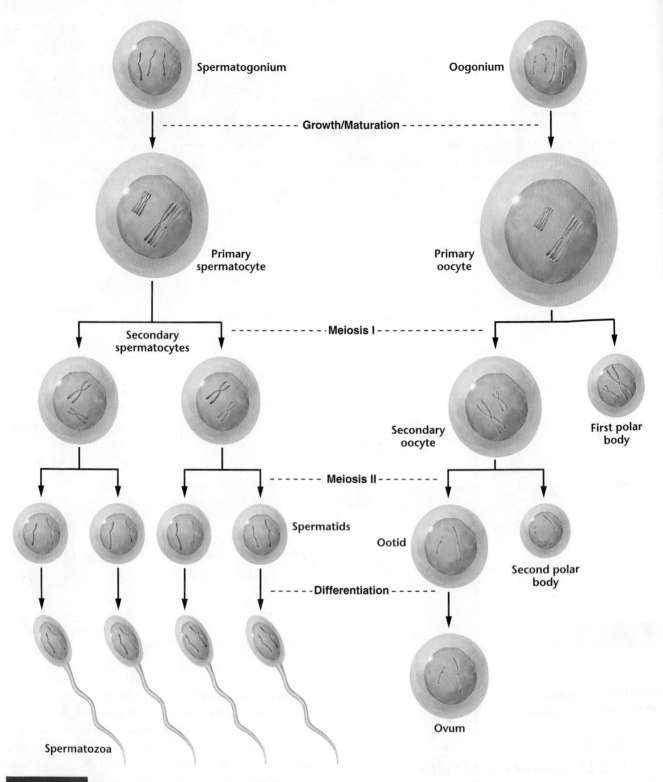

Spermatogenesis and oogenesis in animal cells.

Spermatogenesis may be continuous or may occur periodically in mature male animals; its onset is determined by the species' reproductive cycles. Animals that reproduce year-round produce sperm continuously, whereas those whose breeding period is confined to a particular season produce sperm only during that time.

In animal **oogenesis**, the formation of **ova** (sing. **ovum**), or eggs, occurs in the ovaries, the female reproductive organs. The daughter cells resulting from the two meiotic divisions of this process receive equal amounts of genetic material, but they do *not* receive equal amounts of cytoplasm. Instead, during each division, almost all the cytoplasm of the **primary oocyte**, itself derived from the **oogonium**, is concentrated in one of the two daughter cells. The concentration of cytoplasm is necessary because a major function of the mature ovum is to nourish the developing embryo following fertilization.

During anaphase I in oogenesis, the tetrads of the primary oocyte separate, and the dyads move toward opposite poles. During telophase I, the dyads at one pole are pinched off with very little surrounding cytoplasm to form the **first polar body**. The first polar body may or may not divide again to produce two small haploid cells. The other daughter cell produced by this first meiotic division contains most of the cytoplasm and is called the **secondary oocyte**. The mature ovum will be produced from the secondary oocyte during the second meiotic division. During this division, the cytoplasm of the secondary oocyte again divides unequally, producing an **ootid** and a **second polar body**. The ootid then differentiates into the mature ovum.

Unlike the divisions of spermatogenesis, the two meiotic divisions of oogenesis may not be continuous. In some animal species, the second division may directly follow the first. In others, including humans, the first division of all oocytes begins in the embryonic ovary but arrests in prophase I. Many years later, meiosis resumes in each oocyte just prior to its ovulation. The second division is completed only after fertilization.

NOW SOLVE THIS

2–3 Examine Figure 2–12, which shows oogenesis in animal cells. Will the genotype of the second polar body (derived from meiosis II) always be identical to that of the ootid? Why or why not?

■ HINT: *This problem involves an understanding of meiosis during oogenesis. The key to its solution is to take into account that crossing over occurred between each pair of homologs during meiosis I.*

2.6

Meiosis Is Critical to Sexual Reproduction in All Diploid Organisms

The process of meiosis is critical to the successful sexual reproduction of all diploid organisms. It is the mechanism by which the diploid amount of genetic information is reduced to the haploid amount. In animals, meiosis leads to the formation of gametes, whereas in plants haploid spores are produced, which in turn lead to the formation of

Each diploid organism stores its genetic information in the form of homologous pairs of chromosomes. Each pair consists of one member derived from the maternal parent and one from the paternal parent. Following meiosis, haploid cells potentially contain either the paternal or the maternal representative of every homologous pair of chromosomes. However, the process of crossing over, which occurs in the first meiotic prophase, further reshuffles the alleles between the maternal and paternal members of each homologous pair, which then segregate and assort independently into gametes. These events result in the great amount of genetic variation present in gametes.

It is important to touch briefly on the significant role that meiosis plays in the life cycles of fungi and plants. In many fungi, the predominant stage of the life cycle consists of haploid vegetative cells. They arise through meiosis and proliferate by mitotic cell division. In multicellular plants, the life cycle alternates between the diploid **sporophyte stage** and the haploid **gametophyte stage** (Figure 2–13). While one or the other predominates in different plant groups during this "alternation of generations," the processes of meiosis and fertilization constitute the "bridges" between the sporophyte and gametophyte stages. Therefore, meiosis is an essential component of the life cycle of plants.

2.7

Electron Microscopy Has Revealed the Physical Structure of Mitotic and Meiotic Chromosomes

Thus far in this chapter, we have focused on mitotic and meiotic chromosomes, emphasizing their behavior during cell division and gamete formation. An interesting question is why chromosomes are invisible during interphase but visible during the various stages of mitosis and meiosis. Studies using electron microscopy clearly show why this is the case.

Recall that, during interphase, only dispersed chromatin fibers are present in the nucleus [Figure 2–14(a)]. Once mitosis begins, however, the fibers coil and fold, condensing into typical mitotic chromosomes [Figure 2–14(b)]. If the fibers comprising a mitotic chromosome are loosened, the areas of greatest spreading reveal individual fibers similar to those seen in interphase chromatin [Figure 2–14(c)]. Very few fiber ends seem to be present, and in some cases, none can be seen. Instead, individual fibers always seem to loop back into the interior. Such fibers are obviously twisted and coiled around one another, forming the regular pattern of folding in the mitotic chromosome. Starting in late telophase of mitosis and continuing during G1 of interphase, chromosomes unwind to form the long fibers characteristic of chromatin, which consist of DNA and associated pro-

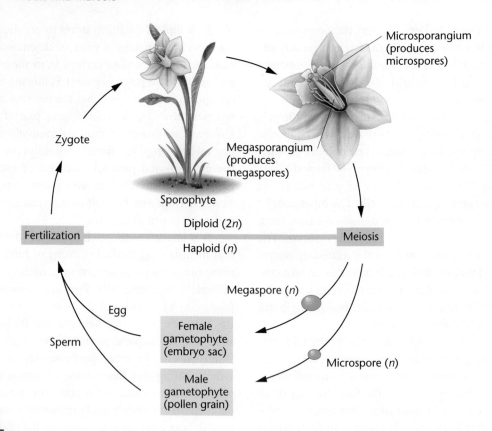

arrangement that DNA can most efficiently function during transcription and replication.

Electron microscopic observations of metaphase chromosomes in varying degrees of coiling led Ernest DuPraw to postulate the **folded-fiber model**, shown in Figure 2–14(c). During metaphase, each chromosome consists of two sister chromatids joined at the centromeric region. Each arm of the chromatid appears to be a single fiber wound much like a skein of yarn. The fiber is composed of tightly coiled double-stranded DNA and protein. An orderly coiling–twisting–condensing process

appears to facilitate the transition of the interphase chromatin into the more condensed mitotic chromosomes. Geneticists believe that during the transition from interphase to prophase, a 5000-fold compaction occurs in the length of DNA within the chromatin fiber! This process must be extremely precise given the highly ordered and consistent appearance of mitotic chromosomes in all eukaryotes. Note particularly in the micrographs the clear distinction between the sister chromatids constituting each chromosome. They are joined only by the common centromere that they share prior to anaphase.

(a)   (b)   (c)

FIGURE 2–14   Comparison of (a) the chromatin fibers characteristic of the interphase nucleus with (b) metaphase chromosomes that are derived from chromatin during mitosis. Part (c) diagrams a mitotic chromosome, showing how chromatin is condensed to produce it. Part (a) is a transmission electron micrograph and part (b) is a scanning electron micrograph.

PubMed: Exploring and Retrieving Biomedical Literature

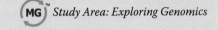

Study Area: Exploring Genomics

In this era of rapidly expanding information on genomics and the biomedical sciences, scientists must be conversant in the use of multiple online databases. These resources provide access to DNA and protein sequences, genomic data, chromosome maps, microarray gene-expression networks, and molecular structures, as well as to the bioinformatics tools necessary for data manipulation. Perhaps the most central database resource is PubMed, an online tool for conducting literature searches and accessing biomedical publications.

PubMed is an Internet-based search system developed by the National Center of Biotechnology Information (NCBI) at the National Library of Medicine. Using PubMed, one can access over 15 million articles in over 4600 biomedical journals. The full text of many of the journals can be obtained electronically through college or university libraries, and some journals (such as *Proceedings of the National Academy of Sciences USA; Genome Biology;* and *Science*) provide free public access to articles within certain time frames.

In this exercise, we will explore PubMed to answer questions about relationships between tubulin, human cancers, and cancer therapies, as well as the genetics of spermatogenesis.

■ Exercise I – Tubulin, Cancer, and Mitosis

In this chapter we were introduced to tubulin and the dynamic behavior of microtubules during the cell cycle. Cancer cells are characterized by continuous and uncontrolled mitotic divisions.

Is it possible that tubulin and microtubules contribute to the development of cancer? Could these important structures be targets for cancer therapies?

1. To begin your search for the answers, access the PubMed site at www.ncbi.nlm.nih.gov/entrez/query. fcgi?DB=pubmed.

2. In the SEARCH box, type "tubulin cancer" and then select the "Go" button to perform the search.

3. Select several research papers and read the abstracts.

To answer the question about tubulin's association with cancer, you may want to limit your search to fewer papers, perhaps those that are review articles. To do this:

1. Select the "Limits" tab near the top of the page.

2. Scroll down the page and select "Review" in the "Type of Article" list.

3. Select "Go" to perform the search.

Explore some of the articles, as abstracts or as full text, available in your library or by free public access. Prepare a brief report or verbally share your experiences with your class. Describe two of the most important things you learned during your exploration and identify the information sources you encountered during the search.

■ Exercise II – Human Disorders of Spermatogenesis

Using the methods described in Exercise I, identify some human disorders associated with defective spermatogenesis. Which human genes are involved in spermatogenesis? How do defects in these genes result in fertility disorders? Prepare a brief written or verbal report on what you have learned and what sources you used to acquire your information.

CASE STUDY | Timing is everything

A man in his early 20s received chemotherapy and radiotherapy as treatment every 60 days for Hodgkin's disease. After unsuccessful attempts to have children, he had his sperm examined at a fertility clinic, upon which multiple chromosomal irregularities were discovered. When examined within 5 days of a treatment, extra chromosomes were often present or one or more chromosomes were completely absent. However, such irregularities were not observed 38 days and later after a treatment.

1. How might a geneticist explain the time-related differences in chromosomal irregularities?
2. Do you think that exposure to chemotherapy and radiotherapy of a spermatogonium would cause more problems than exposure to a secondary spermatocyte?
3. What is the obvious advice that the man received regarding fertility while he remained under treatment?

Summary Points

 For activities, animations, and review quizzes, go to the study area at www.masteringgenetics.com

1. The structure of cells is elaborate and complex, with most components involved directly or indirectly with genetic processes.

2. In diploid organisms, chromosomes exist in homologous pairs, where each member is identical in size, centromere placement, and gene loci. One member of each pair is derived from the maternal parent, and the other from the paternal parent.

3. Mitosis and meiosis are mechanisms by which cells distribute the genetic information contained in their chromosomes to progeny cells in a precise, orderly fashion.

4. Mitosis, which is but one part of the cell cycle, is subdivided into discrete stages that initially depict the condensation of chromatin into the diploid number of chromosomes. Each chromosome first appears as a double structure, consisting of a pair of identical sister chromatids joined at a common centromere. As mitosis proceeds, centromeres split and sister chromatids are pulled apart by spindle fibers and directed toward opposite poles of the cell. Cytoplasmic division then occurs, creating two new cells with the identical genetic information contained in the progenitor cell.

5. Meiosis converts a diploid cell into haploid gametes or spores, making sexual reproduction possible. As a result of chromosome duplication, two subsequent meiotic divisions are required to achieve haploidy, whereby each haploid cell receives one member of each homologous pair of chromosomes.

6. There is a major difference between meiosis in males and in females. Spermatogenesis partitions the cytoplasmic volume equally and produces four haploid sperm cells. Oogenesis, on the other hand, collects the bulk of cytoplasm in one egg cell and reduces the other haploid products to polar bodies. The extra cytoplasm in the egg contributes to zygote development following fertilization.

7. Meiosis results in extensive genetic variation by virtue of the exchange of chromosome segments during crossing over between maternal and paternal chromatids and by virtue of the random separation of maternal and paternal chromatids into gametes. In addition, meiosis plays an important role in the life cycles of fungi and plants, serving as the bridge between alternating generations.

8. Mitotic chromosomes are produced as a result of the coiling and condensation of chromatin fibers of interphase into the characteristic form of chromatids.

INSIGHTS AND SOLUTIONS

This appearance of "Insights and Solutions" begins a feature that will have great value to you as a student. From this point on, "Insights and Solutions" precedes the "Problems and Discussion Questions" at each chapter's end to provide sample problems and solutions that demonstrate approaches you will find useful in genetic analysis. The insights you gain by working through the sample problems will improve your ability to solve the ensuing problems in each chapter.

1. In an organism with a diploid number of $2n = 6$, how many individual chromosomal structures will align on the metaphase plate during (a) mitosis, (b) meiosis I, and (c) meiosis II? Describe each configuration.

Solution: (a) Remember that in mitosis, homologous chromosomes do not synapse, so there will be six double structures, each consisting of a pair of sister chromatids. In other words, the number of structures is equivalent to the diploid number.

(b) In meiosis I, the homologs have synapsed, reducing the number of structures to three. Each is called a tetrad and consists of two pairs of sister chromatids.

(c) In meiosis II, the same number of structures exist (three), but in this case they are called dyads. Each dyad is a pair of sister chromatids. When crossing over has occurred, each chromatid may contain parts of one of its nonsister chromatids, obtained during exchange in prophase I.

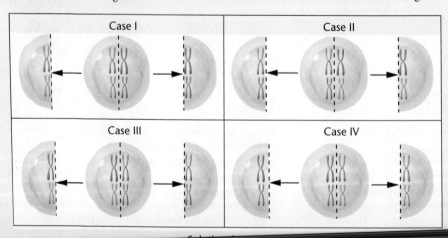

2. Disregarding crossing over, draw all possible alignment configurations that can occur during metaphase for the chromosomes shown in Figure 2–11.

Solution: As shown in the following diagram, four configurations are possible when $n = 2$.

3. For the chromosomes in the previous problem, assume that each of the larger chromosomes has a different allele for a given gene, *A* OR *a*, as shown. Also assume that each of the smaller chromosomes has a different allele for a second gene, *B* OR *b*. Calculate the probability of generating each possible combination of these alleles (*AB*, *Ab*, *aB*, *ab*) following meiosis I.

Solution: As shown in the accompanying diagram:

Case I	AB and ab
Case II	Ab and aB
Case III	aB and Ab
Case IV	ab and AB

Solution for #3

Total: $AB = 2\ (p = 1/4)$

$Ab = 2\ (p = 1/4)$

$aB = 2\ (p = 1/4)$

$ab = 2\ (p = 1/4)$

4. How many different chromosome configurations can occur following meiosis I if three different pairs of chromosomes are present ($n = 3$)?

Solution: If $n = 3$, then eight different configurations would be possible. The formula 2^n, where n equals the haploid number, represents the number of potential alignment patterns. As we will see in the next chapter, these patterns are produced according to the Mendelian postulate of *segregation*, and they serve as the physical basis of another Mendelian postulate called *independent assortment*.

5. Describe the composition of a meiotic tetrad during prophase I, assuming no crossover event has occurred. What impact would a single crossover event have on this structure?

Solution: Such a tetrad contains four chromatids, existing as two pairs. Members of each pair are sister chromatids. They are held together by a common centromere. Members of one pair are maternally derived, whereas members of the other are paternally derived. Maternal and paternal members are called nonsister chromatids. A single crossover event has the effect of exchanging a portion of a maternal and a paternal chromatid, leading to a chiasma, where the two involved chromatids overlap physically in the tetrad. The process of exchange is referred to as crossing over.

Problems and Discussion Questions

MG ™ *For instructor-assigned tutorials and problems, go to www.masteringgentics.com*

HOW DO WE KNOW?

1. In this chapter, we focused on how chromosomes are distributed during cell division, both in dividing somatic cells (mitosis) and in gamete- and spore-forming cells (meiosis). At the same time, we found many opportunities to consider the methods and reasoning by which much of this information was acquired. From the explanations given in the chapter,
   (a) How do we know that chromosomes exist in homologous pairs?
   (b) How do we know that DNA replication occurs during interphase, not early in mitosis?
   (c) How do we know that mitotic chromosomes are derived from chromatin?

2. What role do the following cellular components play in the storage, expression, or transmission of genetic information: (a) chromatin, (b) nucleolus, (c) ribosome, (d) mitochondrion, (e) centriole, (f) centromere?

3. Discuss the concepts of homologous chromosomes, diploidy, and haploidy. What characteristics do two homologous chromosomes share?

4. If two chromosomes of a species are the same length and have similar centromere placements and yet are not homologous, what is different about them?

5. Describe the events that characterize each stage of mitosis.

6. What designations are assigned to chromosomes on the basis of their centromere placement, and where is the centromere located in each case?

7. Contrast telophase in plant and animal mitosis.

8. Describe the phases of the cell cycle and the events that characterize each phase.

9. Contrast the end results of meiosis with those of mitosis.

10. Define and discuss these terms: (a) synapsis, (b) bivalents, (c) chiasmata, (d) crossing over, (e) chromomeres, (f) sister chromatids, (g) tetrads, (h) dyads, (i) monads.

11. Contrast the genetic content and the origin of sister versus non-sister chromatids during their earliest appearance in prophase I of meiosis. How might the genetic content of these change by the time tetrads have aligned at the equatorial plate during metaphase I?

12. Given the end results of the two types of division, why is it necessary for homologs to pair during meiosis and not desirable for them to pair during mitosis?

13. Contrast spermatogenesis and oogenesis. What is the significance of the formation of polar bodies?

14. Explain why meiosis leads to significant genetic variation while mitosis does not.

15. A diploid cell contains three pairs of homologous chromosomes designated C1 and C2, M1 and M2, and S1 and S2. No crossing over occurs. What combinations of chromosomes are possible in (a) daughter cells following mitosis? (b) cells undergoing the first meiotic metaphase? (c) haploid cells following both divisions of meiosis?

16. Considering the preceding problem, predict the number of different haploid cells that could be produced by meiosis if a fourth chromosome pair (W1 and W2) were added.

17. During oogenesis in an animal species with a haploid number of 6, one dyad undergoes nondisjunction during meiosis II. Following the second meiotic division, this dyad ends up intact in the ovum. How many chromosomes are present in (a) the mature ovum and (b) the second polar body? (c) Following fertilization by a normal sperm, what chromosome condition is created?

18. What is the probability that, in an organism with a haploid number of 10, a sperm will be formed that contains all 10 chromosomes whose centromeres were derived from maternal homologs?

19. During the first meiotic prophase, (a) when does crossing over occur; (b) when does synapsis occur; (c) during which stage are the chromosomes least condensed; and (d) when are chiasmata first visible?

20. Describe the role of meiosis in the life cycle of a vascular plant.

21. Contrast the chromatin fiber with the mitotic chromosome. How are the two structures related?

22. Describe the "folded-fiber" model of the mitotic chromosome.

23. You are given a metaphase chromosome preparation (a slide) from an unknown organism that contains 12 chromosomes. Two that are clearly smaller than the rest appear identical in length and centromere placement. Describe all that you can about these chromosomes.

24. If one follows 50 primary oocytes in an animal through their various stages of oogenesis, how many secondary oocytes would be formed? How many first polar bodies would be formed? How many ootids would be formed? If one follows 50 primary spermatocytes in an animal through their various stages of spermatogenesis, how many secondary spermatocytes would be formed? How many spermatids would be formed?

25. The nuclear DNA content of a single sperm cell in *Drosophila melanogaster* is approximately 0.18 picogram. What would be the expected nuclear DNA content of a primary spermatocyte in *Drosophila*? What would be the expected nuclear DNA content of a somatic cell (non-sex cell) in the G1 phase? What would be the expected nuclear DNA content of a somatic cell at metaphase?

Extra-Spicy Problems

 For instructor-assigned tutorials and problems, go to www.masteringgentics.com

As part of the "Problems and Discussion Questions" section in this and each subsequent chapter, we shall present a number of "Extra-Spicy" genetics problems. We have chosen to set these apart in order to identify problems that are particularly challenging. You may be asked to examine and assess actual data, to design genetics experiments, or to engage in cooperative learning. Like genetic varieties of peppers, some of these experiences are just spicy and some are very hot. We hope that you will enjoy the challenges that they pose.

For Questions 26–31, consider a diploid cell that contains three pairs of chromosomes designated AA, BB, and CC. Each pair contains a maternal and a paternal member (e.g., A^m and A^P). Using these designations, demonstrate your understanding of mitosis and meiosis by drawing chromatid combinations as requested. Be sure to indicate when chromatids are paired as a result of replication and/or synapsis. You may wish to use a large piece of brown manila wrapping paper or a cut-up paper grocery bag for this project and to work in partnership with another student. We recommend cooperative learning as an efficacious way to develop the skills you will need for solving the problems presented throughout this text.

26. In mitosis, what chromatid combination(s) will be present during metaphase? What combination(s) will be present at each pole at the completion of anaphase?

27. During meiosis I, assuming no crossing over, what chromatid combination(s) will be present at the completion of prophase? Draw all possible alignments of chromatids as migration begins during early anaphase.

28. Are there any possible combinations present during prophase of meiosis II other than those that you drew in Problem 27? If so, draw them.

29. Draw all possible combinations of chromatids during the early phases of anaphase in meiosis II.

30. Assume that during meiosis I none of the *C* chromosomes disjoin at metaphase, but they separate into dyads (instead of monads) during meiosis II. How would this change the alignments that you constructed during the anaphase stages in meiosis I and II? Draw them.

31. Assume that each gamete resulting from Problem 30 fuses, in fertilization, with a normal haploid gamete. What combinations will result? What percentage of zygotes will be diploid,

containing one paternal and one maternal member of each chromosome pair?

32. A species of cereal rye (*Secale cereale*) has a chromosome number of 14, while a species of Canadian wild rye (*Elymus canadensis*) has a chromosome number of 28. Sterile hybrids can be produced by crossing *Secale* with *Elymus*.

    (a) What would be the expected chromosome number in the somatic cells of the hybrids?

    (b) Given that none of the chromosomes pair at meiosis I in the sterile hybrid (Hang and Franckowlak, 1984), speculate on the anaphase I separation patterns of these chromosomes.

33. An interesting procedure has been applied for assessing the chromosomal balance of potential secondary oocytes for use in human *in vitro* fertilization. Using fluorescence *in situ* hybridization (FISH), Kuliev and Verlinsky (2004) were able to identify individual chromosomes in first polar bodies and thereby infer the chromosomal makeup of "sister" oocytes.

    (a) Assume that when examining a first polar body you saw that it had one copy (dyad) of each chromosome but two dyads of chromosome 21. What would you expect to be the chromosomal 21 complement in the secondary oocyte? What consequences are likely in the resulting zygote, if the secondary oocyte was fertilized?

    (b) Assume that you were examining a first polar body and noted that it had one copy (dyad) of each chromosome except chromosome 21. Chromosome 21 was completely absent. What would you expect to be the chromosome 21 complement (only with respect to chromosome 21) in the secondary oocyte? What consequences are likely in the resulting zygote if the secondary oocyte was fertilized?

    (c) Kuliev and Verlinsky state that there was a relatively high number of separation errors at meiosis I. In these cases the centromere underwent a premature division, occurring at meiosis I rather than meiosis II. Regarding chromosome 21, what would you expect to be the chromosome 21 complement in the secondary oocyte in which you saw a single chromatid (monad) for chromosome 21 in the first polar body? If this secondary oocyte was involved in fertilization, what would be the expected consequences?

Gregor Johann Mendel, who in 1866 put forward the major postulates of transmission genetics as a result of experiments with the garden pea.

3

Mendelian Genetics

CHAPTER CONCEPTS

- Inheritance is governed by information stored in discrete factors called genes.

- Genes are transmitted from generation to generation on vehicles called chromosomes.

- Chromosomes, which exist in pairs in diploid organisms, provide the basis of biparental inheritance.

- During gamete formation, chromosomes are distributed according to postulates first described by Gregor Mendel, based on his nineteenth-century research with the garden pea.

- Mendelian postulates prescribe that homologous chromosomes segregate from one another and assort independently with other segregating homologs during gamete formation.

- Genetic ratios, expressed as probabilities, are subject to chance deviation and may be evaluated statistically.

- The analysis of pedigrees allows predictions concerning the genetic nature of human traits.

Although inheritance of biological traits has been recognized for thousands of years, the first significant insights into how it takes place only occurred about 145 years ago. In 1866, Gregor Johann Mendel published the results of a series of experiments that would lay the foundation for the formal discipline of genetics. Mendel's work went largely unnoticed until the turn of the twentieth century, but eventually, the concept of the gene as a distinct hereditary unit was established. Since then, the ways in which genes, as segments of chromosomes, are transmitted to offspring and control traits have been clarified. Research continued unabated throughout the twentieth century and into the present—indeed, studies in genetics, most recently at the molecular level, have remained at the forefront of biological research since the early 1900s.

When Mendel began his studies of inheritance using *Pisum sativum*, the garden pea, chromosomes and the role and mechanism of meiosis were totally unknown. Nevertheless, he determined that discrete *units of inheritance* exist and predicted their behavior in the formation of gametes. Subsequent investigators, with access to cytological data, were able to relate their own observations of chromosome behavior during meiosis and Mendel's principles of inheritance. Once this correlation was recognized, Mendel's postulates were accepted as the basis for the study of what is known as **transmission genetics**, how genes are transmitted from parents to offspring. These principles were derived directly from Mendel's experimentation. Even today, they serve as the cornerstone of the study of inheritance. In this chapter, we focus on the development of Mendel's principles.

3.1
Mendel Used a Model Experimental Approach to Study Patterns of Inheritance

Johann Mendel was born in 1822 to a peasant family in the Central European village of Heinzendorf. An excellent student in high school, he studied philosophy for several years afterward and in 1843, taking the name Gregor, was admitted to the Augustinian Monastery of St. Thomas in Brno, now part of the Czech Republic. In 1849, he was relieved of pastoral duties, and from 1851 to 1853, he attended the University of Vienna, where he studied physics and botany. He returned to Brno in 1854, where he taught physics and natural science for the next 16 years. Mendel received support from the monastery for his studies and research throughout his life.

In 1856, Mendel performed his first set of hybridization experiments with the garden pea, launching the research phase of his career. His experiments continued until 1868, when he was elected abbot of the monastery. Although he retained his interest in genetics, his new responsibilities demanded most of his time. In 1884, Mendel died of a kidney disorder. The local newspaper paid him the following tribute:

> "His death deprives the poor of a benefactor, and mankind at large of a man of the noblest character, one who was a warm friend, a promoter of the natural sciences, and an exemplary priest."

Mendel first reported the results of some simple genetic crosses between certain strains of the garden pea in 1865. Although his was not the first attempt to provide experimental evidence pertaining to inheritance, Mendel's success where others had failed can be attributed, at least in part, to his elegant experimental design and analysis.

Mendel showed remarkable insight into the methodology necessary for good experimental biology. First, he chose an organism that was easy to grow and to hybridize artificially. The pea plant is self-fertilizing in nature, but it is easy to cross-breed experimentally. It reproduces well and grows to maturity in a single season. Mendel followed seven visible features (we refer to them as characters, or characteristics), each represented by two contrasting properties, or **traits** (Figure 3–1). For the character stem height, for example, he experimented with the traits *tall* and *dwarf*. He selected six other visibly contrasting pairs of traits involving seed shape and color, pod shape and color, and flower color and position. From local seed merchants, Mendel obtained true-breeding strains, those in which each trait appeared unchanged generation after generation in self-fertilizing plants.

There were several other reasons for Mendel's success. In addition to his choice of a suitable organism, he restricted his examination to one or very few pairs of contrasting traits in each experiment. He also kept accurate quantitative records, a necessity in genetic experiments. From the analysis of his data, Mendel derived certain postulates that have become the principles of transmission genetics.

The results of Mendel's experiments went unappreciated until the turn of the century, well after his death. However, once Mendel's publications were rediscovered by geneticists investigating the function and behavior of chromosomes, the implications of his postulates were immediately apparent. He had discovered the basis for the transmission of hereditary traits!

3.2
The Monohybrid Cross Reveals How One Trait Is Transmitted from Generation to Generation

Mendel's simplest crosses involved only one pair of contrasting traits. Each such experiment is called a **monohybrid cross**. A monohybrid cross is made by mating true-breeding

Character	Contrasting traits		F$_1$ results	F$_2$ results	F$_2$ ratio
Seed shape	round/wrinkled		all round	5474 round 1850 wrinkled	2.96:1
Seed color	yellow/green		all yellow	6022 yellow 2001 green	3.01:1
Pod shape	full/constricted		all full	882 full 299 constricted	2.95:1
Pod color	green/yellow		all green	428 green 152 yellow	2.82:1
Flower color	violet/white		all violet	705 violet 224 white	3.15:1
Flower position	axial/terminal		all axial	651 axial 207 terminal	3.14:1
Stem height	tall/dwarf		all tall	787 tall 277 dwarf	2.84:1

FIGURE 3–1   Seven pairs of contrasting traits and the results of Mendel's seven monohybrid crosses of the garden pea (*Pisum sativum*). In each case, pollen derived from plants exhibiting one trait was used to fertilize the ova of plants exhibiting the other trait. In the F$_1$ generation, one of the two traits was exhibited by all plants. The contrasting trait reappeared in approximately 1/4 of the F$_2$ plants.

individuals from two parent strains, each exhibiting one of the two contrasting forms of the character under study. Initially, we examine the first generation of offspring of such a cross, and then we consider the offspring of **selfing**, that is, of self-fertilization of individuals from this first generation. The original parents constitute the **P$_1$**, or **parental generation**; their offspring are the **F$_1$**, or **first filial generation**; the individuals resulting from the selfed F$_1$ generation are the **F$_2$**, or **second filial generation**; and so on.

The cross between true-breeding pea plants with tall stems and dwarf stems is representative of Mendel's monohybrid crosses. *Tall* and *dwarf* are contrasting traits of the character of stem height. Unless tall or dwarf plants are crossed together or with another strain, they will undergo self-fertilization and breed true, producing their respective traits generation after generation. However, when Mendel crossed tall plants with dwarf plants, the resulting F$_1$ generation consisted of only tall plants. When members of the F$_1$ generation were selfed, Mendel observed that 787 of 1064 F$_2$ plants were tall, while 277 of 1064 were dwarf. Note that in this cross (Figure 3–1), the dwarf trait disappeared in the F$_1$ generation, only to reappear in the F$_2$ generation. These observations were important in Mendel's analysis of monohybrid crosses.

Genetic data are usually expressed and analyzed as ratios. In this particular example, many identical P$_1$ crosses were made and many F$_1$ plants—all tall—were produced. As noted, of the 1064 F$_2$ offspring, 787 were tall and 277 were dwarf—a ratio of approximately 2.8:1.0, or about 3:1.

Mendel made similar crosses between pea plants exhibiting each of the other pairs of contrasting traits; the results of these crosses are shown in Figure 3–1. In every case, the outcome was similar to the tall/dwarf cross just described. For the character of interest, all F$_1$ offspring had the same trait exhibited by one of the parents, but in the F$_2$ offspring, an approximate ratio of 3:1 was obtained. That is, three-fourths looked like the F$_1$ plants, while one-fourth exhibited the contrasting trait, which had disappeared in the F$_1$ generation.

We note one further aspect of Mendel's monohybrid crosses. In each cross, the F$_1$ and F$_2$ patterns of inheritance were similar regardless of which P$_1$ plant served as the source of pollen (sperm) and which served as the source of the ovum (egg). The crosses could be made either way—pollination of dwarf plants by tall plants, or vice versa. Crosses made in both these ways are called **reciprocal crosses**. Therefore, the results of Mendel's monohybrid crosses were not sex-dependent.

To explain these results, Mendel proposed the existence of particulate *unit factors* for each trait. He suggested that these factors serve as the basic units of heredity and are passed unchanged from generation to generation, determining various traits expressed by each individual plant. Using these general ideas, Mendel proceeded to hypothesize precisely how such factors could account for the results of the monohybrid crosses.

Mendel's First Three Postulates

Using the consistent pattern of results in the monohybrid crosses, Mendel derived the following three postulates, or principles, of inheritance.

1. UNIT FACTORS IN PAIRS
Genetic characters are controlled by unit factors existing in pairs in individual organisms.

In the monohybrid cross involving tall and dwarf stems, a specific unit factor exists for each trait. Each diploid individual receives one factor from each parent. Because the factors occur in pairs, three combinations are possible: two factors for tall stems, two factors for dwarf stems, or one of each factor. Every individual possesses one of these three combinations, which determines stem height.

2. DOMINANCE/RECESSIVENESS
When two unlike unit factors responsible for a single character are present in a single individual, one unit factor is dominant to the other, which is said to be recessive.

In each monohybrid cross, the trait expressed in the F_1 generation is controlled by the dominant unit factor. The trait not expressed is controlled by the recessive unit factor. The terms *dominant* and *recessive* are also used to designate traits. In this case, tall stems are said to be dominant over recessive dwarf stems.

3. SEGREGATION
During the formation of gametes, the paired unit factors separate, or segregate, randomly so that each gamete receives one or the other with equal likelihood.

If an individual contains a pair of like unit factors (e.g., both specific for tall), then all its gametes receive one of that same kind of unit factor (in this case, tall). If an individual contains unlike unit factors (e.g., one for tall and one for dwarf), then each gamete has a 50 percent probability of receiving either the tall or the dwarf unit factor.

These postulates provide a suitable explanation for the results of the monohybrid crosses. Let's use the tall/dwarf cross to illustrate. Mendel reasoned that P_1 tall plants contained identical paired unit factors, as did the P_1 dwarf

plants. The gametes of tall plants all receive one tall unit factor as a result of **segregation**. Similarly, the gametes of dwarf plants all receive one dwarf unit factor. Following fertilization, all F_1 plants receive one unit factor from each parent—a tall factor from one and a dwarf factor from the other—reestablishing the paired relationship, but because tall is dominant to dwarf, all F_1 plants are tall.

When F_1 plants form gametes, the postulate of segregation demands that each gamete randomly receives either the tall *or* dwarf unit factor. Following random fertilization events during F_1 selfing, four F_2 combinations will result with equal frequency:

1. tall/tall

2. tall/dwarf

3. dwarf/tall

4. dwarf/dwarf

Combinations (1) and (4) will clearly result in tall and dwarf plants, respectively. According to the postulate of dominance/recessiveness, combinations (2) and (3) will both yield tall plants. Therefore, the F_2 is predicted to consist of 3/4 tall and 1/4 dwarf, or a ratio of 3:1. This is approximately what Mendel observed in his cross between tall and dwarf plants. A similar pattern was observed in each of the other monohybrid crosses (Figure 3–1).

Modern Genetic Terminology

To analyze the monohybrid cross and Mendel's first three postulates, we must first introduce several new terms as well as a symbol convention for the unit factors. Traits such as tall or dwarf are physical expressions of the information contained in unit factors. The physical expression of a trait is the **phenotype** of the individual. Mendel's unit factors represent units of inheritance called **genes** by modern geneticists. For any given character, such as plant height, the phenotype is determined by alternative forms of a single gene, called **alleles**. For example, the unit factors representing tall and dwarf are alleles determining the height of the pea plant.

Geneticists have several different systems for using symbols to represent genes. In Chapter 4, we will review a number of these conventions, but for now, we will adopt one to use consistently throughout this chapter. According to this convention, the first letter of the recessive trait symbolizes the character in question; in lowercase italic, it designates the allele for the recessive trait, and in uppercase italic, it designates the allele for the dominant trait. Thus for Mendel's pea plants, we use *d* for the *d*warf allele and *D* for the tall allele. When alleles are written in pairs to represent the two unit factors present in any individual (*DD*, *Dd*, or *dd*), the resulting symbol is called the **genotype**. The genotype designates the genetic makeup of an individual for the trait or traits it describes, whether the individual is haploid or diploid. By reading the genotype, we know the phenotype of the individual: *DD* and *Dd* are tall, and *dd* is dwarf. When both alleles are the same (*DD* or *dd*), the individual is **homozygous** for the trait, or a **homozygote**; when the alleles are different (*Dd*), we use the terms **heterozygous** and **heterozygote**. These symbols and terms are used in Figure 3–2 to describe the monohybrid cross.

Mendel's Analytical Approach

What led Mendel to deduce that unit factors exist in pairs? Because there were two contrasting traits for each of the characters he chose, it seemed logical that two distinct factors must exist. However, why does one of the two traits or phenotypes disappear in the F_1 generation? Observation of the F_2 generation helps to answer this question. The recessive trait and its unit factor do not actually disappear in the F_1; they are merely hidden or masked, only to reappear in one-fourth of the F_2 offspring. Therefore, Mendel concluded that one unit factor for tall and one for dwarf were transmitted to each F_1 individual, but that because the tall factor or allele is dominant to the dwarf factor or allele, all F_1 plants are tall. Given this information, we can ask how Mendel explained the 3:1 F_2 ratio. As shown in Figure 3–2, Mendel deduced that the tall and dwarf alleles of the F_1 heterozygote segregate randomly into gametes. If fertilization is random, this ratio is predicted. If a large population of offspring is generated, the outcome of such a cross should reflect the 3:1 ratio.

Because he operated without the hindsight that modern geneticists enjoy, Mendel's analytical reasoning must be considered a truly outstanding scientific achievement. On the basis of rather simple but precisely executed breeding experiments, he not only proposed that discrete particulate units of heredity exist, but he also explained how they are transmitted from one generation to the next.

Punnett Squares

The genotypes and phenotypes resulting from combining gametes during fertilization can be easily visualized by constructing a diagram called a **Punnett square**, named after the person who first devised this approach, Reginald C. Punnett. Figure 3–3 illustrates this method of analysis for our $F_1 \times F_1$ monohybrid cross. Each of the possible gametes is assigned a column or a row; the

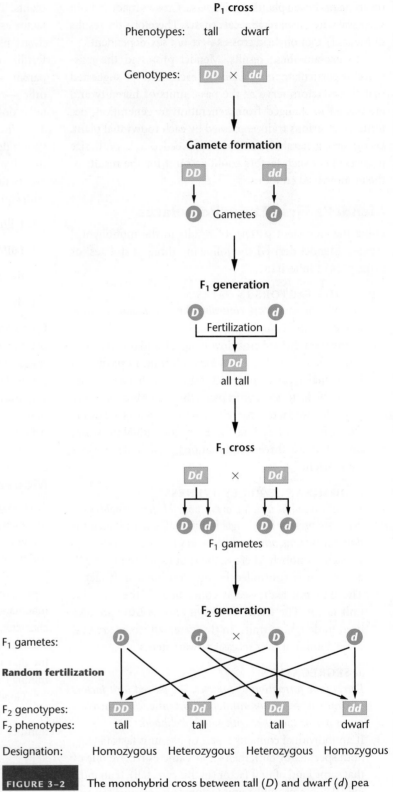

FIGURE 3–2   The monohybrid cross between tall (*D*) and dwarf (*d*) pea plants. Individuals are shown in rectangles, and gametes are shown in circles.

vertical columns represent those of the female parent, and the horizontal rows represent those of the male parent. After assigning the gametes to the rows and columns, we predict the new generation by entering the male and female gametic information into each box and thus producing every possible resulting genotype. By filling out the Punnett square, we are listing all possible random fertilization events. The genotypes and phenotypes of all potential offspring are ascertained by reading the combinations in the boxes.

The Punnett square method is particularly useful when you are first learning about genetics and how to solve genetics problems. Note the ease with which the 3:1 phenotypic ratio and the 1:2:1 genotypic ratio may be derived for the F_2 generation in Figure 3–3.

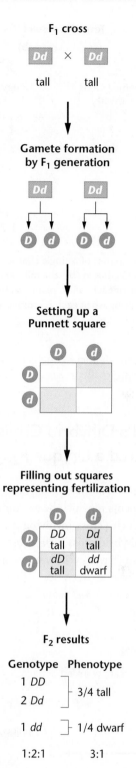

FIGURE 3–3 A Punnett square generating the F_2 ratio of the $F_1 \times F_1$ cross shown in Figure 3–2.

NOW SOLVE THIS

3–1 Pigeons may exhibit a checkered or plain color pattern. In a series of controlled matings, the following data were obtained.

	F₁ Progeny	
P₁ Cross	Checkered	Plain
(a) checkered × checkered	36	0
(b) checkered × plain	38	0
(c) plain × plain	0	35

Then F_1 offspring were selectively mated with the following results. (The P_1 cross giving rise to each F_1 pigeon is indicated in parentheses.)

	F₂ Progeny	
F₁ × F₁ Crosses	Checkered	Plain
(d) checkered (a) × plain (c)	34	0
(e) checkered (b) × plain (c)	17	14
(f) checkered (b) × checkered (b)	28	9
(g) checkered (a) × checkered (b)	39	0

How are the checkered and plain patterns inherited? Select and assign symbols for the genes involved, and determine the genotypes of the parents and offspring in each cross.

■ HINT: *This problem involves an understanding of how traits are inherited and transmitted to offspring. The key to its solution is first to determine whether there is more than one gene pair involved by converting the data to ratios that are characteristic of Mendelian crosses. In the case of this problem, you should consider whether any of the F_2 ratios match Mendel's 3:1 monohybrid ratio.*

The Testcross: One Character

Tall plants produced in the F_2 generation are predicted to have either the DD or the Dd genotype. You might ask if there is a way to distinguish the genotype. Mendel devised a rather simple method that is still used today to discover the genotype of plants and animals: the **testcross**. The organism expressing the dominant phenotype but having an unknown genotype is crossed with a known *homozygous recessive individual*. For example, as shown in **Figure 3–4(a)**, if a tall plant of genotype DD is testcrossed with a dwarf plant, which must have the dd genotype, all offspring will

be tall phenotypically and Dd genotypically. However, as shown in Figure 3–4(b), if a tall plant is Dd and is crossed with a dwarf plant (dd), then one-half of the offspring will be tall (Dd) and the other half will be dwarf (dd). Therefore, a 1:1 tall/dwarf ratio demonstrates the heterozygous nature of the tall plant of unknown genotype. The results of the

Testcross results

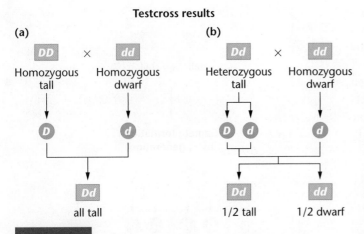

FIGURE 3-4   Testcross of a single character. In (a), the tall parent is homozygous, but in (b), the tall parent is heterozygous. The genotype of each tall P₁ plant can be determined by examining the offspring when each is crossed with the homozygous recessive dwarf plant.

testcross reinforced Mendel's conclusion that separate unit factors control traits.

3.3
Mendel's Dihybrid Cross Generated a Unique F_2 Ratio

As a natural extension of the monohybrid cross, Mendel also designed experiments in which he examined two characters simultaneously. Such a cross, involving two pairs of contrasting traits, is a **dihybrid cross**, or a *two-factor cross*. For example, if pea plants having yellow seeds that are round were bred with those having green seeds that are wrinkled, the results shown in Figure 3–5 would occur: the F_1 offspring would all be yellow and round. It is therefore apparent that

yellow is dominant to green and that round is dominant to wrinkled. When the F_1 individuals are selfed, approximately 9/16 of the F_2 plants express the yellow and round traits, 3/16 express yellow and wrinkled, 3/16 express green and round, and 1/16 express green and wrinkled.

A variation of this cross is also shown in Figure 3–5. Instead of crossing one P_1 parent with both dominant traits (yellow, round) to one with both recessive traits (green, wrinkled), plants with yellow, wrinkled seeds are crossed with those with green, round seeds. In spite of the change in the P_1 phenotypes, both the F_1 and F_2 results remain unchanged. Why this is so will become clear in the next section.

Mendel's Fourth Postulate: Independent Assortment

We can most easily understand the results of a dihybrid cross if we consider it theoretically as consisting of two monohybrid crosses conducted separately. Think of the two sets of traits as being inherited independently of each other; that is, the chance of any plant having yellow or green seeds is not at all influenced by the chance that this plant will have round or wrinkled seeds. Thus, because yellow is dominant to green, all F_1 plants in the first theoretical cross would have yellow seeds. In the second theoretical cross, all F_1 plants would have round seeds because round is dominant to wrinkled. When Mendel examined the F_1 plants of the dihybrid cross, all were yellow and round, as our theoretical cross suggests.

The predicted F_2 results of the first cross are 3/4 yellow and 1/4 green. Similarly, the second cross would yield 3/4 round and 1/4 wrinkled. Figure 3–5 shows that in the dihybrid cross, 12/16 F_2 plants are yellow, while 4/16 are green, exhibiting the expected 3:1 (3/4:1/4) ratio. Similarly, 12/16

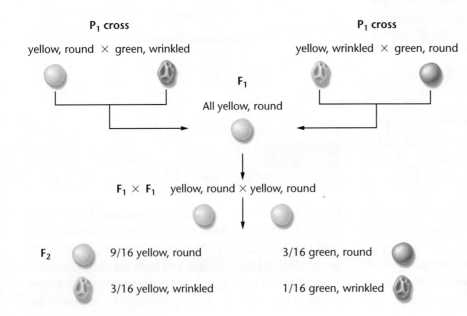

FIGURE 3-5 F_1 and F_2 results of Mendel's dihybrid crosses in which the plants on the top left with yellow, round seeds are crossed with plants having green, wrinkled seeds, and the plants on the top right with yellow, wrinkled seeds are crossed with plants having green, round seeds.

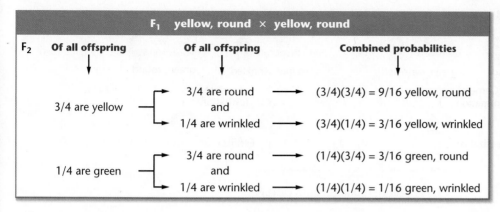

FIGURE 3–6   Computation of the combined probabilities of each F_2 phenotype for two independently inherited characters. The probability of each plant being yellow or green is independent of the probability of it bearing round or wrinkled seeds.

of all F_2 plants have round seeds, while 4/16 have wrinkled seeds, again revealing the 3:1 ratio.

These numbers demonstrate that the two pairs of contrasting traits are inherited independently, so we can predict the frequencies of all possible F_2 phenotypes by applying the **product law** of probabilities: *When two independent events occur simultaneously, the probability of the two outcomes occurring in combination is equal to the product of their individual probabilities of occurrence.* For example, the probability of an F_2 plant having yellow and round seeds is (3/4)(3/4), or 9/16, because 3/4 of all F_2 plants should be yellow and 3/4 of all F_2 plants should be round.

In a like manner, the probabilities of the other three F_2 phenotypes can be calculated: yellow (3/4) and wrinkled (1/4) are predicted to be present together 3/16 of the time; green (1/4) and round (3/4) are predicted 3/16 of the time; and green (1/4) and wrinkled (1/4) are predicted 1/16 of the time. These calculations are shown in **Figure 3–6**.

It is now apparent why the F_1 and F_2 results are identical whether the initial cross is yellow, round plants bred with green, wrinkled plants, or whether yellow, wrinkled plants are bred with green, round plants. In both crosses, the F_1 genotype of all offspring is identical. As a result, the F_2 generation is also identical in both crosses.

On the basis of similar results in numerous dihybrid crosses, Mendel proposed a fourth postulate:

4. INDEPENDENT ASSORTMENT
During gamete formation, segregating pairs of unit factors assort independently of each other.
This postulate stipulates that segregation of any pair of unit factors occurs independently of all others. As a result of random segregation, each gamete receives one member of every pair of unit factors. For one pair, whichever unit factor is received does not influence the outcome of segregation of any other pair. Thus, according to the postulate of **independent assortment**, all possible combinations of gametes should be formed in equal frequency.

The Punnett square in **Figure 3–7** shows how independent assortment works in the formation of the F_2 generation. Examine the formation of gametes by the F_1 plants; segregation

prescribes that every gamete receives either a G or g allele and a W or w allele. Independent assortment stipulates that all four combinations (GW, Gw, gW, and gw) will be formed with equal probabilities.

In every $F_1 \times F_1$ fertilization event, each zygote has an equal probability of receiving one of the four combinations from each parent. If many offspring are produced, 9/16 have yellow, round seeds, 3/16 have yellow, wrinkled seeds, 3/16 have green, round seeds, and 1/16 have green, wrinkled seeds, yielding what is designated as **Mendel's 9:3:3:1 dihybrid ratio**. This is an ideal ratio based on probability events involving segregation, independent assortment, and random fertilization. Because of deviation due strictly to chance, particularly if small numbers of offspring are produced, actual results are highly unlikely to match the ideal ratio.

The Testcross: Two Characters

The testcross may also be applied to individuals that express two dominant traits but whose genotypes are unknown. For example, the expression of the yellow, round seed phenotype in the F_2 generation just described may result from the $GGWW$, $GGWw$, $GgWW$, or $GgWw$ genotypes. If an F_2 yellow, round plant is crossed with the homozygous recessive green, wrinkled plant ($ggww$), analysis of the offspring will indicate the exact genotype of that yellow, round plant. Each of the above genotypes results in a different set of gametes and, in a testcross, a different set of phenotypes in the resulting offspring. You should work out the results of each of these four crosses to be sure that you understand this concept.

3.4

The Trihybrid Cross Demonstrates That Mendel's Principles Apply to Inheritance of Multiple Traits

Thus far, we have considered inheritance of up to two pairs of contrasting traits. Mendel demonstrated that the processes of segregation and independent assortment also apply to three pairs of contrasting traits, in what is called a **trihybrid cross**, or *three-factor cross*.

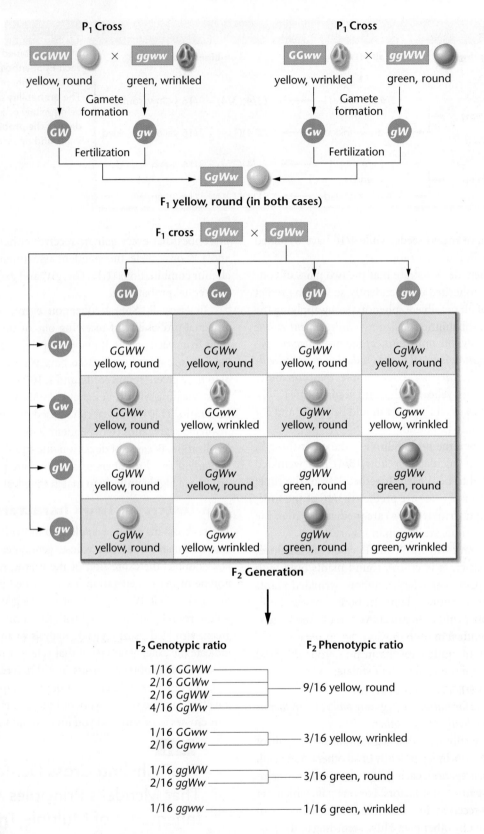

FIGURE 3–7 Analysis of the dihybrid crosses shown in Figure 3–5. The F_1 heterozygous plants are self-fertilized to produce an F_2 generation, which is computed using a Punnett square. Both the phenotypic and genotypic F_2 ratios are shown.

How Mendel's Peas Become Wrinkled: A Molecular Explanation

Only recently, well over a hundred years after Mendel used wrinkled peas in his groundbreaking hybridization experiments, have we come to find out how the *wrinkled* gene makes peas wrinkled. The wild-type allele of the gene encodes a protein called *starch-branching enzyme (SBEI)*. This enzyme catalyzes the formation of highly branched starch molecules as the seed matures.

Wrinkled peas, which result from the homozygous presence of the mutant form of the gene, lack the activity of this enzyme. As a consequence, the production of branch points is inhibited during the synthesis of starch within the seed, which in turn leads to the accumulation of more sucrose and a higher water content while the seed develops. Osmotic pressure inside the seed rises, causing the seed to lose water, ultimately resulting in a wrinkled appearance at maturity. In contrast, developing seeds that bear at least one copy of the normal gene (being either homozygous or heterozygous for the dominant allele) synthesize starch and achieve an osmotic balance that minimizes the loss of water. The end result for them is a smooth-textured outer coat.

Cloning and analysis of the *SBEI* gene has provided new insight into the relationships between genotypes and phenotypes. Interestingly, the mutant gene contains a foreign sequence of some 800 base pairs that disrupts the normal coding sequence. This foreign segment closely resembles sequences called **transposable elements** that have been discovered to have the ability to move from place to place in the genome of certain organisms. Transposable elements have been found in maize (corn), parsley, snapdragons, and fruit flies, among many other organisms.

Wrinkled and round garden peas, the phenotypic traits in one of Mendel's monohybrid crosses.

Although a trihybrid cross is somewhat more complex than a dihybrid cross, its results are easily calculated if the principles of segregation and independent assortment are followed. For example, consider the cross shown in Figure 3–8 where the gene pairs of theoretical contrasting traits are represented by the symbols *A, a, B, b, C,* and *c*. In the cross between *AABBCC* and *aabbcc* individuals, all F_1 individuals are heterozygous for all three gene pairs. Their genotype, *AaBbCc*, results in the phenotypic expression of the dominant *A, B,* and *C* traits. When F_1 individuals serve as parents, each produces eight different gametes in equal frequencies. At this point, we could construct a Punnett square with 64 separate boxes and read out the phenotypes—but such a method is cumbersome in a cross involving so many factors. Therefore, another method has been devised to calculate the predicted ratio.

The Forked-Line Method, or Branch Diagram

It is much less difficult to consider each contrasting pair of traits separately and then to combine these results by using the **forked-line method**, first shown in Figure 3–6. This method, also called a **branch diagram**, relies on the simple application of the laws of probability established for the dihybrid cross. Each gene pair is assumed to behave independently during gamete formation.

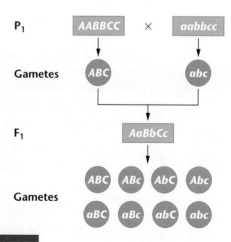

Trihybrid gamete formation

FIGURE 3–8 Formation of P_1 and F_1 gametes in a trihybrid cross.

3–2 Considering the Mendelian traits round versus wrinkled and yellow versus green, consider the crosses below and determine the genotypes of the parental plants by analyzing the phenotypes of their offspring.

Parental Plants	Offspring
(a) round, yellow × round, yellow	3/4 round, yellow
	1/4 wrinkled, yellow
	6/16 wrinkled, yellow
(b) wrinkled, yellow × round, yellow	2/16 wrinkled, green
	6/16 round, yellow
	2/16 round, green
(c) round, yellow × round, yellow	9/16 round, yellow
	3/16 round, green
	3/16 wrinkled, yellow
	1/16 wrinkled, green
(d) round, yellow × wrinkled, green	1/4 round, yellow
	1/4 round, green
	1/4 wrinkled, yellow
	1/4 wrinkled, green

■ HINT: *This problem involves an understanding of Mendelian postulates, including independent assortment. The key to its solution is in each case, to determine everything that you know for certain. This reduces the problem to its bare essentials and clarifies what remains to be figured out. For example, the wrinkled, yellow plant in case (b) must be homozygous for the recessive wrinkled alleles and bear at least one dominant allele for the yellow trait. Having established this, you need only determine the remaining allele for cotyledon color.*

When the monohybrid cross $AA \times aa$ is made, we know that:

1. All F_1 individuals have the genotype Aa and express the phenotype represented by the A allele, which is called the A phenotype in the discussion that follows.

2. The F_2 generation consists of individuals with either the A phenotype or the a phenotype in the ratio of 3:1.

The same generalizations can be made for the $BB \times bb$ and $CC \times cc$ crosses. Thus, in the F_2 generation, 3/4 of all organisms will express phenotype A, 3/4 will express B, and 3/4 will express C. Similarly, 1/4 of all organisms will express a, 1/4 will express b, and 1/4 will express c. The proportions of organisms that express each phenotypic combination can be predicted by assuming that fertilization, following the independent assortment of these three gene pairs during gamete formation, is a random process. We apply the product law of probabilities once again. Figure 3–9 uses the forked-line method to calculate the phenotypic proportions of the F_2 generation. They fall into the trihybrid ratio of 27:9:9:9:3:3:3:1. The same method can be used to solve crosses involving any number of gene pairs, *provided that all gene pairs assort independently from each other.* We shall see later that gene pairs do not always assort with complete independence. However, it appeared to be true for all of Mendel's characters.

Note that in Figure 3–9, only phenotypic ratios of the F_2 generation have been derived. It is possible to generate genotypic ratios as well. To do so, we again consider the A/a, B/b, and C/c gene pairs separately. For example, for the A/a pair, the F_1 cross is $Aa \times Aa$. Phenotypically, an F_2 ratio of 3/4 A:1/4 a is produced. Genotypically, however, the F_2 ratio is different—1/4 AA:1/2 Aa:1/4 aa will result. Using Figure 3–9 as a model, we would enter these genotypic frequencies in the leftmost column of the diagram. Each would be connected by three lines to 1/4 BB, 1/2 Bb, and 1/4 bb,

Generation of F₂ trihybrid phenotypes

A or a	B or b	C or c	Combined proportion	
3/4 A	3/4 B	3/4 C	(3/4)(3/4)(3/4) ABC = 27/64	ABC
		1/4 c	(3/4)(3/4)(1/4) ABc = 9/64	ABc
	1/4 b	3/4 C	(3/4)(1/4)(3/4) AbC = 9/64	AbC
		1/4 c	(3/4)(1/4)(1/4) Abc = 3/64	Abc
1/4 a	3/4 B	3/4 C	(1/4)(3/4)(3/4) aBC = 9/64	aBC
		1/4 c	(1/4)(3/4)(1/4) aBc = 3/64	aBc
	1/4 b	3/4 C	(1/4)(1/4)(3/4) abC = 3/64	abC
		1/4 c	(1/4)(1/4)(1/4) abc = 1/64	abc

FIGURE 3–9 Generation of the F_2 trihybrid phenotypic ratio using the forked-line method. This method is based on the expected probability of occurrence of each phenotype.

respectively. From each of these nine designations, three more lines would extend to the 1/4 *CC*, 1/2 *Cc*, and 1/4 *cc* genotypes. On the right side of the completed diagram, 27 genotypes and their frequencies of occurrence would appear.

In crosses involving two or more gene pairs, the calculation of gametes and genotypic and phenotypic results is quite complex. Several simple mathematical rules will enable you to check the accuracy of various steps required in working these problems. First, you must determine the number of different *heterozygous* gene pairs (n) involved in the cross—for example, where $AaBb \times AaBb$ represents the cross, $n = 2$; for $AaBbCc \times AaBcCc$, $n = 3$; for $AaBBCcDd \times AaBBCcDd$, $n = 3$ (because the *B* genes are not heterozygous). Once n is determined, 2^n is the number of different gametes that can be formed by each parent; 3^n is the number of different genotypes that result following fertilization; and 2^n is the number of different phenotypes that are produced from these genotypes. Table 3.1 summarizes these rules, which may be applied to crosses involving any number of genes, *provided that they assort independently from one another.*

TABLE 3.1

Simple Mathematical Rules Useful in Working Genetics Problems

Crosses between Organisms Heterozygous for Genes Exhibiting Independent Assortment			
Number of Heterozygous Gene Pairs	Number of Different Types of Gametes Formed	Number of Different Genotypes Produced	Number of Different Phenotypes Produced*
n	2^n	3^n	2^n
1	2	3	2
2	4	9	4
3	8	27	8
4	16	81	16

*The fourth column assumes a simple dominant–recessive relationship in each gene pair.

NOW SOLVE THIS

3–3 Using the forked-line, or branch diagram, method, determine the genotypic and phenotypic ratios of these trihybrid crosses: (a) $AaBbCc \times AaBBCC$, (b) $AaBBCc \times aaBBCc$, and (c) $AaBbCc \times AaBbCc$.

■ HINT: *This problem asks you to use the forked-line method to quickly determine genetic outcomes. The key to its solution is to consider each gene pair separately. First predict the outcome of the A/a genes and write these down. Then, for each of those outcomes, write predictions for the B/b genes. Finally, for each of those outcomes, write predictions for the C/c genes. At that point, you will be ready to determine the proportionate ratios of all the different possible combinations.*

Mendel's Work Was Rediscovered in the Early Twentieth Century

Mendel initiated his work in 1856, presented it to the Brünn Society of Natural Science in 1865, and published it the following year. While his findings were often cited and discussed, their significance went unappreciated for about 35 years. Many explanations have been proposed for this delay.

First, Mendel's adherence to mathematical analysis of probability events was quite unusual for biological studies in those days. Perhaps it seemed foreign to his contemporaries. More important, his conclusions did not fit well with existing hypotheses concerning the cause of variation among organisms. The topic of natural variation intrigued students of evolutionary theory. This group, stimulated by the proposal developed by Charles Darwin and Alfred Russel Wallace, subscribed to the theory of **continuous variation**, which held that offspring were a blend of their parents' phenotypes. As we mentioned earlier, Mendel theorized that variation was due to a dominance–recessive relationship between discrete or particulate units, resulting in **discontinuous variation**. For example, note that the F_2 flowers in Figure 3–1 are either white or violet, never something intermediate. Mendel proposed that the F_2 offspring of a dihybrid cross are expressing traits produced by new combinations of previously existing unit factors. As a result, Mendel's hypotheses did not fit well with the evolutionists' preconceptions about causes of variation.

It is also likely that Mendel's contemporaries failed to realize that Mendel's postulates explained *how* variation was transmitted to offspring. Instead, they may have attempted to interpret his work in a way that addressed the issue of *why* certain phenotypes survive preferentially. It was this latter question that had been addressed in the theory of natural selection, but it was not addressed by Mendel. The collective vision of Mendel's scientific colleagues may have been obscured by the impact of Darwin's extraordinary theory of organic evolution.

The Chromosomal Theory of Inheritance

In the latter part of the nineteenth century, a remarkable observation set the scene for the recognition of Mendel's work: Walter Flemming's discovery of chromosomes in the nuclei of salamander cells. In 1879, Flemming described the behavior of these thread-like structures during cell division. As a result of his findings and the work of many other cytologists, the presence of discrete units within the nucleus soon became an integral part of scientists' ideas about inheritance.

In the early twentieth century, hybridization experiments similar to Mendel's were performed independently by three botanists, Hugo de Vries, Karl Correns, and Erich Tschermak. De Vries's work demonstrated the principle of segregation in several plant species. Apparently, he searched the existing literature and found that Mendel's work had anticipated his own conclusions! Correns and Tschermak also reached conclusions similar to those of Mendel.

About the same time, two cytologists, Walter Sutton and Theodor Boveri, independently published papers linking their discoveries of the behavior of chromosomes during meiosis to the Mendelian principles of segregation and independent assortment. They pointed out that the separation of chromosomes during meiosis could serve as the cytological basis of these two postulates. Although they thought that Mendel's unit factors were probably chromosomes rather than genes on chromosomes, their findings reestablished the importance of Mendel's work and led to many ensuing genetic investigations. Sutton and Boveri are credited with initiating the **chromosomal theory of inheritance**, the idea that the genetic material in living organisms is contained in chromosomes, which was developed during the next two decades. As we will see in subsequent chapters, work by Thomas H. Morgan, Alfred H. Sturtevant, Calvin Bridges, and others established beyond a reasonable doubt that Sutton's and Boveri's hypothesis was correct.

Unit Factors, Genes, and Homologous Chromosomes

Because the correlation between Sutton's and Boveri's observations and Mendelian principles serves as the foundation for the modern description of transmission genetics, we will examine this correlation in some depth before moving on to other topics.

As we know, each species possesses a specific number of chromosomes in each somatic cell nucleus. For diploid organisms, this number is called the **diploid number (2n)** and is characteristic of that species. During the formation of gametes (meiosis), the number is precisely halved (n), and when two gametes combine during fertilization, the diploid number is reestablished. During meiosis, however, the chromosome number is not reduced in a random manner. It was apparent to early cytologists that the diploid number of chromosomes is composed of homologous pairs identifiable by their morphological appearance and behavior. The gametes contain one member of each pair—thus the chromosome complement of a gamete is quite specific, and the number of chromosomes in each gamete is equal to the haploid number.

With this basic information, we can see the correlation between the behavior of unit factors and chromosomes and

genes. Figure 3–10 shows three of Mendel's postulates and the chromosomal explanation of each. Unit factors are really genes located on homologous pairs of chromosomes [Figure 3–10(a)]. Members of each pair of homologs separate, or segregate, during gamete formation [Figure 3–10(b)]. In the figure, two different alignments are possible, both of which are shown.

To illustrate the principle of independent assortment, it is important to distinguish between members of any given homologous pair of chromosomes. One member of each pair is derived from the **maternal parent**, whereas the other comes from the **paternal parent**. (We represent the different parental origins with different colors.) As shown in Figure 3–10(c), following independent segregation of each pair of homologs, each gamete receives one member from each pair of chromosomes. All possible combinations are formed with equal probability. If we add the symbols used in Mendel's dihybrid cross (G, g and W, w) to the diagram, we can see why equal numbers of the four types of gametes are formed. The independent behavior of Mendel's pairs of unit factors (G and W in this example) is due to their presence on separate pairs of homologous chromosomes.

Observations of the phenotypic diversity of living organisms make it logical to assume that there are many more genes than chromosomes. Therefore, each homolog must carry genetic information for more than one trait. The currently accepted concept is that a chromosome is composed of a large number of linearly ordered, information-containing genes. Mendel's paired unit factors (which determine tall or dwarf stems, for example) actually constitute a pair of genes located on one pair of homologous chromosomes. The location on a given chromosome where any particular gene occurs is called its **locus** (pl. loci). The different alleles of a given gene (for example, G and g) contain slightly different genetic information (green or yellow) that determines the same character (seed color in this case). Although we have examined only genes with two alternative alleles, most genes have more than two allelic forms. We conclude this section by reviewing the criteria necessary to classify two chromosomes as a homologous pair:

1. During mitosis and meiosis, when chromosomes are visible in their characteristic shapes, both members of a homologous pair are the same size and exhibit identical centromere locations. The sex chromosomes (e.g., the X and the Y chromosomes in mammals) are an exception.

2. During early stages of meiosis, homologous chromosomes form pairs, or synapse.

3. Although it is not generally visible under the microscope, homologs contain the identical linear order of gene loci.

(a) Unit factors in pairs (first meiotic prophase)

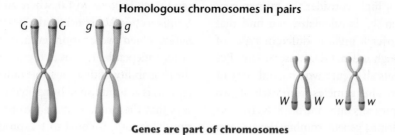

Homologous chromosomes in pairs

Genes are part of chromosomes

(b) Segregation of unit factors during gamete formation (first meiotic anaphase)

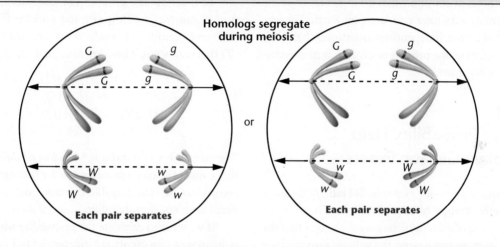

Homologs segregate
during meiosis

or

Each pair separates   Each pair separates

(c) Independent assortment of segregating unit factors (following many meiotic events)

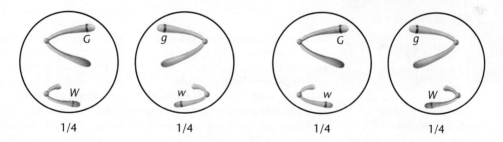

Nonhomologous chromosomes assort independently

1/4   1/4   1/4   1/4

All possible gametic combinations are formed with equal probability

FIGURE 3–10 Illustrated correlation between the Mendelian postulates of (a) unit factors in pairs, (b) segregation, and (c) independent assortment, showing the presence of genes located on homologous chromosomes and their behavior during meiosis.

3.6

Independent Assortment Leads to Extensive Genetic Variation

One consequence of independent assortment is the production by an individual of genetically dissimilar gametes. Genetic variation results because the two members of any homologous pair of chromosomes are rarely, if ever, genetically identical. As the maternal and paternal members of all pairs are distributed to gametes through independent assortment, all possible chromosome combinations are produced, leading to extensive genetic diversity.

We have seen that the number of possible gametes, each with different chromosome compositions, is 2^n, where n equals the haploid number. Thus, if a species has a haploid number of 4, then 2^4, or 16, different gamete combinations can be formed as a result of independent assortment. Although this number is not high, consider the human species, where $n = 23$. When 2^{23} is calculated, we find that in excess of 8×10^6, or over 8 million, different types of gametes are possible through independent assortment. Because fertilization represents an event involving only one of approximately 8×10^6 possible gametes from each of two parents, each offspring represents only one of $(8 \times 10^6)^2$ or one of only 64×10^{12} potential genetic combinations. Given that this probability is less than one in one trillion, it is no wonder that, except for identical twins, each member of the human species exhibits a distinctive set of traits—this number of combinations of chromosomes is far greater than the number of humans who have ever lived on Earth! Genetic variation resulting from independent assortment has been extremely important to the process of evolution in all sexually reproducing organisms.

<hr>

3.7

Laws of Probability Help to Explain Genetic Events

Recall that genetic ratios—for example, 3/4 tall:1/4 dwarf—are most properly thought of as probabilities. These values predict the outcome of each fertilization event, such that the probability of each zygote having the genetic potential for

becoming tall is 3/4, whereas the potential for its being a dwarf is 1/4. Probabilities range from 0.0, where an event *is certain not to occur,* to 1.0, where an event *is certain to occur.* In this section, we consider the relation of probability to genetics. When two or more events with known probabilities occur independently but at the same time, we can calculate the probability of their possible outcomes occurring together. This is accomplished by applying the **product law**, which says that the probability of two or more events occurring simultaneously is equal to the product of their individual probabilities (see Section 3.3). Two or more events are independent of one another if the outcome of each one does not affect the outcome of any of the others under consideration.

To illustrate the product law, consider the possible results if you toss a penny (P) and a nickel (N) at the same time and examine all combinations of heads (H) and tails (T) that can occur. There are four possible outcomes:

$$(P_H{:}N_H) = (1/2)(1/2) = 1/4$$
$$(P_T{:}N_H) = (1/2)(1/2) = 1/4$$
$$(P_H{:}N_T) = (1/2)(1/2) = 1/4$$
$$(P_T{:}N_T) = (1/2)(1/2) = 1/4$$

The probability of obtaining a head or a tail in the toss of either coin is 1/2 and is unrelated to the outcome for the other coin. Thus, all four possible combinations are predicted to occur with equal probability.

If we want to calculate the probability when the possible outcomes of two events are independent of one another but

Tay–Sachs Disease: The Molecular Basis of a Recessive Disorder in Humans

An interesting question involving Mendelian traits centers around how mutant genes result in mutant phenotypes. Insights are gained by considering a modern explanation of the gene that causes **Tay–Sachs disease** (TSD), a devastating inherited recessive disorder involving unalterable destruction of the central nervous system. Infants with TSD are unaffected at birth and appear to develop normally until they are about 6 months old. Then, a progressive loss of mental and physical abilities occurs. Afflicted infants eventually become blind, deaf, mentally retarded, and paralyzed, often

within only a year or two, seldom living beyond age 5. Typical of rare autosomal recessive disorders, two unaffected heterozygous parents, who most often have no immediate family history of the disorder, have a probability of one in four of having a Tay–Sachs child.

We know that proteins are the end products of the expression of most all genes. The protein product involved in TSD has been identified, and we now have a clear understanding of the underlying molecular basis of the disorder. TSD results from the loss of activity of a single enzyme **hexosaminidase A (Hex-A)**. Hex-A, normally found in lysosomes within cells, is needed to break down the ganglioside GM2, a lipid component of nerve cell membranes. Without functional Hex-A, gangliosides accumulate

within neurons in the brain and cause deterioration of the nervous system. Heterozygous carriers of TSD with one normal copy of the gene produce only about 50 percent of the normal amount of Hex-A, but they show no symptoms of the disorder. The observation that the activity of only one gene (one wild-type allele) is sufficient for the normal development and function of the nervous system explains and illustrates the molecular basis of recessive mutations. Only when both genes are disrupted by mutation is the mutant phenotype evident. The responsible gene is located on chromosome 15 and codes for the alpha subunit of the Hex-A enzyme. More than 50 different mutations within the gene have been identified that lead to TSD phenotypes.

can be accomplished in more than one way, we can apply the **sum law**. For example, what is the probability of tossing our penny and nickel and obtaining one head and one tail? In such a case, we do not care whether it is the penny or the nickel that comes up heads, provided that the other coin has the alternative outcome. As we saw above, there are two ways in which the desired outcome can be accomplished, each with a probability of 1/4. The sum law states that the probability of obtaining any single outcome, where that outcome can be achieved by two or more events, is equal to the sum of the individual probabilities of all such events. Thus, according to the sum law, the overall probability in our example is equal to

$$(1/4) + (1/4) = 1/2$$

One-half of all two-coin tosses are predicted to yield the desired outcome.

These simple probability laws will be useful throughout our discussions of transmission genetics and for solving genetics problems. In fact, we already applied the product law when we used the forked-line method to calculate the phenotypic results of Mendel's dihybrid and trihybrid crosses. When we wish to know the results of a cross, we need only calculate the probability of each possible outcome. The results of this calculation then allow us to predict the proportion of offspring expressing each phenotype or each genotype.

An important point to remember when you deal with probability is that predictions of possible outcomes are based on large sample sizes. If we predict that 9/16 of the offspring of a dihybrid cross will express both dominant traits, it is very unlikely that, in a small sample, exactly 9 of every 16 will express this phenotype. Instead, our prediction is that, of a large number of offspring, approximately 9/16 will do so. The deviation from the predicted ratio in smaller sample sizes is attributed to chance, a subject we examine in our discussion of statistics in the next section. As you shall see, the impact of deviation due strictly to chance diminishes as the sample size increases.

The Binomial Theorem

Probability calculations using the **binomial theorem** can be used to analyze cases where there are alternative ways to achieve a combination of events. For families of any size, we can calculate the probability of any combination of male and female children. For example, what is the probability that in a family with four children two will be male and two will be female? This question is complex because each birth is an independent event, and multiple birth orders can achieve the same overall outcome.

The expression of the binomial theorem is

$$(a + b)^n = 1$$

where a and b are the respective probabilities of the two alternative outcomes and n equals the number of trials.

n	Binomial	Expanded Binomial
1	$(a + b)^1$	$a + b$
2	$(a + b)^2$	$a^2 + 2ab + b^2$
3	$(a + b)^3$	$a^3 + 3a^2b + 3ab^2 + b^3$
4	$(a + b)^4$	$a^4 + 4a^3b + 6a^2b^2 + 4ab^3 + b^4$
5	$(a + b)^5$	$a^5 + 5a^4b + 10a^3b^2 + 10a^2b^3 + 5ab^4 + b^5$
etc.		etc.

As the value of n increases and the expanded binomial becomes more complex, Pascal's triangle, shown in Table 3.2, is a useful way to determine the numerical coefficient of each term in the expanded equation. Starting with the third line from the top of this triangle, each number is the sum of the two numbers immediately above it.

To expand any binomial, the various exponents of a and b (e.g., a^3b^2) are determined using the pattern

$$(a + b)^n = a^n, a^{n-1}b, a^{n-2}b^2, a^{n-3}b^3, \ldots, b^n$$

Using these methods for setting up the expression, we find that the expansion of $(a + b)^7$ is

$$a^7 + 7a^6b + 21a^5b^2 + 35a^4b^3 + \cdots + b^7$$

Let's now return to our original question: *What is the probability that in a family with four children two will be male and two will be female?*

First, assign initial probabilities to each outcome:

$$a = \text{male} = 1/2$$
$$b = \text{female} = 1/2$$

Then write out the expanded binomial for the value of $n = 4$,

$$(a + b)^4 = a^4 + 4a^3b + 6a^2b^2 + 4ab^3 + b^4$$

Each term represents a possible outcome, with the exponent of a representing the number of males and the exponent

TABLE 3.2

Pascal's Triangle

n	Numerical Coefficients
	1
1	1 1
2	1 2 1
3	1 3 3 1
4	1 4 6 4 1
5	1 5 10 10 5 1
6	1 6 15 20 15 6 1
7	1 7 21 35 35 21 7 1
etc.	etc.

*Notice that all numbers other than the 1's are equal to the sum of the two numbers directly above them.

of b representing the number of females. Therefore, the term describing the outcome of two males and two females—the expression of the probability (p) we are looking for—is

$$p = 6a^2b^2$$
$$= 6(1/2)^2(1/2)^2$$
$$= 6(1/2)^4$$
$$= 6(1/16)$$
$$= 6/16$$
$$p = 3/8$$

Thus, the probability of families of four children having two boys and two girls is 3/8. Of all families with four children, 3 out of 8 are predicted to have two boys and two girls.

Before examining one other example, we should note that if you prefer not to use Pascal's triangle, a formula can be used to determine the numerical coefficient for any set of exponents,

$$n! \, / \, (s!t!)$$

where

$$n = \text{the total number of events}$$
$$s = \text{the number of times outcome } a \text{ occurs}$$
$$t = \text{the number of times outcome } b \text{ occurs}$$

Therefore, $n = s + t$

The symbol ! denotes a factorial, which is the product of all the positive integers from 1 through some positive integer. For example,

$$5! = (5)(4)(3)(2)(1) = 120$$

Note that in factorials, $0! = 1$.

Using the formula, let's determine the probability that in a family with seven children, five will be males and two females. In this case, $n = 7$, $s = 5$, and $t = 2$. We begin by setting up our equation to find the term for five events having outcome a and two events having outcome b:

$$p = \frac{n!}{s!t!}a^sb^t$$
$$= \frac{7!}{5!2!}(1/2)^5(1/2)^2$$
$$= \frac{(7) \cdot (6) \cdot (5) \cdot (4) \cdot (3) \cdot (2) \cdot (1)}{(5) \cdot (4) \cdot (3) \cdot (2) \cdot (1) \cdot (2) \cdot (1)}(1/2)^7$$
$$= \frac{(7) \cdot (6)}{(2) \cdot (1)}(1/2)^7$$
$$= \frac{42}{2}(1/2)^7$$
$$= 21(1/2)^7$$
$$= 21(1/128)$$
$$p = 21/128$$

Of families with seven children, on the average, 21/128 are predicted to have five males and two females.

Calculations using the binomial theorem have various applications in genetics, including the analysis of polygenic traits (Chapter 23) and studies of population equilibrium (Chapter 25).

Chi-Square Analysis Evaluates the Influence of Chance on Genetic Data

Mendel's 3:1 monohybrid and 9:3:3:1 dihybrid ratios are hypothetical predictions based on the following assumptions: (1) each allele is dominant or recessive, (2) segregation is unimpeded, (3) independent assortment occurs, and (4) fertilization is random. The final two assumptions are influenced by chance events and therefore are subject to random fluctuation. This concept of **chance deviation** is most easily illustrated by tossing a single coin numerous times and recording the number of heads and tails observed. In each toss, there is a probability of 1/2 that a head will occur and a probability of 1/2 that a tail will occur. Therefore, the expected ratio of many tosses is 1/2:1/2, or 1:1. If a coin is tossed 1000 times, usually *about* 500 heads and 500 tails will be observed. Any reasonable fluctuation from this hypothetical ratio (e.g., 486 heads and 514 tails) is attributed to chance.

As the total number of tosses is reduced, the impact of chance deviation increases. For example, if a coin is tossed only four times, you would not be too surprised if all four tosses resulted in only heads or only tails. For 1000 tosses, however, 1000 heads or 1000 tails would be most unexpected. In fact, you might believe that such a result would be impossible. Actually, all heads or all tails in 1000 tosses can be predicted to occur with a probability of $(1/2)^{1000}$. Since $(1/2)^{20}$ is less than one in a million times, an event occurring with a probability as small as $(1/2)^{1000}$ is virtually impossible. Two major points to keep in mind when predicting or analyzing genetic outcomes are:

1. The outcomes of independent assortment and fertilization, like coin tossing, are subject to random fluctuations from their predicted occurrences as a result of chance deviation.

2. As the sample size increases, the average deviation from the expected results decreases. Therefore, a larger sample size diminishes the impact of chance deviation on the final outcome.

Chi-Square Calculations and the Null Hypothesis

In genetics, being able to evaluate observed deviation is a crucial skill. When we assume that data will fit a given ratio such as 1:1, 3:1, or 9:3:3:1, we establish what is called the **null**

TABLE 3.3

Chi-Square Analysis

(a) Monohybrid

Cross Expected Ratio	Observed (o)	Expected (e)	Deviation ($o - e$)	Deviation (d^2)	d^2/e
3/4	740	3/4(1000) = 750	740−750 = −10	$(-10)^2 = 100$	100/750 = 0.13
1/4	260	1/4(1000) = 250	260−250 = +10	$(+10)^2 = 100$	100/250 = 0.40
	Total = 1000				$\chi^2 = 0.53$
					$p = 0.48$

(b) Dihybrid

Cross Expected Ratio	o	e	$(o - e)$	d^2	d^2/e
9/16	587	567	+20	400	0.71
3/16	197	189	+8	64	0.34
3/16	168	189	−21	441	2.33
1/16	56	63	−7	49	0.78
	Total = 1008				$\chi^2 = 4.16$
					$p = 0.26$

hypothesis (H_0). It is so named because the hypothesis assumes that there is *no real difference* between the *measured values* (or ratio) and the *predicted values* (or ratio). Any apparent difference can be attributed purely to chance. The validity of the null hypothesis for a given set of data is measured using statistical analysis. Depending on the results of this analysis, the null hypothesis may either (1) *be rejected* or (2) *fail to be rejected*. If it is rejected, the observed deviation from the expected result is judged not to be attributable to chance alone. In this case, the null hypothesis and the underlying assumptions leading to it must be reexamined. If the null hypothesis fails to be rejected, any observed deviations are attributed to chance.

One of the simplest statistical tests for assessing the goodness of fit of the null hypothesis is **chi-square (χ^2) analysis.** This test takes into account the observed deviation in each component of a ratio (from what was expected) as well as the sample size and reduces them to a single numerical value. The value for χ^2 is then used to estimate how frequently the observed deviation can be expected to occur strictly as a result of chance. The formula used in chi-square analysis is

$$\chi^2 = \Sigma \frac{(o - e)^2}{e}$$

where o is the observed value for a given category, e is the expected value for that category, and Σ (the Greek letter sigma) represents the sum of the calculated values for each category in the ratio. Because ($o - e$) is the deviation (d) in each case, the equation reduces to

$$\chi^2 = \Sigma \frac{d^2}{e}$$

Table 3.3(a) shows the steps in the χ^2 calculation for the F_2 results of a hypothetical monohybrid cross. To analyze the data obtained from this cross, work from left to right across the table, verifying the calculations as appropriate. Note that regardless of whether the deviation d is positive or negative, d^2 always becomes positive after the number is squared. In Table 3.3(b) F_2 results of a hypothetical dihybrid cross are analyzed. Make sure that you understand how each number was calculated in this example.

The final step in chi-square analysis is to interpret the χ^2 value. To do so, you must initially determine a value called the **degrees of freedom (df)**, which is equal to $n - 1$, where n is the number of different categories into which the data are divided, in other words, the number of possible outcomes. For the 3:1 ratio, $n = 2$, so $df = 1$. For the 9:3:3:1 ratio, $n = 4$ and $df = 3$. Degrees of freedom must be taken into account because the greater the number of categories, the more deviation is expected as a result of chance.

Once you have determined the degrees of freedom, you can interpret the χ^2 value in terms of a corresponding **probability value (p)**. Since this calculation is complex, we usually take the p value from a standard table or graph. Figure 3–11 shows a wide range of χ^2 values and the corresponding p values for various degrees of freedom in both a graph and a table. Let's use the graph to explain how to determine the p value. The caption for Figure 3–11(b) explains how to use the table.

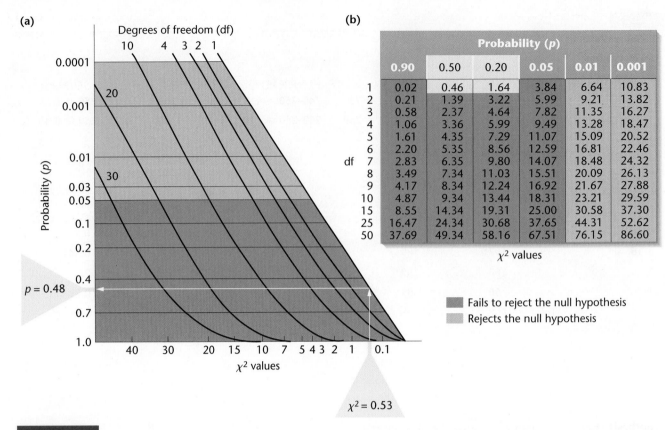

(a)

(b)

FIGURE 3-11   (a) Graph for converting χ^2 values to p values. (b) Table of χ^2 values for selected values of df and p. χ^2 values that lead to a p value of 0.05 or greater (darker blue areas) justify failure to reject the null hypothesis. Values leading to a p value of less than 0.05 (lighter blue areas) justify rejecting the null hypothesis. For example, the table in part (b) shows that for $\chi^2 = 0.53$ with 1 degree of freedom, the corresponding p value is between 0.20 and 0.50. The graph in (a) gives a more precise p value of 0.48 by interpolation. Thus, we fail to reject the null hypothesis.

To determine p using the graph, execute the following steps:

1. Locate the χ^2 value on the abscissa (the horizontal axis, or x-axis).

2. Draw a vertical line from this point up to the line on the graph representing the appropriate df.

3. From there, extend a horizontal line to the left until it intersects the ordinate (the vertical axis, or y-axis).

4. Estimate, by interpolation, the corresponding p value.

We used these steps for the monohybrid cross in Table 3.3(a) to estimate the p value of 0.48, as shown in Figure 3–11(a). Now try this method to see if you can determine the p value for the dihybrid cross [Table 3.3(b)]. Since the χ^2 value is 4.16 and $df = 3$, an approximate p value is 0.26. Checking this result in the table confirms that p values for both the monohybrid and dihybrid crosses are between 0.20 and 0.50.

Interpreting Probability Values

So far, we have been concerned with calculating χ^2 values and determining the corresponding p values. These steps

bring us to the most important aspect of chi-square analysis: understanding the meaning of the p value. It is simplest to think of the p value as a percentage. Let's use the example of the dihybrid cross in Table 3.3(b) where $p = 0.26$, which can be thought of as 26 percent. In our example, the p value indicates that if we repeat the same experiment many times, 26 percent of the trials would be expected to exhibit chance deviation as great as or greater than that seen in the initial trial. Conversely, 74 percent of the repeats would show less deviation than initially observed as a result of chance. Thus, the p value reveals that a null hypothesis (concerning the 9:3:3:1 ratio, in this case) is never proved or disproved absolutely. Instead, a relative standard is set that we use to either *reject* or *fail to reject* the null hypothesis. This standard is most often a p value of 0.05. When applied to chi-square analysis, a p value less than 0.05 means that the observed deviation in the set of results will be obtained by chance alone less than 5 percent of the time. Such a p value indicates that the difference between the observed and predicted results is substantial and requires us to reject the null hypothesis.

On the other hand, p values of 0.05 or greater (0.05 to 1.0) indicate that the observed deviation will be obtained by chance alone 5 percent or more of the time. This conclusion allows us not to reject the null hypothesis (when we are using $p = 0.05$ as our standard). Thus, with its p value of 0.26, the null hypothesis that independent assortment accounts for the results fails to be rejected. Therefore, the observed deviation can be reasonably attributed to chance.

A final note is relevant here concerning the case where the null hypothesis is rejected, that is, where $p < 0.05$. Suppose we had tested a dataset to assess a possible 9:3:3:1 ratio, as in Table 3.3(b), but we rejected the null hypothesis based on our calculation. What are alternative interpretations of the data? Researchers will reassess the assumptions that underlie the null hypothesis. In our dyhibrid cross, we assumed that segregation operates faithfully for both gene pairs. We also assumed that fertilization is random and that the viability of all gametes is equal regardless of genotype—that is, all gametes are equally likely to participate in fertilization. Finally, we assumed that, following fertilization, all preadult stages and adult offspring are equally viable, regardless of their genotype. If any of these assumptions is incorrect, then the original hypothesis is not necessarily invalid.

An example will clarify this point. Suppose our null hypothesis is that a dihybrid cross between fruit flies will result in 3/16 mutant wingless flies. However, perhaps fewer of the mutant embryos are able to survive their preadult development or young adulthood compared to flies whose genotype gives rise to wings. As a result, when the data are gathered, there will be fewer than 3/16 wingless flies. Rejection of the null hypothesis is not in itself cause for us to reject the validity of the postulates of segregation and independent assortment, because other factors we are unaware of may also be affecting the outcome.

NOW SOLVE THIS

3–4 In one of Mendel's dihybrid crosses, he observed 315 round yellow, 108 round green, 101 wrinkled yellow, and 32 wrinkled green F_2 plants. Analyze these data using the χ^2 test to see if
  (a) they fit a 9:3:3:1 ratio.
  (b) the round:wrinkled data fit a 3:1 ratio.
  (c) the yellow:green data fit a 3:1 ratio.

■ HINT: *This problem involves an understanding of χ^2 analysis, as used to determine whether a specific dataset fits certain ratios. The key to its solution is to first determine the expected outcome for each predicted ratio. Then, following a stepwise approach, determine the deviation in each case, and then calculate and interpret each χ^2 value.*

3.9

Pedigrees Reveal Patterns of Inheritance of Human Traits

We now explore how to determine the mode of inheritance of phenotypes in humans, where experimental matings are not made and where relatively few offspring are available for study. The traditional way to study inheritance has been to construct a family tree, indicating the presence or absence of the trait in question for each member of each generation. Such a family tree is called a **pedigree**. By analyzing a pedigree, we may be able to predict how the trait under study is inherited—for example, is it due to a dominant or recessive allele? When many pedigrees for the same trait are studied, we can often ascertain the mode of inheritance.

Pedigree Conventions

Figure 3–12 illustrates some of the conventions geneticists follow in constructing pedigrees. Circles represent females and squares designate males. If the sex of an individual is unknown, a diamond is used. Parents are generally connected to each other by a single horizontal line, and vertical lines lead to their offspring. If the parents are related—that is, **consanguineous**—such as first cousins, they are connected by a double line. Offspring are called **sibs** (short for **siblings**) and are connected by a horizontal **sibship line**. Sibs are placed in birth order from left to right and are labeled with Arabic numerals. Parents also receive an Arabic number

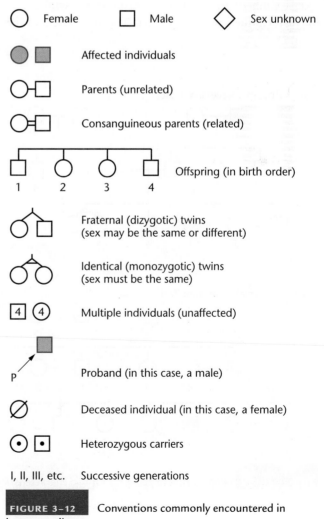

FIGURE 3–12 Conventions commonly encountered in human pedigrees.

designation. Each generation is indicated by a Roman numeral. When a pedigree traces only a single trait, the circles, squares, and diamonds are shaded if the phenotype being considered is expressed and unshaded if not. In some pedigrees, those individuals that fail to express a recessive trait but are known with certainty to be heterozygous carriers have a shaded dot within their unshaded circle or square. If an individual is deceased and the phenotype is unknown, a diagonal line is placed over the circle or square.

Twins are indicated by diagonal lines stemming from a vertical line connected to the sibship line. For identical, or **monozygotic**, twins, the diagonal lines are linked by a horizontal line. Fraternal, or **dizygotic**, twins lack this connecting line. A number within one of the symbols represents that number of sibs of the same sex and of the same or unknown phenotypes. The individual whose phenotype first brought attention to the family is called the **proband** and is indicated by an arrow connected to the designation **p**. This term applies to either a male or a female.

Pedigree Analysis

In Figure 3–13, two pedigrees are shown. The first is a representative pedigree for a trait that demonstrates autosomal recessive inheritance, such as **albinism**, where synthesis of the pigment melanin in obstructed. The male parent of the

first generation (I-1) is affected. Characteristic of a situation in which a parent has a rare recessive trait, the trait "disappears" in the offspring of the next generation. Assuming recessiveness, we might predict that the unaffected female parent (I-2) is a homozygous normal individual because none of the offspring show the disorder. Had she been heterozygous, one-half of the offspring would be expected to exhibit albinism, but none do. However, such a small sample (three offspring) prevents our knowing for certain.

Further evidence supports the prediction of a recessive trait. If albinism were inherited as a dominant trait, individual II-3 would have to express the disorder in order to pass it to his offspring (III-3 and III-4), but he does not. Inspection of the offspring constituting the third generation (row III) provides still further support for the hypothesis that albinism is a recessive trait. If it is, parents II-3 and II-4 are both heterozygous, and approximately one-fourth of their offspring should be affected. Two of the six offspring do show albinism. This deviation from the expected ratio is not unexpected in crosses with few offspring. Once we are confident that albinism is inherited as an autosomal recessive trait, we could portray the II-3 and II-4 individuals with a shaded dot within their larger square and circle. Finally, we can note that, characteristic of pedigrees for autosomal traits, both males and females are affected with equal

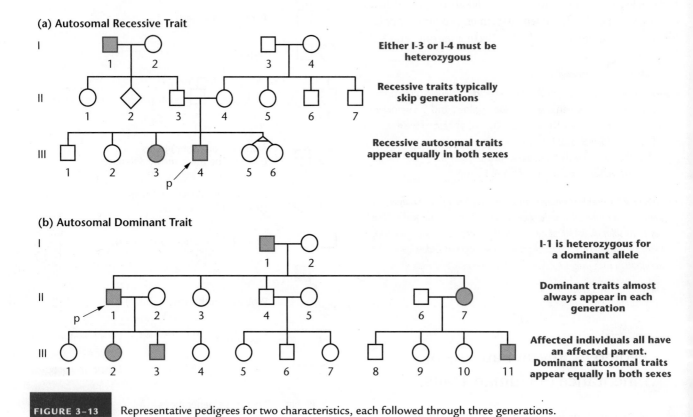

FIGURE 3–13   Representative pedigrees for two characteristics, each followed through three generations.

probability. In Chapter 4, we will examine a pedigree representing a gene located on the sex-determining X chromosome. We will see certain patterns characteristic of the transmission of X-linked traits, such as that these traits are more prevalent in male offspring and are never passed from affected fathers to their sons.

The second pedigree illustrates the pattern of inheritance for a trait such as Huntington disease, which is caused by an autosomal dominant allele. The key to identifying a pedigree that reflects a dominant trait is that all affected offspring will have a parent that also expresses the trait. It is also possible, by chance, that none of the offspring will inherit the dominant allele. If so, the trait will cease to exist in future generations. Like recessive traits, provided that the gene is autosomal, both males and females are equally affected.

When a given autosomal dominant disease is rare within the population, and most are, then it is highly unlikely that affected individuals will inherit a copy of the mutant gene from both parents. Therefore, in most cases, affected individuals are heterozygous for the dominant allele. As a result, approximately one-half of the offspring inherit it. This is borne out in the second pedigree in Figure 3–13. Furthermore, if a mutation is dominant, and a single copy is sufficient to produce a mutant phenotype, homozygotes are likely to be even more severely affected, perhaps even failing to survive. An illustration of this is the dominant gene for **familial hypercholesterolemia**. Heterozygotes display a defect in their receptors for low-density lipoproteins, the so-called LDLs (known popularly as "bad cholesterol"). As a result, too little cholesterol is taken up by cells from the blood, and elevated plasma levels of LDLs result. Without intervention, such heterozygous individuals usually have heart attacks during the fourth decade of their life, or before. While heterozygotes have LDL levels about double that of a normal individual, rare homozygotes have been detected. They lack LDL receptors altogether, and their LDL levels are nearly ten times above the normal range. They are likely to have a heart attack very early in life, even before age 5, and almost inevitably before they reach the age of 20.

Pedigree analysis of many traits has historically been an extremely valuable research technique in human genetic studies. However, the approach does not usually provide the certainty of the conclusions obtained through experimental crosses yielding large numbers of offspring. Nevertheless, when many independent pedigrees of the same trait or disorder are analyzed, consistent conclusions can often be drawn. Table 3.4 lists numerous human traits and classifies them according to their recessive or dominant expression.

TABLE 3.4

Representative Recessive and Dominant Human Traits

Recessive Traits	Dominant Traits
Albinism	Achondroplasia
Alkaptonuria	Brachydactyly
Ataxia telangiectasia	Congenital stationary night blindness
Color blindness	Ehler–Danlos syndrome
Cystic fibrosis	Hypotrichosis
Duchenne muscular dystrophy	Huntington disease
Galactosemia	Hypercholesterolemia
Hemophilia	Marfan syndrome
Lesch–Nyhan syndrome	Myotonic dystrophy
Phenylketonuria	Neurofibromatosis
Sickle-cell anemia	Phenylthiocarbamide tasting
Tay-Sachs disease	Porphyria (some forms)

NOW SOLVE THIS

3–5 The following pedigree is for myopia (nearsightedness) in humans.

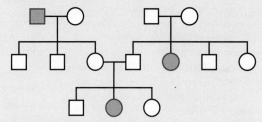

Predict whether the disorder is inherited as the result of a dominant or recessive trait. Determine the most probable genotype for each individual based on your prediction.

■ HINT: *This problem involves the analysis of a pedigree to determine the mode of inheritance of myopia. The key to its solution is to identify whether or not there are individuals who express the trait but neither of whose parents also express the trait. Such an observation is a powerful clue and allows you to rule out one mode of inheritance.*

Online Mendelian Inheritance in Man

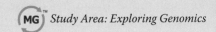

Study Area: Exploring Genomics

The Online Mendelian Inheritance in Man (OMIM) database is a catalog of human genes and human genetic disorders that are inherited in a Mendelian manner. Genetic disorders that arise from major chromosomal aberrations, such as monosomy or trisomy (the loss of a chromosome or the presence of a superfluous chromosome, respectively), are not included. The OMIM database is a daily-updated version of the book *Mendelian Inheritance in Man*, edited by Dr. Victor McKusick of Johns Hopkins University. Scientists use OMIM as an important information source to accompany the sequence data generated by the **Human Genome Project**.

The OMIM entries will give you links to a wealth of information, including DNA and protein sequences, chromosomal maps, disease descriptions, and relevant scientific publications. In this exercise, you will explore OMIM to answer questions about the recessive human disease sickle-cell anemia and other Mendelian inherited disorders.

■ Exercise I – Sickle-cell Anemia

In this chapter, you were introduced to recessive and dominant human traits. You will now discover more about sickle-cell anemia as an autosomal recessive disease by exploring the OMIM database.

1. To begin the search, access the OMIM site at: www.ncbi.nlm.nih.gov/entrez/query.fcgi?db=OMIM&itool=toolbar.

2. In the "SEARCH" box, type "sickle-cell anemia" and click on the "Go" button to perform the search.

3. Select the first entry (#603903).

4. Examine the list of subject headings in the left-hand column and read some of the information about sickle-cell anemia.

5. Select one or two references at the bottom of the page and follow them to their abstracts in PubMed.

6. Using the information in this entry, answer the following questions:

a. Which gene is mutated in individuals with sickle-cell anemia?

b. What are the major symptoms of this disorder?

c. What was the first published scientific description of sickle-cell anemia?

d. Describe two other features of this disorder that you learned from the OMIM database and state where in the database you found this information.

■ Exercise II – Other Recessive or Dominant Disorders

Select another human disorder that is inherited as either a dominant or recessive trait and investigate its features, following the general procedure presented above. Follow links from OMIM to other databases if you choose.

Describe several interesting pieces of information you acquired during your exploration and cite the information sources you encountered during the search.

CASE STUDY | To test or not to test

Thomas first discovered a potentially devastating piece of family history when he learned the medical diagnosis for his brother's increasing dementia, muscular rigidity, and frequency of seizures. His brother, at age 49, was diagnosed with Huntington disease (HD), a dominantly inherited condition that typically begins with such symptoms around the age of 45 and leads to death in one's early 60s. As depressing as the news was to Thomas, it helped explain his father's suicide. Thomas, 38, now wonders what his chances are of carrying the gene for HD, leading he and his wife to discuss the pros and cons of him undergoing genetic testing. Thomas and his wife have two teenage children, a boy and a girl.

1. What role might a genetic counselor play in this real-life scenario?

2. How might the preparation and analysis of a pedigree help explain the dilemma facing Thomas and his family?

3. If Thomas decides to go ahead with the genetic test, what should be the role of the health insurance industry in such cases?

4. If Thomas tests positive for HD, and you were one of his children, would you want to be tested?

Summary Points

For activities, animations, and review quizzes, go to the study area at www.masteringgenetics.com

1. Mendel's postulates help describe the basis for the inheritance of phenotypic traits. Based on the analysis of numerous monohybrid crosses, he hypothesized that unit factors exist in pairs and exhibit a dominant/recessive relationship in determining the expression of traits. He further postulated that unit factors segregate during gamete formation, such that each gamete receives one or the other factor, with equal probability.

2. Mendel's postulate of independent assortment, based initially on his analysis of dihybrid crosses, states that each pair of unit factors segregates independently of other such pairs. As a result, all possible combinations of gametes are formed with equal probability.

3. Both the Punnett square and the forked-line method are used to predict the probabilities of phenotypes or genotypes from crosses involving two or more gene pairs. The forked-line method is less complex, but just as accurate as the Punnett square.

4. The discovery of chromosomes in the late 1800s, along with subsequent studies of their behavior during meiosis, led to the rebirth of Mendel's work, linking his unit factors to chromosomes.

5. Genetic ratios are expressed as probabilities. Thus, deriving outcomes of genetic crosses requires an understanding of the laws of probability.

6. Chi-square analysis allows us to assess the null hypothesis, which states that there is no real difference between the expected and observed values. As such, it tests the probability of whether observed variations can be attributed to chance deviation.

7. Pedigree analysis is a method for studying the inheritance pattern of human traits over several generations, providing the basis for predicting the mode of inheritance of characteristics and disorders in the absence of extensive genetic crossing and large numbers of offspring.

INSIGHTS AND SOLUTIONS

As a student, you will be asked to demonstrate your knowledge of transmission genetics by solving various problems. Success at this task requires not only comprehension of theory but also its application to more practical genetic situations. Most students find problem solving in genetics to be both challenging and rewarding. This section is designed to provide basic insights into the reasoning essential to this process.

Genetics problems are in many ways similar to word problems in algebra. The approach to solving them is identical: (1) analyze the problem carefully; (2) translate words into symbols and define each symbol precisely; and (3) choose and apply a specific technique to solve the problem. The first two steps are the most critical. The third step is largely mechanical.

The simplest problems state all necessary information about a P_1 generation and ask you to find the expected ratios of the F_1 and F_2 genotypes and/or phenotypes. Always follow these steps when you encounter this type of problem:

(a) Determine insofar as possible the genotypes of the individuals in the P_1 generation.

(b) Determine what gametes may be formed by the P_1 parents.

(c) Recombine the gametes by the Punnett square or the forked-line method, or if the situation is very simple, by inspection. From the genotypes of the F_1 generation, determine the phenotypes. Read the F_1 phenotypes.

(d) Repeat the process to obtain information about the F_2 generation.

Determining the genotypes from the given information requires that you understand the basic theory of transmission genetics. Consider this problem: *A recessive mutant allele, black, causes a very dark body in Drosophila when homozygous. The normal wild-type color is described as gray. What F_1 phenotypic ratio is predicted when a black female is crossed to a gray male whose father was black?*

To work out this problem, you must understand dominance and recessiveness, as well as the principle of segregation. Furthermore, you must use the information about the male parent's father. Here is one logical approach to solving this problem:

The female parent is black, so she must be homozygous for the mutant allele (bb). The male parent is gray and must therefore have at least one dominant allele (B). His father was black (bb), and he received one of the chromosomes bearing these alleles, so the male parent must be heterozygous (Bb).

From this point, solving the problem is simple:

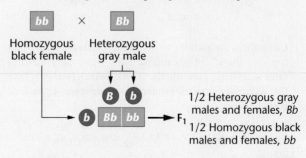

Apply the approach we just studied to the following problems.

1. Mendel found that full pea pods are dominant over constricted pods, while round seeds are dominant over wrinkled seeds. One

of his crosses was between full, round plants and constricted, wrinkled plants. From this cross, he obtained an F_1 generation that was all full and round. In the F_2 generation, Mendel obtained his classic 9:3:3:1 ratio. Using this information, determine the expected F_1 and F_2 results of a cross between homozygous constricted, round and full, wrinkled plants.

Solution: First, assign gene symbols to each pair of contrasting traits. Use the lowercase first letter of each recessive trait to designate that trait, and use the same letter in uppercase to designate the dominant trait. Thus, C and c indicate full and constricted pods, respectively, and W and w indicate the round and wrinkled phenotypes, respectively.

Determine the genotypes of the P_1 generation, form the gametes, combine them in the F_1 generation, and read off the phenotype(s):

P_1:	$ccWW$	$\times$	$CCww$
	constricted, round		full, wrinkled
	$\downarrow$		$\downarrow$
Gametes:	cW		Cw

F_1: $CcWw$

full, round

You can immediately see that the F_1 generation expresses both dominant phenotypes and is heterozygous for both gene pairs. Thus, you expect that the F_2 generation will yield the classic Mendelian ratio of 9:3:3:1. Let's work it out anyway, just to confirm this expectation, using the forked-line method. Both gene pairs are heterozygous and can be expected to assort independently, so we can predict the F_2 outcomes from each gene pair separately and then proceed with the forked-line method.

The F_2 offspring should exhibit the individual traits in the following proportions:

$Cc \times Cc$ $Ww \times Ww$
$\downarrow$ $\downarrow$

$\left.\begin{array}{l} CC \\ Cc \\ cC \end{array}\right\}$ full $\left.\begin{array}{l} WW \\ Ww \\ wW \end{array}\right\}$ round

cc constricted ww wrinkled

1. Using these proportions to complete a forked-line diagram confirms the 9:3:3:1 phenotypic ratio. (Remember that this ratio represents proportions of 9/16:3/16:3/16:1/16.) Note that we are applying the product law as we compute the final probabilities:

```
              ——3/4 round    (3/4)(3/4)   9/16 full, round
3/4 full
              ——1/4 wrinkled  (3/4)(1/4)  3/16 full, wrinkled

              ——3/4 round    (1/4)(3/4)   3/16 constricted, round
1/4 constricted
              ——1/4 wrinkled  (1/4)(1/4)  1/16 constricted, wrinkled
```

2. In another cross, involving parent plants of unknown genotype and phenotype, the following offspring were obtained.

3/8   full, round
3/8   full, wrinkled
1/8   constricted, round
1/8   constricted, wrinkled

Determine the genotypes and phenotypes of the parents.

Solution: This problem is more difficult and requires keener insight because you must work backward to arrive at the answer. The best approach is to consider the outcomes of pod shape separately from those of seed texture.

Of all the plants, $3/8 + 3/8 = 3/4$ are full and $1/8 + 1/8 = 1/4$ are constricted. Of the various genotypic combinations that can serve as parents, which will give rise to a ratio of 3/4:1/4? This ratio is identical to Mendel's monohybrid F_2 results, and we can propose that both unknown parents share the same genetic characteristic as the monohybrid F_1 parents: they must both be heterozygous for the genes controlling pod shape and thus are Cc.

Before we accept this hypothesis, let's consider the possible genotypic combinations that control seed texture. If we consider this characteristic alone, we can see that the traits are expressed in a ratio of $3/8 + 1/8 = 1/2$ round: $3/8 + 1/8 = 1/2$ wrinkled. To generate such a ratio, the parents cannot both be heterozygous or their offspring would yield a 3/4:1/4 phenotypic ratio. They cannot both be homozygous or all offspring would express a single phenotype. Thus, we are left with testing the hypothesis that one parent is homozygous and one is heterozygous for the alleles controlling texture. The potential case of $WW \times Ww$ does not work because it would also yield only a single phenotype. This leaves us with the potential case of $ww \times Ww$. Offspring in such a mating will yield $1/2$ Ww (round): $1/2$ ww (wrinkled), exactly the outcome we are seeking.

Now, let's combine our hypotheses and predict the outcome of the cross. In our solution, we use a dash (–) to indicate that the second allele may be dominant or recessive, since we are only predicting phenotypes.

```
           ——1/2 Ww → 3/8 C–Ww full, round
3/4 C–
           ——1/2 ww → 3/8 C–ww full, wrinkled

           ——1/2 Ww → 1/8 ccWw constricted, round
1/4 cc
           ——1/2 ww → 1/8 ccww constricted, wrinkled
```

As you can see, this cross produces offspring in proportions that match our initial information, and we have solved the problem. Note that, in the solution, we have used genotypes in the forked-line method, in contrast to the use of phenotypes in Solution 1.

3. In the laboratory, a genetics student crossed flies with normal long wings with flies expressing the *dumpy* mutation (truncated wings), which she believed was a recessive trait. In the F_1 generation, all flies had long wings. The following results were obtained in the F_2 generation:

792 long-winged flies

208 dumpy-winged flies

The student tested the hypothesis that the dumpy wing is inherited as a recessive trait using χ^2 analysis of the F_2 data.

(a) What ratio was hypothesized?
(b) Did the analysis support the hypothesis?
(c) What do the data suggest about the *dumpy* mutation?

Solution:

(a) The student hypothesized that the F_2 data (792:208) fit Mendel's 3:1 monohybrid ratio for recessive genes.

(b) The initial step in χ^2 analysis is to calculate the expected results (e) for a ratio of 3:1. Then we can compute deviation $o - e$ (d) and the remaining numbers.

Ratio	o	e	d	d^2	d^2/e
3/4	792	750	42	1764	2.35
1/4	208	250	-42	1764	7.06

Total = 1000

$$\chi^2 = \Sigma \frac{d^2}{e}$$
$$= 2.35 + 7.06$$
$$= 9.41$$

We consult Figure 3–11 to determine the probability (p) and to decide whether the deviations can be attributed to chance. There are two possible outcomes ($n = 2$), so the degrees of freedom (df) $= n - 1$, or 1. The table in Figure 3–11(b) shows that p is a value between 0.01 and 0.001; the graph in Figure 3–11(a) gives an estimate of about 0.001. Since $p < 0.05$, we reject the null hypothesis. The data do not fit a 3:1 ratio.

(c) When the student hypothesized that Mendel's 3:1 ratio was a valid expression of the monohybrid cross, she was tacitly making numerous assumptions. Examining these underlying assumptions may explain why the null hypothesis was rejected. For one thing, she assumed that all the genotypes resulting from the cross were equally viable—that genotypes yielding long wings are equally likely to survive from fertilization through adulthood as the genotype yielding dumpy wings. Further study would reveal that dumpy-winged flies are somewhat less viable than normal flies. As a result, we would expect *less* than 1/4 of the total offspring to express dumpy wings. This observation is borne out in the data, although we have not proven that this is true.

4. If two parents, both heterozygous carriers of the autosomal recessive gene causing cystic fibrosis, have five children, what is the probability that exactly three will be normal?

Solution: This is an opportunity to use the binomial theorem. To do so requires two facts you already possess: the probability of having a normal child during each pregnancy is

$$p_a = \text{normal} = 3/4$$

and the probability of having an afflicted child is

$$p_b = \text{afflicted} = 1/4$$

Insert these into the formula

$$\frac{n!}{s!t!}a^s b^t$$

where $n = 5, s = 3$, and $t = 2$

$$p = \frac{(5) \cdot (4) \cdot (3) \cdot (2) \cdot (1)}{(3) \cdot (2) \cdot (1) \cdot (2) \cdot (1)}(3/4)^3(1/4)^2$$
$$= \frac{(5) \cdot (4)}{(2) \cdot (1)}(3/4)^3(1/4)^2$$
$$= 10(27/64) \cdot (1/16)$$
$$= 10(27/1024)$$
$$= 270/1024$$
$$p = {\sim}0.26$$

Problems and Discussion Questions

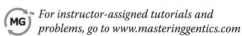 *For instructor-assigned tutorials and problems, go to www.masteringgentics.com*

When working out genetics problems in this and succeeding chapters, always assume that members of the P_1 generation are homozygous, unless the information or data you are given require you to do otherwise.

HOW DO WE KNOW?

1. In this chapter, we focused on the Mendelian postulates, probability, and pedigree analysis. We also considered some of the methods and reasoning by which these ideas, concepts, and techniques were developed. On the basis of these discussions, what answers would you propose to the following questions:

(a) How was Mendel able to derive postulates concerning the behavior of "unit factors" during gamete formation, when he could not directly observe them?

(b) How do we know whether an organism expressing a dominant trait is homozygous or heterozygous?

(c) In analyzing genetic data, how do we know whether deviation from the expected ratio is due to chance rather than to another, independent factor?

(d) Since experimental crosses are not performed in humans, how do we know how traits are inherited?

2. In a cross between a black and a white guinea pig, all members of the F_1 generation are black. The F_2 generation is made up of approximately 3/4 black and 1/4 white guinea pigs.

(a) Diagram this cross, showing the genotypes and phenotypes.

(b) What will the offspring be like if two F_2 white guinea pigs are mated?

(c) Two different matings were made between black members of the F_2 generation, with the following results.

Cross	Offspring
Cross 1	All black
Cross 2	3/4 black, 1/4 white

Diagram each of the crosses.

3. Albinism in humans is inherited as a simple recessive trait. For the following families, determine the genotypes of the parents and offspring. (When two alternative genotypes are possible, list both.)

(a) Two normal parents have five children, four normal and one albino.

(b) A normal male and an albino female have six children, all normal.

(c) A normal male and an albino female have six children, three normal and three albino.

(d) Construct a pedigree of the families in (b) and (c). Assume that one of the normal children in (b) and one of the albino children in (c) become the parents of eight children. Add these children to the pedigree, predicting their phenotypes (normal or albino).

4. Which of Mendel's postulates are illustrated by the pedigree in Problem 3? List and define these postulates.

5. Discuss how Mendel's monohybrid results served as the basis for all but one of his postulates. Which postulate was not based on these results? Why?

6. What advantages were provided by Mendel's choice of the garden pea in his experiments?

7. Mendel crossed peas having round seeds and yellow cotyledons (seed leaves) with peas having wrinkled seeds and green cotyledons. All the F_1 plants had round seeds with yellow cotyledons. Diagram this cross through the F_2 generation, using both the Punnett square and forked-line, or branch diagram, methods.

8. Based on the preceding cross, what is the probability that an organism in the F_2 generation will have round seeds and green cotyledons *and* be true breeding?

9. Which of Mendel's postulates can only be demonstrated in crosses involving at least two pairs of traits? State the postulate.

10. Correlate Mendel's four postulates with what is now known about homologous chromosomes, genes, alleles, and the process of meiosis.

11. What is the basis for homology among chromosomes?

12. In *Drosophila*, *gray* body color is dominant to *ebony* body color, while *long* wings are dominant to *vestigial* wings. Assuming that the P_1 individuals are homozygous, work the following crosses through the F_2 generation, and determine the genotypic and phenotypic ratios for each generation.

(a) gray, long $\times$ ebony, vestigial

(b) gray, vestigial $\times$ ebony, long

(c) gray, long $\times$ gray, vestigial

13. How many different types of gametes can be formed by individuals of the following genotypes: (a) *AaBb*, (b) *AaBB*, (c) *AaBbCc*, (d) *AaBBcc*, (e) *AaBbcc*, and (f) *AaBbCcDdEe*? What are the gametes in each case?

14. Mendel crossed peas having green seeds with peas having yellow seeds. The F_1 generation produced only yellow seeds. In the F_2, the progeny consisted of 6022 plants with yellow seeds and 2001 plants with green seeds. Of the F_2 yellow-seeded plants,

519 were self-fertilized with the following results: 166 bred true for yellow and 353 produced an F_3 ratio of 3/4 yellow: 1/4 green. Explain these results by diagramming the crosses.

15. In a study of black guinea pigs and white guinea pigs, 100 black animals were crossed with 100 white animals, and each cross was carried to an F_2 generation. In 94 of the crosses, all the F_1 offspring were black and an F_2 ratio of 3 black:1 white was obtained. In the other 6 cases, half of the F_1 animals were black and the other half were white. Why? Predict the results of crossing the black and white F_1 guinea pigs from the 6 exceptional cases.

16. Mendel crossed peas having round green seeds with peas having wrinkled yellow seeds. All F_1 plants had seeds that were round and yellow. Predict the results of testcrossing these F_1 plants.

17. Thalassemia is an inherited anemic disorder in humans. Affected individuals exhibit either a minor anemia or a major anemia. Assuming that only a single gene pair and two alleles are involved in the inheritance of these conditions, is thalassemia a dominant or recessive disorder?

18. The following are F_2 results of two of Mendel's monohybrid crosses.

(a)	full pods	882
	constricted pods	299
(b)	violet flowers	705
	white flowers	224

For each cross, state a null hypothesis to be tested using χ^2 analysis. Calculate the χ^2 value and determine the p value for both. Interpret the p values. Can the deviation in each case be attributed to chance or not? Which of the two crosses shows a greater amount of deviation?

19. In assessing data that fell into two phenotypic classes, a geneticist observed values of 250:150. She decided to perform a χ^2 analysis by using the following two different null hypotheses: (a) the data fit a 3:1 ratio, and (b) the data fit a 1:1 ratio. Calculate the χ^2 values for each hypothesis. What can be concluded about each hypothesis?

20. The basis for rejecting any null hypothesis is arbitrary. The researcher can set more or less stringent standards by deciding to raise or lower the p value used to reject or not reject the hypothesis. In the case of the chi-square analysis of genetic crosses, would the use of a standard of $p = 0.10$ be more or less stringent about not rejecting the null hypothesis? Explain.

21. Consider the following pedigree.

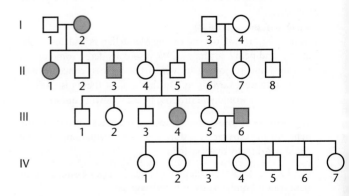

Predict the mode of inheritance of the trait of interest and the most probable genotype of each individual. Assume that the alleles *A* and *a* control the expression.

22. Draw all possible conclusions concerning the mode of inheritance of the trait portrayed in each of the following limited pedigrees. (Each of the 4 cases is based on a different trait.)

(a)

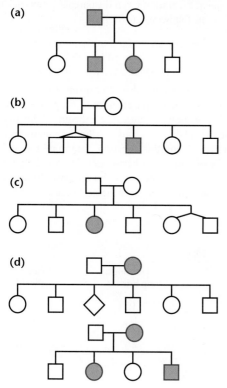

(b)

(c)

(d)

23. In a family of five children, what is the probability that

   (a) all are males?

   (b) three are males and two are females?

   (c) two are males and three are females?

   (d) all are the same sex?

   Assume that the probability of a male child is equal to the probability of a female child ($p = 1/2$).

24. In a family of eight children, where both parents are heterozygous for albinism, what mathematical expression predicts the probability that six are normal and two are albinos?

25. For decades scientists have been perplexed by different circumstances surrounding families with rare, early-onset auditory neuropathy (deafness). In some families, parents and grandparents of the proband have normal hearing, while in other families, a number of affected (deaf) family members are scattered throughout the pedigree, appearing in every generation. Assuming a genetic cause for each case, offer a reasonable explanation for the genetic origin of such deafness in the two types of families.

26. A "wrongful birth" case was recently brought before a court in which a child with Smith–Lemli–Opitz syndrome was born to apparently healthy parents. This syndrome is characterized by a cluster of birth defects including cleft palate, and an array of problems with the reproductive and urinary organs. Originally considered by their physician as having a nongenetic basis, the parents decided to have another child, who was also born with Smith–Lemli–Opitz syndrome. In the role of a genetic counselor, instruct the court about what occurred, including the probability of the parents having two affected offspring, knowing that the disorder is inherited as a recessive trait.

Extra-Spicy Problems

 For instructor-assigned tutorials and problems, go to www.masteringgentics.com

27. Two true-breeding pea plants were crossed. One parent is round, terminal, violet, constricted, while the other expresses the respective contrasting phenotypes of wrinkled, axial, white, full. The four pairs of contrasting traits are controlled by four genes, each located on a separate chromosome. In the F_1 only round, axial, violet, and full were expressed. In the F_2, all possible combinations of these traits were expressed in ratios consistent with Mendelian inheritance.

(a) What conclusion about the inheritance of the traits can be drawn based on the F_1 results?

(b) In the F_2 results, which phenotype appeared most frequently? Write a mathematical expression that predicts the probability of occurrence of this phenotype.

(c) Which F_2 phenotype is expected to occur least frequently? Write a mathematical expression that predicts this probability.

(d) In the F_2 generation, how often is either of the P_1 phenotypes likely to occur?

(e) If the F_1 plants were testcrossed, how many different phenotypes would be produced? How does this number compare with the number of different phenotypes in the F_2 generation just discussed?

28. Tay–Sachs disease (TSD) is an inborn error of metabolism that results in death, often by the age of 2. You are a genetic counselor interviewing a phenotypically normal couple who tell you the male had a female first cousin (on his father's side) who died from TSD and the female had a maternal uncle with TSD. There are no other known cases in either of the families, and none of the matings have been between related individuals. Assume that this trait is very rare.

(a) Draw a pedigree of the families of this couple, showing the relevant individuals.

(b) Calculate the probability that both the male and female are carriers for TSD.

(c) What is the probability that neither of them is a carrier?

(d) What is the probability that one of them is a carrier and the other is not? [*Hint:* The *p* values in (b), (c), and (d) should equal 1.]

29. *Datura stramonium* (the Jimsonweed) expresses flower colors of purple and white and pod textures of smooth and spiny. The results of two crosses in which the parents were not necessarily true breeding were observed to be

white spiny $\times$ white spiny $\rightarrow$ 3/4 white spiny:
1/4 white smooth
purple smooth $\times$ purple smooth $\rightarrow$ 3/4 purple smooth:
1/4 white smooth

(a) Based on these results, put forward a hypothesis for the inheritance of the purple/white and smooth/spiny traits.
(b) Assuming that true-breeding strains of all combinations of traits are available, what single cross could you execute and carry to an F_2 generation that will prove or disprove your hypothesis? Assuming your hypothesis is correct, what results of this cross will support it?

30. The wild-type (normal) fruit fly, *Drosophila melanogaster*, has straight wings and long bristles. Mutant strains have been isolated that have either curled wings or short bristles. The genes representing these two mutant traits are located on separate chromosomes. Carefully examine the data from the following five crosses shown below (running across both columns).

(a) Identify each mutation as either dominant or recessive. In each case, indicate which crosses support your answer.
(b) Assign gene symbols and, for each cross, determine the genotypes of the parents.

31. An alternative to using the expanded binomial equation and Pascal's triangle in determining probabilities of phenotypes in a subsequent generation when the parents' genotypes are known is to use the following equation:

$$\frac{n!}{s!t!}a^s b^t$$

where n is the total number of offspring, s is the number of offspring in one phenotypic category, t is the number of offspring in the other phenotypic category, a is the probability of occurrence of the first phenotype, and b is the probability of the second phenotype. Using this equation, determine the probability of a family of 5 offspring having exactly 2 children afflicted with sickle-cell anemia (an autosomal recessive disease) when both parents are heterozygous for the sickle-cell allele.

Cross	Progeny			
	straight wings, long bristles	straight wings, short bristles	curled wings, long bristles	curled wings, short bristles
1. straight, short $\times$ straight, short	30	90	10	30
2. straight, long $\times$ straight, long	120	0	40	0
3. curled, long $\times$ straight, short	40	40	40	40
4. straight, short $\times$ straight, short	40	120	0	0
5. curled, short $\times$ straight, short	20	60	20	60

32. To assess Mendel's law of segregation using tomatoes, a true-breeding tall variety (*SS*) is crossed with a true-breeding short variety (*ss*). The heterozygous F_1 tall plants (*Ss*) were crossed to produce two sets of F_2 data, as follows.

Set I	Set II
30 tall	300 tall
5 short	50 short

(a) Using the χ^2 test, analyze the results for both datasets. Calculate χ^2 values and estimate the p values in both cases.
(b) From the above analysis, what can you conclude about the importance of generating large datasets in experimental conditions?

33. When examining Sutton's drawings of chromosomes of the grasshopper, *Brachystola magna*, Eleanor Carothers (1913) noted a pair of unlike chromosomes—one large dyad and one small dyad—making up a tetrad in each of 300 primary spermatocytes. In addition, an accessory chromosome (unpaired and later called the X chromosome) was identified in females, such that males had 23 chromosomes and females had 24 chromosomes. Carothers found that the larger dyad in each unlike pair went to the same pole as the accessory chromosome in 154 anaphases, while the smaller dyad went with the accessory chromosome in the remaining 146 anaphases. (a) How do these findings relate to Mendel's postulates, and (b) how do they support the chromosome theory of heredity?

34. *Dentinogenesis imperfecta* is a tooth disorder involving the production of dentin sialophosphoprotein, a bone-like component of the protective middle layer of teeth. The trait is inherited as an autosomal dominant allele located on chromosome 4 in humans and occurs in about 1 in 6000 to 8000 people. Assume that a man with *dentinogenesis imperfecta,* whose father had the disease but whose mother had normal teeth, married a woman with normal teeth. They have six children. What is the probability that their first child will be a male with *dentinogenesis imperfecta*? What is the probability that three of their six children will have the disease?

Labrador retriever puppies expressing brown (chocolate), golden (yellow), and black coat colors, traits controlled by two gene pairs.

4

Extensions of Mendelian Genetics

CHAPTER CONCEPT

- While alleles are transmitted from parent to offspring according to Mendelian principles, they often do not display the clear-cut dominant/recessive relationship observed by Mendel.

- In many cases, in a departure from Mendelian genetics, two or more genes are known to influence the phenotype of a single characteristic.

- Still another exception to Mendelian inheritance occurs when genes are located on the X chromosome, because one of the sexes receives only one copy of that chromosome, eliminating the possibility of heterozygosity.

- Phenotypes are often the combined result of genetics and the environment within which genes are expressed.

- The result of the various exceptions to Mendelian principles is the occurrence of phenotypic ratios that differ from those produced by standard monohybrid, dihybrid, and trihybrid crosses.

W̲ e discussed the fundamental principles of transmission genetics. We saw that genes are present on homologous chromosomes and that these chromosomes segregate from each other and assort independently from other segregating chromosomes during gamete formation. These two postulates are the basic principles of gene transmission from parent to offspring. Once an offspring has received the total set of genes, it is the expression of genes that determines the organism's phenotype. When gene expression does not adhere to a simple dominant/recessive mode, or when more than one pair of genes influences the expression of a single character, the classic 3:1 and 9:3:3:1 F_2 ratios are usually modified. In this and the next several chapters, we consider more complex modes of inheritance. In spite of the greater complexity of these situations, the fundamental principles set down by Mendel still hold.

In this chapter, we restrict our initial discussion to the inheritance of traits controlled by only one set of genes. In diploid organisms, which have homologous pairs of chromosomes, two copies of each gene influence such traits. The copies need not be identical since alternative forms of genes, *alleles*, occur within populations. How alleles influence phenotypes will be our primary focus. We will then consider **gene interaction**, a situation in which a single phenotype is affected by more than one set of genes. Numerous examples will be presented to illustrate a variety of heritable patterns observed in such situations.

Thus far, we have restricted our discussion to chromosomes other than the X and Y pair. By examining cases where genes are present on the X chromosome, illustrating **X-linkage**, we will see yet another modification of Mendelian ratios. Our discussion of modified ratios also includes the consideration of sex-limited and sex-influenced inheritance, cases where the sex of the individual, but not necessarily the genes on the X chromosome, influences the phenotype. We conclude the chapter by showing how a given phenotype often varies depending on the overall environment in which a gene, a cell, or an organism finds itself. This discussion points out that phenotypic expression depends on more than just the genotype of an organism.

4.1

Alleles Alter Phenotypes in Different Ways

Following the rediscovery of Mendel's work in the early 1900s, research focused on the many ways in which genes influence an individual's phenotype. This course of investigation, stemming from Mendel's findings, is called neo-Mendelian genetics (*neo* from the Greek word meaning *since* or *new*).

Each type of inheritance described in this chapter was investigated when observations of genetic data did not conform precisely to the expected Mendelian ratios. Hypotheses that modified and extended the Mendelian principles were proposed and tested with specifically designed crosses. The explanations proffered to account for these observations were constructed in accordance with the principle that a phenotype is under the influence of one or more genes located at specific loci on one or more pairs of homologous chromosomes.

To understand the various modes of inheritance, we must first consider the potential function of an **allele**. An allele is an alternative form of a gene. The allele that occurs most frequently in a population, the one that we arbitrarily designate as normal, is called the *wild-type allele*. This is often, but not always, dominant. Wild-type alleles are responsible for the corresponding wild-type phenotype and are the standards against which all other mutations occurring at a particular locus are compared.

A mutant allele contains modified genetic information and often specifies an altered gene product. For example, in human populations, there are many known alleles of the gene encoding the β chain of human hemoglobin. All such alleles store information necessary for the synthesis of the β chain polypeptide, but each allele specifies a slightly different form of the same molecule. Once the allele's product has been manufactured, the product's function may or may not be altered.

The process of mutation is the source of alleles. For a new allele to be recognized by observation of an organism, the allele must cause a change in the phenotype. A new phenotype results from a change in functional activity of the cellular product specified by that gene. Often, the mutation causes the diminution or the loss of the specific wild-type function. For example, if a gene is responsible for the synthesis of a specific enzyme, a mutation in that gene may ultimately change the conformation of this enzyme and reduce or eliminate its affinity for the substrate. Such a mutation is designated as a **loss-of-function mutation**. If the loss is complete, the mutation has resulted in what is called a **null allele**.

Conversely, other mutations may enhance the function of the wild-type product. Most often when this occurs, it is the result of increasing the quantity of the gene product. For example, the mutation may be affecting the regulation of transcription of the gene under consideration. Such mutations, designated **gain-of-function mutations**, most often result in dominant alleles, since one copy of the mutation in a diploid organism is sufficient to alter the normal phenotype. Examples of gain-of-function mutations include the genetic conversion of *proto-oncogenes*, which regulate the cell cycle, to *oncogenes*, where regulation is overridden by excess gene product. The result is the creation of a cancerous cell. Another example is a mutation that alters the sensitivity of a receptor, whereby an inhibitory signal molecule is unable to quell a particular biochemical response. In a sense, the function of the gene product is always turned on. Having introduced the concepts of gain- and loss-of-function mutations, we should note the possibility that a mutation will create an allele that produces no detectable change in function. In this case, the mutation would not be immediately apparent since no phenotypic variation would be evident. However, such a mutation could be detected if the DNA sequence of the gene was examined directly. These are sometimes referred to as **neutral mutations** since the gene product presents no change to either the phenotype or the evolutionary fitness of the organism.

Finally, we note that while a phenotypic trait may be affected by a single mutation in one gene, traits are often influenced by many gene products. For example, enzymatic reactions are most often part of complex metabolic pathways leading to the synthesis of an end product, such as an amino acid. Mutations in any of a pathway's reactions can have a common effect—the failure to synthesize the end product. Therefore, phenotypic traits related to the end product are often influenced by more than one gene. Such is the case in *Drosophila* eye color mutations. Eye color results from the synthesis and deposition of a brown and a bright red pigment in the facets of the compound eye. This causes the wild-type eye color to appear brick red. There are a series of recessive loss-of-function mutations that interrupt the multistep pathway leading to the synthesis of the brown pigment. While these mutations represent genes located on different chromosomes, they all result in the same phenotype: a bright red eye whose color is due to the absence of the brown pigment. Examples are the mutations *vermilion, cinnabar,* and *scarlet,* which are indistinguishable phenotypically.

In each of the many crosses discussed in the next few chapters, only one or a few gene pairs are involved. Keep in mind that in each cross, all genes that are not under consideration are assumed to have no effect on the inheritance patterns described.

4.2

Geneticists Use a Variety of Symbols for Alleles

We learned a standard convention used to symbolize alleles for very simple Mendelian traits. The initial letter of the name of a recessive trait, lowercased and italicized, denotes the recessive allele, and the same letter in uppercase refers to the dominant allele. Thus, in the case of tall and dwarf, where dwarf is recessive, *D* and *d* represent the alleles responsible for these respective traits. Mendel used upper- and lowercase letters such as these to symbolize his unit factors.

Another useful system was developed in genetic studies of the fruit fly *Drosophila melanogaster* to discriminate between wild-type and mutant traits. This system uses the initial letter, or a combination of several letters, from the name of the mutant trait. If the trait is recessive, lowercase is used; if it is dominant, uppercase is used. The contrasting wild-type trait is denoted by the same letters, but with a superscript +. For example, *ebony* is a recessive body color mutation in *Drosophila*. The normal wild-type body color is gray. Using this system, we denote *ebony* by the symbol e, while gray is denoted by e^+. The responsible locus may be occupied by either the wild-type allele (e^+) or the mutant allele (e). A diploid fly may thus exhibit one of three possible genotypes (the two phenotypes are indicated parenthetically):

e^+/e^+	**gray homozygote (wild type)**
e^+/e	**gray heterozygote (wild type)**
e/e	**ebony homozygote (mutant)**

The slash between the letters indicates that the two allele designations represent the same locus on two homologous chromosomes. If we instead consider a mutant allele that is dominant to the normal wild-type allele, such as *Wrinkled* wing in *Drosophila*, the three possible genotypes are Wr/Wr, Wr/Wr^+, and Wr^+/Wr^+. The initial two genotypes express the mutant wrinkled-wing phenotype.

One advantage of this system is that further abbreviation can be used when convenient: The wild-type allele may simply be denoted by the + symbol. With *ebony* as an example, the designations of the three possible genotypes become

+/+	**gray homozygote (wild type)**
+/e	**gray heterozygote (wild type)**
e/e	**ebony homozygote (mutant)**

Another variation is utilized when no dominance exists between alleles (a situation we will explore in Section 4.3). We simply use uppercase letters and superscripts to denote alternative alleles (e.g., R^1 and R^2, L^M and L^N, and I^A and I^B).

Many diverse systems of genetic nomenclature are used to identify genes in various organisms. Usually, the symbol selected reflects the function of the gene or even a disorder caused by a mutant gene. For example, in yeast, *cdk* is the abbreviation for the *cyclin-dependent kinase* gene, whose product is involved in the cell-cycle regulation mechanism discussed in Chapter 2. In bacteria, *leu⁻* refers to a mutation that interrupts the biosynthesis of the amino acid leucine, and the wild-type gene is designated *leu⁺*. The symbol *dnaA* represents a bacterial gene involved in DNA replication (and DnaA, without italics, is the protein made by that gene). In humans, italicized capital letters are used to name genes: *BRCA1* represents one of the genes associated with susceptibility to *breast cancer*. Although these different systems may seem complex, they are useful ways to symbolize genes.

4.3

Neither Allele Is Dominant in Incomplete, or Partial, Dominance

Unlike the Mendelian crosses reported in Chapter 3, a cross between parents with contrasting traits may sometimes generate offspring with an intermediate phenotype. For example, if a four-o'clock or a snapdragon plant with red flowers is crossed with a white-flowered plant, the offspring have pink flowers. Because some red pigment is produced in the F_1 intermediate-colored plant, neither the red nor white flower color is dominant. Such a situation is known as **incomplete**, or **partial, dominance**.

If the phenotype is under the control of a single gene and two alleles, where neither is dominant, the results of the F_1 (pink) × F_1 (pink) cross can be predicted. The resulting F_2 generation shown in Figure 4–1 confirms the hypothesis that only one pair of alleles determines these phenotypes. The

genotypic ratio (1:2:1) of the F_2 generation is identical to that of Mendel's monohybrid cross. However, because neither allele is dominant, the phenotypic ratio is identical to the genotypic ratio (in contrast to the 3:1 phenotypic ratio of a Mendelian monohybrid cross). Note that because neither allele is recessive, we have chosen not to use upper- and lowercase letters as symbols. Instead, we denote the red and white alleles as R^1 and R^2. We could have chosen W^1 and W^2 or still other designations such as C^W and C^R, where C indicates "color" and the W and R superscripts indicate white and red, respectively.

How are we to interpret lack of dominance whereby an intermediate phenotype characterizes heterozygotes? The most accurate way is to consider gene expression in a quantitative way. In the case of flower color above, the mutation causing white flowers is most likely one where complete "loss of function" occurs. In this case, it is likely that the gene product of the wild-type allele (R^1) is an enzyme that participates in a reaction leading to the synthesis of a red pigment. The mutant allele (R^2) produces an enzyme that cannot catalyze the reaction leading to pigment. The end result is that the heterozygote produces only about half the pigment of the red-flowered plant and the phenotype is pink.

Clear-cut cases of incomplete dominance are relatively rare. However, even when one allele seems to have complete dominance over the other, careful examination of the gene product, rather than the phenotype, often reveals an intermediate level of gene expression. An example is the human biochemical disorder **Tay–Sachs disease**, in which homozygous recessive individuals are severely affected with a fatal lipid-storage disorder and neonates die during their first one to three years of life. Recall the discussion of this human malady (Chapter 3, p. 56). In afflicted individuals, there is almost no activity of the enzyme **hexosaminidase A**, an enzyme normally involved in lipid metabolism. Heterozygotes, with only a single copy of the mutant gene, are phenotypically normal, but with only about 50 percent of the enzyme activity found in homozygous normal individuals. Fortunately, this level of enzyme activity is adequate to achieve normal biochemical function. This situation is not uncommon in enzyme disorders and illustrates the concept of the **threshold effect**, whereby normal phenotypic expression occurs anytime a certain level of gene product is attained. Most often, and in particular in Tay–Sachs disease, the threshold is less than 50 percent.

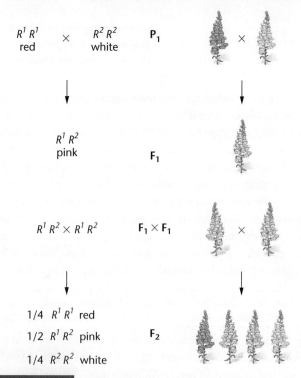

$R^1 R^1$ red   × $R^2 R^2$ white P_1

$R^1 R^2$ pink F_1

$R^1 R^2 \times R^1 R^2$ $F_1 \times F_1$

1/4 $R^1 R^1$ red
1/2 $R^1 R^2$ pink F_2
1/4 $R^2 R^2$ white

FIGURE 4–1   Incomplete dominance shown in the flower color of snapdragons.

4.4

In Codominance, the Influence of Both Alleles in a Heterozygote Is Clearly Evident

If two alleles of a single gene are responsible for producing two distinct, detectable gene products, a situation different from incomplete dominance or dominance/recessiveness arises. In this case, the joint expression of both alleles in a heterozygote is called **codominance**. The **MN blood group** in humans illustrates this phenomenon. Karl Landsteiner and Philip Levin discovered a glycoprotein molecule found on the surface of red blood cells that acts as a native antigen, providing biochemical and immunological identity to individuals. In the human population, two forms of this glycoprotein exist, designated M and N; an individual may exhibit either one or both of them.

The MN system is under the control of a locus found on chromosome 4, with two alleles designated L^M and L^N. Because humans are diploid, three combinations are possible, each resulting in a distinct blood type:

Genotype	Phenotype
$L^M L^M$	M
$L^M L^N$	MN
$L^N L^N$	N

As predicted, a mating between two heterozygous MN parents may produce children of all three blood types, as follows:

$$L^M L^N \times L^M L^N$$
$$\downarrow$$
$$1/4 \; L^M L^M$$
$$1/2 \; L^M L^N$$
$$1/4 \; L^N L^N$$

Once again, the genotypic ratio 1:2:1 is upheld.

Codominant inheritance is characterized by *distinct expression of the gene products of both alleles*. This characteristic distinguishes codominance from incomplete dominance, where heterozygotes express an intermediate, blended, phenotype. For codominance to be studied, both products must be phenotypically detectable. We shall see another example of codominance when we examine the ABO blood-type system.

4.5

Multiple Alleles of a Gene May Exist in a Population

The information stored in any gene is extensive, and mutations can modify this information in many ways. Each change produces a different allele. Therefore, for any gene, the number of alleles within members of a population need not be restricted to two. When three or more alleles of the same gene—which we designate as **multiple alleles**—are present in a population, the resulting mode of inheritance may be unique. It is important to realize that *multiple alleles can be studied only in populations*. Any individual diploid organism has, at most, two homologous gene loci that may be occupied by different alleles of the same gene. However,

among members of a species, numerous alternative forms of the same gene can exist.

The ABO Blood Groups

The simplest case of multiple alleles occurs when three alternative alleles of one gene exist. This situation is illustrated in the inheritance of the **ABO blood groups** in humans, discovered by Karl Landsteiner in the early 1900s. The ABO system, like the MN blood types, is characterized by the presence of antigens on the surface of red blood cells. The A and B antigens are distinct from the MN antigens and are under the control of a different gene, located on chromosome 9. As in the MN system, one combination of alleles in the ABO system exhibits a codominant mode of inheritance.

The ABO phenotype of any individual is ascertained by mixing a blood sample with an antiserum containing type A or type B antibodies. If an antigen is present on the surface of the person's red blood cells, it will react with the corresponding antibody and cause clumping, or agglutination, of the red blood cells. When an individual is tested in this way, one of four phenotypes may be revealed. Each individual has either the A antigen (A phenotype), the B antigen (B phenotype), the A and B antigens (AB phenotype), or neither antigen (O phenotype).

In 1924, it was hypothesized that these phenotypes were inherited as the result of three alleles of a single gene. This hypothesis was based on studies of the blood types of many different families. Although different designations can be used, we will use the symbols I^A, I^B, and i to distinguish these three alleles. The I designation stands for *isoagglutinogen*, another term for antigen. If we assume that the I^A and I^B alleles are responsible for the production of their respective A and B antigens and that i is an allele that does not produce any detectable A or B antigens, we can list the various genotypic possibilities and assign the appropriate phenotype to each:

Genotype	Antigen	Phenotype
$I^A I^A$	A ⎫	A
$I^A i$	A ⎭	
$I^B I^B$	B ⎫	B
$I^B i$	B ⎭	
$I^A I^B$	A, B	AB
$i\,i$	Neither	O

In these assignments, the I^A and I^B alleles are dominant to the i allele, but codominant to each other.

The A and B Antigens

The biochemical basis of the ABO blood type system has now been carefully worked out. The A and B antigens are actually carbohydrate groups (sugars) that are bound to lipid molecules (fatty acids) protruding from the membrane of the red blood cell. The specificity of the A and B antigens is based on the terminal sugar of the carbohydrate group.

Almost all individuals possess what is called the **H substance**, to which one or two terminal sugars are added. As shown in Figure 4–2, the H substance itself contains three sugar molecules—galactose (Gal), *N*-acetylglucosamine (AcGluNH), and fucose—chemically linked together. The I^A allele is responsible for an enzyme that can add the terminal sugar *N*-acetylgalactosamine (AcGalNH) to the H substance. The I^B allele is responsible for a modified enzyme that cannot add *N*-acetylgalactosamine, but instead can add a terminal galactose. Heterozygotes ($I^A I^B$) add either one or the other sugar at the many sites (substrates) available on the surface of the red blood cell, illustrating the biochemical basis of codominance in individuals of the AB blood type. Finally, persons of type O ($i^O i^O$) cannot add either terminal sugar; these persons have only the H substance protruding from the surface of their red blood cells.

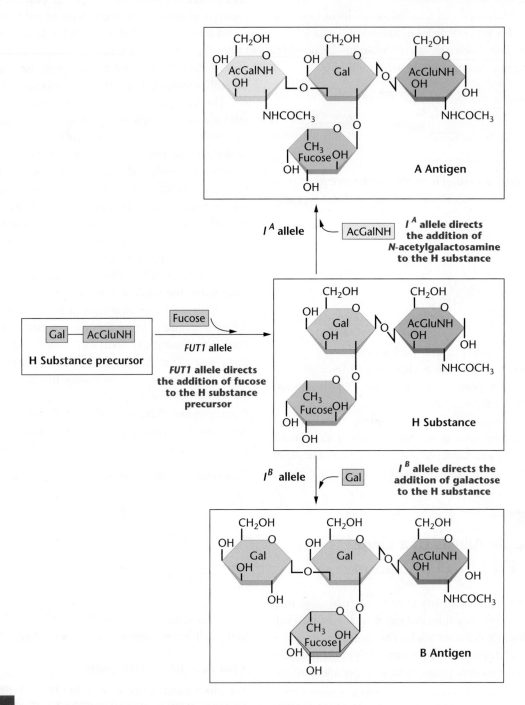

FIGURE 4–2 The biochemical basis of the ABO blood groups. The wild-type *FUT1* allele, present in almost all humans, directs the conversion of a precursor molecule to the H substance by adding a molecule of fucose to it. The I^A and I^B alleles are then able to direct the addition of terminal sugar residues to the H substance. The i^O allele is unable to direct either of these terminal additions. Failure to produce the H substance results in the Bombay phenotype, in which individuals are type O regardless of the presence of an i^A or i^B allele. Gal: galactose; AcGluNH: *N*-acetylglucosamine; AcGalNH: *N*-acetylgalactosamine.

The molecular genetic basis of the mutations leading to the i^A, i^B, and i^O alleles has also been clarified. We will describe it when we discuss mutation and mutagenesis (Ch. 16).

The Bombay Phenotype

In 1952, a very unusual situation provided information concerning the genetic basis of the H substance. A woman in Bombay displayed a unique genetic history inconsistent with her blood type. In need of a transfusion, she was found to lack both the A and B antigens and was thus typed as O. However, as shown in the partial pedigree in Figure 4–3, one of her parents was type AB, and she herself was the obvious donor of an I^B allele to two of her offspring. Thus, she was genetically type B but functionally type O!

This woman was subsequently shown to be homozygous for a rare recessive mutation in a gene designated *FUT1* (encoding an enzyme, fucosyl transferase), which prevented her from synthesizing the complete H substance. In this mutation, the terminal portion of the carbohydrate chain protruding from the red cell membrane lacks fucose, normally added by the enzyme. In the absence of fucose, the enzymes specified by the I^A and I^B alleles apparently are unable to recognize the incomplete H substance as a proper substrate. Thus, neither the terminal galactose nor *N*-acetylgalactosamine can be added, even though the appropriate enzymes capable of doing so are present and functional. As a result, the ABO system genotype cannot be expressed in individuals homozygous for the mutant form of the *FUT1* gene; even though they may have the I^A and/or the I^B alleles, neither antigen is added to the cell surface, and they are functionally type O. To distinguish them from the rest of the population, they are said to demonstrate the **Bombay phenotype**. The frequency of the mutant *FUT1* allele is exceedingly low. Hence, the vast majority of the human population can synthesize the H substance.

The *white* Locus in *Drosophila*

Many other phenotypes in plants and animals are influenced by multiple allelic inheritance. In *Drosophila*, many alleles are present at practically every locus. The recessive mutation that causes white eyes, discovered by Thomas H. Morgan and Calvin Bridges in 1912, is one of over 100

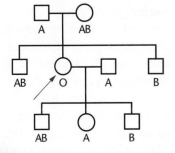

FIGURE 4–3 A partial pedigree of a woman with the Bombay phenotype. Functionally, her ABO blood group behaves as type O. Genetically, she is type B.

TABLE 4.1		
Some of the Alleles Present at the *white* Locus of *Drosophila*		
Allele	**Name**	**Eye Color**
w	white	pure white
w^a	white-apricot	yellowish orange
w^{bf}	white-buff	light buff
w^{bl}	white-blood	yellowish ruby
w^{cf}	white-coffee	deep ruby
w^e	white-eosin	yellowish pink
w^{mo}	white-mottled orange	light mottled orange
w^{sat}	white-satsuma	deep ruby
w^{sp}	white-spotted	fine grain, yellow mottling
w^t	white-tinged	light pink

alleles that can occupy this locus. In this allelic series, eye colors range from complete absence of pigment in the *white* allele to deep ruby in the *white-satsuma* allele, orange in the *white-apricot* allele, and a buff color in the *white-buff* allele. These alleles are designated w, w^{sat}, w^a, and w^{bf}, respectively. In each case, the total amount of pigment in these mutant eyes is reduced to less than 20 percent of that found in the brick-red wild-type eye. Table 4.1 lists these and other *white* alleles and their color phenotypes.

It is interesting to note the biological basis of the original *white* mutation in *Drosophila*. Given what we know about eye color in this organism, it might be logical to presume that the mutant allele somehow interrupts the biochemical synthesis of pigments making up the brick-red eye of the wild-type fly. However, it is now clear that the product of the *white* locus is a protein that is involved in transporting pigments into the ommatidia (the individual units) comprising the compound eye. While flies expressing the *white* mutation can synthesize eye pigments normally, they cannot transport them into these structural units of the eye, thus rendering the white phenotype.

4.6

Lethal Alleles Represent Essential Genes

Many gene products are essential to an organism's survival. Mutations resulting in the synthesis of a gene product that is nonfunctional can often be tolerated in the heterozygous state; that is, one wild-type allele may be sufficient to produce enough of the essential product to allow survival. However, such a mutation behaves as a **recessive lethal allele**, and homozygous recessive individuals will not survive. The time of death will depend on when the product is essential. In mammals, for example, this might occur during development, early childhood, or even adulthood.

4–1 In the guinea pig, one locus involved in the control of coat color may be occupied by any of four alleles: C (full color), c^k (sepia), c^d (cream), or c^a (albino), with an order of dominance of: $C > c^k > c^d > c^a$. (C is dominant to all others, c^k is dominant to c^d and c^a, but not C, etc.) In the following crosses, determine the parental genotypes and predict the phenotypic ratios that would result:

(a) sepia × cream, where both guinea pigs had an albino parent

(b) sepia × cream, where the sepia guinea pig had an albino parent and the cream guinea pig had two sepia parents

(c) sepia × cream, where the sepia guinea pig had two full-color parents and the cream guinea pig had two sepia parents

(d) sepia × cream, where the sepia guinea pig had a full-color parent and an albino parent and the cream guinea pig had two full-color parents

■ HINT: *This problem involves an understanding of multiple alleles. The key to its solution is to note particularly the hierarchy of dominance of the various alleles. Remember also that even though there can be more than two alleles in a population, an individual can have at most two of these. Thus, the allelic distribution into gametes adheres to the principle of segregation.*

In some cases, the allele responsible for a lethal effect when homozygous may also result in a distinctive mutant phenotype when present heterozygously. *It is behaving as a recessive lethal allele but is dominant with respect to the phenotype.* For example, a mutation that causes a yellow coat in mice was discovered in the early part of this century. The yellow coat varies from the normal agouti (wild-type) coat phenotype, as shown in Figure 4–4. Crosses between the various combinations of the two strains yield unusual results:

Crosses				
(A) agouti	×	agouti	⟶	all agouti
(B) yellow	×	yellow	⟶	2/3 yellow: 1/3 agouti
(C) agouti	×	yellow	⟶	1/2 yellow: 1/2 agouti

These results are explained on the basis of a single pair of alleles. With regard to coat color, the mutant *yellow* allele (A^Y) is dominant to the wild-type *agouti* allele (A), so heterozygous mice will have yellow coats. However, the *yellow* allele is also a homozygous recessive lethal. When present in two copies, the mice die before birth. Thus, there are no homozygous yellow mice. The genetic basis for these three crosses is shown in Figure 4–4.

In other cases, a mutation may behave as a **dominant lethal allele**. In such cases, the presence of just one copy of the allele results in the death of the individual. In humans, a disorder called **Huntington disease** is due to a dominant autosomal allele H, where the onset of the disease in heterozygotes (Hh) is delayed, usually well into adulthood. Affected individuals then undergo gradual nervous and motor degeneration until they die. This lethal disorder is particularly tragic because it has such a late onset, typically at about age 40. By that time, the affected individual may have produced a family, and each of their children has a 50 percent probability of inheriting the lethal allele, transmitting the allele to his or her offspring, and eventually developing the disorder. The American folk singer and composer Woody Guthrie (father of modern-day folk singer Arlo Guthrie) died from this disease at age 39.

Dominant lethal alleles are rarely observed. For these alleles to exist in a population, the affected individuals must reproduce before the lethal allele is expressed, as can occur in Huntington disease. If all affected individuals die before reaching reproductive age, the mutant gene will not be passed to future generations, and the mutation will disappear from the population unless it arises again as a result of a new mutation.

4.7

Combinations of Two Gene Pairs with Two Modes of Inheritance Modify the 9:3:3:1 Ratio

Each example discussed so far modifies Mendel's 3:1 F_2 monohybrid ratio. Therefore, combining any two of these modes of inheritance in a dihybrid cross will also modify the classical 9:3:3:1 dihybrid ratio. Having established the foundation of the modes of inheritance of incomplete dominance, codominance, multiple alleles, and lethal alleles, we can now deal with the situation of two modes of inheritance occurring simultaneously. Mendel's principle of independent assortment applies to these situations, provided that the genes controlling each character are not located on the same chromosome—in other words, that they do not demonstrate what is called *genetic linkage*.

Consider, for example, a mating between two humans who are both heterozygous for the autosomal recessive gene that causes albinism and who are both of blood type AB. What is the probability of a particular phenotypic combination occurring in each of their children? Albinism is inherited in the simple Mendelian fashion, and the blood types are determined by the series of three multiple alleles, i^A, i^B, and i^O. The solution to this problem is diagrammed in Figure 4–5 on p. 80, using the forked-line method. This dihybrid cross does not yield four phenotypes in the classical 9:3:3:1 ratio.

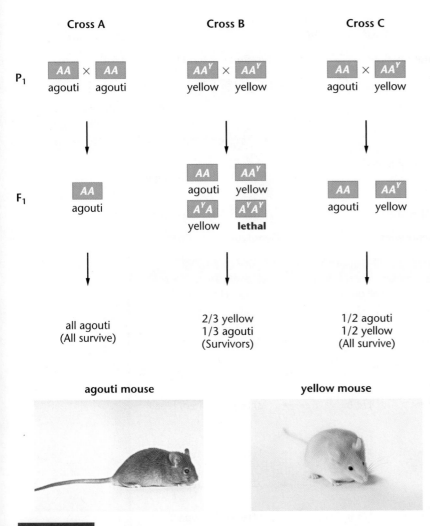

Cross A   Cross B   Cross C

P_1

Cross A: AA (agouti) × AA (agouti)

Cross B: AA^Y (yellow) × AA^Y (yellow)

Cross C: AA (agouti) × AA^Y (yellow)

F_1

Cross A: AA (agouti)

Cross B: AA (agouti), AA^Y (yellow), A^YA (yellow), A^YA^Y (**lethal**)

Cross C: AA (agouti), AA^Y (yellow)

Cross A: all agouti (All survive)

Cross B: 2/3 yellow 1/3 agouti (Survivors)

Cross C: 1/2 agouti 1/2 yellow (All survive)

agouti mouse   **yellow mouse**

FIGURE 4–4   Inheritance patterns in three crosses involving the normal wild-type *agouti* allele (*A*) and the mutant *yellow* allele (A^Y) in the mouse. Note that the mutant allele behaves dominantly to the normal allele in controlling coat color, but it also behaves as a homozygous recessive lethal allele. Mice with the genotype $A^Y A^Y$ do not survive.

Instead, six phenotypes occur in a 3:6:3:1:2:1 ratio, establishing the expected probability for each phenotype. This is just one of the many variants of modified ratios that are possible when different modes of inheritance are combined.

4.8

Phenotypes Are Often Affected by More Than One Gene

Soon after Mendel's work was rediscovered, experimentation revealed that in many cases a given phenotype is affected by more than one gene. This was a significant discovery because it revealed that genetic influence on the phenotype is often much more complex than the situations Mendel encountered in his crosses with the garden pea. Instead of single genes controlling the development of individual parts of a plant or animal body, it soon became clear that phenotypic characters such as eye color, hair color, or fruit shape can be influenced by many different genes and their products.

The term **gene interaction** is often used to express the idea that several genes influence a particular characteristic. This does not mean, however, that two or more genes or their products necessarily interact directly with one another to influence a particular phenotype. Rather, the term means that the

The Molecular Basis of Dominance and Recessiveness: The Agouti Gene

Molecular analysis of the gene resulting in the agouti and yellow mice has provided insight into how a mutation can be both dominant for one phenotypic effect (hair color) and recessive for another (embryonic development). The A^Y allele is a classic example of a gain-of-function mutation. Animals homozygous for the wild-type *A* allele have yellow pigment deposited as a band on the otherwise black hair shaft, resulting in the agouti phenotype (see Figure 4–4). Heterozygotes deposit yellow pigment along the entire length of hair shafts as a result of the deletion of the regulatory region preceding the DNA coding region of the A^Y allele. Without any means to regulate its expression, one copy of the A^Y allele is always turned on in heterozygotes, resulting in the gain of function leading to the dominant effect.

The homozygous lethal effect has also been explained by molecular analysis of the mutant gene. The extensive deletion of genetic material that produced the A^Y allele actually extends into the coding region of an adjacent gene (*Merc*), rendering it nonfunctional. It is this gene that is critical to embryonic development, and the loss of its function in A^Y/A^Y homozygotes is what causes lethality. Heterozygotes exceed the threshold level of the wild-type Merc gene product and thus survive.

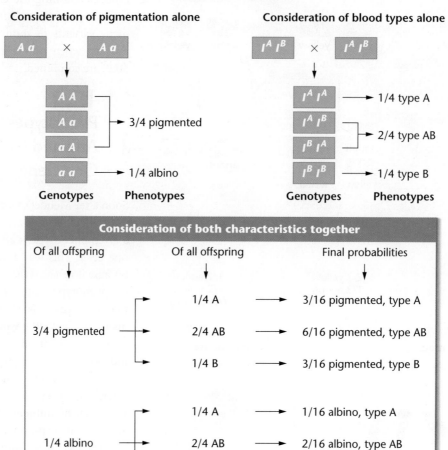

FIGURE 4–5 Calculation of the probabilities in a mating involving the ABO blood type and albinism in humans, using the forked-line method.

cellular function of numerous gene products contributes to the development of a common phenotype. For example, the development of an organ such as the eye of an insect is exceedingly complex and leads to a structure with multiple phenotypic manifestations, for example, to an eye having a specific size, shape, texture, and color. The development of the eye is a complex cascade of developmental events leading to that organ's formation. This process illustrates the developmental concept of **epigenesis**, whereby each step of development increases the complexity of the organ or feature of interest and is under the control and influence of many genes.

An enlightening example of epigenesis and multiple gene interaction involves the formation of the inner ear in mammals, allowing organisms to detect and interpret sound. The structure and function of the inner ear is exceedingly complex. Its formation includes not only distinctive anatomical features to capture, funnel, and transmit external sound toward and through the middle ear, but also to convert sound waves into nerve impulses within the inner ear. Thus, the ear forms as a result of a cascade of intricate developmental events influenced by many genes. Mutations that interrupt many of the steps of ear development lead to a common phenotype: **hereditary deafness**. In a sense, these many genes "interact" to produce a common phenotype. In such situations, the mutant phenotype is described as a **heterogeneous trait**, reflecting the many genes involved. In humans, while a few common alleles are responsible for the vast majority of cases of hereditary deafness, over 50 genes are involved in the development of the ability to discern sound.

Epistasis

Some of the best examples of gene interaction are those showing the phenomenon of **epistasis**, where the expression of one gene pair masks or modifies the effect of another gene pair. Sometimes the genes involved influence the same general phenotypic characteristic in an antagonistic manner, which leads to masking. In other cases, however, the genes involved exert their influence on one another in a complementary, or cooperative, fashion.

For example, the homozygous presence of a recessive allele may prevent or override the expression of other alleles at a second locus (or several other loci). In this case, the alleles at the first locus are said to be *epistatic* to those at the second locus, and the alleles at the second locus are *hypostatic* to those at the first locus. As we will see, there are several variations on this theme. In another example, a single dominant allele at the first locus may be epistatic to the expression of the alleles at a second gene locus. In a third example, two gene pairs may *complement* one another such

that at least one dominant allele in each pair is required to express a particular phenotype.

The Bombay phenotype discussed earlier is an example of the homozygous recessive condition at one locus masking the expression of a second locus. There we established that the homozygous presence of the mutant form of the *FUT1* gene masks the expression of the I^A and I^B alleles. Only individuals containing at least one wild-type *FUT1* allele can form the A or B antigen. As a result, individuals whose genotypes include the I^A or I^B allele and who have no wild-type *FUT1* allele are of the type O phenotype, regardless of their potential to make either antigen. An example of the outcome of matings between individuals heterozygous at both loci is illustrated in Figure 4–6. If many such individuals have children, the phenotypic ratio of 3 A: 6 AB: 3 B: 4 O is expected in their offspring.

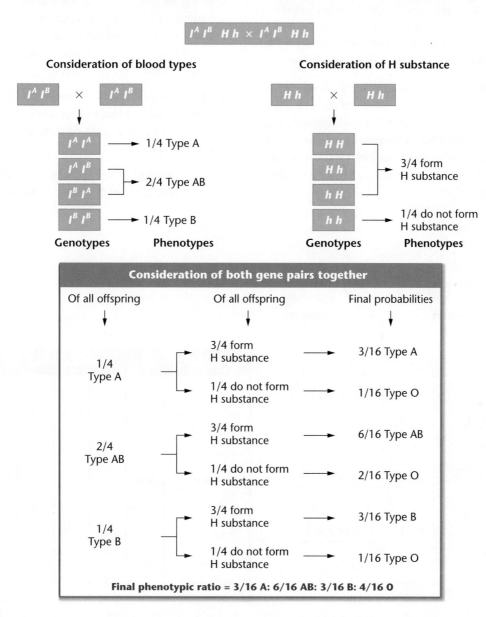

FIGURE 4–6 The outcome of a mating between individuals heterozygous at two genes determining their ABO blood type. Final phenotypes are calculated by considering each gene separately and then combining the results using the forked-line method.

It is important to note two things when examining this cross and the predicted phenotypic ratio:

1. A key distinction exists between this cross and the modified dihybrid cross shown in Figure 4–5: *only one characteristic—blood type—is being followed.* In the modified dihybrid cross in Figure 4–5, blood type *and* skin pigmentation are followed as separate phenotypic characteristics.

2. Even though only a single character was followed, the phenotypic ratio comes out in sixteenths. If we knew nothing about the H substance and the gene controlling it, we could still be confident (because the proportions are in sixteenths) that a second gene pair, other than that controlling the A and B antigens, was involved in the phenotypic expression. *When a single character is being studied, a ratio that is expressed in 16 parts (e.g., 3:6:3:4) suggests that two gene pairs are "interacting" in the expression of the phenotype under consideration.*

The study of gene interaction reveals a number of inheritance patterns that are modifications of the Mendelian dihybrid F_2 ratio (9:3:3:1). In several of the subsequent examples, epistasis has the effect of combining one or more of the four phenotypic categories in various ways. The generation of these four groups is reviewed in Figure 4–7, along with several modified ratios.

As we discuss these and other examples (see Figure 4–8), we will make several assumptions and adopt certain conventions:

1. In each case, distinct phenotypic classes are produced, each clearly discernible from all others. Such traits illustrate discontinuous variation, where phenotypic categories are discrete and qualitatively different from one another.

2. The genes considered in each cross are on different chromosomes and therefore assort independently of one another during gamete formation. To allow you to easily compare the results of different crosses, we designated alleles as *A*, *a* and *B*, *b* in each case.

3. When we assume that complete dominance exists within a gene pair, such that *AA* and *Aa* or *BB* and *Bb* are equivalent in their genetic effects, we use the designations *A*– or *B*– for both combinations, where the dash (–) indicates that either allele may be present without consequence to the phenotype.

4. All P_1 crosses involve homozygous individuals (e.g., *AABB* × *aabb*, *AAbb* × *aaBB*, or *aaBB* × *AAbb*). Therefore, each F_1 generation consists of only heterozygotes of genotype *AaBb*.

5. In each example, the F_2 generation produced from these heterozygous parents is our main focus of analysis.

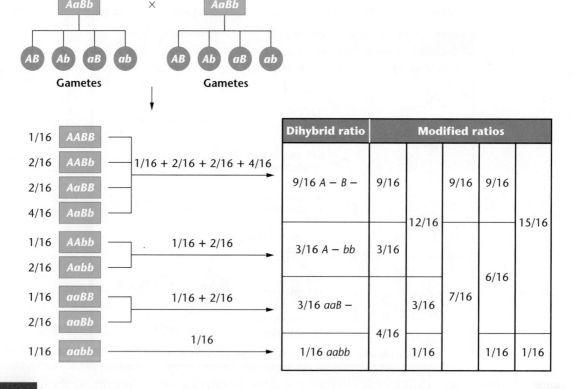

FIGURE 4–7   Generation of various modified dihybrid ratios from the nine unique genotypes produced in a cross between individuals heterozygous at two genes.

Case	Organism	Character	F₂ Phenotypes				Modified ratio
			9/16	3/16	3/16	1/16	
1	Mouse	Coat color	agouti	albino	black	albino	9:3:4
2	Squash	Color	white		yellow	green	12:3:1
3	Pea	Flower color	purple	white			9:7
4	Squash	Fruit shape	disc	sphere		long	9:6:1
5	Chicken	Color	white		colored	white	13:3
6	Mouse	Color	white-spotted	white	colored	white-spotted	10:3:3
7	Shepherd's purse	Seed capsule	triangular			ovoid	15:1
8	Flour beetle	Color	6/16 sooty and 3/16 red	black	jet	black	6:3:3:4

FIGURE 4–8   The basis of modified dihybrid F₂ phenotypic ratios resulting from crosses between doubly heterozygous F₁ individuals. The four groupings of the F₂ genotypes shown in Figure 4–7 and across the top of this figure are combined in various ways to produce these ratios.

When two genes are involved (Figure 4–7), the F₂ genotypes fall into four categories: 9/16 A–B–, 3/16 A–bb, 3/16 aaB–, and 1/16 aabb. Because of dominance, all genotypes in each category are equivalent in their effect on the phenotype.

Case 1 is the inheritance of coat color in mice (Figure 4–8). Normal wild-type coat color is agouti, a grayish pattern formed by alternating bands of pigment on each hair (see Figure 4–4). Agouti is dominant to black (nonagouti) hair, which results from the homozygous expression of a recessive mutation that we designate a. Thus, A– results in agouti, whereas aa yields black coat color. When a recessive mutation, b, at a separate locus is homozygous, it eliminates pigmentation altogether, yielding albino mice (bb), regardless of the genotype at the a locus. Thus, in a cross between agouti (AABB) and albino (aabb) parents, members of the F₁ are all AaBb and have agouti coat color. In the F₂ progeny of a cross between two F₁ double heterozygotes, the following genotypes and phenotypes are observed:

F₁: AaBb × AaBb

F₂ Ratio	Genotype	Phenotype	Final Phenotypic Ratio
9/16	A–B–	agouti	9/16 agouti
3/16	A–bb	albino	3/16 black
3/16	aa B–	black	4/16 albino
1/16	aa bb	albino	

We can envision gene interaction yielding the observed 9:3:4 F₂ ratio as a two-step process:

	Gene B		Gene A	
Precursor molecule (colorless)	→ B–	Black pigment	→ A–	Agouti pattern

In the presence of a B allele, black pigment can be made from a colorless substance. In the presence of an A allele, the black pigment is deposited during the development of hair in a pattern that produces the agouti phenotype. If the aa genotype occurs, all of the hair remains black. If the bb genotype occurs, no black pigment is produced, regardless of the presence of the A or a alleles, and the mouse is albino. Therefore, the bb genotype masks or suppresses the expression of the A allele. As a result, this is referred to as *recessive epistasis*.

A second type of epistasis, called *dominant epistasis*, occurs when a dominant allele at one genetic locus masks the expression of the alleles of a second locus. For instance, Case 2 of Figure 4–8 deals with the inheritance of fruit color in summer squash. Here, the dominant allele A results in white fruit color regardless of the genotype at a second locus, B. In the absence of a dominant A allele (the aa genotype), BB or Bb results in yellow color, while bb results in green color. Therefore, if two white-colored double heterozygotes

(*AaBb*) are crossed, this type of epistasis generates an interesting phenotypic ratio:

$$F_1: AaBb \times AaBb$$
↓

F₂ Ratio	Genotype	Phenotype	Final Phenotypic Ratio
9/16	A– B–	white	12/16 white
3/16	A– bb	white	
3/16	aa B–	yellow	3/16 yellow
1/16	aa bb	green	1/16 green

Of the offspring, 9/16 are *A–B–* and are thus white. The 3/16 bearing the genotypes *A–bb* are also white. Of the remaining squash, 3/16 are yellow (*aaB–*), while 1/16 are green (*aabb*). Thus, the modified phenotypic ratio of 12:3:1 occurs.

Our third example (Case 3 of Figure 4–8), first discovered by William Bateson and Reginald Punnett (of Punnett square fame), is demonstrated in a cross between two true-breeding strains of white-flowered sweet peas. Unexpectedly, the results of this cross yield all purple F₁ plants, and the F₂ plants occur in a ratio of 9/16 purple to 7/16 white. The proposed explanation suggests that the presence of at least one dominant allele of each of two gene pairs is essential in order for flowers to be purple. Thus, this cross represents a case of *complementary gene interaction*. All other genotype combinations yield white flowers because the homozygous condition of either recessive allele masks the expression of the dominant allele at the other locus.

The cross is shown as follows:

$$P_1: AAbb \times aaBB$$
white   white
↓

$$F_1: All\ AaBb\ (purple)$$

F₂ Ratio	Genotype	Phenotype	Final Phenotypic Ratio
9/16	A– B–	purple	9/16 purple
3/16	A– bb	white	
3/16	aa B–	white	7/16 white
1/16	aa bb	white	

We can now envision how two gene pairs might yield such results:

	Gene A			Gene B	
Precursor	↓	Intermediate	↓	Final	
substance	→	product	→	product	
(colorless)	A–	(colorless)	B–	(purple)	

At least one dominant allele from each pair of genes is necessary to ensure both biochemical conversions to the final product, yielding purple flowers. In the preceding

cross, this will occur in 9/16 of the F₂ offspring. All other plants (7/16) have flowers that remain white.

These three examples illustrate in a simple way how the products of two genes interact to influence the development of a common phenotype. In other instances, more than two genes and their products are involved in controlling phenotypic expression.

Novel Phenotypes

Other cases of gene interaction yield novel, or new, phenotypes in the F₂ generation, in addition to producing modified dihybrid ratios. Case 4 in Figure 4–8 depicts the inheritance of fruit shape in the summer squash *Cucurbita pepo*. When plants with disc-shaped fruit (*AABB*) are crossed with plants with long fruit (*aabb*), the F₁ generation all have disc fruit. However, in the F₂ progeny, fruit with a novel shape—sphere—appear, as well as fruit exhibiting the parental phenotypes. A variety of fruit shapes are shown in Figure 4–9.

The F₂ generation, with a modified 9:6:1 ratio, is generated as follows:

$$F_1: AaBb \times AaBb$$
disc   disc
↓

F₂ Ratio	Genotype	Phenotype	Final Phenotypic Ratio
9/16	A– B–	disc	9/16 disc
3/16	A– bb	sphere	6/16 sphere
3/16	aa B–	sphere	
1/16	aa bb	long	1/16 long

In this example of gene interaction, both gene pairs influence fruit shape equally. A dominant allele at either locus ensures a sphere-shaped fruit. In the absence of dominant alleles, the fruit is long. However, if both dominant alleles (*A* and *B*) are present, the fruit displays a flattened, disc shape.

FIGURE 4–9   Summer squash exhibiting various fruit-shape phenotypes, including disc, long, and sphere.

4–2 In some plants a red flower pigment, cyanidin, is synthesized from a colorless precursor. The addition of a hydroxyl group (OH^-) to the cyanidin molecule causes it to become purple. In a cross between two randomly selected purple varieties, the following results were obtained:

    94 purple
    31 red
    43 white

How many genes are involved in the determination of these flower colors? Which genotypic combinations produce which phenotypes? Diagram the purple × purple cross.

■ HINT: *This problem describes a plant in which flower color, a single characteristic, can take on one of three variations. The key to its solution is to first analyze the raw data and convert the numbers to a meaningful ratio. This will guide you in determining how many gene pairs are involved. Then you can group the genotypes in a way that corresponds to the phenotypic ratio.*

Another interesting example of an unexpected phenotype arising in the F$_2$ generation is the inheritance of eye color in *Drosophila melanogaster*. As mentioned earlier, the wild-type eye color is brick red. When two autosomal recessive mutants, *brown* and *scarlet*, are crossed, the F$_1$ generation consists of flies with wild-type eye color. In the F$_2$ generation, wild, scarlet, brown, and white-eyed flies are found in a 9:3:3:1 ratio. While this ratio is numerically the same as Mendel's dihybrid ratio, the *Drosophila* cross involves only one character: eye color. This is an important distinction to make when modified dihybrid ratios resulting from gene interaction are studied.

The *Drosophila* cross is an excellent example of gene interaction because the biochemical basis of eye color in this organism has been determined (Figure 4–10). *Drosophila*, as a typical arthropod, has compound eyes made up of hundreds of individual visual units called ommatidia. The wild-type eye color is due to the deposition and mixing of two separate pigment groups in each ommatidium—the bright-red **drosopterins** and the brown **xanthommatins**. Each type of pigment is produced by a separate biosynthetic pathway. Each step of each pathway is catalyzed by a separate enzyme and is thus under the control of a separate gene. As shown

FIGURE 4–10   A theoretical explanation of the biochemical basis of the four eye color phenotypes produced in a cross between *Drosophila* with brown eyes and scarlet eyes. In the presence of at least one wild-type bw^+ allele, an enzyme is produced that converts substance b to c, and the pigment drosopterin is synthesized. In the presence of at least one wild-type st^+ allele, substance e is converted to f, and the pigment xanthommatin is synthesized. The homozygous presence of the recessive st or bw mutant allele blocks the synthesis of the respective pigment molecule. Either one, both, or neither of these pathways can be blocked, depending on the genotype.

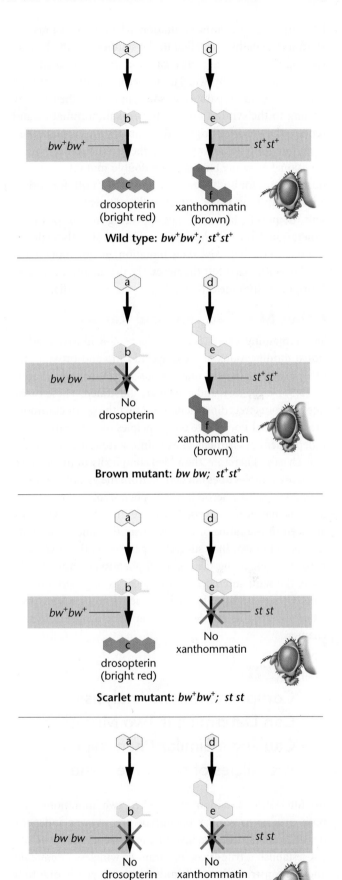

in Figure 4–10, the *brown* mutation, when homozygous, interrupts the pathway leading to the synthesis of the bright-red pigments. Because only xanthommatin pigments are present, the eye is brown. The *scarlet* mutation, affecting a gene located on a separate autosome, interrupts the pathway leading to the synthesis of the brown xanthommatins and renders the eye color bright red in homozygous mutant flies. Each mutation apparently causes the production of a nonfunctional enzyme. Flies that are double mutants and thus homozygous for both *brown* and *scarlet* lack both functional enzymes and can make neither of the pigments; they represent the novel white-eyed flies appearing in 1/16 of the F_2 generation. Note that the absence of pigment in these flies is not due to the X-linked *white* mutation, in which pigments can be synthesized but the necessary precursors cannot be transported into the cells making up the ommatidia.

Other Modified Dihybrid Ratios

The remaining cases (5–8) in Figure 4–8 illustrate additional modifications of the dihybrid ratio and provide still other examples of gene interactions. As you will note, ratios of 13:3, 10:3:3; 15:1, and 6:3:3:4 are illustrated. These cases, like the four preceding them, have two things in common. First, we need not violate the principles of segregation and independent assortment to explain the inheritance pattern of each case. Therefore, the added complexity of inheritance in these examples does not detract from the validity of Mendel's conclusions. Second, the F_2 phenotypic ratio in each example has been expressed in sixteenths. When sixteenths are seen in the ratios of crosses where the inheritance pattern is unknown, they suggest to geneticists that two gene pairs are controlling the observed phenotypes. You should make the same inference in your analysis of genetics problems. Other insights into solving genetics problems are provided in "Insights and Solutions" at the conclusion of this chapter.

4.9

Complementation Analysis Can Determine If Two Mutations Causing a Similar Phenotype Are Alleles of the Same Gene

An interesting situation arises when two mutations that both produce a similar phenotype are isolated independently. Suppose that two investigators independently isolate and establish a true-breeding strain of wingless *Drosophila* and demonstrate that each mutant phenotype is due to a recessive mutation. We might assume that both strains contain mutations in the same gene. However, since we know

that many genes are involved in the formation of wings, we must consider the possibility that mutations in any one of them might inhibit wing formation during development. This is the case with any *heterogeneous trait*, a concept introduced earlier in this chapter in our discussion of hereditary deafness. An analytical procedure called **complementation analysis** allows us to determine whether two independently isolated mutations are in the same gene—that is, whether they are alleles—or whether they represent mutations in separate genes.

To repeat, our analysis seeks to answer this simple question: *Are two mutations that yield similar phenotypes present in the same gene or in two different genes?* To find the answer, we cross the two mutant strains and analyze the F_1 generation. The two possible alternative outcomes and their interpretations are shown in Figure 4–11. To discuss these possibilities (Case 1 and Case 2), we designate one of the mutations m^a and the other m^b.

> **Case 1.** *All offspring develop normal wings.*
> **Interpretation:** The two recessive mutations are in separate genes and are not alleles of one another. Following the cross, all F_1 flies are heterozygous for both genes. Since each mutation is in a separate gene and each F_1 fly is heterozygous at both loci, the normal products of both genes are produced (by the one normal copy of each gene), and wings develop. Under such circumstances, the genes complement one another in restoration of the wild-type phenotype, and complementation is said to occur because the two mutations are in different genes.
>
> **Case 2.** *All offspring fail to develop wings.*
> **Interpretation:** The two mutations affect the same gene and are alleles of one another. Complementation does not occur. Since the two mutations affect the same gene, the F_1 flies are homozygous for the two mutant alleles (the m^a allele and the m^b allele). No normal product of the gene is produced, and in the absence of this essential product, wings do not form.

Complementation analysis, as originally devised by the *Drosophila* geneticist Edward B. Lewis, may be used to screen any number of individual mutations that result in the same phenotype. Such an analysis may reveal that only a single gene is involved or that two or more genes are involved. All mutations determined to be present in any single gene are said to fall into the same **complementation group**, and they will complement mutations in all other groups. When large numbers of mutations affecting the same trait are available and studied using complementation analysis, it is possible to predict the total number of genes involved in the determination of that trait.

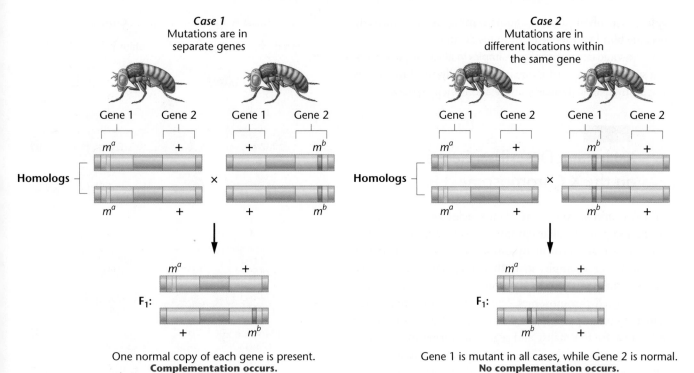

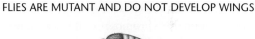

FIGURE 4–11 Complementation analysis of alternative outcomes of two wingless mutations in *Drosophila* (*m^a* and *m^b*). In Case 1, the mutations are not alleles of the same gene, while in Case 2, the mutations are alleles of the same gene.

4.10

Expression of a Single Gene May Have Multiple Effects

While the previous sections have focused on the effects of two or more genes on a single characteristic, the converse situation, where expression of a single gene has multiple phenotypic effects, is also quite common. This phenomenon, which often becomes apparent when phenotypes are examined carefully, is referred to as **pleiotropy**. Many excellent examples can be drawn from human disorders, and we will review two such cases to illustrate this point.

The first disorder is **Marfan syndrome**, a human malady resulting from an autosomal dominant mutation in the gene encoding the connective tissue protein *fibrillin*. Because this protein is widespread in many tissues in the body, one would expect multiple effects of such a defect. In fact, fibrillin is important to the structural integrity of the lens of the eye, to the lining of vessels such as the aorta, and to bones, among other tissues. As a result, the phenotype associated with Marfan syndrome includes lens dislocation,

increased risk of aortic aneurysm, and lengthened long bones in limbs. This disorder is of historical interest in that speculation abounds that Abraham Lincoln was afflicted.

A second example involves another human autosomal dominant disorder, **porphyria variegata**. Afflicted individuals cannot adequately metabolize the porphyrin component of hemoglobin when this respiratory pigment is broken down as red blood cells are replaced. The accumulation of excess porphyrins is immediately evident in the urine, which takes on a deep red color. However, this phenotypic characteristic is merely diagnostic. The severe features of the disorder are due to the toxicity of the buildup of porphyrins in the body, particularly in the brain. Complete phenotypic characterization includes abdominal pain, muscular weakness, fever, a racing pulse, insomnia, headaches, vision problems (that can lead to blindness), delirium, and ultimately convulsions. As you can see, deciding which phenotypic trait best characterizes the disorder is impossible.

Like Marfan syndrome, porphyria variegata is also of historical significance. George III, King of England during the American Revolution, is believed to have suffered from episodes involving all of the above symptoms. He ultimately became blind and senile prior to his death.

We could cite many other examples to illustrate pleiotropy, but suffice it to say that if one looks carefully, most mutations display more than a single manifestation when expressed.

4.11

X-Linkage Describes Genes on the X Chromosome

In many animals and some plant species, one of the sexes contains a pair of unlike chromosomes that are involved in sex determination. In many cases, these are designated as X and Y. For example, in both *Drosophila* and humans, males contain an X and a Y chromosome, whereas females contain two X chromosomes. The Y chromosome must contain a region of pairing homology with the X chromosome if the two are to synapse and segregate during meiosis, but a major portion of the Y chromosome in humans as well as other species is considered to be relatively inert genetically. While we now recognize a number of male-specific genes on the human Y chromosome, it lacks copies of most genes present on the X chromosome. As a result, genes present on the X chromosome exhibit patterns of inheritance that are very different from those seen with autosomal genes. The term **X-linkage** is used to describe these situations.

In the following discussion, we will focus on inheritance patterns resulting from genes present on the X but absent from the Y chromosome. This situation results in a modification of Mendelian ratios, the central theme of this chapter.

X-Linkage in *Drosophila*

One of the first cases of X-linkage was documented in 1910 by Thomas H. Morgan during his studies of the *white* eye mutation in *Drosophila* (Figure 4–12). The normal wild-type red eye color is dominant to white eye color.

Morgan's work established that the inheritance pattern of the white-eye trait was clearly related to the sex of the parent carrying the mutant allele. Unlike the outcome of the typical Mendelian monohybrid cross where F_1 and F_2 data were similar regardless of which P_1 parent exhibited the recessive mutant trait, reciprocal crosses between white-eyed and red-eyed flies did not yield identical results. Morgan's analysis led to the conclusion that the *white* locus is present on the X chromosome rather than on one of the autosomes. Both the gene and the trait are said to be X-linked.

Results of reciprocal crosses between white-eyed and red-eyed flies are shown in Figure 4–12. The obvious differences in phenotypic ratios in both the F_1 and F_2 generations are dependent on whether or not the P_1 white-eyed parent was male or female.

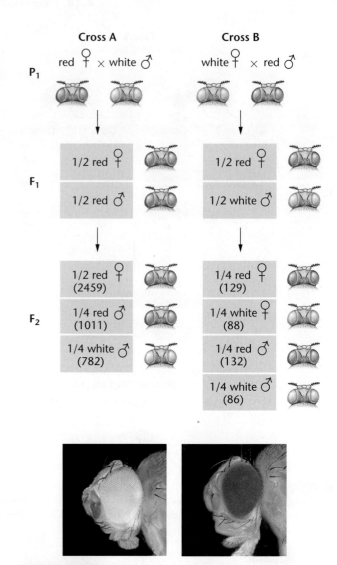

FIGURE 4–12   The F_1 and F_2 results of T. H. Morgan's reciprocal crosses involving the X-linked *white* mutation in *Drosophila melanogaster*. The actual data are shown in parentheses. The photographs show white eye and the brick-red wild-type eye color.

Morgan was able to correlate these observations with the difference found in the sex-chromosome composition of male and female *Drosophila*. He hypothesized that the recessive allele for white eye is found on the X chromosome, but its corresponding locus is absent from the Y chromosome. Females thus have two available gene loci, one on each X chromosome, whereas males have only one available locus, on their single X chromosome.

Morgan's interpretation of X-linked inheritance, shown in Figure 4–13, provides a suitable theoretical explanation for his results. Since the Y chromosome lacks homology with almost all genes on the X chromosome, these alleles present on the X chromosome of the males will be directly expressed in the phenotype. Males cannot be either homozygous or heterozygous for X-linked genes; instead, their condition—possession of only one copy of a gene in an otherwise diploid cell—is referred to as **hemizygosity**. The individual is said to

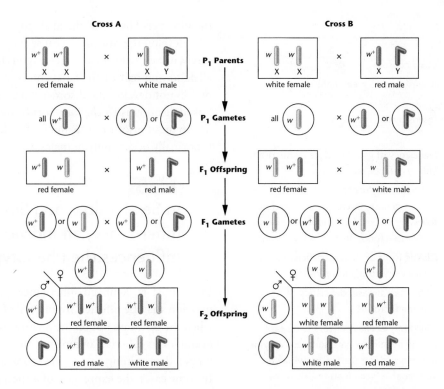

Cross A   **Cross B**

FIGURE 4–13   The chromosomal explanation of the results of the X-linked crosses shown in Figure 4–12.

be **hemizygous**. One result of X-linkage is the *crisscross pattern of inheritance*, in which phenotypic traits controlled by recessive X-linked genes are passed from homozygous mothers to all sons. This pattern occurs because females exhibiting a recessive trait must contain the mutant allele on both X chromosomes. Because male offspring receive one of their mother's two X chromosomes and are hemizygous for all alleles present on that X, all sons will express the same recessive X-linked traits as their mother.

Morgan's work has taken on great historical significance. By 1910, the correlation between Mendel's work and the behavior of chromosomes during meiosis had provided the basis for the **chromosome theory of inheritance**. Morgan's work, and subsequently that of his student, Calvin Bridges, around 1920, provided direct evidence that genes are transmitted on specific chromosomes, and is considered the first solid experimental evidence in support of this theory.

X-Linkage in Humans

In humans, many genes and the respective traits controlled by them are recognized as being linked to the X chromosome (see Table 4.2). These X-linked traits can be easily identified in a pedigree because of the crisscross pattern of inheritance. A pedigree for one form of human **color blindness** is shown in Figure 4–14. The mother in generation I passes the trait to all her sons but to none of her daughters. If the offspring in generation II marry normal individuals, the color-blind sons will produce all normal male and female offspring (III-1, 2, and 3); the normal-visioned daughters will produce normal-visioned

female offspring (III-4, 6, and 7), as well as color-blind (III-8) and normal-visioned (III-5) male offspring.

The way in which X-linked genes are transmitted causes unusual circumstances associated with recessive X-linked

NOW SOLVE THIS

4–3 Below are three pedigrees. For each trait, consider whether it is or is not consistent with X-linked recessive inheritance. In a sentence or two, indicate why or why not.

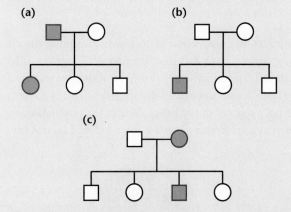

(a)   (b)

(c)

■ HINT: *This problem involves potential X-linked recessive traits as analyzed in pedigrees. The key to its solution is to focus on hemizygosity, where an X-linked recessive allele is always expressed in males, but never passed from a father to his sons. Homozygous females, on the other hand, pass the trait to all sons, but not to their daughters unless the father is also affected.*

TABLE 4.2

Human X-Linked Traits

Condition	Characteristics
Color blindness, deutan type	Insensitivity to green light
Color blindness, protan type	Insensitivity to red light
Fabry's disease	Deficiency of galactosidase A; heart and kidney defects, early death
G-6-PD deficiency	Deficiency of glucose-6-phosphate dehydrogenase; severe anemic reaction following intake of primaquines in drugs and certain foods, including fava beans
Hemophilia A	Classic form of clotting deficiency; deficiency of clotting factor VIII
Hemophilia B	Christmas disease; deficiency of clotting factor IX
Hunter syndrome	Mucopolysaccharide storage disease resulting from iduronate sulfatase enzyme deficiency; short stature, claw-like fingers, coarse facial features, slow mental deterioration, and deafness
Ichthyosis	Deficiency of steroid sulfatase enzyme; scaly dry skin, particularly on extremities
Lesch–Nyhan syndrome	Deficiency of hypoxanthine-guanine phosphoribosyltransferase enzyme (HPRT) leading to motor and mental retardation, self-mutilation, and early death
Duchenne muscular dystrophy	Progressive, life-shortening disorder characterized by muscle degeneration and weakness; sometimes associated with mental retardation; deficiency of the protein dystrophin

disorders, in comparison to recessive autosomal disorders. For example, if an X-linked disorder debilitates or is lethal to the affected individual prior to reproductive maturation, the disorder occurs exclusively in males. This is so because the only sources of the lethal allele in the population are in heterozygous females who are "carriers" and do not express the disorder. They pass the allele to one-half of their sons, who develop the disorder because they are hemizygous but rarely, if ever, reproduce. Heterozygous females also pass the allele to one-half of their daughters, who become carriers but do not develop the disorder. An example of such an X-linked disorder is *Duchenne muscular dystrophy*. The disease has an onset prior to age 6 and is often lethal around age 20. It normally occurs only in males.

4.12

In Sex-Limited and Sex-Influenced Inheritance, an Individual's Sex Influences the Phenotype

In contrast to X-linked inheritance, patterns of gene expression may be affected by the sex of an individual even when the genes are not on the X chromosome. In numerous examples in different organisms, the sex of the individual plays a determining role in the expression of a phenotype. In some cases, the expression of a specific phenotype is absolutely limited to one sex; in others, the sex of an individual influences the expression of a phenotype that is not limited to one sex or the other. This distinction differentiates **sex-limited inheritance** from **sex-influenced inheritance**.

In both types of inheritance, autosomal genes are responsible for the existence of contrasting phenotypes, but the expression of these genes is dependent on the hormone constitution of the individual. Thus, the heterozygous genotype may exhibit one phenotype in males and the contrasting one in females. In domestic fowl, for example, tail and neck plumage is often distinctly different in males and females (Figure 4–15), demonstrating *sex-limited inheritance*. Cock feathering is longer, more curved, and pointed, whereas hen feathering is shorter and less curved. Inheritance of these feather phenotypes is controlled by a single pair of autosomal alleles whose expression is modified by the individual's sex hormones. As shown in the following chart, hen feathering is due to a dominant allele, *H*, but regardless of the homozygous presence of the recessive *h* allele, all females remain hen-feathered. Only in males does the *hh* genotype result in cock feathering.

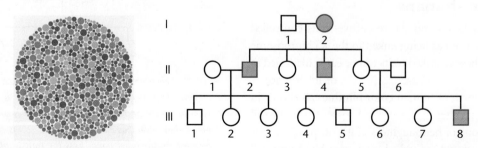

FIGURE 4–14   A human pedigree of the X-linked color-blindness trait. The photograph is of an Ishihara color-blindness chart, which tests for red–green color blindness. Red–green color-blind individuals see a 3 rather than the 8 visualized by those with normal color vision.

FIGURE 4–15 Hen feathering (left) and cock feathering (right) in domestic fowl. The hen's feathers are shorter and less curved.

Genotype	Phenotype	
	♀	♂
HH	Hen-feathered	Hen-feathered
Hh	Hen-feathered	Hen-feathered
hh	Hen-feathered	Cock-feathered

In certain breeds of fowl, the hen feathering or cock feathering allele has become fixed in the population. In the Leghorn breed, all individuals are of the *hh* genotype; as a result, males always differ from females in their plumage. Seabright bantams are all *HH*, showing no sexual distinction in feathering phenotypes.

Another example of sex-limited inheritance involves the autosomal genes responsible for milk yield in dairy cattle. Regardless of the overall genotype that influences the quantity of milk production, those genes are obviously expressed only in females.

Cases of *sex-influenced inheritance* include pattern baldness in humans, horn formation in certain breeds of sheep (e.g., Dorsett Horn sheep), and certain coat patterns in cattle. In such cases, autosomal genes are responsible for the contrasting phenotypes, and while the trait may be displayed by both males and females, the expression of these genes is dependent on the hormone constitution of the individual. Thus, the heterozygous genotype exhibits one phenotype in one sex and the contrasting one in the other. For example, pattern baldness in humans, where the hair is very thin or absent on the top of the head (Figure 4–16), is inherited in the following way:

Genotype	Phenotype	
	♀	♂
BB	Bald	Bald
Bb	Not bald	Bald
bb	Not bald	Not bald

FIGURE 4–16 Pattern baldness, a sex-influenced autosomal trait in humans.

Females can display pattern baldness, but this phenotype is much more prevalent in males. When females do inherit the *BB* genotype, the phenotype is less pronounced than in males and is expressed later in life.

4.13
Genetic Background and the Environment May Alter Phenotypic Expression

In the final section of this chapter we consider *phenotypic expression*. We assumed that the genotype of an organism is always directly expressed in its phenotype (Chapters 2 and 3). For example, pea plants homozygous for the recessive *d* allele (*dd*) will always be dwarf. We discussed gene expression as though the genes operate in a closed system in which the presence or absence of functional products directly determines the collective phenotype of an individual. The situation is actually much more complex. Most gene products function within the internal milieu of the cell, and cells interact with one another in various ways. Furthermore, the organism exists under diverse environmental influences. Thus, gene expression and the resultant phenotype are often modified through the interaction between an individual's particular genotype and the external environment. In this final section of this chapter, we will deal with some of the variables that are known to modify gene expression.

Penetrance and Expressivity

Some mutant genotypes are always expressed as a distinct phenotype, whereas others produce a proportion of individuals whose phenotypes cannot be distinguished from normal (wild type). The degree of expression of a particular trait can be studied quantitatively by determining the *penetrance* and *expressivity* of the genotype under investigation.

The percentage of individuals that show at least some degree of expression of a mutant genotype defines the **penetrance** of the mutation. For example, the phenotypic expression of many of the mutant alleles found in *Drosophila* can overlap with wild-type expression. If 15 percent of flies with a given mutant genotype show the wild-type appearance, the mutant gene is said to have a penetrance of 85 percent.

By contrast, **expressivity** reflects the *range of expression* of the mutant genotype. Flies homozygous for the recessive mutant gene *eyeless* exhibit phenotypes that range from the presence of normal eyes to a partial reduction in size to the complete absence of one or both eyes (Figure 4–17). Although the average reduction of eye size is one-fourth to one-half, expressivity ranges from complete loss of both eyes to completely normal eyes.

Examples such as the expression of the *eyeless* gene have provided the basis for experiments to determine the causes of phenotypic variation. If the laboratory environment is held constant and extensive variation is still observed, other genes may be influencing or modifying the phenotype. On the other hand, if the genetic background is not the cause of the phenotypic variation, environmental factors such as temperature, humidity, and nutrition may be involved. In the case of the *eyeless* phenotype, experiments have shown that both genetic background and environmental factors influence its expression.

Genetic Background: Position Effects

Although it is difficult to assess the specific effect of the **genetic background** and the expression of a gene responsible for determining a potential phenotype, one effect of genetic background has been well characterized, called the **position effect**. In such instances, the physical location of a gene in relation to other genetic material may influence its expression. For example, if a region of a chromosome is relocated or rearranged (called a translocation or inversion event), normal expression of genes in that chromosomal region may be modified. This is particularly true if the gene is relocated to or near certain areas of the chromosome that are condensed and genetically inert, referred to as **heterochromatin**.

An example of a position effect involves female *Drosophila* heterozygous for the X-linked recessive eye color mutant *white* (*w*). The w^+/w genotype normally results in a wild-type brick-red eye color. However, if the region of the X chromosome containing the wild-type w^+ allele is translocated so that it is close to a heterochromatic region, expression of the w^+ allele is modified. Instead of having a red color, the eyes are variegated, or mottled with red and white patches (Figure 4–18). Therefore, following translocation, the dominant effect of the normal w^+ allele is intermittent. A similar position effect is produced if a heterochromatic

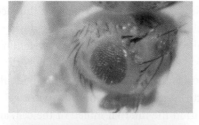

FIGURE 4–17 Variable expressivity as shown in flies homozygous for the *eyeless* mutation in *Drosophila*. Gradations in phenotype range from wild type to partial reduction to eyeless.

(a)

(b)

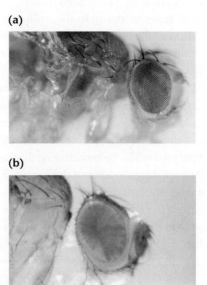

FIGURE 4–18 Position effect, as illustrated in the eye phenotype in two female *Drosophila* heterozygous for the gene *white*. (a) Normal dominant phenotype showing brick-red eye color. (b) Variegated color of an eye caused by translocation of the *white* gene to another location in the genome.

region is relocated next to the *white* locus on the X chromosome. Apparently, heterochromatic regions inhibit the expression of adjacent genes. Loci in many other organisms also exhibit position effects, providing proof that alteration of the normal arrangement of genetic information can modify its expression.

Temperature Effects—An Introduction to Conditional Mutations

Chemical activity depends on the kinetic energy of the reacting substances, which in turn depends on the surrounding temperature. We can thus expect temperature to influence phenotypes. An example is seen in the evening primrose, which produces red flowers when grown at 23°C and white flowers when grown at 18°C. An even more striking example is seen in Siamese cats and Himalayan rabbits, which exhibit dark fur in certain regions where their body temperature is slightly cooler, particularly the nose, ears, and paws (Figure 4–19). In these cases, it appears that the enzyme normally responsible for pigment production is functional only at the lower temperatures present in the extremities, but it loses its catalytic function at the slightly higher temperatures found throughout the rest of the body.

Mutations whose expression is affected by temperature, called **temperature-sensitive mutations**, are examples of **conditional mutations**, whereby phenotypic expression is determined by environmental conditions. Examples of temperature-sensitive mutations are known in viruses and a variety of organisms, including bacteria, fungi, and *Drosophila*. In extreme cases, an organism carrying a mutant allele may express a mutant phenotype when grown at one temperature but express the wild-type phenotype when reared at another temperature. This type of temperature effect is useful in studying mutations that interrupt essential processes during development and are thus normally detrimental or lethal. For example, if bacterial viruses are cultured under *permissive conditions* of 25°C, the mutant gene product is functional, infection proceeds normally, and new viruses are produced and can be studied. However, if bacterial viruses carrying temperature-sensitive mutations infect bacteria cultured at 42°C—the *restrictive condition*—infection progresses up to the point where the essential gene product is required (e.g., for viral assembly) and then arrests. Temperature-sensitive mutations are easily induced and isolated in viruses, and have added immensely to the study of viral genetics.

Nutritional Effects

Another category of phenotypes that are not always a direct reflection of the organism's genotype consists of **nutritional mutations**. In microorganisms, mutations that prevent synthesis of nutrient molecules are quite common, such as when an enzyme essential to a biosynthetic pathway becomes inactive. A microorganism bearing such a mutation is called an **auxotroph**. If the end product of a biochemical pathway can no longer be synthesized, and if that molecule is essential to normal growth and development, the mutation prevents growth and may be lethal. For example, if the bread mold *Neurospora* can no longer synthesize the amino acid leucine, proteins cannot be synthesized. If leucine is present in the growth medium, the detrimental effect is overcome. Nutritional mutants have been crucial to genetic studies in bacteria and also served as the basis for George Beadle and Edward Tatum's proposal, in the early 1940s, that one gene functions to produce one enzyme. (See Chapter 14.)

A slightly different set of circumstances exists in humans. The ingestion of certain dietary substances that normal individuals may consume without harm can adversely affect individuals with abnormal genetic constitutions. Often, a mutation may prevent an individual from metabolizing some substance commonly found in normal diets. For example, those afflicted with the genetic disorder **phenylketonuria** cannot metabolize the amino acid phenylalanine. Those with **galactosemia** cannot metabolize galactose. Those with **lactose intolerance** cannot metabolize lactose. However, if the dietary intake of the involved molecule is drastically reduced or eliminated, the associated phenotype may be ameliorated.

Onset of Genetic Expression

Not all genetic traits become apparent at the same time during an organism's life span. In most cases, the age at which a mutant gene

(a)   (b)

FIGURE 4–19 (a) A Himalayan rabbit. (b) A Siamese cat. Both show dark fur color on the snout, ears, and paws. These patches are due to the effect of a temperature-sensitive allele responsible for pigment production.

exerts a noticeable phenotype depends on events during the normal sequence of growth and development. In humans, the prenatal, infant, preadult, and adult phases require different genetic information. As a result, many severe inherited disorders are not manifested until after birth. For example, as we saw in Chapter 3, **Tay–Sachs disease**, inherited as an autosomal recessive, is a lethal lipid-metabolism disease involving an abnormal enzyme, hexosaminidase A. Newborns appear to be phenotypically normal for the first few months. Then, developmental retardation, paralysis, and blindness ensue, and most affected children die around the age of 3.

The **Lesch–Nyhan syndrome**, inherited as an X-linked recessive disease, is characterized by abnormal nucleic acid metabolism (inability to salvage nitrogenous purine bases), leading to the accumulation of uric acid in blood and tissues, mental retardation, palsy, and self-mutilation of the lips and fingers. The disorder is due to a mutation in the gene encoding hypoxanthine-guanine phosphoribosyl transferase (HPRT). Newborns are normal for six to eight months prior to the onset of the first symptoms.

Still another example is **Duchenne muscular dystrophy (DMD)**, an X-linked recessive disorder associated with progressive muscular wasting. It is not usually diagnosed until a child is 3 to 5 years old. Even with modern medical intervention, the disease is often fatal in the early 20s.

Perhaps the most delayed and highly variable age of onset for a genetic disorder in humans is seen in **Huntington disease**. Inherited as an autosomal dominant disorder, Huntington disease affects the frontal lobes of the cerebral cortex, where progressive cell death occurs over a period of more than a decade. Brain deterioration is accompanied by spastic uncontrolled movements, intellectual deterioration, and ultimately death. While onset of these symptoms has been reported at all ages, they are most often initially observed between ages 30 and 50, with a mean onset age of 38 years.

These examples support the concept that gene products may play more essential roles at certain times during the life cycle of an organism. One may be able to tolerate the impact of a mutant gene for a considerable period of time without noticeable effect. At some point, however, a mutant phenotype is manifested. Perhaps this is the result of the internal physiological environment of an organism changing during development and with age.

Genetic Anticipation

Interest in studying the genetic onset of phenotypic expression has intensified with the discovery of heritable disorders that *exhibit a progressively earlier age of onset and an increased severity of the disorder in each successive generation.* This phenomenon is referred to as **genetic anticipation**.

Myotonic dystrophy (DM), the most common type of adult muscular dystrophy, clearly illustrates genetic anticipation. Individuals afflicted with this autosomal dominant disorder exhibit extreme variation in the severity of symptoms.

Mildly affected individuals develop cataracts as adults, but have little or no muscular weakness. Severely affected individuals demonstrate more extensive weakness, as well as myotonia (muscle hyperexcitability) and in some cases mental retardation. In its most extreme form, the disease is fatal just after birth. A great deal of excitement was generated in 1989, when C. J. Howeler and colleagues confirmed the correlation of increased severity and earlier onset with successive generations of inheritance. The researchers studied 61 parent–child pairs, and in 60 of the cases, age of onset was earlier and more severe in the child than in his or her affected parent.

In 1992, an explanation was put forward to explain both the molecular cause of the mutation responsible for DM and the basis of genetic anticipation in the disorder. As we will see in Chapter 15, a three-nucleotide DNA sequence of the DM gene is repeated a variable number of times and is unstable. Normal individuals have about 5 to 35 copies of this sequence; affected individuals have between 80 and >2500 copies. Those with a greater number of repeats are more severely affected. The most remarkable observation was that, in successive generations of DM individuals, the size of the repeated segment increases. We now know that the RNA transcribed from mutant DM genes is the culprit in the disorder and alters the expression of still other genes. We will return to this topic (Chapter 15). Several other inherited human disorders, including the fragile-X syndrome, Kennedy disease, and Huntington disease, also reveal an association between the size of specific regions of the responsible gene and disease severity.

Genomic (Parental) Imprinting and Gene Silencing

A final example involving genetic background involves what is called **genomic**, or **parental, imprinting**, whereby the process of selective *gene silencing* occurs during early development, impacting on subsequent phenotypic expression. Examples involve cases where genes or regions of a chromosome are imprinted on one homolog but not the other. The impact of silencing depends on the parental origin of the genes or regions that are involved. Such silencing leads to the direct phenotypic expression of the allele(s) on the homolog that is not silenced. Thus, the imprinting step, the critical issue in understanding this phenomenon, is thought to occur before or during gamete formation, leading to differentially marked genes (or chromosome regions) in sperm-forming versus egg-forming tissues.

The first example of genomic imprinting was discovered in 1991, in three specific mouse genes. One is the gene encoding insulin-like growth factor II (*Igf2*). A mouse that carries two normal alleles of this gene is normal in size, whereas a mouse that carries two mutant alleles lacks the growth factor and is dwarf. The size of a heterozygous mouse—one allele normal and one mutant—depends on the parental origin of the wild-type allele. The mouse is normal in size if the normal allele comes from the father, but it is dwarf if the normal allele came from the mother. From this, we can deduce that the normal *Igf2* gene is imprinted and thus silenced during egg production, but it functions normally when it has passed through sperm-producing tissue in males. The imprint is inherited in the sense that the *Igf2* gene in all progeny cells formed during development remain silenced. Imprinting in the next generation then depends on whether the gene passes through sperm-producing or egg-forming tissue.

An example in humans involves two distinct genetic disorders thought to be caused by differential imprinting of the same region of the long arm of chromosome 15 (15q1). In both cases, the disorders are due to an identical deletion of this region in one member of the chromosome 15 pair. The first disorder, **Prader–Willi syndrome (PWS)**, results when the paternal segment is deleted and an undeleted maternal chromosome remains. If the maternal segment is deleted and an undeleted paternal chromosome remains, an entirely different disorder, **Angelman syndrome (AS)**, results.

These two conditions exhibit different phenotypes. PWS entails mental retardation, a severe eating disorder marked by an uncontrollable appetite, obesity, diabetes, and growth retardation. Angelman syndrome also involves mental retardation, but involuntary muscle contractions (chorea) and seizures characterize the disorder. We can conclude that the involved region of chromosome 15 is imprinted differently in male and female gametes and that both an undeleted maternal and a paternal region are required for normal development.

Although numerous questions remain unanswered regarding genomic imprinting, it is now clear that many genes are subject to this process. More than 50 have been identified in mammals thus far. It appears that regions of chromosomes rather than specific genes are imprinted. This phenomenon is an example of the more general topic of **epigenetics**, where genetic expression is *not* the direct result of the information stored in the nucleotide sequence of DNA. Instead, the DNA is altered in a way that affects its expression. These changes are stable in the sense that they are transmitted during cell division to progeny cells, and often through gametes to future generations.

The precise molecular mechanism of imprinting and other epigenetic events is still a matter for conjecture, but it seems certain that **DNA methylation** is involved. In most eukaryotes, methyl groups can be added to the carbon atom at position 5 in cytosine (see Chapter 10) as a result of the activity of the enzyme DNA methyltransferase. Methyl groups are added when the dinucleotide CpG or groups of CpG units (called CpG islands) are present along a DNA chain.

DNA methylation is a reasonable mechanism for establishing a molecular imprint, since there is evidence that a high level of methylation can inhibit gene activity and that active genes (or their regulatory sequences) are often undermethylated. This phenomenon is a fascinating topic. We will encounter other examples throughout the text, and return to more comprehensive coverage of epigenetics in Special Topics in Modern Genetics (p. 493) later in this book.

GENETICS, TECHNOLOGY, AND SOCIETY

Improving the Genetic Fate of Purebred Dogs

For dog lovers, nothing is quite so heartbreaking as watching a dog slowly go blind, struggling to adapt to a life of perpetual darkness. That's what happens in progressive retinal atrophy (PRA), a group of inherited disorders first described in Gordon setters in 1909. Since then, PRA has been detected in more than 100 other breeds of dogs, including Irish setters, border collies, Norwegian elkhounds, toy poodles, miniature schnauzers, cocker spaniels, and Siberian huskies.

The products of many genes are required for the development and maintenance of healthy retinas, and a defect in any one of these genes may cause retinal dysfunction. Decades of research have led to the identification of five such genes (*PDE6A*, *PDE6B*, *PRCD*, *rhodopsin*, and *PRGR*), and more may be discovered. Different mutant alleles are present in different breeds, and each allele is associated with a different form of PRA that varies slightly in its clinical symptoms and rate of progression. Mutations of *PDE6A*, *PDE6B*, and *PRCD* genes are inherited in a recessive pattern, mutations of the *rhodopsin* gene (such as those found in Mastiffs) are dominant, and *PRGR* mutations (in Siberian huskies and Samoyeds) are X-linked.

PRA is almost ten times more common in certain purebred dogs than in mixed breeds. The development of distinct breeds of dogs has involved intensive selection for desirable attributes, such as a particular size, shape, color, or behavior. Many desired characteristics are determined by recessive alleles. The fastest way to increase the homozygosity of these alleles is to mate close relatives, which are likely to carry the same alleles. For example, dogs may be mated to a cousin or a grandparent. Some breeders, in an attempt to profit from impressive pedigrees, also produce hundreds of offspring from individual dogs that have won major prizes at dog shows. This "popular sire effect," as it has been termed, further increases the homozygosity of alleles in purebred dogs.

Unfortunately, the generations of inbreeding that have established favorable characteristics in purebreds have also increased the homozygosity of certain harmful

Continued

Genetics, Technology, and Society, continued

recessive alleles, resulting in a high incidence of inherited diseases. More than 300 genetic diseases have been characterized in purebred dogs, and many breeds have a predisposition to more than 20 of them. According to researchers at Cornell University, purebred dogs suffer the highest incidence of inherited disease of any animal: 25 percent of the 20 million purebred dogs in America are affected with one genetic ailment or another.

Fortunately, advances in canine genetics are beginning to provide new tools to increase the health of purebred dogs. As of 2007, genetic tests are available to detect 30 different retinal diseases in dogs. Tests for PRA are now being used to identify heterozygous carriers of *PRCD* mutations. These carriers show no symptoms of PRA but, if mated with other carriers, pass the trait on to about 25 percent of their offspring. Eliminating PRA carriers from breeding programs has almost eradicated this condition from Portuguese Water Dogs and has greatly reduced its prevalence in other breeds.

Scientists will be able to identify more genes underlying canine inherited diseases thanks to the completion of the Dog Genome Project in 2005. In addition,

new therapies that correct gene-based defects will emerge.

The Dog Genome Project may have benefits for humans beyond the reduction of disease in their canine companions. Eighty-five percent of the genes in the dog genome have equivalents in humans, and over 300 diseases affecting dogs also affect humans, including heart disease, epilepsy, allergies, and cancer. The identification of a disease-causing gene in dogs can be a shortcut to the isolation of the corresponding gene in humans. By contributing to the cure of human diseases, dogs may prove to be "man's best friend" in an entirely new way.

Your Turn

Take time, individually or in groups, to answer the following questions. Investigate the references and links, to help you understand some of the issues surrounding the genetics of purebred dogs.

1. What are some of the limitations of genetic tests, especially as they apply to purebred dog genetic diseases?

This topic is discussed on the OptiGen website (http://www.optigen.com). *OptiGen is a*

company that offers gene tests for all known forms of PRA in dogs and is developing tests for other inherited disorders. From their TESTS list, select prcd-PRA, and visit the link "Benefits and Limitations of All Genetic Tests."

2. Which human disease is similar to PRA in the Siberian husky?

To learn more about these genes and diseases, visit the "Inherited Diseases in Dogs" database (http://server.vet.cam.ac.uk) *and search the database for Progressive Retinal Atrophy in the Siberian husky. Once there, follow OMIM reference link to learn about the human version of PRA in the Siberian husky.*

3. Recently, commercial laboratories have cloned dogs for research purposes and for people who want their beloved pet to return. Do you approve of cloning pet dogs? Why or why not? Do you think that a cloned dog would be identical to the original dog?

To learn about a recent pet dog cloning, read a Manchester Guardian article entitled "Pet cloning service bears five baby Boogers." (http://www. guardian.co.uk/science/2008/aug/05/ genetics.korea)

CASE STUDY | But he isn't deaf

Researching their family histories, a deaf couple learns that each of them has relatives through several generations who are deaf. They also learn that one form of deafness can be inherited as an autosomal recessive trait. They plan to have children, and based on the above information, they assume that all of their children may be deaf. To their surprise, their first child has normal hearing. The couple turns to you as a geneticist to help explain this situation.

1. Is it likely that these parents inherited their deafness as an autosomal recessive trait?
2. If two deaf parents have a hearing child, what conclusions can be drawn about the genetic control of deafness?
3. Is it likely that a future child will be deaf?

Summary Points

 For activities, animations, and review quizzes, go to the study area at www.masteringgenetics.com

1. Since Mendel's work was rediscovered, transmission genetics has been expanded to include many alternative modes of inheritance, including the study of incomplete dominance, codominance, multiple alleles, and lethal alleles.
2. Mendel's classic F_2 ratio is often modified in instances when gene interaction controls phenotypic variation. Many such instances involve epistasis, whereby the expression of one gene influences or inhibits the expression of another gene.
3. Complementation analysis determines whether independently isolated mutations that produce similar phenotypes are alleles of one another, or whether they represent separate genes.
4. Pleiotropy refers to multiple phenotypic effects caused by a single mutation.
5. Genes located on the X chromosome result in a characteristic mode of genetic transmission referred to as X-linkage, displaying

so-called criss-cross inheritance, whereby affected mothers pass X-linked traits to all of their sons.
6. Sex-limited and sex-influenced inheritance occurs when the sex of the organism affects the phenotype controlled by a gene located on an autosome.
7. Phenotypic expression is not always the direct reflection of the genotype. Variable expressivity may be observed, or a percentage of organisms may not express the expected phenotype at all, the basis of the penetrance of a mutant allele. In addition, the phenotype can be modified by genetic background, temperature, and nutrition. Finally, the onset of expression of a gene may vary during the lifetime of an organism, and even depend on whether the mutant allele is transmitted by the male or female parent, the basis of genomic imprinting.

INSIGHTS AND SOLUTIONS

Genetic problems take on added complexity if they involve two independent characters and multiple alleles, incomplete dominance, or epistasis. The most difficult types of problems are those that pioneering geneticists faced during laboratory or field studies. They had to determine the mode of inheritance by working backward from the observations of offspring to parents of unknown genotype.

1. Consider the problem of comb-shape inheritance in chickens, where walnut, rose, pea, and single are observed as distinct phenotypes. These variations are shown in the accompanying photographs. Considering the following data, determine how comb shape is inherited and what genotypes are present in the P_1 generation of each cross.

Cross 1:	single × single	$\longrightarrow$	all single
Cross 2:	walnut × walnut	$\longrightarrow$	all walnut
Cross 3:	rose × pea	$\longrightarrow$	all walnut
Cross 4:	F_1 × F_1 of Cross 3		
	walnut × walnut	$\longrightarrow$	93 walnut
			28 rose
			32 pea
			10 single

Solution: At first glance, this problem appears quite difficult. However, working systematically and breaking the analysis into steps simplifies it. To start, look at the data carefully for any useful information. Once you identify something that is clearly helpful, follow an empirical approach; that is, formulate a hypothesis and test it against the given data. Look for a pattern of inheritance that is consistent with all cases.

This problem gives two immediately useful facts. First, in cross 1, P_1 singles breed true. Second, while P_1 walnut breeds true in cross 2, a walnut phenotype is also produced in cross 3 between rose and pea. When these F_1 walnuts are mated in cross 4, all four comb shapes are produced in a ratio that approximates 9:3:3:1. This observation immediately suggests a cross involving two gene pairs, because the resulting data display the same ratio as in Mendel's dihybrid crosses. Since only one character is involved (comb shape), epistasis may be occurring. This could serve as your working hypothesis, and you must now propose how the two gene pairs "interact" to produce each phenotype.

If you call the allele pairs *A, a* and *B, b*, you might predict that because walnut represents 9/16 of the offspring in cross 4, *A–B–* will produce walnut. (Recall that *A–* and *B–* mean *AA* or *Aa* and *BB* or *Bb*, respectively.) You might also hypothesize that in cross 2, the genotypes are *AABB × AABB* where walnut bred true.

The phenotype representing 1/16 of the offspring of cross 4 is single; therefore you could predict that the single phenotype is the result of the *aabb* genotype. This is consistent with cross 1.

Now you have only to determine the genotypes for rose and pea. The most logical prediction is that at least one

dominant *A* or *B* allele combined with the double recessive condition of the other allele pair accounts for these phenotypes. For example,

$$A\text{–}bb \longrightarrow \text{rose}$$
$$aaB\text{–} \longrightarrow \text{pea}$$

If *AAbb* (rose) is crossed with *aaBB* (pea) in cross 3, all offspring would be *AaBb* (walnut). This is consistent with the data, and you need now look at only cross 4. We predict these walnut genotypes to be *AaBb* (as above), and from the cross

$$AaBb \text{ (walnut)} \times AaBb \text{ (walnut)}$$

we expect

9/16	*A–B–* (walnut)
3/16	*A–bb* (rose)
3/16	*aaB–* (pea)
1/16	*aabb* (single)

Our prediction is consistent with the data given. The initial hypothesis of the interaction of two gene pairs proves consistent throughout, and the problem is solved.

This problem demonstrates the usefulness of a basic theoretical knowledge of transmission genetics. With such knowledge, you can search for clues that will enable you to proceed in a stepwise fashion toward a solution. Mastering problem-solving requires practice, but can give you a great deal of satisfaction. Apply the same general approach to the following problems.

Walnut

Pea

Rose

Single

2. In radishes, flower color may be red, purple, or white. The edible portion of the radish may be long or oval. When only flower color is studied, no dominance is evident, and red × white crosses yield all purple. If these F_1 purples are interbred, the F_2

generation consists of 1/4 red: 1/2 purple: 1/4 white. Regarding radish shape, long is dominant to oval in a normal Mendelian fashion.

(a) Determine the F_1 and F_2 phenotypes from a cross between a true-breeding red, long radish and a radish that is white and oval. Be sure to define all gene symbols at the start.

(b) A red oval plant was crossed with a plant of unknown genotype and phenotype, yielding the following offspring:

103 red long: 101 red oval
98 purple long: 100 purple oval

Determine the genotype and phenotype of the unknown plant.

Solution: First, establish gene symbols:

RR = red $O-$ = long
Rr = purple oo = oval
rr = white

(a) This is a modified dihybrid cross where the gene pair controlling color exhibits incomplete dominance. Shape is controlled conventionally.

P_1: *RROO*   ×   *rroo*
(red long)   (white oval)

F_1: all *RrOo* (purple long)

$F_1 \times F_1$: *RrOo* × *RrOo*

F_2:

1/4 RR ─┬─ 3/4 O−   3/16 RR O−   red long
        └─ 1/4 oo   1/16 RR oo   red oval

2/4 Rr ─┬─ 3/4 O−   6/16 Rr O−   purple long
        └─ 1/4 oo   2/16 Rr oo   purple oval

1/4 rr ─┬─ 3/4 O−   3/16 rr O−   white long
        └─ 1/4 oo   1/16 rr oo   white oval

Note that to generate the F_2 results, we have used the forked-line method. First, we consider the outcome of crossing F_1 parents for the color genes ($Rr \times Rr$). Then the outcome of shape is considered ($Oo \times Oo$).

(b) The two characters appear to be inherited independently, so consider them separately. The data indicate a 1/4:1/4:1/4:1/4 proportion. First, consider color:

P_1:   red × ???   (unknown)
F_1:   204 red   (1/2)
         198 purple   (1/2)

Because the red parent must be *RR*, the unknown must have a genotype of *Rr* to produce these results. Thus it is purple. Now, consider shape:

P_1:   oval × ???   (unknown)
F_1:   201 long   (1/2)
         201 oval   (1/2)

Since the oval plant must be *oo*, the unknown plant must have a genotype of *Oo* to produce these results. Thus it is long. The unknown plant is

RrOo purple long

3. In humans, red–green color blindness is inherited as an X-linked recessive trait. A woman with normal vision whose father is color-blind marries a male who has normal vision. Predict the color vision of their male and female offspring.

Solution: The female is heterozygous, since she inherited an X chromosome with the mutant allele from her father. Her husband is normal. Therefore, the parental genotypes are

$Cc \times C\!\uparrow$ ($\uparrow$ represents the Y chromosome)

All female offspring are normal (*CC* or *Cc*). One-half of the male children will be color-blind ($c\!\uparrow$), and the other half will have normal vision ($C\!\uparrow$).

4. Consider the two very limited unrelated pedigrees shown here. Of the four combinations of X-linked recessive, X-linked dominant, autosomal recessive, and autosomal dominant, which modes of inheritance can be absolutely ruled out in each case?

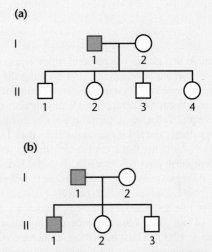

(a)

(b)

Solution: For both pedigrees, X-linked recessive and autosomal recessive remain possible, provided that the maternal parent is heterozygous in pedigree (b). Autosomal dominance seems at first glance unlikely in pedigree (a), since at least half of the offspring should express a dominant trait expressed by one of their parents. However, while it is true that if the affected parent carries an autosomal dominant gene heterozygously, each offspring has a 50 percent chance of inheriting and expressing the mutant gene, the sample size of four offspring is too small to rule this possibility out. In pedigree (b), autosomal dominance is clearly possible. In both cases, one can rule out X-linked dominance because the female offspring would inherit and express the dominant allele, and they do not express the trait in either pedigree.

Problems and Discussion Questions

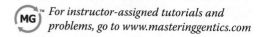 *For instructor-assigned tutorials and problems, go to www.masteringgentics.com*

HOW DO WE KNOW?

1. In this chapter, we focused on extensions and modifications of Mendelian principles and ratios. In the process, we encountered many opportunities to consider how this information was acquired. On the basis of these discussions, what answers would you propose to the following fundamental questions?
   (a) How were early geneticists able to ascertain inheritance patterns that did not fit typical Mendelian ratios?
   (b) How did geneticists determine that inheritance of some phenotypic characteristics involves the interactions of two or more gene pairs? How were they able to determine how many gene pairs were involved?
   (c) How do we know that specific genes are located on the sex-determining chromosomes rather than on autosomes?
   (d) For genes whose expression seems to be tied to the sex of individuals, how do we know whether a gene is X-linked in contrast to exhibiting sex-limited or sex-influenced inheritance?

2. In shorthorn cattle, coat color may be red, white, or roan. Roan is an intermediate phenotype expressed as a mixture of red and white hairs. The following data were obtained from various crosses:

red	$\times$	red	$\rightarrow$	all red
white	$\times$	white	$\rightarrow$	all white
red	$\times$	white	$\rightarrow$	all roan
roan	$\times$	roan	$\rightarrow$	1/4 red:1/2 roan:1/4 white

How is coat color inherited? What are the genotypes of parents and offspring for each cross?

3. Contrast incomplete dominance and codominance. Define the phenomenon of epistasis in the context of the concept of gene interaction.

4. In foxes, two alleles of a single gene, P and p, may result in lethality (PP), platinum coat (Pp), or silver coat (pp). What ratio is obtained when platinum foxes are interbred? Is the P allele behaving dominantly or recessively in causing (a) lethality; (b) platinum coat color?

5. In mice, a short-tailed mutant was discovered. When it was crossed to a normal long-tailed mouse, 4 offspring were short-tailed and 3 were long-tailed. Two short-tailed mice from the F_1 generation were selected and crossed. They produced 6 short-tailed and 3 long-tailed mice. These genetic experiments were repeated three times with approximately the same results. What genetic ratios are illustrated? Hypothesize the mode of inheritance and diagram the crosses.

6. List all possible genotypes for the A, B, AB, and O phenotypes. Is the mode of inheritance of the ABO blood types representative of dominance? of recessiveness? of codominance?

7. With regard to the ABO blood types in humans, determine the genotype of the male parent and female parent shown here:

   Male parent: Blood type B; mother type O
   Female parent: Blood type A; father type B

   Predict the blood types of the offspring that this couple may have and the expected proportion of each.

8. In a disputed parentage case, the child is blood type O, while the mother is blood type A. What blood type would exclude a male from being the father? Would the other blood types prove that a particular male was the father?

9. The A and B antigens in humans may be found in water-soluble form in secretions, including saliva, of some individuals (*Se/Se* and *Se/se*) but not in others (*se/se*). The population thus contains "secretors" and "nonsecretors."
   (a) Determine the proportion of various phenotypes (blood type and ability to secrete) in matings between individuals that are blood type AB and type O, both of whom are *Se/se*.
   (b) How will the results of such matings change if both parents are heterozygous for the gene controlling the synthesis of the H substance (*Hh*)?

10. In chickens, a condition referred to as "creeper" exists whereby the bird has very short legs and wings, and appears to be creeping when it walks. If creepers are bred to normal chickens, one-half of the offspring are normal and one-half are creepers. Creepers never breed true. If bred together, they yield two-thirds creepers and one-third normal. Propose an explanation for the inheritance of this condition.

11. In rabbits, a series of multiple alleles controls coat color in the following way: C is dominant to all other alleles and causes full color. The chinchilla phenotype is due to the c^{ch} allele, which is dominant to all alleles other than C. The c^h allele, dominant only to c^a (albino), results in the Himalayan coat color. Thus, the order of dominance is $C > c^{ch} > c^h > c^a$. For each of the following three cases, the phenotypes of the P_1 generations of two crosses are shown, as well as the phenotype of *one member* of the F_1 generation.

	P_1 Phenotypes		F_1 Phenotypes
(a)	Himalayan $\times$ Himalayan	$\longrightarrow$	albino
		$\times$	$\rightarrow$??
	full color $\times$ albino	$\longrightarrow$	chinchilla
(b)	albino $\times$ chinchilla	$\longrightarrow$	albino
		$\times$	$\rightarrow$??
	full color $\times$ albino	$\longrightarrow$	full color
(c)	chinchilla $\times$ albino	$\longrightarrow$	Himalayan
		$\times$	$\rightarrow$??
	full color $\times$ albino	$\longrightarrow$	Himalayan

For each case, determine the genotypes of the P_1 generation and the F_1 offspring, and predict the results of making each indicated cross between F_1 individuals.

12. Three gene pairs located on separate autosomes determine flower color and shape as well as plant height. The first pair exhibits incomplete dominance, where the color can be red, pink (the heterozygote), or white. The second pair leads to personate (dominant) or peloric (recessive) flower shape, while the third gene pair produces either the dominant tall trait or the recessive dwarf trait. Homozygous plants that are red, personate, and tall are crossed to those that are white, peloric, and dwarf. Determine the F_1 genotype(s) and phenotype(s). If the F_1 plants are

interbred, what proportion of the offspring will exhibit the same phenotype as the F_1 plants?

personate   peloric

13. As in Problem 12, flower color may be red, white, or pink, and flower shape may be personate or peloric. For the following crosses, determine the P_1 and F_1 genotypes:

(a) red, peloric × white, personate
↓
F_1: all pink, personate

(b) red, personate × white, peloric
↓
F_1: all pink, personate

(c) pink, personate × red, peloric → F_1 $\begin{cases} \text{1/4 red, personate} \\ \text{1/4 red, peloric} \\ \text{1/4 pink, peloric} \\ \text{1/4 pink, personate} \end{cases}$

(d) pink, personate × white, peloric → F_1 $\begin{cases} \text{1/4 white, personate} \\ \text{1/4 white, peloric} \\ \text{1/4 pink, personate} \\ \text{1/4 pink, peloric} \end{cases}$

(e) What phenotypic ratios would result from crossing the F_1 of (a) to the F_1 of (b)?

14. Horses can be cremello (a light cream color), chestnut (a brownish color), or palomino (a golden color with white in the horse's tail and mane). Of these phenotypes, only palominos never breed true.

cremello × palomino ⟶ 1/2 cremello
1/2 palomino

chestnut × palomino ⟶ 1/2 chestnut
1/2 palomino

palomino × palomino ⟶ 1/4 chestnut
1/2 palomino
1/4 cremello

(a) From the results given above, determine the mode of inheritance by assigning gene symbols and indicating which genotypes yield which phenotypes.
(b) Predict the F_1 and F_2 results of many initial matings between cremello and chestnut horses.

Chestnut

Palomino

Cremello

15. With reference to the eye color phenotypes produced by the recessive, autosomal, unlinked *brown* and *scarlet* loci in *Drosophila* (see Figure 4–10), predict the F_1 and F_2 results of the following P_1 crosses. (Recall that when both the *brown* and *scarlet* alleles are homozygous, no pigment is produced, and the eyes are white.)
(a) wild type × white
(b) wild type × scarlet
(c) brown   × white

16. Pigment in mouse fur is only produced when the *C* allele is present. Individuals of the *cc* genotype are white. If color is present, it may be determined by the *A, a* alleles. *AA* or *Aa* results in agouti color, while *aa* results in black coats.
(a) What F_1 and F_2 genotypic and phenotypic ratios are obtained from a cross between *AACC* and *aacc* mice?
(b) In three crosses between agouti females whose genotypes were unknown and males of the *aacc* genotype, the following phenotypic ratios were obtained:

(1) 8 agouti	(2) 9 agouti	(3) 4 agouti
8 white	10 black	5 black
		10 white

What are the genotypes of these female parents?

17. In rats, the following genotypes of two independently assorting autosomal genes determine coat color:

A–B–	(gray)
A–bb	(yellow)
aaB–	(black)
aabb	(cream)

A third gene pair on a separate autosome determines whether or not any color will be produced. The *CC* and *Cc* genotypes allow color according to the expression of the *A* and *B* alleles. However, the *cc* genotype results in albino rats regardless of

the *A* and *B* alleles present. Determine the F_1 phenotypic ratio of the following crosses:
(a) *AAbbCC* × *aaBBcc*
(b) *AaBBCC* × *AABbcc*
(c) *AaBbCc* × *AaBbcc*
(d) *AaBBCc* × *AaBBCc*
(e) *AABbCc* × *AABbcc*

18. Given the inheritance pattern of coat color in rats described in Problem 17, predict the genotype and phenotype of the parents who produced the following offspring:
(a) 9/16 gray: 3/16 yellow: 3/16 black: 1/16 cream
(b) 9/16 gray: 3/16 yellow: 4/16 albino
(c) 27/64 gray: 16/64 albino: 9/64 yellow: 9/64 black: 3/64 cream
(d) 3/8 black: 3/8 cream: 2/8 albino
(e) 3/8 black: 4/8 albino: 1/8 cream

19. In a species of the cat family, eye color can be gray, blue, green, or brown, and each trait is true breeding. In separate crosses involving homozygous parents, the following data were obtained:

Cross	P_1	F_1	F_2
A	green × gray	all green	3/4 green: 1/4 gray
B	green × brown	all green	3/4 green: 1/4 brown
C	gray × brown	all green	9/16 green: 3/16 brown
			3/16 gray: 1/16 blue

(a) Analyze the data. How many genes are involved? Define gene symbols and indicate which genotypes yield each phenotype.
(b) In a cross between a gray-eyed cat and one of unknown genotype and phenotype, the F_1 generation was not observed. However, the F_2 resulted in the same F_2 ratio as in cross C. Determine the genotypes and phenotypes of the unknown P_1 and F_1 cats.

20. In a plant, a tall variety was crossed with a dwarf variety. All F_1 plants were tall. When $F_1 \times F_1$ plants were interbred, 9/16 of the F_2 were tall and 7/16 were dwarf.
(a) Explain the inheritance of height by indicating the number of gene pairs involved and by designating which genotypes yield tall and which yield dwarf. (Use dashes where appropriate.)
(b) What proportion of the F_2 plants will be true breeding if self-fertilized? List these genotypes.

21. In a unique species of plants, flowers may be yellow, blue, red, or mauve. All colors may be true breeding. If plants with blue flowers are crossed to red-flowered plants, all F_1 plants have yellow flowers. When these produced an F_2 generation, the following ratio was observed:

9/16 yellow: 3/16 blue: 3/16 red: 1/16 mauve

In still another cross using true-breeding parents, yellow-flowered plants are crossed with mauve-flowered plants. Again, all F_1 plants had yellow flowers and the F_2 showed a 9:3:3:1 ratio, as just shown.
(a) Describe the inheritance of flower color by defining gene symbols and designating which genotypes give rise to each of the four phenotypes.
(b) Determine the F_1 and F_2 results of a cross between true-breeding red and true-breeding mauve-flowered plants.

22. Five human matings (1–5), identified by both maternal and paternal phenotypes for ABO and MN blood-group antigen status, are shown on the left side of the following table:

Parental Phenotypes				Offspring	
(1)	A, M	×	A, N	(a)	A, N
(2)	B, M	×	B, M	(b)	O, N
(3)	O, N	×	B, N	(c)	O, MN
(4)	AB, M	×	O, N	(d)	B, M
(5)	AB, MN	×	AB, MN	(e)	B, MN

Each mating resulted in one of the five offspring shown in the right-hand column (a–e). Match each offspring with one correct set of parents, using each parental set only once. Is there more than one set of correct answers?

23. A husband and wife have normal vision, although both of their fathers are red–green color-blind, an inherited X-linked recessive condition. What is the probability that their first child will be (a) a normal son? (b) a normal daughter? (c) a color-blind son? (d) a color-blind daughter?

24. In humans, the ABO blood type is under the control of autosomal multiple alleles. Color blindness is a recessive X-linked trait. If two parents who are both type A and have normal vision produce a son who is color-blind and is type O, what is the probability that their next child will be a female who has normal vision and is type O?

25. In *Drosophila*, an X-linked recessive mutation, *scalloped* (*sd*), causes irregular wing margins. Diagram the F_1 and F_2 results if (a) a scalloped female is crossed with a normal male; (b) a scalloped male is crossed with a normal female. Compare these results with those that would be obtained if the *scalloped* gene were autosomal.

26. Another recessive mutation in *Drosophila*, *ebony* (*e*), is on an autosome (chromosome 3) and causes darkening of the body compared with wild-type flies. What phenotypic F_1 and F_2 male and female ratios will result if a scalloped-winged female with normal body color is crossed with a normal-winged ebony male? Work out this problem by both the Punnett square method and the forked-line method.

27. In *Drosophila*, the X-linked recessive mutation *vermilion* (*v*) causes bright red eyes, in contrast to the brick-red eyes of wild type. A separate autosomal recessive mutation, *suppressor of vermilion* (*su-v*), causes flies homozygous or hemizygous for *v* to have wild-type eyes. In the absence of *vermilion* alleles, *su-v* has no effect on eye color. Determine the F_1 and F_2 phenotypic ratios from a cross between a female with wild-type alleles at the *vermilion* locus, but who is homozygous for *su-v*, with a *vermilion* male who has wild-type alleles at the *su-v* locus.

28. While *vermilion* is X-linked in *Drosophila* and causes the eye color to be bright red, *brown* is an autosomal recessive mutation that causes the eye to be brown. Flies carrying both mutations lose all pigmentation and are white-eyed. Predict the F_1 and F_2 results of the following crosses:
(a) vermilion females × brown males
(b) brown females × vermilion males
(c) white females × wild-type males

29. In a cross in *Drosophila* involving the X-linked recessive eye mutation *white* and the autosomally linked recessive eye mutation *sepia* (resulting in a dark eye), predict the F_1 and F_2

results of crossing true-breeding parents of the following phenotypes:

(a) white females × sepia males
(b) sepia females × white males

Note that white is epistatic to the expression of sepia.

30. Consider the following three pedigrees, all involving a single human trait:

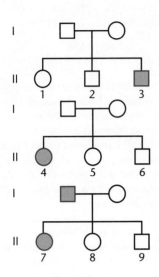

(a) Which combination of conditions, if any, can be excluded?
dominant and X-linked
dominant and autosomal
recessive and X-linked
recessive and autosomal

(b) For each combination that you excluded, indicate the single individual in generation II (e.g., II-1, II-2) that was most instrumental in your decision to exclude it. If none were excluded, answer "none apply."

(c) Given your conclusions in part (a), indicate the genotype of the following individuals:

II-1, II-6, II-9

If more than one possibility applies, list all possibilities. Use the symbols *A* and *a* for the genotypes.

31. In goats, the development of the beard is due to a recessive gene. The following cross involving true-breeding goats was made and carried to the F_2 generation:

P$_1$: bearded female × beardless male
↓
F$_1$: all bearded males and beardless females

$$F_1 \times F_1 \rightarrow \begin{cases} 1/8 \text{ beardless males} \\ 3/8 \text{ bearded males} \\ 3/8 \text{ beardless females} \\ 1/8 \text{ bearded females} \end{cases}$$

Offer an explanation for the inheritance and expression of this trait, diagramming the cross. Propose one or more crosses to test your hypothesis.

32. Predict the F_1 and F_2 results of crossing a male fowl that is cock-feathered with a true-breeding hen-feathered female fowl. Recall that these traits are sex limited.

33. Two mothers give birth to sons at the same time at a busy urban hospital. The son of mother 1 is afflicted with hemophilia, a disease caused by an X-linked recessive allele. Neither parent has the disease. Mother 2 has a normal son, despite the fact that the father has hemophilia. Several years later, couple 1 sues the hospital, claiming that these two newborns were swapped in the nursery following their birth. As a genetic counselor, you are called to testify. What information can you provide the jury concerning the allegation?

34. Discuss the topic of phenotypic expression and the many factors that impinge on it.

35. Contrast penetrance and expressivity as the terms relate to phenotypic expression.

Extra-Spicy Problems

 For instructor-assigned tutorials and problems, go to www.masteringgentics.com

36. Labrador retrievers may be black, brown (chocolate), or golden (yellow) in color (see chapter-opening photo on p. 71). While each color may breed true, many different outcomes are seen when numerous litters are examined from a variety of matings where the parents are not necessarily true breeding. Following are just some of the many possibilities.

(a)	black	×	brown	→	all black
(b)	black	×	brown	→	1/2 black
					1/2 brown
(c)	black	×	brown	→	3/4 black
					1/4 golden
(d)	black	×	golden	→	all black

(e)	black	×	golden	→	4/8 golden
					3/8 black
					1/8 brown
(f)	black	×	golden	→	2/4 golden
					1/4 black
					1/4 brown
(g)	brown	×	brown	→	3/4 brown
					1/4 golden
(h)	black	×	black	→	9/16 black
					4/16 golden
					3/16 brown

Propose a mode of inheritance that is consistent with these data, and indicate the corresponding genotypes of the parents in each mating. Indicate as well the genotypes of dogs that breed true for each color.

37. A true-breeding purple-leafed plant isolated from one side of El Yunque, the rain forest in Puerto Rico, was crossed to a true-breeding white variety found on the other side. The F_1 offspring were all purple. A large number of $F_1 \times F_1$ crosses produced the following results:

purple: 4219   white: 5781   (Total = 10,000)

Propose an explanation for the inheritance of leaf color. As a geneticist, how might you go about testing your hypothesis? Describe the genetic experiments that you would conduct.

38. In Dexter and Kerry cattle, animals may be polled (hornless) or horned. The Dexter animals have short legs, whereas the Kerry animals have long legs. When many offspring were obtained from matings between polled Kerrys and horned Dexters, half were found to be polled Dexters and half polled Kerrys. When these two types of F_1 cattle were mated to one another, the following F_2 data were obtained:

3/8 polled Dexters
3/8 polled Kerrys
1/8 horned Dexters
1/8 horned Kerrys

A geneticist was puzzled by these data and interviewed farmers who had bred these cattle for decades. She learned that Kerrys were true breeding. Dexters, on the other hand, were not true breeding and never produced as many offspring as Kerrys. Provide a genetic explanation for these observations.

39. A geneticist from an alien planet that prohibits genetic research brought with him to Earth two pure-breeding lines of frogs. One line croaks by *uttering*. "rib-it rib-it" and has purple eyes. The other line croaks more softly by *muttering* "knee-deep knee-deep" and has green eyes. With a newfound freedom of inquiry, the geneticist mated the two types of frogs, producing F_1 frogs that were all utterers and had blue eyes. A large F_2 generation then yielded the following ratios:

27/64 blue-eyed, rib-it utterer
12/64 green-eyed, rib-it utterer
9/64 blue-eyed, knee-deep mutterer
9/64 purple-eyed, rib-it utterer
4/64 green-eyed, knee-deep mutterer
3/64 purple-eyed, knee-deep mutterer

(a) How many total gene pairs are involved in the inheritance of both traits? Support your answer.
(b) Of these, how many are controlling eye color? How can you tell? How many are controlling croaking?
(c) Assign gene symbols for all phenotypes and indicate the genotypes of the P_1 and F_1 frogs.
(d) Indicate the genotypes of the six F_2 phenotypes.
(e) After years of experiments, the geneticist isolated pure-breeding strains of all six F_2 phenotypes. Indicate the F_1 and F_2 phenotypic ratios of the following cross using these pure-breeding strains:

blue-eyed, "knee-deep" mutterer × purple-eyed, "rib-it" utterer

(f) One set of crosses with his true-breeding lines initially caused the geneticist some confusion. When he crossed true-breeding purple-eyed, "knee-deep" mutterers with true-breeding green-eyed, "knee-deep" mutterers, he often got different results. In some matings, all offspring were blue-eyed, "knee-deep" mutterers, but in other matings all offspring were purple-eyed, "knee-deep" mutterers. In still a third mating, 1/2 blue-eyed, "knee-deep" mutterers and 1/2 purple-eyed, "knee-deep" mutterers were observed. Explain why the results differed.

(g) In another experiment, the geneticist crossed two purple-eyed, "rib-it" *utterers* together with the results shown here:

9/16 purple-eyed, "rib-it" utterer
3/16 purple-eyed, "knee-deep" mutterer
3/16 green-eyed, "rib-it" utterer
1/16 green-eyed, "knee-deep" mutterer

What were the genotypes of the two parents?

40. The following pedigree is characteristic of an inherited condition known as male precocious puberty, where affected males show signs of puberty by age 4. Propose a genetic explanation of this phenotype.

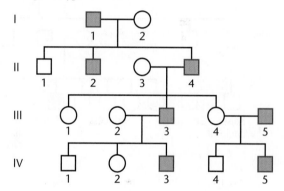

41. Students taking a genetics exam were expected to answer the following question by converting data to a "meaningful ratio" and then solving the problem. The instructor assumed that the final ratio would reflect two gene pairs, and most correct answers did. Here is the exam question:

"Flowers may be white, orange, or brown. When plants with white flowers are crossed with plants with brown flowers, all the F_1 flowers are white. For F_2 flowers, the following data were obtained:

48	white
12	orange
4	brown

Convert the F_2 data to a meaningful ratio that allows you to explain the inheritance of color. Determine the number of genes involved and the genotypes that yield each phenotype."

(a) Solve the problem for two gene pairs. What is the final F_2 ratio?
(b) A number of students failed to reduce the ratio for two gene pairs as described above and solved the problem using three gene pairs. When examined carefully, their solution

was deemed a valid response by the instructor. Solve the problem using three gene pairs.

(c) We now have a dilemma. The data are consistent with two alternative mechanisms of inheritance. Propose an experiment that executes crosses involving the original parents that would distinguish between the two solutions proposed by the students. Explain how this experiment would resolve the dilemma.

42. In four o'clock plants, many flower colors are observed. In a cross involving two true-breeding strains, one crimson and the other white, all of the F_1 generation were rose color. In the F_2, four new phenotypes appeared along with the P_1 and F_1 parental colors. The following ratio was obtained:

1/16 crimson	4/16 rose
2/16 orange	2/16 pale yellow
1/16 yellow	4/16 white
2/16 magenta	

Propose an explanation for the inheritance of these flower colors.

43. Proto-oncogenes stimulate cells to progress through the cell cycle and begin mitosis. In cells that stop dividing, transcription of proto-oncogenes is inhibited by regulatory molecules. As is typical of all genes, proto-oncogenes contain a regulatory DNA region followed by a coding DNA region that specifies the amino acid sequence of the gene product. Consider two types of mutation in a proto-oncogene, one in the regulatory region that eliminates transcriptional control and the other in the coding region that renders the gene product inactive. Characterize both of these mutant alleles as either gain-of-function or loss-of-function mutations and indicate whether each would be dominant or recessive.

44. Below is a partial pedigree of hemophilia in the British Royal Family descended from Queen Victoria, who is believed to be the original "carrier" in this pedigree. Analyze the pedigree and indicate which females are also certain to be carriers. What is the probability that Princess Irene is a carrier?

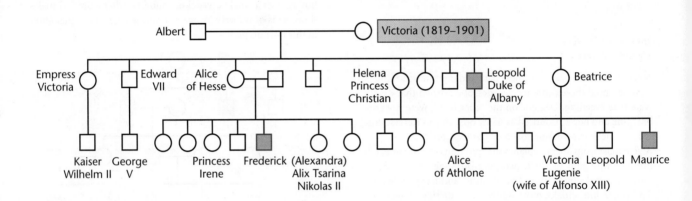

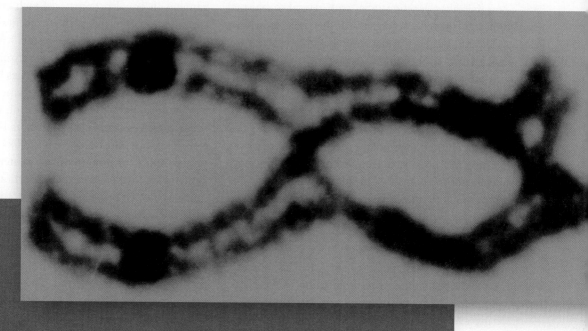

Chiasmata present between synapsed homologs during the first meiotic prophase.

5

Chromosome Mapping in Eukaryotes

CHAPTER CONCEPTS

- Chromosomes in eukaryotes contain large numbers of genes, whose locations are fixed along the length of the chromosomes.

- Unless separated by crossing over, alleles on the same chromosome segregate as a unit during gamete formation.

- Crossing over between homologs during meiosis creates recombinant gametes with different combinations of alleles that enhance genetic variation.

- Crossing over between homologs serves as the basis for the construction of chromosome maps. The greater the distance between two genes on a chromosome, the higher the frequency of crossing over is between them.

- Recombination also occurs between mitotic chromosomes and between sister chromatids.

- Linkage analysis and mapping can be performed for haploid organisms as well as diploid organisms.

Walter Sutton, along with Theodor Boveri, was instrumental in uniting the fields of cytology and genetics. As early as 1903, Sutton pointed out the likelihood that there must be many more "unit factors" than chromosomes in most organisms. Soon thereafter, genetic studies with several organisms revealed that certain genes segregate as if they were somehow joined or linked together. Further investigations showed that such genes are part of the same chromosome, and they may indeed be transmitted as a single unit. We now know that most chromosomes contain a very large number of genes. Those that are part of the same chromosome are said to be *linked* and to demonstrate **linkage** in genetic crosses.

Because the chromosome, not the gene, is the unit of transmission during meiosis, linked genes are not free to undergo independent assortment. Instead, the alleles at all loci of one chromosome should, in theory, be transmitted as a unit during gamete formation. However, in many instances this does not occur. As we saw during the first meiotic prophase, when homologs are paired, or synapsed, a reciprocal exchange of chromosome segments may take place (Chapter 2). This **crossing over** results in the reshuffling, or **recombination**, of the alleles between homologs and always occurs during the tetrad stage.

Crossing over is currently viewed as an actual physical breaking and rejoining process that occurs during meiosis. You can see an example in the micrograph that opens this chapter. The exchange of chromosome segments provides an enormous potential for genetic variation in the gametes formed by any individual. This type of variation, in combination with that resulting from independent assortment, ensures that all offspring will contain a diverse mixture of maternal and paternal alleles.

The frequency of crossing over between any two loci on a single chromosome is proportional to the distance between them, known as the **interlocus distance**. Thus, depending on which loci are being considered, the percentage of recombinant gametes varies. This correlation allows us to construct **chromosome maps**, which indicate the relative locations of genes on the chromosomes.

In this chapter, we will discuss linkage, crossing over, and chromosome mapping in more detail. We will also consider a variety of other topics involving the exchange of genetic information, concluding the chapter with the rather intriguing question of why Mendel, who studied seven genes in an organism with seven chromosomes, did not encounter linkage. Or did he?

5.1

Genes Linked on the Same Chromosome Segregate Together

A simplified overview of the major theme of this chapter is given in Figure 5–1, which contrasts the meiotic consequences of (a) independent assortment, (b) linkage *without*

crossing over, and (c) linkage *with* crossing over. In Figure 5–1(a) we see the results of independent assortment of two pairs of chromosomes, each containing one heterozygous gene pair. No linkage is exhibited. When these same two chromosomes are observed in a large number of meiotic events, they are seen to form four genetically different gametes in equal proportions, each containing a different combination of alleles of the two genes.

Now let's compare these results with what occurs if the same genes are linked on the same chromosome. If no crossing over occurs between the two genes [Figure 5–1(b)], only two genetically different kinds of gametes are formed. Each gamete receives the alleles present on one homolog or the other, which is transmitted intact as the result of segregation. This case illustrates *complete linkage*, which produces only **parental**, or **noncrossover, gametes**. The two parental gametes are formed in equal proportions. Though complete linkage between two genes seldom occurs, it is useful to consider the theoretical consequences of this concept.

Figure 5–1(c) shows the results when crossing over occurs between two linked genes. As you can see, this crossover involves only two nonsister chromatids of the four chromatids present in the tetrad. This exchange generates two new allele combinations, called **recombinant**, or **crossover, gametes**. The two chromatids not involved in the exchange result in noncrossover gametes, like those in Figure 5–1(b). Importantly, the frequency with which crossing over occurs between any two linked genes is proportional to the distance separating the respective loci along the chromosome. As the distance between the two genes increases, the proportion of recombinant gametes increases and that of the parental gametes decreases. In theory, two randomly selected genes can be so close to each other that crossover events are too infrequent to be easily detected. As shown in Figure 5–1(b), this complete linkage produces only parental gametes. On the other hand, if a small, but distinct, distance separates two genes, few recombinant and many parental gametes will be formed.

As we will discuss again later in this chapter, when the loci of two linked genes are far apart, the number of recombinant gametes approaches, but does not exceed, 50 percent. If 50 percent recombinants occur, the result is a 1:1:1:1 ratio of the four types (two parental and two recombinant gametes). In this case, transmission of two linked genes is indistinguishable from that of two unlinked, independently assorting genes. That is, the proportion of the four possible genotypes would be identical, as shown in Figure 5–1(a) and Figure 5–1(c).

The Linkage Ratio

If complete linkage exists between two genes because of their close proximity, and organisms heterozygous at both loci are mated, a unique F_2 phenotypic ratio results, which

(a) **Independent assortment: Two genes on two different homologous pairs of chromosomes**

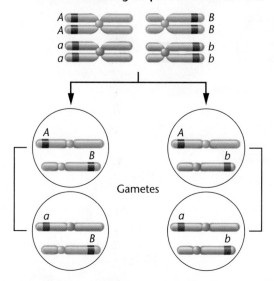

Gametes

(b) **Linkage: Two genes on a single pair of homologs; no exchange occurs**

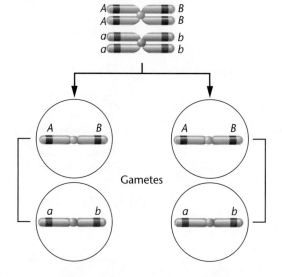

Gametes

(c) **Linkage: Two genes on a single pair of homologs; exchange occurs between two nonsister chromatids**

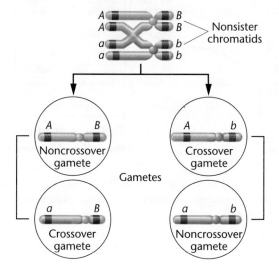

Gametes

we designate the **linkage ratio**. To illustrate this ratio, let's consider a cross involving the closely linked, recessive, mutant genes *heavy* wing vein (*hv*) and *brown* eye (*bw*) in *Drosophila melanogaster* (Figure 5–2). The normal, wild-type alleles hv^+ and bw^+ are both dominant and result in thin wing veins and red eyes, respectively.

In this cross, flies with normal thin wing veins and mutant brown eyes are mated to flies with mutant heavy wing veins and normal red eyes. In more concise terms, heavy-veined flies are crossed with brown-eyed flies. Linked genes are represented by placing their allele designations (the genetic symbols established in Chapter 4) above and below a single or double horizontal line. Those above the line are located at loci on one homolog, and those below the line are located at the homologous loci on the other homolog. Thus, we represent the P_1 generation as follows:

$$P_1: \frac{hv^+\ bw}{hv^+\ bw} \times \frac{hv\ bw^+}{hv\ bw^+}$$

thin, brown   heavy, red

Because these genes are located on an autosome, no designation of male or female is necessary.

In the F_1 generation, each fly receives one chromosome of each pair from each parent. All flies are heterozygous for both gene pairs and exhibit the dominant traits of thin veins and red eyes:

$$F_1 = \frac{hv^+\ bw}{hv\ bw^+}$$

thin, red

As shown in Figure 5–2(a), when the F_1 generation is interbred, each F_1 individual forms only parental gametes because of complete linkage. Following fertilization, the F_2 generation is produced in a 1:2:1 phenotypic and genotypic ratio. One-fourth of this generation shows thin wing veins and brown eyes; one-half shows both wild-type traits, namely, thin veins and red eyes; and one-fourth will show heavy wing veins and red eyes. Therefore, the ratio is 1 heavy: 2 wild: 1 brown. Such a 1:2:1 ratio is characteristic of complete linkage. Complete linkage is usually observed only when genes are very close together and the number of progeny is relatively small.

FIGURE 5–1 Results of gamete formation when two heterozygous genes are (a) on two different pairs of chromosomes; (b) on the same pair of homologs, but with no exchange occurring between them; and (c) on the same pair of homologs, but with an exchange occurring between two nonsister chromatids. Note in this and the following figures that members of homologous pairs of chromosomes are shown in two different colors. This convention was established in Chapter 2 (see, for example, Figure 2–7 and Figure 2–11).

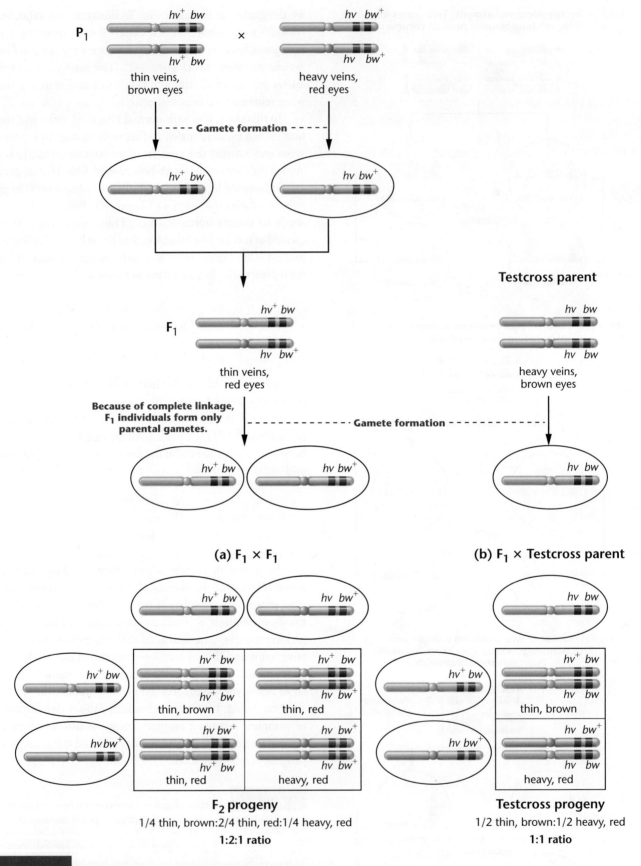

(a) F₁ × F₁

(b) F₁ × Testcross parent

F₂ progeny
1/4 thin, brown:2/4 thin, red:1/4 heavy, red
1:2:1 ratio

Testcross progeny
1/2 thin, brown:1/2 heavy, red
1:1 ratio

FIGURE 5-2   Results of a cross involving two genes located on the same chromosome and demonstrating complete linkage. (a) The F₂ results of the cross. (b) The results of a testcross involving the F₁ progeny.

Figure 5–2(b) demonstrates the results of a testcross with the F_1 flies. Such a cross produces a 1:1 ratio of thin, brown and heavy, red flies. Had the genes controlling these traits been incompletely linked or located on separate autosomes, the testcross would have produced four phenotypes rather than two.

When large numbers of mutant genes in any given species are investigated, genes located on the same chromosome show evidence of linkage to one another. As a result, **linkage groups** can be identified, one for each chromosome. In theory, the number of linkage groups should correspond to the haploid number of chromosomes. In diploid organisms in which large numbers of mutant genes are available for genetic study, this correlation has been confirmed.

NOW SOLVE THIS

5–1 Consider two hypothetical recessive autosomal genes *a* and *b,* where a heterozygote is testcrossed to a double-homozygous mutant. Predict the phenotypic ratios under the following conditions:

(a) *a* and *b* are located on separate autosomes.

(b) *a* and *b* are linked on the same autosome but are so far apart that a crossover always occurs between them.

(c) *a* and *b* are linked on the same autosome but are so close together that a crossover almost never occurs.

■ HINT: *This problem involves an understanding of linkage, crossing over, and independent assortment. The key to its solution is to be aware that results are indistinguishable when two genes are unlinked compared to the case where they are linked but so far apart that crossing over always intervenes between them during meiosis.*

5.2

Crossing Over Serves as the Basis for Determining the Distance between Genes in Chromosome Mapping

It is highly improbable that two randomly selected genes linked on the same chromosome will be so close to one another along the chromosome that they demonstrate complete linkage. Instead, crosses involving two such genes will almost always produce a percentage of offspring resulting from recombinant gametes. The percentage will vary depending on the distance between the two genes along the chromosome. This phenomenon was first explained in 1911 by two *Drosophila* geneticists, Thomas H. Morgan and his undergraduate student, Alfred H. Sturtevant.

Morgan and Crossing Over

As you may recall from our discussion in Chapter 4, Morgan was the first to discover the phenomenon of X-linkage.

In his studies, he investigated numerous *Drosophila* mutations located on the X chromosome. His original analysis, based on crosses involving only one gene on the X chromosome, led to the discovery of X-linked inheritance. However, when he made crosses involving two X-linked genes, his results were initially puzzling. For example, female flies expressing the mutant *yellow* body (*y*) and *white* eyes (*w*) alleles were crossed with wild-type males (gray body and red eyes). The F_1 females were wild type, while the F_1 males expressed both mutant traits. In the F_2 the vast majority of the total offspring showed the expected parental phenotypes—yellow-bodied, white-eyed flies and wild-type flies (gray-bodied, red-eyed). The remaining flies, less than 1.0 percent, were either yellow-bodied with red eyes or gray-bodied with white eyes. It was as if the two mutant alleles had somehow separated from each other on the homolog during gamete formation in the F_1 female flies. This outcome is illustrated in cross A of Figure 5–3, using data later compiled by Sturtevant.

When Morgan studied other X-linked genes, the same basic pattern was observed, but the proportion of F_2 phenotypes differed. For example, when he crossed *white*-eye, *miniature*-wing mutants with wild-type flies, only 65.5 percent of all the F_2 flies showed the parental phenotypes, while 34.5 percent of the offspring appeared as if the mutant genes had been separated during gamete formation. This is illustrated in cross B of Figure 5–3, again using data subsequently compiled by Sturtevant.

Morgan was faced with two questions: (1) What was the source of gene separation and (2) why did the frequency of the apparent separation vary depending on the genes being studied? The answer Morgan proposed for the first question was based on his knowledge of earlier cytological observations made by F. A. Janssens and others. Janssens had observed that synapsed homologous chromosomes in meiosis wrapped around each other, creating **chiasmata** (sing. chiasma), X-shaped intersections where points of overlap are evident (see the photo on p. 105). Morgan proposed that these chiasmata could represent points of genetic exchange.

Regarding the crosses shown in Figure 5–3, Morgan postulated that if an exchange of chromosome material occurs during gamete formation, at a chiasma between the mutant genes on the two X chromosomes of the F_1 females, the unique phenotypes will occur. He suggested that such exchanges led to recombinant gametes in both the *yellow–white* cross and the *white–miniature* cross, as compared to the parental gametes that underwent no exchange. On the basis of this and other experimentation, Morgan concluded that linked genes are arranged in a linear sequence along the chromosome and that a variable frequency of exchange occurs between any two genes during gamete formation.

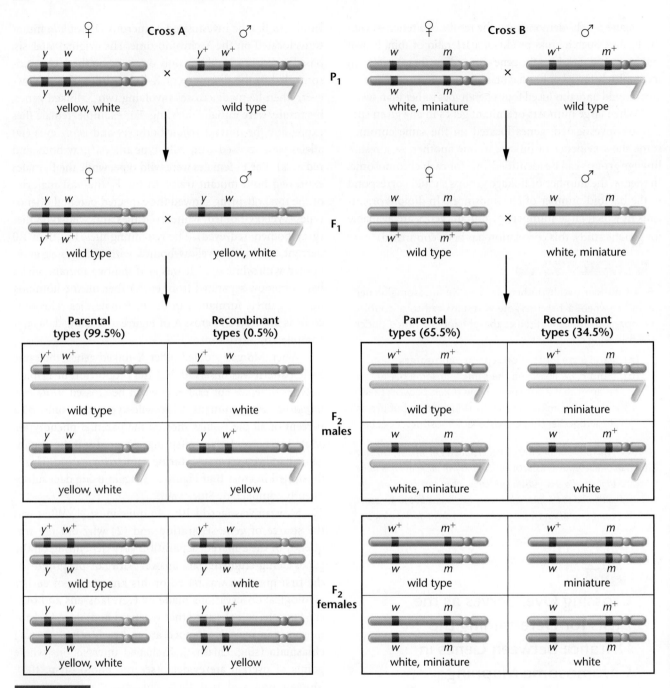

FIGURE 5–3   The F_1 and F_2 results of crosses involving the *yellow* (*y*), *white* (*w*) mutations (cross A), and the *white*, *miniature* (*m*) mutations (cross B), as compiled by Sturtevant. In cross A, 0.5 percent of the F_2 flies (males and females) demonstrate recombinant phenotypes, which express either *white* or *yellow*. In cross B, 34.5 percent of the F_2 flies (males and females) demonstrate recombinant phenotypes, which are either *miniature* or *white* mutants.

In answer to the second question, Morgan proposed that two genes located relatively close to each other along a chromosome are less likely to have a chiasma form between them than if the two genes are farther apart on the chromosome. Therefore, the closer two genes are, the less likely that a genetic exchange will occur between them. Morgan was the first to propose the term *crossing over* to describe the physical exchange leading to recombination.

Sturtevant and Mapping

Morgan's student, Alfred H. Sturtevant, was the first to realize that his mentor's proposal could be used to map the sequence of linked genes. According to Sturtevant,

"In a conversation with Morgan . . . I suddenly realized that the variations in strength of linkage, already attributed by Morgan to differences in the spatial separation of the genes, offered the possibility of determining sequences in the linear dimension of a chromosome. I went home and spent most of the night (to the neglect of my undergraduate homework) in producing the first chromosomal map."

Sturtevant, in a paper published in 1913, compiled data from numerous crosses made by Morgan and other geneticists involving recombination between the genes represented by the *yellow, white,* and *miniature* mutants. A subset of these data is shown in Figure 5–3. The frequencies of recombination between each pair of these three genes are as follows:

(1) *yellow, white*	0.5%
(2) *white, miniature*	34.5%
(3) *yellow, miniature*	35.4%

Because the sum of (1) and (2) approximately equals (3), Sturtevant suggested that the recombination frequencies between linked genes are additive. On this basis, he predicted that the order of the genes on the X chromosome is *yellow–white–miniature.* In arriving at this conclusion, he reasoned as follows: The *yellow* and *white* genes are apparently close to each other because the recombination frequency is low. However, both of these genes are quite far from the *miniature* gene because the *white–miniature* and *yellow–miniature* combinations show larger recombination frequencies. Because *miniature* shows more recombination with *yellow* than with *white* (35.4 percent vs. 34.5 percent), it follows that *white* is located between the other two genes, not outside of them.

Sturtevant knew from Morgan's work that the frequency of exchange could be used as an estimate of the distance between two genes or loci along the chromosome. He constructed a chromosome map of the three genes on the X chromosome, setting one map unit (mu) equal to 1 percent recombination between two genes.[*] The distance between *yellow* and *white* is thus 0.5 mu, and the distance between *yellow* and *miniature* is 35.4 mu. It follows that the distance between *white* and *miniature* should be 35.4 − 0.5 = 34.9 mu. This estimate is close to the actual frequency of recombination between *white* and *miniature* (34.5 mu). The map for these three genes is shown in Figure 5–4. The fact that these numbers do not add up perfectly is due to normal variation that one would expect between crosses, leading to the minor imprecisions encountered in independently conducted mapping experiments.

In addition to these three genes, Sturtevant considered crosses involving two other genes on the X chromosome and

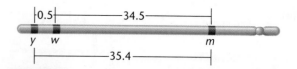

FIGURE 5–4 A map of the *yellow (y), white (w),* and *miniature (m)* genes on the X chromosome of *Drosophila melanogaster.* Each number represents the percentage of recombinant offspring produced in one of three crosses, each involving two different genes.

produced a more extensive map that included all five genes. He and a colleague, Calvin Bridges, soon began a search for autosomal linkage in *Drosophila.* By 1923, they had clearly shown that linkage and crossing over are not restricted to X-linked genes but could also be demonstrated with autosomes. During this work, they made another interesting observation. In *Drosophila,* crossing over was shown to occur only in females. The fact that no crossing over occurs in males made genetic mapping much less complex to analyze in *Drosophila.* While crossing over does occur in both sexes in most other organisms, crossing over in males is often observed to occur less frequently than in females. For example, in humans, such recombination occurs only about 60 percent as often in males compared to females.

Although many refinements have been added to chromosome mapping since Sturtevant's initial work, his basic principles are accepted as correct. These principles are used to produce detailed chromosome maps of organisms for which large numbers of linked mutant genes are known. Sturtevant's findings are also historically significant to the broader field of genetics. In 1910, the **chromosomal theory of inheritance** was still widely disputed—even Morgan was skeptical of this theory before he conducted his experiments. Research has now firmly established that chromosomes contain genes in a linear order and that these genes are the equivalent of Mendel's unit factors.

Single Crossovers

Why should the relative distance between two loci influence the amount of crossing over and recombination observed between them? During meiosis, a limited number of crossover events occur in each tetrad. These recombinant events occur randomly along the length of the tetrad. Therefore, the closer that two loci reside along the axis of the chromosome, the less likely that any **single crossover** event will occur between them. The same reasoning suggests that the farther apart two linked loci, the more likely a random crossover event will occur in between them.

In Figure 5–5(a), a single crossover occurs between two nonsister chromatids, but not between the two loci being

---

[*]In honor of Morgan's work, map units are often referred to as centi-Morgans (cM).

(a)

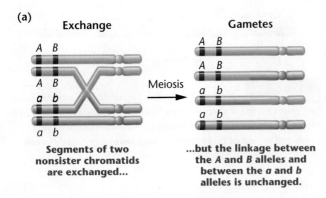

Segments of two nonsister chromatids are exchanged...

...but the linkage between the *A* and *B* alleles and between the *a* and *b* alleles is unchanged.

(b)

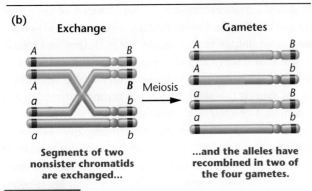

Segments of two nonsister chromatids are exchanged...

...and the alleles have recombined in two of the four gametes.

FIGURE 5–5   Two examples of a single crossover between two nonsister chromatids and the gametes subsequently produced. In (a) the exchange does not alter the linkage arrangement between the alleles of the two genes, only parental gametes are formed, and the exchange goes undetected. In (b) the exchange separates the alleles, resulting in recombinant gametes, which are detectable.

studied; therefore, the crossover is undetected because no recombinant gametes are produced for the two traits of interest. In Figure 5–5(b), where the two loci under study are quite far apart, the crossover does occur between them, yielding gametes in which the traits of interest are recombined.

When a single crossover occurs between two nonsister chromatids, the other two chromatids of the tetrad are not involved in the exchange and enter the gamete unchanged. Even if a single crossover occurs 100 percent of the time between two linked genes, recombination is subsequently observed in only 50 percent of the potential gametes formed. This concept is diagrammed in Figure 5–6. Theoretically, if we assume only single exchanges between a given pair of loci and observe 20 percent recombinant gametes, we will conclude that crossing over actually occurs between these two loci in 40 percent of the tetrads. The general rule is that, under these conditions, the percentage of tetrads involved in an exchange between two genes is twice as great as the percentage of recombinant gametes produced. Therefore, the theoretical limit of observed recombination due to crossing over is 50 percent.

When two linked genes are more than 50 map units apart, a crossover can theoretically be expected to occur between them in 100 percent of the tetrads. If this prediction were achieved, each tetrad would yield equal proportions of the four gametes shown in Figure 5–6, just as if the genes were on different chromosomes and assorting independently. For a variety of reasons, this theoretical limit is seldom achieved.

5.3

Determining the Gene Sequence during Mapping Requires the Analysis of Multiple Crossovers

The study of single crossovers between two linked genes provides a basis for determining the *distance* between them. However, when many linked genes are studied, their *sequence* along the chromosome is more difficult to determine. Fortunately, the discovery that multiple crossovers occur between the chromatids of a tetrad has facilitated the

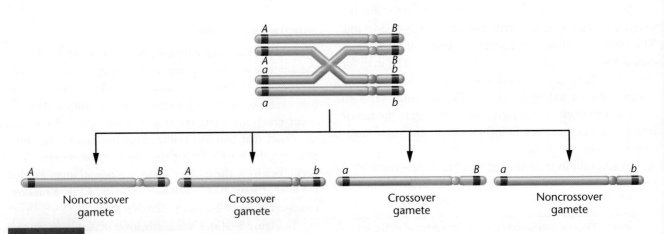

FIGURE 5–6   The consequences of a single exchange between two nonsister chromatids occurring in the tetrad stage. Two noncrossover (parental) and two crossover (recombinant) gametes are produced.

process of producing more extensive chromosome maps. As we shall see next, when three or more linked genes are investigated simultaneously, it is possible to determine first the sequence of genes and then the distances between them.

Multiple Exchanges

It is possible that in a single tetrad, two, three, or more exchanges will occur between nonsister chromatids as a result of several crossing over events. Double exchanges of genetic material result from **double crossovers (DCOs)**, as shown in Figure 5–7. To study a double exchange, three gene pairs must be investigated, each heterozygous for two alleles. Before we determine the frequency of recombination among all three loci, let's review some simple probability calculations.

As we have seen, the probability of a single exchange occurring in between the A and B or the B and C genes is related directly to the distance between the respective loci.

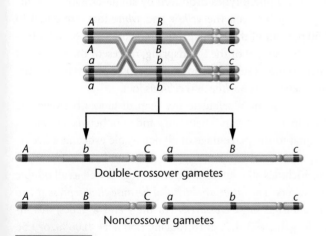

Double-crossover gametes

Noncrossover gametes

FIGURE 5–7 Consequences of a double exchange occurring between two nonsister chromatids. Because the exchanges involve only two chromatids, two noncrossover gametes and two double-crossover gametes are produced. The Chapter Opening photograph on p. 105 illustrates two chiasmata present in a tetrad isolated during the first meiotic prophase stage.

The closer A is to B and B is to C, the less likely it is that a single exchange will occur in between either of the two sets of loci. In the case of a double crossover, two separate and independent events or exchanges must occur simultaneously. The mathematical probability of two independent events occurring simultaneously is equal to the product of the individual probabilities. This is the *product law* introduced in Chapter 3.

Suppose that crossover gametes resulting from single exchanges are recovered 20 percent of the time ($p = 0.20$) between A and B, and 30 percent of the time ($p = 0.30$) between B and C. The probability of recovering a double-crossover gamete arising from two exchanges (between A and B and between B and C) is predicted to be $(0.20)(0.30) = 0.06$, or 6 percent. It is apparent from this calculation that the expected frequency of double-crossover gametes is always expected to be much lower than that of either single-crossover class of gametes.

If three genes are relatively close together along one chromosome, the expected frequency of double-crossover gametes is extremely low. For example, suppose that the A–B distance in Figure 5–7 is 3 mu and the B–C distance is 2 mu. The expected double-crossover frequency is $(0.03)(0.02) = 0.0006$, or 0.06 percent. This translates to only 6 events in 10,000. Thus in a mapping experiment where closely linked genes are involved, very large numbers of offspring are required to detect double-crossover events. In this example, it is unlikely that a double crossover will be observed even if 1000 offspring are examined. Thus, it is evident that if four or five genes are being mapped, even fewer triple and quadruple crossovers can be expected to occur.

Three-Point Mapping in *Drosophila*

The information presented in the previous section enables us to map three or more linked genes in a single cross. To illustrate the mapping process in its entirety, we examine two situations involving three linked genes in two quite different organisms.

To execute a successful mapping cross, three criteria must be met:

1. The genotype of the organism producing the crossover gametes must be heterozygous at all loci under consideration. If homozygosity occurred at any locus, all gametes produced would contain the same allele, precluding mapping analysis.

2. The cross must be constructed so that the genotypes of all gametes can be accurately determined by observing the phenotypes of the resulting offspring. This is necessary because the gametes and their genotypes can never be observed directly. To overcome this problem, each phenotypic class must reflect the genotype of the gametes of the parents producing it.

3. A sufficient number of offspring must be produced in the mapping experiment to recover a representative sample of all crossover classes.

These criteria are met in the three-point mapping cross of *Drosophila melanogaster* shown in Figure 5–8. In this cross three X-linked recessive mutant genes—*yellow* body color, *white* eye color, and *echinus* eye shape—are considered. To diagram the cross, *we must assume some theoretical sequence, even though we do not yet know if it is correct.* In Figure 5–8, we initially assume the sequence of the three genes to be *y–w–ec*. If this is incorrect, our analysis shall demonstrate it and reveal the correct sequence.

In the P_1 generation, males hemizygous for all three wild-type alleles are crossed to females that are homozygous for all three recessive mutant alleles. Therefore, the P_1 males are wild type with respect to body color, eye color, and eye shape. They are said to have a *wild-type phenotype.* The females, on the other hand, exhibit the three mutant traits: yellow body color, white eyes, and echinus eye shape.

This cross produces an F_1 generation consisting of females that are heterozygous at all three loci and males that, because of the Y chromosome, are hemizygous for the three mutant alleles. Phenotypically, all F_1 females are wild type, while all F_1 males are yellow, white, and echinus. The genotype of the F_1 females fulfills the first criterion for constructing a map of the three linked genes; that is, it is heterozygous at the three loci and may serve as the source of recombinant gametes generated by crossing over. Note that, because of the genotypes of the P_1 parents, all three of the mutant alleles are on one homolog and all three wild-type alleles are on the other homolog. With other parents, *other arrangements would be possible that could produce a heterozygous genotype.* For example, a heterozygous female could have the *y* and *ec* mutant alleles on one homolog and the *w* allele on the other. This would occur if one of her parents was *yellow, echinus* and the other parent was *white.*

In our cross, the second criterion is met as a result of the gametes formed by the F_1 males. Every gamete contains either an X chromosome bearing the three mutant alleles or a Y chromosome, which does not contain any of the three loci being considered. Whichever type participates in fertilization, the genotype of the gamete produced by the F_1 female will be expressed phenotypically in the F_2 female and male offspring derived from it. As a result, all noncrossover and crossover gametes produced by the F_1 female parent can be determined by observing the F_2 phenotypes.

With these two criteria met, we can construct a chromosome map from the crosses illustrated in Figure 5–8. First, we must determine which F_2 phenotypes correspond to the various noncrossover and crossover categories. To determine the **noncrossover** F_2 phenotypes, we must identify individuals derived from the parental gametes formed by the F_1 female. Each such gamete contains *an X chromosome unaffected by crossing over.* As a result of segregation, approximately equal proportions of the two types of gametes, and subsequently their F_2 phenotypes, are produced. Because they derive from a heterozygote, the genotypes of the two parental gametes and the F_2 phenotypes complement one another. For example, if one is wild type, the other is mutant for all three genes. This is the case in the cross being considered. In other situations, if one chromosome shows one mutant allele, the second chromosome shows the other two mutant alleles, and so on. These are therefore called **reciprocal classes** of gametes and phenotypes.

The two noncrossover phenotypes are most easily recognized because *they occur in the greatest proportion of offspring.* Figure 5–8 shows that gametes (1) and (2) are present in the greatest numbers. Therefore, flies that are yellow, white, and echinus and those that are normal, or wild type, for all three characters constitute the noncrossover category and represent 94.44 percent of the F_2 offspring.

The second category that can be easily detected is represented by the double-crossover phenotypes. Because of their low probability of occurrence, *they must be present in the least numbers.* Remember that this group represents two independent but simultaneous single-crossover events. Two reciprocal phenotypes can be identified: gamete 7, which shows the mutant traits yellow and echinus, but normal eye color; and gamete 8, which shows the mutant trait white, but normal body color and eye shape. Together these double-crossover phenotypes constitute only 0.06 percent of the F_2 offspring.

The remaining four phenotypic classes fall into two categories resulting from single crossovers. Gametes 3 and 4, reciprocal phenotypes produced by single-crossover events occurring between the *yellow* and *white* loci, are equal to 1.50 percent of the F_2 offspring. Gametes 5 and 6, constituting 4.00 percent of the F_2 offspring, represent the reciprocal phenotypes resulting from single-crossover events occurring between the *white* and *echinus* loci.

We can now calculate the map distances between the three loci. The distance between *y* and *w*, or between *w* and *ec*, is equal to the percentage of all detectable exchanges occurring between them. For any two genes under consideration, this includes all related single crossovers as well as all double crossovers. *The latter are included because they represent two simultaneous single crossovers.* For the *y* and *w* genes, this includes gametes 3, 4, 7, and 8, totaling 1.50% + 0.06%, or 1.56 mu. Similarly, the distance between *w* and *ec* is equal to the percentage of offspring resulting from an exchange between these two loci: gametes 5, 6, 7, and 8, totaling 4.00% + 0.06%, or 4.06 mu. The map of these three loci on the X chromosome is shown at the bottom of Figure 5–8.

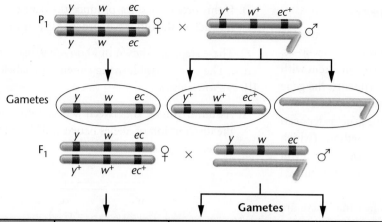

Origin of female gametes	Gametes	Gametes		F₂ phenotype	Observed Number	Category, total, and percentage
NCO	① *y w ec*	*y w ec*	*y w ec*	*y w ec*	4685	**Non-crossover**
	② *y⁺ w⁺ ec⁺*	*y⁺ w⁺ ec⁺*	*y⁺ w⁺ ec⁺*	*y⁺ w⁺ ec⁺*	4759	9444 94.44%
SCO	③ *y w⁺ ec⁺*	*y w⁺ ec⁺*	*y w⁺ ec⁺*	*y w⁺ ec⁺*	80	**Single crossover between y and w**
	④ *y⁺ w ec*	*y⁺ w ec*	*y⁺ w ec*	*y⁺ w ec*	70	150 1.50%
SCO	⑤ *y w ec⁺*	*y w ec⁺*	*y w ec⁺*	*y w ec⁺*	193	**Single crossover between w and ec**
	⑥ *y⁺ w⁺ ec*	*y⁺ w⁺ ec*	*y⁺ w⁺ ec*	*y⁺ w⁺ ec*	207	400 4.00%
DCO	⑦ *y w⁺ ec*	*y w⁺ ec*	*y w⁺ ec*	*y w⁺ ec*	3	**Double crossover between y and w and between w and ec**
	⑧ *y⁺ w ec⁺*	*y⁺ w ec⁺*	*y⁺ w ec⁺*	*y⁺ w ec⁺*	3	6 0.06%

y w ec Map of *y, w,* and *ec* loci
├─1.56─┼──4.06──┤

FIGURE 5–8 A three-point mapping cross involving the *yellow* (*y* or *y⁺*), *white* (*w* or *w⁺*), and *echinus* (*ec* or *ec⁺*) genes in *Drosophila melanogaster*. NCO, SCO, and DCO refer to noncrossover, single-crossover, and double-crossover groups, respectively. Centromeres are not drawn on the chromosomes, and only two nonsister chromatids are initially shown in the left-hand column.

Determining the Gene Sequence

In the preceding example, we assumed that the sequence (or order) of the three genes along the chromosome was *y–w–ec*. Our analysis established that the sequence is consistent with the data. However, in most mapping experiments, the gene sequence is not known, and this constitutes another variable in the analysis. In our example, had the gene order been unknown, we could have used one of two methods (which we will study next) to determine it. In your own work, you should select one of these methods and use it consistently.

Method I This method is based on the fact that there are only three possible arrangements, each containing a different one of the three genes between the other two:

(I)   *w–y–ec*   (*y* is in the middle)

(II)   *y–ec–w*   (*ec* is in the middle)

(III)   *y–w–ec*   (*w* is in the middle)

Use the following steps during your analysis to determine the gene order:

1. Assuming any of the three orders, first determine the *arrangement of alleles* along each homolog of the heterozygous parent giving rise to noncrossover and crossover gametes (the F₁ female in our example).

2. Determine whether a double-crossover event occurring within that arrangement will produce the *observed double-crossover phenotypes*. Remember that these phenotypes occur least frequently and are easily identified.

3. If this order does not produce the correct phenotypes, try each of the other two orders. One must work!

These steps are shown in **Figure 5–9**, using our *y–w–ec* cross. The three possible arrangements are labeled I, II, and III, as shown above.

1. Assuming that *y* is between *w* and *ec* (arrangement I), the distribution of alleles between the homologs of the F₁ heterozygote is:

$$\frac{w \qquad y \qquad ec}{w^+ \qquad y^+ \qquad ec^+}$$

We know this because of the way in which the P₁ generation was crossed: The P₁ female contributes an X chromosome bearing the *w*, *y*, and *ec* alleles, while the P₁ male contributed an X chromosome bearing the w^+, y^+, and ec^+ alleles.

2. A double crossover within that arrangement yields the following gametes:

$$\underline{w \qquad y^+ \qquad ec} \quad \text{and} \quad \underline{w^+ \qquad y \qquad ec^+}$$

Following fertilization, if *y* is in the middle, the F₂ double-crossover phenotypes will correspond to these gametic genotypes, yielding offspring that express the white, echinus phenotype and offspring that express the yellow phenotype. Instead, determination of the actual double crossovers reveals them to be yellow, echinus flies and white flies. *Therefore, our assumed order is incorrect.*

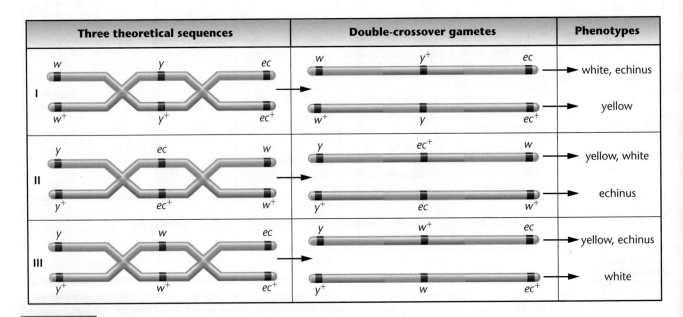

Three theoretical sequences	Double-crossover gametes	Phenotypes
I		white, echinus / yellow
II		yellow, white / echinus
III		yellow, echinus / white

FIGURE 5–9   The three possible sequences of the *white, yellow,* and *echinus* genes, the results of a double crossover in each case, and the resulting phenotypes produced in a testcross. For simplicity, the two noncrossover chromatids of each tetrad are omitted.

3. If we consider arrangement II, with the ec/ec^+ alleles in the middle, or arrangement III, with the w/w^+ alleles in the middle:

$$(\text{II}) \quad \frac{y \quad\;\; ec \quad\;\; w}{y^+ \quad ec^+ \quad w^+} \quad \text{or} \quad (\text{III}) \quad \frac{y \quad\;\; w \quad\;\; ec}{y^+ \quad w^+ \quad ec^+}$$

we see that arrangement II again provides *predicted* double-crossover phenotypes that do not correspond to the actual (observed) double-crossover phenotypes. The predicted phenotypes are yellow, white flies and echinus flies in the F_2 generation. *Therefore, this order is also incorrect.* However, arrangement III produces the observed phenotypes—yellow, echinus flies and white flies. *Therefore, this arrangement, with the w gene in the middle, is correct.*

To summarize Method I: First, determine the arrangement of alleles on the homologs of the heterozygote yielding the crossover gametes by identifying the reciprocal non-crossover phenotypes. Then, test each of the three possible orders to determine which one yields the observed double-crossover phenotypes—*the one that does so represents the correct order.* This method is summarized in Figure 5–9.

Method II Method II also begins by determining the arrangement of alleles along each homolog of the heterozygous parent. In addition, it requires one further assumption:

Following a double-crossover event, the allele in the middle position will fall between the outside, or flanking, alleles that were present on the opposite parental homolog.

To illustrate, assume order I, *w–y–ec*, in the following arrangement:

$$\frac{w \quad\;\; y \quad\;\; ec}{w^+ \quad y^+ \quad ec^+}$$

Following a double-crossover event, the y and y^+ alleles would be switched to this arrangement:

$$\frac{w \quad\;\; y^+ \quad ec}{w^+ \quad y \quad\;\; ec^+}$$

After segregation, two gametes would be formed:

$$\underline{w \quad\;\; y^+ \quad ec} \quad \text{and} \quad \underline{w^+ \quad y \quad\;\; ec^+}$$

Because the genotype of the gamete will be expressed directly in the phenotype following fertilization, the double-crossover phenotypes will be:

white, echinus flies and yellow flies

Note that the *yellow* allele, assumed to be in the middle, is now associated with the two outside markers of the other homolog, w^+ and ec^+. However, these predicted phenotypes do not coincide with the observed double-crossover phenotypes. Therefore, the *yellow* gene is not in the middle.

This same reasoning can be applied to the assumption that the *echinus* gene or the *white* gene is in the middle. In the former case, we will reach a negative conclusion. If we assume that the *white* gene is in the middle, the *predicted* and *actual* double crossovers coincide. Therefore, we conclude that the *white* gene is located between the *yellow* and *echinus* genes.

To summarize Method II, determine the arrangement of alleles on the homologs of the heterozygote yielding crossover gametes. Then examine the actual double-crossover phenotypes and identify the single allele that has been switched so that it is now no longer associated with its original neighboring alleles. That allele will be the one located between the other two in the sequence.

In our example y, ec, and w are on one homolog in the F_1 heterozygote, and y^+, ec^+, and w^+ are on the other. In the F_2 double-crossover classes, it is w and w^+ that have been switched. The w allele is now associated with y^+ and ec^+, while the w^+ allele is now associated with the y and ec alleles. Therefore, the *white* gene is in the middle, and the *yellow* and *echinus* genes are the flanking markers.

A Mapping Problem in Maize

Having established the basic principles of chromosome mapping, we will now consider a related problem in maize (corn). This analysis differs from the preceding example in two ways. First, the previous mapping cross involved X-linked genes. Here, we consider autosomal genes. Second, in the discussion of this cross, we will change our use of symbols, as first suggested in Chapter 4. Instead of using the gene symbols and superscripts (e.g., bm^+, v^+, and pr^+), we simply use + to denote each wild-type allele. This system is easier to manipulate but requires a better understanding of mapping procedures.

When we look at three autosomally linked genes in maize, our experimental cross must still meet the same three criteria we established for the X-linked genes in *Drosophila*: (1) One parent must be heterozygous for all traits under consideration; (2) the gametic genotypes produced by the heterozygote must be apparent from observing the phenotypes of the offspring; and (3) a sufficient sample size must be available for complete analysis.

In maize, the recessive mutant genes *bm* (*brown* midrib), *v* (*virescent* seedling), and *pr* (*purple* aleurone) are linked on chromosome 5. Assume that a female plant is known to be heterozygous for all three traits, but we do not know: (1) the arrangement of the mutant alleles on the maternal and paternal homologs of this heterozygote; (2) the sequence of genes; or (3) the map distances between the genes. What genotype must the male plant have to allow successful

(a) Some possible allele arrangements and gene sequences in a heterozygous female

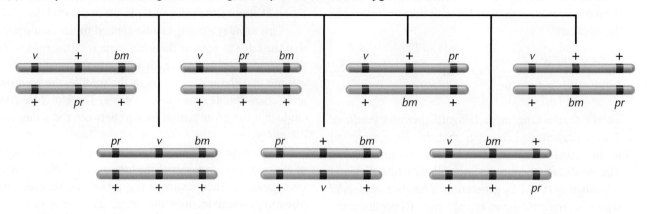

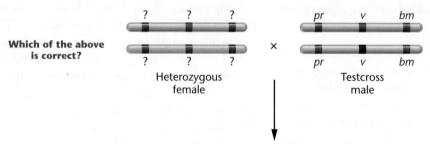

Which of the above is correct?

Heterozygous female   ×   Testcross male

(b) Actual results of mapping cross*

Phenotypes of offspring	Number	Total and percentage	Exchange classification
+ v bm	230	467 42.1%	Noncrossover (NCO)
pr + +	237		
+ + bm	82	161 14.5%	Single crossover (SCO)
pr v +	79		
+ v +	200	395 35.6%	Single crossover (SCO)
pr + bm	195		
pr v bm	44	86 7.8%	Double crossover (DCO)
+ + +	42		

* The sequence pr – v – bm may or may not be correct.

FIGURE 5–10 (a) Some possible allele arrangements and gene sequences in a heterozygous female. The data from a three-point mapping cross, depicted in (b), where the female is testcrossed, provide the basis for determining which combination of arrangement and sequence is correct. [See Figure 5–11(d).]

mapping? To meet the second criterion, the male must be homozygous for all three recessive mutant alleles. Otherwise, offspring of this cross showing a given phenotype might represent more than one genotype, making accurate mapping impossible. Note that this is equivalent to performing a testcross.

Figure 5–10 diagrams this cross. As shown, we know neither the *arrangement of alleles* nor the *sequence of loci* in

the heterozygous female. Several possibilities are shown, but we have yet to determine which is correct. We don't know the sequence in the testcross male parent either, so we must designate it randomly. Note that we initially placed *v* in the middle. *This may or may not be correct.*

The offspring have been arranged in groups of two, representing each pair of reciprocal phenotypic classes. The four reciprocal classes are derived from no crossing over (NCO), each of two possible single-crossover events (SCO), and a double-crossover event (DCO).

To solve this problem, refer to Figures 5–10 and 5–11 as you consider the following questions:

1. *What is the correct heterozygous arrangement of alleles in the female parent?*

   Determine the two noncrossover classes, those that occur with the highest frequency. In this case, they are $+ \ v \ bm$ and $pr + +$. Therefore, the alleles on the homologs of the female parent must be distributed as shown in Figure 5–11(a). These homologs segregate into gametes, unaffected by any recombination event.

Any other arrangement of alleles will not yield the observed noncrossover classes. (Remember that $+ \ v \ bm$ is equivalent to $pr^+ \ v \ bm$ and that $pr \ + \ +$ is equivalent to $pr \ v^+ \ bm^+$.)

2. *What is the correct sequence of genes?*

   To answer this question, we will first use the approach described in Method I. We know, based on the answer to question 1, that the correct arrangement of alleles is

   $$\frac{+ \qquad v \qquad bm}{pr \qquad + \qquad +}$$

   But is the gene sequence correct? That is, will a double-crossover event yield the observed double-crossover

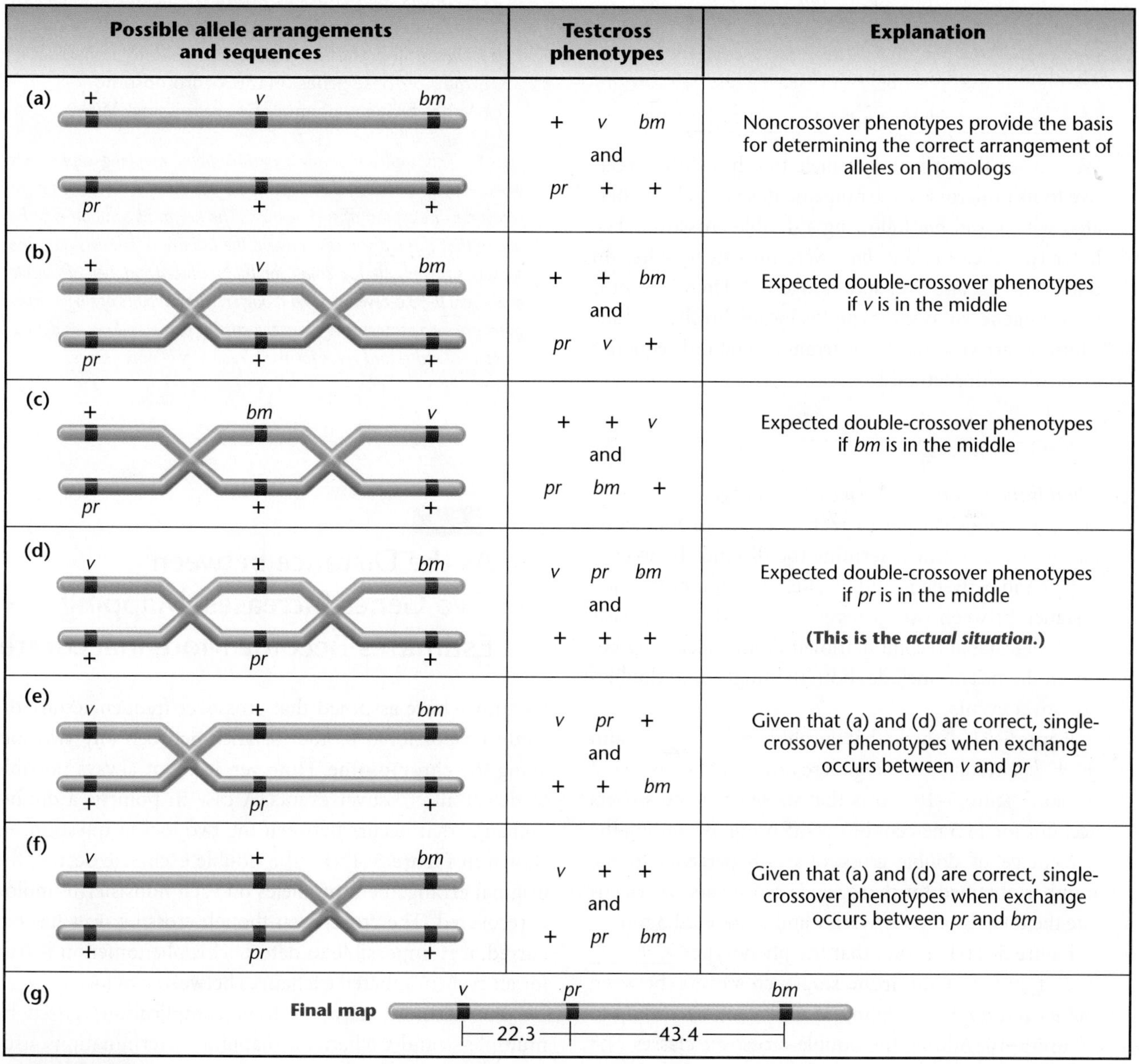

Possible allele arrangements and sequences		Testcross phenotypes	Explanation
(a)	+ v bm pr + +	+ v bm and pr + +	Noncrossover phenotypes provide the basis for determining the correct arrangement of alleles on homologs
(b)	+ v bm pr + +	+ + bm and pr v +	Expected double-crossover phenotypes if *v* is in the middle
(c)	+ bm v pr + +	+ + v and pr bm +	Expected double-crossover phenotypes if *bm* is in the middle
(d)	v + bm + pr +	v pr bm and + + +	Expected double-crossover phenotypes if *pr* is in the middle **(This is the *actual situation*.)**
(e)	v + bm + pr +	v pr + and + + bm	Given that (a) and (d) are correct, single-crossover phenotypes when exchange occurs between *v* and *pr*
(f)	v + bm + pr +	v + + and + pr bm	Given that (a) and (d) are correct, single-crossover phenotypes when exchange occurs between *pr* and *bm*
(g)	**Final map** v — 22.3 — pr — 43.4 — bm		

FIGURE 5–11   Producing a map of the three genes in the cross in Figure 5–10, where neither the arrangement of alleles nor the sequence of genes in the heterozygous female parent is known.

phenotypes following fertilization? *Observation shows that it will not* [Figure 5–11(b)]. Now try the other two orders [Figure 5–11(c) and 5–11(d)], *keeping the same allelic arrangement:*

$$\frac{+ \quad bm \quad v}{pr \quad + \quad +} \text{ or } \frac{v \quad + \quad bm}{+ \quad pr \quad +}$$

Only the order on the right yields the observed double-crossover gametes [Figure 5–11(d)]. Therefore, the *pr* gene is in the middle.

The same conclusion is reached if we use Method II to analyze the problem. In this case, no assumption of gene sequence is necessary. The arrangement of alleles along homologs in the heterozygous parent is

$$\frac{+ \quad v \quad bm}{pr \quad + \quad +}$$

The double-crossover gametes are also known:

$$\underline{pr \quad v \quad bm} \text{ and } \underline{+ \quad + \quad +}$$

We can see that it is the *pr* allele that has shifted relative to its noncrossover arrangement, so as to be associated with *v* and *bm* following a double crossover. The latter two alleles (*v* and *bm*) were present together on one homolog, and they stayed together. Therefore, *pr* is the odd gene, so to speak, and is located in the middle. Thus, we arrive at the same arrangement and sequence as we did with Method I:

$$\frac{v \quad + \quad bm}{+ \quad pr \quad +}$$

3. *What is the distance between each pair of genes?*
Having established the correct sequence of loci as *v–pr–bm*, we can now determine the distance between *v* and *pr* and between *pr* and *bm*. Remember that the map distance between two genes is calculated on the basis of all detectable recombinational events occurring between them. This includes both the single- and double-crossover events.

Figure 5–11(e) shows that the phenotypes *v pr +* and *+ + bm* result from single crossovers between *v* and *pr,* and Figure 5–10 shows that those single crossovers account for 14.5 percent of the offspring. By adding the percentage of double crossovers (7.8 percent) to the number obtained for those single crossovers, we calculate the total distance between *v* and *pr* to be 22.3 mu.

Figure 5–11(f) shows that the phenotypes *v + +* and *+ pr bm* result from single crossovers between the *pr* and *bm* loci, totaling 35.6 percent, according to Figure 5–10. Adding the double-crossover classes (7.8 percent), we compute the distance between *pr* and *bm* as 43.4 mu. The final map for all three genes in this example is shown in Figure 5–11(g).

5–3 In *Drosophila*, a heterozygous female for the X-linked recessive traits *a*, *b*, and *c* was crossed to a male that phenotypically expressed *a*, *b*, and *c*. The offspring occurred in the following phenotypic ratios.

+	*b*	*c*	460
a	+	+	450
a	*b*	*c*	32
+	+	+	38
a	+	*c*	11
+	*b*	+	9

No other phenotypes were observed.

(a) What is the genotypic arrangement of the alleles of these genes on the X chromosome of the female?
(b) Determine the correct sequence and construct a map of these genes on the X chromosome.
(c) What progeny phenotypes are missing? Why?

■ HINT: *This problem involves a three-point mapping experiment where only six phenotypic categories are observed, even though eight categories are typical of such a cross. The key to its solution is to be aware that if the distances between the loci are relatively small, the sample size may be too small for the predicted number of double crossovers to be recovered, even though reciprocal pairs of single crossovers are seen. You should write the missing gametes down as double crossovers and record zeros for their frequency of appearance.*

5.4

As the Distance between Two Genes Increases, Mapping Estimates Become More Inaccurate

So far, we have assumed that crossover frequencies are directly proportional to the distance between any two loci along the chromosome. However, it is not always possible to detect all crossover events. A case in point is a double exchange that occurs between the two loci in question. As shown in Figure 5–12(a), if a double exchange occurs, the original arrangement of alleles on each nonsister homolog is recovered. Therefore, even though crossing over has occurred, it is impossible to detect. This phenomenon is true for all even-numbered exchanges between two loci.

Furthermore, as a result of complications posed by multiple-strand exchanges, mapping determinations usually underestimate the actual distance between two genes. The farther apart two genes are, the greater the probability that undetected crossovers will occur. While the discrepancy

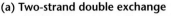

(a) Two-strand double exchange

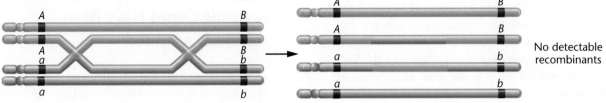

No detectable recombinants

(b)

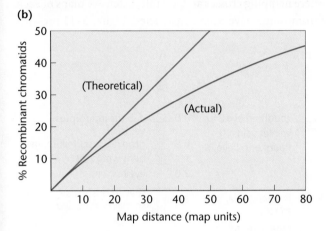

FIGURE 5–12 (a) A double crossover is undetected because no rearrangement of alleles occurs. (b) The theoretical and actual percentage of recombinant chromatids versus map distance. The straight line shows the theoretical relationship if a direct correlation between recombination and map distance exists. The curved line is the actual relationship derived from studies of *Drosophila, Neurospora,* and *Zea mays.*

is minimal for two genes relatively close together, the degree of inaccuracy increases as the distance increases, as shown in the graph of recombination frequency versus map distance in Figure 5–12(b). There, the theoretical frequency where a direct correlation between recombination and map distance exists is contrasted with the actual frequency observed as the distance between two genes increases. The most accurate maps are constructed from experiments in which genes are relatively close together.

Interference and the Coefficient of Coincidence

As review of the product law in Section 5.3 would indicate, the expected frequency of multiple exchanges, such as double crossovers, can be predicted once the distance between genes is established. For example, in the maize cross of the previous section, the distance between *v* and *pr* is 22.3 mu, and the distance between *pr* and *bm* is 43.4 mu. If the two single crossovers that make up a double crossover occur independently of one another, we can calculate the expected frequency of double crossovers (DCO_{exp}) as follows:

$$DCO_{exp} = (0.223) \times (0.434) = 0.097 = 9.7\%$$

Often in mapping experiments, the observed DCO frequency is less than the expected number of DCOs. In the maize cross, for example, only 7.8 percent of the DCOs are observed when 9.7 percent are expected. **Interference** (*I*), the inhibition of further crossover events by a crossover

event in a nearby region of the chromosome, causes this reduction.

To quantify the disparities that result from interference, we calculate the **coefficient of coincidence** (*C*):

$$C = \frac{\text{Observed DCO}}{\text{Expected DCO}}$$

In the maize cross, we have

$$C = \frac{0.078}{0.097} = 0.804$$

Once we have found *C*, we can quantify interference (*I*) by using this simple equation

$$I = 1 - C$$

In the maize cross, we have

$$I = 1.000 - 0.804 = 0.196$$

If interference is complete and no double crossovers occur, then *I* = 1.0. If fewer DCOs than expected occur, *I* is a positive number and **positive interference** has occurred. If more DCOs than expected occur, *I* is a negative number and **negative interference** has occurred. In this example, *I* is a positive number (0.196), indicating that 19.6 percent fewer double crossovers occurred than expected.

Positive interference is most often observed in eukaryotic systems. In *C. elegans,* for example, only one crossover event per chromosome is observed, and intereference along each chromosome is complete (*C* = 1.0). In other organisms, the closer genes are to one another along the chromosome, the more positive interference occurs. Interference in *Drosophila* is often complete within a distance of 10 map units. This observation suggests that physical constraints preventing the formation of closely spaced chiasmata contribute to interference. The interpretation is consistent with the finding that interference decreases as the genes in question are located farther apart. In the maize cross illustrated in Figures 5–10 and 5–11, the three genes are relatively far apart, and 80 percent of the expected double crossovers are observed.

5.5

Drosophila Genes Have Been Extensively Mapped

In organisms such as fruit flies, maize, and the mouse, where large numbers of mutants have been discovered and where mapping crosses are possible, extensive maps of each chromosome have been constructed. Figure 5–13 presents

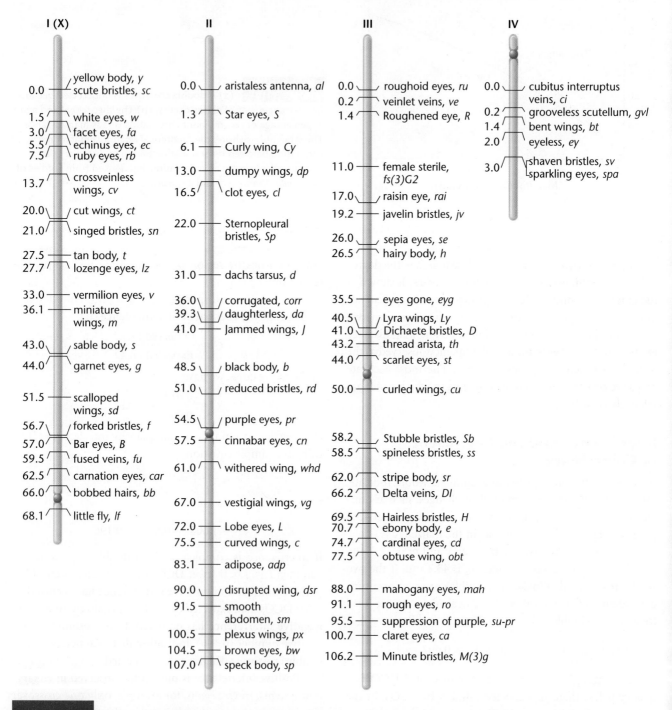

FIGURE 5–13   A partial genetic map of the four chromosomes of *Drosophila melanogaster*. The circle on each chromosome represents the position of the centromere. Chromosome I is the X chromosome. Chromosome IV is not drawn to scale; that is, it is relatively smaller than indicated.

partial maps of the four chromosomes of *Drosophila melanogaster*. Virtually every morphological feature of the fruit fly has been subjected to mutation. Each locus affected by mutation is first localized to one of the four chromosomes, or linkage groups, and then mapped in relation to other genes present on that chromosome. As you can see, the genetic map of the X chromosome is somewhat shorter than that of autosome II or III. In comparison to these three, autosome IV is miniscule. Cytological evidence has shown that the relative lengths of the genetic maps correlate roughly with the relative physical lengths of these chromosomes.

5.6

Lod Score Analysis and Somatic Cell Hybridization Were Historically Important in Creating Human Chromosome Maps

In humans, genetic experiments involving carefully planned crosses and large numbers of offspring are neither ethical nor feasible, so the earliest linkage studies were based on pedigree analysis. These studies attempted to establish whether certain traits were X-linked or autosomal. As we showed in Chapter 4, traits determined by genes located on the X chromosome result in characteristic pedigrees; thus, such genes were easier to identify. For autosomal traits, geneticists tried to distinguish clearly whether pairs of traits demonstrated linkage or independent assortment. When extensive pedigrees are available, it is possible to conclude that two genes under consideration are closely linked (i.e., rarely separated by crossing over) from the fact that the two traits segregate together. This approach established linkage between the genes encoding the **Rh antigens** and the gene responsible for the phenotype referred to as **elliptocytosis**, where the shape of erythrocytes is oval. It was hoped that from these kinds of observations a human gene map could be created.

A difficulty arises, however, when two genes of interest are separated on a chromosome to the degree that recombinant gametes are formed, obscuring linkage in a pedigree. In these cases, an approach relying on probability calculations, called the **lod score method**, helps to demonstrate linkage. First devised by J. B. S. Haldane and C. A. Smith in 1947 and refined by Newton Morton in 1955, the lod score (standing for *log* of the *odds* favoring linkage) assesses the probability that a particular pedigree (or several pedigrees for the same traits of interest) involving two traits reflects genetic linkage between them. First, the probability is calculated that the family (pedigree) data concerning two traits conform to transmission without linkage—that is, the traits appear to

be independently assorting. Then the probability is calculated that the identical family data for these same traits result from linkage with a specified recombination frequency. These probability calculations factor in the statistical significance at the $p = 0.05$ level. The ratio of these probability values is then calculated and converted to the logarithm of this value, which reflects the "odds" for, and against, linkage. Traditionally, a value of 3.0 or higher strongly indicates linkage, whereas a value of -2.0 or less argues strongly against linkage. Values between -2.0 and 3.0 are inconclusive.

The lod score method represented an important advance in assigning human genes to specific chromosomes and in constructing preliminary human chromosome maps. However, its accuracy is limited by the extent of the pedigree, and the initial results were discouraging—both because of this limitation and because of the relatively high haploid number of human chromosomes (23). By 1960, very little autosomal linkage information had become available. Today, however, in contrast to its restricted impact when originally developed, this elegant technique has become important in human linkage analysis, owing to the discovery of countless molecular *DNA markers* along every human chromosome. The discovery of these markers, which behave just as genes do in occupying a particular locus along a chromosome, was a result of recombinant DNA techniques (Chapter 20) and genomic analysis (Chapter 21). Any human trait may now be tested for linkage with such markers. We will return to a consideration of DNA markers in Section 5.7.

In the 1960s, a new technique, **somatic cell hybridization**, proved to be an immense aid in assigning human genes to their respective chromosomes. This technique, first discovered by Georges Barski, relies on the fact that two cells in culture can be induced to fuse into a single hybrid cell. Barsky used two mouse-cell lines, but it soon became evident that cells from different organisms could also be fused. When fusion occurs, an initial cell type called a **heterokaryon** is produced. The hybrid cell contains two nuclei in a common cytoplasm. Using the proper techniques, we can fuse human and mouse cells, for example, and isolate the hybrids from the parental cells.

As the heterokaryons are cultured *in vitro*, two interesting changes occur. Eventually, the nuclei fuse together, creating a **synkaryon**. Then, as culturing is continued for many generations, chromosomes from one of the two parental species are gradually lost. In the case of the human–mouse hybrid, human chromosomes are lost randomly until eventually the synkaryon has a full complement of mouse chromosomes and only a few human chromosomes. It is the preferential loss of human chromosomes (rather than mouse chromosomes) that makes possible the assignment of human genes to the chromosomes on which they reside.

The experimental rationale is straightforward. If a specific human gene product is synthesized in a synkaryon containing

Hybrid cell lines	Human chromosomes present								Gene products expressed			
	1	**2**	**3**	**4**	**5**	**6**	**7**	**8**	**A**	**B**	**C**	**D**
23	⬤	⬤	⬤	⬤					−	+	−	+
34	⬤	⬤			⬤	⬤			+	−	−	+
41	⬤		⬤		⬤		⬤		+	+	−	+

FIGURE 5–14   A hypothetical grid of data used in synteny testing to assign genes to their appropriate human chromosomes. Three somatic hybrid cell lines, designated 23, 34, and 41, have each been scored for the presence, or absence, of human chromosomes 1 through 8, as well as for their ability to produce the hypothetical human gene products A, B, C, and D.

three human chromosomes, for example, then the gene responsible for that product must reside on one of the three human chromosomes remaining in the hybrid cell. On the other hand, if the human gene product is not synthesized in the synkaryon, the responsible gene cannot be present on any of the remaining three human chromosomes. Ideally, one would have a panel of 23 hybrid cell lines, each with a different human chromosome, allowing the immediate assignment to a particular chromosome of any human gene for which the product could be characterized.

In practice, a panel of cell lines each of which contains several remaining human chromosomes is most often used. The correlation of the presence or absence of each chromosome with the presence or absence of each gene product is called **synteny testing**. Consider, for example, the hypothetical data provided in Figure 5–14, where four gene products (A, B, C, and D) are tested in relationship to eight human chromosomes. Let us carefully analyze the results to locate the gene that produces product A.

1. Product A is not produced by cell line 23, but chromosomes 1, 2, 3, and 4 are present in cell line 23. Therefore, we can rule out the presence of gene *A* on those four chromosomes and conclude that it might be on chromosome 5, 6, 7, or 8.

2. Product A is produced by cell line 34, which contains chromosomes 5 and 6, but not 7 and 8. Therefore, gene *A* is on chromosome 5 or 6, but cannot be on 7 or 8 because they are absent, even though product A is produced.

3. Product A is also produced by cell line 41, which contains chromosome 5 but not chromosome 6. Therefore, gene *A* is on chromosome 5, according to this analysis.

Using a similar approach, we can assign gene *B* to chromosome 3. Perform the analysis for yourself to demonstrate that this is correct.

Gene *C* presents a unique situation. The data indicate that it is not present on chromosomes 1–7. While it might be on chromosome 8, no direct evidence supports this conclusion. Other panels are needed. We leave gene *D* for you to analyze. Upon what chromosome does it reside?

By using the approach just described, researchers were able to assign literally hundreds of human genes to one chromosome or another. To map genes for which the products have yet to be discovered, researchers have had to rely on other approaches. For example, by combining recombinant DNA technology with pedigree analysis, it was possible to assign the genes responsible for Huntington disease, cystic fibrosis, and neurofibromatosis to their respective chromosomes 4, 7, and 17. Modern genomic analysis has expanded our knowledge of the mapping location of countless other human traits, as described in the next section.

5.7

Chromosome Mapping Is Now Possible Using DNA Markers and Annotated Computer Databases

Although traditional methods based on recombination analysis have produced detailed chromosomal maps in several organisms, such maps in other organisms (including humans) that do not lend themselves to such studies are greatly limited. Fortunately, the development of technology allowing direct analysis of DNA has greatly enhanced mapping in those organisms. We will address this topic using humans as an example.

Progress has initially relied on the discovery of **DNA markers** (mentioned earlier) that have been identified during recombinant DNA and genomic studies. These markers are short segments of DNA whose sequence and location are known, making them useful *landmarks* for mapping purposes. The analysis of human genes in relation to these markers has extended our knowledge of the location within the genome of countless genes, which is the ultimate goal of mapping.

The earliest examples are the DNA markers referred to as **restriction fragment length polymorphisms (RFLPs)** (see Chapter 22) and **microsatellites** (see Chapter 12). RFLPS are polymorphic sites generated when specific DNA sequences are recognized and cut by restriction enzymes. Microsatellites are short repetitive sequences that are found throughout the genome, and they vary in the number of repeats at any given site. For example, the two-nucleotide sequence CA is repeated 5–50 times per site [$(CA)_n$] and appears throughout the genome approximately every 10,000 bases, on average. Microsatellites may be identified not only by the number of repeats but by the DNA sequences that flank them. More recently, variation in single nucleotides (called **single-nucleotide polymorphisms** or **SNPs**) has been utilized. Found throughout the genome, up to several million of these variations may be screened for an association with a disease or trait of interest, thus providing geneticists with a means to identify and locate related genes.

Cystic fibrosis offers an early example of a gene located by using DNA markers. It is a life-shortening autosomal recessive exocrine disorder resulting in excessive, thick mucus that impedes the function of organs such as the lung and pancreas. After scientists established that the gene causing this disorder is located on chromosome 7, they were then able to pinpoint its exact location on the long arm (the q arm) of that chromosome.

Several years ago (June 2007), using SNPs as DNA markers, associations between 24 genomic locations were established with seven common human diseases: *Type 1* (insulin dependent) and *Type 2 diabetes, Crohn's disease* (inflammatory bowel disease), *hypertension, coronary artery disease, bipolar* (manic-depressive) *disorder*, and *rheumatoid arthritis*. In each case, an inherited susceptibility effect was mapped to a specific location on a specific chromosome within the genome. In some cases, this either confirmed or led to the identification of a specific gene involved in the cause of the disease. In other cases, new genes will no doubt soon be discovered as a result of the identification of their location. We will return to this topic in much greater detail in Chapters 21 and 22.

The many Human Genome Project databases that have been completed now make it possible to map genes along a human chromosome in base-pair distances rather than recombination frequency. This distinguishes what is referred to as a *physical map* of the genome from the *genetic maps* described above. Distances can then be determined relative to other genes and to features such as the DNA markers discussed. Through this approach geneticists will soon be able to construct chromosome maps for individuals that designate specific allele combinations at each gene site.

5.8
Crossing Over Involves a Physical Exchange between Chromatids

Once genetic mapping techniques had been developed, they were used to study the relationship between the chiasmata observed in meiotic prophase I and crossing over. For example, are chiasmata visible manifestations of crossover events? If so, then crossing over in higher organisms appears to be the result of an actual physical exchange between homologous chromosomes. That this is the case was demonstrated independently in the 1930s by Harriet Creighton and Barbara McClintock in *Zea mays* (maize) and by Curt Stern in *Drosophila*.

Because the experiments are similar, we will consider only one of them, the work with maize. Creighton and McClintock studied two linked genes on chromosome 9 of the maize plant. At one locus, the alleles *colorless* (c) and *colored* (C) control endosperm coloration (the endosperm is the nutritive tissue inside the corn kernel). At the other locus, the alleles *starchy* (Wx) and *waxy* (wx) control the carbohydrate characteristics of the endosperm. The maize plant studied was heterozygous at both loci. The key to this experiment is that one of the homologs contained two unique cytological markers. The markers consisted of a densely stained knob at one end of the chromosome and a translocated piece of another chromosome (8) at the other end. The arrangements of these alleles and markers could be detected cytologically and are shown in **Figure 5–15**.

Creighton and McClintock crossed this plant to one homozygous for the *colorless* allele (c) and heterozygous for the *waxy/starchy* alleles. They obtained a variety of different phenotypes in the offspring, but they were most interested in one that occurred as a result of a crossover involving the chromosome with the unique cytological markers. They examined the chromosomes of this plant, having a colorless, waxy phenotype (Case I in Figure 5–15), for the presence of the cytological markers. If genetic crossing over was accompanied by a physical exchange between homologs, the translocated chromosome would still be present, but the knob would not. This was the case! In a second plant (Case II), the phenotype colored, starchy should result from either nonrecombinant gametes or crossing over. Some of the cases then ought to contain chromosomes with the dense knob but not the translocated chromosome. This condition was also found, and the conclusion that a physical exchange had taken place was again supported. Along with Curt Stern's findings in *Drosophila*, this work clearly established that crossing over has a cytological basis.

Once we have introduced the chemical structure and replication of DNA (Chapters 10 and 11), we will return to the topic of crossing over in Chapter 11 to examine how

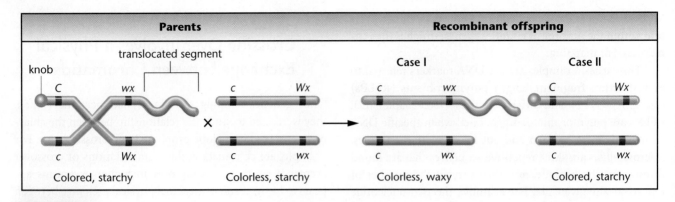

FIGURE 5–15   The phenotypes and chromosome compositions of parents and recombinant offspring in Creighton and McClintock's experiment in maize. The knob and translocated segment served as cytological markers, which established that crossing over involves an actual exchange of chromosome arms.

breakage and reunion occur between the strands of DNA making up chromatids. This discussion will provide a better understanding of genetic recombination.

5.9

Exchanges Also Occur between Sister Chromatids

Considering that crossing over occurs between synapsed homologs in meiosis, we might ask whether such a physical exchange occurs between sister chromatids that are aligned together during mitosis. Each individual chromosome in prophase and metaphase of mitosis consists of two identical sister chromatids, joined at a common centromere. A number of experiments have demonstrated that reciprocal exchanges similar to crossing over do occur between sister chromatids. While these **sister chromatid exchanges** (**SCEs**) do not produce new allelic combinations, evidence is accumulating that attaches significance to these events.

Identification and study of SCEs are facilitated by several unique staining techniques. In one approach, cells are allowed to replicate for two generations in the presence of the thymidine analog bromodeoxyuridine (BrdU). Following two rounds of replication, each pair of sister chromatids has one member with one strand of DNA "labeled" with BrdU, and the other member with both strands labeled with BrdU. Using a differential stain, chromatids with the analog in both strands stain *less brightly* than chromatids with BrdU in only one strand. As a result, any SCEs are readily detectable. In Figure 5–16, numerous instances of SCE events are clearly evident. Because of their patterns of alternating patches, these sister chromatids are sometimes referred to as **harlequin chromosomes**.

The significance of SCEs is still uncertain, but several observations have led to great interest in this phenomenon.

We know, for example, that agents that induce chromosome damage (e.g., viruses, X rays, ultraviolet light, and certain chemical mutagens) also increase the frequency of SCEs. Further, the frequency of SCEs is elevated in **Bloom syndrome**, a human disorder caused by a mutation in the *BLM* gene on chromosome 15. This rare, recessively inherited disease is characterized by prenatal and postnatal retardation of growth, a great sensitivity of the facial skin to

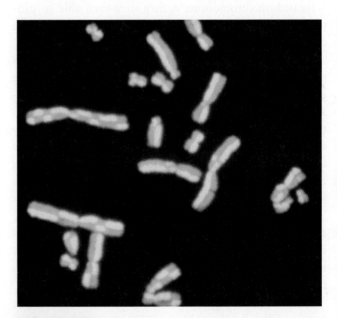

FIGURE 5–16   Demonstration of sister chromatid exchanges (SCEs) in mitotic chromosomes. Sometimes called harlequin chromosomes because of the alternating patterns they exhibit, sister chromatids containing the thymidine analog BrdU are seen to fluoresce *less* brightly where they contain the analog in both DNA strands than when they contain the analog in only one strand. These chromosomes were stained with 33258-Hoechst reagent and acridine orange and then viewed under fluorescence microscopy.

the sun, immune deficiency, a predisposition to malignant and benign tumors, and abnormal behavior patterns. The chromosomes from cultured leukocytes, bone marrow cells, and fibroblasts derived from homozygotes are very fragile and unstable when compared with those derived from homozygous and heterozygous normal individuals. Increased breaks and rearrangements between nonhomologous chromosomes are observed in addition to excessive amounts of sister chromatid exchanges. Work by James German and colleagues suggests that the *BLM* gene encodes an enzyme called **DNA helicase**, which is best known for its role in DNA replication (see Chapter 11).

The mechanisms of exchange between nonhomologous chromosomes and between sister chromatids may prove to share common features because the frequency of both events increases substantially in individuals with certain genetic disorders. These findings suggest that further study of sister chromatid exchange may contribute to the understanding of recombination mechanisms and to the relative stability of normal and genetically abnormal chromosomes. We shall encounter still another demonstration of SCEs (Chapter 11) when we consider replication of DNA (see Figure 11–5).

Linkage and Mapping Studies Can Be Performed in Haploid Organisms

We now turn to yet another extension of transmission genetics: linkage analysis and chromosome mapping in *haploid* eukaryotes. As we shall see, even though analysis of the location of genes relative to one another in the genome of haploid organisms may seem a bit more complex than it is in diploid organisms, the underlying principles are the same. In fact, many basic principles of inheritance were established through the study of haploid fungi.

Many of the single-celled eukaryotes are haploid during the vegetative stages of their life cycle. The alga *Chlamydomonas* and the mold *Neurospora* demonstrate this genetic condition. These organisms do form reproductive cells that fuse during fertilization, forming a diploid zygote; however, the zygote soon undergoes meiosis and reestablishes haploidy. The haploid meiotic products are the progenitors of the subsequent members of the vegetative phase of the life cycle. Figure 5–17 illustrates this type of cycle in the green

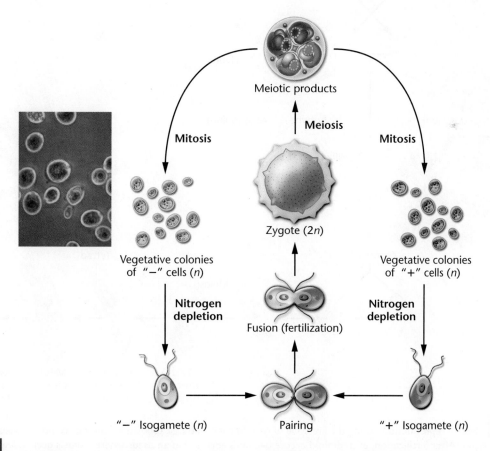

Meiotic products

Mitosis   Meiosis   Mitosis

Zygote (2*n*)

Vegetative colonies   Vegetative colonies
of "−" cells (*n*)   of "+" cells (*n*)

Nitrogen   Nitrogen
depletion   depletion

Fusion (fertilization)

"−" Isogamete (*n*)   Pairing   "+" Isogamete (*n*)

FIGURE 5–17   The life cycle of *Chlamydomonas*. The diploid zygote (in the center) undergoes meiosis, producing "+" or "−" haploid cells that undergo mitosis, yielding vegetative colonies. Unfavorable conditions stimulate them to form isogametes, which fuse in fertilization, producing a zygote that repeats the cycle. Vegetative colonies are illustrated photographically.

alga *Chlamydomonas*. Even though the haploid cells that fuse during fertilization look identical, and are thus called **isogametes**, a chemical identity that distinguishes two distinct types exists on their surface. As a result, all strains are either "+" or "−," and fertilization occurs only between unlike cells.

To perform genetic experiments with haploid organisms, researchers isolate genetic strains of different genotypes and cross them with one another. Following fertilization and meiosis, the meiotic products remain close together and can be analyzed. Such is the case in *Chlamydomonas* as well as in the fungus *Neurospora*, which we shall use as an example in the ensuing discussion. Following fertilization in *Neurospora* (Figure 5–18), meiosis occurs in a saclike structure called the **ascus** (pl. asci), within which the initial set of haploid products, called a **tetrad**, are retained. *The term tetrad has a different meaning here than when it is used to describe the four-stranded chromosome configuration characteristic of meiotic prophase I in diploids.*

Following meiosis in *Neurospora*, each cell in the ascus divides mitotically, producing eight haploid **ascospores**. These can be dissected and examined morphologically or tested to determine their genotypes and phenotypes. Because the arrangement of the eight cells reflects the *sequence* of their formation following meiosis, the tetrad is "ordered" and we can do *ordered tetrad analysis*. This process is critical to our subsequent discussion.

Gene-to-Centromere Mapping

When the ascospore pattern for a single pair of alleles $(a/+)$ is analyzed in *Neurospora*, as diagrammed in Figure 5–19, the data can be used to calculate the map distance between that gene locus and the centromere. This process is sometimes referred to as **mapping the centromere**. It is accomplished by experimentally determining the frequency of recombination using tetrad data. Note in Figure 5–19 that once the four meiotic products of the tetrad are formed, a mitotic division occurs, resulting in eight ordered products (ascospores).

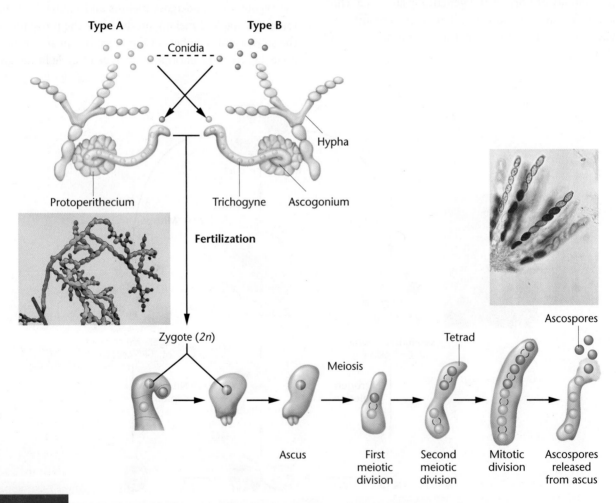

FIGURE 5–18   Sexual reproduction during the life cycle of *Neurospora* is initiated following fusion of conidia (asexual spores) of opposite mating types. After fertilization, each diploid zygote becomes enclosed in an ascus where meiosis occurs, leading to four haploid cells, two of each mating type. A mitotic division then occurs, and the eight haploid ascospores are later released. Upon germination, the cycle may be repeated. The photographs show the vegetative stage of the organism and several asci that may form in a single structure, even though we have illustrated the events occurring in only one ascus.

Condition	Four-strand stage	Chromosomes following meiosis	Chromosomes following mitotic division	Ascospores in ascus
(a) No crossover	*a* *a* + +	*a* *a* + +	*a* *a* *a* *a* + + + +	*a* *a* *a* *a* + + + +
		First-division segregation		
(b) One form of crossover in four-strand stage	*a* *a* + +	*a* + *a* +	*a* *a* + + *a* *a* + +	*a* *a* + + *a* *a* + +
		Second-division segregation		
(c) An alternate crossover in four-strand stage	*a* *a* + +	+ *a* *a* +	+ + *a* *a* *a* *a* + +	+ + *a* *a* *a* *a* + +
		Second-division segregation		

FIGURE 5–19 Three ways in which different ascospore patterns can be generated in *Neurospora*. Analysis of these patterns can serve as the basis of gene-to-centromere mapping. The photograph shows a variety of ascospore arrangements within *Neurospora* asci.

If no crossover event occurs between the gene under study and the centromere, the pattern of ascospores (contained within an ascus) appears as shown in Figure 5–19(a) (*aaaa++++*).*

This pattern represents **first-division segregation**, because the two alleles are separated during the first meiotic division. However, crossover events will alter the pattern, as shown in Figure 5–19(b) (*aa++aa++*) and 5–19(c)

(*++aaaa++*). Two other recombinant patterns also occur, depending on the chromatid orientation during the second meiotic division: (*++aa++aa*) and (*aa++++aa*). All four patterns, resulting from a crossover event between the *a* gene and the centromere, reflect **second-division segregation**, because the two alleles are not separated until the second meiotic division. Since the mitotic division simply replicates the patterns (increasing the 4 ascospores to 8), ordered tetrad data are usually condensed to reflect the genotypes of the four ascospore pairs, and six unique combinations are possible.

The pattern (++++aaaa*) can also be formed, but it is indistinguishable from (*aaaa++++*).

First-Division Segregation			
(1)	a	a	+	+
(2)	+	+	a	a

Second-Division Segregation			
(3)	a	+	a	+
(4)	+	a	+	a
(5)	+	a	a	+
(6)	a	+	+	a

To calculate the distance between the gene and the centromere, data must be tabulated from a large number of asci resulting from a controlled cross. We then use these data to calculate the distance (d):

$$d = \frac{1/2 \text{ (second division segregant asci)}}{\text{total asci scored}} \times 100$$

The distance (d) reflects the percentage of recombination and is only half the number of second-division segregant asci. This is because crossing over occurs in only two of the four chromatids during meiosis.

To illustrate, we use a for albino and $+$ for wild type in *Neurospora*. In crosses between the two genetic types, suppose the following data are observed:

65 first-division segregants

70 second-division segregants

Thus, the distance between a and the centromere is

$$d = \frac{(1/2)(70)}{135} = 0.259 \times 100 = 25.9$$

or about 26 mu.

As the distance increases to 50 units, all asci should theoretically reflect second-division segregation. However, numerous factors prevent it in actuality. As in diploid organisms, accuracy is greatest when the gene and the centromere are relatively close together.

As we will discuss in the next section, we can also analyze haploid organisms in order to distinguish between linkage and independent assortment of two genes. Once linkage is established, mapping distances between gene loci are calculated. As a result, detailed maps of organisms such as *Saccharomyces, Neurospora,* and *Chlamydomonas* are now available.

Ordered versus Unordered Tetrad Analysis

In our previous discussion, we assumed that the genotype of each ascospore and its position in the tetrad can be determined. To perform such an **ordered tetrad analysis**, individual asci must be dissected, and each ascospore must be

tracked as it germinates. This is a tedious process, but it is essential for two types of analysis:

1. To distinguish between first-division segregation and second-division segregation of alleles in meiosis.

2. To determine whether or not recombination events are reciprocal. Such information is essential for "mapping the centromere," as we have just discussed. Thus, ordered tetrad analysis must be performed in order to map the distance between a gene and the centromere.

Ordered tetrad analysis has revealed that recombination events are not always reciprocal, particularly when the genes under study are closely linked. This observation has led to the investigation of the phenomenon called *gene conversion*. Because its discussion requires a background in DNA structure and analysis, we will return to this topic in Chapter 11.

Much less tedious than ordered tetrad analysis is to isolate individual asci, allow them to mature, and then determine the genotypes of each ascospore in no particular order. This approach is referred to as **unordered tetrad analysis**. As we shall see in the next section, such an analysis can be used to discover whether or not two genes are linked on the same chromosome and, if so, to determine the map distance between them.

Linkage and Mapping

To show how analysis of genetic data derived from haploid organisms can be used to distinguish between linkage and independent assortment of two genes, and then allows mapping distances to be calculated between gene loci, we shall consider tetrad analysis in the alga *Chlamydomonas*. Except that the four meiotic products are not ordered and *do not* undergo a mitotic division following the completion of meiosis, the general principles discussed for *Neurospora* also apply to *Chlamydomonas*.

To compare independent assortment and linkage, imagine two mutant alleles, a and b, representing two distinct loci in *Chlamydomonas*. Suppose that 100 tetrads derived from the cross $ab \times ++$ yield the tetrad data shown in Table 5.1. As you can see, all tetrads produce one of three patterns. For example, all tetrads in category I produce two $++$ cells and two ab cells and are designated as **parental ditypes (P)**. Category II tetrads produce two $a+$ cells and two $+b$ cells and are called **nonparental ditypes (NP)**. Category III tetrads produce four cells that each have one of the four possible genotypes and are thus termed **tetratypes (T)**.

These data support the hypothesis that the genes represented by the a and b alleles are located on separate chromosomes. To understand why, you should refer to Figure 5–20. In parts (a) and (b) of that figure, the origin of parental (P) and nonparental (NP) ditypes is demonstrated for two unlinked genes. According to the Mendelian principle of

TABLE 5.1

Tetrad Analysis in *Chlamydomonas*

Category	I	II	III
Tetrad type	Parental (P)	Nonparental (NP)	Tetratypes (T)
Genotypes present	+ + + + a b a b	a + a + + b + b	+ + a + + b a b
Number of tetrads	43	43	14

independent assortment of unlinked genes, approximately equal proportions of these tetrad types are predicted. Thus, when the parental ditypes are equal to the nonparental ditypes, the two genes are not linked. The data in Table 5.1

confirm this prediction. Because independent assortment has occurred, it can be concluded that the two genes are located on separate chromosomes.

The origin of category III, the tetratypes, is diagrammed in Figure 5–20(c,d). The genotypes of tetrads in this category can be generated in two possible ways. Both involve a crossover event between one of the genes and the centromere. In Figure 5–20(c), the exchange involves one of the two chromosomes and occurs between gene *a* and the centromere; in Figure 5–20(d), the other chromosome is involved, and the exchange occurs between gene *b* and the centromere.

Production of tetratype tetrads does not alter the final ratio of the four genotypes present in all meiotic products. If the genotypes from 100 tetrads (which yield 400 cells) are computed, 100 of each genotype are found. This 1:1:1:1 ratio is predicted according to independent assortment.

Now consider the case where the genes *a* and *b* are linked (Figure 5–21). The same categories of tetrads will be

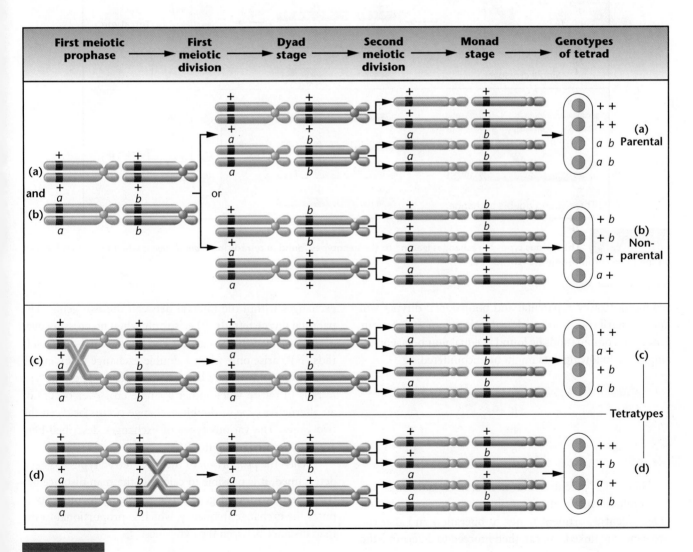

FIGURE 5–20   The origin of various genotypes found in tetrads in *Chlamydomonas* when two genes located on separate chromosomes are considered.

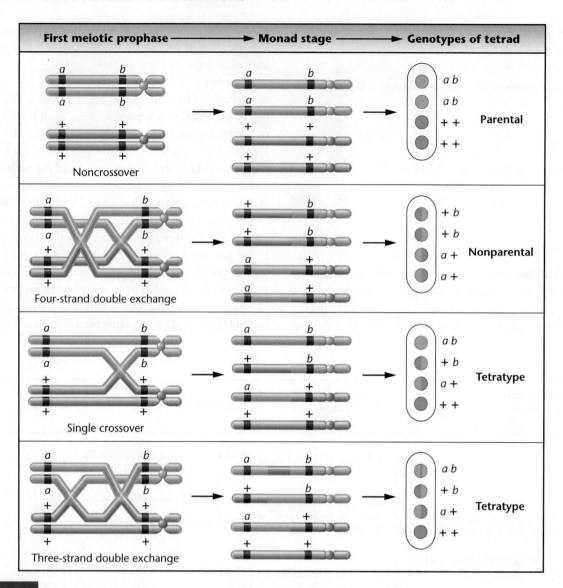

FIGURE 5-21   The various types of exchanges leading to the genotypes found in tetrads in *Chlamydomonas* when two genes located on the same chromosome are considered.

produced. However, parental and nonparental ditypes will not necessarily occur in equal proportions; nor will the four genotypic combinations be found in equal numbers. For example, the following data might be encountered:

Category I	Category II	Category III
P	NP	T
64	6	30

Since the parental and nonparental categories are not produced in equal proportions, we can conclude that independent assortment is not in operation and that the two genes are linked. We can then proceed to determine the map distance between them.

In the analysis of these data, we are concerned with the determination of which tetrad types represent genetic exchanges within the interval between the two genes. The parental ditype tetrads (P) arise only when no crossing over occurs between the two genes. The nonparental ditype tetrads (NP) arise only when a double exchange involving all four chromatids occurs between two genes. The tetratype tetrads (T) arise when either a single crossover occurs or an alternative type of double exchange occurs between the two genes. The various types of exchanges described here are diagrammed in Figure 5–21.

When the proportion of the three tetrad types has been determined, it is possible to calculate the map distance between the two linked genes. The following formula computes the exchange frequency, which is proportional to the map distance between the two genes:

$$\text{exchange frequency (\%)} = \frac{\text{NP} + 1/2(\text{T})}{\text{total number of tetrads}} \times 100$$

In this formula, NP represents the nonparental tetrads; all meiotic products represent an exchange. The tetratype tetrads are represented by T; assuming only single exchanges, half of the meiotic products represent exchanges. The sum of the scored tetrads that fall into these categories is then divided by the total number of tetrads examined (P + NP + T). Multiplying that result by 100 converts it to a percentage, which is directly equivalent to the map distance between the genes.

In our example, the calculation reveals that genes a and b are separated by 21 mu:

$$\frac{6 + 1/2(30)}{100} = \frac{6 + 15}{100} = \frac{21}{100} = 0.21 \times 100 = 21\%$$

Although we have considered linkage analysis and mapping of only two genes at a time, such studies often involve three or more genes. In these cases, both gene sequence and map distances can be determined.

5.11
Did Mendel Encounter Linkage?

We conclude this chapter by examining a modern-day interpretation of the experiments that form the cornerstone of transmission genetics—Mendel's crosses with garden peas.

Some observers believe that Mendel had extremely good fortune in his classic experiments with the garden pea. In their view, he did not encounter any linkage relationships between the seven mutant characters in his crosses. Had Mendel obtained highly variable data characteristic of linkage and crossing over, these observers say, he might not have succeeded in recognizing the basic patterns of inheritance and interpreting them correctly.

The article by Stig Blixt, reprinted in its entirety in the box that follows, demonstrates the inadequacy of this hypothesis. As we shall see, some of Mendel's genes were indeed linked. We shall leave it to Stig Blixt to enlighten you as to why Mendel did not detect linkage.

Why Didn't Gregor Mendel Find Linkage?

It is quite often said that Mendel was very fortunate not to run into the complication of linkage during his experiments. He used seven genes, and the pea has only seven chromosomes. Some have said that had he taken just one more, he would have had problems. This, however, is a gross oversimplification. The actual situation, most probably, is that Mendel worked with three genes in chromosome 4, two genes in chromosome 1, and one gene in each of chromosomes 5 and 7. (See Table 1.)

It seems at first glance that, out of the 21 dihybrid combinations Mendel theoretically could have studied, no less than four (that is, a–i, v–fa, v–le, fa–le) ought to have resulted in linkages. As found, however, in hundreds of crosses and shown by the genetic map of the pea, a and i in chromosome 1 are so distantly located on the chromosome that no linkage is normally detected. The same is true for v and le on the one hand, and fa on the other, in chromosome 4. This leaves v–le, which ought to have shown linkage.

Mendel, however, does not seem to have published this particular combination and thus, presumably, never made the appropriate cross to obtain both genes segregating simultaneously. It is therefore not so astonishing that Mendel did not run into the complication of linkage, although he did not avoid it by choosing one gene from each chromosome.

Stig Blixt

Weibullsholm Plant Breeding Institute, Landskrona, Sweden, and Centro de Energia Nuclear naAgricultura, Piracicaba, SP, Brazil.

Source: Reprinted by permission from Macmillan Publishers Ltd: Nature, "Molecular Structure of Nucleic Acids: A Structure for Deoxyribose Nucleic Acid" by F.H.C. Crick and J.D. Watson, *Nature*, Vol. 171, No. 4356, pp. 737–38. Copyright 1953. www.nature.com

TABLE 1

Relationship between Modern Genetic Terminology and Character Pairs Used by Mendel

Character Pair Used by Mendel	Alleles in Modern Terminology	Located in Chromosome
Seed color, yellow–green	*I–i*	1
Seed coat and flowers, colored–white	*A–a*	1
Mature pods, smooth expanded–wrinkled indented	*V–v*	4
Inflorescences, from leaf axis–umbellate in top of plant	*Fa–fa*	4
Plant height, 0.5–1 m	*Le–le*	4
Unripe pods, green yellow	*Gp–gp*	5
Mature seeds, smooth wrinkled	*R–r*	7

Human Chromosome Maps on the Internet

Study Area: Exploring Genomics

In this chapter we discussed how recombination data can be analyzed to develop chromosome maps based on linkage. Although linkage analysis and chromosome mapping continue to be important approaches in genetics, chromosome maps are increasingly being developed for many species using genomics techniques to sequence entire chromosomes. As a result of the Human Genome Project, maps of human chromosomes are now freely available on the Internet. With the click of a mouse you can have immediate access to an incredible wealth of information. In this exercise we will explore the **National Center for Biotechnology Information (NCBI) Genes and Disease** Web site to learn more about human chromosome maps.

■ NCBI Genes and Disease

The NCBI Web site is an outstanding resource for genome data. Here we explore the Genes and Disease site, which presents human chromosome maps that show the locations of specific disease genes.

1. Access the Genes and Disease site at http://www.ncbi.nlm.nih.gov/books/ NBK22183/

2. Under contents, click on "chromosome maps" to see a page with an image of a karyotype of human chromosomes. Click on a chromosome in the chromosome map image, scroll down the page to view a chromosome or click on a chromosome listed on the right side of the page. For example, click on chromosome 7. Notice that above the chromosome image displays the number of genes on the chromosome and the number of base pairs the chromosome contains.

3. Look again at chromosome 7. At first you might think there are only five disease genes on this chromosome because the initial view shows only selected disease genes. However, if you click the "MapViewer" link for the chromosome (just above the drawing), you will see detailed information about the chromosome, including a complete "Master Map" of the genes it contains, including the symbols used in naming genes:

   Gene Symbols: Clicking on the gene symbols takes you to the **NCBI Entrez Gene database**, a searchable tool for information on genes in the NCBI database. Links: The items in the "Links"

column provide access to OMIM (Online Mendelian Inheritance in Man, discussed in the "Exploring Genomics" feature for Chapter 3) data for a particular gene, as well as to protein information (*pn*) and lists of homologous genes (*hm*; these are other genes that have similar sequences).

4. Click on the links in MapViewer to learn more about a gene of interest.

5. Scan the chromosome maps in MapViewer until you see one of the genes listed as a "hypothetical gene or protein."

   a. What does it mean if a gene or protein is referred to as hypothetical?

   b. What information do you think genome scientists use to assign a gene locus for a gene encoding a hypothetical protein?

Visit the **NCBI Map Viewer** homepage (www.ncbi.nlm.nih.gov/mapview) for an excellent database containing chromosome maps for a wide variety of different organisms. Search this database to learn more about chromosome maps for an organism you are interested in.

CASE STUDY | Links to autism

As parents of an autistic child, entering a research study seemed to be a way of not only educating themselves about their son's condition, but also of furthering research into this complex, behaviorally defined disorder. Researchers told the couple that there is a strong genetic influence for autism since the concordance rate in identical twins is about 75 percent and only about 5 percent in fraternal twins. In addition, researchers have identified interactions among at least ten candidate genes distributed among chromosomes 3, 7, 15, and 17. Generally unaware of the principles of basic genetics, the couple asked a number of interesting questions. How would you respond to them?

1. What is a "candidate" gene?

2. Since several candidate genes must be on the same chromosome, will they always be transmitted as a block of harmful genetic information to future offspring?

3. With such an apparently complex genetic condition, what is the likelihood that our next child will also be autistic?

4. Is prenatal diagnosis possible during future pregnancies?

Summary Points

 For activities, animations, and review quizzes, go to the study area at www.masteringgenetics.com

1. Genes located on the same chromosome are said to be linked. Alleles of linked genes located close together on the same homolog are usually transmitted together during gamete formation.

2. Crossover frequency between linked genes during gamete formation is proportional to the distance between genes, providing the experimental basis for mapping the location of genes relative to one another along the chromosome.

3. Determining the sequence of genes in a three-point mapping experiment requires the analysis of the double-crossover gametes, as reflected in the phenotype of the offspring receiving those gametes.

4. Interference describes the extent to which a crossover in one region of a chromosome influences the occurrence of a crossover in an adjacent region of the chromosome and is quantified by calculating the coefficient of coincidence (C).

5. Human linkage studies, initially relying on pedigree and lod score analysis, and subsequently on somatic cell hybridization techniques, are now enhanced by the use of newly discovered molecular DNA markers.

6. Linkage analysis and chromosome mapping are possible in haploid eukaryotes, relying on the direct analysis of meiotic products such as ascospores in *Neurospora*. Such studies also provide the basis of distinguishing between linkage and independent assortment.

7. Cytological investigations of both maize and *Drosophila* reveal that crossing over involves a physical exchange of segments between nonsister chromatids.

8. Recombination events are known to occur between sister chromatids in mitosis and are referred to as sister chromatid exchanges (SCEs).

INSIGHTS AND SOLUTIONS

1. In a series of two-point mapping crosses involving three genes linked on chromosome III in *Drosophila*, the following distances were calculated:

$$cd-sr \; 13 \; mu$$
$$cd-ro \; 16 \; mu$$

(a) Why can't we determine the sequence and construct a map of these three genes?

(b) What mapping data will resolve the issue?

(c) Can we tell which of the sequences shown here is correct?

$$ro \xrightarrow{\;16\;} cd \xrightarrow{\;13\;} sr$$

or

$$sr \xrightarrow{\;13\;} cd \xrightarrow{\;16\;} ro$$

Solution:

(a) It is impossible to do so because there are two possibilities based on these limited data:

Case 1: $cd \xrightarrow{\;13\;} sr \xrightarrow{\;3\;} ro$

or

Case 2: $ro \xrightarrow{\;16\;} cd \xrightarrow{\;13\;} sr$

(b) The map distance is determined by crossing over between *ro* and *sr*. If case 1 is correct, it should be 3 mu, and if case 2 is correct, it should be 29 mu. In fact, this distance is 29 mu, demonstrating that case 2 is correct.

(c) No; based on the mapping data, they are equivalent.

2. In *Drosophila*, *Lyra (Ly)* and *Stubble (Sb)* are dominant mutations located at loci 40 and 58, respectively, on chromosome III. A recessive mutation with bright red eyes was discovered and shown also to be on chromosome III. A map

is obtained by crossing a female who is heterozygous for all three mutations to a male homozygous for the bright red mutation (which we refer to here as *br*). The data in the table are generated. Determine the location of the *br* mutation on chromosome III. By referring to Figure 5–14, predict what mutation has been discovered. How could you be sure?

Phenotype	Number
(1) *Ly Sb br*	404
(2) + + +	422
(3) *Ly* + +	18
(4) + *Sb br*	16
(5) *Ly* + *br*	75
(6) + *Sb* +	59
(7) *Ly Sb* +	4
(8) + + *br*	2
Total	= 1000

Solution: First determine the distribution of the alleles between the homologs of the heterozygous crossover parent (the female in this case). To do this, locate the most frequent reciprocal phenotypes, which arise from the noncrossover gametes. These are phenotypes 1 and 2. Each phenotype represents the alleles on one of the homologs. Therefore, the distribution of alleles is

Second, determine the correct *sequence* of the three loci along the chromosome. This is done by determining which sequence yields the observed double-crossover phenotypes that are the least frequent reciprocal phenotypes (7 and 8). If the sequence is correct as written, then a double crossover, depicted here,

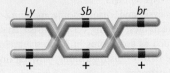

would yield *Ly + br* and *+ Sb +* as phenotypes. Inspection shows that these categories (5 and 6) are actually single crossovers, not double crossovers. Therefore, the sequence, as written, is incorrect. There are only two other possible sequences. Either the *Ly* gene (Case A below) or the *br* gene (Case B below) is in the middle between the other two genes.

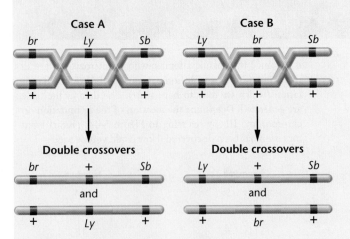

Comparison with the actual data shows that case B is correct. The double-crossover gametes 7 and 8 yield flies that express *Ly* and *Sb* but not *br*, or express *br* but not *Ly* and *Sb*. Therefore, the correct *arrangement* and *sequence* are as follows:

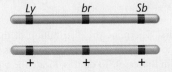

Once the sequence is found, determine the location of *br* relative to *Ly* and *Sb*. A single crossover between *Ly* and *br*, as shown here,

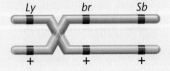

yields flies that are *Ly + +* and *+ br Sb* (phenotypes 3 and 4). Therefore, the distance between the *Ly* and *br* loci is equal to

$$\frac{18 + 16 + 4 + 2}{1000} = \frac{40}{1000} = 0.04 = 4 \text{ mu}$$

Remember that, because we need to know the frequency of all crossovers between *Ly* and *br*, we must add in the double crossovers, since they represent two single crossovers occurring simultaneously. Similarly, the distance between the *br* and *Sb* loci is derived mainly from single crossovers between them.

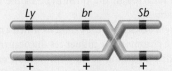

This event yields *Ly br +* and *+ + Sb* phenotypes (phenotypes 5 and 6). Therefore, the distance equals

$$\frac{75 + 59 + 4 + 2}{1000} = \frac{140}{1000} = 0.14 = 14 \text{ mu}$$

The final map shows that *br* is located at locus 44, since *Lyra* and *Stubble* are known:

Inspection of Figure 5–13 reveals that the mutation *scarlet*, which produces bright red eyes, is known to sit at locus 44, so it is reasonable to hypothesize that the bright red eye mutation is an allele of *scarlet*. To test this hypothesis, we could cross females of our bright red mutant with known *scarlet* males. If the two mutations are alleles, no complementation will occur, and all progeny will reveal a bright red mutant eye phenotype. If complementation occurs, all progeny will show normal brick-red (wild-type) eyes, since the bright red mutation and *scarlet* are at different loci. (They are probably very close together.) In such a case, all progeny will be heterozygous at both the bright eye and the *scarlet* loci and will not express either mutation because they are both recessive. This cross represents what is called an *allelism test*.

3. In rabbits, black color (*B*) is dominant to brown (*b*), while full color (*C*) is dominant to chinchilla (*c^ch*). The genes controlling these traits are linked. Rabbits that are heterozygous for both traits and express black, full color were crossed with rabbits that express brown, chinchilla, with the following results:

   31 brown, chinchilla

   34 black, full color

   16 brown, full color

   19 black, chinchilla

Determine the arrangement of alleles in the heterozygous parents and the map distance between the two genes.

Solution: This is a two-point mapping problem. The two most prevalent reciprocal phenotypes are the noncrossovers, and the less frequent reciprocal phenotypes arise from a single crossover. The distribution of alleles is derived from the noncrossover phenotypes because they enter gametes intact.

The single crossovers give rise to 35/100 offspring (35 percent). Therefore, the distance between the two genes is 35 mu.

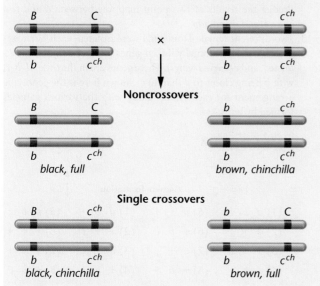

Noncrossovers

black, full   *brown, chinchilla*

Single crossovers

black, chinchilla   *brown, full*

4. In a cross in *Neurospora* where one parent expresses the mutant allele *a* and the other expresses a wild-type phenotype (+), the following data were obtained in the analysis of ascospores:

Ascus Types						
	1	2	3	4	5	6
	+	*a*	*a*	+	*a*	+
	+	*a*	*a*	+	*a*	+
	+	*a*	+	*a*	+	*a*
	+	*a*	+	*a*	+	*a*
	a	+	*a*	+	+	*a*
	a	+	*a*	+	+	*a*
	a	+	+	*a*	*a*	+
	a	+	+	*a*	*a*	+
	39	33	5	4	9	10	Total = 100

(Sequence of ascospores in ascus — leftmost column label)

Calculate the gene-to-centromere distance.

Solution: Ascus types 1 and 2 represent first-division segregation (fds), where no crossing over occurred between the *a* locus and the centromere. All others (3–6) represent second-division segregation (sds). By applying the formula

$$\text{distance} = \frac{1/2 \text{ sds}}{\text{total asci}}$$

we obtain the following result:

$$d = 1/2(5 + 4 + 9 + 10)/100$$
$$= 1/2(28)/100$$
$$= 0.14$$
$$= 14 \text{ mu}$$

Problems and Discussion Questions

 For instructor-assigned tutorials and problems, go to www.masteringgentics.com

HOW DO WE KNOW?

1. In this chapter, we focused on linkage, chromosomal mapping, and many associated phenomena. In the process, we found many opportunities to consider the methods and reasoning by which much of this information was acquired. From the explanations given in the chapter, what answers would you propose to the following fundamental questions?

(a) How was it established experimentally that the frequency of recombination (crossing over) between two genes is related to the distance between them along the chromosome?

(b) How do we know that specific genes are linked on a single chromosome, in contrast to being located on separate chromosomes?

(c) How do we know that crossing over results from a physical exchange between chromatids?

(d) How do we know that sister chromatids undergo recombination during mitosis?

(e) When designed matings cannot be conducted in an organism (for example, in humans), how do we learn that genes are linked, and how do we map them?

2. What is the significance of crossing over (which leads to genetic recombination) to the process of evolution?

3. Describe the cytological observation that suggests that crossing over occurs during the first meiotic prophase.

4. Why does more crossing over occur between two distantly linked genes than between two genes that are very close together on the same chromosome?

5. Explain why a 50 percent recovery of single-crossover products is the upper limit, even when crossing over *always* occurs between two linked genes?

6. Why are double-crossover events expected less frequently than single-crossover events?

7. What is the proposed basis for positive interference?

8. What two essential criteria must be met in order to execute a successful mapping cross?

9. The genes *dumpy* (*dp*), *clot* (*cl*), and *apterous* (*ap*) are linked on chromosome II of *Drosophila*. In a series of two-point mapping crosses, the following genetic distances were determined. What is the sequence of the three genes?

dp–ap	42
dp–cl	3
ap–cl	39

10. Colored aleurone in the kernels of corn is due to the dominant allele *R*. The recessive allele *r*, when homozygous, produces colorless aleurone. The plant color (not the kernel color) is controlled by another gene with two alleles, *Y* and *y*. The dominant *Y* allele results in green color, whereas the homozygous presence of the recessive *y* allele causes the plant to appear yellow. In a testcross between a plant of unknown genotype and phenotype and a plant that is homozygous recessive for both traits, the following progeny were obtained:

Colored, green	88
Colored, yellow	12
Colorless, green	8
Colorless, yellow	92

Explain how these results were obtained by determining the exact genotype and phenotype of the unknown plant, including the precise arrangement of the alleles on the homologs.

11. In the cross shown here, involving two linked genes, *ebony* (*e*) and *claret* (*ca*), in *Drosophila*, where crossing over does not occur in males, offspring were produced in a 2 + : 1 *ca* : 1 *e* phenotypic ratio:

♀		♂
$\dfrac{e \quad ca^+}{e^+ \quad ca}$	×	$\dfrac{e \quad ca^+}{e^+ \quad ca}$

These genes are 30 units apart on chromosome III. What did crossing over in the female contribute to these phenotypes?

12. In a series of two-point mapping crosses involving five genes located on chromosome II in *Drosophila*, the following recombinant (single-crossover) frequencies were observed:

pr–adp	29%
pr–vg	13
pr–c	21
pr–b	6
adp–b	35
adp–c	8
adp–vg	16
vg–b	19
vg–c	8
c–b	27

(a) Given that the *adp* gene is near the end of chromosome II (locus 83), construct a map of these genes.

(b) In another set of experiments, a sixth gene, *d*, was tested against *b* and *pr*:

d–b	17%
d–pr	23%

Predict the results of two-point mapping between *d* and *c*, *d* and *vg*, and *d* and *adp*.

13. Two different female *Drosophila* were isolated, each heterozygous for the autosomally linked genes *b* (*black body*), *d* (*dachs tarsus*), and *c* (*curved wings*). These genes are in the order *d–b–c*, with *b* being closer to *d* than to *c*. Shown here is the genotypic arrangement for each female along with the various gametes formed by both:

Female A		Female B	
$\dfrac{d\ b\ +}{+\ +\ c}$		$\dfrac{d\ +\ +}{+\ b\ c}$	
↓ Gamete formation		↓	
(1) *d b c*	(5) *d + +*	(1) *d b +*	(5) *d b c*
(2) + + +	(6) + *b c*	(2) + + *c*	(6) + + +
(3) + + *c*	(7) *d + c*	(3) *d + c*	(7) *d + +*
(4) *d b +*	(8) + *b +*	(4) + *b +*	(8) + *b C*

Identify which categories are noncrossovers (NCOs), single crossovers (SCOs), and double crossovers (DCOs) in each case. Then, indicate the relative frequency in which each will be produced.

14. In *Drosophila*, a cross was made between females—all expressing the three X-linked recessive traits *scute* bristles (*sc*), *sable* body (*s*), and *vermilion* eyes (*v*)—and wild-type males. In the F_1, all females were wild type, while all males expressed all three mutant traits. The cross was carried to the F_2 generation, and 1000 offspring were counted, with the results shown in the following table.

Phenotype			Offspring
sc	*s*	*v*	314
+	+	+	280
+	*s*	*v*	150
sc	+	+	156
sc	+	*v*	46
+	*s*	+	30
sc	*s*	+	10
+	+	*v*	14

No determination of sex was made in the data.

(a) Using proper nomenclature, determine the genotypes of the P_1 and F_1 parents.
(b) Determine the sequence of the three genes and the map distances between them.
(c) Are there more or fewer double crossovers than expected?
(d) Calculate the coefficient of coincidence. Does it represent positive or negative interference?

15. Another cross in *Drosophila* involved the recessive, X-linked genes *yellow* (*y*), *white* (*w*), and *cut* (*ct*). A yellow-bodied, white-eyed female with normal wings was crossed to a male whose eyes and body were normal but whose wings were cut. The F_1 females were wild type for all three traits, while the F_1 males expressed the yellow-body and white-eye traits. The cross was carried to an F_2 progeny, and only male offspring were tallied. On the basis of the data shown here, a genetic map was constructed.

Phenotype			Male Offspring
y	+	ct	9
+	w	+	6
y	w	ct	90
+	+	+	95
+	+	ct	424
y	w	+	376
y	+	+	0
+	w	ct	0

(a) Diagram the genotypes of the F_1 parents.
(b) Construct a map, assuming that *white* is at locus 1.5 on the X chromosome.
(c) Were any double-crossover offspring expected?
(d) Could the F_2 female offspring be used to construct the map? Why or why not?

16. In *Drosophila*, *Dichaete* (*D*) is a mutation on chromosome III with a dominant effect on wing shape. It is lethal when homozygous. The genes *ebony* body (*e*) and *pink* eye (*p*) are recessive mutations on chromosome III. Flies from a *Dichaete* stock were crossed to homozygous ebony, pink flies, and the F_1 progeny, with a Dichaete phenotype, were backcrossed to the ebony, pink homozygotes. Using the results of this backcross shown in the table,

(a) Diagram this cross, showing the genotypes of the parents and offspring of both crosses.
(b) What is the sequence and interlocus distance between these three genes?

Phenotype	Number
Dichaete	401
ebony, pink	389
Dichaete, ebony	84
pink	96
Dichaete, pink	2
ebony	3
Dichaete, ebony, pink	12
wild type	13

17. *Drosophila* females homozygous for the third chromosomal genes *pink* and *ebony* (the same genes from Problem 16) were crossed with males homozygous for the second chromosomal gene *dumpy*. Because these genes are recessive, all offspring were wild type (normal). F_1 females were testcrossed to triply recessive males. If we assume that the two linked genes, *pink* and *ebony*, are 20 mu apart, predict the results of this cross. If the reciprocal cross were made (F_1 males—where no crossing over occurs—with triply recessive females), how would the results vary, if at all?

18. In *Drosophila*, two mutations, *Stubble* (*Sb*) and *curled* (*cu*), are linked on chromosome III. *Stubble* is a dominant gene that is lethal in a homozygous state, and *curled* is a recessive gene. If a female of the genotype

$$\frac{Sb \quad\quad cu}{+ \quad\quad +}$$

is to be mated to detect recombinants among her offspring, what male genotype would you choose as a mate?

19. If the cross described in Problem 18 were made, and if *Sb* and *cu* are 8.2 map units apart on chromosome III, and if 1000 offspring were recovered, what would be the outcome of the cross, assuming that equal numbers of males and females were observed.

20. Are mitotic recombinations and sister chromatid exchanges effective in producing genetic variability in an individual? in the offspring of individuals?

21. What possible conclusions can be drawn from the observations that in male *Drosophila*, no crossing over occurs, and that during meiosis, synaptonemal complexes are not seen in males but are observed in females where crossing over occurs?

22. An organism of the genotype *AaBbCc* was testcrossed to a triply recessive organism (*aabbcc*). The genotypes of the progeny are presented in the following table.

20	*AaBbCc*	20	*AaBbcc*
20	*aabbCc*	20	*aabbcc*
5	*AabbCc*	5	*Aabbcc*
5	*aaBbCc*	5	*aaBbcc*

(a) If these three genes were all assorting independently, how many genotypic and phenotypic classes would result in the offspring, and in what proportion, assuming simple dominance and recessiveness in each gene pair?
(b) Answer part (a) again, assuming the three genes are so tightly linked on a single chromosome that no crossover gametes were recovered in the sample of offspring.
(c) What can you conclude from the *actual* data about the location of the three genes in relation to one another?

23. Based on our discussion of the potential inaccuracy of mapping (see Figure 5–12), would you revise your answer to Problem 22? If so, how?

24. Traditional gene mapping has been applied successfully to a variety of organisms including yeast, fungi, maize, and *Drosophila*. However, human gene mapping has only recently shared a similar spotlight. What factors have delayed the application of traditional gene-mapping techniques in humans?

25. DNA markers have greatly enhanced the mapping of genes in humans. What are DNA markers, and what advantage do they confer?

26. In a certain plant, fruit is either red or yellow, and fruit shape is either oval or long. Red and oval are the dominant traits.

Two plants, both heterozygous for these traits, were testcrossed, with the following results.

Phenotype	Progeny	
	Plant A	Plant B
red, long	46	4
yellow, oval	44	6
red, oval	5	43
yellow, long	5	47
	100	100

Determine the location of the genes relative to one another and the genotypes of the two parental plants.

27. Two plants in a cross were each heterozygous for two gene pairs (Ab/aB) whose loci are linked and 25 mu apart. Assuming that crossing over occurs during the formation of both male and female gametes and that the A and B alleles are dominant, determine the phenotypic ratio of their offspring.

28. In a cross in *Neurospora* involving two alleles, B and b, the following tetrad patterns were observed. Calculate the distance between the gene and the centromere.

Tetrad Pattern	Number
BBbb	36
bbBB	44
BbBb	4
bBbB	6
BbbB	3
bBBb	7

29. In *Neurospora*, the cross $a+ \times +b$ yielded only two types of ordered tetrads in approximately equal numbers. What can be concluded?

	Spore Pair			
	1–2	3–4	5–6	7–8
Tetrad Type 1	a +	a +	+ b	+ b
Tetrad Type 2	+ +	+ +	a b	a b

30. Here are two sets of data derived from crosses in *Chlamydomonas* involving three genes represented by the mutant alleles *a*, *b*, and *c*:

Genes	Cross	P	NP	T
1	a and b	36	36	28
2	b and c	79	3	18
3	a and c	?	?	?

Determine as much as you can concerning the arrangement of these three genes relative to one another. Assuming that *a* and *c* are linked and are 38 mu apart and that 100 tetrads are produced, describe the expected results of cross 3.

31. In *Chlamydomonas*, a cross $ab \times ++$ yielded the following unordered tetrad data where *a* and *b* are linked:

(1)	+ + + + a b a b	38	(3)	a + a + + b + b	6	(5)	a b + + + b a +	2
(2)	+ + a b + + a b	5	(4)	a b a + + b + +	17	(6)	a b + b a + + +	3

(a) Identify the tetrads representing parental ditypes (P), nonparental ditypes (NP), and tetratypes (T).
(b) Explain the origin of tetrad (2).
(c) Determine the map distance between *a* and *b*.

32. The following results are ordered tetrad pairs from a cross between strain *cd* and strain $++$:

Tetrad Class						
1	2	3	4	5	6	7
c +	c +	c d	+ d	c +	c d	c +
c +	c d	c d	c +	+ +	+ +	+ d
+ d	+ +	+ +	c +	c d	c d	c d
+ d	+ d	+ +	+ d	+ d	+ +	+ +
1	17	41	1	5	3	1

They are summarized by tetrad classes.

(a) Name the ascus type of each class from 1 to 7 (P, NP, or T).
(b) The data support the conclusion that the *c* and *d* loci are linked. State the evidence in support of this conclusion.
(c) Calculate the gene–centromere distance for each locus.
(d) Calculate the distance between the two linked loci.
(e) Draw a chromosome map, including the centromere, and explain the discrepancy between the distances determined by the two different methods in parts (c) and (d).
(f) Describe the arrangement of crossovers needed to produce the ascus class 6.

33. In a cross in *Chlamydomonas*, $AB \times ab$, 211 unordered asci were recovered:

10	AB	Ab	aB	ab
102	Ab	aB	Ab	aB
99	AB	AB	ab	Ab

(a) Correlate each of the three tetrad types in the problem with their appropriate tetrad designations (names).
(b) Are genes *A* and *B* linked?
(c) If they are linked, determine the map distance between the two genes. If they are unlinked, provide the maximum information you can about why you drew this conclusion.

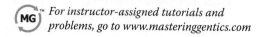

Extra-Spicy Problems

For instructor-assigned tutorials and problems, go to www.masteringgentics.com

34. A number of human–mouse somatic cell hybrid clones were examined for the expression of specific human genes and the presence of human chromosomes. The results are summarized in the following table. Assign each gene to the chromosome on which it is located.

	Hybrid Cell Clone					
	A	B	C	D	E	F
Genes expressed						
ENO1 (enolase-1)	−	+	−	+	+	−
MDH1 (malate dehydrogenase-1)	+	+	−	+	−	+
PEPS (peptidase S)	+	−	+	−	−	−
PGM1 (phosphoglucomutase-1)	−	+	−	+	+	−
Chromosomes (present or absent)						
1	−	+	−	+	+	−
2	+	+	−	+	−	+
3	+	+	−	−	+	−
4	+	−	+	−	−	−
5	−	+	+	+	+	+

35. A female of genotype

$$\frac{a \quad b \quad c}{+ \quad + \quad +}$$

produces 100 meiotic tetrads. Of these, 68 show no crossover events. Of the remaining 32, 20 show a crossover between *a* and *b*, 10 show a crossover between *b* and *c*, and 2 show a double crossover between *a* and *b* and between *b* and *c*. Of the 400 gametes produced, how many of each of the 8 different genotypes will be produced? Assuming the order *a–b–c* and the allele arrangement previously shown, what is the map distance between these loci?

36. In laboratory class, a genetics student was assigned to study an unknown mutation in *Drosophila* that had a whitish eye. He crossed females from his true-breeding mutant stock to wild-type (brick-red-eyed) males, recovering all wild-type F_1 flies. In the F_2 generation, the following offspring were recovered in the following proportions:

wild type	5/8
bright red	1/8
brown eye	1/8
white eye	1/8

The student was stumped until the instructor suggested that perhaps the whitish eye in the original stock was the result of homozygosity for a mutation causing brown eyes *and* a mutation causing bright red eyes, illustrating gene interaction (see Chapter 4). After much thought, the student was able to analyze the data, explain the results, and learn several things about the location of the two genes relative to one another. One key to his understanding was that crossing over occurs in *Drosophila*

females but not in males. Based on his analysis, what did the student learn about the two genes?

37. *Drosophila melanogaster* has one pair of sex chromosomes (XX or XY) and three pairs of autosomes, referred to as chromosomes II, III, and IV. A genetics student discovered a male fly with very short (*sh*) legs. Using this male, the student was able to establish a pure breeding stock of this mutant and found that it was recessive. She then incorporated the mutant into a stock containing the recessive gene *black* (*b*, body color located on chromosome II) and the recessive gene *pink* (*p*, eye color located on chromosome III). A female from the homozygous black, pink, short stock was then mated to a wild-type male. The F_1 males of this cross were all wild type and were then backcrossed to the homozygous *b, p, sh* females. The F_2 results appeared as shown in the following table. No other phenotypes were observed.

	Wild	Pink*	Black, Short*	Black, Pink, Short
Females	63	58	55	69
Males	59	65	51	60

*Other trait or traits are wild type.

(a) Based on these results, the student was able to assign *short* to a linkage group (a chromosome). Which one was it? Include your step-by-step reasoning.
(b) The student repeated the experiment, making the reciprocal cross, F_1 females backcrossed to homozygous *b, p, sh* males. She observed that 85 percent of the offspring fell into the given classes, but that 15 percent of the offspring were equally divided among *b + p, b + +, + sh p,* and *+ sh +* phenotypic males and females. How can these results be explained, and what information can be derived from the data?

38. In *Drosophila*, a female fly is heterozygous for three mutations, *Bar* eyes (*B*), *miniature* wings (*m*), and *ebony* body (*e*). Note that *Bar* is a dominant mutation. The fly is crossed to a male with normal eyes, miniature wings, and ebony body. The results of the cross are as follows.

111 miniature	101 Bar, ebony
29 wild type	31 Bar, miniature, ebony
117 Bar	35 ebony
26 Bar, miniature	115 miniature, ebony

Interpret the results of this cross. If you conclude that linkage is involved between any of the genes, determine the map distance(s) between them.

39. The gene controlling the Xg blood group alleles (Xg^+ and $Xg^−$) and the gene controlling a newly described form of inherited recessive muscle weakness called *episodic muscle weakness* (*EMWX*) (Ryan et al., 1999) are closely linked on the X chromosome in humans at position Xp22.3 (the tip of the short arm). A male with EMWX who is $Xg^−$ marries a woman who is Xg^+, and they have eight daughters and one son all of whom are normal for

muscle function, the male being Xg^+ and all the daughters being heterozygous at both the *EMWX* and *Xg* loci. Following is a table that lists three of the daughters with the phenotypes of their husbands and children.

(a) Create a pedigree that represents all data stated above and in the following table.
(b) For each of the offspring, indicate whether or not a crossover was required to produce the phenotypes that are given.

	Husband's Phenotype	Offspring's Sex	Offspring's Phenotype
Daughter 1:	Xg^+	female	Xg^+
		male	EMWX, Xg^+
Daughter 2:	Xg^-	male	Xg^-
		female	Xg^+
		male	EMWX, Xg^-
Daughter 3:	Xg^-	male	EMWX, Xg^-
		male	Xg^+
		male	Xg^-
		male	EMWX, Xg^+
		male	Xg^-
		male	EMWX, Xg^-
		female	Xg^+
		female	Xg^-
		female	Xg^+

40. Because of the relatively high frequency of meiotic errors that lead to developmental abnormalities in humans, many research efforts have focused on identifying correlations between error frequency and chromosome morphology and behavior. Tease et al. (2002) studied human fetal oocytes of chromosomes 21, 18, and 13 using an immunocytological approach that allowed a direct estimate of the frequency and position of meiotic recombination. Below is a summary of information (modified from Tease et al., 2002) that compares recombination frequency with the frequency of trisomy for chromosomes 21, 18, and 13. (*Note:* You may want to read appropriate portions of Chapter 8 for descriptions of these trisomic conditions.)

Trisomic	Mean Recombination Frequency	Live-born Frequency
Chromosome 21	1.23	1/700
Chromosome 18	2.36	1/3000–1/8000
Chromosome 13	2.50	1/5000–1/19000

(a) What conclusions can be drawn from these data in terms of recombination and nondisjunction frequencies? How might recombination frequencies influence trisomic frequencies?
(b) Other studies indicate that the number of crossovers per oocyte is somewhat constant, and it has been suggested that positive chromosomal interference acts to spread out a limited number of crossovers among as many chromosomes as possible. Considering information in part (a), speculate on the selective advantage positive chromosomal interference might confer.

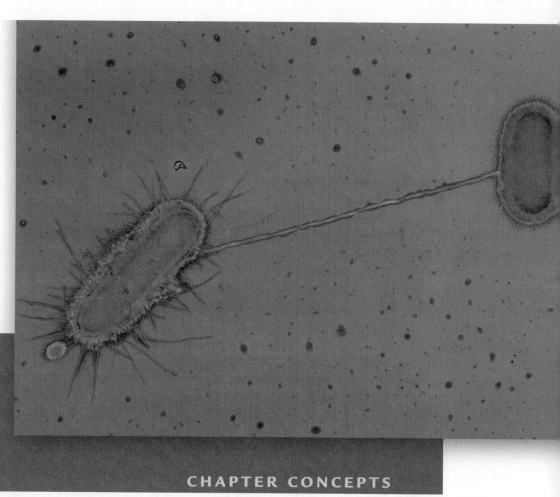

Transmission electron micrograph of conjugating *E. coli.*

6

Genetic Analysis and Mapping in Bacteria and Bacteriophages

CHAPTER CONCEPTS

- Bacterial genomes are most often contained in a single circular chromosome.

- Bacteria have developed numerous ways of exchanging and recombining genetic information between individual cells, including conjugation, transformation, and transduction.

- The ability to undergo conjugation and to transfer the bacterial chromosome from one cell to another is governed by genetic information contained in the DNA of a "fertility," or F, factor.

- The F factor can exist autonomously in the bacterial cytoplasm as a plasmid, or it can integrate into the bacterial chromosome, where it facilitates the transfer of the host chromosome to the recipient cell, leading to genetic recombination.

- Bacteriophages are viruses that have bacteria as their hosts. Viral DNA is injected into the host cell, where it replicates and directs the reproduction of the bacteriophage and the lysis of the bacterium.

- During bacteriophage infection, replication of the phage DNA may be followed by its recombination, which may serve as the basis for intergenic and intragenic mapping.

- Rarely, following infection, bacteriophage DNA integrates into the host chromosome, becoming a prophage, where it is replicated along with the bacterial DNA.

n this chapter, we shift from consideration of transmission genetics and mapping in eukaryotes to discussion of the analysis of genetic recombination and mapping in bacteria and bacteriophages, viruses that have bacteria as their host. As we focus on these topics, it will become clear that complex processes have evolved in bacteria and bacteriophages that transfer genetic information between individual cells within populations. These processes provide geneticists with the basis for chromosome mapping.

The study of bacteria and bacteriophages has been essential to the accumulation of knowledge in many areas of genetic study. For example, much of what we know about the expression and regulation of genetic information was initially derived from experimental work with them. Furthermore, as we shall see (Chapter 20), our extensive knowledge of bacteria and their resident plasmids has served as the basis for their widespread use in DNA cloning and other recombinant DNA procedures.

The value of bacteria and their viruses as research organisms in genetics is based on two important characteristics that they display. First, they have extremely short reproductive cycles. Literally hundreds of generations, amounting to billions of genetically identical bacteria or phages, can be produced in short periods of time. Second, they can also be studied in pure culture. That is, a single species or mutant strain of bacteria or one type of virus can with ease be isolated and investigated independently of other similar organisms. As a result, they have been indispensable to the progress made in genetics over the past half century.

<table>
<tr><td>6.1</td></tr>
</table>

Bacteria Mutate Spontaneously and Grow at an Exponential Rate

Genetic studies using bacteria depend on our ability to study mutations in these organisms. It has long been known that genetically homogeneous cultures of bacteria occasionally give rise to cells exhibiting heritable variation, particularly with respect to growth under unique environmental conditions. Prior to 1943, the source of this variation was hotly debated. The majority of bacteriologists believed that environmental factors induced changes in certain bacteria, leading to their survival or adaptation to the new conditions. For example, strains of *E. coli* are known to be sensitive to infection by the bacteriophage T1. Infection by the bacteriophage T1 leads to reproduction of the virus at the expense of the bacterial host, from which new phages are released as the host cell is disrupted, or lysed. If a plate of *E. coli* is uniformly sprayed with T1, almost all cells are lysed. Rare *E. coli* cells, however, survive infection and are not lysed. If these cells are isolated and established in pure culture, all their descendants are resistant to T1 infection. The **adaptation**

hypothesis, put forth to explain this type of observation, implies that the interaction of the phage and bacterium is essential to the acquisition of immunity. In other words, exposure to the phage "induces" resistance in the bacteria.

On the other hand, the occurrence of **spontaneous mutations**, which occur regardless of the presence or absence of bacteriophage T1, suggested an alternative model to explain the origin of resistance in *E. coli*. In 1943, Salvador Luria and Max Delbruck presented the first convincing evidence that bacteria, like eukaryotic organisms, are capable of spontaneous mutation. Their experiment, referred to as the **fluctuation test**, marks the initiation of modern bacterial genetic study. We will explore this discovery in Chapter 16. Mutant cells that arise spontaneously in otherwise pure cultures can be isolated and established independently from the parent strain by the use of selection techniques. *Selection* refers to culturing the organism under conditions where only the desired mutant grows well, while the wild type does not grow. With carefully designed selection, mutations for almost any desired characteristic can now be isolated. Because bacteria and viruses usually contain only one copy of a single chromosome, and are therefore haploid, all mutations are expressed directly in the descendants of mutant cells, adding to the ease with which these microorganisms can be studied.

Bacteria are grown either in a liquid culture medium or in a petri dish on a semisolid agar surface. If the nutrient components of the growth medium are very simple and consist only of an organic carbon source (such as glucose or lactose) and various inorganic ions, including Na^+, K^+, Mg^{++}, Ca^{++}, and NH_4^+ present as inorganic salts, it is called **minimal medium**. To grow on such a medium, a bacterium must be able to synthesize all essential organic compounds (e.g., amino acids, purines, pyrimidines, sugars, vitamins, and fatty acids). A bacterium that can accomplish this remarkable biosynthetic feat—one that the human body cannot duplicate—is a **prototroph**. It is said to be wild type for all growth requirements. On the other hand, if a bacterium loses, through mutation, the ability to synthesize one or more organic components, it is an **auxotroph**. For example, a bacterium that loses the ability to make histidine is designated as a *his⁻* auxotroph, in contrast to its prototrophic *his⁺* counterpart. For the *his⁻* bacterium to grow, this amino acid must be added as a supplement to the minimal medium. Medium that has been extensively supplemented is called *complete medium*.

To study bacterial growth quantitatively, an inoculum of bacteria—a small amount of a bacteria-containing solution, for example, 0.1 or 1.0 mL—is placed in liquid culture medium. A graph of the characteristic growth pattern for a bacteria culture is shown in Figure 6–1. Initially, during the **lag phase**, growth is slow. Then, a period of rapid growth, called the **logarithmic (log) phase**, ensues. During this phase, cells divide continually with a fixed time interval between cell divisions, resulting in exponential growth. When a cell density of about 10^9 cells/mL of culture medium is reached, nutrients become limiting and

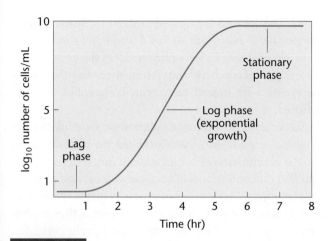

FIGURE 6–1 Typical bacterial population growth curve showing the initial lag phase, the subsequent log phase where exponential growth occurs, and the stationary phase that occurs when nutrients are exhausted. Eventually, all cells will die.

cells cease dividing; at this point, the cells enter the **stationary phase**. The doubling time during the log phase can be as short as 20 minutes. Thus, an initial inoculum of a few thousand cells added to the culture easily achieves maximum cell density during an overnight incubation.

Cells grown in liquid medium can be quantified by transferring them to the semisolid medium of a petri dish. Following incubation and many divisions, each cell gives rise to a colony visible on the surface of the medium. By counting colonies, it is possible to estimate the number of bacteria present in the original culture. If the number of colonies is too great to count, then successive dilutions (in a technique called *serial dilution*) of the original liquid culture are made and plated, until the colony number is reduced to the point where it can be counted (Figure 6–2). This technique allows the number of bacteria present in the original culture to be calculated.

For example, let's assume that the three petri dishes in Figure 6–2 represent dilutions of the liquid culture by 10^{-3}, 10^{-4}, and 10^{-5} (from left to right).* We need only select the dish in which the number of colonies can be counted accurately. Because each colony presumably arose from a single bacterium, the number of colonies times the dilution factor represents the number of bacteria in each milliliter (mL) of the initial inoculum before it was diluted. In Figure 6–2, the rightmost dish contains 12 colonies. The dilution factor for a 10^{-5} dilution is 10^5. Therefore, the initial number of bacteria was 12×10^5 per mL.

Genetic Recombination Occurs in Bacteria

Development of techniques that allowed the identification and study of bacterial mutations led to detailed investigations of the transfer of genetic information between individual organisms. As we shall see, as with meiotic crossing over in eukaryotes, the process of **genetic recombination** in bacteria provided the basis for the development of chromosome mapping methodology. It is important to note at the outset of our discussion that the term *genetic recombination*, as applied to bacteria, refers to the *replacement* of one or more genes present in the chromosome of one cell with those from the chromosome of a genetically distinct cell. While this is somewhat different from our use of the term in eukaryotes—where it describes crossing over *resulting in a reciprocal exchange*—the overall effect is the same: Genetic information is transferred and results in an altered genotype.

We will discuss three processes that result in the transfer of genetic information from one bacterium to another: *conjugation, transformation,* and *transduction.* Collectively, knowledge of these processes has helped us understand the origin of genetic variation between members of the same bacterial species, and in some cases, between members of different species. When transfer of genetic information occurs

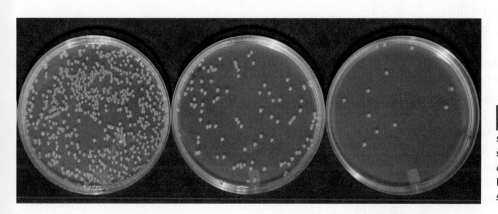

FIGURE 6–2 Results of the serial dilution technique and subsequent culture of bacteria. Each dilution varies by a factor of 10. Each colony was derived from a single bacterial cell.

---

*10^{-5} represents a 1:100,000 dilution.

between members of the same species, the term **vertical gene transfer** applies. When transfer occurs between members of related, but distinct bacterial species, the term **horizontal gene transfer** is used. The horizontal gene transfer process has played a significant role in the evolution of bacteria. Often, the genes discovered to be involved in horizontal transfer are those that also confer survival advantages to the recipient species. For example, one species may transfer antibiotic resistance genes to another species. Or genes conferring enhanced pathogenicity may be transferred. Thus, the potential for such transfer is a major concern in the medical community. In addition, horizontal gene transfer has been a major factor in the process of speciation in bacteria. Many, if not most, bacterial species have been the recipient of genes from other species.

Conjugation in Bacteria: The Discovery of F^+ and F^- Strains

Studies of bacterial recombination began in 1946, when Joshua Lederberg and Edward Tatum showed that bacteria undergo **conjugation**, a process by which genetic information from one bacterium is transferred to and recombined with that of another bacterium. Their initial experiments were performed with two multiple auxotrophs (nutritional mutants) of *E. coli* strain K12. As shown in Figure 6–3, strain A required methionine (met) and biotin (bio) in order to grow, whereas strain B required threonine (thr), leucine (leu), and thiamine (thi). Neither strain would grow on minimal medium. The two strains were first grown separately in supplemented media, and then cells from both were mixed and grown together for several more generations. They were then plated on minimal medium. Any cells that grew on minimal medium were prototrophs. It is highly improbable that any of the cells containing two or three mutant genes would undergo spontaneous mutation simultaneously at two or three independent locations to become wild-type cells. Therefore, the researchers assumed that any prototrophs recovered must have arisen as a result of some form of genetic exchange and recombination between the two mutant strains.

In this experiment, prototrophs were recovered at a rate of $1/10^7$ (or 10^{-7}) cells

plated. The controls for this experiment consisted of separate plating of cells from strains A and B on minimal medium. No prototrophs were recovered. On the basis of these observations, Lederberg and Tatum proposed that, while the events were indeed rare, genetic recombination had occurred.

Lederberg and Tatum's findings were soon followed by numerous experiments that elucidated the physical nature and the genetic basis of conjugation. It quickly became evident that different strains of bacteria were capable of effecting a unidirectional transfer of genetic material. When cells serve as donors of parts of their chromosomes, they are designated as F^+ **cells** (F for "fertility"). Recipient bacteria, designated as F^- **cells**, receive the donor chromosome material

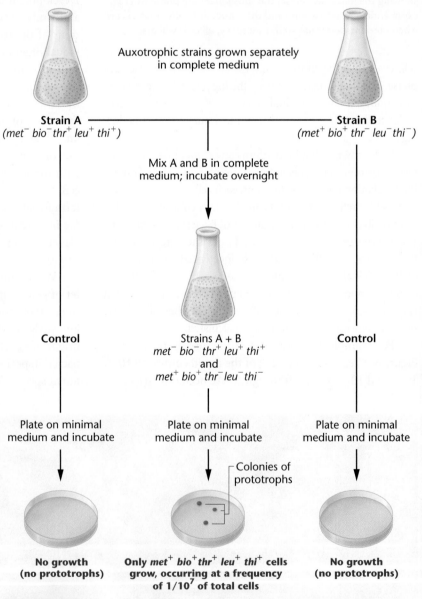

Auptrophic strains grown separately in complete medium

Strain A
(met^- bio^- thr^+ leu^+ thi^+)

Strain B
(met^+ bio^+ thr^- leu^- thi^-)

Mix A and B in complete medium; incubate overnight

Control

Strains A + B
met^- bio^- thr^+ leu^+ thi^+
and
met^+ bio^+ thr^- leu^- thi^-

Control

Plate on minimal medium and incubate

Plate on minimal medium and incubate

Plate on minimal medium and incubate

Colonies of prototrophs

No growth (no prototrophs)

Only met^+ bio^+ thr^+ leu^+ thi^+ cells grow, occurring at a frequency of $1/10^7$ of total cells

No growth (no prototrophs)

FIGURE 6–3   Production of prototrophs as a result of genetic recombination between two auxotrophic strains. Neither auxotrophic strain will grow on minimal medium, but prototrophs do, suggesting that genetic recombination has occurred.

(now known to be DNA) and recombine it with part of their own chromosome.

Experimentation subsequently established that cell-to-cell contact is essential for chromosome transfer. Support for this concept was provided by Bernard Davis, who designed the Davis U-tube for growing F$^+$ and F$^-$ cells (Figure 6–4). At the base of the tube is a sintered glass filter with a pore size that allows passage of the liquid medium but is too small to allow passage of bacteria. The F$^+$ cells are placed on one side of the filter and F$^-$ cells on the other side. The medium passes back and forth across the filter so that it is shared by both sets of bacterial cells during incubation. When Davis plated samples from both sides of the tube on minimal medium, no prototrophs were found, and he logically concluded that *physical contact between cells of the two strains is essential to genetic recombination*. We now know that this physical interaction is the initial stage of the process of conjugation and is mediated by a structure called the **F pilus** (or **sex pilus**; pl. pili), a six to nine nm tubular extension of the cell (see the Chapter Opening photograph on p. 143). Bacteria often have many pili of different types performing different cellular functions, but all pili are involved in some way with adhesion (the binding together of cells). After contact has been initiated between mating pairs, chromosome transfer is possible.

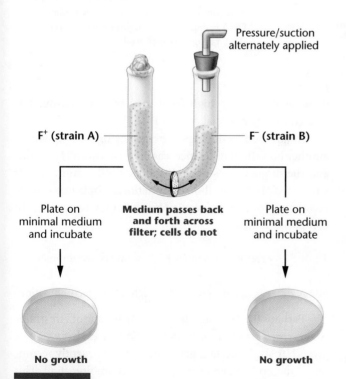

FIGURE 6–4 When strain A and strain B auxotrophs are grown in a common medium but separated by a filter, as in this Davis U-tube apparatus, no genetic recombination occurs and no prototrophs are produced.

Later evidence established that F$^+$ cells contain a **fertility factor (F factor)** that confers the ability to donate part of their chromosome during conjugation. Experiments by Joshua and Esther Lederberg and by William Hayes and Luca Cavalli-Sforza showed that certain environmental conditions eliminate the F factor from otherwise fertile cells. However, if these "infertile" cells are then grown with fertile donor cells, the F factor is regained. These findings led to the hypothesis that the F factor is a mobile element, a conclusion further supported by the observation that, after conjugation, recipient F$^-$ cells always become F$^+$. Thus, in addition to the rare cases of gene transfer (genetic recombination) that result from conjugation, the F factor itself is passed to *all* recipient cells. Accordingly, the initial cross of Lederberg and Tatum (Figure 6–3) can be described as follows:

Strain A	Strain B
F$^+$	F$^-$
×	
(DONOR)	(RECIPIENT)

Characterization of the F factor confirmed these conclusions. Like the bacterial chromosome, though distinct from it, the F factor has been shown to consist of a circular, double-stranded DNA molecule, equivalent in size to about 2 percent of the bacterial chromosome (about 100,000 nucleotide pairs). There are as many as 40 genes contained within the F factor. Many are *tra* genes, whose products are involved in the *tra*nsfer of genetic information, including the genes essential to the formation of the sex pilus.

Geneticists believe that transfer of the F factor during conjugation involves separation of the two strands of its double helix and movement of one of the two strands into the recipient cell. The other strand remains in the donor cell. Both strands, one moving across the conjugation tube and one remaining in the donor cell, are replicated. The result is that both the donor and the recipient cells become F$^+$. This process is diagrammed in Figure 6–5.

To summarize, an *E. coli* cell may or may not contain the F factor. When it is present, the cell is able to form a sex pilus and potentially serve as a donor of genetic information. During conjugation, a copy of the F factor is almost always transferred from the F$^+$ cell to the F$^-$ recipient, converting the recipient to the F$^+$ state. The question remained as to exactly why such a low proportion of these matings (10^{-7}) also results in genetic recombination. Also, it was unclear what the transfer of the F factor had to do with the transfer and recombination of particular genes. The answers to these questions awaited further experimentation.

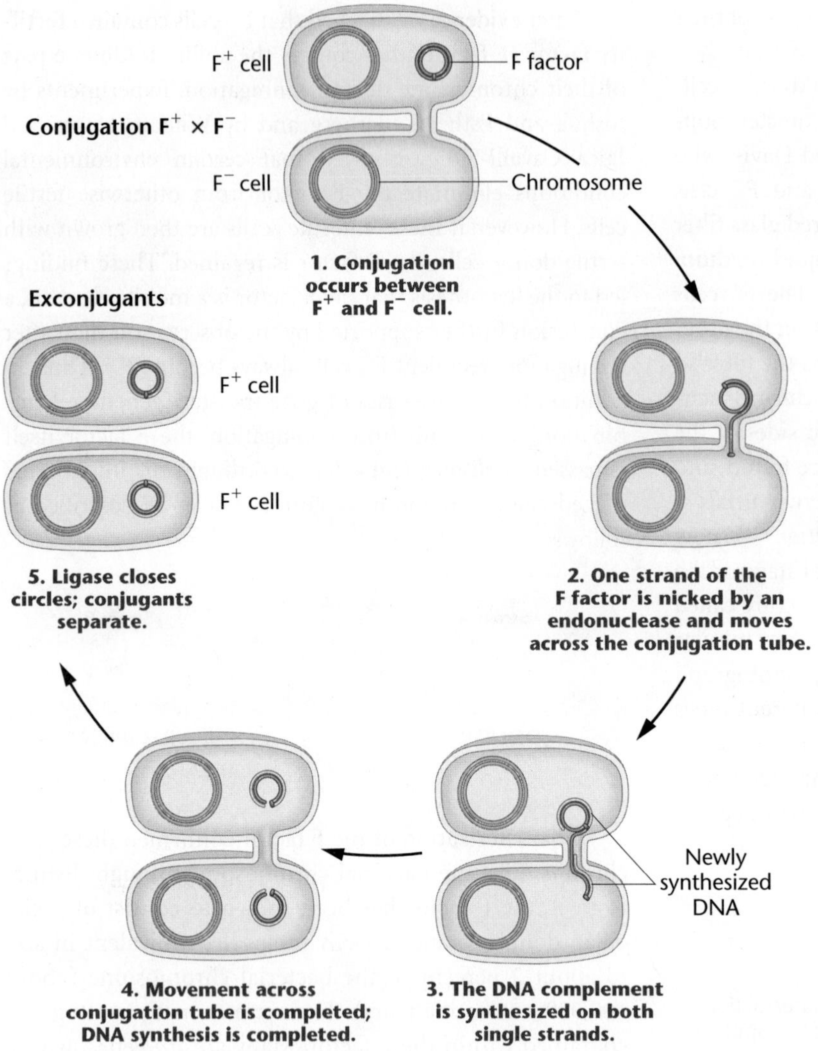

FIGURE 6–5   An $F^+ \times F^-$ mating, demonstrating how the recipient F^- cell is converted to F^+. During conjugation, the DNA of the F factor is replicated, with one new copy entering the recipient cell, converting it to F^+. The bars drawn on the F factors indicate their clockwise rotation during replication. Newly replicated DNA is depicted by a lighter shade of blue as the F factor is transferred.

As you soon shall see, the F factor is in reality an autonomous genetic unit referred to as a *plasmid*. However, in covering the history of its discovery, in this chapter we will continue to refer to it as a "factor."

Hfr Bacteria and Chromosome Mapping

Subsequent discoveries not only clarified how genetic recombination occurs but also defined a mechanism by which the *E. coli* chromosome could be mapped. Let's address chromosome mapping first.

In 1950, Cavalli-Sforza treated an F^+ strain of *E. coli* K12 with nitrogen mustard, a potent chemical known to induce mutations. From these treated cells, he recovered a genetically altered strain of donor bacteria that underwent recombination at a rate of $1/10^4$ (or 10^{-4}), 1000 times more frequently than the original F^+ strains. In 1953, William Hayes isolated another strain that demonstrated a similarly elevated frequency of recombination. Both strains were

designated **Hfr**, for **high-frequency recombination**. Hfr cells constitute a special class of F^+ cells.

In addition to the higher frequency of recombination, another important difference was noted between Hfr strains and the original F^+ strains. If a donor cell is from an Hfr strain, recipient cells, though sometimes displaying genetic recombination, *never become Hfr*; thus they remain F^-. In comparison, then,

$F^+ \times F^- \rightarrow$ recipient becomes F^+ (low rate of recombination)

$\text{Hfr} \times F^- \rightarrow$ recipient remains F^- (high rate of recombination)

Perhaps the most significant characteristic of Hfr strains is the *specific nature of recombination* in each case. In a given Hfr strain, certain genes are more frequently recombined than others, and some do not recombine at all. This *nonrandom* pattern of gene transfer was shown to vary among Hfr strains. Although these results were puzzling,

Hayes interpreted them to mean that some physiological alteration of the F factor had occurred to produce Hfr strains of *E. coli*.

In the mid-1950s, experimentation by Ellie Wollman and François Jacob explained the differences between Hfr cells and F$^+$ cells and showed how Hfr strains would allow genetic mapping of the *E. coli* chromosome. In Wollman's and Jacob's experiments, Hfr and antibiotic-resistant F$^-$ strains with suitable marker genes were mixed, and recombination of these genes was assayed at different times. Specifically, a culture containing a mixture of an Hfr and an F$^-$ strain was incubated, and samples were removed at intervals and placed in a blender. The shear forces created in the blender separated conjugating bacteria so that the transfer of the chromosome was terminated. Then the sampled cells were grown on medium containing the antibiotic, so that only recipient cells would be recovered. These cells were subsequently tested for the transfer of specific genes.

This process, called the **interrupted mating technique**, demonstrated that, depending on the specific Hfr strain, certain genes are transferred and recombined sooner than others. The graph in **Figure 6–6** illustrates this point. During the first 8 minutes after the two strains were mixed, no genetic recombination was detected. At about 10 minutes,

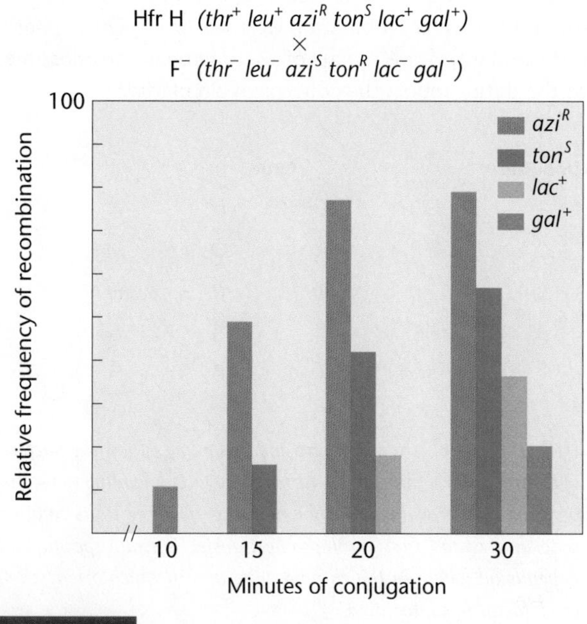

recombination of the *azi*^{R} gene could be detected, but no transfer of the *ton*^{S}, *lac*$^+$, or *gal*$^+$ genes was noted. By 15 minutes, 50 percent of the recombinants were *azi*^{R} and 15 percent were also *ton*^{S}; but none was *lac*$^+$ or *gal*$^+$. Within 20 minutes, the *lac*$^+$ gene was found among the recombinants; and within 25 minutes, *gal*$^+$ was also beginning to be transferred. Wollman and Jacob had demonstrated an *ordered transfer of genes* that correlated with the length of time conjugation proceeded.

It appeared that the chromosome of the Hfr bacterium was transferred linearly, so that the gene order and distance between genes, as measured in minutes, could be predicted from experiments such as Wollman and Jacob's (**Figure 6–7**). This information, sometimes referred to as **time mapping**, served as the basis for the first genetic map of the *E. coli* chromosome. Minutes in bacterial mapping provide a measure similar to map units in eukaryotes.

Wollman and Jacob repeated the same type of experiment with other Hfr strains, obtaining similar results but with one important difference. Although genes were always transferred linearly with time, as in their original experiment, the order in which genes entered the recipient seemed to vary from Hfr strain to Hfr strain [**Figure 6–8(a)**]. Nevertheless, when the researchers reexamined the entry rate of genes, and thus the different genetic maps for each strain, a distinct pattern emerged. The major difference between each strain was simply the point of the origin (*O*)—the first part of the donor chromosome to enter the recipient—and the direction in which entry proceeded from that point [Figure 6–8(b)].

To explain these results, Wollman and Jacob postulated that the *E. coli* chromosome is circular (a closed circle, with no free ends). If the point of origin (*O*) varies from strain to strain, a different sequence of genes will be transferred in each case. But what determines *O*? They proposed that, in various Hfr strains, the F factor

Hfr H *(thr$^+$ leu$^+$ azi^R ton^S lac$^+$ gal$^+$)*
×
F$^-$ *(thr$^-$ leu$^-$ azi^S ton^R lac$^-$ gal$^-$)*

Relative frequency of recombination

100

azi^R
ton^S
lac$^+$
gal$^+$

10   15   20   30

Minutes of conjugation

FIGURE 6–6 The progressive transfer during conjugation of various genes from a specific Hfr strain of *E. coli* to an F$^-$ strain. Certain genes (*azi* and *ton*) transfer sooner than others and recombine more frequently. Others (*lac* and *gal*) transfer later, and recombinants are found at a lower frequency. Still others (*thr* and *leu*) are always transferred and were used in the initial screen for recombinants but are not shown here.

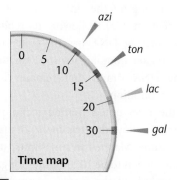

Time map

FIGURE 6–7 A time map of the genes studied in the experiment depicted in Figure 6–6.

(a)

Hfr strain	(earliest)			Order of transfer				(latest)
H	thr –	leu –	azi –	ton –	pro –	lac –	gal –	thi
1	leu –	thr –	thi –	gal –	lac –	pro –	ton –	azi
2	pro –	ton –	azi –	leu –	thr –	thi –	gal –	lac
7	ton –	azi –	leu –	thr –	thi –	gal –	lac –	pro

(b)

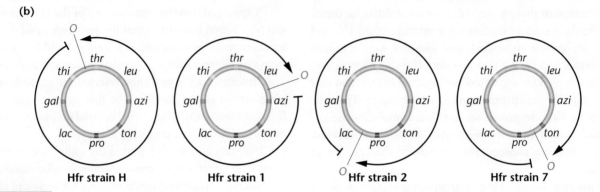

Hfr strain H   Hfr strain 1   Hfr strain 2   Hfr strain 7

FIGURE 6–8   (a) The order of gene transfer in four Hfr strains, suggesting that the *E. coli* chromosome is circular. (b) The point where transfer originates (*O*) is identified in each strain. The origin is the point of integration of the F factor into the chromosome; the direction of transfer is determined by the orientation of the F factor as it integrates. The arrowheads indicate the points of initial transfer.

integrates into the chromosome at different points, and its position determines the *O* site. One such case of integration is shown in step 1 of Figure 6–9. During conjugation between an Hfr and an F⁻ cell, the position of the F factor determines the initial point of transfer (steps 2 and 3). Those genes adjacent to *O* are transferred first, and the F factor becomes the last part that can be transferred (step 4). However, conjugation rarely, if ever, lasts long enough to allow the entire chromosome to pass across the conjugation tube (step 5). *This proposal explains why recipient cells, when mated with Hfr cells, remain F⁻.*

Figure 6–9 also depicts the way in which the two strands making up a donor's DNA molecule behave during transfer, allowing for the entry of one strand of DNA into the recipient (step 3). Following its replication in the recipient, the entering DNA has the potential to recombine with the region homologous to it on the host chromosome. The DNA strand that remains in the donor also undergoes replication.

Use of the interrupted mating technique with different Hfr strains allowed researchers to map the entire *E. coli* chromosome. Mapped in time units, strain K12 (or *E. coli* K12) was shown to be 100 minutes long. While modern genome analysis of the *E. coli* chromosome has now established the presence of just over 4000 protein-coding sequences, this original mapping procedure established the location of approximately 1000 genes.

NOW SOLVE THIS

6–1 When the interrupted mating technique was used with five different strains of Hfr bacteria, the following orders of gene entry and recombination were observed. On the basis of these data, draw a map of the bacterial chromosome. Do the data support the concept of circularity?

Hfr Strain			Order		
1	T	C	H	R	O
2	H	R	O	M	B
3	M	O	R	H	C
4	M	B	A	K	T
5	C	T	K	A	B

■ HINT: *This problem involves an understanding of how the bacterial chromosome is transferred during conjugation, leading to recombination and providing data for mapping. The key to its solution is to understand that chromosome transfer is strain-specific and depends on where in the chromosome, and in which orientation, the F factor has integrated.*

Recombination in F⁺ × F⁻ Matings: A Reexamination

The preceding model helped geneticists better understand how genetic recombination occurs during the F⁺ × F⁻

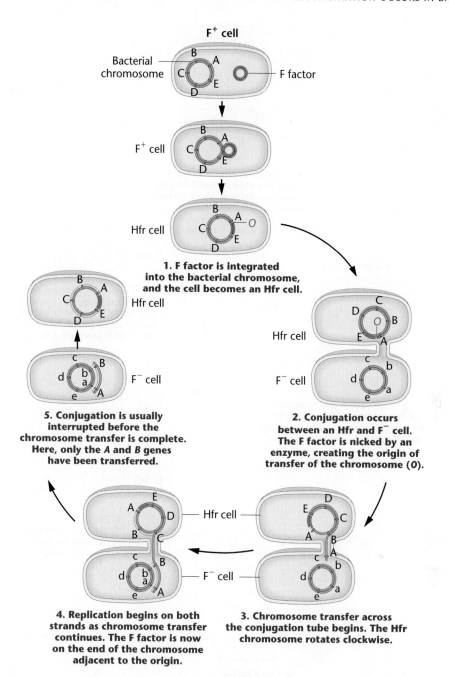

1. F factor is integrated into the bacterial chromosome, and the cell becomes an Hfr cell.

2. Conjugation occurs between an Hfr and F⁻ cell. The F factor is nicked by an enzyme, creating the origin of transfer of the chromosome (O).

3. Chromosome transfer across the conjugation tube begins. The Hfr chromosome rotates clockwise.

4. Replication begins on both strands as chromosome transfer continues. The F factor is now on the end of the chromosome adjacent to the origin.

5. Conjugation is usually interrupted before the chromosome transfer is complete. Here, only the A and B genes have been transferred.

FIGURE 6–9 Conversion of F⁺ to an Hfr state occurs by integration of the F factor into the bacterial chromosome. The point of integration determines the origin (O) of transfer. During conjugation, an enzyme nicks the F factor, now integrated into the host chromosome, initiating the transfer of the chromosome at that point. Conjugation is usually interrupted prior to complete transfer. Here, only the A and B genes are transferred to the F⁻ cell; they may recombine with the host chromosome. Newly replicated DNA of the chromosome is depicted by a lighter shade of orange.

matings. Recall that recombination occurs much less frequently in them than in Hfr × F⁻ matings and that random gene transfer is involved. The current belief is that when F⁺ and F⁻ cells are mixed, conjugation occurs readily, and each F⁻ cell involved in conjugation with an F⁺ cell receives a copy of the F factor, *but no genetic recombination occurs*. However, at an extremely low frequency in a population of F⁺ cells, the F factor integrates spontaneously into a random point in the bacterial chromosome, converting

that F⁺ cell to the Hfr state as shown in Figure 6–9. Therefore, in F⁺ × F⁻ matings, the extremely low frequency of genetic recombination (10^{-7}) is attributed to the rare, newly formed Hfr cells, which then undergo conjugation with F⁻ cells. Because the point of integration of the F factor is random, the genes transferred by any newly formed Hfr donor *will also appear to be random within the larger* F⁺/F⁻ *population*. The recipient bacterium will appear as a recombinant but will, in fact, remain F⁻. If it subsequently

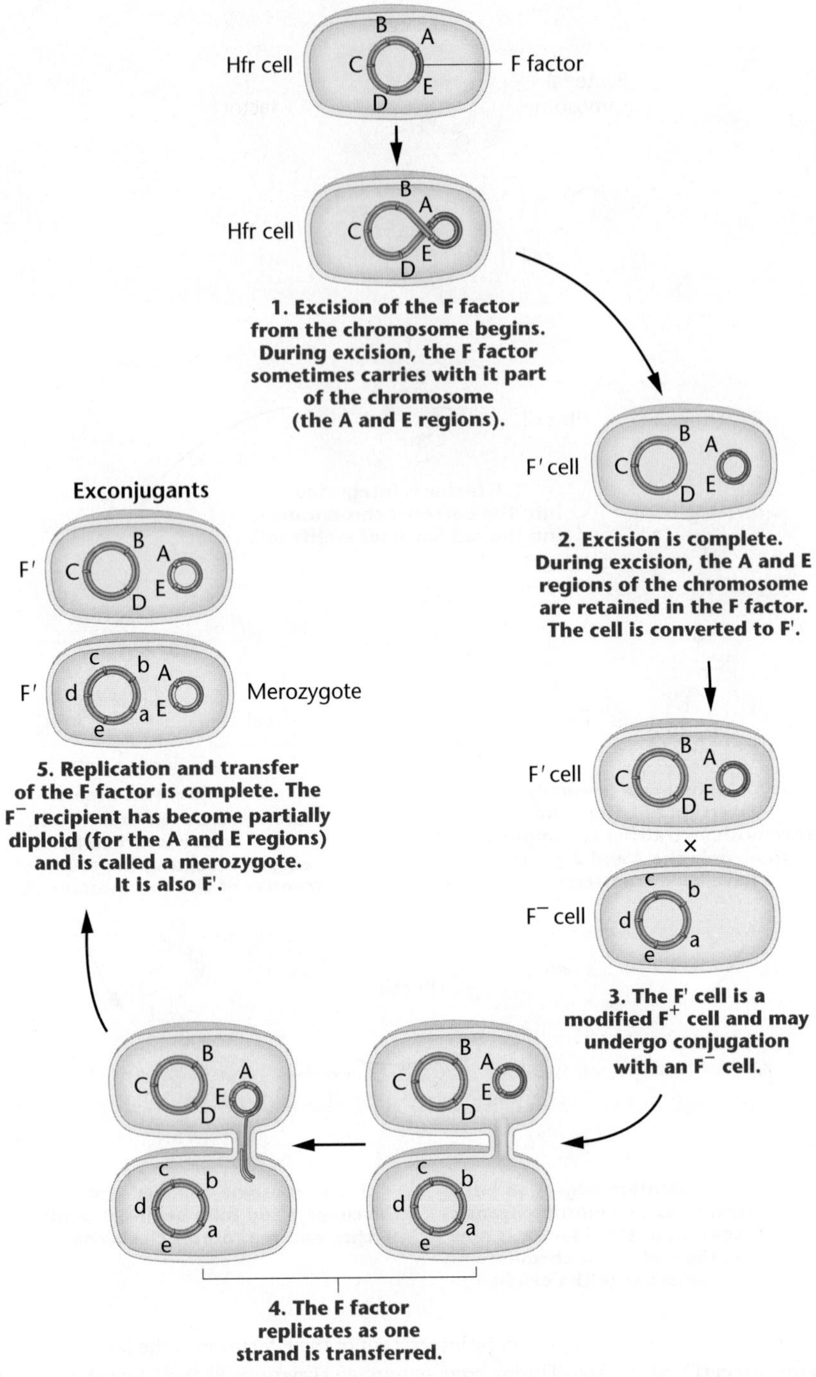

FIGURE 6–10   Conversion of an Hfr bacterium to F′ and its subsequent mating with an F⁻ cell. The conversion occurs when the F factor loses its integrated status. During excision from the chromosome, the F factor may carry with it one or more chromosomal genes (in this case, A and E). Following conjugation, the recipient cell becomes partially diploid and is called a merozygote. It also behaves as an F⁺ donor cell.

undergoes conjugation with an F⁺ cell, it will be converted to F⁺.

The F′ State and Merozygotes

In 1959, during experiments with Hfr strains of *E. coli*, Edward Adelberg discovered that the F factor could lose its integrated status, causing the cell to revert to the F⁺ state (Figure 6–10, step 1). When this occurs, the F factor frequently carries several

adjacent bacterial genes along with it (step 2). Adelberg designated this condition **F′** to distinguish it from F⁺ and Hfr. F′, like Hfr, is thus another special case of F⁺.

The presence of bacterial genes within a cytoplasmic F factor creates an interesting situation. An F′ bacterium behaves like an F⁺ cell by initiating conjugation with F⁻ cells (Figure 6–10, step 3). When this occurs, the F factor, containing chromosomal genes, is transferred to the F⁻ cell (step 4). As a result, whatever chromosomal genes are part of the F factor are now present as duplicates in the recipient cell (step 5) because the recipient still has a complete chromosome. This creates a partially diploid cell called a **merozygote**. Pure cultures of F′ merozygotes can be established. They have been extremely useful in the study of genetic regulation in bacteria, as we will discuss in Chapter 16.

6.3
Rec Proteins Are Essential to Bacterial Recombination

Once researchers established that a unidirectional transfer of DNA occurs between bacteria, they became interested in determining how the actual recombination event occurs in the recipient cell. Just how does the donor DNA replace the homologous region in the recipient chromosome? As with many systems, the biochemical mechanism by which recombination occurs was deciphered through genetic studies. Major insights were gained as a result of the isolation of a group of mutations that impaired the process of recombination and led to the discovery of *rec* (for recombination) genes.

The first relevant observation in this case involved a series of mutant genes labeled *recA*, *recB*, *recC*, and *recD*. The first mutant gene, *recA*, diminished genetic recombination in bacteria 1000-fold, nearly eliminating it altogether; each of the other *rec* mutations reduced recombination by about 100 times. Clearly, the normal wild-type products of these genes play an essential role in the process of recombination.

Researchers looked for, and subsequently isolated, several functional gene products present in normal cells but missing in *rec* mutant cells and showed that they played a role in genetic recombination. The first product is called the **RecA protein**.* This protein plays an important role in recombination involving either a single-stranded DNA molecule or the linear end of a double-stranded DNA molecule that has unwound. As it turns out, **single-strand displacement** is a common form of recombination in many bacterial species. When double-stranded DNA enters a recipient cell, one strand is often degraded, leaving the complementary strand as the only source of recombination. This strand must find its homologous region along the host chromosome, and once it does, RecA facilitates recombination.

The second related gene product is a more complex protein called the **RecBCD protein**, an enzyme consisting of polypeptide subunits encoded by three other *rec* genes. This protein is important when double-stranded DNA serves as the source of genetic recombination. RecBCD unwinds the helix, facilitating

recombination that involves RecA. These discoveries have extended our knowledge of the process of recombination considerably and underscore the value of isolating mutations, establishing their phenotypes, and determining the biological role of the normal, wild-type genes. The model of recombination based on the *rec* discoveries also applies to eukaryotes: eukaryotic proteins similar to RecA have been isolated and studied. We will return to this topic in Chapter 11 where we will discuss two models of DNA recombination.

6.4
The F Factor Is an Example of a Plasmid

The preceding sections introduced the extrachromosomal heredity unit called the F factor that bacteria require for conjugation. When it exists autonomously in the bacterial cytoplasm, it is composed of a double-stranded closed circle of DNA. These characteristics place the F factor in the more general category of genetic structures called **plasmids** [Figure 6–11(a)]. Plasmids often exist in multiple copies in the cytoplasm; each may contain one or more genes and often quite a few. Their replication depends on the same enzymes that replicate the chromosome of the host cell, and they are distributed to daughter cells along with the host chromosome during cell division. Most often, the cell has multiple copies of each type of plasmid it possesses. Many plasmids are confined to the cytoplasm of the bacterial cell. Others, such as the F factor, can integrate into the host chromosome. Those plasmids that can exist autonomously or can integrate into the chromosome are further designated as **episomes**.

Plasmids can be classified according to the genetic information specified by their DNA. The F factor plasmid confers fertility and contains genes essential for sex pilus formation, on which conjugation and subsequent genetic recombination depend. Other examples of plasmids include the R and the Col plasmids.

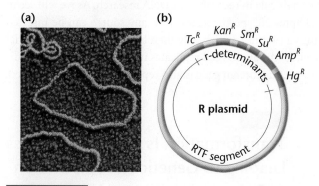

(a) **(b)**

FIGURE 6–11 (a) Electron micrograph of plasmids isolated from *E. coli*. (b) An R plasmid containing a resistance transfer factor (RTF) and multiple r-determinants (Tc, tetracycline; Kan, kanamycin; Sm, streptomycin; Su, sulfonamide; Amp, ampicillin; and Hg, mercury).

---

*Note that the names of bacterial genes use lowercase letters and are italicized, while the names of the corresponding gene products begin with capital letters and are not italicized. For example, the *recA* gene encodes the RecA protein.

Most **R plasmids** consist of two components: the **resistance transfer factor (RTF)** and one or more **r-determinants** [Figure 6–11(b)]. The RTF encodes genetic information essential to transferring the plasmid between bacteria, and the r-determinants are genes conferring resistance to antibiotics or heavy metals such as mercury. While RTFs are quite similar in a variety of plasmids from different bacterial species, there is wide variation in r-determinants, each of which is specific for resistance to one class of antibiotic. Determinants with resistance to tetracycline, streptomycin, ampicillin, sulfanilamide, kanamycin, or chloramphenicol are the most frequently encountered. Sometimes plasmids contain many r-determinants, conferring resistance to several antibiotics [Figure 6–11(b)]. Bacteria bearing such plasmids are of great medical significance, not only because of their multiple resistance but also because of the ease with which the plasmids may be transferred to other pathogenic bacteria, rendering those bacteria resistant to a wide range of antibiotics.

The first known case of such a plasmid occurred in Japan in the 1950s in the bacterium *Shigella*, which causes dysentery. In hospitals, bacteria were isolated that were resistant to as many as five of the above antibiotics. Obviously, this phenomenon represents a major health threat. Fortunately, a bacterial cell sometimes contains r-determinant plasmids but no RTF. Although such a cell is resistant, it cannot transfer the genetic information for resistance to recipient cells. The most commonly studied plasmids, however, contain the RTF as well as one or more r-determinants.

The **Col plasmid**, ColE1 (derived from *E. coli*), is clearly distinct from R plasmids. It encodes one or more proteins that are highly toxic to bacterial strains that do not harbor the same plasmid. These proteins, called **colicins**, can kill neighboring bacteria, and bacteria that carry the plasmid are said to be *colicinogenic*. Present in 10 to 20 copies per cell, the Col plasmid also contains a gene encoding an immunity protein that protects the host cell from the toxin. Unlike an R plasmid, the Col plasmid is not usually transmissible to other cells.

Interest in plasmids has increased dramatically because of their role in recombinant DNA research. As we will see in Chapter 20, specific genes from any source can be inserted into a plasmid, which may then be inserted into a bacterial cell. As the altered cell replicates its DNA and undergoes division, the foreign gene is also replicated, thus being cloned.

6.5

Transformation Is a Second Process Leading to Genetic Recombination in Bacteria

Transformation provides another mechanism for recombining genetic information in some bacteria. Small pieces of extracellular (exogenous) DNA are taken up by a living bacterium, potentially leading to a stable genetic change in the recipient cell. We discuss transformation in this chapter because in those bacterial species in which it occurs, the process can be used to map bacterial genes, though in a more limited way than conjugation. As we will see in Chapter 10, the process of transformation was also instrumental in proving that DNA is the genetic material. Further, in recombinant DNA studies (Chapter 20), transformation, albeit an artificial version enhanced by *electroporation* (electric current), is instrumental in gene cloning.

The Transformation Process

The process of transformation (Figure 6–12) consists of numerous steps that achieve two basic outcomes: (1) entry of foreign DNA into a recipient cell; and (2) recombination between the foreign DNA and its homologous region in the recipient chromosome. While completion of both outcomes is required for genetic recombination, the first step of transformation can occur without the second step, resulting in the addition of foreign DNA to the bacterial cytoplasm but not to its chromosome.

In a population of bacterial cells, only those in a particular physiological state of **competence** take up DNA. Studies have shown that various kinds of bacteria readily undergo transformation naturally (e.g., *Haemophilus influenzae, Bacillus subtilis, Shigella paradysenteriae, Streptococcus pneumoniae,* and *E. coli*). Others can be induced in the laboratory to become competent. Entry of DNA is thought to occur at a limited number of receptor sites on the surface of a competent bacterial cell (Figure 6–12, step 1). Passage into the cell is thought to be an active process that requires energy and specific transport molecules. This model is supported by the fact that substances that inhibit energy production or protein synthesis also inhibit transformation.

Soon after entry, one of the two strands of the double helix is digested by nucleases, leaving only a single strand to participate in transformation (Figure 6–12, steps 2 and 3). The surviving DNA strand aligns with the complementary region of the bacterial chromosome. In a process involving several enzymes, this segment of DNA replaces its counterpart in the chromosome (step 4), which is excised and degraded.

For recombination to be detected, the transforming DNA must be derived from a different strain of bacteria that bears some distinguishing genetic variation, such as a mutation. Once this is integrated into the chromosome, the recombinant region contains one host strand (present originally) and one mutant strand. Because these strands are from different sources, the region is referred to as a **heteroduplex**, which usually contains some mismatch of base sequence. This mismatch activates a repair process (see Chapter 15). Following repair and one round of DNA

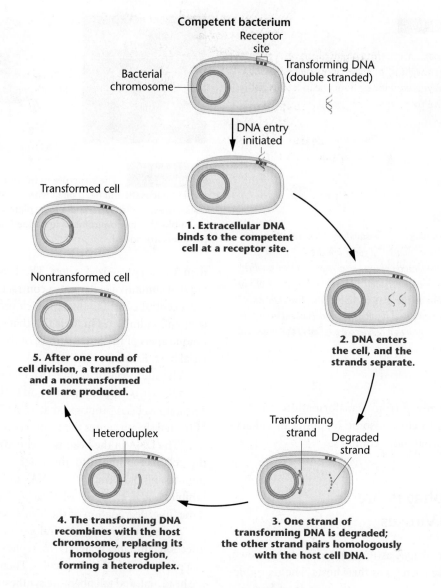

Competent bacterium

Receptor site

Transforming DNA (double stranded)

Bacterial chromosome

DNA entry initiated

1. Extracellular DNA binds to the competent cell at a receptor site.

Transformed cell

Nontransformed cell

5. After one round of cell division, a transformed and a nontransformed cell are produced.

Heteroduplex

4. The transforming DNA recombines with the host chromosome, replacing its homologous region, forming a heteroduplex.

2. DNA enters the cell, and the strands separate.

Transforming strand

Degraded strand

3. One strand of transforming DNA is degraded; the other strand pairs homologously with the host cell DNA.

FIGURE 6–12 Proposed steps for transformation of a bacterial cell by exogenous DNA. Only one of the two strands of the entering DNA is involved in the transformation event, which is completed following cell division.

replication, one chromosome is restored to its original DNA sequence, identical to that of the original recipient cell, and the other contains the properly aligned mutant gene. Following cell division, one nontransformed cell (nonmutant) and one transformed cell (mutant) are produced (step 5).

Transformation and Linked Genes

In early transformation studies, the most effective exogenous DNA was a size containing 10,000–20,000 nucleotide pairs, a length sufficient to encode several genes.* Genes adjacent to or very close to one another on the bacterial chromosome can be carried on a single segment of this size. Consequently, a

single transfer event can result in the **cotransformation** of several genes simultaneously. Genes that are close enough to each other to be cotransformed are *linked*. In contrast to *linkage groups* in eukaryotes, which consist of all genes on a single chromosome, note that here *linkage* refers to the proximity of genes that permits cotransformation (i.e., the genes are next to, or close to, one another).

If two genes are not linked, simultaneous transformation occurs only as a result of two independent events involving two distinct segments of DNA. As with double crossovers in eukaryotes, the probability of two independent events occurring simultaneously is equal to the product of the individual probabilities. Thus, the frequency of two unlinked genes being transformed simultaneously is much lower than if they are linked. Under certain conditions, relative distances between linked

_____

*Today, we know that a 2000 nucleotide pair length of DNA is highly effective in gene cloning experiments.

NOW SOLVE THIS

6–2 In a transformation experiment involving a recipient bacterial strain of genotype $a^- b^-$, the following results were obtained. What can you conclude about the location of the a and b genes relative to each other?

	Transformants (%)		
Transforming DNA	$a^+ b^-$	$a^- b^+$	$a^+ b^+$
$a^+ b^+$	3.1	1.2	0.04
$a^+ b^-$ and $a^- b^+$	2.4	1.4	0.03

■ HINT: *This problem involves an understanding of how transformation can be used to determine if bacterial genes are closely "linked". You are asked to predict the location of two genes relative to one another. The key to its solution is to understand that cotransformation (of two genes) occurs according to the laws of probability. Two "unlinked" genes are transformed only as a result of two independent events. In such a case, the probability of that occurrence is equal to the product of the individual probabilities.*

genes can be determined from transformation data in a manner analogous to chromosome mapping in eukaryotes, though somewhat more complex.

6.6

Bacteriophages Are Bacterial Viruses

Bacteriophages, or **phages** as they are commonly known, are viruses that have bacteria as their hosts. The reproduction of phages can lead to still another mode of bacterial genetic recombination, called transduction. To understand this process, we first must consider the genetics of bacteriophages, which themselves can undergo recombination.

A great deal of genetic research has been done using bacteriophages as a model system. In this section, we will first examine the structure and life cycle of one type of bacteriophage. We then discuss how these phages are studied during their infection of bacteria. Finally, we contrast two possible modes of behavior once initial phage infection occurs. This information is background for our discussion of *transduction* and *bacteriophage recombination*.

Phage T4: Structure and Life Cycle

Bacteriophage T4 is one of a group of related bacterial viruses referred to as T-even phages. It exhibits the intricate structure shown in Figure 6–13. Its genetic material, DNA, is contained within an icosahedral (referring to a polyhedron with 20 faces) protein coat, making up the head of the virus. The DNA is sufficient in quantity to encode more

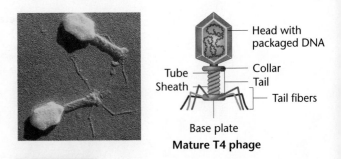

Mature T4 phage

FIGURE 6–13 The structure of bacteriophage T4 which includes an icosahedral head filled with DNA; a tail consisting of a collar, tube, and sheath; and a base plate with tail fibers. During assembly, the tail components are added to the head and then tail fibers are added.

than 150 average-sized genes. The head is connected to a tail that contains a collar and a contractile sheath surrounding a central core. Tail fibers, which protrude from the tail, contain binding sites in their tips that specifically recognize unique areas of the outer surface of the cell wall of the bacterial host, *E. coli*.

The life cycle of phage T4 (Figure 6–14) is initiated when the virus binds by adsorption to the bacterial host cell. Then, it has been proposed that an ATP-driven contraction of the tail sheath causes the central core to penetrate the cell wall. The DNA in the head is extruded, and it moves across the cell membrane into the bacterial cytoplasm. Within minutes, all bacterial DNA, RNA, and protein synthesis is inhibited, and synthesis of viral molecules begins. At the same time, degradation of the host DNA is initiated.

A period of intensive viral gene activity characterizes infection. Initially, phage DNA replication occurs, leading to a pool of viral DNA molecules. Then, the components of the head, tail, and tail fibers are synthesized. The assembly of mature viruses is a complex process that has been well studied by William Wood, Robert Edgar, and others. Three sequential pathways take part: (1) DNA packaging as the viral heads are assembled, (2) tail assembly, and (3) tail-fiber assembly. Once DNA is packaged into the head, that structure combines with the tail components, to which tail fibers are added. Total construction is a combination of self-assembly and enzyme-directed processes.

When approximately 200 new viruses are constructed, the bacterial cell is ruptured by the action of lysozyme (a phage gene product), and the mature phages are released from the host cell. This step during infection is referred to as **lysis**, and it completes what is referred to as the **lytic cycle**. The 200 new phages infect other available bacterial cells, and the process repeats itself over and over again.

The Plaque Assay

Bacteriophages and other viruses have played a critical role in our understanding of molecular genetics. During infection

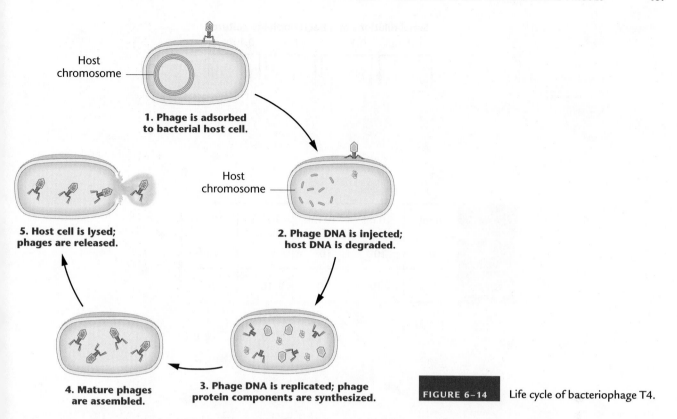

Host chromosome

1. Phage is adsorbed to bacterial host cell.

Host chromosome

2. Phage DNA is injected; host DNA is degraded.

5. Host cell is lysed; phages are released.

3. Phage DNA is replicated; phage protein components are synthesized.

4. Mature phages are assembled.

FIGURE 6–14 Life cycle of bacteriophage T4.

of bacteria, enormous quantities of bacteriophages may be obtained for investigation. Often, more than 10^{10} viruses are produced per milliliter of culture medium. Many genetic studies have relied on our ability to determine the number of phages produced following infection under specific culture conditions. The **plaque assay**, routinely used for such determinations, is invaluable in quantitative analysis during mutational and recombinational studies of bacteriophages.

This assay is illustrated in Figure 6–15, where actual plaque morphology is also shown. A serial dilution of the original virally infected bacterial culture is performed. Then, a 0.1-mL sample (an aliquot, meaning a fractional portion) from a dilution is added to a small volume of melted nutrient agar (about 3 mL) into which a few drops of a healthy bacterial culture have been added. The solution is then poured evenly over a base of solid nutrient agar in a petri dish and allowed to solidify before incubation. A clear area called a **plaque** occurs wherever a single virus initially infected one bacterium in the culture (the lawn) that has grown up during incubation. The plaque represents clones of the single infecting bacteriophage, created as reproduction cycles are repeated. If the dilution factor is too low, the plaques will be plentiful, and they may fuse, lysing the entire lawn of bacteria. This has occurred in the 10^{-3} dilution in Figure 6–15. However, if the dilution factor is increased appropriately, plaques can be counted, and the density of viruses in the initial culture can be estimated,

$$\text{initial phage density} = (\text{plaque number/mL}) \times (\text{dilution factor})$$

Figure 6–15 shows that 23 phage plaques were derived from the 0.1-mL aliquot of the 10^{-5} dilution. Therefore, we estimate a density of 230 phages/mL *at this dilution* (since the initial aliquot was 0.1 mL). The initial phage density in the undiluted sample, given that 23 plaques were observed from 0.1 mL of the 10^{-5} dilution, is calculated as

$$\text{initial phage density} = (230/\text{mL}) \times (10^5) = (230 \times 10^5)/\text{mL}$$

Because this figure is derived from the 10^{-5} dilution, we can also estimate that there would be only 0.23 phage/0.1 mL in the 10^{-7} dilution. Thus, if 0.1 mL from this tube were assayed, we would predict that no phage particles would be present. This prediction is borne out in Figure 6–15, where an intact lawn of bacteria lacking any plaques is depicted. The dilution factor is simply too great.

Use of the plaque assay has been invaluable in mutational and recombinational studies of bacteriophages. We will apply this technique more directly later in this chapter when we discuss Seymour Benzer's elegant genetic analysis of a single gene in phage T4.

Lysogeny

Infection of a bacterium by a virus does not always result in viral reproduction and lysis. As early as the 1920s, it was known that a virus can enter a bacterial cell and coexist with it. The precise molecular basis of this relationship is now well understood. Upon entry, the viral DNA is integrated into the bacterial chromosome instead of replicating in the

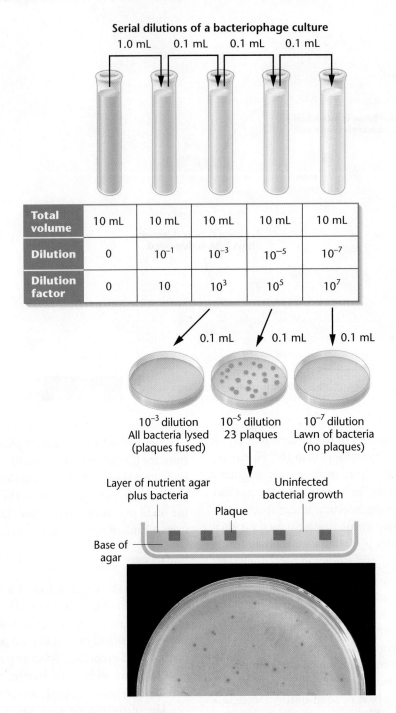

FIGURE 6–15   A plaque assay for bacteriophage analysis. First, serial dilutions are made of a bacterial culture infected with bacterio-phages. Then, three of the dilutions (10^{-3}, 10^{-5}, and 10^{-7}) are analyzed using the plaque assay technique. Each plaque represents the initial infection of one bacterial cell by one bacteriophage. In the 10^{-3} dilution, so many phages are present that all bacteria are lysed. In the 10^{-5} dilution, 23 plaques are produced. In the 10^{-7} dilution, the dilution factor is so great that no phages are present in the 0.1-mL sample, and thus no plaques form. From the 0.1-mL sample of the 10^{-5} dilution, the original bacteriophage density is calculated to be $23 \times 10 \times 10^5$ phages/mL (230×10^5). The photograph shows phage T2 plaques on lawns of *E. coli*.

bacterial cytoplasm; this integration characterizes the developmental stage referred to as **lysogeny**. Subsequently, each time the bacterial chromosome is replicated, the viral DNA is also replicated and passed to daughter bacterial cells following division. No new viruses are produced, and no lysis of the bacterial cell occurs. However, under certain stimuli, such as chemical or ultraviolet-light treatment, the viral DNA loses its integrated status and initiates replication, phage reproduction, and lysis of the bacterium.

Several terms are used in describing this relationship. The viral DNA integrated into the bacterial chromosome is called a **prophage**. Viruses that can either lyse the cell or behave as a prophage are called **temperate phages**. Those that can only lyse the cell are referred to as **virulent phages**.

A bacterium harboring a prophage has been **lysogenized** and is said to be **lysogenic**; that is, it is capable of being lysed as a result of induced viral reproduction. The viral DNA (like the F factor discussed earlier) is classified as an *episome*, meaning a genetic molecule that can replicate either in the cytoplasm of a cell or as part of its chromosome.

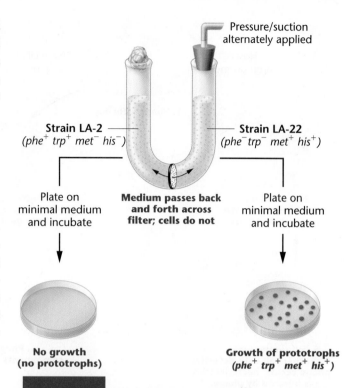

FIGURE 6–16 The Lederberg–Zinder experiment using *Salmonella*. After placing two auxotrophic strains on opposite sides of a Davis U-tube, Lederberg and Zinder recovered prototrophs from the side with the LA-22 strain but not from the side containing the LA-2 strain.

6.7

Transduction Is Virus-Mediated Bacterial DNA Transfer

In 1952, Norton Zinder and Joshua Lederberg were investigating possible recombination in the bacterium *Salmonella typhimurium*. Although they recovered prototrophs from mixed cultures of two different auxotrophic strains, subsequent investigations showed that recombination was not due to the presence of an F factor and conjugation, as in *E. coli*. What they discovered was a process of bacterial recombination mediated by bacteriophages and now called **transduction**.

The Lederberg–Zinder Experiment

Lederberg and Zinder mixed the *Salmonella* auxotrophic strains LA-22 and LA-2 together, and when the mixture was plated on minimal medium, they recovered prototrophic cells. The LA-22 strain was unable to synthesize the amino acids phenylalanine and tryptophan ($phe^- trp^-$), and LA-2 could not synthesize the amino acids methionine and histidine ($met^- his^-$). Prototrophs ($phe^+ trp^+ met^+ his^+$) were recovered at a rate of about $1/10^5$ (or 10^{-5}) cells.

Although these observations at first suggested that the recombination was the type observed earlier in conjugative strains of *E. coli*, experiments using the Davis U-tube soon showed otherwise (Figure 6–16). The two auxotrophic strains were separated by a sintered glass filter, thus preventing contact between the strains while allowing them to grow in a common medium. Surprisingly, when samples were removed from both sides of the filter and plated independently on minimal medium, prototrophs *were* recovered, but only from the side of the tube containing LA-22 bacteria. Recall that if conjugation were responsible, the Davis U-tube should have *prevented* recombination altogether (see Figure 6–4).

Since LA-2 cells appeared to be the source of the new genetic information (phe^+ and trp^+), how that information crossed the filter from the LA-2 cells to the LA-22 cells, allowing recombination to occur, was a mystery. The unknown source was designated simply as a **filterable agent (FA)**.

Three observations were used to identify the FA:

1. The FA was produced by the LA-2 cells only when they were grown in association with LA-22 cells. If LA-2 cells were grown independently in a culture medium that was later added to LA-22 cells, recombination did not occur. Therefore, the LA-22 cells played some role in the production of FA by LA-2 cells but did so only when the two strains were sharing a common growth medium.

2. The addition of DNase, which enzymatically digests DNA, did not render the FA ineffective. Therefore, the FA is not exogenous DNA, ruling out transformation.

3. The FA could not pass across the filter of the Davis U-tube when the pore size was reduced below the size of bacteriophages.

Aided by these observations and aware that temperate phages could lysogenize *Salmonella*, researchers proposed that the genetic recombination event was mediated by bacteriophage P22, present initially as a prophage in the chromosome of the LA-22 *Salmonella* cells. They hypothesized that P22 prophages sometimes enter the vegetative, or lytic, phase, reproduce, and are released by the LA-22 cells. Such P22 phages, being much smaller than a bacterium, then cross the filter of the U-tube and subsequently infect and lyse some of the LA-2 cells. In the process of lysis of LA-2, the P22 phages occasionally package a region of the LA-2 chromosome in their heads. If this region contains the phe^+ and trp^+ genes, and if the phages subsequently pass back across the filter and infect LA-22 cells, these newly lysogenized cells will behave as prototrophs. This process of transduction,

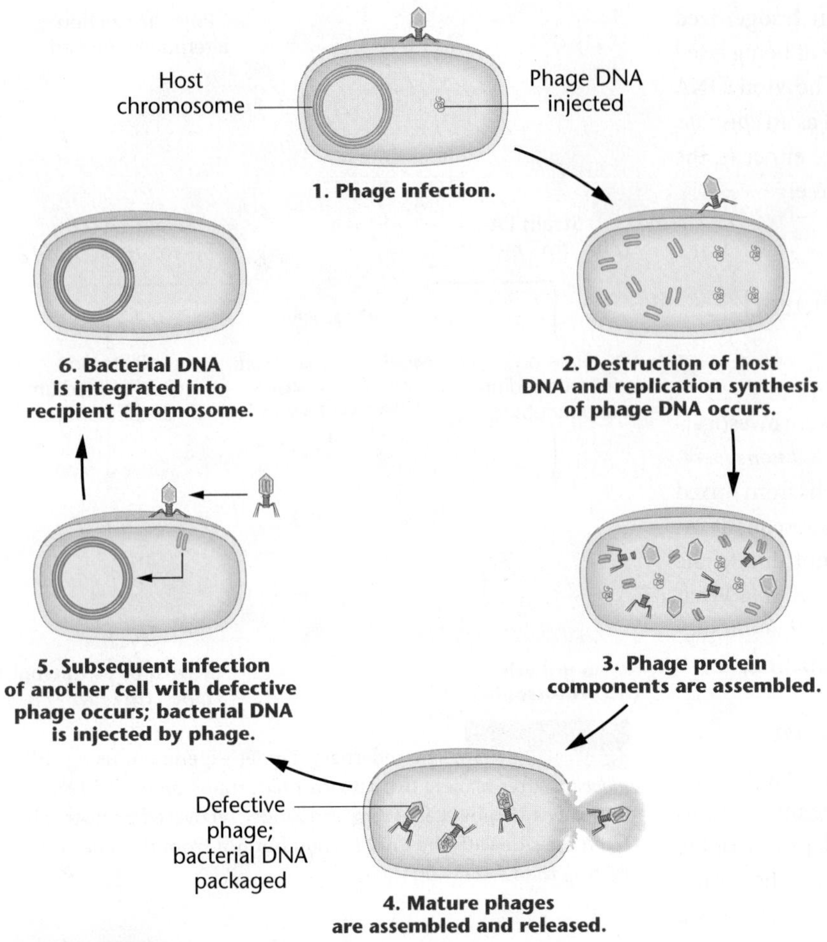

FIGURE 6–17   Generalized transduction.

1. Phage infection.

2. Destruction of host DNA and replication synthesis of phage DNA occurs.

3. Phage protein components are assembled.

4. Mature phages are assembled and released.

Defective phage; bacterial DNA packaged

5. Subsequent infection of another cell with defective phage occurs; bacterial DNA is injected by phage.

6. Bacterial DNA is integrated into recipient chromosome.

Host chromosome

Phage DNA injected

whereby bacterial recombination is mediated by bacteriophage P22, is diagrammed in Figure 6–17.

The Nature of Transduction

Further studies have revealed the existence of transducing phages in other species of bacteria. For example, *E. coli* can be transduced by phage P1, and *B. subtilis* and *Pseudomonas aeruginosa* can be transduced by phages SPO1 and F116, respectively. The details of several different modes of transduction have also been established. Even though the initial discovery of transduction involved a temperate phage and a lysogenized bacterium, the same process can occur during the normal lytic cycle. Sometimes a small piece of bacterial DNA is packaged *along with* the viral chromosome, *or instead of it*, so that the transducing phage either contains both viral and bacterial DNA, or just bacterial DNA. In either case, only a few bacterial genes are present in the transducing phage, although the phage head is capable of enclosing up to 1 percent of the bacterial chromosome. In either case, the ability to infect a host cell is unrelated to the type of DNA in the phage head, making transduction possible.

When bacterial rather than viral DNA is injected into the host cell, it either remains in the cytoplasm or recombines with the homologous region of the bacterial

chromosome. If the bacterial DNA remains in the cytoplasm, it does not replicate but is transmitted to one progeny cell following each division. When this happens, only a single cell, partially diploid for the transduced genes, is produced—a phenomenon called *abortive transduction*. If the bacterial DNA recombines with its homologous region of the bacterial chromosome, *complete transduction* occurs, whereby the transduced genes become a permanent part of the chromosome, which is passed to all daughter cells.

Both abortive and complete transduction are subclasses of the broader category of *generalized transduction*, which is characterized by the random nature of DNA fragments and genes that are transduced. Each fragment of the bacterial chromosome has a finite but small chance of being packaged in the phage head and subsequently recombined during transduction. Most cases of generalized transduction are of the abortive type; some data suggest that complete transduction occurs 10 to 20 times less frequently. In contrast to generalized transduction, *specialized transduction* occurs when transfer of bacterial DNA is not random, but instead, only strain-specific genes are transduced. This occurs when the DNA representing a temperate bacteriophage breaks out of the host chromosome, bringing with it bacterial DNA on either of its ends.

Transduction and Mapping

Like transformation, transduction has been used in linkage and mapping studies of the bacterial chromosome. The fragment of bacterial DNA involved in a transduction event may be large enough to include several genes. As a result, two genes that are close to one another along the bacterial chromosome (i.e., are linked) can be transduced simultaneously, a process called **cotransduction**. If two genes are not close enough to one another along the chromosome to be included on a single DNA fragment, two independent transduction events must occur to carry them into a single cell. Since this occurs with a much lower probability than cotransduction, linkage can be determined by comparing the frequency of specific simultaneous recombinations.

By concentrating on two or three linked genes, transduction studies can also determine the precise order of these genes. The closer linked genes are to each other, the greater the frequency of cotransduction. Mapping studies can be done on three closely aligned genes, predicated on the same rationale that underlies other mapping techniques.

6.8

Bacteriophages Undergo Intergenic Recombination

Around 1947, several research teams demonstrated that genetic recombination can be detected in bacteriophages. This led to the discovery that gene mapping can be performed in these viruses. Such studies relied on finding numerous phage mutations that could be visualized or assayed. As in bacteria and eukaryotes, these mutations allow genes to be identified and followed in mapping experiments. Before considering recombination and mapping in these bacterial viruses, we briefly introduce several of the mutations that were studied.

Bacteriophage Mutations

Phage mutations often affect the morphology of the plaques formed following lysis of bacterial cells. For example, in 1946, Alfred Hershey observed unusual T2 plaques on plates of *E. coli* strain B. Normal T2 plaques are small and have a clear center surrounded by a diffuse (nearly invisible) halo. In contrast, the unusual plaques were larger and possessed a distinctive outer perimeter (compare the lighter plaques in Figure 6–18). When the viruses were isolated from these plaques and replated on *E. coli* B cells, the resulting plaque appearance was identical. Thus, the plaque phenotype was an inherited trait resulting from the reproduction of mutant phages. Hershey named the mutant *rapid lysis* (*r*) because the plaques' larger size was thought to be due to a more rapid or more efficient life cycle of the phage. We now know that, in wild-type phages, reproduction is inhibited once a particular-sized plaque has been formed. The *r* mutant T2 phages overcome this inhibition, producing larger plaques.

Salvador Luria discovered another bacteriophage mutation, *host range* (*h*). This mutation extends the range of bacterial hosts that the phage can infect. Although wild-type T2 phages can infect *E. coli* B (a unique strain), they normally cannot attach or be adsorbed to the surface of *E. coli* B-2 (a different strain). The *h* mutation, however, confers the ability to adsorb to and subsequently infect *E. coli* B-2. When grown on a mixture of *E. coli* B and B-2, the *h* plaque has a center that appears much darker than that of the *h*⁺ plaque (Figure 6–18).

Table 6.1 lists other types of mutations that have been isolated and studied in the T-even series of bacteriophages (e.g., T2, T4, T6). These mutations are important to the study of genetic phenomena in bacteriophages.

Mapping in Bacteriophages

Genetic recombination in bacteriophages was discovered during **mixed infection experiments**, in which two distinct mutant strains were allowed to *simultaneously* infect the same bacterial culture. These studies were designed so that the number of viral particles sufficiently exceeded the number of bacterial cells to ensure simultaneous infection

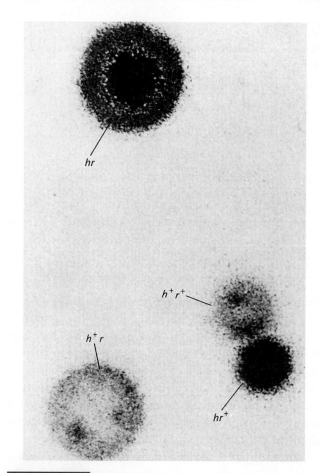

hr

$h^+ r^+$

$h^+ r$

hr^+

FIGURE 6–18   Plaque morphology phenotypes observed following simultaneous infection of *E. coli* by two strains of phage T2, $h^+ r$ and hr^+. In addition to the parental genotypes, recombinant plaques *hr* and $h^+ r^+$ are shown.

of most cells by both viral strains. If two loci are involved, recombination is referred to as **intergenic**.

For example, in one study using the T2/*E. coli* system, the parental viruses were of either the $h^+ r$ (wild-type host range, rapid lysis) or the hr^+ (extended host range, normal lysis) genotype. If no recombination occurred, these two parental genotypes would be the only expected phage progeny. However, the recombinants $h^+ r^+$ and *hr* were detected

TABLE 6.1

Some Mutant Types of T-Even Phages

Name	Description
minute	Forms small plaques
turbid	Forms turbid plaques on *E. coli* B
star	Forms irregular plaques
UV-sensitive	Alters UV sensitivity
acriflavin-resistant	Forms plaques on acriflavin agar
osmotic shock	Withstands rapid dilution into distilled water
lysozyme	Does not produce lysozyme
amber	Grows in *E. coli* K12 but not B
temperature-sensitive	Grows at 25°C but not at 42°C

TABLE 6.2

TABLE 6.2

Results of a Cross Involving the *h* and *r* Genes in Phage T2
($hr^+ \times h^+r$)

Genotype	Plaques	Designation
$h\,r^+$	42	Parental progeny
h^+r	34	76%
h^+r^+	12	Recombinants
$h\,r$	12	24%

Source: Data derived from Hershey and Rotman (1949).

in addition to the parental genotypes (see Figure 6–18). As with eukaryotes, the percentage of recombinant plaques divided by the total number of plaques reflects the relative distance between the genes:

recombinational frequency = $(h^+r^+ + hr)$/total plaques $\times$ 100

Sample data for the *h* and *r* loci are shown in Table 6.2.

Similar recombinational studies have been conducted with numerous mutant genes in a variety of bacteriophages. Data are analyzed in much the same way as in eukaryotic mapping experiments. Two- and three-point mapping crosses are possible, and the percentage of recombinants in the total number of phage progeny is calculated. This value is proportional to the relative distance between two genes along the DNA molecule constituting the chromosome.

Investigations into phage recombination support a model similar to that of eukaryotic crossing over—a breakage and reunion process between the viral chromosomes. A fairly clear picture of the dynamics of viral recombination has emerged. Following the early phase of infection, the chromosomes of the phages begin replication. As this stage progresses, a pool of chromosomes accumulates in the bacterial cytoplasm. If double infection by phages of two genotypes has occurred, then the pool of chromosomes initially consists of the two parental types. Genetic exchange between these two types will occur before, during, and after replication, producing recombinant chromosomes.

In the case of the h^+r and hr^+ example discussed here, recombinant h^+r^+ and hr chromosomes are produced. Each of these chromosomes can undergo replication, with new replicates undergoing exchange with each other and with parental chromosomes. Furthermore, recombination is not restricted to exchange between two chromosomes—three or more may be involved simultaneously. As phage development progresses, chromosomes are randomly removed from the pool and packed into the phage head, forming mature phage particles. Thus, a variety of parental and recombinant genotypes are represented in progeny phages.

As we will see in the next section, powerful selection systems have made it possible to detect *intragenic* recombination in viruses, where exchanges occur at points within a single gene, as opposed to *intergenic* recombination, where exchanges occur at points located between genes. Such studies have led to what has been called the fine-structure analysis of the gene.

6.9
Intragenic Recombination Occurs in Phage T4

We conclude this chapter with an account of an ingenious example of genetic analysis. In the early 1950s, Seymour Benzer undertook a detailed examination of a single locus, *rII*, in phage T4. Benzer successfully designed experiments to recover the extremely rare genetic recombinants arising as a result of intragenic exchange. Such recombination is equivalent to eukaryotic crossing over, but in this case, within a gene rather than at a point between two genes. Benzer demonstrated that such recombination occurs between the DNA of individual bacteriophages during simultaneous infection of the host bacterium *E. coli*.

The end result of Benzer's work was the production of a detailed map of the *rII* locus. Because of the extremely detailed information provided by his analysis, and because these experiments occurred decades before DNA-sequencing techniques were developed, the insights concerning the internal structure of the gene were particularly noteworthy.

The *rII* Locus of Phage T4

The primary requirement in genetic analysis is the isolation of a large number of mutations in the gene being investigated. Mutants at the *rII* locus produce distinctive plaques when plated on *E. coli* strain B, allowing their easy identification. Figure 6–18 illustrates mutant *r* plaques compared to their wild-type r^+ counterparts in the related T2 phage. Benzer's approach was to isolate many independent *rII* mutants—he eventually obtained about 20,000—and to perform recombinational studies so as to produce a genetic map of this locus. Benzer assumed that most of these mutations, because they were randomly isolated, would represent different locations within the *rII* locus and would thus provide an ample basis for mapping studies.

The key to Benzer's analysis was that *rII* mutant phages, though capable of infecting and lysing *E. coli* B, could not successfully lyse a second related strain, *E. coli* K12(λ).[*] Wild-type phages, by contrast, could lyse both the B and the K12 strains. Benzer reasoned that these conditions provided the potential for a highly sensitive screening system. If phages from any two different mutant strains were allowed to simultaneously infect *E. coli* B, exchanges between the two mutant sites within the locus would produce rare wild-type recombinants (Figure 6–19). If the progeny phage population, which contained more than 99.9 percent *rII* phages and

---

[*]The inclusion of "(λ)" in the designation of K12 indicates that this bacterial strain is lysogenized by phage λ. This, in fact, is the reason that *rII* mutants cannot lyse such bacteria. In future discussions, this strain will simply be abbreviated as *E. coli* K12.

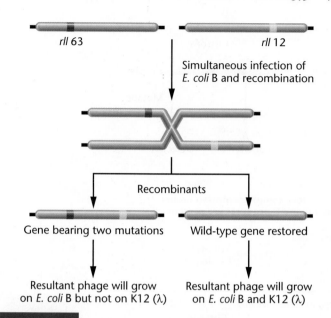

rII 63   rII 12

Simultaneous infection of
E. coli B and recombination

Recombinants

Gene bearing two mutations   Wild-type gene restored

Resultant phage will grow   Resultant phage will grow
on E. coli B but not on K12 (λ)   on E. coli B and K12 (λ)

FIGURE 6–19 Illustration of intragenic recombination between two mutations in the *rII* locus of phage T4. The result is the production of a wild-type phage, which will grow on both *E. coli* B and K12, and of a phage that has incorporated both mutations into the *rII* locus. The latter will grow on *E. coli* B but not on *E. coli* K12.

less than 0.1 percent wild-type phages, were then allowed to infect strain K12, the wild-type recombinants would successfully reproduce and produce wild-type plaques. *This is the critical step in recovering and quantifying rare recombinants.*

By using serial dilution techniques, Benzer was able to determine the total number of mutant *rII* phages produced on *E. coli* B and the total number of recombinant wild-type phages that would lyse *E. coli* K12. These data provided the basis for calculating the frequency of recombination, a value proportional to the distance within the gene between the two mutations being studied. As we will see, this experimental design was extraordinarily sensitive. Remarkably, it was possible for Benzer to detect as few as one recombinant wild-type phage among 100 million mutant phages. When information from many such experiments is combined, a detailed map of the locus is possible.

Before we discuss this mapping, we need to describe an important discovery Benzer made during the early development of his screen—a discovery that led to the development of a technique used widely in genetics labs today, the **complementation assay** you learned about in Chapter 4.

Complementation by *rII* Mutations

Before Benzer was able to initiate these intragenic recombination studies, he had to resolve a problem encountered during the early stages of his experimentation. While doing a control study in which K12 bacteria were simultaneously infected with pairs of different *rII* mutant strains, Benzer sometimes found that certain pairs of the *rII* mutant strains lysed the K12 bacteria. This was initially quite puzzling, since only the wild-type *rII*

was supposed to be capable of lysing K12 bacteria. How could two mutant strains of *rII*, each of which was thought to contain a defect in the same gene, show a wild-type function?

Benzer reasoned that, during simultaneous infection, each mutant strain provided something that the other lacked, thus restoring wild-type function. This phenomenon, which he called **complementation**, is illustrated in Figure 6–20(a). When many pairs of mutations were tested, each mutation fell into one of two possible **complementation groups**, A or B. Those that failed to complement one another were placed in the same complementation group, while those that did complement one another were each assigned to a different complementation group. Benzer coined the term **cistron**, which he defined as the smallest functional genetic unit, to describe a complementation group. In modern terminology, we know that a cistron represents a gene.

We now know that Benzer's A and B cistrons represent two separate genes in what we originally referred to as the *rII* locus (because of the initial assumption that it was a single gene). Complementation occurs when K12 bacteria are infected with two *rII* mutants, one with a mutation in the A gene and one with a mutation in the B gene. Therefore, there is a source of both wild-type gene products, since the A mutant provides wild-type B and the B mutant provides wild-type A. We can also explain why two strains that fail to complement, say two A-cistron mutants, are actually mutations in the same gene. In this case, if two A-cistron mutants are combined, there will be an immediate source of the wild-type B product, but no immediate source of the wild-type A product [Figure 6–20(b)].

Once Benzer was able to place all *rII* mutations in either the A or the B cistron, he was set to return to his intragenic recombination studies, testing mutations in the A cistron against each other and testing mutations in the B cistron against each other.

NOW SOLVE THIS

6–3 In complementation studies of the *rII* locus of phage T4, three groups of three different mutations were tested. For each group, only two combinations were tested. On the basis of each set of data (shown here), predict the results of the third experiment for each group.

Group A	Group B	Group C
d × *e*–lysis	*g* × *b*–no lysis	*j* × *k*–lysis
d × *f*–no lysis	*g* × *i*–no lysis	*j* × *l*–lysis
e × *f*–?	*b* × *i*–?	*k* × *l*–?

■ HINT: *This problem involves an understanding of why complementation occurs during simultaneous infection of a bacterial cell by two bacteriophage strains, each with a different mutation within the rII locus. The key to its solution is to be aware that if each mutation alters a different genetic product, then each strain will provide the product that the other is missing, thus leading to complementation.*

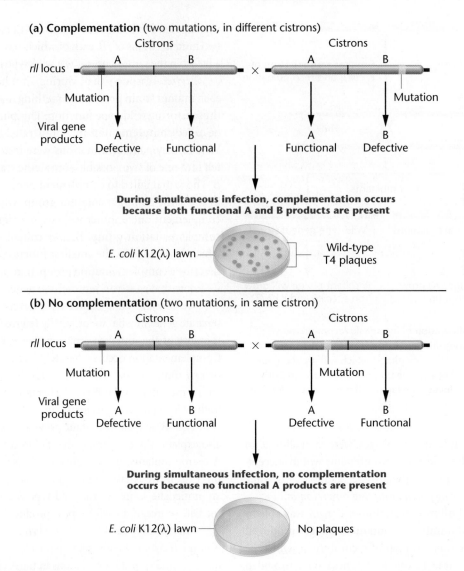

FIGURE 6–20 Comparison of two pairs of *rII* mutations. (a) In one case, they complement one another. (b) In the other case, they do not complement one another. Complementation occurs when each mutation is in a separate cistron. Failure to complement occurs when the two mutations are in the same cistron.

Recombinational Analysis

Of the approximately 20,000 *rII* mutations, roughly half fell into each cistron. Benzer set about mapping the mutations within each one. For example, if two *rII* A mutants (i.e., two phage strains with different mutations in the A cistron) were first allowed to infect *E. coli* B in a liquid culture, and if a recombination event occurred between the mutational sites in the A cistron, then wild-type progeny viruses would be produced at low frequency. If samples of the progeny viruses from such an experiment were then plated on *E. coli* K12, only the wild-type recombinants would lyse the bacteria and produce plaques. The total number of nonrecombinant progeny viruses would be determined by plating samples on *E. coli* B.

This experimental protocol is illustrated in Figure 6–21. The percentage of recombinants can be determined by counting the plaques at the appropriate dilution in each case. As in eukaryotic mapping experiments, the frequency of recombination is an estimate of the distance between the two mutations within the cistron. For example, if the number of recombinants is equal to 4×10^3/mL, and the total number of progeny is 8×10^9/mL, then the frequency of recombination between the two mutants is

$$2\left(\frac{4 \times 10^3}{8 \times 10^9}\right) = 2(0.5 \times 10^{-6})$$
$$= 10^{-6}$$
$$= 0.000001$$

Multiplying by 2 is necessary because each recombinant event yields two reciprocal products, only one of which—the wild type—is detected.

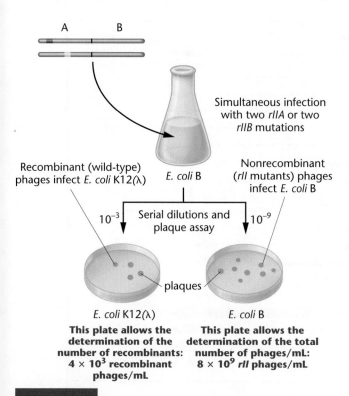

Simultaneous infection with two *rIIA* or two *rIIB* mutations

Recombinant (wild-type) phages infect *E. coli* K12(λ)

E. coli B

Nonrecombinant (*rII* mutants) phages infect *E. coli* B

10^{-3}

Serial dilutions and plaque assay

10^{-9}

plaques

E. coli K12(λ)

E. coli B

This plate allows the determination of the number of recombinants: 4×10^3 recombinant phages/mL

This plate allows the determination of the total number of phages/mL: 8×10^9 *rII* phages/mL

FIGURE 6–21 The experimental protocol for recombination studies between pairs of mutations in the same cistron. In this figure, all phage infecting *E. coli* B (in the flask) contain one of two mutations in the A cistron, as shown in the depiction of their chromosomes to the left of the flask.

Deletion Testing of the *rII* Locus

Although the system for assessing recombination frequencies described earlier allowed for mapping mutations within each cistron, testing 1000 mutants two at a time in all combinations would have required millions of experiments. Fortunately, Benzer was able to overcome this obstacle when he devised an analytical approach referred to as **deletion testing**. He discovered that some of the *rII* mutations were, in reality, deletions of small parts of both cistrons. That is, the genetic changes giving rise to the *rII* properties were not a characteristic of point mutations. Most importantly, when a deletion mutation was tested using simultaneous infection by two phage strains, one having the deletion mutation and the other having a point mutation located in the deleted part of the same cistron, the test never yielded wild-type recombinants. The reason is illustrated in Figure 6–22. Because the deleted area is lacking the area of DNA containing the point mutation, no recombination is possible. Thus, a method was available that could roughly, but quickly, localize any mutation, provided it was contained within a region covered by a deletion.

Deletion testing could thus provide data for the initial localization of each mutation. For example, as shown in Figure 6–23, seven overlapping deletions spanning various regions of the A cistron were used for the initial screening of point mutations in that cistron. Depending on whether the viral chromosome bearing a point mutation does or does not undergo recombination with the chromosome bearing a deletion, each point mutation can be assigned to a specific area of the cistron. Further deletions within each of the seven areas can be used to localize, or map, each *rII* point mutation more precisely. Remember that, in each case, a point mutation is localized in the area of a deletion when it fails to give rise to any wild-type recombinants.

The *rII* Gene Map

After several years of work, Benzer produced a genetic map of the two cistrons composing the *rII* locus of phage T4 (Figure 6–24). From the 20,000 mutations analyzed, 307 distinct sites within this locus were mapped in relation to one another. Areas containing many mutations, designated as **hot spots**, were apparently more susceptible to mutation than were areas in which only one or a few mutations were found. In addition, Benzer discovered areas within the cistrons in which no mutations were localized. He estimated that as many as 200 recombinational units had not been localized by his studies.

The significance of Benzer's work is his application of genetic analysis to what had previously been considered an abstract unit—the gene. Benzer had demonstrated in 1955 that a gene is not an indivisible particle, but instead consists of mutational and recombinational units that are arranged in a specific order. Today, we know these are nucleotides composing DNA. His analysis, performed prior to the detailed molecular studies of the gene in the 1960s, is considered a classic example of genetic experimentation.

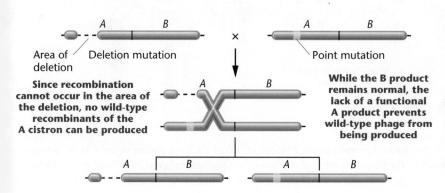

A

B

×

A

B

Area of deletion

Deletion mutation

Point mutation

Since recombination cannot occur in the area of the deletion, no wild-type recombinants of the A cistron can be produced

A

B

While the B product remains normal, the lack of a functional A product prevents wild-type phage from being produced

A

B

A

B

FIGURE 6–22 Demonstration that recombination between a phage chromosome with a deletion in the A cistron and another phage with a point mutation overlapped by that deletion cannot yield a chromosome with wild-type A and B cistrons.

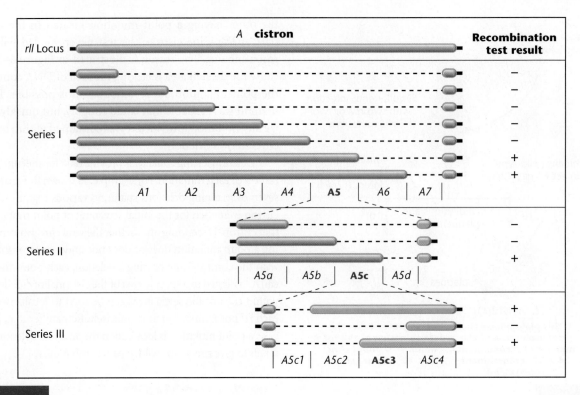

FIGURE 6–23   Three series of overlapping deletions in the *A* cistron of the *rII* locus used to localize the position of an unknown *rII* mutation. For example, if a mutant strain tested against each deletion (dashed areas) in Series I for the production of recombinant wild-type progeny shows the results at the right (+ or −), the mutation must be in segment *A5*. In Series II, the mutation is further narrowed to segment *A5c*, and in Series III to segment *A5c3*.

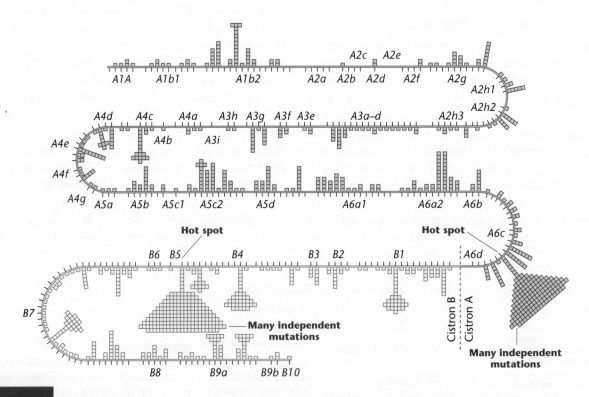

FIGURE 6–24   A partial map of mutations in the *A* and *B* cistrons of the *rII* locus of phage *T4*. Each square represents an independently isolated mutation. Note the two areas in which the largest number of mutations are present, referred to as "hot spots" (*A6cd* and *B5*).

GENETICS, TECHNOLOGY, AND SOCIETY

From Cholera Genes to Edible Vaccines

Using an expanding toolbox of molecular genetic tools, scientists are tackling some of the most serious bacterial diseases affecting our species. Our ability to clone bacterial genes and transfer them into other organisms is leading directly to exciting new treatments. The story of edible vaccines for the treatment of cholera illustrates how genetic engineering is being applied to control a serious human disease.

Cholera is caused by *Vibrio cholerae*, a curved, rod-shaped bacterium found in rivers and oceans. Most genetic strains of *V. cholerae* are harmless; only a few are pathogenic. Infection occurs when a person drinks water or eats food contaminated with pathogenic *V. cholerae*. Once in the digestive system, these bacteria colonize the small intestine and produce proteins called enterotoxins that invade the mucosal cells lining the intestine. This triggers a massive secretion of water and dissolved salts resulting in violent diarrhea, severe dehydration, muscle cramps, lethargy, and often death. The enterotoxin consists of two polypeptides, called the A and B subunits, encoded by two separate genes.

Cholera remains a leading cause of human deaths throughout the Third World, where basic sanitation is lacking and water supplies are often contaminated. For example, in July 1994, 70,000 cases of cholera leading to 12,000 fatalities were reported among the Rwandans crowded into refugee camps in Goma, Zaire. And after an absence of over 100 years, cholera reappeared in Latin America in 1991, spreading from Peru to Mexico and claiming more than 10,000 lives. Following the 2010 earthquakes in Haiti, a severe cholera outbreak spread through the country, claiming more than 4000 lives.

A new gene-based technology is emerging to attack cholera. This technology centers on genetically engineered plants that act as vaccines. Scientists introduce a cloned gene—such as a gene encoding a bacterial protein—into the plant genome. The transgenic plant produces the new gene product, and immunity is acquired when an animal eats the plant. The gene product in the plant acts as an antigen, stimulating the production of antibodies to protect against bacterial infection or the effects of their toxins. Since the B subunit of the cholera enterotoxin binds to intestinal cells, research has focused on using this polypeptide as the antigen, with the hope that antibodies against it will prevent toxin binding and render the bacteria harmless.

Leading the efforts to develop an edible vaccine are Charles Arntzen and associates at Cornell University. To test the system, they are using the B subunit of an *E. coli* enterotoxin, which is similar in structure and immunological properties to the cholera protein. Their first step was to obtain the DNA clone of the gene encoding the B subunit and to attach it to a promoter that would induce transcription in all tissues of the plant. Second, the researchers introduced the hybrid gene into potato plants by means of *Agrobacterium*-mediated transformation. The engineered plants expressed their new gene and produced the enterotoxin B subunit. Third, they fed mice a few grams of the genetically engineered tubers. Arntzen's group found that the mice produced specific antibodies against the B subunit and secreted them into the small intestine. When they fed purified enterotoxin to the mice, the mice were protected from its effects and did not develop the symptoms of cholera. In clinical trials conducted using humans in 1998, almost all of the volunteers developed an immune response, and none experienced adverse side effects.

Arntzen's experiments have served as models for other research efforts involving edible vaccines. Currently, scientists are developing edible vaccines against bacterial diseases such as anthrax and tetanus, as well as viral diseases such as rabies, AIDS, and measles.

Your Turn

Take time, individually or in groups, to answer the following questions. Investigate the references and links to help you understand some of the issues that surround the development and uses of edible vaccines.

1. What are the latest research developments on edible vaccines for cholera?

   A source of information is the PubMed Web site (http://www.ncbi.nlm.nih.gov/sites/entrez?db=PubMed), *as described in the "Exploring Genomics" feature in Chapter 2.*

2. Several oral vaccines against cholera are currently available. Given the availability of these vaccines, why do you think that scientists are also developing edible vaccines for cholera? Which vaccine type would you choose for vaccinating populations at risk for cholera?

   Read about these cholera vaccines on the World Health Organization Web site at http://www.who.int/topics/cholera/vaccines/current/en/index.html

3. Cholera is spread through ingestion of contaminated water and food. Despite its severity, cholera patients can be effectively treated by oral rehydration. Cholera becomes a major health problem only when proper sanitation and medical treatment are lacking. Given these facts, how much research funding do you think we should spend to develop cholera vaccines, relative to funds spent on improved sanitation, water treatment, and education about treatments?

   A discussion of cholera prevention and treatment can be found at http://en.wikipedia.org/wiki/Cholera

4. One of the problems associated with edible vaccines is the public's concern about genetically modified organisms (GMOs). Attitudes vary from outright moral opposition to GMOs to concern about the potential environmental hazards associated with growing transgenic plants. How do you feel about the use of GMOs? What do you think are the most valid arguments for and against them, and why?

   Scientists also debate these issues. One such debate is presented in the article Arntzen, C. J., et al., 2003, GM crops: Science, politics and communication, Nature Rev Genet 4: 839–843.

CASE STUDY | To treat or not to treat

A 4-month-old infant had been running a moderate fever for 36 hours, and a nervous mother made a call to her pediatrician. Examination and testing revealed no outward signs of infection or cause of the fever. The anxious mother asked the pediatrician about antibiotics, but the pediatrician recommended watching the infant carefully for two days before making a decision. He explained that decades of rampant use of antibiotics in medicine and agriculture had caused a worldwide surge in bacteria that are now resistant to such drugs. He also said that the reproductive behavior of bacteria allows them to exchange antibiotic resistance traits with a wide range of other disease-causing bacteria, and that many strains are now resistant to multiple antibiotics. The physician's information raises several interesting questions.

1. Was the physician correct in saying that bacteria can share resistance?

2. Where do bacteria carry antibiotic resistance genes, and how are they exchanged?

3. If the infant was given an antibiotic as a precaution, how might it contribute to the production of resistant bacteria?

4. Aside from hospitals, where else would infants and children come in contact with antibiotic-resistant strains of bacteria? Does the presence of such bacteria in the body always mean an infection?

Summary Points

 For activities, animations, and review quizzes, go to the study area at www.masteringgenetics.com

1. Genetic recombination in bacteria takes place in three ways: conjugation, transformation, and transduction.

2. Conjugation may be initiated by a bacterium housing a plasmid called the F factor in its cytoplasm, making it a donor cell. Following conjugation, the recipient cell receives a copy of the F factor and is converted to the F$^+$ status.

3. When the F factor is integrated from the cytoplasm into the chromosome, the cell remains as a donor and is referred to as an Hfr cell. Upon mating, the donor chromosome moves unidirectionally into the recipient, initiating recombination and providing the basis for time mapping of the bacterial chromosome.

4. Plasmids, such as the F factor, are autonomously replicating DNA molecules found in the bacterial cytoplasm, sometimes containing unique genes conferring antibiotic resistance as well as the genes necessary for plasmid transfer during conjugation.

5. Transformation in bacteria, which does not require cell-to-cell contact, involves exogenous DNA that enters a recipient bacterium and recombines with the host's chromosome. Linkage mapping of closely aligned genes is possible during the analysis of transformation.

6. Bacteriophages, viruses that infect bacteria, demonstrate a well-defined life cycle where they reproduce within the host cell and can be studied using the plaque assay.

7. Bacteriophages can be lytic, meaning they infect the host cell, reproduce, and then lyse it, or in contrast, they can lysogenize the host cell, where they infect it and integrate their DNA into the host chromosome, but do not reproduce.

8. Transduction is virus-mediated bacterial DNA recombination. When a lysogenized bacterium subsequently reenters the lytic cycle, the new bacteriophages serve as vehicles for the transfer of host (bacterial) DNA.

9. Various mutant phenotypes, including mutations in plaque morphology and host range, have been studied in bacteriophages and have served as the basis for mapping in these viruses.

10. Transduction is also used for bacterial linkage and mapping studies.

11. Various mutant phenotypes, including mutations in plaque morphology and host range, have been studied in bacteriophages. These have served as the basis for investigating genetic exchange and mapping in these viruses.

12. Genetic analysis of the *rII* locus in bacteriophage T4 allowed Seymour Benzer to study intragenic recombination. By isolating *rII* mutants and performing complementation analysis, recombinational studies, and deletion mapping, Benzer was able to locate and map more than 300 distinct sites within the two cistrons of the *rII* locus.

INSIGHTS AND SOLUTIONS

1. Time mapping is performed in a cross involving the genes *his*, *leu*, *mal*, and *xyl*. The recipient cells were auxotrophic for all four genes. After 25 minutes, mating was interrupted with the following results in recipient cells. Diagram the positions of these genes relative to the origin (*O*) of the F factor and to one another.

   (a) 90% were xyl^+

   (b) 80% were mal^+

   (c) 20% were his^+

   (d) none were leu^+

   Solution: The *xyl* gene was transferred most frequently, which shows it is closest to *O* (very close). The *mal* gene is next closest and reasonably near *xyl*, followed by the more distant *his* gene. The *leu* gene is far beyond these three, since no recombinants are recovered that include it. The diagram shows these relative locations along a piece of the circular chromosome.

2. Three strains of bacteria, each bearing a separate mutation, a^-, b^-, or c^-, are the sources of donor DNA in a transformation experiment. Recipient cells are wild type for those genes but express the mutation d^-.

   (a) Based on the following data, and assuming that the location of the *d* gene precedes the *a*, *b*, and *c* genes, propose a linkage map for the four genes.

DNA Donor	Recipient	Transformants	Frequency of Transformants
a^-d^+	a^+d^-	a^+d^+	0.21
b^-d^+	b^+d^-	b^+d^+	0.18
c^-d^+	c^+d^-	c^+d^+	0.63

   (b) If the donor DNA were wild type and the recipient cells were either a^-b^-, a^-c^-, or b^-c^-, which of the crosses would be expected to produce the greatest number of wild-type transformants?

   Solution: (a) These data reflect the relative distances between the *a*, *b*, and *c* genes, individually, and the *d* gene. The *a* and *b* genes are about the same distance from the *d* gene and are thus tightly linked to one another. The *c* gene is more distant. Assuming that the *d* gene precedes the others, the map looks like this:

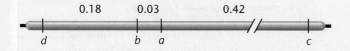

   (b) Because the a and b genes are closely linked, they most likely cotransform in a single event. Thus, recipient cells $a^-\ b^-$ are most likely to convert to wild type.

3. For his fine-structure analysis of the *rII* locus in phage T4, Benzer was able to perform complementation testing of any pair of mutations once it was clear that the locus contained two cistrons. Complementation was assayed by simultaneously infecting *E. coli* K12 with two phage strains, each with an independent mutation, neither of which could alone lyse K12. From the data that follow, determine which mutations are in which cistron, assuming that mutation 1 (M-1) is in the A cistron and mutation 2 (M-2) is in the B cistron. Are there any cases where the mutation cannot be properly assigned?

Test Pair	Results*
1, 2	+
1, 3	−
1, 4	−
1, 5	+
2, 3	−
2, 4	+
2, 5	−

   *+ or – indicates complementation or the failure of complementation, respectively.

   Solution: M-1 and M-5 complement one another and, therefore, are not in the same cistron. Thus, M-5 must be in the B cistron. M-2 and M-4 complement one another. By the same reasoning, M-4 is not with M-2 and, therefore, is in the A cistron. M-3 fails to complement either M-1 or M-2, and so it would seem to be in both cistrons. One explanation is that the physical cause of M-3 somehow overlaps both the A and the B cistrons. It might be a double mutation with one sequence change in each cistron. It might also be a deletion that overlaps both cistrons and thus could not complement either M-1 or M-2.

4. Another mutation, M-6, was tested with the results shown here:

Test Pair	Results
1, 6	+
2, 6	−
3, 6	−
4, 6	+
5, 6	−

   Draw all possible conclusions about M-6.

   Solution: These results are consistent with assigning M-6 to the B cistron.

5. Recombination testing was then performed for M-2, M-5, and M-6 so as to map the B cistron. Recombination analysis using both *E. coli* B and K12 showed that recombination occurred between M-2 and M-5 and between M-5 and M-6, but not between M-2 and M-6. Why not?

    Solution: Either M-2 and M-6 represent identical mutations, or one of them may be a deletion that overlaps the other but does not overlap M-5. Furthermore, the data cannot rule out the possibility that both are deletions.

6. In recombination studies of the *rII* locus in phage T4, what is the significance of the value determined by calculating phage growth in the K12 versus the B strains of *E. coli*

following simultaneous infection in *E. coli* B? Which value is always greater?

Solution: When plaque analysis is performed on *E. coli* B, in which the wild-type and mutant phages are both lytic, the total number of phages per milliliter can be determined. Because almost all cells are *rII* mutants of one type or another, this value is much larger than the value obtained with K12. To avoid total lysis of the plate, extensive dilution is necessary. In K12, *rII* mutations will not grow, but wild-type phages will. Because wild-type phages are the rare recombinants, there are relatively few of them and extensive dilution is not required.

Problems and Discussion Questions

 For instructor-assigned tutorials and problems, go to www.masteringgentics.com

HOW DO WE KNOW?

1. In this chapter, we have focused on genetic systems present in bacteria and on the viruses that use bacteria as hosts (bacteriophages). In particular, we discussed mechanisms by which bacteria and their phages undergo genetic recombination, which allows geneticists to map bacterial and bacteriophage chromosomes. In the process, we found many opportunities to consider how this information was acquired. From the explanations given in the chapter, what answers would you propose to the following questions?

    (a) How do we know that genes exist in bacteria and bacteriophages?

    (b) How do we know that bacteria undergo genetic recombination, allowing the transfer of genes from one organism to another?

    (c) How do we know whether or not genetic recombination between bacteria involves cell-to-cell contact?

    (d) How do we know that bacteriophages recombine genetic material through transduction and that cell-to-cell contact is not essential for transduction to occur?

    (e) How do we know that intergenic exchange occurs in bacteriophages?

    (f) How do we know that in bacteriophage T4 the *rII* locus is subdivided into two regions, or cistrons?

2. Distinguish among the three modes of recombination in bacteria.

3. With respect to F^+ and F^- bacterial matings, answer the following questions:

    (a) How was it established that physical contact between cells was necessary?

    (b) How was it established that chromosome transfer was unidirectional?

    (c) What is the genetic basis for a bacterium's being F^+?

4. List all major differences between (a) the $F^+ \times F^-$ and the Hfr $\times$ F^- bacterial crosses; and (b) the F^+, F^- Hfr, and F' bacteria.

5. Describe the basis for chromosome mapping in the Hfr $\times$ F^- crosses.

6. In general, when recombination experiments are conducted with bacteria, participating bacteria are mixed in complete medium, then transferred to a minimal growth medium. Why isn't the protocol reversed; minimal medium first, complete medium second?

7. Why are the recombinants produced from an Hfr $\times$ F^- cross rarely, if ever, F^+?

8. Describe the origin of F' bacteria and merozygotes.

9. In a transformation experiment, donor DNA was obtained from a prototroph bacterial strain ($a^+b^+c^+$), and the recipient was a triple auxotroph ($a^-b^-c^-$). What general conclusions can you draw about the linkage relationships among the three genes from the following transformant classes that were recovered?

a^+ b^- c^-		180
a^- b^+ c^-		150
a^+ b^+ c^-		210
a^- b^- c^+		179
a^+ b^- c^+		2
a^- b^+ c^+		1
a^+ b^+ c^+		3

10. Describe the role of heteroduplex formation during transformation.

11. Explain the observations that led Zinder and Lederberg to conclude that the prototrophs recovered in their transduction experiments were not the result of F^+ mediated conjugation.

12. Define plaque, lysogeny, and prophage.

13. Differentiate between generalized and specialized transduction.

14. Two theoretical genetic strains of a virus ($a^-b^-c^-$ and $a^+b^+c^+$) were used to simultaneously infect a culture of host bacteria. Of 10,000 plaques scored, the following genotypes

were observed. Determine the genetic map of these three genes on the viral chromosome. Decide whether interference was positive or negative.

$a^+\ b^+\ c^+$	4100		$a^-\ b^+\ c^-$	160
$a^-\ b^-\ c^-$	3990		$a^+\ b^-\ c^+$	140
$a^+\ b^-\ c^-$	740		$a^-\ b^-\ c^+$	90
$a^-\ b^+\ c^+$	670		$a^+\ b^+\ c^-$	110

15. The bacteriophage genome consists of many genes encoding proteins that make up the head, collar, tail, and tail fibers. When these genes are transcribed following phage infection, how are these proteins synthesized, since the phage genome lacks genes essential to ribosome structure?

16. If a single bacteriophage infects one *E. coli* cell present on a lawn of bacteria and, upon lysis, yields 200 viable viruses, how many phages will exist in a single plaque if three more lytic cycles occur?

17. A phage-infected bacterial culture was subjected to a series of dilutions, and a plaque assay was performed in each case, with the results shown in the following table. What conclusion can be drawn in the case of each dilution, assuming that 0.1 mL was used in each plaque assay?

	Dilution Factor	Assay Results
(a)	10^4	All bacteria lysed
(b)	10^5	14 plaques
(c)	10^6	0 plaques

18. In recombination studies of the *rII* locus in phage T4, what is the significance of the value determined by calculating phage growth in the K12 versus the B strains of *E. coli* following simultaneous infection in *E. coli* B? Which value is always greater?

19. In an analysis of other *rII* mutants, complementation testing yielded the following results:

Mutants	Results (+ / − lysis)
1, 2	+
1, 3	+
1, 4	−
1, 5	−

(a) Predict the results of testing 2 and 3, 2 and 4, and 3 and 4 together.
(b) If further testing yielded the following results, what would you conclude about mutant 5?

Mutants	Results
2, 5	−
3, 5	−
4, 5	−

20. Using mutants 2 and 3 from the previous problem, following mixed infection on *E. coli* B, progeny viruses were plated in a series of dilutions on both *E. coli* B and K12 with the following results. What is the recombination frequency between the two mutants?

Strain Plated	Dilution	Plaques
E. coli B	10^{-5}	2
E. coli K12	10^{-1}	5

21. Another mutation, 6, was tested in relation to mutations 1 through 5 from the previous problems. In initial testing, mutant 6 complemented mutants 2 and 3. In recombination testing with 1, 4, and 5, mutant 6 yielded recombinants with 1 and 5, but not with 4. What can you conclude about mutation 6?

Extra-Spicy Problems

 For instructor-assigned tutorials and problems, go to www.masteringgentics.com

22. During the analysis of seven *rII* mutations in phage T4, mutants 1, 2, and 6 were in cistron A, while mutants 3, 4, and 5 were in cistron B. Of these, mutant 4 was a deletion overlapping mutant 5. The remainder were point mutations. Nothing was known about mutant 7. Predict the results of complementation (+ or −) between 1 and 2; 1 and 3; 2 and 4; and 4 and 5.

23. In studies of recombination between mutants 1 and 2 from the previous problem, the results shown in the following table were obtained.

Strain	Dilution	Plaques	Phenotypes
E. coli B	10^{-7}	4	*r*
E. coli K12	10^{-2}	8	+

(a) Calculate the recombination frequency.
(b) When mutant 6 was tested for recombination with mutant 1, the data were the same as those shown above for strain B, but not for K12. The researcher lost the K12 data, but remembered that recombination was ten times more frequent than when

mutants 1 and 2 were tested. What were the lost values (dilution and colony numbers)?
(c) Mutant 7 (Problem 22) failed to complement any of the other mutants (1–6). Define the nature of mutant 7.

24. In *Bacillus subtilis*, linkage analysis of two mutant genes affecting the synthesis of two amino acids, tryptophan (trp_2^-) and tyrosine (tyr_1^-), was performed using transformation. Examine the following data and draw all possible conclusions regarding linkage. What is the purpose of Part B of the experiment? [Reference: E. Nester, M. Schafer, and J. Lederberg (1963).]

Donor DNA	Recipient Cell	Transformants	No.
A. $trp_2^+\ tyr_1^+$	$trp_2^-\ tyr_1^-$	$trp^+\ tyr^-$	196
		$trp^-\ tyr^+$	328
		$trp^+\ tyr^+$	367
B. $trp_2^+\ tyr_1^-$ and $trp_2^-\ tyr_1^+$	$trp_2^-\ tyr_1^-$	$trp^+\ tyr^-$	190
		$trp^-\ tyr^+$	256
		$trp^+\ tyr^+$	2

25. An Hfr strain is used to map three genes in an interrupted mating experiment. The cross is $Hfr/a^+b^+c^+rif \times F^-/a^-b^-c^-rif^r$. (No map order is implied in the listing of the alleles; rif^r is resistance to the antibiotic rifampicin.) The a^+ gene is required for the biosynthesis of nutrient A, the b^+ gene for nutrient B, and c^+ for nutrient C. The minus alleles are auxotrophs for these nutrients. The cross is initiated at time = 0, and at various times, the mating mixture is plated on three types of medium. Each plate contains minimal medium (MM) plus rifampicin plus specific supplements that are indicated in the following table. (The results for each time interval are shown as the number of colonies growing on each plate.)

	Time of Interruption			
	5 min	10 min	15 min	20 min
Nutrients A and B	0	0	4	21
Nutrients B and C	0	5	23	40
Nutrients A and C	4	25	60	82

(a) What is the purpose of rifampicin in the experiment?
(b) Based on these data, determine the approximate location on the chromosome of the a, b, and c genes relative to one another and to the F factor.
(c) Can the location of the rif gene be determined in this experiment? If not, design an experiment to determine the location of rif relative to the F factor and to gene b.

26. A plaque assay is performed beginning with 1 mL of a solution containing bacteriophages. This solution is serially diluted three times by combining 0.1 mL of each sequential dilution with 9.9 mL of liquid medium. Then 0.1 mL of the final dilution is plated in the plaque assay and yields 17 plaques. What is the initial density of bacteriophages in the original 1 mL?

27. In a cotransformation experiment, using various combinations of genes two at a time, the following data were produced. Determine which genes are "linked" to which others.

Successful Cotransformation	Unsuccessful Cotransformation
a and d; b and c;	a and b; a and c; a and f;
b and f	d and b; d and c; d and f
	a and e; b and e; c and e
	d and e; f and e

28. For the experiment in Problem 27, another gene, g, was studied. It demonstrated positive cotransformation when tested with gene f. Predict the results of testing gene g with genes a, b, c, d, and e.

29. Bacterial conjugation, mediated mainly by conjugative plasmids such as F, represents a potential health threat through the sharing of genes for pathogenicity or antibiotic resistance. Given that more than 400 different species of bacteria coinhabit a healthy human gut and more than 200 coinhabit human skin, Francisco Dionisio [*Genetics* (2002) 162:1525–1532] investigated the ability of plasmids to undergo between-species conjugal transfer. The following data are presented for various species of the enterobacterial genus *Escherichia*. The data are presented as "log base 10" values; for example, -2.0 would be equivalent to 10^{-2} as a rate of transfer. Assume that all differences between values presented are statistically significant.

(a) What general conclusion(s) can be drawn from these data?
(b) In what species is within-species transfer most likely? In what species pair is between-species transfer most likely?
(c) What is the significance of these findings in terms of human health?

	Donor			
Recipient	E. chrysanthemi	E. blattae	E. fergusonii	E. coli
E. chrysanthemi	−2.4	−4.7	−5.8	−3.7
E. blattae	−2.0	−3.4	−5.2	−3.4
E. fergusonii	−3.4	−5.0	−5.8	−4.2
E. coli	−1.7	−3.7	−5.3	−3.5

30. A study was conducted in an attempt to determine which functional regions of a particular conjugative transfer gene (*tra1*) are involved in the transfer of plasmid R27 in *Salmonella enterica*. The R27 plasmid is of significant clinical interest because it is capable of encoding multiple-antibiotic resistance to typhoid fever. To identify functional regions responsible for conjugal transfer, an analysis by Lawley et al. (2002. *J. Bacteriol.* 184:2173–2180) was conducted in which particular regions of the *tra1* gene were mutated and tested for their impact on conjugation. Shown here is a map of the regions tested and believed to be involved in conjugative transfer of the plasmid. Similar coloring indicates related function. Numbers correspond to each functional region subjected to mutation analysis.

Accompanying the map is a table showing the effects of these mutations on R27 conjugation.

Effects of Mutations in Functional Regions of Transfer Region 1 (*tra1*) on R27 Conjugation

R27 Mutation in Region	Conjugative Transfer	Relative Conjugation Frequency (%)
1	+	100
2	+	100
3	−	0
4	+	100
5	−	0
6	−	0
7	+	12
8	−	0
9	−	0
10	−	0
11	+	13
12	−	0
13	−	0
14	−	0

(a) Given the data, do all functional regions appear to influence conjugative transfer?

(b) Which regions appear to have the most impact on conjugation?

(c) Which regions appear to have a limited impact on conjugation?

(d) What general conclusions might one draw from these data?

31. Influenza (the flu) is responsible for approximately 250,000 to 500,000 deaths annually, but periodically its toll has been much higher. For example, the 1918 flu pandemic killed approximately 30 million people worldwide and is considered the worst spread of a deadly illness in recorded history. With highly virulent flu strains emerging periodically, it is little wonder that the scientific community is actively studying influenza biology. In 2007, the National Institute of Allergy and Infectious Diseases completed sequencing of 2035 human and avian influenza virus strains. Influenza strains undergo recombination as described in this chapter, and they have a high mutation rate owing to the error-prone replication of their genome (which consists of RNA rather than DNA). In addition, they are capable of chromosome reassortment in which various combinations of their eight chromosomes (or portions thereof) can be packaged into progeny viruses when two or more strains infect the same cell. The end result is that we can make vaccines, but they must change annually, and even then, we can only guess at what specific viral strains will be prevalent in any given year. Based on the above information, consider the following questions:

(a) Of what evolutionary value to influenza viruses are high mutation and recombination rates coupled with chromosome reassortment?

(b) Why can't humans combat influenza just as they do mumps, measles, or chicken pox?

(c) Why are vaccines available for many viral diseases but not influenza?

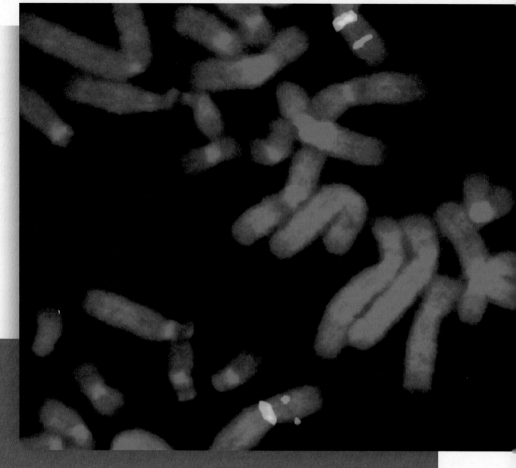

Human X chromosomes highlighted using fluorescence *in situ* hybridization (FISH), a method in which specific probes bind to specific sequences of DNA. The probe producing green fluorescence binds to the DNA of the X chromosome centromeres. The probe producing red fluorescence binds to the DNA sequence of the X-linked Duchenne muscular dystrophy (DMD) gene.

7

Sex Determination and Sex Chromosomes

CHAPTER CONCEPTS

- A variety of mechanisms have evolved that result in sexual differentiation, leading to sexual dimorphism and greatly enhancing the production of genetic variation within species.

- Often, specific genes, usually on a single chromosome, cause maleness or femaleness during development.

- In humans, the presence of extra X or Y chromosomes beyond the diploid number may be tolerated but often leads to syndromes demonstrating distinctive phenotypes.

- While segregation of sex-determining chromosomes should theoretically lead to a one-to-one sex ratio of males to females, in humans the actual ratio greatly favors males at conception.

- In mammals, females inherit two X chromosomes compared to one in males, but the extra genetic information in females is compensated for by random inactivation of one of the X chromosomes early in development.

- In some reptilian species, temperature during incubation of eggs determines the sex of offspring.

n the biological world, a wide range of reproductive modes and life cycles are observed. Some organisms are entirely asexual, displaying no evidence of sexual reproduction. Some organisms alternate between short periods of sexual reproduction and prolonged periods of asexual reproduction. In most diploid eukaryotes, however, sexual reproduction is the only natural mechanism for producing new members of the species. Orderly transmission of genetic material from parents to offspring, and the resultant phenotypic variability, relies on the processes of segregation and independent assortment that occur during meiosis. Meiosis produces haploid gametes so that, following fertilization, the resulting offspring maintain the diploid number of chromosomes characteristic of their kind. Thus, meiosis ensures genetic constancy within members of the same species.

These events, seen in the perpetuation of all sexually reproducing organisms, depend ultimately on an efficient union of gametes during fertilization. In turn, successful fertilization depends on some form of sexual differentiation in the reproductive organisms. Even though it is not overtly evident, this differentiation occurs in such diverse organisms as bacteria, archaea, and unicellular eukaryotes such as algae. In many animal species, including humans, the differentiation of the sexes is more evident as phenotypic dimorphism of males and females. The ancient symbol for iron and for Mars, depicting a shield and spear ($\male$), and the ancient symbol for copper and for Venus, depicting a mirror ($\female$), have also come to symbolize maleness and femaleness, respectively.

Dissimilar, or **heteromorphic chromosomes**, such as the XY pair in mammals, characterize one sex or the other in a wide range of species, resulting in their label as **sex chromosomes**. Nevertheless, in many species, genes rather than chromosomes ultimately serve as the underlying basis of **sex determination**. As we will see, some of these genes are present on sex chromosomes, but others are autosomal. Extensive investigation has revealed variation in sex-chromosome systems—even in closely related organisms—suggesting that mechanisms controlling sex determination have undergone rapid evolution many times in the history of life.

In this chapter, we review some representative modes of sexual differentiation by examining the life cycles of three model organisms often studied in genetics: the green alga *Chlamydomonas*; the maize plant, *Zea mays*; and the nematode (roundworm), *Caenorhabditis elegans*. These organisms contrast the different roles that sexual differentiation plays in the lives of diverse organisms. Then, we delve more deeply into what is known about the genetic basis for the determination of sexual differences, with a particular emphasis on two organisms: our own species, representative of mammals; and *Drosophila*, on which pioneering sex-determining studies were performed.

Life Cycles Depend on Sexual Differentiation

In describing sexual dimorphism (differences between males and females) in multicellular animals, biologists distinguish between **primary sexual differentiation**, which involves only the gonads, where gametes are produced, and **secondary sexual differentiation**, which involves the overall appearance of the organism, including clear differences in such organs as mammary glands and external genitalia as well as in nonreproductive organs. In plants and animals, the terms **unisexual, dioecious**, and **gonochoric** are equivalent; they all refer to an individual containing only male *or* only female reproductive organs. Conversely, the terms **bisexual, monoecious**, and **hermaphroditic** refer to individuals containing both male *and* female reproductive organs, a common occurrence in both the plant and animal kingdoms. These organisms can produce both eggs and sperm. The term **intersex** is usually reserved for individuals of an intermediate sexual condition, most of whom are sterile.

Chlamydomonas

The life cycle of the green alga *Chlamydomonas* (Cla may da moan us), shown in **Figure 7–1**, is representative of organisms exhibiting only infrequent periods of sexual reproduction. Such organisms spend most of their life cycle in a haploid phase, asexually producing daughter cells by mitotic divisions. However, under unfavorable nutrient conditions, such as nitrogen depletion, certain daughter cells function as gametes, joining together in fertilization. Following fertilization, a diploid zygote, which may withstand the unfavorable environment, is formed. When conditions change for the better, meiosis ensues and haploid vegetative cells are again produced.

In such species, there is little visible difference between the haploid vegetative cells that reproduce asexually and the haploid gametes that are involved in sexual reproduction. Moreover, the two gametes that fuse during mating are not usually morphologically distinguishable from one another, which is why they are called **isogametes** (*iso*- means equal, or uniform). Species producing them are said to be **isogamous**.

In 1954, Ruth Sager and Sam Granick demonstrated that gametes in *Chlamydomonas* could be subdivided into two **mating types**. Working with clones derived from single haploid cells, they showed that cells from a given clone mate with cells from some but not all other clones. When they tested the mating abilities of large numbers of clones, all could be placed into one of two mating categories, either mt^+ or mt^- cells. "Plus" cells mate only with "minus" cells, and vice versa. Following fertilization, which involves fusion

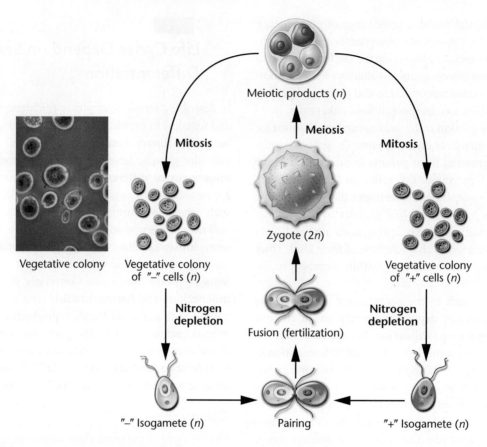

Meiotic products (*n*)

Mitosis

Meiosis

Mitosis

Vegetative colony

Vegetative colony of *"–"* cells (*n*)

Zygote (2*n*)

Vegetative colony of *"+"* cells (*n*)

Nitrogen depletion

Nitrogen depletion

Fusion (fertilization)

"–" Isogamete (*n*)

Pairing

"+" Isogamete (*n*)

FIGURE 7–1   The life cycle of *Chlamydomonas*. Unfavorable conditions stimulate the formation of isogametes of opposite mating types that may fuse in fertilization. The resulting zygote undergoes meiosis, producing two haploid cells of each mating type. The photograph shows vegetative cells of this green alga.

of entire cells, and subsequently, meiosis, the four haploid cells, or **zoospores**, produced (see the top of Figure 7–1) were found to consist of two plus types and two minus types.

Further experimentation established that plus and minus cells differ chemically. When extracts are prepared from cloned *Chlamydomonas* cells (or their flagella) of one type and then added to cells of the opposite mating type, clumping, or agglutination, occurs. No such agglutination occurs if the extracts are added to cells of the mating type from which they were derived. These observations suggest that despite the morphological similarities between isogametes, they are differentiated biochemically. Therefore, in this alga, a primitive means of sex differentiation exists, even though there is no morphological indication that such differentiation has occurred. Further research has pinpointed the *mt* locus to *Chlamydomonas* chromosome VI and has identified the gene that mediates the expression of the *mt⁻* mating type, which is essential for cell fusion in response to nitrogen depletion.

Zea mays

The life cycles of many plants alternate between the *haploid gametophyte stage* and the *diploid sporophyte stage* (see Figure 2–13). The processes of meiosis and fertilization

link the two phases during the life cycle. The relative amount of time spent in the two phases varies between the major plant groups. In some nonseed plants, such as mosses, the haploid gametophyte phase and the morphological structures representing this stage predominate. The reverse is true in seed plants.

Maize (*Zea mays*), familiar to you as corn, exemplifies a *monoecious* seed plant, meaning a plant in which the sporophyte phase and the morphological structures representing that phase predominate during the life cycle. Both male and female structures are present on the adult plant. Thus, sex determination occurs differently in different tissues of the same organism, as shown in the life cycle of this plant (Figure 7–2). The stamens, which collectively constitute the tassel, produce diploid microspore mother cells, each of which undergoes meiosis and gives rise to four haploid microspores. Each haploid microspore in turn develops into a mature male microgametophyte—the pollen grain—which contains two haploid sperm nuclei.

Equivalent female diploid cells, known as megaspore mother cells, are produced in the pistil of the sporophyte. Following meiosis, only one of the four haploid megaspores survives. It usually divides mitotically three times, producing

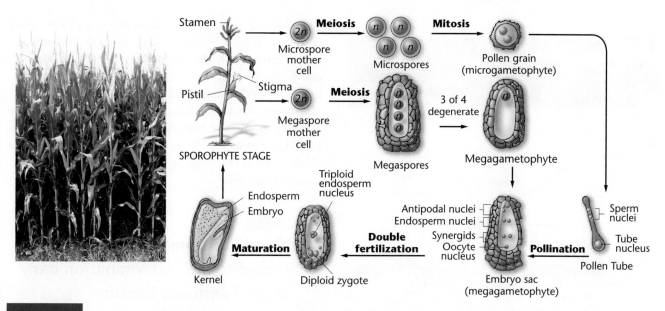

FIGURE 7–2 The life cycle of maize (*Zea mays*). The diploid sporophyte bears stamens and pistils that give rise to haploid microspores and megaspores, which develop into the pollen grain and the embryo sac that ultimately house the sperm and oocyte, respectively. Following fertilization, the embryo develops within the kernel and is nourished by the endosperm. Germination of the kernel gives rise to a new sporophyte (the mature corn plant), and the cycle repeats itself.

a total of eight haploid nuclei enclosed in the embryo sac. Two of these nuclei unite near the center of the embryo sac, becoming the endosperm nuclei. At the micropyle end of the sac, where the sperm enters, sit three other nuclei: the oocyte nucleus and two synergids. The remaining three, antipodal nuclei, cluster at the opposite end of the embryo sac.

Pollination occurs when pollen grains make contact with the silks (or stigma) of the pistil and develop long pollen tubes that grow toward the embryo sac. When contact is made at the micropyle, the two sperm nuclei enter the embryo sac. One sperm nucleus unites with the haploid oocyte nucleus, and the other sperm nucleus unites with two endosperm nuclei. This process, known as *double fertilization*, creates the diploid zygote nucleus and the triploid endosperm nucleus, respectively. Each ear of corn can contain as many as 1000 of these structures, each of which develops into a single kernel. Each kernel, if allowed to germinate, gives rise to a new plant, the *sporophyte*.

The mechanism of sex determination and differentiation in a monoecious plant such as *Zea mays*, where the tissues that form the male and female gametes have the same genetic constitution, was difficult to unravel at first. However, the discovery of a number of mutant genes that disrupt normal tassel and pistil formation supports the concept that normal products of these genes play an important role in sex determination by affecting the differentiation of male or female tissue in several ways.

For example, mutant genes altering the sexual development of florets provide valuable information. When homozygous, mutations classified as *tassel seed* (*ts1 and ts2*) convert male tassels to female pistils. The wild-type form of these genes normally eliminate the cells forming pistils by inducing their cell death. Thus, a single gene can cause a normally monoecious plant to become exclusively female. On the other hand, the recessive mutations *silkless* (*sk*) and *barren stalk* (*ba*) interfere with the development of the pistil, resulting in plants with only male-functioning reproductive organs, provided that the plant has wild-type *ts* genes.

Data gathered from studies of these and other mutants suggest that the products of many wild-type alleles of these genes interact in controlling sex determination. Following sexual differentiation of florets into either male or female structures, male or female gametes are produced.

Caenorhabditis elegans

The nematode worm *Caenorhabditis elegans* [*C. elegans*, for short; Figure 7–3(a)] is a popular organism in genetic studies, particularly for investigating the genetic control of development. Its usefulness is based on the fact that adults consist of 959 somatic cells, the precise lineage of which can be traced back to specific embryonic origins.

There are two sexual phenotypes in these worms: males, which have only testes, and hermaphrodites, which during larval development form two gonads, which subsequently produce both sperm and eggs. The eggs that are produced are fertilized by the stored sperm in a process of self-fertilization.

The outcome of this process is quite interesting [**Figure 7–3(b)**]. The vast majority of organisms that result

(a)

(b)

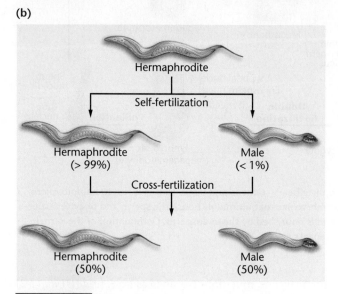

Hermaphrodite

Self-fertilization

Hermaphrodite   Male
(> 99%)   (< 1%)

Cross-fertilization

Hermaphrodite   Male
(50%)   (50%)

FIGURE 7–3   (a) Photomicrograph of a hermaphroditic nematode, *C. elegans*; (b) the outcomes of self-fertilization in a hermaphrodite, and a mating of a hermaphrodite and a male worm.

are hermaphrodites, like the parental worm; less than 1 percent of the offspring are males. As adults, males can mate with hermaphrodites, producing about half male and half hermaphrodite offspring.

The signal that determines maleness in contrast to hermaphroditic development is provided by the expression of

NOW SOLVE THIS

7–1 The marine echiurid worm *Bonellia viridis* is an extreme example of environmental influence on sex determination. Undifferentiated larvae either remain free-swimming and differentiate into females, or they settle on the proboscis of an adult female and become males. If larvae that have been on a female proboscis for a short period are removed and placed in seawater, they develop as intersexes. If larvae are forced to develop in an aquarium where pieces of proboscises have been placed, they develop into males. Contrast this mode of sexual differentiation with that of mammals. Suggest further experimentation to elucidate the mechanism of sex determination in *B. viridis*.

■ HINT: *This problem asks you to analyze experimental findings related to sex determination. The key to its solution is to devise further testing with the goal of isolating the unknown factor affecting sex determination and testing it experimentally.*

genes located on both the X chromosome and autosomes. *C. elegans* lacks a Y chromosome altogether—hermaphrodites have two X chromosomes, while males have only one X chromosome. It is believed that the ratio of X chromosomes to the number of sets of autosomes ultimately determines the sex of these worms. A ratio of 1.0 (two X chromosomes and two copies of each autosome) results in hermaphrodites, and a ratio of 0.5 results in males. The absence of a heteromorphic Y chromosome is not uncommon in organisms.

X and Y Chromosomes Were First Linked to Sex Determination Early in the Twentieth Century

How sex is determined has long intrigued geneticists. In 1891, H. Henking identified a nuclear structure in the sperm of certain insects, which he labeled the X-body. Several years later, Clarence McClung showed that some of the sperm in grasshoppers contain an unusual genetic structure, which he called a *heterochromosome*, but the remainder of the sperm lack this structure. He mistakenly associated the presence of the heterochromosome with the production of male progeny. In 1906, Edmund B. Wilson clarified Henking and McClung's findings when he demonstrated that female somatic cells in the butterfly *Protenor* contain 14 chromosomes, including two X chromosomes. During oogenesis, an even reduction occurs, producing gametes with seven chromosomes, including one X chromosome. Male somatic cells, on the other hand, contain only 13 chromosomes, including one X chromosome. During spermatogenesis, gametes are produced containing either six chromosomes, without an X, or seven chromosomes, one of which is an X. Fertilization by X-bearing sperm results in female offspring, and fertilization by X-deficient sperm results in male offspring [Figure 7–4(**a**)].

The presence or absence of the X chromosome in male gametes provides an efficient mechanism for sex determination in this species and also produces a 1:1 sex ratio in the resulting offspring. This mechanism, now called the **XX/XO**, or ***Protenor*, mode of sex determination**, depends on the random distribution of the X chromosome into one-half of the male gametes during segregation. As we saw earlier, *C. elegans* exhibits this system of sex determination.

Wilson also experimented with the milkweed bug *Lygaeus turcicus*, in which both sexes have 14 chromosomes. Twelve of these are autosomes (A). In addition, the females have two X chromosomes, while the males have only a single X and a smaller heterochromosome labeled the **Y chromosome**. Females in this species produce only

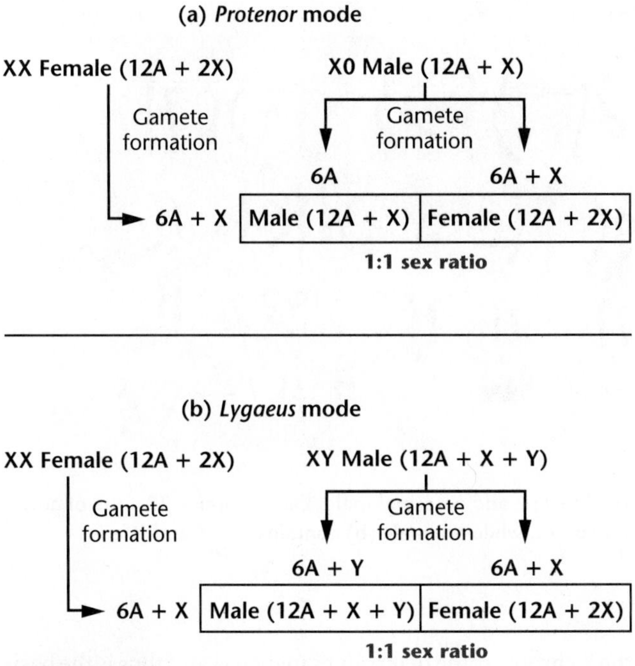

(a) *Protenor* **mode**

XX Female (12A + 2X)   X0 Male (12A + X)

Gamete formation   Gamete formation

6A   6A + X

6A + X | Male (12A + X) | Female (12A + 2X) |

1:1 sex ratio

(b) *Lygaeus* **mode**

XX Female (12A + 2X)   XY Male (12A + X + Y)

Gamete formation   Gamete formation

6A + Y   6A + X

6A + X | Male (12A + X + Y) | Female (12A + 2X) |

1:1 sex ratio

FIGURE 7–4 (a) The *Protenor* mode of sex determination where the heterogametic sex (the male in this example) is X0 and produces gametes with or without the X chromosome; (b) the *Lygaeus* mode of sex determination, where the heterogametic sex (again, the male in this example) is XY and produces gametes with either an X or a Y chromosome. In both cases, the chromosome composition of the offspring determines its sex.

gametes of the (6A + X) constitution, but males produce two types of gametes in equal proportions, (6A + X) and (6A + Y). Therefore, following random fertilization, equal numbers of male and female progeny will be produced with distinct chromosome complements. This mode of sex determination is called the *Lygaeus,* or **XX/XY, mode of sex determination** [Figure 7–4(b)].

In *Protenor* and *Lygaeus* insects, males produce unlike gametes. As a result, they are described as the **heterogametic sex**, and in effect, their gametes ultimately determine the sex of the progeny in those species. In such cases, the female, who has like sex chromosomes, is the **homogametic sex**, producing uniform gametes with regard to chromosome numbers and types.

The male is not always the heterogametic sex. In some organisms, the female produces unlike gametes, exhibiting either the *Protenor* XX/XO or *Lygaeus* XX/XY mode of sex determination. Examples include certain moths and butterflies, some fish, reptiles, amphibians, at least one species of plants (*Fragaria orientalis*), and most birds. To immediately distinguish situations in which the female is the heterogametic sex, some geneticists use the notation **ZZ/ZW**, where ZZ is the homogametic male and ZW is the heterogametic female, instead of the XX/XY notation. For example, chickens are so denoted.

The Y Chromosome Determines Maleness in Humans

The first attempt to understand sex determination in our own species occurred almost 100 years ago and involved the visual examination of chromosomes in dividing cells. Efforts were made to accurately determine the diploid chromosome number of humans, but because of the relatively large number of chromosomes, this proved to be quite difficult. In 1912, H. von Winiwarter counted 47 chromosomes in a spermatogonial metaphase preparation. It was believed that the sex-determining mechanism in humans was based on the presence of an extra chromosome in females, who were thought to have 48 chromosomes. However, in the 1920s, Theophilus Painter counted between 45 and 48 chromosomes in cells of testicular tissue and also discovered the small Y chromosome, which is now known to occur only in males. In his original paper, Painter favored 46 as the diploid number in humans, but he later concluded incorrectly that 48 was the chromosome number in both males and females.

For 30 years, this number was accepted. Then, in 1956, Joe Hin Tjio and Albert Levan discovered a better way to prepare chromosomes for viewing. This improved technique led to a strikingly clear demonstration of metaphase stages showing that 46 was indeed the human diploid number. Later that same year, C. E. Ford and John L. Hamerton, also working with testicular tissue, confirmed this finding. The familiar karyotypes of humans are shown in Figure 7–5.

Of the normal 23 pairs of human chromosomes, one pair was shown to vary in configuration in males and females. These two chromosomes were designated the X and Y sex chromosomes. The human female has two X chromosomes, and the human male has one X and one Y chromosome.

We might believe that this observation is sufficient to conclude that the Y chromosome determines maleness. However, several other interpretations are possible. The Y could play no role in sex determination; the presence of two X chromosomes could cause femaleness; or maleness could result from the lack of a second X chromosome. The evidence that clarified which explanation was correct came from study of the effects of human sex-chromosome variations, described below. As such investigations revealed, the Y chromosome does indeed determine maleness in humans.

Klinefelter and Turner Syndromes

In about 1940, scientists identified two human abnormalities characterized by aberrant sexual development, **Klinefelter syndrome (47,XXY)** and **Turner syndrome (45,X).***

---

*Although the possessive form of the names of eponymous syndromes is sometimes used (e.g., Klinefelter's syndrome), the current preference is to use the nonpossessive form.

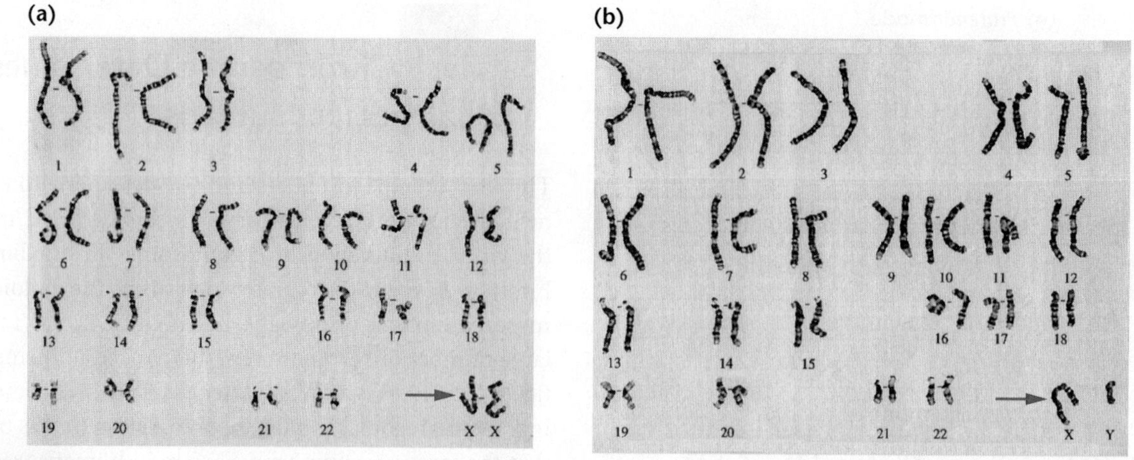

(a)   (b)

FIGURE 7–5   The traditional human karyotypes derived from a normal female and a normal male. Each contains 22 pairs of autosomes and two sex chromosomes. The female (a) contains two X chromosomes, while the male (b) contains one X and one Y chromosome.

Individuals with Klinefelter syndrome are generally tall and have long arms and legs and large hands and feet. They usually have genitalia and internal ducts that are male, but their testes are rudimentary and fail to produce sperm. At the same time, feminine sexual development is not entirely suppressed. Slight enlargement of the breasts (gynecomastia) is common, and the hips are often rounded. This ambiguous sexual development, referred to as intersexuality, can lead to abnormal social development. Intelligence is often below the normal range as well.

In Turner syndrome, the affected individual has female external genitalia and internal ducts, but the ovaries are rudimentary. Other characteristic abnormalities include short stature (usually under 5 feet), cognitive impairment, skin folds on the back of the neck, and underdeveloped breasts. A broad, shieldlike chest is sometimes noted. In 1959, the karyotypes of individuals with these syndromes were determined to be abnormal with respect to the sex chromosomes. Individuals with Klinefelter syndrome have more than one X chromosome. Most often they have an XXY complement in addition to 44 autosomes [Figure 7–6(a)], which is why people with this karyotype are designated 47,XXY. Individuals with Turner syndrome most often have only 45 chromosomes, including just a single X chromosome; thus, they are designated 45,X [**Figure 7–6(b)**]. Note the convention used in designating these chromosome compositions: the number states the total number of chromosomes present, and the symbols after the comma indicate the deviation from the normal diploid content. Both conditions result from **nondisjunction**, the failure of the X chromosomes to segregate properly during meiosis (nondisjunction is described in Chapter 8 and illustrated in Figure 8–1).

These Klinefelter and Turner karyotypes and their corresponding sexual phenotypes led scientists to conclude that the Y chromosome determines maleness and thus is the basis for phenotypic sex determination in humans. In its absence, the person's sex is female, even if only a single X chromosome is present. The presence of the Y chromosome in the presence of two X chromosomes characteristic of Klinefelter syndrome is sufficient to determine maleness, even though male development is not complete. Similarly, in the absence of a Y chromosome, as in the case of individuals with Turner syndrome, no masculinization occurs. Note that we cannot conclude anything regarding sex determination under circumstances where a Y chromosome is present without an X because Y-containing human embryos lacking an X chromosome (designated 45,Y) do not survive.

Klinefelter syndrome occurs in about 1 of every 660 male births. The karyotypes **48,XXXY, 48,XXYY, 49,XXXXY**, and **49,XXXYY** are similar phenotypically to 47,XXY, but manifestations are often more severe in individuals with a greater number of X chromosomes.

Turner syndrome can also result from karyotypes other than 45,X, including individuals called **mosaics**, whose somatic cells display two different genetic cell lines, each exhibiting a different karyotype. Such cell lines result from a mitotic error during early development, the most common chromosome combinations being **45,X/46,XY** and **45,X/46,XX**. Thus, an embryo that began life with a normal karyotype can give rise to an individual whose cells show a mixture of karyotypes and who exhibits varying aspects of this syndrome.

Turner syndrome is observed in about 1 in 2000 female births, a frequency much lower than that for Klinefelter syndrome. One explanation for this difference is the observation that the majority of 45,X fetuses die *in utero* and are aborted spontaneously. Thus, a similar frequency of the two syndromes may occur at conception.

(a)

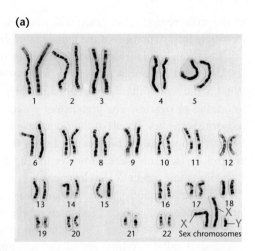

(b)

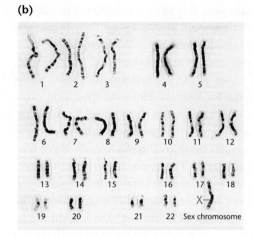

FIGURE 7–6 The karyotypes of individuals with (a) Klinefelter syndrome (47,XXY) and (b) Turner syndrome (45,X).

47,XXX Syndrome

The abnormal presence of three X chromosomes along with a normal set of autosomes (**47,XXX**) results in female differentiation. The highly variable syndrome that accompanies this genotype, often called **triplo-X**, occurs in about 1 of 1000 female births. Frequently, 47,XXX women are perfectly normal and may remain unaware of their abnormality in chromosome number unless a karyotype is done. In other cases, underdeveloped secondary sex characteristics, sterility, delayed development of language and motor skills, and mental retardation may occur. In rare instances, **48,XXXX** (tetra-X) and **49,XXXXX** (penta-X) karyotypes have been reported. The syndromes associated with these karyotypes are similar to but more pronounced than the 47,XXX syndrome. Thus, in many cases, the presence of additional X chromosomes appears to disrupt the delicate balance of genetic information essential to normal female development.

47,XYY Condition

Another human condition involving the sex chromosomes is **47,XYY**. Studies of this condition, where the only deviation from diploidy is the presence of an additional Y chromosome in an otherwise normal male karyotype, were initiated in 1965 by Patricia Jacobs. She discovered that 9 of 315 males in a Scottish maximum security prison had the 47,XYY karyotype. These males were significantly above average in height and had been incarcerated as a result of antisocial (nonviolent) criminal acts. Of the nine males studied, seven were of subnormal intelligence, and all suffered personality disorders. Several other studies produced similar findings.

The possible correlation between this chromosome composition and criminal behavior piqued considerable interest, and extensive investigation of the phenotype and frequency of the 47,XYY condition in both criminal and noncriminal populations ensued. Above-average height (usually over 6 feet) and subnormal intelligence have been generally substantiated, and the frequency of males displaying this karyotype is indeed higher among people in penal and mental institutions than among unincarcerated populations (see Table 7.1). A particularly relevant question involves the characteristics displayed by XYY males who are not incarcerated. The only nearly constant association is that such individuals are over 6 feet tall.

TABLE 7.1

Frequency of XYY Individuals in Various Settings

Setting	Restriction	Number Studied	Number XYY	Frequency XYY
Control population	Newborns	28,366	29	0.10%
Mental–penal	No height restriction	4,239	82	1.93
Penal	No height restriction	5,805	26	0.44
Mental	No height restriction	2,562	8	0.31
Mental–penal	Height restriction	1,048	48	4.61
Penal	Height restriction	1,683	31	1.84
Mental	Height restriction	649	9	1.38

Source: Compiled from data presented in Hook, 1973, Tables 1–8. Copyright 1973 by the American Association for the Advancement of Science.

A study addressing this issue was initiated to identify 47,XYY individuals at birth and to follow their behavioral patterns during preadult and adult development. By 1974, the two investigators, Stanley Walzer and Park Gerald, had identified about 20 XYY newborns in 15,000 births at Boston Hospital for Women. However, they soon came under great pressure to abandon their research. Those opposed to the study argued that the investigation could not be justified and might cause great harm to individuals who displayed this karyotype. The opponents argued that (1) no association between the additional Y chromosome and abnormal behavior had been previously established in the population at large, and (2) "labeling" the individuals in the study might create a self-fulfilling prophecy. That is, as a result of participation in the study, parents, relatives, and friends might treat individuals identified as 47,XYY differently, ultimately producing the expected antisocial behavior. Despite the support of a government funding agency and the faculty at Harvard Medical School, Walzer and Gerald abandoned the investigation in 1975.

Since Walzer and Gerald's work, it has become clear that many XYY males are present in the population who do not exhibit antisocial behavior and who lead normal lives. Therefore, we must conclude that there is a high, but not constant, correlation between the extra Y chromosome and the predisposition of these males to exhibit behavioral problems.

Sexual Differentiation in Humans

Once researchers had established that, in humans, it is the Y chromosome that houses genetic information necessary for maleness, they attempted to pinpoint a specific gene or genes capable of providing the "signal" responsible for sex determination. Before we delve into this topic, it is useful to consider how sexual differentiation occurs in order to better comprehend how humans develop into sexually dimorphic males and females. During early development, every human embryo undergoes a period when it is potentially hermaphroditic. By the fifth week of gestation, gonadal primordia (the tissues that will form the gonad) arise as a pair of **gonadal (genital) ridges** associated with each embryonic kidney. The embryo is potentially hermaphroditic because at this stage its gonadal phenotype is sexually indifferent—male or female reproductive structures cannot be distinguished, and the gonadal ridge tissue can develop to form male or female gonads. As development progresses, primordial germ cells migrate to these ridges, where an outer cortex and inner medulla form (*cortex* and *medulla* are the outer and inner tissues of an organ, respectively). The cortex is capable of developing into an ovary, while the medulla may develop into a testis. In addition, two sets of undifferentiated ducts called the Wolffian and Müllerian ducts exist in each embryo. Wolffian ducts differentiate into other organs of the male reproductive tract, while Müllerian ducts differentiate into structures of the female reproductive tract.

Because gonadal ridges can form either ovaries or testes, they are commonly referred to as **bipotential gonads**. What is the switch that triggers gonadal ridge development into testes or ovaries? The presence or absence of a Y chromosome is the key. If cells of the ridge have an XY constitution, development of the medulla into a testis is initiated around the seventh week. However, in the absence of the Y chromosome, no male development occurs, the cortex of the ridge subsequently forms ovarian tissue, and the Müllerian duct forms oviducts (Fallopian tubes), uterus, cervix, and portions of the vagina. Depending on which pathway is initiated, parallel development of the appropriate male or female duct system then occurs, and the other duct system degenerates. If testes differentiation is initiated, the embryonic testicular tissue secretes hormones that are essential for continued male sexual differentiation. As we will discuss in the next section, presence of a Y chromosome and development of the testes also inhibit formation of female reproductive organs.

In females, as the 12th week of fetal development approaches, oogonia within the ovaries begin meiosis, and primary oocytes can be detected. By the 25th week of gestation, all oocytes become arrested in meiosis and remain dormant until puberty is reached some 10 to 15 years later. In males, on the other hand, primary spermatocytes are not produced until puberty is reached (refer to Figure 2–12).

The Y Chromosome and Male Development

The human Y chromosome, unlike the X, was long thought to be mostly blank genetically. It is now known that this is not true, even though the Y chromosome contains far fewer genes than does the X. Data from the Human Genome Project indicate that the Y chromosome has at least 75 genes, compared to 900–1400 genes on the X. Current analysis of these genes and regions with potential genetic function reveals that some have homologous counterparts on the X chromosome and others do not. For example, present on both ends of the Y chromosome are so-called **pseudoautosomal regions (PARs)** that share homology with regions on the X chromosome and synapse and recombine with it during meiosis. The presence of such a pairing region is critical to segregation of the X and Y chromosomes during male gametogenesis. The remainder of the chromosome, about 95 percent of it, does not synapse or recombine with the X chromosome. As a result, it was originally referred to as the *nonrecombining region of the Y* (*NRY*). More recently, researchers have designated this region as the **male-specific region of the Y (MSY)**. As you will see, some portions of the MSY share homology with genes on the X chromosome, and others do not.

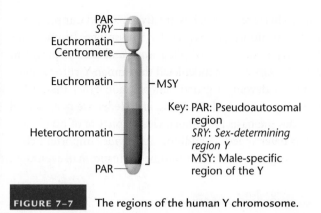

Key: PAR: Pseudoautosomal region
SRY: Sex-determining region Y
MSY: Male-specific region of the Y

FIGURE 7-7 The regions of the human Y chromosome.

The human Y chromosome is diagrammed in Figure 7–7. The MSY is divided about equally between *euchromatic* regions, containing functional genes, and *heterochromatic* regions, lacking genes. Within euchromatin, adjacent to the PAR of the short arm of the Y chromosome, is a critical gene that controls male sexual development, called the *sex-determining region Y (SRY)*. In humans, the absence of a Y chromosome almost always leads to female development; thus, this gene is absent from the X chromosome. At 6 to 8 weeks of development, the *SRY* gene becomes active in XY embryos. *SRY* encodes a protein that causes the undifferentiated gonadal tissue of the embryo to form testes. This protein is called the **testis-determining factor (TDF)**.*

SRY (or a closely related version) is present in all mammals thus far examined, indicative of its essential function throughout this diverse group of animals. Our ability to identify the presence or absence of DNA sequences in rare individuals whose sex-chromosome composition does not correspond to their sexual phenotype has provided evidence that *SRY* is the gene responsible for male sex determination. For example, there are human males who have two X and no Y chromosomes. Often, attached to one of their X chromosomes is the region of the Y that contains *SRY*. There are also females who have one X and one Y chromosome. Their Y is almost always missing the *SRY* gene.

Further support of this conclusion involves experiments using **transgenic mice**. These animals are produced from fertilized eggs injected with foreign DNA that is subsequently incorporated into the genetic composition of the developing embryo. In normal mice, a chromosome region designated *Sry* has been identified that is comparable to *SRY* in humans. When mouse DNA containing *Sry* is injected into normal XX mouse eggs, most of the offspring develop into males.

The question of how the product of this gene triggers development of embryonic gonadal tissue into testes rather than ovaries has been under investigation for a number of years. TDF is believed to function as a *transcription factor*, a DNA-binding protein that interacts directly with regulatory sequences of other genes to stimulate their expression. Thus, while TDF behaves as a master switch that controls other genes downstream in the process of sexual differentiation, identifying TDF target genes has been difficult. To date, *Sox9* in mice is one such gene. Another potential target for activation by TDF that has been extensively studied is the gene for **Müllerian inhibiting substance (MIS**, also called Müllerian inhibiting hormone, MIH, or anti-Müllerian hormone). Cells of the developing testes secrete MIS. As its name suggests, MIS protein causes regression (atrophy) of cells in the Müllerian duct. Degeneration of the duct prevents formation of the female reproductive tract.

Other *autosomal genes* are part of a cascade of genetic expression initiated by *SRY*. Examples include the human *SOX9* gene, which when activated by *SRY*, leads to the differentiation of cells that form the seminiferous tubules that contain male germ cells. In the mouse fibroblast growth factor 9 (*Fgf9*) is upregulated in XY gonads. Testis development is completely blocked in gonads lacking *Fgf9*, and signs of ovarian development occur. Another gene, *SF1*, is involved in the regulation of enzymes affecting steroid metabolism. In mice, this gene is initially active in both the male and female bisexual genital ridge, persisting until the point in development when testis formation is apparent. At that time, its expression persists in males but is extinguished in females. Recent work using mice have suggested that testicular development may be actively repressed throughout the life of females by downregulating expression of specific genes. This is based on experiments showing that, in adult female mice, deletion of a gene *Foxl2*, which encodes a transcription factor, leads to transdifferentiation of the ovary into the testis. Establishment of the link between these various genes and sex determination has brought us closer to a complete understanding of how males and females arise in humans, but much work remains to be done.

Findings by David Page and his many colleagues have now provided a reasonably complete picture of the MSY region of the human Y chromosome. Page has spearheaded the detailed study of the Y chromosome for the past several decades. The MSY consists of about 23 million base pairs (23 Mb) and can be divided into three regions. The first region is the *X-transposed region*. It comprises about 15 percent of the MSY and was originally derived from the X chromosome in the course of human evolution (about 3 to 4 million years ago). The X-transposed region is 99 percent identical to region Xq21 of the modern human

---

*It is interesting to note that in chickens, a similar gene has recently been identified. Called *DMRT1*, it is located on the Z chromosome. This gene is the subject of Problem 35 in the Problems section at the end of the chapter.

X chromosome. Two genes, both with X chromosome homologs, are present in this region.

Research by Page and others has also revealed that sequences called **palindromes**—sequences of base pairs that read the same but in the opposite direction on complementary strands—are present throughout the MSY. Recombination between palindromes on sister chromatids of the Y during replication is a mechanism used to repair mutations in the Y. This discovery has fascinating implications concerning how the Y chromosome may maintain its size and structure.

In early 2010, Page and colleagues demonstrated the first comprehensive comparison of the Y chromosome structure from two species. One interesting finding was that the MSY of the human Y chromosome is very different in sequence structure than the MSY from chimpanzees. The study indicates that rapid evolution has occurred since separation of these species over 6 million years ago—a surprise given that primate sex chromosomes have been in existence for hundreds of millions of years. Over 30 percent of the chimpanzee MSY sequence has no homologous sequence in the human MSY. The chimpanzee MSY has lost many protein-coding genes compared to common ancestors but contains twice the number of palindromic sequences as the human MSY.

The second area of the MSY is designated the *X-degenerative region*. Comprising about 20 percent of the MSY, this region contains DNA sequences that are even more distantly related to those present on the X chromosome. The X-degenerative region contains 27 single-copy genes and a number of *pseudogenes* (genes whose sequences have degenerated sufficiently during evolution to render them nonfunctional). As with the genes present in the X-transposed region, all share some homology with counterparts on the X chromosome. One of these is the *SRY* gene, discussed earlier. Other X-degenerative genes that encode protein products are expressed ubiquitously in all tissues in the body, but *SRY* is expressed only in the testes.

The third area, the *ampliconic region*, contains about 30 percent of the MSY, including most of the genes closely associated with the development of testes. These genes lack counterparts on the X chromosome, and their expression is limited to the testes. There are 60 transcription units (genes that yield a product) divided among nine gene families in this region, most represented by multiple copies. Members of each family have nearly identical (>98 percent) DNA sequences. Each repeat unit is an **amplicon** and is contained within seven segments scattered across the euchromatic regions of both the short and long arms of the Y chromosome. Genes in the ampliconic region encode proteins specific to the development and function of the testes, and the products of many of these genes are directly related to fertility in males. It is currently believed that a great deal of male sterility in our population can be linked to mutations in these

genes. This recent work has greatly expanded our picture of the genetic information carried by this unique chromosome. It clearly refutes the so-called *wasteland theory*, prevalent only 20 years ago, that depicted the human Y chromosome as almost devoid of genetic information other than a few genes that cause maleness. The knowledge we have gained provides the basis for a much clearer picture of how maleness is determined. In addition, it provides important clues to the origin of the Y chromosome during human evolution.

NOW SOLVE THIS

7–2 Campomelic dysplasia (CMD1) is a congenital human syndrome featuring malformation of bone and cartilage. It is caused by an autosomal dominant mutation of a gene located on chromosome 17. Consider the following observations in sequence, and in each case, draw whatever appropriate conclusions are warranted.

(a) Of those with the syndrome who are karyotypically 46,XY, approximately 75 percent are sex reversed, exhibiting a wide range of female characteristics.

(b) The nonmutant form of the gene, called *SOX9*, is expressed in the developing gonad of the XY male, but not the XX female.

(c) The *SOX9* gene shares 71 percent amino acid coding sequence homology with the Y-linked *SRY* gene.

(d) CMD1 patients who exhibit a 46,XX karyotype develop as females, with no gonadal abnormalities.

■ HINT: *This problem asks you to apply the information presented in this chapter to a real-life example. The key to its solution is knowing that some genes are activated and produce their normal product as a result of expression of products of other genes found on different chromosomes—in this case, perhaps one that is on the Y chromosome.*

7.4

The Ratio of Males to Females in Humans Is Not 1.0

The presence of heteromorphic sex chromosomes in one sex of a species but not the other provides a potential mechanism for producing equal proportions of male and female offspring. This potential depends on the segregation of the X and Y (or Z and W) chromosomes during meiosis, such that half of the gametes of the heterogametic sex receive one of the chromosomes and half receive the other one. As we learned in the previous section, small pseudoautosomal regions of pairing homology do exist at both ends of the human X and Y chromosomes, suggesting that the X and Y chromosomes do synapse and then segregate into different gametes. Provided that both types of gametes are equally successful in fertilization and that the two sexes are equally viable during fetal and embryonic development, a 1:1 ratio of male and female offspring should result.

The actual proportion of male to female offspring, referred to as the **sex ratio**, has been assessed in two ways. The **primary sex ratio** reflects the proportion of males to females conceived in a population. The **secondary sex ratio** reflects the proportion of each sex that is born. The secondary sex ratio is much easier to determine but has the disadvantage of not accounting for any disproportionate embryonic or fetal mortality.

When the secondary sex ratio in the human population was determined in 1969 by using worldwide census data, it was found not to equal 1.0. For example, in the Caucasian population in the United States, the secondary ratio was a little less than 1.06, indicating that about 106 males were born for each 100 females. (In 1995, this ratio dropped to slightly less than 1.05.) In the African-American population in the United States, the ratio was 1.025. In other countries the excess of male births is even greater than is reflected in these values. For example, in Korea, the secondary sex ratio was 1.15.

Despite these ratios, it is possible that the primary sex ratio is 1.0 and is later altered between conception and birth. For the secondary ratio to exceed 1.0, then, prenatal female mortality would have to be greater than prenatal male mortality. However, this hypothesis has been examined and shown to be false. In fact, just the opposite occurs. In a Carnegie Institute study, reported in 1948, the sex of approximately 6000 embryos and fetuses recovered from miscarriages and abortions was determined, and fetal mortality was actually higher in males. On the basis of the data derived from that study, the primary sex ratio in U.S. Caucasians was estimated to be 1.079. It is now believed that the figure is much higher—between 1.20 and 1.60, suggesting that many more males than females are conceived in the human population.

It is not clear why such a radical departure from the expected primary sex ratio of 1.0 occurs. To come up with a suitable explanation, researchers must examine the assumptions on which the theoretical ratio is based:

1. Because of segregation, males produce equal numbers of X- and Y-bearing sperm.

2. Each type of sperm has equivalent viability and motility in the female reproductive tract.

3. The egg surface is equally receptive to both X- and Y-bearing sperm.

No direct experimental evidence contradicts any of these assumptions; however, the human Y chromosome is smaller than the X chromosome and therefore of less mass. Thus, it has been speculated that Y-bearing sperm are more motile than X-bearing sperm. If this is true, then the probability of a fertilization event leading to a male zygote is increased, providing one possible explanation for the observed primary ratio.

7.5

Dosage Compensation Prevents Excessive Expression of X-Linked Genes in Mammals

The presence of two X chromosomes in normal human females and only one X in normal human males is unique compared with the equal numbers of autosomes present in the cells of both sexes. On theoretical grounds alone, it is possible to speculate that this disparity should create a "genetic dosage" difference between males and females, with attendant problems, for all X-linked genes. There is the potential for females to produce twice as much of each product of all X-linked genes. The additional X chromosomes in both males and females exhibiting the various syndromes discussed earlier in this chapter are thought to compound this dosage problem. Embryonic development depends on proper timing and precisely regulated levels of gene expression. Otherwise, disease phenotypes or embryonic lethality can occur. In this section, we will describe research findings regarding X-linked gene expression that demonstrate a genetic mechanism of **dosage compensation** that balances the dose of X chromosome gene expression in females and males.

Barr Bodies

Murray L. Barr and Ewart G. Bertram's experiments with female cats, as well as Keith Moore and Barr's subsequent study in humans, demonstrate a genetic mechanism in mammals that compensates for X chromosome dosage disparities. Barr and Bertram observed a darkly staining body in interphase nerve cells of female cats that was absent in similar cells of males. In humans, this body can be easily demonstrated in female cells derived from the buccal mucosa (cheek cells) or in fibroblasts (undifferentiated connective tissue cells), but not in similar male cells (**Figure 7–8**). This highly condensed structure,

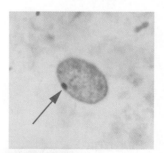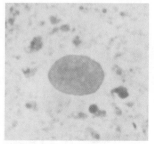

FIGURE 7–8 Photomicrographs comparing cheek epithelial cell nuclei from a male that fails to reveal Barr bodies (right) with a nucleus from a female that demonstrates a Barr body (indicated by the arrow in the left image). This structure, also called a sex chromatin body, represents an inactivated X chromosome.

about 1 μm in diameter, lies against the nuclear envelope of interphase cells. It stains positively in the Feulgen reaction, a cytochemical test for DNA.

Current experimental evidence demonstrates that this body, called a **sex chromatin body**, or simply a **Barr body**, is an inactivated X chromosome. Susumu Ohno was the first to suggest that the Barr body arises from one of the two X chromosomes. This hypothesis is attractive because it provides a possible mechanism for dosage compensation. If one of the two X chromosomes is inactive in the cells of females, the dosage of genetic information that can be expressed in males and females will be equivalent. Convincing, though indirect, evidence for this hypothesis comes from the study of the sex-chromosome syndromes described earlier in this chapter. Regardless of how many X chromosomes a somatic cell possesses, all but one of them appear to be inactivated and can be seen as Barr bodies. For example, no Barr body is seen in the somatic cells of Turner 45,X females; one is seen in Klinefelter 47,XXY males; two in 47,XXX females; three in 48,XXXX females; and so on (Figure 7–9). Therefore, the number of Barr bodies follows an $N - 1$ rule, where N is the total number of X chromosomes present.

Although this apparent inactivation of all but one X chromosome increases our understanding of dosage compensation, it further complicates our perception of other matters. For example, because one of the two X chromosomes is inactivated in normal human females, why then is the Turner 45,X individual not entirely normal?

Why aren't females with the triplo-X and tetra-X karyotypes (47,XXX and 48,XXXX) completely unaffected by the additional X chromosomes? Furthermore, in Klinefelter syndrome (47,XXY), X chromosome inactivation effectively renders the person 46,XY. Why aren't these males unaffected by the extra X chromosome in their nuclei?

One possible explanation is that chromosome inactivation does not normally occur in the very early stages of development of those cells destined to form gonadal tissues. Another possible explanation is that not all of each X chromosome forming a Barr body is inactivated. Recent studies have indeed demonstrated that as many as 15 percent of the human X-chromosomal genes actually escape inactivation. Clearly, then, not every gene on the X requires inactivation. In either case, excessive expression of certain X-linked genes might still occur at critical times during development despite apparent inactivation of superfluous X chromosomes.

The Lyon Hypothesis

In mammalian females, one X chromosome is of maternal origin, and the other is of paternal origin. Which one is inactivated? Is the inactivation random? Is the same chromosome inactive in all somatic cells? In 1961, Mary Lyon and Liane Russell independently proposed a hypothesis that answers these questions. They postulated that the inactivation of X chromosomes occurs randomly in somatic cells at a point early in embryonic development, most likely sometime during the blastocyst stage of development. Once inactivation has occurred, all descendant cells have the same X chromosome inactivated as their initial progenitor cell.

This explanation, which has come to be called the **Lyon hypothesis**, was initially based on observations of female mice heterozygous for X-linked coat color genes. The pigmentation of these heterozygous females was mottled, with large patches expressing the color allele on one X and other patches expressing the allele on the other X. This is the phenotypic pattern that would be expected if different X chromosomes were inactive in adjacent patches of cells. Similar mosaic patterns occur in the black and yellow-orange patches of female tortoiseshell and calico cats (Figure 7–10). Such X-linked coat color patterns do not occur in male cats because all their cells contain the single maternal X chromosome and are therefore hemizygous for only one X-linked coat color allele.

The most direct evidence in support of the Lyon hypothesis comes from studies of gene expression in clones of human fibroblast cells. Individual cells are isolated following biopsy and cultured *in vitro*. A culture of cells derived from a single cell is called a **clone**. The synthesis of the enzyme glucose-6-phosphate dehydrogenase (G6PD) is

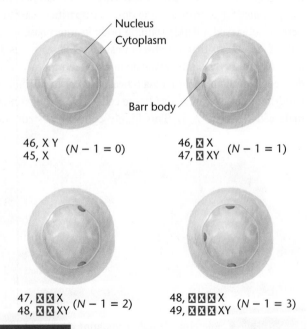

46, X Y
45, X $(N - 1 = 0)$

46, ⊠ X
47, ⊠ XY $(N - 1 = 1)$

47, ⊠ ⊠ X
48, ⊠ ⊠ XY $(N - 1 = 2)$

48, ⊠ ⊠ ⊠ X
49, ⊠ ⊠ ⊠ XY $(N - 1 = 3)$

FIGURE 7–9   Occurrence of Barr bodies in various human karyotypes, where all X chromosomes except one ($N - 1$) are inactivated.

(a)

(b)

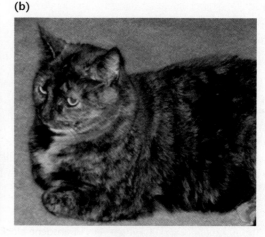

FIGURE 7–10   (a) The random distribution of orange and black patches in a calico cat illustrates the Lyon hypothesis. The white patches are due to another gene, distinguishing calico cats from tortoiseshell cats (b), which lack the white patches.

controlled by an X-linked gene. Numerous mutant alleles of this gene have been detected, and their gene products can be differentiated from the wild-type enzyme by their migration pattern in an electrophoretic field.

Fibroblasts have been taken from females heterozygous for different allelic forms of *G6PD* and studied. The Lyon hypothesis predicts that if inactivation of an X chromosome occurs randomly early in development, and thereafter all progeny cells have the same X chromosome inactivated as their progenitor, such a female should show two types of clones, each containing only one electrophoretic form of *G6PD*, in approximately equal proportions.

In 1963, Ronald Davidson and colleagues performed an experiment involving 14 clones from a single heterozygous female. Seven showed only one form of the enzyme, and 7 showed only the other form. Most important was the finding that none of the 14 clones showed both forms of the enzyme. Studies of *G6PD* mutants thus provide strong support for the random permanent inactivation of either the maternal or paternal X chromosome.

The Lyon hypothesis is generally accepted as valid; in fact, the inactivation of an X chromosome into a Barr body is sometimes referred to as **lyonization**. One extension of the hypothesis is that mammalian females are mosaics for all heterozygous X-linked alleles—some areas of the body express only the maternally derived alleles, and others express only the paternally derived alleles. An especially interesting example involves **red-green color blindness**, an X-linked recessive disorder. In humans, hemizygous males are fully color-blind in all retinal cells. However, heterozygous females display mosaic retinas, with patches of defective color perception and surrounding areas with normal color perception. In this example, random inactivation of one or the other X chromosome early in the development of heterozygous females has led to these phenotypes.

NOW SOLVE THIS

7–3 CC (Carbon Copy), the first cat produced from a clone, was created from an ovarian cell taken from her genetic donor, Rainbow, a calico cat. The diploid nucleus from the cell was extracted and then injected into an enucleated egg. The resulting zygote was then allowed to develop in a petri dish, and the cloned embryo was implanted in the uterus of a surrogate mother cat, who gave birth to CC. CC's surrogate mother was a tabby (see the photo on page 196 at the end of this chapter). Geneticists were very interested in the outcome of cloning a calico cat, because they were not certain if the cat would have patches of orange and black, just orange, or just black. Taking into account the Lyon hypothesis, explain the basis of the uncertainty. Would you expect CC to appear identical to Rainbow? Explain why or why not.

■ HINT: *This problem involves an understanding of the Lyon hypothesis. The key to its solution is to realize that the donor nucleus was from a differentiated ovarian cell of an adult female cat, which itself had inactivated one of its X chromosomes.*

The Mechanism of Inactivation

The least understood aspect of the Lyon hypothesis is the mechanism of X chromosome inactivation. Somehow, either DNA, the attached histone proteins, or both DNA and histone proteins, are chemically modified, silencing most genes that are part of that chromosome. Once silenced, a memory is created that keeps the same homolog inactivated following chromosome replications and cell divisions. Such a process, whereby expression of genes on one homolog, but not the other, is affected, is referred to as **imprinting**. This term also applies to a number of other examples in which genetic information is modified and gene expression is

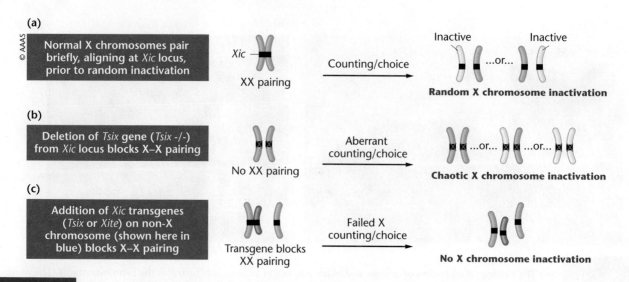

FIGURE 7–11 (a) Transient pairing of X chromosomes may be required for initiating X-inactivation. (b) Deleting the *Tsix* gene of the *Xic* locus prevents X–X pairing and leads to chaotic X-inactivation. (c) Blocking X–X pairing by addition of *Xic*-containing transgenes blocks X–X pairing and prevents X-inactivation.

repressed. Collectively, such events are part of the growing field of **epigenetics** (see Chapter 26).

Ongoing investigations are beginning to clarify the mechanism of inactivation. A region of the mammalian X chromosome is the major control unit. This region, located on the proximal end of the p arm in humans, is called the **X inactivation center (*Xic*)**, and its genetic expression *occurs only on the X chromosome that is inactivated*. The *Xic* is about 1 Mb (10^6 base pairs) in length and is known to contain several putative regulatory units and four genes. One of these, ***X-inactive specific transcript (XIST)***, is now known to be a critical gene for X-inactivation.

Several interesting observations have been made regarding the RNA that is transcribed from the *XIST* gene, many coming from experiments that used the equivalent gene in the mouse (*Xist*). First, the RNA product is quite large and does not encode a protein, and thus is not translated. The RNA products of *Xist* spread over and coat the X chromosome *bearing the gene that produced them*. Two other noncoding genes at the *Xic* locus, *Tsix* (an antisense partner of *Xist*) and *Xite*, are also believed to play important roles in X-inactivation.

A second observation is that transcription of *Xist* initially occurs at low levels on all X chromosomes. As the inactivation process begins, however, transcription continues, and is enhanced, only on the X chromosome that becomes inactivated. In 1996, a research group led by Neil Brockdorff and Graeme Penny provided convincing evidence that transcription of *Xist* is the critical event in chromosome inactivation. These researchers introduced a targeted deletion (7 kb) into this gene, disrupting its sequence. As a result, the chromosome bearing the deletion lost its ability to become inactivated. Interesting questions remain regarding imprinting and inactivation. For example, in cells with more than

two X chromosomes, what sort of "counting" mechanism exists that designates all but one X chromosome to be inactivated? Studies by Jeannie T. Lee and colleagues suggest that maternal and paternal X chromosomes must first pair briefly and align at their *Xic* loci as a mechanism for counting the number of X chromosomes prior to X-inactivation [Figure 7–11(a)]. Using mouse embryonic stem cells, Lee's group deleted the *Tsix* gene contained in the *Xic* locus. This deletion blocked X–X pairing and resulted in chaotic inactivation of 0, 1, or 2 X chromosomes [**Figure 7–11(b)**]. Lee and colleagues provided further evidence for the role of *Xic* locus in chromosome counting by adding copies of genetically engineered non-X chromosomes containing multiple copies of *Tsix* or *Xite*. (These are referred to as **transgenes** because they are artificially introduced into the organism.) This experimental procedure effectively blocked X–X pairing and prevented X chromosome inactivation [**Figure 7–11(c)**].

Other genes and protein products are being examined for their role in X chromosome pairing and counting. Recent studies by Lee and colleagues have provided evidence that the inactivated X must associate with regions at the periphery of the nucleus to maintain a state of silenced gene expression. Indeed, in a majority of human female somatic cells the inactivated X, present as a Barr body, is observed attached to the nuclear envelope.

Many questions remain. What "blocks" the *Xic* locus of the active chromosome, preventing further transcription of *Xist*? How does imprinting impart a memory such that inactivation of the same X chromosome or chromosomes is subsequently maintained in progeny cells, as the Lyon hypothesis calls for? Whatever the answers to these questions, scientists have taken exciting steps toward understanding how dosage compensation is accomplished in mammals.

The Ratio of X Chromosomes to Sets of Autosomes Determines Sex in *Drosophila*

Because males and females in *Drosophila melanogaster* (and other *Drosophila* species) have the same general sex-chromosome composition as humans (males are XY and females are XX), we might assume that the Y chromosome also causes maleness in these flies. However, the elegant work of Calvin Bridges in 1916 showed this not to be true. His studies of flies with quite varied chromosome compositions led him to the conclusion that the Y chromosome is not involved in sex determination in this organism. Instead, Bridges proposed that the X chromosomes and autosomes together play a critical role in sex determination. Recall that in the nematode *C. elegans*, which lacks a Y chromosome, both the sex chromosomes and autosomes are critical to sex determination.

Bridges' work can be divided into two phases: (1) A study of offspring resulting from nondisjunction of the X chromosomes during meiosis in females and (2) subsequent work with progeny of females containing three copies of each chromosome, called triploid (3n) females. As we have seen previously in this chapter (and as you will see in Figure 8–1), nondisjunction is the failure of paired chromosomes to segregate or separate during the anaphase stage of the first or second meiotic divisions. The result is the production of two types of abnormal gametes, one of which contains an extra chromosome (n + 1) and the other of which lacks a chromosome (n − 1). Fertilization of such aberrant gametes with a haploid gamete produces 2n + 1 or 2n − 1 zygotes. As in humans, if nondisjunction involves the X chromosome, in addition to the normal complement of autosomes, both an XXY and an X0 sex-chromosome composition may result. (The "0" signifies that neither a second X nor a Y chromosome is present, as occurs in X0 genotypes of individuals with Turner syndrome.)

Contrary to what was later discovered in humans, Bridges found that the XXY flies were normal females and the X0 flies were sterile males. The presence of the Y chromosome in the XXY flies did not cause maleness, and its absence in the X0 flies did not produce femaleness. From these data, he concluded that the Y chromosome in *Drosophila* lacks male-determining factors, but since the X0 males were sterile, it does contain genetic information essential to male fertility. Recent work has shown that the Y chromosome in *Drosophila* contains only about 20 protein-coding genes but mutation of these genes has significant impacts on regulating expression of hundreds of genes on other chromosomes, including genes on the X chromosome.

Bridges was able to clarify the mode of sex determination in *Drosophila* by studying the progeny of triploid females (3n), which have three copies each of the haploid complement of chromosomes. *Drosophila* has a haploid number of 4, thereby possessing three pairs of autosomes in addition to its pair of sex chromosomes. Triploid females apparently originate from rare diploid eggs fertilized by normal haploid sperm. Triploid females have heavy-set bodies, coarse bristles, and coarse eyes, and they may be fertile. Because of the odd number of each chromosome (3), during meiosis, a variety of different chromosome complements are distributed into gametes that give rise to offspring with a variety of abnormal chromosome constitutions. Correlations between the sexual morphology and chromosome composition, along with Bridges' interpretation, are shown in **Figure 7–12**.

Bridges realized that the critical factor in determining sex is *the ratio of X chromosomes to the number of haploid sets of autosomes (A) present.* Normal (2X:2A) and triploid (3X:3A) females each have a ratio equal to 1.0, and both are fertile. As the ratio exceeds unity (3X:2A, or 1.5,

Normal diploid male

(IV)
(II)
(III)
(I)
XY

2 sets of autosomes
+
X   Y

Chromosome formulation	Ratio of X chromosomes to autosome sets	Sexual morphology
3X/2A	1.5	Metafemale
3X/3A	1.0	Female
2X/2A	1.0	Female
3X/4A	0.75	Intersex
2X/3A	0.67	Intersex
X/2A	0.50	Male
XY/2A	0.50	Male
XY/3A	0.33	Metamale

FIGURE 7–12 Chromosome compositions, the corresponding ratios of X chromosomes to sets of autosomes, and the resultant sexual morphology seen in *Drosophila melanogaster*. The normal diploid male chromosome composition is shown as a reference on the left (XY/2A). The rows representing normally occurring females and males are lightly shaded.

for example), what was once called a *superfemale* is produced. Because such females are most often inviable, they are now more appropriately called **metafemales**.

Normal (XY:2A) and sterile (X0:2A) males each have a ratio of 1:2, or 0.5. When the ratio decreases to 1:3, or 0.33, as in the case of an XY:3A male, infertile **metamales** result. Other flies recovered by Bridges in these studies had an X:A ratio intermediate between 0.5 and 1.0. These flies were generally larger, and they exhibited a variety of morphological abnormalities and rudimentary bisexual gonads and genitalia. They were invariably sterile and expressed both male and female morphology, thus being designated as **intersexes**.

Bridges' results indicate that in *Drosophila*, factors that cause a fly to develop into a male are not located on the sex chromosomes but are instead found on the autosomes. Some female-determining factors, however, are located on the X chromosomes. Thus, with respect to primary sex determination, male gametes containing one of each autosome plus a Y chromosome result in male offspring not because of the presence of the Y but because they fail to contribute an X chromosome. This mode of sex determination is explained by the **genic balance theory**. Bridges proposed that a threshold for maleness is reached when the X:A ratio is 1:2 (X:2A), but that the presence of an additional X (XX:2A) alters the balance and results in female differentiation.

Numerous mutant genes have been identified that are involved in sex determination in *Drosophila*. Recessive mutations in the autosomal gene, transformer (tra), discovered over 75 years ago by Alfred H. Sturtevant, clearly demonstrated that a single autosomal gene could have a profound impact on sex determination. Females homozygous for *tra* are transformed into sterile males, but homozygous males are unaffected.

More recently, another gene, *Sex-lethal (Sxl)*, has been shown to play a critical role, serving as a "master switch" in sex determination. Activation of the X-linked *Sxl* gene, which relies on a ratio of X chromosomes to sets of autosomes that equals 1.0, is essential to female development. In the absence of activation—as when, for example, the X:A ratio is 0.5—male development occurs. It is interesting to note that mutations that inactivate the *Sxl* gene, as originally studied in 1960 by Hermann J. Muller, kill female embryos but have no effect on male embryos, consistent with the role of the gene. Although it is not yet exactly clear how this ratio influences the *Sxl* locus, we do have some insights into the question. The *Sxl* locus is part of a hierarchy of gene expression and exerts control over other genes, including *tra* (discussed in the previous paragraph) and *dsx* (doublesex). The wild-type allele of *tra* is activated by the product of *Sxl* only in females and in turn influences the expression of *dsx*.

Depending on how the initial RNA transcript of *dsx* is processed (spliced, as explained below), the resultant dsx protein activates either male- or female-specific genes required for sexual differentiation. Each step in this regulatory cascade requires a form of processing called **RNA splicing**, in which portions of the RNA are removed and the remaining fragments are "spliced" back together prior to translation into a protein. In the case of the *Sxl* gene, the RNA transcript may be spliced in different ways, a phenomenon called **alternative splicing**. A different RNA transcript is produced in females than in males. In potential females, the transcript encodes a functional protein and initiates a cascade of regulatory gene expression, ultimately leading to female differentiation. In potential males, the transcript encodes a nonfunctional protein, leading to a different pattern of gene activity, whereby male differentiation occurs. In Chapter 17, alternative splicing is again addressed as one of the mechanisms involved in the regulation of genetic expression in eukaryotes.

Dosage Compensation in *Drosophila*

Since *Drosophila* females contain two copies of X-linked genes, whereas males contain only one copy, a dosage problem exists as it does in mammals such as humans and mice. However, the mechanism of dosage compensation in *Drosophila* differs considerably from that in mammals, since X chromosome inactivation is not observed. Instead, male X-linked genes are transcribed at twice the level of the comparable genes in females. Interestingly, if groups of X-linked genes are moved (translocated) to autosomes, dosage compensation still affects them, even though they are no longer part of the X chromosome.

As in mammals, considerable gains have been made recently in understanding the process of dosage compensation in *Drosophila*. At least four autosomal genes are known to be involved, under the same master-switch gene, *Sxl*, that induces female differentiation during sex determination. Mutations in any of these genes severely reduce the increased expression of X-linked genes in males, causing lethality.

Evidence supporting a mechanism of increased genetic activity in males is now available. The well-accepted model proposes that one of the autosomal genes, *mle (maleless)*, encodes a protein that binds to numerous sites along the X chromosome, causing enhancement of genetic expression. The products of three other autosomal genes also participate in and are required for *mle* binding. In addition, proteins called male-specific lethals (MSLs) have been shown to bind to gene-rich regions of the X to increase gene expression in male flies. Collectively, this cluster of gene-activating proteins is called the **dosage compensation complex (DCC)**. **Figure 7–13** illustrates the presence of these proteins bound to the X chromosome in *Drosophila* in contrast to their failure to bind to the autosomes. The location of DCC proteins may be identified when fluorescent antibodies against these proteins are added to preparations of the large polytene chromosomes characteristic of salivary gland cells in fly larvae (see Chapter 12).

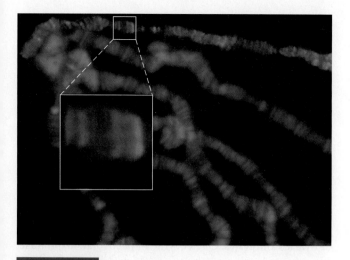

FIGURE 7–13   Fluorescent antibodies against proteins in the dosage compensation complex (DCC) bind only to the X chromosome in *Drosophila* polytene chromosome preparations, providing evidence concerning the role of the DCC in increasing the expression of X-linked genes.

FIGURE 7–14   A bilateral gynandromorph of *Drosophila me-lanogaster* formed following the loss of one X chromosome in one of the two cells during the first mitotic division. The left side of the fly, composed of male cells containing a single X, expresses the mutant *white*-eye and *miniature*-wing alleles. The right side is composed of female cells containing two X chromosomes heterozygous for the two recessive alleles.

This model predicts that the master-switch *Sxl* gene plays an important role during dosage compensation. In XY flies, *Sxl* is inactive; therefore, the autosomal genes are activated, causing enhanced X chromosome activity. On the other hand, *Sxl* is active in XX females and functions to inactivate one or more of the male-specific autosomal genes, perhaps *mle*. By dampening the activity of these autosomal genes, it ensures that they will not serve to double the expression of X-linked genes in females, which would further compound the dosage problem.

Tom Cline has proposed that, before the aforementioned dosage compensation mechanism is activated, *Sxl* acts as a sensor for the expression of several other X-linked genes. In a way, *Sxl* counts X chromosomes. When it registers the dose of their expression as being high—for example, as the result of the presence of two X chromosomes—the *Sxl* gene product is modified to dampen the expression of the autosomal genes. Although this model may yet be revised or refined, it is useful for guiding future research.

Clearly, an entirely different mechanism of dosage compensation exists in *Drosophila* (and probably many related organisms) than that seen in mammals. The development of elaborate mechanisms to equalize the expression of X-linked genes demonstrates the critical importance of level of gene expression. A delicate balance of gene products is necessary to maintain normal development of both males and females.

Drosophila Mosaics

Our knowledge of sex determination and our discussion of X-linkage in *Drosophila* (Chapter 4) helps us to understand the unusual appearance of a unique fruit fly, shown in

Figure 7–14. This fly was recovered from a stock in which all other females were heterozygous for the X-linked genes *white* eye (*w*) and *miniature* wing (*m*). It is a **bilateral gynandromorph**, which means that one-half of its body (the left half) has developed as a male and the other half (the right half) as a female.

We can account for the presence of both sexes in a single fly in the following way. If a female zygote (heterozygous for *white* eye and *miniature* wing) were to lose one of its X chromosomes during the first mitotic division, the two cells would be of the XX and X0 constitution, respectively. Thus, one cell would be female and the other would be male. Each of these cells is responsible for producing all progeny cells that make up either the right half or the left half of the body during embryogenesis.

In the case of the bilateral gynandromorph, the original cell of X0 constitution apparently produced only identical progeny cells and gave rise to the left half of the fly, which, because of its chromosomal constitution, was male. Since the male half demonstrated the *white, miniature* phenotype, the X chromosome bearing the w^+, m^+ alleles was lost, while the *w, m*-bearing homolog was retained. All cells on the right side of the body were derived from the original XX cell, leading to female development. These cells, which remained heterozygous for both mutant genes, expressed the wild-type eye–wing phenotypes.

Depending on the orientation of the spindle during the first mitotic division, gynandromorphs can be produced that have the "line" demarcating male versus female development along or across any axis of the fly's body.

7.7

Temperature Variation Controls Sex Determination in Reptiles

We conclude this chapter by discussing several cases involving reptiles, in which the environment—specifically temperature—has a profound influence on sex determination. In contrast to **genotypic sex determination (GSD)**, in which sex is determined genetically (as is true of all examples thus far presented in the chapter), the cases that we will now discuss are categorized as **temperature-dependent sex determination (TSD)**. As we shall see, the investigations leading to this information may well have come closer to revealing the true nature of the underlying basis of sex determination than any findings previously discussed.

In many species of reptiles, GSD is involved at conception based on sex-chromosome composition, as is the case in many organisms already considered in this chapter. For example, in many snakes, including vipers, a ZZ/ZW mode is in effect, in which the female is the heterogamous sex (ZW). However, in boas and pythons, it is impossible to distinguish one sex chromosome from the other in either sex. In many lizards, both the XX/XY and ZZ/ZW systems are found, depending on the species.

In still other reptilian species, however, TSD is the norm, including all crocodiles, most turtles, and some lizards, where sex determination is achieved according to the incubation temperature of eggs during a critical period of embryonic development. Three distinct patterns of TSD emerge (cases I–III in Figure 7–15). In case I, low temperatures yield 100 percent females, and high temperatures yield 100 percent males. Just the opposite occurs in case II. In case III, low *and* high temperatures yield 100 percent females, while intermediate temperatures yield various proportions of males. The third pattern is seen in various species of crocodiles, turtles, and lizards, although other members of these groups are known to exhibit the other patterns.

Two observations are noteworthy. First, in all three patterns, certain temperatures result in both male and female offspring; second, this pivotal temperature (T_p) range is fairly narrow, usually spanning less than 5°C, and sometimes only 1°C. The central question raised by these observations is this: What are the metabolic or physiological parameters affected by temperature that lead to the differentiation of one sex or the other?

The answer is thought to involve steroids (mainly estrogens) and the enzymes involved in their synthesis. Studies clearly demonstrate that the effects of temperature on estrogens, androgens, and inhibitors of the enzymes controlling their synthesis are involved in the sexual differentiation of ovaries and testes. One enzyme in particular, **aromatase**, converts androgens (male hormones such as testosterone) to estrogens (female hormones such as estradiol). The activity of this enzyme is correlated with the pathway of reactions that occurs during gonadal differentiation activity and is high in developing ovaries and low in developing testes. Researchers in this field, including Claude Pieau and colleagues, have proposed that a thermosensitive factor mediates the transcription of the reptilian aromatase gene, leading to temperature-dependent sex determination. Several other genes are likely to be involved in this mediation.

The involvement of sex steroids in gonadal differentiation has also been documented in birds, fishes, and amphibians. Thus, sex-determining mechanisms involving estrogens seem to be characteristic of nonmammalian vertebrates. The regulation of such systems, while temperature dependent in many reptiles, appears to be controlled by sex chromosomes (XX/XY or ZZ/ZW) in many of these other organisms. A final intriguing thought on this matter is that the product of *SRY*, a key component in mammalian sex determination, has been shown to bind *in vitro* to a regulatory portion of the aromatase gene, suggesting a mechanism whereby it could act as a repressor of ovarian development.

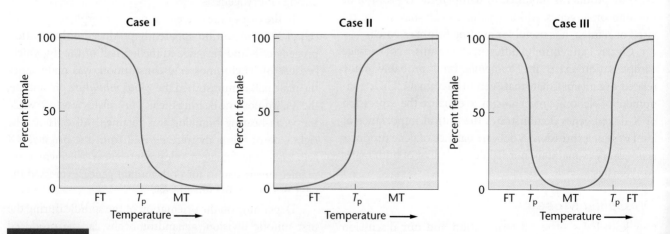

FIGURE 7–15   Three different patterns of temperature-dependent sex determination (TSD) in reptiles, as described in the text. The relative pivotal temperature T_p is crucial to sex determination during a critical point during embryonic development. FT = Female-determining temperature; MT = male-determining temperature.

GENETICS, TECHNOLOGY, AND SOCIETY

A Question of Gender: Sex Selection in Humans

Throughout history, people have attempted to influence the gender of their unborn offspring by following varied and sometimes bizarre procedures. In medieval Europe, prospective parents would place a hammer under the bed to help them conceive a boy, or a pair of scissors to conceive a girl. Other practices were based on the ancient belief that semen from the right testicle created male offspring and that from the left testicle created females. As late as the eighteenth century, European men might tie off or remove their left testicle to increase the chances of getting a male heir. In some cultures, efforts to control the sex of offspring has had a darker outcome—female infanticide. In ancient Greece, the murder of female infants was so common that the male:female ratio in some areas approached 4:1. In some parts of rural India, hundreds of families admitted to female infanticide as late as the 1990s. In 1997, the World Health Organization reported population data showing that about 50 million women were "missing" in China, likely because of selective abortion of female fetuses and institutionalized neglect of female children.

In recent times, sex-specific abortion has replaced much of the traditional female infanticide. For a fee, some companies offer amniocentesis and ultrasound tests for prenatal sex determination. Studies in India estimate that hundreds of thousands of fetuses are aborted each year because they are female. As a result of sex-selective abortion, the female:male ratio in India was 927:1000 in 1991. In some northern states, the ratio was as low as 600:1000.

In Western industrial countries, new genetics and reproductive technologies offer parents ways to select their children's gender prior to implantation of the embryo in the uterus—called *preimplantation gender selection (PGS)*. Following *in vitro* fertilization, embryos are biopsied and assessed for gender. Only sex-selected embryos are then implanted. The

simplest method involves separating X and Y chromosome-bearing spermatozoa based on their DNA content. Because of the difference in size of the X and Y chromosomes, X-bearing sperm contain 2.8 to 3.0 percent more DNA than Y-bearing sperm. Sperm samples are treated with a fluorescent DNA stain, then passed through a laser beam in a Fluorescence-Activated Cell Sorter machine that separates the sperm into two fractions based on the intensity of their DNA-fluorescence. The sorted sperm are then used for standard intrauterine insemination.

The emerging PGS methods raise a number of legal and ethical issues. Some people feel that prospective parents have the legal right to use sex-selection techniques as part of their fundamental procreative liberty. Proponents state that PGS will reduce the suffering of many families. For example, people at risk for transmitting X-linked diseases such as hemophilia or Duchenne muscular dystrophy can now enhance their chance of conceiving a female child, who will not express the disease.

The majority of people who undertake PGS, however, do so for nonmedical reasons—to "balance" their families. A possible argument in favor of this use is that the ability to intentionally select the sex of an offspring may reduce overpopulation and economic burdens for families who would repeatedly reproduce to get the desired gender. By the same token, PGS may reduce the number of abortions. It is also possible that PGS may increase the happiness of both parents and children, as the children would be more "wanted."

On the other hand, some argue that PGS serves neither the individual nor the common good. They argue that PGS is inherently sexist, having its basis in the idea that one sex is superior to the other, and leads to an increase in linking a child's worth to gender. Other critics fear that social approval of PGS will open the door to other genetic manipulations of children's characteristics. It is difficult to

predict the full effects that PGS will bring to the world. But the gender-selection genie is now out of the bottle and is unwilling to return.

Your Turn

Take time, individually or in groups, to answer the following questions. Investigate the references and links to help you understand some of the issues that surround the topic of gender selection.

1. What do you think are valid arguments for and against the use of PGS?

2. A generally accepted moral and legal concept is that of reproductive autonomy—the freedom to make individual reproductive decisions without external interference. Are there circumstances under which reproductive autonomy should be restricted?

The above questions, and others, are explored in a series of articles in the American Journal of Bioethics, *Volume 1 (2001). See the article by J. A. Robertson on pages 2–9, for a summary of the moral and legal issues surrounding PGS.*

3. What do you think are the reasons that some societies practice female infanticide and prefer the birth of male children?

For a discussion of this topic, visit the "Gendercide Watch" Web site http://www.gendercide.org.

4. If safe and efficient methods of PGS were available to you, do you think that you would use them to help you with family planning? Under what circumstances might you use them?

The Genetics and IVF Institute (Fairfax, Virginia) is presently using PGS techniques based on sperm sorting, in an FDA-approved clinical trial. As of 2008, over 1000 human pregnancies have resulted, with an approximately 80 percent success rate. Read about these methods on their Web site: http://www.microsort.net.

CASE STUDY | Doggone it!

A dog breeder discovers one of her male puppies has abnormal genitalia. After a visit to the veterinary clinic at a nearby university, the breeder learns that the dog's karyotype lacks a Y chromosome, but instead has an XX chromosome pair, with one of the X chromosomes slightly larger than usual (being mammals, male dogs are normally XY and females are XX). The veterinarian tells her that in other breeds, some females display an XY chromosome pair, with the Y chromosome being slightly shorter than normal. These observations raise several interesting questions:

1. Can you offer a chromosomal explanation of these two cases?

2. How could such cases be used to locate the gene(s) responsible for maleness?

3. Suppose you discover a female dog with a normal-sized XY chromosome pair. What kind of mutation might be involved in this case?

4. Suppose you discover a female dog with only a single X chromosome. What predictions might you make about the sex organs and reproductive capacity of this dog?

Summary Points

 For activities, animations, and review quizzes, go to the study area at www.masteringgenetics.com

1. Sexual reproduction depends on the differentiation of male and female structures responsible for the production of male and female gametes, which in turn is controlled by specific genes, most often housed on specific sex chromosomes.

2. Specific sex chromosomes contain genetic information that controls sex determination and sexual differentiation.

3. The presence or absence of a Y chromosome that contains an intact *SRY* gene is responsible for causing maleness in humans and other mammals.

4. In humans, while many more males than females are conceived, and although fewer male than female embryos and fetuses survive *in utero*, there are nevertheless more males than females born.

5. In mammals, female somatic cells randomly inactivate one of two X chromosomes during early embryonic development, a process important for balancing the expression of X chromosome linked genes in males and females.

6. The Lyon hypothesis states that early in development, inactivation of either the maternal or paternal X chromosome occurs in each cell, and that all progeny cells subsequently inactivate the same chromosome. Mammalian females thus develop as mosaics for the expression of heterozygous X-linked alleles.

7. Many reptiles show temperature-dependent effects on sex determination. Although specific sex chromosomes determine genotypic sex in many reptiles, temperature effects on genes involved in sexual determination affect whether an embryo develops a male or female phenotype.

INSIGHTS AND SOLUTIONS

1. In *Drosophila*, the X chromosomes may become attached to one another ($\widehat{XX}$) such that they always segregate together. Some flies thus contain a set of attached X chromosomes plus a Y chromosome.

   (a) What sex would such a fly be? Explain why this is so.

   (b) Given the answer to part (a), predict the sex of the offspring that would occur in a cross between this fly and a normal one of the opposite sex.

   (c) If the offspring described in part (b) are allowed to interbreed, what will be the outcome?

   Solution:

   (a) The fly will be a female. The ratio of X chromosomes to sets of autosomes—which determines sex in *Drosophila*—will be 1.0, leading to normal female development. The Y chromosome has no influence on sex determination in *Drosophila*.

   (b) All progeny flies will have two sets of autosomes along with one of the following sex-chromosome compositions:

   (1) $\widehat{XX}X \rightarrow$ a metafemale with 3 X's (called a trisomic)

   (2) $\widehat{XX}Y \rightarrow$ a female like her mother

   (3) XY $\rightarrow$ a normal male

   (4) YY $\rightarrow$ no development occurs

   (c) A stock will be created that maintains attached-X females generation after generation.

2. The Xg cell-surface antigen is coded for by a gene located on the X chromosome. No equivalent gene exists on the Y chromosome. Two codominant alleles of this gene have been identified: *Xg1* and *Xg2*. A woman of genotype *Xg2/Xg2* bears children with a man of genotype *Xg1/Y*, and they produce a son with Klinefelter syndrome of genotype *Xg1/Xg2/Y*. Using proper genetic terminology, briefly explain how this individual was generated. In which parent and in which meiotic division did the mistake occur?

   Solution: Because the son with Klinefelter syndrome is *Xg1/Xg2/Y*, he must have received both the *Xg1* allele and the Y chromosome from his father. Therefore, nondisjunction must have occurred during meiosis I in the father.

Problems and Discussion Questions

MG *For instructor-assigned tutorials and problems, go to www.masteringgentics.com*

1. In this chapter, we focused on sex differentiation, sex chromosomes, and genetic mechanisms involved in sex determination. At the same time, we found many opportunities to consider the methods and reasoning by which much of this information was acquired. From the explanations given in the chapter, what answers would you propose to the following fundamental questions?

   (a) How do we know that specific genes in maize play a role in sexual differentiation?

   (b) How do we know whether or not a heteromorphic chromosome such as the Y chromosome plays a crucial role in the determination of sex?

   (c) How do we know that in humans the X chromosomes play no role in human sex determination, while the Y chromosome causes maleness and its absence causes femaleness?

   (d) How did we learn that, although the sex ratio at birth in humans favors males slightly, the sex ratio at conception favors them much more?

   (e) How do we know that *Drosophila* utilizes a different sex-determination mechanism than mammals, even though it has the same sex-chromosome compositions in males and females?

   (f) How do we know that X chromosomal inactivation of either the paternal or maternal homolog is a random event during early development in mammalian females?

2. As related to sex determination, what is meant by

   (a) homomorphic and heteromorphic chromosomes?

   (b) isogamous and heterogamous organisms?

3. Contrast the life cycle of a plant such as *Zea mays* with an animal such as *C. elegans*.

4. Discuss the role of sexual differentiation in the life cycles of *Chlamydomonas, Zea mays,* and *C. elegans*.

5. Distinguish between the concepts of sexual differentiation and sex determination.

6. Contrast the *Protenor* and *Lygaeus* modes of sex determination.

7. Describe the major difference between sex determination in *Drosophila* and in humans.

8. How do mammals, including humans, solve the "dosage problem" caused by the presence of an X and Y chromosome in one sex and two X chromosomes in the other sex?

9. The phenotype of an early-stage human embryo is considered sexually indifferent. Explain why this is so even though the embryo's genotypic sex is already fixed.

10. What specific observations (evidence) support the conclusions about sex determination in *Drosophila* and humans?

11. Describe how nondisjunction in human female gametes can give rise to Klinefelter and Turner syndrome offspring following fertilization by a normal male gamete.

12. An insect species is discovered in which the heterogametic sex is unknown. An X-linked recessive mutation for *reduced wing* (*rw*) is discovered. Contrast the F_1 and F_2 generations from a cross between a female with reduced wings and a male with normal-sized wings when

    (a) the female is the heterogametic sex.

    (b) the male is the heterogametic sex.

13. When cows have twin calves of unlike sex (fraternal twins), the female twin is usually sterile and has masculinized reproductive organs. This calf is referred to as a freemartin. In cows, twins may share a common placenta and thus fetal circulation. Predict why a freemartin develops.

14. An attached-X female fly, $\overline{XX}\,Y$ (see the "Insights and Solutions" box), expresses the recessive X-linked *white*-eye mutation. It is crossed to a male fly that expresses the X-linked recessive *miniature*-wing mutation. Determine the outcome of this cross in terms of sex, eye color, and wing size of the offspring.

15. Assume that on rare occasions the attached X chromosomes in female gametes become unattached. Based on the parental phenotypes in Problem 14, what outcomes in the F_1 generation would indicate that this has occurred during female meiosis?

16. It has been suggested that any male-determining genes contained on the Y chromosome in humans cannot be located in the limited region that synapses with the X chromosome during meiosis. What might be the outcome if such genes were located in this region?

17. What is a Barr body, and where is it found in a cell?

18. Indicate the expected number of Barr bodies in interphase cells of individuals with Klinefelter syndrome; Turner syndrome; and karyotypes 47,XYY, 47,XXX, and 48,XXXX.

19. Define the Lyon hypothesis.

20. Can the Lyon hypothesis be tested in a human female who is homozygous for one allele of the X-linked *G6PD* gene? Why, or why not?

21. Predict the potential effect of the Lyon hypothesis on the retina of a human female heterozygous for the X-linked red-green color-blindness trait.

22. Cat breeders are aware that kittens expressing the X-linked calico coat pattern and tortoiseshell pattern are almost invariably females. Why are they certain of this?

23. What does the apparent need for dosage compensation mechanisms suggest about the expression of genetic information in normal diploid individuals?

24. How does X chromosome dosage compensation in *Drosophila* differ from that process in humans?

25. What type of evidence supports the conclusion that the primary sex ratio in humans is as high as 1.20 to 1.60?

26. Devise as many hypotheses as you can that might explain why so many more human male conceptions than human female conceptions occur.

27. In mice, the *Sry* gene (see Section 7.3) is located on the Y chromosome very close to one of the pseudoautosomal regions that pairs with the X chromosome during male meiosis. Given this information, propose a model to explain the generation of unusual males who have two X chromosomes (with an *Sry*-containing piece of the Y chromosome attached to one X chromosome).

28. The genes encoding the red- and green-color-detecting proteins of the human eye are located next to one another on the X chromosome and probably evolved from a common ancestral pigment gene. The two proteins demonstrate 76 percent homology in their amino acid sequences. A normal-visioned woman (with both genes present on each of her two X chromosomes) has a red-color-blind son who was shown to have one copy of the green-detecting gene and no copies of the red-detecting gene. Devise an explanation for these observations at the chromosomal level (involving meiosis).

29. What is the role of the enzyme aromatase in sexual differentiation in reptiles?

Extra-Spicy Problems

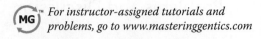

For instructor-assigned tutorials and problems, go to www.masteringgentics.com

30. In mice, the X-linked dominant mutation *Testicular feminization (Tfm)* eliminates the normal response to the testicular hormone testosterone during sexual differentiation. An XY mouse bearing the *Tfm* allele on the X chromosome develops testes, but no further male differentiation occurs—the external genitalia of such an animal are female. From this information, what might you conclude about the role of the *Tfm* gene product and the X and Y chromosomes in sex determination and sexual differentiation in mammals? Can you devise an experiment, assuming you can "genetically engineer" the chromosomes of mice, to test and confirm your explanation?

31. In the wasp *Bracon hebetor*, a form of parthenogenesis (the development of unfertilized eggs into progeny) resulting in haploid organisms is not uncommon. All haploids are males. When offspring arise from fertilization, females almost invariably result. P. W. Whiting has shown that an X-linked gene with nine multiple alleles (X_a, X_b, etc.) controls sex determination. Any homozygous or hemizygous condition results in males, and any heterozygous condition results in females. If an X_a/X_b female mates with an X_a male and lays 50 percent fertilized and 50 percent unfertilized eggs, what proportion of male and female offspring will result?

32. Shown below are two graphs that plot the percentage of fertilized eggs containing males against the atmospheric temperature during early development in (a) snapping turtles and (b) most lizards. Interpret these data as they relate to the effect of temperature on sex determination.

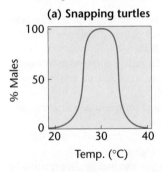
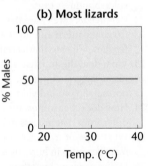

33. When the cloned cat Carbon Copy (CC) was born (see the Now Solve This question on p. 187), she had black patches and white patches, but completely lacked any orange patches. The knowledgeable students of genetics were not surprised at this outcome. Starting with the somatic ovarian cell used as the source of the nucleus in the cloning process, explain how this outcome occurred.

Carbon Copy with her surrogate mother.

34. In a number of organisms, including *Drosophila* and butterflies, genes that alter the sex ratio have been described. In the pest species *Musca domesticus* (the house fly), *Aedes aegypti* (the mosquito that is the vector for yellow fever), and *Culex pipiens* (the mosquito vector for filariasis and some viral diseases), scientists are especially interested in such genes. Sex in *Culex* is determined by a single gene pair, *Mm* being male and *mm* being female. Males homozygous for the recessive gene *dd* never produce many female offspring. The *dd* combination in males causes fragmentation of the *m*-bearing dyad during the first meiotic division, hence its failure to complete spermatogenesis.

(a) Account for this sex-ratio distortion by drawing labeled chromosome arrangements in primary and secondary spermatocytes for each of the following genotypes: *Mm Dd* and *Mm dd*. How do meiotic products differ between *Dd* and *dd* genotypes? Note that the diploid chromosome number is 6 in *Culex pipiens* and both *D* and *M* loci are linked on the same chromosome.

(b) How might a sex-ratio distorter such as *dd* be used to control pest population numbers?

35. In chickens, a key gene involved in sex determination has recently been identified. Called *DMRT1*, it is located on the Z chromosome, and is absent on the W chromosome. Like *SRY* in humans, it is male determining. Unlike *SRY* in humans, however, female chickens (ZW) have a single copy while males (ZZ) have two copies of the gene. Nevertheless, it is transcribed only in the developing testis. Working in the laboratory of Andrew Sinclair (a co-discoverer of the human *SRY* gene), Craig Smith and colleagues were able to "knock down" expression of *DMRT1* in ZZ embryos using RNA interference techniques (see Chapter 17). In such cases, the developing gonads look more like ovaries than testes [*Nature* 461: 267 (2009)]. What conclusions can you draw about the role that the *DMRT1* gene plays in chickens in contrast to the role the *SRY* gene plays in humans?

36. The paradigm in vertebrates is that once sex determination occurs and testes or ovaries are formed, that secondary sexual differentiation (male vs. female characteristics) is dependent on male or female hormones that are produced. Recently, D. Zhao and colleagues studied three chickens that were bilateral gynandromorphs, with the right side of the body being clearly female and the left side of the body clearly male [*Nature* 464: 237 (2010)]. Propose experimental questions that can be investigated using these chickens to test this paradigm. What alternative interpretation contrasts with the paradigm?

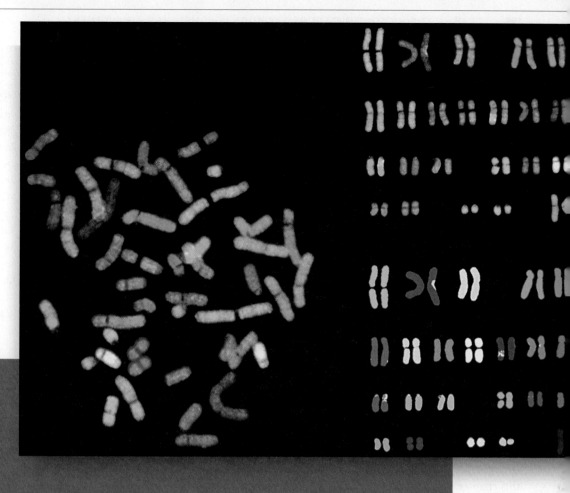

Spectral karyotyping of human chromosomes, utilizing differentially labeled "painting" probes.

8

Chromosome Mutations: Variation in Number and Arrangement

CHAPTER CONCEPTS

- The failure of chromosomes to properly separate during meiosis results in variation in the chromosome content of gametes and subsequently in offspring arising from such gametes.

- Plants often tolerate an abnormal genetic content, but, as a result, they often manifest unique phenotypes. Such genetic variation has been an important factor in the evolution of plants.

- In animals, genetic information is in a delicate equilibrium whereby the gain or loss of a chromosome, or part of a chromosome, in an otherwise diploid organism often leads to lethality or to an abnormal phenotype.

- The rearrangement of genetic information within the genome of a diploid organism may be tolerated by that organism but may affect the viability of gametes and the phenotypes of organisms arising from those gametes.

- Chromosomes in humans contain fragile sites—regions susceptible to breakage, which leads to abnormal phenotypes.

n previous chapters, we have emphasized how mutations and the resulting alleles affect an organism's phenotype and how traits are passed from parents to offspring according to Mendelian principles. In this chapter, we look at phenotypic variation that results from more substantial changes than alterations of individual genes—modifications at the level of the chromosome.

Although most members of diploid species normally contain precisely two haploid chromosome sets, many known cases vary from this pattern. Modifications include a change in the total number of chromosomes, the deletion or duplication of genes or segments of a chromosome, and rearrangements of the genetic material either within or among chromosomes. Taken together, such changes are called **chromosome mutations** or **chromosome aberrations**, to distinguish them from gene mutations. Because the chromosome is the unit of genetic transmission, according to Mendelian laws, chromosome aberrations are passed to offspring in a predictable manner, resulting in many unique genetic outcomes.

Because the genetic component of an organism is delicately balanced, even minor alterations of either content or location of genetic information within the genome can result in some form of phenotypic variation. More substantial changes may be lethal, particularly in animals. Throughout this chapter, we consider many types of chromosomal aberrations, the phenotypic consequences for the organism that harbors an aberration, and the impact of the aberration on the offspring of an affected individual. We will also discuss the role of chromosome aberrations in the evolutionary process.

8.1

Variation in Chromosome Number: Terminology and Origin

Variation in chromosome number ranges from the addition or loss of one or more chromosomes to the addition of one or more haploid sets of chromosomes. Before we embark on our discussion, it is useful to clarify the terminology that describes such changes. In the general condition known as **aneuploidy**, an organism gains or loses one or more chromosomes but not a complete set. The loss of a single chromosome from an otherwise diploid genome is called *monosomy*. The gain of one chromosome results in *trisomy*. These changes are contrasted with the condition of **euploidy**, where complete haploid sets of chromosomes are present. If more than two sets are present, the term **polyploidy** applies. Organisms with three sets are specifically *triploid*, those with four sets are *tetraploid*, and so on. Table 8.1 provides an organizational framework for you to follow as we discuss

TABLE 8.1

Terminology for Variation in Chromosome Numbers

Term	Explanation
Aneuploidy	$2n \pm x$ chromosomes
Monosomy	$2n - 1$
Disomy	$2n$
Trisomy	$2n + 1$
Tetrasomy, pentasomy, etc.	$2n + 2$, $2n + 3$, etc.
Euploidy	Multiples of n
Diploidy	$2n$
Polyploidy	$3n$, $4n$, $5n$, . . .
Triploidy	$3n$
Tetraploidy, pentaploidy, etc.	$4n$, $5n$, etc.
Autopolyploidy	Multiples of the same genome
Allopolyploidy (amphidiploidy)	Multiples of closely related genomes

each of these categories of aneuploid and euploid variation and the subsets within them.

As we consider cases that include the gain or loss of chromosomes, it is useful to examine how such aberrations originate. For instance, how do the syndromes arise where the number of sex-determining chromosomes in humans is altered, as described in Chapter 7? As you may recall, the gain (47,XXY) or loss (45,X) of an X chromosome from an otherwise diploid genome affects the phenotype, resulting in **Klinefelter syndrome** or **Turner syndrome**, respectively (see Figure 7–6). Human females may contain extra X chromosomes (e.g., 47,XXX, 48,XXXX), and some males contain an extra Y chromosome (47,XYY).

Such chromosomal variation originates as a random error during the production of gametes, a phenomenon referred to as **nondisjunction**, whereby paired homologs fail to disjoin during segregation. This process disrupts the normal distribution of chromosomes into gametes. The results

NOW SOLVE THIS

8–1 A human female with Turner syndrome (47,X) also expresses the X-linked trait hemophilia, as did her father. Which of her parents underwent nondisjunction during meiosis, giving rise to the gamete responsible for the syndrome?

■ HINT: *This problem involves an understanding of how nondisjunction leads to aneuploidy. The key to its solution is first to review Turner syndrome, discussed above and in more detail in Chapter 7, then to factor in that she expresses hemophilia, and finally, to consider which parent contributed a gamete with an X chromosome that underwent normal meiosis.*

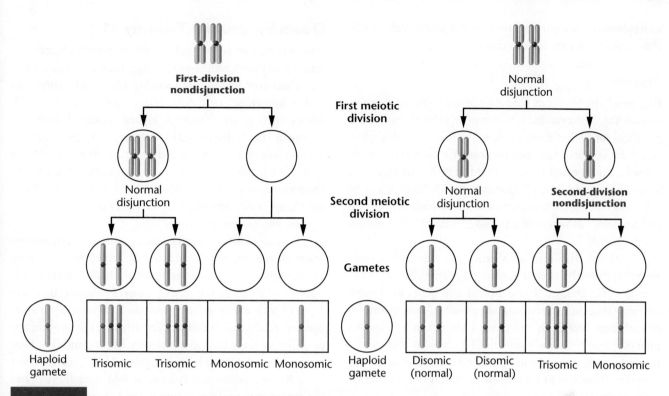

FIGURE 8-1 Nondisjunction during the first and second meiotic divisions. In both cases, some of the gametes that are formed either contain two members of a specific chromosome or lack that chromosome. After fertilization by a gamete with a normal haploid content, monosomic, disomic (normal), or trisomic zygotes are produced.

of nondisjunction during meiosis I and meiosis II for a single chromosome of a diploid organism are shown in Figure 8–1. As you can see, abnormal gametes can form containing either two members of the affected chromosome or none at all. Fertilizing these with a normal haploid gamete produces a zygote with either three members (trisomy) or only one member (monosomy) of this chromosome. Nondisjunction leads to a variety of aneuploid conditions in humans and other organisms.

8.2

Monosomy and Trisomy Result in a Variety of Phenotypic Effects

We turn now to a consideration of variations in the number of autosomes and the genetic consequence of such changes. The most common examples of *aneuploidy*, where an organism has a chromosome number other than an exact multiple of the haploid set, are cases in which a single chromosome is either added to, or lost from, a normal diploid set.

Monosomy

The loss of one chromosome produces a $2n - 1$ complement called **monosomy**. Although monosomy for the X chromosome occurs in humans, as we have seen in 45,X

Turner syndrome, monosomy for any of the autosomes is not usually tolerated in humans or other animals. In *Drosophila*, flies that are monosomic for the very small chromosome IV (containing less than 5 percent of the organism's genes) develop more slowly, exhibit reduced body size, and have impaired viability. Monosomy for the larger autosomal chromosomes II and III is apparently lethal because such flies have never been recovered.

The failure of monosomic individuals to survive is at first quite puzzling, since at least a single copy of every gene is present in the remaining homolog. However, if just one of those genes is represented by a lethal allele, the unpaired chromosome condition will result in the death of the organism. This will occur because monosomy unmasks recessive lethals that are otherwise tolerated in heterozygotes carrying the corresponding wild-type alleles. Another possible cause of lethality of aneuploidy is that a single copy of a recessive gene may be insufficient to provide adequate function for sustaining the organism, a phenomenon called **haploinsufficiency**.

Aneuploidy is better tolerated in the plant kingdom. Monosomy for autosomal chromosomes has been observed in maize, tobacco, the evening primrose (*Oenothera*), and the jimson weed (*Datura*), among many other plants. Nevertheless, such monosomic plants are usually less viable than their diploid derivatives. Haploid pollen grains, which undergo extensive development before participating in fertilization, are particularly sensitive to the lack of one chromosome and are seldom viable.

Trisomy

In general, the effects of **trisomy** ($2n + 1$) parallel those of monosomy. However, the addition of an extra chromosome produces somewhat more viable individuals in both animal and plant species than does the loss of a chromosome. In animals, this is often true, provided that the chromosome involved is relatively small. However, the addition of a large autosome to the diploid complement in both *Drosophila* and humans has severe effects and is usually lethal during development.

In plants, trisomic individuals are viable, but their phenotype may be altered. A classical example involves the jimson weed, *Datura*, whose diploid number is 24. Twelve primary trisomic conditions are possible, and examples of each one have been recovered. Each trisomy alters the phenotype of the plant's capsule (Figure 8–2) sufficiently to produce a unique phenotype. These capsule phenotypes were first thought to be caused by mutations in one or more genes.

Still another example is seen in the rice plant (*Oryza sativa*), which has a haploid number of 12. Trisomic strains for each chromosome have been isolated and studied—the plants of 11 strains can be distinguished from one another and from wild-type plants. Trisomics for the longer chromosomes are the most distinctive, and the plants grow more slowly. This is in keeping with the belief that larger chromosomes cause greater genetic imbalance than smaller ones. Leaf structure, foliage, stems, grain morphology, and plant height also vary among the various trisomies.

FIGURE 8-2   The capsule of the fruit of the jimson weed, *Datura stramonium*, the phenotype of which is uniquely altered by each of the possible 12 trisomic conditions.

Down Syndrome: Trisomy 21

The only human autosomal trisomy in which a significant number of individuals survive longer than a year past birth was discovered in 1866 by Langdon Down. The condition is now known to result from trisomy of chromosome 21, one of the G group*(Figure 8–3), and is called **Down syndrome** or simply **trisomy 21** (designated 47,21+). This trisomy is found in approximately 1 infant in every 800 live births. While this might seem to be a rare, improbable event, there are approximately 4000–5000 such births annually in the United States, and there are currently over 250,000 individuals with Down syndrome.

Typical of other conditions classified as syndromes, many phenotypic characteristics *may* be present in trisomy 21, but any single affected individual usually exhibits only a subset of these. In the case of Down syndrome, there are 12 to 14 such characteristics, with each individual, on average, expressing 6 to 8 of them. Nevertheless, the outward appearance of these individuals is very similar, and they bear a striking resemblance to one another. This is, for the most part, due to a prominent epicanthic fold in each eye** and the typically flat face and round head. People with Down syndrome are also characteristically short and may have a protruding, furrowed tongue (which causes the mouth to remain partially open) and short, broad hands with characteristic palm and fingerprint patterns. Physical, psychomotor, and mental development are retarded, and poor muscle tone is characteristic. While life expectancy is shortened to an average of about 50 years, individuals are known to survive into their 60s.

In the way of further illustrating the impact of just one additional chromosome in an otherwise diploid genome, children afflicted with Down syndrome are prone to respiratory disease and heart malformations, and they show an incidence of leukemia approximately 20 times higher than that of the normal population. In addition, death in older Down syndrome adults is frequently due to Alzheimer disease, the onset of which occurs at a much earlier age than in the normal population.

Because Down syndrome is common in our population, a comprehensive understanding of the underlying genetic basis has long been a research goal. Investigations have given rise to the idea that a critical region of chromosome 21 contains the genes that are dosage sensitive in this trisomy

---

* On the basis of size and centromere placement, human autosomal chromosomes are divided into seven groups: A (1–3), B (4–5), C (6–12), D (13–15), E (16–18), F (19–20), and G (21–22).

** The epicanthic fold, or epicanthus, is a skin fold of the upper eyelid, extending from the nose to the inner side of the eyebrow. It covers and appears to lower the inner corner of the eye, giving the eye a slanted, or almond-shaped, appearance. The epicanthus is a prominent normal component of the eyes in many Asian groups.

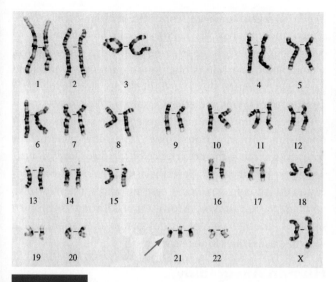

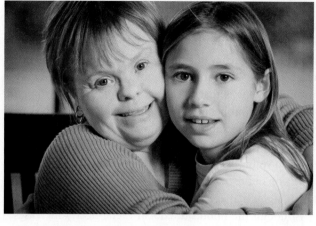

FIGURE 8-3 The karyotype and a photograph of a child with Down syndrome (hugging her unaffected sister on the right). In the karyotype, three members of the G-group chromosome 21 are present, creating the 47,21+ condition.

and responsible for the many phenotypes associated with the syndrome. This hypothetical portion of the chromosome has been called the **Down syndrome critical region (DSCR)**. A mouse model was created in 2004 that is trisomic for the DSCR, although some mice do not exhibit the characteristics of the syndrome. Nevertheless, this remains an important investigative approach.

Current studies of the DSCR region in both humans and mice have led to several interesting findings. Several research investigations have now led us to believe that the presence of three copies of the genes present in this region are necessary, but themselves not sufficient for the cognitive deficiencies characteristic of the syndrome. Another finding involves still another important observation about Down syndrome—that such individuals have a decreased risk of developing a number of cancers involving solid tumors, including lung cancer and melanoma. This health benefit has been correlated with the presence of an extra copy of the *DSCR1* gene, which encodes a protein that suppresses *vascular endothelial growth factor* (*VEGF*). This suppression, in turn, blocks the process of angiogenesis. As a result, the overexpression of this gene inhibits tumors from forming proper vascularization, diminishing their growth. A 14-year study published in 2002 involving 17,800 Down syndrome individuals revealed an approximate 10 percent reduction in cancer mortality in contrast to a control population. No doubt, further information will be forthcoming from the study of the DSCR region.

The Origin of the Extra 21st Chromosome in Down Syndrome

Most frequently, this trisomic condition occurs through nondisjunction of chromosome 21 during meiosis. Failure of paired homologs to disjoin during either anaphase I or

II may lead to gametes with the $n + 1$ chromosome composition. About 75 percent of these errors leading to Down syndrome are attributed to nondisjunction during meiosis I. Subsequent fertilization with a normal gamete creates the trisomic condition.

Chromosome analysis has shown that, while the additional chromosome may be derived from either the mother or father, the ovum is the source in about 95 percent of 47,21+ trisomy cases. Before the development of techniques using polymorphic markers to distinguish paternal from maternal homologs, this conclusion was supported by the more indirect evidence derived from studies of the age of mothers giving birth to infants afflicted with Down syndrome. Figure 8–4 shows the relationship between the incidence of

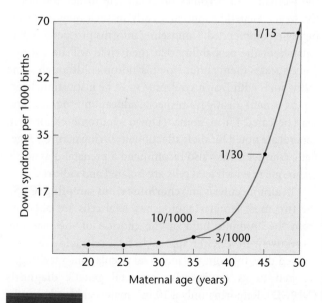

FIGURE 8-4 Incidence of Down syndrome births related to maternal age.

Down syndrome births and maternal age, illustrating the dramatic increase as the age of the mother increases. While the frequency is about 1 in 1000 at maternal age 30, a tenfold increase to a frequency of 1 in 100 is noted at age 40. The frequency increases still further to about 1 in 30 at age 45. A very alarming statistic is that as the age of childbearing women exceeds 45, the probability of a Down syndrome birth continues to increase substantially. In spite of these statistics, substantially more than half of Down syndrome births occur to women younger than 35 years, because the overwhelming proportion of pregnancies in the general population involve women under that age.

Although the nondisjunctional event clearly increases with age, we do not know with certainty why this is so. However, one observation may be relevant. Meiosis is initiated in all the eggs of a human female when she is still a fetus, until the point where the homologs synapse and recombination has begun. Then oocyte development is arrested in meiosis I. Thus, all primary oocytes have been formed by birth. When ovulation begins at puberty, meiosis is reinitiated in one egg during each ovulatory cycle and continues into meiosis II. The process is once again arrested after ovulation and is not completed unless fertilization occurs.

The end result of this progression is that each ovum that is released has been arrested in meiosis I for about a month longer than the one released during the preceding cycle. As a consequence, women 30 or 40 years old produce ova that are significantly older and that have been arrested longer than those they ovulated 10 or 20 years previously. In spite of the logic underlying this hypothesis explaining the cause of the increased incidence of Down syndrome as women age, it remains difficult to prove directly.

These statistics obviously pose a serious problem for the woman who becomes pregnant late in her reproductive years. Genetic counseling early in such pregnancies is highly recommended. Counseling informs prospective parents about the probability that their child will be affected and educates them about Down syndrome. Although some individuals with Down syndrome must be institutionalized, others benefit greatly from special education programs and may be cared for at home. (Down syndrome children in general are noted for their affectionate, loving nature.) A genetic counselor may also recommend a prenatal diagnostic technique in which fetal cells are isolated and cultured.

In **amniocentesis** and **chorionic villus sampling (CVS)**, the two most familiar approaches, fetal cells are obtained from the amniotic fluid or the chorion of the placenta, respectively. In a newer approach, fetal cells and DNA are derived directly from the maternal circulation, a technique referred to as **noninvasive prenatal genetic diagnosis (NIPGD)**. Requiring only a 10 mL maternal blood sample, this procedure will become increasingly more common because it poses no risk to the fetus.

With regard to Down syndrome, after fetal cells are obtained and cultured, the karyotype can be determined by cytogenetic analysis. If the fetus is diagnosed as being affected, a therapeutic abortion is one option currently available to parents. Obviously, this is a difficult decision involving a number of religious and ethical issues.

Since Down syndrome is caused by a random error—nondisjunction of chromosome 21 during maternal or paternal meiosis—the occurrence of the disorder is *not* expected to be inherited. Nevertheless, Down syndrome occasionally runs in families. These instances, referred to as *familial Down syndrome*, involve a translocation of chromosome 21, another type of chromosomal aberration, which we will discuss later in the chapter.

Human Aneuploidy

Besides Down syndrome, only two human trisomies, and no monosomies, survive to term: **Patau** and **Edwards syndromes** (47,13+ and 47,18+, respectively). Even so, these individuals manifest severe malformations and early lethality. Figure 8–5 illustrates the abnormal karyotype and the many defects characterizing Patau infants.

The above observation leads us to ask whether many other aneuploid conditions arise but that the affected

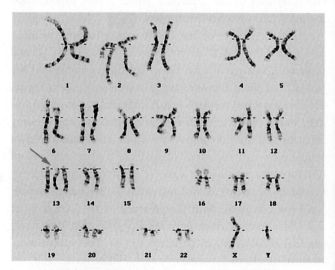

Mental retardation	Microcephaly
Growth failure	Cleft lip and palate
Low-set, deformed ears	Polydactyly
Deafness	Deformed finger nails
Atrial septal defect	Kidney cysts
Ventricular septal defect	Double ureter
Abnormal polymorphonuclear granulocytes	Umbilical hernia
	Developmental uterine abnormalities
	Cryptorchidism

FIGURE 8–5   The karyotype and phenotypic description of an infant with Patau syndrome, where three members of the D-group chromosome 13 are present, creating the 47,13+ condition.

fetuses do not survive to term. That this is the case has been confirmed by karyotypic analysis of spontaneously aborted fetuses. These studies reveal two striking statistics: (1) Approximately 20 percent of all conceptions terminate in spontaneous abortion (some estimates are considerably higher); and (2) about 30 percent of all spontaneously aborted fetuses demonstrate some form of chromosomal imbalance. This suggests that at least 6 percent (0.20×0.30) of conceptions contain an abnormal chromosome complement. A large percentage of fetuses demonstrating chromosomal abnormalities are aneuploids.

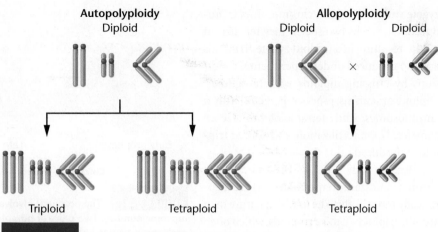

FIGURE 8–6   Contrasting chromosome origins of an autopolyploid versus an allopolyploid karyotype.

An extensive review of this subject by David H. Carr has revealed that a significant percentage of aborted fetuses are trisomic for one of the chromosome groups. Trisomies for every human chromosome have been recovered. Interestingly, the monosomy with highest incidence among abortuses is the 45,X condition, which produces an infant with Turner syndrome, if the fetus survives to term. Autosomal monosomies are seldom found, however, even though nondisjunction should produce $n - 1$ gametes with a frequency equal to $n + 1$ gametes. This finding suggests that gametes lacking a single chromosome are functionally impaired to a serious degree or that the embryo dies so early in its development that recovery occurs infrequently. We discussed the potential causes of monosomic lethality earlier in this chapter. Carr's study also found various forms of polyploidy and other miscellaneous chromosomal anomalies.

These observations support the hypothesis that normal embryonic development requires a precise diploid complement of chromosomes to maintain the delicate equilibrium in the expression of genetic information. The prenatal mortality of most aneuploids provides a barrier against the introduction of these genetic anomalies into the human population.

8.3

Polyploidy, in Which More Than Two Haploid Sets of Chromosomes Are Present, Is Prevalent in Plants

The term *polyploidy* describes instances in which more than two multiples of the haploid chromosome set are found. The naming of polyploids is based on the number of sets of chromosomes found: A triploid has $3n$ chromosomes; a tetraploid has $4n$; a pentaploid, $5n$; and so forth (Table 8.1).

Several general statements can be made about polyploidy. This condition is relatively infrequent in many animal species but is well known in lizards, amphibians, and fish, and is much more common in plant species. Usually, odd numbers of chromosome sets are not reliably maintained from generation to generation because a polyploid organism with an uneven number of homologs often does not produce genetically balanced gametes. For this reason, triploids, pentaploids, and so on, are not usually found in plant species that depend solely on sexual reproduction for propagation.

Polyploidy originates in two ways: (1) The addition of one or more extra sets of chromosomes, identical to the normal haploid complement of the same species, resulting in **autopolyploidy**; or (2) the combination of chromosome sets from different species occurring as a consequence of hybridization, resulting in **allopolyploidy** (from the Greek word *allo,* meaning "other" or "different"). The distinction between auto- and allopolyploidy is based on the genetic origin of the extra chromosome sets, as shown in Figure 8–6.

In our discussion of polyploidy, we use the following symbols to clarify the origin of additional chromosome sets. For example, if A represents the haploid set of chromosomes of any organism, then

$$A = a_1 + a_2 + a_3 + a_4 + \cdots + a_n$$

where a_1, a_2, and so on, are individual chromosomes and n is the haploid number. A normal diploid organism is represented simply as AA.

Autopolyploidy

In autopolyploidy, each additional set of chromosomes is identical to the parent species. Therefore, triploids are represented as AAA, tetraploids are $AAAA$, and so forth.

Autotriploids arise in several ways. A failure of all chromosomes to segregate during meiotic divisions can produce a diploid gamete. If such a gamete is fertilized by a haploid gamete, a zygote with three sets of chromosomes is produced. Or, rarely, two sperm may fertilize an ovum, resulting in a triploid zygote. Triploids are also produced under experimental conditions by crossing diploids with tetraploids. Diploid organisms produce gametes with n chromosomes, while tetraploids produce $2n$ gametes. Upon fertilization, the desired triploid is produced.

Because they have an even number of chromosomes, **autotetraploids** ($4n$) are theoretically more likely to be found in nature than are autotriploids. Unlike triploids, which often produce genetically unbalanced gametes with odd numbers of chromosomes, tetraploids are more likely to produce balanced gametes when involved in sexual reproduction.

How polyploidy arises naturally is of great interest to geneticists. In theory, if chromosomes have replicated, but the parent cell never divides and instead reenters interphase, the chromosome number will be doubled. That this very likely occurs is supported by the observation that tetraploid cells can be produced experimentally from diploid cells. This is accomplished by applying cold or heat shock to meiotic cells or by applying colchicine to somatic cells undergoing mitosis. **Colchicine**, an alkaloid derived from the autumn crocus, interferes with spindle formation, and thus replicated chromosomes cannot separate at anaphase and do not migrate to the poles. When colchicine is removed, the cell can reenter interphase. When the paired sister chromatids separate and uncoil, the nucleus contains twice the diploid number of chromosomes and is therefore $4n$. This process is shown in Figure 8–7.

In general, autopolyploids are larger than their diploid relatives. This increase seems to be due to larger cell size rather than greater cell number. Although autopolyploids do not contain new or unique information compared with their diploid relatives, the flower and fruit of plants are often increased in size, making such varieties of greater horticultural or commercial value. Economically important triploid plants include several potato species of the genus *Solanum*, Winesap apples, commercial bananas, seedless watermelons, and the cultivated tiger lily *Lilium tigrinum*. These plants are propagated asexually. Diploid bananas contain hard seeds, but the commercial triploid "seedless" variety has edible seeds. Tetraploid alfalfa, coffee, peanuts, and McIntosh apples are also of economic value because they are either larger or grow more vigorously than do their diploid or triploid counterparts. Many of the most popular varieties of hosta plant are tetraploid. In each case, leaves are thicker and larger, the foliage is more vivid, and the plant grows more vigorously. The commercial strawberry is an octoploid.

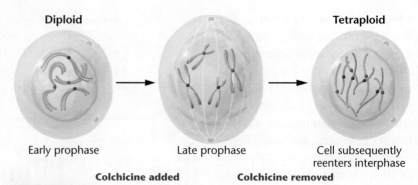

Diploid | | **Tetraploid**

Early prophase — Late prophase — Cell subsequently reenters interphase

Colchicine added   Colchicine removed

FIGURE 8–7   The potential involvement of colchicine in doubling the chromosome number. Two pairs of homologous chromosomes are shown. While each chromosome had replicated its DNA earlier during interphase, the chromosomes do not appear as double structures until late prophase. When anaphase fails to occur normally, the chromosome number doubles if the cell reenters interphase.

We have long been curious about how cells with increased ploidy values, where no new genes are present, express different phenotypes from their diploid counterparts. Our current ability to examine gene expression using modern biotechnology has provided some interesting insights. For example, Gerald Fink and his colleagues have been able to create strains of the yeast *Saccharomyces cerevisiae* with one, two, three, or four copies of the genome. Thus, each strain contains identical genes (they are said to be isogenic) but different ploidy values. These scientists then examined the expression levels of all genes during the entire cell cycle of the organism. Using the rather stringent standards of a tenfold increase or decrease of gene expression, Fink and coworkers proceeded to identify ten cases where, as ploidy increased, gene expression was increased at least tenfold and seven cases where it was reduced by a similar level.

One of these genes provides insights into how polyploid cells become larger than their haploid or diploid counterparts. In polyploid yeast, two **G1 cyclins**, Cln1, and Pcl1, are repressed when ploidy increases, while the size of the yeast cells increases. This is explained by the observation that G1 cyclins facilitate the cell's movement through G1, which is delayed when expression of these genes is repressed. The polyploid cell stays in the G1 phase longer and, on average, grows to a larger size before it moves beyond the G1 stage of the cell cycle. Yeast cells also show different morphology as ploidy increases. Several of the other genes, repressed as ploidy increases, have been linked to cytoskeletal dynamics that account for the morphological changes.

Allopolyploidy

Polyploidy can also result from hybridizing two closely related species. If a haploid ovum from a species with chromosome sets AA is fertilized by sperm from a species with sets BB, the resulting hybrid is AB, where $A = a_1, a_2, a_3, \ldots a_n$ and $B = b_1, b_2, b_3, \ldots b_n$. The hybrid plant may be sterile

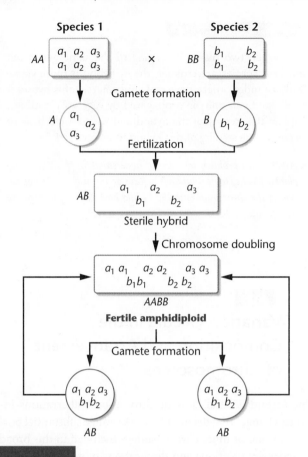

FIGURE 8–8 The origin and propagation of an amphidiploid. Species 1 contains genome A consisting of three distinct chromosomes, a_1, a_2, and a_3. Species 2 contains genome B consisting of two distinct chromosomes, b_1 and b_2. Following fertilization between members of the two species and chromosome doubling, a fertile amphidiploid containing two complete diploid genomes ($AABB$) is formed.

because of its inability to produce viable gametes. Most often, this occurs when some or all of the a and b chromosomes are not homologous and therefore cannot synapse in meiosis. As a result, unbalanced genetic conditions result. If, however, the new AB genetic combination undergoes a natural or an induced chromosomal doubling, two copies of all a chromosomes and two copies of all b chromosomes will be present, and they will pair during meiosis. As a result, a fertile $AABB$ tetraploid is produced. These events are shown in Figure 8–8. Since this polyploid contains the equivalent of four haploid genomes derived from separate species, such an organism is called an **allotetraploid**. When both original species are known, an equivalent term, **amphidiploid**, is preferred in describing the allotetraploid.

Amphidiploid plants are often found in nature. Their reproductive success is based on their potential for forming balanced gametes. Since two homologs of each specific chromosome are present, meiosis occurs normally (Figure 8–8) and fertilization successfully propagates the plant sexually. This discussion assumes the simplest situation, where none

of the chromosomes in set A are homologous to those in set B. In amphidiploids, formed from closely related species, some homology between a and b chromosomes is likely. Allopolyploids are rare in most animals because mating behavior is most often species-specific, and thus the initial step in hybridization is unlikely to occur.

A classical example of amphidiploidy in plants is the cultivated species of American cotton, *Gossypium* (Figure 8–9). This species has 26 pairs of chromosomes: 13 are large and 13 are much smaller. When it was discovered that Old World cotton had only 13 pairs of large chromosomes, allopolyploidy was suspected. After an examination of wild American cotton revealed 13 pairs of small chromosomes, this speculation was strengthened. J. O. Beasley reconstructed the origin of cultivated cotton experimentally by crossing the Old World strain with the wild American strain and then treating the hybrid with colchicine to double the chromosome number. The result of these treatments was a fertile amphidiploid variety of cotton. It contained 26 pairs of chromosomes as well as characteristics similar to the cultivated variety.

Amphidiploids often exhibit traits of both parental species. An interesting example, but one with no practical economic importance, is that of the hybrid formed between the radish *Raphanus sativus* and the cabbage *Brassica oleracea*. Both species have a haploid number $n = 9$. The initial hybrid consists of nine *Raphanus* and nine *Brassica* chromosomes ($9R + 9B$). Although hybrids are almost always sterile, some fertile amphidiploids ($18R + 18B$) have been produced. Unfortunately, the root of this plant is more like the cabbage and its shoot more like the radish; had the converse occurred, the hybrid might have been of economic importance.

A much more successful commercial hybridization uses the grasses wheat and rye. Wheat (genus *Triticum*) has a basic haploid genome of seven chromosomes. In addition to normal diploids ($2n = 14$), cultivated autopolyploids exist,

FIGURE 8–9 The pods of the amphidiploid form of *Gossypium,* the cultivated American cotton plant.

including tetraploid ($4n = 28$) and hexaploid ($6n = 42$) species. Rye (genus *Secale*) also has a genome consisting of seven chromosomes. The only cultivated species is the diploid plant ($2n = 14$).

Using the technique outlined in Figure 8–8, geneticists have produced various hybrids. When tetraploid wheat is crossed with diploid rye and the F_1 plants are treated with colchicine, a hexaploid variety ($6n = 42$) is obtained; the hybrid, designated *Triticale,* represents a new genus. Fertile hybrid varieties derived from various wheat and rye species can be crossed or backcrossed. These crosses have created many variations of the genus *Triticale.* The hybrid plants demonstrate characteristics of both wheat and rye. For example, certain hybrids combine the high-protein content of wheat with rye's high content of the amino acid lysine. (The lysine content is low in wheat and thus is a limiting nutritional factor.) Wheat is considered to be a high-yielding grain, whereas rye is noted for its versatility of growth in unfavorable environments. *Triticale* species, which combine both traits, have the potential of significantly increasing grain production. Programs designed to improve crops through hybridization have long been under way in several developing countries.

Endopolyploidy

Endopolyploidy is the condition in which only certain cells in an otherwise diploid organism are polyploid. In such cells, the set of chromosomes replicates repeatedly without nuclear division. Numerous examples of naturally occurring endopolyploidy have been observed. For example, vertebrate liver cell nuclei, including human ones, often contain $4n$, $8n$, or $16n$ chromosome sets. The stem and parenchymal tissue of apical regions of flowering plants are also often endopolyploid. Cells lining the gut of mosquito larvae attain a $16n$ ploidy, but during the pupal stages, such cells undergo very quick reduction divisions, giving rise to smaller diploid cells. In the water strider *Gerris,* wide variations in chromosome numbers are found in different tissues, with as many as 1024 to 2048 copies of each chromosome in the salivary gland cells. Since the diploid number in this organism is 22, the nuclei of these cells may contain more than 40,000 chromosomes.

Although the role of endopolyploidy is not clear, the proliferation of chromosome copies often occurs in cells where high levels of certain gene products are required. In fact, it is well established that certain genes whose product is in high demand in *every* cell exist naturally in multiple copies in the genome. Ribosomal and transfer RNA genes are examples of multiple-copy genes. In certain cells of organisms, where even this condition may not allow for a sufficient amount of a particular gene product, it may be necessary to replicate the entire genome, allowing an even greater rate of expression of that gene.

NOW SOLVE THIS

8–2 When two plants belonging to the same genus but different species are crossed, the F_1 hybrid is more viable and has more ornate flowers. Unfortunately, this hybrid is sterile and can only be propagated by vegetative cuttings. Explain the sterility of the hybrid and what would have to occur for the sterility of this hybrid to be reversed.

■ HINT: *This problem involves an understanding of allopolyploid plants. The key to its solution is to focus on the origin and composition of the chromosomes in the F_1 and how they might might be manipulated.*

8.4

Variation Occurs in the Composition and Arrangement of Chromosomes

The second general class of chromosome aberrations includes changes that delete, add, or rearrange substantial portions of one or more chromosomes. Included in this broad category are deletions and duplications of genes or part of a chromosome and rearrangements of genetic material in which a chromosome segment is inverted, exchanged with a segment of a nonhomologous chromosome, or merely transferred to another chromosome. Exchanges and transfers are called *translocations,* in which the locations of genes are altered within the genome. These types of chromosome alterations are illustrated in Figure 8–10.

In most instances, these structural changes are due to one or more breaks along the axis of a chromosome, followed by either the loss or rearrangement of genetic material. Chromosomes can break spontaneously, but the rate of breakage may increase in cells exposed to chemicals or radiation. Although the actual ends of chromosomes, known as telomeres, do not readily fuse with newly created ends of "broken" chromosomes or with other telomeres, the ends produced at points of breakage are "sticky" and can rejoin other broken ends. If breakage and rejoining do not reestablish the original relationship and if the alteration occurs in germ cells, the gametes will contain the structural rearrangement, which is heritable.

If the aberration is found in one homolog, but not the other, the individual is said to be *heterozygous for the aberration.* In such cases, unusual but characteristic pairing configurations are formed during meiotic synapsis. These patterns are useful in identifying the type of change that has occurred. If no loss or gain of genetic material occurs, individuals bearing the aberration "heterozygously" are

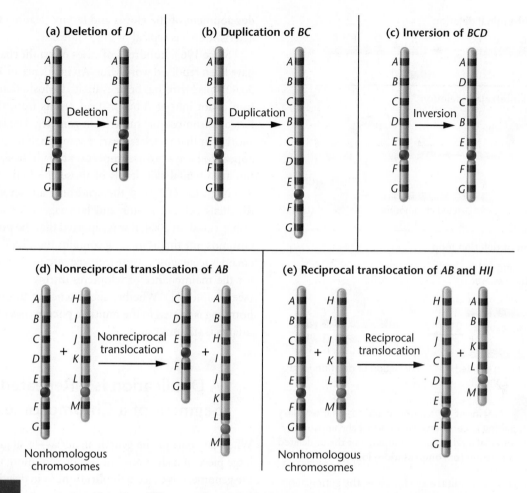

(a) Deletion of *D*

(b) Duplication of *BC*

(c) Inversion of *BCD*

(d) Nonreciprocal translocation of *AB*

Nonhomologous chromosomes

(e) Reciprocal translocation of *AB* and *HIJ*

Nonhomologous chromosomes

FIGURE 8–10 An overview of the five different types of gain, loss, or rearrangement of chromosome segments.

likely to be unaffected phenotypically. However, the unusual pairing arrangements often lead to gametes that are duplicated or deficient for some chromosomal regions. When this occurs, the offspring of "carriers" of certain aberrations have an increased probability of demonstrating phenotypic changes.

A Deletion Is a Missing Region of a Chromosome

When a chromosome breaks in one or more places and a portion of it is lost, the missing piece is called a **deletion** (or a **deficiency**). The deletion can occur either near one end or within the interior of the chromosome. These are **terminal** and **intercalary deletions**, respectively [Figure 8–11(a) and (b)]. The portion of the chromosome that retains the centromere region is usually maintained when the cell divides, whereas the segment without the centromere is eventually lost in progeny cells following mitosis or meiosis. For synapsis to occur between a chromosome with a large intercalary

deletion and a normal homolog, the unpaired region of the normal homolog must "buckle out" into a **deletion**, or **compensation, loop** [Figure 8–11(c)].

If only a small part of a chromosome is deleted, the organism might survive. However, a deletion of a portion of a chromosome need not be very great before the effects become severe. We see an example of this in the following discussion of the cri du chat syndrome in humans. If even more genetic information is lost as a result of a deletion, the aberration is often lethal, in which case the chromosome mutation never becomes available for study.

Cri du Chat Syndrome in Humans

In humans, the **cri du chat syndrome** results from the deletion of a small terminal portion of chromosome 5. It might be considered a case of *partial monosomy*, but since the region that is missing is so small, it is better referred to as a **segmental deletion**. This syndrome was first reported by Jérôme LeJeune in 1963, when he described the clinical symptoms, including an eerie cry similar to the meowing of a cat, after which the syndrome is named. This syndrome is associated with the loss of a small, variable part of the short

(a) Origin of terminal deletion

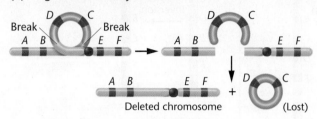

Break

A +

(Lost)

(b) Origin of intercalary deletion

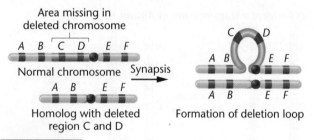

(c) Formation of deletion loop

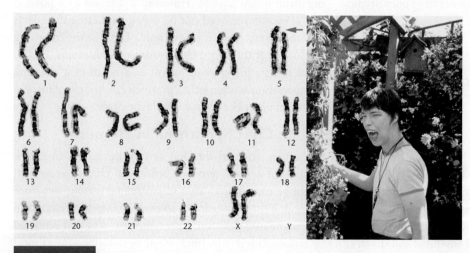

FIGURE 8–11   Origins of (a) a terminal and (b) an intercalary deletion. In (c), pairing occurs between a normal chromosome and one with an intercalary deletion by looping out the undeleted portion to form a deletion (or compensation) loop.

arm of chromosome 5 (Figure 8–12). Thus, the genetic constitution may be designated as 46,5p−, meaning that the individual has all 46 chromosomes but that some or all of the p arm (the petite, or short, arm) of one member of the chromosome 5 pair is missing.

Infants with this syndrome may exhibit anatomic malformations, including gastrointestinal and cardiac complications, and they are often mentally retarded. Abnormal

development of the glottis and larynx (leading to the characteristic cry) is typical of this syndrome.

Since 1963, hundreds of cases of cri du chat syndrome have been reported worldwide. An incidence of 1 in 25,000–50,000 live births has been estimated. Most often, the condition is not inherited but instead results from the sporadic loss of chromosomal material in gametes. The length of the short arm that is deleted varies somewhat; longer deletions appear to have a greater impact on the physical, psychomotor, and mental skill levels of those children who survive. Although the effects of the syndrome are severe, most individuals achieve motor and language skills and may be home-cared. In 2004, it was reported that the portion of the chromosome that is missing contains the *TERT* gene, which encodes *telomerase reverse transcriptase*, an enzyme essential for the maintenance of telomeres during DNA replication (see Chapter 11). Whether the absence of this gene on one homolog is related to the multiple phenotypes of cri du chat infants is still unknown.

8.6

A Duplication Is a Repeated Segment of a Chromosome

When any part of the genetic material—a single locus or a large piece of a chromosome—is present more than once in the genome, it is called a **duplication**. As in deletions, pairing in heterozygotes can produce a compensation loop. Duplications may arise as the result of unequal crossing over between synapsed chromosomes during meiosis (Figure 8–13) or through a replication error prior to meiosis. In the former case, both a duplication and a deletion are produced.

We consider three interesting aspects of duplications. First, they may result in gene redundancy. Second, as with deletions, duplications may produce phenotypic variation. Third, according to one convincing theory, duplications have also been an important source of genetic variability during evolution.

Gene Redundancy and Amplification— Ribosomal RNA Genes

Duplication of chromosomal segments has the potential to amplify the number of copies of individual genes. This has clearly been the case with the gene encoding ribosomal RNA, which is needed in abundance in the ribosomes of all cells to support protein synthesis. We might

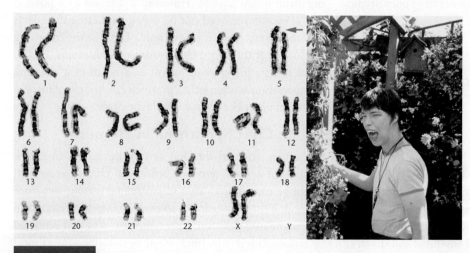

FIGURE 8–12   A representative karyotype and a photograph of a child exhibiting cri du chat syndrome (46,5p−). In the karyotype, the arrow identifies the absence of a small piece of the short arm of one member of the chromosome 5 homologs.

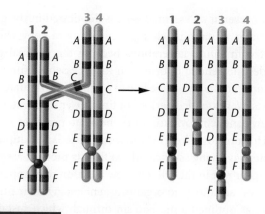

FIGURE 8–13 The origin of duplicated and deficient regions of chromosomes as a result of unequal crossing over. The tetrad on the left is mispaired during synapsis. A single crossover between chromatids 2 and 3 results in the deficient (chromosome 2) and duplicated (chromosome 3) chromosomal regions shown on the right. The two chromosomes uninvolved in the crossover event remain normal in gene sequence and content.

hypothesize that a single copy of the gene encoding rRNA is inadequate in many cells that demonstrate intense metabolic activity. Studies using the technique of molecular hybridization, which enables us to determine the percentage of the genome that codes for specific RNA sequences, show that our hypothesis is correct. Indeed, multiple copies of genes code for rRNA. Such DNA is called **rDNA**, and the general phenomenon is referred to as **gene redundancy**. For example, in the common intestinal bacterium *Escherichia coli* (*E. coli*), about 0.7 percent of the haploid genome consists of rDNA—the equivalent of seven copies of the gene. In *Drosophila melanogaster,* 0.3 percent of the haploid genome, equivalent to 130 gene copies, consists of rDNA. Although the presence of multiple copies of the same gene is not restricted to those coding for rRNA, we will focus on those genes in this section.

Interestingly, in some cells, particularly oocytes, even the normal redundancy of rDNA is insufficient to provide adequate amounts of rRNA needed to construct ribosomes. Oocytes store abundant nutrients, including huge quantities of ribosomes, for use by the embryo during early development. More ribosomes are included in oocytes than in any other cell type. By considering how the amphibian *Xenopus laevis* acquires this abundance of ribosomes, we shall see a second way in which the amount of rRNA is increased. This phenomenon is called **gene amplification**.

The genes that code for rRNA are located in an area of the chromosome known as the **nucleolar organizer region (NOR)**. The NOR is intimately associated with the

nucleolus, which is a processing center for ribosome production. Molecular hybridization analysis has shown that each NOR in the frog *Xenopus* contains the equivalent of 400 redundant gene copies coding for rRNA. Even this number of genes is apparently inadequate to synthesize the vast amount of ribosomes that must accumulate in the amphibian oocyte to support development following fertilization.

To further amplify the number of rRNA genes, the rDNA is selectively replicated, and each new set of genes is released from its template. Because each new copy is equivalent to one NOR, multiple small nucleoli are formed in the oocyte. As many as 1500 of these "micronucleoli" have been observed in a single *Xenopus* oocyte. If we multiply the number of micronucleoli (1500) by the number of gene copies in each NOR (400), we see that amplification in *Xenopus* oocytes can result in over half a million gene copies! If each copy is transcribed only 20 times during the maturation of the oocyte, in theory, sufficient copies of rRNA are produced to result in well over 12 million ribosomes.

The *Bar* Mutation in *Drosophila*

Duplications can cause phenotypic variation that might at first appear to be caused by a simple gene mutation. The *Bar*-eye phenotype in *Drosophila* (Figure 8–14) is a classic example. Instead of the normal oval-eye shape, *Bar*-eyed flies have narrow, slitlike eyes. This phenotype is inherited in the same way as a dominant X-linked mutation.

In the early 1920s, Alfred H. Sturtevant and Thomas H. Morgan discovered and investigated this "mutation." Normal wild-type females (B^+/B^+) have about 800 facets in each eye. Heterozygous females (B/B^+) have about 350 facets, while homozygous females (B/B) average only about 70 facets. Females were occasionally recovered with even fewer facets and were designated as *double Bar* (B^D/B^+).

About 10 years later, Calvin Bridges and Herman J. Muller compared the polytene X chromosome banding pattern of the *Bar* fly with that of the wild-type fly. These chromosomes contain specific banding patterns that have been well categorized into regions. Their studies revealed that one copy of the region designated as 16A is present on both X-chromosomes of wild-type flies but that this region was

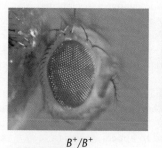
B^+/B^+

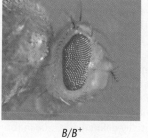
B/B^+

B/B

FIGURE 8–14 *Bar*-eye phenotypes in contrast to the wild-type eye in *Drosophila*.

duplicated in *Bar* flies and triplicated in *double Bar* flies. These observations provided evidence that the *Bar* phenotype is not the result of a simple chemical change in the gene but is instead a duplication.

The Role of Gene Duplication in Evolution

During the study of evolution, it is intriguing to speculate on the possible mechanisms of genetic variation. The origin of unique gene products present in more recently evolved organisms but absent in ancestral forms is a topic of particular interest. In other words, how do "new" genes arise? As we will see below, the process of gene duplication is hypothesized to be the major source of new genes, as proposed in 1970 by Susumu Ohno in his provocative monograph, *Evolution by Gene Duplication*. Ohno's thesis is based on the supposition that the products of many genes, present as only a single copy in the genome, are indispensable to the survival of members of any species during evolution. Therefore, unique genes are not free to accumulate mutations sufficient to alter their primary function and give rise to new genes.

However, if an essential gene is duplicated in the germ line, major mutational changes in this extra copy will be tolerated in future generations because the original gene provides the genetic information for its essential function. The duplicated copy will be free to acquire many mutational changes over extended periods of time. Over short intervals, the new genetic information may be of no practical advantage. However, over long evolutionary periods, the duplicated gene may change sufficiently so that its product assumes a divergent role in the cell. The new function may impart an "adaptive" advantage to organisms, enhancing their fitness. Ohno has outlined a mechanism through which sustained genetic variability may have originated.

Ohno's thesis is supported by the discovery of genes that have a substantial amount of their DNA sequence in common, but whose gene products are distinct. The genes encoding the digestive enzymes *trypsin* and *chymotrypsin* are examples, as are those that encode the respiratory molecules *myoglobin* and the various forms of *hemoglobin*. We conclude that the genes arose from a common ancestral gene through duplication. During evolution, the related genes diverged sufficiently that their products became unique.

Copy Number Variants (CNVs)— Duplications and Deletions at the Molecular Level

Genomic investigations that focus on the duplication or deletion of DNA sequences in humans are providing new insights into our understanding of duplications and deletions. These variations, most often involving thousands of base pairs, may prove to play crucial roles in many of our individual attributes, such as our sensitivity to drugs and susceptibility to disease. These differences, because they represent *quantitative differences in the number of large DNA sequences,* are termed **copy number variants (CNVs)**, and are found in both coding and noncoding regions of the genome.

In 2004, two research groups independently described the presence of CNVs in the genomes of healthy individuals with no known genetic disorders. CNVs were initially defined as regions of DNA at least 1 kb in length (1000 base pairs) that display at least 90 percent sequence identity. This initial study revealed 50 loci consisting of CNVs, and in 2005, several other groups began sifting through the genome in search of CNVs, defining almost 300 additional sites. The current number of CNV sites, when scaled down to sequences of DNA of at least 500 base pairs, was in 2010 estimated at over 10,000 regions, representing a substantial proportion of the total genetic variability within humans.

Current CNV studies have focused on finding associations with human diseases. CNVs appear to have both positive and negative associations with many diseases in which the genetic basis is not yet fully understood. For example, an association has been reported between CNVs and autism, a neurodevelopmental disorder that impairs communication, behavior, and social interaction. Interestingly, a mutant CNV site has been found to appear *de novo* (anew) in 10 percent of so-called sporadic cases of autism, where unaffected parents lack the CNV mutation. This is in contrast to only 2 percent of affected individuals where the disease appears to be familial (run in the family). Similarly, a higher than average copy number of the gene *CCL3L1* imparts an HIV-suppressive effect during viral infection, diminishing the progression to AIDS. Another research group has associated specific mutant CNV sites with certain subset populations of individuals with lung cancer—the greater number of copies of the *EGFR* (*Epidermal Growth Factor Receptor*) gene, the more responsive are patients with non–small-cell lung cancer to treatment. Finally, the greater the reduction in copy number of the gene designated *DEFB*, the greater the risk of developing Crohn's disease, a condition affecting the colon. Relevant to this chapter, these findings reveal that duplications and deletions are no longer restricted to textbook examples of these chromosomal mutations.

Other support includes the presence of **gene families**—groups of contiguous genes whose products perform the same function. Again, members of a family show DNA sequence homology sufficient to conclude that they share a common origin. Examples are the various types of human hemoglobin polypeptide chains, as well as the immunologically important T-cell receptors and antigens encoded by the major histocompatibility complex.

8.7

Inversions Rearrange the Linear Gene Sequence

The **inversion**, another class of structural variation, is a type of chromosomal aberration in which a segment of a chromosome is turned around 180 degrees within a chromosome. An inversion does not involve a loss of genetic information but simply rearranges the linear gene sequence. An inversion requires breaks at two points along the length of the chromosome and subsequent reinsertion of the inverted segment. Figure 8–15 illustrates how an inversion might arise. By forming a chromosomal loop prior to breakage, the newly created "sticky ends" are brought close together and rejoined.

The inverted segment may be short or quite long and may or may not include the centromere. If the centromere is not part of the rearranged chromosome segment, it is a **paracentric inversion**. If the centromere is part of the inverted segment, it is described as a **pericentric inversion**, which is the type shown in Figure 8–15.

Although inversions appear to have a minimal impact on the individuals bearing them, their consequences are of great interest to geneticists. Organisms that are heterozygous for inversions may produce aberrant gametes that have a major impact on their offspring.

Consequences of Inversions during Gamete Formation

If only one member of a homologous pair of chromosomes has an inverted segment, normal linear synapsis during meiosis is not possible. Organisms with one inverted chromosome and one noninverted homolog are called **inversion heterozygotes.** Pairing between two such chromosomes in meiosis is accomplished only if they form an **inversion loop** (Figure 8–16).

If crossing over does not occur within the inverted segment of the inversion loop, the homologs will segregate, which results in two normal and two inverted chromatids that are distributed into gametes. However, if crossing over does occur within the inversion loop, abnormal chromatids are produced. The effect of a single crossover (SCO) event within a paracentric inversion is diagrammed in Figure 8–16(a).

In any meiotic tetrad, a single crossover between non-sister chromatids produces two parental chromatids and two recombinant chromatids. When the crossover occurs within a paracentric inversion, however, one recombinant **dicentric chromatid** (two centromeres) and one recombinant **acentric chromatid** (lacking a centromere) are produced. Both contain duplications and deletions of chromosome segments as well. During anaphase, an acentric chromatid moves randomly to one pole or the other or may be lost, while a dicentric chromatid is pulled in two directions. This polarized movement produces *dicentric bridges* that are cytologically recognizable. A dicentric chromatid usually breaks at some point so that part of the chromatid goes into one gamete and part into another gamete during the reduction divisions. Therefore, gametes containing either recombinant chromatid are deficient in genetic material. When such a gamete participates in fertilization, the zygote most often develops abnormally, if at all.

A similar chromosomal imbalance is produced as a result of a crossover event between a chromatid bearing a pericentric inversion and its noninverted homolog, as shown in Figure 8–16(b). The recombinant chromatids that are directly involved in the exchange have duplications and deletions. In plants, gametes receiving such aberrant chromatids fail to develop normally, leading to aborted pollen or ovules. Thus, lethality occurs prior to fertilization, and inviable seeds result. In animals, the gametes have developed prior to the meiotic error, so fertilization is more likely to occur in spite of the chromosome error. However, the end result is the production of inviable embryos following fertilization. In both cases, viability is reduced.

Because offspring bearing crossover gametes are inviable and not recovered, it *appears* as if the inversion suppresses crossing over. Actually, in inversion heterozygotes, the inversion has the effect of *suppressing the recovery of crossover products* when chromosome exchange occurs within the inverted region.

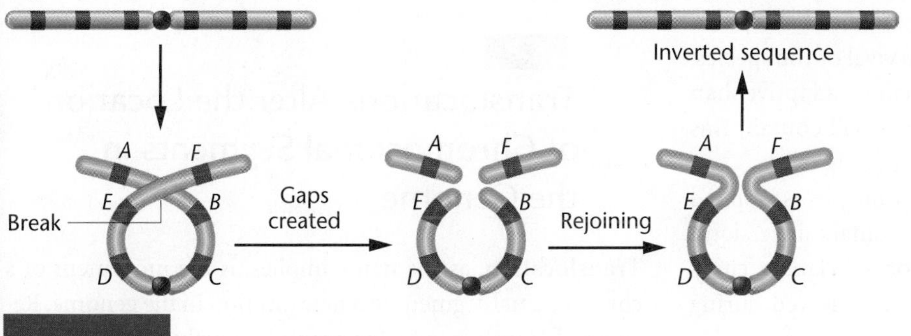

FIGURE 8–15   One possible origin of a pericentric inversion.

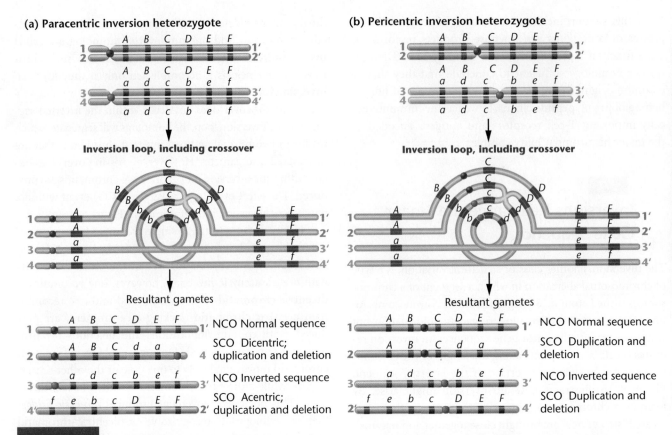

FIGURE 8–16   (a) The effects of a single crossover (SCO) within an inversion loop in a paracentric inversion heterozygote, where two altered chromosomes are produced, one acentric and one dicentric. Both chromosomes also contain duplicated and deficient regions. (b) The effects of a crossover in a pericentric inversion heterozygote, where two altered chromosomes are produced, both with duplicated and deficient regions.

If crossing over always occurred within a paracentric or pericentric inversion, 50 percent of the gametes would be ineffective. The viability of the resulting zygotes is therefore greatly diminished. Furthermore, up to one-half of the viable gametes have the inverted chromosome, and the inversion will be perpetuated within the species. The cycle will be repeated continuously during meiosis in future generations.

Evolutionary Advantages of Inversions

Because recovery of crossover products is suppressed in inversion heterozygotes, groups of specific alleles at adjacent loci within inversions may be preserved from generation to generation. If the alleles of the involved genes confer a survival advantage on organisms maintaining them, the inversion is beneficial to the evolutionary survival of the species. For example, if a set of alleles *ABcDef* is more adaptive than sets *AbCdeF* or *abcdEF*, effective gametes will contain this favorable set of genes, undisrupted by crossing over.

In laboratory studies, the same principle is applied using **balancer chromosomes**, which contain inversions. When an organism is heterozygous for a balancer chromosome, desired sequences of alleles are preserved during experimental work.

NOW SOLVE THIS

8–3 What is the effect of a rare double crossover (a) within a chromosome segment that is heterozygous for a pericentric inversion; and (b) within a segment that is heterozygous for a paracentric inversion?

■ HINT: *This problem involves an understanding of how homologs synapse in the presence of a heterozygous inversion, as well as the distinction between pericentric and paracentric inversions. The key to its solution is to draw out the tetrad and follow the chromatids undergoing a double crossover.*

8.8

Translocations Alter the Location of Chromosomal Segments in the Genome

Translocation, as the name implies, is the movement of a chromosomal segment to a new location in the genome. Reciprocal translocation, for example, involves the exchange

(a) Possible origin of a reciprocal translocation between two nonhomologous chromosomes

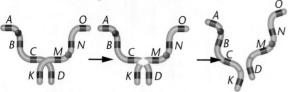

(b) Synapsis of translocation heterozygote

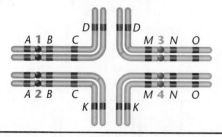

(c) Two possible segregation patterns leading to gamete formation

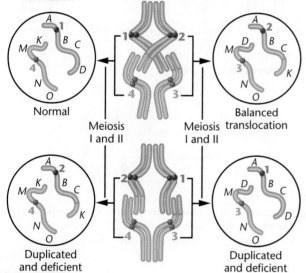

FIGURE 8–17 (a) Possible origin of a reciprocal transloca-tion. (b) Synaptic configuration formed during meiosis in an individual that is heterozygous for the translocation. (c) Two possible segregation patterns, one of which leads to a normal and a balanced gamete (called alternate segregation) and one that leads to gametes containing duplications and deficiencies (called adjacent segregation-1).

of segments between two nonhomologous chromosomes. The least complex way for this event to occur is for two nonhomologous chromosome arms to come close to each other so that an exchange is facilitated. Figure 8–17(a) shows a simple reciprocal translocation in which only two breaks are required. If the exchange includes internal chro-mosome segments, four breaks are required, two on each chromosome.

The genetic consequences of reciprocal translocations are, in several instances, similar to those of inversions. For example, genetic information is not lost or gained. Rather,

there is only a rearrangement of genetic material. The pres-ence of a translocation does not, therefore, directly alter the viability of individuals bearing it.

Homologs that are heterozygous for a reciprocal trans-location undergo unorthodox synapsis during meiosis. As shown in Figure 8–17(b), pairing results in a cruciform, or crosslike, configuration. As with inversions, some of the gametes produced are genetically unbalanced as a result of an unusual alignment during meiosis. In the case of trans-locations, however, aberrant gametes are not necessarily the result of crossing over. To see how unbalanced gametes are produced, focus on the homologous centromeres in Figure 8–17(b) and (c). According to the principle of independent assortment, the chromosome containing centromere 1 mi-grates randomly toward one pole of the spindle during the first meiotic anaphase; it travels along with *either* the chro-mosome having centromere 3 *or* the chromosome having centromere 4. The chromosome with centromere 2 moves to the other pole, along with *either* the chromosome con-taining centromere 3 *or* centromere 4. This results in four potential meiotic products. The 1,4 combination contains chromosomes that are not involved in the translocation and are normal. The 2,3 combination, however, contains the translocated chromosomes, but these contain a com-plete complement of genetic information and are balanced. When this result is achieved [the top configuration in Figure 8–17(c)], the segregation pattern at the first meiotic divi-sion is referred to as **alternate segregation**. A second pattern [the bottom configuration in Figure 8–17(c)] produces the other two potential products, the 1,3 and 2,4 combinations, which contain chromosomes displaying duplicated and de-leted (deficient) segments. This pattern is called **adjacent segregation-1**. Note that a third type of arrangement, where homologous centromeres segregate to the same pole during meiosis (called *adjacent segregation-2*), has not been includ-ed in this figure. This type of segregation has an outcome similar to adjacent segregation-1, with meiotic products containing genetically unbalanced duplicated and deleted chromosomal material.

When genetically unbalanced gametes participate in fer-tilization in animals, the resultant offspring do not usually survive. Fewer than 50 percent of the progeny of parents het-erozygous for a reciprocal translocation survive. This condi-tion in a parent is called **semisterility**, and its impact on the reproductive fitness of organisms plays a role in evolution. In humans, such an unbalanced condition results in partial monosomy or trisomy, leading to a variety of birth defects.

Translocations in Humans: Familial Down Syndrome

Research conducted since 1959 has revealed numerous translocations in members of the human population.

One common type of translocation involves breaks at the extreme ends of the short arms of two nonhomologous acrocentric chromosomes. These small segments are lost, and the larger segments fuse at their centromeric region. This type of translocation produces a new, large submetacentric or metacentric chromosome, often called a **Robertsonian translocation**.

One such translocation accounts for cases in which Down syndrome is familial (inherited). Earlier in this chapter, we pointed out that most instances of Down syndrome are due to trisomy 21. This chromosome composition results from nondisjunction during meiosis in one parent. Trisomy accounts for over 95 percent of all cases of Down syndrome. In such instances, the chance of the same parents producing a second affected child is extremely low. However, in the remaining families with a Down child, the syndrome occurs in a much higher frequency over several generations—it "runs in families."

Cytogenetic studies of the parents and their offspring from these unusual cases explain the cause of **familial Down syndrome**. Analysis reveals that one of the parents contains a **14/21, D/G translocation** (Figure 8–18). That is, one parent has the majority of the G-group chromosome 21 translocated to one end of the D-group chromosome 14. This individual is phenotypically normal, even though he or she has only 45 chromosomes and is referred to as a **balanced translocation carrier**. During meiosis in such an individual, one of the gametes contains two copies of chromosome 21—a normal chromosome and a second copy translocated to chromosome 14. When such a gamete is fertilized by a standard haploid gamete, the resulting zygote has 46 chromosomes but three copies of chromosome 21. These individuals exhibit Down syndrome. Other potential surviving offspring contain either the standard diploid genome (without a translocation) or the balanced translocation like the parent. Both cases result in normal individuals. In the fourth case, a monosomic individual is produced, which is lethal. Although not illustrated, adjacent-2 segregation is also thought to occur, but rarely. Such gametes are unbalanced, and upon fertilization, lethality occurs.

The above findings have allowed geneticists to resolve the seeming paradox of an inherited trisomic phenotype in an individual with an apparent diploid number of chromosomes. It is interesting to note that the "carrier," who has 45 chromosomes and exhibits a normal phenotype, does not contain the complete diploid amount of genetic material.

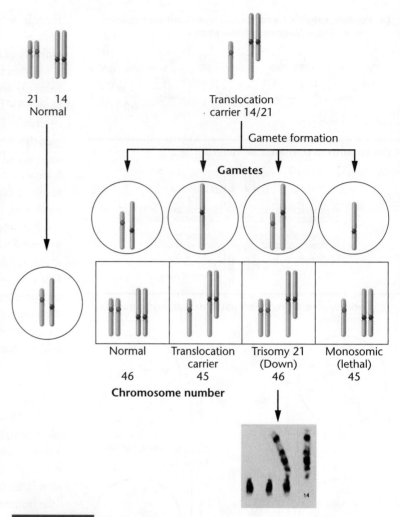

FIGURE 8–18 Chromosomal involvement and translocation in familial Down syndrome. The photograph shows the relevant chromosomes from a trisomy 21 offspring produced by a translocation carrier parent.

A small region is lost from both chromosomes 14 and 21 during the translocation event. This occurs because the ends of both chromosomes have broken off prior to their fusion. These specific regions are known to be two of many chromosomal locations housing multiple copies of the genes encoding rRNA, the major component of ribosomes. Despite the loss of up to 20 percent of these genes, the carrier is unaffected.

8.9

Fragile Sites in Human Chromosomes Are Susceptible to Breakage

We conclude this chapter with a brief discussion of the results of an intriguing discovery made around 1970 during observations of metaphase chromosomes prepared following human cell culture. In cells derived from certain individuals, a specific area along one of the chromosomes failed to stain, giving the appearance of a gap. In other individuals whose chromosomes displayed such morphology, the gaps appeared at other positions within the set of chromosomes. Such areas eventually became known as **fragile sites**, since they appeared to be susceptible to chromosome breakage when cultured in the absence of certain chemicals such as folic acid, which is normally present in the culture medium.

The cause of the fragility at these sites is unknown. However, since they represent points along the chromosome that are susceptible to breakage, these sites may indicate regions where the chromatin is not tightly coiled. Note that even though almost all studies of fragile sites have been carried out *in vitro* using mitotically dividing cells, interest in them increased when clear associations were established between several of these sites and a corresponding altered phenotype, including mental retardation and cancer.

Fragile-X Syndrome (Martin-Bell Syndrome)

While most fragile sites do not appear to be associated with any clinical syndrome, individuals bearing a folate-sensitive site on the X chromosome (Figure 8–19) exhibit the **fragile-X syndrome** (or *Martin-Bell syndrome*), the most common form of inherited mental retardation. This syndrome affects about 1 in 4000 males and 1 in 8000 females. Since affected females usually carry only one fragile X chromosome the disorder is considered a dominant trait. Fortunately, penetrance is not complete, and the trait is fully expressed in only about 30 percent of fragile X-bearing females and 80 percent of fragile X-bearing males. In addition to mental retardation, affected males have characteristic long, narrow faces with protruding chins, enlarged ears, and increased testicular size.

A gene that spans the fragile site is now known to be responsible for this syndrome. This gene, *FMR1*, is one of a growing number where a sequence of three nucleotides is repeated many times, expanding the size of the gene. Such **trinucleotide repeats** are also characteristic of other human disorders, including Huntington disease and myotonic dystrophy. In *FMR1*, the trinucleotide sequence CGG is repeated in an untranslated area adjacent to the coding sequence of the gene (called the "upstream" region). The number of repeats varies within the human population, and a high number correlates directly with expression of fragile-X syndrome. Normal individuals have between 6 and 54 repeats, whereas those with 55 to 230 repeats are unaffected, but are considered "carriers" of the disorder. More than 230 repeats lead to expression of the syndrome.

It is thought that once the gene contains this increased number of repeats it becomes chemically modified so that the bases within and around the repeats are methylated, which inactivates the gene. The normal product of the gene is an RNA-binding protein, FMRP, known to be expressed in the brain. Evidence is now accumulating that directly links the absence of the protein in the brain with the cognitive defects associated with the syndrome.

From a genetic standpoint, an interesting aspect of fragile-X syndrome is the instability of the CGG repeats. An individual with 6 to 54 repeats transmits a gene containing the same number of copies to his or her offspring. However, carrier individuals with 55 to 230 repeats, though not at risk to develop the syndrome, may transmit to their offspring a gene with an increased number of repeats. The number of repeats increases in future generations, demonstrating the phenomenon known as **genetic anticipation**. Once the threshold of 230 repeats is exceeded, retardation becomes more severe in each successive generation as the number of trinucleotide repeats increases. Interestingly, expansion from the carrier status (55 to 230 repeats) to the syndrome status (over 230 repeats) occurs only during the transmission of the gene by the maternal parent, not by the paternal parent. Thus, a "carrier" male may transmit a stable chromosome to his daughter, who may subsequently transmit an unstable chromosome with an increased number of repeats to her offspring. Their grandfather was the source of the original chromosome.

The Link between Fragile Sites and Cancer

While the study of the fragile-X syndrome first brought unstable chromosome regions to the attention of geneticists, a link between an autosomal fragile site and lung cancer was reported in 1996 by Carlo Croce, Kay Huebner, and

FIGURE 8–19 A human fragile X chromosome. The "gap" region (near the bottom of the chromosome) is associated with the fragile-X syndrome.

their colleagues. They have subsequently postulated that the defect is associated with the formation of a variety of different tumor types. Croce and Huebner first showed that the *FHIT* gene (standing for *f*ragile *hi*stidine *t*riad), located within the well-defined fragile site designated as *FRA3B* on the p arm of chromosome 3, is often altered or missing in cells taken from tumors of individuals with lung cancer. More extensive studies have now revealed that the normal protein product of this gene is absent in cells of many other cancers, including those of the esophagus, breast, cervix, liver, kidney, pancreas, colon, and stomach. Genes such as *FHIT* that are located within fragile regions undoubtedly have an increased susceptibility to mutations and deletions.

More recently, Muller Fabbri and Kay Huebner, working with others in Croce's lab, have identified and studied another fragile site, with most interesting results. Found within the *FRA16D* site on chromosome 16 is the *WWOX* gene. Like the *FHIT* gene, it has been implicated in a range of human cancers. In particular, like *FHIT*, it has been found to be either lost or genetically silenced in the large majority of lung tumors, as well as in cancer tissue of the breast, ovary, prostate, bladder, esophagus, and pancreas. When the gene is present but silent, its DNA is thought to be heavily methylated, rendering it inactive. Furthermore, the active gene is also thought to behave as a *tumor suppressor*, providing a surveillance function by recognizing cancer cells and inducing apoptosis, effectively eliminating them before malignant tumors can be initiated.

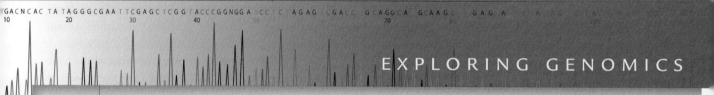

EXPLORING GENOMICS

Atlas of Genetics and Cytogenetics in Oncology and Haematology

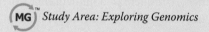 *Study Area: Exploring Genomics*

In this chapter, we discussed how variations in chromosome number and alterations in chromosome structure can affect the chromosome content of gametes to create genetic alterations in offspring and abnormal phenotypes. The **Atlas of Genetics and Cytogenetics in Oncology and Haematology** is a peer-reviewed, online journal and database of cytogenetics that specializes in cataloging chromosome abnormalities and genes involved in different cancers. Hematology is the study of blood. Much of the information presented in the atlas has been provided from clinical studies of patients with blood cancers, such as different forms of leukemia (cancer of white blood cells), that have revealed variations in chromosome number and chromosome structural defects (duplications, deletions, and translocations). In this exercise we explore the Atlas of Genetics site to learn more about human chromosome abnormalities.

■ **Exercise I – Exploring Chromosome 9**

1. Access the Atlas of Genetics and Cytogenetics in Oncology and Haematology at http://atlasgeneticsoncology.org.

2. Notice that the homepage lists database entries in several ways, including an alphabetical listing of cancer genes, a listing by chromosome, and a catalog of case reports. We will explore each of these features here.

3. Under "Entities by Chromosome" click on chromosome 9. Explore the many links to abnormalities of chromosome 9 involved in different leukemias. Notice that many of these are translocations. For example, t9;12 (q34;p13) symbolizes a translocation between chromosome 9, band 34 of the q-arm, and chromosome 12, band 13 of the p-arm.

4. Find the link to "Familial melanoma," review the information on this

condition, and then address the following items:

a. Describe this disease condition.

b. What locus on chromosome 9 is implicated in this disease? What gene is found at this locus? What is the function of this gene? (Use the disease gene links from the familial melanoma page or search the "Genes" feature of this database to learn about the gene's function.)

■ **Exercise II – Case Reports: Rare Examples of Chromosome Alterations**

The "Case reports" section of the atlas (see the link at the top of the page) provides reports on examples of rare conditions caused by chromosome abnormalities that have been observed in different patients. Many of these case reports contain excellent examples of data generated using the cytogenetic techniques discussed in this chapter and elsewhere in the book.

1. Find the link to the case report by Shambhu Roy, Sonal Bakshi, Shailesh Patel, and colleagues. Explore the information presented in their report and then answer the following questions:

   a. What chromosome abnormality is reported by this group?

   b. What techniques were used to diagnose this condition, and what tissue samples were used for the diagnosis?

   c. What is the disease condition associated with this patient?

   d. Why is this disease referred to as "a typical CML"? (Refer to Chapter 19 and Figure 19–4 to find the answer to this question.)

   e. Name another genetic condition associated with extra copies of chromosome 21.

2. Explore more case reports presented in the atlas to learn about other rare examples of chromosome abnormalities that have been observed by scientists and physicians around the world.

CASE STUDY | Fish tales

Aquatic vegetation overgrowth, usually controlled by dredging or herbicides, represents a significant issue in maintaining private and public waterways. In 1963, diploid grass carp were introduced in Arkansas to consume vegetation, but they reproduced prodigiously and spread to eventually become a hazard to aquatic ecosystems in 35 states. In the 1980s, many states adopted triploid grass carp as an alternative because of their high, but not absolute, sterility level and their longevity of seven to ten years. Today, most states require permits for vegetation control by triploid carp, requiring their containment in the body of water to which they are introduced. Genetic modifications of organisms to achieve specific outcomes will certainly become more common in the future and raise several interesting questions.

1. Taking triploid carp as an example, what controversies may emerge as similar modified species become available for widespread use?

2. If you were a state employee in charge of a specific waterway, what questions would you ask before you approved the introduction of a laboratory-produced, polyploid species into your waterway?

3. Why would the creation and use of a tetraploid carp species be less desirable in the above situation?

Summary Points

 For activities, animations, and review quizzes, go to the study area at www.masteringgenetics.com

1. Alterations of the precise diploid content of chromosomes are referred to as chromosomal aberrations or chromosomal mutations.

2. Studies of monosomic and trisomic disorders are increasing our understanding of the delicate genetic balance that is essential for normal development.

3. When more than two haploid sets of chromosomes are present, these may be derived from the same or different species, the basis of autopolyploidy and allotetraploidy, respectively.

4. Deletions or duplications of segments of a gene or a chromosome may be the source of mutant phenotypes, such as cri du chat syndrome in humans and Bar eyes in *Drosophila*, while duplications can be particularly important as a source of amplified or new genes.

5. Inversions and translocations may initially cause little or no loss of genetic information or deleterious effects. However, heterozygous combinations of the involved chromosome segments may result in genetically abnormal gametes following meiosis, with lethality or inviability often ensuing.

6. Fragile sites in human mitotic chromosomes have sparked research interest because one such site on the X chromosome is associated with the most common form of inherited mental retardation, while other autosomal sites have been linked to various forms of cancer.

INSIGHTS AND SOLUTIONS

1. In a cross using maize that involves three genes, *a*, *b*, and *c*, a heterozygote (*abc*/+++) is testcrossed to *abc*/*abc*. Even though the three genes are separated along the chromosome, thus predicting that crossover gametes and the resultant phenotypes should be observed, only two phenotypes are recovered: *abc* and +++. In addition, the cross produced significantly fewer viable plants than expected. Can you propose why no other phenotypes were recovered and why the viability was reduced?

Solution: One of the two chromosomes may contain an inversion that overlaps all three genes, effectively precluding the recovery of any "crossover" offspring. If this is a paracentric inversion and the genes are clearly separated (assuring that a significant number of crossovers occurs between them), then numerous acentric and dicentric chromosomes will form, resulting in the observed reduction in viability.

2. A male *Drosophila* from a wild-type stock is discovered to have only seven chromosomes, whereas normally $2n = 8$. Close examination reveals that one member of chromosome IV (the smallest chromosome) is attached to (translocated to) the distal end of chromosome II and is missing its centromere, thus accounting for the reduction in chromosome number.

(a) Diagram all members of chromosomes II and IV during synapsis in meiosis I.

Solution:

(b) If this male mates with a female with a normal chromosome composition who is homozygous for the recessive chromosome IV mutation *eyeless* (*ey*), what chromosome compositions will occur in the offspring regarding chromosomes II and IV?

Solution:

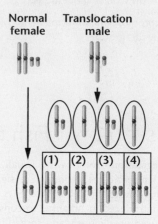

(c) Referring to the diagram in the solution to part (b), what phenotypic ratio will result regarding the presence of eyes, assuming all abnormal chromosome compositions survive?

Solution:

1. normal (heterozygous)

2. eyeless (monosomic, contains chromosome IV from mother)

3. normal (heterozygous; trisomic and may die)

4. normal (heterozygous; balanced translocation)

The final ratio is 3/4 normal: 1/4 eyeless.

Problems and Discussion Questions

 For instructor-assigned tutorials and problems, go to www.masteringgentics.com

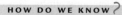

HOW DO WE KNOW?

1. In this chapter, we have focused on chromosomal mutations resulting from a change in number or arrangement of chromosomes. In our discussions, we found many opportunities to consider the methods and reasoning by which much of this information was acquired. From the explanations given in the chapter, what answers would you propose to the following fundamental questions?

(a) How do we know that the extra chromosome causing Down syndrome is usually maternal in origin?

(b) How do we know that human aneuploidy for each of the 22 autosomes occurs at conception, even though most often human aneuploids do not survive embryonic or fetal development and thus are never observed at birth?

(c) How do we know that specific mutant phenotypes are due to changes in chromosome number or structure?

(d) How do we know that the mutant Bar-eye phenotype in *Drosophila* is due to a duplicated gene region rather than to a change in the nucleotide sequence of a gene?

2. For a species with a diploid number of 18, indicate how many chromosomes will be present in the somatic nuclei of individuals that are haploid, triploid, tetraploid, trisomic, and monosomic.

3. Define these pairs of terms, and distinguish between them.

    aneuploidy/euploidy

    monosomy/trisomy

    Patau syndrome/Edwards syndrome

    autopolyploidy/allopolyploidy

    autotetraploid/amphidiploid

    paracentric inversion/pericentric inversion

4. Contrast the relative survival times of individuals with Down, Patau, and Edwards syndromes. Speculate as to why such differences exist.

5. What evidence suggests that Down syndrome is more often the result of nondisjunction during oogenesis rather than during spermatogenesis?

6. What evidence indicates that humans with aneuploid karyotypes occur at conception but are usually inviable?

7. Contrast the fertility of an allotetraploid with an autotriploid and an autotetraploid.

8. Describe the origin of cultivated American cotton.

9. Predict how the synaptic configurations of homologous pairs of chromosomes might appear when one member is normal and the other member has sustained a deletion or duplication.

10. Inversions are said to "suppress crossing over." Is this terminology technically correct? If not, restate the description accurately.

11. Contrast the genetic composition of gametes derived from tetrads of inversion heterozygotes where crossing over occurs within a paracentric versus a pericentric inversion.

12. Human adult hemoglobin is a tetramer containing two alpha (α) and two beta (β) polypeptide chains. The α gene cluster on chromosome 16 and the β gene cluster on chromosome 11 share amino acid similarities such that 61 of the amino acids of the α-globin polypeptide (141 amino acids long) are shared in identical sequence with the β-globin polypeptide (146 amino acids long). How might one explain the existence of two polypeptides with partially shared function and structure on two different chromosomes?

13. Discuss Ohno's hypothesis on the role of gene duplication in the process of evolution. What evidence supports this hypothesis?

14. What roles have inversions and translocations played in the evolutionary process?

15. The primrose, *Primula kewensis,* has 36 chromosomes that are similar in appearance to the chromosomes in two related species, *P. floribunda* ($2n = 18$) and *P. verticillata* ($2n = 18$). How could *P. kewensis* arise from these species? How would you describe *P. kewensis* in genetic terms?

16. Certain varieties of chrysanthemums contain 18, 36, 54, 72, and 90 chromosomes; all are multiples of a basic set of nine chromosomes. How would you describe these varieties genetically? What feature do the karyotypes of each variety share? A variety with 27 chromosomes has been discovered, but it is sterile. Why?

17. *Drosophila* may be monosomic for chromosome 4, yet remain fertile. Contrast the F_1 and F_2 results of the following crosses involving the recessive chromosome 4 trait, bent bristles: (a)

monosomic IV, bent bristles $\times$ diploid, normal bristles; (b) monosomic IV, normal bristles $\times$ diploid, bent bristles.

18. Mendelian ratios are modified in crosses involving autotetraploids. Assume that one plant expresses the dominant trait green seeds and is homozygous (*WWWW*). This plant is crossed to one with white seeds that is also homozygous (*wwww*). If only one dominant allele is sufficient to produce green seeds, predict the F_1 and F_2 results of such a cross. Assume that synapsis between chromosome pairs is random during meiosis.

19. Having correctly established the F_2 ratio in Problem 18, predict the F_2 ratio of a "dihybrid" cross involving two independently assorting characteristics (e.g., $P_1 = WWWWAAAA \times wwwwaaaa$).

20. The mutations called *bobbed* in *Drosophila* result from variable reductions (deletions) in the number of amplified genes coding for rRNA. Researchers trying to maintain *bobbed* stocks have often documented their tendency to revert to wild type in successive generations. Propose a mechanism based on meiotic recombination which could account for this reversion phenomenon. Why would wild-type flies become more prevalent in *Drosophila* cultures?

21. The outcome of a single crossover between nonsister chromatids in the inversion loop of an inversion heterozygote varies depending on whether the inversion is of the paracentric or pericentric type. What differences are expected?

22. A couple planning their family are aware that through the past three generations on the husband's side a substantial number of stillbirths have occurred and several malformed babies were born who died early in childhood. The wife has studied genetics and urges her husband to visit a genetic counseling clinic, where a complete karyotype-banding analysis is performed. Although the tests show that he has a normal complement of 46 chromosomes, banding analysis reveals that one member of the chromosome 1 pair (in group A) contains an inversion covering 70 percent of its length. The homolog of chromosome 1 and all other chromosomes show the normal banding sequence.

(a) How would you explain the high incidence of past stillbirths?

(b) What can you predict about the probability of abnormality/normality of their future children?

(c) Would you advise the woman that she will have to bring each pregnancy to term to determine whether the fetus is normal? If not, what else can you suggest?

Extra-Spicy Problems

 For instructor-assigned tutorials and problems, go to www.masteringgentics.com

23. In a cross in *Drosophila*, a female heterozygous for the autosomally linked genes *a, b, c, d,* and *e* (*abcde*/+++++) was testcrossed with a male homozygous for all recessive alleles. Even though the distance between each of the loci was at least 3 map units, only four phenotypes were recovered, yielding the following data:

Phenotype	No. of Flies
+ + + + +	440
a b c d e	460
+ + + + e	48
a b c d +	52
	Total = 1000

Why are many expected crossover phenotypes missing? Can any of these loci be mapped from the data given here? If so, determine map distances.

24. A woman who sought genetic counseling is found to be heterozygous for a chromosomal rearrangement between the second and third chromosomes. Her chromosomes, compared to those in a normal karyotype, are diagrammed here:

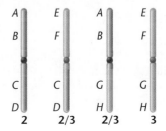

(a) What kind of chromosomal aberration is shown?

(b) Using a drawing, demonstrate how these chromosomes would pair during meiosis. Be sure to label the different segments of the chromosomes.

(c) This woman is phenotypically normal. Does this surprise you? Why or why not? Under what circumstances might you expect a phenotypic effect of such a rearrangement?

25. The woman in Problem 24 has had two miscarriages. She has come to you, an established genetic counselor, with these questions: Is there a genetic explanation of her frequent miscarriages? Should she abandon her attempts to have a child of her own? If not, what is the chance that she could have a normal child? Provide an informed response to her concerns.

26. In a recent cytogenetic study on 1021 cases of Down syndrome, 46 were the result of translocations, the most frequent of which was symbolized as t(14;21). What does this symbol represent, and how many chromosomes would you expect to be present in t(14;21) Down syndrome individuals?

27. A boy with Klinefelter syndrome (47,XXY) is born to a mother who is phenotypically normal and a father who has the X-linked skin condition called anhidrotic ectodermal dysplasia. The mother's skin is completely normal with no signs of the skin abnormality. In contrast, her son has patches of normal skin and patches of abnormal skin.

(a) Which parent contributed the abnormal gamete?

(b) Using the appropriate genetic terminology, describe the meiotic mistake that occurred. Be sure to indicate in which division the mistake occurred.

(c) Using the appropriate genetic terminology, explain the son's skin phenotype.

28. To investigate the origin of nondisjunction, 200 human oocytes that had failed to be fertilized during *in vitro* fertilization procedures were subsequently examined (Angel, R. 1997. *Am. J. Hum. Genet.* 61: 23–32). These oocytes had completed meiosis I and were arrested in metaphase II (MII). The majority (67 percent) had a normal MII-metaphase complement, showing 23 chromosomes, each consisting of two sister chromatids joined at a common centromere. The remaining oocytes all had abnormal chromosome compositions. Surprisingly, when trisomy was considered, none of the abnormal oocytes had 24 chromosomes.

(a) Interpret these results in regard to the origin of trisomy, as it relates to nondisjunction, and when it occurs. Why are the results surprising?

(b) A large number of the abnormal oocytes contained 22 1/2 chromosomes; that is, 22 chromosomes plus a single chromatid representing the 1/2 chromosome. What chromosome compositions will result in the zygote if such oocytes proceed through meiosis and are fertilized by normal sperm?

(c) How could the complement of 22 1/2 chromosomes arise? Provide a drawing that includes several pairs of MII chromosomes.

(d) Do your answers support or dispute the generally accepted theory regarding nondisjunction and trisomy, as outlined in Figure 8–1?

29. In a human genetic study, a family with five phenotypically normal children was investigated. Two were "homozygous" for a Robertsonian translocation between chromosomes 19 and 20 (they contained two identical copies of the fused chromosome). These children have only 44 chromosomes but a complete genetic complement. Three of the children were "heterozygous" for the translocation and contained 45 chromosomes, with one translocated chromosome plus a normal copy of both chromosomes 19 and 20. Two other pregnancies resulted in stillbirths. It was later discovered that the parents were first cousins. Based on this information, determine the chromosome compositions of the parents. What led to the stillbirths? Why was the discovery that the parents were first cousins a key piece of information in understanding the genetics of this family?

30. A 3-year-old child exhibited some early indication of Turner syndrome, which results from a 45,X chromosome composition. Karyotypic analysis demonstrated two cell types: 46,XX (normal) and 45,X. Propose a mechanism for the origin of this mosaicism.

31. A normal female is discovered with 45 chromosomes, one of which exhibits a Robertsonian translocation containing most of chromosomes 18 and 21. Discuss the possible outcomes in her offspring when her husband contains a normal karyotype.

32. Prader-Willi and Angelman syndromes (see Chapter 4) are human conditions that result in part through the process of genomic imprinting, where a particular gene is "marked" during gamete formation in the parent of origin. In the case of these two syndromes, a portion of the long arm of chromosome 15 (15q11 to 15q13) is imprinted. Another condition, Beckwith-Wiedemann syndrome (accelerated growth and increased risk of cancer), is associated with abnormalities of imprinted genes on the short arm of chromosome 11. All three of the above syndromes occur when imprinted sequences become abnormally exposed by some genetic or chromosomal event. One such event is uniparental disomy (UPD), in which a person receives two homologs of one chromosome (or part of a chromosome) from one parent and no homolog from the other. In many cases, UPD is without consequence; however, if chromosome 11 or 15 is involved, then, coupled with genomic imprinting, complications can arise. Listed below are possible origins of UPD. Provide a diagram and explanation for each.

1. *Trisomic rescue:* loss of a chromosome in a trisomic zygote
2. *Monosomic rescue:* gain of a chromosome in a monosomic zygote
3. *Gamete complementation:* fertilization of a gamete with two copies of a chromosome by a gamete with no copies of that chromosome
4. *Isochromosome formation:* a chromosome that contains two copies of an arm (15q-15q, for example) rather than one.

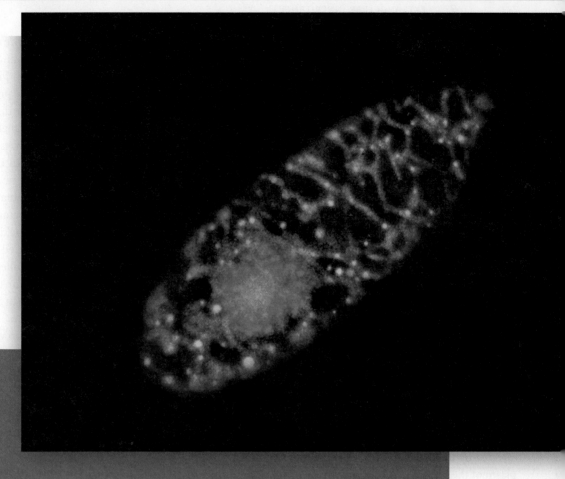

The distribution of the multiple small mitochondrial DNA molecules (bright orange-yellow spots) and the nuclear genome (red) in an *Euglena gracilis* cell.

9

Extranuclear Inheritance

CHAPTER CONCEPTS

- Extranuclear inheritance occurs when phenotypes result from genetic influence other than the biparental transmission of genes located on chromosomes housed within the nucleus.

- Organelle heredity, an example of extranuclear inheritance, is due to the transmission of genetic information contained in mitochondria or chloroplasts, most often from only one parent.

- Traits determined by mitochondrial DNA are most often transmitted uniparentally through the maternal gamete, while traits determined by chloroplast DNA may be transmitted uniparentally or biparentally.

- Maternal effect, the expression of the maternal nuclear genotype during gametogenesis and during early development, may have a strong influence on the phenotype of an organism.

Throughout the history of genetics, occasional reports have challenged the basic tenet of Mendelian transmission genetics—that the phenotype is transmitted by nuclear genes located on the chromosomes of both parents. Observations have revealed inheritance patterns that fail to reflect Mendelian principles, and some indicate an apparent extranuclear influence on the phenotype. Before the role of DNA in genetics was firmly established, such observations were commonly regarded with skepticism. However, with the increasing knowledge of molecular genetics and the discovery of DNA in mitochondria and chloroplasts, extranuclear inheritance is now recognized as an important aspect of genetics.

There are several varieties of extranuclear inheritance. One major type, referred to above, is also described as **organelle heredity**. In this type of inheritance, DNA contained in mitochondria or chloroplasts determines certain phenotypic characteristics of the offspring. Examples are often recognized on the basis of the uniparental transmission of these organelles, usually from the female parent through the egg to progeny. A second type, called **infectious heredity**, results from a symbiotic or parasitic association with a microorganism. In such cases, an inherited phenotype is affected by the presence of the microorganism in the cytoplasm of the host cells. A third variety involves the **maternal effect** on the phenotype, whereby nuclear gene products are stored in the egg and then transmitted through the ooplasm to offspring. These gene products are distributed to cells of the developing embryo and influence its phenotype.

The common element in all of these examples is the transmission of genetic information to offspring through the cytoplasm rather than through the nucleus, most often from only one of the parents. This shall constitute our definition of **extranuclear inheritance**.

9.1

Organelle Heredity Involves DNA in Chloroplasts and Mitochondria

In this section and the next, we will examine examples of inheritance patterns arising from chloroplast and mitochondrial function. Such patterns are now appropriately called **organelle heredity**. Before DNA was discovered in these organelles, the exact mechanism of transmission of the traits we are about to discuss was not clear, except that their inheritance appeared to be linked to something in the cytoplasm rather than to genes in the nucleus. Most often (but not in all cases), the traits appeared to be transmitted from the maternal parent through the ooplasm, causing the results of reciprocal crosses to vary.

Analysis of the inheritance patterns resulting from mutant alleles in chloroplasts and mitochondria has been difficult for two major reasons. First, the function of these organelles is dependent on gene products from both nuclear and organelle DNA, making the discovery of the genetic origin of mutations affecting organelle function difficult. Second, large numbers of these organelles are contributed to each progeny cell following cell division. If only one or a few of the organelles acquire a new mutation or contain an existing one in a cell with a population of mostly normal organelles, the corresponding mutant phenotype may not be revealed, since the organelles lacking the mutation perform the wild-type function for the cell. Such variation in the genetic content of organelles is called **heteroplasmy**. Analysis is thus much more complex for traits controlled by genes encoded by organelle DNA than for Mendelian characters controlled by nuclear genes.

We will begin our discussion with several of the classical examples that ultimately called attention to organelle heredity. After that, we will discuss information concerning the DNA and the resultant genetic function in each organelle.

Chloroplasts: Variegation in Four O'Clock Plants

In 1908, Carl Correns (one of the rediscoverers of Mendel's work) provided the earliest example of inheritance linked to chloroplast transmission. Correns discovered a variant of the four o'clock plant, *Mirabilis jalapa*, in which some branches had white leaves, some had green, and some had variegated leaves. The completely white leaves and the white areas in variegated leaves lack chlorophyll that otherwise provides green color. Chlorophyll is the light-absorbing pigment made within chloroplasts.

Correns was curious about how inheritance of this phenotypic trait occurred. Inheritance in all possible combinations of crosses is strictly determined by the phenotype of the ovule source Figure 9–1. For example, if the seeds (representing the progeny) were derived from ovules on branches with green leaves, all progeny plants bore only green leaves, regardless of the phenotype of the source of pollen. Correns concluded that inheritance was transmitted through the cytoplasm of the maternal parent because the pollen, which contributes little or no cytoplasm to the zygote, had no apparent influence on the progeny phenotypes.

Since leaf coloration is a function of the chloroplast, genetic information either contained in that organelle or somehow present in the cytoplasm and influencing the chloroplast must be responsible for the inheritance pattern. It now seems certain that the genetic "defect" that eliminates the green chlorophyll in the white patches on leaves is a mutation in the DNA housed in the chloroplast.

Source of Pollen	Location of Ovule		
	White branch	Green branch	Variegated branch
White branch	White	Green	White, green, or variegated
Green branch	White	Green	White, green, or variegated
Variegated branch	White	Green	White, green, or variegated

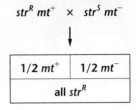

$str^R \ mt^+ \quad \times \quad str^S \ mt^-$

1/2 mt^+	1/2 mt^-
all str^R	

$str^S \ mt^+ \quad \times \quad str^R \ mt^-$

1/2 mt^+	1/2 mt^-
all str^S	

FIGURE 9–1   Offspring of crosses involving leaves from various branches of variegated four o'clock plants. The photograph illustrates variegation in the leaves of the four o'clock plant.

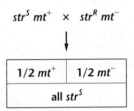

FIGURE 9–2   The results of reciprocal crosses between streptomycin-resistant (str^R) and streptomycin-sensitive (str^S) strains in the green alga *Chlamydomonas* (shown in the photograph).

Chloroplast Mutations in *Chlamydomonas*

The unicellular green alga *Chlamydomonas reinhardtii* has provided an excellent system for the investigation of plastid inheritance. This haploid eukaryotic organism (Figure 5–20) has a single large chloroplast containing about 75 copies of a circular double-stranded DNA molecule. Matings that reestablish diploidy are immediately followed by meiosis, and the various stages of the life cycle are easily studied in culture in the laboratory. The first known cytoplasmic mutation, streptomycin resistance (str^R) in *Chlamydomonas*, was reported in 1954 by Ruth Sager. Although *Chlamydomonas*'s two mating types, mt^+ and mt^-, appear to make equal cytoplasmic contributions to the zygote, Sager determined that the str^R phenotype is transmitted only through the mt^+ parent (Figure 9–2). Reciprocal crosses between sensitive and resistant strains yield different results depending on the genotype of the mt^+ parent, which is expressed in all offspring. As shown in the figure, one-half of the offspring are mt^+ and one-half of them are mt^-, indicating that mating type is controlled by a nuclear gene that segregates in a Mendelian fashion.

Since Sager's discovery, a number of other *Chlamydomonas* mutations (including resistance to, or dependence on, a variety of bacterial antibiotics) that show a similar uniparental inheritance pattern have been discovered. These mutations have all been linked to the transmission of the chloroplast, and their study has extended our knowledge of chloroplast inheritance.

Following fertilization, which involves the fusion of two cells of opposite mating type, the single chloroplasts of the

NOW SOLVE THIS

9–1 *Chlamydomonas*, a eukaryotic green alga, may be sensitive to the antibiotic erythromycin, which inhibits protein synthesis in prokaryotes. There are two mating types in this alga, mt^+ and mt^-. If an mt^+ cell sensitive to the antibiotic is crossed with an mt^- cell that is resistant, all progeny cells are sensitive. The reciprocal cross (mt^+ resistant and mt^- sensitive) yields all resistant progeny cells. Assuming that the mutation for resistance is in the chloroplast DNA, what can you conclude from the results of these crosses?

■ HINT: *This problem involves an understanding of the cytoplasmic transmission of organelles in unicellular algae. The key to its solution is to consider the results you would expect from two possibilities: that inheritance of the trait is uniparental or that inheritance is biparental.*

FIGURE 9–3 Micrograph illustrating the growth of the bread mold *Neurospora crassa*.

two mating types fuse. After the resulting zygote has undergone meiosis and haploid cells are produced, it is apparent that the genetic information in the chloroplast of progeny cells is derived only from the *mt*⁺ parent. The genetic information originally present within the *mt*⁻ chloroplast has degenerated.

The inheritance of phenotypes influenced by mitochondria is also uniparental in *Chlamydomonas*. However, studies of the transmission of several cases of antibiotic resistance governed by mitochondria have shown that it is the *mt*⁻ parent that transmits the mitochondrial genetic information to progeny cells. This is just the opposite of what occurs with chloroplast-derived phenotypes, such as *str*ᴿ. The significance of inheriting one organelle from one parent and the other organelle from the other parent is not yet established.

Mitochondrial Mutations: Early Studies in *Neurospora* and Yeast

As alluded to earlier, mutations affecting mitochondrial function have also been discovered and studied, revealing that mitochondria, too, contain a distinctive genetic system. As with chloroplasts, mitochondrial mutations are transmitted through the cytoplasm during reproduction. In 1952, Mary B. Mitchell and Herschel K. Mitchell studied the pink bread mold *Neurospora crassa* (Figure 9–3). They discovered a slow-growing mutant strain and named it *poky*. (It is also designated *mi-1*, for *maternal inheritance*.) Slow growth is associated with impaired mitochondrial function, specifically caused by the absence of several cytochrome proteins essential for electron transport. In the absence of cytochromes, aerobic respiration leading to ATP synthesis is curtailed.

Results of genetic crosses between wild-type and *poky* strains suggest that the trait is maternally inherited. If one mating type is *poky* and the other is wild type, all progeny colonies are *poky*, yet the reciprocal cross produces normal wild-type colonies.

Another study of mitochondrial mutations has been performed with the yeast *Saccharomyces cerevisiae*. The first such mutation, described by Boris Ephrussi and his coworkers in 1956, was named *petite* because of the small size of the yeast colonies (Figure 9–4). Many independent *petite* mutations have since been discovered and studied, and all have a common characteristic: a deficiency in cellular respiration involving abnormal electron transport, as performed by mitochondria. This organism is a *facultative anaerobe* (an organism that can function both with and without the presence of oxygen), so in the absence of oxygen it can grow by fermenting glucose through glycolysis. Thus, it may survive the loss of mitochondrial function by generating energy anaerobically.

The complex genetics of *petite* mutations is diagrammed in Figure 9–5. A small proportion of these mutants are the result of nuclear mutations in genes whose products are transported to and function in mitochondria. They exhibit Mendelian inheritance and are thus called **segregational petites**. The remaining mutants demonstrate cytoplasmic transmission, indicating alterations in the DNA of the mitochondria. They produce one of two effects in matings. **Neutral petites**, when crossed to wild type, yield meiotic products (called ascospores) that give rise only to wild-type, or normal, colonies. The same pattern continues if progeny of such crosses are backcrossed to neutral *petites*. The majority of "neutrals" lack mtDNA completely or have lost a substantial portion of it, so for their offspring to be normal, the neutrals must also be inheriting mitochondria capable of aerobic respiration from the normal parent following reproduction. This establishes that in yeast, mitochondria are inherited from both parental cells. The functional

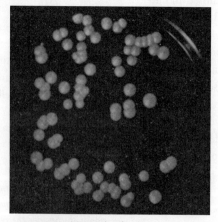

Normal colonies

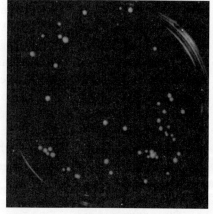

Petite colonies

FIGURE 9–4 A comparison of normal versus petite colonies in the yeast *Saccharomyces cerevisiae*.

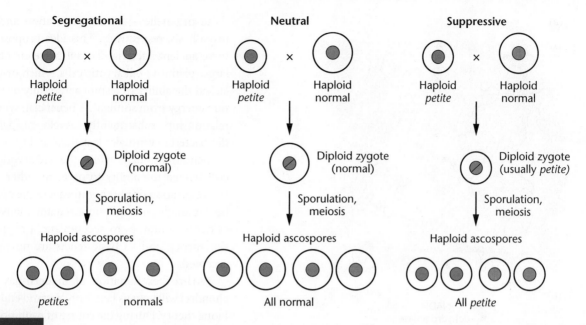

FIGURE 9–5 The outcome of crosses involving the three types of *petite* mutations affecting mitochondrial function in the yeast *Saccharomyces cerevisiae*.

mitochondria from the normal parent are replicated in offspring and support aerobic respiration.

The third mutational type, ***suppressive petites***, provides different results. Crosses between mutant and wild type give rise to diploid zygotes, which after meiosis, yield haploid cells that all express the *petite* phenotype. Assuming that the offspring have received mitochondria from both parents, the *petite* cells behave as what is called a **dominant-negative mutation**, which somehow suppresses the function of the wild-type mitochondria.

Two major hypotheses concerning the organelle DNA have been advanced to explain this suppressiveness. One explanation suggests that the mutant (or deleted) DNA in the mitochondria (mtDNA) replicates more rapidly, resulting in the mutant mitochondria "taking over" or dominating the phenotype by numbers alone. The second explanation suggests that recombination occurs between the mutant and wild-type mtDNA, introducing errors into or disrupting the normal mtDNA. It is not yet clear which one, if either, of these explanations is correct.

9.2
Knowledge of Mitochondrial and Chloroplast DNA Helps Explain Organelle Heredity

That both mitochondria and chloroplasts contain their own DNA and a system for expressing genetic information was first suggested by the discovery of mutations and the resultant inheritance patterns in plants, yeast, and other fungi, as already discussed. Because both mitochondria and chloroplasts are inherited through the maternal cytoplasm in most organisms, and because each of the above-mentioned examples of mutations could be linked hypothetically to the altered function of either chloroplasts or mitochondria, geneticists set out to look for more direct evidence of DNA in these organelles. Not only was unique DNA found to be a normal component of both mitochondria and chloroplasts, but careful examination of the nature of this genetic information would provide essential clues as to the evolutionary origin of these organelles.

Organelle DNA and the Endosymbiotic Theory

Electron microscopists not only documented the presence of DNA in mitochondria and chloroplasts, but they also saw

---

NOW SOLVE THIS

9–2 In aerobically cultured yeast, a *petite* mutant is isolated. To determine the type of mutation causing this phenotype, the *petite* and wild-type strains are crossed. Such a cross has three potential outcomes.

(a) all wild type
(b) some *petite*: some wild type
(c) all *petite*

For each set of results, what conclusion about the type of *petite* mutation is justified?

■ HINT: *This problem involves the understanding that the petite phenotype is related to mitochondrial function and to how mitochondria are inherited. The key to its solution is to remember that in yeast, inheritance of mitochondria is biparental.*

(a)

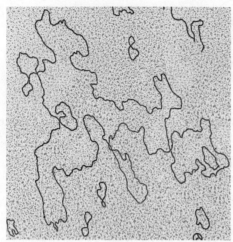

(b)

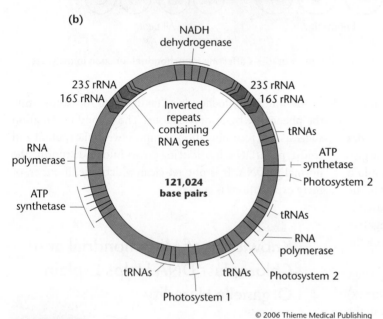

© 2006 Thieme Medical Publishing

FIGURE 9–6 Illustrations of chloroplast DNA. (a) Electron micrograph of chloroplast DNA derived from lettuce. (b) Diagram illustrating the arrangement of many of the genes encoded by cpDNA of the moss, *Marchantia polymorpha*. Photosystems 1 and 2 are groups of genes with photosynthetic functions.

these organelles—aerobic respiration and photosynthesis, respectively. This idea proposes that these ancient bacteria-like cells were engulfed by larger primitive eukaryotic cells, which originally lacked the ability to respire aerobically or to capture energy from sunlight. A beneficial, symbiotic relationship subsequently developed, whereby the bacteria eventually lost their ability to function autonomously, while the eukaryotic host cells gained the ability to perform either oxidative respiration or photosynthesis, as the case may be. Although some questions remain unanswered, evidence continues to accumulate in support of this theory, and its basic tenets are now widely accepted.

A brief examination of modern-day mitochondria will help us better understand endosymbiotic theory. During the course of evolution subsequent to the invasion event, distinct branches of diverse eukaryotic organisms arose. As the evolution of the host cells progressed, the companion bacteria also underwent their own independent changes. The primary alteration was the transfer of many of the genes from the invading bacterium to the nucleus of the host. The *products* of these genes, though still functioning in the organelle, are nevertheless now encoded and transcribed in the nucleus and translated in the cytoplasm prior to their transport into the organelle. The amount of DNA remaining today in the typical mitochondrial genome is minuscule compared with that in the free-living bacteria from which it was derived. The most gene-rich organelles now have fewer than 10 percent of the genes present in the smallest bacterium known.

Similar changes have characterized the evolution of chloroplasts. In the subsequent sections, we will explore in some detail what is known about modern-day chloroplasts and mitochondria.

Molecular Organization and Gene Products of Chloroplast DNA

The details of the autonomous genetic system of chloroplasts have now been worked out, providing further support of the endosymbiotic theory. The chloroplast, responsible for photosynthesis, contains both DNA (as a source of genetic information) and a complete protein-synthesizing apparatus. The molecular components of the chloroplast's translation apparatus are derived from both nuclear and organelle genetic information.

Chloroplast DNA (cpDNA), shown in Figure 9–6(a), is fairly uniform in size among different organisms, ranging

that it exists there in a form quite unlike the form seen in the nucleus of the eukaryotic cells that house these organelles (Figures 9–6 and 9–7). The DNA in chloroplasts and mitochondria looks remarkably similar to the DNA seen in bacteria. This similarity, along with the observation of the presence of a unique genetic system capable of organelle-specific transcription and translation, led Lynn Margulis and others to the postulate known as the **endosymbiotic theory**. Basically, the theory states that mitochondria and chloroplasts arose independently about 2 billion years ago from free-living protobacteria (primitive bacteria). Progenitors possessed the abilities now attributed to

(a)

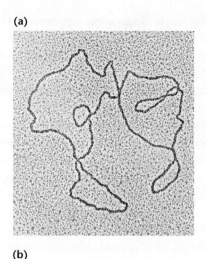

(b)

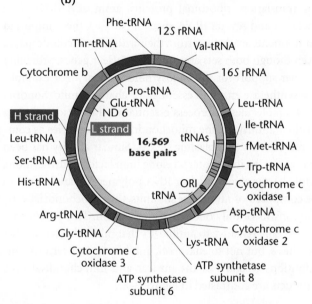

© 2006 Thieme Medical Publishing

FIGURE 9–7 Examples of mitochondrial DNA. (a) Electron micrograph of mitochondrial DNA derived from the frog *Xenopus laevis*. (b) Diagram illustrating the arrangement of many of the genes encoded by human mtDNA.

between 100 and 225 kb in length. It shares many similarities to DNA found in prokaryotic cells. It is circular and double stranded, and it is free of the associated proteins characteristic of eukaryotic DNA. Compared with nuclear DNA from the same organism, it invariably shows a different density and base composition.

The size of cpDNA is much larger than that of mtDNA. To some extent, this can be accounted for by a larger number of genes. However, most of the difference appears to be due to the presence in cpDNA of many long noncoding nucleotide sequences both between and within genes, the latter being introns (noncoding DNA sequences, also characteristic of eukaryotic nuclear DNA, as you'll see in Chapter 13). Duplications of many DNA sequences are also present. Since such noncoding sequences vary in different plants, they indicate

that independent evolution occurred in chloroplasts following their initial invasion of a primitive eukaryotic-like cell.

In the green alga *Chlamydomonas*, there are about 75 copies of the chloroplast DNA molecule per organelle. In higher plants, such as the sweet pea, multiple copies of the DNA molecule are also present in each organelle, but the molecule (134 kb) is considerably smaller than that in *Chlamydomonas* (195 kb). Interestingly, genetic recombination between the multiple copies of DNA within chloroplasts has been documented in some organisms.

Numerous gene products encoded by chloroplast DNA function during translation within the organelle. Figure 9–6(b) illustrates some of the genes that are present on cpDNA of the moss, which is representative of a variety of higher plants. Two sets each of the genes coding for the ribosomal RNAs—16S, and 23S rRNA—are present (S refers to the Svedberg coefficient, which is described in Chapter 10 and is related to the molecule's size and shape). In addition, cpDNA encodes numerous transfer RNAs (tRNAs), as well as many ribosomal proteins specific to the chloroplast ribosomes. In the liverwort, whose cpDNA was the first to be sequenced, there are genes encoding 30 tRNAs, RNA polymerase, multiple rRNAs, and numerous ribosomal proteins. The variations in the gene products encoded in the cpDNA of different plants again attest to the independent evolution that occurred within chloroplasts.

Chloroplast ribosomes differ significantly from those present in the cytoplasm and encoded by nuclear genes. They have a Svedberg coefficient slightly less than 70S, which characterizes bacterial ribosomes. Even though some chloroplast ribosomal proteins are encoded by chloroplast DNA and some by nuclear DNA, most, if not all, such proteins are chemically distinct from their counterparts present in cytoplasmic ribosomes. Both observations provide direct support for the endosymbiotic theory.

Still other chloroplast genes specific to the photosynthetic function have been identified [Figure 9–6(b)]. For example, in the moss, there are 92 chloroplast genes encoding proteins that are part of the thylakoid membrane, a cellular component integral to the light-dependent reactions of photosynthesis. Mutations in these genes may inactivate photosynthesis. A typical distribution of genes between the nucleus and the chloroplast is illustrated by one of the major photosynthetic enzymes, ribulose-1-5-bisphosphate carboxylase (known as *Rubisco*). This enzyme has its small subunit encoded by a nuclear gene, whereas the large subunit is encoded by cpDNA.

Molecular Organization and Gene Products of Mitochondrial DNA

Extensive information is also available concerning the structure and gene products of mitochondrial DNA (mtDNA). In most eukaryotes, mtDNA exists as a double-stranded,

TABLE 9.1

The Size of mtDNA in Different Organisms

Organisms	Size (kb)
Homo sapiens (human)	16.6
Mus musculus (mouse)	16.2
Xenopus laevis (frog)	18.4
Drosophila melanogaster (fruit fly)	18.4
Saccharomyces cerevisiae (yeast)	75.0
Pisum sativum (pea)	110.0
Arabidopsis thaliana (mustard plant)	367.0

closed circle [Figure 9–7(a)] that, like cpDNA, is free of the chromosomal proteins characteristic of eukaryotic chromosomal DNA. An exception is found in some ciliated protozoans, in which the DNA is linear.

In size, mtDNA is much smaller than cpDNA and varies greatly among organisms, as demonstrated in Table 9.1. In a variety of animals, including humans, mtDNA consists of about 16,000 to 18,000 bp (16 to 18 kb). However, yeast (*Saccharomyces*) mtDNA consists of 75 kb. Plants typically exceed this amount—367 kb is present in mitochondria in the mustard plant, *Arabidopsis*. Vertebrates have 5 to 10 such DNA molecules per organelle, while plants have 20 to 40 copies per organelle.

There are several other noteworthy aspects of mtDNA. With only rare exceptions, introns are absent from mitochondrial genes, and gene repetitions are seldom present. Nor is there usually much in the way of intergenic spacer DNA. This is particularly true in species whose mtDNA is fairly small in size, such as humans. In *Saccharomyces*, with a much larger mtDNA molecule, much of the excess DNA is accounted for by introns and intergenic spacer DNA. As will be discussed in Chapter 13, the expression of mitochondrial genes uses several modifications of the otherwise standard genetic code. Also of interest is the fact that replication in mitochondria is dependent on enzymes encoded by nuclear DNA.

Human mtDNA [Figure 9–7(b)] encodes two ribosomal RNAs (rRNAs), 22 transfer RNAs (tRNAs), and 13 polypeptides essential to the oxidative respiration functions of the organelle. For instance, mitochondrial-encoded gene products are present in all of the protein complexes of the electron transport chain found in the inner membrane of mitochondria. In most cases, these polypeptides are part of multichain proteins, many of which also contain subunits that are encoded in the nucleus, synthesized in the cytoplasm, and then transported into the organelle. Thus, the protein-synthesizing apparatus and the molecular components for cellular respiration are jointly derived from nuclear and mitochondrial genes.

Another interesting observation is that in vertebrate mtDNA, the two strands vary in density, as revealed by centrifugation. This provides researchers with a way to isolate the strands for study, designating one heavy (H) and the other light (L). While most of the mitochondrial genes are encoded by the H strand, several are encoded by the complementary L strand.

As might be predicted by the endosymbiotic theory, ribosomes found in the organelle differ from those present in the neighboring cytoplasm. Mitochondrial ribosomes of different species vary considerably in their Svedberg coefficients, ranging from 55S to 80S, while cytoplasmic ribosomes are uniformly 80S. The majority of proteins that function in mitochondria are encoded by nuclear genes. In fact, over 1000 nuclear-coded gene products are essential to biological activity in the organelle. They include, for example, DNA and RNA polymerases, initiation and elongation factors essential for translation, ribosomal proteins, aminoacyl tRNA synthetases, and several tRNA species. Notably, these imported components are distinct from their cytoplasmic counterparts, even though both sets are coded by nuclear genes, providing further support for the endosymbiotic theory. For example, the synthetase enzymes essential for charging mitochondrial tRNA molecules (a process essential to translation) show a distinct affinity for the mitochondrial tRNA species as compared with the cytoplasmic tRNAs. Similar affinity has been shown for the initiation and elongation factors. Furthermore, while bacterial and nuclear RNA polymerases are known to be composed of numerous subunits, the mitochondrial variety consists of only one polypeptide chain. This polymerase is generally sensitive to antibiotics that inhibit bacterial RNA synthesis, but not to eukaryotic inhibitors. The various contributions of some of the nuclear and mitochondrial gene products are contrasted in Figure 9–8.

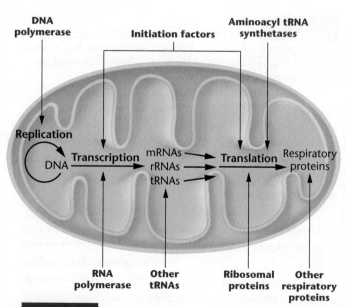

FIGURE 9–8   Gene products that are essential to mitochondrial function. Those shown entering the organelle are derived from the cytoplasm and encoded by nuclear genes.

9–3 DNA in human mitochondria encode 22 different tRNA molecules. However, 32 different tRNA molecules are required for translation of proteins within mitochondria. Explain.

■ HINT: *This problem involves understanding the origin of mitochondria in eukaryotes. The key to its solution is to consider the endosymbiotic theory and its ramifications.*

9.3

Mutations in Mitochondrial DNA Cause Human Disorders

The DNA found in human mitochondria has been completely sequenced and contains 16,569 base pairs. As mentioned earlier, mtDNA gene products include 13 of over 70 proteins required for aerobic cellular respiration. Because a cell's energy supply is largely dependent on aerobic cellular respiration to generate ATP, disruption of any mitochondrial gene by mutation may potentially have a severe impact on that organism. We have seen this in our previous discussion of the *petite* mutations in yeast, which would be lethal were it not for this organism's ability to respire anaerobically. In fact, mtDNA is particularly vulnerable to mutations, for two possible reasons. First, mtDNA does not have the structural protection from mutations provided by histone proteins present in nuclear DNA. Second, mitochondria concentrate highly mutagenic **reactive oxygen species (ROS)** generated by cell respiration. In such a confined space, ROS are toxic to the contents of the organelle and are known to damage proteins, lipids, and mtDNA. Ultimately, this increases the frequency of point mutations and deletions in mitochondria.

The number of copies of mtDNA in human cells can range from several hundred in somatic cells to approximately 100,000 copies of mtDNA in an oocyte. Fortunately, a zygote receives a large number of organelles through the egg, so if only one organelle or a few of them contain a mutation, its impact is greatly diluted by the many mitochondria that lack the mutation and function normally. However, during early development, cell division disperses the initial population of mitochondria present in the zygote, and in the newly formed cells, these organelles reproduce autonomously. Therefore, if a deleterious mutation arises or is present in the initial population of organelles, adults will have cells with a variable mixture of both normal and abnormal organelles. This variation in the genetic content of organelles, as indicated earlier in the chapter, is called *heteroplasmy*.

In order for a human disorder to be attributable to genetically altered mitochondria, several criteria must be met:

1. Inheritance must exhibit a maternal rather than a Mendelian pattern.

2. The disorder must reflect a deficiency in the bioenergetic function of the organelle.

3. There must be a mutation in one or more of the mitochondrial genes.

Thus far, several disorders in humans are known to demonstrate these characteristics. For example, **myoclonic epilepsy and ragged-red fiber disease (MERRF)** demonstrates a pattern of inheritance consistent with maternal transmission. Only the offspring of affected mothers inherit the disorder; the offspring of affected fathers are normal. Individuals with this rare disorder express ataxia (lack of muscular coordination), deafness, dementia, and epileptic seizures. The disease is so named because of the presence of "ragged-red" skeletal muscle fibers that exhibit blotchy red patches resulting from the proliferation of aberrant mitochondria (**Figure 9–9**). Brain function, which has a high energy demand, is affected in this disorder, leading to the neurological symptoms described.

Analysis of mtDNA from patients with MERRF has revealed a mutation in one of the 22 mitochondrial genes encoding a transfer RNA. Specifically, the gene encoding $tRNA^{Lys}$ (the tRNA that delivers lysine during translation) contains an A-to-G transition within its sequence. This genetic alteration apparently interferes with the capacity for translation within the organelle, which in turn leads to the various manifestations of the disorder.

The cells of affected individuals exhibit heteroplasmy, containing a mixture of normal and abnormal mitochondria. Different patients contain different proportions of the two, and even different cells from the same patient exhibit various levels of abnormal mitochondria. Were it not for heteroplasmy, the mutation would very likely be lethal, testifying to the essential nature of mitochondrial function and its reliance on the genes encoded by mtDNA within the organelle.

A second disorder, **Leber's hereditary optic neuropathy (LHON)**, also exhibits maternal inheritance as well as mtDNA lesions. The disorder is characterized by sudden bilateral blindness. The average age of vision loss is 27, but onset is quite variable. Four mutations have been identified, all of which disrupt normal oxidative phosphorylation, the final pathway of respiration in cells. More than 50 percent of cases are due to a mutation at a specific position in the mitochondrial gene encoding a subunit of NADH dehydrogenase. This mutation is transmitted maternally through the mitochondria to all offspring. Noteworthy is the observation that in many instances of LHON, there is no family history; a significant number of cases are "sporadic," resulting from newly arisen mutations.

(a)

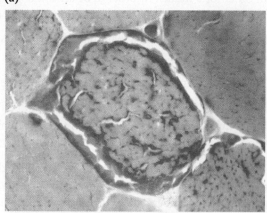

(b)

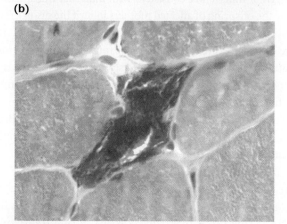

FIGURE 9–9   Ragged-red fibers in skeletal muscle cells from patients with the mitochondrial disease MERRF. (a) The muscle fiber has mild proliferation of mitochondria. (See red rim and speckled cytoplasm.) (b) Marked proliferation in which mitochondria have replaced most cellular structures.

Individuals severely affected by a third disorder, **Kearns–Sayre syndrome (KSS)**, lose their vision, experience hearing loss, and display heart conditions. The genetic basis of KSS involves deletions at various positions within mtDNA. Many KSS patients are symptom-free as children but display progressive symptoms as adults. The proportion of mtDNAs that reveal deletion mutations increases as the severity of symptoms increases.

Mitochondria, Human Health, and Aging

The study of hereditary mitochondrial-based disorders provides insights into the critical importance of this organelle during normal development. In fact, mitochondrial dysfunction seems to be implicated in most all major disease conditions, including Type II (late-onset) diabetes, atherosclerosis, neurodegenerative diseases such as Parkinson, Alzheimer, and Huntington disease, schizophrenia and bipolar disorders, and a variety of cancers. It is becoming evident, for example, that mutations in mtDNA are present in such human malignancies as skin, colorectal, liver, breast, pancreatic, lung, prostate, and bladder cancers. Genetic tests for detecting mutations in the mtDNA genome that may serve as early-stage disease markers have been developed. For example, mtDNA mutations in skin cells have been detected as a biomarker of cumulative exposure of ultraviolet light and development of skin cancer. However, it is still unclear whether mtDNA mutations are causative effects contributing to development of malignant tumors or whether they are the consequences of tumor formation. Nonetheless, there is an interesting link between mtDNA mutations and cancer, including data suggesting that many chemical carcinogens have significant mutation effects on mtDNA.

The study of hereditary mitochondrial-based disorders has also suggested a link between the progressive decline of mitochondrial function and the aging process. It has been hypothesized that the accumulation of sporadic mutations in mtDNA leads to an increased prevalence of defective mitochondria (and the concomitant decrease in the supply of ATP) in cells over a lifetime. This condition in turn plays a significant role in aging. It has been suggested that cells require a threshold level of ATP production resulting from oxidative phosphorylation for normal function. When the level drops below this threshold, the aging process is accelerated.

Many studies have now documented that aging tissue contain mitochondria with increased levels of DNA damage. As with cancer, the major question is whether such changes are simply biomarkers of the aging process or whether they lead to a decline in physiological function that contributes significantly to aging. In support of the latter hypothesis, one study links age-related muscle fiber atrophy in rats to deletions in mtDNA and electron transport abnormalities. Such deletions appear to be present in the fibers of atrophied muscle, but are absent from fibers in regions of normal tissue. Another study using genetically altered mice is even more convincing. Such mice have a nuclear gene altered that diminishes proofreading during the replication of mtDNA. These mice display reduced fidelity and accumulate mutations over time at a much higher rate than is normal. Such mice also show many characteristics of premature aging, as observed by loss and graying of hair, reduction in bone density and muscle mass, decline in fertility, anemia, and reduced life span.

These, and other studies, continue to speak to the importance of generating adequate ATP as a result of oxidative phosphorylation under the direction of DNA in mitochondria. As cells undergo genetic damage, which appears to be a natural phenomenon, their function declines, which may be

9–4 Given the maternal origin of mitochondria, from mother to offspring through the egg, and coupled with the relatively high mutation rate of mitochondrial DNA, one would expect that a high percentage of mutated, nonfunctional mitochondria would accumulate in cells of adults. One could envision that as a result, a mitochondrial meltdown would eventually occur as generations ensue. Such is not the case, however, and data indicate that over generations, mutations in mtDNA tend to be reduced in frequency in a given lineage. How might this phenomenon be explained?

■ HINT: *This problem involves an understanding of heteroplasmy in mitochondria. The key to its solution is to consider factors that drive evolution, particularly natural selection, and to apply this concept to the presence of many mitochondria in each cell.*

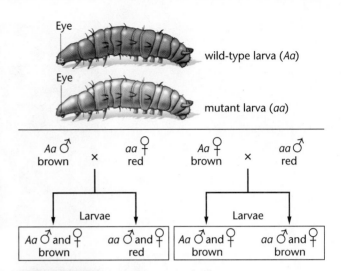

FIGURE 9–10 Maternal influence in the inheritance of eye pigment in the meal moth *Ephestia kuehniella*. Multiple light receptor structures (eyes) are present on each side of the anterior portion of larvae.

an underlying factor in aging as well as in the progression of age-related disorders.

9.4

In Maternal Effect, the Maternal Genotype Has a Strong Influence during Early Development

In **maternal effect**, also referred to as *maternal influence*, an offspring's phenotype for a particular trait is under the control of nuclear gene products present in the egg. This is in contrast to biparental inheritance, where both parents transmit information on genes in the nucleus that determines the offspring's phenotype. In cases of maternal effect, the nuclear genes of the female gamete are transcribed, and the genetic products (either proteins or untranslated RNAs) accumulate in the egg cytoplasm. After fertilization, these products are distributed among newly formed cells and influence the patterns or traits established during early development. Three examples will illustrate such an influence of the maternal genome on particular traits.

Ephestia Pigmentation

A very straightforward illustration of a maternal effect is seen in the Mediterranean meal moth, *Ephestia kuehniella*. The wild-type larva of this moth has a pigmented skin and brown eyes as a result of the dominant gene *A*. The pigment is derived from a precursor molecule, kynurenine, which is in turn a derivative of the amino acid tryptophan. A mutation, *a*, interrupts the synthesis of kynurenine and, when homozygous, may result in red eyes and little pigmentation in larvae. However, as illustrated in Figure 9–10, results of the cross

Aa × *aa* depend on which parent carries the dominant gene. When the male is the heterozygous parent, a 1:1 brown- to red-eyed ratio is observed in larvae, as predicted by Mendelian segregation. When the female is heterozygous for the *A* gene, however, all larvae are pigmented and have brown eyes, in spite of half of them being *aa*. As these larvae develop into adults, one-half of them gradually develop red eyes, reestablishing the 1:1 ratio.

One explanation for these results is that the *Aa* oocytes synthesize kynurenine or an enzyme necessary for its synthesis and accumulate it in the ooplasm prior to the completion of meiosis. Even in *aa* progeny, if the mothers were *Aa*, this pigment is distributed in the cytoplasm of the cells of the developing larvae; thus, they develop pigmentation and brown eyes. In these progeny, however, the pigment is eventually diluted among many cells and depleted, resulting in the conversion to red eyes as adults. The *Ephestia* example demonstrates the maternal effect in which a cytoplasmically stored nuclear gene product influences the larval phenotype and, at least temporarily, overrides the genotype of the progeny.

Limnaea Coiling

Shell coiling in the snail *Limnaea peregra* is an excellent example of maternal effect on a permanent rather than a transitory phenotype. Some strains of this snail have left-handed, or sinistrally, coiled shells (*dd*), while others have right-handed, or dextrally, coiled shells (*DD* or *Dd*). These snails are hermaphroditic and may undergo either cross- or self-fertilization, providing a variety of types of matings.

Figure 9–11 illustrates the results of reciprocal crosses between true-breeding snails. As you can see, these crosses yield different outcomes, even though both are between sinistral and dextral organisms and produce all

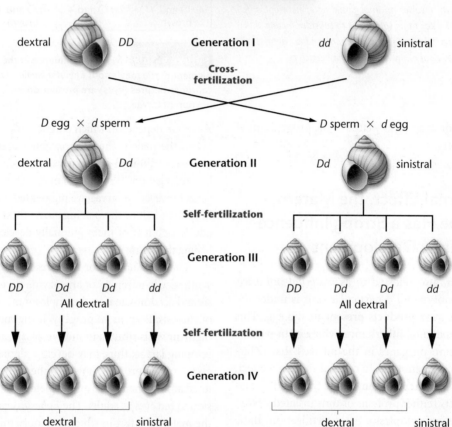

dextral   *DD*   **Generation I**   *dd*   sinistral

Cross-fertilization

D egg × *d* sperm   *D* sperm × *d* egg

dextral   *Dd*   **Generation II**   *Dd*   sinistral

Self-fertilization

Generation III

DD *Dd* *Dd* *dd*   *DD* *Dd* *Dd* *dd*

All dextral   All dextral

Self-fertilization

Generation IV

dextral   sinistral   dextral   sinistral

FIGURE 9–11   Inheritance of coiling in the snail *Limnaea peregra*. Coiling is either dextral (right handed) or sinistral (left handed). A maternal effect is evident in generations II and III, where the genotype of the maternal parent, rather than the offspring's own genotype, controls the phenotype of the offspring. The photograph illustrates a mixture of right- versus left-handed coiled snails.

heterozygous offspring. Examination of the progeny reveals that their phenotypes depend on the genotypes of the female parents. If we adopt that conclusion as a working hypothesis, we can test it by examining the offspring in subsequent generations of self-fertilization events. In each case, the hypothesis is upheld. Ovum donors that are *DD* or *Dd* produce only dextrally coiled progeny. Maternal parents that are *dd* produce only sinistrally coiled progeny. The coiling pattern of the progeny snails is determined by *the genotype of the parent producing the egg, regardless of the phenotype of that parent.*

Investigation of the developmental events in *Limnaea* reveals that the orientation of the spindle in the first cleavage division after fertilization determines the direction of coiling. Spindle orientation appears to be controlled by maternal genes acting on the developing eggs in the ovary. The orientation of the spindle, in turn, influences cell divisions following fertilization and establishes the permanent adult coiling pattern. The dextral allele (*D*) produces an active gene product that causes right-handed coiling. If ooplasm from dextral eggs is injected into uncleaved sinistral eggs, they cleave in a dextral pattern. However, in the converse experiment, sinistral ooplasm has no effect when injected into dextral eggs. Apparently, the sinistral allele is the result of a classic recessive mutation that encodes an inactive gene product.

We can conclude, then, that females that are either *DD* or *Dd* produce oocytes that synthesize the *D* gene product, which is stored in the ooplasm. Even if the oocyte contains only the *d* allele following meiosis and is fertilized by a *d*-bearing sperm, the resulting *dd* snail will be dextrally coiled (right handed).

Embryonic Development in *Drosophila*

A more recently documented example of maternal effect involves various genes that control embryonic development in *Drosophila melanogaster*. The genetic control of embryonic development in *Drosophila*, discussed in greater detail in Chapter 18, is a fascinating story. The protein products of the maternal-effect genes function to activate other genes,

which may in turn activate still other genes. This cascade of gene activity leads to a normal embryo whose subsequent development yields a normal adult fly. The extensive work by Edward B. Lewis, Christiane Nüsslein-Volhard, and Eric Wieschaus (who shared the 1995 Nobel Prize for Physiology or Medicine for their findings) has clarified how these and other genes function. Genes that illustrate maternal effect have products that are synthesized by the developing egg and stored in the oocyte prior to fertilization. Following fertilization, these products create molecular gradients that determine spatial organization as development proceeds.

For example, the gene *bicoid* (bcd^+) plays an important role in specifying the development of the anterior portion of the fly. The RNA transcribed by this gene is deposited anteriorly in the egg (Figure 9–12), and upon translation, forms a gradient highest at the anterior end and gradually diluted posteriorly. Embryos derived from mothers who

---

GENETICS, TECHNOLOGY, AND SOCIETY

Mitochondrial DNA and the Mystery of the Romanovs

In 1917, after more than 300 years of Romanov Imperial Russian rule and a violent Bolshevik revolution, the last Romanov ruler abdicated his crown. Tsar Nicholas II, his wife Tsarina Alexandra, and their five children—Olga, Tatiana, Maria, Anastasia, and Alexei—were arrested and sent into exile in the town of Ekaterinburg in western Siberia. They were accompanied by their family doctor and three loyal servants.

Just after midnight on July 17, 1918, the group was awakened and brought downstairs to a cellar room, where they were told that they would be moved to a safer location. Instead, eleven men with revolvers entered the room and shot the family and their staff. To complete the execution, the bodies were bayoneted and their faces smashed with rifle butts. The corpses were then hauled away and thrown down a mineshaft. Two days later, the bodies were removed from the mineshaft and buried in shallow graves in the woods. There they rested for more than 60 years.

An air of mystery soon developed around the demise of the Romanovs. Because the burial location was unknown and there were few eyewitness accounts

of the assassination, speculation arose that some—or perhaps all—of the family may have survived. Romantic mythologies developed over the possible survival of the youngest daughter Anastasia (17) and the heir to the throne, Alexei (13), both of whom were said to have escaped from Russia and awaited the restoration of their imperial birthrights. Since 1918, more than 200 people have claimed to be the Grand Duchess Anastasia.

The solution to the Romanov mystery began in 1979, when a Siberian geologist discovered four skulls in a shallow grave near Ekaterinburg. In the summer of 1991, after the establishment of *glasnost* in the former Soviet Union, exhumation of the site began. Altogether, almost 1000 bone fragments were recovered, which were reassembled into nine skeletons—five females and four males. Based on measurements of the bones and computer-assisted superimposition of the skulls onto photographs, the remains were tentatively identified as belonging to the murdered Romanovs and their four staff members. But there were two missing bodies—those of one daughter and the boy Alexei.

The next step in authenticating the remains involved DNA analysis. The goals were to establish family relationships among the remains, and then to determine, by comparisons with living relatives, whether the family group was in fact the Romanovs. The scientists extracted small amounts of DNA from bone fragments from each skeleton. Genomic DNA typing confirmed the familial relationships between an adult male and female (the presumed parents) and three young females. DNA typing from four of the adult skeletons revealed that they were not related to each other or to the family group.

After nuclear DNA typing, the scientists analyzed mitochondrial DNA (mtDNA). Since mitochondria are transmitted strictly from mother to offspring, mtDNA sequences can be used to trace maternal lineages without the complicating effects of meiotic crossing over. Scientists determined the nucleotide sequences of two highly variable regions of mtDNA from all nine bone samples. The sequences from the presumed Tsarina Alexandra were an exact match with those from Prince Philip

(continued)

are homozygous for the mutant allele (bcd^-/bcd^-) fail to develop anterior areas that normally give rise to the head and thorax of the adult fly. Embryos whose mothers contain at least one wild-type allele (bcd^+) develop normally, even if the genotype of the embryo is homozygous for the mutation. Consistent with the concept of *maternal effect*, the *genotype of the female parent, not the genotype of the embryo* determines the phenotype of the offspring. Nüsslein-Volhard, and Wieschaus, using large-scale mutant screens, discovered many other maternal-effect genes critical to normal *Drosophila* development, influencing not only anterior and posterior morphogenesis, but also the expression of zygotic (nuclear) genes that control segmentation in this arthropod. When we return to our discussion of this general topic in Chapter 18, we will see examples of other genes illustrating maternal effect that influence both anterior and posterior morphogenesis. Many of these genes function to regulate the spatial expression of "zygotic" (nuclear) genes that influence other aspects of development and that behave genetically in the conventional Mendelian fashion.

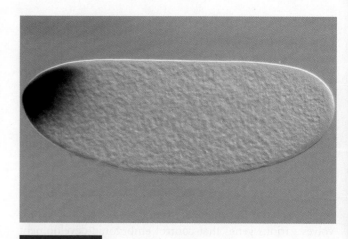

FIGURE 9–12   A gradient of *bicoid* mRNA as concentrated in anterior region of the *Drosophila* embryo during early development.

Genetics, Technology, and Society, continued

of England, who is her grandnephew, verifying her identity. However, identification was more complicated for the Tsar.

The Tsar's mtDNA sequences were compared with those from two living relatives—the Countess Xenia Cheremeteff-Sfiri (his great-grandniece) and James George Alexander Bannerman Carnegie, third Duke of Fife (descended from a line of women stretching back to the Tsar's grandmother). These comparisons produced a surprise. At position 16,169 of the mtDNA, the Tsar seemed to have either one or another base, a C or a T. The Countess and the Duke, in contrast, had only a T at this position. The conclusion was that Tsar Nicholas had two different populations of mitochondria in his cells, each with a different base at position 16,169 of its DNA. This condition, called heteroplasmy, is now believed to occur in 10 to 20 percent of humans. To clarify the discrepancy between the Tsar's mtDNA and that of his living relatives, the Russian government granted a request to analyze the remains of the Tsar's younger brother, Grand Duke Georgij Romanov, who died in 1899 of tuberculosis. The Grand Duke's mtDNA contained the same heteroplasmic variant at position 16,169. It was concluded that the Ekaterinburg bones were the doomed imperial family. With years of controversy finally resolved,

the authenticated remains of Tsar Nicholas II, Tsarina Alexandra, and three of their daughters were buried in the Saint Peter and Paul Cathedral in Saint Petersburg on July 17, 1998, 80 years to the day after they were murdered.

Take time, individually or in groups, to answer the following questions. Investigate the references and links, to help you understand some of the technologies and issues surrounding the identification of the Romanov family.

1. In 2007, a second grave was discovered near the mass grave of the Romanovs. This one contained the remains of an adolescent male and female, consistent with accounts from eyewitnesses in 1918 that Alexei and one of his sisters had been buried nearby after unsuccessful attempts to burn their bodies. Were these the remains of the last two Romanovs?

Read about the forensic DNA analyses involved in identifying these remains in Coble, M. D., et al. 2009. Mystery solved: the iden-

tification of the two missing Romanov children using DNA analysis. PLoS ONE 4: 1–9.

2. One of the most famous people claiming to be Grand Duchess Anastasia was Anna Anderson, who died in the United States in 1984. What types of DNA analyses were used to verify the true identity of Ms. Anderson, and what were the conclusions from these studies?

Begin your search by reading Stoneking, M., et al. 1995. Establishing the identity of Anna Anderson Manahan. Nature Genet. 9: 9–10.

3. Alexei Romanov suffered from hemophilia, inherited through his mother from Queen Victoria of England. Historians consider that his condition contributed to the Russian Revolution and the demise of the Romanovs. Explain how this came about and the genetic basis for his disease.

Two interesting accounts of the "Royal Disease" and how Alexei's hemophilia contributed to history can be found in Rogaev, E. I., et al. 2009. Genotype analysis identifies the cause of the "Royal Disease." Science 326: 817, and in Alexei Nikolaevich, Tsarevich of Russia at http://en.wikipedia.org/wiki/Alexei_Nikolaevich,_Tsarevich_of_Russia

CASE STUDY | A twin difference

Recent news of a professional baseball player's "mysterious cell disease" being traced to a mitochondrial disorder (Rocco Baldelli of the Tampa Bay Rays) has brought not only personal clarification to one set of identical twin brothers, but also national awareness of an array of similar disorders that affect thousands of children and adults regardless of age, gender, race, or position in society. For 14 years the brothers competed on equal terms until one began to suffer loss of visual acuity and muscularity. Diagnosis of a mitochondrial disorder explained the cause and forecasted supportive treatment, but also raised some important questions.

1. The parents and other relatives of the twins are apparently healthy, so how could this genetic disorder have arisen?

2. How could one identical twin have an inherited mitochondrial disorder and not the other?

3. Will the unaffected brother be guaranteed continued good health?

Summary Points

For activities, animations, and review quizzes, go to the study area at www.masteringgenetics.com

1. Patterns of inheritance sometimes vary from those expected from the biparental transmission of nuclear genes. Often, phenotypes appear to result from genetic information transmitted through the ooplasm of the egg.

2. Organelle heredity is based on the genotypes of chloroplast and mitochondrial DNA, as these organelles are transmitted to offspring. Chloroplast mutations affect the photosynthetic capabilities of plants, whereas mitochondrial mutations affect cells highly dependent on energy (ATP) generated through cellular respiration. The resulting mutants display phenotypes related to the loss of function of these organelles.

3. Both chloroplasts and mitochondria first appeared in primitive eukaryotic cells some 2 billion years ago, originating as invading protobacteria, which then coevolved with the host cell according to the endosymbiotic theory. Evidence in support of this endosymbiotic theory is extensive and centers around many observations involving the DNA and genetic machinery present in modern-day chloroplasts and mitochondria.

4. Mutations in human mtDNA are the underlying causes of a range of heritable genetic disorders in humans.

5. Maternal-effect patterns result when nuclear gene products expressed by the maternal genotype of the egg influence early development. *Ephestia* pigmentation, coiling in snails, and gene expression during early development in *Drosophila* are examples.

INSIGHTS AND SOLUTIONS

1. Analyze the following hypothetical pedigree, determine the most consistent interpretation of how the trait is inherited, and point out any inconsistencies:

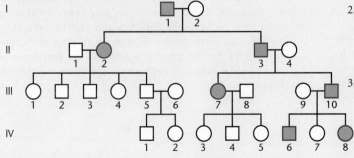

2. Can the explanation in Solution 1 be attributed to a gene on the Y chromosome? Defend your answer.

Solution: The trait is passed from all affected male parents to all but one offspring, but it is *never* passed maternally. Individual IV-7 (a female) is the only exception.

2. Can the explanation in Solution 1 be attributed to a gene on the Y chromosome? Defend your answer.

Solution: No, because male parents pass the trait to their daughters as well as to their sons.

3. Is the above pedigree an example of a paternal effect or of paternal inheritance?

Solution: It has all the characteristics of paternal inheritance because males pass the trait to almost all of their offspring. To assess whether the trait is due to a paternal effect (resulting from a nuclear gene in the male gamete), analysis of further matings would be needed.

Problems and Discussion Questions

MG™ *For instructor-assigned tutorials and problems, go to www.masteringgentics.com*

HOW DO WE KNOW?

1. In this chapter, we focused on extranuclear inheritance and how traits can be determined by genetic information contained in mitochondria and chloroplasts, and we discussed how expression of maternal genotypes can affect the phenotype of an organism. At the same time, we found many opportunities to consider the methods and reasoning by which much of this information was acquired. From the explanations given in the chapter, what answers would you propose to the following fundamental questions?

   (a) How was it established that particular phenotypes are inherited as a result of genetic information present in the chloroplast rather than in the nucleus?
   (b) How did the discovery of three categories of *petite* mutations in yeast lead researchers to postulate extranuclear inheritance of colony size?
   (c) What observations support the endosymbiotic theory?
   (d) What key observations in crosses between dextrally and sinistrally coiled snails support the explanation that this phenotype is the result of maternal-effect inheritance?
   (e) What findings demonstrate a maternal effect as the basis of a mode of inheritance?

2. What genetic criteria distinguish a case of extranuclear inheritance from a case of Mendelian autosomal inheritance? from a case of X-linked inheritance?

3. Streptomycin resistance in *Chlamydomonas* may result from a mutation in either a chloroplast gene or a nuclear gene. What phenotypic results would occur in a cross between a member of an mt^+ strain resistant in both genes and a member of a strain sensitive to the antibiotic? What results would occur in the reciprocal cross?

4. A plant may have green, white, or green-and-white (variegated) leaves on its branches, owing to a mutation in the chloroplast that prevents color from developing. Predict the results of the following crosses:

Ovule Source		Pollen Source
(a) Green branch	×	White branch
(b) White branch	×	Green branch
(c) Variegated branch	×	Green branch
(d) Green branch	×	Variegated branch

5. In diploid yeast strains, sporulation and subsequent meiosis can produce haploid ascospores, which may fuse to reestablish diploid cells. When ascospores from a *segregational petite* strain fuse with those of a normal wild-type strain, the diploid

zygotes are all normal. Following meiosis, ascospores are *petite* and normal. Is the *segregational petite* phenotype inherited as a dominant or a recessive trait?

6. Predict the results of a cross between ascospores from a *segregational petite* strain and a *neutral petite* strain. Indicate the phenotype of the zygote and the ascospores it may subsequently produce.

7. In *Limnaea*, what results would you expect in a cross between a *Dd* dextrally coiled and a *Dd* sinistrally coiled snail, assuming cross-fertilization occurs as shown in Figure 9–11? What results would occur if the *Dd* dextral produced only eggs and the *Dd* sinistral produced only sperm?

8. In a cross of *Limnaea*, the snail contributing the eggs was dextral but of unknown genotype. Both the genotype and the phenotype of the other snail are unknown. All F_1 offspring exhibited dextral coiling. Ten of the F_1 snails were allowed to undergo self-fertilization. One-half produced only dextrally coiled offspring, whereas the other half produced only sinistrally coiled offspring. What were the genotypes of the original parents?

9. In *Drosophila subobscura*, the presence of a recessive gene called *grandchildless* (*gs*) causes the offspring of homozygous females, but not those of homozygous males, to be sterile. Can you offer an explanation as to why females and not males are affected by the mutant gene?

10. A male mouse from a true-breeding strain of hyperactive animals is crossed with a female mouse from a true-breeding strain of lethargic animals. (These are both hypothetical strains.) All the progeny are lethargic. In the F_2 generation, all offspring are lethargic. What is the best genetic explanation for these observations? Propose a cross to test your explanation.

11. Consider the case where a mutation occurs that disrupts translation in a single human mitochondrion found in the oocyte participating in fertilization. What is the likely impact of this mutation on the offspring arising from this oocyte?

12. What is the endosymbiotic theory, and why is this theory relevant to the study of extranuclear DNA in eukaryotic organelles?

13. In the Problems and Discussion Questions in Chapter 7, question 33, we described CC, the cat created by nuclear transfer cloning, whereby a diploid nucleus from one cell is injected into an enucleated egg cell to create an embryo. Cattle, sheep, rats, dogs, and several other species have been cloned using nuclei from somatic cells. Embryos and adults produced by this approach often show a number of different mitochondrial defects. Explain possible reasons for the prevalence of mitochondrial defects in embryos created by nuclear transfer cloning.

Extra-Spicy Problems

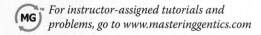

For instructor-assigned tutorials and problems, go to www.masteringgentics.com

14. The specification of the anterior–posterior axis in *Drosophila* embryos is initially controlled by various gene products that are synthesized and stored in the mature egg following oogenesis. Mutations in these genes result in abnormalities of the axis during embryogenesis. These mutations illustrate *maternal effect*. How do such mutations vary from those produced by organelle heredity? Devise a set of parallel crosses and expected outcomes involving mutant genes that contrast maternal effect and organelle heredity.

15. The maternal-effect mutation *bicoid* (*bcd*) is recessive. In the absence of the bicoid protein product, embryogenesis is not completed. Consider a cross between a female heterozygous for the *bicoid* alleles (bcd^+/bcd^-) and a male homozygous for the mutation (bcd^-/bcd^-).
    (a) How is it possible for a male homozygous for the mutation to exist?
    (b) Predict the outcome (normal vs. failed embryogenesis) in the F_1 and F_2 generations of the cross described.

16. Shown here is a pedigree for a hypothetical human disorder:

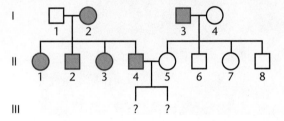

    Analyze the pedigree and propose a genetic explanation for the nature of its inheritance. Consistent with your explanation, predict the outcome of a mating between individuals II-4 and II-5.

17. Extrachromosomally inherited traits are widespread among arthropods. In the two-spotted ladybird beetle, *Adalia bipunctata*, a male-killing trait has been discovered in which certain strains of females display a distorted sex ratio that favors female offspring (Werren, J., et al. 1994. *J. Bacteriol.* 176: 388–394). Unaffected strains show a normal one-to-one sex ratio. Two key observations are that affected strains can be cured by antibiotics, and that in addition to their normal 18*S* and 28*S* rRNA, 16*S* rDNA can be detected by PCR (polymerase chain reaction) analysis. Of the modes of extranuclear inheritance described in the text (organelle heredity, infectious heredity, and maternal effect), which is most likely to be causing this altered sex ratio in *Adalia*?

18. Researchers examined a family with an interesting distribution of Leigh syndrome symptoms. In this disorder, individuals may show a progressive loss of motor function (ataxia, A) with peripheral neuropathy (PN, meaning impairment of the peripheral nerves). A mitochondrial DNA (mtDNA) mutation that reduces ATPase activity was identified in various tissues of affected individuals. The accompanying table summarizes the presence of symptoms in an extended family.

Person	Condition	Percent Mitochondria with Mutation
Proband	A and PN	>90%
Brother	A and PN	>90%
Brother	Asymptomatic	17%
Mother	PN	86%
Maternal uncle	PN	85%
Maternal cousin	A and PN	90%
Maternal cousin	A and PN	91%
Maternal grandmother	Asymptomatic	56%

    (a) Develop a pedigree that summarizes the information presented in the table.
    (b) Provide an explanation for the pattern of inheritance of the disease. What term describes this pattern?
    (c) How can some individuals in the same family show such variation in symptoms? What term, as related to organelle heredity, describes such variation?
    (d) In what way does a condition caused by mtDNA differ in expression and transmission from a mutation that causes albinism?

19. Mutations in mitochondrial DNA appear to be responsible for a number of neurological disorders including myoclonic epilepsy with ragged-red fibers disease, Leber's hereditary optic neuropathy, and Kearns-Sayre syndrome. In each case, the disease phenotype is expressed when the ratio of mutant to wild-type mitochondria exceeds a threshold peculiar to each disease, but usually in the 60 to 95 percent range.
    (a) Given that these are debilitating conditions, why has no cure been developed? Can you suggest a general approach that might be used to treat, or perhaps even cure, these disorders?
    (b) Compared with the vast number of mitochondria in an embryo, the number of mitochondria in an ovum is relatively small. Might such an ooplasmic mitochondrial bottleneck present an opportunity for therapy or cure? Explain.

10

DNA Structure and Analysis

CHAPTER CONCEPTS

- Except in some viruses, DNA serves as the genetic material in all living organisms on Earth.

- According to the Watson–Crick model, DNA exists in the form of a right-handed double helix.

- The strands of the double helix are antiparallel and held together by hydrogen bonding between complementary nitrogenous bases.

- The structure of DNA provides the means of storing and expressing genetic information.

- RNA has many similarities to DNA but exists mostly as a single-stranded molecule.

- In some viruses, RNA serves as the genetic material.

- Many techniques have been developed that facilitate the analysis of nucleic acids, most based on detection of the complementarity of nitrogenous bases.

Up to this point in the text, we have described chromosomes as containing genes that control phenotypic traits that are transmitted through gametes to future offspring. Logically, genes must contain some sort of information that, when passed to a new generation, influences the form and characteristics of each individual. We refer to that information as the **genetic material**. Logic also suggests that this same information in some way directs the many complex processes that lead to an organism's adult form.

Until 1944, it was not clear what chemical component of the chromosome makes up genes and constitutes the genetic material. Because chromosomes were known to have both a nucleic acid and a protein component, both were candidates. In 1944, however, direct experimental evidence emerged showing that the nucleic acid DNA serves as the informational basis for the process of heredity.

Once the importance of DNA to genetic processes was realized, work was intensified with the hope of discerning not only the structure of this molecule but also the relationship of its structure to its function. Between 1944 and 1953, many scientists sought information that might answer the most significant and intriguing question in the history of biology: How does DNA serve as the genetic basis for living processes? Researchers believed the answer must depend strongly on the chemical structure of the DNA molecule, given the complex but orderly functions ascribed to it.

These efforts were rewarded in 1953, when James Watson and Francis Crick put forth their hypothesis for the double-helical nature of DNA. The assumption that the molecule's functions would be easier to clarify once its general structure was determined proved to be correct. In this chapter, we first review the evidence that DNA is the genetic material and then discuss the elucidation of its structure. We conclude the chapter with a discussion of various analytical techniques useful during the investigation of the nucleic acids, DNA and RNA.

10.1

The Genetic Material Must Exhibit Four Characteristics

For a molecule to serve as the genetic material, it must exhibit four crucial characteristics: **replication, storage of information, expression of information**, and **variation by mutation**. *Replication* of the genetic material is one facet of the cell cycle and as such is a fundamental property of all living organisms. Once the genetic material of cells replicates and is doubled in amount, it must then be partitioned equally—through mitosis—into daughter cells. The genetic material is also replicated during the formation of gametes, but is partitioned so that each cell gets only one-half of the original amount of genetic material—the process of *meiosis*, discussed in Chapter 2. Although the products of mitosis and meiosis are different, these processes are both part of the more general phenomenon of cellular reproduction.

Storage of information requires the molecule to act as a repository of genetic information that may or may not be expressed by the cell in which it resides. It is clear that while most cells contain a complete copy of the organism's genome, at any point in time they express only a part of this genetic potential. For example, in bacteria many genes "turn on" in response to specific environmental cues and "turn off" when conditions change. In vertebrates, skin cells may display active melanin genes but never activate their hemoglobin genes; in contrast, digestive cells activate many genes specific to their function but do not activate their melanin genes.

Inherent in the concept of storage is the need for the genetic material to be able to encode the vast variety of gene products found among the countless forms of life on our planet. The chemical language of the genetic material must have the capability of storing such diverse information and transmitting it to progeny cells and organisms.

Expression of the stored genetic information is a complex process that is the underlying basis for the concept of **information flow** within the cell (Figure 10–1). The initial event in this flow of information is the **transcription** of DNA, in which three main types of RNA molecules are synthesized: messenger RNA (mRNA), transfer RNA (tRNA), and ribosomal RNA (rRNA). Of these, mRNAs are translated into proteins, by means of a process mediated by the tRNA and rRNA. Each mRNA is the product of a

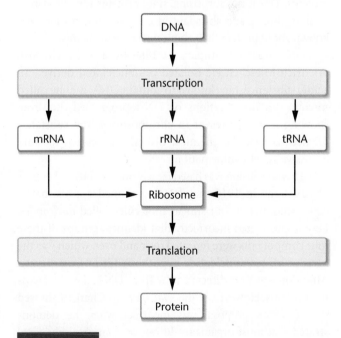

FIGURE 10–1   Simplified diagram of information flow (the central dogma) from DNA to RNA to produce the proteins within cells.

specific gene and leads to the synthesis of a different protein. In **translation**, the chemical information in mRNA directs the construction of a chain of amino acids, called a polypeptide, which then folds into a protein. Collectively, these processes serve as the foundation for the **central dogma of molecular genetics**: "DNA makes RNA, which makes proteins."

The genetic material is also the source of *variability* among organisms, through the process of mutation. If a mutation—a change in the chemical composition of DNA—occurs, the alteration is reflected during transcription and translation, affecting the specific protein. If a mutation is present in a gamete, it may be passed to future generations and, with time, become distributed in the population. Genetic variation, which also includes alterations of chromosome number and rearrangements within and between chromosomes (as discussed in Chapter 8), provides the raw material for the process of evolution.

10.2

Until 1944, Observations Favored Protein as the Genetic Material

The idea that genetic material is physically transmitted from parent to offspring has been accepted for as long as the concept of inheritance has existed. Beginning in the late nineteenth century, research into the structure of biomolecules progressed considerably, setting the stage for describing the genetic material in chemical terms. Although proteins and nucleic acid were both considered major candidates for the role of genetic material, until the 1940s many geneticists favored proteins. This is not surprising, since proteins were known to be both diverse and abundant in cells, and much more was known about protein than about nucleic acid chemistry.

DNA was first studied in 1869 by a Swiss chemist, Friedrich Miescher. He isolated cell nuclei and derived an acidic substance, now known to contain DNA, that he called **nuclein**. As investigations of DNA progressed, however, showing it to be present in chromosomes, the substance seemed to lack the chemical diversity necessary to store extensive genetic information.

This conclusion was based largely on Phoebus A. Levene's observations in 1910 that DNA contained approximately equal amounts of four similar molecules called *nucleotides*. Levene postulated incorrectly that identical groups of these four components were repeated over and over, which was the basis of his **tetranucleotide hypothesis** for DNA structure. Attention was thus directed away from DNA, thereby favoring proteins. However, in the 1940s, Erwin Chargaff showed that Levene's proposal was incorrect when he demonstrated that most organisms do not contain precisely equal proportions of the four nucleotides. We shall see later that the structure of DNA accounts for Chargaff's observations.

10.3

Evidence Favoring DNA as the Genetic Material Was First Obtained during the Study of Bacteria and Bacteriophages

The 1944 publication by Oswald Avery, Colin MacLeod, and Maclyn McCarty concerning the chemical nature of a "transforming principle" in bacteria was the initial event leading to the acceptance of DNA as the genetic material. Their work, along with subsequent findings of other research teams, constituted the first direct experimental proof that DNA, and not protein, is the biomolecule responsible for heredity. It marked the beginning of the *era of molecular genetics*, a period of discovery in biology that made biotechnology feasible and has moved us closer to an understanding of the basis of life. The impact of their initial findings on future research and thinking paralleled that of the publication of Darwin's theory of evolution and the subsequent rediscovery of Mendel's postulates of transmission genetics. Together, these events constitute three great revolutions in biology.

Transformation: Early Studies

The research that provided the foundation for Avery, MacLeod, and McCarty's work was initiated in 1927 by Frederick Griffith, a medical officer in the British Ministry of Health. He performed experiments with several different strains of the bacterium *Diplococcus pneumoniae*.* Some were *virulent*, that is, infectious, strains that cause pneumonia in certain vertebrates (notably humans and mice), whereas others were *avirulent*, or noninfectious strains, which do not cause illness.

The difference in virulence depends on the presence of a polysaccharide capsule; virulent strains have this capsule, whereas avirulent strains do not. The nonencapsulated bacteria are readily engulfed and destroyed by phagocytic cells in the host animal's circulatory system. Virulent bacteria, which possess the polysaccharide coat, are not easily engulfed; they multiply and cause pneumonia.

The presence or absence of the capsule causes a visible difference between colonies of virulent and avirulent strains. Encapsulated bacteria form smooth, shiny-surfaced colonies (*S*) when grown on an agar culture plate; nonencapsulated strains produce rough colonies (*R*) (Figure 10–2). Thus, virulent and avirulent strains are easily distinguished by standard microbiological culture techniques.

Each strain of *Diplococcus* may be one of dozens of different types called *serotypes* that differ in the precise chemical structure of the polysaccharide constituent of the thick,

---

*This organism is now named *Streptococcus pneumoniae*.

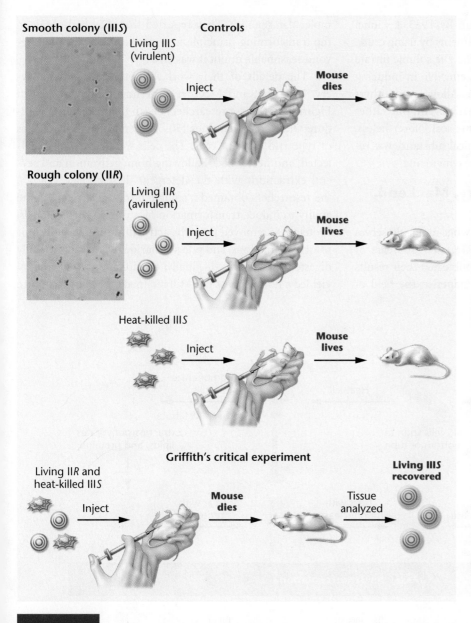

FIGURE 10–2 Griffith's transformation experiment. The photographs show bacterial colonies containing cells with capsules (type IIIS) and without capsules (type IIR).

slimy capsule. Serotypes are identified by immunological techniques and are usually designated by Roman numerals. In the United States, types I and II are the most common in causing pneumonia. Griffith used types IIR and IIIS in his critical experiments that led to new concepts about the genetic material. Table 10.1 summarizes the characteristics of Griffith's two strains.

TABLE 10.1

Strains of *Diplococcus pneumonia* Used by Frederick Griffith in His Original Transformation Experiments

Serotype	Colony Morphology	Capsule	Virulence
IIR	Rough	Absent	Avirulent
IIIS	Smooth	Present	Virulent

Griffith knew from the work of others that only living virulent cells would produce pneumonia in mice. If heat-killed virulent bacteria are injected into mice, no pneumonia results, just as living avirulent bacteria fail to produce the disease. Griffith's critical experiment (Figure 10–2) involved an injection into mice of living IIR (avirulent) cells combined with heat-killed IIIS (virulent) cells. Since neither cell type caused death in mice when injected alone, Griffith expected that the double injection would not kill the mice. But, after five days, all of the mice that received both types of cells were dead. Paradoxically, analysis of their blood revealed a large number of living type IIIS (virulent) bacteria.

As far as could be determined, these IIIS bacteria were identical to the IIIS strain from which the heat-killed cell preparation had been made. The control mice, injected only with living avirulent IIR bacteria for this set of experiments, did not develop pneumonia and remained healthy. This ruled out the possibility that the avirulent IIR cells simply changed (or mutated) to virulent IIIS cells in the absence of the heat-killed IIIS bacteria. Instead, some type of interaction had taken place between living IIR and heat-killed IIIS cells.

Griffith concluded that the heat-killed IIIS bacteria somehow converted live avirulent IIR cells into virulent IIIS cells. Calling the phenomenon **transformation**, he suggested that the *transforming principle* might be some part of the polysaccharide capsule or a compound required for capsule synthesis, although the capsule alone did not cause pneumonia. To use Griffith's term, the transforming principle from the dead IIIS cells served as a "pabulum"—that is, a nutrient source—for the IIR cells.

Griffith's work led other physicians and bacteriologists to research the phenomenon of transformation. By 1931, M. Henry Dawson at the Rockefeller Institute had confirmed Griffith's observations and extended his work one step further. Dawson and his coworkers showed that transformation could occur *in vitro* (in a test tube). When heat-killed IIIS cells were incubated with living IIR cells, living IIIS cells were recovered. Therefore, injection into mice was not necessary for transformation to occur. By 1933, J. Lionel Alloway had refined the *in vitro* experiments by using crude extracts of IIIS cells and living IIR cells. The soluble filtrate from the heat-killed IIIS cells was as effective in inducing transformation as were the intact cells. Alloway and others did not view transformation as a genetic event, but rather as a physiological modification of some sort. Nevertheless, the experimental evidence that a chemical substance was responsible for transformation was quite convincing.

Transformation: The Avery, MacLeod, and McCarty Experiment

The critical question, of course, was what molecule serves as the transforming principle? In 1944, after 10 years of work, Avery, MacLeod, and McCarty published their results in what is now regarded as a classic paper in the field of

molecular genetics. They reported that they had obtained the transforming principle in a purified state and that beyond reasonable doubt it was DNA.

The details of their work, sometimes called the Avery, MacLeod, and McCarty experiment, are outlined in Figure 10–3. These researchers began their isolation procedure with large quantities (50–75 liters) of liquid cultures of type IIIS virulent cells. The cells were centrifuged, collected, and heat killed. Following homogenization and several extractions with the detergent deoxycholate (DOC), the researchers obtained a soluble filtrate that retained the ability to induce transformation of type IIR avirulent cells. Protein was removed from the active filtrate by several chloroform extractions, and polysaccharides were enzymatically digested and removed. Finally, precipitation with ethanol yielded a fibrous mass that still retained the ability to induce

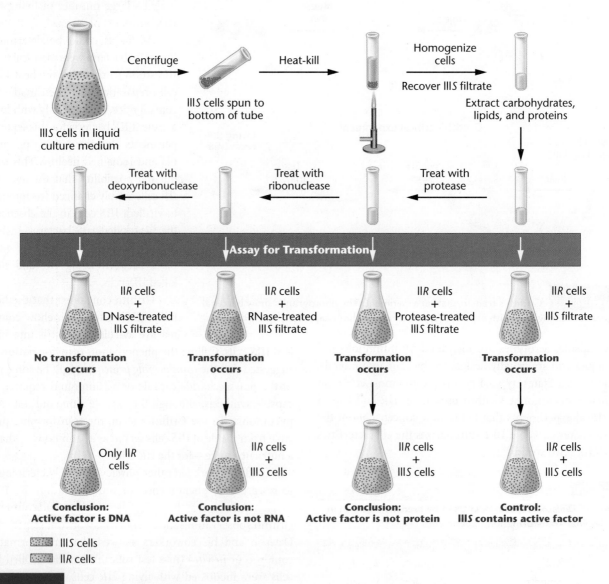

FIGURE 10–3   Summary of Avery, MacLeod, and McCarty's experiment demonstrating that DNA is the transforming principle.

transformation of type IIR avirulent cells. From the original 75-liter sample, the procedure yielded 10 to 25 mg of this "active factor."

Further testing clearly established that the transforming principle was DNA. The fibrous mass was first analyzed for its nitrogen: phosphorus ratio, which was shown to coincide with the ratio of "sodium desoxyribonucleate," the chemical name then used to describe DNA. To solidify their findings, Avery, MacLeod, and McCarty sought to eliminate, to the greatest extent possible, all probable contaminants from their final product. Thus, it was treated with the proteolytic enzymes trypsin and chymotrypsin and then with an RNA-digesting enzyme, called **ribonuclease** (**RNase**). Such treatments destroyed any remaining activity of proteins and RNA. Nevertheless, transforming activity still remained. Chemical testing of the final product gave strong positive reactions for DNA. The final confirmation came with experiments using crude samples of the DNA-digesting enzyme **deoxyribonuclease** (**DNase**), which was isolated from dog and rabbit sera. Digestion with this enzyme destroyed the transforming activity of the filtrate—thus Avery and his coworkers were certain that the active transforming principle in these experiments was DNA.

The great amount of work involved in this research, the confirmation and reconfirmation of the conclusions drawn, and the unambiguous logic of the experimental design are truly impressive. Avery, MacLeod, and McCarty's conclusion in the 1944 publication was, however, very simply stated: "The evidence presented supports the belief that a nucleic acid of the desoxyribose* type is the fundamental unit of the transforming principle of *Pneumococcus* Type III."

Avery and his colleagues recognized the genetic and biochemical implications of their work. They observed that "nucleic acids of this type must be regarded not merely as structurally important but as functionally active in determining the biochemical activities and specific characteristics of pneumococcal cells." This suggested that the transforming principle interacts with the IIR cell and gives rise to a coordinated series of enzymatic reactions culminating in the synthesis of the type IIIS capsular polysaccharide. Avery, MacLeod, and McCarty emphasized that, once transformation occurs, the capsular polysaccharide is produced in successive generations. Transformation is therefore heritable, and the process affects the genetic material.

Immediately after publication of the report, several investigators turned to, or intensified, their studies of transformation in order to clarify the role of DNA in genetic mechanisms. In particular, the work of Rollin Hotchkiss was instrumental in confirming that the critical factor in transformation was DNA and not protein. In 1949, in a separate study, Harriet Taylor isolated an **extremely rough** (*ER*)

mutant strain from a rough (*R*) strain. This *ER* strain produced colonies that were more irregular than the *R* strain. The DNA from *R* accomplished the transformation of *ER* to *R*. Thus, the *R* strain, which served as the recipient in the Avery experiments, was shown also to be able to serve as the DNA donor in transformation.

Transformation has now been shown to occur in *Haemophilus influenzae*, *Bacillus subtilis*, *Shigella paradysenteriae*, and *Escherichia coli*, among many other microorganisms. Transformation of numerous genetic traits other than colony morphology has also been demonstrated, including traits involving resistance to antibiotics. These observations further strengthened the belief that transformation by DNA is primarily a genetic event rather than simply a physiological change. We will pursue this idea again in the "Insights and Solutions" section at the end of this chapter.

The Hershey–Chase Experiment

The second major piece of evidence supporting DNA as the genetic material was provided during the study of the bacterium *Escherichia coli* and one of its infecting viruses, **bacteriophage T2**. Often referred to simply as a **phage**, the virus consists of a protein coat surrounding a core of DNA. Electron micrographs reveal that the phage's external structure is composed of a hexagonal head plus a tail. Figure 10–4 shows as much of the life cycle as was known in 1952 for a T-even bacteriophage such as T2. Briefly, the phage adsorbs to the bacterial cell, and some genetic component of the phage enters the bacterial cell. Following infection, the viral component "commandeers" the cellular machinery of the host and causes viral reproduction. In a reasonably short time, many new phages are constructed and the bacterial cell is lysed, releasing the progeny viruses. This process is referred to as the **lytic cycle**.

In 1952, Alfred Hershey and Martha Chase published the results of experiments designed to clarify the events leading to phage reproduction. Several of the experiments clearly established the independent functions of phage protein and nucleic acid in the reproduction process associated with the bacterial cell. Hershey and Chase knew from existing data that:

1. T2 phages consist of approximately 50 percent protein and 50 percent DNA.

2. Infection is initiated by adsorption of the phage by its tail fibers to the bacterial cell.

3. The production of new viruses occurs within the bacterial cell.

It appeared that some molecular component of the phage—DNA or protein (or both)—entered the bacterial cell and directed viral reproduction. Which was it?

Hershey and Chase used the radioisotopes ^{32}P and ^{35}S to follow the molecular components of phages during infection. Because DNA contains phosphorus (P) but not sulfur, ^{32}P

---

*Desoxyribose is now spelled deoxyribose.

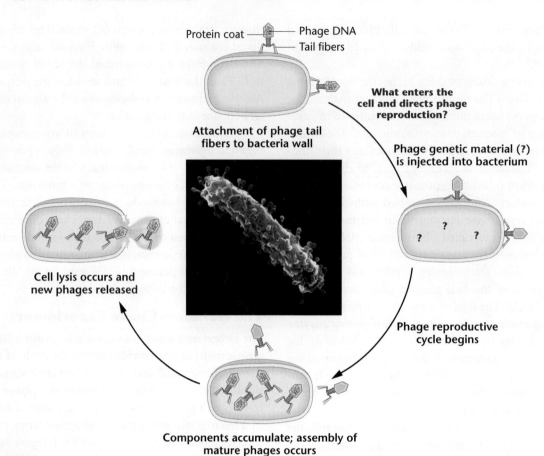

Protein coat — Phage DNA
— Tail fibers

**Attachment of phage tail
fibers to bacteria wall**

**What enters the
cell and directs phage
reproduction?**

**Phage genetic material (?)
is injected into bacterium**

**Cell lysis occurs and
new phages released**

**Phage reproductive
cycle begins**

**Components accumulate; assembly of
mature phages occurs**

FIGURE 10–4   Life cycle of a T-even bacteriophage, as known in 1952. The electron micrograph shows an *E. coli* cell during infection by numerous T2 phages (shown in blue).

effectively labels DNA; because proteins contain sulfur (S) but not phosphorus, ^{35}S labels protein. *This is a key feature of the experiment.* If *E. coli* cells are first grown in the presence of ^{32}P *or* ^{35}S and then infected with T2 viruses, the progeny phages will have *either* a radioactively labeled DNA core *or* a radioactively labeled protein coat, respectively. These labeled phages can be isolated and used to infect unlabeled bacteria (Figure 10–5).

When labeled phages and unlabeled bacteria were mixed, an adsorption complex was formed as the phages attached their tail fibers to the bacterial wall. These complexes were isolated and subjected to a high shear force in a blender. The force stripped off the attached phages so that the phages and bacteria could be analyzed separately. Centrifugation separated the lighter phage particles from the heavier bacterial cells (Figure 10–5). By tracing the radioisotopes, Hershey and Chase were able to demonstrate that most of the ^{32}P-labeled DNA had been transferred into the bacterial cell following adsorption; on the other hand, almost all of the ^{35}S-labeled protein remained outside the bacterial cell and was recovered in the phage "ghosts" (empty phage coats) after the blender treatment. Following this separation, the bacterial cells, which now contained viral DNA, were eventually lysed as new phages were produced. These progeny phages contained ^{32}P, but not ^{35}S.

Hershey and Chase interpreted these results as indicating that the protein of the phage coat remains outside the host cell and is not involved in directing the production of new phages. On the other hand, and most important, phage DNA enters the host cell and directs phage reproduction. Hershey and Chase had demonstrated that the genetic material in phage T2 is DNA, not protein.

These experiments, along with those of Avery and his colleagues, provided convincing evidence that DNA was the molecule responsible for heredity. This conclusion has since served as the cornerstone of the field of molecular genetics.

NOW SOLVE THIS

10–1 Would an experiment similar to that performed by Hershey and Chase work if the basic design were applied to the phenomenon of transformation? Explain why or why not.

■HINT: *This problem involves an understanding of the protocol of the Hershey–Chase experiment as applied to the investigation of transformation. The key to its solution is to remember that in transformation, exogenous DNA enters the soon-to-be transformed cell and that no cell-to-cell contact is involved in the process.*

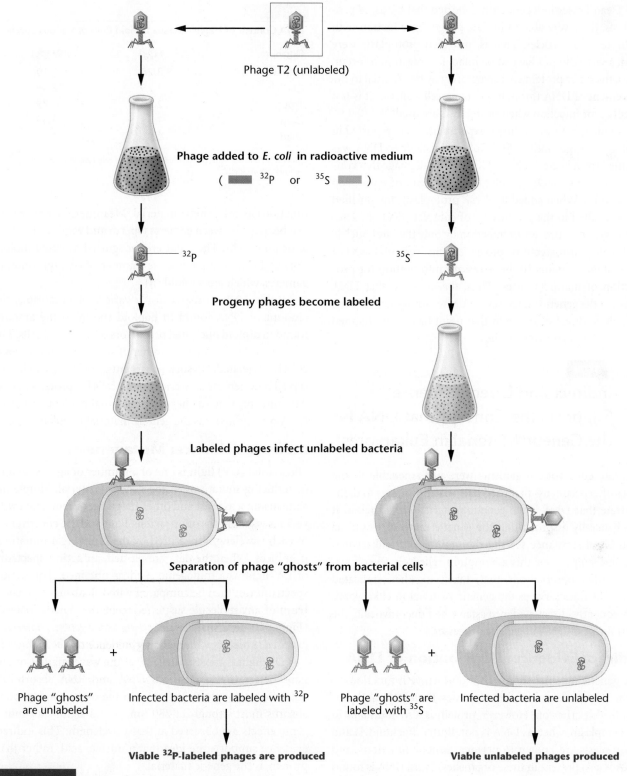

Phage T2 (unlabeled)

Phage added to *E. coli* in radioactive medium

(■ 32P or 35S ■)

32P

35S

Progeny phages become labeled

Labeled phages infect unlabeled bacteria

Separation of phage "ghosts" from bacterial cells

Phage "ghosts" are unlabeled

Infected bacteria are labeled with 32P

Phage "ghosts" are labeled with 35S

Infected bacteria are unlabeled

Viable 32P-labeled phages are produced

Viable unlabeled phages produced

FIGURE 10–5 Summary of the Hershey–Chase experiment demonstrating that DNA, and not protein, is responsible for directing the reproduction of phage T2 during the infection of *E. coli*.

Transfection Experiments

During the eight years following publication of the Hershey–Chase experiment, additional research using bacterial viruses provided even more solid proof that DNA is the genetic material. In 1957, several reports demonstrated that if *E. coli* is treated with the enzyme lysozyme, the outer wall of the cell can be removed without destroying the bacterium. Enzymatically treated cells are naked, so to speak, and contain only the cell membrane as their outer boundary. Such structures are called **protoplasts** (or **spheroplasts**). John Spizizen and Dean Fraser independently reported that by using protoplasts, they were able to initiate phage reproduction with disrupted T2 particles. That is, provided protoplasts were used, a virus did not have to be intact for infection to occur. Thus, the outer protein coat structure may be essential to the movement of DNA through the intact cell wall, but it is not essential for infection when protoplasts are used.

Similar, but more refined, experiments were reported in 1960 by George Guthrie and Robert Sinsheimer. DNA was purified from bacteriophage ϕX174, a small phage that contains a single-stranded circular DNA molecule of some 5386 nucleotides. When added to *E. coli* protoplasts, the purified DNA resulted in the production of complete ϕX174 bacteriophages. This process of infection by only the viral nucleic acid, called **transfection**, proves conclusively that ϕX174 DNA alone contains all the necessary information for production of mature viruses. Thus, the evidence that DNA serves as the genetic material was further strengthened, even though all direct evidence to that point had been obtained from bacterial and viral studies.

10.4

Indirect and Direct Evidence Supports the Concept that DNA Is the Genetic Material in Eukaryotes

In 1950, eukaryotic organisms were not amenable to the types of experiments that used bacteria and viruses to demonstrate that DNA is the genetic material. Nevertheless, it was generally assumed that the genetic material would be a universal substance serving the same role in eukaryotes. Initially, support for this assumption relied on several circumstantial observations that, taken together, indicated that DNA does serve as the genetic material in eukaryotes. Subsequently, direct evidence established unequivocally the central role of DNA in genetic processes.

Indirect Evidence: Distribution of DNA

The genetic material should be found where it functions—in the nucleus as part of chromosomes. Both DNA and protein fit this criterion. However, protein is also abundant in the cytoplasm, whereas DNA is not. Both mitochondria and chloroplasts are known to perform genetic functions, and DNA is also present in these organelles. Thus, DNA is found only where primary genetic functions occur. Protein, on the other hand, is found everywhere in the cell. These observations are consistent with the interpretation favoring DNA over proteins as the genetic material.

Because it had earlier been established that chromosomes within the nucleus contain the genetic material, a correlation was expected to exist between the ploidy (n, 2n, etc.) of a cell and the quantity of the substance that

TABLE 10.2

DNA Content of Haploid versus Diploid Cells of Various Species*

Organism	n (pg)	2n (pg)
Human	3.25	7.30
Chicken	1.26	2.49
Trout	2.67	5.79
Carp	1.65	3.49
Shad	0.91	1.97

*Sperm (n) and nucleated precursors to red blood cells (2n) were used to contrast ploidy levels.

functions as the genetic material. Meaningful comparisons can be made between gametes (sperm and eggs) and somatic or body cells. The latter are recognized as being diploid (2n) and containing twice the number of chromosomes as gametes, which are haploid (n).

Table 10.2 compares, for a variety of organisms, the amount of DNA found in haploid sperm to the amount found in diploid nucleated precursors of red blood cells. The amount of DNA and the number of sets of chromosomes is closely correlated. No such consistent correlation can be observed between gametes and diploid cells for proteins. These data thus provide further circumstantial evidence favoring DNA over proteins as the genetic material of eukaryotes.

Indirect Evidence: Mutagenesis

Ultraviolet (UV) light is one of a number of agents capable of inducing mutations in the genetic material. Simple organisms such as yeast and other fungi can be irradiated with various wavelengths of ultraviolet light and the effectiveness of each wavelength measured by the number of mutations it induces. When the data are plotted, an **action spectrum** of UV light as a mutagenic agent is obtained. This action spectrum can then be compared with the **absorption spectrum** of any molecule suspected to be the genetic material (Figure 10–6). *The molecule serving as the genetic material is expected to absorb at the wavelength(s) found to be mutagenic.*

UV light is most mutagenic at the wavelength λ of 260 nanometers (nm), and both DNA and RNA absorb UV light most strongly at 260 nm. On the other hand, protein absorbs most strongly at 280 nm, yet no significant mutagenic effects are observed at that wavelength. This indirect evidence supports the idea that a nucleic acid, rather than protein, is the genetic material.

Direct Evidence: Recombinant DNA Studies

Although the circumstantial evidence just described does not constitute direct proof that DNA is the genetic material in eukaryotes, those observations spurred researchers to forge ahead using this supposition as the underlying hypothesis. Today, there is no doubt of its validity; DNA *is* the genetic material

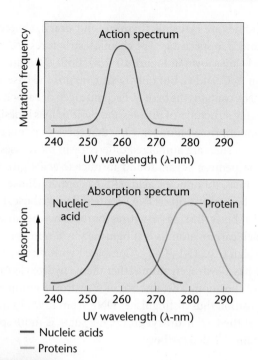

FIGURE 10–6 Comparison of the action spectrum (which determines the most effective mutagenic UV wavelength) and the absorption spectrum (which shows the range of wavelength where nucleic acids and proteins absorb UV light).

in all eukaryotes. The strongest evidence is provided by molecular analysis utilizing **recombinant DNA technology**. In this procedure, segments of eukaryotic DNA corresponding to specific genes are isolated and spliced into bacterial DNA. The resulting complex can be inserted into a bacterial cell, and then its genetic expression is monitored. If a eukaryotic gene is introduced, the subsequent production of the corresponding eukaryotic protein product demonstrates directly that the eukaryotic DNA is now present and functional in the bacterial cell. This has been shown to be the case in countless instances. For example, the products of the human genes specifying insulin and interferon are produced by bacteria after the human genes that encode these proteins are inserted. As the bacterium divides, the eukaryotic DNA replicates along with the bacterial DNA and is distributed to the daughter cells, which also express the human genes by creating the corresponding proteins.

The availability of vast amounts of DNA coding for specific genes, derived from recombinant DNA research, has led to other direct evidence that DNA serves as the genetic material. Work in the laboratory of Beatrice Mintz has demonstrated that DNA encoding the human *β-globin* gene, when microinjected into a fertilized mouse egg, is later found to be present and expressed in adult mouse tissue and transmitted to and expressed in that mouse's progeny. These mice are examples of what are called **transgenic animals**. Other work has introduced rat DNA encoding a growth hormone into fertilized mouse eggs. About one-third of the resultant mice grew to twice their normal size, indicating

that foreign DNA was present and functional. Subsequent generations of mice inherited this genetic information and also grew to a large size. This clearly demonstrates that DNA meets the requirement of expression of genetic information in eukaryotes. Later, we will see exactly how DNA is stored, replicated, expressed, and mutated.

10.5

RNA Serves as the Genetic Material in Some Viruses

Some viruses contain an RNA core rather than a DNA core. In these viruses, it appears that RNA serves as the genetic material—an exception to the general rule that DNA performs this function. In 1956, it was demonstrated that when purified RNA from **tobacco mosaic virus** (**TMV**) was spread on tobacco leaves, the characteristic lesions caused by viral infection subsequently appeared. Thus, it was concluded that RNA is the genetic material of this virus.

In 1965 and 1966, Norman Pace and Sol Spiegelman demonstrated that RNA from the phage Qβ can be isolated and replicated *in vitro*. Replication depends on an enzyme, **RNA replicase**, which is isolated from host *E. coli* cells following normal infection. When the RNA replicated *in vitro* is added to *E. coli* protoplasts, infection and viral multiplication (*transfection*) occur. Thus, RNA synthesized in a test tube serves as the genetic material in these phages by directing the production of all the components necessary for viral reproduction. While many viruses, such the T2 virus used by Hershey and Chase, use DNA as their hereditary material, another group of RNA-containing viruses bears mention. These are the **retroviruses**, which replicate in an unusual way. Their RNA serves as a template for the synthesis of the complementary DNA molecule. The process, **reverse transcription**, occurs under the direction of an RNA-dependent DNA polymerase enzyme called **reverse transcriptase**. This DNA intermediate can be incorporated into the genome of the host cell, and when the host DNA is transcribed, copies of the original retroviral RNA chromosomes are produced. Retroviruses include the human immunodeficiency virus (HIV), which causes AIDS, as well as several RNA tumor viruses.

10.6

Knowledge of Nucleic Acid Chemistry Is Essential to the Understanding of DNA Structure

Having established the critical importance of DNA and RNA in genetic processes, we will now take a brief look at the chemical structures of these molecules. As we shall see, the structural components of DNA and RNA are very similar.

This chemical similarity is important in the coordinated functions played by these molecules during gene expression. Like the other major groups of organic biomolecules (proteins, carbohydrates, and lipids), nucleic acid chemistry is based on a variety of similar building blocks that are polymerized into chains of varying lengths.

Nucleotides: Building Blocks of Nucleic Acids

DNA is a nucleic acid, and **nucleotides** are the building blocks of all nucleic acid molecules. Sometimes called mononucleotides, these structural units consist of three essential components: a **nitrogenous base**, a **pentose sugar** (a 5-carbon sugar), and a **phosphate group**. There are two kinds of nitrogenous bases: the nine-member double-ring **purines** and the six-member single-ring **pyrimidines**.

Two types of purines and three types of pyrimidines are commonly found in nucleic acids. The two purines are **adenine** and **guanine**, abbreviated **A** and **G**. The three

pyrimidines are **cytosine**, **thymine**, and **uracil**, abbreviated **C, T,** and **U**, respectively. The chemical structures of A, G, C, T, and U are shown in Figure 10–7(a). Both DNA and RNA contain A, C, and G, but only DNA contains the base T and only RNA contains the base U. Each nitrogen or carbon atom of the ring structures of purines and pyrimidines is designated by an unprimed number. Note that corresponding atoms in the two rings are numbered differently in most cases.

The pentose sugars found in nucleic acids give them their names. Ribonucleic acids (RNA) contain **ribose**, while deoxyribonucleic acids (DNA) contain **deoxyribose**. Figure 10–7(b) shows the ring structures for these two pentose sugars. Each carbon atom is distinguished by a number with a prime sign (e.g., C-1′, C-2′). Compared with ribose, deoxyribose has a hydrogen atom rather than a hydroxyl group at the C-2′ position. The absence of a hydroxyl group at the C-2′ position thus distinguishes DNA from RNA. In the absence of the C-2′ hydroxyl group, the sugar is more specifically named **2-deoxyribose**.

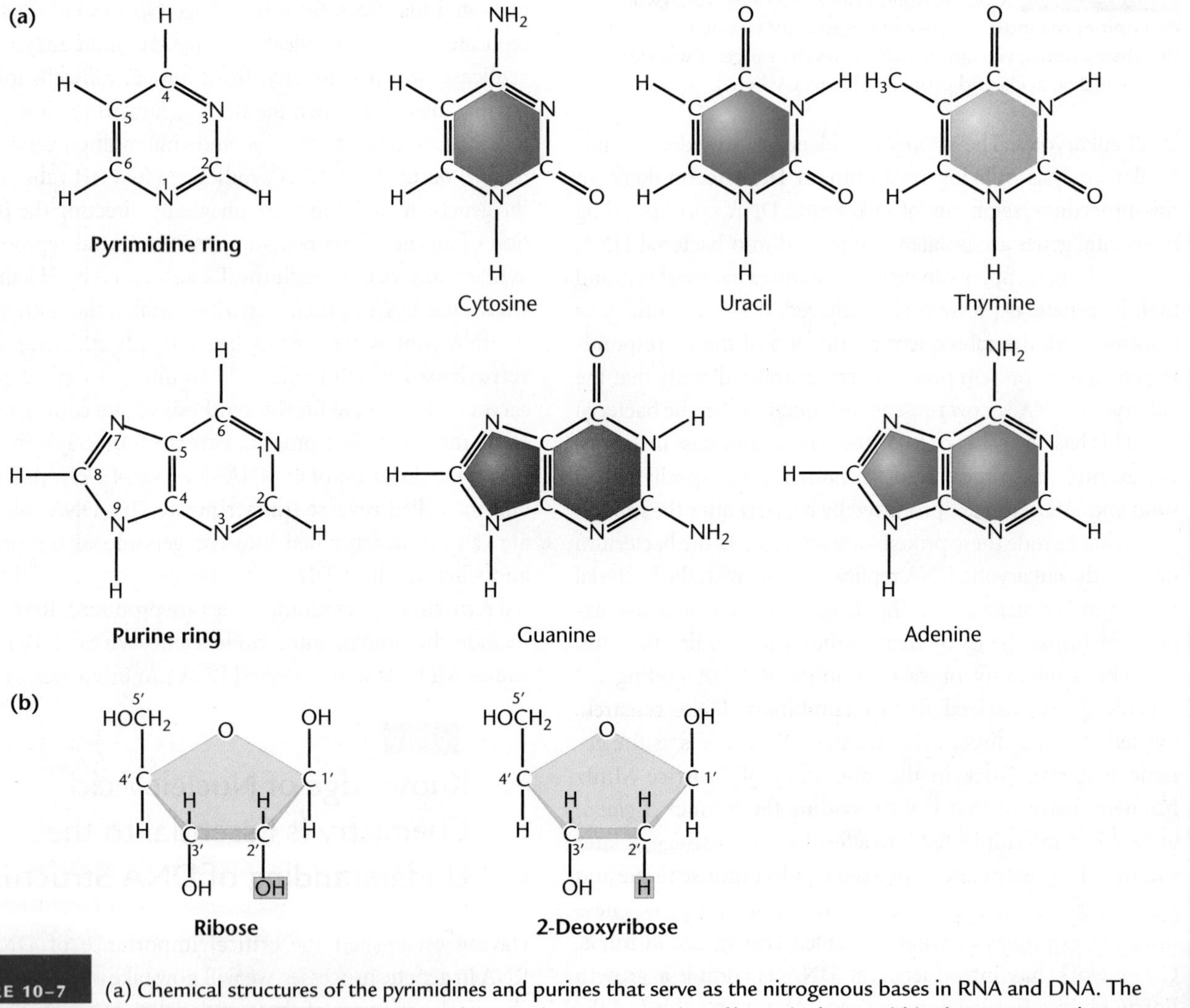

FIGURE 10–7   (a) Chemical structures of the pyrimidines and purines that serve as the nitrogenous bases in RNA and DNA. The convention for numbering carbon and nitrogen atoms making up the two categories of bases is shown within the structures that appear on the left. (b) Chemical ring structures of ribose and 2-deoxyribose, which serve as the pentose sugars in RNA and DNA, respectively.

If a molecule is composed of a purine or pyrimidine base and a ribose or deoxyribose sugar, the chemical unit is called a **nucleoside**. If a phosphate group is added to the nucleoside, the molecule is now called a **nucleotide**. Nucleosides and nucleotides are named according to the specific nitrogenous base (A, T, G, C, or U) that is part of the molecule. The structures of a nucleoside and a nucleotide and the nomenclature used in naming nucleosides and nucleotides are given in Figure 10–8.

The bonding between components of a nucleotide is highly specific. The C-1′ atom of the sugar is involved in the chemical linkage to the nitrogenous base. If the base is a purine, the N-9 atom is covalently bonded to the sugar; if the base is a pyrimidine, the N-1 atom bonds to the sugar. In deoxyribonucleotides, the phosphate group may be bonded to the C-2′, C-3′, or C-5′ atom of the sugar. The C-5′ phosphate configuration is shown in Figure 10–8. It is by far the prevalent form in biological systems and the one found in DNA and RNA.

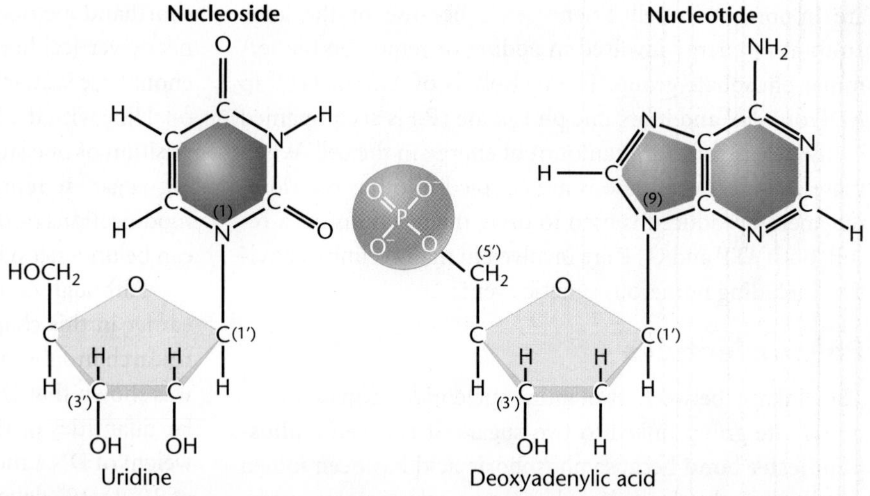

Ribonucleosides	Ribonucleotides
Adenosine	Adenylic acid
Cytidine	Cytidylic acid
Guanosine	Guanylic acid
Uridine	Uridylic acid
Deoxyribonucleosides	**Deoxyribonucleotides**
Deoxyadenosine	Deoxyadenylic acid
Deoxycytidine	Deoxycytidylic acid
Deoxyguanosine	Deoxyguanylic acid
Deoxythymidine	Deoxythymidylic acid

FIGURE 10–8 Structures and names of the nucleosides and nucleotides of RNA and DNA.

Nucleoside Diphosphates and Triphosphates

Nucleotides are also described by the term **nucleoside monophosphate (NMP)**. The addition of one or two phosphate groups results in **nucleoside diphosphates (NDPs)** and **triphosphates (NTPs)**, respectively, as shown in Figure 10–9. The triphosphate form is significant because it serves as the precursor molecule during nucleic acid synthesis within the cell (see Chapter 11). In addition, **adenosine triphosphate (ATP)** and **guanosine triphosphate (GTP)**

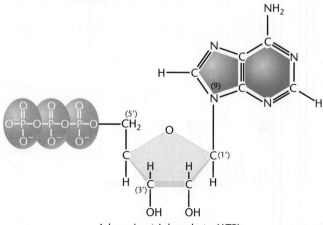

Deoxynucleoside diphosphate (dNDP)

Deoxythymidine diphosphate (dTDP)

Nucleoside triphosphate (NTP)

Adenosine triphosphate (ATP)

FIGURE 10–9 Structures of nucleoside diphosphates and triphosphates. Deoxythymidine diphosphate and adenosine triphosphate are diagrammed here.

are important in cell bioenergetics because of the large amount of energy involved in adding or removing the terminal phosphate group. The hydrolysis of ATP or GTP to ADP or GDP and inorganic phosphate (P_i) is accompanied by the release of a large amount of energy in the cell. When these chemical conversions are coupled to other reactions, the energy produced is used to drive the reactions. As a result, both ATP and GTP are involved in many cellular activities, including numerous genetic events.

Polynucleotides

The linkage between two mononucleotides consists of a phosphate group linked to two sugars. It is called a **phosphodiester bond** because phosphoric acid has been joined to two alcohols (the hydroxyl groups on the two sugars) by an ester linkage on both sides. Figure 10–10(a) shows the phosphodiester bond in DNA. The same bond is found in RNA. Each structure has a **C-5′ end** and a **C-3′ end**. Two joined nucleotides form a **dinucleotide**; three nucleotides, a **trinucleotide**; and so forth. Short chains consisting of up to approximately 30 nucleotides linked together are called **oligonucleotides**; longer chains are called **polynucleotides**.

Because drawing polynucleotide structures, as shown in Figure 10–10(a), is time consuming and complex, a schematic shorthand method has been devised [Figure 10–10(b)]. The nearly vertical lines represent the pentose sugar; the nitrogenous base is attached at the top, in the C-1′ position. A diagonal line with the P in the middle of it is attached to the C-3′ position of one sugar and the C-5′ position of the neighboring sugar; it represents the phosphodiester bond. Several modifications of this shorthand method are in use, and they can be understood in terms of these guidelines.

Although Levene's tetranucleotide hypothesis (described earlier in this chapter) was generally accepted before 1940, research in subsequent decades revealed it to be incorrect. It was shown that DNA does not necessarily contain equimolar quantities of the four bases. In addition, the molecular weight of DNA molecules was determined to be in the range of 10^6 to 10^9 daltons, far in excess of that of a tetranucleotide. The current view of DNA is that it consists of exceedingly long polynucleotide chains.

Long polynucleotide chains account for the large molecular weight of DNA and explain its most important property—storage of vast quantities of genetic information. If each nucleotide position in this long chain can be occupied by any one of four nucleotides, extraordinary variation is possible. For example, a polynucleotide only 1000 nucleotides in length can be arranged 4^{1000} different ways, each one different from all other possible sequences. This potential variation in molecular structure is essential if DNA is to store the vast amounts of chemical information necessary to direct cellular activities.

10.7

The Structure of DNA Holds the Key to Understanding Its Function

The previous sections in this chapter have established that DNA is the genetic material in all organisms (with certain viruses being the exception) and have provided details as to the basic chemical components making up nucleic acids. What remained to be deciphered was the precise structure of DNA. That is, how are polynucleotide chains organized into DNA, which serves as the genetic material? Is DNA composed of a single chain or more than one? If the latter is the case, how do the chains relate chemically to one another? Do the chains branch? And more important, how does the structure of this molecule relate to the various genetic functions served by DNA (i.e., storage, expression, replication, and mutation)?

From 1940 to 1953, many scientists were interested in solving the structure of DNA. Among others, Erwin Chargaff, Maurice Wilkins, Rosalind Franklin, Linus Pauling, Francis Crick, and James Watson sought information that might answer what many consider to be the most significant and intriguing question in the history of biology: *How does*

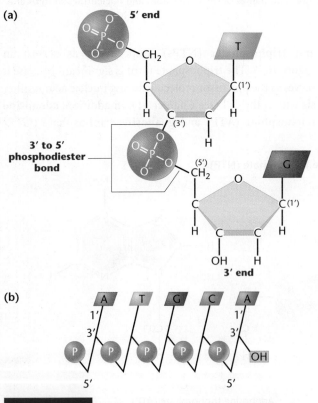

FIGURE 10–10   (a) Linkage of two nucleotides by the formation of a C-3′-C-5′ (3′-5′) phosphodiester bond, producing a dinucleotide. (b) Shorthand notation for a polynucleotide chain.

DNA serve as the genetic basis for life? The answer was believed to depend strongly on the chemical structure and organization of the DNA molecule, given the complex but orderly functions ascribed to it.

In 1953, James Watson and Francis Crick proposed that the structure of DNA is in the form of a double helix. Their model was described in a short paper published in the journal *Nature*. (The article is reprinted in its entirety on page 255.) In a sense, this publication was the finish of a highly competitive scientific race. Watson's book *The Double Helix* recounts the human side of the scientific drama that eventually led to the elucidation of DNA structure.

The data available to Watson and Crick, crucial to the development of their proposal, came primarily from two sources: (1) base composition analysis of hydrolyzed samples of DNA and (2) X-ray diffraction studies of DNA. Watson and Crick's analytical success can be attributed to their focus on building a model that conformed to the existing data. If the correct solution to the structure of DNA is viewed as a puzzle, Watson and Crick, working at the Cavendish Laboratory in Cambridge, England, were the first to fit the pieces together successfully.

Before learning the details of this far-reaching discovery, you may find some of the background on James Watson and Francis Crick to be of interest. Watson began his undergraduate studies at the University of Chicago at age 15, and was originally interested in ornithology. He then pursued his Ph.D. at Indiana University, where he studied viruses. He was only 24 years old in 1953, when he and Crick proposed the double-helix theory. Crick, now considered one of the great theoretical biologists of our time, had studied undergraduate physics at University College, London, and went on to perform military research during World War II. At the time of his collaboration with Watson, he was 35 years old and was performing X-ray diffraction studies of polypeptides and proteins as a graduate student. Immediately after they made their major discovery, Crick is reputed to have walked into the Eagle Pub in Cambridge, where the two frequently lunched, and announced for all to hear, "We have discovered the secret of life." It turns out that more than 50 years later, many scientists would quite agree!

Base-Composition Studies

Between 1949 and 1953, Erwin Chargaff and his colleagues used chromatographic methods to separate the four bases in DNA samples from various organisms. Quantitative methods were then used to determine the amounts of the four bases from each source. Table 10.3(a) lists some of Chargaff's original data. Parts (b) and (c) of the table show

TABLE 10.3

DNA Base-Composition Data

(a) Chargaff's Data*

Organism/Source	Molar Proportions[a]			
	1	2	3	4
	A	T	G	C
Ox thymus	26	25	21	16
Ox spleen	25	24	20	15
Yeast	24	25	14	13
Avian tubercle bacilli	12	11	28	26
Human sperm	29	31	18	18

(c) G + C Content in Several Organisms

Organism	%G + C
Phage T2	36.0
Drosophila	45.0
Maize	49.1
Euglena	53.5
Neurospora	53.7

(b) Base Compositions of DNAs from Various Sources

Organism	Base Composition				Base Ratio		Combined Base Ratios	
	1	2	3	4	5	6	7	8
	A	T	G	C	A/T	G/C	(A + G)/(C + T)	(A + T)/(C + G)
Human	30.9	29.4	19.9	19.8	1.05	1.00	1.04	1.52
Sea urchin	32.8	32.1	17.7	17.3	1.02	1.02	1.02	1.58
E. coli	24.7	23.6	26.0	25.7	1.04	1.01	1.03	0.93
Sarcina lutea	13.4	12.4	37.1	37.1	1.08	1.00	1.04	0.35
T7 bacteriophage	26.0	26.0	24.0	24.0	1.00	1.00	1.00	1.08

*Source: From Chargaff, 1950.

[a]Moles of nitrogenous constituent per mole of P. (Often, the recovery was less than 100 percent.)

more recently derived base-composition information that reinforces Chargaff's findings. As we shall see, Chargaff's data were critical to the success of Watson and Crick as they devised the double helical model of DNA. On the basis of these data, the following conclusions may be drawn:

1. As shown in Table 10–3(b), the amount of adenine residues is proportional to the amount of thymine residues in DNA (columns 1, 2, and 5). Also, the amount of guanine residues is proportional to the amount of cytosine residues (columns 3, 4, and 6).

2. Based on this proportionality, the sum of the purines (A + G) equals the sum of the pyrimidines (C + T) as shown in column 7.

3. The percentage of (G + C) does not necessarily equal the percentage of (A + T). As you can see, this ratio varies greatly among organisms, as shown in column 8 and in Part (c) of Table 10.3.

These conclusions indicate a definite pattern of base composition in DNA molecules. The data provided the initial clue to the DNA puzzle. In addition, they directly refute Levene's tetranucleotide hypothesis, which stated that all four bases are present in equal amounts.

X-Ray Diffraction Analysis

When fibers of a DNA molecule are subjected to X-ray bombardment, the X rays scatter (diffract) in a pattern that depends on the molecule's atomic structure. The pattern of diffraction can be captured as spots on photographic film and analyzed for clues to the overall shape of and regularities within the molecule. This process, **X-ray diffraction analysis**, was applied successfully to the study of protein structure by Linus Pauling and other chemists. The technique had been attempted on DNA as early as 1938 by William Astbury. By 1947, he had detected a periodicity of 3.4 angstroms (3.4-Å) repetitions* within the structure of the molecule, which suggested to him that the bases were stacked like coins on top of one another.

Between 1950 and 1953, Rosalind Franklin, working in the laboratory of Maurice Wilkins, obtained improved X-ray data from more purified samples of DNA (Figure 10–11). Her work confirmed the 3.4 Å periodicity seen by Astbury and suggested that the structure of DNA was some sort of helix. However, she did not propose a definitive model. Pauling had analyzed the work of Astbury and others and incorrectly proposed that DNA was a triple helix.

The Watson–Crick Model

Watson and Crick published their analysis of DNA structure in 1953. By building models based on the above-mentioned

*Today, measurement in nanometers (nm) is favored (1 nm = 10 Å).

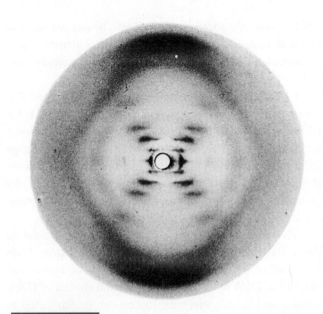

FIGURE 10–11   X-ray diffraction photograph by Rosalind Franklin using the B form of purified DNA fibers. The strong arcs on the periphery represent closely spaced aspects of the molecule, allowing scientists to estimate the periodicity of the nitrogenous bases, which are 3.4 Å apart. The inner cross pattern of spots reveals the grosser aspects of the molecule, indicating its helical nature.

parameters, they arrived at the double-helical form of DNA shown in Figure 10–12(a).

This model has the following major features:

1. Two long polynucleotide chains are coiled around a central axis, forming a right-handed double helix.

2. The two chains are **antiparallel**; that is, their C-5′-to-C-3′ orientations run in opposite directions.

3. The bases of both chains are flat structures lying perpendicular to the axis; they are "stacked" on one another, 3.4 Å (0.34 nm) apart, on the inside of the double helix.

4. The nitrogenous bases of opposite chains are *paired* as the result of the formation of hydrogen bonds; in DNA, only A═T and G≡C pairs occur.

5. Each complete turn of the helix is 34 Å (3.4 nm) long; thus, each turn of the helix is the length of a series of 10 base pairs.

6. A larger **major groove** alternating with a smaller **minor groove** winds along the length of the molecule.

7. The double helix has a diameter of 20 Å (2.0 nm).

The nature of the base pairing (point 4 above) is the model's most significant feature in terms of explaining its genetic functions. Before we discuss it, however, several other important features warrant emphasis. First, the antiparallel arrangement of the two chains is a key part of the double-helix

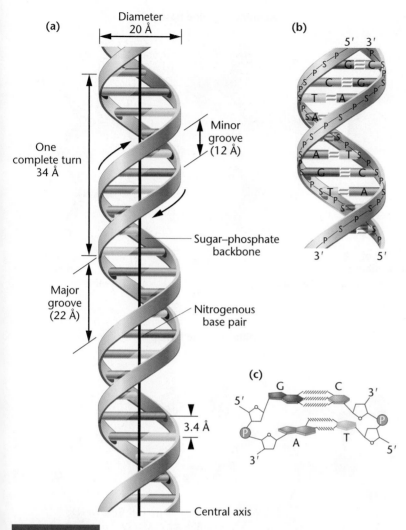

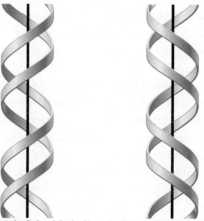

FIGURE 10–12 (a) The DNA double helix as proposed by Watson and Crick. The ribbon-like strands represent the sugar-phosphate backbones, and the horizontal rungs depict the nitrogenous base pairs, of which there are 10 per complete turn. The major and minor grooves are apparent. A solid vertical line shows the central axis. (b) A detailed view depicting the bases, sugars, phosphates, and hydrogen bonds of the helix. (c) A demonstration of the anti-parallel arrangement of the chains and the horizontal stacking of the bases.

The key to the model proposed by Watson and Crick is the specificity of base pairing. Chargaff's data suggested that A was equal in amount to T and that G was equal in amount to C. Watson and Crick realized that pairing A with T and C with G would account for these proportions, and that such pairing could occur as a result of hydrogen bonds between base pairs [Figure 10–12(b)], which would also provide the chemical stability necessary to hold the two chains together. Arrangement of the components in this way produces the major and minor grooves along the molecule's length. Furthermore, a purine (A or G) opposite a pyrimidine (T or C) on each "rung of the spiral staircase" in the proposed helix accounts for the 20 Å (2-nm) diameter suggested by X-ray diffraction studies.

The specific $A=T$ and $G\equiv C$ base pairing is described as **complementarity** and results from the chemical affinity that produces the hydrogen bonds in each pair of bases. As we will see, complementarity is very important in the processes of DNA replication and gene expression.

Two questions are particularly worthy of discussion. First, why aren't other base pairs possible? Watson and Crick discounted the pairing of A with G or of C with T because these would represent purine–purine and pyrimidine–pyrimidine pairings, respectively. Such pairings would lead to aberrant diameters of, in one case, more than and, in the other case, less than 20 Å because of the respective sizes of the purine and pyrimidine rings. In addition, the three-dimensional configurations that would be formed by such

model. While one chain runs in the 5′-to-3′ orientation (what seems right side up to us), the other chain goes in the 3′-to-5′ orientation (and thus appears upside down). This is indicated in Figure 10–12(b) and (c). Given the bond angles in the structures of the various nucleotide components, the double helix could not be constructed easily if both chains ran parallel to one another.

Second, the right-handed nature of the helix modeled by Watson and Crick is best appreciated by comparison with its left-handed counterpart, which is a mirror image, as shown in Figure 10–13. The conformation in space of the right-handed helix is most consistent with the data that were available to Watson and Crick, although an alternative form of DNA (Z-DNA) does exist as a left-handed helix, as we discuss below.

Right-handed double helix   Left-handed double helix

FIGURE 10–13 The right- and left-handed helical forms of DNA. Note that they are mirror images of one another.

pairings would not produce an alignment that allows sufficient hydrogen-bond formation. It is for this reason that A═C and G≡T pairings were also discounted, even though those pairs would each consist of one purine and one pyrimidine.

The second question concerns the properties of hydrogen bonds. Just what is the nature of such a bond, and is it strong enough to stabilize the helix? A **hydrogen bond** is a very weak electrostatic attraction between a covalently bonded hydrogen atom and an atom with an unshared electron pair. The hydrogen atom assumes a partial positive charge, while the unshared electron pair—characteristic of covalently bonded oxygen and nitrogen atoms—assumes a partial negative charge. These opposite charges are responsible for the weak chemical attraction that is the basis of the hydrogen bond. As oriented in the double helix, adenine forms two hydrogen bonds with thymine, and guanine forms three hydrogen bonds with cytosine (Figure 10–14). Although two or three hydrogen bonds taken alone are energetically very weak, thousands of bonds in tandem (as found in long polynucleotide chains) provide great stability to the helix.

Another stabilizing factor is the arrangement of sugars and bases along the axis. In the Watson–Crick model, the hydrophobic ("water-fearing") nitrogenous bases are stacked almost horizontally on the interior of the axis and are thus shielded from the watery environment that surrounds the molecule within the cell. The hydrophilic ("water-loving") sugar-phosphate backbones are on the outside of the axis, where both components may interact with water. These molecular arrangements provide significant chemical stabilization to the helix.

A more recent and accurate analysis of the form of DNA that served as the basis for the Watson–Crick model has revealed a minor structural difference between the substance and the model. A precise measurement of the number of base pairs per turn has demonstrated a value of 10.4, rather than the 10.0 predicted by Watson and Crick. In the classic model, each base pair is rotated 36° around the helical axis relative to the adjacent base pair, but the new finding requires a rotation of 34.6°. This results in slightly more than 10 base pairs per 360° turn.

The Watson–Crick model had an instant effect on the emerging discipline of molecular biology. Even in their initial 1953 article in *Nature*, the authors observed, "It has not escaped our notice that the specific pairing we have postulated immediately suggests a possible copying mechanism for the genetic material." Two months later, Watson and Crick pursued this idea in a second article in *Nature*, suggesting a specific mechanism of replication of *DNA—the semiconservative mode of replication* (described in Chapter 11). The second article also alluded to two new concepts: (1) the storage of genetic information in the sequence of the bases and (2) the mutations or genetic changes that would result from an alteration of bases. These ideas have received vast amounts of experimental support since 1953 and are now universally accepted.

Adenine-thymine base pair

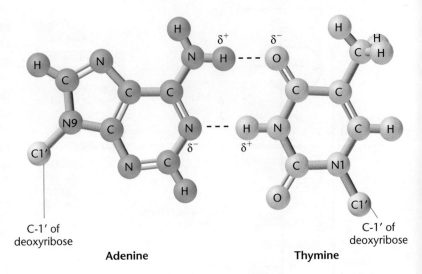

C-1' of deoxyribose

Adenine

C-1' of deoxyribose

Thymine

Guanine-cytosine base pair

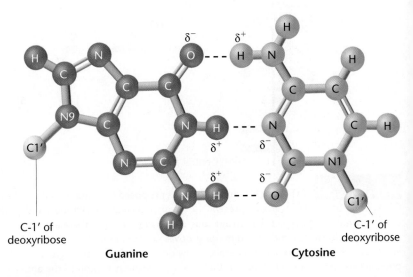

C-1' of deoxyribose

Guanine

C-1' of deoxyribose

Cytosine

- - - **Hydrogen bond**

FIGURE 10–14   Ball-and-stick models of A═T and G≡C base pairs. The dashes (– – –) represent the hydrogen bonds that form between bases.

Molecular Structure of Nucleic Acids: A Structure for Deoxyribose Nucleic Acid

We wish to suggest a structure for the salt of deoxyribose nucleic acid (DNA). This structure has novel features which are of considerable biological interest. A structure for nucleic acid has already been proposed by Pauling and Corey.[1] They kindly made their manuscript available to us in advance of publication. Their model consists of three intertwined chains, with the phosphates near the fibre axis, and the bases on the outside. In our opinion, this structure is unsatisfactory for two reasons: (1) We believe that the material which gives the X-ray diagrams is the salt, not the free acid. Without the acidic hydrogen atoms it is not clear what forces would hold the structure together, especially as the negatively charged phosphates near the axis will repel each other. (2) Some of the van der Waals distances appear to be too small.

Another three-chain structure has also been suggested by Fraser (in the press). In his model the phosphates are on the outside and the bases on the inside, linked together by hydrogen bonds. This structure as described is rather ill-defined, and for this reason we shall not comment on it.

We wish to put forward a radically different structure for the salt of deoxyribose nucleic acid. This structure has two helical chains each coiled round the same axis. We have made the usual chemical assumptions, namely, that each chain consists of phosphate diester groups joining β-D-deoxyribofuranose residues with 3′, 5′ linkages. The two chains (but not their bases) are related by a dyad perpendicular to the fibre axis. Both chains follow right-handed helices, but owing to the dyad the sequences of the atoms in the two chains run in opposite directions. Each chain loosely resembles Furberg's[2] model No. 1; that is, the bases are on the inside of the helix and the phosphates on the outside. The configuration of the sugar and the atoms near it is close to Furberg's "standard configuration," the sugar being roughly perpendicular to the attached base. There is a residue on each chain every 3.4 Å in the z-direction. We have assumed an angle of 36° between adjacent residues in the same chain, so that the structure repeats after 10 residues on each chain, that is, after 34 Å the distance of a phosphorus atom from the fibre axis is 10 Å. As the phosphates are on the outside, cations have easy access to them.

The structure is an open one, and its water content is rather high. At lower water contents we would expect the bases to tilt so that the structure could become more compact.

The novel feature of the structure is the manner in which the two chains are held together by the purine and pyrimidine bases. The planes of the bases are perpendicular to the fibre axis. They are joined together in pairs, a single base from one chain being hydrogen-bonded to a single base from the other chain, so that the two lie side by side with identical z-coordinates. One of the pair must be a purine and the other a pyrimidine for bonding to occur. The hydrogen bonds are made as follows: purine position 1 to pyrimidine position 1; purine position 6 to pyrimidine position 6.

If it is assumed that the bases only occur in the structure in the most plausible tautomeric forms (that is, with the keto rather than the enol configurations) it is found that only specific pairs of bases can bond together. These pairs are: adenine (purine) with thymine (pyrimidine), and guanine (purine) with cytosine (pyrimidine).

In other words, if an adenine forms one member of a pair, on either chain, then on these assumptions the other member must be thymine; similarly for guanine and cytosine. The sequence of bases on a single chain does not appear to be restricted in any way. However, if only specific pairs of bases can be formed, it follows that if the sequence of bases on one chain is given, then the sequence on the other chain is automatically determined.

It has been found experimentally[3,4] that the ratio of the amounts of adenine to thymine, and the ratio of guanine to cytosine, are always very close to unity for deoxyribose nucleic acid.

It is probably impossible to build this structure with a ribose sugar in place of the deoxyribose, as the extra oxygen atom would make too close a van der Waals contact.

The previously published X-ray data[5,6] on deoxyribose nucleic acid are insufficient for a rigorous test of our structure. So far as we can tell, it is roughly compatible with the experimental data, but it must be regarded as unproved until it has been checked against more exact results. Some of these are given in the following communications. We were not aware of the details of the results presented there when we devised our structure, which rests mainly though not entirely on published experimental data and stereochemical arguments.

It has not escaped our notice that the specific pairing we have postulated immediately suggests a possible copying mechanism for the genetic material. Full details of the structure, including the conditions assumed in building it, together with a set of co-ordinates for the atoms, will be published elsewhere.

We are much indebted to Dr. Jerry Donohue for constant advice and criticism, especially on interatomic distances. We have also been stimulated by a knowledge of the general nature of the unpublished experimental results and ideas of Dr. M. H. F. Wilkins, Dr. R. E. Franklin and their co-workers at King's College, London. One of us (J. D. W.) has been aided by a fellowship from the National Foundation for Infantile Paralysis.

—J. D. **Watson**
—F.H.C. **Crick**

Medical Research Council Unit for the Study of the Molecular Structure of Biological Systems, Cavendish Laboratory, Cambridge, England.

[1] Pauling, L., and Corey, R. B., *Nature*, 171, 346 (1953); *Proc. U.S. Nat. Acad. Sci.*, 39, 84 (1953).
[2] Furberg, S., *Acta Chem. Scand.*, 6, 634 (1952).
[3] Chargaff, E. For references see Zamenhof, S., Brawerman, G., and Chargaff, E., *Biochim. et Biophys. Acta*, 9, 402 (1952).
[4] Wyatt, G. R., *J. Gen. Physiol.*, 36, 201 (1952).
[5] Astbury, W. T., *Symp. Soc. Exp. Biol. 1, Nucleic Acid*, 66 (Camb. Univ. Press, 1947).
[6] Wilkins, M.H.F., and Randall, J. T., *Biochim. et Biophys. Acta*, 10, 192 (1953).

Watson and Crick's synthesis of ideas was highly significant with regard to subsequent studies of genetics and biology. The nature of the gene and its role in genetic mechanisms could now be viewed and studied in biochemical terms. Recognition of their work, along with that of Wilkins, led to all three receiving the Nobel Prize in Physiology and Medicine in 1962. Unfortunately, Rosalind Franklin had died in 1958 at the age of 37, making her contributions ineligible for consideration, since the award is not given posthumously. The Nobel Prize was to be one of many such awards bestowed for work in the field of molecular genetics.

10.8
Alternative Forms of DNA Exist

Under different conditions of isolation, different conformations of DNA are seen. At the time when Watson and Crick performed their analysis, two forms—**A-DNA** and **B-DNA**—were known. Watson and Crick's analysis was based on Rosalind Franklin's X-ray studies of the B form, which is seen under aqueous, low-salt conditions and is believed to be the biologically significant conformation.

While DNA studies around 1950 relied on the use of X-ray diffraction, more recent investigations have been performed using **single-crystal X-ray analysis**. The earlier studies achieved resolution of about 5 Å, but single crystals diffract X rays at about 1-Å intervals, near atomic resolution. As a result, every atom is "visible," and much greater structural detail is available.

With this modern technique, A-DNA, which is prevalent under high-salt or dehydration conditions, has now been scrutinized. In comparison to B-DNA, A-DNA is slightly more compact, with 9 base pairs in each complete turn of the helix, which is 23 Å (2.3 nm) in diameter (Figure 10–15). While it is also a right-handed helix, the orientation of the bases is somewhat different—they are tilted and displaced laterally in relation to the axis of the helix. As a result, the appearance of the major and minor grooves is modified. It seems doubtful that A-DNA occurs *in vivo* (under physiological conditions).

Still other forms of DNA right-handed helices have been discovered when investigated under various laboratory conditions. These have been designated C-, D-, E-, and most recently P-DNA. **C-DNA** is found under even greater dehydration conditions than those observed during the isolation of A- and B-DNA. It has only 9.3 base pairs per turn and is, thus, less compact. Its helical diameter is 19 Å. Like A-DNA, C-DNA does not have its base pairs lying flat; rather, they are tilted relative to the axis of the helix. Two other forms, **D-DNA** and **E-DNA**, occur in helices lacking guanine in their base composition. They have even fewer base pairs per turn: 8 and 7, respectively. And most recently, Jean-François Allemand and colleagues have shown that if DNA is artificially stretched, still another conformation is assumed, called **P-DNA** (named for Linus Pauling). Contrasted with the B form of DNA, P-DNA is quite

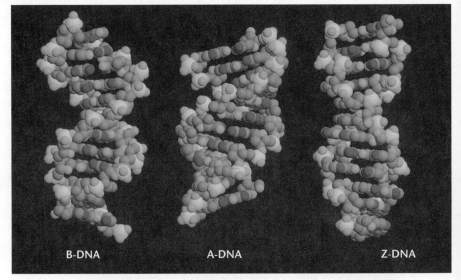

B-DNA   A-DNA   Z-DNA

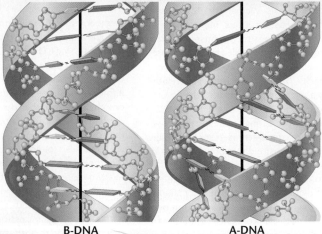

B-DNA   A-DNA

FIGURE 10–15   The top half of the figure shows computer-generated space-filling models of B-DNA, A-DNA, and Z-DNA. Below the photograph is an artist's depiction illustrating the orientation of the base pairs of B-DNA and A-DNA. Note that in B-DNA the base pairs are perpendicular to the helix, while they are tilted and pulled away from the helix in A-DNA.

interesting, since it is longer and narrower, and the phosphate groups, found on the outside of B-DNA, are located inside the molecule. The nitrogenous bases, present inside the helix in B-DNA, are found closer to the external surface in P-DNA, and there are fewer hydrogen bonds formed as a result. There are 2.62 bases per turn, in contrast to the 10.4 per turn in B-DNA.

Finally, still another form of DNA, called **Z-DNA**, was discovered by Andrew Wang, Alexander Rich, and their colleagues in 1979, when they examined a small synthetic DNA oligonucleotide containing only G≡C base pairs. Z-DNA takes on the rather remarkable configuration of a *left-handed double helix* (Figure 10–15). Like A- and B-DNA, Z-DNA consists of two antiparallel chains held together by Watson–Crick base pairs. Beyond these characteristics, Z-DNA is quite different. The left-handed helix is 18 Å (1.8 nm) in diameter, contains 12 base pairs per turn, and has a zigzag conformation (hence its name). The major groove present in B-DNA is nearly eliminated in Z-DNA. Speculation abounds over the possibility that regions of Z-DNA exist in the chromosomes of living organisms. The unique helical arrangement has the potential to provide an important recognition site for interaction with DNA-binding molecules. However, the extent to which Z-DNA occurs *in vivo* is still not clear.

The interest in alternative forms of DNA, such as Z and P, stems from the belief that DNA might have to assume a structure other than the B-form for some of its genetic functions. During both replication and transcription (when its RNA complement is synthesized during gene expression), the strands of the helix must separate and become accessible to large enzymes, as well as to a variety of other proteins involved in these processes. It is possible that changes in the shape of the DNA facilitate these functions. However, verification of the biological significance of alternative forms awaits further study.

NOW SOLVE THIS

10–2 In sea urchin DNA, which is double stranded, 17.5 percent of the bases were shown to be cytosine (C). What percentages of the other three bases are expected to be present in this DNA?

■ HINT: *This problem asks you to extrapolate from one measurement involving a unique DNA molecule to three other values characterizing the molecule. The key to its solution is to understand the base-pairing rules in the Watson–Crick model of DNA.*

10.9

The Structure of RNA Is Chemically Similar to DNA, but Single Stranded

The structure of RNA molecules resembles DNA, with several important exceptions. Although RNA also has nucleotides linked into polynucleotide chains, the sugar ribose replaces deoxyribose, and the nitrogenous base uracil replaces thymine. Another important difference is that most RNA is single stranded, although there are two important exceptions. First, RNA molecules sometimes fold back on themselves to form double-stranded regions of complementary base pairs. Second, some animal viruses that have RNA as their genetic material contain double-stranded helices.

As established earlier (see Figure 10–1), three major classes of cellular RNA molecules function during the expression of genetic information: **ribosomal RNA (rRNA), messenger RNA (mRNA)**, and **transfer RNA (tRNA)**. These molecules all originate as complementary copies of one of the two strands of DNA segments during the process of transcription. That is, their nucleotide sequence is complementary to the deoxyribonucleotide sequence of DNA that served as the template for their synthesis. Because uracil replaces thymine in RNA, uracil is complementary to adenine during transcription and during RNA base pairing.

Table 10.4 characterizes these major forms of RNA as found in prokaryotic and eukaryotic cells. Different RNAs are distinguished according to their sedimentation behavior in a centrifugal field and by their size (the number of nucleotides each contains). Sedimentation behavior depends on a

TABLE 10.4

Principal Classes of RNA

Class	% of Total RNA*	Svedberg Coefficient	Eukaryotic (E) or Prokaryotic (P)	Number of Nucleotides
Ribosomal (rRNA)	80	5	P and E	120
		5.8	E	160
		16	P	1542
		18	E	1874
		23	P	2904
		28	E	4718
Transfer (tRNA)	15	4	P and E	75–90
Messenger (mRNA)	5	varies	P and E	100–10,000

*In E. coli.

molecule's density, mass, and shape, and its measure is called the **Svedberg coefficient** (*S*). While higher *S* values almost always designate molecules of greater molecular weight, the correlation is not direct; that is, a twofold increase in molecular weight does not lead to a twofold increase in *S*. This is because, in addition to a molecule's mass, the size and shape of the molecule also affect its rate of sedimentation (*S*). As you can see in Table 10.4, RNA molecules come in a wide range of sizes.

Ribosomal RNA usually constitutes about 80 percent of all RNA in an *E. coli* cell. Ribosomal RNAs are important structural components of **ribosomes**, which function as nonspecific workbenches where proteins are synthesized during translation. The various forms of rRNA found in prokaryotes and eukaryotes differ distinctly in size.

Messenger RNA molecules carry genetic information from the DNA of the gene to the ribosome. The mRNA molecules vary considerably in size, reflecting the range in the sizes of the proteins encoded by the mRNA as well as the different sizes of the genes serving as the templates for transcription of mRNA. While Table 10.4 shows that about 5 percent of RNA is mRNA in *E. coli*, this percentage varies from cell to cell and even at different times in the life of the same cell.

Transfer RNA, accounting for up to 15 percent of the RNA in a typical cell, is the smallest class (in terms of average molecule size) of these RNA molecules and carries amino acids to the ribosome during translation. Because more than one tRNA molecule interacts simultaneously with the ribosome, the molecule's smaller size facilitates these interactions.

We will discuss the functions of these three classes of RNA in much greater detail in Chapters 13 and 14. These RNAs represent the major forms of the molecule involved in genetic expression, but other unique RNAs exist that perform various roles, especially in eukaryotes. For example, **telomerase RNA** is involved in DNA replication at the ends of chromosomes (Chapter 11), **small nuclear RNA (snRNA)** participates in processing mRNAs (Chapter 13), and **antisense RNA, microRNA (miRNA)**, and **short**

interfering RNA (siRNA) are involved in gene regulation (Chapter 17).

10.10

Many Analytical Techniques Have Been Useful during the Investigation of DNA and RNA

Since 1953, the role of DNA as the genetic material and the role of RNA in transcription and translation have been clarified through detailed analysis of nucleic acids. In this section, we consider several fundamental methods that have been particularly important for accomplishing this analysis. Many of them make use of the unique nature of the hydrogen bond that is so integral to the structure of nucleic acids. In Chapter 20 we will consider still other techniques that have been developed for manipulating DNA in various ways, including cloning.

Absorption of Ultraviolet Light

Nucleic acids absorb ultraviolet (UV) light most strongly at wavelengths of 254 to 260 nm (Figure 10–6), due to the interaction between UV light and the ring systems of the purines and pyrimidines. In aqueous solution, peak absorption by DNA and RNA occurs at 260 nm. Thus, UV light can be used in the localization, isolation, and characterization of molecules that contain nitrogenous bases (i.e., nucleosides, nucleotides, and polynucleotides).

Ultraviolet analysis is used in conjunction with many standard procedures that separate and identify molecules. As we shall see in the next section, the use of UV absorption is critical to the isolation of nucleic acids following their separation.

Sedimentation Behavior

Nucleic acid mixtures can be separated into different components by several possible centrifugation procedures (Figure 10–16). For example, the mixture can be loaded on top of a solution in which a concentration gradient has been formed from top to bottom. Then the tube holding the two is spun at high speeds in an ultracentrifuge so that the molecules of the mixture migrate downward through the concentration gradient, with each kind of molecule moving at a different rate. After centrifugation is stopped, successive fractions are eluted from the tube and measured spectrophotometrically for absorption at 260 nm. In this way, the position of a nucleic acid fraction along the gradient can be determined and the fraction isolated and studied further.

These gradient centrifugations rely on the sedimentation behavior of molecules in solution. Two major types of gradient centrifugation techniques are employed in the analysis of nucleic acids: *sedimentation equilibrium* and *sedimentation velocity*. Both require the use of high-speed

NOW SOLVE THIS

10–3 German measles results from an infection of the rubella virus, which can cause a multitude of health problems in newborns. What conclusions can you reach from a nucleic acid analysis that reveals an A + G/U + C ratio of 1.13?

■ HINT: *This problem asks you to analyze information about the chemical composition of a nucleic acid serving as the genetic material of a virus. The key to its solution is to apply your knowledge of nucleic acid chemistry, in particular your understanding of base-pairing.*

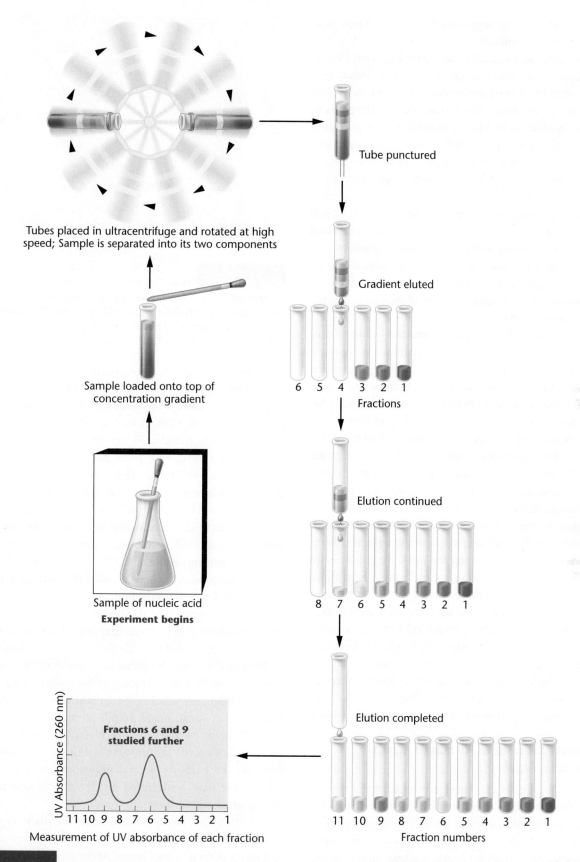

Tubes placed in ultracentrifuge and rotated at high speed; Sample is separated into its two components

Tube punctured

Gradient eluted

Sample loaded onto top of concentration gradient

6 5 4 3 2 1
Fractions

Sample of nucleic acid
Experiment begins

Elution continued

8 7 6 5 4 3 2 1

UV Absorbance (260 nm)

Fractions 6 and 9 studied further

Elution completed

11 10 9 8 7 6 5 4 3 2 1

11 10 9 8 7 6 5 4 3 2 1
Measurement of UV absorbance of each fraction

11 10 9 8 7 6 5 4 3 2 1
Fraction numbers

FIGURE 10–16   Separation of a mixture of two types of nucleic acid by gradient centrifugation. To fractionate the gradient, successive samples are eluted from the bottom of the tube. Each is measured for absorbance of ultraviolet light at 260 nm, producing a profile of the sample in graphic form.

centrifugation to create large centrifugal forces upon molecules in a gradient solution.

In **sedimentation equilibrium centrifugation** (sometimes called density gradient centrifugation), a density gradient is created that overlaps the densities of the individual components of a mixture of molecules. Usually, the gradient is made of a solution of a heavy metal salt, such as cesium chloride (CsCl). During centrifugation, the molecules migrate until they reach a point of neutral buoyant density. At this point, the centrifugal force on them is equal and opposite to the upward diffusion force, and no further migration occurs. If DNAs of different densities are present, they will separate as the molecules of each density reach equilibrium with the corresponding density of CsCl. The gradient may then be fractionated and the components isolated (Figure 10–16). When properly executed, this technique provides high resolution in separating mixtures of molecules varying only slightly in density.

The second technique, **sedimentation velocity centrifugation**, employs an analytical centrifuge that uses ultraviolet absorption optics to monitor the migration of the molecules during centrifugation and determine the "velocity of sedimentation." As mentioned earlier, this velocity has been standardized in units called Svedberg coefficients (S).

In this technique, the molecules are loaded on top of the gradient, and the gravitational forces created by centrifugation drive them toward the bottom of the tube. Two forces work against this downward movement: (1) the viscosity of the solution creates a frictional resistance, and (2) part of the force of diffusion is directed upward. Under these conditions, the key variables are the mass and shape of the molecules being examined. In general, the greater the mass, the greater is the sedimentation velocity. However, the molecule's shape affects the frictional resistance. Therefore, two molecules of equal mass, but different in shape, will sediment at different rates.

One use of the sedimentation velocity technique is the determination of **molecular weight (MW)**. If certain physical-chemical properties of a molecule under study are also known, the MW can be calculated based on the sedimentation velocity. As noted earlier, the S values increase with molecular weight, but they are not directly proportional to it.

Denaturation and Renaturation of Nucleic Acids

Heat or other stresses can cause a complex molecule like DNA to **denature**, or lose its function due to the unfolding of its three-dimensional structure. In the denaturation of double-stranded DNA, the hydrogen bonds of the duplex (double-stranded) structure break, the helix unwinds, and the strands separate. However, no covalent bonds break. The viscosity of the DNA decreases, and both the UV absorption and the buoyant density increase. In laboratory studies, denaturation may be caused intentionally either by heating

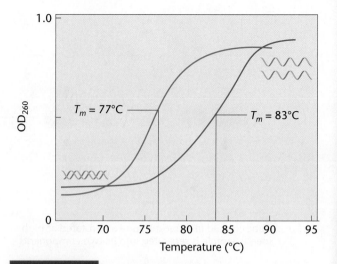

FIGURE 10–17   A melting profile shows the increase in UV absorption versus temperature (the hyperchromic effect) for two DNA molecules with different $G \equiv C$ contents. The molecule with a melting point (T_m) of 83°C has a greater $G \equiv C$ content than the molecule with a T_m of 77°C.

or chemical treatment. (Denaturation as a result of heating is sometimes referred to as *melting*.) In such studies, the increase in UV absorption of heated DNA in solution, called the **hyperchromic shift**, is the easiest change to measure. This effect is illustrated in Figure 10–17.

Because $G \equiv C$ base pairs have one more hydrogen bond than do $A = T$ pairs (see Figure 10–14), they are more stable under heat treatment. Thus, DNA with a greater proportion of $G \equiv C$ pairs than $A = T$ pairs requires higher temperatures to denature completely. When absorption at 260 nm is monitored and plotted against temperature during heating, a **melting profile** of the DNA is obtained. The midpoint of this profile, or curve, is called the **melting temperature (T_m)** and represents the point at which 50 percent of the strands are unwound, or denatured (Figure 10–17). When the curve plateaus at its maximum optical density, denaturation is complete, and only single strands exist. Analysis of melting profiles provides a characterization of DNA and an alternative method of estimating its base composition.

One might ask whether the denaturation process can be reversed; that is, can single strands of nucleic acids re-form a double helix, provided that each strand's complement is present? Not only is the answer yes, but such reassociation provides the basis for several important analytical techniques that have contributed much valuable information during genetic experimentation.

If DNA that has been denatured thermally is cooled slowly, random collisions between complementary strands will result in their reassociation. At the proper temperature, hydrogen bonds will re-form, securing pairs of strands into duplex structures. With time during cooling, more and more duplexes will form. Depending on the conditions, a complete match (that is, perfect complementarity) is not essential for duplex formation, provided there are stretches of base pairing on the two reassociating strands.

Molecular Hybridization

The denaturation and renaturation of nucleic acids is the basis for one of the most powerful and useful techniques in molecular genetics—**molecular hybridization**. This technique derives its name from the fact that single strands need not originate from the same nucleic acid source in order to combine to form duplex structures. For example, if DNA strands are isolated from two distinct organisms and a reasonable degree of base complementarity exists between them, under the proper temperature conditions, double-stranded molecular hybrids will form during renaturation. Such hybridization may also occur between single strands of DNA and RNA. A case in point is when RNA and the DNA from which it has been transcribed are mixed

together (Figure 10–18). The RNA will find and attach to its single-stranded DNA complement. In this example, DNA molecules are heated, causing strand separation, and then the strands are slowly cooled in the presence of single-stranded RNA. If the RNA has been transcribed from the DNA used in the experiment, and is therefore complementary to it, molecular hybridization will occur, creating a DNA:RNA duplex. Several methods are available for monitoring the amount of double-stranded molecules produced following strand separation. In early studies, radioisotopes were utilized to "tag" one of the strands and monitor its presence in the hybrid duplexes that formed.

In the 1960s, molecular hybridization techniques contributed to our increased understanding of transcriptional events occurring at the gene level. Refinements of this process have occurred continually, helping to advance the study

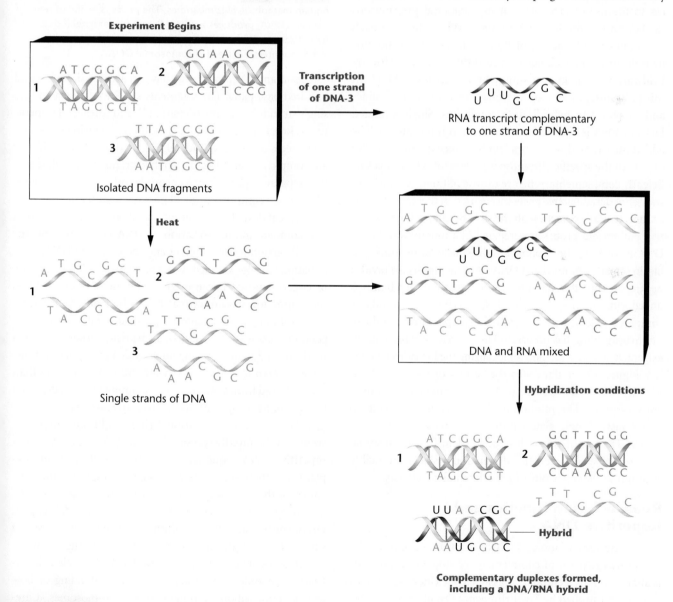

FIGURE 10–18   Diagrammatic representation of the process of molecular hybridization between DNA fragments and RNA that has been transcribed on strand of one of the DNA fragments.

of both molecular evolution and the organization of DNA in chromosomes. Hybridization can occur in solution or when DNA is bound either to a gel or to a specialized binding filter. Such filters are used in a variety of **DNA blotting procedures**, whereby hybridization serves as a way to "probe" for complementary nucleic acid sequences. Blotting is used routinely in modern genomic analysis. In addition, hybridization will occur even when DNA is present in tissue affixed to a slide, as in the FISH procedure (discussed in the next section), or when affixed to a glass chip, the basis of **DNA microarray analysis** (discussed in Chapter 24). Microarray analysis allows mass screening for a specific DNA sequence from among thousands of cloned genes in a single assay.

Fluorescent *in situ* Hybridization (FISH)

A refinement in the molecular hybridization technique has led to the use of DNA present in cytological preparations as the "target" for hybrid formation. When this approach is combined with the use of fluorescent probes to monitor hybridization, the technique is called **fluorescent *in situ* hybridization**, or is known simply by the acronym **FISH**. In this procedure, mitotic or interphase cells are fixed to slides and subjected to hybridization conditions. Single-stranded DNA or RNA is added, and hybridization is monitored. The added nucleic acid serves as a "probe," since it will hybridize only with the specific chromosomal areas for which it is sufficiently complementary. Before the use of fluorescent probes was refined, radioactive probes were used in these *in situ* procedures to allow detection on the slide by autoradiography.

Fluorescent probes are prepared in a unique way. When DNA is used, it is first coupled to the small organic molecule biotin (creating biotinylated DNA). Another molecule (avidin or streptavidin) that has a high binding affinity for biotin is linked with a fluorescent molecule such as fluorescein, and this complex is reacted with the cytological preparation after the *in situ* hybridization has occurred. This procedure represents an extremely sensitive method for localizing the hybridized DNA.

Figure 10–19 illustrates the results of using FISH to identify the DNA specific to the centromeres of human chromosomes. The resolution of FISH is great enough to detect just a single gene within an entire set of chromosomes. The use of this technique to identify chromosomal locations of specific genetic information has been a valuable addition to the repertoire of experimental geneticists.

Reassociation Kinetics and Repetitive DNA

In one extension of molecular hybridization procedures, the *rate of reassociation* of complementary single DNA strands is analyzed. This technique, **reassociation kinetics**, was first refined and studied by Roy Britten and David Kohne.

The DNA used in such studies is first fragmented into small pieces by shearing forces introduced during its isolation.

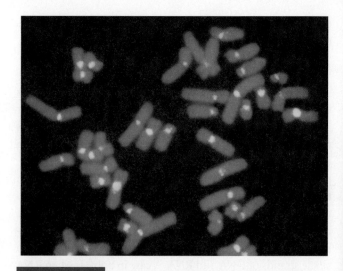

FIGURE 10–19   Fluorescent *in situ* hybridization (FISH) of human metaphase chromosomes. The probe, specific to centromeric DNA, produces a yellow fluorescence signal indicating hybridization. The red fluorescence is produced by propidium iodide counterstaining of chromosomal DNA.

The resultant DNA fragments have an average size of several hundred base pairs. The fragments are then dissociated into single strands by heating (denatured), and when the temperature is lowered, reassociation is monitored. During reassociation, pieces of single-stranded DNA randomly collide. If they are complementary, a stable double strand is formed; if not, they separate and are free to encounter other DNA fragments. The process continues until all possible matches are made.

A great deal of information can be obtained from studies that compare the reassociation of DNA of different organisms. As genome size increases and there is more DNA, reassociation time is extended. Reassociation occurs more slowly in larger genomes because with random collisions it takes more time for all correct matches to be made. When reassociation kinetics in eukaryotic organisms with much larger genome sizes was first studied, a surprising observation was made. Rather than requiring an extended reassociation time, *some* eukaryotic DNA reassociated even more rapidly than those derived from bacteria. The remaining DNA, as expected because of its complexity, took longer to reassociate.

Based on this observation, Britten and Kohne hypothesized that the rapidly reassociating fraction might represent **repetitive DNA sequences**. This interpretation would explain why these segments reassociate so rapidly—multiple copies of the same sequence are much more likely to make matches, thus reassociating more quickly than single copies. On the other hand, the remaining DNA segments consist of **unique DNA sequences**, present only once in the genome.

It is now known that repetitive DNA is prevalent in eukaryotic genomes and is key to our understanding of how genetic information is organized in chromosomes. Careful study has shown that various levels of repetition exist. In some cases, short DNA sequences are repeated over a million times. In other cases, longer sequences are repeated only a few times, or intermediate levels of sequence redundancy are present. We will return to this topic in Chapter 12, where we will discuss the organization of DNA in genes and chromosomes. For now, we will simply point out that the discovery of repetitive DNA was one of the first clues that much of the DNA in eukaryotes is not contained in genes that encode proteins. We will develop and elaborate on this concept as we proceed with our coverage of the molecular basis of heredity.

Electrophoresis of Nucleic Acids

We conclude the chapter by considering **electrophoresis**, a technique that has made essential contributions to the analysis of nucleic acids. Electrophoresis, also useful in protein studies, can be applied to the separation of different-sized fragments of DNA and RNA chains and is invaluable in current research investigations in molecular genetics.

In general, electrophoresis separates, or *resolves*, molecules in a mixture by causing them to migrate under the influence of an electric field. A sample to be analyzed is placed on a porous substance (a piece of filter paper or a semisolid gel) that in turn is placed in a solution that conducts electricity. If two molecules have approximately the same shape and mass, the one with the greatest net charge will migrate more rapidly toward the electrode of opposite polarity.

As electrophoretic technology developed from its initial application (which was protein separation), researchers discovered that using gels of varying pore sizes significantly improved the method's resolution. This advance is particularly useful for mixtures of molecules with a similar charge-mass ratio but different sizes. For example, two polynucleotide chains of different lengths (say, 10 vs. 20 nucleotides) are both negatively charged because of the phosphate groups of the nucleotides. Thus, both chains move to the positively charged pole (the anode), and because they have the same charge-mass ratio, the electric field moves them at similar speeds. Consequently, the separation between the two chains is minimal. However, using a porous medium such as a **polyacrylamide gel** or an **agarose gel**, which can be prepared with various pore sizes, enables us to separate the two molecules. *Smaller molecules migrate at a faster rate through the gel than larger molecules* (Figure 10–20). The key to separation is the porous gel matrix, which restricts migration of larger molecules more than it restricts smaller molecules. The resolving power of this method is so great that polynucleotides that vary by even one nucleotide in length may be separated. Once electrophoresis is complete, bands representing the variously sized molecules are identified either by autoradiography (if a component of the molecule is radioactive) or by use of a fluorescent dye that binds to nucleic acids.

Electrophoretic separation of nucleic acids is at the heart of a variety of commonly used research techniques discussed later in the text (Chapters 20 and 21). Of particular note, as you will see in those discussions, are the various "blotting" techniques (e.g., Southern blots and Northern blots), as well as DNA sequencing methods.

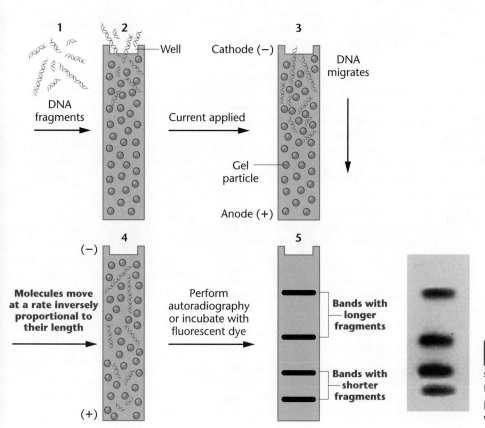

FIGURE 10–20 Electrophoretic separation of a mixture of DNA fragments that vary in length. The photograph shows an agarose gel with DNA bands.

ACTGACNCAC TA TAGGGCGAA T TCGAGC TCG G T ACCCGGNGG A TCC C AGAG CGACC GCAGGC A GCAAG C GAG A
  10   20   30   40   50   60   70

EXPLORING GENOMICS

Introduction to Bioinformatics: BLAST

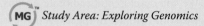 *Study Area: Exploring Genomics*

In this chapter, we focused on the structural details of DNA, the genetic material for living organisms. In Chapter 20, you will learn how scientists can clone and sequence DNA—a routine technique in molecular biology and genetics laboratories. The explosion of DNA and protein sequence data that has occurred in the last 15 years has launched the field of *bioinformatics,* an interdisciplinary science that applies mathematics and computing technology to develop hardware and software for storing, sharing, comparing, and analyzing nucleic acid and protein sequence data.

A large number of sequence databases that make use of bioinformatics have been developed. An example is **GenBank** (www.ncbi. nlm.nih.gov/Genbank/index.html), which is the National Institutes of Health sequence database. This global resource, with access to databases in Europe and Japan, currently contains more than 148 billion base pairs of sequence data!

In the Exploring Genomics exercises for Chapter 5, you were introduced to the National Center for Biotechnology Information (NCBI) Genes and Disease site. Now we will use an NCBI application called **BLAST, Basic Local Alignment Search Tool**. BLAST is an invaluable program for searching through GenBank and other databases to find DNA- and protein-sequence similarities between cloned substances. It has many additional functions that we will explore in other exercises.

■ Exercise I – Introduction to BLAST

1. Access BLAST from the NCBI Web site at www.ncbi.nlm.nih.gov/BLAST.

2. Click on "nucleotide blast." This feature allows you to search DNA databases to look for a similarity between a sequence you enter and other sequences in the database. Do a nucleotide search with the following sequence:

   CCAGAGTCCAGCTGCTGCT CATA CTACTGATACTGCTGGG

3. Imagine that this sequence is a short part of a gene you cloned in your laboratory. You want to know if this gene or others with similar sequences have been discovered. Enter this sequence into the "Enter Query Sequence" text box at the top of the page. Near the bottom of the page, under the "Program Selection" category, choose "blastn"; then click on the "BLAST" button at the bottom of the page to run the search. It may take several minutes for results to be available because BLAST is using powerful algorithms to scroll through billions of bases of sequence data! A new page will appear with the results of your search.

4. Near the top of this page you will see a table showing significant matches to the sequence you searched with (called the query sequence). BLAST determines significant matches based on statistical measures that consider the length of the query sequence, the number of matches with sequences in the database, and other factors. Significant *alignments,* regions of significant similarity in the query and subject sequences, typically have E values less than 1.0.

5. The top part of the table lists matches to transcripts (mRNA sequences), and the lower part lists matches to genomic DNA sequences, in order of highest to lowest number of matches. Use the "Links" column to the far right of this table to explore gene and chromosome databases relevant to the matched sequences.

6. Alignments are indicated by horizontal lines. BLAST adjusts for gaps in the sequences, that is, for areas that may not align precisely because of missing bases in otherwise similar sequences. Scroll below the table to see the aligned sequences from this search, and then answer the following questions:

   (a) What were the top three matches to your query sequence?

   (b) For each alignment, BLAST also indicates the percent *identity* and the number of gaps in the match between the query and subject sequences (shown in the column under "Max ident"). What was the percent identity for the top three matches? What percentage of each aligned sequence showed gaps indicating sequence differences?

   (c) Click on the links for the first matched sequence (far-right column). These will take you to a wealth of information, including the size of the sequence; the species it was derived from; a PubMed-linked chronology of research publications pertaining to this sequence; the complete sequence; and if the sequence encodes a polypeptide, the predicted amino acid sequence coded by the gene. Skim through the information presented for this gene. What is the gene's function?

7. A BLAST search can also be done by entering the *accession number* for a sequence, which is a unique identifying number assigned to a sequence before it can be put into a database. For example, search with the accession number NM_007305. What did you find?

8. Run a BLAST search using the sequences or accession numbers listed below. In each case, after entering the accession number or sequence in the "Enter Query Sequence" box, go to the "Choose Search Set" box and click on the "Others" button for database. Then go to the "Program Selection" box and click "megablast" before running your search. These features will allow you to align the query sequence with similar genes from a number of other species. When each search is completed, explore the information BLAST provides so that you can identify and learn about the gene encoded by the sequence.

   (a) NM_001006650. What is the top sequence that aligns with the query sequence of this accession number and shows 100 percent sequence identity?

   (b) DQ991619. What gene is encoded by this sequence?

   (c) NC_007596. What living animal has a sequence similar to this one?

CASE STUDY | Zigs and zags of the smallpox virus

Smallpox, a once highly lethal contagious disease, has been eradicated worldwide. However, research continues with stored samples of variola, the smallpox virus, because it is a potential weapon in bioterrorism. Human cells protect themselves from the variola virus (and other viruses) by activating genes that encode protective proteins. It has recently been discovered that in response to variola, human cells create small transitory stretches of Z-DNA at sites that regulate these genes. The smallpox virus can bypass this cellular defense mechanism by specifically targeting the segments of Z-DNA and inhibiting the synthesis of the protective proteins. This discovery raises some interesting questions:

1. What is unique about Z-DNA that might make it a specific target during viral infection?
2. How might the virus target host-cell Z-DNA formation to block the synthesis of antiviral proteins?
3. To study the interaction between viral proteins and Z-DNA, how could Z DNA-forming DNA be synthesized in the lab?
4. How could this research lead to the development of drugs to combat infection by variola and related viruses?

Summary Points

 For activities, animations, and review quizzes, go to the study area at www.masteringgenetics.com

1. Although both proteins and nucleic acids were initially considered as possible candidates for genetic material, proteins were initially favored.
2. By 1952, transformation studies and experiments using bacteria infected with bacteriophages strongly suggested that DNA is the genetic material in bacteria and most viruses.
3. Although initially only indirect observations supported the hypothesis that DNA controls inheritance in eukaryotes, subsequent studies involving recombinant DNA techniques and transgenic mice provided direct experimental evidence that the eukaryotic genetic material is DNA.
4. RNA serves as the genetic material in some bacteriophages as well as some plant and animal viruses.
5. As proposed by Watson and Crick, DNA exists in the form of a right-handed double helix composed of two long antiparallel polynucleotide chains held together by hydrogen bonds formed between complementary, nitrogenous base pairs.

6. The second category of nucleic acids important in genetic function is RNA, which is similar to DNA with the exceptions that it is usually single stranded, the sugar ribose replaces the deoxyribose, and the pyrimidine uracil replaces thymine.
7. Various methods of analysis of nucleic acids, particularly molecular hybridization and electrophoresis, have led to studies essential to our understanding of genetic mechanisms.
8. Among the techniques used to study double-stranded DNA, reassociation kinetics analysis enabled geneticists to postulate the existence of repetitive DNA in eukaryotes, where certain nucleotide sequences are present many times in the genome.

INSIGHTS AND SOLUTIONS

The current chapter, in contrast to preceding chapters, does not emphasize genetic problem solving. Instead, it recounts some of the initial experimental analyses that launched the era of molecular genetics. Accordingly, our "Insights and Solutions" section shifts its emphasis to experimental rationale and analytical thinking, an approach that will continue to be used in later chapters whenever appropriate.

1. (a) Based strictly on your scrutiny of the transformation data of Avery, MacLeod, and McCarty, what objection might be made to the conclusion that DNA is the genetic material? What other conclusion might be considered? (b) What observations, including later ones, argue against this objection?

   Solution: (a) Based solely on their results, it may be concluded that DNA is essential for transformation. However, DNA might have been a substance that caused capsular formation by *directly* converting nonencapsulated cells to cells with a capsule. That is, DNA may simply have played a catalytic role in capsular synthesis, leading to cells displaying smooth type III colonies.

   (b) First, transformed cells pass the trait onto their progeny cells, thus supporting the conclusion that DNA is responsible for heredity, not for the direct production of polysaccharide coats. Second, subsequent transformation studies over a period of five years showed that other traits, such as antibiotic resistance, could be transformed. Therefore, the transforming factor has a broad general effect, not one specific to polysaccharide synthesis. This observation is more in keeping with the conclusion that DNA is the genetic material.

2. If RNA were the universal genetic material, how would it have affected the Avery experiment and the Hershey–Chase experiment?

   Solution: In the Avery experiment, digestion of the soluble filtrate with RNase, rather than DNase, would have eliminated transformation. Had this occurred, Avery and his colleagues would have concluded that RNA was the transforming factor. Hershey and Chase would have obtained identical results, since ^{32}P would also label RNA but not protein. Had they been using a bacteriophage with RNA as its nucleic acid, and had they known this, they would have concluded that RNA was responsible for directing the reproduction of their bacteriophage.

3. A quest to isolate an important disease-causing organism was successful, and molecular biologists were hard at work analyzing the results. The organism contained as its genetic material a remarkable nucleic acid with a base composition of A = 21 percent, C = 29 percent, G = 29 percent, U = 21 percent. When heated, it showed a major hyperchromic shift, and when the reassociation kinetics were studied, the nucleic acid of this organism reannealed more slowly than that of phage T4 and *E. coli*. T4 contains 10^5 nucleotide pairs.

Analyze this information carefully, and draw *all* possible conclusions about the genetic material of this organism, based strictly on the preceding observations. As a test of your model, make one prediction that if upheld would strengthen your hypothesis about the nature of this molecule.

Solution: First of all, because of the presence of uracil (U), the molecule appears to be RNA. In contrast to normal RNA, this one has base ratios of A/U = G/C = 1, suggesting that the molecule may be a double helix. The hyperchromic shift and reassociation kinetics support this hypothesis. In the kinetic study, since none of the nucleic acid segments reannealed more rapidly than bacterial or viral nucleic acid, there is no repetitive sequence RNA. Furthermore, the total length of unique-sequence DNA is greater than that of either phage T4 (10^5 nucleotide pairs) or *E. coli*. A prediction might be made concerning the sugars. Our model suggests that ribose rather than deoxyribose should be present. If so, this observation would support the hypothesis that RNA is the genetic material in this organism.

Problems and Discussion Questions

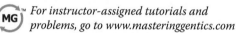 *For instructor-assigned tutorials and problems, go to www.masteringgentics.com*

HOW DO WE KNOW?

1. In this chapter, we first focused on the information that showed DNA to be the genetic material and then discussed the structure of DNA as proposed by Watson and Crick. We concluded the chapter by describing various techniques developed to study DNA. Along the way, we found many opportunities to consider the methods and reasoning by which much of this information was acquired. From the explanations given in the chapter, what answers would you propose to the following fundamental questions:

(a) How were scientists able to determine that DNA, and not some other molecule, serves as the genetic material in bacteria and bacteriophages?

(b) How do we know that DNA also serves as the genetic material in eukaryotes such as humans?

(c) How was it determined that the structure of DNA is a double helix with the two strands held together by hydrogen bonds formed between complementary nitrogenous bases?

(d) How do we know that G pairs with C and that A pairs with T as complementary base pairs are formed?

(e) How do we know that repetitive DNA sequences exist in eukaryotes?

2. The functions ascribed to the genetic material are replication, expression, storage, and mutation. What does each of these terms mean in the context of genetics?

3. Discuss the reasons proteins were generally favored over DNA as the genetic material before 1940. What was the role of the tetranucleotide hypothesis in this controversy?

4. Contrast the various contributions made to an understanding of transformation by Griffith, by Avery and his colleagues, and by Taylor.

5. When Avery and his colleagues had obtained what was concluded to be the transforming factor from the IIIS virulent cells, they treated the fraction with proteases, RNase, and DNase, followed in each case by the assay for retention or loss of transforming ability. What were the purpose and results of these experiments? What conclusions were drawn?

6. Why were ^{32}P and ^{35}S chosen for use in the Hershey–Chase experiment? Discuss the rationale and conclusions of this experiment.

7. Does the design of the Hershey–Chase experiment distinguish between DNA and RNA as the molecule serving as the genetic material? Why or why not?

8. What observations are consistent with the conclusion that DNA serves as the genetic material in eukaryotes? List and discuss them.

9. What are the exceptions to the general rule that DNA is the genetic material in all organisms? What evidence supports these exceptions?

10. Draw the chemical structure of the three components of a nucleotide, and then link the three together. What atoms are removed from the structures when the linkages are formed?

11. How are the carbon and nitrogen atoms of the sugars, purines, and pyrimidines numbered?

12. Adenine may also be named 6-amino purine. How would you name the other four nitrogenous bases, using this alternative system? (O is indicated by "oxy-," and CH_3 by "methyl.")

13. Draw the chemical structure of a dinucleotide composed of A and G. Opposite this structure, draw the dinucleotide composed of T and C in an antiparallel (or upside-down) fashion. Form the possible hydrogen bonds.

14. Describe the various characteristics of the Watson–Crick double-helix model for DNA.

15. What evidence did Watson and Crick have at their disposal in 1953? What was their approach in arriving at the structure of DNA?

16. What might Watson and Crick have concluded had Chargaff's data from a single source indicated the following?

	A	T	G	C
%	29	19	21	31

Why would this conclusion be contradictory to Wilkins's and Franklin's data?

17. How do covalent bonds differ from hydrogen bonds? Define base complementarity.
18. List three main differences between DNA and RNA.
19. What are the three major types of RNA molecules? How is each related to the concept of information flow?
20. What component of the nucleotide is responsible for the absorption of ultraviolet light? How is this technique important in the analysis of nucleic acids?
21. Distinguish between sedimentation velocity and sedimentation equilibrium centrifugation (density gradient centrifugation).
22. What is the physical state of DNA following denaturation?
23. What is the hyperchromic effect? How is it measured? What does T_m imply?
24. Why is T_m related to base composition?
25. What is the chemical basis of molecular hybridization?
26. What did the Watson–Crick model suggest about the replication of DNA?
27. A genetics student was asked to draw the chemical structure of an adenine- and thymine-containing dinucleotide derived from DNA. His answer is shown here:

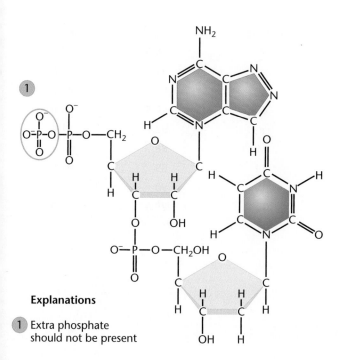

Explanations

1 Extra phosphate should not be present

The student made more than six major errors. One of them is circled, numbered 1, and explained. Find five others. Circle them, number them 2 through 6, and briefly explain each in the manner of the example given.

28. Considering the information in this chapter on B- and Z-DNA and right- and left-handed helices, carefully analyze structures (a) and (b) below and draw conclusions about their helical nature. Which is right handed and which is left handed?

(a)    **(b)**

29. One of the most common spontaneous lesions that occurs in DNA under physiological conditions is the hydrolysis of the amino group of cytosine, converting the cytosine to uracil. What would be the effect on DNA structure of a uracil group replacing cytosine?
30. In some organisms, cytosine is methylated at carbon 5 of the pyrimidine ring after it is incorporated into DNA. If a 5-methyl cytosine molecule is then hydrolyzed, as described in Problem 29, what base will be generated?
31. Because of its rapid turnaround time, fluorescent *in situ* hybridization (FISH) is commonly used in hospitals and laboratories as an aneuploid screen of cells retrieved from amniocentesis and chorionic villus sampling (CVS). Chromosomes 13, 18, 21, X, and Y (see Chapter 8) are typically screened for aneuploidy in this way. Explain how FISH might be accomplished using amniotic or CVS samples and why the above chromosomes have been chosen for screening.
32. Assume that you are interested in separating short (25–40 nucleotides) DNA molecules from a pool of longer molecules in the 900–1000 nucleotide range. You have two recipes for making your polyacrylamide gels: one recipe uses 12 percent acrylamide and would be considered a "hard gel," while the other uses 4 percent acrylamide and would be considered a loose gel. Which recipe would you consider using and why?

Extra-Spicy Problems

 For instructor-assigned tutorials and problems, go to www.masteringgentics.com

33. A primitive eukaryote was discovered that displayed a unique nucleic acid as its genetic material. Analysis provided the following information:
(a) The general X-ray diffraction pattern is similar to that of DNA, but with somewhat different dimensions and more irregularity.
(b) A major hyperchromic shift is evident upon heating and monitoring UV absorption at 260 nm.
(c) Base-composition analysis reveals four bases in the following proportions:

Adenine	=	8%
Guanine	=	37%
Xanthine	=	37%
Hypoxanthine	=	18%

(d) About 75 percent of the sugars are deoxyribose, while 25 percent are ribose.
Postulate a model for the structure of this molecule that is consistent with the foregoing observations.

34. *Newsdate: March 1, 2030.* A unique creature has been discovered during exploration of outer space. Recently, its genetic material has been isolated and analyzed. This material is similar in some ways to DNA in its chemical makeup. It contains in abundance the 4-carbon sugar erythrose and a molar equivalent of phosphate groups. In addition, it contains six nitrogenous bases: adenine (A), guanine (G), thymine (T), cytosine (C), hypoxanthine (H), and xanthine (X). These bases exist in the following relative proportions:

$$A = T = H \text{ and } C = G = X$$

X-ray diffraction studies have established a regularity in the molecule and a constant diameter of about 30 Å. Together, these data have suggested a model for the structure of this molecule.

(a) Propose a general model of this molecule. Describe it briefly.
(b) What base-pairing properties must exist for H and for X in the model?
(c) Given the constant diameter of 30 Å, do you think that *either* (i) both H and X are purines or both pyrimidines, *or* (ii) one is a purine and one is a pyrimidine?

35. You are provided with DNA samples from two newly discovered bacterial viruses. Based on the various analytical techniques discussed in this chapter, construct a research protocol that would be useful in characterizing and contrasting the DNA of both viruses. For each technique that you include in the protocol, indicate the type of information you hope to obtain.

36. During gel electrophoresis, DNA molecules can easily be separated according to size because all DNA molecules have the same charge-to-mass ratio and the same shape (long rod). Would you expect RNA molecules to behave in the same manner as DNA during gel electrophoresis? Why or why not?

37. Electrophoresis is an extremely useful procedure when applied to analysis of nucleic acids as it can resolve molecules of different sizes with relative ease and accuracy. Large molecules migrate more slowly than small molecules in agarose and polyacrylamide gels. However, the fact that nucleic acids of the same length may exist in a variety of conformations can often complicate the interpretation of electrophoretic separations. For instance, when a single species of a bacterial plasmid is isolated from cells, the individual plasmids may exist in three forms (depending on the genotype of their host and conditions of isolation): superhelical/supercoiled (form I), nicked/open circle (form II), and linear (form III). Form I is compact and very tightly coiled, with both DNA strands continuous. Form II exists as a loose circle because one of the two DNA strands has been broken, thus releasing the supercoil. All three have the same mass, but each will migrate at a different rate through a gel. Based on your understanding of gel composition and DNA migration, predict the relative rates of migration of the various DNA structures mentioned above.

38. Following is a table (modified from Kropinski, 1973) that presents the T_m and chemical composition ($\%G \equiv C$) of DNA from certain bacteriophages. From these data develop a graph that presents $\%G \equiv C$ (ordinate) and T_m (abscissa). What is the relationship between T_m and $\%G \equiv C$ for these samples? What might be the molecular basis of this relationship?

Phage	T_m	$\%G \equiv C$
α	86.5	44.0
κ	91.5	53.8
λ	89	49.2
ϕ 80	90.5	53.0
χ	92.1	57.4
Mu-1	88	51.4
T1	89	48.0
T3	90	49.6
T7	89.5	48.0

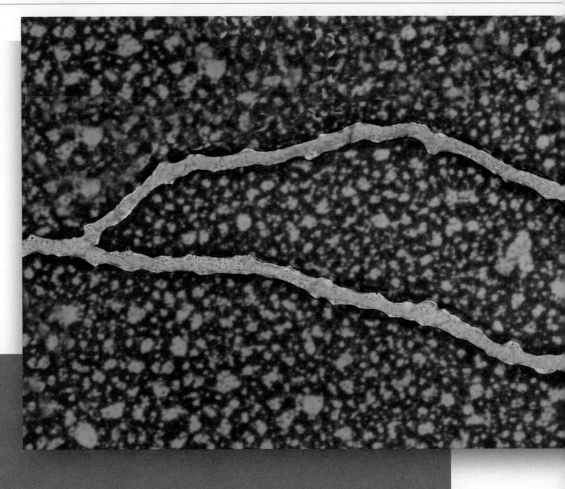

Transmission electron micrograph of human DNA from a HeLa cell, illustrating a replication fork characteristic of active DNA replication.

11

DNA Replication and Recombination

CHAPTER CONCEPTS

- Genetic continuity between parental and progeny cells is maintained by semiconservative replication of DNA, as predicted by the Watson–Crick model.

- Semiconservative replication uses each strand of the parent double helix as a template, and each newly replicated double helix includes one "old" and one "new" strand of DNA.

- DNA synthesis is a complex but orderly process, occurring under the direction of a myriad of enzymes and other proteins.

- DNA synthesis involves the polymerization of nucleotides into polynucleotide chains.

- DNA synthesis is similar in prokaryotes and eukaryotes, but more complex in eukaryotes.

- In eukaryotes, DNA synthesis at the ends of chromosomes (telomeres) poses a special problem, overcome by a unique RNA-containing enzyme, telomerase.

- Genetic recombination, an important process leading to the exchange of segments between DNA molecules, occurs under the direction of a group of enzymes.

F ollowing Watson and Crick's proposal for the structure of DNA, scientists focused their attention on how this molecule is replicated. Replication is an essential function of the genetic material and must be executed precisely if genetic continuity between cells is to be maintained following cell division. It is an enormous, complex task. Consider for a moment that more than 3×10^9 (3 billion) base pairs exist within the human genome. To duplicate faithfully the DNA of just one of these chromosomes requires a mechanism of extreme precision. Even an error rate of only 10^6 (one in a million) will still create 3000 errors (obviously an excessive number) during each replication cycle of the genome. Although it is not error free, and much of evolution would not have occurred if it were, an extremely accurate system of DNA replication has evolved in all organisms.

As Watson and Crick noted in the concluding paragraph of their 1953 paper (reprinted on page 255, their proposed model of the double helix provided the initial insight into how replication occurs. Called *semiconservative replication,* this mode of DNA duplication was soon to receive strong support from numerous studies of viruses, prokaryotes, and eukaryotes. Once the general *mode* of replication was clarified, research to determine the precise details of *DNA synthesis* intensified. What has since been discovered is that numerous enzymes and other proteins are needed to copy a DNA helix. Because of the complexity of the chemical events during synthesis, this subject remains an extremely active area of research.

In this chapter, we will discuss the general mode of replication, as well as the specific details of DNA synthesis. The research leading to such knowledge is another link in our understanding of life processes at the molecular level.

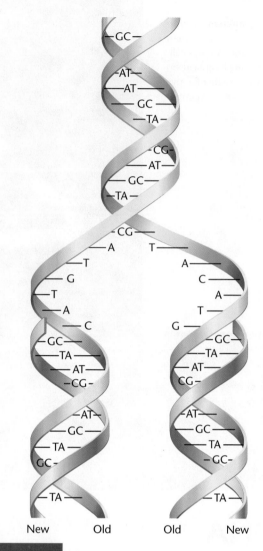

FIGURE 11–1   Generalized model of semiconservative replication of DNA. New synthesis is shown in blue.

DNA Is Reproduced by Semiconservative Replication

Watson and Crick recognized that, because of the arrangement and nature of the nitrogenous bases, each strand of a DNA double helix could serve as a template for the synthesis of its complement (Figure 11–1). They proposed that, if the helix were unwound, each nucleotide along the two parent strands would have an affinity for its complementary nucleotide. As we learned in Chapter 10, the complementarity is due to the potential hydrogen bonds that can be formed. If thymidylic acid (T) were present, it would "attract" adenylic acid (A); if guanidylic acid (G) were present, it would attract cytidylic acid (C); likewise, A would attract T, and C would attract G. If these nucleotides were then covalently linked into polynucleotide chains along both templates, the result would be the production of two identical double strands of

DNA. Each replicated DNA molecule would consist of one "old" and one "new" strand, hence the reason for the name **semiconservative replication**.

Two other theoretical modes of replication are possible that also rely on the parental strands as a template (Figure 11–2). In **conservative replication**, complementary polynucleotide chains are synthesized as described earlier. Following synthesis, however, the two newly created strands then come together and the parental strands reassociate. The original helix is thus "conserved."

In the second alternative mode, called **dispersive replication**, the parental strands are dispersed into two new double helices following replication. Hence, each strand consists of both old and new DNA. This mode would involve cleavage of the parental strands during replication. It is the most complex of the three possibilities and is therefore considered to be least likely to occur. It could not, however, be ruled out as an experimental model. Figure 11–2 shows

Conservative   Semiconservative   Dispersive

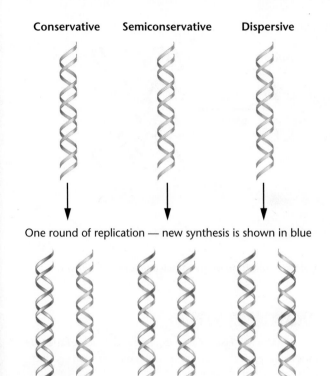

One round of replication — new synthesis is shown in blue

FIGURE 11–2 Results of one round of replication of DNA for each of the three possible modes by which replication could be accomplished.

the theoretical results of a single round of replication by each of the three different modes.

The Meselson–Stahl Experiment

In 1958, Matthew Meselson and Franklin Stahl published the results of an experiment providing strong evidence that semiconservative replication is the mode used by bacterial cells to produce new DNA molecules. They grew *E. coli* cells for many generations in a medium that had $^{15}NH_4Cl$ (ammonium chloride) as the only nitrogen source. A "heavy" isotope of nitrogen, ^{15}N contains one more neutron than the naturally occurring ^{14}N isotope; thus, molecules containing ^{15}N are more dense than those containing ^{14}N. Unlike radioactive isotopes, ^{15}N is stable. After many generations in this medium, almost all nitrogen-containing molecules in the *E. coli* cells, including the nitrogenous bases of DNA, contained the heavier isotope.

Critical to the success of this experiment, DNA containing ^{15}N can be distinguished from DNA containing ^{14}N. The experimental procedure involves the use of a technique referred to as **sedimentation equilibrium centrifugation**, or as it is also called, *buoyant density gradient centrifugation*. Samples are forced by centrifugation through a density gradient of a heavy metal salt, such as cesium chloride. Molecules of DNA will reach equilibrium when their density equals the density of the gradient medium. In this case, ^{15}N-DNA will reach this point at a position closer to the bottom of the tube than will ^{14}N-DNA.

In this experiment (Figure 11–3), uniformly labeled ^{15}N cells were transferred to a medium containing only $^{14}NH_4Cl$. Thus, all "new" synthesis of DNA during replication contained only the "lighter" isotope of nitrogen. The time of transfer to the new medium was taken as time zero ($t = 0$). The *E. coli* cells were allowed to replicate over several generations, with cell samples removed after each

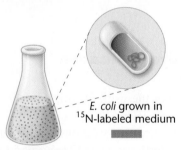

E. coli grown in ^{15}N-labeled medium

E. coli DNA becomes uniformly labeled with ^{15}N in nitrogenous bases

FIGURE 11–3 The Meselson–Stahl experiment.

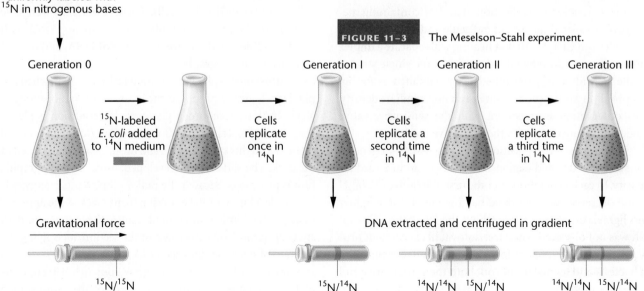

Generation 0 — ^{15}N-labeled *E. coli* added to ^{14}N medium — Cells replicate once in ^{14}N — Generation I — Cells replicate a second time in ^{14}N — Generation II — Cells replicate a third time in ^{14}N — Generation III

Gravitational force — DNA extracted and centrifuged in gradient

$^{15}N/^{15}N$ $^{15}N/^{14}N$ $^{14}N/^{14}N$ $^{15}N/^{14}N$ $^{14}N/^{14}N$ $^{15}N/^{14}N$

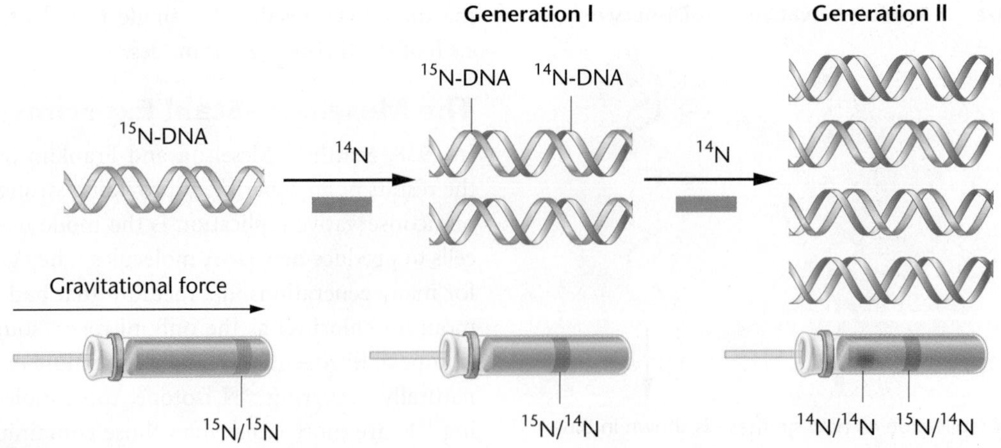

FIGURE 11–4   The expected results of two generations of semiconservative replication in the Meselson–Stahl experiment.

replication cycle. DNA was isolated from each sample and subjected to sedimentation equilibrium centrifugation.

After one generation, the isolated DNA was present in only a single band of intermediate density—the expected result for semiconservative replication in which each replicated molecule was composed of one new ^{14}N-strand and one old ^{14}N-strand (Figure 11–4). This result was not consistent with the prediction of conservative replication, in which two distinct bands would occur; thus this mode may be rejected.

After two cell divisions, DNA samples showed two density bands—one intermediate band and one lighter band corresponding to the ^{14}N position in the gradient. Similar results occurred after a third generation, except that the proportion of the lighter band increased. This was again consistent with the interpretation that replication is semiconservative.

You may have realized that a molecule exhibiting intermediate density is also consistent with dispersive replication. However, Meselson and Stahl ruled out this mode of replication on the basis of two observations. First, after the first generation of replication in an ^{14}N-containing medium, they isolated the hybrid molecule and heat denatured it. Recall from Chapter 10 that heating will separate a duplex into single strands. When the densities of the single strands of the hybrid were determined, they exhibited *either* an ^{15}N profile *or* an ^{14}N profile, but *not* an intermediate density. This observation is consistent with the semiconservative mode but inconsistent with the dispersive mode.

Furthermore, if replication were dispersive, *all* generations after $t = 0$ would demonstrate DNA of an intermediate density. In each generation after the first, the ratio of $^{15}N/^{14}N$ would decrease and the hybrid band would become lighter and lighter, eventually approaching the ^{14}N band. This result was not observed. The Meselson–Stahl experiment provided conclusive support for semiconservative replication in bacteria and tended to rule out both the conservative and dispersive modes.

11–1 In the Meselson–Stahl experiment, which of the three modes of replication could be ruled out after one round of replication? after two rounds?

■ HINT: *This problem involves an understanding of the nature of the experiment as well as the difference between the three possible modes of replication. The key to its solution is to determine which mode will not create "hybrid" helices after one round of replication.*

Semiconservative Replication in Eukaryotes

In 1957, the year before the work of Meselson and Stahl was published, J. Herbert Taylor, Philip Woods, and Walter Hughes presented evidence that semiconservative replication also occurs in eukaryotic organisms. They experimented with root tips of the broad bean *Vicia faba*, which are an excellent source of dividing cells. These researchers were able to monitor the process of replication by labeling DNA with ^{3}H-thymidine, a radioactive precursor of DNA, and by performing autoradiography.

Autoradiography is a common technique that, when applied cytologically, pinpoints the location of a radioisotope in a cell. In this procedure, a photographic emulsion is placed over a histological preparation containing cellular material (root tips, in this experiment), and the preparation is stored in the dark. The slide is then developed, much as photographic film is processed. Because the radioisotope emits energy, following development the emulsion turns black at the approximate point of emission. The end result is the presence of dark spots or "grains" on the surface of the section, identifying the location of newly synthesized DNA within the cell.

Taylor and his colleagues grew root tips for approximately one generation in the presence of the radioisotope and then placed them in unlabeled medium in which cell division continued. At the conclusion of each generation, they arrested the cultures at metaphase by adding colchicine (a chemical derived from the crocus plant that poisons the spindle fibers) and then examined the chromosomes by autoradiography. They found radioactive thymidine only in association with chromatids that contained newly synthesized DNA. Figure 11–5 illustrates the replication of a

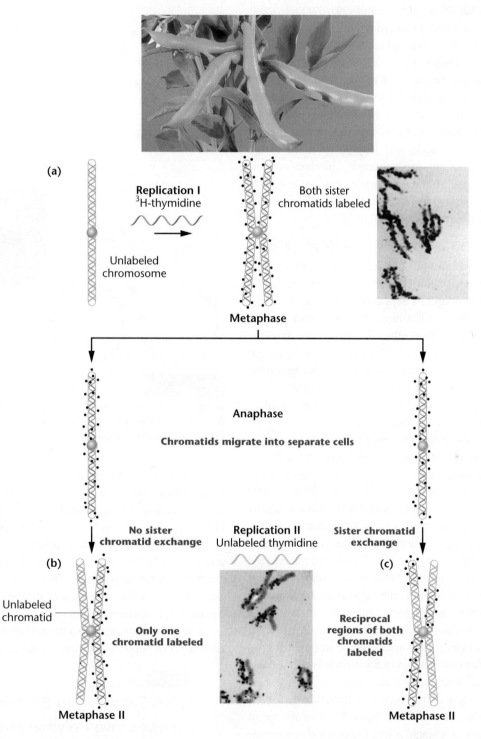

FIGURE 11–5 The Taylor–Woods–Hughes experiment, demonstrating the semiconservative mode of replication of DNA in root tips of *Vicia faba*. A portion of the plant is shown in the top photograph. (a) An unlabeled chromosome proceeds through the cell cycle in the presence of ³H-thymidine. As it enters mitosis, both sister chromatids of the chromosome are labeled, as shown, by auto-radiography. After a second round of replication (b), this time in the absence of ³H-thymidine, only one chromatid of each chromosome is expected to be surrounded by grains. Except where a reciprocal exchange has occurred between sister chromatids (c), the expectation was upheld. The micrographs are of the actual autoradiograms obtained in the experiment.

single chromosome over two division cycles, including the distribution of grains.

These results are compatible with the semiconservative mode of replication. After the first replication cycle in the presence of the isotope, both sister chromatids show radioactivity, indicating that each chromatid contains one new radioactive DNA strand and one old unlabeled strand. After the second replication cycle, *which takes place in unlabeled medium*, only one of the two sister chromatids of each chromosome should be radioactive because half of the parent strands are unlabeled. With only the minor exceptions of *sister chromatid exchanges* (discussed in Chapter 5), this result was observed.

Together, the Meselson–Stahl experiment and the experiment by Taylor, Woods, and Hughes soon led to the general acceptance of the semiconservative mode of replication. Later studies with other organisms reached the same conclusion and also strongly supported Watson and Crick's proposal for the double-helix model of DNA.

Origins, Forks, and Units of Replication

To enhance our understanding of semiconservative replication, let's briefly consider a number of relevant issues. The first concerns the **origin of replication**. Where along the chromosome is DNA replication initiated? Is there only a single origin, or does DNA synthesis begin at more than one point? Is any given point of origin random, or is it located at a specific region along the chromosome? Second, once replication begins, does it proceed in a single direction or in both directions away from the origin? In other words, is replication *unidirectional* or *bidirectional*?

To address these issues, we need to introduce two terms. First, at each point along the chromosome where replication is occurring, the strands of the helix are unwound, creating what is called a **replication fork**. Such a fork will initially appear at the point of origin of synthesis and then move along the DNA duplex as replication proceeds. If replication is bidirectional, two such forks will be present, migrating in opposite directions away from the origin. The second term refers to the length of DNA that is replicated following one initiation event at a single origin. This is a unit referred to as the **replicon**.

The evidence is clear regarding the origin and direction of replication. John Cairns tracked replication in *E. coli*, using radioactive precursors of DNA synthesis and autoradiography. He was able to demonstrate that in *E. coli* there is only a single region, called *oriC*, where replication is initiated. The presence of only a single origin is characteristic of bacteria, which have only one circular chromosome. Since DNA synthesis in bacteria originates at a single point, the entire chromosome constitutes one replicon. In *E. coli*, the replicon consists of the entire genome of 4.6 Mb (4.6 million base pairs).

Figure 11–6 illustrates Cairns's interpretation of DNA replication in *E. coli*. This interpretation (and the accompanying micrograph) does not answer the question of unidirectional

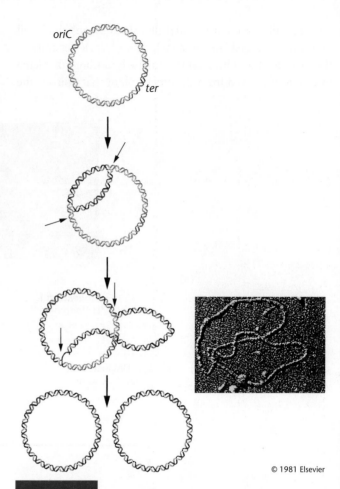

© 1981 Elsevier

FIGURE 11–6   Bidirectional replication of the *E. coli* chromosome. The thin black arrows identify the advancing replication forks. The micrograph is of a bacterial chromosome in the process of replication, comparable to the figure next to it.

versus bidirectional synthesis. However, other results, derived from studies of bacteriophage lambda, demonstrated that replication is bidirectional, moving away from *oriC* in both directions. Figure 11–6 therefore interprets Cairns's work with that understanding. Bidirectional replication creates two replication forks that migrate farther and farther apart as replication proceeds. These forks eventually merge, as semiconservative replication of the entire chromosome is completed, at a termination region, called *ter*.

Later in this chapter we will see that in eukaryotes, each chromosome contains multiple points of origin.

11.2

DNA Synthesis in Bacteria Involves Five Polymerases, as Well as Other Enzymes

To say that replication is semiconservative and bidirectional describes the overall *pattern* of DNA duplication and the association of finished strands with one another once synthesis

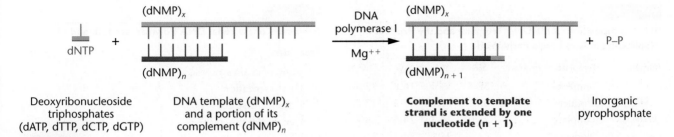

Deoxyribonucleoside triphosphates (dATP, dTTP, dCTP, dGTP)

DNA template (dNMP)$_x$ and a portion of its complement (dNMP)$_n$

Complement to template strand is extended by one nucleotide (n + 1)

Inorganic pyrophosphate

FIGURE 11-7   The chemical reaction catalyzed by DNA polymerase I. During each step, a single nucleotide is added to the growing complement of the DNA template, using a nucleoside triphosphate as the substrate. The release of inorganic pyrophosphate drives the reaction energetically.

is completed. However, it says little about the more complex issue of how the actual *synthesis* of long complementary polynucleotide chains occurs on a DNA template. Like most questions in molecular biology, this one was first studied using microorganisms. Research on DNA synthesis began about the same time as the Meselson–Stahl work, and the topic is still an active area of investigation. What is most apparent in this research is the tremendous complexity of the biological synthesis of DNA.

DNA Polymerase I

Studies of the enzymology of DNA replication were first reported by Arthur Kornberg and colleagues in 1957. They isolated an enzyme from *E. coli* that was able to direct DNA synthesis in a cell-free (*in vitro*) system. The enzyme is called **DNA polymerase I**, because it was the first of several similar enzymes to be isolated.

Kornberg determined that there were two major requirements for *in vitro* DNA synthesis under the direction

of DNA polymerase I: (1) all four deoxyribonucleoside triphosphates (dNTPs) and (2) template DNA. If any one of the four deoxyribonucleoside triphosphates was omitted from the reaction, no measurable synthesis occurred. If derivatives of these precursor molecules other than the nucleoside triphosphate were used (nucleotides or nucleoside diphosphates), synthesis also did not occur. If no template DNA was added, synthesis of DNA occurred but was reduced greatly.

Most of the synthesis directed by Kornberg's enzyme appeared to be exactly the type required for semiconservative replication. The reaction is summarized in Figure 11–7, which depicts the addition of a single nucleotide. The enzyme has since been shown to consist of a single polypeptide containing 928 amino acids.

The way in which each nucleotide is added to the growing chain is a function of the specificity of DNA polymerase I. As shown in Figure 11–8, the precursor dNTP contains the three phosphate groups attached to the 5'-carbon of

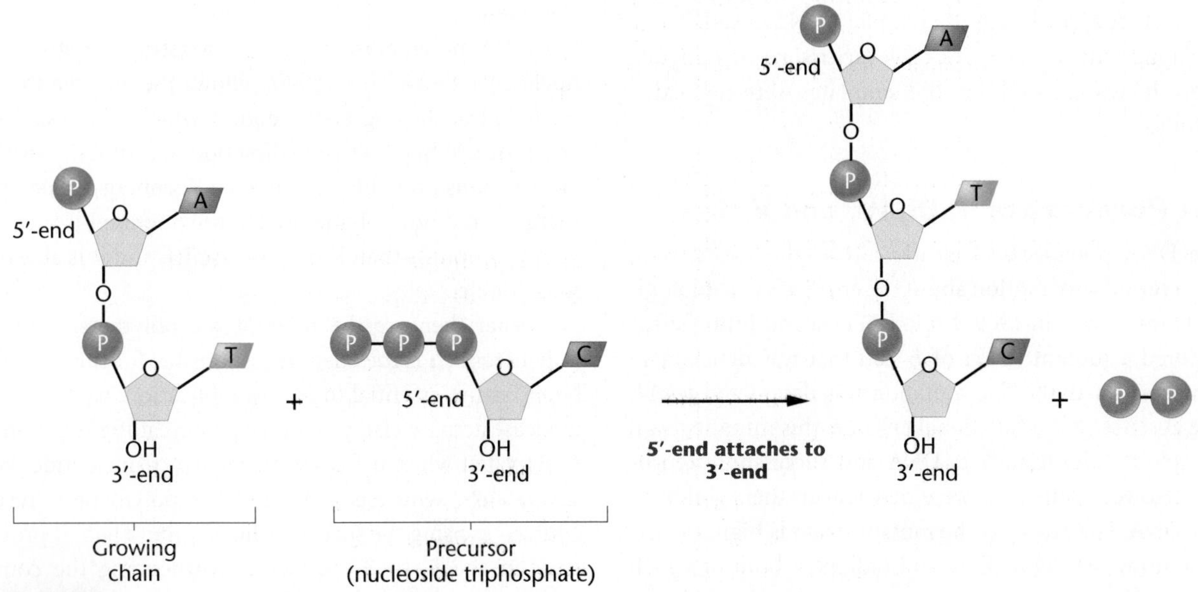

Growing chain

Precursor (nucleoside triphosphate)

5'-end attaches to 3'-end

FIGURE 11-8   Demonstration of 5' to 3' synthesis of DNA.

Base Composition of the DNA Template and the Product of Replication in Kornberg's Early Work

Organism	Template or Product	%A	%T	%G	%C
T2	Template	32.7	33.0	16.8	17.5
	Product	33.2	32.1	17.2	17.5
E. coli	Template	25.0	24.3	24.5	26.2
	Product	26.1	25.1	24.3	24.5
Calf	Template	28.9	26.7	22.8	21.6
	Product	28.7	27.7	21.8	21.8

Source: Kornberg (1960).

Properties of Bacterial DNA Polymerases I, II, and III

Properties	I	II	III
Initiation of chain synthesis	−	−	−
5′–3′ polymerization	+	+	+
3′–5′ exonuclease activity	+	+	+
5′–3′ exonuclease activity	+	−	−
Molecules of polymerase/cell	400	?	15

deoxyribose. As the two terminal phosphates are cleaved during synthesis, the remaining phosphate attached to the 5′-carbon is covalently linked to the 3′-OH group of the deoxyribose to which it is added. Thus, **chain elongation** occurs in the **5′ to 3′ direction** by the addition of one nucleotide at a time to the growing 3′ end. Each step provides a newly exposed 3′-OH group that can participate in the next addition of a nucleotide as DNA synthesis proceeds.

Having isolated DNA polymerase I and demonstrated its catalytic activity, Kornberg next sought to demonstrate the accuracy, or fidelity, with which the enzyme replicated the DNA template. Because technology for ascertaining the nucleotide sequences of the template and newly synthesized strand was not yet available in 1957, he initially had to rely on several indirect methods.

One of Kornberg's approaches was to compare the nitrogenous base compositions of the DNA template with those of the recovered DNA product. Table 11.1 shows Kornberg's base-composition analysis of three DNA templates. Within experimental error, the base composition of each product agreed with the template DNAs used. These data, along with other types of comparisons of template and product, suggested that the templates were replicated faithfully.

DNA Polymerase II, III, IV, and V

While DNA polymerase I clearly directs the synthesis of DNA, a serious reservation about the enzyme's true biological role was raised in 1969. Paula DeLucia and John Cairns discovered a mutant strain of E. coli that was deficient in polymerase I activity. The mutation was designated *polA1*. In the absence of the functional enzyme, this mutant strain of E. coli still duplicated its DNA and successfully reproduced. However, the cells were deficient in their ability to repair DNA. For example, the mutant strain is highly sensitive to ultraviolet light (UV) and radiation, both of which damage DNA and are mutagenic. Nonmutant bacteria are able to repair a great deal of UV-induced damage.

These observations led to two conclusions:

1. At least one other enzyme that is responsible for replicating DNA *in vivo* is present in *E. coli* cells.

2. DNA polymerase I serves a secondary function *in vivo*, now believed to be critical to the maintenance of fidelity of DNA synthesis.

To date, four other unique DNA polymerases have been isolated from cells lacking polymerase I activity and from normal cells that contain polymerase I. Table 11.2 contrasts several characteristics of DNA polymerase I with **DNA polymerase II** and **III**. Although none of the three can *initiate* DNA synthesis on a template, all three can *elongate* an existing DNA strand, called a **primer**.

All the DNA polymerase enzymes are large proteins exhibiting a molecular weight in excess of 100,000 Daltons (Da). All three possess 3′ to 5′ exonuclease activity, which means that they have the potential to polymerize in one direction and then pause, reverse their direction, and excise nucleotides just added. As we will discuss later in the chapter, this activity provides a capacity to proofread newly synthesized DNA and to remove and replace incorrect nucleotides.

DNA polymerase I also demonstrates 5′ to 3′ exonuclease activity. This activity allows the enzyme to excise nucleotides, starting at the end at which synthesis begins and proceeding in the same direction of synthesis. Two final observations probably explain why Kornberg isolated polymerase I and not polymerase III: polymerase I is present in greater amounts than is polymerase III, and it is also much more stable.

What then are the roles of the polymerases *in vivo*? Polymerase III is the enzyme responsible for the 5′ to 3′ polymerization essential to *in vivo* replication. Its 3′ to 5′ exonuclease activity also provides a proofreading function that is activated when it inserts an incorrect nucleotide. When this occurs, synthesis stalls and the polymerase "reverses course," excising the incorrect nucleotide. Then, it proceeds back in the 5′ to 3′ direction, synthesizing the complement of the template strand. Polymerase I is believed to be responsible for removing the primer, as well as for the synthesis

TABLE 11.3

Subunits of the DNA Polymerase III Holoenzyme

Subunit	Function	Groupings
α	5'–3' polymerization	Core enzyme: Elongates polynucleotide chain and proofreads
ϵ	3'–5' exonuclease	
θ	Core assembly	
γ		
δ		
δ'	Loads enzyme on template (serves as clamp loader)	γ complex
χ		
ψ		
β	Sliding clamp structure (processivity factor)	
τ	Dimerizes core complex	

that fills gaps produced after this removal. Its exonuclease activities also allow for its participation in DNA repair. Polymerase II, as well as **polymerase IV** and **V**, are involved in various aspects of repair of DNA that has been damaged by external forces, such as ultraviolet light. Polymerase II is encoded by a gene activated by disruption of DNA synthesis at the replication fork.

We end this section by emphasizing the complexity of the DNA polymerase III molecule. In contrast to DNA polymerase I, which is but a single polypeptide, the active form of DNA polymerase III is a **holoenzyme**—a complex enzyme made up of multiple subunits. Polymerase III consists of ten kinds of polypeptide subunits (Table 11.3) and has a molecular weight of 900,000 Da. The largest subunit, α, has a molecular weight of 140,000 Da and, along with subunits ε and θ, constitutes a **core enzyme** responsible for the polymerization activity. The α subunit is responsible for nucleotide polymerization on the template strands, whereas the ε subunit of the core enzyme possesses the 3' to 5' exonuclease activity. A single DNA polymerase III holoenzyme contains, along with other components, two core enzymes combined into a dimer.

A second group of five subunits (γ, δ, δ', χ, and ψ) forms what is called the γ complex, which is involved in "loading" the enzyme onto the template at the replication fork. This enzymatic function requires energy and is dependent on the hydrolysis of ATP. The β subunit serves as a "clamp" and prevents the core enzyme from falling off the template during polymerization. Finally, the τ subunit functions to dimerize two core polymerases, facilitating simultaneous synthesis of both strands of the helix at the replication fork. The holoenzyme and several other proteins at the replication fork together form a huge complex (nearly as large as a ribosome) known as the **replisome**. We consider the function of DNA polymerase III in more detail later in this chapter.

Many Complex Issues Must Be Resolved during DNA Replication

We have thus far established that in bacteria and viruses replication is semiconservative and bidirectional along a single replicon. We also know that synthesis is catalyzed by DNA polymerase III and occurs in the 5' to 3' direction. Bidirectional synthesis creates two replication forks that move in opposite directions away from the origin of synthesis. As we can see from the following list, many issues remain to be resolved in order to provide a comprehensive understanding of DNA replication:

1. The helix must undergo localized unwinding, and the resulting "open" configuration must be stabilized so that synthesis may proceed along both strands.

2. As unwinding and subsequent DNA synthesis proceed, increased coiling creates tension further down the helix, which must be reduced.

3. A primer of some sort must be synthesized so that polymerization can commence under the direction of DNA polymerase III. Surprisingly, RNA, not DNA, serves as the primer.

4. Once the RNA primers have been synthesized, DNA polymerase III begins to synthesize the DNA complement of both strands of the parent molecule. Because the two strands are antiparallel to one another, continuous synthesis in the direction that the replication fork moves is possible along only one of the two strands. On the other strand, synthesis must be discontinuous and thus involves a somewhat different process.

5. The RNA primers must be removed prior to completion of replication. The gaps that are temporarily created must be filled with DNA complementary to the template at each location.

6. The newly synthesized DNA strand that fills each temporary gap must be joined to the adjacent strand of DNA.

7. While DNA polymerases accurately insert complementary bases during replication, they are not perfect, and, occasionally, incorrect nucleotides are added to the growing strand. A proofreading mechanism that also corrects errors is an integral process during DNA synthesis.

As we consider these points, examine Figures 11–9, 11–10, 11–11, and 11–12 to see how each issue is resolved. Figure 11–13 summarizes the model of DNA synthesis.

Unwinding the DNA Helix

As discussed earlier, there is a single point of origin along the circular chromosome of most bacteria and viruses at which DNA synthesis is initiated. This region of the *E. coli* chromosome has been particularly well studied. Called *oriC*, it consists of 245 nucleotide pairs characterized by repeating sequences of 9 and 13 bases (called **9mers** and **13mers**). As shown in Figure 11–9, one particular protein, called **DnaA** (because it is encoded by the gene called *dnaA*), is responsible for the initial step in unwinding the helix. A number of subunits of the DnaA protein bind to each of several

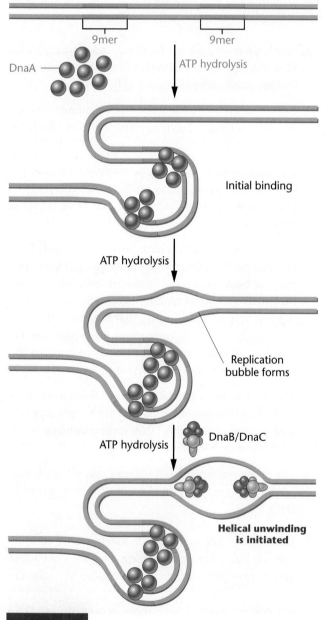

FIGURE 11–9   Helical unwinding of DNA during replication as accomplished by DnaA, DnaB, and DnaC proteins. Initial binding of many monomers of DnaA occurs at DNA sites containing repeating sequences of 9 nucleotides, called 9mers. Not illustrated are 13mers, which are also involved.

9mers. This step facilitates the subsequent binding of **DnaB** and **DnaC** proteins that further open and destabilize the helix. Proteins such as these, which require the energy supplied by the hydrolysis of ATP in order to break hydrogen bonds and denature the double helix, are called **helicases**. Other proteins, called **single-stranded binding proteins** (**SSBPs**), stabilize this open conformation.

As unwinding proceeds, a coiling tension is created ahead of the replication fork, often producing **supercoiling**. In circular molecules, supercoiling may take the form of added twists and turns of the DNA, much like the coiling you can create in a rubber band by stretching it out and then twisting one end. Such supercoiling can be relaxed by **DNA gyrase**, a member of a larger group of enzymes referred to as **DNA topoisomerases**. The gyrase makes either single- or double-stranded "cuts" and also catalyzes localized movements that have the effect of "undoing" the twists and knots created during supercoiling. The strands are then resealed. These various reactions are driven by the energy released during ATP hydrolysis.

Together, the DNA, the polymerase complex, and associated enzymes make up an array of molecules that participate in DNA synthesis and are part of what we have previously called the *replisome*.

Initiation of DNA Synthesis Using an RNA Primer

Once a small portion of the helix is unwound, what else is needed to initiate synthesis? As we have seen, DNA polymerase III requires a primer with a free 3′-hydroxyl group in order to elongate a polynucleotide chain. Since none is available in a circular chromosome, this absence prompted researchers to investigate how the first nucleotide could be added. It is now clear that RNA serves as the primer that initiates DNA synthesis.

A short segment of RNA (about 10 to 12 nucleotides long), complementary to DNA, is first synthesized on the DNA template. Synthesis of the RNA is directed by a form of RNA polymerase called **primase**, which does not require a free 3′ end to initiate synthesis. It is to this short segment of RNA that DNA polymerase III begins to add deoxyribonucleotides, initiating DNA synthesis. A conceptual diagram of initiation on a DNA template is shown in Figure 11–10. Later, the RNA primer is clipped out and replaced with DNA. This is thought to occur under the direction of DNA polymerase I. Recognized in viruses, bacteria, and several eukaryotic organisms, RNA priming is a universal phenomenon during the initiation of DNA synthesis.

Continuous and Discontinuous DNA Synthesis

We must now revisit the fact that the two strands of a double helix are **antiparallel** to each other—that is, one runs in the

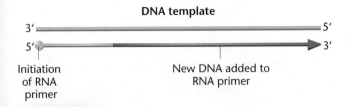

DNA template

3′ ━━━━━━━━━━━━━━━━━━ 5′
5′ ●━━━━━━━━━━━━━━━━━▶ 3′

Initiation   New DNA added to
of RNA   RNA primer
primer

FIGURE 11–10 The initiation of DNA synthesis. A complementary RNA primer is first synthesized, to which DNA is added. All synthesis is in the 5′ to 3′ direction. Eventually, the RNA primer is replaced with DNA under the direction of DNA polymerase I.

5′–3′ direction, while the other has the opposite 3′–5′ polarity. Because DNA polymerase III synthesizes DNA in only the 5′–3′ direction, synthesis along an advancing replication fork occurs in one direction on one strand and in the opposite direction on the other.

As a result, as the strands unwind and the replication fork progresses down the helix (Figure 11–11), only one strand can

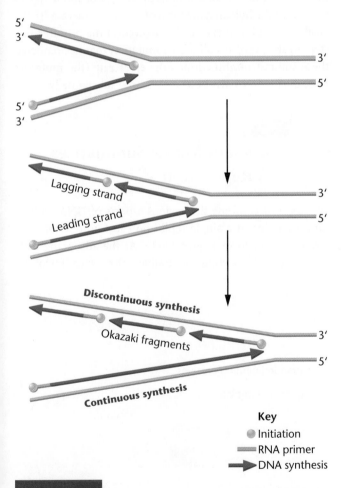

Key
● Initiation
━━ RNA primer
▶ DNA synthesis

FIGURE 11–11 Opposite polarity of synthesis along the two strands of DNA is necessary because they run antiparallel to one another, and because DNA polymerase III synthesizes in only one direction (5′ to 3′). On the lagging strand, synthesis must be discontinuous, resulting in the production of Okazaki fragments. On the leading strand, synthesis is continuous. RNA primers are used to initiate synthesis on both strands.

serve as a template for **continuous DNA synthesis**. This newly synthesized DNA is called the **leading strand**. As the fork progresses, many points of initiation are necessary on the opposite DNA template, resulting in **discontinuous DNA synthesis** of the **lagging strand**.*

Evidence supporting the occurrence of discontinuous DNA synthesis was first provided by Reiji and Tuneko Okazaki. They discovered that when bacteriophage DNA is replicated in *E. coli,* some of the newly formed DNA that is hydrogen bonded to the template strand is present as small fragments containing 1000 to 2000 nucleotides. RNA primers are part of each such fragment. These pieces, now called **Okazaki fragments**, are converted into longer and longer DNA strands of higher molecular weight as synthesis proceeds.

Discontinuous synthesis of DNA requires enzymes that both remove the RNA primers and unite the Okazaki fragments into the lagging strand. As we have noted, DNA polymerase I removes the primers and replaces the missing nucleotides. Joining the fragments is the work of **DNA ligase**, which is capable of catalyzing the formation of the phosphodiester bond that seals the nick between the discontinuously synthesized strands. The evidence that DNA ligase performs this function during DNA synthesis is strengthened by the observation of a ligase-deficient mutant strain (*lig*) of *E. coli,* in which a large number of unjoined Okazaki fragments accumulate.

Concurrent Synthesis Occurs on the Leading and Lagging Strands

Given the model just discussed, we might ask how DNA polymerase III synthesizes DNA on both the leading and lagging strands. Can both strands be replicated simultaneously at the same replication fork, or are the events distinct, involving two separate copies of the enzyme? Evidence suggests that both strands can be replicated simultaneously. As Figure 11–12 illustrates, if the lagging strand forms a loop, nucleotide polymerization can occur on both template strands under the direction of a dimer of the enzyme. After the synthesis of 1000 to 2000 nucleotides, the monomer of the enzyme on the lagging strand will encounter a completed Okazaki fragment, at which point it releases the lagging strand. A new loop is then formed with the lagging template strand, and the process is repeated. Looping inverts the orientation of the template but not the direction of actual synthesis on the lagging strand, which is always in the 5′ to 3′ direction.

Another important feature of the holoenzyme that facilitates synthesis at the replication fork is a dimer of the β subunit that forms a clamplike structure around the newly

─────────────

*Because DNA synthesis is continuous on one strand and discontinuous on the other, the term **semidiscontinuous synthesis** is sometimes used to describe the overall process.

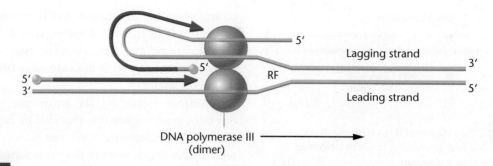

5'

Lagging strand

3'

RF

5'
3'

5'

5'

Leading strand

DNA polymerase III
(dimer)

FIGURE 11–12   Illustration of how concurrent DNA synthesis may be achieved on both the leading and lagging strands at a single replication fork (RF). The lagging template strand is "looped" in order to invert the physical direction of synthesis, but not the biochemical direction. The enzyme functions as a dimer, with each core enzyme achieving synthesis on one or the other strand.

formed DNA duplex. This β-subunit clamp prevents the **core enzyme** (the α, ε, and θ subunits that are responsible for catalysis of nucleotide addition) from falling off the template as polymerization proceeds. Because the entire holoenzyme moves along the parent duplex, advancing the replication fork, the θ-subunit dimer is often referred to as a *sliding clamp*.

Proofreading and Error Correction Occurs during DNA Replication

The immediate purpose of DNA replication is the synthesis of a new strand that is precisely complementary to the template strand at each nucleotide position. Although the action of DNA polymerases is very accurate, synthesis is not perfect and a noncomplementary nucleotide is occasionally inserted erroneously. To compensate for such inaccuracies, the DNA polymerases all possess 3' to 5' exonuclease activity. This property imparts the potential for them to detect and excise a mismatched nucleotide (in the 3' to 5' direction).

Once the mismatched nucleotide is removed, 5' to 3' synthesis can again proceed. This process, called **proofreading**, increases the fidelity of synthesis by a factor of about 100. In the case of the holoenzyme form of DNA polymerase III, the epsilon (ε) subunit is directly involved in the proofreading step. In strains of *E. coli* with a mutation that has rendered the ε subunit nonfunctional, the error rate (the mutation rate) during DNA synthesis is increased substantially.

11.4

A Coherent Model Summarizes DNA Replication

We can now combine the various aspects of DNA replication occurring at a single replication fork into a coherent model, as shown in Figure 11–13. At the advancing fork, a helicase is unwinding the double helix. Once unwound,

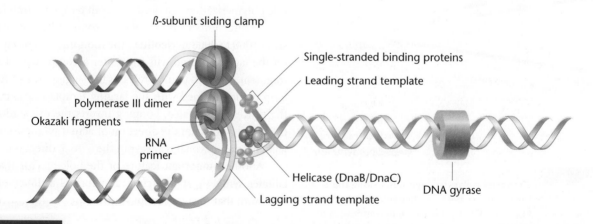

ß-subunit sliding clamp

Single-stranded binding proteins

Leading strand template

Polymerase III dimer

Okazaki fragments

RNA
primer

Helicase (DnaB/DnaC)

DNA gyrase

Lagging strand template

FIGURE 11–13   Summary of DNA synthesis at a single replication fork. Various enzymes and proteins essential to the process are shown.

single-stranded binding proteins associate with the strands, preventing the re-formation of the helix. In advance of the replication fork, DNA gyrase functions to diminish the tension created as the helix supercoils. Each half of the dimeric polymerase is a core enzyme bound to one of the template strands by a β-subunit sliding clamp. Continuous synthesis occurs on the leading strand, while the lagging strand must loop around in order for simultaneous (concurrent) synthesis to occur on both strands. Not shown in the figure, but essential to replication on the lagging strand, is the action of DNA polymerase I and DNA ligase, which together replace the RNA primers with DNA and join the Okazaki fragments, respectively.

Because the investigation of DNA synthesis is still an extremely active area of research, this model will no doubt be extended in the future. In the meantime, it provides a summary of DNA synthesis against which genetic phenomena can be interpreted.

NOW SOLVE THIS

11–2 An alien organism was investigated. When DNA replication was studied, a unique feature was apparent: No Okazaki fragments were observed. Create a model of DNA that is consistent with this observation.

■ HINT: *This problem involves an understanding of the process of DNA synthesis in prokaryotes, as depicted in Figure 11–13. The key to its solution is to consider why Okazaki fragments are observed during DNA synthesis and how their formation relates to DNA structure, as described in the Watson–Crick model.*

11.5

Replication Is Controlled by a Variety of Genes

Much of what we know about DNA replication in viruses and bacteria is based on genetic analysis of the process. For example, we have already discussed studies involving the *polA1* mutation, which revealed that DNA polymerase I is not the major enzyme responsible for replication. Many other mutations interrupt or seriously impair some aspect of replication, such as the ligase-deficient and the proofreading-deficient mutations mentioned previously. Because such mutations are lethal ones, genetic analysis frequently uses **conditional mutations**, which are expressed under one condition but not under a different condition. For example, a **temperature-sensitive mutation** may not be expressed at a particular *permissive* temperature. When mutant cells are grown at a *restrictive* temperature, the

TABLE 11.4

Some of the Various *E. coli* Genes and Their Products or Role in Replication

Gene	Product or Role
polA	DNA polymerase I
polB	DNA polymerase II
dnaE, N, Q, X, Z	DNA polymerase III subunits
dnaG	Primase
dnaA, I, P	Initiation
dnaB, C	Helicase at *oriC*
gyrA, B	Gyrase subunits
lig	DNA ligase
rep	DNA helicase
ssb	Single-stranded binding proteins
rpoB	RNA polymerase subunit

mutant phenotype is expressed and can be studied. By examining the effect of the loss of function associated with the mutation, the investigation of such temperature-sensitive mutants can provide insight into the product and the associated function of the normal, nonmutant gene.

As shown in Table 11.4, a variety of genes in *E. coli* specify the subunits of the DNA polymerases and encode products involved in specification of the origin of synthesis, helix unwinding and stabilization, initiation and priming, relaxation of supercoiling, repair, and ligation. The discovery of such a large group of genes attests to the complexity of the process of replication, even in the relatively simple prokaryote. Given the enormous quantity of DNA that must be unerringly replicated in a very brief time, this level of complexity is not unexpected. As we will see in the next section, the process is even more involved and therefore more difficult to investigate in eukaryotes.

11.6

Eukaryotic DNA Replication Is Similar to Replication in Prokaryotes, But Is More Complex

Eukaryotic DNA replication shares many features with replication in bacteria. In both systems, double-stranded DNA is unwound at replication origins, replication forks are formed, and bidirectional DNA synthesis creates leading and lagging strands from single-stranded DNA templates under the direction of DNA polymerase. Eukaryotic polymerases have the same fundamental requirements for DNA synthesis as do bacterial polymerases: four deoxyribonucleoside triphosphates, a template, and a primer. However, eukaryotic DNA replication is more complex, due to several features of eukaryotic DNA. Eukaryotic cells contain much more DNA, this DNA is complexed with nucleosomes, and eukaryotic chromosomes are linear rather than circular. In this section, we will describe some of the ways in which eukaryotes deal with this added complexity.

Initiation at Multiple Replication Origins

The most obvious difference between eukaryotic and prokaryotic DNA replication is that eukaryotic replication must deal with greater amounts of DNA. For example, yeast cells contain 3 times as much DNA, and *Drosophila* cells contain 40 times as much as *E. coli* cells. In addition, eukaryotic DNA polymerases synthesize DNA at a rate 25 times slower (about 2000 nucleotides per minute) than that in prokaryotes. Under these conditions, replication from a single origin on a typical eukaryotic chromosome would take days to complete. However, replication of entire eukaryotic genomes is usually accomplished in a matter of minutes to hours.

To facilitate the rapid synthesis of large quantities of DNA, eukaryotic chromosomes contain multiple replication origins. Yeast genomes contain between 250 and 400 origins, and mammalian genomes have as many as 25,000. Multiple origins are visible under the electron microscope as "replication bubbles" that form as the DNA helix opens up, each bubble providing two potential replication forks (Figure 11–14). Origins in yeast, called **autonomously replicating sequences (ARSs)**, consist of approximately 120 base pairs containing a **consensus sequence** (meaning a sequence that is the same, or nearly the same, in all yeast ARSs) of 11 base pairs. Origins in mammalian cells appear to be unrelated to specific sequence motifs and may be defined more by chromatin structure over a 6–55 kb region.

Eukaryotic replication origins not only act as sites of replication initiation, but also control the timing of DNA

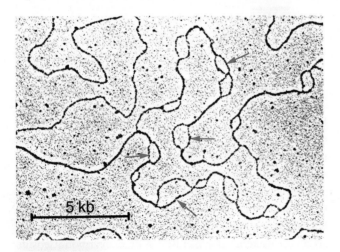

FIGURE 11–14   A demonstration of the multiple origins of replication along a eukaryotic chromosome. Each origin is apparent as a replication bubble along the axis of the chromosome. Arrows identify some of these replication bubbles.

replication. These regulatory functions are carried out by a complex of more than 20 proteins, called the **prereplication complex (pre-RC)**, which assembles at replication origins. In the early G1 phase of the cell cycle, replication origins are recognized by a six-protein complex known as an **origin recognition complex (ORC)**, which tags the origin as a site of initiation. Throughout the G1 phase of the cell cycle, other proteins associate with the ORC to form the pre-RC. The presence of a pre-RC at an origin "licenses" that origin for replication. Once DNA polymerases initiate synthesis at the origin, the pre-RC is disrupted and does not reassemble again until the G1 phase of the next cell cycle. This is an important mechanism because it distinguishes segments of DNA that have completed replication from segments of unreplicated DNA, thus maintaining orderly and efficient replication. It ensures that replication occurs only once along each stretch of DNA during each cell cycle.

The initiation of DNA replication is also regulated at the pre-RC. A number of cell-cycle kinases that phosphorylate replication proteins, along with helicases that unwind DNA, associate with the pre-RC and are essential for initiation. The kinases are activated in S phase, at which time they phosphorylate other proteins that trigger the initiation of DNA replication. The end result is the unwinding of DNA at the replication forks, the stabilization of single-stranded DNA, the association of DNA polymerases with the origins, and the initiation of DNA synthesis.

Multiple Eukaryotic DNA Polymerases

To accommodate the increased number of replicons, eukaryotic cells contain many more DNA polymerase molecules than do bacterial cells. For example, a single *E. coli* cell contains about 15 molecules of DNA polymerase III, but mammalian cells contain tens of thousands of DNA polymerase molecules.

Eukaryotes also utilize a larger number of different DNA polymerase types than do prokaryotes. The human genome contains genes that encode at least 14 different DNA polymerases, only three of which are involved in the majority of nuclear genome DNA replication. The nomenclature and characteristics of these DNA polymerases are summarized in Table 11.5.

Pol α, δ, and ε are the major forms of the enzyme involved in initiation and elongation during eukaryotic nuclear DNA synthesis, so we will concentrate our discussion on these. Two of the four subunits of the **Pol α enzyme** synthesize RNA primers on both the leading and lagging strands. After the RNA primer reaches a length of about 10 ribonucleotides, another subunit adds 20 to 30 complementary deoxyribonucleotides. Pol α is said to possess low **processivity**, a term that refers to the strength of the association between the enzyme and its substrate, and thus the length of DNA that

TABLE 11.5

Properties of Eukaryotic DNA Polymerases

DNA Polymerase	Subunits	3'–5' Exonuclease	Function
α (alpha)	4	No	RNA/DNA primers, initiation of DNA synthesis
δ (delta)	4	Yes	Lagging strand synthesis, DNA repair, proofreading
ε (epsilon)	4	Yes	Leading strand synthesis, proofreading
γ (gamma)	2	Yes	Mitochondrial DNA replication and repair
β (beta)	1	No	Base-excision DNA repair
η (eta)	1	No	Translesion DNA synthesis
ζ (zeta)	2	No	Translesion DNA synthesis
κ (kappa)	1	No	Translesion DNA synthesis
ι (iota)	1	No	Translesion DNA synthesis
θ (theta)	1	No	DNA repair
λ (lambda)	1	No	DNA repair
μ (mu)	1	No	DNA repair
υ (nu)	1	No	Unknown
Rev 1	1	No	DNA repair

is synthesized before the enzyme dissociates from the template. Once the primer is in place, an event known as **polymerase switching** occurs, whereby Pol α dissociates from the template and is replaced by Pol δ and ε. These enzymes extend the primers on opposite strands of DNA, possess much greater processivity, and exhibit 3' to 5' exonuclease activity, thus having the potential to proofread. Pol ε synthesizes DNA on the leading strand, and Pol δ synthesizes the lagging strand. Both Pol δ and ε participate in other DNA synthesizing events in the cell, including several types of DNA repair and recombination. All three DNA polymerases are essential for viability.

As in prokaryotic DNA replication, the final stages in eukaryotic DNA replication involve replacing the RNA primers with DNA and ligating the Okazaki fragments on the lagging strand. In eukaryotes, the Okazaki fragments are about ten times smaller (100 to 150 nucleotides) than in prokaryotes.

Included in the remainder of DNA-replicating enzymes is Pol γ, which is found exclusively in mitochondria, synthesizing the DNA present in that organelle. DNA polymerases are involved in DNA repair and replication through regions of the DNA template that contain damage or distortions (called **translesion synthesis**, or **TLS**). Although these translesion DNA polymerases are less faithful in copying DNA and thus make more errors than Pol α, δ, and γ, they are able to bypass the distortions, leaving behind gaps that may then be repaired.

Replication through Chromatin

One of the major differences between prokaryotic and eukaryotic DNA is that eukaryotic DNA is complexed with DNA-binding proteins, existing in the cell as *chromatin*.

As we will discuss in Chapter 12, chromatin consists of regularly repeating units called nucleosomes, each of which consists of about 200 base pairs of DNA complexed with eight histone protein molecules (**Figure 11–15**). Before polymerases can begin synthesis, nucleosomes and other DNA-binding proteins must be stripped away or otherwise modified to allow the passage of replication proteins. As DNA synthesis proceeds, the histones and nonhistone proteins must rapidly reassociate with the newly formed duplexes, reestablishing the characteristic nucleosome pattern.

In order to re-create nucleosomal chromatin on replicated DNA, the synthesis of new histone proteins is tightly coupled to DNA synthesis during the S phase of the cell cycle. Research data suggest that nucleosomes are disrupted just ahead of the replication fork and that the preexisting histone proteins can assemble with newly synthesized histone proteins into new nucleosomes. The new nucleosomes are assembled behind the replication fork, onto the two daughter strands of DNA. The assembly of new nucleosomes is carried out by **chromatin assembly factors (CAFs)** that move along with the replication fork.

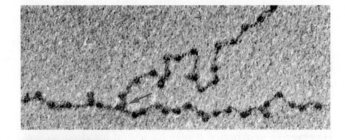

FIGURE 11–15   An electron micrograph of a eukaryotic replicating fork demonstrating the presence of histone-protein-containing nucleosomes on both branches.

The Ends of Linear Chromosomes Are Problematic during Replication

A final difference between prokaryotic and eukaryotic DNA synthesis stems from the structural differences in their chromosomes. Unlike the closed, circular DNA of bacteria and most bacteriophages, eukaryotic chromosomes are linear. During replication, a special problem arises at the "ends" of these linear molecules.

Eukaryotic chromosomes end in distinctive sequences called **telomeres** that help preserve the integrity and stability of the chromosome. Telomeres are necessary because the double-stranded "ends" of DNA molecules at the termini of linear chromosomes potentially resemble the **double-stranded breaks (DSBs)** that can occur when a chromosome becomes fragmented internally. In such cases, the double-stranded loose ends can fuse to other such ends; if they don't fuse, they are vulnerable to degradation by nucleases. Either outcome can lead to problems. Telomeres are believed to create inert chromosome ends, protecting intact eukaryotic chromosomes from improper fusion or degradation.

Telomere Structure

We could speculate that there must be something unique about the DNA sequence or the proteins that bind to it that confers this protective property to telomeres. Indeed, this has been shown to be the case. First discovered by Elizabeth Blackburn and Joe Gall in their study of micronuclei—the smaller of two nuclei in the ciliated protozoan *Tetrahymena*—the DNA at the protozoan's chromosome ends consists of the short tandem repeating sequence TTGGGG. This sequence is present many times on one of the two helical strands making up each telomere. This strand is referred to as the G-rich strand, in contrast to its complementary strand, the so-called C-rich strand, which displays the repeated sequence AACCCC. In a similar way, all vertebrates contain the sequence TTAGGG at the ends of G-rich strands, repeated several thousand times in somatic cells. Since each linear chromosome ends with two helical DNA strands running antiparallel to one another, one strand has a 3'-ending and the other has a 5'-ending. It is the 3'-strand that is the G-rich one. This has special significance during telomere replication.

But first, let's describe how this tandemly repeated DNA confers inertness to the chromosome ends. One model is based on the discovery that the 3'-ending G-rich strand extends as an overhang, lacking a complement, and thus forms a single-stranded tail at the terminus of each telomere. In *Tetrahymena,* this tail is only 12 to 16 nucleotides long. However, in vertebrates, it may be several hundred nucleotides long. The final conformation of these tails has been correlated with chromosome inertness. Though not considered

complementary in the same way as A-T and G-C base pairs are, G-containing nucleotides are nevertheless capable of base pairing with one another when several are aligned opposite another G-rich sequence. Thus, the G-rich single-stranded tails are capable of looping back on themselves, forming multiple G-G hydrogen bonds to create what are referred to as **G-quartets**. The resulting loops at the chromosome ends (called **t-loops**) are much like those created when you tie your shoelaces into a bow. It is believed that these structures, in combination with specific proteins that bind to them, essentially close off the ends of chromosomes and make them inert.

Replication at the Telomere

Now let's consider the problem that semiconservative replication poses at the end of a double-stranded DNA molecule. Although 5' to 3' synthesis on the leading-strand template may proceed to the end, a difficulty arises on the lagging strand once the final RNA primer is removed (Figure 11–16).

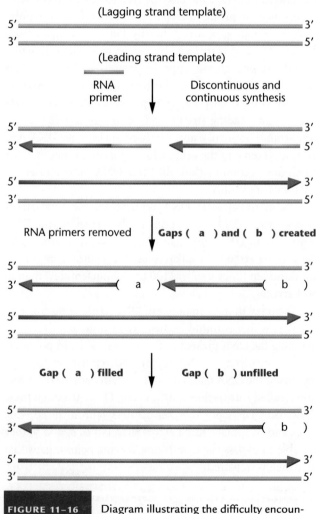

FIGURE 11–16   Diagram illustrating the difficulty encountered during the replication of the ends of linear chromosomes. A gap (b) is left following synthesis on the lagging strand.

Normally, the newly created gap would be filled in starting with the addition of a nucleotide to the adjacent 3'-OH group [the group to the right of gap (a) in Figure 11–16]. However, since the final gap [gap (b) in Figure 11–16] is at the end of the strand being synthesized, there is no Okazaki fragment present to provide the needed 3'-OH group. Thus, in the situation depicted in Figure 11–16, a gap remains on the lagging strand produced in each successive round of synthesis, shortening the double-stranded end of the chromosome by the length of the RNA primer. With each round of replication, the shortening becomes more severe in each daughter cell, *eventually extending beyond the telomere and potentially deleting gene-coding regions.*

A unique eukaryotic enzyme called **telomerase**, first discovered by Elizabeth Blackburn and Carol Greider in studies of *Tetrahymena,* has helped us understand the solution to the problem of telomere shortening. As noted earlier, telomeric DNA in eukaryotes is always found to consist of many short, repeated nucleotide sequences, with the G-rich strand overhanging in the form of a single-stranded tail. In *Tetrahymena* the tail contains several repeats of the sequence 5'-TTGGGG-3'. As we will see, telomerase is capable of adding several more repeats of this six-nucleotide sequence to the 3'-end of the G-rich strand (using 5'–3' synthesis). Detailed investigation by Blackburn and Greider of how the *Tetrahymena* telomerase enzyme accomplishes this synthesis yielded an extraordinary finding. The enzyme is highly unusual in that it is a *ribonucleoprotein,* containing within its molecular structure a short piece of RNA that is essential to its catalytic activity. The RNA component serves as both a "guide" (to proper attachment of the enzyme to the telomere) and a "template" for the synthesis of its DNA complement, the latter being a process called **reverse transcription**. In *Tetrahymena,* the RNA contains the sequence AACCCCAAC, within which is found the complement of the repeating telomeric DNA sequence that must be synthesized (TTGGGG).

Figure 11–17 shows one model of how researchers envision the enzyme working. Part of the RNA sequence of the enzyme (shown in green) base-pairs with the ending sequence of the single-stranded overhanging DNA, while the remainder of the RNA extends beyond the overhang. Next, reverse transcription of this extending RNA sequence—synthesizing DNA on an RNA template—extends the length of the G-rich lagging strand. It is believed that the enzyme is then translocated toward the (newly formed) end of the

(a) Telomerase binds to 3′ G-rich tail

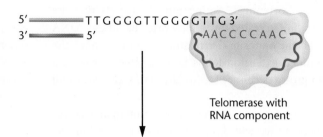

Telomerase with
RNA component

(b) Telomeric DNA is synthesized on G-rich tail

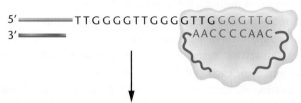

(c) Telomerase is translocated and steps (a) and (b) are repeated

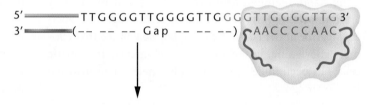

(d) Telomerase released; primase and DNA polymerase fill gap

Gap   Primer

(e) Primer removed; gap sealed by DNA ligase

Gap sealed

FIGURE 11–17   The predicted solution to the problem posed in Figure 11–16. The enzyme telomerase (with its RNA component shown in green) directs synthesis of repeated TTGGGG sequences, resulting in the formation of an extended 3'-overhang. This facilitates DNA synthesis on the opposite strand, filling in the gap that would otherwise have been created on the ends of linear chromosomes during each replication cycle.

strand, and the same events are repeated, continuing the extension process.

At this point, if conventional DNA synthesis then ensues using the overhang as a template and involving primase, DNA polymerase, and DNA ligase, most of the original gap is filled [Figure 11–17 (d) and (e)]. When the primase is removed a small gap remains. However, it is now found well beyond the original end of the chromosome, thus preventing any shortening. Another model suggests that the DNA extension, created by telomerase, facilitates DNA synthesis on the opposite C-rich strand. In this model, the single-stranded extension loops back on itself, providing the 3′-OH group necessary for initiation of synthesis to fill the gap.

Telomerase function has now been found in all eukaryotes studied. As we will discuss in Chapter 12, telomeric DNA sequences have been highly conserved throughout evolution, reflecting the critical function of telomeres. As mentioned earlier, in humans, the telomeric DNA sequence on the lagging strand that is repeated is 5′-TTAGGG-3′, differing from *Tetrahymena* by only one nucleotide.

In most eukaryotic somatic cells, telomerase is not active, and thus, with each cell division, the telomeres of each chromosome do shorten. After many divisions, the telomere may be seriously eroded, causing the cell to lose the capacity for further division. Malignant cells, on the other hand, maintain telomerase activity and in this way are immortalized. In the "Genetics, Technology, and Society" feature at the end of this chapter, we will see that telomere shortening, in the absence of telomerase in somatic cells, has been linked to a molecular mechanism involved in cellular aging.

11.8

DNA Recombination, Like DNA Replication, Is Directed by Specific Enzymes

We now turn to a topic previously discussed in Chapter 5—genetic recombination. There, we pointed out that the process of crossing over between homologs depends on the breakage and rejoining of the DNA strands, and results in the exchange of genetic information between DNA molecules. Now that we have discussed the chemistry and replication of DNA, we can consider how recombination occurs at the molecular level. In general, our discussion pertains to genetic exchange between any two homologous double-stranded DNA molecules, whether they are viral or bacterial chromosomes or eukaryotic homologs during meiosis. Genetic exchange at equivalent positions along two chromosomes with substantial DNA sequence homology is referred to as **general**, or **homologous recombination**.

Several models attempt to explain homologous recombination, but they all have certain features in common. First, all are based on proposals first put forth independently by Robin Holliday and Harold L. K. Whitehouse in 1964. Second, they all depend on the complementarity between DNA strands to explain the precision of the exchange. Finally, each model relies on a series of enzymatic processes in order to accomplish genetic recombination.

One such model is shown in Figure 11–18. It begins with two paired DNA duplexes, or homologs [Step (a)], in each of which an endonuclease introduces a single-stranded nick at an identical position [Step (b)]. The internal strand endings produced by these cuts are then displaced and subsequently pair with their complements on the opposite duplex [Step (c)]. Next, a ligase seals the loose ends [Step (d)], creating hybrid duplexes called **heteroduplex DNA molecules**, held together by a cross-bridge structure. The position of this cross bridge can then move down the chromosome by a process referred to as **branch migration** [Step (e)], which occurs as a result of a zipper-like action as hydrogen bonds are broken and then re-formed between complementary bases of the displaced strands of each duplex. This migration yields an increased length of heteroduplex DNA on both homologs.

If the duplexes bend [Step (f)] and the bottom portion shown in the figure rotates 180° [Step (g)], an intermediate planar structure called a χ (chi) form—or **Holliday structure**—is created. If the two strands on opposite homologs previously uninvolved in the exchange are now nicked by an endonuclease [Step (h)] and ligation occurs as in Step (i), two recombinant duplexes are created. Note that the arrangement of alleles is altered as a result of this recombination.

Whereas the model above involves *single-stranded breaks*, other recombination models have been proposed that involve *double-stranded breaks* in one of the DNA double helices. In these models, endonucleases remove nucleotides at the breakpoint, creating 3′ overhangs on each strand. One of the broken strands invades the intact double helix of the other homolog, and both strands line up with the intact homolog. DNA repair synthesis then fills all gaps, and two Holliday junctions are formed. Endonuclease cleavages and ligations finalize the exchange. The end result is the same as our original model: genetic exchange occurs during crossing over in meiotic tetrads. A similar mechanism is thought to occur when cells repair double-stranded breaks in chromosomes. Such damage can occur from numerous causes, including the energy of ionizing radiation. We discuss this topic again in Chapter 15.

As with DNA replication, the processes involved in DNA recombination require the activities of numerous enzymes and other proteins. Mutations in genes encoding these proteins may cause defects in recombination, as well as in DNA repair and replication. One of the key proteins involved in *E. coli* recombination is the **RecA protein**. This molecule

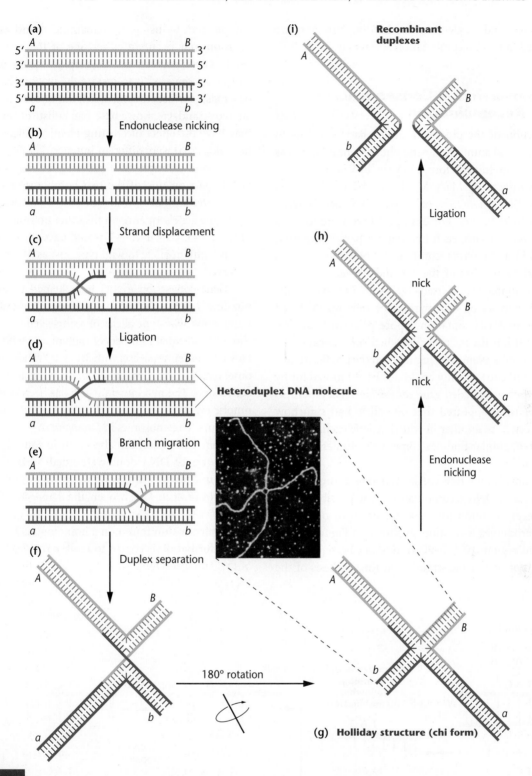

FIGURE 11–18   Model depicting how genetic recombination can occur as a result of the breakage and rejoining of heterologous DNA strands. Each stage is described in the text. The electron micrograph shows DNA in a χ-form structure similar to the diagram in (g); the DNA is an extended Holliday structure, derived from the *Col*E1 plasmid of *E. coli*. *David Dressler, Oxford University, England.*

promotes the exchange of reciprocal single-stranded DNA molecules as occurs in Step (c) of the model. RecA also enhances the hydrogen-bond formation during strand displacement, thus initiating heteroduplex formation. The

RecB, RecC, and **RecD proteins** can cleave DNA strands and help unwind the duplex. Other proteins are involved in branch migration and resolution of Holliday structures. DNA replication proteins, such as DNA polymerases, DNA ligase, gyrases, and single-stranded binding proteins, are also involved in DNA recombination and repair.

Gene Conversion, a Consequence of DNA Recombination

A modification of the preceding model has helped us to better understand a unique genetic phenomenon known as **gene conversion**. Initially found in yeast by Carl Lindegren and in *Neurospora* by Mary Mitchell, gene conversion is characterized by a *nonreciprocal* genetic exchange between two closely linked genes. For example, if we were to cross two *Neurospora* strains, each bearing a separate mutation ($a + \times + b$), a *reciprocal* recombination between the genes would yield spore pairs of the ++ and the *ab* genotypes. However, a nonreciprocal exchange yields one pair without the other. Working with pyridoxine mutants, Mitchell observed several asci-containing spore pairs with the ++ genotype, but not the reciprocal product (*ab*). Because the frequency of these events was higher than the predicted mutation rate and consequently could not be accounted for by mutation, they were called *gene conversions*. They were so named because it appeared that one allele had somehow been "converted" to another in which genetic exchange had also occurred. Similar findings come from studies of other fungi as well.

Gene conversion is now considered to be a consequence of the process of DNA recombination. One possible explanation interprets conversion as a mismatch of base pairs during heteroduplex formation, as shown in Figure 11–19. Mismatched regions of hybrid strands can be repaired by the excision of one of the strands and the synthesis of the

complement by using the remaining strand as a template. Excision may occur in either one of the strands, yielding two possible "corrections." One repairs the mismatched base pair and "converts" it to restore the original sequence. The other also corrects the mismatch but does so by copying the altered strand, creating a base-pair substitution. Conversion may have the effect of creating identical alleles on the two homologs that were different initially.

In our example in Figure 11–19, suppose that the G≡C pair on one of the two homologs was responsible for the mutant allele, while the A═T pair was part of the wild-type gene sequence on the other homolog. Conversion of the G≡C pair to A═T would have the effect of changing the mutant allele to wild type, just as Mitchell originally observed.

Gene-conversion events have helped to explain other puzzling genetic phenomena in fungi. For example, when mutant and wild-type alleles of a single gene are crossed, asci should yield equal numbers of mutant and wild-type spores. However, exceptional asci with 3:1 or 1:3 ratios are sometimes observed. These ratios can be understood in terms of gene conversion. The phenomenon also has been detected during mitotic events in fungi, as well as during the study of unique compound chromosomes in *Drosophila*.

Gene conversion can also occur in somatic cells during the process of DNA double-stranded break repair, as discussed in Chapter 15. When the backbones of both strands of a DNA molecule are broken, the damaged strands are degraded at the break sites and repaired by gene conversion using information found on a homologous DNA molecule. This information is copied to replace the degraded portion of the damaged DNA.

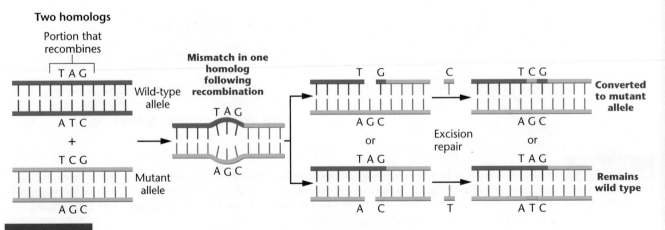

FIGURE 11–19   A proposed mechanism that accounts for the phenomenon of gene conversion during recombination in meiosis. A base-pair mismatch occurs in a recombining homolog (owing to the presence of a mutant allele) during heteroduplex formation. During excision repair, one of the two mismatched bases is removed and the complement is synthesized. In one case, the mutant base pair is preserved. When it is subsequently included in a recombinant spore, the mutant genotype will be maintained. In the other case, the mutant base pair is converted to the wild-type sequence. When included in a recombinant spore, the wild-type genotype will be expressed, leading to a nonreciprocal exchange ratio.

Telomeres: The Key to Immortality?

Humans, like all multicellular organisms, grow old and die. As we age, our immune systems become less efficient, wound healing is impaired, and tissues lose resilience. It has always been a mystery why we go through these age-related declines and why each species has a characteristic finite life span. Why do we grow old? Can we reverse this march to mortality? Some recent discoveries suggest that the answers to these questions may lie at the ends of our chromosomes.

The study of human aging begins with a study of human cells growing in culture dishes. Like the organisms from which the cells are taken, cells in culture have a finite life span. This *replicative senescence* was first noted by Leonard Hayflick in the 1960s. He reported that normal human fibroblasts lose their ability to grow and divide after about 50 cell divisions. These senescent cells remain metabolically active but can no longer proliferate. Eventually, they die. Although we don't know whether cellular senescence directly causes organismal aging, the evidence is suggestive. For example, cells derived from young people undergo more divisions than those from older people; cells from short-lived species stop growing after fewer divisions than those from longer-lived species; and cells from patients with premature aging syndromes undergo fewer divisions than those from normal patients.

Another characteristic of aging cells involves their telomeres. In most mammalian somatic cells, telomeres become shorter with each DNA replication because DNA polymerase cannot synthesize new DNA at the ends of each parent strand. However, as discussed in detail in this chapter, cells that undergo extensive proliferation, like embryonic cells, germ cells, and adult stem cells, maintain their telomere length by using *telomerase*—a remarkable RNA-containing enzyme that adds telomeric DNA sequences onto the ends of linear chromosomes. However, most somatic cells in adult organisms do not proliferate and do not contain active telomerase.

Could we gain perpetual youth and vitality by increasing our telomere lengths? Studies suggest that it may be possible to reverse senescence by artificially increasing the amount of telomerase in our cells. When investigators introduced cloned telomerase genes into normal human cells in culture, telomeres lengthened, and the cells continued to grow past their typical senescence point. These studies suggest that some of the atrophy of tissues that accompanies old age could be reversed by activating telomerase genes. However, before we use telomerase to achieve immortality, we need to consider a potential serious side effect: cancer.

Although normal cells shorten their telomeres and undergo senescence after a specific number of cell divisions, cancer cells do not. More than 80 percent of human tumor cells contain telomerase activity, maintain telomeres, and achieve immortality. Those that do not contain active telomerase use a less well understood mechanism known as ALT (for "alternative lengthening of telomeres").

These observations have motivated scientists to devise new cancer therapies based on the idea that agents that inhibit telomerase might destroy cancer cells by allowing telomeres to shorten, thereby forcing the cells into senescence. Because most normal human cells do not express telomerase, such a therapy might target tumor cells and be less toxic than most current anticancer drugs. Many such anti-telomerase drugs are currently under development, and some are in clinical trials.

Will a deeper understanding of telomeres allow us to both arrest cancers *and* reverse the descent into old age? Time will tell.

Your Turn

Take time, individually or in groups, to answer the following questions. Investigate the references and links to help you understand some of the research on telomeres, aging, and cancer.

1. How might our knowledge about telomeres and telomerase be applied to anti-aging strategies? Are such strategies or therapies being developed?

   Sources of information can be obtained by using the PubMed Web site (http://www.ncbi.nlm.nih.gov/sites/entrez?db=PubMed).

2. One anti-telomerase drug, called GRN163L, is being developed by Geron Corporation as a treatment for cancer. How does GRN163L work? What is the current status of GRN163L clinical trials? What are some possible side effects for anti-telomerase drugs?

   Read about this drug and its clinical trials on the Geron Web site at http://www.geron.com. *Search on PubMed for scientific papers dealing with GRN163L's anticancer effects.*

3. People suffering from chronic stress appear to have more health problems and to age prematurely. Is there any evidence that chronic stress, poor health, and telomere length are linked? How might stress affect telomere length or vice versa?

   Some recent papers suggest how these phenomena may be linked. One such paper is Epel, E. S., et al. 2004. Accelerated telomere shortening in response to life stress. *Proc. Natl. Acad. Sci. USA* 101(49): 17312–17315.

4. In 2006, the Lasker Award for Basic Medical Research was awarded to Drs. Elizabeth Blackburn, Carol Greider, and Jack Szostak, who subsequently were awarded the 2009 Nobel Prize in Physiology or Medicine. How did the intersections of people, ideas, and good fortune lead to their discovery of telomerase and its role in aging and cancer? What is the future for this research?

   Listen to interviews with these scientists, in which they tell their stories about their research and where they see the field going, at http://www.laskerfoundation.org/2006videoawards.

CASE STUDY | At loose ends

A researcher was asked if his work on the genetic control of human telomere replication was related to any genetic disorders. He replied that one might think that any mutations involving replication would be lethal during early development, and thus unavailable for study. But, in fact, a rare human genetic disorder affecting telomeres is known. This disorder, dyskeratosis congenita (DKC), is associated with mutations in the protein subunits of telomerase, the enzyme responsible for replicating the ends of eukaryotic chromosomes. Initial symptoms appear in tissues derived from rapidly dividing cells, including the skin, nails, and bone marrow, and first affect children between the ages of 5 and 15 years.

This disorder raises several interesting questions.

1. How could such individuals survive?
2. Why are the tissues derived from rapidly dividing cells initially affected?
3. Is this disorder likely to impact the life span?
4. Would you predict that mutations in the RNA component of telomerase might also cause DKC?

Summary Points

 For activities, animations, and review quizzes, go to the study area at www.masteringgenetics.com

1. In 1958, Meselson and Stahl resolved the question of which of three potential modes of replication is utilized by *E. coli* during the duplication of DNA in favor of semiconservative replication, showing that newly synthesized DNA consists of one old strand and one new strand.

2. Taylor, Woods, and Hughes demonstrated semiconservative replication in eukaryotes using the root tips of the broad bean as the source of dividing cells.

3. Arthur Kornberg isolated the enzyme DNA polymerase I from *E. coli* and showed that it is capable of directing *in vitro* DNA synthesis, provided that a template and precursor nucleoside triphosphates are supplied.

4. The discovery of the *polA1* mutant strain of *E. coli*, capable of DNA replication despite its lack of polymerase I activity, cast doubt on the enzyme's hypothesized *in vivo* replicative function. Polymerase III has been identified as the enzyme responsible for DNA replication *in vivo*.

5. During the initiation of DNA synthesis, the double helix unwinds, forming a replication fork at which synthesis begins.

Proteins stabilize the unwound helix and assist in relaxing the coiling tension created ahead of the replication fork.

6. DNA synthesis is initiated at specific sites along each template strand by the enzyme primase, resulting in short segments of RNA that provide suitable 3' ends upon which DNA polymerase III can begin polymerization.

7. Concurrent DNA synthesis occurs continuously on the leading strand and discontinuously on the opposite lagging strand, resulting in short Okazaki fragments that are later joined by DNA ligase.

8. DNA replication in eukaryotes is more complex than replication in prokaryotes, using multiple replication origins, multiple forms of DNA polymerases, and factors that disrupt and assemble nucleosomal chromatin.

9. Replication at the ends of linear chromosomes in eukaryotes poses a special problem that can be solved by the presence of telomeres and by a unique RNA-containing enzyme called telomerase.

10. Homologous recombination between DNA molecules relies on precise alignment of homologs and the actions of a series of enzymes that can cut, realign, and reseal DNA strands.

INSIGHTS AND SOLUTIONS

1. Predict the theoretical results of conservative and dispersive replication of DNA under the conditions of the Meselson–Stahl experiment. Follow the results through two generations of replication after cells have been shifted to a ^{14}N-containing medium, using the following sedimentation pattern.

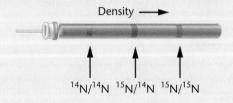

Density ⟶

^{14}N/^{14}N ^{15}N/^{14}N ^{15}N/^{15}N

Solution:

Conservative replication

Generation I   Generation II

Dispersive replication

Generation I   Generation II

2. Mutations in the *dnaA* gene of *E. coli* are lethal and can only be studied following the isolation of conditional, temperature-sensitive mutations. Such mutant strains grow nicely and

replicate their DNA at the permissive temperature of 18°C, but they do not grow or replicate their DNA at the restrictive temperature of 37°C. Two observations were useful in determining the function of the DnaA protein product. First, *in vitro* studies using DNA templates that have unwound do not require the DnaA protein. Second, if intact cells are grown at 18°C and are then shifted to 37°C, DNA synthesis continues at this temperature until one round of replication is completed and then stops. What do these observations suggest about the role of the *dnaA* gene product?

Solution: At 18°C (the permissive temperature), the mutation is not expressed and DNA synthesis begins. Following the shift to the restrictive temperature, the already initiated DNA synthesis continues, but no new synthesis can begin. Because the DnaA protein is not required for synthesis of unwound DNA, these observations suggest that, *in vivo*, the DnaA protein plays an essential role in DNA synthesis by interacting with the intact helix and somehow facilitating the localized denaturation necessary for synthesis to proceed.

Problems and Discussion Questions

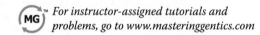 *For instructor-assigned tutorials and problems, go to www.masteringgentics.com*

HOW DO WE KNOW?

1. In this chapter, we focused on how DNA is replicated and synthesized. We also discussed recombination at the DNA level and the phenomenon of gene conversion. Along the way, we encountered many opportunities to consider how this information was acquired. On the basis of these discussions, what answers would you propose to the following fundamental questions?
   (a) What is the experimental basis for concluding that DNA replicates semiconservatively in both prokaryotes and eukaryotes?
   (b) How was it demonstrated that DNA synthesis occurs under the direction of DNA polymerase III and not polymerase I?
   (c) How do we know that *in vivo* DNA synthesis occurs in the 5′ to 3′ direction?
   (d) How do we know that DNA synthesis is discontinuous on one of the two template strands?
   (e) What observations reveal that a "telomere problem" exists during eukaryotic DNA replication, and how did we learn of the solution to this problem?

2. Compare conservative, semiconservative, and dispersive modes of DNA replication.

3. Describe the role of ^{15}N in the Meselson–Stahl experiment.

4. Predict the results of the experiment by Taylor, Woods, and Hughes if replication were (a) conservative and (b) dispersive.

5. Reconsider Problem 30 in Chapter 10. In the model you proposed, could the molecule be replicated semiconservatively? Why? Would other modes of replication work?

6. What are the requirements for *in vitro* synthesis of DNA under the direction of DNA polymerase I?

7. In Kornberg's initial experiments, it was rumored that he grew *E. coli* in Anheuser-Busch beer vats. (Kornberg was working at Washington University in St. Louis.) Why do you think this might have been helpful to the experiment?

8. How did Kornberg assess the fidelity of DNA polymerase I in copying a DNA template?

9. Which characteristics of DNA polymerase I raised doubts that its *in vivo* function is the synthesis of DNA leading to complete replication?

10. Kornberg showed that nucleotides are added to the 3′ end of each growing DNA strand. In what way does an exposed 3′-OH group participate in strand elongation?

11. What was the significance of the *polA1* mutation?

12. Summarize and compare the properties of DNA polymerase I, II, and III.

13. List and describe the function of the ten subunits constituting DNA polymerase III. Distinguish between the holoenzyme and the core enzyme.

14. Distinguish between (a) unidirectional and bidirectional synthesis, and (b) continuous and discontinuous synthesis of DNA.

15. List the proteins that unwind DNA during *in vivo* DNA synthesis. How do they function?

16. Define and indicate the significance of (a) Okazaki fragments, (b) DNA ligase, and (c) primer RNA during DNA replication.

17. Outline the current model for DNA synthesis.

18. Why is DNA synthesis expected to be more complex in eukaryotes than in bacteria? How is DNA synthesis similar in the two types of organisms?

19. Suppose that *E. coli* synthesizes DNA at a rate of 100,000 nucleotides per minute and takes 40 minutes to replicate its chromosome. (a) How many base pairs are present in the entire *E. coli* chromosome? (b) What is the physical length of the chromosome in its helical configuration—that is, what is the circumference of the chromosome if it were opened into a circle?

20. Several temperature-sensitive mutant strains of *E. coli* display the following characteristics. Predict what enzyme or function is being affected by each mutation.
   (a) Newly synthesized DNA contains many mismatched base pairs.
   (b) Okazaki fragments accumulate, and DNA synthesis is never completed.
   (c) No initiation occurs.
   (d) Synthesis is very slow.
   (e) Supercoiled strands remain after replication, which is never completed.

21. While many commonly used antibiotics interfere with protein synthesis or cell wall formation, clorobiocin, one of several antibiotics in the aminocoumarin class, inhibits the activity of bacterial DNA gyrase. Similar drugs have been tested as treatments for human cancer. How might such drugs be effective against bacteria as well as cancer?

22. Define gene conversion, and describe how this phenomenon is related to genetic recombination.

23. Many of the gene products involved in DNA synthesis were initially defined by studying mutant *E. coli* strains that could not synthesize DNA.

(a) The *dnaE* gene encodes the α subunit of DNA polymerase III. What effect is expected from a mutation in this gene? How could the mutant strain be maintained?

(b) The *dnaQ* gene encodes the ε subunit of DNA polymerase. What effect is expected from a mutation in this gene?

24. In 1994, telomerase activity was discovered in human cancer cell lines. Although telomerase is not active in human somatic tissue, this discovery indicated that humans do contain the genes for telomerase proteins and telomerase RNA. Since inappropriate activation of telomerase can cause cancer, why do you think the genes coding for this enzyme have been maintained in the human genome throughout evolution? Are there any types of human body cells where telomerase activation would be advantageous or even necessary? Explain.

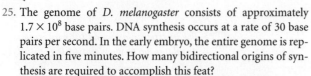

Extra-Spicy Problems

 For instructor-assigned tutorials and problems, go to www.masteringgentics.com

25. The genome of *D. melanogaster* consists of approximately 1.7×10^8 base pairs. DNA synthesis occurs at a rate of 30 base pairs per second. In the early embryo, the entire genome is replicated in five minutes. How many bidirectional origins of synthesis are required to accomplish this feat?

26. Assume a hypothetical organism in which DNA replication is conservative. Design an experiment similar to that of Taylor, Woods, and Hughes that will unequivocally establish this fact. Using the format established in Figure 11–5, draw sister chromatids and illustrate the expected results establishing this mode of replication.

27. DNA polymerases in all organisms add only 5′ nucleotides to the 3′ end of a growing DNA strand, never to the 5′ end. One possible reason for this is the fact that most DNA polymerases have a proofreading function that would not be *energetically* possible if DNA synthesis occurred in the 3′ to 5′ direction.

(a) Sketch the reaction that DNA polymerase would have to catalyze if DNA synthesis occurred in the 3′ to 5′ direction.

(b) Consider the information in your sketch and speculate as to why proofreading would be problematic.

28. An alien organism was investigated that demonstrated the "telomere problem" during DNA synthesis, but on only one end of each chromosome. Create a model of DNA that is compatible with this observation. Is this organism a prokaryote or eukaryote?

29. Assume that the sequence of bases given in this problem is present on one nucleotide chain of a DNA duplex and that the chain has opened up at a replication fork. Synthesis of an RNA primer occurs on this template starting at the base that is underlined.

(a) If the RNA primer consists of eight nucleotides, what is its base sequence?

(b) In the intact RNA primer, which nucleotide has a free 3′-OH terminus?

$$3′ .... GGCTACC\underline{T}GGATTCA ...... 5′$$

30. Given the following diagram, assume that the phase G1 chromosome on the left underwent one round of replication in 3′-thymidine and that the metaphase chromosome on the right had both chromatids labeled. Which of the replicative models (conservative, dispersive, semiconservative) could be eliminated by this observation?

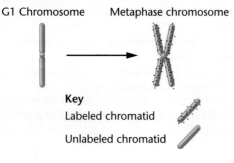

31. Reiji and Tuneko Okazaki conducted a now classic experiment in 1968 in which they discovered a population of short fragments synthesized during DNA replication. They introduced a short pulse of ³H-thymidine into a culture of *E. coli* and extracted DNA from the cells at various intervals. In analyzing the DNA after centrifugation in denaturing gradients, they noticed that as the interval between the time of ³H-thymidine introduction and the time of centrifugation increased, the proportion of short strands decreased and more labeled DNA was found in larger strands. What would account for this observation?

32. The following table (data adapted from Khodursky et al., 2000) presents the percentage of DNA synthesis after 15 minutes from initiation for four strains of *Escherichia coli* grown under permissive (30°C) and restrictive (42°C) temperatures and various concentrations of the gyrase inhibitor novobiocin. The strains have the following characteristics and genotypes: wild type, temperature-sensitive gyrase mutation (gyr^{ts}), novobiocin resistant (gyr^r), and the double mutant ($gyr^{ts,r}$). Based on data contained in the table, assign the appropriate genotypes to the strains labeled *A*, *B*, *C*, and *D*.

Temperature °C	30			42		
Novobiocin (g/ml)	0	5	40	0	5	40
Bacterial strain						
A	100	56	9	40	26	1
B	100	100	100	41	43	40
C	100	58	4	100	56	3
D	100	100	100	100	100	100

33. Consider the drawing of a dinucleotide below.

(a) Is it DNA or RNA?

(b) Is the arrow closest to the 5′ or the 3′ end?

(c) Suppose that the molecule was cleaved with the enzyme spleen diesterase, which breaks the covalent bond connecting the phosphate to C-5′. After cleavage, to which nucleoside is the phosphate now attached (A or T)?

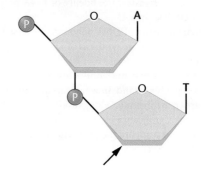

34. To gauge the fidelity of DNA synthesis, Arthur Kornberg and colleagues devised the nearest-neighbor frequency test, which determines the frequency with which any two bases occur adjacent to each other along the polynucleotide chain (*J. Biol. Chem.* 236: 864–875). This test relies on the enzyme spleen phosphodiesterase (see the previous problem). As we saw in Figure 11–8, DNA is synthesized by polymerization of 5′-nucleotides—that is, each nucleotide is added with the phosphate on the C-5′ of deoxyribose. However, as shown in the accompanying figure, the phosphodiesterase enzyme cleaves DNA between the phosphate and the C-5′ atom, thereby producing 3′-nucleotides. In this test, the phosphates on only one of the four nucleotide precursors of DNA (cytidylic acid, for example) are made radioactive with ^{32}P, and DNA is synthesized. Then the DNA is subjected to enzymatic cleavage, in which the radioactive phosphate is transferred to the base that is the "nearest neighbor" on the 5′ side of all cytidylic acid nucleotides.

Following four separate experiments, in each of which a different one of the four nucleotide types is radioactive, the frequency of all 16 possible nearest neighbors can be calculated. When Kornberg applied the nearest-neighbor frequency test to the DNA template and resultant product from a variety of experiments, he found general agreement between the nearest-neighbor frequencies of the two.

Analysis of nearest-neighbor data led Kornberg to conclude that the two strands of the double helix are in opposite polarity to one another. Demonstrate this approach by determining the outcome of such an analysis if the strands of DNA shown here are (a) antiparallel versus (b) parallel:

(a)

5′ ———————————————————————→ 3′

p – G – p – C – p – T – p – T – p – A – p – C – p – A

    – C – p – G – p – A – p – A – p – T – p – G – p – T – p

3′ ←——————————————————————— 5′

vs.

(b)

5′ ———————————————————————→ 3′

p – G – p – C – p – T – p – T – p – A – p – C – p – A

p – C – p – G – p – A – p – A – p – T – p – G – p – T

5′ ———————————————————————→ 3′

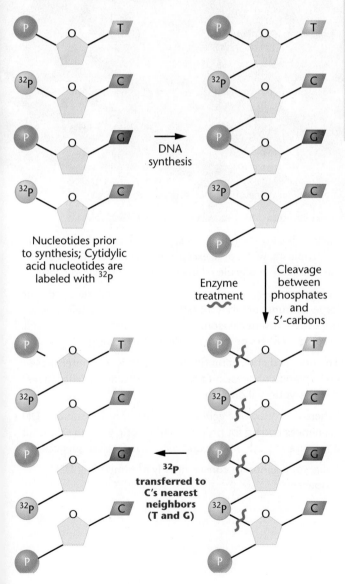

Nucleotides prior to synthesis; Cytidylic acid nucleotides are labeled with ^{32}P

DNA synthesis

Enzyme treatment

Cleavage between phosphates and 5′-carbons

^{32}P transferred to C's nearest neighbors (T and G)

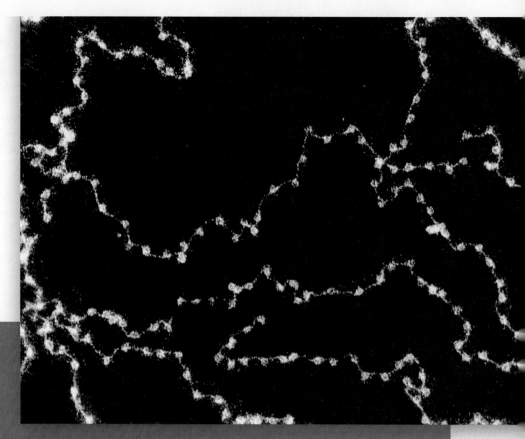

Chromatin fibers spilling out of a chicken erythrocyte, viewed using electron microscopy.

12

DNA Organization in Chromosomes

CHAPTER CONCEPTS

- Genetic information in viruses, bacteria, mitochondria, and chloroplasts, with some exceptions, is contained in a short, circular DNA molecule relatively free of associated proteins.

- Eukaryotic cells, in contrast to viruses and bacteria, contain large amounts of DNA that during most of the cell cycle is organized into nucleosomes and is present as either uncoiled chromatin fibers or more condensed structures.

- The uncoiled chromatin fibers characteristic of interphase coil up and condense into chromosomes during the stages of eukaryotic cell division.

- Whereas prokaryotic genomes consist of mostly unique DNA sequences coding for proteins, eukaryotic genomes contain a mixture of both unique and repetitive DNA sequences.

- Eukaryotic genomes consist mostly of noncoding DNA sequences.

nce geneticists understood that DNA houses genetic information, they focused their energies on discovering how DNA is organized into genes and how these basic units of genetic function are organized into chromosomes. In short, the next major questions they tackled had to do with how the genetic material is organized within the genome. These issues have a bearing on many areas of genetic inquiry. For example, the ways in which genomic organization varies has a major bearing on the regulation of genetic expression.

In this chapter, we focus on the various ways DNA is organized into chromatin, which in turn is organized into chromosomes. These structures have been studied using numerous approaches, including biochemical analysis as well as visualization by light microscopy and electron microscopy. In the first part of the chapter, after surveying what we know about chromosomes in viruses and bacteria, we examine two specialized eukaryotic structures called polytene and lampbrush chromosomes. Then, we turn to a consideration of how eukaryotic chromosomes are organized at the molecular level—for example, how DNA is complexed with proteins to form chromatin and how the chromatin fibers characteristic of interphase are condensed into chromosome structures visible during mitosis and meiosis. We conclude the chapter by examining certain aspects of DNA sequence organization in eukaryotic genomes.

12.1
Viral and Bacterial Chromosomes Are Relatively Simple DNA Molecules

The chromosomes of viruses and bacteria are much less complicated than those in eukaryotes. They usually consist of a single nucleic acid molecule quite different from the multiple chromosomes constituting the genome of higher forms. Prokaryotic chromosomes are largely devoid of associated proteins and contain relatively less genetic information. These characteristics have greatly simplified genetic analysis in prokaryotic organisms, and we now have a fairly comprehensive view of the structure of their chromosomes.

The chromosomes of viruses consist of a nucleic acid molecule—either DNA or RNA—that can be either single or double stranded. They can exist as circular structures (covalently closed circles), or they can take the form of linear molecules. The single-stranded DNA of the ϕX174 **bacteriophage** and the double-stranded DNA of the polyoma virus are closed circles housed within the protein coat of the mature viruses. The **bacteriophage lambda** (λ), on the other hand, possesses a linear double-stranded DNA molecule prior to infection, but it closes to form a ring upon infection of the host cell. Still other viruses, such as the T-even series of bacteriophages, have linear double-stranded chromosomes of DNA that do not form circles inside the bacterial host. Thus, circularity is not an absolute requirement for replication in viruses.

Viral nucleic acid molecules have been visualized with the electron microscope. Figure 12–1 shows a mature

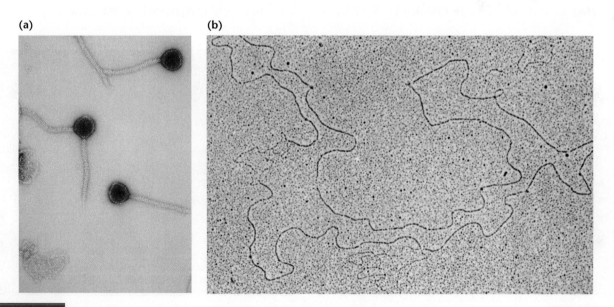

(a) **(b)**

FIGURE 12–1   Electron micrographs of (a) phage λ and (b) the DNA isolated from it. The chromosome is 17 μm long. The phages are magnified about five times more than the DNA.

TABLE 12.1

The Genetic Material of Representative Viruses and Bacteria

Source		Nucleic Acid			Overall Size of Viral Head or Bacterial Cell (μm)
		Type	SS or DS*	Length (μm)	
Viruses	ϕX174	DNA	SS	2.0	0.025 × 0.025
	Tobacco mosaic virus	RNA	SS	3.3	0.30 × 0.02
	Lambda phage	DNA	DS	17.0	0.07 × 0.07
	T2 phage	DNA	DS	52.0	0.07 × 0.10
Bacteria	*Haemophilus influenzae*	DNA	DS	832.0	1.00 × 0.30
	Escherichia coli	DNA	DS	1200.0	2.00 × 0.50

*SS = single-stranded; DS = double-stranded

bacteriophage λ and its double-stranded DNA molecule in the circular configuration. One constant feature shared by viruses, bacteria, and eukaryotic cells is the ability to package an exceedingly long DNA molecule into a relatively small volume. In λ, the DNA is 17 μm long and must fit into the phage head, which is less than 0.1 μm on any side. Table 12.1 compares the length of the chromosomes of several viruses with the size of their head structure. In each case, a similar packaging feat must be accomplished. Compare the dimensions given for phage T2 with the micrograph of both the DNA and the viral particle shown in Figure 12–2. Seldom does the space available in the head of a virus exceed the chromosome volume by more than a factor of two. In many cases, almost all space is filled, indicating nearly perfect packing. Once packed within the head, the virus's genetic material is functionally inert until it is released into a host cell.

Bacterial chromosomes are also relatively simple in form. They always consist of a double-stranded DNA molecule, compacted into a structure sometimes referred to as the **nucleoid**. *Escherichia coli,* the most extensively studied bacterium, has a large, circular chromosome measuring approximately 1200 μm (1.2 mm) in length. When the cell is gently lysed and the chromosome released, the chromosome can be visualized under the electron microscope (Figure 12–3).

DNA in bacterial chromosomes is found to be associated with several types of **DNA-binding proteins**. Two, called **HU** and **H1 proteins**, are small but abundant in the cell and contain a high percentage of positively charged amino acids that can bond ionically to the negative charges of the phosphate groups in DNA. Although these proteins are structurally similar to molecules called *histones* that are associated with eukaryotic DNA (as described in Section 12.4), they are not involved in compacting DNA in a similar way. Unlike the tightly packed chromosome present in the head of a virus, the bacterial chromosome is *not* functionally inert, and can be readily replicated and transcribed.

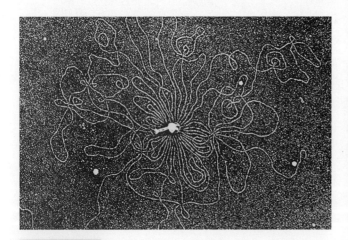

FIGURE 12–2   Electron micrograph of bacteriophage T2, which has had its DNA released by osmotic shock. The chromosome is 52 μm long.

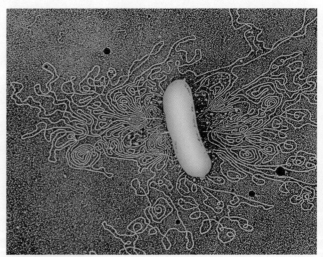

FIGURE 12–3   Electron micrograph of the bacterium *Escherichia coli,* which has had its DNA released by osmotic shock. The chromosome is 1200 μm long.

NOW SOLVE THIS

12–1 In bacteriophages and bacteria, the DNA is almost always organized into circular (closed loops) chromosomes. Phage λ is an exception, maintaining its DNA in a linear chromosome within the viral particle. However, as soon as this DNA is injected into a host cell, it circularizes before replication begins. What advantage exists in replicating circular DNA molecules compared to linear molecules, characteristic of eukaryotic chromosomes?

■ HINT: *This problem involves an understanding of eukaryotic DNA replication, as discussed in Chapter 11. The key to its solution is to consider why the enzyme telomerase is essential in eukaryotic DNA replication, and why bacterial and viral chromosomes can be replicated without encountering the "telomere" problem.*

12.2

Supercoiling Facilitates Compaction of the DNA of Viral and Bacterial Chromosomes

One major insight into the way DNA is organized and packaged in viral and bacterial chromosomes has come from the discovery of **supercoiled DNA**, which is particularly characteristic of closed-circular molecules. Supercoiled DNA was first proposed as a result of a study of double-stranded DNA molecules derived from the polyoma virus, which causes tumors in mice. In 1963, it was observed that when the polyoma DNA was subjected to high-speed centrifugation, it was resolved into three distinct components, each of different density and compactness. The one that was least compact, and thus least dense, demonstrated a decreased sedimentation velocity; the other two fractions each showed greater velocities owing to their greater compaction and density. All three were of identical molecular weight.

In 1965, Jerome Vinograd proposed an explanation for these observations. He postulated that the two fractions of greatest sedimentation velocity both consisted of circular DNA molecules, whereas the fraction of lower sedimentation velocity contained linear DNA molecules. Closed-circular molecules are more compact and sediment more rapidly than do the same molecules in linear form.

Vinograd proposed further that the denser of the two fractions of circular molecules consisted of covalently closed DNA helices that are slightly *underwound* in comparison to the less dense circular molecules. Energetic forces stabilizing the double helix resist this underwinding, causing the molecule as a whole to **supercoil**, that is, to contort in a certain way, in order to retain normal base pairing. Vinograd proposed that it is the supercoiled shape that causes tighter packing and thus the increase in sedimentation velocity.

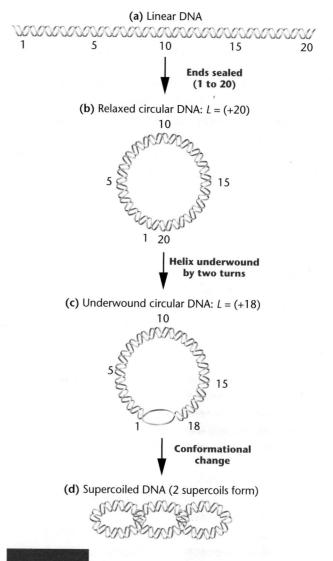

FIGURE 12–4 Depictions of the transformations leading to the supercoiling of circular DNA. *L* signifies linking number.

The transitions just described are illustrated in Figure 12–4. Consider a double-stranded linear molecule existing in the normal Watson–Crick right-handed helix [Figure 12–4(a)]. This helix contains 20 complete turns, which means the **linking number** (*L*) of this molecule is 20, or *L* = 20. Suppose we were to change this linear molecule into a closed circle by bringing the opposite ends together and joining them [Figure 12–4(b)]. If the closed circle still has a linking number of 20 (if we haven't introduced or eliminated any turns when joining the ends), we define the molecule as being *energetically relaxed*. Now suppose that the circle were subsequently cut open, underwound by two full turns, and then resealed [Figure 12–4(c)]. Such a structure, in which *L* has been changed to 18, would be *energetically strained* and, as a result, would change its form to relieve the strain.

In order to assume a more energetically favorable conformation, an underwound molecule will form supercoils in the direction opposite to that of the underwinding. In our case

[Figure 12–4(d)], two negative supercoils are introduced spontaneously, reestablishing, in total, the original number of turns in the helix. Use of the term *negative* refers to the fact that, by definition, the supercoils are left-handed (whereas the helix is right-handed). The end result is the formation of a more compact structure with enhanced physical stability.

In most closed-circular DNA molecules in bacteria and their phages, the DNA helix is slightly underwound [as in Figure 12–4(c)]. For example, the virus SV40 contains 5200 base pairs. In energetically relaxed DNA, 10.4 base pairs occupy each complete turn of the helix, and the linking number can be calculated as

$$L = \frac{5200}{10.4} = 500$$

However, analysis of circular SV40 DNA reveals that it is underwound by 25 turns, so L is equal to only 475. Predictably, 25 negative supercoils are observed in the molecule. In *E. coli,* an even larger number of supercoils is observed, greatly facilitating chromosome condensation in the nucleoid region.

Two otherwise identical molecules that differ only in their linking number are said to be **topoisomers** of one another. But how can a molecule convert from one topoisomer to the other if there are no free ends, as is the case in closed circles of DNA? Biologically, this may be accomplished by any one of a group of enzymes that cut one or both of the strands and wind or unwind the helix before resealing the ends.

Appropriately, these enzymes are called **topoisomerases**. First discovered by Martin Gellert and James Wang, these catalytic molecules are known as either type I or type II, depending on whether they cleave one or both strands in the helix, respectively. In *E. coli,* topoisomerase I serves to reduce the number of negative supercoils in a closed-circular DNA molecule. Topoisomerase II introduces negative supercoils into DNA. This latter enzyme is thought to bind to DNA, twist it, cleave both strands, and then pass them through the loop that it has created. Once the phosphodiester bonds are re-formed, the linking number is decreased and one or more supercoils form spontaneously.

Supercoiled DNA and topoisomerases are also found in eukaryotes. While the chromosomes in these organisms are not usually circular, supercoils can occur when areas of DNA are embedded in a lattice of proteins associated with the chromatin fibers. This association creates "anchored" ends, providing the stability for the maintenance of supercoils once they are introduced by topoisomerases. In both prokaryotes and eukaryotes, DNA replication and transcription create supercoils downstream as the double helix unwinds and becomes accessible to the appropriate enzyme.

Topoisomerases may play still other genetic roles involving eukaryotic DNA conformational changes. Interestingly, these enzymes are involved in separating (decatenating) the DNA of sister chromatids following replication.

12.3

Specialized Chromosomes Reveal Variations in the Organization of DNA

We now consider two cases of genetic organization that demonstrate specialized forms that eukaryotic chromosomes can take. Both types—*polytene chromosomes* and *lampbrush chromosomes*—are so large that their organization was discerned using light microscopy long before we understood how mitotic chromosomes form from interphase chromatin. The study of these chromosomes provided many of our initial insights into the arrangement and function of the genetic information. It is important to know that polytene and lampbrush chromosomes are unusual and not typically found in most eukaryotic cells, but the study of their structure has revealed many common themes of chromosome organization.

Polytene Chromosomes

Giant **polytene chromosomes** are found in various tissues (salivary, midgut, rectal, and malpighian excretory tubules) in the larvae of some flies, as well as in several species of protozoans and plants. They were first observed by E. G. Balbiani in 1881. The large amount of information obtained from studies of these genetic structures provided a model system for subsequent investigations of chromosomes. What is particularly intriguing about polytene chromosomes is that they can be seen in the nuclei of interphase cells.

Each polytene chromosome is 200 to 600 μm long, and when they are observed under the light microscope, they exhibit a linear series of alternating bands and interbands (Figure 12–5 and 12–6). The banding pattern is distinctive for each chromosome in any given species. Individual bands are sometimes called **chromomeres**, a more generalized

FIGURE 12–5   Polytene chromosomes derived from larval salivary gland cells of *Drosophila.*

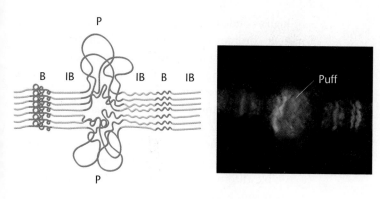

FIGURE 12–6 Photograph of a puff within a polytene chromosome. The diagram depicts the uncoiling of strands within a band region (B) to produce a puff (P) in a polytene chromosome. Each band (B) represents a chromomere. Interband regions (IBs) are also labeled.

term describing lateral condensations of material along the axis of a chromosome.

Extensive study using electron microscopy and radioactive tracers led to an explanation for the unusual appearance of these chromosomes. First, polytene chromosomes represent paired homologs. This in itself is highly unusual, since they are present in somatic cells, where, in most organisms, chromosomal material is normally dispersed as chromatin and homologs are not paired. Second, their large size and distinctive appearance result from their being composed of large numbers of identical DNA strands. The DNA of these paired homologs undergoes many rounds of replication, *but without strand separation or cytoplasmic division.* As replication proceeds, chromosomes are created, having 1000 to 5000 DNA strands that remain in precise parallel alignment with one another. Apparently, the parallel register of so many DNA strands gives rise to the distinctive band pattern along the axis of the chromosome.

The presence of bands on polytene chromosomes was initially interpreted as the visible manifestation of individual genes. The discovery that the strands present in bands undergo localized uncoiling during genetic activity further strengthened this view. Each such uncoiling event results in a bulge called a **puff**, so labeled because of its appearance under the microscope (Figure 12–6). That puffs are visible manifestations of a high level of gene activity (transcription that produces RNA) is evidenced by their high rate of incorporation of radioactively labeled RNA precursors, as assayed by autoradiography. Bands that are not extended into puffs incorporate fewer radioactive precursors or none at all.

The study of bands during development in insects such as *Drosophila* and the midge fly *Chironomus* reveals differential gene activity. A characteristic pattern of band formation that is equated with gene activation is observed as development proceeds. Despite attempts to resolve the issue, it is not yet clear how many genes are contained in each band. However, we do know that in *Drosophila*, which contains about 15,000 genes, there are approximately 5000 bands. Interestingly, a band may contain up to 10^7 base pairs of DNA, enough DNA to encode 50 to 100 average-size genes.

NOW SOLVE THIS

12–2 After salivary gland cells from *Drosophila* are isolated and cultured in the presence of radioactive thymidylic acid, autoradiography is performed, revealing polytene chromosomes. Predict the distribution of the grains along the chromosomes.

■ HINT: *This problem involves an understanding of the organization of DNA in polytene chromosomes. The key to its solution is to be aware that* ^{3}H-*thymidine, as a molecular tracer, will only be incorporated into DNA during its replication.*

Lampbrush Chromosomes

Another specialized chromosome that has given us insight into chromosomal structure is the **lampbrush chromosome**, so named because it resembles the brushes used to clean kerosene lamp chimneys in the nineteenth century. Lampbrush chromosomes were first discovered in 1882 by Walther Flemming in salamander oocytes, and then seen in 1892 by J. Ruckert in shark oocytes. They are now known to be characteristic of most vertebrate oocytes, as well as the spermatocytes of some insects. Therefore, they are meiotic chromosomes. Most of the experimental work on them has been done with material taken from amphibian oocytes.

These unique chromosomes are easily isolated from oocytes in the diplotene stage of the first prophase of meiosis, where they are active in directing the metabolic activities of the developing cell. The homologs are seen as synapsed pairs held together by chiasmata. However, instead of condensing, as most meiotic chromosomes do, lampbrush chromosomes are often extended to lengths of 500 to 800 μm. Later, in meiosis, they revert to their normal length of 15 to 20 μm. Based on these observations, lampbrush chromosomes are interpreted as being extended, uncoiled versions of the normal meiotic chromosomes.

The two views of lampbrush chromosomes in **Figure 12–7** provide significant insights into their morphology. Part (a) shows the meiotic configuration under the light microscope. The linear axis of each horizontal structure seen in the figure contains a large number of condensed areas,

(a)

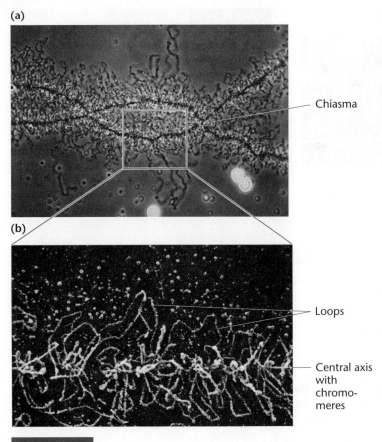

Chiasma

(b)

Loops

Central axis with chromomeres

FIGURE 12–7 Lampbrush chromosomes derived from amphibian oocytes. (a) A photomicrograph. (b) A scanning electron micrograph.

which, as with polytene chromosomes, are referred to as *chromomeres.* Emanating from each chromomere is a pair of lateral loops, giving the chromosome its distinctive appearance. In part (b), the scanning electron micrograph (SEM) shows many adjacent pairs of loops in detail along one of the axes. As with bands in polytene chromosomes, much more DNA is present in each loop than is needed to encode a single gene. Such an SEM provides a clear view of the chromomeres and the chromosomal fibers emanating from them. Each chromosomal loop is thought to be composed of one DNA double helix, while the central axis is made up of two DNA helices. This hypothesis is consistent with the belief that each meiotic chromosome is composed of a pair of sister chromatids. Studies using radioactive RNA precursors reveal that the loops are active in the synthesis of RNA. The lampbrush loops, in a manner similar to puffs in polytene chromosomes, represent DNA that has been reeled out from the central chromomere axis during transcription.

12.4

DNA Is Organized into Chromatin in Eukaryotes

We now turn our attention to the way DNA is organized in eukaryotic chromosomes, which are most clearly visible as

highly condensed structures during mitosis. However, after chromosome separation and cell division, cells enter the interphase stage of the cell cycle, at which time the components of the chromosome uncoil and decondense into a form referred to as **chromatin**. While in interphase, the chromatin is dispersed throughout the nucleus. As the cell cycle progresses, cells may replicate their DNA and reenter mitosis, whereupon chromatin coils and condenses back into visible chromosomes once again. This condensation represents a length contraction of some 10,000 times for each chromatin fiber.

The organization of DNA during the transitions just described is much more intricate and complex than in viruses or bacteria, which never exhibit a complex process similar to mitosis. This is due to the greater amount of DNA per chromosome in eukaryotes, as well as the presence of a large number of proteins associated with eukaryotic DNA. For example, while DNA in the *E. coli* chromosome is 1200 μm long, the DNA in each human chromosome ranges from 19,000 to 73,000 μm in length. In a single human nucleus, all 46 chromosomes contain sufficient DNA to extend almost 2 meters. This genetic material, along with its associated proteins, is contained within a nucleus that usually measures about 5 to 10 μm in diameter.

Such intricacy parallels the structural and biochemical diversity of the many types of cells present in a multicellular eukaryotic organism. Different cells assume specific functions based on highly specific biochemical activity. Although all cells carry a full genetic complement, different cells activate different sets of genes. Clearly, then, a highly ordered regulatory system must exist to govern the use of the genetic information. Such a system must in some way be imposed on or related to the molecular structure of the genetic material.

Chromatin Structure and Nucleosomes

As we have seen, the genetic material of viruses and bacteria consists of strands of DNA or RNA relatively devoid of proteins. In contrast, eukaryotic chromatin has a substantial amount of protein associated with the chromosomal DNA in all phases of the cell cycle. The associated proteins can be categorized as either positively charged **histones** or less positively charged *nonhistone proteins.* Of these two groups, the histones play the most essential structural role. Histones contain large amounts of the positively charged amino acids lysine and arginine, making it possible for them to bond electrostatically to the negatively charged phosphate groups of nucleotides. Recall that a similar interaction has been proposed for several bacterial proteins. The five main types of histones are shown in Table 12.2.

TABLE 12.2

TABLE 12.2

Categories and Properties of Histone Proteins

Histone Type	Lysine-Arginine Content	Molecular Weight (Da)
H1	Lysine-rich	23,000
H2A	Slightly lysine-rich	14,000
H2B	Slightly lysine-rich	13,800
H3	Arginine-rich	15,300
H4	Arginine-rich	11,300

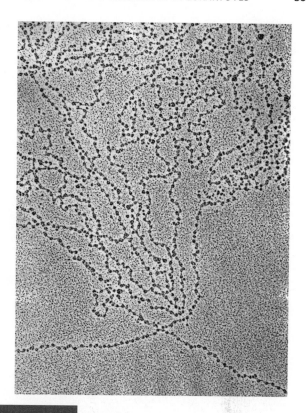

FIGURE 12–8   An electron micrograph revealing nucleosomes appearing as "beads on a string" along chromatin strands derived from *Drosophila melanogaster*.

The general model for chromatin structure is based on the assumption that chromatin fibers, composed of DNA and protein, undergo extensive coiling and folding as they are condensed within the cell nucleus. Moreover, X-ray diffraction studies confirm that histones play an important role in chromatin structure. Chromatin produces regularly spaced diffraction rings, suggesting that repeating structural units occur along the chromatin axis. If the histone molecules are chemically removed from chromatin, the regularity of this diffraction pattern is disrupted.

A basic model for chromatin structure was worked out in the mid-1970s. The following observations were particularly relevant to the development of this model:

1. Digestion of chromatin by certain endonucleases, such as micrococcal nuclease, yields DNA fragments that are approximately 200 base pairs in length or multiples thereof. This enzymatic digestion is not random, for if it were, we would expect a wide range of fragment sizes. Thus, chromatin consists of some type of repeating unit, each of which protects the DNA from enzymatic cleavage except where any two units are joined. It is the area between units that is attacked and cleaved by the endonuclease.

2. Electron microscopic observations of chromatin have revealed that chromatin fibers are composed of linear arrays of spherical particles (**Figure 12–8**). Discovered by Ada and Donald Olins, the particles occur regularly along the axis of a chromatin strand and resemble beads on a string. These particles, initially referred to as *v-bodies* (*v* is the Greek letter nu), are now called **nucleosomes**. These findings conform to the above observation that suggests the existence of repeating units.

3. Studies of the chemical association between histone molecules and DNA in the nucleosomes of chromatin show that histones H2A, H2B, H3, and H4 occur as two types of tetramers, $(H2A)_2 \cdot (H2B)_2$ and $(H3)_2 \cdot (H4)_2$. Roger Kornberg predicted that each repeating nucleosome unit consists of one of each tetramer (creating an octomer) in association with about 200 base pairs of DNA. Such a structure is consistent with previous observations and provides the basis for a model that explains the interaction of histones and DNA in chromatin.

4. When nuclease digestion time is extended, some of the 200 base pairs of DNA are removed from the nucleosome, creating what is called a **nucleosome core particle** consisting of 147 base pairs. The DNA lost in the prolonged digestion is responsible for linking nucleosomes together. This **linker DNA** is associated with the fifth histone, H1.

On the basis of this information, as well as on X-ray and neutron-scattering analyses of crystallized core particles by John T. Finch, Aaron Klug, and others, a detailed model of the nucleosome was put forward in 1984, providing a basis for predicting chromatin structure and its condensation into chromosomes. In this model, illustrated in **Figure 12–9**, a 147-bp length of the 2-nm-diameter DNA molecule coils around an octamer of histones in a left-handed superhelix that completes about 1.7 turns per nucleosome. Each nucleosome, ellipsoidal in shape, measures about 11 nm at its longest point [Figure 12–9a]. Significantly, the formation of the nucleosome represents the first level of packing, whereby the DNA helix is reduced to about one-third of its original length by winding around the histones.

In the nucleus, the chromatin fiber seldom, if ever, exists in the extended form described in the previous paragraph (that is, as an extended chain of nucleosomes). Instead, the 11-nm-diameter fiber is further packed into a thicker, 30-nm-diameter structure that was initially called a *solenoid* [Figure 12–9b]. This thicker structure, which is dependent on the presence of histone H1, consists of numerous nucleosomes coiled around

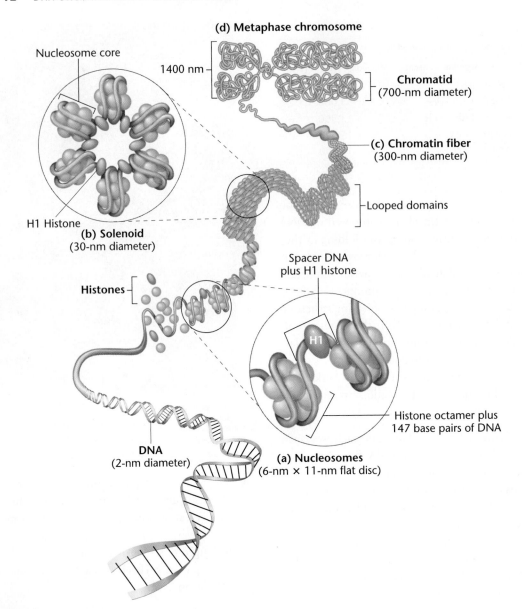

(d) Metaphase chromosome

1400 nm

Chromatid
(700-nm diameter)

Nucleosome core

(c) Chromatin fiber
(300-nm diameter)

Looped domains

H1 Histone

(b) Solenoid
(30-nm diameter)

Spacer DNA
plus H1 histone

Histones

H1

Histone octamer plus
147 base pairs of DNA

DNA
(2-nm diameter)

(a) Nucleosomes
(6-nm × 11-nm flat disc)

FIGURE 12–9   General model of the association of histones and DNA to form nucleosomes, illustrating the way in which each thickness of fiber may be coiled into a more condensed structure, ultimately producing a metaphase chromosome.

and stacked upon one another, creating a second level of packing. This provides a six-fold increase in compaction of the DNA. It is this structure that is characteristic of an uncoiled chromatin fiber in interphase of the cell cycle. In the transition to the mitotic chromosome, still further compaction must occur. The 30-nm structures are folded into a series of *looped domains*, which further condense the chromatin fiber into a structure that is 300 nm in diameter [Figure 12–9c]. These *coiled chromatin fibers* are then compacted into the chromosome arms that constitute a chromatid, one of the longitudinal subunits of the metaphase chromosome [Figure 12–9d]. While Figure 12–9 shows the chromatid arms to be 700 nm in diameter, this value undoubtedly varies among different organisms. At a value of 700 nm, a pair of sister chromatids comprising a chromosome measures about 1400 nm.

The importance of the organization of DNA into chromatin and of chromatin into mitotic chromosomes can be illustrated by considering that a human cell stores its genetic material in a nucleus about 5 to 10 μm in diameter. The haploid genome contains more than 3 billion base pairs of DNA distributed among 23 chromosomes. The diploid cell contains twice that amount. At 0.34 nm per base pair, this amounts to an enormous length of DNA (as stated earlier, almost 2 meters)! One estimate is that the DNA inside a typical human nucleus is complexed with roughly 25×10^6 nucleosomes.

In the overall transition from a fully extended DNA helix to the extremely condensed status of the mitotic chromosome, a packing ratio (the ratio of DNA length to the length of the structure containing it) of about 500 to 1 must be achieved. In fact, our model accounts for a ratio of only

12–3 If a human nucleus is 10 μm in diameter, and it must hold as much as 2 m of DNA, which is complexed into nucleosomes that during full extension are 11 nm in diameter, what percentage of the volume of the nucleus is occupied by the genetic material?

■ HINT: *This problem asks you to make some numerical calculations in order to see just how "filled" the eukaryotic nucleus is with a diploid amount of DNA. The key to its solution is the use of the formula V = (4/3)πr³, which calculates the volume of a sphere.*

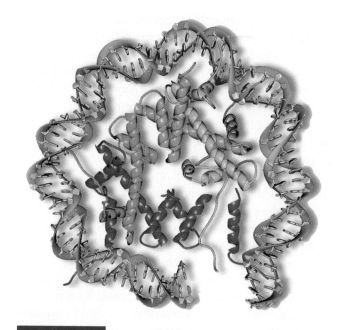

about 50 to 1. Obviously, the larger fiber can be further bent, coiled, and packed to achieve even greater condensation during the formation of a mitotic chromosome.

Chromatin Remodeling

As with many significant endeavors in genetics, the study of nucleosomes has answered some important questions but at the same time has opened up new ones. For example, in the preceding discussion, we established that histone proteins play an important structural role in packaging DNA into the nucleosomes that make up chromatin. While this discovery helped solve the structural problem of how the huge amount of DNA is organized within the eukaryotic nucleus, it brought another problem to the fore: *In the chromatin fiber, complexed with histones into nucleosomes, which may be further folded into several more levels of compaction, the DNA is inaccessible to interaction with important nonhistone proteins.* For example, the various proteins that function in enzymatic and regulatory roles during the processes of replication and transcription must interact directly with DNA. To accommodate these protein–DNA interactions, chromatin must be induced to change its structure, a process now referred to as **chromatin remodeling**. To allow replication and gene expression, chromatin must relax its compact structure and expose regions of DNA to regulatory proteins, but there must also be a mechanism for reversing the process during periods of inactivity.

Insights into how different states of chromatin structure might be achieved began to emerge in 1997, when Timothy Richmond and members of his research team were able to significantly improve the level of resolution in X-ray diffraction studies of nucleosome crystals, from 7 Å in the 1984 studies to 2.8 Å in the 1997 studies. One model based on their work is shown in **Figure 12–10**. At this resolution, most atoms are visible, thus revealing the subtle twists and turns of the superhelix of DNA encircling the histones. Recall that the double-helical ribbon in the figure represents 147 bp of DNA surrounding four pairs of histone proteins.

FIGURE 12–10 The nucleosome core particle derived from X-ray crystal analysis at 2.8 Å resolution. The double-helical DNA surrounds four pairs of histones.

This configuration is essentially repeated over and over in the chromatin fiber and is the principal packaging unit of DNA in the eukaryotic nucleus.

By 2003, Richmond and colleagues achieved a resolution of 1.9 Å that revealed the details of the location of each histone entity within the nucleosome. Of particular relevance to the discussion of chromatin remodeling is the observation that there are unstructured **histone tails** that are *not* packed into the folded histone domains within the core of the nucleosomes but instead protrude from it. For example, tails devoid of any secondary structure extending from histones H3 and H2B protrude through the minor-groove channels of the DNA helix. You should look carefully at Figure 12–10 and locate examples of such tails. Other tails of histone H4 appear to make a connection with adjacent nucleosomes. The significance of histone tails is that they provide potential targets along the chromatin fiber for a variety of chemical modifications that may be linked to genetic functions, including chromatin remodeling and the possible regulation of gene expression.

Several of these potential chemical modifications are now recognized as important to genetic function. One of the most well-studied histone modifications involves **acetylation** by the action of the enzyme *histone acetyltransferase (HAT)*. The addition of an acetyl group to the positively charged amino group present on the side chain of the amino acid lysine effectively changes the net charge of the protein by neutralizing the positive charge. Lysine is in abundance in histones, and geneticists have known for some time that acetylation is linked to gene activation. It appears that high levels of acetylation open up, or remodel, the chromatin fiber, an effect that increases in regions of active genes and decreases in inactive regions. In a well-studied example, the inactivation of the X chromosome in mammals, forming a Barr body (Chapter 7), histone H4 is known to be greatly underacetylated.

Two other important chemical modifications are the **methylation** and **phosphorylation** of amino acids that are part of histones. These chemical processes result from the action of enzymes called *methyltransferases* and *kinases*, respectively. Methyl groups can be added to both arginine and lysine residues in histones, and this change has been correlated with gene activity. Phosphate groups can be added to the hydroxyl groups of the amino acids serine and histidine, introducing a negative charge on the protein. During the cell cycle, increased phosphorylation, particularly of histone H3, is known to occur at characteristic times. Such chemical modification is believed to be related to the cycle of chromatin unfolding and condensation that occurs during and after DNA replication. It is important to note that the above chemical modifications (acetylation, methylation, and phosphorylation) are all reversible, under the direction of specific enzymes.

Interestingly, while methylation of histones *within nucleosomes* is often positively correlated with gene activity in eukaryotes, methylation of the nitrogenous base cytosine *within polynucleotide chains of DNA*, forming **5-methyl cytosine**, is usually negatively correlated with gene activity. Methylation of cytosine occurs most often when the nucleotide cytidylic acid is next to the nucleotide guanylic acid, forming what is called a **CpG island**. We must conclude, then, that methylation can have a positive or a negative impact on gene activity.

The research described above has extended our knowledge of nucleosomes and chromatin organization and serves here as a general introduction to the concept of chromatin remodeling. A great deal more work must be done, however, to elucidate the specific involvement of chromatin remodeling during genetic processes. In particular, the way in which the modifications are influenced by regulatory molecules within cells will provide important insights into the mechanisms of gene expression. What is clear is that the dynamic forms in which chromatin exists are vitally important to the way that all genetic processes directly involving DNA are executed. We will return to a more detailed discussion of the role of chromatin remodeling when we consider the regulation of eukaryotic gene expression in Chapter 17. In addition, chromatin remodeling is an important topic in the discussion of **epigenetics**, the study of modifications of an organism's genetic and phenotypic expression that are *not* attributable to alteration of the DNA sequence making up a gene. Epigenetics is the subject of one of the new *Special Topics in Modern Genetics* chapters that follows Chapter 19.

Heterochromatin

Although we know that the DNA of the eukaryotic chromosome consists of one continuous double-helical fiber along its entire length, we also know that the whole chromosome is not structurally uniform from end to end. In the early part of the twentieth century, it was observed that some parts of the chromosome remain condensed and stain deeply during interphase, while most parts are uncoiled and do not stain. In 1928, the terms **euchromatin** and **heterochromatin** were coined to describe the parts of chromosomes that are uncoiled and those that remain condensed, respectively.

Subsequent investigation revealed a number of characteristics that distinguish heterochromatin from euchromatin. Heterochromatic areas are genetically inactive because they either lack genes or contain genes that are repressed. Also, heterochromatin replicates later during the S phase of the cell cycle than does euchromatin. The discovery of heterochromatin provided the first clues that parts of eukaryotic chromosomes do not always encode proteins. For example, one particular heterochromatic region of the chromosome, the *telomere*, is thought to be involved in maintenance of the chromosome's structural integrity, and another region, the *centromere*, is involved in chromosome movement during cell division.

The presence of heterochromatin is unique to and characteristic of the genetic material of eukaryotes. In some cases, whole chromosomes are heterochromatic. A case in point is the mammalian Y chromosome, much of which is genetically inert. And, as we discussed in Chapter 7, the inactivated X chromosome in mammalian females is condensed into an inert heterochromatic Barr body. In some species, such as mealy bugs, the chromosomes of one entire haploid set are heterochromatic.

When certain heterochromatic areas from one chromosome are translocated to a new site on the same or another nonhomologous chromosome, genetically active areas sometimes become genetically inert if they now lie adjacent to the translocated heterochromatin. As we saw in Chapter 4, this influence on existing euchromatin is one example of what is more generally referred to as a **position effect**. That is, the position of a gene or group of genes relative to all other genetic material may affect their expression.

12.5

Chromosome Banding Differentiates Regions along the Mitotic Chromosome

Until about 1970, mitotic chromosomes viewed under the light microscope could be distinguished only by their relative sizes and the positions of their centromeres. Unfortunately, even in organisms with a low haploid number, two or more chromosomes are often visually indistinguishable from one another. Since that time, however, new cytological procedures made possible differential staining along the longitudinal axis of mitotic chromosomes. Such methods are now referred to as **chromosome-banding techniques**, because the staining patterns resemble the bands of polytene chromosomes.

One of the first chromosome-banding techniques was devised by Mary Lou Pardue and Joe Gall. They found that if chromosome preparations from mice were heat denatured and then treated with Giemsa stain, a unique staining pattern emerged: Only the centromeric regions of mitotic chromosomes took up the stain! The staining pattern was thus referred to as **C-banding**. Relevant to our immediate discussion, this cytological technique identifies a specific area of the chromosome composed of heterochromatin. A micrograph of the human karyotype treated in this way is shown in **Figure 12–11**. Mouse chromosomes are all telocentric, thus localizing the stain at the end of each chromosome.

Other chromosome-banding techniques were developed at about the same time. The most useful of these techniques produces a staining pattern differentially along the length of each chromosome. This method, producing **G-bands** (**Figure 12–12**), involves the digestion of the mitotic chromosomes with the proteolytic enzyme trypsin, followed by Giemsa staining. The differential staining reactions reflect the heterogeneity and complexity of the chromosome along its length.

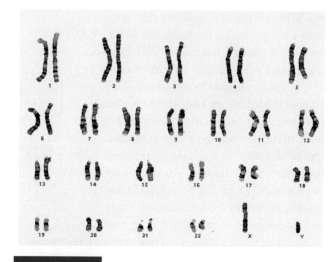

FIGURE 12–12 G-banded karyotype of a normal human male. Chromosomes were derived from cells in metaphase.

In 1971 a uniform nomenclature for human chromosome-banding patterns was established based on G-banding. **Figure 12–13** illustrates the application of this nomenclature to the X chromosome. On the left of the chromosome are the various organizational levels of banding of the p and q arms that can be identified; the resulting designation for each of the specific regions is shown on the right side.

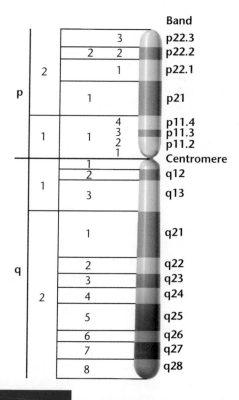

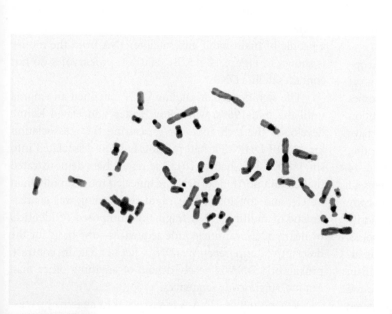

FIGURE 12–11 A human mitotic chromosome preparation processed to demonstrate C-banding. Only the centromeres stain.

FIGURE 12–13 The regions of the human X chromosome distinguished by its banding pattern. The designations on the right identify specific bands.

Although the molecular mechanisms involved in producing the various banding patterns are not well understood, the bands have played an important role in cytogenetic analysis, particularly in humans. The pattern of banding on each chromosome is unique, allowing a distinction to be made even between those chromosomes that are identical in size and centromere placement (e.g., human chromosomes 4 and 5 and 21 and 22). So precise is the banding pattern of each chromosome that homologs can be distinguished from one another, and when a segment of one chromosome has been translocated to another chromosome, its origin can be determined with great precision.

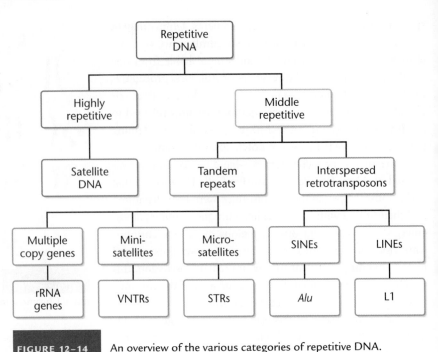

FIGURE 12–14   An overview of the various categories of repetitive DNA.

12.6

Eukaryotic Genomes Demonstrate Complex Sequence Organization Characterized by Repetitive DNA

Thus far, we have examined the general structure of chromosomes in bacteriophages, bacteria, and eukaryotes. We now begin an examination of what we know about the organization of DNA sequences within the chromosomes making up an organism's genome, placing our emphasis on eukaryotes. Once we establish certain general aspects of this organization, we will focus on how genes themselves are organized within chromosomes (see Chapter 21).

We learned in Chapter 10 that, in addition to single copies of unique DNA sequences that make up genes, a great deal of the DNA sequencing within eukaryotic chromosomes is repetitive in nature and that various levels of repetition occur within the genomes of organisms. Many studies have now provided insights into **repetitive DNA**, demonstrating various classes of sequences and organization. Figure 12–14 schematizes these categories. Some functional genes are present in more than one copy (they are referred to as **multiple-copy genes**) and so are repetitive in nature. However, the majority of repetitive sequences do not encode proteins. Nevertheless, many are transcribed, and the resultant RNA of some is involved in chromatin remodeling. We will explore three main categories of repetitive sequences: (1) heterochromatin found to be associated with centromeres and making up telomeres; (2) tandem repeats of both short and long DNA sequences; and (3) transposable sequences that are interspersed throughout the genome of eukaryotes.

Satellite DNA

The nucleotide composition (e.g., the percentage of $G{\equiv}C$ versus $A{=}T$ pairs) of the DNA of a particular species is reflected in the DNA's density, which can be measured with sedimentation equilibrium centrifugation (introduced in Chapter 10). When eukaryotic DNA is analyzed in this way, a graph describes its composition as a single main peak, representing a single main band, of fairly uniform density. However, one or more additional peaks indicate the presence of DNA that differs slightly in density. This component, called **satellite DNA**, makes up a variable proportion of the total DNA, depending on the species. For example, a profile of main-band and satellite DNA from the mouse is shown in Figure 12–15. By contrast, prokaryotes do not contain satellite DNA.

The significance of satellite DNA remained an enigma until the mid-1960s, when Roy Britten and David Kohne developed the technique for measuring the reassociation kinetics of DNA that had previously been dissociated into single strands (Chapter 10). The researchers demonstrated that certain portions of DNA reannealed more rapidly than others, and concluded that rapid reannealing was characteristic of multiple DNA fragments composed of identical or nearly identical nucleotide sequences—the basis for the descriptive term *repetitive DNA*. Recall that, in contrast, prokaryotic DNA is nearly devoid of anything other than unique, single-copy sequences.

When satellite DNA was subjected to analysis by reassociation kinetics (see Chapter 10), it fell into the category of *highly repetitive DNA,* consisting of short sequences repeated a large number of times. Further evidence suggested

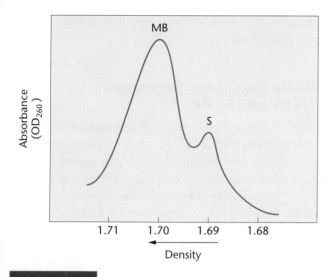

FIGURE 12–15 Separation of main-band (MB) and satellite (S) DNA from the mouse by using ultracentrifugation in a CsCl gradient.

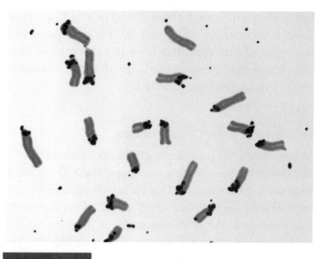

FIGURE 12–16 *In situ* hybridization between a radioactive probe representing mouse satellite DNA and mouse mitotic chromosomes. The grains in the autoradiograph are concentrated in the chromosome regions (the centromeres), revealing them to be the location of satellite DNA sequences.

that these sequences are present as tandem (meaning adjacent) repeats clustered in very specific chromosomal areas known to be heterochromatic—the regions flanking centromeres. This was discovered in 1969 when several researchers, including Mary Lou Pardue and Joe Gall, applied the technique of *in situ* **molecular hybridization** to the study of satellite DNA. This technique involves molecular hybridization between an isolated fraction of labeled DNA or RNA probes and the DNA contained in the chromosomes of a cytological preparation. While fluorescent probes are standard today, radioactive probes were used by Pardue and Gall. Following the hybridization procedure, autoradiography was performed to locate the chromosome areas complementary to the fraction of DNA or RNA.

Pardue and Gall demonstrated that radioactive molecular probes made from mouse satellite DNA hybridize with DNA of centromeric regions of mouse mitotic chromosomes (**Figure 12–16**). Several conclusions were drawn: Satellite DNA differs from main-band DNA in its molecular composition, as established by buoyant density studies. It is composed of short repetitive sequences. Finally, satellite DNA is found in the heterochromatic centromeric regions of chromosomes.

Centromeric DNA Sequences

The separation of homologs during mitosis and meiosis depends on **centromeres**, described cytologically in the late nineteenth century as the *primary constrictions* along eukaryotic chromosomes (see Chapter 2). In this role,

it is believed that the DNA sequence contained within the centromere is critical to this role. Careful analysis has confirmed this belief. The minimal region of the centromere that supports the function of chromosomal segregation is designated the **CEN region**. Within this heterochromatic region of the chromosome, the DNA binds a platform of proteins, which in multicellular organisms includes the **kinetochore** that binds to the spindle fiber during division (see Figure 12–8).

The CEN regions of the yeast *Saccharomyces cerevisiae* were the first to be studied. Each centromere serves an identical function, so it is not surprising that CENs from different chromosomes were found to be remarkably similar in their organization. The CEN region of yeast chromosomes consists of about 120 bp. Mutational analysis suggests that portions near the 3′ end of this DNA region are most critical to centromere function since mutations in them, but not those nearer the 5′ end, disrupt centromere function. Thus, the DNA of this region appears to be essential to the eventual binding to the spindle fiber.

Centromere sequences of multicellular eukaryotes are much more extensive than in yeast and vary considerably in size. For example, in *Drosophila* the CEN region is found embedded within some 200 to 600 kb of DNA, much of which is highly repetitive (recall from our prior discussion that highly repetitive satellite DNA is localized in the centromere regions of mice). In humans, one of the most recognized satellite DNA sequences is the **alphoid family**, found mainly in the centromere regions.

Alphoid sequences, each about 170 bp in length, are present in tandem arrays of up to 1 million base pairs. It is now believed such repetitive DNA in eukaryotes is transcribed and that the RNA that is produced is ultimately involved in kinetochore function.

Telomeric DNA Sequences

We now return to a topic that was introduced in Chapter 11, the **telomere**—the structure that "caps" the ends of linear eukaryotic chromosomes. Our earlier discussion focused on replication of telomeric DNA as well as on the model describing how the DNA sequences in these structures function to maintain the stability of chromosomes. The heterochromatic cap structure renders chromosome ends inert in interactions with other chromosome ends and with enzymes that use double-stranded DNA ends as substrates (such as repair enzymes). As with centromeres, the analysis of telomeres was first approached by investigating the smaller chromosomes of simple eukaryotes, such as protozoans and yeast. The idea that all telomeres of all chromosomes in related organisms might share a common nucleotide sequence has now been borne out.

Telomeric DNA sequences consist of short tandem repeats. It is this group of repetitive sequences that contributes to the stability and integrity of the chromosome. In the ciliate *Tetrahymena,* more than 50 tandem repeats of the hexanucleotide sequence 5'-TTGGGG-3' occur. In all vertebrates, including humans, the sequence 5'-TTAGGGG-3' is repeated many times. The number of copies making up a telomere varies in different organisms, and there may be as many as 1000 repeats in some species.

A central question concerns the role of these repeat DNA sequences in telomere function. Recent findings have established that the sequences are transcribed and that the RNA product, called **TERRA (telomere repeat-containing RNA)**, is an integral component of the telomere, contributing to its heterochromatic nature by facilitating methylation of the histone H3K9. In addition, these sequences have been shown to regulate telomerase, the RNA-containing enzyme that replicates the telomere (see Chapter 11). This is possible because TERRA sequences are complementary to those of the RNA component of telomerase, which provides the template for the synthesis of telomeric DNA. Serving as a telomerase ligand, TERRA acts as an inhibitor of telomerase.

Learning about telomerase regulation is of great interest, since in multicellular organisms, including humans, telomerase is active in germ-line cells but is inactive in somatic cells. And, in human cancer cells, which have become immortalized, the transition to malignancy appears to require the activation of telomerase in order to overcome the normal senescence associated with chromosome shortening.

Middle Repetitive Sequences: VNTRs and STRs

A brief look at still another prominent category of repetitive DNA sheds additional light on the organization of the eukaryotic genome. In addition to highly repetitive DNA, which constitutes about 5 percent of the human genome (and 10 percent of the mouse genome), a second category, **middle** (or **moderately**) **repetitive DNA**, recognized by C_0t analysis, is fairly well characterized. Because we now know a great deal about the human genome, we will use our own species to illustrate this category of DNA in genome organization.

Although middle repetitive DNA does include some duplicated genes (such as those encoding ribosomal RNA), most prominent in this category are either noncoding tandemly repeated sequences or noncoding interspersed sequences. No function has been ascribed to these components of the genome. An example is DNA described as **variable number tandem repeats (VNTRs)**. These repeating DNA sequences may be 15 to 100 bp long and are found within and between genes. Many such clusters are dispersed throughout the genome, and they are often referred to as **minisatellites**.

The number of tandem copies of each specific sequence at each location varies from one individual to the next, creating localized regions of 1000 to 20,000 bp (1–20 kb) in length. As we will see in Chapter 21, the variation in size (length) of these regions between individual humans was originally the basis for the forensic technique referred to as **DNA fingerprinting**.

Another group of tandemly repeated sequences consists of di-, tri-, tetra-, and pentanucleotides, also referred to as **microsatellites** or **short tandem repeats (STRs)**. Like VNTRs, they are dispersed throughout the genome and vary among individuals in the number of repeats present at any site. For example, in humans, the most common microsatellite is the dinucleotide $(CA)_n$, where n equals the number of repeats. Most commonly, n is between 5 and 50. These clusters have served as useful molecular markers for genome analysis.

Repetitive Transposed Sequences: SINEs and LINEs

Still another category of repetitive DNA consists of sequences that are interspersed individually throughout the genome, rather than being tandemly repeated. They can be either short or long, and many have the added distinction of being **transposable sequences**, which are mobile and can potentially move to different locations within the genome. A large portion of the human genome is composed of such sequences. Transposable sequences are discussed in more detail in Chapter 15.

For example, **short interspersed elements**, called **SINEs**, are less than 500 base pairs long and may be present 500,000 times or more in the human genome. The best characterized human SINE is a set of closely related sequences called the *Alu* **family** (the name is based on the presence of DNA sequences recognized by the restriction endonuclease *Alu* I). Members of this DNA family, also found in other mammals, are 200 to 300 base pairs long and are dispersed rather uniformly throughout the genome, both between and within genes. In humans, this family encompasses more than 5 percent of the entire genome.

Alu sequences are of particular interest because some members of the *Alu* family are transcribed into RNA, although the specific role of this RNA is not certain. Even so, the consequence of *Alu* sequences is their potential for transposition within the genome, which is related to chromosome rearrangements during evolution. *Alu* sequences are thought to have arisen from an RNA element whose DNA complement was dispersed throughout the genome as a result of the activity of reverse transcriptase (an enzyme that synthesizes DNA on an RNA template).

The group of **long interspersed elements (LINEs)** represents yet another category of repetitive transposable DNA sequences. LINEs are usually about 6 kb in length and in the human genome are present approximately 850,000 times. The most prominent example in humans is the **L1 family**. Members of this sequence family are about 6400 base pairs long and are present more than 500,000 times. Their 5′-end is highly variable, and their role within the genome has yet to be defined.

The general mechanism for transposition of L1 elements is now clear. The L1 DNA sequence is first transcribed into an RNA molecule. The RNA then serves as the template for synthesis of the DNA complement using the enzyme reverse transcriptase. This enzyme is encoded by a portion of the L1 sequence. The new L1 copy then integrates into the DNA of the chromosome at a new site. Because of the similarity of this transposition mechanism to that used by retroviruses, LINEs are referred to as **retrotransposons**.

SINEs and LINEs represent a significant portion of human DNA. SINEs constitute about 13 percent of the human genome, whereas LINEs constitute up to 21 percent. Within both types of elements, repeating sequences of DNA are present in combination with unique sequences.

Middle Repetitive Multiple-Copy Genes

In some cases, middle repetitive DNA includes functional genes present tandemly in multiple copies. For example, many copies exist of the genes encoding ribosomal RNA. *Drosophila* has 120 copies per haploid genome. Single genetic units encode a large precursor molecule that is processed into the 5.8S, 18S, and 28S rRNA components. In humans, multiple copies of this gene are clustered on the p arm of the acrocentric chromosomes 13, 14, 15, 21, and 22. Multiple copies of the genes encoding 5S rRNA are transcribed separately from multiple clusters found together on the terminal portion of the p arm of chromosome 1.

12.7

The Vast Majority of a Eukaryotic Genome Does Not Encode Functional Genes

Given the preceding information concerning various forms of repetitive DNA in eukaryotes, it is of interest to pose an important question: *What proportion of the eukaryotic genome actually encodes functional genes?*

We have seen that, taken together, the various forms of highly repetitive and moderately repetitive DNA comprise a substantial portion of the human genome. In addition to repetitive DNA, a large amount of the DNA consists of single-copy sequences as defined by C_0t analysis that appear to be noncoding. Included are many instances of what we call **pseudogenes**. These are DNA sequences representing evolutionary vestiges of duplicated copies of genes that have undergone significant mutational alteration. As a result, although they show some homology to their parent gene, they are usually not transcribed because of insertions and deletions throughout their structure.

While the proportion of the genome consisting of repetitive DNA varies among organisms, one feature seems to be shared: *Only a very small part of the genome actually codes for proteins.* For example, the 20,000 to 30,000 genes encoding proteins in sea urchin occupy less than 10 percent of the genome. In *Drosophila*, only 5 to 10 percent of the genome is occupied by genes coding for proteins. In humans, it appears that the coding regions of the estimated 20,000 functional genes occupy only about 2 percent of the total DNA sequence making up the genome.

Study of the various forms of repetitive DNA has significantly enhanced our understanding of genome organization. In the next chapter, we will explore the organization of genes within chromosomes.

Database of Genomic Variants: Structural Variations in the Human Genome

MG *Study Area: Exploring Genomics*

In this chapter, we focused on structural details of chromosomes and DNA sequence organization in chromosomes. As discussed in Chapter 6, we have learned that large segments of DNA and a number of genes can vary greatly in copy number due to duplications, creating **copy number variations (CNVs).** Many studies are underway to identify and map CNVs and to find possible disease conditions associated with them.

To date, approximately 2000 CNVs have been identified in the human genome, and estimates suggest there may be thousands more within human populations. In this Exploring Genomics exercise we will visit the **Database of Genomic Variants (DGV)**, which provides a quickly expanding summary of structural variations in the human genome including CNVs.

- **Exercise I – Database of Genomic Variants**

1. Access the DGV at http://projects .tcag.ca/variation. Click the "About the Project" tab to learn more about the purpose of the DGV.

2. Information in the DGV is easily viewed by clicking on a chromosome of interest using the "View Data by Chromosome" feature or by clicking on a chromosome using the "View Data by Genome" feature.

3. Click on a chromosome of interest to you using the "View Data by Chromosome" feature. A table will appear showing several columns of data including:

- Locus: Shows the locus for the CNV, including the base pairs that span the variation.

- Landmark: Shows different variations of CNVs for a particular locus.

- Variation ID: Provides a unique identifying number for each variation.

- Variation Type: Listed as "copy number." Most variations in this database are CNVs. Variations known to be insertions or deletions based on relatively small changes (a few bases) are labeled "InDel." Inversions labeled "Inv."

- Cytoband: Indicates the chromosomal banding location for the variation.

- Position (Mb): Shows the relative location in megabases (Mb) on the chromosome.

- Known Genes in the Locus: Lists genes that are located in a particular CNV.

4. Let's analyze a particular group of CNVs. Many CNVs are unlikely to affect phenotype because they involve large areas of non–protein-coding or nonregulatory sequences. But gene-containing CNVs have been identified, including variants containing genes associated with Alzheimer's disease, Parkinson's disease, and other conditions.

Defensin (*DEF*) genes are part of a large family of highly duplicated genes. To learn more about *DEF* genes and CNVs, use the Keyword Search box to search for *DEF* genes (click "No" for the exact match button). A results page for the search will appear with a listing of CNVs. Click on one of the links shown.

The top part of each report has graphs indicating the position of each CNV on a chromosome. Scroll down to the bottom of this page until you see "Known Genes." Click on the different *DEF* genes listed in the known genes category, which will take you to the National Center for Biotechnology Information (NCBI) Entrez site with a wealth of information about these genes so that you can answer the following questions. Do this for several *DEF*-containing CNVs on different chromosomes to find the information you will need for your answers.

On what chromosome(s) did you find CNVs containing *DEF* genes?

(a) What did you learn about the function of *DEF* gene products? What do DEF proteins do?

(b) Variations in *DEF* genotypes and *DEF* gene expression in humans have been implicated in a number of different human disease conditions. Give examples of the kinds of disorders affected by variations in *DEF* genotypes.

(c) Explore the DGV to search a chromosome of interest to you and learn more about CNVs that have been mapped to that chromosome. Try the "View Data by Genome" feature that will show you maps of each chromosome indicating different variations. For CNVs (shown in blue), clicking on the CNV will take you to its locus on the chromosome.

CASE STUDY | Art inspires learning

A genetics student visiting a museum saw a painting by Goya showing a woman with a newborn baby in her lap that had very short arms and legs along with some facial abnormalities. Wondering whether this condition might be a genetic disorder, the student went online, learning that the baby might have Roberts syndrome (RBS), a rare autosomal recessive trait. She read that cells in RBS have mitotic errors, including the premature separation of centromeres and other heterochromatic regions of homologs in metaphase instead of anaphase. As a result, metaphase chromosomes have a rigid, or "railroad track" appearance. RBS has been shown to be caused by mutant alleles of the *ESCO2* gene, which functions during cell division.

The student wrote a list of questions to investigate in an attempt to better understand this condition. How would you answer these questions?

1. What do centromeres and other heterochromatic regions have in common that might cause this appearance?
2. What might be the role of the protein encoded by *ESCO2*, which in mutant form could cause these changes in mitotic chromosomes?
3. How could premature separation of centromeres cause the problems seen in RBS?

Summary Points

For activities, animations, and review quizzes, go to the study area at www.masteringgenetics.com

1. Bacteriophage and bacterial chromosomes, in contrast to those of eukaryotes, are largely devoid of associated proteins, are of much smaller size, and, with some exceptions, consist of circular DNA.
2. Polytene and lampbrush chromosomes are examples of specialized structures that extended our knowledge of genetic organization and function well in advance of the technology available to the modern-day molecular biologist.
3. Eukaryotic chromatin is a nucleoprotein organized into repeating units called nucleosomes, which are composed of about 200 base pairs of DNA, an octamer of four types of histones, plus one linker histone.
4. Nucleosomes provide a mechanism for compaction of chromatin within the nucleus. Several forms of chemical modification, for example, acetylation and methylation, may alter the level of compaction, a process referred to as chromatin remodeling, which is critical to replication and transcription of DNA.

5. Heterochromatin, prematurely condensed in interphase and for the most part genetically inert, is illustrated by centromeric and telomeric regions of eukaryotic chromosomes, the Y chromosome, and the Barr body.
6. Chromosome banding techniques provide a way to subdivide and identify specific regions of mitotic chromosomes.
7. Eukaryotic genomes demonstrate complex sequence organization characterized by numerous categories of repetitive DNA, consisting of either tandem repeats clustered in various regions of the genome or single sequences repeatedly interspersed at random throughout the genome.
8. The vast majority of the DNA in most eukaryotic genomes does not encode functional genes. In humans, for example, only about 2 percent of the genome is used to encode the 20,000 genes found there.

INSIGHTS AND SOLUTIONS

A previously undiscovered single-celled organism was found living at a great depth on the ocean floor. Its nucleus contains only a single, linear chromosome consisting of 7×10^6 nucleotide pairs of DNA coalesced with three types of histonelike proteins. Consider the following questions:

1. A short micrococcal nuclease digestion yielded DNA fractions consisting of 700, 1400, and 2100 base pairs. Predict what these fractions represent. What conclusions can be drawn?

   Solution: The chromatin fiber may consist of a nucleosome variant containing 700 base pairs of DNA. The 1400- and 2100-bp fractions represent two and three of these nucleosomes, respectively, linked together. Enzymatic digestion may have been incomplete, leading to the latter two fractions.

2. The analysis of individual nucleosomes revealed that each unit contained one copy of each protein and that the short

linker DNA had no protein bound to it. If the entire chromosome consists of nucleosomes (discounting any linker DNA), how many are there, and how many total proteins are needed to form them?

Solution: Since the chromosome contains 7×10^6 base pairs of DNA, the number of nucleosomes, each containing 7×10^2 base pairs, is equal to

$$7 \times 10^6 / 7 \times 10^2 = 10^4 \text{ nucleosomes}$$

The chromosome thus contains 10^4 copies of each of the three proteins, for a total of 3×10^4 proteins.

3. Analysis then revealed the organism's DNA to be a double helix similar to the Watson–Crick model, but containing 20 base pairs per complete turn of the right-handed helix. The physical size of the nucleosome was exactly double the

Insights and Solutions, continued

volume occupied by the nucleosome found in any other known eukaryote, and the nucleosome's axis length was greater by a factor of two. Compare the degree of compaction (the number of turns per nucleosome) of this organism's nucleosome with that found in other eukaryotes.

Solution: The unique organism compacts a length of DNA consisting of 35 complete turns of the helix (700 base pairs per nucleosome/20 base pairs per turn) into each nucleosome. The normal eukaryote compacts a length of DNA consisting of 20 complete turns of the helix (200 base pairs per nucleosome/10 base pairs per turn) into a nucleosome half the volume of that in the unique organism. The degree of compaction is therefore less in the unique organism.

4. No further coiling or compaction of this unique chromosome occurs in the newly discovered organism. Compare this situation with that of a eukaryotic chromosome. Do you think an interphase human chromosome 7×10^6 base pairs in length would be a shorter or longer chromatin fiber?

Solution: The eukaryotic chromosome contains still another level of condensation in the form of solenoids, which are coils consisting of nucleosomes connected with linker DNA. Solenoids condense the eukaryotic fiber by still another factor of five. The length of the unique chromosome is compacted into 10^4 nucleosomes, each containing an axis length twice that of the eukaryotic fiber. The eukaryotic fiber consists of $7 \times 10^6/2 \times 10^2 = 3.5 \times 10^4$ nucleosomes, 3.5 times more than the unique organism. However, they are compacted by the factor of five in each solenoid. Therefore, the chromosome of the unique organism is a longer chromatin fiber.

Problems and Discussion Questions

 For instructor-assigned tutorials and problems, go to www.masteringgentics.com

HOW DO WE KNOW?

1. In this chapter, we focused on how DNA is organized at the chromosomal level. Along the way, we found many opportunities to consider the methods and reasoning by which much of this information was acquired. From the explanations given in the chapter, what answers would you propose to the following fundamental questions:

    (a) How do we know that viral and bacterial chromosomes most often consist of circular DNA molecules devoid of protein?

    (b) What is the experimental basis for concluding that puffs in polytene chromosomes and loops in lampbrush chromosomes are areas of intense transcription of RNA?

    (c) How did we learn that eukaryotic chromatin exists in the form of repeating nucleosomes, each consisting of about 200 base pairs and an octamer of histones?

    (d) How do we know that satellite DNA consists of repetitive sequences and has been derived from regions of the centromere?

2. Contrast the size of the single chromosome in bacteriophage λ and T2 with that of *E. coli*. How does this relate to the relative size and complexity of phages and bacteria?

3. Describe the structure of giant polytene chromosomes and how they arise.

4. What genetic process is occurring in a puff of a polytene chromosome? How do we know this experimentally?

5. Describe the structure of LINE sequences. Why are LINEs referred to as retrotransposons?

6. During what genetic process are lampbrush chromosomes present in vertebrates?

7. Why might we predict that the organization of eukaryotic genetic material will be more complex than that of viruses or bacteria?

8. Describe the sequence of research findings that led to the development of the model of chromatin structure.

9. Describe the molecular composition and arrangement of the components in the nucleosome.

10. Describe the transitions that occur as nucleosomes are coiled and folded, ultimately forming a chromatid.

11. Provide a comprehensive definition of heterochromatin and list as many examples as you can.

12. Mammals contain a diploid genome consisting of at least 10^9 bp. If this amount of DNA is present as chromatin fibers, where each group of 200 bp of DNA is combined with 9 histones into a nucleosome and each group of 6 nucleosomes is combined into a solenoid, achieving a final packing ratio of 50, determine (a) the total number of nucleosomes in all fibers, (b) the total number of histone molecules combined with DNA in the diploid genome, and (c) the combined length of all fibers.

13. Assume that a viral DNA molecule is a 50-μm-long circular strand with a uniform 20-Å diameter. If this molecule is contained in a viral head that is a 0.08-μm-diameter sphere, will the DNA molecule fit into the viral head, assuming complete flexibility of the molecule? Justify your answer mathematically.

14. How many base pairs are in a molecule of phage T2 DNA 52 μm long?

15. Examples of histone modifications are acetylation (by histone acetyltransferase, or HAT), which is often linked to gene activation, and deacetylation (by histone deacetylases, or HDACs), which often leads to gene silencing typical of heterochromatin. Such heterochromatinization is initiated from a nucleation site and spreads bidirectionally until encountering boundaries that delimit the silenced areas. Recall from Chapter 4 the brief discussion of position effect, where repositioning of the w^+ allele in *Drosophila* by translocation or inversion near heterochromatin produces intermittent w^+ activity. In the heterozygous state (w^+/w), a variegated eye is produced, with white and red patches. How might one explain position-effect variegation in terms of histone acetylation and/or deacetylation?

16. In light of indications that there are at least 19,000 pseudogenes in the human genome and perhaps as few as 21,000 protein-coding genes, some researchers suggest that as assays become more refined, the number of pseudogenes may outnumber protein-coding genes (Gerstein and Zheng, 2006).

    (a) Why would it take more time to locate pseudogenes than protein-coding genes?

    (b) Most pseudogenes are damaged remains of previously functional genes and are considered to be nonfunctional.

However, it has been suggested that some pseudogenes do function. In light of the various ways in which DNA is organized in chromosomes, how might pseudogenes function?

17. Variable number tandem repeats (VNTRs) are repeating DNA sequences of about 15 to 100 bp in length, found both within and between genes. Why are they commonly used in forensics?

18. A number of recent studies have determined that disease pathogenesis, whether it be related to viruses, cancer, aging, or a host of other causes, is often associated with specific changes in DNA methylation. If such patterns are to be considered as biomarkers for disease diagnosis what requisite criteria would you consider essential to their use?

19. It has been shown that infectious agents such as viruses often exert a dramatic effect on their host cell's genome architecture.

In many cases, viruses induce methylation of host DNA sequences in order to enhance their infectivity. What specific host gene functions would you consider as strong candidates for such methylation by infecting viruses?

20. Cancer can be defined as an abnormal proliferation of cells that defy the normal regulatory controls observed by normal cells. Recently, histone deacetylation therapies have been attempted in the treatment of certain cancers (reviewed by Delcuve et al., 2009). Specifically, the FDA has approved histone deacetylation (HDAC) inhibitors for the treatment of cutaneous T-cell lymphoma. Explain why histone acetylation might be associated with cancer and what the rationale is for the use of HDAC inhibitors in the treatment of certain forms of cancer?

Extra-Spicy Problems

 For instructor-assigned tutorials and problems, go to www.masteringgentics.com

21. In a study of *Drosophila*, two normally active genes, w^+ (wild-type allele of the *white*-eye gene) and *hsp*26 (a heat-shock gene), were introduced (using a plasmid vector) into euchromatic and heterochromatic chromosomal regions, and the relative activity of each gene was assessed (Sun et al., 2002). An approximation of the resulting data is shown in the following table. Which characteristic or characteristics of heterochromatin are supported by the experimental data?

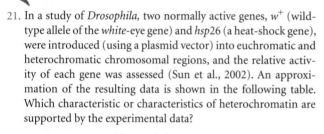

Gene	Activity (relative percentage)	
	Euchromatin	Heterochromatin
*hsp*26	100%	31%
w^+	100%	8%

22. While much remains to be learned about the role of nucleosomes and chromatin structure and function, recent research indicates that *in vivo* chemical modification of histones is associated with changes in gene activity. One study determined that acetylation of H3 and H4 is associated with 21.1 percent and 13.8 percent increases in yeast gene activity, respectively, and that yeast heterochromatin is hypomethylated relative to the genome average (Bernstein et al., 2000). Speculate on the significance of these findings in terms of nucleosome–DNA interactions and gene activity.

23. An article entitled "Nucleosome Positioning at the Replication Fork" states, "both the 'old' randomly segregated nucleosomes as well as the 'new' assembled histone octamers rapidly position themselves (within seconds) on the newly replicated DNA strands" (Lucchini et al., 2002). Given this statement, how would one compare the distribution of nucleosomes and DNA in newly replicated chromatin? How could one experimentally test the distribution of nucleosomes on newly replicated chromosomes?

24. The human genome contains approximately 10^6 copies of an *Alu* sequence, one of the best-studied classes of short interspersed elements (SINEs), per haploid genome. Individual *Alus* share a 282-nucleotide consensus sequence followed by a 3'-adenine-rich tail region (Schmid, 1998). Given that there are approximately 3×10^9 base pairs per human haploid genome, about how many base pairs are spaced between each *Alu* sequence?

25. Following is a diagram of the general structure of the bacteriophage λ chromosome. Speculate on the mechanism by which it forms a closed ring upon infection of the host cell.

5′ GGGCGGCGACCT—double-stranded region—3′
          3′—double-stranded region—CCCGCCGCTGGA 5′

26. Tandemly repeated DNA sequences with a repeat sequence of one to six base pairs—for example, $(GACA)_n$—are called microsatellites and are common in eukaryotes. A particular subset of such sequences, the trinucleotide repeat, is of great interest because of the role such repeats play in human neurodegenerative disorders (Huntington disease, myotonic dystrophy, spinal-bulbar muscular atrophy, spinocerebellar ataxia, and fragile-X syndrome). Following are data (modified from Toth et al., 2000) regarding the location of microsatellites within and between genes. What general conclusions can be drawn from these data?

Percentage of Microsatellite DNA Sequences within Genes and between Genes		
Taxonomic Group	**Within Genes**	**Between Genes**
Primates	7.4	92.6
Rodents	33.7	66.3
Arthropods	46.7	53.3
Yeasts	77.0	23.0
Other fungi	66.7	33.3

27. More information from the research effort in Problem 26 produced data regarding the pattern of the length of such repeats within genes. Each value in the following table represents the number of times a microsatellite of a particular sequence length, one to six bases long, is found within genes. For instance, in primates, a dinucleotide sequence (GC, for example) is found 10 times, while a trinucleotide is found 1126 times. In fungi, a repeat motif composed of 6 nucleotides (GACACC, for example) is found 219 times, whereas a tetranucleotide repeat (GACA, for example) is found only 2 times. Analyze and interpret these data by indicating what general pattern is

apparent for the distribution of various microsatellite lengths within genes. Of what significance might this general pattern be?

Distribution of Microsatellites by Unit Length within Genes

Taxonomic Group	Length of Repeated Motif (bp)					
	1	2	3	4	5	6
Primates	49	10	1126	29	57	244
Rodents	62	70	1557	63	116	620
Arthropods	12	34	1566	0	21	591
Yeasts	36	19	706	7	52	330
Other fungi	9	4	381	2	35	219

28. Microsatellites are currently exploited as markers for paternity testing. A sample paternity test is shown in the following table in which ten microsatellite markers were used to test samples from a mother, her child, and an alleged father. The name of the microsatellite locus is given in the left-hand column, and the genotype of each individual is recorded as the number of repeats he or she carries at that locus. For example, at locus D9S302, the mother carries 30 repeats on one of her chromosomes and 31 on the other. In cases where an individual carries the same number of repeats on both chromosomes, only a single number is recorded. (Some of the numbers are followed by a decimal point, for example, 20.2, to indicate a partial repeat in addition to the complete repeats.) Assuming that these markers are inherited in a simple Mendelian fashion, can the alleged father be excluded as the source of the sperm that produced the child? Why or why not? Explain.

Microsatellite Locus-Chromosome Location	Mother	Child	Alleged Father
D9S302-9q31-q33	30	31	32
	31	32	33
D22S883-22pter-22qter	17	20.2	20.2
	22	22	
D18S535-18q12.2-q12.3	12	13	11
	14	14	13
D7SI 804-7pter-7qter	27	26	26
	30	30	27
D3S2387-3p24.2.3pter	23	24	20.2
	25.2	25.2	24
D4S2386-4pter-qter	12	12	12
			16
D5S1719-5pter-5qter	11	10.3	10
	11.3	11	10.3
CSF1PO-5q33.3.q34	11	11	10
		12	12
FESFPS-15q25-15qter	11	12	10
	12	13	13
TH01-11p15.5	7	7	7
			8

29. At the end of the short arm of human chromosome 16 (16p), several genes associated with disease are present, including thalassemia and polycystic kidney disease. When that region of chromosome 16 was sequenced, gene-coding regions were found to be very close to the telomere-associated sequences. Could there be a possible link between the location of these genes and the presence of the telomere-associated sequences? What further information concerning the disease genes would be useful in your analysis?

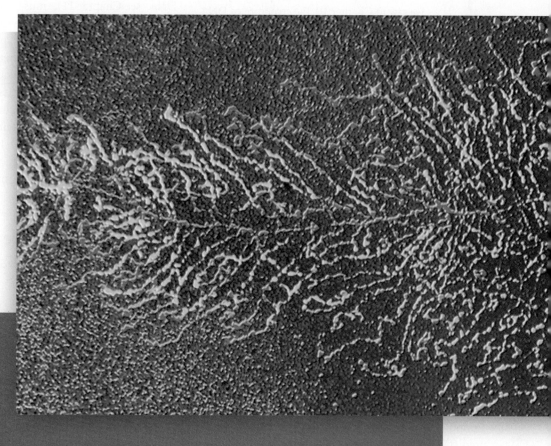

Electron micrograph of a segment of DNA undergoing transcription.

13

The Genetic Code and Transcription

CHAPTER CONCEPTS

- Genetic information is stored in DNA by means of a triplet code that is nearly universal to all living things on Earth.

- The genetic code is initially transferred from DNA to RNA, in the process of transcription.

- Once transferred to RNA, the genetic code exists as triplet codons, which are sets of three nucleotides in which each nucleotide is one of the four kinds of ribonucleotides composing RNA.

- RNA's four ribonucleotides, analogous to an alphabet of four "letters," can be arranged into 64 different three-letter sequences. Most of the triplets in RNA encode one of the 20 amino acids present in proteins, which are the end products of most genes.

- Several codons act as signals that initiate or terminate protein synthesis.

- In eukaryotes, the process of transcription is similar to, but more complex than, that in prokaryotes and in the bacteriophages that infect them.

As we saw in Chapter 10, the structure of DNA consists of a linear sequence of deoxyribonucleotides. This sequence ultimately dictates the components of proteins, the end products of most genes. A central issue is how information stored as a nucleic acid can be decoded into a protein. Figure 13–1 provides a simple overview of how this transfer of information, resulting ultimately in gene expression, occurs. In the first step, information present on one of the two strands of DNA (the template strand) is transferred into an RNA complement through the process of transcription. Once synthesized, this RNA acts as a "messenger" molecule, transporting the coded information out of the nucleus—hence its name, *messenger RNA (mRNA)*. The mRNAs then associate with ribosomes, where decoding into proteins occurs.

In this chapter, we will focus on the initial phases of gene expression by addressing two major questions. First, how is genetic information encoded? Second, how does the transfer from DNA to RNA occur, thus explaining the process of transcription? As you will see, ingenious analytical research has established that the genetic code is written in units of three letters—triplets of ribonucleotides in mRNA that reflect the stored information in genes. Most of the triplet code words direct the incorporation of a specific amino acid into a protein as it is synthesized. As we can predict based on the complexity of the replication of DNA (see Chapter 11), transcription is also a complex process dependent on a major polymerase enzyme and a cast of supporting proteins. We will explore what is known about transcription in bacteria and then contrast this prokaryotic model with transcription in eukaryotes.

We will continue our discussion of gene expression by addressing how translation occurs and then describing the structure and function of proteins (Chapter 14). Together, these chapters provide a comprehensive picture of molecular genetics, which serves as the most basic foundation for the understanding of living organisms.

13.1

The Genetic Code Uses Ribonucleotide Bases as "Letters"

Before we consider the various analytical approaches that led to our current understanding of the genetic code, let's summarize the general features that characterize it:

1. The genetic code is written in linear form, using as "letters" the ribonucleotide bases that compose mRNA molecules. The ribonucleotide sequence is derived from the complementary nucleotide bases in DNA.

2. Each "word" within the mRNA consists of three ribonucleotide letters, thus referred to as a **triplet code**. With several exceptions, each group of *three* ribonucleotides, called a **codon**, specifies *one* amino acid.

3. The code is **unambiguous**—each triplet specifies only a single amino acid.

4. The code is **degenerate**, meaning that a given amino acid can be specified by more than one triplet codon. This is the case for 18 of the 20 amino acids.

5. The code contains one "start" and three "stop" signals, triplets that **initiate** and **terminate** translation, respectively.

6. No internal punctuation (analogous, for example, to a comma) is used in the code. Thus, the code is said to be **commaless**. Once translation of mRNA begins, the codons are read one after the other with no breaks between them (until a stop signal is reached).

7. The code is **nonoverlapping**. After translation commences, any single ribonucleotide within the mRNA is part of only one triplet.

8. The sequence of codons in a gene is colinear, with the sequence of amino acids making up the encoded protein.

9. The code is nearly **universal**. With only minor exceptions, a single coding dictionary is used by almost all viruses, prokaryotes, archaea, and eukaryotes.

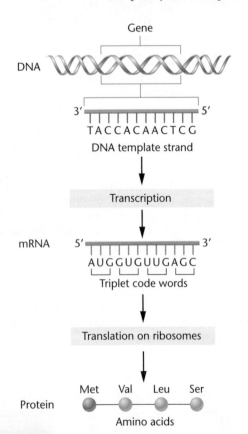

FIGURE 13–1 Flowchart illustrating how genetic information encoded in DNA produces protein.

Early Studies Established the Basic Operational Patterns of the Code

In the late 1950s, before it became clear that mRNA serves as an intermediate in the transfer of genetic information from DNA to proteins, researchers thought that DNA might encode protein synthesis more directly. Because ribosomes had already been identified, the initial thinking was that information in DNA was transferred in the nucleus to the RNA of the ribosome, which served as the template for protein synthesis in the cytoplasm. This concept soon became untenable, as accumulating evidence indicated that there was an unstable intermediate template. The RNA of ribosomes, on the other hand, was extremely stable. In 1961, François Jacob and Jacques Monod postulated the existence of **messenger RNA (mRNA)**. After mRNA was discovered, it soon became clear that, even though genetic information is stored in DNA, the code that is translated into proteins resides in RNA. The central question then was how only four letters—the four nucleotides—could specify 20 words, the amino acids? Scientists considered the possibility that DNA codes could overlap in such a way that each nucleotide would be part of more than one code word, but accumulating evidence was inconsistent with the constraints such a code would place on the resultant amino acid sequences. We will address some of this evidence below.

The Triplet Nature of the Code

In the early 1960s, Sydney Brenner argued on theoretical grounds that the code must be a triplet, since three-letter words represent the minimal length needed for using four letters to specify 20 amino acids. A code of four nucleotides combined into two-letter words, for example, provides only 16 unique code words (4^2). A triplet code yields 64 words (4^3) and therefore is sufficient for the 20 needed but far less cumbersome than a four-letter code (4^4), which would specify 256 words.

The ingenious experimental work of Francis Crick, Leslie Barnett, Brenner, and R. J. Watts-Tobin provided the first solid evidence for a triplet code. These researchers induced insertion and deletion mutations in the *rII* locus of phage T4 (see Chapter 6). Wild type phage cause lysis and plaque formation in strains B and K12 of *E. coli*. However, *rII* mutations prevent this infection process in strain K12, while allowing infection in strain B. Crick and his colleagues used the acridine dye *proflavin* to induce the mutations. This mutagenic agent intercalates within the double helix of DNA, causing the insertion or deletion of one or more nucleotides during replication. As shown in Figure 13–2(a), an insertion of a single nucleotide causes the **reading frame** (the contiguous sequence of nucleotides encoding a polypeptide) to shift,

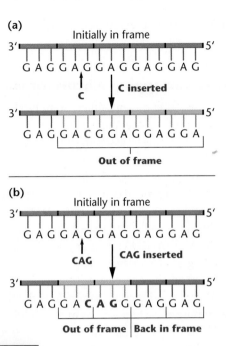

FIGURE 13–2   The effect of frameshift mutations on a DNA sequence repeating the triplet sequence GAG. (a) The insertion of a single nucleotide shifts all subsequent triplets out of the reading frame. (b) The insertion of three nucleotides changes only two triplets, after which the original reading frame is reestablished.

changing all subsequent downstream codons (codons to the right of the insertion in the figure). Such mutations are referred to as **frameshift mutations**. Upon translation, the amino acid sequence of the encoded protein will be altered radically starting at the point where the mutation occurred. When these mutations are present at the *rII* locus, T4 will not reproduce on *E. coli* K12.

Crick and his colleagues reasoned that if phages with these induced frameshift mutations were treated again with proflavin, still other insertions or deletions would occur. A second change might result in a revertant phage, which would display wild-type behavior and successfully infect *E. coli* K12. For example, if the original mutant contained an insertion (+), a second event causing a deletion (−) close to the insertion would restore the original reading frame. In the same way, an event resulting in an insertion (+) might correct an original deletion (−).

By studying many mutations of this type, these researchers were able to compare combinations of various frameshift mutations within the same DNA sequence. They found that various combinations of one plus (+) and one minus (−) caused reversion to wild-type behavior. Still other observations shed light on the number of nucleotides constituting the genetic code. When two pluses or two minuses occurred in the same sequence, the correct reading frame *was not* reestablished. This argued against a doublet (two-letter) code. However, when three pluses [Figure 13–2(b)] or three minuses were present together, the original frame *was* reestablished. These observati.ons strongly supported the triplet nature of the code.

The Nonoverlapping Nature of the Code

Work by Sydney Brenner and others soon established that the code could not be overlapping for any one transcript. For example, beginning by assuming a triplet code, Brenner deduced the restrictions that might be expected if the code were overlapping. He imagined theoretical nucleotide sequences encoding a protein consisting of three amino acids. If the nucleotide sequence were, say, GTACA, and parts of the central codon, TAC, were shared by the outer codons, GTA and ACA, then only certain amino acids (those with codons ending in TA or beginning with AC) should be found adjacent to the one encoded by the central codon. This consideration led Brenner to conclude that if the code were overlapping, the compositions of tripeptide sequences within proteins should be somewhat limited.

For example, when any particular central amino acid is considered, only 16 combinations (2^4) of three amino acids (tripeptide sequences) would theoretically be possible. Looking at the amino acid sequences of proteins that were known at that time, Brenner failed to find such restrictions in tripeptide sequences. For any central amino acid, he found many more than 16 different tripeptides. This observation led him to conclude that the code is not overlapping.

A second major argument against an overlapping code concerned the effect of a single-nucleotide change. With an overlapping code, two adjacent amino acids would be affected by such a point mutation. However, mutations in the genes coding for the protein coat of tobacco mosaic virus (TMV), human hemoglobin, and the bacterial enzyme tryptophan synthetase invariably revealed only single amino acid changes.

The third argument against an overlapping code was presented by Francis Crick in 1957, when he predicted that DNA would not serve as a direct template for the formation of proteins. Crick reasoned that any affinity between nucleotides and an amino acid would require hydrogen bonding. Chemically, however, such specific affinities seemed unlikely. Instead, Crick proposed that there must be an "adaptor molecule" that could covalently bind to the amino acid, yet also be capable of hydrogen bonding to a nucleotide sequence. With an overlapping code, various adaptors would somehow have to overlap one another at nucleotide sites during translation, making the translation process overly complex, in Crick's opinion, and possibly inefficient. As we will see later in this chapter, Crick's prediction was correct—*transfer RNA* (*tRNA*) serves as the adaptor in protein synthesis, and the ribosome accommodates two tRNA molecules at a time.

Crick's and Brenner's arguments, taken together, strongly suggested that, during translation, the genetic code is nonoverlapping. Without exception, this concept has been upheld.

The Commaless and Degenerate Nature of the Code

Between 1958 and 1960, information relating to the genetic code continued to accumulate. In addition to his adaptor proposal, Crick hypothesized, on the basis of genetic evidence, that the code would be *commaless*—that is, he believed no internal punctuation would occur along the reading frame. Crick also speculated that only 20 of the 64 possible codons would specify an amino acid and that the remaining 44 would carry no coding assignment.

Was Crick wrong with respect to the 44 "blank" codes? Or is the code degenerate, meaning that more than one codon specifies the same amino acid? Crick's frameshift studies (discussed earlier in this chapter) suggested that, in fact, contrary to his earlier proposal, the code is degenerate. For the cases in which wild-type function is restored—that is, (+) with (−); (++) with (−−); and (+++) with (−−−)—the original frame of reading is also restored. However, in between the insertion and deletion, there may be numerous codons that would still be out of frame. If 44 of the 64 possible codons were blank, referred to as **nonsense codons**, and did not specify an amino acid, at least one blank codon would very likely occur in the string of nucleotides still out of frame. If such a nonsense codon was encountered during protein synthesis, the process would probably stop or be terminated at that point. If so, the product of the *rII* locus would not be made, and restoration would not occur. Because the various mutant combinations were able to reproduce on *E. coli* K12, Crick and his colleagues concluded that, in all likelihood, most, if not all, of the remaining 44 triplets were not blank. It followed that the genetic code was degenerate. As we shall see, this reasoning proved to be correct.

13.3

Studies by Nirenberg, Matthaei, and Others Led to Deciphering of the Code

In 1961, Marshall Nirenberg and J. Heinrich Matthaei became the first to characterize specific coding sequences, laying a cornerstone for the complete analysis of the genetic code. Their success, as well as that of others who made important contributions in deciphering the code, was dependent on the use of two experimental tools, an *in vitro* (*cell-free in a test tube*) *protein-synthesizing system* and the enzyme **polynucleotide phosphorylase**, which allowed the production of synthetic mRNAs. These mRNAs served as templates for polypeptide synthesis in the cell-free system.

Synthesizing Polypeptides in a Cell-Free System

In the cell-free protein-synthesizing system, amino acids are incorporated into polypeptide chains. The process begins with an *in vitro* mixture containing all the essential factors for protein synthesis in the cell: ribosomes, tRNAs, amino acids, and other molecules essential to translation (see Chapter 14). To allow scientists to follow (or trace) the progress of protein synthesis, one or more of the amino acids must be radioactive. Finally, an mRNA must be added, to serve as the template to be translated.

In 1961, mRNA had yet to be isolated. However, use of the enzyme polynucleotide phosphorylase allowed the artificial synthesis of RNA templates, which could be added to the cell-free system. This enzyme, isolated from bacteria, catalyzes the reaction shown in Figure 13–3. Discovered in 1955 by Marianne Grunberg-Manago and Severo Ochoa, the enzyme functions metabolically in bacterial cells to degrade RNA. However, *in vitro*, in the presence of high concentrations of ribonucleoside diphosphates, the reaction can be "forced" in the opposite direction, to synthesize RNA, as illustrated in the figure.

In contrast to RNA polymerase (which constructs mRNA *in vivo*), polynucleotide phosphorylase does not require a DNA template. As a result, the order in which ribonucleotides are added is random, depending on the relative concentration of the four ribonucleoside diphosphates present in the reaction mixture. The probability of the insertion of a specific ribonucleotide is proportional to the availability of that molecule relative to other available ribonucleotides. *This point is absolutely critical to understanding the work of Nirenberg and others in the ensuing discussion.*

Together, the cell-free system for protein synthesis and the availability of synthetic mRNAs provided a means of deciphering the ribonucleotide composition of various codons encoding specific amino acids.

Homopolymer Codes

For their initial experiments, Nirenberg and Matthaei synthesized **RNA homopolymers**, RNA molecules containing

TABLE 13.1

Incorporation of ^{14}C-Phenylalanine into Protein

Artificial mRNA	Radioactivity (counts/min)
None	44
Poly U	39,800
Poly A	50
Poly C	38

Source: After Nirenberg and Matthaei (1961).

only one type of ribonucleotide, and used them for synthesizing polypeptides *in vitro*. In other words, the mRNA they used in their cell-free protein-synthesizing system was either UUUUUU . . ., AAAAAA . . ., CCCCCC . . ., or GGGGGG. . . . They tested each of these types of mRNA to see which, if any, amino acids were consequently incorporated into newly synthesized proteins. The method they used was to conduct many experimental syntheses with each homopolymer. They always made all 20 amino acids available, but for each experiment they attached a radioactive label to a different amino acid and thus could tell when that amino acid had been incorporated into the resulting polypeptide.

For example, in experiments using ^{14}C-phenylalanine (Table 13.1), Nirenberg and Matthaei concluded that the RNA homopolymer UUUUU . . . (polyuridylic acid) directed the incorporation of only phenylalanine into the peptide homopolymer polyphenylalanine. Assuming the validity of a triplet code, they made the first specific codon assignment: UUU codes for phenylalanine. Using similar experiments, they quickly found that AAA codes for lysine and CCC codes for proline. Poly G was not a functional template, probably because the molecule folds back on itself. Thus, the assignment for GGG had to await other approaches.

Note that the specific triplet codon assignments were possible only because homopolymers were used. In this method, only the general nucleotide composition of the template is known, not the specific order of the nucleotides in each triplet, but since three identical letters can have only one possible sequence (e.g., UUU), three of the actual codons for phenylalanine, lysine, and proline could be identified.

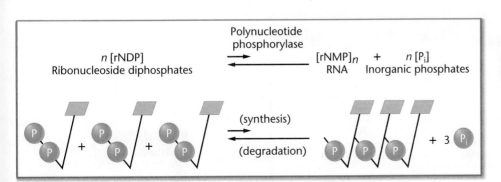

$$ n \, [\text{rNDP}] \quad \xrightarrow[]{\text{Polynucleotide phosphorylase}} \quad [\text{rNMP}]_n \; + \; n \, [\text{P}_i] $$

Ribonucleoside diphosphates → RNA   Inorganic phosphates

(synthesis) / (degradation)

+ 3 P$_i$

FIGURE 13–3 The reaction catalyzed by the enzyme polynucleotide phosphorylase. Note that the equilibrium of the reaction favors the degradation of RNA but that the reaction can be "forced" in the direction favoring synthesis.

Mixed Copolymers

With the initial success of these techniques, Nirenberg and Matthaei, and Ochoa and coworkers turned to the use of **RNA heteropolymers**. In their next experiments, two or more different ribonucleoside diphosphates were used in combination to form the artificial message. The researchers reasoned that if they knew the relative proportion of each type of ribonucleoside diphosphate in their synthetic mRNA, they could predict the frequency of each of the possible triplet codons it contained. If they then added the mRNA to the cell-free system and ascertained the percentage of each amino acid present in the resulting polypeptide, they could analyze the results and predict the *composition* of the triplets that had specified those particular amino acids.

This approach is illustrated in Figure 13–4. Suppose that only A and C are used for synthesizing the mRNA, in a ratio of 1A:5C. The insertion of a ribonucleotide at any position along the RNA molecule during its synthesis is determined by the ratio of A:C. Therefore, there is a 1/6 possibility for an A and a 5/6 chance for a C to occupy each position. On this basis, we can calculate the frequency of any given triplet appearing in the message.

For AAA, the frequency is $(1/6)^3$ or about 0.4 percent. For AAC, ACA, and CAA, the frequencies are identical—that is, $(1/6)^3 (5/6)$ or about 2.3 percent for each triplet. Together, all three 2A:1C triplets account for 6.9 percent of the total three-letter sequences. In the same way, each of three 1A:2C triplets accounts for $(1/6) (5/6)^2$ or 11.6 percent (or a total of 34.8 percent); CCC is represented by $(5/6)^3$, or 57.9 percent of the triplets.

By examining the percentages of the different amino acids incorporated into the polypeptide synthesized under the direction of this message, we can propose probable base compositions for each of those amino acids (Figure 13–4). Because proline appears 69 percent of the time, we could propose that proline is encoded by CCC (57.9 percent) and also by one of the codons consisting of 2C:1A (11.6 percent). Histidine, at 14 percent, is probably coded by one 2C:1A codon (11.6 percent) and one 1C:2A codon (2.3 percent). Threonine, at 12 percent, is likely coded by only one 2C:1A codon. Asparagine and glutamine each appear to be coded by one of the 1C:2A codons, and lysine appears to be coded by AAA.

Using as many as all four ribonucleotides to construct the mRNA, the researchers conducted many similar experiments. Although the determination by this means of the *composition* of triplet code words corresponding to all 20 amino acids represented a very significant breakthrough, the *specific sequences* of triplets were still unknown—other approaches were still needed.

The Triplet Binding Assay

It was not long before more advanced techniques for discovering codons were developed. In 1964, Nirenberg and

Possible compositions	Possible triplets	Probability of occurrence of any triplet	Final %
3A	AAA	$(1/6)^3 = 1/216 = 0.4\%$	0.4
1C:2A	AAC ACA CAA	$(5/6)(1/6)^2 = 5/216 = 2.3\%$	3 x 2.3 = 6.9
2C:1A	ACC CAC CCA	$(5/6)^2(1/6) = 25/216 = 11.6\%$	3 x 11.6 = 34.8
3C	CCC	$(5/6)^3 = 125/216 = 57.9\%$	57.9
			100.0

Chemical synthesis of message ↓

CCCCCCCCACCCCCCAACCACCCCCACCCCCACCCAA ——— RNA

Translation of message ↓

Percentage of amino acids in protein		Probable base-composition assignments
Lysine	<1	AAA
Glutamine	2	1C:2A
Asparagine	2	1C:2A
Threonine	12	2C:1A
Histidine	14	2C:1A, 1C:2A
Proline	69	CCC, 2C:1A

FIGURE 13–4   Results and interpretation of a mixed copolymer experiment in which a ratio of 1A:5C is used (1/6A:5/6C).

NOW SOLVE THIS

13–1 In a mixed copolymer experiment using polynucleotide phosphorylase, 3/4G:1/4C was used to form the synthetic message. The amino acid composition of the resulting protein was determined to be:

Glycine	36/64	(56 percent)
Alanine	12/64	(19 percent)
Arginine	12/64	(19 percent)
Proline	4/64	(6 percent)

From this information,

(a) indicate the percentage (or fraction) of the time each possible codon will occur in the message.
(b) determine one consistent base-composition assignment for the amino acids present.
(c) in view of the wobble hypothesis, predict as many specific codon assignments as possible.

■ HINT: *This problem asks you to analyze a mixed copolymer experiment and to predict codon composition assignments for the amino acids encoded by the synthetic message. The key to its solution is to first calculate the proportion of each triplet codon in the synthetic RNA and then match these to the proportions of amino acids that are synthesized.*

Philip Leder developed the **triplet binding assay**, leading to specific assignments of triplet codons. This technique took advantage of the observation that ribosomes, when presented *in vitro* with an RNA sequence as short as three ribonucleotides, will bind to it and form a complex similar to what is found *in vivo*. The triplet RNA sequence acts like a codon in mRNA, attracting a tRNA molecule containing a complementary sequence (Figure 13–5). Such a triplet sequence in tRNA, that is, complementary to a codon of mRNA, is known as an **anticodon**.

Although it was not yet feasible to chemically synthesize long stretches of RNA, specific triplet sequences could be synthesized in the laboratory to serve as templates. All that was needed was a method to determine which tRNA–amino acid was bound to the triplet RNA-ribosome complex. The test system Nirenberg and Leder devised was quite simple. The amino acid to be tested was made radioactive and combined with its cognate tRNA, creating a "charged" tRNA. Because codon compositions (though not exact sequences) were known, it was possible to narrow the decision as to which amino acids should be tested for each specific triplet.

The radioactively charged tRNA, the RNA triplet, and ribosomes were incubated together on a nitrocellulose filter, which retains ribosomes (because of their larger size) but not the other, smaller components, such as charged tRNA. If radioactivity is not retained on the filter, an incorrect amino acid has been tested. If radioactivity remains on the filter, it does so because the charged tRNA has bound to the RNA triplet associated with the ribosome, which itself remains on the filter. In such a case, a specific codon assignment can be made.

Work proceeded in several laboratories, and in many cases clear-cut, unambiguous results were obtained. For example, Table 13.2 shows 26 triplet codons assigned to ten different amino acids. However, in some cases the triplet binding was inefficient, and assignments were not possible. Eventually, about 50 of the 64 triplets were assigned. These specific assignments led to two major conclusions. First, the genetic code is degenerate—that is, one amino acid may be specified by more than one triplet. Second, the code is also *unambiguous*—that is, a single codon specifies only one amino acid. As we will see later in this chapter, these conclusions have been upheld with only minor exceptions. The triplet binding technique was a major innovation in the effort to decipher the genetic code.

Repeating Copolymers

Yet another innovative technique for deciphering the genetic code was developed in the early 1960s by Gobind Khorana, who was able to chemically synthesize long RNA molecules

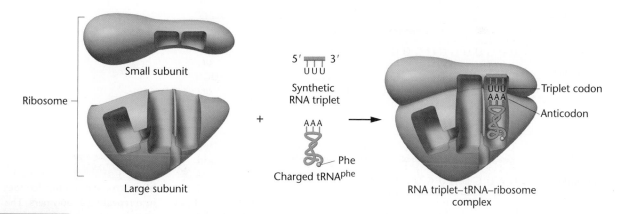

FIGURE 13–5   The behavior of the components during the triplet-binding assay. When the UUU triplet is positioned in the ribosome, it acts as a codon, attracting the complementary AAA anticodon of the charged tRNA^{phe}.

TABLE 13.2

Amino Acid Assignments to Specific Trinucleotides Derived from the Triplet-Binding Assay

Trinucleotides	Amino Acid
AAA AAG	Lysine
AUG	Methionine
AUU AUG AUA	Isoleucine
CCU CCC CCG CCA	Proline
CUC CUA CUG CUU	Leucine
GAA GAG	Glutamic acid
UCU UCC UCA UCG	Serine
UGU UGC	Cysteine
UUA UUG	Leucine
UUU UUC	Phenylalanine

consisting of short sequences repeated many times. First, he created the individual short sequences (e.g., di-, tri-, and tetra-nucleotides); then he replicated them many times and finally joined them enzymatically to form the long polynucleotides. In Figure 13–6, a dinucleotide made in this way is converted to an mRNA with two repeating triplet codons. A trinucleotide is converted into an mRNA that can be read as containing one of three potential repeating triplets, depending on the point at which initiation occurs. Finally, a tetranucleotide creates a message with four repeating triplet sequences.

When these synthetic mRNAs were added to a cell-free system, the predicted proportions of amino acids were found to be incorporated in the resulting polypeptides. When these data were combined with data drawn from mixed copolymer and triplet binding experiments, specific assignments were possible.

One example of specific assignments made in this way will illustrate the value of Khorana's approach. Consider the following experiments in concert with one another: (1) The repeating *trinucleotide sequence* UUCUUCUUC . . . can be read as three possible repeating triplets—UUC, UCU, and CUU—depending on the initiation point. When placed in a cell-free translation system, three different polypeptide homopolymers—containing either phenylalanine (phe), serine (ser), or leucine (leu)—are produced. Thus, we know that each of the three triplets encodes one of the three amino acids, but we do not know which codes which; (2) On the other hand, the *repeating dinucleotide sequence* UCUCUCUC . . . produces the triplets UCU and CUC and, when used in an experiment, leads to the incorporation of leucine and serine into a polypeptide. Thus, the triplets UCU and CUC specify leucine and serine, but we still do not know which triplet specifies which amino acid. However, when considering both sets of results in concert, we can conclude that UCU, which is common to both experiments, must encode either leucine or serine but not phenylalanine. Thus, either CUU *or* UUC encodes leucine *or* serine, while the other encodes phenylalanine; (3) To derive more specific information, we can examine the results of using the repeating tetranucleotide sequence UUAC, which produces the triplets UUA, UAC, ACU, and CUU. The CUU triplet is one of the two in which we are interested. Three amino acids are incorporated by this experiment: leucine, threonine, and tyrosine. Because CUU must specify only serine or leucine, and because, of these two, only leucine appears in the resulting polypeptide, we may conclude that CUU specifies leucine. Once this assignment is established, we can logically determine

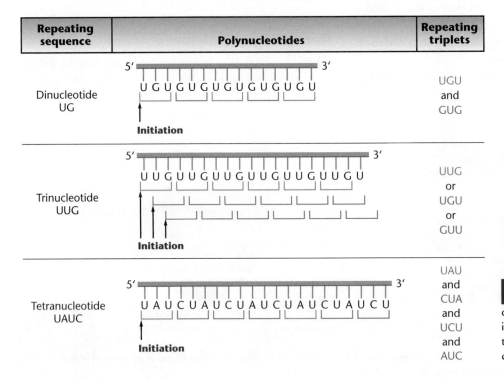

Repeating sequence	Polynucleotides	Repeating triplets
Dinucleotide UG	5′ U G U G U G U G U G U G U 3′ Initiation	UGU and GUG
Trinucleotide UUG	5′ U U G U U G U U G U U G U U G U 3′ Initiation	UUG or UGU or GUU
Tetranucleotide UAUC	5′ U A U C U A U C U A U C U A U C U A U C U 3′ Initiation	UAU and CUA and UCU and AUC

FIGURE 13–6 The conversion of di-, tri-, and tetranucleotides into repeating copolymers. The triplet codons produced in each case are shown.

TABLE 13.3

Amino Acids Incorporated Using Repeated Synthetic Copolymers of RNA

Repeating Copolymer	Codons Produced	Amino Acids in Resulting Polypeptides
UG	UGU	Cysteine
	GUG	Valine
AC	ACA	Threonine
	CAC	Histidine
UUC	UUC	Phenylalanine
	UCU	Serine
	CUU	Leucine
AUC	AUC	Isoleucine
	UCA	Serine
	CAU	Histidine
UAUC	UAU	Tyrosine
	CUA	Leucine
	UCU	Serine
	AUC	Isoleucine
GAUA	GAU	None
	AGA	None
	UAG	None
	AUA	None

all others. Of the two triplet pairs remaining (UUC and UCU from the first experiment *and* UCU and CUC from the second experiment), whichever triplet is common to both must encode serine. This is UCU. By elimination, UUC is determined to encode phenylalanine and CUC is determined to encode leucine. Thus, through painstaking logical analysis, four specific triplets encoding three different amino acids have been assigned from these experiments.

From these interpretations, Khorana reaffirmed the identity of triplets that had already been deciphered and filled in gaps left from other approaches. A number of examples are shown in Table 13.3. For example, the use of

NOW SOLVE THIS

13–2 When repeating copolymers are used to form synthetic mRNAs, dinucleotides produce a single type of polypeptide that contains only two different amino acids. On the other hand, using a trinucleotide sequence produces three different polypeptides, each consisting of only a single amino acid. Why? What will be produced when a repeating tetranucleotide is used?

■ HINT: *This problem asks you to consider different outcomes of repeating copolymer experiments. The key to its solution is to be aware that when using a repeating copolymer of RNA, translation can be initiated at different ribonucleotides. You must simply determine the number of triplet codons produced by initiation at each of the different ribonucleotides.*

two tetranucleotide sequences, GAUA and GUAA, suggested that at least two triplets were *termination codons*. Khorana reached this conclusion because neither of these repeating sequences directed the incorporation of more than a few amino acids into a polypeptide, too few to detect. There are no triplets common to both messages, and both seemed to contain at least one triplet that terminates protein synthesis. Of the possible triplets in the poly–(GAUA) sequence, shown in Table 13.3, UAG was later shown to be a termination codon.

13.4

The Coding Dictionary Reveals Several Interesting Patterns among the 64 Codons

The various techniques applied to decipher the genetic code have yielded a dictionary of 61 triplet codons assigned to amino acids. The remaining three codons are termination signals, not specifying any amino acid.

Degeneracy and the Wobble Hypothesis

A general pattern of triplet codon assignments becomes apparent when we look at the genetic coding dictionary. **Figure 13–7** displays the assignments in a particularly revealing form first suggested by Francis Crick.

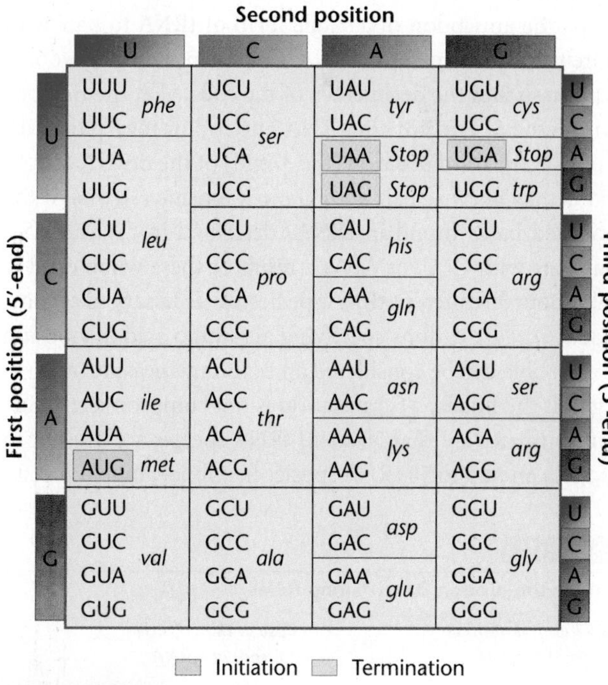

Second position

	U	C	A	G	
U	UUU / UUC phe	UCU / UCC ser	UAU / UAC tyr	UGU / UGC cys	U / C
	UUA / UUG	UCA / UCG	UAA Stop / UAG Stop	UGA Stop / UGG trp	A / G
C	CUU / CUC leu	CCU / CCC pro	CAU / CAC his	CGU / CGC arg	U / C
	CUA / CUG	CCA / CCG	CAA / CAG gln	CGA / CGG	A / G
A	AUU / AUC ile	ACU / ACC thr	AAU / AAC asn	AGU / AGC ser	U / C
	AUA / AUG met	ACA / ACG	AAA / AAG lys	AGA / AGG arg	A / G
G	GUU / GUC val	GCU / GCC ala	GAU / GAC asp	GGU / GGC gly	U / C
	GUA / GUG	GCA / GCG	GAA / GAG glu	GGA / GGG	A / G

First position (5'-end) — Third position (3'-end)

▢ Initiation   ▢ Termination

FIGURE 13–7 The coding dictionary. AUG encodes methionine, which initiates most polypeptide chains. All other amino acids except tryptophan, which is encoded only by UGG, are represented by two to six triplets. The triplets UAA, UAG, and UGA are termination signals and do not encode any amino acids.

Most evident is that the code is degenerate, as the early researchers predicted. That is, almost all amino acids are specified by two, three, or four different codons. Three amino acids (serine, arginine, and leucine) are each encoded by six different codons. Only tryptophan and methionine are encoded by single codons.

Also evident is the *pattern* of degeneracy. Most often sets of codons specifying the same amino acid are grouped, such that the first two letters are the same, with only the third differing. For example, as you can see in Figure 13–7, the codons for phenylalanine (UUU and UUC in the top left corner of the coding table) differ only by their third letter. Either U or C in the third position specifies phenylalanine. Four codons specify valine (GUU, GUC, GUA, and GUG, in the bottom left corner), and they differ only by their third letter. In this case, all four letters in the third position specify valine.

Crick observed this pattern in the degeneracy throughout the code, and in 1966, he postulated the **wobble hypothesis**. Crick's hypothesis predicted that the initial two ribonucleotides of triplet codes are often more critical than the third member in attracting the correct tRNA. He postulated that hydrogen bonding at the third position of the codon–anticodon interaction would be *less* spatially constrained and need not adhere as strictly to the established base-pairing rules. The wobble hypothesis proposes a more flexible set of base-pairing rules at the third position of the codon (Table 13.4).

This relaxed base-pairing requirement, or "wobble," allows the anticodon of a single form of tRNA to pair with more than one triplet in mRNA. Consistent with the wobble hypothesis and the degeneracy of the code, U at the first position (the 5′ end) of the tRNA anticodon may pair with A or G at the third position (the 3′ end) of the mRNA codon, and G may likewise pair with U or C. Inosine (I), one of the modified bases found in tRNA (described in Chapter 15), may pair with C, U, or A. As a result of these wobble rules, only about 30 different tRNA species are necessary to accommodate the 61 codons specifying an amino acid. If nothing else, wobble can be considered an economy measure, assuming that the fidelity of translation is not compromised. Current estimates are that 30 to 40 tRNA species are present in bacteria and up to 50 tRNA species in animal and plant cells.

TABLE 13.4

Anticodon–Codon Base-Pairing Rules

Base at first position (5′ end) of tRNA	Base at third position (3′ end) of mRNA
A	U
C	G
G	C or U
U	A or G
I	A, U, or C

The Ordered Nature of the Code

Still another observation has been made concerning the pattern of codon sequences and their corresponding amino acids, leading to the description referred to as the **ordered genetic code**. By this is meant that *chemically similar amino acids* often share one or two "middle" bases in the different triplets encoding them. For example, either U or C is often present in the second position of triplets that specify hydrophobic amino acids, including valine and alanine, among others. Two codons (AAA and AAG) specify the positively charged amino acid lysine. If only the middle letter of these codons is changed from A to G (AGA and AGG), the positively charged amino acid arginine is specified. Hydrophilic amino acids, such as serine or threonine are specified by triplet codons with G or C in the second position.

The chemical properties of amino acids will be discussed in more detail in Chapter 14. The end result of an "ordered" code is that it buffers the potential effect of mutation on protein function. While many mutations of the second base of triplet codons result in a change of one amino acid to another, the change is often to an amino acid with similar chemical properties. In such cases, protein function may not be noticeably altered.

Initiation, Termination, and Suppression

In contrast to the *in vitro* experiments discussed earlier, initiation of protein synthesis *in vivo* is a highly specific process. In bacteria, the initial amino acid inserted into all polypeptide chains is a modified form of methionine—**N-formylmethionine (fmet)**. Only one codon, AUG, codes for methionine, and it is sometimes called the **initiator codon**. However, when AUG appears internally in mRNA, rather than at an initiating position, unformylated methionine is inserted into the polypeptide chain. Rarely, another codon, GUG, specifies methionine during initiation, though it is not clear why this happens, since GUG normally encodes valine.

In bacteria, either the formyl group is removed from the initial methionine upon completion of synthesis of a protein, or the entire formylmethionine residue is removed. In eukaryotes, unformylated methionine is the initial amino acid of polypeptide synthesis. As in bacteria, this initial methionine residue may be cleared from the polypeptide.

As mentioned in the preceding section, three other codons (UAG, UAA, and UGA) serve as **termination codons**, punctuation signals that do not code for any amino acid.[*] They are not recognized by a tRNA molecule, and translation terminates when they are encountered. Mutations that produce any of the three codons internally in a gene also

---

[*]Historically, the terms *amber* (UAG), *ochre* (UAA), and *opal* (UGA) were used to distinguish mutations producing any of the three termination codons.

result in termination. In that case, only a partial polypeptide is synthesized, since it is prematurely released from the ribosome. When such a change occurs in the DNA, it is called a **nonsense mutation**.

The Genetic Code Has Been Confirmed in Studies of Phage MS2

The various aspects of the genetic code discussed so far yield a fairly complete picture. The code is triplet in nature, degenerate, unambiguous, and commaless, although it contains punctuation in the form of start-and-stop signals. These individual principles have been confirmed by the detailed analysis of the RNA-containing bacteriophage MS2 by Walter Fiers and his coworkers.

MS2 is a bacteriophage that infects *E. coli*. Its nucleic acid (RNA) contains only about 3500 ribonucleotides, making up only three genes. These genes specify a coat protein, an RNA-directed replicase, and a maturation protein (the A protein). This simple system of a small genome and few gene products allowed Fiers and his colleagues to sequence the genes and their products. The amino acid sequence of the coat protein was completed in 1970, and the nucleotide sequence of the gene and a number of nucleotides on each end of it were reported in 1972.

When the chemical constitutions of this gene and its encoded protein are compared, they are found to exhibit **colinearity** with one another. That is, the linear sequence of triplet codons formed by the nucleotides corresponds precisely with the linear sequence of amino acids in the protein. Furthermore, the codon for the first amino acid is AUG, the common initiator codon; the codon for the last amino acid is followed by two consecutive termination codons, UAA and UAG. We will return to discuss this concept again in Chapter 14.

By 1976, the other two genes of MS2 and their protein products were sequenced. The analysis clearly showed that the genetic code in this virus was identical to that established in bacterial systems. Other evidence suggests that the code is also identical in eukaryotes, thus providing confirmation of what seemed to be a universal code.

The Genetic Code Is Nearly Universal

Between 1960 and 1978, it was generally assumed that the genetic code would be found to be universal, applying equally to viruses, bacteria, archaea, and eukaryotes. Certainly, the nature of mRNA and the translation machinery seemed to be very similar in these organisms. For example, cell-free systems derived from bacteria could translate eukaryotic mRNAs. Poly U stimulates synthesis of polyphenylalanine in cell-free systems when the components are derived from eukaryotes. Many recent studies involving recombinant DNA technology (Chapter 20) reveal that eukaryotic genes can be inserted into bacterial cells, which are then transcribed and translated. Within eukaryotes, mRNAs from mice and rabbits have been injected into amphibian eggs and efficiently translated. For the many eukaryotic genes that have been sequenced, notably those for hemoglobin molecules, the amino acid sequence of the encoded proteins adheres to the coding dictionary established from bacterial studies.

However, several 1979 reports on the coding properties of DNA derived from mitochondria (mtDNA) of yeast and humans undermined the hypothesis of the universality of the genetic language. Since then, mtDNA has been examined in many other organisms.

Cloned mtDNA fragments were sequenced and compared with the amino acid sequences of various mitochondrial proteins, revealing several exceptions to the coding dictionary (Table 13.5). Most surprising was that the codon UGA, normally specifying termination, specifies the insertion of tryptophan during translation in yeast and human mitochondria. In human mitochondria, AUA, which normally specifies isoleucine, directs the internal insertion of methionine. In yeast mitochondria, threonine is inserted instead of leucine when CUA is encountered in mRNA.

In 1985, several other exceptions to the standard coding dictionary were discovered in the bacterium *Mycoplasma capricolum,* and in the nuclear genes of the protozoan ciliates *Paramecium, Tetrahymena,* and *Stylonychia.* For example, as shown in Table 13.5, one alteration converts the termination codon (UGA) to tryptophan. Several other code alterations convert a termination codon to glutamine. These changes are significant because they are

TABLE 13.5

Exceptions to the Universal Code

Codon	Normal Code Word	Altered Code Word	Source
UGA	Termination	Trp	Human and yeast mitochondria; *Mycoplasma*
CUA	Leu	Thr	Yeast mitochondria
AUA	Ile	Met	Human mitochondria
AGA	Arg	Termination	Human mitochondria
AGG	Arg	Termination	Human mitochondria
UAA	Termination	Gln	*Paramecium, Tetrahymena, and Stylonychia*
UAG	Termination	Gln	*Paramecium*

seen in both a prokaryote and several eukaryotes, that is, in distinct species that have evolved separately over a long period of time.

Note the pattern apparent in several of the altered codon assignments: the change in coding capacity involves only a shift in recognition of the third, or wobble, position. For example, AUA specifies isoleucine in the cytoplasm and methionine in the mitochondrion, but in the cytoplasm, methionine is specified by AUG. In a similar example, UGA calls for termination in the cytoplasm, but for tryptophan in the mitochondrion; in the cytoplasm, tryptophan is specified by UGG. It has been suggested that such changes in codon recognition may represent an evolutionary trend toward reducing the number of tRNAs needed in mitochondria; only 22 tRNA species are encoded in human mitochondria, for example. However, until other cases are discovered, the differences must be considered to be exceptions to the previously established general coding rules.

13.7
Different Initiation Points Create Overlapping Genes

Earlier we stated that the genetic code is nonoverlapping, meaning that each ribonucleotide in the code *for a given polypeptide* is part of only one codon. However, this characteristic of the code does not rule out the possibility that a single mRNA may have multiple initiation points for translation. If so, these points could theoretically create several different reading frames within the same mRNA, thus specifying more than one polypeptide. This concept of **overlapping genes** is illustrated in Figure 13–8(a). Note that gene is also referred to as an **ORF**, or **open reading frame**, which is

defined as a DNA sequence that produces an RNA that has a start-and-stop codon, between which is a series of triplet codons specifying the amino acids making up a polypeptide. Thus, we sometimes also refer to *overlapping ORFs.*

That this might actually occur in some viruses was suspected when phage ϕX174 was carefully investigated. The circular DNA chromosome consists of 5386 nucleotides, which should encode a maximum of 1795 amino acids, sufficient for five or six proteins. However, this small virus in fact synthesizes nine proteins consisting of more than 2300 amino acids. A comparison of the nucleotide sequence of the DNA and the amino acid sequences of the polypeptides synthesized has clarified the apparent paradox. At least four instances of multiple initiation have been discovered, creating overlapping genes [Figure 13–8(b)].

The sequences specifying the K and B polypeptides are initiated in separate reading frames within the sequence specifying the A polypeptide. The *K* gene sequence overlaps into the adjacent sequence specifying the C polypeptide. The E sequence is out of frame with, but initiated within, that of the D polypeptide. Finally, the A′ sequence, while in frame with the A sequence, is initiated in the middle of the A sequence. They both terminate at the identical point. In all, seven different polypeptides are created from a DNA sequence that might otherwise have specified only three (A, C, and D).

A similar situation has been observed in other viruses and bacteria. The employment of overlapping reading frames optimizes the limited amount of DNA present. However, such an approach to storing information has a distinct disadvantage in that a single mutation may affect more than one protein and thus increase the chances that the change will be deleterious or lethal. In the case we just discussed [Figure 13–8(b)], a single mutation in the middle of the *B* gene could potentially affect three other proteins (the A, A′, and K proteins). It may be for this reason that, while present, overlapping genes are not the rule in other organisms.

More recently, genomic analysis has made it possible to examine extensive DNA sequence information in eukaryotes, revealing that the concept of overlapping genes also applies to these organisms, including mammals. For example, in one study of mouse and human DNA, over 1200 such cases were revealed. In the humans, head to tail and head to head overlap was observed as well as cases where one gene is fully embedded in the sequence of another larger gene. In the case of three human genes, *MUTYH*, *TOE1*, and *TESK2*, *TOE1* contains overlapping coding regions with *MUTYH* at the 5′ end and with *TESK2* at the 3′ end.

(a)

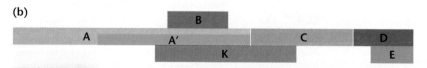

(b)

FIGURE 13–8 Illustration of the concept of overlapping genes. (a) Translation initiated at two different AUG positions out of frame with one another will give rise to two distinct amino acid sequences. (b) The relative positions of the sequences encoding seven polypeptides of the phage ϕX174.

Transcription Synthesizes RNA on a DNA Template

Even while the genetic code was being studied, it was quite clear that proteins were the end products of many genes. Hence, while some geneticists were attempting to elucidate the code, other research efforts were directed toward the nature of genetic expression. The central question was how DNA, a nucleic acid, is able to specify a protein composed of amino acids.

The complex, multistep process begins with the transfer of genetic information stored in DNA to RNA. The process by which RNA molecules are synthesized on a DNA template is called **transcription**. It results in an mRNA molecule complementary to the gene sequence of one of the two strands of the double helix. Each triplet codon in the mRNA is, in turn, complementary to the anticodon region of its corresponding tRNA, which inserts the correct amino acid into the polypeptide chain during translation. The significance of transcription is enormous, for it is the initial step in the process of *information flow* within the cell. The idea that RNA is involved as an intermediate molecule in the process of information flow between DNA and protein is suggested by the following observations:

1. DNA is, for the most part, associated with chromosomes in the nucleus of the eukaryotic cell. However, protein synthesis occurs in association with ribosomes located outside the nucleus, in the cytoplasm. Therefore, DNA does not appear to participate directly in protein synthesis.

2. RNA is synthesized in the nucleus of eukaryotic cells, in which DNA is found, and is chemically similar to DNA.

3. Following its synthesis, most RNA migrates to the cytoplasm, in which protein synthesis (translation) occurs.

4. The amount of RNA is generally proportional to the amount of protein in a cell.

Collectively, these observations suggested that genetic information, stored in DNA, is transferred to an RNA intermediate, which directs the synthesis of the proteins. As with most new ideas in molecular genetics, the initial supporting experimental evidence for an RNA intermediate was based on studies of bacteria and bacteriophages.

Studies with Bacteria and Phages Provided Evidence for the Existence of mRNA

In two papers published in 1956 and 1958, Elliot Volkin and his colleagues reported their analysis of RNA produced

TABLE 13.6

Base Compositions (in mole percents) of RNA Produced Immediately Following Infection of *E. coli* by the Bacteriophages T2 and T7 in Contrast to the Composition of RNA of Uninfected *E. coli*

Source	Adenine	Thymine	Uracil	Cytosine	Guanine
Post-infection RNA in T2-infected cells	33	—	32	18	18
T2 DNA	32	32	—	17*	18
Post-infection RNA in T7-infected cells	27	—	28	24	22
T7 DNA	26	26	—	24	22
E. coli RNA	23	—	22	18	17

*5-hydroxymethyl cytosine.
Source: Volkin and Astrachan (1956); Volkin, Astrachan, and Countryman (1958).

immediately after bacteriophage infection of *E. coli*. Using the isotope ^{32}P to follow newly synthesized RNA, they found that its base composition closely resembled that of the phage DNA, but was different from that of bacterial RNA (Table 13.6). This newly synthesized RNA was unstable (short lived); however, its production was shown to precede the synthesis of new phage proteins. Thus, Volkin and his coworkers considered the possibility that synthesis of RNA is a preliminary step in the process of protein synthesis.

Although ribosomes were known to participate in protein synthesis, their role in this process was not clear. As we noted earlier, one possibility was that each ribosome is specific for the protein synthesized in association with it. That is, perhaps genetic information in DNA is transferred to the RNA of a ribosome (rRNA) during the latter's synthesis so that each ribosome is restricted to the production of a particular protein. The alternative hypothesis was that ribosomes are nonspecific "workbenches" for protein synthesis and that specific genetic information rests with a messenger RNA.

In an elegant experiment using the *E. coli*–phage system, the results of which were reported in 1961, Sydney Brenner, François Jacob, and Matthew Meselson clarified this question. They labeled uninfected *E. coli* ribosomes with heavy isotopes and then allowed phage infection to occur in the presence of radioactive RNA precursors. By following these components during translation, the researchers demonstrated that the synthesis of phage proteins (under the direction of newly synthesized RNA) occurred on bacterial ribosomes that were present prior to infection. The ribosomes appeared to be nonspecific, strengthening the case that another type of RNA serves as an intermediary in the process of protein synthesis.

That same year, Sol Spiegelman and his colleagues reached the same conclusion when they isolated ^{32}P-labeled RNA following the infection of bacteria and used it in molecular hybridization studies. They tried hybridizing this

RNA to the DNA of both phages and bacteria in separate experiments. The RNA hybridized only with the phage DNA, showing that it was complementary in base sequence to the viral genetic information.

The results of these experiments agree with the concept of a messenger RNA (mRNA) being made on a DNA template and then directing the synthesis of specific proteins in association with ribosomes. This concept was formally proposed by François Jacob and Jacques Monod in 1961 as part of a model for gene regulation in bacteria. Since then, mRNA has been isolated and thoroughly studied. There is no longer any question about its role in genetic processes.

13.10

RNA Polymerase Directs RNA Synthesis

To prove that RNA can be synthesized on a DNA template, it was necessary to demonstrate that there is an enzyme capable of directing this synthesis. By 1959, several investigators, including Samuel Weiss, had independently discovered such a molecule in rat liver. Called **RNA polymerase**, it has the same general substrate requirements as does DNA polymerase, the major exception being that the substrate nucleotides contain the ribose rather than the deoxyribose form of the sugar. Unlike DNA polymerase, no primer is required to initiate synthesis. The overall reaction summarizing the synthesis of RNA on a DNA template can be expressed as

$$n(\text{NTP}) \xrightarrow[\text{enzyme}]{\text{DNA}} (\text{NMP})_n + n(\text{PP}_i)$$

As the equation reveals, nucleoside triphosphates (NTPs) serve as substrates for the enzyme, which catalyzes the polymerization of nucleoside monophosphates (NMPs), or nucleotides, into a polynucleotide chain $(\text{NMP})n$. Nucleotides are linked during synthesis by 5' to 3' phosphodiester bonds (see Figure 10–12). The energy created by cleaving the triphosphate precursor into the monophosphate form drives the reaction, and inorganic phosphates (PP_i) are produced.

A second equation summarizes the sequential addition of each ribonucleotide as the process of transcription progresses:

$$(\text{NMP})_n + \text{NTP} \xrightarrow[\text{enzyme}]{\text{DNA}} (\text{NMP})_{n+1} + \text{PP}_i$$

As this equation shows, each step of transcription involves the addition of one ribonucleotide (NMP) to the growing polyribonucleotide chain $(\text{NMP})_{n+1}$, using a nucleoside triphosphate (NTP) as the precursor.

RNA polymerase from *E. coli* has been extensively characterized and shown to consist of subunits designated α, β, β', ω, and σ. The complex, active form of the enzyme, the **holoenzyme**, contains the subunits $\alpha_2\beta\beta'\sigma$ and has a molecular weight of almost 500,000 Da. While there is some variation in the subunit composition of other bacteria, it is the β and β' polypeptides that provide the catalytic mechanism and active site for transcription. As we will see, the σ (**sigma**) **factor** [Figure 13–9(a)] plays a regulatory function in the initiation of RNA transcription.

While there is but a single form of the enzyme in *E. coli*, there are several different σ factors, creating variations of

(a) Transcription components

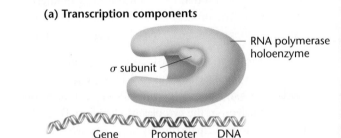

(b) Template binding and initiation of transcription

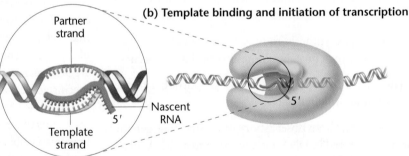

(c) Chain elongation

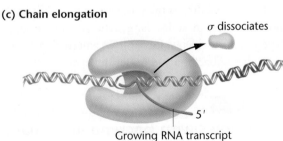

FIGURE 13–9   The early stages of transcription in prokaryotes, showing (a) the components of the process; (b) template binding at the −10 site involving the sigma subunit of RNA polymerase and subsequent initiation of RNA synthesis; and (c) chain elongation, after the σ subunit has dissociated from the transcription complex and the enzyme moves along the DNA template.

the polymerase holoenzyme. On the other hand, eukaryotes display three distinct forms of RNA polymerase, each consisting of a greater number of polypeptide subunits than in bacteria. In this section, we will discuss the process of transcription in prokaryotes. We will return to a discussion of eukaryotic transcription later in this chapter.

NOW SOLVE THIS

13–3 The following represent deoxyribonucleotide sequences in the template strand of DNA:

Sequence 1:	5'-CTTTTTTGCCAT-3'
Sequence 2:	5'-ACATCAATAACT-3'
Sequence 3:	5'-TACAAGGGTTCT-3'

(a) For each strand, determine the mRNA sequence that would be derived from transcription.

(b) Using Figure 13–7, determine the amino acid sequence that is encoded by these mRNAs.

(c) For Sequence 1, what is the sequence of the partner DNA strand?

■ HINT: *This problem asks you to consider the outcome of the transfer of complementary information from DNA to RNA and to determine the amino acids encoded by this information. The key to its solution is to remember that in RNA, uracil is complementary to adenine, and that while DNA stores genetic information in the cell, the code that is translated is contained in the RNA complementary to the template strand of DNA making up a gene.*

Promoters, Template Binding, and the σ Subunit

Transcription results in the synthesis of a single-stranded RNA molecule complementary to a region along only one of the two strands of the DNA double helix. When discussing transcription, we can call the DNA strand that is transcribed the *template strand* and its complement the *partner strand*.

The initial step in prokaryotic gene transcription is referred to as **template binding** [Figure 13–9(b)]. In bacteria, the site of this initial binding is established when the RNA polymerase σ subunit recognizes specific DNA sequences called **promoters**. These sequences are located in the 5' region, upstream from the point of initial transcription of a gene. It is believed that the enzyme "explores" a length of DNA until it encounters a promoter region and binds there to about 60 nucleotide pairs along the helix, 40 of which are upstream from the point of initial transcription. Once this occurs, the helix is denatured, or unwound, locally, making the template strand of the DNA accessible to the action of

the enzyme. The point at which transcription actually begins is called the **transcription start site**.

Because the interaction of promoters with RNA polymerase governs the efficiency of transcription—by regulating the initiation of transcription—the importance of promoter sequences cannot be overemphasized. The nature of the binding between promoter and polymerase is at the heart of discussions concerning genetic regulation, the subject of Chapters 16 and 17. While those chapters present more detailed information concerning promoter–enzyme interactions, we must address three points here.

The first point is the concept of **consensus sequences** of DNA, sequences that are similar (homologous) in different genes of the same organism or in one or more genes of related organisms. Their conservation during evolution attests to the critical nature of their role in biological processes. Two consensus sequences have been found in bacterial promoters. One, TATAAT, is located 10 nucleotides upstream from the site of initial transcription (the −10 region, or **Pribnow box**). The other, TTGACA, is located 35 nucleotides upstream (the −35 region). Mutations in either region diminish transcription, often severely.

Sequences such as these, in regions adjacent to the gene itself, are said to be *cis*-acting elements. The term *cis*, drawn from organic chemistry nomenclature, means "next to" or on the same side as other functional groups, in contrast to being *trans* to or "across from," them. In molecular genetics, then, *cis*-elements are those that are located on the same DNA molecule. In contrast, *trans*-acting factors are molecules that bind to these DNA elements.

The second point is that the degree of RNA polymerase binding to different promoters varies greatly, causing variable gene expression. Currently, this is attributed to sequence variation in the promoters. In bacteria, both strong promoters and weak promoters have been discovered, causing a variation in time of initiation from once every 1 to 2 seconds to as little as once every 10 to 20 minutes. Mutations in promoter sequences may severely reduce the initiation of gene expression.

A final general point to be made involves the σ subunit in bacteria. The major form is designated σ^{70} based on its molecular weight of 70 kilodaltons (kDa). The promoters of most bacterial genes recognize this form; however, several alternative forms of RNA polymerase in *E. coli* have unique σ subunits associated with them (e.g., σ^{32}, σ^{54}, σ^{S}, and σ^{E}). Each form recognizes different promoter sequences, which in turn provides specificity to the initiation of transcription.

Initiation, Elongation, and Termination of RNA Synthesis

Once RNA polymerase has recognized and bound to the promoter, DNA is converted from its double-stranded form to an open structure, exposing the template strand. The enzyme then proceeds to **initiate** RNA synthesis, whereby the first 5′-ribonucleoside triphosphate, which is complementary to the first nucleotide, is inserted at the start site. As we noted, no primer is required. Subsequent ribonucleotide complements are inserted and linked together by phosphodiester bonds as RNA polymerization proceeds. This process continues in a 5′ to 3′ direction (in terms of the nascent RNA), creating a temporary 8-bp DNA/RNA duplex whose chains run antiparallel to one another [Figure 13–9(b)].

After these ribonucleotides have been added to the growing RNA chain, the σ subunit dissociates from the holoenzyme, and **chain elongation** proceeds under the direction of the core enzyme [Figure 13–9(c)]. In *E. coli*, this process proceeds at the rate of about 50 nucleotides/second at 37°C.

The enzyme traverses the entire gene until eventually it encounters a specific nucleotide sequence that acts as a termination signal. Such termination sequences, about 40 base pairs in length, are extremely important in prokaryotes because of the close proximity of the end of one gene to the upstream sequences of the adjacent gene. An interesting aspect of termination in bacteria is that the termination sequence alluded to above is actually transcribed into RNA. The unique sequence of nucleotides in this termination region causes the newly formed transcript to fold back on itself, forming what is called a **hairpin secondary structure**, held together by hydrogen bonds. The hairpin is important to termination. In some cases, the termination of synthesis is also dependent on the **termination factor, rho** (ρ). Rho is a large hexameric protein that physically interacts with the growing RNA transcript, facilitating termination of transcription.

When termination is achieved, the transcribed RNA molecule is released from the DNA template, and the core polymerase enzyme dissociates. The synthesized RNA molecule is precisely complementary to the DNA sequence of the template strand of the gene. Wherever an A, T, C, or G residue was encountered there, a corresponding U, A, G, or C residue, respectively, was incorporated into the RNA molecule. These RNA molecules ultimately provide the information leading to the synthesis of all proteins in the cell.

In bacteria, groups of genes whose protein products are involved in the same metabolic pathway are often clustered together along the chromosome. In many such cases, the genes are contiguous, and all but the last gene lack the encoded signals for termination of transcription. The result is that during transcription, a large mRNA is produced, encoding more than one protein. Since genes in bacteriophages are sometimes referred to as *complementation groups* and are called *cistrons* (see Chapter 6), the RNA is called a **polycistronic mRNA**. The products of genes transcribed in this fashion are usually all needed by the cell at the same time,

so this is an efficient way to transcribe and subsequently translate the needed genetic information. In eukaryotes, **monocistronic mRNAs** are the rule, although an increasing number of exceptions are being reported.

13.11

Transcription in Eukaryotes Differs from Prokaryotic Transcription in Several Ways

Much of our knowledge of transcription has been derived from studies of prokaryotes. Most of the general aspects of the mechanics of these processes are similar in eukaryotes, but there are several notable differences:

1. Transcription in eukaryotes occurs within the nucleus under the direction of three separate forms of RNA polymerase. Unlike the prokaryotic process, in eukaryotes the RNA transcript is not free to associate with ribosomes prior to the completion of transcription. For the mRNA to be translated, it must move out of the nucleus into the cytoplasm.

2. Initiation of transcription of eukaryotic genes requires the compact chromatin fiber, characterized by nucleosome coiling, to be uncoiled and the DNA to be made accessible to RNA polymerase and other regulatory proteins. This transition, referred to as *chromatin remodeling*, reflects the dynamics involved in the conformational change that occurs as the DNA helix is opened (Chapter 12).

3. Initiation and regulation of transcription entail a more extensive interaction between *cis*-acting DNA sequences and *trans*-acting protein factors involved in stimulating and initiating transcription. Eukaryotic RNA polymerases, for example, rely on *transcription factors* (*TFs*) to scan and bind to DNA. In addition to promoters, other control units, called *enhancers* and *silencers*, may be located in the 5′ regulatory region upstream from the initiation point, but they have also been found within the gene or even in the 3′ downstream region, beyond the coding sequence.

4. Alteration of the primary RNA transcript to produce mature eukaryotic mRNA involves many complex stages referred to generally as "processing." An initial processing step involves the addition of a 5′ cap and a 3′ tail to most transcripts destined to become mRNAs. The initial (or primary) transcripts are most often much larger than those that are eventually translated into protein. Sometimes called **pre-mRNAs**, these primary transcripts are found only in the nucleus and referred to collectively as **heterogeneous nuclear RNA (hnRNA)**.

Only about 25 percent of hnRNA molecules are converted to mRNA. From those that are converted, substantial amounts of the ribonucleotide sequence are excised, and the remaining segments are spliced back together prior to nuclear export and translation. This phenomenon has given rise to the concepts of *split genes* and *splicing* in eukaryotes (discussed in Section 13.12).

In the remainder of this chapter we will look at the basic details of transcription in eukaryotic cells. The process of transcription is highly regulated, determining which DNA sequences are copied into RNA and when and how frequently they are transcribed. We will return to topics directly related to regulation of eukaryotic transcription in Chapter 17.

Initiation of Transcription in Eukaryotes

Eukaryotic RNA polymerase exists in three distinct forms. While the three forms of the enzyme share certain polypeptide subunits, each nevertheless transcribes different types of genes, as indicated in Table 13.7. Each enzyme is larger and more complex than the single prokaryotic polymerase. For example, in yeast, the holoenzyme consists of two large subunits and 10 smaller subunits.

In regard to the initial template-binding step and promoter regions, most is known about **RNA polymerase II (RNP II)**, which is responsible for the transcription of a wide range of genes in eukaryotes. The activity of RNP II is dependent on both *cis*-acting elements surrounding the gene itself and a number of *trans*-acting transcription factors that bind to these DNA elements (we will return to the topic of transcription factors below). At least four *cis*-acting DNA elements regulate the initiation of transcription by RNP II. The first of these elements, the **core-promoter**, determines where RNP II binds to the DNA and where it begins copying the DNA into RNA. The other three types of regulatory DNA sequences, called **proximal-promoter elements, enhancers**, and **silencers**, influence the efficiency or the rate of transcription initiation by RNP II from the core promoter element. Recall that in prokaryotes, the DNA sequence recognized by RNA polymerase is also called the promoter. In eukaryotes, however, transcriptional initiation is controlled by a larger number of *cis*-acting DNA elements.

In many eukaryotic genes, a *cis*-acting core-promoter element is the **Goldberg–Hogness**, or **TATA, box**. Located about 30 nucleotide pairs upstream (−30) from the start point of transcription, TATA boxes share a consensus sequence TATAA/TAAR, where R indicates any purine nucleotide. The sequence and function of TATA boxes are analogous to those found in the −10 promoter region of prokaryotic genes. However, recall that in prokaryotes, RNA polymerase binds directly to the −10 promoter region. As we will see below, this is not the case in eukaryotes. A wide range of core-promoter and proximal-promoter elements are also found within eukaryotic gene-regulatory regions, and each can have an effect on the efficiency of transcription initiation from the start site. Many of these elements will be discussed in more detail in Chapter 17.

Although eukaryotic promoter elements can determine the site and general efficiency of initiation, other elements, known as enhancers and silencers, have more dramatic effects on eukaryotic gene transcription. As their names suggest, enhancers increase transcription levels and silencers decrease them. The locations of these elements can vary from immediately upstream from a promoter to downstream, within, or kilobases away, from a gene. Thus they can modulate transcription from a distance. Enhancers and silencers often act to increase or decrease transcription in response to a cell's requirement for a gene product, or at a particular time during development or place within the organism. Each eukaryotic gene has its own unique arrangement of proximal-promoter, enhancer, and silencer elements.

Complementing the *cis*-acting regulatory sequences are various *trans*-acting factors that facilitate RNP II binding and, therefore, the initiation of transcription. These proteins are referred to as **transcription factors**. There are two broad categories of transcription factors: the **general transcription factors** (GTFs) that are absolutely required for all RNP II–mediated transcription, and the **transcriptional activators and repressors** that influence the efficiency or the rate of RNP II transcription initiation. The general transcription factors are essential because RNP II cannot bind directly to eukaryotic core-promoter sites and initiate transcription without their presence. The general transcription factors involved with human RNP II binding are well characterized and designated **TFIIA, TFIIB**, and so on. One of these, **TFIID**, binds directly to the TATA-box sequence. Once initial binding of TFIID to DNA occurs, the other general transcription factors, along with RNP II, bind sequentially to TFIID, forming an extensive **pre-initiation complex**.

The specific transcription factors (activators and repressors, above) bind to enhancer and silencer elements and regulate transcription initiation by aiding or preventing the assembly of pre-initiation complexes. They appear to supplant the role of the σ factor seen in the prokaryotic enzyme and are important in eukaryotic gene regulation. We will consider the roles of general and specific transcription

TABLE 13.7

RNA Polymerases in Eukaryotes

Form	Product	Location
I	rRNA	Nucleolus
II	mRNA, snRNA	Nucleoplasm
III	5S rRNA, tRNA	Nucleoplasm

factors in eukaryotic gene regulation, as well as the various DNA elements that bind these factors (Chapter 17).

Recent Discoveries Concerning RNA Polymerase Function

It is of great interest to learn more about RNA polymerase in eukaryotes because of the difficult task it faces during transcription along the chromatin fiber. The enzyme is confronted with its substrate DNA that, unlike in bacteria, is wrapped around histone proteins organized into nucleosomes. Recall that these structures have compacted the double-stranded DNA, which is a barrier to transcription (Chapter 12). Thus, the enzyme must open up the molecule and separate (denature) the two strands so that the template strand may pass through its active site during RNA synthesis. The ability to crystallize large nucleic acid–protein structures and perform X-ray diffraction analysis at a resolution below 5 Å has shed light on this issue, particularly the work of Roger Kornberg and colleagues, using RNP II isolated from yeast. It is useful to note here that achieving a resolution below 2.8 Å allows the visualization of each amino acid of every protein in the complex!

Kornberg's discovery provides a highly detailed account of the most critical processes of transcription. RNP II in yeast contains two large subunits and ten smaller ones, forming a huge three-dimensional complex with a molecular weight of about 500 kDa. The promoter region of the DNA duplex that is to be transcribed enters a positively charged cleft between the two large subunits of the enzyme. The subunit assemblage in fact resembles a pair of jaws that can open and partially close. Prior to association with DNA, the cleft is open; once associated with DNA, the cleft partially closes, securing the duplex during the initiation of transcription. The part of the enzyme that is critical for this transition is about 50 kDa in size and is called the *clamp*.

Once secured by the clamp, the strands of a small duplex region of DNA separate at a position within the enzyme referred to as the *active center,* and complementary RNA synthesis is initiated on the DNA template strand. However, the entire complex remains unstable, and transcription usually terminates following the incorporation of only a few ribonucleotides. It is not clear why, but this so-called *abortive transcription* is repeated a number of times before a stable DNA:RNA hybrid containing a transcript of 11 ribonucleotides is formed. Once this occurs, abortive transcription is overcome, a stable complex is achieved, and elongation of the RNA transcript proceeds in earnest. Transcription at this point is said to have achieved a level of highly processive RNA polymerization.

As transcription proceeds, the enzyme moves along the DNA, and at any given time, about 40 base pairs of DNA and 18 residues of the growing RNA chain are part of the enzyme complex. The RNA synthesized earliest runs through a groove in the enzyme and exits under a structure at the top and back designated as the *lid.* Another area, called the *pore,* has been identified at the bottom of the enzyme. It serves as the point through which RNA precursors gain entry into the complex.

Eventually, as transcription proceeds, the portion of the DNA that signals termination is encountered, and the complex once again becomes unstable, much as it was during the earlier state of abortive transcription. The clamp opens, and both DNA and RNA are released from the enzyme as transcription is terminated. This completes the cycle that constitutes transcription. In sum, an unstable complex is formed during the initiation of transcription, stability is established once elongation manages to create a duplex of sufficient size, elongation proceeds, and then instability again characterizes termination of transcription.

Clearly, Kornberg's findings extended our knowledge of transcription considerably. For this work, he was awarded the Nobel Prize in Chemistry in 2006. As you think back on this cycle, try to visualize the process mentally, from the time the DNA first associates with the enzyme until the transcript is released from the large molecular complex. If these images are clear to you, you no doubt have acquired a firm understanding of transcription in eukaryotes, which is more complex than in prokaryotes.

Heterogeneous Nuclear RNA and Its Processing: Caps and Tails

While in bacteria the base sequence of DNA is transcribed into an mRNA that is immediately and directly translated into the amino acid sequence as dictated by the genetic code, eukaryotic RNA transcripts require significant alteration before they are transported to the cytoplasm and translated. By 1970, accumulating evidence showed that eukaryotic mRNA is transcribed initially as a precursor molecule much larger than that which is translated into protein. This notion was based on the observation by James Darnell and his coworkers of the large **heterogeneous nuclear RNA (hnRNA)** in mammalian nuclei that contained nucleotide sequences common to the smaller mRNA molecules present in the cytoplasm. They proposed that the initial transcript of a gene results in a large RNA molecule that must be processed in the nucleus before it appears in the cytoplasm as a mature mRNA molecule. The various processing steps, discussed in the sections that follow, are summarized in **Figure 13–10.**

An important **posttranscriptional modification** of eukaryotic RNA transcripts destined to become mRNAs occurs at the 5′ end of these molecules, where a **7-methylguanosine (7-mG) cap** is added. The cap is added even

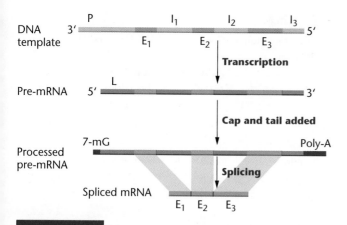

FIGURE 13–10   Posttranscriptional RNA processing in eukaryotes. Transcription produces a pre-mRNA containing a leader sequence (L), several introns (I), and several exons (E), as identified in the DNA template strand. This is processed by the addition of a 5′7-mG cap and a 3′-poly-A tail. The introns are then spliced out and the exons joined to create the mature mRNA. While the above figure depicts these steps sequentially, in some eukaryotic transcripts, splicing actually occurs in introns before transcription is complete and the polyA tail has been added, leading to the concept of *cotranscriptional splicing.*

before synthesis of the initial transcript is complete and appears to be important to subsequent processing within the nucleus. The cap also protects the 5′ end of the molecule from nuclease attack. Subsequently, it may be involved in the transport of mature mRNAs across the nuclear membrane into the cytoplasm and in the initiation of translation of the mRNA into protein. The cap is fairly complex and is distinguished by the unique 5′-5′ bonding that connects it to the initial ribonucleotide of the RNA. Some eukaryotes also acquire a methyl group (CH_3) at the 2′-carbon of the ribose sugars of the first two ribonucleotides of the RNA.

Further insights into the processing of RNA transcripts during the maturation of mRNA came from the discovery that both pre-RNAs and mRNAs contain at their 3′ end a stretch of as many as 250 adenylic acid residues. This **poly-A sequence** is added after the 3′ end of the initial transcript is cleaved enzymatically at a position some 10 to 35 ribonucleotides from a highly conserved AAUAAA sequence. Poly A has now been found at the 3′ end of almost all mRNAs studied in a variety of eukaryotic organisms. In fact, poly-A tails have also been detected in some prokaryotic mRNAs. The exceptions in eukaryotes seem to be the RNAs that encode the histone proteins.

While the AAUAAA sequence is not found on all eukaryotic transcripts, it appears to be essential to those that have it. If the sequence is changed as a result of a mutation, those transcripts that would normally have it cannot add the poly-A tail. In the absence of this tail, these RNA transcripts are rapidly degraded. Both the 5′ cap and the 3′ poly-A tail

are critical if an mRNA transcript is to be transported to the cytoplasm and translated.

The Coding Regions of Eukaryotic Genes Are Interrupted by Intervening Sequences

One of the most exciting breakthroughs in the history of molecular genetics occurred in 1977, when Susan Berget, Philip Sharp, and Richard Roberts presented direct evidence that the genes of animal viruses contain *internal* nucleotide sequences that are not expressed in the amino acid sequence of the proteins they encode. These internal DNA sequences are represented in initial RNA transcripts, but they are removed before the mature mRNA is translated (Figure 13–10). Such nucleotide segments are called **intervening sequences**, and the genes that contain them are **split genes**. DNA sequences that are not represented in the final mRNA product are also called **introns** ("int" for intervening), and those retained and expressed are called **exons** ("ex" for expressed). Splicing involves the removal of the corresponding ribonucleotide sequences representing introns as a result of an excision process and the rejoining of the regions representing exons.

Similar discoveries were soon made in many other eukaryotic genes. Two approaches have been most fruitful for this purpose. The first involves the molecular hybridization of purified, functionally mature mRNAs with DNA containing the genes from which the RNA was originally transcribed. Hybridization between nucleic acids that are not perfectly complementary results in **heteroduplexes**, in which introns present in the DNA but absent in the mRNA loop out and remain unpaired. Such structures can be visualized with the electron microscope, as shown in Figure 13–11. The chicken ovalbumin complex shown in the figure is a heteroduplex with seven loops (A through G), representing seven introns whose sequences are present in the DNA but not in the final mRNA.

The second approach provides more specific information. It involves a direct comparison of nucleotide sequences of DNA with those of mRNA and their correlation with amino acid sequences. Such an approach allows the precise identification of all intervening sequences.

Thus far, most eukaryotic genes have been shown to contain introns (Figure 13–12). One of the first so identified was the **β-globin gene** in mice and rabbits, studied independently by Philip Leder and Richard Flavell. The mouse gene contains an intron 550 nucleotides long, beginning immediately after the codon specifying the 104th amino acid. In the rabbit, there is an intron of 580 base pairs near the codon for the 110th amino acid. In addition, another intron of about 120 nucleotides exists earlier in both genes.

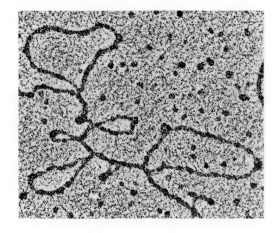

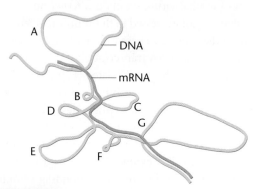

FIGURE 13–11 An electron micrograph and an interpretive drawing of the hybrid molecule (heteroduplex) formed between the template DNA strand of the chicken ovalbumin gene and the mature ovalbumin mRNA. Seven DNA introns, A–G, produce unpaired loops.

Similar introns have been found in the β-globin gene in all mammals examined.

The **ovalbumin gene** of chickens has been extensively characterized by Bert O'Malley in the United States and Pierre Chambon in France. As shown in Figure 13–12, the gene contains seven introns. In fact, the majority of the gene's DNA sequence is composed of introns and is thus "silent." The initial RNA transcript is nearly three times the length of the

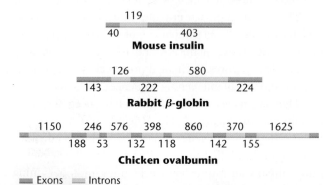

FIGURE 13–12 Intervening sequences in various eukaryotic genes. The numbers indicate the number of nucleotides present in various intron and exon regions.

TABLE 13.8

Contrasting Human Gene Size, mRNA Size, and Number of Introns

Gene	Gene Size (kb)	mRNA Size (kb)	Number of Introns
Insulin	1.7	0.4	2
Collagen [*pro-α-2(1)*]	38.0	5.0	50
Albumin	25.0	2.1	14
Phenylalanine hydroxylase	90.0	2.4	12
Dystrophin	2000.0	17.0	50

mature mRNA. Compare the ovalbumin gene in Figures 13–11 and 13–12. Can you match the unpaired loops in Figure 13–11 with the order of introns specified in Figure 13–12?

The list of genes containing intervening sequences is long. In fact, few eukaryotic genes seem to lack introns. An extreme example of the number of introns in a single gene is provided by the gene coding for one of the subunits of collagen, the major connective tissue protein in vertebrates. The *pro-α-2(1) collagen* gene contains 50 introns. The precision of cutting and splicing that occur must be extraordinary if errors are not to be introduced into the mature mRNA. Equally noteworthy is the difference between the size of a typical gene and the size of the final mRNA transcribed from it once introns are removed. As shown in Table 13.8, only about 15 percent of the collagen gene consists of exons that finally appear in mRNA. For other proteins, an even more extreme picture emerges. Only about 8 percent of the albumin gene remains to be translated, and in the largest human gene known, dystrophin (which is the protein product absent in Duchenne muscular dystrophy), less than 1 percent of the gene sequence is retained in the mRNA. Two other human genes are also contrasted in Table 13.8.

Although the vast majority of eukaryotic genes examined thus far contain introns, there are several exceptions. Notably, the genes coding for histones and for interferon appear to contain no introns. It is not clear why or how the genes encoding these molecules have been maintained throughout evolution without acquiring the extraneous information characteristic of almost all other genes.

Splicing Mechanisms: Self-Splicing RNAs

The discovery of split genes led to intensive attempts to elucidate the mechanism by which introns of RNA are excised and exons are spliced back together. A great deal of progress has already been made, relying heavily on *in vitro* studies. Interestingly, it appears that somewhat different mechanisms exist for different types of RNA, as well as for RNAs produced in mitochondria and chloroplasts.

We might envision the simplest possible mechanism for removing an intron to be as illustrated in Figure 13–10. After an endonucleolytic "cut" is made at each end of an intron, the intron is removed, and the terminal ends of the adjacent exons are ligated by an enzyme (in short, the intron is snipped out, and the exon ends are rejoined). This is apparently what happens to the introns present in transfer RNAs (tRNAs) in bacteria. A specific endonuclease recognizes the intron termini and excises the intervening sequences. Then RNA ligase seals the exon ends to complete each splicing event. However, in the studies of all other RNAs—tRNA in higher eukaryotes and rRNAs and pre-mRNAs in all eukaryotes—precise excision of introns is much more complex and a much more interesting story.

Introns in eukaryotes can be categorized into several groups based on their splicing mechanisms. Group I, represented by introns that are part of the primary transcript of rRNAs, require no additional components for intron excision; the intron itself is the source of the enzymatic activity necessary for removal. This amazing discovery was made in 1982 by Thomas Cech and his colleagues during a study of the ciliate protozoan *Tetrahymena*. RNAs that are capable of catalytic activity are referred to as **ribozymes**. The self-excision process for Group I introns serves to illustrate this concept and is shown in **Figure 13–13**. Chemically, two nucleophilic reactions take place—that is, reactions caused by the presence of electron-rich chemical species (in this case, they are transesterification reactions). The first is an interaction between guanosine, which acts as a cofactor in the reaction, and the primary transcript [Figure 13–13(a)]. The 3′-OH group of guanosine is transferred to the nucleotide adjacent to the 5′ end of the intron [Figures 13–13(b) and 13–13(c)]. The second reaction involves the interaction of the newly acquired 3′-OH group on the left-hand exon and the phosphate on the 3′ end of the intron [Figure 13–13(c)]. The intron is spliced out and the two exon regions are ligated, leading to the mature RNA [Figure 13–13(d)].

Self-excision of group I introns, as described above, is now known to apply to pre-rRNAs from other protozoans besides *Tetrahymena*. Self-excision also seems to govern the removal of introns from the primary mRNA and tRNA transcripts produced in mitochondria and chloroplasts. These are referred to as group II introns. As in group I molecules, splicing here involves two autocatalytic reactions leading to the excision of introns. However, guanosine is not involved as a cofactor with group II introns.

Splicing Mechanisms: The Spliceosome

Introns are a major component of nuclear-derived pre-mRNA transcripts of eukaryotes. Compared to the group I and group II introns discussed above, those in nuclear-derived

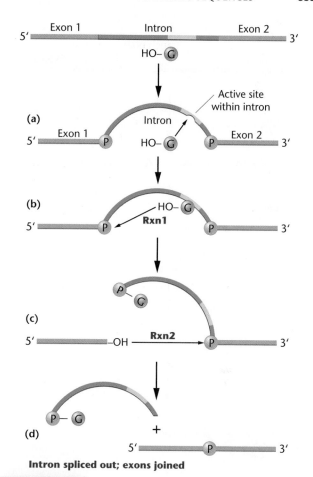

FIGURE 13–13   Splicing mechanism for removal of group I introns from the initial rRNA transcript (pre-rRNA). The process is one of self-excision involving two transesterification reactions.

mRNA can be much larger—up to 20,000 nucleotides—and they are more plentiful. Their removal appears to require a much more complex mechanism. Nevertheless, we now have a good handle on the process.

Interestingly, the splicing reactions are mediated by a huge molecular complex called a **spliceosome**, which has now been identified in extracts of yeast as well as in mammalian cells. This structure is very large, 40S in yeast and 60S in mammals, being the same size as ribosomal subunits! One set of essential components of spliceosomes is a unique set of **small nuclear RNAs (snRNAs)**. These RNAs are usually 100 to 200 nucleotides long or less and are complexed with proteins to form **small nuclear ribonucleoproteins (snRNPs or snurps)**. Because they are rich in uridine residues, the snRNAs have been arbitrarily designated U1, U2, . . ., U6.

Figure 13–14 depicts a model illustrating the steps involved in the removal of one intron. Keep in mind that while this figure shows separate components, the process involves the huge spliceosome that envelopes the RNA being spliced. The nucleotide sequences near the ends of the intron begin

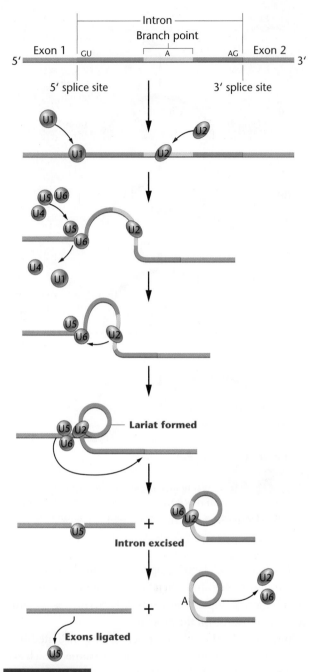

FIGURE 13–14   A model of the splicing mechanism for removal of an intron from nuclear-derived pre-mRNA. Excision is dependent on various snRNAs (U1, U2, . . ., U6) that combine with proteins to form snRNPs (snurps), which function as part of a large structure referred to as the spliceosome. The lariat structure in the intermediate stage is characteristic of this mechanism.

at the 5′ end with a GU dinucleotide sequence, called the *donor sequence,* and terminate at the 3′ end with an AG dinucleotide, called the *acceptor sequence.* These, as well as other consensus sequences shared by introns, attract specific snRNAs of the spliceosome. For example, the snRNA U1 bears a nucleotide sequence that is complementary to the 5′-splice donor sequence end of the intron. Base pairing resulting from this homology promotes the binding that represents the initial step in the formation of the spliceosome. After the other snRNPs (U2, U4, U5, and U6) are added, splicing commences. As with group I splicing, two *transesterification reactions* occur. The first involves the interaction of the 3′-OH group from an adenine (A) residue present within the **branch point** region of the intron. The A residue attacks the 5′-splice site, cutting the RNA chain. In a subsequent step involving several other snRNPs, an intermediate structure is formed and the second reaction ensues, linking the cut 5′ end of the intron to the A. This results in the formation of a characteristic loop structure called a *lariat,* which contains the excised intron. The exons are then ligated and the snRNPs are released.

The processing involved in splicing, which occurs within the nucleus, represents a potential regulatory step in gene expression in eukaryotes. For instance, several cases are known wherein introns present in pre-mRNAs *derived from the same gene* are spliced *in more than one way,* thereby yielding different collections of exons in the mature mRNA. Such **alternative splicing** yields a group of similar but nonidentical mRNAs that, upon translation, result in a series of related proteins called **isoforms**. Many examples have been encountered in organisms ranging from viruses to *Drosophila* to humans. Alternative splicing of pre-mRNAs represents a way of producing related proteins from a single gene, increasing the number of gene products that can be derived from an organism's genome. We will return to this topic in Chapter 17 in our discussion of the regulation of gene expression in eukaryotes.

13.13

RNA Editing May Modify the Final Transcript

In the late 1980s, still another unexpected form of post-transcriptional RNA processing was discovered in several organisms. In this form, referred to as **RNA editing**, the nucleotide sequence of a pre-mRNA is actually changed prior to translation. As a result, the ribonucleotide sequence of the mature RNA differs from the sequence encoded in the exons of the DNA from which the RNA was transcribed.

Although other variations exist, there are two main types of RNA editing: **substitution editing**, in which the identities of individual nucleotide bases are altered; and **insertion/deletion editing**, in which nucleotides are added to or subtracted from the total number of bases. Substitution editing is used in some nuclear-derived eukaryotic RNAs and is prevalent in mitochondrial and chloroplast RNAs transcribed in plants. *Physarum polycephalum,* a slime mold, uses both substitution and insertion/deletion editing for its mitochondrial mRNAs.

Trypanosoma, a parasite that causes African sleeping sickness, and its relatives use extensive insertion/deletion editing in mitochondrial RNAs. The uridines added to an individual transcript can make up more than 60 percent of the coding sequence, usually forming the initiation codon and bringing the rest of the sequence into the proper reading frame. Insertion/deletion editing in trypanosomes is directed by **gRNA (guide RNA)** templates, which are also transcribed from the mitochondrial genome. These small RNAs share a high degree of complementarity to the edited region of the final mRNAs. They base-pair with the pre-edited mRNAs to direct the editing machinery to make the correct changes.

The best-studied examples of substitutional editing occur in mammalian nuclear-encoded mRNA transcripts. One such example is the protein *apolipoprotein B (apo B)*, which exists in both a long and a short form that are encoded by the same gene. In human intestinal cells, apo B mRNA is edited by a single C-to-U change, which converts a CAA glutamine codon into a UAA stop codon and terminates the polypeptide at approximately half its genomically encoded length. The editing is performed by a complex of proteins that bind to a "mooring sequence" on the mRNA transcript just downstream of the editing site. A second example involves the subunits constituting the *glutamate receptor channels (GluR)* in mammalian brain tissue. In this case, adenosine (A) to inosine (I) editing occurs in pre-mRNAs prior to their translation, during which I is read as guanosine (G). A family of three ADAR (*a*denosine *d*eaminase *a*cting on *R*NA) enzymes is believed to be responsible for the editing of various sites within the glutamate channel subunits. The double-stranded RNAs required for editing by the ADAR enzymes are provided by intron/exon pairing of the GluR mRNA transcripts. The editing changes alter the physiological parameters (solute permeability and desensitization response time) of the receptors containing the subunits.

The importance of RNA editing resulting from the action of ADARs is most apparent in situations where these enzymes have lost their functional capacity as a result of mutation. In several investigations, the loss of function was shown to have a lethal impact in mice. In one study, embryos heterozygous for a defective *ADAR1* gene died during embryonic development as a result of a defective hematopoietic system. In another study, mice with two defective copies of *ADAR2* progressed through development normally but were prone to epileptic seizures and died while still in the weaning stage. Their tissues contained the unedited version of one of the GluR products. The defect leading to death was believed to be in the brain. Heterozygotes for the mutation were normal.

Findings such as these in mammals have established that RNA editing provides still another important mechanism of posttranscriptional modification, and that this process is not restricted to small or asexually reproducing genomes, such as those in mitochondria. Several new examples of RNA editing have been found each year since its discovery, and the trend is likely to continue. These discoveries, too, have important implications for the regulation of genetic expression.

13.14
Transcription Has Been Visualized by Electron Microscopy

We conclude this chapter by presenting a striking visual demonstration of the transcription process based on the electron microscope studies of Oscar Miller, Jr., Barbara Hamkalo, and Charles Thomas. Their combined work has captured the transcription process in both prokaryotes and eukaryotes. **Figure 13–15** shows a micrograph and

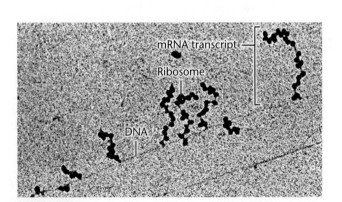

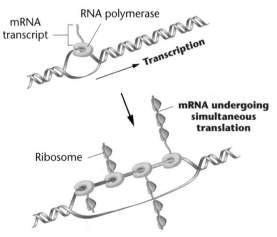

FIGURE 13–15 Electron micrograph and interpretive drawing of simultaneous transcription of genes in *E. coli*. As each transcript is forming, ribosomes attach, initiating simultaneous translation along each strand. *O.L. Miller, Jr., Barbara A. Hamkalo, C.A. Thomas, Jr. Science 169: 392–395, 1970 by the American Association for the Advancement of Science. F:2.*

interpretive drawings of transcription in *E. coli*. In the micrograph, multiple strands of RNA are seen to emanate from different points along a central DNA template. Many RNA strands result because numerous transcription events are occurring simultaneously along each gene. Progressively longer RNA strands are found farther downstream from the point of initiation of transcription along a given gene, whereas the shortest strands are closest to the point of initiation.

An interesting picture emerges from study of the *E. coli* micrograph. Because prokaryotes lack nuclei, cytoplasmic ribosomes are not separated physically from the chromosome. As a result, ribosomes are free to attach to *partially* transcribed mRNA molecules and initiate translation. The longer the RNA strand, the greater the number of ribosomes attached to it. As we will see in Chapter 14, these structures are called **polyribosomes**. Visualization of transcription confirms many of the predictions scientists had made from the biochemical analysis of this process.

CASE STUDY | A drug that sometimes works

A 30-year-old woman with β-thalassemia, a recessively inherited genetic disorder caused by absence of the hemoglobin β chain, had been treated with blood transfusions since the age of 7. However, in spite of the transfusions, her health was declining. As an alternative treatment, her physician administered 5-azacytidine to induce transcription of the fetal β hemoglobin chain to replace her missing β chain. This drug activates gene transcription by removing methyl groups from DNA. Addition of methyl groups silences genes. However, the physician expressed concern that approximately 40 percent of all human genes are normally silenced by methylation. Nevertheless, after several weeks of 5-azacytidine treatment, the patient's condition improved dramatically. Although the treatment was successful, use of this drug raises several important questions.

1. Why was her physician concerned that a high percentage of human genes are transcriptionally silenced by methylation?

2. What genes might raise the greatest concern?

3. What criteria would you use when deciding to administer a drug such as 5-azacytidine?

Summary Points

 For activities, animations, and review quizzes, go to the study area at www.masteringgenetics.com

1. Early studies of the genetic code revealed it to be triplet in nature and to be nonoverlapping, commaless, and degenerate.

2. The use of RNA homopolymers and mixed copolymers in a cell-free protein synthesizing system allowed the determination of the composition, but not the sequence, of triplet codons designating specific amino acids.

3. Use of the triplet-binding assay and of repeating copolymers allowed the determination of the specific sequences of triplet codons designating specific amino acids.

4. The complete coding dictionary reveals that of the 64 possible triplet codons, 61 encode the 20 amino acids found in proteins, while three triplets terminate translation. One of these 61 is the initiation codon and specifies methionine.

5. Confirmation for the coding dictionary, including codons for initiation and termination, was obtained by comparing the complete nucleotide sequences of phage MS2 with the amino acid sequence of the corresponding proteins. Other findings support the belief that, with only minor exceptions, the code is universal for all organisms.

6. Transcription—the initial step in gene expression—is the synthesis, under the direction of RNA polymerase, of a strand of RNA complementary to a DNA template.

7. Like DNA replication, the processes of transcription can be subdivided into the stages of initiation, elongation, and termination. Also like DNA replication, transcription relies on base-pairing affinities between complementary nucleotides.

8. Initiation of transcription is dependent on an upstream (5′) DNA region, called the promoter, that represents the initial binding site for RNA polymerase. Promoters contain specific DNA sequences, such as the TATA box, that are essential to polymerase binding.

9. The process of creating the initial transcript during transcription is more complex in eukaryotes than in prokaryotes, including the addition of a 5′ 7-mG cap and a 3′ poly-A tail, to the pre-mRNA.

10. The primary transcript in eukaryotes reflects the presence of intervening sequences, or introns, present in DNA, which must be spliced out to create the mature mRNA.

GENETICS, TECHNOLOGY, AND SOCIETY

Nucleic Acid-Based Gene Silencing: Attacking the Messenger

Standard chemotherapies for diseases such as cancer and AIDS are often accompanied by toxic side effects. Conventional therapeutic drugs affect both normal and diseased cells, with diseased or infected cells being only slightly more susceptible than the patient's normal cells. Scientists have long wished for a magic bullet that could seek out and destroy viruses or diseased cells, leaving normal cells alive and healthy. Over the last decade, a group of promising candidates has emerged, collectively described as *nucleic acid-based gene-silencing* drugs.

The two chief nucleic acid-based therapies currently being investigated are *antisense oligonucleotides* (*ASOs*) and *RNA interference* (*RNAi*). Both have been developed through an understanding of the molecular biology of gene expression: First, a single-stranded messenger RNA (mRNA) is copied from the template strand of the duplex DNA molecule; and second, the mRNA is complexed with ribosomes, and its coded information is translated into the amino acid sequence of a polypeptide.

Normally, a gene is transcribed into RNA from only one strand of the DNA duplex. The resulting RNA is known as *sense RNA*. However, it is possible for the other DNA strand to be copied into RNA, and this RNA, produced by transcription of the "wrong" strand of DNA, is called *antisense RNA*. When present together, sense and antisense RNA strands can form double-stranded duplex structures, the formation of which may affect the sense RNA in several ways.

In ASO technologies, scientists design single-stranded antisense DNA oligonucleotides (about 20 nucleotides long) of known sequence and then synthesize large amounts of these antisense nucleic acids *in vitro*. It is theoretically possible to treat cells with these synthetic antisense oligonucleotides, so that they enter the cells and bind to precise target mRNAs.

The binding of antisense DNA to sense mRNA may physically block its translation. Alternatively, the degradation of the RNA may result. In either case, gene expression is blocked.

The antisense approach is exciting because of its potential specificity. Because an ASO has a sequence that specifically binds to a particular sense RNA, it should be possible to inhibit synthesis of the specific protein encoded by the sense RNA. If the protein is necessary for virus reproduction or cancer cell growth (but is not necessary in normal cells), the antisense oligonucleotide should have only therapeutic effects.

One ASO drug, which targets cytomegalovirus infection, is currently on the market. Others, which are designed to counteract cancers, Crohn's disease, HIV-1, and Hepatitis C, are in Phase II clinical trials.

The second gene-silencing approach, called RNAi technology (discussed in more detail in Chapter 17), uses short double-stranded RNA molecules (~20–25 nucleotides long) with sequences complementary to specific mRNAs within cells. These are known as *short interfering RNAs* (siRNAs). These siRNA molecules may be synthesized *in vitro* or may be transcribed within a cell from cloned vectors that are introduced into cells.

Once within a cell's cytoplasm, the siRNA associates with an enzyme complex called an *RNA-induced silencing complex* (*RISC*), which is found within the cell's cytoplasm. One strand of the siRNA is cleaved within the RISC, and the other strand binds to a target mRNA that contains the complementary RNA sequence. When RISC and the siRNA are bound to the target mRNA, the RISC may cleave the target mRNA or may interfere with its translation.

RNAi clinical trials are being conducted to study its use in combating the eye disease macular degeneration, with encouraging results. Other areas of high interest for RNAi-based treatments are cancers, diseases of the nervous system, and viral infections such as hepatitis B and HIV-1.

Your Turn

Take time, individually or in groups, to answer the following questions. Investigate the references and links to help you discuss some of the issues that surround the development and uses of antisense therapies.

1. What are some of the challenges in the use of ASOs as therapeutics? Do siRNAs share these challenges?

   A balanced discussion of antisense oligonucleotide drugs is presented in: Lebedeva, I. and Stein, C.A. 2001. Antisense oligonucleotides: Promise and reality. *Annu. Rev. Pharmacol.* 41:403–419. *RNAi drugs are critiqued in* Dykxhoorn, D. M. and Lieberman, J. 2006. Running interference: Prospects and obstacles to using small interfering RNAs as small molecule drugs. *Annu. Rev. Biomed. Eng.* 8:377–402.

2. Have any RNAi-based therapeutic drugs reached the market? What clinical trials for RNAi drugs are currently in progress?

   Information about clinical trials can be found at http://www.ClinicalTrials.gov. *A number of biotechnology companies, including Sirna Therapeutics, Alnylam Pharmaceuticals, and Opko Health, are developing RNAi-based therapeutics. Their Web sites also contain information about the RNAi drug pipelines and clinical trials.*

3. Studies in model organisms show that RNAi is effective in silencing genes involved in a wide range of infections and diseases. What do you think is the most promising use of RNAi as a therapeutic?

   To read about animal studies using siRNA gene silencing, see Dykxhoorn, D.M., et al. 2006. The silent treatment: siRNAs as small molecule drugs. *Gene Therapy* 13:541–552.

INSIGHTS AND SOLUTIONS

1. Calculate how many triplet codons would be possible had evolution seized on six bases (three complementary base pairs) rather than four bases with which to construct DNA. Would six bases accommodate a two-letter code, assuming 20 amino acids and start-and-stop codons?

Solution: Six bases taken three at a time would produce $(6)^3$ or 216, triplet codes. If the code was a doublet, there would be $(6)^2$ or 36, two-letter codes, more than enough to accommodate 20 amino acids and start-and-stop signals.

2. In a heteropolymer experiment using $1/2C : 1/4A : 1/4G$, how many different triplets will occur in the synthetic RNA molecule? How often will the most frequent triplet occur?

Solution: There will be $(3)^3$ or 27, triplets produced. The most frequent will be CCC, present $(1/2)^3$ or 1/8 of the time.

3. In a regular copolymer experiment, in which UUAC is repeated over and over, how many different triplets will occur in the synthetic RNA, and how many amino acids will occur in the polypeptide when this RNA is translated? (Consult Figure 13–7.)

Solution: The synthetic RNA will repeat four triplets—UUA, CUU, ACU, and UAC—over and over. Because both UUA and CUU encode leucine, while ACU and UAC encode threonine and tyrosine, respectively, the polypeptides synthesized under the directions of such an RNA contain three amino acids in the repeating sequence leu-leu-thr-tyr.

4. Actinomycin D inhibits DNA-dependent RNA synthesis. This antibiotic is added to a bacterial culture in which a specific protein is being monitored. Compared to a control culture, into which no antibiotic is added, translation of the protein declines over a period of 20 minutes, until no further protein is made. Explain these results.

Solution: The mRNA, which is the basis for the translation of the protein, has a lifetime of about 20 minutes. When actinomycin D is added, transcription is inhibited, and no new mRNAs are made. Those already present support the translation of the protein for up to 20 minutes.

Problems and Discussion Questions

 For instructor-assigned tutorials and problems, go to www.masteringgentics.com

HOW DO WE KNOW?

1. In this chapter, we focused on the genetic code and the transcription of genetic information stored in DNA into complementary RNA molecules. Along the way, we found many opportunities to consider the methods and reasoning by which much of this information was acquired. From the explanations given in the chapter, what answers would you propose to the following fundamental questions:
   (a) Why did geneticists believe, even before direct experimental evidence was obtained, that the genetic code would turn out to be composed of triplet sequences and be nonoverlapping? Experimentally, how were these suppositions shown to be correct?
   (b) What experimental evidence provided the initial insights into the *compositions* of codons encoding specific amino acids?
   (c) How were the specific sequences of triplet codes determined experimentally?
   (d) How were the experimentally derived triplet codon assignments verified in studies using bacteriophage MS2?
   (e) What evidence do we have that the expression of the information encoded in DNA involves an RNA intermediate?
   (f) How do we know that the initial transcript of a eukaryotic gene contains noncoding sequences that must be removed before accurate translation into proteins can occur?

2. Early proposals regarding the genetic code considered the possibility that DNA served directly as the template for polypeptide synthesis. (See Gamow, 1954, in Selected Readings.)

In eukaryotes, what difficulties would such a system pose? What observations and theoretical considerations argue against such a proposal?

3. In their studies of frameshift mutations, Crick, Barnett, Brenner, and Watts-Tobin found that either three "pluses" or three "minuses" restored the correct reading frame. (a) Assuming the code is a triplet, what effect would the addition or loss of six nucleotides have on the reading frame? (b) If the code were a sextuplet (consisting of six nucleotides), would the reading frame be restored by the addition or loss of three, six, or nine nucleotides?

4. The mRNA formed from the repeating tetranucleotide UUAC incorporates only three amino acids, but the use of UAUC incorporates four amino acids. Why?

5. In studies using repeating copolymers, AC . . . incorporates threonine and histidine, and CAACAA . . . incorporates glutamine, asparagine, and threonine. What triplet code can definitely be assigned to threonine?

6. In a coding experiment using repeating copolymers (as demonstrated in Table 13.3), the following data were obtained:

Copolymer	Codons Produced	Amino Acids in Polypeptide
AG	AGA, GAG	Arg, Glu
AAG	AGA, AAG, GAA	Lys, Arg, Glu

AGG is known to code for arginine. Taking into account the wobble hypothesis, assign each of the four codons produced in the experiment to its correct amino acid.

7. In the triplet-binding technique, radioactivity remains on the filter when the amino acid corresponding to the codon is labeled. Explain the rationale for this technique.

8. When the amino acid sequences of insulin isolated from different organisms were determined, some differences were noted. For example, alanine was substituted for threonine, serine was substituted for glycine, and valine was substituted for isoleucine at corresponding positions in the protein. List the single-base changes that could occur in codons of the genetic code to produce these amino acid changes.

9. In studies of the amino acid sequence of wild-type and mutant forms of tryptophan synthetase in *E. coli,* the following changes have been observed:

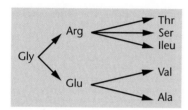

Determine a set of triplet codes in which only a single-nucleotide change produces each amino acid change.

10. Why doesn't polynucleotide phosphorylase (Ochoa's enzyme) synthesize RNA *in vivo*?

11. Refer to Table 13.1. Can you hypothesize why a mixture of Poly U + Poly A would not stimulate incorporation of ^{14}C-phenylalanine into protein?

12. Predict the amino acid sequence produced during translation by the following short hypothetical mRNA sequences (note that the second sequence was formed from the first by a deletion of only one nucleotide):

Sequence 1: 5′-AUGCCGGAUUAUAGUUGA-3′

Sequence 2: 5′-AUGCCGGAUUAAGUUGA-3′

What type of mutation gave rise to Sequence 2?

13. A short RNA molecule was isolated that demonstrated a hyperchromic shift, indicating secondary structure. Its sequence was determined to be

5′-AGGCGCCGACUCUACU-3′

(a) Propose a two-dimensional model for this molecule.

(b) What DNA sequence would give rise to this RNA molecule through transcription?

(c) If the molecule were a tRNA fragment containing a CGA anticodon, what would the corresponding codon be?

(d) If the molecule were an internal part of a message, what amino acid sequence would result from it following translation? (Refer to the code chart in Figure 13–7.)

14. A glycine residue is in position 210 of the tryptophan synthetase enzyme of wild-type *E. coli*. If the codon specifying glycine is GGA, how many single-base substitutions will result in an amino acid substitution at position 210? What are they? How many will result if the wild-type codon is GGU?

15. Refer to Figure 13–7 to respond to the following:

(a) Shown here is a hypothetical viral mRNA sequence:

5′-AUGCAUACCUAUGAGACCCUUGGA-3′

Assuming that it could arise from overlapping genes, how many different polypeptide sequences can be produced? What are the sequences?

(b) A base-substitution mutation that altered the sequence in (a) eliminated the synthesis of all but one polypeptide. The altered sequence is shown here:

5′-AUGCAUACCUAUGUGACCCUUGGA-3′

Determine why.

16. Most proteins have more leucine than histidine residues, but more histidine than tryptophan residues. Correlate the number of codons for these three amino acids with this information.

17. Define the process of transcription. Where does this process fit into the central dogma of molecular genetics (DNA makes RNA makes protein)?

18. What was the initial evidence for the existence of mRNA?

19. Describe the structure of RNA polymerase in bacteria. What is the core enzyme? What is the role of the σ subunit?

20. Write a paragraph describing the abbreviated chemical reactions that summarize RNA polymerase-directed transcription.

21. Messenger RNA molecules are very difficult to isolate in prokaryotes because they are rather quickly degraded in the cell. Can you suggest a reason why this occurs? Eukaryotic mRNAs are more stable and exist longer in the cell than do prokaryotic mRNAs. Is this an advantage or a disadvantage for a pancreatic cell making large quantities of insulin?

22. Present an overview of various forms of posttranscriptional processing in eukaryotes. For each, provide an example.

23. One form of posttranscriptional modification of most eukaryotic RNA transcripts is the addition of a poly-A sequence at the 3′ end. The absence of a poly-A sequence leads to rapid degradation of the transcript. Poly-A sequences of various lengths are also added to many prokaryotic RNA transcripts where, instead of promoting stability, they enhance degradation. In both cases, RNA secondary structures, stabilizing proteins, or degrading enzymes interact with poly-A sequences. Considering the activities of RNAs, what might be general functions of 3′-polyadenylation?

24. Describe the role of two forms of RNA editing that lead to changes in the size and sequence of pre-mRNAs. Briefly describe several examples of each form of editing, including their impact on respective protein products.

25. Substitution RNA editing is known to involve either C-to-U or A-to-I conversions. What common chemical event accounts for each?

Extra-Spicy Problems

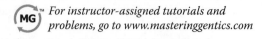

For instructor-assigned tutorials and problems, go to www.masteringgentics.com

26. It has been suggested that the present-day triplet genetic code evolved from a doublet code when there were fewer amino acids available for primitive protein synthesis.

    (a) Can you find any support for the doublet code notion in the existing coding dictionary?

    (b) The amino acids Ala, Val, Gly, Asp, and Glu are all early members of biosynthetic pathways (Taylor and Coates, 1989) and are more evolutionarily conserved than other amino acids (Brooks and Fresco, 2003). They therefore probably represent "early" amino acids. Of what significance is this information in terms of the evolution of the genetic code? Also, which base, of the first two, would likely have been the more significant in originally specifying these amino acids?

    (c) As determined by comparisons of ancient and recently evolved proteins, cysteine, tyrosine, and phenylalanine appear to be late-arriving amino acids. In addition, they are considered to have been absent in the abiotic earth (Miller, 1987). All three of these amino acids have only two codons each, while many others, earlier in origin, have more. Is this mere coincidence, or might there be some underlying explanation?

27. In a mixed copolymer experiment, messages were created with either 4/5C:1/5A or 4/5A:1/5C. These messages yielded proteins with the following amino acid compositions.

4/5C:1/5A		4/5A:1/5C	
Proline	63.0 percent	Proline	3.5 percent
Histidine	13.0 percent	Histidine	3.0 percent
Threonine	16.0 percent	Threonine	16.6 percent
Glutamine	3.0 percent	Glutamine	13.0 percent
Asparagine	3.0 percent	Asparagine	13.0 percent
Lysine	0.5 percent	Lysine	50.0 percent
	98.5 percent		99.1 percent

Using these data, predict the most specific coding composition for each amino acid.

28. Shown here are the amino acid sequences of the wild-type and three mutant forms of a short protein. Use this information to answer the following questions:

Wild-type:	Met-Trp-Tyr-Arg-Gly-Ser-Pro-Thr
Mutant 1:	Met-Trp
Mutant 2:	Met-Trp-His-Arg-Gly-Ser-Pro-Thr
Mutant 3:	Met-Cys-Ile-Val-Val-Val-Gln-His

    (a) Using Figure 13–7, predict the type of mutation that led to each altered protein.

    (b) For each mutant protein, determine the specific ribonucleotide change that led to its synthesis.

    (c) The wild-type RNA consists of nine triplets. What is the role of the ninth triplet?

    (d) Of the first eight wild-type triplets, which, if any, can you determine specifically from an analysis of the mutant proteins? In each case, explain why or why not.

    (e) Another mutation (Mutant 4) is isolated. Its amino acid sequence is unchanged, but the mutant cells produce abnormally low amounts of the wild-type proteins. As specifically as you can, predict where this mutation exists in the gene.

29. The genetic code is degenerate. Amino acids are encoded by either 1, 2, 3, 4, or 6 triplet codons. (See Figure 13–7.) An interesting question is whether the number of triplet codes for a given amino acid is in any way correlated with the frequency with which that amino acid appears in proteins. That is, is the genetic code optimized for its intended use? Some approximations of the frequency of appearance of nine amino acids in proteins in *E. coli* are

Amino Acid	Percentage
Met	2
Cys	2
Gln	5
Pro	5
Arg	5
Ile	6
Glu	7
Ala	8
Leu	10

    (a) Determine how many triplets encode each amino acid.

    (b) Devise a way to graphically compare the two sets of information (data).

    (c) Analyze your data to determine what, if any, correlations can be drawn between the relative frequency of amino acids making up proteins and the number of codons for each. Write a paragraph that states your specific and general conclusions.

    (d) How would you proceed with your analysis if you wanted to pursue this problem further?

30. As described in Chapter 12, *Alu* elements proliferate in the human genome by a process called retrotransposition, in which the *Alu* DNA sequence is transcribed into RNA, copied into double-stranded DNA, and then inserted back into the genome at a site distant from that of its "parent" *Alu* gene.

    Clearly, this has been an extremely efficient process, since *Alu* genes have proliferated to about 10^6 copies in the human genome! This efficiency is largely due to the fact that *Alu* elements, like many small structural RNAs, carry their promoter sequences *within the transcribed region of the gene,* rather than 5′ to the transcription start site. If *Alu* elements carried promoters upstream of the transcription site, as do protein-coding genes, what would happen once they were retrotransposed? Would a retrotransposed *Alu* gene be able to proliferate? Explain.

31. M. Klemke et al. (2001) discovered an interesting coding phenomenon in which an exon within a neurologic hormone receptor gene in mammals appears to produce two different protein entities (XLαs, ALEX). Below is the DNA sequence of the exon's 5′ end derived from a rat. The lowercase letters represent the initial coding portion for the XLαs protein, and the uppercase letters indicate the portion where the ALEX entity is initiated. (For simplicity, and to correspond with the RNA

coding dictionary, it is customary to represent the noncoding, nontemplate strand of the DNA segment.)

5'-gtcccaaccatgcccaccgatcttccgcctgcttctgaagATGCGGGCCCAG

(a) Convert the noncoding DNA sequence to the coding RNA sequence.

(b) Locate the initiator codon within the XLαs segment.

(c) Locate the initiator codon within the ALEX segment. Are the two initiator codons in frame?

(d) Provide the amino acid sequence for each coding sequence. In the region of overlap, are the two amino acid sequences the same?

(e) Are there any evolutionary advantages to having the same DNA sequence code for two protein products? Are there any disadvantages?

32. The concept of consensus sequences of DNA was defined in this chapter as sequences that are similar (homologous) in different genes of the same organism or in genes of different organisms. Examples were the Pribnow box and the −35 region in prokaryotes and the TATA-box region in eukaryotes. One study found that among 73 isolates from the virus HIV-Type 1C (a major contributor to the AIDS epidemic), a GGGNNNNNCC consensus sequence exists (where N equals any nitrogenous base) in the promoter–enhancer region of the NF-κB transcription factor, a *cis*-acting element that is critical for initiating HIV transcription in human macrophages (Novitsky et al., 2002). The authors contend that finding this and other conserved sequences may be of value in designing an AIDS vaccine. What advantages would knowing these consensus sequences confer? Are there disadvantages as a vaccine is designed?

33. Recent observations indicate that alternative splicing is a common way for eukaryotes to expand their repertoire of gene functions. Studies indicate that approximately 50 percent of human genes exhibit alternative splicing and approximately 15 percent of disease-causing mutations involve aberrant alternative splicing. Different tissues show remarkably different frequencies of alternative splicing, with the brain accounting for approximately 18 percent of such events (Xu et al., 2002. *Nuc. Acids Res.* 30: 3754–3766).

(a) Define alternative splicing and speculate on the evolutionary strategy alternative splicing offers to organisms.

(b) Why might some tissues engage in more alternative splicing than others?

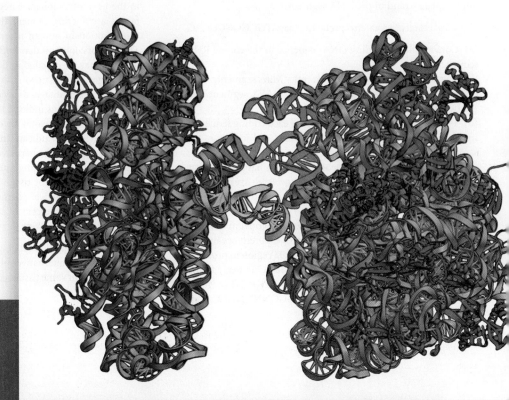

Crystal structure of a *Thermus thermophilus* 70S ribosome containing three bound transfer RNAs, bridging the two subunits. *Image provided by Dr. Albion Baucom (baucom@biology. ucsc.edu). Reprinted from the front cover of* Science, *Vol. 292, May 4, 2001.*

14

Translation and Proteins

CHAPTER CONCEPTS

- The ribonucleotide sequence of messenger RNA (mRNA) reflects genetic information stored in the DNA of genes and corresponds to the amino acid sequences in proteins encoded by those genes.

- The process of translation decodes the information in mRNA, leading to the synthesis of polypeptide chains.

- Translation involves the interactions of mRNA, tRNA, ribosomes, and a variety of translation factors essential to the initiation, elongation, and termination of the polypeptide chain.

- Proteins, the final product of most genes, achieve a three-dimensional conformation that arises from the primary amino acid sequences of the polypeptide chains making up each protein.

- The function of any protein is closely tied to its three-dimensional structure, which can be disrupted by mutation.

n Chapter 13, we established that a genetic code stores information in the form of triplet codons in DNA and that this information is initially expressed, through the process of transcription, as a messenger RNA complementary to the template strand of the DNA helix. However, the final product of gene expression, in most instances, is a polypeptide chain consisting of a linear series of amino acids whose sequence has been prescribed by the genetic code. In this chapter, we will examine how the information present in mRNA is translated to create polypeptides, which then fold into protein molecules. We will also review the evidence confirming that proteins are the end products of genes and discuss briefly the various levels of protein structure, diversity, and function. This information extends our understanding of gene expression and provides an important foundation for interpreting how the mutations that arise in DNA can result in the diverse phenotypic effects observed in organisms.

14.1

Translation of mRNA Depends on Ribosomes and Transfer RNAs

Translation of mRNA is the biological polymerization of amino acids into polypeptide chains. This process, alluded to in our earlier discussion of the genetic code, occurs only in association with **ribosomes**, which serve as nonspecific workbenches. The central question in translation is how triplet codons of mRNA direct specific amino acids into their correct position in the polypeptide. That question was answered once **transfer RNA (tRNA)** was discovered. This class of molecules adapts genetic information present as specific triplet codons in mRNA to their corresponding amino acids. As noted in Chapter 13, the requirement for some sort of "adaptor" was postulated by Francis Crick in 1957.

In association with a ribosome, mRNA presents a triplet codon that calls for a specific amino acid. A specific tRNA molecule contains within its nucleotide sequence three consecutive ribonucleotides complementary to the codon, called the **anticodon**, which can base-pair with the codon. Another region of this tRNA is covalently bonded to the codon's corresponding amino acid.

Hydrogen bonding of tRNAs to mRNA holds amino acids in proximity to each other so that a peptide bond can be formed between them. The process occurs over and over as mRNA runs through the ribosome, and amino acids are polymerized into a polypeptide. Before looking more closely at this process, we will first consider the structures of the ribosome and transfer RNA.

Ribosomal Structure

Because of its essential role in the expression of genetic information, the ribosome has been analyzed extensively. One bacterial cell contains about 10,000 ribosomes, and a eukaryotic cell contains many times more. Electron microscopy has revealed that the bacterial ribosome is about 25 μm at its largest diameter and consists of two subunits, one large and one small. Both subunits consist of one or more molecules of rRNA and an array of **ribosomal proteins**. When the two subunits are associated with each other in a single ribosome, the structure is sometimes called a **monosome**.

The main differences between prokaryotic and eukaryotic ribosomes are summarized in Figure 14–1. The subunit and rRNA components are most easily isolated and characterized on the basis of their sedimentation behavior in sucrose gradients (their rate of migration when centrifuged, abbreviated S for Svedberg, as introduced in Chapter 10). In prokaryotes, the monosome is a $70S$ particle, and in eukaryotes it is approximately $80S$. Sedimentation coefficients, which reflect differences in the rate of migration of different-sized particles and molecules, are not additive. For example, the prokaryotic $70S$ monosome consists of a $50S$ and a $30S$ subunit, and the eukaryotic $80S$ monosome consists of a $60S$ and a $40S$ subunit.

The larger subunit in prokaryotes consists of a $23S$ rRNA molecule, a $5S$ rRNA molecule, and 31 ribosomal proteins. In the eukaryotic equivalent, a $28S$ rRNA molecule is accompanied by a $5.8S$ and a $5S$ rRNA molecule and 46 proteins. The smaller prokaryotic subunits consist of a $16S$ rRNA component and 21 proteins. In the eukaryotic equivalent, an $18S$ rRNA component and about 33 proteins are found. The approximate molecular weights (in daltons, or Da) and number of nucleotides of these components are also shown in Figure 14–1.

It is now clear that the RNA components of the ribosome perform all-important catalytic functions associated with translation. The many ribosomal proteins, whose functions were long a mystery, are thought to promote the binding of the various molecules involved in translation and, in general, to fine-tune the process. This conclusion is based on the observation that some of the catalytic functions in ribosomes still occur in experiments involving "ribosomal protein-depleted" ribosomes.

Molecular hybridization studies have established the degree of redundancy of the genes coding for the rRNA components. The *E. coli* genome contains seven copies of a single sequence that encodes all three components—$23S$, $16S$, and $5S$. The initial transcript of each set of these genes produces a $30S$ RNA molecule that is enzymatically cleaved into these smaller components. The coupling of the genetic information encoding these three rRNA components ensures that, following multiple transcription events, equal quantities of all three will be present as ribosomes are assembled.

Prokaryotes Monosome 70S (2.5 × 10^6 Da)				**Eukaryotes** Monosome 80S (4.2 × 10^6 Da)			

Large subunit		**Small subunit**		**Large subunit**		**Small subunit**	
50S	1.6 × 10^6 Da	30S	0.9 × 10^6 Da	60S	2.8 × 10^6 Da	40S	1.4 × 10^6 Da

| 23S rRNA
(2904 nucleotides)
+
31 proteins
+
5S rRNA
(120 nucleotides) | 16S rRNA
(1541 nucleotides)
+
21 proteins | 28S rRNA
(4718 nucleotides)
+
46 proteins
+
5S rRNA   5.8S rRNA
(120   +   (160
nucleotides)   nucleotides) | 18S rRNA
(1874 nucleotides)
+
33 proteins |

FIGURE 14–1   A comparison of the components in prokaryotic and eukaryotic ribosomes.

In eukaryotes, many more copies of a sequence encoding the 28S, 18S, and 5.8S components are present. In *Drosophila,* approximately 120 copies per haploid genome are each transcribed into a molecule of about 45S. This is processed into the 28S, 18S, and 5.8S rRNA species. These species are homologous to the three rRNA components of *E. coli.* In *Xenopus laevis,* more than 500 copies of the 34S component are present per haploid genome. In mammalian cells, the initial transcript is 45S. The rRNA genes, called **rDNA**, are part of the moderately repetitive DNA fraction and are present in clusters at various chromosomal sites.

Each cluster in eukaryotes consists of **tandem repeats**, with each unit separated by a noncoding **spacer DNA** sequence. In humans, these gene clusters have been localized near the ends of chromosomes 13, 14, 15, 21, and 22. The unique 5S rRNA component of eukaryotes is not part of this larger transcript. Instead, genes coding for the 5S ribosomal component are distinct and located separately. In humans, a gene cluster encoding 5S rRNA has been located on chromosome 1.

Despite their detailed knowledge of the structure and genetic origin of the ribosomal components, a complete understanding of the function of these components has eluded geneticists. This is not surprising; the ribosome is the largest and perhaps the most intricate of all cellular structures. For example, the bacterial monosome has a combined molecular weight of 2.5 million Da!

tRNA Structure

Because of their small size and stability in the cell, transfer RNAs (tRNAs) have been investigated extensively and are the best characterized RNA molecules. They are composed of only 75 to 90 nucleotides, displaying a nearly identical structure in bacteria and eukaryotes. In both types of organisms, tRNAs are transcribed from DNA as larger precursors, which are cleaved into mature 4S tRNA molecules. In *E. coli,* for example, tRNA^{Tyr} (the superscript identifies the specific tRNA by the amino acid that binds to it, called its *cognate amino acid*) is composed of 77 nucleotides, yet its precursor contains 126 nucleotides.

In 1965, Robert Holley and his colleagues reported the complete sequence of tRNA^{Ala} isolated from yeast. Of great interest was their finding that a number of nucleotides are unique to tRNA, each containing a so-called *modified base.* As illustrated in Figure 14–2, each of these nucleotides contains a modification of one of the four nitrogenous bases expected in RNA (G, C, A, and U). Shown are inosinic acid (which contains the purine hypoxanthine), ribothymidylic acid, pseudouridine, and several others. These modified structures are created *after* transcription of tRNA, illustrating the more general concept of **posttranscriptional modification**. In this case, the unmodified base is inserted during transcription of tRNA, and subsequently, enzymatic reactions catalyze the chemical modifications to the base. You should compare the structure of these with the five standard

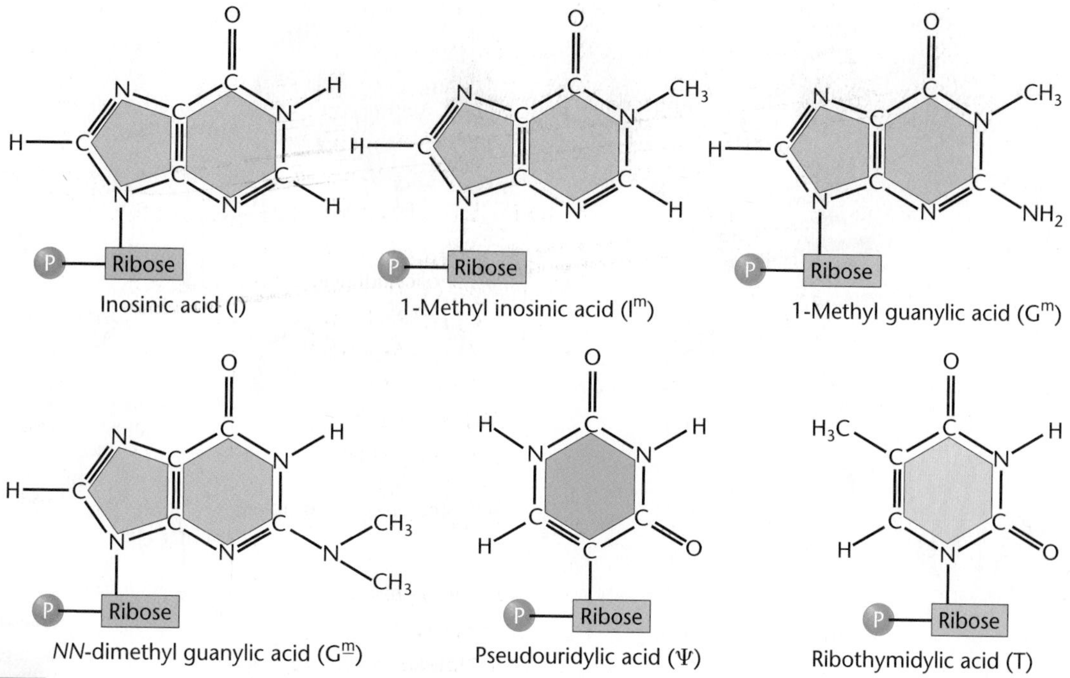

Inosinic acid (I)

1-Methyl inosinic acid (Iᵐ)

1-Methyl guanylic acid (Gᵐ)

NN-dimethyl guanylic acid (G$\underline{\underline{m}}$)

Pseudouridylic acid (Ψ)

Ribothymidylic acid (T)

FIGURE 14–2   Ribonucleotides containing unusual nitrogenous bases are found in transfer RNA.

nucleotides introduced and illustrated in Chapter 10. While it is still not clear why such modified bases are created, it is believed that their presence enhances hydrogen bonding efficiency during translation.

Holley's sequence analysis led him to propose the two-dimensional **cloverleaf model of tRNA**. It had been known that tRNA has a characteristic secondary structure created by base pairing. Holley discovered that he could arrange the linear sequence in such a way that several stretches of base pairing would result. His arrangement, with its series of paired stems and unpaired loops, resembled the shape of a cloverleaf. The loops consistently contained modified bases that did not generally form base pairs. Holley's model is shown in Figure 14–3.

The triplets GCU, GCC, and GCA specify alanine; therefore, Holley looked for an anticodon sequence complementary to one of these codons in his tRNAᴬˡᵃ molecule. He found it in the form of CGI (the 3′ to 5′ direction) in one loop of the cloverleaf. The nitrogenous base I (inosinic acid) can form hydrogen bonds with U, C, or A, the third members of the alanine triplets. Thus, the **anticodon loop** was established.

Studies of other tRNA species revealed many constant features. First, at the 3′ end, all tRNAs contain the sequence (. . . pCpCpA-3′). At this end of the molecule, the amino acid is covalently joined to the terminal adenosine residue. All tRNAs contain the nucleotide (5′-pG. . .) at the other end of the molecule. In addition, the lengths of the analogous stems

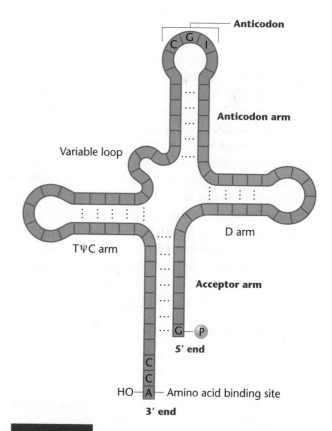

FIGURE 14–3   Holley's two-dimensional cloverleaf model of transfer RNA. Hydrogen bonds are designated by dots (···).

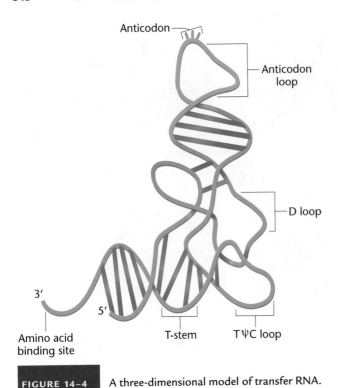

FIGURE 14–4   A three-dimensional model of transfer RNA.

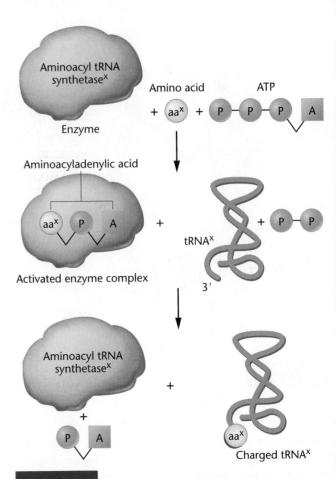

FIGURE 14–5   Steps involved in charging tRNA. The superscript x denotes that only the corresponding specific tRNA and specific aminoacyl tRNA synthetase enzyme are involved in the charging process for each amino acid.

and loops in tRNA molecules are very similar. Each tRNA examined also contains an anticodon complementary to the known codon for the tRNA's cognate amino acid, and all anticodon loops are present in the same position of the cloverleaf.

Because the cloverleaf model was predicted strictly on the basis of nucleotide sequence, there was great interest in the three-dimensional structure that would be revealed by X-ray crystallographic examination of tRNA. By 1974, Alexander Rich and his colleagues in the United States, and J. Roberts, B. Clark, Aaron Klug, and their colleagues in England had succeeded in crystallizing tRNA and performing X-ray crystallography at a resolution of 3 Å. At such resolution, the pattern formed by individual nucleotides is discernible.

As a result of these studies, a complete three-dimensional model of tRNA is now available (Figure 14–4). Both the anticodon loop and the 3′-acceptor region (to which the amino acid is covalently linked) have been located. Geneticists speculate that the shapes of the intervening loops are recognized by the enzymes responsible for attaching amino acids to tRNAs—a subject to which we now turn our attention.

Charging tRNA

Before translation can proceed, the tRNA molecules must be chemically linked to their respective amino acids. This activation process, called **charging**, or *aminoacylation,* occurs under the direction of enzymes called **aminoacyl tRNA synthetases**. Because there are 20 different amino acids, there must be at least 20 different tRNA molecules and as many different enzymes. In theory, because there are 61 triplets that encode amino acids, there could be 61 specific

tRNAs and enzymes. However, because of the ability of the third member of a triplet code to "wobble," the minimum number of different tRNAs required is only 31. In actuality, there are more than this number. It is also believed that there are only 20 synthetases, one for each amino acid, regardless of the greater number of corresponding tRNAs.

The charging process is outlined in Figure 14–5. In the initial step, the amino acid is converted to an activated form, reacting with ATP to create an **aminoacyladenylic acid**. A covalent linkage is formed between the 5′-phosphate group of ATP and the carboxyl end of the amino acid. This reaction occurs in association with the synthetase enzyme, forming a complex that then reacts with a specific tRNA molecule. During the next step, the amino acid is transferred to the appropriate tRNA and bonded covalently to the adenine residue at the 3′ end. The charged tRNA may then participate directly in protein synthesis. Aminoacyl tRNA synthetases are highly specific enzymes because they recognize only one amino acid and only the tRNAs corresponding to that amino acid, called **isoaccepting tRNAs**. This is a crucial point if fidelity of translation is to be maintained.

14–1 In 1962, F. Chapeville and others reported an experiment in which they isolated radioactive ^{14}C-cysteinyl-tRNA^{Cys} (charged tRNA^{Cys} + cysteine). They then removed the sulfur group from the cysteine, creating alanyl-tRNA^{Cys} (charged tRNA^{Cys} + alanine). When alanyl-tRNA^{Cys} was added to a synthetic mRNA calling for cysteine, but not alanine, a polypeptide chain was synthesized containing alanine. What can you conclude from this experiment?

■ HINT: *This problem is concerned with establishing whether tRNA or the amino acid added to the tRNA during charging is responsible for attracting the charged tRNA to mRNA during translation. The key to its solution is the observation that in this experiment, when the triplet codon in mRNA calls for cysteine, alanine is inserted during translation, even though it is the "incorrect" amino acid.*

14.2

Translation of mRNA Can Be Divided into Three Steps

Much like transcription, the process of translation can best be described by breaking it into discrete phases. We will consider three such phases, each with its own set of illustrations (Figures 14–6, 14–7, and 14–8), but keep in mind that translation is a dynamic, continuous process. As you read the following discussion, keep track of the step-by-step events depicted in the figures. Many of the protein factors and their roles in translation are presented in Table 14.1.

Initiation

Initiation of prokaryotic translation is depicted in Figure 14–6. Recall that the ribosome serves as a nonspecific workbench for the translation process. Ribosomes, when they are not involved in translation, are dissociated into their large and small subunits. Initiation of translation in *E. coli* involves the small ribosomal subunit, an mRNA molecule, a specific charged initiator tRNA, GTP, Mg^{2+}, and three proteinaceous **initiation factors (IFs)** that enhance the binding affinity of the various translational components. In prokaryotes, the initiation codon of mRNA—AUG—calls for the modified amino acid **N-formylmethionine (f-met)**.

The small ribosomal subunit binds to several initiation factors, and this complex in turn binds to mRNA (Step 1). In bacteria, this binding involves a sequence of up to six ribonucleotides (AGGAGG, not shown in Figure 14–6) that *precedes* the initial AUG start codon of mRNA. This sequence—containing only purines and called the **Shine–Dalgarno sequence**—base-pairs with a region of the 16S rRNA of the small ribosomal subunit, facilitating initiation.

Another initiation protein then enhances the binding of charged formylmethionyl tRNA to the small subunit in response to the AUG triplet (Step 2). This step "sets" the reading frame so that all subsequent groups of three ribonucleotides are translated accurately. The aggregate represents the **initiation complex**, which then combines with the large ribosomal subunit. In this process, a molecule of GTP is hydrolyzed, providing the required energy, and the initiation factors are released (Step 3).

Elongation

The second phase of translation, elongation, is depicted in Figure 14–7. Once both subunits of the ribosome are assembled with the mRNA, binding sites for two charged tRNA molecules are formed. These are the **P (peptidyl) site** and the **A (aminoacyl) site**. The charged initiator tRNA binds to the P site, provided that the AUG codon of mRNA is in the corresponding position of the small subunit.

The lengthening of the growing polypeptide chain by one amino acid is called **elongation**. The sequence of the

TABLE 14.1

Various Protein Factors Involved during Translation in *E. coli*

Process	Factor	Role
Initiation of translation	IF1	Stabilizes 30S subunit
	IF2	Binds f-met-tRNA to 30S-mRNA complex; binds to GTP and stimulates hydrolysis
	IF3	Binds 30S subunit to mRNA; dissociates monosomes into subunits following termination
Elongation of polypeptide	EF-Tu	Binds GTP; brings aminoacyl-tRNA to the A site of the ribosome
	EF-Ts	Generates active EF-Tu
	EF-G	Stimulates translocation; GTP-dependent
Termination of translation and release of polypeptide	RF1	Catalyzes release of the polypeptide chain from tRNA and dissociation of the translocation complex; specific for UAA and UAG termination codons
	RF2	Behaves like RF1; specific for UGA and UAA codons
	RF3	Stimulates RF1 and RF2

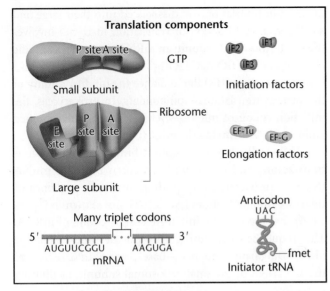

Translation components

FIGURE 14–6   Initiation of translation. The separate components are depicted at the left of the figure.

to this, the covalent bond between the tRNA occupying the P site and its cognate amino acid is hydrolyzed (broken). The newly formed dipeptide remains attached to the end of the tRNA still residing in the A site (Step 2). These reactions were initially believed to be catalyzed by an enzyme called **peptidyl transferase**, embedded in but never isolated from the large subunit of the ribosome. However, it is now clear that this catalytic activity is a function of the $23S$ rRNA of the large subunit, perhaps in conjunction with one or more of the ribosomal proteins. In such a case, as we saw with splicing of pre-mRNAs in Chapter 13, we refer to the complex as a **ribozyme**, recognizing the catalytic role that RNA plays in the process.

Before elongation can be repeated, the tRNA attached to the P site, which is now uncharged, must be released from the large subunit. The uncharged tRNA moves briefly into a third site on the ribosome called the **E (exit) site**. The entire **mRNA–tRNA–aa$_2$–aa$_1$** complex then shifts in the direction of the P site by a distance of three nucleotides (Step 3). This event, called *translocation,* requires several protein elongation factors (EFs). While it was originally thought that the energy derived from hydrolysis of GTP was essential for translocation, the energy produced is now thought to lock the proper structures in place during each step of elongation. The result is that the third codon of mRNA has now moved into the A site and is ready to accept its specific charged tRNA (Step 4). One simple way to distinguish the two sites in your mind is to remember that, *following the shift,* the P site (*P* for peptide) contains a tRNA attached to a peptide chain, whereas the A site (*A* for amino acid) contains a tRNA with an amino acid attached.

The sequence of elongation and translocation is repeated over and over (Steps 4 and 5). An additional amino acid is added to the growing polypeptide chain each time the mRNA advances by three nucleotides through the ribosome. Once a polypeptide chain of reasonable size is assembled (about 30 amino acids), it begins to emerge from the base of the large subunit, as illustrated in Step 6. The large subunit contains a tunnel through which the elongating polypeptide emerges.

As we have seen, the role of the small subunit during elongation is to "decode" the codons in the mRNA, while the role of the large subunit is peptide-bond synthesis. The efficiency of the process is remarkably high: The observed error rate is only about 10^{-4}. At this rate, an incorrect amino acid will occur only once in every 20 polypeptides of an average length of 500 amino acids! Elongation in *E. coli* proceeds at a rate of about 15 amino acids per second at 37°C.

Initiation of Translation

1. mRNA binds to small subunit along with initiation factors (IF1, 2, 3)

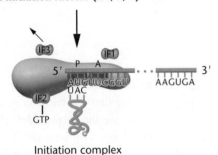

Initiation complex

2. Initiator tRNA^{fmet} binds to mRNA codon in P site; IF3 released

3. Large subunit binds to complex; IF1 and IF2 released; EF-Tu binds to tRNA, facilitating entry into A site

second triplet in mRNA dictates which charged tRNA molecule will become positioned at the A site (Step 1). Once it is present, an enzymatic reaction occurs within the large subunit of the ribosome, catalyzing the formation of the peptide bond that links the two amino acids together. Just prior

Termination

Termination, the third phase of translation, is depicted in Figure 14–8. The process is signaled by the presence of any one of the three possible triplet codons appearing in the

Elongation during Translation

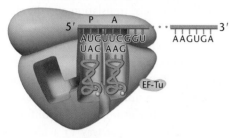

1. Second charged tRNA has entered A site, facilitated by EF-Tu; first elongation step commences

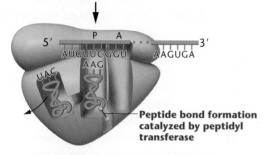

2. Peptide bond forms; uncharged tRNA moves to the E site and subsequently out of the ribosome; the mRNA has been translocated three bases to the left, causing the tRNA bearing the dipeptide to shift into the P site.

— Peptide bond formation catalyzed by peptidyl transferase

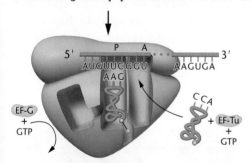

3. The first elongation step is complete, facilitated by EF-G. The third charged tRNA is ready to enter the A site.

A site: UAG, UAA, or UGA. These codons do not specify an amino acid, nor do they call for a tRNA in the A site. They are called **stop codons, termination codons**, or **nonsense codons**. Often, several consecutive several consecutive stop codons are part of an mRNA. When one such termination stop codon is encountered, the finished polypeptide is still attached to the terminal tRNA at the P site, and the A site is empty. The termination codon signals the action of a **GTP-dependent release factor**, which stimulates steps leading to the release of the polypeptide chain from the terminal tRNA and subsequently from the translation complex (Step 1). Then, the tRNA is released from the ribosome, which then dissociates into its subunits (Step 2). If a termination codon should appear in the middle of an mRNA molecule as a result of mutation, the same process occurs, and the polypeptide chain is prematurely terminated.

Polyribosomes

As elongation proceeds and the initial portion of an mRNA molecule has passed through the ribosome, this portion of mRNA is free to associate with another small subunit to form a second initiation complex. The process can be repeated several times with a single mRNA and results in what are called **polyribosomes**, or just **polysomes**.

After cells are gently lysed in the laboratory, polyribosomes can be isolated from them and analyzed. The photos in Figure 14–9 show these complexes as seen under the electron microscope. In Figure 14–9(a), you can see the thin lines of mRNA between the individual ribosomes. The micrograph in Figure 14–9(b) is even more remarkable, for it shows the polypeptide chains emerging from the ribosomes during translation. The formation of polysome complexes represents an efficient use of the components available for protein synthesis during a unit of time. Using the analogy of a tape and a tape recorder, in polysome complexes, we

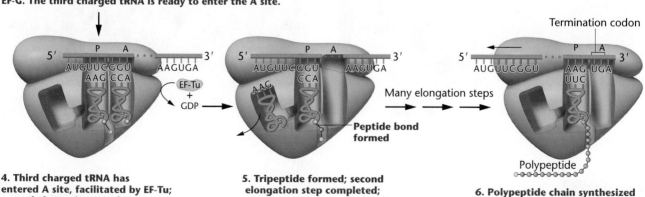

4. Third charged tRNA has entered A site, facilitated by EF-Tu; second elongation step begins

5. Tripeptide formed; second elongation step completed; uncharged tRNA moves to E site

— Peptide bond formed

Many elongation steps

Termination codon

Polypeptide

6. Polypeptide chain synthesized and exiting ribosome

FIGURE 14–7 Elongation of the growing polypeptide chain during translation.

Termination of Translation

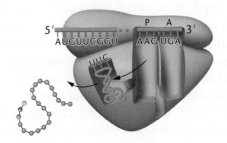

1. tRNA and polypeptide chain released

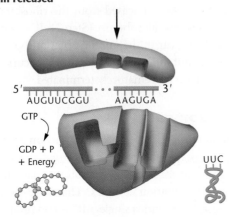

2. GTP-dependent termination factors stimulate the release of tRNA and the dissociation of the ribosomal subunits. The polypeptide folds into a protein.

FIGURE 14–8 Termination of the process of translation.

would thread and play one tape (mRNA) simultaneously through several recorders (the ribosomes). At any given moment, each recorder would be playing a different part of the song (the polypeptide being synthesized in each ribosome).

14.3

High-Resolution Studies Have Revealed Many Details about the Functional Prokaryotic Ribosome

Our knowledge of the process of translation and the structure of the ribosome, as described in the previous sections, is based primarily on biochemical and genetic observations, in addition to the visualization of ribosomes under the electron microscope. To confirm and refine this information, the next step is to examine the ribosome at even higher levels of resolution. For example, X-ray diffraction analysis of ribosome crystals is one way to achieve this. However, because of its tremendous size and the complexity of molecular interactions occurring in the functional ribosome, it was extremely difficult to obtain the crystals necessary to perform X-ray diffraction studies. Nevertheless, great strides have been made over the past decade. First, the individual ribosomal subunits were crystallized and examined in several laboratories, most prominently that of V. Ramakrishnan. Then, the crystal structure of the intact 70S ribosome, complete with associated mRNA and tRNAs, was examined by Harry Noller and colleagues. In essence, the entire translational complex was seen at the atomic level. Both Ramakrishnan and Noller derived the ribosomes from the bacterium *Thermus thermophilus*.

Many noteworthy observations have been made from these investigations. For example, the sizes and shapes of the subunits, measured at atomic dimensions, are in agreement with earlier estimates based on high-resolution electron microscopy. Furthermore, the shape of the ribosome changes

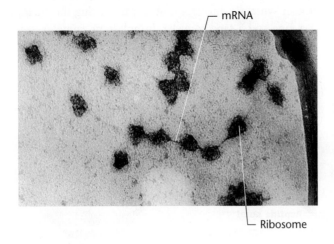

(a)

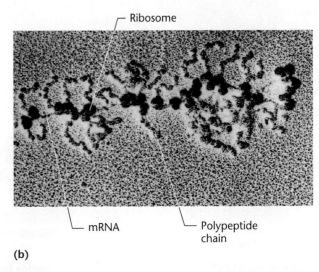

(b)

FIGURE 14–9 Polyribosomes as seen under the electron microscope. (a) Examples derived from rabbit reticulocytes engaged in the translation of hemoglobin mRNA. (b) Examples taken from salivary gland cells of the midgefly, *Chironomus thummi*. Note that the nascent polypeptide chains are apparent as they emerge from each ribosome. Their length increases as translation proceeds from left to right along the mRNA.

during different functional states, attesting to the dynamic nature of the process of translation. A great deal has also been learned about the prominence and location of the RNA components of the subunits. For example, about one-third of the 16S RNA is responsible for producing a flat projection, referred to as the *platform,* within the smaller 30S subunit, and it modulates movement of the mRNA–tRNA complex during translocation. One of the models based on Noller's findings is shown in the opening photograph of this chapter (p. 344).

Crystallographic analysis also supports the concept that RNA is the real "player" in the ribosome during translation. The interface between the two subunits, considered to be the location in the ribosome where polymerization of amino acids occurs, is composed almost exclusively of RNA. In contrast, the numerous ribosomal proteins are found mostly on the periphery of the ribosome. These observations confirm what has been predicted on genetic grounds—the catalytic steps that join amino acids during translation occur under the direction of RNA, not proteins.

Another interesting finding involves the actual location of the various sites predicted to house tRNAs during translation. All three sites (A, P, and E), have been identified in X-ray diffraction studies, and in each case, the RNA of the ribosome makes direct contact with the various loops and domains of the tRNA molecule. This observation supports the hypotheses that had been developed concerning the roles of the different regions of tRNA and helps us understand why the distinctive three-dimensional conformation that is characteristic of all tRNA molecules has been preserved throughout evolution.

Still another noteworthy observation is that the intervals between the A, P, and E sites are at least 20 Å, and perhaps as much as 50 Å, wide, thus defining the atomic distance that the tRNA molecules must shift during each translocation event. This is considered a fairly large distance relative to the size of the tRNAs themselves. Further analysis has led to the identification of molecular (RNA–protein) bridges existing between the three sites and apparently involved in the translocation events. Other such bridges are present at other key locations and have been related to ribosome function. These observations provide us with a much more complete picture of the dynamic changes that must occur within the ribosome during translation. A final observation takes us back almost 50 years, to when Francis Crick proposed the **wobble hypothesis**, as introduced in Chapter 13. The Ramakrishnan group has identified the precise location along the 16S rRNA of the 30S subunit involved in the decoding step that connects mRNA to the proper tRNA. At this location, two particular nucleotides of the 16S rRNA actually flip out and probe the codon:anticodon region, and are believed to check for accuracy of base pairing during this interaction. According to the wobble hypothesis, the stringency of this step is high for the first two base pairs but less so for the third (or wobble) base pair.

As our knowledge of the translation process in prokaryotes has continued to grow, a remarkable study was reported in 2010 by Niels Fischer and colleagues. Using a unique high-resolution approach—the technique of **time-resolved single particle cryo-electron microscopy (cryo-EM)**—the 70S *E. coli* ribosome was captured and examined while in the process of translation at a resolution of 5.5 Å. In this work, over two million images were obtained and computationally analyzed, establishing a temporal snapshot of the trajectories of tRNA during the process of translocation. This research team examined how tRNA is translocated during elongation of the polypeptide chain. They demonstrated that the trajectories are coupled with dynamic conformational changes in the components of the ribosome. Surprisingly, the work has revealed that during translation, the ribosome behaves as a complex molecular machine *powered by Brownian movement driven by thermal energy.* That is, the energetic requirement for achieving the various conformational changes essential to translocation are inherent to the ribosome itself.

Numerous questions about ribosome structure and function still remain. In particular, the precise role of the many ribosomal proteins is yet to be clarified. Nevertheless, the models that are emerging based on the work of Noller, Ramakrishnan, Fisher, and their many colleagues provide us with a much better understanding of the mechanism of translation.

14.4

Translation Is More Complex in Eukaryotes

The general features of the model we just discussed were initially derived from investigations of the translation process in bacteria. As we have seen (Figure 14–1), one main difference between prokaryotes and eukaryotes is that in the latter, translation occurs on larger ribosomes whose rRNA and protein components are more complex. Interestingly, prokaryotic and eukaryotic rRNAs do share what is called a *core sequence,* but in eukaryotes, they are lengthened by the addition of *expansion sequences (ES)*, which presumably impart added functionality. Another significant distinction is that whereas transcription and translation are coupled in prokaryotes, in eukaryotes these two processes are separated both spatially and temporally. In eukaryotic cells transcription occurs in the nucleus and translation in the cytoplasm. This separation provides multiple opportunities for regulation of genetic expression in eukaryotic cells.

A number of aspects of the initiation of translation vary in eukaryotes. Three differences center on the mRNA that is being translated. First, the 5′ end of mRNA is capped with a 7-methylguanosine (7-mG) residue at maturation (see Chapter 13). The presence of the cap, absent in prokaryotes, is essential for efficient initiation of translation. A second difference is that many mRNAs contain a purine (A or G) three bases upstream from the AUG initiator codon, which is followed by a G (A/GNNAUGG). Named after its discoverer, Marilyn Kozak, its presence in eukaryotes is considered to increase the efficiency of translation by interacting with the initator tRNA. This **Kozak sequence** is considered analogous to the *Shine-Dalgarno* sequence found in the upstream region of prokaryotic mRNAs.

Third, eukaryotic mRNAs require the posttranscriptional addition of a poly-A tail on their 3′ end; that is, they are *polyadenylated*. In the absence of poly A, these potential messages are rapidly degraded in the cytoplasm. Interestingly, histone mRNAs serve as an exception and are not polyadenylated. Still another difference related to initiation of translation is that in eukaryotes the amino acid formylmethionine is not required as it is in prokaryotes. However, the AUG triplet, which encodes methionine, is essential to the formation of the translational complex, and a unique transfer RNA ($tRNA_i^{Met}$) is used during initiation.

Still other differences are noteworthy. Eukaryotic mRNAs are much longer lived than are their prokaryotic counterparts. Most exist for hours rather than minutes prior to degradation by nucleases in the cell; thus they remain available much longer to orchestrate protein synthesis. And, during translation, protein factors similar to those in prokaryotes guide the initiation, elongation, and termination of translation in eukaryotes. Many of these eukaryotic factors are clearly homologous to their counterparts in prokaryotes. However, a greater number of factors are usually required during each step, and some are more complex than in prokaryotes. Finally, recall that in eukaryotes many, but not all, of the cell's ribosomes are found in association with the membranes that make up the endoplasmic reticulum (forming the rough ER). Such membranes are absent from the cytoplasm of prokaryotic cells. This association in eukaryotes facilitates the secretion of newly synthesized proteins from the ribosomes directly into the channels of the endoplasmic reticulum. Recent studies using electron microscopy have established how this occurs. A *tunnel* in the large subunit of the ribosome begins near the point where the two subunits interface and exits near the back of the large subunit. The location of the tunnel within the large subunit is the basis for the belief that it provides the conduit for the movement of the newly synthesized polypeptide chain out of the ribosome. In studies in yeast, newly synthesized polypeptides enter the ER through a membrane channel formed by a specific protein,

Sec61. This channel is perfectly aligned with the exit point of the ribosomal tunnel. In prokaryotes, the polypeptides are released by the ribosome directly into the cytoplasm.

We conclude this section by noting that at the end of 2010, after years of work, the crystal structure of the highly complex 80S eukaryotic ribosome has now been visualized by Marat Yusupov and colleagues at the remarkable resolution of 4.15 Å. This research, focusing on the yeast ribosome, is beginning to reveal at the atomic level how many of the eukaryote-specific aspects of translation are achieved. For one, it has been established that as the mRNA shifts by three nucleotides during translation (the process of translocation), a 4°–5° CW rotation of the small subunit occurs. As expected, translation is indeed a dynamic process as the subunits shift with relationship to one another as the mRNA proceeds through the ribosome. Further analysis at such a resolution will no doubt provide greater insights into the interactions of mRNA and tRNA with the components of the ribosome, and thus a better understanding of the process in eukaryotes.

14.5

The Initial Insight That Proteins Are Important in Heredity Was Provided by the Study of Inborn Errors of Metabolism

Now, let's consider how we know that proteins are the end products of genetic expression. The first insight into the role of proteins in genetic processes was provided by observations made by Sir Archibald Garrod and William Bateson early in the twentieth century. Garrod was born into an English family of medical scientists. His father was a physician with a strong interest in the chemical basis of rheumatoid arthritis, and his eldest brother was a leading zoologist in London. As a practicing physician, Garrod himself became interested in several human disorders that seemed to be inherited. Although he also studied albinism and cystinuria, we will describe his investigation of the disorder **alkaptonuria**. Individuals afflicted with this disorder have a disruption in an important metabolic pathway (Figure 14–10). As a result, they cannot metabolize the alkapton 2,5-dihydroxyphenylacetic acid, also known as homogentisic acid. Homogentisic acid thus accumulates in their cells and tissues and is excreted in the urine. The molecule's oxidation products are black and easily detectable in the diapers of newborns. The unmetabolized products tend to accumulate in cartilaginous areas, causing a darkening of the ears and nose. In joints, this deposition leads to a benign arthritic condition. Alkaptonuria is a rare but not serious disease that persists throughout an individual's life.

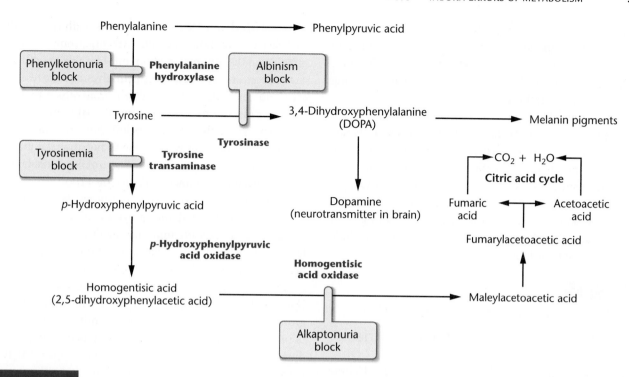

FIGURE 14–10 Metabolic pathway involving phenylalanine and tyrosine. Various metabolic blocks resulting from mutations lead to the disorders phenylketonuria, alkaptonuria, albinism, and tyrosinemia.

Garrod studied alkaptonuria by looking for patterns of inheritance of this benign trait. Eventually he concluded that it was genetic in nature. Of 32 known cases, he ascertained that 19 were confined to 7 families, with one family having four affected siblings. In several instances, the parents were unaffected but known to be related as first cousins, and therefore **consanguine**, a term describing relatives having a common recent ancestor. Parents who are so related have a higher probability than unrelated parents of producing offspring that express recessive traits, because such parents are both more likely to be heterozygous for some of the same recessive traits (see Chapter 25). Garrod concluded that this inherited condition was the result of an alternative mode of metabolism, thus implying that hereditary information controls chemical reactions in the body. While *genes* and *enzymes* were not familiar terms during Garrod's time, he used the corresponding concepts of *unit factors* and *ferments*. Garrod published his initial observations in 1902.

Only a few geneticists, including Bateson, were familiar with and made reference to Garrod's work. Garrod's ideas fit nicely with Bateson's belief that inherited conditions were caused by the lack of some critical substance. In 1909, Bateson published *Mendel's Principles of Heredity,* in which he linked ferments with heredity. However, for almost 30 years, most geneticists failed to see the relationship between genes and enzymes. Garrod and Bateson, like Mendel, were ahead of their time.

Phenylketonuria

The inherited human metabolic disorder **phenylketonuria (PKU)** results when another reaction in the pathway shown in Figure 14–10 is blocked. Described first in 1934, this disorder can result in mental retardation and is transmitted as an autosomal recessive disease. Afflicted individuals are unable to convert the amino acid phenylalanine to the amino acid tyrosine. These molecules differ by only a single hydroxyl group (OH), present in tyrosine but absent in phenylalanine. The reaction is catalyzed by the enzyme **phenylalanine hydroxylase**, which is inactive in affected individuals and active at a level of about 30 percent in heterozygotes. The enzyme functions in the liver. While the normal blood level of phenylalanine is about 1 mg/100 mL, people with phenylketonuria show a level as high as 50 mg/100 mL.

As phenylalanine accumulates, it may be converted to phenylpyruvic acid and, subsequently, to other derivatives. These are less efficiently resorbed by the kidney and tend to spill into the urine more quickly than phenylalanine. Both phenylalanine and its derivatives enter the cerebrospinal fluid, resulting in elevated levels in the brain. The presence of these substances during early development is thought to cause mental retardation.

Phenylketonuria occurs in approximately 1 in 11,000 births, and newborns are routinely screened for PKU throughout the United States. When the condition is detected in the analysis of an infant's blood, a strict dietary regimen is instituted in time to prevent retardation. A low-phenylalanine diet can reduce by-products such as phenylpyruvic acid, and the development of abnormalities characterizing the disease can be diminished.

Our knowledge of inherited metabolic disorders such as alkaptonuria and phenylketonuria has caused a revolution in medical thinking and practice. Human disease, once thought to be solely attributable to the action of invading microorganisms, viruses, or parasites, clearly can have a genetic basis. We know now that hundreds of medical conditions are caused by errors in metabolism resulting from mutant genes. These human biochemical disorders include all classes of organic biomolecules.

14.6
Studies of *Neurospora* Led to the One-Gene:One-Enzyme Hypothesis

In two separate investigations beginning in 1933, George Beadle provided the first convincing experimental evidence that genes are directly responsible for the synthesis of enzymes. The first investigation, conducted in collaboration with Boris Ephrussi, involved *Drosophila* eye pigments. Together, Beadle and Ephrussi confirmed that mutant genes that altered the eye color of fruit flies could be linked to biochemical errors that, in all likelihood, involved the loss of enzyme function. Encouraged by these findings, Beadle then joined with Edward Tatum to investigate nutritional mutations in the pink bread mold *Neurospora crassa*. This investigation led to the **one-gene:one-enzyme hypothesis**.

Analysis of *Neurospora* Mutants by Beadle and Tatum

In the early 1940s, Beadle and Tatum chose to work with *Neurospora* because much was known about its biochemistry, and mutations could be induced and isolated with relative ease. By inducing mutations, they produced strains that had genetic blocks of reactions essential to the growth of the organism.

Beadle and Tatum knew that this mold could manufacture nearly every biomolecule necessary for normal development. For example, using rudimentary carbon and nitrogen sources, the organism can synthesize nine water-soluble vitamins, 20 amino acids, numerous carotenoid pigments, and all essential purines and pyrimidines. Beadle and Tatum irradiated asexual conidia (spores) with X rays to increase the frequency of mutations and then grew the spores on "complete" medium containing all the necessary growth factors (e.g., vitamins and amino acids). Under such growth conditions, a mutant strain unable to grow on minimal medium would be able to grow by ingesting the supplements present in the enriched, complete medium. All the cultures were then transferred to minimal medium. Any organisms capable of growing on the minimal medium must be able to synthesize all the necessary growth factors themselves, and the researchers could conclude that the cultures from which those organisms came did not contain a nutritional mutation. If no growth occurred, then it was concluded that the culture that had not been able to grow contained a nutritional mutation. The next task was to determine the type of nutritional mutation. The results are shown in Figure 14–11(a).

Many thousands of individual spores derived by this procedure were isolated and grown on complete medium. In subsequent tests on minimal medium, many cultures failed to grow, indicating that a nutritional mutation had been induced. To identify the mutant type, the mutant strains were tested on a series of different incomplete media [Figure 14–11(b) and 14–11(c)], each containing different groups of supplements, and subsequently on media containing single vitamins, amino acids, purines, or pyrimidines as supplements, until one specific supplement that permitted growth was found. Beadle and Tatum reasoned that the supplement that restored growth was the molecule that the mutant strain could not synthesize.

The first mutant strain isolated required vitamin B_6 (pyridoxine) in the medium, and the second one required vitamin B_1 (thiamine). Using the same procedure, Beadle and Tatum eventually isolated and studied hundreds of mutants deficient in the ability to synthesize other vitamins, amino acids, nucleotides, or other substances.

The findings derived from testing more than 80,000 spores convinced Beadle and Tatum that genetics and biochemistry have much in common. It seemed likely that each nutritional mutation caused the loss of the enzymatic activity that facilitates an essential reaction in wild-type organisms. It also appeared that a mutation could be found for nearly any enzymatically controlled reaction. Beadle and Tatum had thus provided sound experimental evidence for the hypothesis that *one gene specifies one enzyme*, an idea alluded to more than 30 years earlier by Garrod and Bateson. With modifications, this concept was to become another major principle of genetics.

Genes and Enzymes: Analysis of Biochemical Pathways

The one-gene:one-enzyme concept and its attendant research methods have been used over the years to work out many details of metabolism in *Neurospora, Escherichia coli*, and a number of other microorganisms. One of the first metabolic pathways to be investigated in detail was that leading to the synthesis of the amino acid arginine in *Neurospora*. By studying seven mutant strains, each requiring arginine for growth (*arg*⁻), Adrian Srb and Norman Horowitz ascertained a partial biochemical pathway that leads to

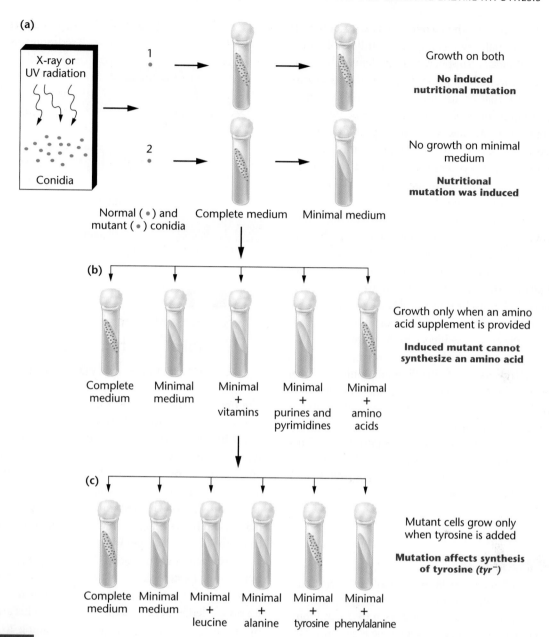

FIGURE 14–11 Induction, isolation, and characterization of a nutritional auxotrophic mutation in *Neurospora*. (a) Most conidia (blue) are not affected, but one conidium (shown in red) contains a mutation (b and c). The precise nature of the mutation is established and found to involve the biosynthesis of tyrosine.

the synthesis of the amino acid. Their work demonstrates how genetic analysis can be used to establish biochemical information.

Srb and Horowitz tested each mutant strain's ability to reestablish growth if either citrulline or ornithine, two compounds with close chemical similarity to arginine, was used as a supplement to minimal medium. If either was able to substitute for arginine, they reasoned that it must be involved in the biosynthetic pathway of arginine. The researchers found that both molecules could be substituted in one or more strains.

Of the seven mutant strains, four of them (*arg 4* through *arg 7*) grew if supplied with either citrulline, ornithine, or arginine. Two of them (*arg 2* and *arg 3*) grew if supplied with citrulline or arginine. One strain (*arg 1*) would grow only if arginine were supplied; neither citrulline nor ornithine could substitute for it. From these experimental observations, the following pathway and metabolic blocks for each mutation were deduced:

$$\text{Precursor} \xrightarrow[\text{Enzyme A}]{arg\,4-7} \text{Ornithine} \xrightarrow[\text{Enzyme B}]{arg\,2-3} \text{Citrulline} \xrightarrow[\text{Enzyme C}]{arg\,1} \text{Arginine}$$

The logic supporting these conclusions is as follows: If mutants *arg 4* through *arg 7* can grow regardless of which of the three molecules is supplied as a supplement to minimal medium, the mutations preventing growth must cause a metabolic block that occurs *prior* to the involvement of ornithine, citrulline, or arginine in the pathway. When any one of these three molecules is added, *its presence bypasses the block.* As a result, both citrulline and ornithine appear to be involved in the biosynthesis of arginine. However, the sequence of their participation in the pathway cannot be determined on the basis of these data.

On the other hand, both the *arg 2* and the *arg 3* mutations grow if supplied with citrulline, but not if they are supplied with only ornithine. Therefore, ornithine must be synthesized in the pathway *prior to the block.* Its presence will not overcome the block. Citrulline, however, *does overcome the block,* so it must be synthesized beyond the point of blockage. Therefore, the conversion of ornithine to citrulline represents the correct sequence in the pathway.

Finally, we can conclude that *arg 1* represents a mutation preventing the conversion of citrulline to arginine. Neither ornithine nor citrulline can overcome the metabolic block in this mutation because both participate earlier in the pathway.

Taken together, the analysis as described above supports the sequence of biosynthesis shown in Figure 14–12. Since Srb and Horowitz's experiments in 1944, the detailed pathway has been worked out, and the enzymes controlling each step have been characterized.

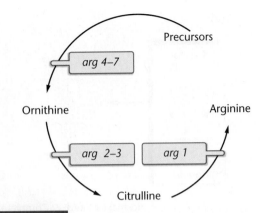

FIGURE 14–12   Abbreviated pathway describing the biosynthesis of arginine in *Neurospora.*

14.7

Studies of Human Hemoglobin Established That One Gene Encodes One Polypeptide

The one-gene:one-enzyme concept developed in the early 1940s was not immediately accepted by all geneticists. This is not surprising, since it was not yet clear how mutant enzymes could cause variation in the many different kinds of phenotypic traits. For example, *Drosophila* mutants demonstrated altered eye size, wing shape, wing vein pattern, and so on. Plants exhibited mutant varieties of seed texture, height, and fruit size. How an inactive mutant enzyme could result in such phenotypes was puzzling to many geneticists.

Two factors soon modified the one-gene: one-enzyme hypothesis. First, while *nearly all enzymes are proteins, not all proteins are enzymes.* As the study of biochemical genetics proceeded, it became clear that all proteins are specified by the information stored in genes, leading to the more accurate phraseology **one-gene: one-protein hypothesis.** Second, proteins often have a subunit structure consisting of two or more polypeptide chains. This is the basis of the quaternary structure of proteins, which we will discuss later in the chapter. Because each distinct polypeptide chain is encoded by a separate gene, a more modern statement of Beadle and Tatum's basic principle is **one-gene: one-polypeptide chain hypothesis.** The need for these modifications of the original hypothesis became apparent during the analysis of hemoglobin structure in individuals afflicted with sickle-cell anemia.

Sickle-Cell Anemia

The first direct evidence that genes specify proteins other than enzymes came from the work on mutant hemoglobin molecules derived from humans afflicted with the disorder

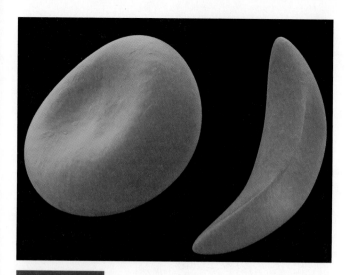

A comparison of a normal erythrocyte (left) with one derived from a patient with sickle-cell anemia (right).

sickle-cell anemia. Affected individuals have erythrocytes that, under low oxygen tension, become elongated and curved because of the polymerization of hemoglobin. The sickle shape of these erythrocytes is in contrast to the biconcave disc shape characteristic in unaffected individuals (Figure 14–13). Those with the disease suffer attacks when red blood cells aggregate in the venous side of capillary systems, where oxygen tension is very low. As a result, a variety of tissues are deprived of oxygen and suffer severe damage. When this occurs, an individual is said to experience a *sickle-cell crisis*. If untreated, a crisis may be fatal. The kidneys, muscles, joints, brain, gastrointestinal tract, and lungs can be affected.

In addition to suffering crises, these individuals are anemic because their erythrocytes are destroyed more rapidly than normal red blood cells. Compensatory physiological mechanisms include increased red-cell production by bone marrow and accentuated heart action. These mechanisms lead to abnormal bone size and shape as well as dilation of the heart.

In 1949, James Neel and E. A. Beet demonstrated that the disease is inherited as a Mendelian trait. Pedigree analysis revealed three genotypes and phenotypes controlled by a single pair of alleles, Hb^A and Hb^S. Unaffected and affected individuals result from the homozygous genotypes Hb^A/Hb^A and Hb^S/Hb^S, respectively. The red blood cells of heterozygotes, who exhibit the **sickle-cell trait** but not the disease, undergo much less sickling because more than half of their hemoglobin is normal. Though largely unaffected, such heterozygotes are "carriers" of the defective gene, which is transmitted on average to 50 percent of their offspring.

In the same year, Linus Pauling and his coworkers provided the first insight into the molecular basis of sickle-cell

anemia. They showed that hemoglobins isolated from diseased and normal individuals differ in their rates of electrophoretic migration. In electrophoresis (described in Chapter 10), charged molecules migrate in an electric field. If the net charge of two molecules is different, their rates of migration will be different. Hence, Pauling and his colleagues concluded that a chemical difference exists between normal and sickle-cell hemoglobin. The two molecules are now designated **HbA** and **HbS**, respectively.

Figure 14–14(a) illustrates the migration pattern of hemoglobin derived from individuals of all three possible genotypes when subjected to **starch gel electrophoresis**. The gel provides the supporting medium for the molecules during migration. In this experiment, samples were placed at a point of origin between the cathode (–) and the anode (+), and an electric current was applied. The migration pattern revealed that all molecules moved toward the anode, indicating a net negative charge. However, HbA migrated farther than HbS, suggesting that its net negative charge was greater. The electrophoretic pattern of hemoglobin derived from individuals who were carriers revealed the presence of both HbA and HbS, and confirmed their heterozygous genotype.

Pauling's findings suggested two possibilities. It was known that hemoglobin consists of four nonproteinaceous, iron-containing *heme groups* and a *globin portion* that contains four polypeptide chains. The alteration in net charge in HbS had to be due, theoretically, to a chemical change in one of these components.

Work carried out between 1954 and 1957 by Vernon Ingram resolved this question. He demonstrated that the chemical change occurs in the primary structure of the globin portion of the hemoglobin molecule. Using the **fingerprinting technique** shown in Figure 14–14(b), Ingram showed that HbS differs in amino acid composition compared to HbA. Human adult hemoglobin contains two identical α chains of 141 amino acids and two identical β chains of 146 amino acids in its quaternary structure.

The fingerprinting technique involves the enzymatic digestion of the protein into peptide fragments. The mixture is then placed on absorbent paper and exposed to an electric field, where migration occurs according to net charge. The paper is next turned at a right angle to its first exposure and placed in a solvent, in which chromatographic action causes the migration of the peptides in the second direction. The end result is a two-dimensional separation of the peptide fragments into a distinctive pattern of spots, or a "fingerprint." Ingram's work revealed that HbS and HbA differed by only a single peptide fragment [Figure 14–14(b)]. Further analysis then revealed a single amino acid change: Valine was substituted for glutamic acid at the sixth position of the β chain, accounting for the peptide difference [Figure 14–14(c)].

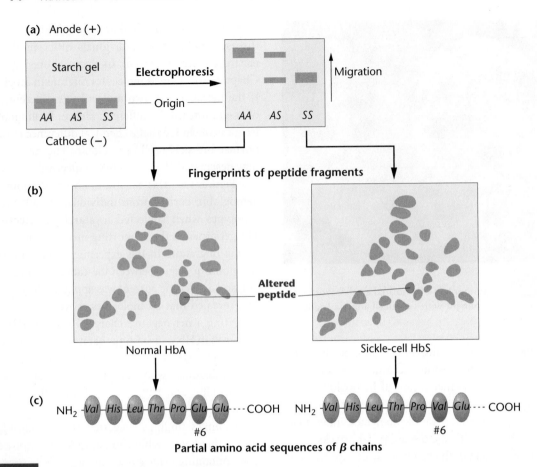

FIGURE 14-14 Investigation of hemoglobin derived from $Hb^A Hb^A$, $Hb^A Hb^S$, and $Hb^S Hb^S$ individuals using electrophoresis, finger-printing, and amino acid analysis. Hemoglobin from individuals with sickle-cell anemia ($Hb^S Hb^S$): (a) migrates differently in an electrophoretic field; (b) shows an altered peptide in fingerprint analysis; and (c) shows an altered amino acid, valine, at the sixth position in the β chain. During electrophoresis, heterozygotes ($Hb^A Hb^S$) are shown to have both forms of hemoglobin.

The significance of this discovery has been multifaceted. It clearly establishes that a single gene provides the genetic information for a single polypeptide chain. Studies of HbS also demonstrate that a mutation can affect the phenotype by directing a single amino acid substitution. Also, by providing the explanation for sickle-cell anemia, the concept of inherited **molecular disease** was firmly established. Finally, this work led to a thorough study of human hemoglobins, which has provided valuable genetic insights.

In the United States, sickle-cell anemia is found almost exclusively in the African-American population. It affects about one in every 625 African-American infants. Currently, about 50,000 to 75,000 individuals are afflicted. In about 1 of every 145 African-American married couples, both partners are heterozygous carriers. In these cases, each of their children has a 25 percent chance of having the disease.

Human Hemoglobins

Having introduced human hemoglobins in a historical context, we now end the discussion by providing an update of what is currently known about these molecules in our species. Molecular analysis reveals that an individual human produces different types of hemoglobin molecules at different stages of the life cycle. All are tetramers consisting of different combinations of seven distinct polypeptide chains, each encoded by a separate gene. The expression of these various genes is developmentally regulated.

Almost all adult hemoglobin consists of **HbA**, which contains two α and two β **chains**. Recall that the mutation in sickle-cell anemia involves the β chain. HbA represents about 98 percent of all hemoglobin found in an adult's erythrocytes after the age of six months. The remaining 2 percent consists of **HbA$_2$**, a minor adult component. This molecule contains two α chains and two **delta (δ) chains**. The δ chain is very similar to the β chain, consisting of 146 amino acids.

During embryonic and fetal development, quite a different set of hemoglobins is found. The earliest set to develop is called **Gower 1**, containing two **zeta (ζ) chains**, which are most similar to α chains, and two **epsilon (ε) chains**, which are similar to β chains. By eight weeks' gestation, the embryonic form is gradually replaced by still another hemoglobin molecule with still different chains. This molecule is called **HbF**, or **fetal hemoglobin**, and consists of two α

TABLE 14.2

Chain Compositions of Human Hemoglobins from Conception to Adulthood

Hemoglobin Type	Chain Composition
Embryonic-Gower 1	$\zeta_2 \, \varepsilon_2$
Fetal-HbF	$\alpha_2 \, ^G\gamma_2$
	$\alpha_2 \, ^A\gamma_2$
Adult-HbA	$\alpha_2 \, \beta_2$
Minor adult-HbA$_2$	$\alpha_2 \, \delta_2$

chains and two **gamma (γ) chains**. There are two types of γ chains, designated $^G\gamma$ and $^A\gamma$. Both are similar to β chains and differ from each other by only a single amino acid. These persist until birth, after which HbF is, again gradually, replaced with HbA and HbA$_2$. The nomenclature and sequence of appearance of the five tetramers we have described are summarized in Table 14.2.

14.8

The Nucleotide Sequence of a Gene and the Amino Acid Sequence of the Corresponding Protein Exhibit Colinearity

Once it was established that genes specify the synthesis of polypeptide chains, the next logical question was how the genetic information contained in the nucleotide sequence of a gene can be transferred to the amino acid sequence of a polypeptide chain? It seemed most likely that a colinear relationship would exist between the two molecules. That is, the order of nucleotides in the DNA of a gene would correlate directly with the order of amino acids in the corresponding polypeptide—the concept of **colinearity**.

The initial experimental evidence in support of this concept was derived from Charles Yanofsky's studies of the *trpA* gene that encodes the A subunit of the enzyme *tryptophan synthetase* in *E. coli*. Yanofsky isolated many independent mutants that had lost the activity of the enzyme. He then mapped the various mutations, establishing their location with respect to one another within the gene. He also determined where the amino acid substitution had occurred in each mutant protein. When the two sets of data were compared, the colinear relationship was apparent. The location of each mutation in the *trpA* gene correlated with the position of the altered amino acid in the A polypeptide of tryptophan synthetase. This comparison is illustrated in Figure 14–15.

14.9

Variation in Protein Structure Provides the Basis of Biological Diversity

In contrast to nucleic acids, which store and express genetic information, proteins, as end products of genetic expression, are more closely aligned with biological function. It is the variation in biological function that provides the basis of diversity between cell types and between organisms. What is it about proteins that enable them to perform or control enormous numbers of complex and important cellular activities in an organism? As we will see, the secret of the complexity of protein function lies in the incredible structural diversity of proteins.

At the outset of our discussion, we should differentiate between **polypeptides** and **proteins**. Polypeptides, most simply, are precursors of proteins. Thus, as the amino acid polymer is assembled on and then released from the ribosome during translation, it is called a *polypeptide*. Once a polypeptide subsequently folds up and assumes a functional three-dimensional conformation, it is called a *protein*. In most cases, several polypeptides combine during this process to

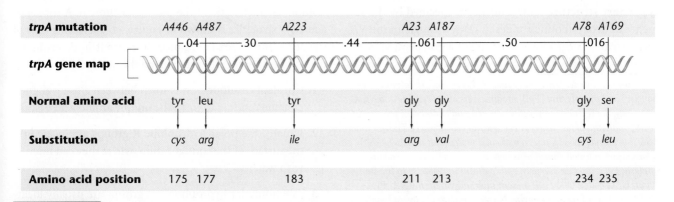

FIGURE 14–15 Demonstration of colinearity between the genetic map of various *trpA* mutations in *E. coli* and the affected amino acids in the protein product. The numbers shown between mutations represent linkage distances.

produce an even higher order of protein structure. It is its three-dimensional conformation in space that is essential to a protein's specific function and that distinguishes it from other proteins.

Like nucleic acids, the polypeptide chains comprising proteins are linear, nonbranched polymers. In the vast majority of organisms, there are 20 amino acids that serve as the subunits (the building blocks) of proteins.* Each amino acid has a **carboxyl group**, an **amino group**, and a **radical (R) group** (a side chain that determines the type of amino acid) bound covalently to a **central carbon (C) atom.** Figure 14–16 shows the 20 R groups that define the 20 amino acids in proteins. The R groups are varied in structure and can be divided into four main classes: (1) **nonpolar (hydrophobic)**, (2) **polar (hydrophilic)**, (3) **positively charged**, and (4) **negatively charged**. Because polypeptides are often long polymers, and because each unit in the polymer may be any 1 of 20 amino acids, each with unique chemical properties, enormous variation in the molecule's final conformation and chemical activity is possible. For example, if an average polypeptide is composed of 200 amino acids (a molecular weight of about 20,000 Da), 20^{200} different molecules, each with a unique sequence, can be created from the 20 different building blocks.

Around 1900, German chemist Emil Fischer determined the manner in which the amino acids are bonded together. He showed that the amino group of one amino acid can react with the carboxyl group of another amino acid in a dehydration (condensation) reaction, releasing a molecule of H_2O. The resulting covalent bond is called a **peptide bond** (Figure 14–17). Two amino acids linked together constitute a *dipeptide,* three a *tripeptide,* and so on. Once ten or more amino acids are linked by peptide bonds, the chain is referred to as a *polypeptide.* Generally, no matter how long a polypeptide is, it will have an amino group at one end (the N-terminus) and a carboxyl group at the other end (the C-terminus).

Four levels of protein structure are recognized: primary, secondary, tertiary, and quaternary. The sequence of amino acids in the linear backbone of the polypeptide constitutes its **primary structure**. This sequence is specified by the sequence of deoxyribonucleotides in DNA through an mRNA intermediate. The primary structure of a polypeptide helps determine the specific characteristics of the higher orders of organization as a protein is formed.

Secondary structures are certain regular or repeating configurations in space assumed by amino acids lying close to one another in the polypeptide chain. In 1951, Linus Pauling and Robert Corey predicted, on theoretical grounds, an **α helix** as one type of secondary structure. The α-helix model [Figure 14–18(a)] has since been confirmed by X-ray crystallographic studies. It is rodlike and has the greatest possible theoretical stability. The helix is composed of a spiral chain of amino acids stabilized by hydrogen bonds.

The side chains (the R groups) of amino acids extend outward from the helix, and each amino acid residue occupies a distance of 1.5 Å in the length of the helix. There are 3.6 residues per turn. Although left-handed helices are theoretically possible, all α helices seen in proteins are right-handed.

Also in 1951, Pauling and Corey proposed another secondary structure, the **β-pleated sheet**. In this model, a single polypeptide chain folds back on itself, or several chains run in either parallel or antiparallel fashion next to one another. Each such structure is stabilized by hydrogen bonds formed between certain atoms on adjacent chains [Figure 14–18(b)]. In the zigzagging plane formation that results, amino acids in adjacent rows are 3.5 Å apart.

As a general rule, most proteins exhibit a mixture of α-helix and β-pleated sheet structures. Globular proteins, most of which are round in shape and water soluble, usually contain a β-pleated sheet structure at their core, as well as many areas with α helices. The more rigid structural proteins, many of which are water insoluble, rely on more extensive β-pleated sheet regions for their rigidity. For example, **fibroin**, the protein made by the silk moth, depends extensively on this form of secondary structure.

The secondary structure describes the folding and interactions of amino acids in certain parts of a polypeptide chain, but the **tertiary structure** defines the three-dimensional spatial conformation of the chain as a whole. Each polypeptide twists and turns and loops around itself in a very specific fashion, characteristic of the particular protein. A model of a tertiary structure is shown in Figure 14–19. At this level of structure, three factors are most important in determining the conformation and in stabilizing the molecule:

1. Covalent disulfide bonds form between closely aligned cysteine residues to form the unique amino acid cystine.

2. Usually, the polar hydrophilic R groups are located on the surface of the configuration, where they can interact with water.

3. The nonpolar hydrophobic R groups are usually located on the inside of the molecule, where they interact with one another, avoiding interaction with water.

---

*Two other amino acids are exceptions to this rule of 20: **selenocysteine (Sec)** and **pyrrolysine (Pyl)**, modified versions of cysteine and lysine, respectively. Sec is found in the active sites of a small number of proteins in the three domains of life (archaea, bacteria, and eukaryotes). The insertion of Sec into polypeptides requires a molecular process that recodes UGA codons, which normally function as stop signals, to serve as Sec codons. Similarly, Pyl is inserted into polypeptide chains in response to the stop codon, UAG codon when it is present internally in an mRNA. Pyl is found in several methane-loving bacteria and, as discovered more recently, in a bacterium that lives symbiotically under the skin of an annelid worm.*

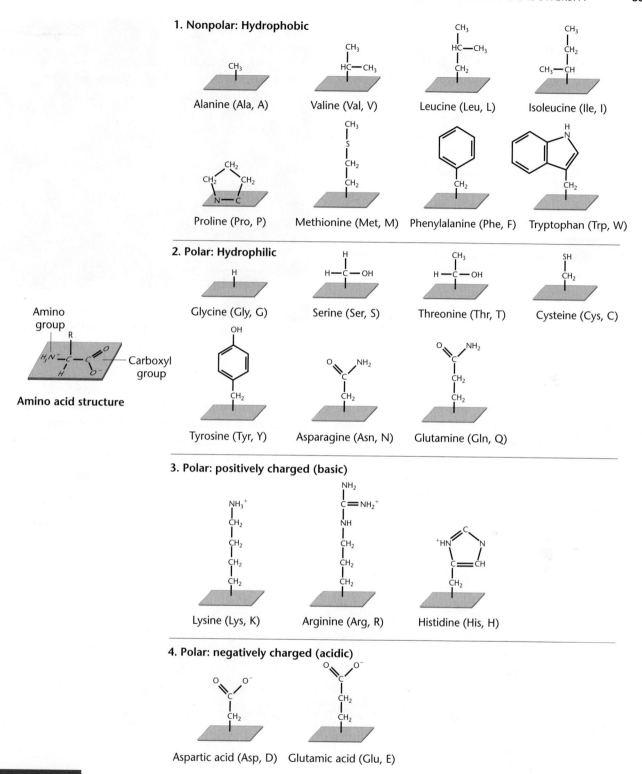

FIGURE 14–16 Chemical structures and designations of the 20 amino acids encoded by living organisms, divided into four major categories. Each amino acid has two abbreviations in universal use; for example, alanine is designated either Ala or A.

It is important to emphasize that the three-dimensional conformation achieved by any protein is a product of the *primary structure* of the polypeptide. Thus, the genetic code need only specify the sequence of amino acids in order ultimately to produce the final configuration of proteins. The effects of the three stabilizing factors depend on the location of each amino acid relative to all others in the chain. As folding occurs, the most thermodynamically stable conformation results. This level of organization is extremely important because the specific function of any protein is directly related to its tertiary structure.

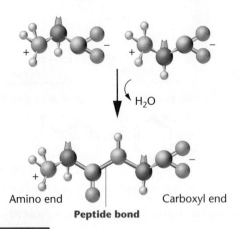

Peptide bond formation between two amino acids, resulting from a dehydration reaction.

Amino end   Carboxyl end
Peptide bond

H_2O

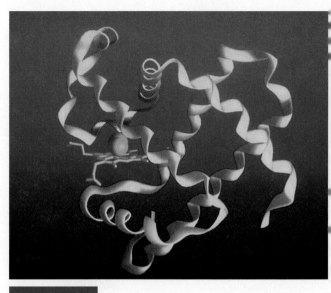

The tertiary level of protein structure for the respiratory pigment myoglobin. The bound oxygen atom is shown in red.

The concept of **quaternary structure** of proteins applies only to those composed of more than one polypeptide chain and refers to the position of the various chains in relation to one another. This type of protein is *oligomeric,* and each chain in it is a *protomer,* or, less formally, a *subunit.* Protomers have conformations that facilitate their fitting together in a specific complementary fashion. Hemoglobin, an oligomeric protein consisting of four protomers (two α and two β chains), has been studied in great detail. Its quaternary structure is shown in Figure 14–20. Most enzymes, including DNA and RNA polymerase, demonstrate quaternary structure.

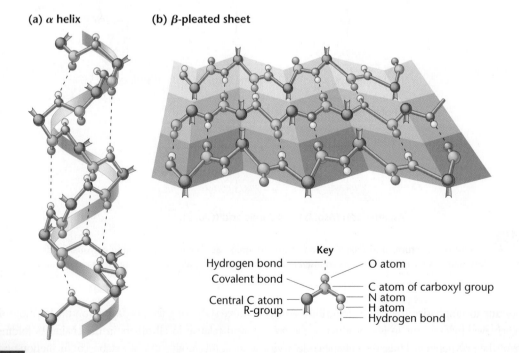

(a) α **helix**   (b) β-**pleated sheet**

Key

Hydrogen bond ———   O atom

Covalent bond ———   C atom of carboxyl group

Central C atom ———   N atom

R-group ———   H atom

Hydrogen bond

(a) The right-handed α helix, which represents one form of secondary structure of a polypeptide chain. (b) The β-pleated sheet, an alternative form of secondary structure of polypeptide chains. To maintain clarity, not all atoms are shown.

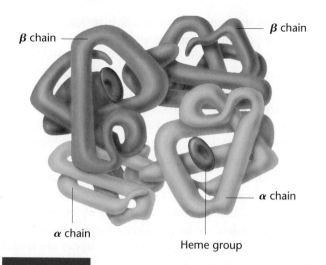

FIGURE 14–20   The quaternary level of protein structure as seen in hemoglobin. Four chains (two α and two β) interact with four heme groups to form the functional molecule.

NOW SOLVE THIS

14–3 HbS results from the substitution of valine for glutamic acid at the number 6 position in the β chain of human hemoglobin. HbC is the result of a change at the same position in the β chain, but in this case lysine replaces glutamic acid. Return to the genetic code table (Figure 13–7) and determine whether single-nucleotide changes can account for these mutations. Then view Figure 14–16 and examine the R groups in the amino acids glutamic acid, valine, and lysine. Describe the chemical differences between the three amino acids. Predict how the changes might alter the structure of the molecule and lead to altered hemoglobin function.

■ HINT: *This problem asks you to consider the potential impact of several amino acid substitutions that result from mutations in one of the genes encoding one of the chains making up human hemoglobin. The key to its solution is to consider and compare the structure of the three amino acids (glutamic acid, lysine, and valine) and their net charge.*

14.10

Posttranslational Modification Alters the Final Protein Product

Before we turn to a discussion of protein function, it is important to point out that polypeptide chains, like RNA transcripts, are often modified once they have been synthesized. This additional processing is broadly described as **posttranslational modification**. Although many of these alterations are detailed biochemical transformations and beyond the scope of our discussion, you should be aware that they

occur and that they are critical to the functional capability of the final protein product. Several examples of posttranslational modification are as follows:

1. *The N-terminus amino acid is usually removed or modified.* For example, either the formyl group or the entire formylmethionine residue in bacterial polypeptides is usually removed enzymatically. In eukaryotic polypeptide chains, the amino group of the initial methionine residue is often removed, and the amino group of the N-terminal residue may be modified (acetylated).

2. *Individual amino acid residues are sometimes modified.* For example, phosphates may be added to the hydroxyl groups of certain amino acids, such as tyrosine. Modifications such as these create negatively charged residues that may form an ionic bond with other molecules. The process of phosphorylation is extremely important in regulating many cellular activities and is a result of the action of enzymes called *kinases*. At other amino acid residues, methyl groups or acetyl groups may be added enzymatically, which can affect the function of the modified polypeptide chain.

3. *Carbohydrate side chains are sometimes attached.* These are added covalently, producing *glycoproteins*, an important category of cell-surface molecules, such as those specifying the antigens in the ABO blood-type system in humans.

4. *Polypeptide chains may be trimmed.* For example, the insulin gene is first translated into a longer molecule that is enzymatically trimmed to insulin's final 51-amino acid form.

5. *Signal sequences are removed.* At the N-terminal end of some proteins is a sequence of up to 30 amino acids that plays an important role in directing the protein to the location in the cell in which it functions. This is called a **signal sequence**, and it determines the final destination of a protein in the cell. The process is called **protein targeting**. For example, proteins whose fate involves secretion or proteins that are to become part of the plasma membrane are dependent on specific sequences for their initial transport into the lumen of the endoplasmic reticulum. While the signal sequence of various proteins with a common destination might differ in their primary amino acid sequence, they share many chemical properties. For example, those destined for secretion all contain a string of up to 15 hydrophobic amino acids preceded by a positively charged amino acid at the N-terminus of the signal sequence. Once the polypeptides are transported, but before they achieve their functional status as proteins, the signal sequence is enzymatically removed from them.

6. *Polypeptide chains are often complexed with metals.* The tertiary and quaternary levels of protein structure often include and are dependent on metal atoms. The functional protein is thus a molecular complex that includes both polypeptide chains and metal atoms. Hemoglobin, which contains four iron atoms along with its four polypeptide chains, is a good example.

Protein Folding and Misfolding

The posttranslational modifications described above are obviously important in achieving the functional status specific to any given protein. Because the final three-dimensional structure of the molecule is directly responsible for its specific function, how polypeptide chains ultimately fold into their final conformations is also an important topic. It was long thought that **protein folding** was a spontaneous process whereby a linear molecule exiting the ribosome achieved a three-dimensional, thermodynamically stable conformation based solely on the combined chemical properties inherent in the amino acid sequence. This indeed is the case for many proteins. However, numerous studies have shown that for other proteins, correct folding is dependent on members of a family of molecules called **chaperones**. Chaperones are themselves proteins (sometimes called *molecular chaperones* or *chaperonins*) that function by mediating the folding process by excluding the formation of alternative, incorrect patterns. While they may bind to the protein in question, like enzymes, they do not become part of the final product. Initially discovered in *Drosophila*, in which they are called **heat-shock proteins**, chaperones are ubiquitous, having now been discovered in all organisms. They are even present in mitochondria and chloroplasts.

In eukaryotic cells, chaperones are particularly important when translation occurs on membrane-bound ribosomes, where the newly translated polypeptide is extruded into the lumen of the endoplasmic reticulum. Even in their presence, misfolding may still occur, and one more system of "quality control" exists. As misfolded proteins are transported out of the endoplasmic reticulum to the cytoplasm, they are "tagged" by another class of small proteins called **ubiquitins**. A *polyubiquitin–protein complex* is formed that moves to a cellular structure called the **proteasome**, within which the ubiquitins are released and the misfolded proteins are degraded by proteases.

Protein folding is a critically important process, not only because misfolded proteins may be nonfunctional, but also because improperly folded proteins can accumulate and be detrimental to cells and the organisms that contain them. For example, a group of transmissible brain disorders in mammals—**scrapie** in sheep, **bovine spongiform encephalopathy** (**mad cow disease**) in cattle, and **Creutzfeldt–Jakob disease** in humans—are caused by the presence in the brain of **prions**, which are aggregates of a misfolded protein. The misfolded protein (called PrP^{Sc}) is an altered version of a normal cellular protein (called PrP^C) synthesized in neurons and found in the brains of all adult animals. The difference between PrP^C and PrP^{Sc} lies in their secondary protein structures. Normal, noninfectious PrP^C folds into an α helix, whereas infectious PrP^{Sc} folds into a β-pleated sheet. When an abnormal PrP^{Sc} molecule contacts a PrP^C molecule, the normal protein refolds into the abnormal conformation. The process continues as a chain reaction, with potentially devastating results—the formation of prion particles that eventually destroy the brain. Hence, this group of disorders can be considered diseases of secondary protein structure. Currently, many laboratories are studying protein folding and misfolding, particularly as related to genetics. Numerous inherited human disorders are caused by misfolded proteins that form abnormal aggregates. **Sickle-cell anemia**, discussed earlier in this chapter, is a case in point, where the β chains of hemoglobin are altered as the result of a single amino acid change, causing the molecules to aggregate within erythrocytes, with devastating results. An autosomal dominant inherited form of **Creutzfeldt–Jakob disease** is known in which the mutation alters the PrP amino acid sequence, leading to prion formation. And various progressive neurodegenerative diseases such as **Huntington disease**, **Alzheimer disease**, and **Parkinson disease** are linked to the formation of abnormal protein aggregate in the brain. Huntington disease is inherited as an autosomal dominant trait, whereas less clearly defined genetic components are associated with Alzheimer and Parkinson diseases.

Proteins Function in Many Diverse Roles

The essence of life on Earth resides at the level of diversity of cellular function. While DNA and RNA serve as vehicles for storing and expressing genetic information, proteins are the *means* of cellular function. And it is the capability of cells to assume diverse structures and functions that distinguishes most eukaryotes from less evolutionarily advanced organisms such as bacteria. Therefore, an introductory understanding of protein function is critical to a complete view of genetic processes.

Proteins are the most abundant macromolecules found in cells. As the end products of genes, they play many diverse roles. For example, the respiratory pigments **hemoglobin** and **myoglobin** transport oxygen, which is essential for cellular metabolism. **Collagen** and **keratin** are structural proteins associated with the skin, connective tissue, and hair of organisms. **Actin** and **myosin** are contractile proteins, found in abundance in muscle tissue, while **tubulin** is the basis of the function of microtubules in mitotic and meiotic spindles.

Still other examples are the **immunoglobulins**, which function in the immune system of vertebrates; **transport proteins**, involved in the movement of molecules across membranes; some of the **hormones** and their **receptors**, which regulate various types of chemical activity; **histones**, which bind to DNA in eukaryotic organisms; and **transcription factors** that regulate gene expression.

Nevertheless, the most diverse and extensive group of proteins (in terms of function) are the **enzymes**, to which we have referred throughout this chapter. Enzymes specialize in catalyzing chemical reactions within living cells. Like all catalysts, they increase the rate at which a chemical reaction reaches equilibrium, but they do not alter the endpoint of the chemical equilibrium. Their remarkable, highly specific catalytic properties largely determine the metabolic capacity of any cell type and provide the underlying basis of what we refer to as **biochemistry**. The specific functions of many enzymes involved in the genetic and cellular processes of cells are described throughout the text.

Biological catalysis is a process whereby enzymes lower the **energy of activation** for a given reaction (Figure 14–21). The energy of activation is the increased kinetic energy state that molecules must usually reach before they react with one another. This state can be attained as a result of elevated temperatures, but enzymes allow biological reactions to occur at lower, physiological temperatures. Thus, enzymes make life as we know it possible.

The catalytic properties of an enzyme are determined by the chemical configuration of the molecule's **active site**. This site is associated with a crevice, a cleft, or a pit on the surface of the enzyme that binds the reactants, or substrates, and facilitates their interaction. Enzymatically catalyzed

reactions control metabolic activities in the cell. Each reaction is either *catabolic* or *anabolic*. **Catabolism** is the degradation of large molecules into smaller, simpler ones with the release of chemical energy. **Anabolism** is the synthetic phase of metabolism, building the various components that make up nucleic acids, proteins, lipids, and carbohydrates.

14.12
Proteins Are Made Up of One or More Functional Domains

We conclude this chapter by briefly discussing the important discovery that distinct regions made up of specific amino acid sequences are associated with unique functions in protein molecules. Such sequences, usually between 50 and 300 amino acids, constitute **protein domains** and represent modular portions of the protein that fold into stable, unique conformations independently of the rest of the molecule. Different domains impart different functional capabilities. Some proteins contain only a single domain, while others contain two or more.

The significance of domains resides in the tertiary structures of proteins. Each domain can contain a mixture of secondary structures, including α helices and β-pleated sheets. The unique conformation of a given domain imparts a specific function to the protein. For example, a domain may serve as the catalytic site of an enzyme, or it may impart an ability to bind to a specific ligand. Thus, discussions of proteins may mention *catalytic domains, DNA-binding domains,* and so on. In short, a protein must be seen as being composed of a series of structural and functional modules. Obviously, the presence of multiple domains in a single protein increases the versatility of each molecule and adds to its functional complexity.

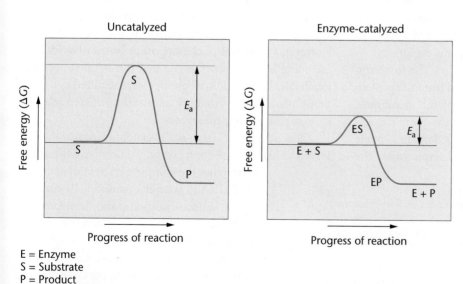

E = Enzyme
S = Substrate
P = Product

FIGURE 14–21 Energy requirements of an uncatalyzed versus an enzymatically catalyzed chemical reaction. The energy of activation (E_a) necessary to initiate the reaction is substantially lower as a result of catalysis.

Exon Shuffling

An interesting hypothesis concerning the genetic origin of protein domains was put forward by Walter Gilbert in 1977. Gilbert suggested that the functional regions of genes in higher organisms consist of collections of exons originally present in ancestral genes and brought together through recombination during the course of evolution. Referring to the process as **exon shuffling**, Gilbert proposed that exons, like protein domains, are also modular, and that during evolution exons may have been reshuffled between genes in eukaryotes, with the result that different genes have similar domains.

Several observations lend support to this proposal. Most exons are fairly small, averaging about 150 base pairs and encoding about 50 amino acids, consistent with the sizes of many functional domains in proteins. Second, recombinational events that during evolution may lead to exon shuffling would be expected to occur within areas of genes represented by introns. Because introns are free to accumulate mutations without harm to the organism, recombinational events would tend to further randomize their nucleotide sequences. Over extended evolutionary periods, sequence diversity would increase. This is, in fact, what is observed: Introns range from 50 to 20,000 bases in length and exhibit fairly random base sequences.

Since 1977, a serious research effort has been aimed at analyzing gene structure. In 1985, more direct evidence in favor of Gilbert's proposal of exon modules was presented. For example, the human gene encoding the membrane receptor for low-density lipoproteins (LDLs) was isolated and sequenced. The *LDL receptor protein* is essential to the transport of plasma cholesterol into the cell. It mediates endocytosis and is expected to have numerous functional domains. These include domains capable of binding specifically to the LDL substrates and interacting with other proteins located at different depths within the membrane as the LDL is being transported across it. In addition, the receptor molecule is modified posttranslationally by the addition of a carbohydrate; a domain must exist that links to this carbohydrate. Given these functional constraints, one would predict that the LDL receptor polypeptide would contain several distinct domains.

Detailed analysis of the gene encoding the LDL receptor supports the concept of exon modules and their shuffling during evolution. The gene is quite large—45,000 base pairs—and contains 18 exons, which in turn contain only slightly less than 2600 nucleotides. These exons code the various functional domains of the protein *and* appear to have been recruited from other genes during evolution.

Figure 14–22 shows these relationships. The first exon encodes a signal sequence that is removed from the protein before the LDL receptor becomes part of the membrane. The next five exons, collectively, represent the domain specifying the binding site for cholesterol. This domain is made up of a 40-amino acid sequence repeated seven times. The next domain, encoded by eight exons, consists of a sequence of 400 amino acids bearing a striking homology to the peptide-hormone *epidermal growth factor* (*EGF*) in mice (a similar sequence is also found in three blood-clotting proteins). This region contains three repetitive sequences of 40 amino acids. The fifteenth exon specifies the domain for the posttranslational addition of the carbohydrate, while the next two specify regions of the protein that are integrated into the membrane, anchoring the receptor to specific sites called *coated pits* on the cell surface.

These observations concerning the LDL exons are fairly compelling in support of the theory of exon shuffling during evolution. Certainly, there is no disagreement concerning the concept of protein domains being responsible for specific molecular interactions.

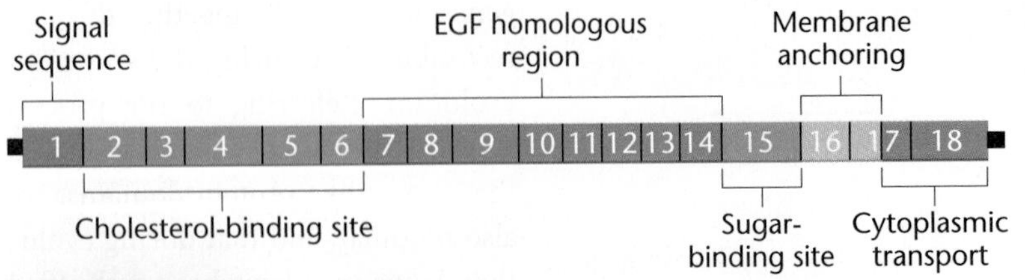

The Origin of Protein Domains

What remains controversial and evocative in the exon-shuffling theory is the question of when introns first appeared on the evolutionary scene. In 1978, W. Ford Doolittle proposed that these intervening sequences were part of the genome of the most primitive ancestors of modern-day eukaryotes. In support of this "intron-early" idea, Gilbert has argued that if similarities in intron DNA sequences are found in identical positions within genes shared by distantly related eukaryotes (such as humans, chickens, and corn), they must have also been present in primitive ancestral genomes.

If Doolittle's proposal is correct, why are introns absent in most prokaryotes and infrequent in yeast? Gilbert argues that they were present at one point during evolution, but as the genome of these primitive organisms evolved, they were lost. The loss resulted from strong selection pressure to streamline chromosomes so as to minimize the energy expenditure on replication and gene expression. Furthermore, streamlining led to fewer errors in mRNA production. However, supporters of the opposing "intron-late" school, including Jeffrey Palmer, argue that introns first appeared much later in evolution, when they were acquired by a single group of eukaryotes that are ancestral to modern-day eukaryotes but not to prokaryotes.

The advent of large-scale genome sequencing has served to increase the controversy. We now have the ability to decipher the complete nucleotide sequence of all DNA in specific organisms. By comparing the amino acid coding and the noncoding DNA sequences from evolutionarily spaced species, geneticists hope to gain insight into how these sequences evolved. Currently, no indisputable evidence in support of either the "intron-early" or the "intron-late" theory has been found, and the question has remained difficult to resolve.

FIGURE 14–22 The 18 exons making up the gene encoding the LDL receptor protein are organized into five functional domains and one signal sequence.

AC TA TAGGGCGAA TTCGAGCTCGG TACCCGGNGG A TCCTC AGAG CGACC GCAGG A G AAG GAG A
 20   30   40   50   70

EXPLORING GENOMICS

Translation Tools and Swiss-Prot for Studying Protein Sequences

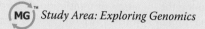 *Study Area: Exploring Genomics*

Many of the databases and bioinformatics programs we have used for Exploring Genomics exercises have focused on manipulating and analyzing DNA and RNA sequences. However, scientists working on various aspects of translation and protein structure and function also have a wide range of bioinformatics tools and databases at their disposal via the Internet. Many of these sites were developed through *proteomics,* the study of all the proteins expressed in a cell or tissue. We will discuss proteomics in more detail in Chapter 21.

In this Exploring Genomics exercise, we will use a program from **ExPASy (Expert Protein Analysis System)** to translate a segment of a gene into a possible polypeptide. We will then explore databases for learning more about this polypeptide.

■ Exercise I – Translating a Nucleotide Sequence and Analyzing a Polypeptide

ExPASy (Expert Protein Analysis System), hosted by the Swiss Institute of Bioinformatics, provides a wealth of resources for studies in proteomics. In this exercise we will use a program from ExPASy called **Translate Tool** to translate a nucleotide sequence to a polypeptide sequence. Although many other programs are available on the Web for this purpose, ExPASy is one of the more student-friendly tools. Translate Tool allows you to make a predicted polypeptide sequence from a cloned gene and then look for open reading frames and variations in possible polypeptides.

1. Below is a partial sequence for a human gene based on a complementary DNA (cDNA) sequence. In Chapter 20 you will learn that cDNA sequences are DNA copies complementary to mRNA molecules expressed in a cell. Before you translate this sequence in ExPASy, run a nucleotide–nucleotide BLAST search from the NCBI Web site (http://www.ncbi.nlm.nih.gov/BLAST) to identify the gene corresponding to

this sequence. Refer to the Exploring Genomics exercise in Chapter 10 if you need help with BLAST searches.

ACATTGCTTCTGACACAATTGTGT
TCACTAGCAACCTCAAACAGACAC
CATGGTGCATCTGACTCCTGAGGAG
AAGTCTGCCGTTACTGCCCTGTGGG
GCAAGGTGAACGTGGATGAAGTTG
GTGGTGAGGCCCTGGGCAG

2. Access the ExPASy Translate Tool program at http://us.expasy.org/tools/dna.html. Copy and paste the cDNA sequence into Translate Tool and click "Translate Sequence" to generate possible polypeptide sequences encoded by this cDNA.

3. Review the translation results and then answer the following questions:

   (a) Did Translate Tool provide one or multiple possible polypeptide sequences?

   (b) If the translation results showed multiple polypeptide sequences, what does this mean? Explain.

   (c) Refer to Figure 14–14 in the chapter. Based on this figure, which reading frame generated by Translate Tool appears to be correct?

4. ExPASy also provides access to a wealth of information about this polypeptide by connecting to a large number of different databases such as **UniProt KB/Swiss-Prot**, a protein sequence database maintained by the Swiss Institute for Bioinformatics (SIB) and the European Bioinformatics Institute (EBI), and a database called the Protein Data Bank. UniProt KB/Swiss-Prot is widely used by scientists around the world. Visit UniProt KB/Swiss-Prot (http://ca.expasy.org/sprot/hpi) for a wealth of information on the human genome and proteomics.

5. To learn more about the features of this polypeptide, it is best to work with a complete sequence. To obtain

a complete sequence, click on the reading frame you believe is correct.

6. To retrieve the amino acid sequence for the entire polypeptide, click "methionine (m)" as the first amino acid in the polypeptide and then use the "BLAST" link near the bottom of the next page to run a BLAST search. From the BLAST results page, locate UniProtKB/Swiss-Prot entry P68871; this is an accession number for the protein sequence, similar to the accession numbers assigned to DNA and RNA sequences, and it is the correct match for this sequence. Click on this link to reveal a comprehensive report about the protein. Be sure the identity of this protein agrees with what you discovered in Step 1.

7. Scroll past the references in the Swiss-Prot report; then explore the following features:

   (a) 3D Structure Databases—presents 3D modeling representations showing polypeptide folding arrangements.

   (b) Under 2D Gel Databases, use the Swiss-2D PAGE link to view 2D gels of this polypeptide from different tissue samples. Refer to Figure 21–24 for a representation of 2D gels. When viewing a 2D gel image with this feature, click on links under "Map Locations" to identify specific spots on a gel that correspond to this polypeptide.

   (c) The Family and Domain databases links will take you to a wealth of information about this polypeptide and related polypeptides and proteins.

   (d) Explore the "Other" category and visit the DrugBank link that provides information on drugs that bind to and affect this polypeptide.

   (e) At the very bottom of the page, under "Sequence analysis tools," use the "ProtParam" feature to learn more about predicted secondary structures formed by this polypeptide.

CASE STUDY | Lost in translation

A recessively inherited brain disorder called vanishing white matter (VWM) was first described in the 1990s using magnetic resonance imaging (MRI). Affected individuals show neurological deterioration early in childhood, dying soon after diagnosis, or they may express a slow progressive form of the disease. VWM is caused by mutations in any of the five genes that encode the protein subunits of the translation initiation factor 2B. This factor helps position the ribosome on the mRNA during the initiation of translation. It is known that other cells that rapidly synthesize large amounts of protein, such as insulin-secreting cells, are also affected. Many questions about this disorder remain to be answered.

1. Given the two forms of the disease, discuss the possible nature of VWM mutations. Why do you think that this is a recessive rather than a dominant mutation?
2. Why might some cells in the body be more susceptible to mutations in genes encoding factor 2B than other cells?
3. How could we study the cellular and molecular aspects of VWM?

Summary Points

 For activities, animations, and review quizzes, go to the study area at www.masteringgenetics.com

1. Translation is the synthesis of polypeptide chains under the direction of mRNA in association with ribosomes.
2. Translation depends on tRNA molecules that serve as adaptors between triplet codons in mRNA and the corresponding amino acids.
3. Translation occurs in association with ribosomes, and like transcription, is subdivided into the stages of initiation, elongation, and termination and relies on base-pairing affinities between complementary nucleotides.
4. Inherited metabolic disorders are most often due to the loss of enzyme activity resulting from mutations in genes encoding those proteins.
5. Beadle and Tatum's work with nutritional mutations in *Neurospora* led them to propose that one gene encodes one enzyme.
6. Pauling and Ingram's investigations of hemoglobin from patients with sickle-cell anemia led to the modification of the one-gene:one-enzyme hypothesis to indicate that one gene encodes one polypeptide chain.
7. Proteins, the end products of genes, demonstrate four levels of structural organization that together describe their three-dimensional conformation, which is the basis of each molecule's function.
8. Of the myriad functions performed by proteins, the most influential role belongs to enzymes, which serve as highly specific biological catalysts that play a central role in the production of all classes of molecules in living systems.
9. In eukaryotes, proteins contain one or more functional domains, each prescribed by exon regions interspersed within genes. Specific domains impart specific functional capacities to proteins and appear to have been "shuffled" between genes during evolution.

INSIGHTS AND SOLUTIONS

1. The growth responses in the following chart were obtained by growing four mutant strains of *Neurospora* on different media, each containing one of four related compounds, A, B, C, and D. None of the mutations grows on minimal medium. Draw all possible conclusions from this data.

	Growth Product			
Mutation	A	B	C	D
1	–	–	–	–
2	+	+	–	+
3	+	+	–	–
4	–	+	–	–

Solution: Nothing can be concluded about mutation *1* except that it is lacking some essential factor, perhaps even unrelated to any biochemical pathway in which A, B, C, and D participate. Nor can anything be concluded about compound C. If it is involved in a pathway with the other compounds, it is a product synthesized prior to the synthesis of A, B, and D.

We must now analyze these three compounds and the control of their synthesis by the enzymes encoded by genes *2*, *3*, and *4*. Because product B allows growth in all three cases, it may be considered the "end product"—it bypasses the block in all three instances. Similar reasoning suggests that product A precedes B in the pathway, since A bypasses the block in two of the three steps. Product D precedes B, yielding a partial solution:

$$C(?) \longrightarrow D \longrightarrow A \longrightarrow B$$

Now let's determine which mutations control which steps. Since mutation *2* can be alleviated by products D, B, and A, it must control a step prior to all three products, perhaps the direct conversion to D, although we cannot be certain. Mutation *3* is alleviated by B and A, so its effect must precede theirs in the pathway. Thus, we will assign it a role controlling the conversion of D to A. Likewise, we can provisionally assign mutation *4* to the conversion of A to B, leading to the more complete solution

$$C(?) \xrightarrow{2(?)} D \xrightarrow{3} A \xrightarrow{4} B$$

Problems and Discussion Questions

For instructor-assigned tutorials and problems, go to www.masteringgentics.com

HOW DO WE KNOW?

1. In this chapter, we focused on the translation of mRNA into proteins as well as on protein structure and function. Along the way, we found many opportunities to consider the methods and reasoning by which much of this information was acquired. From the explanations given in the chapter, what answers would you propose to the following fundamental questions:

   (a) What experimentally derived information led to Holley's proposal of the two-dimensional cloverleaf model of tRNA?

   (b) What experimental information verifies that certain codons in mRNA specify chain termination during translation?

   (c) How do we know, based on studies of *Neurospora* nutritional mutations, that one gene specifies one enzyme?

   (d) On what basis have we concluded that proteins are the end products of genetic expression?

   (e) What experimental information directly confirms that the genetic code, as shown in Figure 13–7, is correct?

   (f) How do we know that the structure of a protein is intimately related to the function of that protein?

2. List and describe the role of all of the molecular constituents of a functional polyribosome.

3. Contrast the roles of tRNA and mRNA during translation and list all enzymes that participate in the transcription and translation process.

4. Francis Crick proposed the "adaptor hypothesis" for the function of tRNA. Why did he choose that description?

5. During translation, what molecule bears the codon? the anticodon?

6. The α chain of eukaryotic hemoglobin is composed of 141 amino acids. What is the minimum number of nucleotides in an mRNA coding for this polypeptide chain?

7. Assuming that each nucleotide is 0.34 nm long in the mRNA, how many triplet codes can occupy at one time the space in a ribosome that is 20 nm in diameter?

8. Summarize the steps involved in charging tRNAs with their appropriate amino acids.

9. To carry out its role, each transfer RNA requires at least four specific recognition sites that must be inherent in its tertiary structure. What are they?

10. What are isoaccepting tRNAs? Assuming that there are only 20 different aminoacyl tRNA synthetases but 31 different tRNAs, speculate on parameters that might be used to ensure that each charged tRNA has received the correct amino acid.

11. Discuss the potential difficulties of designing a diet to alleviate the symptoms of phenylketonuria.

12. Phenylketonurics cannot convert phenylalanine to tyrosine. Why don't these individuals exhibit a deficiency of tyrosine?

13. Phenylketonurics are often more lightly pigmented than are normal individuals. Can you suggest a reason why this is so?

14. Early detection and adherence to a strict dietary regime has prevented much of the mental retardation that used to occur in those afflicted with phenylketonuria (PKU). Affected individuals now often lead normal lives and have families. For various reasons, such individuals adhere less rigorously to their diet as they get older. Predict the effect that mothers with PKU who neglect their diets might have on newborns.

15. The synthesis of flower pigments is known to be dependent on enzymatically controlled biosynthetic pathways. For the crosses shown here, postulate the role of mutant genes and their products in producing the observed phenotypes:

   (a) P_1: white strain A $\times$ white strain B
   F_1: all purple
   F_2: 9/16 purple: 7/16 white

   (b) P_1: white $\times$ pink
   F_1: all purple
   F_2: 9/16 purple: 3/16 pink: 4/16 white

16. The study of biochemical mutants in organisms such as *Neurospora* has demonstrated that some pathways are branched. The data shown here illustrate the branched nature of the pathway resulting in the synthesis of thiamine:

Mutation	Growth Supplement			
	Minimal Medium	Pyrimidine	Thiazole	Thiamine
thi-1	−	−	+	+
thi-2	−	+	−	+
thi-3	−	−	−	+

Why don't the data support a linear pathway? Can you postulate a pathway for the synthesis of thiamine in *Neurospora*?

17. Explain why the one-gene:one-enzyme concept is not considered totally accurate today.

18. Why is an alteration of electrophoretic mobility interpreted as a change in the primary structure of the protein under study?

19. Contrast the polypeptide-chain components of each of the hemoglobin molecules found in humans.

20. Using sickle-cell anemia as an example, describe what is meant by a molecular or genetic disease. What are the similarities and dissimilarities between this type of a disorder and a disease caused by an invading microorganism?

21. Contrast the contributions of Pauling and Ingram to our understanding of the genetic basis for sickle-cell anemia.

22. Hemoglobins from two individuals are compared by electrophoresis and by fingerprinting. Electrophoresis reveals no difference in migration, but fingerprinting shows an amino acid difference. How is this possible?

23. HbS results in anemia and resistance to malaria, whereas in those with HbA, the parasite *Plasmodium falciparum* invades red blood cells and causes the disease. Predict whether those with HbC are likely to be anemic and whether they would be resistant to malaria.

24. Shown here are several amino acid substitutions in the α and β chains of human hemoglobin:

Using the code table (Figure 13–7), determine how many of them can occur as a result of a single-nucleotide change.

Hb Type	Normal Amino Acid	Substituted Amino Acid
Hb Toronto	Ala	Asp (α-5)
HbJ Oxford	Gly	Asp (α-15)
Hb Mexico	Gln	Glu (α-54)
Hb Bethesda	Tyr	His (β-145)
Hb Sydney	Val	Ala (β-67)
HbM Saskatoon	His	Tyr (β-63)

25. Certain mutations called *amber* in bacteria and viruses result in premature termination of polypeptide chains during translation. Many *amber* mutations have been detected at different points along the gene coding for a head protein in phage T4.

26. Describe what colinearity means. Of what significance is the concept of colinearity in the study of genetics?

27. Does Yanofsky's work with the *trpA* locus in *E. coli* (discussed in this chapter) constitute more or less direct evidence in support of colinearity than Fiers's work with phage MS2 (discussed in Chapter 13)? Explain.

28. Define and compare the four levels of protein organization.

29. List as many different categories of protein functions as you can. Wherever possible, give an example of each category.

30. How does an enzyme function? Why are enzymes essential for living organisms on Earth?

31. Exon shuffling is a proposal that relates exons in DNA to the repositioning of functional domains in proteins. What evidence exists in support of exon shuffling? Two schools of thought have emerged concerning the origin of exons, "intron-early" and "intron-late." Briefly describe both theories and present support for each.

Extra-Spicy Problems

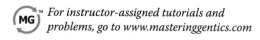

For instructor-assigned tutorials and problems, go to www.masteringgentics.com

32. Three independently assorting genes are known to control the following biochemical pathway that provides the basis for flower color in a hypothetical plant:

$$\text{Colorless} \xrightarrow{A-} \text{yellow} \xrightarrow{B-} \text{green} \xrightarrow{C-} \text{speckled}$$

Three homozygous recessive mutations are also known, each of which interrupts a different one of these steps. Determine the phenotypic results in the F_1 and F_2 generations resulting from the P_1 crosses of true-breeding plants listed here:

(a) speckled (*AABBCC*) × yellow (*AAbbCC*)
(b) yellow (*AAbbCC*) × green (*AABBcc*)
(c) colorless (*aaBBCC*) × green (*AABBcc*)

33. How would the results vary in cross (a) of Problem 32 if genes *A* and *B* were linked with no crossing over between them? How would the results of cross (a) vary if genes *A* and *B* were linked and 20 μm apart?

34. Deep in a previously unexplored South American rain forest, a species of plants was discovered with true-breeding varieties whose flowers were either pink, rose, orange, or purple. A very astute plant geneticist made a single cross, carried to the F_2 generation, as shown:

P_1:	purple × pink
F_1:	all purple
F_2:	27/64 purple
	16/64 pink
	12/64 rose
	9/64 orange

Based solely on these data, he was able to propose both a mode of inheritance for flower pigmentation and a biochemical pathway for the synthesis of these pigments.

Carefully study the data. Create a hypothesis of your own to explain the mode of inheritance. Then propose a biochemical pathway consistent with your hypothesis. How could you test the hypothesis by making other crosses?

35. The emergence of antibiotic-resistant strains of *Enterococci* and transfer of resistant genes to other bacterial pathogens have highlighted the need for new generations of antibiotics to combat serious infections. To grasp the range of potential sites for the action of existing antibiotics, sketch the components of the translation machinery (e.g., see Step 3 of Figure 14–6), and using a series of numbered pointers, indicate the specific location for the action of the antibiotics shown in the following table.

Antibiotic	Action
1. Streptomycin	Binds to 30S ribosomal subunit
2. Chloramphenicol	Inhibits peptidyl transferase of 70S ribosome
3. Tetracycline	Inhibits binding of charged tRNA to ribosome
4. Erythromycin	Binds to free 50S particle and prevents formation of 70S ribosome
5. Kasugamycin	Inhibits binding of tRNA^{fmet}
6. Thiostrepton	Prevents translocation by inhibiting EF-G

36. The development of antibiotic resistance by pathogenic bacteria represents a major health concern. One potential new antibiotic is evernimicin, which was isolated from *Micromonospora carbonaceae*. Evernimicin is an oligosaccharide with antibiotic activity against a broad range of gram-positive pathogenic bacteria. To determine the mode of action of this drug, researchers have analyzed 23S ribosomal DNA mutants that showed reduced sensitivity to evernimicin (e.g., Adrian et al., 2000. *Antimicrob. Ag. and Chemo.* 44: 3101–3106). They discovered two classes of mutants that conferred resistance: In one class, the mutation occurs in 23S rRNA nucleotides 2475–2483; in the other, it occurs in ribosomal protein L16. This suggests that these two ribosomal components are structurally and functionally linked. It turns out that the tRNA anticodon

stem-loop appears to bind to the A site of the ribosome at rRNA bases 2465–2485. This finding conforms to the proposed function of L16, which appears to be involved in attracting the aminoacyl stem of the tRNA to the ribosome at its A site. Using your sketch of the translation machinery from Problem 35 along with this information, designate where the proposed antibacterial action of evernimicin is likely to occur.

37. The flow of genetic information from DNA to protein is mediated by messenger RNA. If you introduce short DNA strands (called antisense oligonucleotides) that are complementary to mRNAs, hydrogen bonding may occur and "label" the DNA/RNA hybrid for ribonuclease-H degradation of the RNA. One study compared the effect of different-length antisense oligonucleotides upon ribonuclease-H–mediated degradation of tumor necrosis factor (*TNFα*) mRNA. *TNFα* exhibits antitumor and proinflammatory activities (Lloyd et al., 2001. *Nuc. Acids Res.* 29: 3664–3673). The following graph indicates the efficacy of various-sized antisense oligonucleotides in causing ribonuclease-H cleavage.

(a) Describe how antisense oligonucleotides interrupt the flow of genetic information in a cell.
(b) What general conclusion can be drawn from the graph?
(c) What factors other than oligonucleotide length are likely to influence antisense efficacy *in vivo*?

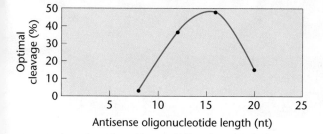

38. The fidelity of translation is dependent on the reliable action of aminoacyl tRNA synthetases that ensure the association of only one type of amino acid with a specific tRNA. Two relatively rare human conditions, Charcot-Marie-Tooth disease 2D (CMT2D) and distal spinal muscular atrophy type V (dSMA-V), are neuropathologies of the peripheral axons that map to a well-defined region of the short arm of chromosome 7, as does the gene for glycyl tRNA synthetase. Families with CMT2D or dSMA-V have been identified with missense mutations in the gene for glycyl tRNA synthetase; and the mutations present in CMT2D and dSMA-V have been found to result in a loss of activity of glycyl tRNA synthetase (Antonellis et al., 2003). A sample pedigree (from Antonellis et al., 2003) is presented here (to maintain anonymity, sexes are not provided).

(a) Considering the following pedigree and the function of aminoacyl tRNA synthetases in general, would you conclude that the genes causing these diseases are dominant or recessive in their action?
(b) Considering the vital role that synthetases play in protein synthesis, speculate as to how individuals might survive with such a defect in translational efficiency.
(c) Why might some tissues (neural) be more affected than others?

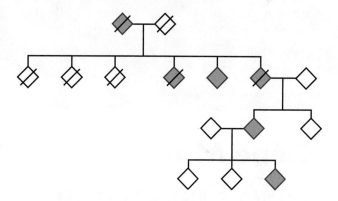

Pigment mutations within an ear of corn, caused by transposition of the *Ds* element.

15

Gene Mutation, DNA Repair, and Transposition

CHAPTER CONCEPTS

- Mutations comprise any change in the base-pair sequence of DNA.
- Mutation is a source of genetic variation and provides the raw material for natural selection. It is also the source of genetic damage that contributes to cell death, genetic diseases, and cancer.
- Mutations have a wide range of effects on organisms depending on the type of base-pair alteration, the location of the mutation within the chromosome, and the function of the affected gene product.
- Mutations can occur spontaneously as a result of natural biological and chemical processes, or they can be induced by external factors, such as chemicals or radiation.
- Single-gene mutations cause a wide variety of human diseases.
- Organisms rely on a number of DNA repair mechanisms to counteract mutations. These mechanisms range from proofreading and correction of replication errors to base excision and homologous recombination repair.
- Mutations in genes whose products control DNA repair lead to genome hypermutability, human DNA repair diseases, and cancers.
- Geneticists induce gene mutations as the first step in classical genetic analysis.
- Transposable elements move into and out of chromosomes, causing chromosome breaks and inducing mutations both within coding regions and in gene-regulatory regions.

The ability of DNA molecules to store, replicate, transmit, and decode information is the basis of genetic function. But equally important are the changes that occur to DNA sequences. Without the variation that arises from changes in DNA sequences, there would be no phenotypic variability, no adaptation to environmental changes, and no evolution. Gene mutations are the source of new alleles and are the origin of genetic variation within populations. On the downside, they are also the source of genetic changes that can lead to cell death, genetic diseases, and cancer.

Mutations also provide the basis for genetic analysis. The phenotypic variations resulting from mutations allow geneticists to identify and study the genes responsible for the modified trait. In genetic investigations, mutations act as identifying "markers" for genes so that they can be followed during their transmission from parents to offspring. Without phenotypic variability, classical genetic analysis would be impossible. For example, if all pea plants displayed a uniform phenotype, Mendel would have had no foundation for his research.

In Chapter 8, we examined mutations in large regions of chromosomes—chromosomal mutations. In contrast, the mutations we will now explore are those occurring primarily in the base-pair sequence of DNA within individual genes—**gene mutations**. We will also describe how the cell defends itself from such mutations using various mechanisms of DNA repair. The chapter also describes how geneticists use mutations to identify genes and analyze gene functions in humans and other organisms.

15.1

Gene Mutations Are Classified in Various Ways

A mutation can be defined as an alteration in DNA sequence. Any base-pair change in any part of a DNA molecule can be considered a mutation. A mutation may comprise a single base-pair substitution, a deletion or insertion of one or more base pairs, or a major alteration in the structure of a chromosome.

Mutations may occur within regions of a gene that code for protein or within noncoding regions of a gene such as introns and regulatory sequences. Mutations may or may not bring about a detectable change in phenotype. The extent to which a mutation changes the characteristics of an organism depends on which type of cell suffers the mutation and the degree to which the mutation alters the function of a gene product or a gene-regulatory region.

Mutations can occur in somatic cells or within germ cells. Those that occur in germ cells are heritable and are the basis for the transmission of genetic diversity and evolution, as well as genetic diseases. Those that occur in somatic cells are not transmitted to the next generation but may lead to altered cellular function or tumors.

Because of the wide range of types and effects of mutations, geneticists classify mutations according to several different schemes. These organizational schemes are not mutually exclusive. In this section, we outline some of the ways in which gene mutations are classified.

Spontaneous and Induced Mutations

Mutations can be classified as either spontaneous or induced, although these two categories overlap to some degree. **Spontaneous mutations** are changes in the nucleotide sequence of genes that appear to have no known cause. No specific agents are associated with their occurrence, and they are generally assumed to be accidental. Many of these mutations arise as a result of normal biological or chemical processes in the organism that alter the structure of nitrogenous bases. Often, spontaneous mutations occur during the enzymatic process of DNA replication, as we discuss later in this chapter.

In contrast to spontaneous mutations, mutations that result from the influence of extraneous factors are considered to be **induced mutations**. Induced mutations may be the result of either natural or artificial agents. For example, radiation from cosmic and mineral sources and ultraviolet radiation from the sun are energy sources to which most organisms are exposed and, as such, may be factors that cause induced mutations.

The earliest demonstration of the artificial induction of mutations occurred in 1927, when Hermann J. Muller reported that X rays could cause mutations in *Drosophila*. In 1928, Lewis J. Stadler reported that X rays had the same effect on barley. In addition to various forms of radiation, numerous natural and synthetic chemical agents are also mutagenic.

Several generalizations can be made regarding spontaneous mutation rates in organisms. The **mutation rate** is defined as the likelihood that a gene will undergo a mutation in a single generation or in forming a single gamete. First, the rate of spontaneous mutation is exceedingly low for all organisms. Second, the rate varies between different organisms. Third, even within the same species, the spontaneous mutation rate varies from gene to gene.

Viral and bacterial genes undergo spontaneous mutation at an average of about 1 in 100 million (10^{-8}) replications or cell divisions (Table 15.1). Maize, *Drosophila*, and humans demonstrate rates several orders of magnitude higher. The genes studied in these groups average between 1 in 1,000,000 (10^{-6}) and 1 in 100,000 (10^{-5}) mutations per gamete formed. Some mouse genes are another order of magnitude higher in their spontaneous mutation rate, 1 in 100,000 to 1 in 10,000 (10^{-5} to 10^{-4}). It is not clear why such large variations occur in mutation rates. The variation between genes in a given organism may be due to inherent

TABLE 15.1

Spontaneous Mutation Rates at Various Loci in Different Organisms

Organism	Character	Locus	Rate*
Bacteriophage T2	Lysis inhibition	$r \rightarrow r^+$	1×10^{-8}
	Host range	$h^+ \rightarrow h$	4×10^{-9}
E. coli	Lactose fermentation	$lac^- \rightarrow lac^+$	2×10^{-7}
	Streptomycin sensitivity	$str\text{-}d \rightarrow str\text{-}s$	1×10^{-8}
Zea mays	Shrunken seeds	$sh^+ \rightarrow sh^-$	1×10^{-6}
	Purple	$pr^+ \rightarrow pr^-$	1×10^{-5}
Drosophila melanogaster	Yellow body	$y^+ \rightarrow y$	1.2×10^{-6}
	White eye	$w^+ \rightarrow w$	4×10^{-5}
Mus musculus	Piebald coat	$s^+ \rightarrow s$	3×10^{-5}
	Brown coat	$b^+ \rightarrow b$	8.5×10^{-4}
Homo sapiens	Hemophilia	$h^+ \rightarrow h$	2×10^{-5}
	Huntington disease	$Hu^+ \rightarrow Hu$	5×10^{-6}

*Rates are expressed per gene replication (T2), per cell division (*E. coli*), or per gamete per generation (*Zea mays, Drosophila melanogaster, Mus musculus,* and *Homo sapiens*).

differences in mutability in different regions of the genome. Some DNA sequences appear to be highly susceptible to mutation and are known as **mutation hot spots**. The variation between organisms may, in part, reflect the relative efficiencies of their DNA proofreading and repair systems. We will discuss these systems later in the chapter.

The Fluctuation Test: Are Mutations Random or Adaptive?

One of the basic concepts in genetics is that mutations occur randomly and are not directed by the environment in which the organism finds itself—that is, mutations arise spontaneously in the absence of selective pressure rather than being the consequence of selective pressure. This concept has been verified by many experiments, including the classic Luria–Delbrück Fluctuation Test.

In 1943, Salvador Luria and Max Delbrück presented the first direct evidence that mutations do not occur as part of an adaptive mechanism, but instead take place spontaneously and randomly. Their experiment, known as the **Luria–Delbrück fluctuation test**, is an example of exquisite analytical and theoretical work.

Luria and Delbrück carried out their experiments using the *E. coli*-T1 system. The T1 bacteriophage is a bacterial virus that infects *E. coli* cells and lyses the infected bacteria. Mutations can occur in *E. coli* cells that make the cells resistant to T1. To begin their experiment, Luria and Delbruck inoculated a large flask and a number of individual culture tubes with a few *E. coli* cells and grew all the cultures for several generations in the absence of any T1 bacteriophage. After growth, each small flask contained about 20 million cells and the large flask was allowed to grow to a higher density. At this point, the bacteria in the smaller individual tubes were spread onto the surface of growth media containing T1

bacteriophage, in a petri dish. Similarly, several portions of the large flask (each portion also containing about 20 million cells) were spread onto T1-containing media in petri plates. After incubation, any individual cells of *E. coli* that were resistant to the T1 bacteriophage grew and formed colonies. All other *E. coli* cells that were not resistant would die. The colonies present on the plates were then counted.

The experimental rationale for distinguishing between the two hypotheses (adaptive versus random mutation) was as follows:

Hypothesis 1: Adaptive Mutations. If mutations occur adaptively, in response to the presence of T1 bacteriophage in the growth medium, every *E. coli* cell that is grown on the T1-containing growth medium would have a constant probability of acquiring a mutation that would give it resistance to T1. Therefore, the number of resistant cells (colonies) would depend only on the number of *E. coli* cells and bacteriophages present on the petri plates. In this case, because a constant number of cells and T1 were present on each plate, there would be a fairly constant number of resistant colonies from plate to plate and from experiment to experiment.

Hypothesis 2: Random Mutations. If mutations occur randomly and are not affected by the presence or absence of bacteriophage, mutations leading to resistance would occur at a low rate at any time during incubation even in the absence of T1 bacteriophage. If a mutation occurred early in the incubation process in the culture tubes, the subsequent growth of the mutant bacteria would produce a large number of resistant cells in the culture. If a mutation occurred later in the incubation process, there would be fewer resistant colonies. The random mutation hypothesis predicts that the number of resistant cells would fluctuate

TABLE 15.2

The Luria–Delbrück Experiment Demonstrating That Mutations Are Spontaneous

Sample No.	Number of T1-Resistant Bacteria	
	Same Culture (Control)	Different Cultures
1	14	6
2	15	5
3	13	10
4	21	8
5	15	24
6	14	13
7	26	165
8	16	15
9	20	6
10	13	10
Mean	16.7	26.2
Variance	15.0	2178.0

Source: After Luria and Delbrück (1943).

from experiment to experiment, and from tube to tube, reflecting the varying times at which the resistance mutations occurred in liquid culture. In contrast, each portion of the large culture flask (containing a stirred and homogeneous mixture of resistant and susceptible cells) would produce a constant number of resistant colonies from plate to plate.

Table 15.2 shows a representative set of data from the Luria–Delbrück experiments. The middle column shows the number of mutant colonies recovered from a series of aliquots derived from the large liquid culture. These data serve as a control, and, as predicted, little fluctuation is observed in these samples. In contrast, the right-hand column shows the number of resistant colonies recovered from each of 10 independently incubated liquid cultures. These data reveal a great deal of fluctuation between independently incubated cultures, supporting the hypothesis that mutations arise randomly, even in the absence of selective pressure, and are inherited in a stable fashion.

Although the concept of spontaneous mutation has been accepted for some time, the possibility that organisms might also be capable of inducing specific mutations as a result of environmental pressures has long intrigued geneticists. Some recent and controversial research has suggested that under some stressful nutritional conditions such as starvation, bacteria may be capable of activating mechanisms that create a hypermutable state in genes that would, when mutated, enhance survival. The conclusions from these studies are still a source of debate, but they keep alive the interest in the possibility of adaptive mutation.

Classification Based on Location of Mutation

Mutations may be classified according to the cell type or chromosomal locations in which they occur. **Somatic mutations** are those occurring in any cell in the body except germ cells.

Autosomal mutations are mutations within genes located on the autosomes, whereas **X-linked and Y-linked mutations** are those within genes located on the X or Y chromosome, respectively.

Mutations arising in somatic cells are not transmitted to future generations. When a recessive autosomal mutation occurs in a somatic cell of a diploid organism, it is unlikely to result in a detectable phenotype. The expression of most such mutations is likely to be masked by expression of the wild-type allele within that cell. Somatic mutations will have a greater impact if they are dominant or, in males, if they are X-linked, since such mutations are most likely to be immediately expressed. Similarly, the impact of dominant or X-linked somatic mutations will be more noticeable if they occur early in development, when a small number of undifferentiated cells replicate to give rise to several differentiated tissues or organs. Dominant mutations that occur in cells of adult tissues are often masked by the activity of thousands upon thousands of nonmutant cells in the same tissue that perform the nonmutant function.

Mutations in germ cells are of greater significance because they may be transmitted to offspring as gametes. They have the potential of being expressed in all cells of an offspring. Inherited dominant autosomal mutations will be expressed phenotypically in the first generation. X-linked recessive mutations arising in the gametes of a **homogametic** female may be expressed in hemizygous male offspring. This will occur provided that the male offspring receives the affected X chromosome. Because of heterozygosity, the occurrence of an autosomal recessive mutation in the gametes of either males or females (even one resulting in a lethal allele) may go unnoticed for many generations, until the resultant allele has become widespread in the population. Usually, the new allele will become evident only when a chance mating brings two copies of it together into the homozygous condition.

Classification Based on Type of Molecular Change

Geneticists often classify gene mutations in terms of the nucleotide changes that constitute the mutation. A change of one base pair to another in a DNA molecule is known as a **point mutation**, or **base substitution** (Figure 15–1). A change of one nucleotide of a triplet within a protein-coding portion of a gene may result in the creation of a new triplet that codes for a different amino acid in the protein product. If this occurs, the mutation is known as a **missense mutation**. A second possible outcome is that the triplet will be changed into a stop codon, resulting in the termination of translation of the protein. This is known as a **nonsense mutation**. If the point mutation alters a codon but does not result in a change in the amino acid at that position in the protein (due to degeneracy of the genetic code), it can be considered a **silent mutation**.

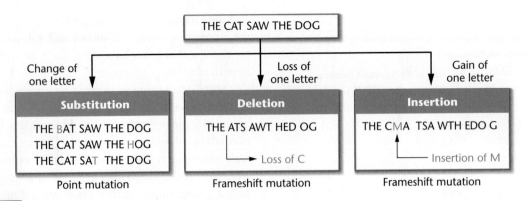

Analogy showing the effects of substitution, deletion, and insertion of one letter in a sentence composed of three-letter words to demonstrate point and frameshift mutations.

You will often see two other terms used to describe base substitutions. If a pyrimidine replaces a pyrimidine or a purine replaces a purine, a **transition** has occurred. If a purine replaces a pyrimidine, or vice versa, a **transversion** has occurred.

Another type of change is the insertion or deletion of one or more nucleotides at any point within the gene. As illustrated in Figure 15–1, the loss or addition of a single nucleotide causes all of the subsequent three-letter codons to be changed. These are called **frameshift mutations** because the frame of triplet reading during translation is altered. A frameshift mutation will occur when any number of bases are added or deleted, except multiples of three, which would reestablish the initial frame of reading. It is possible that one of the many altered triplets will be UAA, UAG, or UGA, the translation termination codons. When one of these triplets is encountered during translation, polypeptide synthesis is terminated at that point. Obviously, the results of frameshift mutations can be very severe, especially if they occur early in the coding sequence.

Classification Based on Phenotypic Effects

Depending on their type and location, mutations can have a wide range of phenotypic effects, from none to severe.

As discussed in Chapter 4, a **loss-of-function mutation** is one that reduces or eliminates the function of the gene product. Any type of mutation, from a point mutation to deletion of the entire gene, may lead to a loss of function. Mutations that result in complete loss of function are known as **null mutations**. Most loss-of-function mutations are recessive; however a dominant effect of a loss-of-function mutation can occur during a situation known as haploinsufficiency. In diploid organisms, **haploinsufficiency** occurs when the single functional copy of the gene does not produce enough gene product to bring about a wild-type phenotype. In humans, Marfan syndrome is an example of a disorder caused by haploinsufficiency—in this case as a result of loss-of-function mutation in one copy of the *FBN1* gene.

A **gain-of-function mutation** results in a gene product with enhanced or new functions. This may be due to a change in the amino acid sequence of the protein that confers a new activity, or it may result from a mutation in a regulatory region of the gene, leading to expression of the gene at higher levels, or the synthesis of the gene product at abnormal times, or places. Most gain-of-function mutations are dominant.

The most easily observed mutations are those affecting a morphological trait. These mutations are known as **visible mutations** and are recognized by their ability to alter a normal or wild-type visible phenotype. For example, all of Mendel's pea characteristics and many genetic variations encountered in *Drosophila* fit this designation, since they cause obvious changes to the morphology of the organism.

Some mutations give rise to nutritional or biochemical effects. In bacteria and fungi, a typical **nutritional mutation** results in a loss of ability to synthesize an amino acid or vitamin. In humans, sickle-cell anemia and hemophilia are examples of diseases resulting from **biochemical mutations**. Although such mutations do not always affect morphological characters, they affect the function of proteins that can affect the well-being and survival of the affected individual.

Still another category consists of mutations that affect the behavior patterns of an organism. The primary effect of **behavioral mutations** is often difficult to analyze. For example, the mating behavior of a fruit fly may be impaired if it cannot beat its wings. However, the defect may be in the flight muscles, the nerves leading to them, or the brain, where the nerve impulses that initiate wing movements originate.

Another group of mutations—**regulatory mutations**—affect the regulation of gene expression. A mutation in a regulatory gene or a gene control region can disrupt normal regulatory processes and inappropriately activate or inactivate expression of a gene. For example, as we will see with the *lac* operon discussed in Chapter 16, a regulatory gene produces a product that controls the transcription of the entire *lac* operon. Mutations within this regulatory gene can lead to the production of a regulatory protein with abnormal effects on the *lac* operon. Our knowledge of genetic regulation has been dependent on the study of such regulatory mutations. Regulatory mutations may also occur in regions such as splice junctions, promoters, or other regulatory regions of a gene that affect many aspects of gene regulation including transcription initiation, mRNA splicing, and mRNA stability.

It is also possible that a mutation may interrupt a process that is essential to the survival of the organism. In this case, it is referred to as a **lethal mutation**. For example, a mutant bacterium that has lost the ability to synthesize an essential amino acid will cease to grow and eventually will die when placed in a medium lacking that amino acid. Various inherited human biochemical disorders are also examples of lethal mutations. For example, Tay–Sachs disease and Huntington disease are caused by mutations that result in lethality, but at different points in the life cycle of humans.

Another interesting class of mutations exert effects on the organism in ways that depend on the environment in which the organism finds itself. Such mutations are called **conditional mutations** because these mutations are present in the genome of an organism but can be detected only under certain conditions. Among the best examples of conditional mutations are **temperature-sensitive mutations**. At a "permissive" temperature, the mutant gene product functions normally, but it loses its function at a different, "restrictive" temperature. Therefore, when the organism is shifted from the permissive to the restrictive temperature, the effect of the mutation becomes apparent. The temperature-sensitive coat color variations in Siamese cats and Himalayan rabbits, discussed in Chapter 4, are striking examples of the effects of conditional mutations.

A **neutral mutation** is a mutation that can occur either in a protein-coding region or in any part of the genome, and whose effect on the genetic fitness of the organism is negligible. For example, a neutral mutation within a gene may change a lysine codon (AAA) to an arginine codon (AGA). The two amino acids are chemically similar; therefore, this change may be insignificant to the function of the protein. Because eukaryotic genomes consist mainly of noncoding regions, the vast majority of mutations are likely to occur in the large portions of the genome that do not contain genes. These may be considered neutral mutations, if they do not affect gene products or gene expression.

NOW SOLVE THIS

15–1 If one spontaneous mutation occurs within a human egg cell genome, and this mutation changes an A to a T, what is the most likely effect of this mutation on the phenotype of an offspring that develops from this mutated egg?

■ HINT: *This problem asks you to predict the effects of a single base-pair mutation on phenotype. The key to its solution involves an understanding of the organization of the human genome as well as the effects of mutations on coding and noncoding regions of genes, and the effects of mutations on development.*

15.2

Spontaneous Mutations Arise from Replication Errors and Base Modifications

In this section, we will outline some of the processes that lead to spontaneous mutations. It is useful to keep in mind, however, that many of the DNA changes that occur during spontaneous mutagenesis also occur, at a higher rate, during induced mutagenesis.

DNA Replication Errors and Slippage

As we learned in Chapter 11, the process of DNA replication is imperfect. Occasionally, DNA polymerases insert incorrect nucleotides during replication of a strand of DNA. Although DNA polymerases can correct most of these replication errors using their inherent 3′ to 5′ exonuclease proofreading capacity, misincorporated nucleotides may persist after replication. If these errors are not detected and corrected by DNA repair mechanisms, they may lead to mutations. Replication errors due to mispairing predominantly lead to point mutations. The fact that bases can take several forms, known as **tautomers**, increases the chance of mispairing during DNA replication, as we explain next.

In addition to mispairing and point mutations, DNA replication can lead to the introduction of small insertions or deletions. These mutations can occur when one strand of the DNA template loops out and becomes displaced during replication, or when DNA polymerase slips or stutters during replication. If a loop occurs in the template strand during replication, DNA polymerase may miss the looped-out nucleotides, and a small deletion in the new strand will be introduced. If DNA polymerase repeatedly introduces nucleotides that are not present in the template strand, an insertion of one or more nucleotides will occur, creating an unpaired loop on the newly synthesized strand. Insertions and deletions may lead to frameshift mutations, or amino acid insertions or deletions in the gene product.

Replication slippage can occur anywhere in the DNA but seems distinctly more common in regions containing repeated sequences. Repeat sequences are hot spots for DNA mutation and in some cases contribute to hereditary diseases, such as fragile-X syndrome and Huntington disease. The hypermutability of repeat sequences in noncoding regions of the genome is the basis for current methods of forensic DNA analysis.

Tautomeric Shifts

Purines and pyrimidines can exist in tautomeric forms—that is, in alternate chemical forms that differ by only a single proton shift in the molecule. The biologically important

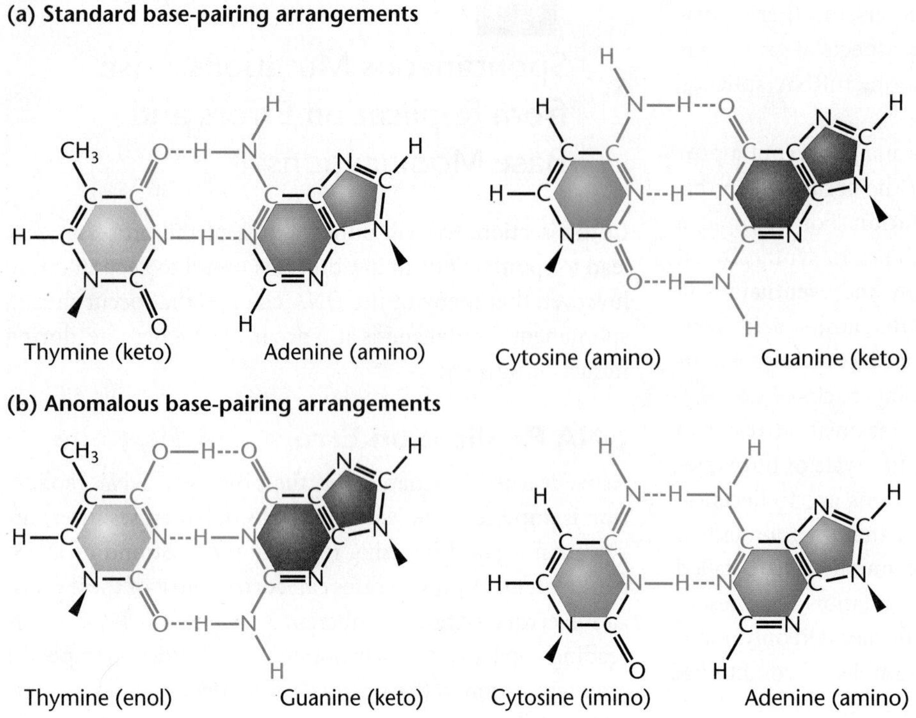

(a) Standard base-pairing arrangements

Thymine (keto)   Adenine (amino)   Cytosine (amino)   Guanine (keto)

(b) Anomalous base-pairing arrangements

Thymine (enol)   Guanine (keto)   Cytosine (imino)   Adenine (amino)

FIGURE 15–2   Standard base-pairing relationships (a) compared with examples of the anomalous base-pairing that occurs as a result of tautomeric shifts (b). The long triangle indicates the point at which the base bonds to the pentose sugar.

tautomers are the keto–enol forms of thymine and guanine and the amino–imino forms of cytosine and adenine. These shifts change the bonding structure of the molecule, allowing hydrogen bonding with noncomplementary bases. Hence, **tautomeric shifts** may lead to permanent base-pair changes and mutations. Figure 15–2 compares normal base-pairing relationships with rare unorthodox pairings. Anomalous $T \equiv G$ and $C = A$ pairs, among others, may be formed.

A mutation occurs during DNA replication when a transiently formed tautomer in the template strand pairs with a noncomplementary base. In the next round of replication, the "mismatched" members of the base pair are separated, and each becomes the template for its normal complementary base. The end result is a point mutation (Figure 15–3).

Depuration and Deamination

Some of the most common causes of spontaneous mutations are two forms of DNA base damage: depurination and deamination. **Depurination** is the loss of one of the nitrogenous bases in an intact double-helical DNA molecule. Most frequently, the base is a purine—either guanine or adenine. These bases may be lost if the glycosidic bond linking the 1′-C of the deoxyribose and the number 9 position of the purine ring is broken, leaving an **apurinic site** on one strand of the DNA. Geneticists estimate that thousands of such spontaneous lesions are formed daily in the DNA of

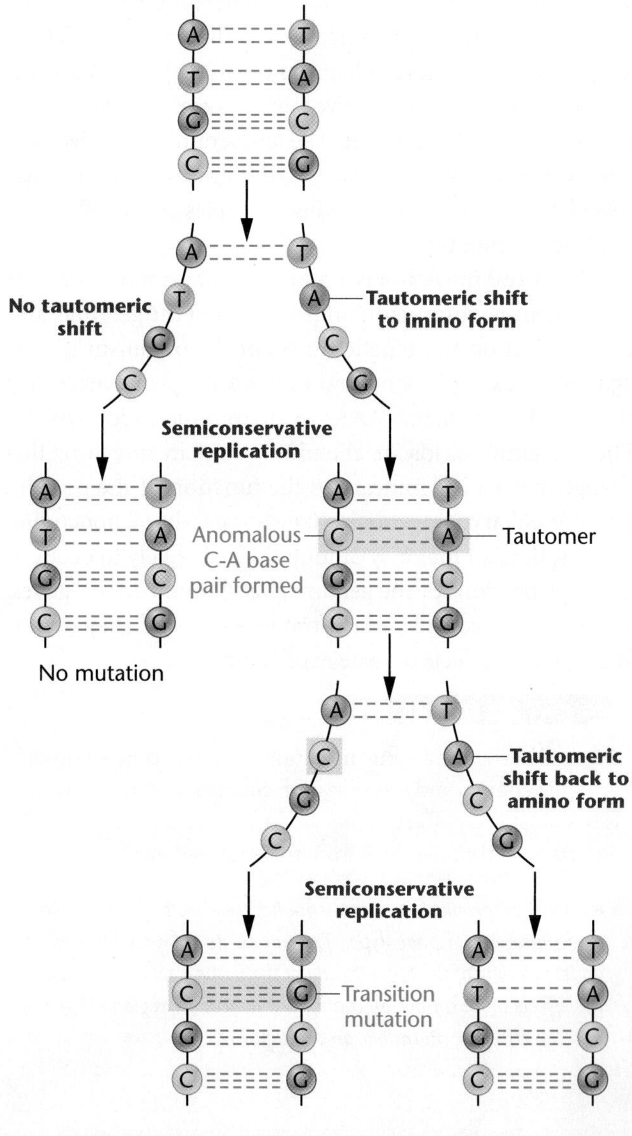

FIGURE 15–3   Formation of an $A = T$ to $G \equiv C$ transition mutation as a result of a tautomeric shift in adenine.

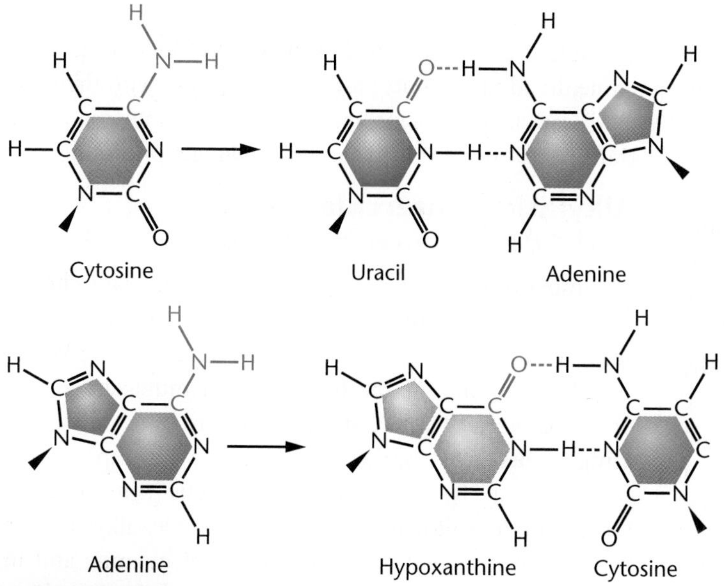

Deamination of cytosine and adenine, leading to new base pairing and mutation. Cytosine is converted to uracil, which base-pairs with adenine. Adenine is converted to hypoxanthine, which base-pairs with cytosine.

mammalian cells in culture. If apurinic sites are not repaired, there will be no base at that position to act as a template during DNA replication. As a result, DNA polymerase may introduce a nucleotide at random at that site.

In **deamination**, an amino group in cytosine or adenine is converted to a keto group (Figure 15–4). In these cases, cytosine is converted to uracil, and adenine is changed to hypoxanthine. The major effect of these changes is an alteration in the base-pairing specificities of these two bases during DNA replication. For example, cytosine normally pairs with guanine. Following its conversion to uracil, which pairs with adenine, the original $G \equiv C$ pair is converted to an $A = U$ pair and then, in the next replication, is converted to an $A = T$ pair. When adenine is deaminated, the original $A = T$ pair is converted to a $G \equiv C$ pair because hypoxanthine pairs naturally with cytosine. Deamination may occur spontaneously or as a result of treatment with chemical mutagens such as nitrous acid (HNO_2).

Oxidative Damage

DNA may also suffer damage from the by-products of normal cellular processes. These by-products include reactive oxygen species (electrophilic oxidants) that are generated during normal aerobic respiration. For example, superoxides (O_2^-), hydroxyl radicals ($\cdot OH$), and hydrogen peroxide (H_2O_2) are created during cellular metabolism and are constant threats to the integrity of DNA. Such **reactive oxidants**, also generated by exposure to high-energy radiation, can produce more than 100 different types of chemical

modifications in DNA, including modifications to bases, loss of bases, and single-stranded breaks.

Transposons

Transposable genetic elements, or **transposons**, are DNA elements that can move within, or between, genomes. These elements are present in the genomes of all organisms, from bacteria to humans, and often comprise large portions of these genomes. Transposons can act as naturally occurring mutagens. If in moving to a new location they insert themselves into the coding region of a gene, they can alter the reading frame or introduce stop codons. If they insert into the regulatory region of a gene, they can disrupt proper expression of the gene. Transposons can also create chromosomal damage, including double-stranded breaks, inversions, and translocations. Transposable genetic elements are described in detail later in this chapter (Section 15.8).

NOW SOLVE THIS

15–2 One of the most famous cases of an X-linked recessive mutation in humans is that of hemophilia found in the descendants of Britain's Queen Victoria. The pedigree of the royal family indicates that Victoria was heterozygous for the trait; however, her father was not affected, and there is no evidence that her mother was a carrier. What are some possible explanations of how the mutation arose? What types of mutations could lead to the disease?

■ HINT: *This problem asks you to determine the sources of new mutations. The key to its solution is to consider the ways in which mutations occur, the types of cells in which they can occur, and how they are inherited.*

15.3

Induced Mutations Arise from DNA Damage Caused by Chemicals and Radiation

Induced mutations are those that increase the rate of mutation above the spontaneous background. All cells on Earth are exposed to a plethora of agents called **mutagens**, which have the potential to damage DNA and cause induced mutations. Some of these agents, such as some fungal toxins, cosmic rays, and ultraviolet light, are natural components of our environment. Others, including some industrial pollutants, medical X rays, and chemicals within tobacco smoke, can be considered as unnatural or human-made additions to our modern world. On the positive side, geneticists harness some mutagens for use in analyzing genes and gene functions, as discussed later in this chapter (Section 15.7). The mechanisms by which some of these natural and un-natural agents lead to mutations are outlined in this section.

Base Analogs

One category of mutagenic chemicals is **base analogs**, compounds that can substitute for purines or pyrimidines during nucleic acid biosynthesis. For example, **5-bromouracil (5-BU)**, a derivative of uracil, behaves as a thymine analog but is halogenated at the number 5 position of the pyrimidine ring. If 5-BU is chemically linked to deoxyribose, the nucleoside analog **bromodeoxyuridine (BrdU)** is formed. Figure 15–5 compares the structure of this analog with that of thymine. The presence of the bromine atom in place of the methyl group increases the probability that a tautomeric shift will occur. If 5-BU is incorporated into DNA in place of thymine and a tautomeric shift to the enol form occurs, 5-BU base-pairs with guanine. After one round of replication, an $A = T$ to $G \equiv C$ transition results. Furthermore, the presence of 5-BU within DNA increases the sensitivity of the molecule to ultraviolet (UV) light, which itself is mutagenic.

There are other base analogs that are mutagenic. For example, **2-amino purine (2-AP)** can act as an analog of adenine. In addition to its base-pairing affinity with thymine, 2-AP can also base-pair with cytosine, leading to possible transitions from $A = T$ to $G \equiv C$ following replication.

Alkylating, Intercalating, and Adduct-Forming Agents

A number of naturally occurring and human-made chemicals alter the structure of DNA and cause mutations. The sulfur-containing mustard gases, discovered during World War I, were some of the first chemical mutagens identified in chemical warfare studies. Mustard gases are **alkylating agents**—that is, they donate an alkyl group, such as CH_3 or CH_3CH_2, to amino or keto groups in nucleotides. Ethylmethane sulfonate (EMS), for example, alkylates the keto groups in the number 6 position of guanine and in the number 4 position of thymine. As with base analogs, base-pairing affinities are altered, and transition mutations result. For example, 6-ethylguanine acts as an analog of adenine and pairs with thymine (Figure 15–6).

Intercalating agents are chemicals that have dimensions and shapes that allow them to wedge between the base pairs of DNA. When bound between base pairs, intercalating agents cause base pairs to distort and DNA strands to unwind. These changes in DNA structure affect many functions including transcription, replication, and repair. Deletions and insertions occur during DNA replication and repair, leading to frameshift mutations.

Some intercalating agents are used as DNA stains. An example is ethidium bromide, a fluorescent compound that is commonly used in molecular biology laboratories to visualize DNA during purifications and gel electrophoresis. The mutagenic characteristics of both ethidium bromide and the ultraviolet light used to visualize its fluorescence mean that this chemical must be used with caution. Other intercalating agents are used for cancer chemotherapy. Examples are doxorubicin, which is used to treat Hodgkin's lymphoma, and dactinomycin, which is used to treat a variety of sarcomas. Because cancer cells undergo DNA replication more frequently than noncancer cells, they are more sensitive than normal cells to the mutagenic and damaging effects of these chemotherapeutic agents.

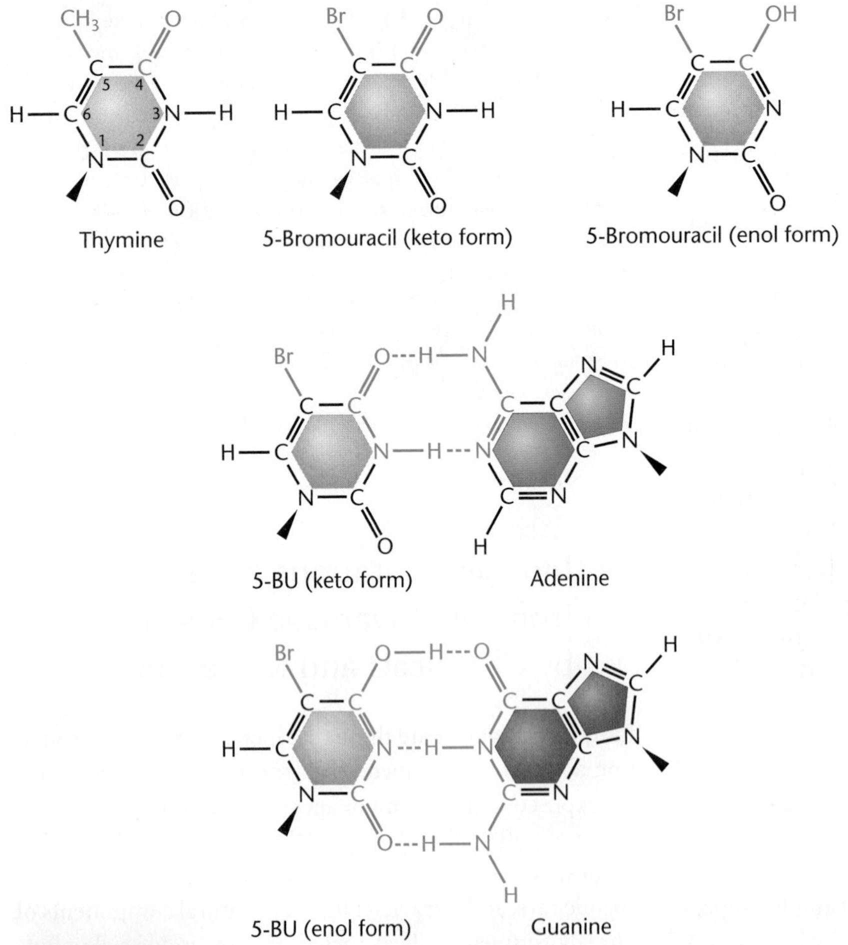

FIGURE 15-5   Similarity of the chemical structure of 5-bromouracil (5-BU) and thymine. In the common keto form, 5-BU base-pairs normally with adenine, behaving as a thymine analog. In the rare enol form, it pairs anomalously with guanine.

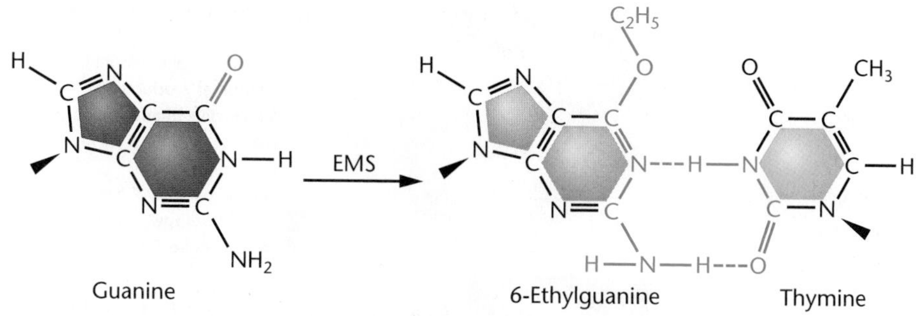

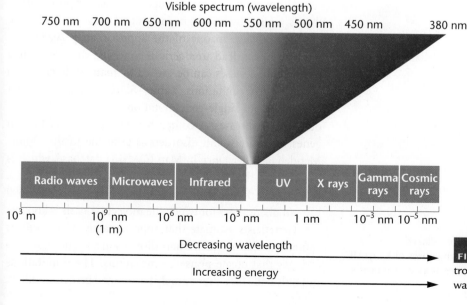

FIGURE 15–6 Conversion of guanine to 6-ethylguanine by the alkylating agent ethylmethane sulfonate (EMS). The 6-ethylguanine base-pairs with thymine.

Guanine

6-Ethylguanine

Thymine

Another group of chemicals that cause mutations are known as adduct-forming agents. A DNA adduct is a substance that covalently binds to DNA, altering its conformation and interfering with replication and repair. Two examples of adduct-forming substances are acetaldehyde (a component of cigarette smoke) and heterocyclic amines (HCAs). HCAs are cancer-causing chemicals that are created during the cooking of meats such as beef, chicken, and fish. HCAs are formed at high temperatures from amino acids and creatine. Many HCAs covalently bind to guanine bases. At least 17 different HCAs have been linked to the development of cancers, such as those of the stomach, colon, and breast.

Ultraviolet Light

All electromagnetic radiation consists of energetic waves that we define by their different wavelengths (Figure 15–7). The full range of wavelengths is referred to as the **electromagnetic spectrum**, and the energy of any radiation in the spectrum varies inversely with its wavelength. Waves in the range of visible light and longer are benign when they interact with most organic molecules. However, waves of shorter length than visible light, being inherently more energetic, have the potential to disrupt organic molecules. As we know, purines and pyrimidines absorb **ultraviolet (UV) radiation** most intensely at a wavelength of about 260 nm. Although Earth's ozone layer absorbs the most dangerous types of UV radiation, sufficient UV radiation can induce thousands of DNA lesions per hour in any cell exposed to this radiation. One major effect of UV radiation on DNA is the creation of **pyrimidine dimers**—chemical species consisting of two identical pyrimidines—particularly ones consisting of two thymine residues (Figure 15–8). The dimers distort the DNA conformation and inhibit normal replication. As a result, errors can be introduced in the base sequence of DNA during replication. When UV-induced dimerization is extensive, it is responsible (at least in part) for the killing effects of UV radiation on cells.

Ionizing Radiation

As noted above, the energy of radiation varies inversely with wavelength. Therefore, **X rays, gamma rays,** and **cosmic rays** are more energetic than UV radiation (Figure 15–7). As a result, they penetrate deeply into tissues, causing ionization of the molecules encountered along the way. Hence, this type of radiation is called **ionizing radiation**.

Visible spectrum (wavelength)

750 nm 700 nm 650 nm 600 nm 550 nm 500 nm 450 nm 380 nm

| Radio waves | Microwaves | Infrared | | UV | X rays | Gamma rays | Cosmic rays |

10^3 m 10^9 nm (1 m) 10^6 nm 10^3 nm   1 nm 10^{-3} nm 10^{-5} nm

Decreasing wavelength

Increasing energy

FIGURE 15–7 The regions of the electromagnetic spectrum and their associated wavelengths.

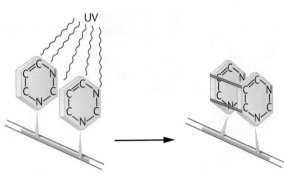

Dimer formed between adjacent thymidine residues along a DNA strand

FIGURE 15–8   Induction of a thymine dimer by UV radiation, leading to distortion of the DNA. The covalent crosslinks occur between the atoms of the pyrimidine ring.

As ionizing radiation penetrates cells, stable molecules and atoms are transformed into **free radicals**—chemical species containing one or more unpaired electrons. Free radicals can directly or indirectly affect the genetic material, altering purines and pyrimidines in DNA, breaking phosphodiester bonds, disrupting the integrity of chromosomes, and producing a variety of chromosomal aberrations, such as deletions, translocations, and chromosomal fragmentation.

Given the capacity of ionizing radiation to cause serious genetic damage, it is important to consider what levels of radiation are mutagenic in humans and what sources of ionizing radiation cause the most damage in everyday life. Figure 15–9 shows a graph of the percentage of induced X-linked recessive lethal mutations versus the dose of X rays administered. There is a linear relationship between X-ray dose and the induction of mutation; for each doubling of the dose, twice as many mutations are induced. Because the

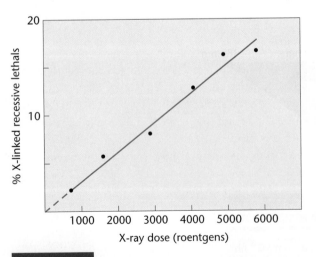

FIGURE 15–9   Plot of the percentage of X-linked recessive mutations induced in *Drosophila* by increasing doses of X rays. If extrapolated, the graph intersects the zero axis as shown by the dashed line.

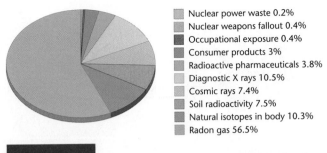

Nuclear power waste 0.2%
Nuclear weapons fallout 0.4%
Occupational exposure 0.4%
Consumer products 3%
Radioactive pharmaceuticals 3.8%
Diagnostic X rays 10.5%
Cosmic rays 7.4%
Soil radioactivity 7.5%
Natural isotopes in body 10.3%
Radon gas 56.5%

FIGURE 15–10   Chart showing average yearly dose of radiation from natural and human-made sources.

line intersects near the zero axis, this graph suggests that even very small doses of radiation are mutagenic.

Although it is often assumed that radiation from artificial sources such as nuclear power plant waste and medical X rays are the most significant sources of radiation exposure for humans, scientific data indicate otherwise. Scientists estimate that less than 20 percent of human radiation exposure arises from human-made sources. Figure 15–10 summarizes the annual radiation exposure for humans residing in the United States. As these data indicate, the greatest radiation exposure comes from radon gas, cosmic rays, and natural soil radioactivity. More than half of human-made radiation exposure comes from medical X rays and radioactive pharmaceuticals.

15.4

Single-Gene Mutations Cause a Wide Range of Human Diseases

Now that the human genome has been sequenced and genome-wide techniques are available to pinpoint disease-associated genes and the mutations within them, scientists are rapidly accumulating information about how specific types of mutations contribute to human disorders.

Although most human genetic diseases are **polygenic**—that is, caused by variations in several genes—even a single base-pair change in one of the approximately 20,000 human genes can lead to a serious inherited disorder. These **monogenic** diseases can be caused by many different types of single-gene mutations. Table 15.3 lists some examples of the types of single-gene mutations that can lead to serious genetic diseases. A comprehensive database of human genes, mutations, and disorders is available in the Online Mendelian Inheritance in Man (OMIM) database, which is described in the "Exploring Genomics" feature in Chapter 3. As of 2009, the OMIM database has catalogued over 2600 human phenotypes for which the molecular basis is known.

Geneticists estimate that approximately 30 percent of mutations that cause human diseases are single base-pair changes that create nonsense mutations. These mutations not only code for a prematurely terminated protein product,

TABLE 15.3

Examples of Human Disorders Caused by Single-Gene Mutations

Type of DNA Mutation	Disorder	Molecular Change
Missense	Achondroplasia	Glycine to Arginine at position 380 of *FGFR2* gene
Nonsense	Marfan syndrome	Tyrosine to STOP codon at position 2113 of *fibrillin-1* gene
Insertion	Familial hypercholesterolemia	Various short insertions throughout the *LDLR* gene
Deletion	Cystic fibrosis	Three base-pair deletion of phenylalanine codon at position 508 of *CFTR* gene
Trinucleotide repeat expansions	Huntington disease	>40 repeats of (CAG) sequence in coding region of *Huntingtin* gene

but also trigger rapid decay of the mRNA. Many more mutations are missense mutations that alter the amino acid sequence of a protein and frameshift mutations that alter the protein sequence and create internal nonsense codons. Other common disease-associated mutations affect the sequences of gene promoters, mRNA splicing signals, and other noncoding sequences that affect transcription, processing, and stability of mRNA or protein. One recent study showed that about 15 percent of all point mutations that cause human genetic diseases result in abnormal mRNA splicing. Approximately 85 percent of these spicing mutations alter the sequence of 5′ and 3′ splice signals. The remainder create new splice sites within the gene. Splicing defects often result in degradation of the abnormal mRNA or creation of abnormal protein products.

Single Base-Pair Mutations and β-Thalassemia

Although some genetic diseases, such as sickle-cell anemia, are caused by one specific base-pair change within a single gene, most are caused by any of a large number of different mutations. The mutation profile associated with β-thalassemia provides an example of how one inherited disease can arise from a large number of possible mutations.

β-thalassemia is an inherited autosomal recessive blood disorder resulting from a reduction or absence of hemoglobin. It is the most common single-gene disease in the world, affecting people worldwide, but especially populations in Mediterranean, North African, Middle Eastern, Central Asian, and Southeast Asian countries.

People with β-thalassemia have varying degrees of anemia—from severe to mild—with symptoms including weakness, delayed development, jaundice, enlarged organs, and often a need for frequent blood transfusions.

Mutations in the *β-globin* gene (*HBB* gene) cause β-thalassemia. The *HBB* gene encodes the 146 amino acid β-globin polypeptide. Two β-globin polypeptides associate with two α-globin polypeptides to form the adult hemoglobin tetramer. The *HBB* gene spans 1.6 kilobases of DNA on the short arm of chromosome 11. It is made up of three exons and two introns.

Scientists have discovered over 250 different mutations in the *HBB* gene that cause β-thalassemia, although most cases worldwide are associated with about 20 of these mutations. Most mutations change a single nucleotide within or surrounding the *HBB* gene, or create small insertions and deletions. In addition, each population affected by β-thalassemia has a unique mix of mutations. For example, the most prevalent mutation in a Sardinian population—a mutation that accounts for more than 95 percent of cases—is a single base-pair change at codon 39, creating a nonsense mutation and premature termination of the β-globin polypeptide. In contrast, a study of β-thalassemia mutations in a Yugoslavian population revealed 14 different mutations, with only three (all in intron 1 splice signals) accounting for 75 percent of cases.

The types of mutations that cause β-thalassemia not only affect the β-globin amino acid sequence (missense, nonsense, and frameshift mutations), but also alter *HBB* transcription efficiency, mRNA splicing and stability, translation, and protein stability.

Table 15.4 provides a summary of the types of single-gene mutations that cause β-thalassemia. More than half of these mutations are single base-pair changes, and the remainder are short insertions, deletions, and duplications.

Mutations Caused by Expandable DNA Repeats

Beginning in about 1990, molecular analyses of the genes responsible for a number of inherited human disorders revealed a remarkable set of observations. Researchers discovered that some mutant genes contain expansion of **trinucleotide repeat sequences**—specific short DNA sequences repeated many times. Normal individuals have fewer than 30 repetitions of these sequences; however, individuals with over 20 different human disorders appear to have abnormally large numbers of repeat sequences—often over 200—within and surrounding specific genes.

Examples of diseases associated with these trinucleotide repeat expansions are fragile-X syndrome (discussed in detail in Chapter 8), myotonic dystrophy, and Huntington disease (discussed in Chapter 4). When trinucleotide repeats such as $(CAG)_n$ occur within a coding region, they can be translated

TABLE 15.4

Types of Mutations in the *HBB* Gene that Cause β-thalassemia

Number of Mutations Known	Gene Region Affected	Description
22	5' upstream region	Single base-pair mutations occur between −101 and −25 upstream from transcription start site. For example, a T−>A transition in the TATA sequence at −30 results in decreased gene transcription and severe disease.
1	mRNA CAP site	Single base-pair mutation (A−>C transversion) at +1 position leads to decreased levels of mRNA.
3	5' untranslated region	Single base-pair mutations at +20, +22, and +33 cause decreases in transcription and translation and mild disease.
7	ATG translation initiation codon	Single base-pair mutations alter the mRNA AUG sequence, resulting in no translation and severe disease.
36	Exons 1, 2, and 3 coding regions	Single base-pair missense and nonsense mutations, and mutations that create abnormal mRNA splice sites. Disease severity varies from mild to extreme.
38	Introns 1 and 2	Single base-pair transitions and transversions that reduce or abolish mRNA splicing, and create abnormal splice sites that affect mRNA stability. Most cause severe disease.
6	Polyadenylation site	Single base-pair changes in the AATAAA sequence reduce the efficiency of mRNA cleavage and polyadenylation, yielding long mRNAs or unstable mRNAs. Disease is mild.
> 100	Throughout and surrounding the *HBB* gene	Short insertions, deletions, and duplications that alter coding sequences, create frameshift stop codons, and alter mRNA splicing.

into long tracks of glutamine. These glutamine tracks may cause the proteins to aggregate abnormally. When the repeats occur outside coding regions, but within the mRNA, it is thought that the mRNAs may act as "toxic" RNAs that bind to important regulatory proteins, sequestering them away from their normal functions in the cell. Another possible consequence of long trinucleotide repeats is that the regions of DNA containing the repeats may become abnormally methylated, leading to silencing of gene transcription.

The mechanisms by which the repeated sequences expand from generation to generation are of great interest. It is thought that expansion may result from either errors during DNA replication or errors during DNA damage repair.

NOW SOLVE THIS

15–3 The cancer drug melphalan is an alkylating agent of the mustard gas family. It acts in two ways: by causing alkylation of guanine bases and by cross linking DNA strands together. Describe two ways in which melphalan might kill cancer cells. What are two ways in which cancer cells could repair the DNA-damaging effects of melphalan?

■ HINT: *This problem asks you to consider the effect of the alkylation of guanine on base pairing during DNA replication. The key to its solution is to consider the effects of mutations on cellular processes that allow cells to grow and divide. In Section 15.5, you will learn about the ways in which cells repair the types of mutations introduced by alkylating agents.*

Whatever the cause may be, the presence of these short and unstable repeat sequences seems to be prevalent in humans and in many other organisms.

15.5

Organisms Use DNA Repair Systems to Counteract Mutations

Living systems have evolved a variety of elaborate repair systems that counteract both spontaneous and induced DNA damage. These **DNA repair** systems are absolutely essential to the maintenance of the genetic integrity of organisms and, as such, to the survival of organisms on Earth. The balance between mutation and repair results in the observed mutation rates of individual genes and organisms. Of foremost interest in humans is the ability of these systems to counteract genetic damage that would otherwise result in genetic diseases and cancer. The link between defective DNA repair and cancer susceptibility is described in detail in Chapter 19.

We now embark on a review of some systems of DNA repair, with the emphasis on the major approaches that organisms use to counteract genetic damage.

Proofreading and Mismatch Repair

Some of the most common types of mutations arise during DNA replication when an incorrect nucleotide is inserted by DNA polymerase. The major DNA synthesizing enzyme in bacteria (**DNA polymerase III**) makes an error approximately once every 100,000 insertions, leading to an error rate of 10^{-5}. Fortunately, DNA polymerase proofreads each step, catching 99 percent of those errors. If an incorrect nucleotide is inserted during polymerization, the enzyme can recognize the error and "reverse" its direction. It then behaves as a 3' to 5' exonuclease, cutting out the incorrect nucleotide and replacing it with the correct one. This improves the efficiency of replication one hundred-fold, creating only 1 mismatch in every 10^7 insertions, for a final error rate of 10^{-7}.

To cope with errors such as base–base mismatches, small insertions, and deletions that remain after proofreading, another mechanism, called **mismatch repair**, may be activated. During mismatch repair, the mismatches are detected, the incorrect nucleotide is removed, and the correct nucleotide is inserted in its place. But how does the repair system recognize which nucleotide is correct (on the template strand) and which nucleotide is incorrect (on the newly synthesized strand)? If the mismatch is recognized but no such discrimination occurs, the excision will be random, and the strand bearing the correct base will be clipped out 50 percent of the time. Hence, strand discrimination is a critical step.

The process of strand discrimination has been elucidated in some bacteria, including *E. coli,* and is based on **DNA methylation**. These bacteria contain an enzyme, **adenine methylase**, which recognizes the DNA sequence

$$5'\text{——GATC——}3'$$
$$3'\text{——CTAG——}5'$$

as a substrate, adding a methyl group to each of the adenine residues during DNA replication.

Following replication, the newly synthesized DNA strand remains temporarily unmethylated, as the adenine methylase lags behind the DNA polymerase. Prior to methylation, the repair enzyme recognizes the mismatch and binds to the unmethylated (newly synthesized) DNA strand. An **endonuclease** enzyme creates a nick in the backbone of the unmethylated DNA strand, either 5' or 3' to the mismatch. An **exonuclease** unwinds and degrades the nicked DNA strand, until the region of the mismatch is reached. Finally, DNA polymerase fills in the gap created by the exonuclease, using the correct DNA strand as a template. DNA ligase then seals the gap.

A series of *E. coli* gene products, MutH, MutL, and MutS, as well as exonucleases, DNA polymerase III and ligase, are involved in mismatch repair. Mutations in the *MutH, MutL,* and *MutS* genes result in bacterial strains deficient in mismatch repair. While the preceding mechanism occurs in *E. coli,* similar mechanisms involving homologous proteins exist in yeast and in mammals.

In humans, mutations in genes that code for DNA mismatch repair proteins (such as the *hMSH2* and *hMLH1,* which are the human equivalents of the *MutS* and *MutL* genes of *E. coli*) are associated with the hereditary nonpolyposis colon

cancer. Mismatch repair defects are commonly found in other cancers, such as leukemias, lymphomas, and tumors of the ovary, prostate, and endometrium. Cells from these cancers show genome-wide increases in the rate of spontaneous mutation. The link between defective mismatch repair and cancer is supported by experiments with mice. Mice that are engineered to have deficiencies in mismatch repair genes accumulate large numbers of mutations and are cancer-prone.

Postreplication Repair and the SOS Repair System

Another DNA repair system, called **postreplication repair**, responds *after* damaged DNA has escaped repair and has failed to be completely replicated. As illustrated in Figure 15–11,

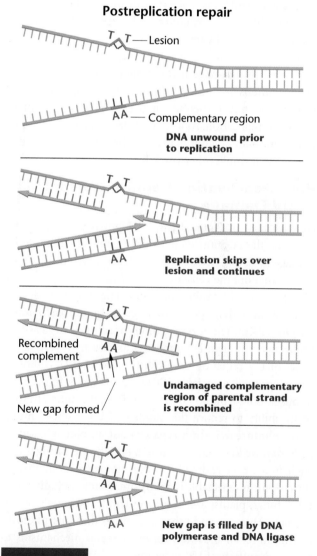

Postreplication repair

— Lesion

— Complementary region

DNA unwound prior to replication

Replication skips over lesion and continues

Recombined complement

New gap formed

Undamaged complementary region of parental strand is recombined

New gap is filled by DNA polymerase and DNA ligase

FIGURE 15–11 Postreplication repair occurs if DNA replication has skipped over a lesion such as a thymine dimer. Through the process of recombination, the correct complementary sequence is recruited from the parental strand and inserted into the gap opposite the lesion. The new gap is filled by DNA polymerase and DNA ligase.

when DNA bearing a lesion of some sort (such as a pyrimidine dimer) is being replicated, DNA polymerase may stall at the lesion and then skip over it, leaving an unreplicated gap on the newly synthesized strand. To correct the gap, the RecA protein directs a recombinational exchange with the corresponding region on the undamaged parental strand of the same polarity (the "donor" strand). When the undamaged segment of the donor strand DNA replaces the gapped segment, a gap is created on the donor strand. The gap can be filled by repair synthesis as replication proceeds. Because a recombinational event is involved in this type of DNA repair, it is considered to be a form of **homologous recombination repair**.

Still another repair pathway, the *E. coli* **SOS repair system**, also responds to damaged DNA, but in a different way. In the presence of a large number of unrepaired DNA mismatches and gaps, bacteria can induce the expression of about 20 genes (including *lexA, recA,* and *uvr*) whose products allow DNA replication to occur even in the presence of these lesions. This type of repair is a last resort to minimize DNA damage, hence its name. During SOS repair, DNA synthesis becomes error-prone, inserting random and possibly incorrect nucleotides in places that would normally stall DNA replication. As a result, SOS repair itself becomes mutagenic—although it may allow the cell to survive DNA damage that would otherwise kill it.

Photoreactivation Repair: Reversal of UV Damage

As illustrated in Figure 15–8, UV light is mutagenic as a result of the creation of pyrimidine dimers. UV-induced damage to *E. coli* DNA can be partially reversed if, following irradiation, the cells are exposed briefly to light in the blue range of the visible spectrum. The process is dependent on the activity of a protein called **photoreactivation enzyme (PRE)**. The enzyme's mode of action is to cleave the bonds between thymine dimers, thus directly reversing the effect of UV radiation on DNA (Figure 15–12). Although the enzyme will associate with a thymine dimer in the dark, it must absorb a photon of light to cleave the dimer. In spite of its ability to reduce the number of UV-induced mutations, **photoreactivation repair** is not absolutely essential in *E. coli*; we know this because a mutation creating a null allele in the gene coding for PRE is not lethal. Nonetheless, the enzyme is detectable in many organisms, including bacteria, fungi, plants, and some vertebrates—though not in humans. Humans and other organisms that lack photoreactivation repair must rely on other repair mechanisms to reverse the effects of UV radiation.

Base and Nucleotide Excision Repair

A number of light-independent DNA repair systems exist in all prokaryotes and eukaryotes. The basic mechanisms

Photoreactivation repair

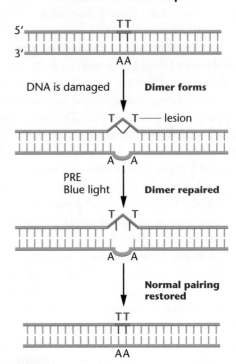

FIGURE 15–12   Damaged DNA repaired by photoreactivation repair. The bond creating the thymine dimer is cleaved by the photoreactivation enzyme (PRE), which must be activated by blue light in the visible spectrum.

involved in these types of repair—collectively referred to as **excision repair** or cut-and-paste mechanisms—consist of the following three steps.

1. The distortion or error present on one of the two strands of the DNA helix is recognized and enzymatically clipped out by an endonuclease. Excisions in the phosphodiester backbone usually include a number of nucleotides adjacent to the error as well, leaving a gap on one strand of the helix.

2. A DNA polymerase fills in the gap by inserting nucleotides complementary to those on the intact strand, which it uses as a replicative template. The enzyme adds these nucleotides to the free 3'-OH end of the clipped DNA. In *E. coli*, this step is usually performed by DNA polymerase I.

3. DNA ligase seals the final "nick" that remains at the 3'-OH end of the last nucleotide inserted, closing the gap.

There are two types of excision repair: base excision repair and nucleotide excision repair. **Base excision repair (BER)** corrects DNA that contains a damaged DNA base. The first step in the BER pathway in *E. coli* involves the recognition of the altered base by an enzyme called **DNA**

Base excision repair

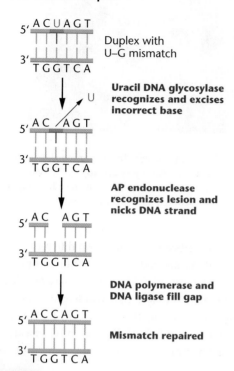

FIGURE 15–13   Base excision repair (BER) accomplished by uracil DNA glycosylase, AP endonuclease, DNA polymerase, and DNA ligase. Uracil is recognized as a noncomplementary base, excised, and replaced with the complementary base (C).

Nucleotide excision repair

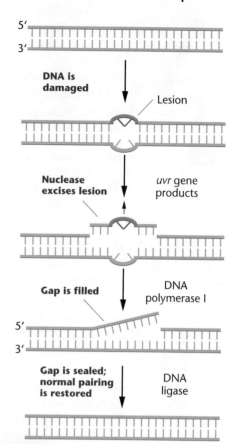

FIGURE 15–14   Nucleotide excision repair (NER) of a UV-induced thymine dimer. During repair, 13 nucleotides are excised in prokaryotes, and 28 nucleotides are excised in eukaryotes.

glycosylase. There are a number of DNA glycosylases, each of which recognizes a specific base (Figure 15–13). For example, the enzyme uracil DNA glycosylase recognizes the presence of uracil in DNA. DNA glycosylases first cut the glycosidic bond between the base and the sugar, creating an **apyrimidinic or apurinic site**. The sugar with the missing base is then recognized by an enzyme called **AP endonuclease**. The AP endonuclease makes a cut in the phosphodiester backbone at the apyrimidinic or apurinic site. Endonucleases then remove the deoxyribose sugar, and the gap is filled by DNA polymerase and DNA ligase.

Although much has been learned about the mechanisms of BER in *E. coli*, BER systems have also been detected in eukaryotes from yeast to humans. Experimental evidence shows that both mouse and human cells that are defective in BER activity are hypersensitive to the killing effects of gamma rays and oxidizing agents.

Nucleotide excision repair (NER) pathways repair "bulky" lesions in DNA that alter or distort the double helix. These lesions include the UV-induced pyrimidine dimers and DNA adducts discussed previously.

The NER pathway (Figure 15–14) was first discovered in *E. coli* by Paul Howard-Flanders and coworkers, who isolated several independent mutants that are sensitive to UV

radiation. One group of genes was designated *uvr* (ultraviolet repair) and included the *uvrA, uvrB,* and *uvrC* mutations. In the NER pathway, the *uvr* gene products are involved in recognizing and clipping out lesions in the DNA. Usually, a specific number of nucleotides is clipped out around both sides of the lesion. In *E. coli,* usually a total of 13 nucleotides is removed, including the lesion. The repair is then completed by DNA polymerase I and DNA ligase, in a manner similar to that occurring in BER. The undamaged strand opposite the lesion is used as a template for the replication, resulting in repair.

Nucleotide Excision Repair and Human Disease

The mechanism of NER in eukaryotes is much more complicated than that in prokaryotes and involves many more proteins, encoded by about 30 genes. Much of what is known about the system in humans has come from detailed studies of individuals with **xeroderma pigmentosum (XP)**, a rare recessive genetic disorder that predisposes individuals

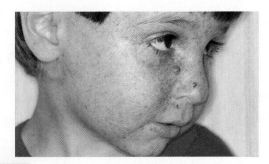
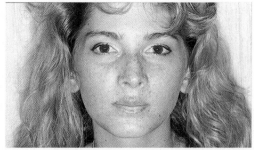

FIGURE 15-15   Two individuals with xeroderma pigmentosum. The 4-year-old boy on the left shows marked skin lesions induced by sunlight. Mottled redness (erythema) and irregular pigment changes in response to cellular injury are apparent. Two nodular cancers are present on his nose. The 18-year-old girl on the right has been carefully protected from sunlight since her diagnosis of xeroderma pigmentosum in infancy. Several cancers have been removed, and she has worked as a successful model.

to severe skin abnormalities, skin cancers, and a wide range of other symptoms including developmental and neurological defects. Patients with XP are extremely sensitive to UV radiation in sunlight. In addition, they have a 2000-fold higher rate of cancer, particularly skin cancer, than the general population. The condition is severe and may be lethal, although early detection and protection from sunlight can arrest it (Figure 15–15).

The repair of UV-induced lesions in XP has been investigated *in vitro,* using human fibroblast cell cultures derived from normal individuals and those with XP. (Fibroblasts are undifferentiated connective tissue cells.) The results of these studies suggest that the XP phenotype is caused by defects in NER pathways and by mutations in more than one gene.

In 1968, James Cleaver showed that cells from XP patients were deficient in DNA synthesis other than that occurring during chromosome replication—a phenomenon known as **unscheduled DNA synthesis**. Unscheduled DNA synthesis is elicited in normal cells by UV radiation. Because this type of synthesis is thought to represent the activity of DNA polymerization during NER, the lack of unscheduled DNA synthesis in XP patients suggested that XP may be a deficiency in NER.

The involvement of multiple genes in NER and XP was further investigated by studies using **somatic cell hybridization**. Fibroblast cells from any two unrelated XP patients, when grown together in tissue culture, can fuse together, forming heterokaryons. A **heterokaryon** is a single cell with two nuclei from different organisms but a common cytoplasm. NER in the heterokaryon can be measured by the level of unscheduled DNA synthesis. If the mutation in each of the two XP cells occurs in the same gene, the heterokaryon, like the cells that fused to form it, will still be unable to undergo NER. This is because there is no normal copy of the relevant gene present in the heterokaryon. However, if NER does occur in the heterokaryon, the mutations in the two XP

cells must have been present in two different genes. Hence, the two mutants are said to demonstrate **complementation**, a concept also discussed in Chapter 4. Complementation occurs because the heterokaryon has at least one normal copy of each gene in the fused cell. By fusing XP cells from a large number of XP patients, researchers were able to determine how many genes contribute to the XP phenotype.

Based on these and other studies, XP patients have been divided into seven complementation groups, indicating that at least seven different genes are involved in nucleotide excision repair in humans. A gene representing each of these complementation groups, *XPA* to *XPG* (Xeroderma Pigmentosum gene *A* to *G*), has now been identified, and a homologous gene for each has been identified in yeast. Approximately 20 percent of XP patients do not fall into any of the seven complementation groups. Cells from these patients often have mutations in the gene coding for DNA polymerase H and are defective in repair DNA synthesis.

As a result of the study of defective genes in XP, a great deal is now known about how NER counteracts DNA damage in normal cells. The first step in humans is recognition of the damaged DNA by proteins encoded by the *XPC, XPE,* and *XPA* genes. These proteins then recruit the remainder of the repair proteins to the site of DNA damage. The *XPB* and *XPD* genes encode helicases, and the *XPF* and *XPG* genes encode nucleases. The excision repair complex containing these and other factors is responsible for the excision of an approximately 28-nucleotide-long fragment from the DNA strand that contains the lesion.

Two other rare autosomal recessive diseases are associated with defects in NER pathways—Cockayne syndrome (CS) and trichothiodystrophy (TTD). The symptoms of CS include developmental and neurological defects, and sensitivity to sunlight, but not an increase in cancers. Patients with CS age prematurely and usually die before the age of 20. Patients with TTD suffer from dwarfism, retardation, brittle hair and skin, and facial deformities. Like CS, these patients are sensitive to sunlight but do not have higher than normal rates of cancer. TTD patients have a median life span of six years.

Both CS and TTD arise from mutations in some of the same genes involved in XP (such as XPB and XPD), as well as other genes that encode proteins involved in NER within transcribed regions of the genome. It is not known why such a wide variety of different symptoms result from mutations in the same genes or DNA repair pathways; however, it may reflect the fact that products of many NER genes are also involved in other essential processes.

Double-Strand Break Repair in Eukaryotes

Thus far, we have discussed repair pathways that deal with damage or errors within one strand of DNA. We conclude our discussion of DNA repair by considering what happens when both strands of the DNA helix are cleaved—as a result of exposure to ionizing radiation, for example. These types of damage are extremely dangerous to cells, leading to chromosome rearrangements, cancer, or cell death. In this section, we will discuss double-strand breaks in eukaryotic cells.

Specialized forms of DNA repair, the DNA **double-strand break repair (DSB repair)** pathways, are activated and are responsible for reattaching two broken DNA strands. Recently, interest in DSB repair has grown because defects in these pathways are associated with X-ray hypersensitivity and immune deficiency. Such defects may also underlie familial disposition to breast and ovarian cancer. Several human disease syndromes, such as Fanconi's anemia and ataxia telangiectasia, result from defects in DSB repair.

One pathway involved in double-strand break repair is **homologous recombination repair**. The first step in this process involves the activity of an enzyme that recognizes the double-strand break, and then digests back the 5′ ends of the broken DNA helix, leaving overhanging 3′ ends (Figure 15–16). One overhanging end searches for a region of sequence complementarity on the sister chromatid and then invades the homologous DNA duplex, aligning the complementary sequences. Once aligned, DNA synthesis proceeds from the 3′ overhanging ends, using the undamaged DNA strands as templates. The interaction of two sister chromatids is necessary because, when both strands of one helix are broken, there is no undamaged parental DNA strand available to use as a source of the complementary template DNA sequence

during repair. After DNA repair synthesis, the resulting heteroduplex molecule is resolved and the two chromatids separate, as previously discussed in Chapter 11.

DSB repair usually occurs during the late S or early G2 phase of the cell cycle, after DNA replication, a time when sister chromatids are available to be used as repair templates. Because an undamaged template is used during repair synthesis, homologous recombination repair is an accurate process.

A second pathway, called **nonhomologous end joining**, also repairs double-strand breaks. However, as the name implies, the mechanism does not recruit a homologous region of DNA during repair. This system is activated in G1,

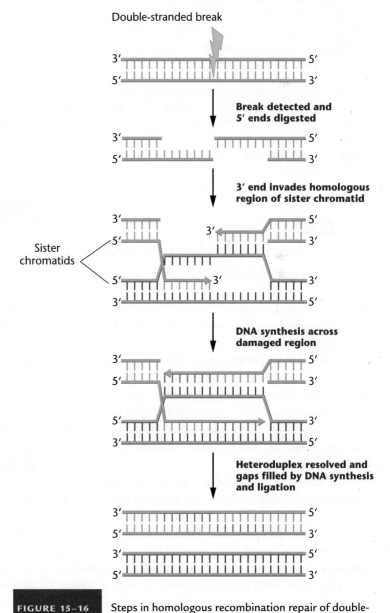

FIGURE 15–16 Steps in homologous recombination repair of double-stranded breaks.

prior to DNA replication. End joining involves a complex of many proteins, including DNA-dependent protein kinase and the breast cancer susceptibility gene product, BRCA1. These and other proteins bind to the free ends of the broken DNA, trim the ends, and ligate them back together. Because some nucleotide sequences are lost in the process of end joining, it is an error-prone repair system. In addition, if more than one chromosome suffers a double-strand break, the wrong ends could be joined together, leading to abnormal chromosome structures, such as those discussed in Chapter 8.

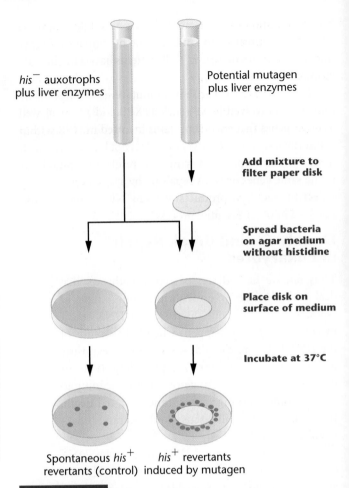

his⁻ auxotrophs plus liver enzymes

Potential mutagen plus liver enzymes

Add mixture to filter paper disk

Spread bacteria on agar medium without histidine

Place disk on surface of medium

Incubate at 37°C

Spontaneous his^+ revertants (control)

his^+ revertants induced by mutagen

FIGURE 15–17   The Ames test, which screens compounds for potential mutagenicity.

15.6

The Ames Test Is Used to Assess the Mutagenicity of Compounds

There is great concern about the possible mutagenic properties of any chemical that enters the human body, whether through the skin, the digestive system, or the respiratory tract. Examples of synthetic chemicals that concern us are those found in air and water pollution, food preservatives, artificial sweeteners, herbicides, pesticides, and pharmaceutical products. Mutagenicity can be tested in various organisms, including fungi, plants, and cultured mammalian cells; however, one of the most common tests, which we describe here, uses bacteria.

The **Ames test** uses a number of different strains of the bacterium *Salmonella typhimurium* that have been selected for their ability to reveal the presence of specific types of mutations. For example, some strains are used to detect base-pair substitutions, and other strains detect various frameshift mutations. Each strain contains a mutation in one of the genes of the histidine operon. The mutant strains are unable to synthesize histidine (*his⁻* strains) and therefore require histidine for growth (Figure 15–17). The assay measures the frequency of reverse mutations that occur within the mutant gene, yielding wild-type bacteria (*his⁺* revertants). These *Salmonella* strains also have an increased sensitivity to mutagens due to the presence of mutations in genes involved in DNA damage repair and synthesis of the lipopolysaccharide barrier that coats bacteria and protects them from external substances.

Many substances entering the human body are relatively innocuous until activated metabolically, usually in the liver, to more chemically reactive products. Thus, the Ames test includes a step in which the test compound is incubated *in vitro* in the presence of a mammalian liver extract. Alternatively, test compounds may be injected into a mouse where they are modified by liver enzymes and then recovered for use in the Ames test.

In the initial use of Ames testing in the 1970s, a large number of known **carcinogens**, or cancer-causing agents, were examined, and more than 80 percent of these were shown to be strong mutagens. This is not surprising, as the transformation of cells to the malignant state occurs as a result of mutations. For example, more than 60 compounds found in cigarette smoke test positive in the Ames test and cause cancer in animal tests. Although a positive response in the Ames test does not prove that a compound is carcinogenic, the Ames test is useful as a preliminary screening device. The Ames test is used extensively during the development of industrial and pharmaceutical chemical compounds.

15.7
Geneticists Use Mutations to Identify Genes and Study Gene Function

In order to dissect the genes and processes that regulate biological functions, geneticists often study the effects of mutations in **model organisms**. A good model organism must be easy to grow, have a short generation time, produce abundant progeny, and be readily mutagenized and crossed. The most extensively used model organisms are bacteria (*E. coli*), budding yeasts (*Saccharomyces cerevisiae*), fruit flies (*Drosophila melanogaster*), nematodes (*Caenorhabditis elegans*), zebrafish (*Danio rerio*), mustard plants (*Arabidopsis thaliana*), and mice (*Mus musculus*).

In this section, we will describe some of the methods that geneticists use to generate and detect gene mutations, as the first steps toward a full genetic analysis.

Inducing Mutations with Radiation, Chemicals, and Transposon Insertion

Geneticists sometimes begin their analyses of genes by examining spontaneous, naturally occurring mutations. However, as mutations are generally rare in nature, researchers need to induce mutations in model organisms in order to increase the chances of detecting a relevant mutant. The goal of mutagenesis is to create one mutation at random in the genome of each individual in the experimental population, so that one gene product is disrupted in each individual, leaving the rest of the genome wild type.

During genetic analyses, researchers use a wide range of different mutagens, depending on the type of mutation desired. For example, ionizing radiation can be used to create chromosome breaks, deletions, and translocations. Though useful for some types of studies, such mutations often have severe effects on the phenotype, making further genetic analysis difficult. In contrast, chemicals such as ethyl methane sulfonate (EMS) and nitrosoguanidine cause single base-pair changes and small deletions and insertions. With these mutagens, a range of mild to severe mutant phenotypes can be generated. Conditional mutations, such as temperature sensitive mutations, are particularly useful for the study of essential gene functions. Such mutations are more likely to arise from single base-pair changes than from larger deletions or insertions. Geneticists also use transposons to create mutations. If a transposon, such as a *Drosophila P* element, inserts into a gene's coding or regulatory regions, it can disrupt the gene's function. Transposons and their mutations are discussed in more detail in the next section.

More recently, genomics and reverse genetic techniques have expanded the methods available for studying mutations, gene function, and human diseases. Once a gene has been identified and cloned, sequence analysis of mutant and normal genes from affected and unaffected individuals may reveal the molecular basis of the disease or mutant phenotype. In addition, knowledge of the mutant gene sequence may open the way to developing specific genetic tests and gene-based therapeutics.

Screening and Selecting for Mutations

The next step in a genetic analysis using model organisms involves detecting those individuals, within the mutagenized population, who display mutations in the gene or genes affecting the phenotype of interest. The most frequently used method to detect mutants is a **genetic screen**. A genetic screen may involve the visual or biochemical examination of large numbers of mutagenized organisms. For example, Mendel formed the scientific basis for transmission genetics by screening thousands of individual pea plants for visible phenotypic features. Similarly, scientists used modern biochemical assays of the proteins in maize endosperm, in order to detect the *opaque-2* mutant strain. This strain contains high levels of lysine, which improves the nutritional value of the plant.

It is easy to see how dominant mutations can be detected during a genetic screen, in either haploid or diploid organisms. As long as the dominant mutation is not lethal at an early stage of development, the phenotype will be visible immediately in any organism that bears a dominant mutation in the relevant gene. The mutant can then be crossed, and heterozygous or homozygous stocks can be maintained.

Most mutations, however, result in a loss of gene function, and most of these loss-of-function mutations are recessive. Organisms that have a haploid phase in their life cycle, such as yeast, have a significant advantage for recessive mutant detection, as the mutant phenotype will be immediately evident in the mutated haploid individual. The mutation can then be crossed and analyzed in diploid phases of the life cycle.

Diploid organisms that are heterozygous for a recessive allele will not show the mutant phenotype. To deal with this situation, geneticists have devised some intricate strategies to detect and recover recessive mutations in diploid organisms. These methods involve crossing mutagenized individuals with special strains bearing a number of visible and lethal gene markers or chromosomes containing specific deletions. After three or more generations, the recessive lethal mutation may be revealed, but the mutant strain can be maintained in a heterozygous state. Although these techniques are cumbersome to perform, they are fairly efficient. Many of the developmental mutations in *Drosophila* that are discussed in Chapter 18 were identified using these techniques.

15.8

Transposable Elements Move within the Genome and May Create Mutations

Transposable elements, also known as **transposons** or "jumping genes," can move or transpose within and between chromosomes, inserting themselves into various locations within the genome.

Transposons are present in the genomes of all organisms from bacteria to humans. Not only are they ubiquitous, but they also comprise large portions of some eukaryotic genomes. For example, almost 50 percent of the human genome is derived from transposable elements. Some organisms with unusually large genomes, such as salamanders and barley, contain hundreds of thousands of copies of various types of transposable elements. Although the function of these elements is unknown, data from human genome sequencing suggest that some genes may have evolved from transposons and that transposons may help to modify and reshape the genome. Transposable elements are also valuable tools in genetic research. Geneticists harness transposons as mutagens, as cloning tags, and as vehicles for introducing foreign DNA into model organisms.

In this section, we discuss transposable elements as naturally occurring mutagens. The movement of transposons from one place in the genome to another has the capacity to disrupt genes and cause mutations, as well as to create chromosomal damage such as double-strand breaks.

Insertion Sequences and Bacterial Transposons

There are two types of transposable elements in bacteria: insertion sequences and bacterial transposons. **Insertion sequences (IS elements)** can move from one location to another and, if they insert into a gene or gene-regulatory region, may cause mutations.

IS elements were first identified during analyses of mutations in the *gal* operon of *E. coli*. Researchers discovered that certain mutations in this operon were due to the presence of several hundred base pairs of extra DNA inserted into the beginning of the operon. Surprisingly, the segment of mutagenic DNA could spontaneously excise from this location, restoring wild-type function to the *gal* operon. Subsequent research revealed that several other DNA elements could behave in a similar fashion, inserting into bacterial chromosomes and affecting gene function.

IS elements are relatively short, not exceeding 2000 bp (2 kb). The first insertion sequence to be characterized in *E. coli*, IS1, is about 800 bp long. Other IS elements such as IS2, 3, 4, and 5 are about 1250 to 1400 bp in length. IS elements are present in multiple copies in bacterial genomes. For example, the *E. coli* chromosome contains five to eight copies of IS1, five copies each of IS2 and IS3, as well as copies of IS elements on plasmids such as F factors.

All IS elements contain two features that are essential for their movement. First, they contain a gene that encodes an enzyme called **transposase**. This enzyme is responsible for making staggered cuts in chromosomal DNA, into which the IS element can insert. Second, the ends of IS elements contain **inverted terminal repeats (ITRs)**. ITRs are short segments of DNA that have the same nucleotide sequence as each other but are oriented in the opposite direction (Figure 15–18). Although Figure 15–18 shows the ITRs to consist of only a few nucleotides, IS ITRs usually contain about 20 to 40 nucleotide pairs. ITRs are essential for transposition and act as recognition sites for the binding of the transposase enzyme.

Bacterial transposons (**Tn elements**) are larger than IS elements and contain protein-coding genes that are unrelated to their transposition. Some Tn elements, such as Tn10, are comprised of a drug-resistance gene flanked by two IS elements present in opposite orientations. The IS elements encode the transposase enzyme that is necessary for transposition of the Tn element. Other types of Tn elements, such as Tn3, have shorter inverted repeat sequences at their ends and encode their transposase enzyme from a transposase gene located in the middle of the Tn element. Like IS elements, Tn elements are mobile in both bacterial

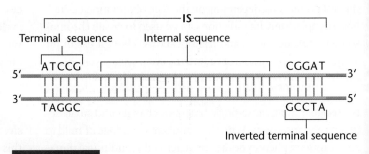

FIGURE 15–18 An insertion sequence (IS), shown in purple. The terminal sequences are perfect inverted repeats of one another.

chromosomes and in plasmids, and can cause mutations if they insert into genes or gene-regulatory regions.

Tn elements are currently of interest because they can introduce multiple drug resistance onto bacterial plasmids. These plasmids, called **R factors**, may contain many Tn elements conferring simultaneous resistance to heavy metals, antibiotics, and other drugs. These elements can move from plasmids onto bacterial chromosomes and can spread multiple drug resistance between different strains of bacteria.

The *Ac–Ds* System in Maize

About 20 years before the discovery of transposons in bacteria, Barbara McClintock discovered mobile genetic elements in corn plants (maize). She did this by analyzing the genetic behavior of two mutations, *Dissociation (Ds)* and *Activator (Ac)*, expressed in either the endosperm or aleurone layers. She then correlated her genetic observations with cytological examinations of the maize chromosomes. Initially, McClintock determined that *Ds* was located on chromosome 9. If *Ac* was also present in the genome, *Ds* induced breakage at a point on the chromosome adjacent to its own location. If chromosome breakage occurred in somatic cells during their development, progeny cells often lost part of the broken chromosome, causing a variety of phenotypic effects.

Subsequent analysis suggested to McClintock that both *Ds* and *Ac* elements sometimes moved to new chromosomal locations. While *Ds* moved only if *Ac* was also present, *Ac* was capable of autonomous movement. Where *Ds* came to reside determined its genetic effects—that is, it might cause chromosome breakage, or it might inhibit expression of a certain gene. In cells in which *Ds* caused a gene mutation, *Ds* might move again, restoring the gene mutation to wild type.

Figure 15–19 illustrates the types of movements and effects brought about by *Ds* and *Ac* elements. In McClintock's original observation, pigment synthesis was restored in cells in which the *Ds* element jumped out of chromosome 9. McClintock concluded that the *Ds* and *Ac* genes were **mobile controlling elements**. We now commonly refer to them as transposable elements, a term coined by another great maize geneticist, Alexander Brink.

Several *Ac* and *Ds* elements have now been analyzed, and the relationship between the two elements has been clarified. The first *Ds* element studied (*Ds9*) is nearly identical to *Ac* except for a 194-bp deletion within the transposase gene. The deletion of part of the transposase gene in the *Ds9* element explains its dependence on the *Ac* element for transposition. Several other *Ds* elements have also been sequenced, and each contains an even larger deletion within the transposase gene. In each case, however, the ITRs are retained.

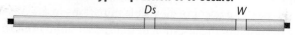

(a) In absence of *Ac*, *Ds* is not transposable.

Wild-type expression of *W* occurs.

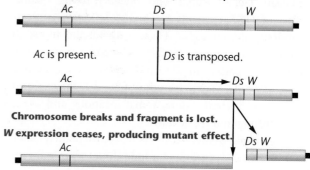

(b) When *Ac* is present, *DS* may be transposed.

Ac is present.

Ds is transposed.

Chromosome breaks and fragment is lost.
W expression ceases, producing mutant effect.

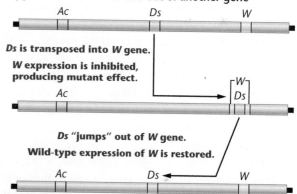

(c) *DS* can move into and out of another gene

Ds is transposed into *W* gene.

W expression is inhibited, producing mutant effect.

Ds "jumps" out of *W* gene.

Wild-type expression of *W* is restored.

FIGURE 15–19 Effects of *Ac* and *Ds* elements on gene expression. (a) If *Ds* is present in the absence of *Ac*, there is normal expression of a distantly located hypothetical gene *W*. (b) In the presence of *Ac*, *Ds* may transpose to a region adjacent to *W*. *Ds* can induce chromosome breakage, which may lead to loss of a chromosome fragment bearing the *W* gene. (c) In the presence of *Ac*, *Ds* may transpose into the *W* gene, disrupting *W*-gene expression. If *Ds* subsequently transposes out of the *W* gene, *W*-gene expression may return to normal.

Although the significance of Barbara McClintock's mobile controlling elements was not fully appreciated following her initial observations, molecular analysis has since verified her conclusions. She was awarded the Nobel Prize in Physiology or Medicine in 1983.

Copia and *P* Elements in *Drosophila*

There are more than 30 families of transposable elements in *Drosophila*, each of which is present in 20 to 50 copies in the genome. Together, these families constitute about

5 percent of the *Drosophila* genome and over half of the middle repetitive DNA of this organism. One study suggests that 50 percent of all visible mutations in *Drosophila* are the result of the insertion of transposons into otherwise wild-type genes.

In 1975, David Hogness and his colleagues David Finnegan, Gerald Rubin, and Michael Young identified a class of DNA elements in *Drosophila melanogaster* that they designated as ***copia***. These elements are transcribed into "copious" amounts of RNA (hence their name). *Copia* elements are present in 10 to 100 copies in the genomes of *Drosophila* cells. Mapping studies show that they are transposable to different chromosomal locations and are dispersed throughout the genome.

Each *copia* element consists of approximately 5000 to 8000 bp of DNA, including a long **direct terminal repeat (DTR)** sequence of 267 bp at each end. Within each DTR is an inverted terminal repeat (ITR) of 17 bp (Figure 15–20). The short ITR sequences are characteristic of *copia* elements. The DTR sequences are found in other transposons in other organisms, but they are not universal.

Insertion of *copia* is dependent on the presence of the ITR sequences and seems to occur preferentially at specific target sites in the genome. The *copia*-like elements demonstrate regulatory effects at the point of their insertion in the chromosome. Certain mutations, including those affecting eye color and segment formation, are due to *copia* insertions within genes. For example, the eye-color mutation *white-apricot*, is caused by an allele of the *white* gene, which contains a *copia* element within the gene. Transposition of the *copia* element out of the *white-apricot* allele can restore the allele to wild type.

Perhaps the most significant *Drosophila* transposable elements are the ***P elements***. These were discovered while studying the phenomenon of **hybrid dysgenesis**, a condition characterized by sterility, elevated mutation rates, and chromosome rearrangements in the offspring of crosses between certain strains of fruit flies. Hybrid dysgenesis is caused by high rates of *P* element transposition in the germ line, in which transposons insert themselves into or near genes, thereby causing mutations. *P* elements range from 0.5 to 2.9 kb long, with 31-bp ITRs. Full-length *P* elements encode at least two proteins, one of which is the transposase enzyme that is required for transposition, and another is a repressor protein that inhibits transposition. The transposase gene is expressed only in the germ line, accounting for the tissue specificity of *P* element transposition. Strains of flies that contain full-length *P* elements inserted into their genomes are resistant to further transpositions due to the presence of the repressor protein encoded by the *P* elements.

Mutations can arise from several kinds of insertional events. If a *P* element inserts into the coding region of a gene, it can terminate transcription of the gene and destroy normal gene expression. If it inserts into the promoter region of a gene, it can affect the level of expression of the gene. Insertions into introns can affect splicing or cause the premature termination of transcription.

Geneticists have harnessed *P* elements as tools for genetic analysis. One of the most useful applications of *P* elements is as vectors to introduce transgenes into *Drosophila*—a technique known as **germ-line transformation**. *P* elements are also used to generate mutations and to clone mutant genes. In addition, researchers are perfecting methods to target *P* element insertions to precise single-chromosomal sites, which should increase the precision of germ-line transformation in the analysis of gene activity.

Transposable Elements in Humans

The human genome, like that of other eukaryotes, is riddled with DNA derived from transposons. Recent genomic sequencing data reveal that approximately half of the human genome is comprised of transposable element DNA. As discussed in Chapter 12, the major families of human transposons are the long interspersed elements and short interspersed elements (**LINEs and SINEs**, respectively). Together, they comprise over 30 percent of the human genome. Other families of transposable elements account for a further 11 percent (Table 15.5). As coding

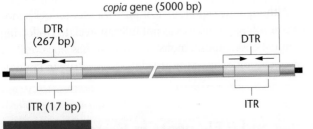

FIGURE 15–20   Structural organization of a copia transposable element in *Drosophila melanogaster,* showing the terminal repeats.

TABLE 15.5

Transposable Elements in the Human Genome

Element Type	Length	Copies in Genome	% of Genome
LINEs	1–6 kb	850,000	21
SINEs	100–500 bp	1,500,000	13
LTR Elements	<5 kb	443,000	8
DNA Elements	80–300 bp	294,000	3
Unclassified	–	3,000	0.1

sequences comprise only about 1 percent of the human genome, there is about 40 to 50 times more transposable element DNA in the human genome than DNA in functional genes.

Although most human transposons appear to be inactive, the potential mobility and mutagenic effects of transposable elements have far-reaching implications for human genetics, as can be seen in a recent example of a transposon "caught in the act." The case involves a male child with hemophilia. One cause of hemophilia is a defect in blood-clotting factor VIII, the product of an X-linked gene. Haig Kazazian and his colleagues found LINEs inserted at two points within the gene. Researchers were interested in determining if one of the mother's X chromosomes also contained this specific LINE. If so, the unaffected mother would be heterozygous and pass the LINE-containing chromosome to her son. The surprising finding was that the LINE sequence was *not* present on either of her X chromosomes but *was* detected on chromosome 22 of both parents. This suggests that this mobile element may have transposed from one chromosome to another in the gamete-forming cells of the mother, prior to being transmitted to the son.

LINE insertions into the human *dystrophin* gene have resulted in at least two separate cases of Duchenne muscular dystrophy. In one case, a transposon inserted into exon 48, and in another case, a transposon inserted into exon 44, both leading to frameshift mutations and premature termination of translation of the dystrophin protein. There are also reports that LINEs have inserted into the *APC* and *c-myc* genes, leading to mutations that may have contributed to the development of some colon and breast cancers. In the latter cases, the transposition had occurred within one or a few somatic cells. As of 2009, researchers have determined that cases of at least 11 human diseases are due to insertions of LINE elements.

SINE insertions are also responsible for more than 30 cases of human disease. In one case, an **Alu element** integrated into the *BRCA2* gene, inactivating this tumor suppressor gene and leading to a familial case of breast cancer. Other genes that have been mutated by *Alu* integrations are the *factor IX* gene (leading to hemophilia B), the *ChE* gene (leading to acholinesterasemia), and the *NF1* gene (leading to neurofibromatosis).

Transposons, Mutations, and Evolution

Transposons can have a wide range of effects on genes. The insertion of a transposon into the coding region of a gene may disrupt the gene's normal translation reading frame or may induce premature termination of translation of the mRNA transcribed from the gene. Many transposons contain their own promoters and enhancers, as well as splice sites and polyadenylation signals. The presence of these transposon regulatory sequences can have effects on nearby genes. The insertion of a transposon containing polyadenylation or transcription termination signals into a gene's intron may bring about termination of the gene's transcription within the transposon. In addition, it can cause aberrant splicing of an RNA transcribed from the gene. Insertions of a transposon into a gene's transcription regulatory region may disrupt the gene's normal regulation or may cause the gene to be expressed differently as a result of the presence of the transposon's own promoter or enhancer sequences. The presence of two or more identical transposons in a genome creates the potential for recombination between the transposons, leading to duplications, deletions, inversions, or chromosome translocations. Any of these rearrangements may bring about phenotypic changes or disease.

New germ-line transpositions are estimated to occur once in every 50 to 100 human births. Most of these do not cause disease or a change in phenotype; however, it is thought that about 0.2 percent of detectable human mutations may be due to transposon insertions. Other organisms appear to suffer more damage due to transposition. About 10 percent of new mouse mutations and 50 percent of *Drosophila* mutations are caused by insertions of transposons in or near genes.

Because of their ability to alter genes and chromosomes, transposons may contribute to the variability that underlies evolution. For example, the Tn elements of bacteria carry antibiotic resistance genes between organisms, conferring a survival advantage to the bacteria under certain conditions. Another example of a transposon's contribution to evolution is provided by *Drosophila* telomeres. LINE-like elements are present at the ends of *Drosophila* chromosomes, and these elements act as telomeres, maintaining the length of *Drosophila* chromosomes over successive cell divisions. Other examples of evolved transposons are the *RAG1* and *RAG2* genes in humans. These genes encode **recombinase** enzymes that are essential to the development of the immune system. These two genes appear to have evolved from transposons.

Transposons may also affect the evolution of genomes by altering gene-expression patterns in ways that are subsequently retained by the host. For example, the human *amylase* gene contains an enhancer that causes the gene to be expressed in the parotid gland. This enhancer evolved from transposon sequences that were inserted into the gene-regulatory region early in primate evolution. Other examples of gene-expression patterns that were affected by the presence of transposon sequences are T-cell-specific expression of the *CD8* gene and placenta-specific expression of the *leptin* and *CYP19* genes.

Sequence Alignment to Identify a Mutation

MG *Study Area: Exploring Genomics*

In this chapter, we examined the causes of different types of mutations and how mutations affect phenotype by altering the structure and function of proteins. The emergence of genomics, bioinformatics, and proteomics as key disciplines in modern genetics has provided geneticists with an unprecedented set of tools for identifying and analyzing mutations in gene and protein sequences.

In this exercise we will use the **ExPASy (Expert Protein Analysis System)** site, which is hosted by the Swiss Institute for Bioinformatics and provides a wealth of resources for studying proteins. Here we will use an ExPASy program called SIM (for "similarity" in sequence) to compare two polypeptide sequences so as to pinpoint a mutation. Once the mutation has been identified, you will learn more about the gene encoding these polypeptides and about a human disease condition associated with this gene.

Exercise I — Identifying a Missense Mutation Affecting a Protein in Humans

1. Begin this exercise by accessing the ExPASy site at http://www.expasy.ch/tools/sim-prot.html. The SIM feature is an algorithm-based software program that allows us to compare multiple polypeptide sequences by looking for amino acid similarity in the sequences.

2. Following are amino acid sequences for polypeptides expressed in two different people.

Person A

MGAPACALALCVAVAIVAGASSESLGTEQ
RVVGRAAEVPGPEPGQQEQLVFGSGDAV
ELSCPPPGGGPMGPTVVVKDGTGLVPSE
RVLVGPQRLQVLNASHEDSGAYSCRQRLT
QRVLCHFSVRVTDAPSSGDDEDGEDEA
EDTGVDTGAPYWTRPERMDKKLLAVPA
ANTVRFRCPAAGNPTPSISWLKNGREFR
GEHRIGGIKLRHQQWSLVMESVVPSDRG
NYTCVVENKFGSIRQTYTLDVLERS
PHRPILQAGLPANQTAVLGSDVEFHC
KVYSDAQPHIQWLKHVEVNGSKVG
PDGTPYVTVLKTAGANTTDKELEVLSLH
NVTFEDAGEYTCLAGNSIGFSHHSAWLVV
LPAEEELVEADEAGSVYAGILSYGVGFFL
FILVVAAVTLCRLRSPPKKGLGSPTVHK
ISRFPLKRQVSLESNASMSSNTPLVRIARL
SSGEGPTLANVSELELPADPKWELSRARL
TLGKPLGEGCFGQVVMAEAIGIDKDRAA
KPVTVAVKMLKDDATDKDLSDLVSEMEM
MKMIGKHKNIINLLGACTQGGPLYVLVEY
AAKGNLREFLRARRPPGLDYSFDTCKPPE
EQLTFKDLVSCAYQVARGMEYLASQKCI
HRDLAARNVLVTEDNVMKIADFGLARD
VHNLDYYKKTTNGRLPVKWMAPEALFD
RVYTHQSDVWSFGVLLWEIFTLGGSPYPG
IPVEELFKLLKEGHRMDKPANCTHDLYMI
MRECWHAAPSQRPTFKQLVEDLDRVLT
VTSTDEYLDLSAPFEQYSPGGQDTPSSSS
GDDSVFAHDLLPPAPPSSGGSRT

Person B

MGAPACALALCVAVAIVAGASSESLGTEQ
RVVGRAAEVPGPEPGQQEQLVFGSGDAV
ELSCPPPGGGPMGPTVVVKDGTGLVPSE
RVLVGPQRLQVLNASHEDSGAYSCRQRLT
QRVLCHFSVRVTDAPSSGDDEDGEDEAE
DTGVDTGAPYWTRPERMDKKLLAVPAA
NTVRFRCPAAGNPTPSISWLKNGREFRGE
HRIGGIKLRHQQWSLVMESVVPSDRGNY
TCVVENKFGSIRQTYTLDVLERSPHRPILQ
AGLPANQTAVLGSDVEFHCKVYSDAQPHI
QWLKHVEVNGSKVGPDGTPYVTVLKTA
GANTTDKELEVLSLHNVTFEDAGEYTCL
AGNSIGFSHHSAWLVVLPAEEELVEADEA
GSVYAGILSYRVGFFLFILVVAAVTLCRLR
SPPKKGLGSPTVHKISRFPLKRQVSLESNA
SMSSNTPLVRIARLSSGEGPTLANVSELEL
PADPKWELSRARLTLGKPLGEGCFGQVV
MAEAIGIDKDRAAKPVTVAVKMLKDDAT
DKDLSDLVSEMEMMKMIGKHKNIINLL
GACTQGGPLYVLVEYAAKGNLREFLRAR

RPPGLDYSFDTCKPPEEQLTFKDLVSCAY
QVARGMEYLASQKCIHRDLAARNVLVT
EDNVMKIADFGLARDVHNLDYYKKTTN
GRLPVKWMAPEALFDRVYTHQSDVWSF
GVLLWEIFTLGGSPYPGIPVEELFKLLKEG
HRMDKPANCTHDLYMIMRECWHAAPSQ
RPTFKQLVEDLDRVLTVTSTDEYLDLSAP
FEQYSPGGQDTPSSSSSGDDSVFAHDLLP
PAPPSSGGSRT

3. Copy and paste each sequence into the "SEQUENCE" text boxes in SIM. (*Hint:* Access these sequences from the Companion Web site so that you can copy and paste the sequence into SIM.) Use the Person A sequence for sequence 1 and the Person B sequence for sequence 2. Click the "User-entered sequence" button for each. Name the sequences Person A and Person B as appropriate. Submit the sequences for comparison and then answer the following questions:

   (a) How many amino acids are in each polypeptide sequence that was analyzed?

   (b) Look carefully at the alignment results. Can you find any differences in amino acid sequence when comparing these two polypeptides? What did you find?

Exercise II — Identifying the Genetic Basis for a Human Genetic Disease Condition

1. Go to the ExPASy home page and find the BLAST link. Run a protein BLAST (blastp) search to identify which polypeptide you have been studying. Explore the BLAST reports for the top three protein sequences that aligned with your query sequence by clicking on the link for each sequence. Pay particular attention to the "Comment" section of each report to help you answer the questions in the following part.

2. Now that you know what gene you are working with, go to PubMed (http://www.ncbi.nlm.nih.gov/entrez/query.fcgi?db=PubMed) and search for a review article from the authors Vajo, Z., Francomano, C. A., and Wilkin, D. J.

3. Answer the following questions:

   (a) What gene codes for the polypeptides have you been studying?

   (b) What is the function of this protein?

   (c) Based on what you learned from the alignment results you analyzed in Exercise I, the BLAST reports, and your PubMed search, what human disease is caused by the mutation you identified in Exercise I? Explain your answer and briefly describe phenotypes associated with the disease.

CASE STUDY | Genetic dwarfism

Seven months pregnant, an expectant mother was undergoing a routine ultrasound. While prior tests had been normal, this one showed that the limbs of the fetus were unusually short. The doctor suspected that the baby might have a genetic form of dwarfism called achondroplasia. He told her that the disorder was due to an autosomal dominant mutation and occurred with a frequency of about 1 in 25,000 births. The expectant mother had studied genetics in college and immediately raised several questions. How would you answer them?

1. How could her baby have a dominantly inherited disorder if there was no history of this condition on either side of the family?
2. Is the mutation more likely to have come from the mother or the father?
3. If this child has achondroplasia, would the chances increase that their next child would also have this disorder?
4. Could this disorder have been caused by X rays or ultrasounds she had earlier in pregnancy?

Summary Points

 For activities, animations, and review quizzes, go to the study area at www.masteringgenetics.com

1. Mutations can be spontaneous or induced, somatic or germ-line, autosomal or sex-linked. Mutations can have many different effects on gene function depending on the type of nucleotide changes that comprise the mutation and the location of those mutations.
2. Spontaneous mutations occur in many ways, ranging from errors during DNA replication to damage caused to DNA bases as a result of normal cellular metabolism.
3. Mutations can be induced by many types of chemicals and radiation. These agents can damage both bases and the sugar-phosphate backbone of DNA molecules.
4. Single-gene mutations in humans cause a wide range of diseases. These mutations may be base-pair changes, insertions, deletions, and expanded DNA repeats.

5. Organisms counteract mutations using DNA repair systems including proofreading, mismatch repair, postreplication repair, photoreactivation repair, SOS repair, base excision repair, nucleotide excision repair, and double-strand break repair.
6. The Ames test allows scientists to estimate the mutagenicity and cancer-causing potential of chemical agents.
7. Geneticists use a range of genetic and biochemical techniques to induce and analyze mutations in model organisms.
8. Transposable elements can move within a genome, creating mutations and altering gene expression. In addition, transposons may contribute to evolution.

INSIGHTS AND SOLUTIONS

1. The base analog 2-amino purine (2-AP) substitutes for adenine during DNA replication, but it may base-pair with cytosine. The base analog 5-bromouracil (5-BU) substitutes for thymidine, but it may base-pair with guanine. Follow the double-stranded trinucleotide sequence shown here through three rounds of replication, assuming that, in the first round, both analogs are present and become incorporated wherever possible. Before the second and third round of replication, any unincorporated base analogs are removed. What final sequences occur?

Solution:

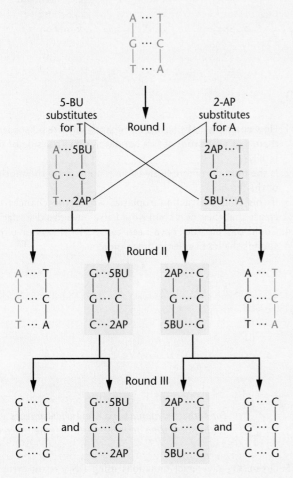

2. A rare dominant mutation expressed at birth was studied in humans. Records showed that six cases were discovered in 40,000 live births. Family histories revealed that in two cases, the mutation was already present in one of the parents. Calculate the spontaneous mutation rate for this mutation. What are some underlying assumptions that may affect our conclusions?

Solution: Only four cases represent a new mutation. Because each live birth represents two gametes, the sample size is from 80,000 meiotic events. The rate is equal to

$$4/80,000 = 1/20,000 = 5 \times 10^{-5}$$

We have assumed that the mutant gene is fully penetrant and is expressed in each individual bearing it. If it is not fully penetrant, our calculation may be an underestimate because one or more mutations may have gone undetected. We have also assumed that the screening was 100 percent accurate. One or more mutant individuals may have been "missed," again leading to an underestimate. Finally, we assumed that the viability of the mutant and nonmutant individuals is equivalent and that they survive equally *in utero*. Therefore, our assumption is that the number of mutant individuals at birth is equal to the number at conception. If this were not true, our calculation would again be an underestimate.

3. Consider the following estimates:

(a) There are 5.5×10^9 humans living on this planet.

(b) Each individual has about 30,000 (0.3×10^5) genes.

(c) The average mutation rate at each locus is 10^{-5}.

How many spontaneous mutations are currently present in the human population? Assuming that these mutations are equally distributed among all genes, how many new mutations have arisen in each gene in the human population?

Solution: First, since each individual is diploid, there are two copies of each gene per person, each arising from a separate gamete. Therefore, the total number of spontaneous mutations is

$$(2 \times 0.3 \times 10^5 \text{ genes/individual})$$
$$\times (5.5 \times 10^9) \text{ individuals} \times (10^{-5} \text{ mutations/gene})$$
$$= (0.6 \times 10^5) \times (5.5 \times 10^9) \times (10^{-5}) \text{ mutations}$$
$$= 3.3 \times 10^9 \text{ mutations in the population.}$$
$$3.3 \times 10^9 \text{ mutations}/0.3 \times 10^5 \text{ genes}$$
$$= 11 \times 10^4 \text{ mutations per gene in the population.}$$

Problems and Discussion Questions

 For instructor-assigned tutorials and problems, go to www.masteringgentics.com

HOW DO WE KNOW?

1. In this chapter, we focused on how gene mutations arise and how cells repair DNA damage. At the same time, we found opportunities to consider the methods and reasoning by which much of this information was acquired. From the explanations given in the chapter,
   (a) How do we know that mutations occur spontaneously?
   (b) How do we know that certain chemicals and wavelengths of radiation induce mutations in DNA?

   (c) How do we know that DNA repair mechanisms detect and correct the majority of spontaneous and induced mutations?
2. Discuss the importance of mutations in genetic studies.
3. What is a spontaneous mutation, and why are spontaneous mutations rare?
4. Why would a mutation in a somatic cell of a multicellular organism escape detection?

5. Most mutations are thought to be deleterious. Why, then, is it reasonable to state that mutations are essential to the evolutionary process?

6. Why is a random mutation more likely to be deleterious than beneficial?

7. Most mutations in a diploid organism are recessive. Why?

8. What is meant by a conditional mutation?

9. Describe a tautomeric shift and how it may lead to a mutation.

10. Contrast and compare the mutagenic effects of deaminating agents, alkylating agents, and base analogs.

11. Acridine dyes induce frameshift mutations. Why are frameshift mutations likely to be more detrimental than point mutations, in which a single pyrimidine or purine has been substituted?

12. Why are X rays more potent mutagens than UV radiation?

13. DNA damage brought on by a variety of natural and artificial agents elicits a wide variety of cellular responses involving numerous signaling pathways. In addition to the activation of DNA repair mechanisms, there can be activation of pathways leading to apoptosis (programmed cell death) and cell-cycle arrest. Why would apoptosis and cell-cycle arrest often be part of a cellular response to DNA damage?

14. Contrast the various types of DNA repair mechanisms known to counteract the effects of UV radiation. What is the role of visible light in repairing UV-induced mutations?

15. Mammography is an accurate screening technique for the early detection of breast cancer in humans. Because this technique uses X rays diagnostically, it has been highly controversial. Can you explain why? What reasons justify the use of X rays for such a medical screening technique?

16. A significant number of mutations in the *HBB* gene that cause human β-thalassemia occur within introns or in upstream noncoding sequences. Explain why mutations in these regions often lead to severe disease, although they may not directly alter the coding regions of the gene.

17. Describe how the Ames test screens for potential environmental mutagens. Why is it thought that a compound that tests positively in the Ames test may also be carcinogenic?

18. What genetic defects result in the disorder xeroderma pigmen-

tosum (XP) in humans? How do these defects create the phenotypes associated with the disorder?

19. Compare several transposable elements in bacteria, maize, *Drosophila*, and humans. What properties do they share?

20. Speculate on how improved living conditions and medical care in the developed nations might affect human mutation rates, both neutral and deleterious.

21. In maize, a *Ds* or *Ac* transposon can cause mutations in genes at or near the site of transposon insertion. It is possible for these elements to transpose away from their original site, causing a reversion of the mutant phenotype. In some cases, however, even more severe phenotypes appear, due to events at or near the mutant allele. What might be happening to the transposon or the nearby gene to create more severe mutations?

22. It is estimated that about 0.2 percent of human mutations are due to transposon insertions and a much higher degree of mutational damage is known to occur in some other organisms. In what way might transposons contribute positively to evolution?

23. In a bacterial culture in which all cells are unable to synthesize leucine (*leu⁻*), a potent mutagen is added, and the cells are allowed to undergo one round of replication. At that point, samples are taken, a series of dilutions is made, and the cells are plated on either minimal medium or minimal medium containing leucine. The first culture condition (minimal medium) allows the growth of only *leu⁺* cells, while the second culture condition (minimal medium with leucine added) allows growth of all cells. The results of the experiment are as follows:

Culture Condition	Dilution	Colonies
Minimal medium	10^{-1}	18
Minimal medium + leucine	10^{-7}	6

What is the rate of mutation at the locus associated with leucine biosynthesis?

Extra-Spicy Problems

 For instructor-assigned tutorials and problems, go to www.masteringgenetics.com

24. Presented here are hypothetical findings from studies of heterokaryons formed from seven human xeroderma pigmentosum cell strains:

	XP1	XP2	XP3	XP4	XP5	XP6	XP7
XP1	—						
XP2	—	—					
XP3	—	—	—				
XP4	+	+	+	—			
XP5	+	+	+	+	—		
XP6	+	+	+	+	—	—	
XP7	+	+	+	+	—	—	—

Note: "+" = complementation; "—" = no complementation

These data are measurements of the occurrence or nonoccurrence of unscheduled DNA synthesis in the fused heterokaryon.

None of the strains alone shows any unscheduled DNA synthesis. Which strains fall into the same complementation groups? How many different groups are revealed based on these data? What can we conclude about the genetic basis of XP from these data?

25. Imagine yourself as one of the team of geneticists who launches a study of the genetic effects of high-energy radiation on the surviving Japanese population immediately following the atom bomb attacks at Hiroshima and Nagasaki in 1945. Demonstrate your insights into both chromosomal and gene mutation by outlining a short-term and long-term study that addresses these radiation effects. Be sure to include strategies for considering the effects on both somatic and germ-line tissues.

26. With the knowledge that radiation causes mutations, many assume that human-made forms of radiation are the major contributors to the mutational load in humans. What evidence suggests otherwise?

27. Human equivalents of bacterial DNA mismatch repair proteins are subject to mutational damage just as are other proteins. What evidence indicates that mutations in human DNA mismatch repair genes are related to certain forms of cancer?

28. Among Betazoids in the world of *Star Trek®*, the ability to read minds is under the control of a gene called *mindreader* (abbreviated *mr*). Most Betazoids can read minds, but rare recessive mutations in the *mr* gene result in two alternative phenotypes: *delayed-receivers* and *insensitives*. Delayed-receivers have some mind-reading ability but perform the task much more slowly than normal Betazoids. Insensitives cannot read minds at all. Betazoid genes do not have introns, so the gene only contains coding DNA. It is 3332 nucleotides in length, and Betazoids use a four-letter genetic code.

The following table shows some data from five unrelated *mr* mutations.

Mutation	Description of Mutation	Phenotype
mr-1	Nonsense mutation in codon 829	Delayed-receiver
mr-2	Missense mutation in codon 52	Delayed-receiver
mr-3	Deletion of nucleotides 83–150	Delayed-receiver
mr-4	Missense mutation in codon 192	Insensitive
mr-5	Deletion of nucleotides 83–93	Insensitive

For each mutation, provide a plausible explanation for why it gives rise to its associated phenotype and not to the other phenotype. For example, hypothesize why the *mr-1* nonsense mutation in codon 829 gives rise to the milder delayed-receiver phenotype rather than the more severe insensitive phenotype. Then repeat this type of analysis for the other mutations. (More than one explanation is possible, so be creative within *plausible* bounds!)

29. Skin cancer carries a lifetime risk nearly equal to that of all other cancers combined. Following is a graph (modified from Kraemer, 1997. *Proc. Natl. Acad. Sci. (USA)* 94: 11–14) depicting the age of onset of skin cancers in patients with or without XP, where the cumulative percentage of skin cancer is plotted against age. The non-XP curve is based on 29,757 cancers surveyed by the National Cancer Institute, and the curve representing those with XP is based on 63 skin cancers from the Xeroderma Pigmentosum Registry.

(a) Provide an overview of the information contained in the graph.

(b) Explain why individuals with XP show such an early age of onset.

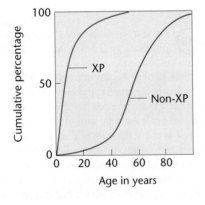

30. The initial discovery of IS elements in bacteria revealed the presence of an element upstream (5′) of three genes controlling galactose metabolism. All three genes were affected simultaneously, although there was only one IS insertion. Offer an explanation as to why this might occur.

31. It has been noted that most transposons in humans and other organisms are located in noncoding regions of the genome—regions such as introns, pseudogenes, and stretches of particular types of repetitive DNA. There are several ways to interpret this observation. Describe two possible interpretations. Which interpretation do you favor? Why?

32. Mutations in the *IL2RG* gene cause approximately 30 percent of severe combined immunodeficiency disorder (SCID) cases. These mutations result in alterations to a protein component of cytokine receptors that are essential for proper development of the immune system. The *IL2RG* gene is composed of eight exons and contains upstream and downstream sequences that are necessary for proper transcription and translation. Below are some of the mutations observed. For each, explain its likely influence on the *IL2RG* gene product (assume its length to be 375 amino acids).

(a) Nonsense mutation in coding regions

(b) Insertion in Exon 1, causing frameshift

(c) Insertion in Exon 7, causing frameshift

(d) Missense mutation

(e) Deletion in Exon 2, causing frameshift

(f) Deletion in Exon 2, in frame

(g) Large deletion covering Exons 2 and 3

33. A variety of neural and muscular disorders are associated with expansions of trinucleotide repeat sequences. The table below lists several disorders, the repeat motifs, their locations, the normal number of repeats, and the number of repeats in the full mutations.

(a) Most disorders attributable to trinucleotide repeats result from expansion of the repeats. Two mechanisms are often proposed to explain repeat expansion: (1) unequal synapsis and crossing over and (2) errors in DNA replication where single-stranded, base-paired loops are formed that conflict with linear replication. Present a simple sketch of each mechanism.

(b) Notice that some of the repeats occur in areas of the gene that are not translated. How can a mutation occur if the alteration is not reflected in an altered amino acid sequence?

(c) In the two cases where the repeat expansions occur in exons, the extent of expansion is considerably less than when the expansion occurs outside exons. Present an explanation for this observation.

Disorder	Repeat	Location	Normal Number	Full Mutation
Fragile X	CCG	5′ untranslated region	6–230	>230
Huntington	CAG	Exon	6–35	36–120
Myotonic dystrophy	CTG	3′ untranslated region	5–37	37–1500
Oculopharyngeal muscular dystrophy	GCG	Exon	6	8–13
Friedreich ataxia	AAG	Intron	20	200–900

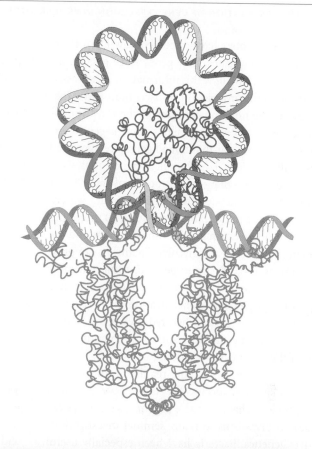

Model showing how the *lac* repressor (red) and catabolite-activating protein (dark blue in center of DNA loop) bind to the *lac* operon promoter, creating a 93-base-pair repression loop in the *lac* regulatory DNA.

16

Regulation of Gene Expression in Prokaryotes

CHAPTER CONCEPTS

- In bacteria, regulation of gene expression is often linked to the metabolic needs of the cell.

- Efficient expression of genetic information in bacteria is dependent on intricate regulatory mechanisms that exert control over transcription.

- Mechanisms that regulate transcription are categorized as exerting either positive or negative control of gene expression.

- Prokaryotic genes that encode proteins with related functions tend to be organized in clusters and are often under coordinated control. Such clusters, including their adjacent regulatory sequences, are called operons.

- Transcription of genes within operons is either inducible or repressible.

- Often, the metabolic end product of a biosynthetic pathway induces or represses gene expression in that pathway.

Previous chapters have discussed how DNA is organized into genes, how genes store genetic information, and how this information is expressed through the processes of transcription and translation. We now consider one of the most fundamental questions in molecular genetics: *How is genetic expression regulated?* It is clear that not all genes are expressed at all times in all situations. For example, detailed analysis of proteins in *E. coli* shows that concentrations of the 4000 or so polypeptide chains encoded by the genome vary widely. Some proteins may be present in as few as 5 to 10 molecules per cell, whereas others, such as ribosomal proteins and the many proteins involved in the glycolytic pathway, are present in as many as 100,000 copies per cell. Although most prokaryotic gene products are present continuously at a basal level (a few copies), the concentration of these products can increase dramatically when required. Clearly, fundamental regulatory mechanisms must exist to control the expression of the genetic information.

In this chapter, we will explore regulation of gene expression in prokaryotes specifically. As we have seen in a number of previous chapters, these organisms served as excellent research organisms in many seminal investigations in molecular genetics. Bacteria have been especially useful research organisms in genetics for a number of reasons. For one thing, they have extremely short reproductive cycles. Literally hundreds of generations, giving rise to billions of genetically identical bacteria, can be produced in overnight cultures. In addition, they can be studied in "pure culture," allowing mutant strains of genetically unique bacteria to be isolated and investigated separately.

Relevant to our current topic, bacteria also serve as an excellent model system for studies involving the induction of genetic transcription in response to changes in environmental conditions. Our focus will be on regulation at the level of the gene. Keep in mind that posttranscriptional regulation also occurs in bacteria. However, we will defer discussion of this level of regulation to Chapter 17, in which we consider eukaryotic regulation as well.

16.1
Prokaryotes Regulate Gene Expression in Response to Environmental Conditions

Regulation of gene expression has been extensively studied in prokaryotes, particularly in *E. coli*. Geneticists have learned that highly efficient genetic mechanisms have evolved in these organisms to turn transcription of specific genes on and off, depending on the cell's metabolic need for the respective gene products. Not only do bacteria respond to changes in their environment, but they also regulate gene activity associated with a variety of nonenvironmentally regulated cellular activities (including the replication, recombination, and repair of their DNA) and with cell division.

The idea that microorganisms regulate the synthesis of their gene products is not a new one. As early as 1900, it was shown that when lactose (a galactose and glucose-containing disaccharide) is present in the growth medium of yeast, the organisms synthesize enzymes required for lactose metabolism. When lactose is absent, the enzymes are not manufactured. Soon thereafter, investigators were able to generalize that bacteria also adapt to their environment, producing certain enzymes only when specific chemical substrates are present. These were thus referred to as **adaptive enzymes** (sometimes also called *facultative*). In contrast, enzymes that are produced continuously, regardless of the chemical makeup of the environment, were called **constitutive enzymes**. Since then, the term *adaptive* has been replaced with the more accurate term **inducible**, reflecting the role of the substrate, which serves as the **inducer** in enzyme production.

More recent investigation has revealed a contrasting system, whereby the presence of a specific molecule inhibits gene expression. Such molecules are usually end products of anabolic biosynthetic pathways. For example, the amino acid tryptophan can be synthesized by bacterial cells. If a sufficient supply of tryptophan is present in the environment or culture medium, then there is no reason for the organism to expend energy in synthesizing the enzymes necessary for tryptophan production. A mechanism has therefore evolved whereby tryptophan plays a role in repressing the transcription of mRNA needed for producing tryptophan-synthesizing enzymes. In contrast to the inducible system controlling lactose metabolism, the system governing tryptophan expression is said to be **repressible**.

Regulation, whether of the inducible or repressible type, may be under either **negative** or **positive control**. Under negative control, genetic expression occurs *unless it is shut off by some form of a regulator molecule*. In contrast, under positive control, transcription occurs *only if a regulator molecule directly stimulates RNA production*. In theory, either type of control or a combination of the two can govern inducible or repressible systems. Our discussion in the ensuing sections of this chapter will help clarify these contrasting systems of regulation. The enzymes involved in lactose digestion and tryptophan synthesis are under negative control.

16.2
Lactose Metabolism in *E. coli* Is Regulated by an Inducible System

Beginning in 1946 with the studies of Jacques Monod and continuing through the next decade with significant contributions by Joshua Lederberg, François Jacob, and

André Lwoff, genetic and biochemical evidence concerning lactose metabolism was amassed. Research provided insights into the way in which the gene activity is repressed when lactose is absent but induced when it is available. In the presence of lactose, the concentration of the enzymes responsible for its metabolism increases rapidly from a few molecules to thousands per cell. The enzymes responsible for lactose metabolism are thus inducible, and lactose serves as the inducer.

In prokaryotes, genes that code for enzymes with related functions (for example, the set of genes involved with lactose metabolism) tend to be organized in clusters on the bacterial chromosome, and transcription of these genes is often under the coordinated control of a single regulatory region. The location of this regulatory region is almost always upstream (5′) of the gene cluster it controls. Because the regulatory region is on the same strand as those genes, we refer to it as a **cis-acting site**. *Cis*-acting regulatory regions bind molecules that control transcription of the gene cluster. Such molecules are called **trans-acting elements**. Events at the regulatory site determine whether the genes are transcribed into mRNA and thus whether the corresponding enzymes or other protein products may be synthesized from the genetic information in the mRNA. Binding of a *trans*-acting element at a *cis*-acting site can regulate the gene cluster either negatively (by turning off transcription) or positively (by turning on transcription of genes in the cluster). In this section, we discuss how transcription of such bacterial gene clusters is coordinately regulated.

The discovery of a regulatory gene and a regulatory site that are part of the gene cluster was paramount to the understanding of how gene expression is controlled in the system. Neither of these regulatory elements encodes enzymes necessary for lactose metabolism—the function of the three genes in the cluster. As illustrated in Figure 16–1, the three structural genes and the adjacent regulatory site constitute the **lactose**, or **lac**, **operon**. Together, the entire gene cluster functions in an integrated fashion to provide a rapid response to the presence or absence of lactose.

Structural Genes

Genes coding for the primary structure of an enzyme are called **structural genes**. There are three structural genes in

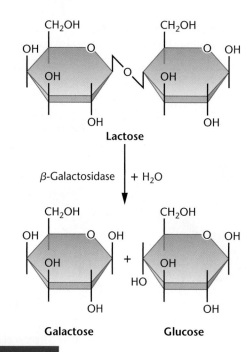

Lactose

β-Galactosidase + H₂O

Galactose   **Glucose**

FIGURE 16–2   The catabolic conversion of the disaccharide lactose into its monosaccharide units, galactose and glucose.

the *lac* operon. The *lacZ* gene encodes **β-galactosidase**, an enzyme whose primary role is to convert the disaccharide lactose to the monosaccharides glucose and galactose (Figure 16–2). This conversion is essential if lactose is to serve as the primary energy source in glycolysis. The second gene, *lacY*, specifies the primary structure of **permease**, an enzyme that facilitates the entry of lactose into the bacterial cell. The third gene, *lacA,* codes for the enzyme **transacetylase**. While its physiological role is still not completely clear, it may be involved in the removal of toxic by-products of lactose digestion from the cell.

To study the genes coding for these three enzymes, researchers isolated numerous mutations that lacked the function of one or the other enzyme. Such *lac⁻* mutants were first isolated and studied by Joshua Lederberg. Mutant cells that fail to produce active β-galactosidase (*lacZ⁻*) or permease (*lacY⁻*) are unable to use lactose as an energy source. Mutations were also found in the transacetylase gene. Mapping studies by Lederberg established that all three genes are

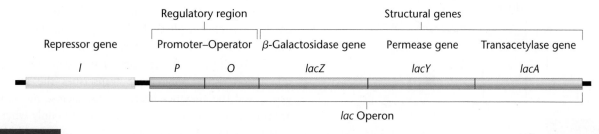

Regulatory region Structural genes

| Repressor gene | Promoter–Operator | β-Galactosidase gene | Permease gene | Transacetylase gene |

I   *P*   *O*   *lacZ*   *lacY*   *lacA*

lac Operon

FIGURE 16–1   A simplified overview of the genes and regulatory units involved in the control of lactose metabolism. (The regions within this stretch of DNA are not drawn to scale.) A more detailed model will be developed later in this chapter. (See Figure 16–10.)

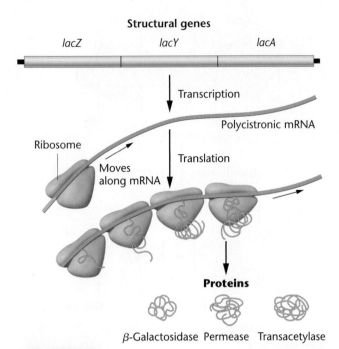

FIGURE 16-3   The structural genes of the *lac* operon are transcribed into a single polycistronic mRNA, which is translated simultaneously by several ribosomes into the three enzymes encoded by the operon.

closely linked or contiguous to one another on the bacterial chromosome, in the order *Z–Y–A* (see Figure 16–1).

Knowledge of their close linkage led to another discovery relevant to what later became known about the regulation of structural genes: All three genes are transcribed as a single unit, resulting in a so-called *polycistronic mRNA* (Figure 16–3; recall that *cistron* refers to the part of a nucleotide sequence coding for a single gene). This results in the coordinate regulation of all three genes, since a single-message RNA is simultaneously translated into all three gene products.

The Discovery of Regulatory Mutations

How does lactose stimulate transcription of the *lac* operon and induce the synthesis of the enzymes for which it codes? A partial answer came from studies using **gratuitous inducers**, chemical analogs of lactose such as the sulfur-containing analog **isopropylthiogalactoside (IPTG)**, shown in Figure 16–4.

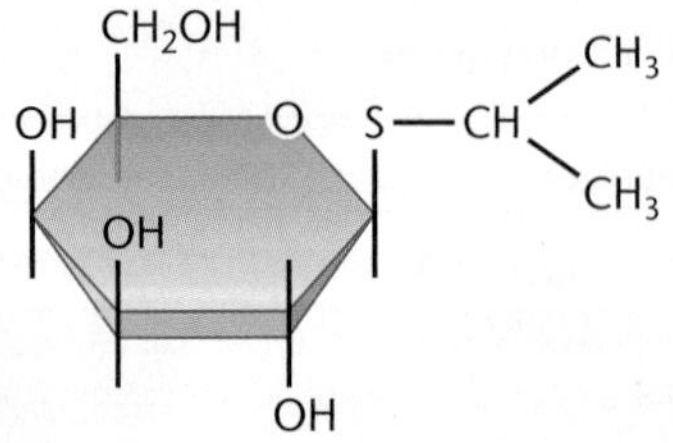

FIGURE 16-4   The gratuitous inducer isopropylthiogalactoside (IPTG).

Gratuitous inducers behave like natural inducers, but they do not serve as substrates for the enzymes that are subsequently synthesized. Their discovery provides strong evidence that the primary induction event does *not* depend on the interaction between the inducer and the enzyme.

What, then, is the role of lactose in induction? The answer to this question required the study of another class of mutations described as **constitutive mutations**. In cells bearing these types of mutations, enzymes are produced regardless of the presence or absence of lactose. Studies of the constitutive mutation *lacI⁻* mapped the mutation to a site on the bacterial chromosome close to, but distinct from, the structural genes *lacZ, lacY,* and *lacA*. This mutation led researchers to discover the *lacI* gene, which is appropriately called a **repressor gene**. A second set of constitutive mutations producing effects identical to those of *lacI⁻* is present in a region immediately adjacent to the structural genes. This class of mutations, designated *lac O^C*, is located in the **operator region** of the operon. In both types of constitutive mutants, the enzymes are produced continually, inducibility is eliminated, and gene regulation has been lost.

The Operon Model: Negative Control

Around 1960, Jacob and Monod proposed a hypothetical mechanism involving negative control that they called the **operon model**, in which a group of genes is regulated and expressed together as a unit. As we saw in Figure 16–1, the *lac* operon they proposed consists of the *Z, Y,* and *A* structural genes, as well as the adjacent sequences of DNA referred to as the *operator region*. They argued that the *lacI* gene regulates the transcription of the structural genes by producing a **repressor molecule**, and that the repressor is **allosteric**, meaning that the molecule reversibly interacts with another molecule, undergoing both a conformational change in three-dimensional shape and a change in chemical activity. Figure 16–5 illustrates the components of the *lac* operon as well as the action of the *lac* repressor in the presence and absence of lactose.

Jacob and Monod suggested that the repressor normally binds to the DNA sequence of the operator region. When it does so, it inhibits the action of RNA polymerase, effectively repressing the transcription of the structural genes [Figure 16–5(b)]. However, when lactose is present, this sugar binds to the repressor and causes an allosteric (conformational) change. The change alters the binding site of the repressor, rendering it incapable of interacting with operator DNA [Figure 16–5(c)]. In the absence of the repressor–operator interaction, RNA polymerase transcribes the structural genes, and the enzymes necessary for lactose metabolism are produced. Because transcription occurs only when the repressor *fails* to bind to the operator region, regulation is said to be under *negative control*.

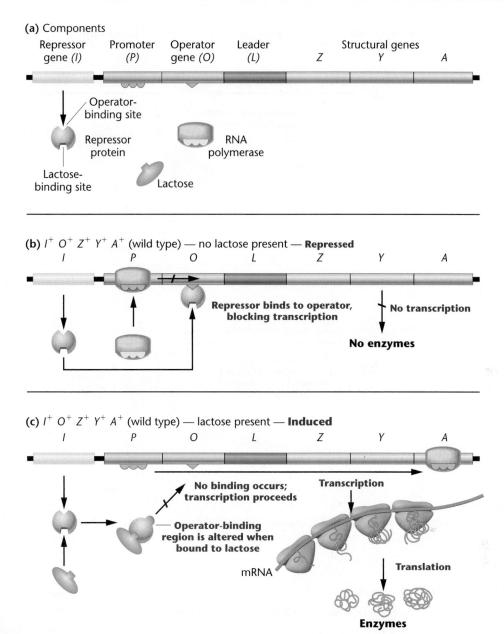

(a) Components

(b) $I^+ \, O^+ \, Z^+ \, Y^+ \, A^+$ (wild type) — no lactose present — **Repressed**

(c) $I^+ \, O^+ \, Z^+ \, Y^+ \, A^+$ (wild type) — lactose present — **Induced**

FIGURE 16–5 The components of the wild-type *lac* operon and the response in the absence and presence of lactose.

To summarize, the operon model invokes a series of molecular interactions between proteins, inducers, and DNA to explain the efficient regulation of structural gene expression. In the absence of lactose, the enzymes encoded by the genes are not needed, and expression of genes encoding these enzymes is repressed. When lactose is present, it indirectly induces the activation of the genes by binding with the repressor.* If all lactose is metabolized, none is available to bind to the repressor, which is again free to bind to operator DNA and to repress transcription.

Both the I^- and O^C constitutive mutations interfere with these molecular interactions, allowing continuous transcription of the structural genes. In the case of the I^- mutant, seen in Figure 16–6(a), the repressor protein is altered or absent and cannot bind to the operator region, so the structural genes are always turned on. In the case of the O^C mutant [Figure 16–6(b)], the nucleotide sequence of the operator DNA is altered and will not bind with a normal repressor molecule. The result is the same: The structural genes are always transcribed.

---

*Technically, the inducer is allolactose, an isomer of lactose. When lactose enters the bacterial cell, some of it is converted to allolactose by the β-galactosidase enzyme.

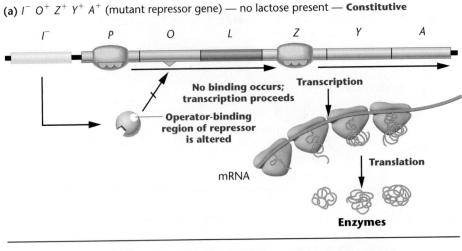

(a) $I^- \; O^+ \; Z^+ \; Y^+ \; A^+$ (mutant repressor gene) — no lactose present — **Constitutive**

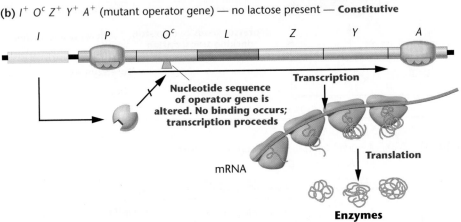

(b) $I^+ \; O^c \; Z^+ \; Y^+ \; A^+$ (mutant operator gene) — no lactose present — **Constitutive**

FIGURE 16–6   The response of the *lac* operon in the absence of lactose when a cell bears either the I^- or the O^C mutation.

Genetic Proof of the Operon Model

The operon model is a good one because it leads to three major predictions that can be tested to determine its validity. The major predictions to be tested are that (1) the I gene produces a diffusible product (that is, a *trans*-acting product); (2) the O region is involved in regulation but does not produce a product (it is *cis*-acting); and (3) the O region must be adjacent to the structural genes in order to regulate transcription.

The creation of partially diploid bacteria allows us to assess these assumptions, particularly those that predict the presence of *trans*-acting regulatory elements. For example, the F plasmid may contain chromosomal genes (Chapter 6), in which case it is designated F′. When an F⁻ cell acquires such a plasmid, it contains its own chromosome plus one or more additional genes present in the plasmid. This host cell is thus a **merozygote**, a cell that is diploid for certain added genes (but not for the rest of the chromosome). The use of such a plasmid makes it possible, for example, to introduce an I^+ gene into a host cell whose genotype is I^-, or to introduce an O^+ region into a host cell of genotype O^C. The Jacob–Monod operon model predicts how regulation should be affected in such cells. Adding an I^+ gene to an I^- cell should restore inducibility, because the normal wild-type repressor, which is a *trans*-acting factor, would be produced by the inserted I^+ gene. In contrast, adding an O^+ region to an O^C cell should have no effect on constitutive enzyme production, since regulation depends on an O^+ region being located immediately adjacent to the structural genes—that is, O^+ is a *cis*-acting regulator.

Results of these experiments are shown in Table 16.1, where Z represents the structural genes (and the inserted genes are listed after the designation F′). In both cases described above, the Jacob–Monod model is upheld (part B of Table 16.1). Part C of the table shows the reverse experiments, where either an I^- gene or an O^C region is added to cells of normal inducible genotypes. As the model predicts, inducibility is maintained in these partial diploids.

Another prediction of the operon model is that certain mutations in the I gene should have the opposite effect of I^-.

TABLE 16.1

A Comparison of Gene Activity (+ or −) in the Presence or Absence of Lactose for Various *E. coli* Genotypes

Genotype	Presence of β-Galactosidase Activity	
	Lactose Present	Lactose Absent
$I^+O^+Z^+$	+	−
A. $I^+O^+Z^-$	−	−
$I^-O^+Z^+$	+	+
$I^+O^cZ^+$	+	+
B. $I^-O^+Z^+/F'I^+$	+	−
$I^+O^cZ^+/F'O^+$	+	+
C. $I^+O^+Z^+/F'I^-$	+	−
$I^+O^+Z^+/F'O^c$	+	−
D. $I^SO^+Z^+$	−	−
$I^SO^+Z^+/F'I^+$	−	−

Note: In parts B to D, most genotypes are partially diploid, containing an F factor plus attached genes (F′).

That is, instead of being constitutive because the repressor can't bind the operator, mutant repressor molecules should be produced that cannot interact with the inducer, lactose. As a result, these repressors would always bind to the operator sequence, and the structural genes would be permanently repressed. In cases like this, the presence of an additional I^+ gene would have little or no effect on repression.

In fact, such a mutation, I^S, was discovered wherein the operon, as predicted, is "superrepressed," as shown in part D of Table 16.1 (and depicted in Figure 16–7). An additional I^+ gene does not effectively relieve repression of gene activity. These observations are consistent with the idea that the repressor contains separate DNA-binding domains and inducer-binding domains.

Isolation of the Repressor

Although Jacob and Monod's operon theory succeeded in explaining many aspects of genetic regulation in prokaryotes,

NOW SOLVE THIS

16–1 Even though the *lac Z, Y,* and *A* structural genes are transcribed as a single polycistronic mRNA, each gene contains the initiation and termination signals essential for translation. Predict what will happen when a cell growing in the presence of lactose contains a deletion of one nucleotide (a) early in the *Z* gene and (b) early in the *A* gene.

■ **HINT:** *This problem requires you to combine your understanding of the genetic expression of the lac operon with that of the genetic code, frameshift mutations, and termination of transcription. The key to its solution is to consider the effect of the loss of one nucleotide within a polycistronic mRNA.*

the nature of the repressor molecule was not known when their landmark paper was published in 1961. While they had assumed that the allosteric repressor was a protein, RNA was also a candidate because the activity of the molecule required the ability to bind to DNA. Despite many attempts to isolate and characterize the hypothetical repressor molecule, no direct chemical evidence was forthcoming. A single *E. coli* cell contains no more than ten or so copies of the *lac* repressor, and direct chemical identification of ten molecules in a population of millions of proteins and RNAs in a single cell presented a tremendous challenge.

In 1966, Walter Gilbert and Benno Müller-Hill reported the isolation of the *lac* repressor in partially purified form. To achieve the isolation, they used a *regulator quantity (I^q)* mutant strain that contains about ten times as much repressor as do wild-type *E. coli* cells. Also instrumental in their success was the use of the gratuitous inducer IPTG, which binds to the repressor, and the technique of **equilibrium dialysis**. In this technique, extracts of I^q cells were placed in a dialysis bag and allowed to attain equilibrium with an external solution of radioactive IPTG, a molecule small enough to diffuse freely in and out of the bag. At equilibrium, the

$I^S\ O^+\ Z^+\ Y^+\ A^+$ (mutant repressor gene) — lactose present — **Repressed**

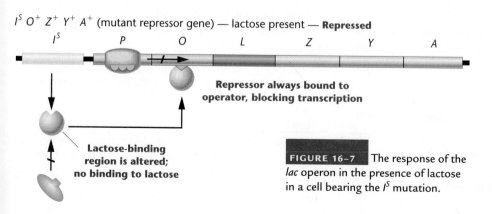

Repressor always bound to operator, blocking transcription

Lactose-binding region is altered; no binding to lactose

FIGURE 16–7 The response of the *lac* operon in the presence of lactose in a cell bearing the I^S mutation.

concentration of radioactive IPTG was higher inside the bag than in the external solution, indicating that an IPTG-binding material was present in the cell extract and was too large to diffuse across the wall of the bag.

Ultimately, the IPTG-binding material was purified and shown to have various characteristics of a protein. In contrast, extracts of I^- constitutive cells having no *lac* repressor *activity* did not exhibit IPTG binding, strongly suggesting that the isolated protein was the repressor molecule.

To confirm this thinking, Gilbert and Müller-Hill grew *E. coli* cells in a medium containing radioactive sulfur and then isolated the IPTG-binding protein, which was labeled in its sulfur-containing amino acids. Next, this protein was mixed with DNA from a strain of phage lambda (λ) carrying the *lacO*$^+$ gene. When the two substances are not mixed together, the DNA sediments at 40*S*, while the IPTG-binding protein sediments at 7*S*. However, when the DNA and protein were mixed and sedimented in a gradient, using ultracentrifugation, the radioactive protein sedimented at the same rate as did DNA, indicating that the protein binds to the DNA. Further experiments showed that the IPTG-binding, or repressor, protein binds only to DNA containing the *lac* region and does not bind to *lac* DNA containing an operator-constitutive O^C mutation.

16.3

The Catabolite-Activating Protein (CAP) Exerts Positive Control over the *lac* Operon

As described in the preceding discussion of the *lac* operon, the role of β-galactosidase is to cleave lactose into its components, glucose and galactose. Then, in order to be used by the cell, the galactose, too, must be converted to glucose. What if the cell found itself in an environment that contained ample amounts of both lactose *and* glucose? Given that glucose is the preferred carbon source for *E. coli,* it would not be energetically efficient for a cell to induce transcription of the *lac* operon, since what it really needs— glucose—is already present. As we will see next, still another molecular component, called the **catabolite-activating protein (CAP)**, is involved in effectively repressing the expression of the *lac* operon when glucose is present. This inhibition is called **catabolite repression**.

To understand CAP and its role in regulation, let's backtrack for a moment to review the system depicted in Figure 16–5. When the *lac* repressor is bound to the inducer (lactose), the *lac* operon is activated, and RNA polymerase transcribes the structural genes. As stated in Chapter 14, transcription is initiated as a result of the binding that occurs

between RNA polymerase and the nucleotide sequence of the **promoter region**, found upstream (5′) from the initial coding sequences. Within the *lac* operon, the promoter is found between the *I* gene and the operator region (*O*) (see Figure 16–1). Careful examination has revealed that polymerase binding is never very efficient unless CAP is also present to facilitate the process.

The mechanism is summarized in Figure 16–8. In the absence of glucose and under inducible conditions, CAP exerts positive control by binding to the CAP site, facilitating RNA-polymerase binding at the promoter, and thus transcription. Therefore, for maximal transcription of the structural genes, the repressor must be bound by lactose (so as not to repress operon expression), *and* CAP must be bound to the CAP-binding site.

This leads to the central question about CAP: How does the presence of glucose inhibit CAP binding? The answer involves still another molecule, **cyclic adenosine monophosphate (cAMP)**, upon which CAP binding is dependent. *In order to bind to the promoter, CAP must be bound to cAMP.* The level of cAMP is itself dependent on an enzyme, **adenyl cyclase**, which catalyzes the conversion of ATP to cAMP (see Figure 16–9).* The role of glucose in catabolite repression is now clear. It inhibits the activity of adenyl cyclase, causing a decline in the level of cAMP in the cell. Under this condition, CAP cannot form the CAP–cAMP complex essential to the positive control of transcription of the *lac* operon.

Like the *lac* repressor, CAP and cAMP–CAP have been examined by X-ray crystallography. CAP is a dimer that inserts into adjacent regions of a specific nucleotide sequence of the DNA making up the *lac* promoter. The cAMP–CAP complex, when bound to DNA, bends it, causing it to assume a new conformation.

Binding studies in solution further clarify the mechanism of gene activation. Alone, neither cAMP–CAP nor RNA polymerase has a strong tendency to bind to *lac* promoter DNA, nor does either molecule have a strong affinity for the other. However, when both are together in the presence of the *lac* promoter DNA, a tightly bound complex is formed, an example of what is called **cooperative binding**. In the case of cAMP–CAP and the *lac* operon, this phenomenon illustrates the high degree of specificity that is involved in the genetic regulation of just one small group of genes.

Regulation of the *lac* operon by catabolite repression results in efficient energy use, because the presence of glucose will

---

* Because of its involvement with cAMP, CAP is also called *cyclic AMP receptor protein (CRP),* and the gene encoding the protein is named *crp.* Since the protein was first named CAP, we will adhere to the initial nomenclature.

(a) Glucose absent

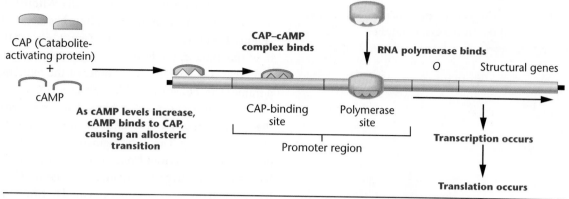

(b) Glucose present

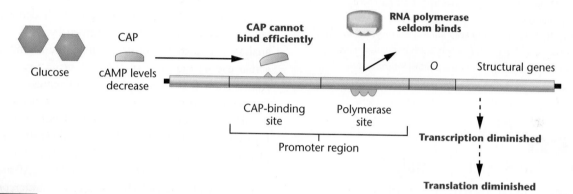

FIGURE 16–8 Catabolite repression. (a) In the absence of glucose, cAMP levels increase, resulting in the formation of a CAP–cAMP complex, which binds to the CAP site of the promoter, stimulating transcription. (b) In the presence of glucose, cAMP levels decrease, CAP–cAMP complexes are not formed, and transcription is not stimulated.

override the need for the metabolism of lactose, should the lactose also be available to the cell. In contrast to the negative regulation conferred by the *lac* repressor, the action of cAMP–CAP constitutes positive regulation. Thus, a combination of positive and negative regulatory mechanisms determines transcription levels of the *lac* operon. Catabolite repression involving CAP has also been observed for other inducible operons, including those controlling the metabolism of galactose and arabinose.

FIGURE 16–9 The formation of cAMP from ATP, catalyzed by adenyl cyclase.

16–2 Predict the level of genetic activity of the *lac* operon as well as the status of the *lac* repressor and the CAP protein under the cellular conditions listed in the accompanying table.

	Lactose	Glucose
(a)	–	–
(b)	+	–
(c)	–	+
(d)	+	+

■ HINT: *This problem asks you to combine your knowledge of the Jacob–Monod model of the regulation of the lac operon with your understanding of how catabolite repression impacts on this model. The key to its solution is to keep in mind that regulation involving lactose is a negative control system, while regulation involving glucose and catabolite repression is a positive control system.*

16.4

Crystal Structure Analysis of Repressor Complexes Has Confirmed the Operon Model

We now have thorough knowledge of the biochemical nature of the regulatory region of the *lac* operon, including the precise locations of its various components relative to one another (Figure 16–10). In 1996, Mitchell Lewis, Ponzy Lu, and their colleagues succeeded in determining the crystal structure of the *lac* repressor, as well as the structure of the repressor bound to the inducer and to operator DNA. As a result, previous information that was based on genetic and biochemical data has now been complemented with the missing structural interpretation. Together, these contributions provide a nearly complete picture of the regulation of the operon.

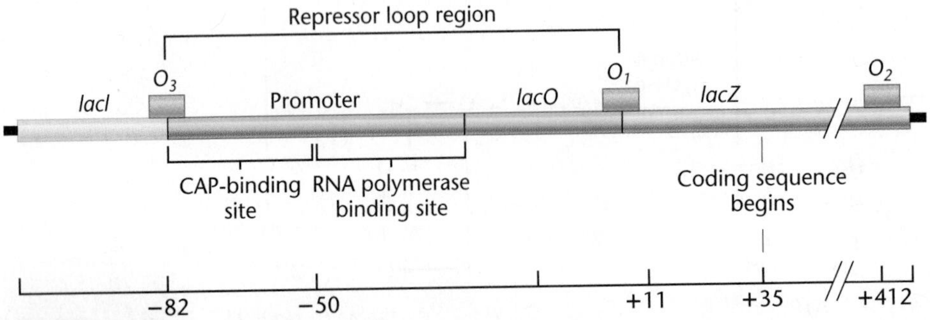

The repressor, as the gene product of the *I* gene, is a monomer consisting of 360 amino acids. Within this monomer, the region of inducer binding has been identified [Figure 16–11(a)]. While dimers [Figure 16–11(b)] can also bind the inducer, the functional repressor is a homotetramer (that is, it contains four copies of the monomer). The tetramer can be cleaved with a protease under controlled conditions to yield five fragments. Four are derived from the N-terminal ends of the tetramer subunits, and they bind to operator DNA. The fifth fragment is the remaining core of the tetramer, derived from the COOH-terminus ends; it binds to lactose and gratuitous inducers such as IPTG. Analysis has revealed that each tetramer can bind to two symmetrical operator DNA helices at a time.

The operator DNA that was previously defined by mutational studies ($lacO^C$) and confirmed by DNA-sequencing analysis is located just upstream from the beginning of the actual coding sequence of the *lacZ* gene. Crystallographic studies show that the actual region of repressor binding of this primary operator, O_1, consists of 21 base pairs. Two other auxiliary operator regions have been identified, as shown in Figure 16–10. One, O_2, is 401 base pairs downstream from the primary operator, within the *lacZ* gene. The other, O_3, is 93 base pairs upstream from O_1, just beyond the CAP site. *In vivo*, all three operators must be bound for maximum repression.

Binding by the repressor simultaneously at two operator sites distorts the conformation of DNA, causing it to bend away from the repressor. When a model is created to demonstrate dual binding of operators O_1 and O_3 [Figure 16–11(c)], the 93 base pairs of DNA that intervene must jut out, forming what is called a **repression loop**. This model positions the promoter region that binds RNA polymerase on the inside of the loop, which prevents access by the polymerase during repression. In addition, the repression loop positions the CAP-binding site in a way that facilitates CAP interaction with RNA polymerase upon subsequent induction. The DNA looping caused by repression in this model is similar to configuration changes that are predicted to occur in eukaryotic systems (see Chapter 17).

Studies have also defined the three-dimensional conformational changes that accompany the interactions with the inducer molecules. Taken together, the crystallographic studies have brought a new level of understanding of the regulatory process occurring within the *lac* operon, confirming the findings and predictions of Jacob and Monod in their model set forth over 40 years ago, which was based strictly on genetic observations.

FIGURE 16–10   The various regulatory regions involved in the control of genetic expression of the *lac* operon, as described in the text. The numbers on the bottom scale represent nucleotide sites upstream and downstream from the initiation of transcription.

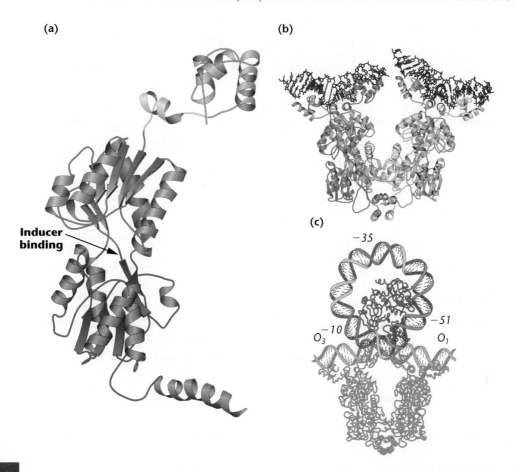

(a)

(b)

(c)

FIGURE 16–11 Models of the *lac* repressor and its binding to operator sites in DNA, as generated from crystal structure analysis. (a) The repressor monomer, showing the inducer-binding site. The DNA-binding region is shown in red. (b) The repressor dimer bound to two 21-base-pair segments of operator DNA (shown in dark blue). (c) The repressor (shown in pink) and CAP (shown in dark blue) bound to the *lac* DNA. Binding to operator regions O_1 and O_3 creates a 93-base-pair repression loop of promoter DNA.

16.5

The Tryptophan (*trp*) Operon in *E. coli* Is a Repressible Gene System

Although the process of induction had been known for some time, it was not until 1953 that Monod and colleagues discovered a repressible operon. Wild-type *E. coli* are capable of producing the enzymes necessary for the biosynthesis of amino acids as well as other essential macromolecules. Focusing his studies on the amino acid tryptophan and the enzyme **tryptophan synthetase**, Monod discovered that if tryptophan is present in sufficient quantity in the growth medium, the enzymes necessary for its synthesis are not produced. It is energetically advantageous for bacteria to repress expression of genes involved in tryptophan synthesis when ample tryptophan is present in the growth medium.

Further investigation showed that a series of enzymes encoded by five contiguous genes on the *E. coli* chromosome are involved in tryptophan synthesis. These genes are part of an operon, and in the presence of tryptophan, all are coordinately repressed, and none of the enzymes is produced. Because of the great similarity between this repression and the induction of enzymes for lactose metabolism, Jacob and Monod proposed a model of gene regulation analogous to the *lac* system (the updated version is shown in Figure 16–12).

To account for repression, Jacob and Monod suggested the presence of a *normally inactive repressor* that alone cannot interact with the operator region of the operon. However, the repressor is an allosteric molecule that can bind to tryptophan. When tryptophan is present, the resultant complex of repressor and tryptophan attains a new conformation that binds to the operator, repressing transcription. Thus, when tryptophan, the end product of this anabolic pathway, is present, the system is repressed and enzymes are not made. Since the regulatory complex inhibits transcription of the operon, this repressible system is under negative control. And as tryptophan participates in repression, it is referred to as a **corepressor** in this regulatory scheme.

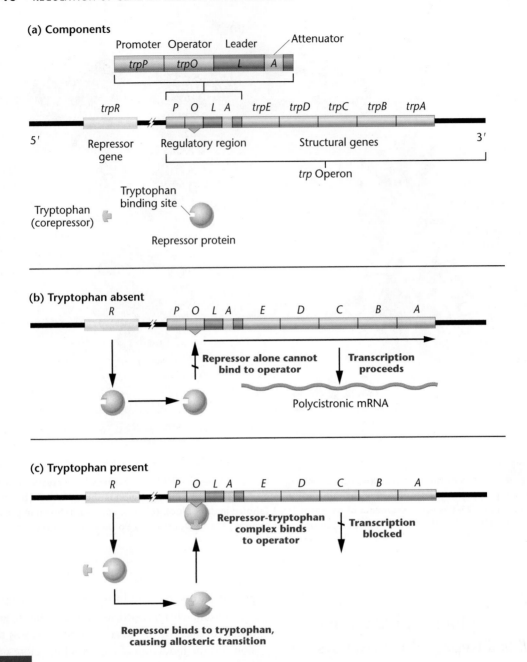

FIGURE 16–12   A repressible operon. (a) The components involved in regulation of the tryptophan operon. (b) In the absence of tryptophan, an inactive repressor is made that cannot bind to the operator (*O*), thus allowing transcription to proceed. (c) When tryptophan is present, it binds to the repressor, causing an allosteric transition to occur. This complex binds to the operator region, leading to repression of the operon.

Evidence for the *trp* Operon

Support for the concept of a repressible operon was soon forth-coming, based primarily on the isolation of two distinct catego-ries of constitutive mutations. The first class, *trpR⁻*, maps at a considerable distance from the structural genes. This locus rep-resents the gene coding for the repressor. Presumably, the muta-tion inhibits either the repressor's interaction with tryptophan or repressor formation entirely. Whichever the case, no repres-sion ever occurs in cells with the *trpR⁻* mutation. As expected,

if the *trpR⁺* gene encodes a functional repressor molecule, the presence of a copy of this gene will restore repressibility.

The second constitutive mutation is analogous to that of the operator of the lactose operon, because it maps im-mediately adjacent to the structural genes. Furthermore, the addition of a wild-type operator gene into mutant cells (as an external element) does not restore repression. This is what would be predicted if the mutant operator no longer interacts with the repressor–tryptophan complex.

The entire *trp* operon has now been well defined, as shown in Figure 16–12. The five contiguous structural genes (*trp E, D, C, B,* and *A*) are transcribed as a polycistronic message directing translation of the enzymes that catalyze the biosynthesis of tryptophan. As in the *lac* operon, a promoter region (*trpP*) represents the binding site for RNA polymerase, and an operator region (*trpO*) binds the repressor. In the absence of binding, transcription is initiated within the *trpP–trpO* region and proceeds along a **leader sequence** 162 nucleotides prior to the first structural gene (*trpE*). Within that leader sequence, still another regulatory site has been demonstrated, called an *attenuator*—the subject of the next section. As we will see, this regulatory unit is an integral part of this operon's control mechanism.

16.6

Attenuation Is a Process Critical to the Regulation of the *trp* Operon in *E. coli*

The preceding section has established that the genes within the *trp* operon are expressed only when the quantity of tryptophan is severely limiting in the cell, but repressed in the presence of tryptophan. However, in the presence of low levels of tryptophan, transcription of the upstream region of the operon may nevertheless be initiated. This discovery was made by Charles Yanofsky, Kevin Bertrand, and their colleagues, who realized that once transcription is initiated, an independent mechanism exists that effectively represses the expression of the operon if tryptophan is present. They named this mechanism **attenuation**, a word derived from the verb *attenuate*, which means "to weaken or impair." After many years of work, this group discovered that following initiation of transcription of the *trp* operon, mRNA synthesis may be terminated at a point about 140 nucleotides along the transcript, prior to encountering the DNA template representing the structural genes. This upstream RNA molecule is referred to as the *5′-leader sequence.* Obviously, in the absence of tryptophan, attenuation is somehow overcome, allowing transcription of the entire operon.

Identification of the site involved in attenuation was made possible by the isolation of various deletion mutations in a 25-nt region 115 to 140 nucleotides into the leader sequence. Such mutations abolish attenuation. This site is referred to as the *attenuator*. An explanation of how attenuation occurs and how it is overcome, as presented by Yanofsky and colleagues, is summarized in Figure 16–13. The operon and the initial DNA sequence that is transcribed is shown in Figure 16–13(a). Transcription initially gives rise to the leader sequence [Figure 16–13(b)], which has the potential to fold into either of two mutually exclusive stem-loop structures described as *hairpins.* In the presence of tryptophan, the hairpin that is formed behaves as a *terminator,* causing transcription to be terminated prematurely. On the other hand, if tryptophan is limiting, the alternative hairpin structure, referred to as the *antiterminator,* is formed. In this case, transcription proceeds beyond the antiterminator region, and the mRNA representing the entire operon is subsequently produced. The hairpins are examples of *secondary RNA structures* and are shown in Figure 16–13(c).

The major issue remains as to how the presence of tryptophan leads to attenuation and how the absence of tryptophan circumvents it. The Yanofsky model, a revelation in the study of regulatory genetics, relies on two key points: (1) Bacterial ribosomes may attach to mRNA molecules and initiate *translation* before *transcription* of a gene (or operon) is complete (see Figure 13–15); and (2) the leader region of mRNA includes several tryptophan codons. Yanofsky proposed that if the tryptophan codons are *translated,* the *terminator hairpin* is formed. However, when tryptophan is very scarce or absent in the cell, inadequate charged $tRNA^{Trp}$ is available and translation stalls at the tryptophan codons. This translational stalling induces the *antiterminator hairpin* to form, allowing *transcription* to proceed throughout the operon.

The above model represents a unique approach to gene regulation, which can be termed *translational regulation of transcription.* The bottom line is that availability of $tRNA^{Trp}$, the formation of which is dependent on tryptophan levels, is critical to this mechanism that fine tunes the regulation of the operon. This model is supported experimentally by a number of genetic studies, whereby various regions of the leader sequence representing regulatory stems have been mutated or deleted. In each case, the predictions of such genetic alterations have been upheld.

Because regulation by attenuation is dependent on simultaneous transcription and translation, a linkage that is impossible in eukaryotes, the process is limited to prokaryotes. It has now been established in other bacterial operons in *E. coli,* including those responsible for the biosynthesis of amino acids, including threonine, histidine, leucine, and phenylalanine. As with the *trp* operon, attenuators in these operons contain multiple codons calling for regulation of the amino acid. When the relevant amino acid is unavailable, translation "stalls," allowing transcription to proceed. For example, the leader sequence in the histidine operon encodes seven contiguous histidine residues.

TRAP and AT Proteins Govern Attenuation in *B. subtilis*

Keep in mind that not all organisms, even those of the same type, solve problems such as gene regulation in exactly the same way. Thus, it is not unusual for geneticists to discover that new strategies have arisen during evolution. Often, a

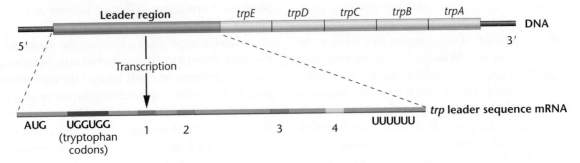

(a) Transcription of *trp* Operon (DNA)

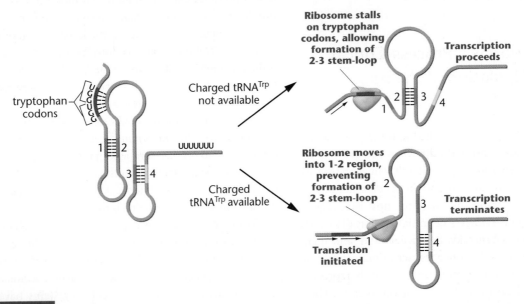

(b) Stem-loop structures in leader RNA sequence

(c) Alternative secondary structures of leader RNA

FIGURE 16–13   Attenuation in the tryptophan operon. (a) Transcription of the *trp* operon, showing the leader region followed by the five *trp* genes. The leader region in the mRNA is enlarged to show the translation start codon (AUG), the two tryptophan codons (UGGUGG), and the string of U residues at the end of the leader. The four regions marked 1, 2, 3, and 4 indicate the location of four sequences that have the potential to form stems by base pairing. (b) Two possible stem-loop structures that can form within the leader sequence mRNA. (c) Two alternative secondary structures that can be formed within the leader mRNA sequence during translation of the mRNA, depending on the availability of tryptophan. The first (at the top) occurs in the absence of tryptophan and thus in the absence of its charged cognate tRNA. As a result, a stem-loop involving regions 2 and 3 forms. During translation, the ribosome stalls at the trp codons prior to region 1, forming an antiterminator conformation. As a result, transcription by RNA polymerase continues. At the bottom, in the presence of tryptophan, and thus its charged cognate tRNA, the ribosome continues translation, moving past region 1 and the tryptophan codons, which inhibits the formation of the antiterminator region 2-3 stem-loop. As a result, the antiterminator signal is disrupted and transcription terminates. It is this process that is referred to as attenuation.

new approach is a variation on a well-established theme. Such is the case with the regulation of the *trp* operon in different bacteria.

As we saw previously, *E. coli*, a Gram-negative bacterium, uses charged tRNA^{Trp} and a terminator hairpin in the leader sequence of the transcript as machinery for attenuating transcription of its *trp* operon. The Gram-positive bacterium *Bacillus subtilis* also uses attenuation and hairpins to regulate its *trp* operon. In fact, *B. subtilis* relies on attenuation as its sole mechanism of regulation because that bacterium lacks a mechanism that represses transcription entirely in this operon, as is present in *E. coli*.

However, the molecular signals that cause attenuation in *B. subtilis* do not use a process of translation and stalling, as *E. coli* does, to induce the hairpin that terminates transcription. Instead, a specific protein, isolated in the 1990s by Charles Yanofsky and coworkers, either binds or does not bind to the attenuator leader sequence, thereby inducing the alternative terminator or antiterminator configurations, respectively.

Attenuation is accomplished in *B. subtilis* in a unique way. The protein, **trp RNA-binding attenuation protein (TRAP)**, binds to tryptophan if it is present in the cell. TRAP consists of 11 subunits, forming a symmetrical quaternary protein structure. Each subunit can bind one molecule of

(a)

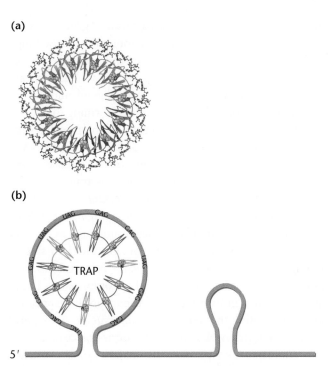

(b)

TRAP

5'

Terminator hairpin
(Tryptophan abundant)

FIGURE 16–14 Model of the *trp* RNA-binding attenuation protein (TRAP). The symmetrical molecule consists of 11 subunits, each capable of binding to one molecule of tryptophan that is embedded within the subunit. (b) The interaction of a tryptophan-bound TRAP molecule with the leader sequence of the *trp* operon of *B. subtilis*. This interaction induces the formation of the terminator hairpin, attenuating expression of the *trp* operon.

tryptophan, incorporating it into a deep pocket within the protein [Figure 16–14(a)]. When fully saturated with tryptophan, this protein can bind to the 5′-leader sequence of the RNA transcript, which contains 11 triplet repeats of either GAG or UAG, each separated by several spacer nucleotides. Each triplet is linked to one of the subunits, which contains a binding pocket for the triplet. When bound to the transcript, the TRAP is encircled by a belt of RNA [Figure 16–14(b)], and this conformation prevents the antiterminator hairpin from forming. Instead, the terminator configuration results, leading to premature termination of transcription and, consequently, attenuation of expression of the operon.

The attenuation strategy of *B. subtilis* involves an interesting mechanism of regulation. Two observations had suggested to Yanofsky and his colleagues that the attenuation strategy in *B. subtilis* might be even more complex than had originally appeared. First, regulation of the *trp* operon in *B. subtilis* is extremely sensitive to a wide range of tryptophan concentrations. Such a fine tuning suggests that something more than the simple on–off mechanism attributed to TRAP might be at work. Second, a mutation in the gene encoding tryptophanyl-tRNA synthetase leads to the overexpression of

the *trp* operon, even in the presence of excess tryptophan. This enzyme is responsible for charging tRNA[Trp]; in its absence, uncharged tRNA[Trp] molecules accumulate, and regulation is interrupted. The interruption of regulation would suggest that uncharged tRNA[Trp] plays an essential role in attenuation, leading Yanofsky and his coworker Angela Valbuzzi to ask how tRNA[Trp] might be involved with TRAP. What they found extends our knowledge of the system of regulation.

They discovered that there exists still another protein, **anti-TRAP (AT)**, which provides a metabolic signal that tRNA[Trp] is uncharged, thereby indicating that tryptophan is very scarce in the cell. Yanofsky and Valbuzzi theorized that uncharged tRNA[Trp] induces a separate operon to express the *AT* gene. The AT protein then associates with TRAP, specifically when it is in the tryptophan-activated state, and inhibits the binding of TRAP to its target leader RNA sequence.

Such a finding not only adds yet another dimension to our understanding of this system of gene regulation, but it also explains the original observation of the mutation in the tRNA[Trp] synthetase gene that leads to overexpression of the *trp* operon. The mutation prevents charging of tRNA[Trp], even in the presence of tryptophan. As uncharged tRNA[Trp] accumulates, it induces AT, which binds to TRAP. The TRAP is then unable to bind to the leader RNA sequence; as a result, the *trp* operon is overinduced.

The preceding description demonstrates the complexity of regulatory mechanisms in bacteria. That such intricate strategies have resulted from the evolutionary process attests to the critical importance of carefully regulating gene expression in bacteria.

NOW SOLVE THIS

16–3 Consider the regulation of the *trp* operon in *B. subtilis*. For each of the following cases, indicate whether the structural genes in the operon are being expressed or not expressed:

(a) TRAP present, tRNA[Trp] abundant, tryptophan abundant.
(b) TRAP present, tRNA[Trp] abundant, tryptophan scarce.
(c) TRAP present, tRNA[Trp] scarce, tryptophan abundant.
(d) TRAP absent, tRNA[Trp] abundant, tryptophan abundant.

In each case, also indicate whether AT is present. Describe the role of TRAP and AT in attenuation.

■ HINT: *This problem concerns the regulation of the B. subtilis trp operon, which involves the TRAP protein and requires you to understand the concept of attenuation. You are asked to predict whether or not the structural genes are expressed under various conditions. The key to its solution is to realize that unlike E. coli, which uses tryptophan as a repressor along with the process of attenuation in regulating the trp operon, B. subtilis relies solely on the process of attenuation and the TRAP protein for regulation of this operon.*

16.7

Riboswitches Utilize Metabolite-sensing RNAs to Regulate Gene Expression

Since the elucidation of attenuation in the *trp* operon, numerous cases of gene regulation that also depend on alternative forms of mRNA secondary structure have been documented. These involve what are called **riboswitches**, which are mRNA sequences (or elements), present in the 5′-untranslated region (5′-UTR) upstream from the coding sequences. These elements are capable of binding with small molecule ligands, such as metabolites, whose synthesis or activity is controlled by the genes encoded by the mRNA. Such binding causes a conformational change in one domain of the riboswitch element, which induces another change at a second RNA domain, most often creating a *terminator structure*. This terminator region interfaces directly with the transcriptional machinery and shuts it down. Thus, the mRNA is able to respond to the environment of the cell and regulate its own expression. Common to all riboswitches, then, is the presence of a *metabolite-sensing RNA sequence* that can adopt one of two secondary structures that allows transcription of that RNA to either proceed or not to proceed. The decision depends on the concentration of a molecular component of the cell whose existence is influenced by the genes encoded by that RNA.

Many classes of riboswitches recognize a broad range of ligands, including amino acids, purines, vitamin cofactors, aminosugars, and metal ions, among others. They are widespread in bacteria. In *Bacillus subtilis,* for example, approximately 5 percent of this bacterium's genes are regulated by riboswitches. They are also found in archaea, fungi, and plants, and may prove to be present in animals as well.

The two important domains within a riboswitch are called the **aptamer**, which binds to the ligand, and the **expression platform**, which is capable of forming the terminator structure. **Figure 16–15** illustrates the principles involved in riboswitch control. The 5′-UTR of an mRNA is shown on the left side of the figure in the absence of the ligand (metabolite). RNA polymerase has transcribed the unbound aptamer domain, and in the *default conformation*, the expression domain adopts an *antiterminator conformation*. Thus, transcription continues through the expression platform and into the coding region. On the right side of the figure, in the presence of the ligand, it binds to the aptamer domain, inducing an alternative conformation in the aptamer, which in turn induces a change in the expression domain, creating the *terminator conformation*. RNA polymerase is effectively blocked and transcription ceases.

While this description fits the majority of riboswitches that have been studied, there are two variations from this model that complete our coverage of the topic. First, it is possible for the default position of a riboswitch to be in the *terminator conformation*. For example, consider the 5′UTRs representing genes encoding aminoacyl tRNA synthetases, the enzymes that charge tRNAs with their cognate amino acid (see Chapter 14). In these cases, uncharged tRNAs serve as the ligand and bind to the aptamer of the 5′UTR, inducing the *antiterminator conformation*. This allows the transcription of the genes, leading to the subsequent translation of the enzymes that are required to charge tRNAs. When charged tRNAs are in abundance and uncharged tRNAs are

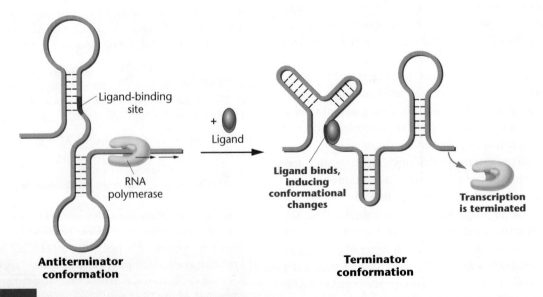

Antiterminator conformation

Terminator conformation

FIGURE 16–15   Illustration of the mechanism of riboswitch regulation of gene expression, where the default position (left) is in the antiterminator conformation and, upon binding by the ligand, adopts the terminator conformation.

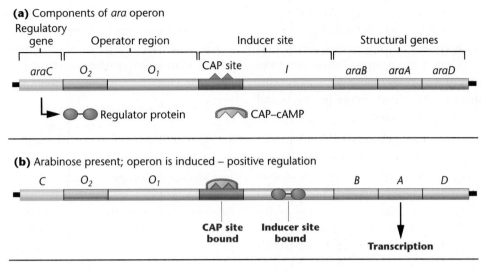

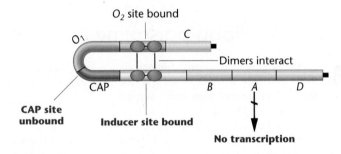

FIGURE 16–16   Genetic regulation of the *ara* operon. The regulatory protein of the *araC* gene acts as either an inducer (in the presence of arabinose) or a repressor (in the absence of arabinose).

not present to bind to the aptamer, the riboswitch, in the default condition, adapts the *terminator conformation,* effectively shutting down transcription. As predicted, charged tRNAs are unable to bind to the aptamer.

Finally, there are examples of bacterial riboswitches where, like attenuation of the *trp* operon discussed earlier, the alternative 5′UTR conformations allow transcription to be completed, but ligand binding to the aptamer induces a conformation that inhibits translation by ribosomes. In such cases, regulation of gene expression has been invoked at the level of translation.

16.8

The *ara* Operon Is Controlled by a Regulator Protein That Exerts Both Positive and Negative Control

We conclude this chapter with a brief discussion of the **arabinose (ara) operon** as analyzed in *E. coli.* This inducible operon is unique because the same regulatory protein is capable of exerting both positive and negative control, and as

a result, at some times induces and at others represses gene expression. Figure 16–16(a) shows the operon components, and Figure 16–16(b) and (c) show, respectively, the conditions under which the operon is active or inactive.

1. The metabolism of the sugar arabinose is governed by the enzymatic products of three structural genes, *ara B, A,* and *D* [Figure 16–16(a)]. Their transcription is controlled by the regulatory protein AraC, encoded by the *araC* gene, which interacts with two regulatory regions, *araI* and *araO₂.*[*] These sites can be bound individually or coordinately by the AraC protein.

2. The *I* region bears that designation because when it is the only region bound by AraC, the system is induced [Figure 16–16(b)]. For this binding to occur, both arabinose and cAMP must be present. Thus, like induction of the *lac* operon, a CAP-binding site is present in the promoter region that modulates catabolite repression in the presence of glucose.

---

[*]An additional operator O_1 region is also present, but is not involved in the regulation of the *ara* structural gene.

3. In the absence of both arabinose and cAMP, the AraC protein binds coordinately to both the *I* site *and* the O_2 site (so named because it was the second operator region to be discovered in this operon). When both *I* and O_2 are bound by AraC, a conformational change occurs in the DNA to form a tight loop [Figure 16–16(c)], and the structural genes are repressed.

4. The O_2 region is located about 200 nucleotides upstream from the *I* region. Binding at either regulatory region involves a dimer of AraC. When both regions are bound, the dimers interact, producing the loop that causes repression. Presumably, this loop inhibits the access of RNA polymerase to the promoter region. The region of DNA that loops out (the stretch

between *I* and O_2) is critical to the formation of the repression complex. Genetic alteration here by insertion or deletion of even a few nucleotides is sufficient to interfere with loop formation and, therefore, repression.

Our observations involving the *ara* operon illustrate the degree of complexity that may be seen in the regulation of a group of related genes. As we mentioned at the beginning of this chapter, the development of regulatory mechanisms has provided evolutionary advantages that allow bacterial systems to adjust to a range of natural environments. Without question, these systems are well equipped genetically not only to survive under varying physiological conditions but to do so with great biochemical efficiency.

CASE STUDY | Food poisoning and bacterial gene expression

At a Midwestern university, several undergraduates reported to the health service with fever, muscle aches, and vomiting. Blood tests showed they had bacterial food poisoning, caused by *Listeria monocytogenes*. One student, a biology major, asked how the bacteria were able to get from the intestine into the bloodstream. The physician explained that *Listeria* cells that express a cell-surface protein, InlA, are able to use receptors on the surface of intestinal cells to gain entry and spread from there to the blood. Once in the blood, the bacteria reach the liver, where they use another cell-surface protein InlB, to enter liver cells, which are the main sites of infection. She further explained that transcriptional levels of the *inlA* and *inlB* genes determine the infective ability of different *Listeria* strains.

Later, the student discussed the incident with his genetics instructor, and together they pondered several questions about

how bacteria are able to hijack the normal functions of receptors on human cells as part of their life cycle.

1. How can we tell if the *inlA* and *inlB* genes are expressed constitutively or are induced in the presence of the human receptors?

2. Might these genes be normally expressed at low levels, but undergo increased transcription once in contact with intestinal epithelial cells? How could this be determined?

3. Are the *inlA* and *inlB* genes part of an operon, or do they use a common promoter? What experiments might show this?

4. How might an understanding of the regulation of these bacterial genes be used to treat infections?

Summary Points

 For activities, animations, and review quizzes, go to the study area at www.masteringgenetics.com

1. Research on the *lac* operon in *E. coli* pioneered our understanding of gene regulation in bacteria.
2. Genes involved in the metabolism of lactose are coordinately regulated by a negative control system that responds to the presence or absence of lactose.
3. The catabolite-activating protein (CAP) exerts positive control over *lac* gene expression by interacting with RNA polymerase at the *lac* promoter and by responding to the levels of cyclic AMP in the bacterial cell.
4. The *lac* repressor of *E. coli* has been isolated and studied. Crystal structure analysis has shown how it interacts with the DNA of the operon as well as with inducers, revealing conformational changes in DNA leading to the formation of a repression loop that inhibits binding between RNA polymerase and the promoter region of the operon.

5. Unlike the inducible *lac* operon, the *trp* operon is repressible. In the presence of tryptophan, the repressor binds to the regulatory region of the *trp* operon and represses transcription initiation.
6. The process of attenuation, which regulates operons based on the presence of an end product, involves alterations to mRNA secondary structure, leading to premature termination of transcription.
7. Riboswitches regulate gene expression by virtue of conformational changes in the secondary structure of the 5'UTR leader region of mRNAs.
8. The *ara* operon in *E. coli* is unique in that the regulator protein exerts both positive and negative control over expression of the genes specifying the enzymes that metabolize the sugar arabinose.

GENETICS, TECHNOLOGY, AND SOCIETY

Quorum Sensing: Social Networking in the Bacterial World

For decades, scientists regarded bacteria as independent single-celled organisms, incapable of cell-to-cell communication. However, recent research is revealing that many bacteria can regulate gene expression and coordinate group behavior through a form of communication termed *quorum sensing*. Through this process, bacteria send and receive chemical signals called autoinducers that relay information about population size. When the population size reaches a "quorum," defined in the business world as the minimum number of members of an organization that must be present to conduct business, the autoinducers regulate gene expression in a way that benefits the group as a whole. Quorum sensing has been described in more than 70 species of bacteria, and its uses range from controlling bioluminescence in marine bacteria to regulating the expression of virulence factors in pathogenic bacteria. Our understanding of quorum sensing has altered our perceptions of prokaryotic gene regulation and is leading to the development of practical applications, including new antibiotic drugs.

Quorum sensing was discovered in the 1960s, during research on the bioluminescent bacterium *Vibrio fischeri* which lives in a symbiotic relationship with the Hawaiian bobtail squid *Euprymna scolopes*. While hunting for food at night, the squid uses light emitted by the *V. fischeri* present in its light organ to illuminate the ocean floor and to counter the shadows created by moonlight that normally act as a beacon for the squid's predators. In return, the bacteria gain a protected, nutrient-rich environment in the squid's light organ. During the day, the bacteria do not glow as a result of the squid's ability to reduce the concentration of bacteria in its light organ, which in turn prevents expression of the bacterial luciferase (*lux*) operon.

What turns on the bacteria's *lux* genes in response to high cell density and off in response to low cell density? In *V. fischeri*, the responsible autoinducer is a secreted homoserine lactone molecule. At a critical population size, these molecules accumulate, are taken up by bacteria within the population, and regulate the *lux* operon by binding directly to transcription factors that stimulate *lux* gene expression.

Since the discovery of quorum sensing in *Vibrio fischeri*, scientists have identified similar microbial communication systems in other bacteria, including significant human pathogens such as pseudomonal, staphylococcal, and streptococcal species. The expression of as many as 15 percent of bacterial genes may be regulated by quorum sensing.

Quorum sensing molecules may also mediate communication among members of different species. In 1994, Bonnie Bassler and her colleagues at Princeton University discovered an autoinducer molecule in the marine bacterium *Vibrio harveyi* that was also present in many diverse types of bacteria. This molecule, autoinducer-2 (AI-2), has the potential to mediate "quorum-sensing cross talk" between species and thus serve as a universal language for bacterial communication. Because the accumulation of AI-2 is proportional to cell number, and because the structure of AI-2 may vary slightly between different species, the current hypothesis is that AI-2 can transmit information about both the cell density and species composition of a bacterial community.

Pathogenic bacteria use quorum sensing to regulate the expression of genes whose products help these bacteria invade a host and avoid immune system detection. For example, *Vibrio cholerae*, the causative agent of cholera, uses AI-2 and an additional species-specific autoinducer to activate the genes controlling the production of cholera toxin. *Pseudomonas aeruginosa*, the Gram-negative bacterium that often affects cystic fibrosis patients, uses quorum sensing to regulate the production of elastase, a protease that disrupts the respiratory epithelium and interferes with ciliary function. *P. aeruginosa* also uses autoinducers to control the production of biofilms, tough protective shells that resist host defenses and make treatment with antibiotics nearly impossible. Other bacteria determine cell density through quorum sensing to delay the production of toxic substances until the colony is large enough to overpower the host's immune system and establish an infection. Because many bacteria rely on quorum sensing to regulate disease-causing genes, therapeutics that block quorum sensing may help combat infections.

Research into these potential therapies is now in progress, and several are now approaching the clinical trial phase. Thus, what began as a fascinating observation in the glowing squid has launched an exciting era of research in bacterial genetics that may one day prove of great clinical significance.

Your Turn

Take time, individually or in groups, to answer the following questions. Investigate the references and links to help you understand the mechanisms and potential uses of quorum sensing in bacteria.

1. Inhibitors of quorum sensing molecules have potential as antibacterial agents. What are some ways in which quorum sensing inhibitors could work to combat bacterial infections? Have any of these therapeutics reached clinical trials?

A recent review of quorum sensing therapeutics can be found in Njoroge, J. and Sperandio, V. 2009. Jamming bacterial communication: New approaches for the treatment of infectious diseases. EMBO Mol. Med. 1(4): 201–210.

2. Regulation of bacterial gene expression by autoinducer molecules involves several different mechanisms. Describe these mechanisms and how each could be used as a target for the control of bacterial infections.

The mechanisms by which autoinducers regulate gene expression are summarized in Asad, S. and Opal, S.M. 2008. Bench-to-bedside review: Quorum sensing and the role of cell-to-cell communication during invasive bacterial infection. Critical Care 12: 236–247.

3. Quorum sensing systems are also capable of detecting and responding to chemical signals given off by host cells. Explain how this works and how this might benefit pathogenic bacteria.

A review article dealing with interkingdom communication and quorum sensing can be found at Wagner, V.E. et al. 2006. Quorum sensing: dynamic response of Pseudomonas aeruginosa to external signals. Trends Microbiol. 14(2): 55–58.

INSIGHTS AND SOLUTIONS

1. A hypothetical operon (*theo*) in *E. coli* contains several structural genes encoding enzymes that are involved sequentially in the biosynthesis of an amino acid. Unlike the *lac* operon, in which the repressor gene is separate from the operon, the gene encoding the regulator molecule is contained within the *theo* operon. When the end product (the amino acid) is present, it combines with the regulator molecule, and this complex binds to the operator, repressing the operon. In the absence of the amino acid, the regulatory molecule fails to bind to the operator, and transcription proceeds.

Categorize and characterize this operon, then consider the following mutations, as well as the situation in which the wild-type gene is present along with the mutant gene in partially diploid cells (F′):

(a) Mutation in the operator region.

(b) Mutation in the promoter region.

(c) Mutation in the regulator gene.

In each case, will the operon be active or inactive in transcription, assuming that the mutation affects the regulation of the *theo* operon? Compare each response with the equivalent situation for the *lac* operon.

Solution: The *theo* operon is repressible and under negative control. When there is no amino acid present in the medium (or the environment), the product of the regulatory gene cannot bind to the operator region, and transcription proceeds under the direction of RNA polymerase. The enzymes necessary for the synthesis of the amino acid are produced, as

is the regulator molecule. If the amino acid *is* present, either initially or after sufficient synthesis has occurred, the amino acid binds to the regulator, forming a complex that interacts with the operator region, causing repression of transcription of the genes within the operon.

The *theo* operon is similar to the tryptophan system, except that the regulator gene is within the operon rather than separate from it. Therefore, in the *theo* operon, the regulator gene is itself regulated by the presence or absence of the amino acid.

(a) As in the *lac* operon, a mutation in the *theo* operator gene inhibits binding with the repressor complex, and transcription occurs constitutively. The presence of an F′ plasmid bearing the wild-type allele would have no effect, since it is not adjacent to the structural genes.

(b) A mutation in the *theo* promoter region would no doubt inhibit binding to RNA polymerase and therefore inhibit transcription. This would also happen in the *lac* operon. A wild-type allele present in an F′ plasmid would have no effect.

(c) A mutation in the *theo* regulator gene, as in the *lac* system, may inhibit either its binding to the repressor or its binding to the operator gene. In both cases, transcription will be constitutive, because the *theo* system is repressible. Both cases result in the failure of the regulator to bind to the operator, allowing transcription to proceed. In the *lac* system, failure to bind the corepressor lactose would permanently repress the system. The addition of a wild-type allele would restore repressibility, provided that this gene was transcribed constitutively.

Problems and Discussion Questions

 For instructor-assigned tutorials and problems, go to www.masteringgenetics.com

HOW DO WE KNOW?

1. In this chapter, we focused on the regulation of gene expression in prokaryotes. Along the way, we found many opportunities to consider the methods and reasoning by which much of this information was acquired. From the explanations given in the chapter, what answers would you propose to the following fundamental questions?
   (a) How do we know that bacteria regulate the expression of certain genes in response to the environment?
   (b) What evidence established that lactose serves as the inducer of a gene whose product is related to lactose metabolism?
   (c) What led researchers to conclude that a repressor molecule regulates the lac operon?
   (d) How do we know that the *lac* repressor is a protein?
   (e) How do we know that the *trp* operon is a repressible control system, in contrast to the *lac* operon, which is an inducible control system?

2. Contrast the need for the enzymes involved in lactose and tryptophan metabolism in bacteria when lactose and tryptophan, respectively, are (a) present and (b) absent.

3. Contrast positive versus negative control of gene expression.

4. Contrast the role of the repressor in an inducible system and in a repressible system.

5. For the *lac* genotypes shown in the accompanying table, predict whether the structural genes (Z) are constitutive, permanently repressed, or inducible in the presence of lactose.

Genotype	Constitutive	Repressed	Inducible
$I^+O^+Z^+$			×
$I^-O^+Z^+$			
$I^-O^cZ^+$			
$I^-O^cZ^+/F'O^+$			
$I^+O^cZ^+/F'O^+$			
$I^sO^+Z^+$			
$I^sO^+Z^+/F'I^+$			

6. For the genotypes and conditions (lactose present or absent) shown in the following table, predict whether functional enzymes, nonfunctional enzymes, or no enzymes are made.

Genotype	Condition	Functional Enzyme Made	Nonfunctional Enzyme Made	No Enzyme Made
$I^+O^+Z^+$	No lactose			×
$I^+O^CZ^+$	Lactose			
$I^-O^+Z^-$	No lactose			
$I^-O^+Z^-$	Lactose			
$I^-O^+Z^+/F'I^+$	No lactose			
$I^+O^CZ^+/F'O^+$	Lactose			
$I^+O^+Z^-/$ $F'I^+O^+Z^+$	Lactose			
$I^-O^+Z^-/$ $F'I^+O^+Z^+$	No lactose			
$I^SO^+Z^+/F'O^+$	No lactose			
$I^+O^CZ^+/$ $F'O^+Z^+$	Lactose			

7. The locations of numerous $lacI^-$ and $lacI^S$ mutations have been determined within the DNA sequence of the *lacI* gene. Among these, $lacI^-$ mutations were found to occur in the 5'-upstream region of the gene, while $lacI^S$ mutations were found to occur farther downstream in the gene. Are the locations of the two types of mutations within the gene consistent with what is known about the function of the repressor that is the product of the *lacI* gene?

8. Describe the experimental rationale that allowed the *lac* repressor to be isolated.

9. What properties demonstrate the *lac* repressor to be a protein? Describe the evidence that it indeed serves as a repressor within the operon system.

10. Predict the effect on the inducibility of the *lac* operon of a mutation that disrupts the function of (a) the *crp* gene, which encodes the CAP protein, and (b) the CAP-binding site within the promoter.

11. Erythritol, a natural sugar abundant in fruits and fermenting foods, is about 65 percent as sweet as table sugar and has about 95 percent less calories. It is "tooth friendly" and generally devoid of negative side effects as a human consumable product. Pathogenic *Brucella* strains that catabolize erythritol contain four closely spaced genes, all involved in erythritol metabolism. One of the four genes (*eryD*) encodes a product that represses the expression of the other three genes. Erythritol catabolism is stimulated by erythritol. Present a simple regulatory model to account for the regulation of erythritol catabolism in *Brucella*. Does this system appear to be under inducible or repressible control?

12. Describe the role of attenuation in the regulation of tryptophan biosynthesis.

13. Attenuation of the *trp* operon was viewed as a relatively inefficient way to achieve genetic regulation when it was first discovered in the 1970s. Since then, however, attenuation has been found to be a relatively common regulatory strategy. Assuming that attenuation is a relatively inefficient way to achieve genetic regulation, what might explain its widespread occurrence?

14. Neelaredoxin is a 15-kDa protein that is a gene product common in anaerobic prokaryotes. It has superoxide-scavenging activity, and it is *constitutively expressed*. In addition, its expression is not further *induced* during its exposure to O_2 or H_2O_2 (Silva, G., et al. 2001. *J. Bacteriol.* 183: 4413–4420). What do the terms *constitutively expressed* and *induced* mean in terms of neelaredoxin synthesis?

15. Milk products such as cheeses and yogurts are dependent on the conversion by various anaerobic bacteria, including several *Lactobacillus* species, of lactose to glucose and galactose, ultimately producing lactic acid. These conversions are dependent on both permease and β-galactosidase as part of the *lac* operon. After selection for rapid fermentation for the production of yogurt, one *Lactobacillus* subspecies lost its ability to regulate *lac* operon expression (Lapierre, L., et al. 2002. *J. Bacteriol.* 184: 928–935). Would you consider it likely that in this subspecies the *lac* operon is "on" or "off"? What genetic events would likely contribute to the loss of regulation as described above?

16. Assume that the structural genes of the *lac* operon have been fused, through recombinant DNA techniques, to the regulatory apparatus of the *ara* operon. If arabinose is provided in a minimal medium to *E. coli* carrying this gene fusion, would you expect β-galactosidase to be produced at induced levels? Explain.

17. Both attenuation of the *trp* operon in *E. coli* and riboswitches in *B. subtilis* rely on changes in the secondary structure of the leader regions of mRNA to regulate gene expression. Compare and contrast the specific mechanisms in these two types of regulation.

18. Keeping in mind the life cycle of bacteriophages discussed in Chapter 6, consider the following problem: During the reproductive cycle of a temperate bacteriophage, the viral DNA inserts into the bacterial chromosome where the resultant prophage behaves much like a Trojan horse. It can remain quiescent, or it can become lytic and initiate a burst of progeny viruses. Several operons maintain the prophage state by interacting with a repressor that keeps the lytic cycle in check. Insults (ultraviolet light, for example) to the bacterial cell lead to a partial breakdown of the repressor, which in turn causes the production of enzymes involved in the lytic cycle. As stated in this simple form, would you consider this system of regulation to be operating under positive or negative control?

Extra-Spicy Problems

 For instructor-assigned tutorials and problems, go to www.masteringgenetics.com

19. Bacterial strategies to evade natural or human-imposed antibiotics are varied and include membrane-bound efflux pumps that export antibiotics from the cell. A review of efflux pumps (Grkovic, S., et al., 2002) states that, because energy is required to drive the pumps, activating them in the absence of the antibiotic has a selective disadvantage. The review also states that a given antibiotic may play a role in the regulation of efflux by interacting with either an activator protein or a repressor protein, depending on the system involved. How might such systems be categorized in terms of *negative control* (*inducible* or *repressible*) or *positive control* (*inducible* or *repressible*)?

20. In a theoretical operon, genes A, B, C, and D represent the repressor gene, the promoter sequence, the operator gene, and the structural gene, *but not necessarily in the order named*. This operon is concerned with the metabolism of a theoretical molecule (tm). From the data provided in the accompanying table, first decide whether the operon is inducible or repressible. Then assign A, B, C, and D to the four parts of the operon. Explain your rationale. (AE = active enzyme; IE = inactive enzyme; NE = no enzyme)

Genotype	tm Present	tm Absent
$A^+B^+C^+D^+$	AE	NE
$A^-B^+C^+D^+$	AE	AE
$A^+B^-C^+D^+$	NE	NE
$A^+B^+C^-D^+$	IE	NE
$A^+B^+C^+D^-$	AE	AE
$A^-B^+C^+D^+/F'A^+B^+C^+D^+$	AE	AE
$A^+B^-C^+D^+/F'A^+B^+C^+D^+$	AE	NE
$A^+B^+C^-D^+/F'A^+B^+C^+D^+$	AE + IE	NE
$A^+B^+C^+D^-/F'A^+B^+C^+D^+$	AE	NE

21. A bacterial operon is responsible for the production of the biosynthetic enzymes needed to make the hypothetical amino acid tisophane (tis). The operon is regulated by a separate gene, R, deletion of which causes the loss of enzyme synthesis. In the wild-type condition, when tis is present, no enzymes are made; in the absence of tis, the enzymes are made. Mutations in the operator gene (O^-) result in repression regardless of the presence of tis. Is the operon under positive or negative control? Propose a model for (a) repression of the genes in the presence of tis in wild-type cells and (b) the mutations.

22. A marine bacterium is isolated and shown to contain an inducible operon whose genetic products metabolize oil when it is encountered in the environment. Investigation demonstrates that the operon is under positive control and that there is a *reg* gene whose product interacts with an operator region (o) to regulate the structural genes, designated *sg*.

    In an attempt to understand how the operon functions, a constitutive mutant strain and several partial diploid strains were isolated and tested with the results shown here:

Host Chromosome	F′ Factor	Phenotype
wild type	none	inducible
wild type	*reg* gene from mutant strain	inducible
wild type	operon from mutant strain	constitutive
mutant strain	*reg* gene from wild type	constitutive

Draw all possible conclusions about the mutation as well as the nature of regulation of the operon. Is the constitutive mutation in the *trans*-acting *reg* element or in the *cis*-acting *o* operator element?

23. The SOS repair genes in *E. coli* (discussed in Chapter 13) are negatively regulated by the *lexA* gene product, called the LexA repressor. When a cell sustains extensive damage to its DNA, the LexA repressor is inactivated by the *recA* gene product (RecA), and transcription of the SOS genes is increased dramatically.

One of the SOS genes is the *uvrA* gene. You are a student studying the function of the *uvrA* gene product in DNA repair. You isolate a mutant strain that shows constitutive expression of the UvrA protein. Naming this mutant strain *uvrA^C*, you construct the following diagram showing the *lexA* and *uvrA* operons:

(a) Describe two different mutations that would result in a *uvrA* constitutive phenotype. Indicate the actual genotypes involved.
(b) Outline a series of genetic experiments that would use partial diploid strains to determine which of the two possible mutations you have isolated.

24. A fellow student considers the issues in Problem 23 and argues that there is a more straightforward, nongenetic experiment that could differentiate between the two types of mutations. The experiment requires no fancy genetics and would allow you to easily assay the products of the other SOS genes. Propose such an experiment.

25. Figure 16–13 depicts numerous critical regions of the leader sequence of mRNA that play important roles during the process of attenuation in the *trp* operon. A closer view of the leader sequence, which begins at about position 30 downstream from the 5′ end, is given below (running across both columns of this page):

AUGAAAGCAAUUUUCGUACUGAAAGGUUGGUGGCGCACUUCCUGAAACGGGCAGUGUAUUCACCAUGCGUAAAGCAAUCAGAUACCCAGCCCGCCUAAUGAGCGGGCUUUUUUUU

Within this molecule are the sequences that cause the formation of the alternative hairpins. It also contains the successive triplets that encode tryptophan, where stalling during translation occurs. Take a large piece of paper (such as manila wrapping paper) and, along with several other students from your genetics class, work through the base sequence to identify the *trp* codons and the parts of the molecule representing the base-pairing regions that form the terminator and antiterminator hairpins shown in Figure 16–13.

26. One of the most prevalent sexually transmitted diseases is caused by the bacterium *Chlamydia trachomatis* and leads to blindness if left untreated. Upon infection, metabolically inert cells differentiate, through gene expression, to become metabolically active cells that divide by binary fission. It has been proposed that release from the inert state is dependent on heat-shock proteins that both activate the reproductive cycle and facilitate the binding of chlamydiae to host cells. Researchers made the following observations regarding the heat-shock regulatory system in *Chlamydia trachomatis*: (1) a regulator protein (call it R) binds to a *cis*-acting DNA element (call it D); (2) R and D function as a repressor–operator pair; (3) R functions as a negative regulator of transcription; (4) D is composed of an inverted-repeat sequence; (5) repression by R is dependent on D being supercoiled (Wilson & Tan, 2002).
(a) Based on this information, devise a model to explain the heat-dependent regulation of metabolism in *Chlamydia trachomatis*.
(b) Some bacteria, like *E. coli*, use a heat-shock sigma factor to regulate heat-shock transcription. Are the above findings in *Chlamydia* compatible with use of a heat-sensitive sigma factor?

27. A reporter gene—that is, a gene whose activity is relatively easy to detect—is commonly used to study promoter activity under a variety of investigative circumstances. Recombinant

DNA technology is used to attach such a reporter gene to the promoter of interest, after which the recombinant molecule is inserted by transformation or transfection into the cell or host organism of choice. Green fluorescent protein (GFP) from the jellyfish *Aequorea victoria* fluoresces green when exposed to blue light. Assume that you wished to study the regulatory apparatus of the *lac* operon and you developed a plasmid (p) that contained all the elements of the *lac* operon except that you replaced the *lac* structural genes with the *GFP* gene (your reporter gene). You used this plasmid to transform *E. coli*. Considering the following genotypes, under which of the following conditions would you expect β-galactosidase production and/or green colonies when examined under blue light?

Genotype	Medium Condition		β-galactosidase	Green Colonies
	Lactose	Glucose		
$I^+O^+Z^+/pI^+O^+GFP$	−	+		
	+	+		
	+	−		
$I^+O^+Z^-/pI^+O^cGFP$	−	+		
	+	−		
$I^+O^+Z^+/pI^sO^cGFP$	−	+		
	+	−		
$I^+O^cZ^+/pI^sO^+GFP$	−	+		
	+	−		

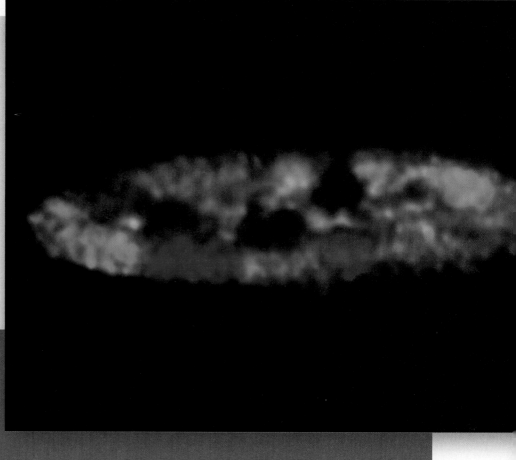

17

Regulation of Gene Expression in Eukaryotes

CHAPTER CONCEPTS

- Expression of genetic information is regulated by mechanisms that exert control over transcription, mRNA processing, export from the nucleus, mRNA stability, translation, and posttranslational modifications.

- Chromatin remodeling, histone modifications, and DNA alterations play important roles in regulating gene expression in eukaryotes.

- Programmed DNA rearrangements contribute to gene regulation in a small number of genes.

- Eukaryotic transcription initiation requires the assembly of transcription regulatory proteins at *cis*-acting DNA sites known as promoters, enhancers, and silencers.

- Transcription activators and repressors stimulate the association of the general transcription factors into pre-initiation complexes at gene promoters.

- Eukaryotic gene expression is also regulated by multiple posttranscriptional mechanisms, including alternative splicing of pre-mRNA, export from the nucleus, control of mRNA stability, translation regulation, and posttranslational processing.

- RNA-induced gene silencing controls gene expression in several ways and is being harnessed for scientific research and medical treatments.

Virtually all cells in a eukaryotic organism contain a complete genome; however, only a subset of genes is expressed in any particular cell type. For example, some white blood cells express genes encoding certain immunoglobulins, allowing these cells to synthesize antibodies that defend the organism from infection and foreign agents. However, skin, kidney, and liver cells do not express immunoglobulin genes. Pancreatic islet cells synthesize and secrete insulin in response to the presence of blood sugars; however, they do not manufacture immunoglobulins. In addition, they do not synthesize insulin when it is not required. Eukaryotic cells, as part of multicellular organisms, do not grow solely in response to the availability of nutrients. Instead, they regulate their growth and division to occur at appropriate places in the body and at appropriate times during development. The loss of gene regulation that controls normal cell growth and division may lead to developmental defects or cancer.

Eukaryotic gene regulation is one of the most rapidly advancing fields in genetic research. Our ever-increasing understanding about the mechanisms that regulate gene expression in eukaryotes is contributing knowledge that is fascinating and often surprising. It is also leading to practical applications, such as potential therapies for many human diseases.

In this chapter, we will explore the regulation of gene expression in eukaryotes. We will outline some of the general features of gene regulation, from DNA rearrangements and chromatin modifications to transcriptional and postranscriptional mechanisms.

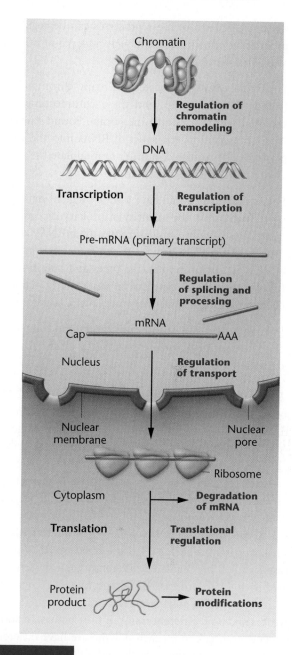

FIGURE 17–1 Regulation can occur at any stage in the expression of genetic material in eukaryotes. All these forms of regulation affect the degree to which a gene is expressed.

17.1

Eukaryotic Gene Regulation Can Occur at Any of the Steps Leading from DNA to Protein Product

In eukaryotes, gene expression is tightly controlled in order to express the required levels of gene products at specific times, in specific cell types, and in response to complex changes in the environment. To achieve this degree of fine tuning, eukaryotes employ a wide range of mechanisms for altering the expression of genes. In contrast to prokaryotic gene regulation, which occurs primarily at the level of transcription initiation, regulation of gene expression in eukaryotes can occur at many different levels. These include alterations in the DNA template, initiation of transcription, mRNA modifications and stability, and synthesis, modification, and stability of the protein product (Figure 17–1). Several features of eukaryotic cells make it possible for them to use more types of gene regulation than are possible in prokaryotic cells:

- Eukaryotic cells contain a much greater amount of DNA than do prokaryotic cells, and this DNA is associated with histones and other proteins to form highly compact chromatin structures within an enclosed nucleus. Eukaryotic cells modify this structural organization in order to influence gene expression.

- The mRNAs of most eukaryotic genes must be spliced, capped, and polyadenylated prior to transport from the nucleus. Each of these processes can be regulated in order to influence the numbers and types of mRNAs available for translation.

- Eukaryotic genes are located on many chromosomes (rather than just one), and these chromosomes are enclosed within a double-membrane-bound nucleus. After transcription, transport of RNAs into the cytoplasm can be regulated in order to modulate the availability of mRNAs for translation.

- Eukaryotic mRNAs can have a wide range of half-lives ($t_{1/2}$). In contrast, the majority of prokaryotic mRNAs decay very rapidly. Rapid turnover of mRNAs allows prokaryotic cells to rapidly respond to environmental changes. In eukaryotes, the types and quantities of mRNAs in each cell type can be more subtly manipulated by altering mRNA decay rates over a larger range.

- In eukaryotes, translation rates can be modulated, as can the way proteins are processed, modified, and degraded.

In the following sections, we examine some of the major ways in which eukaryotic gene expression is regulated. We will limit our discussion to regulation of genes transcribed by RNA polymerase II. As previously described in Chapter 13, eukaryotes use three different RNA polymerases for transcription. RNA polymerase II transcribes genes that encode all mRNAs and some small nuclear RNAs, whereas RNA polymerases I and III transcribe genes that code for ribosomal RNAs, some small nuclear RNAs, and transfer RNAs. The genes transcribed by RNA polymerase I and III are regulated differently than those transcribed by RNA polymerase II. The promoters recognized by each type of RNA polymerase have different nucleotide sequences and bind different transcription factors. In addition, genes transcribed by each polymerase have different transcription termination signals and mRNA processing mechanisms.

17.2

Programmed DNA Rearrangements Regulate Expression of a Small Number of Genes

Except for the occurrence of sequence alterations due to mutations, genomic DNA in most organisms remains remarkably stable. However, despite this genomic stability, some significant examples of gene regulation by DNA rearrangement exist. In Chapter 8, we were introduced to DNA amplification and learned how amplification of specific genes, such as the *rRNA* genes in *Xenopus* oocytes, allows expression

of these genes at times during development when high levels of these gene products are required.

Besides gene amplifications, several other types of programmed DNA rearrangements occur during developmental regulation of some eukaryotic genes. One type of rearrangement results in the creation of new genes from gene fragments located at various positions in the genome. Another type of rearrangement results in a switch in expression of genes as a result of a recombination or DNA copying event. In these cases, these DNA rearrangements are programmed and essential for normal expression of the genes involved. In addition, rearrangements that lead to loss of some DNA sequences occur only in particular types of somatic cells in higher eukaryotes; hence, these changes are not passed on to progeny.

In this section, we present an example of gene regulation by **programmed DNA rearrangements**: the immunoglobulin system in humans.

The Immune System and Antibody Diversity

The immune system protects organisms against infections and the presence of foreign substances that may enter blood or body tissues. It does this by recognizing molecules on the surface of these foreign substances, and by physically binding to them in a kind of lock-and-key configuration—a phenomenon known as **antigen recognition**. **Antigens** are defined as molecules, usually proteins, which bring about an immune response. Although the immune system has many components, we will focus on **humoral immunity**, which involves the production of proteins called **immunoglobulins**, or **antibodies**, that bind directly to antigens.

Immunoglobulins are synthesized by a type of blood cell known as the **B lymphocyte**, or **B cell**, which undergoes development and maturation in the bone marrow. Immunoglobulin molecules consist of four polypeptide chains, held together by disulfide bonds. Two of these polypeptides are identical **light (L) chains**, and two are identical **heavy (H) chains** (Figure 17–2). Together, they form a Y-shaped immunoglobulin structure. Each of the light and heavy chains contains a **constant region** at the C-terminus of the polypeptide and a **variable region** at the N-terminus. The four variable regions, when combined together into the immunoglobulin molecule, form a unique structure that recognizes one specific antigen. Each B cell synthesizes only one type of immunoglobulin.

Mammals are capable of producing hundreds of millions of different types of antibodies in response to the presence of a wide variety of antigens—in fact, more types of antibodies than there are genes in the genome. How can such immense diversity arise from a limited number of genes? The answer lies in the fact that the DNA encoding

(a)

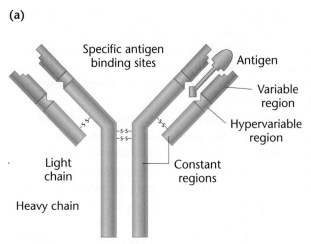

(b)

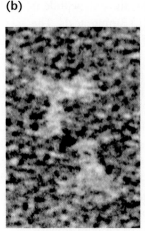

FIGURE 17–2 Structure of immunoglobulin molecules. (a) Immunoglobulin molecules are Y-shaped and consist of four polypeptide chains. Each chain contains a constant region (purple), a variable region (red), and a hypervariable region (blue). The variable and hypervariable regions form an antigen-recognition site that interacts with a specific antigen in a lock-and-key arrangement. The chains are joined by disulfide bonds. (b) This colored scanning electron micrograph reveals the Y-shaped structure of immunoglobulin molecules, which have been negatively stained prior to electron microscopy.

the immunoglobulin genes is not fixed but undergoes multiple programmed changes, including deletions, translocations, and random mutations.

Gene Rearrangements in the κ Light-chain Gene

To illustrate how DNA rearrangements contribute to antibody diversity, we will describe the formation of one type of the immunoglobulin light-chain gene (the κ **light-chain gene**) in humans.

The human κ light-chain gene is assembled during B-cell development, from multiple DNA regions located along chromosome 2. In most somatic cells, as well as in germ cells, these DNA regions are organized as shown in Figure 17–3. There are 70 to 100 different **L (leader)** and **V (variable) regions**, about half of which are functional and can encode the variable (V) portion of the κ light-chain molecule. Each LV region is preceded by transcription initiation sequences. The LV regions are located about 6 kilobases away from the **J (joining) region**, a region that contains five possible J-encoding exons. Adjacent to the J region is an enhancer sequence that stimulates transcription from any promoter located nearby, and a single **C exon** that encodes the constant region of the κ light-chain molecule.

During development of a mature B cell, one of the LV regions (L_2V_2 in Figure 17–3), along with its promoter, is randomly joined by a recombination event to one of the five J regions (J_3 in our example). This recombination event deletes the entire intervening DNA on chromosome 2. After this somatic recombination event, the LV promoter is activated by the nearby presence of the immunoglobulin enhancer, resulting in transcription of the gene. The resulting RNA is spliced to remove the introns between the L and V regions, and between the J and C regions.

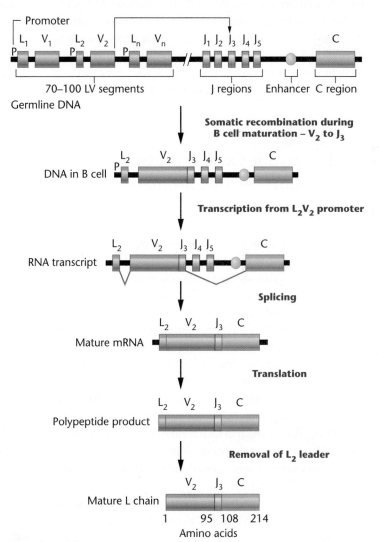

FIGURE 17–3 Assembly of a κ light-chain gene in a B cell, followed by its transcription, mRNA processing, translation, and postranslational modification. The B-cell gene contains one LV region (L_2V_2) recombined with one J region (J_3). The remaining J regions and the enhancer-containing intron are removed from the transcript by intron splicing. Following translation, the leader sequence (L_2) is cleaved off as the mature polypeptide chain crosses the cell membrane.

After translation of the mature mRNA, the polypeptide is processed by removal of the leader (L) amino acids. The rearranged κ gene is then maintained and passed on to all progeny of that B cell.

Antibody diversity results in part from the random recombination of one of 35 to 50 different functional LV regions with any one of five different J regions. Two other mechanisms further increase the level of diversity. The recombination event that joins an LV region to a J region is not precise and can occur anywhere within a region containing several base pairs. Because of this imprecision, the recombination between any particular pair of LV and J regions still shows considerable variation. In addition, V regions are susceptible to high rates of random somatic mutation—**hypermutation**—during B-cell development. Hypermutation introduces even more variation into the LVJ region's sequence. The mechanisms of DNA rearrangements described here for the human κ light-chain genes also occur during the formation of the gene that encodes another type of light-chain gene (called the λ light-chain gene) as well as the gene that encodes the immunoglobulin heavy chains.

17.3

Eukaryotic Gene Expression Is Influenced by Chromatin Modifications

Two structural features of eukaryotic genes distinguish them from the genes of prokaryotes. First, eukaryotic genes are situated on chromosomes that occupy a distinct location within the cell—the nucleus. This sequestering of genetic information in a discrete compartment allows the proteins that directly regulate transcription to be kept apart from those involved with translation and other aspects of cellular metabolism. Second, as described in Chapter 12, eukaryotic DNA is combined with histones and nonhistone proteins to form chromatin. Chromatin's basic structure is characterized by repeating units called nucleosomes that are wound into 30-nm fibers, which in turn form other, even more compact structures. The presence of these chromatin structures is inhibitory to many processes, including transcription, replication, and DNA repair.

In this section, we outline some of the ways in which eukaryotic cells modify chromatin in order to regulate gene expression.

Chromosome Territories and Transcription Factories

The development of chromosome-painting techniques has revealed that the interphase nucleus is not a bag of tangled chromosome arms, but has a highly organized structure. In the interphase nucleus, each chromosome occupies a discrete domain called a **chromosome territory** and stays separate from other chromosomes. Channels between chromosomes contain little or no DNA and are called **interchromosomal domains**.

Research suggests that transcriptionally active genes are located at the edges of chromosome territories next to interchromosomal domain channels. Scientists hypothesize that this organization may bring actively expressed genes into closer association with transcription factors, or with other actively expressed genes, thereby facilitating their coordinated expression.

Another feature within the nucleus—the **transcription factory**—may also contribute to regulating gene expression. Transcription factories are nuclear sites that contain most of the active RNA polymerase and transcription regulatory molecules. By concentrating transcription proteins and actively transcribed genes in specific locations in the nucleus, the cell may enhance the expression of these genes.

Histone Modifications and Nucleosomal Chromatin Remodeling

The ability of the cell to alter the association of DNA with other chromatin components is essential to allow regulatory proteins to access DNA. Hence, chromatin modification is an important step in gene regulation. Chromatin modification appears to be a prerequisite for transcription of some eukaryotic genes, although it can occur simultaneously with transcription of other genes.

Chromatin can be modified in two general ways. The first involves changes to nucleosomes, and the second involves modifications to DNA. In this subsection, we will discuss changes to the nucleosomal component of chromatin. In the next subsection, we present DNA modifications, specifically DNA methylation.

Changes in nucleosome composition can affect gene transcription. For example, most nucleosomes contain the normal histones H2A and H3. Some gene promoter regions may be flanked by nucleosomes containing variant histones, such as H2A.Z and H3.3. These variant nucleosomes help keep promoter regions free of normal repressive nucleosomes, thereby facilitating gene transcription.

A second mechanism of chromatin alteration involves histone modification. One such modification is acetylation, a chemical alteration of the histone component of nucleosomes that is catalyzed by **histone acetyltransferase enzymes (HATs)**. When an acetate group is added to specific basic amino acids on a histone tail, the attraction between the basic histone protein and acidic DNA is lessened. These modifications make promoter regions available for binding to transcription factors that initiate the chain of events leading to gene transcription. In some cases, HATs are recruited to genes by the presence of certain transcription activator proteins that bind to transcription regulatory regions. Of course, what can be opened can also be closed. In that case, **histone deacetylases (HDACs)** remove acetate groups from histone tails. HDACs can be recruited to genes by the presence of certain repressor proteins on regulatory regions. In addition to acetylation, histones can be modified in several other ways, including phosphorylation and methylation.

The third mechanism is chromatin remodeling which involves the repositioning or removal of nucleosomes on DNA, brought about by chromatin remodeling complexes. Repositioned nucleosomes make regions of the chromosome accessible to transcription regulatory proteins, such as transcription activators and RNA polymerase II. One of the best-studied remodeling complexes is the **SWI/SNF** complex. Remodelers such as SWI/SNF can act in several different ways (Figure 17–4). They may loosen the attachment between histones and DNA, resulting in the nucleosome sliding along the DNA and exposing regulatory regions. Alternatively, they may loosen the DNA strand from the nucleosome core, or they may cause reorganization of the internal nucleosome components. In all cases, the DNA is left transiently exposed to association with transcription factors and RNA polymerase.

(a) Alteration of DNA-histone contacts

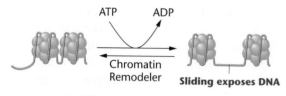

(b) Alteration of the DNA path

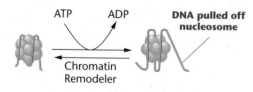

(c) Remodeling of nucleosome core particle

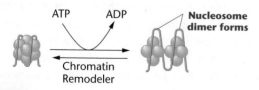

FIGURE 17–4 Three ways by which chromatin remodelers, such as the SWI/SNF complex, alter the association of nucleosomes with DNA. (a) The DNA-histone contacts may be loosened, allowing the nucleosomes to slide along the DNA, exposing DNA regulatory regions. (b) The path of the DNA around a nucleosome core particle may be altered. (c) Components of the core nucleosome particle may be rearranged, resulting in a modified nucleosome structure.

DNA Methylation

Another type of change in chromatin that plays a role in gene regulation is the addition or removal of methyl groups to or from bases in DNA. The DNA of most eukaryotic organisms can be modified after DNA replication by the enzyme-mediated addition of methyl groups to bases and sugars. **DNA methylation** most often occurs at position 5 of cytosine (**5-methylcytosine**), causing the methyl group to protrude into the major groove of the DNA helix. Methylation occurs most often on the cytosine of CG doublets in DNA, usually on both strands:

$$5' - {}^{m}CpG - 3'$$
$$3' - GpC^{m} - 5'$$

Potentially, methylatable CpG sequences are not randomly distributed throughout the genome, but are concentrated in CpG-rich regions, called **CpG islands**, located at the 5′ends of genes, usually in promoter regions. In the genome of eukaryotic species, approximately 5 percent of the cytosine residues are methylated.

Evidence of a role for methylation in eukaryotic gene expression is based on a number of observations. First, an inverse relationship exists between the degree of methylation and the degree of expression. Large transcriptionally inert regions of the genome, such as the inactivated X chromosome in mammalian female cells, are often heavily methylated.

Second, methylation patterns are tissue specific and, once established, are heritable for all cells of that tissue. It appears that proper patterns of DNA methylation are essential for normal mammalian development. Despite this heritability, DNA methylation in specific gene regions can be altered by methylase and demethylase enzymes in order to activate or silence regions of DNA.

Perhaps the most direct evidence of a role for methylation in gene expression comes from studies using base analogs. The nucleoside **5-azacytidine** can be incorporated into DNA in place of cytidine during DNA replication. This analog cannot be methylated, causing the undermethylation of the sites where it is incorporated. The incorporation of 5-azacytidine into DNA changes the pattern of gene expression and stimulates expression of alleles on inactivated X chromosomes. In addition, the presence of 5-azacytidine in DNA can induce the expression of genes that would normally be silent in certain differentiated cells.

How might methylation affect gene regulation? Data from *in vitro* studies suggest that methylation can repress transcription by inhibiting the binding of transcription factors to DNA. Methylated DNA may also recruit repressive chromatin remodeling complexes to gene-regulatory regions.

17–1 Cancer cells often have abnormal patterns of chromatin modifications. In some cancers, the DNA repair genes *MLH1* and *BRCA1* are hypermethylated on their promoter regions. Explain how this abnormal methylation pattern could contribute to cancer.

■ HINT: *This problem involves an understanding of the types of genes that are mutated in cancer cells. The key to its solution is to consider how methylation affects gene expression of cancer-related genes.*

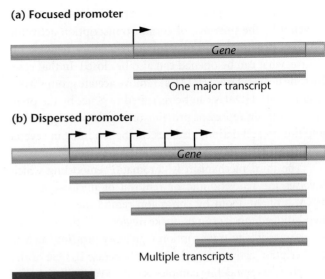

(a) Focused promoter

One major transcript

(b) Dispersed promoter

Multiple transcripts

FIGURE 17–5   Focused and dispersed promoters. Focused promoters (a) specify one specific transcription initiation site. Dispersed promoters (b) specify weak transcription initiation at multiple start site positions over an approximately 100-bp region. Dispersed promoters are common in vertebrates and are associated with housekeeping genes. Transcription start sites and the directions of transcription are indicated with arrows.

17.4

Eukaryotic Transcription Initiation Is Regulated at Specific *Cis*-Acting Sites

Eukaryotic transcription regulation requires the binding of many regulatory factors to specific DNA sequences located in and around genes, as well as to sequences located at great distances. In this section, we will discuss some of the DNA sequences—known as *cis*-acting sequences—that are required for the accurate and regulated transcription of genes transcribed by RNA polymerase II. In particular, we will discuss promoter, enhancer, and silencer elements. As previously defined in Chapter 13, a **cis-acting sequence** is one that is located on the same chromosome as the gene that it regulates. This is in contrast to the actions of **trans-acting factors** (such as DNA-binding proteins), that can regulate a gene on any chromosome.

Promoter Elements

A **promoter** is a region of DNA that recognizes the transcription machinery and binds one or more proteins that regulate transcription initiation. Promoters are necessary in order for transcription to be initiated accurately and at a basal level. Promoters are located immediately adjacent to the genes they regulate. They may be up to several hundred nucleotides in length and specify the site or sites at which transcription begins and the direction of transcription along the DNA. Within promoters are a number of **promoter elements**—short nucleotide sequences that bind specific regulatory factors.

There are two subcategories within eukaryotic promoters. First, the **core promoter** determines the accurate initiation of transcription by RNA polymerase II. Second, **proximal promoter elements** are those that modulate the efficiency of basal levels of transcription.

Recent bioinformatic research has introduced new complexities to our understanding of promoters and how they work. This research reveals that there is a great deal of diversity in eukaryotic promoters, in terms of both their structures and functions. Promoters are now thought to be either *focused* or *dispersed*. **Focused promoters** specify transcription initiation at a single specific nucleotide (the **transcription start site**). In contrast, **dispersed promoters** direct initiation from a number of weak transcription start sites located over a 50- to 100-nucleotide region (Figure 17–5). Focused transcription initiation is the major type of initiation for most genes of lower eukaryotes, but for only about 30 percent of vertebrate genes. Focused promoters are usually associated with genes whose transcription levels are highly regulated, whereas dispersed promoters are associated with genes that are transcribed constitutively.

Little is known about the DNA elements that make up dispersed promoters. These promoters are usually found within CpG islands, suggesting that chromatin modifications may influence initiation from these promoters. In addition, dispersed promoters are free of nucleosomes over a 150- to 200-nucleotide region.

Much more is known about the structure of focused promoters. These promoters are made up of one or more DNA sequence elements, including the Initiator (Inr), TATA box, TFIIB recognition element (BRE), downstream promoter element (DPE), and motif ten element (MTE). The locations and consensus sequences of these elements are

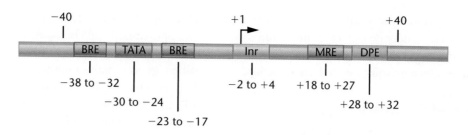

FIGURE 17–6 Core promoter elements found in focused promoters. Core promoter elements are usually located between −40 and +40 nucleotides, relative to the transcription start site, indicated as +1. None of these elements is universal, and a core promoter may contain only one, or several, of these elements. BRE is the TFIIB recognition element, which can be found on either side of the TATA box. TATA is the TATA box, Inr is the initiator element, MTE is the motif ten element, and DPE is the downstream promoter element.

summarized in Figure 17–6. Each of these elements is found in only some core promoters, with no element being a universal component of all focused promoters. The role of these core promoter elements is to bind to specific transcription initiation proteins, and these will be discussed in the next section.

The **Inr element** encompasses the transcription start site, from approximately nucleotides −2 to +4, relative to the start site. In humans, the Inr consensus sequence is YYAN^A/$_T$YY (where Y indicates any pyrimidine nucleotide and N indicates any nucleotide). The transcription start site is the first A residue at +1. The **TATA box** element is located at approximately −30 relative to the transcription start site and has the consensus sequence TATA^A/$_T$AAR (where R indicates any purine nucleotide). Although the TATA box is a common element in prokaryotic and eukaryotic promoters, it is found in only about 15 percent of mammalian gene core promoters. The **BRE** is found in some core promoters at positions either immediately upstream or downstream from the TATA box. The **MTE** and **DPE** sequence motifs are located downstream of the transcription start

site, at approximately +18 to +27 and +28 to +33, respectively. Most core promoters that contain a DPE also contain an Inr element; however, some core promoters have been discovered that also contain TATA boxes. The MTE motif can be present along with the Inr element but can also occur in core promoters with TATA boxes and/or DPE motifs.

Many promoters contain proximal promoter elements located upstream of the TATA and BRE motifs, and these elements act along with the core promoter elements to increase the levels of basal transcription. For example, the CAAT box is a common proximal promoter element. The **CAAT box** has the consensus sequence CAAT or CCAAT and is usually located about 70 to 80 base pairs upstream from the start site. Mutational analysis suggests that CAAT boxes (when present) are critical to the promoter's ability to initiate transcription. Mutations on either side of this element have no effect on transcription, whereas mutations within the CAAT sequence dramatically lower the rate of transcription. Figure 17–7 summarizes the transcriptional effects of mutations in the CAAT box and other promoter elements. The **GC box** is another

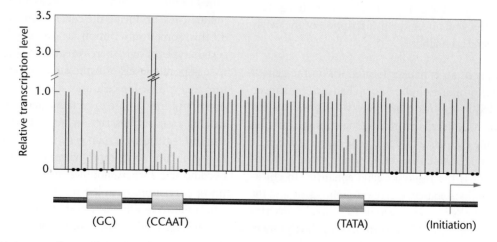

FIGURE 17–7 Summary of the effects on transcription levels of different point mutations in the promoter region of the β-globin gene. Each line represents the level of transcription produced in a separate experiment by a single-nucleotide mutation (relative to wild type) at a particular location. Dots represent nucleotides for which no mutation was obtained. Note that mutations within specific elements of the promoter have the greatest effects on the level of transcription.

element often found in promoter regions and has the consensus sequence GGGCGG. It is located, in one or more copies, at about position −110. The CAAT and GC boxes function somewhat like enhancers, which we will cover in the next section.

Enhancers and Silencers

Transcription of eukaryotic genes is regulated not only at promoter sequences but also at DNA sequences called **enhancers**. Enhancers can be located on either side of a gene, at some distance from the gene, or even within the gene. Like promoters, they are *cis* regulators because they function when located on the same chromosome as the structural genes they regulate. While promoter sequences are essential for basal-level transcription, enhancers are necessary for achieving the maximum level of transcription. In addition, enhancers are responsible for time- and tissue-specific gene expression. Thus, some degree of analogy exists between enhancers and operator regions in prokaryotes; however, enhancers are more complex in both structure and function.

Scientists have studied promoters and enhancers by analyzing the effects that specific mutations have on the transcription of cloned genes introduced into cultured cells. In addition, they have moved these elements into new positions relative to the coding genes to which the elements are attached. These studies have revealed several features that distinguish promoters from enhancers:

1. The position of an enhancer need not be fixed relative to the gene it regulates; it will function whether it is upstream, downstream, or within a gene.

2. The orientation of an enhancer can be inverted without significant effect on its action.

3. If an enhancer is experimentally moved adjacent to a gene elsewhere in the genome, or if an unrelated gene is placed near an enhancer, the transcription of the newly adjacent gene is enhanced.

An example of an enhancer located *within* the gene it regulates is the immunoglobulin heavy-chain gene enhancer, which is located in an intron between two exons. This enhancer is active only in cells expressing the immunoglobulin genes, indicating that tissue-specific gene expression can be modulated through the enhancer. An example of a downstream enhancer is the β-globin gene enhancer. In chickens, an enhancer located between the β-globin gene and the ε-globin gene works in one direction to control transcription of the ε-globin gene during embryonic development and in the opposite direction to regulate expression of the β-globin gene during adult life.

Enhancers are modular and often contain several different short DNA sequences. For example, the enhancer of the SV40 virus (which is transcribed inside a eukaryotic cell)

has a complex structure consisting of two adjacent sequences of approximately 100 bp each, located some 200 bp upstream from a transcriptional start point. Each of the two 100-bp regions contains multiple sequence motifs that contribute to achieving the maximum rate of transcription. If one or the other of these regions is deleted, there is no effect on transcription; but if both are deleted, *in vivo* transcription is greatly reduced.

Another type of *cis*-acting transcription regulatory element, the **silencer**, acts upon eukaryotic genes to repress the level of transcription initiation. Silencers, like enhancers, are *cis*-acting short DNA sequence elements that affect the rate of transcription initiated from an associated promoter. They often act in tissue- or temporal-specific ways to control gene expression.

17.5

Eukaryotic Transcription Initiation Is Regulated by Transcription Factors that Bind to *Cis*-Acting Sites

It is generally accepted that *cis*-acting regulatory sites—including promoters, enhancers, and silencers—influence transcription initiation by acting as binding sites for transcription regulatory proteins. These transcription regulatory proteins, known as **transcription factors**, can have diverse and complicated effects on transcription. Some transcription factors increase the levels of transcription initiation and are known as **activators**, whereas others reduce transcription levels and are known as **repressors**.

The effects of activators and repressors can be finely tuned to the appropriate cell type, as a response to environmental cues, or during the correct time in development. To do this, some transcription factors may be present in only certain types of cells, thereby regulating their target genes for tissue-specific levels of expression. Some transcription factors are expressed in cells only at certain times during development or in response to certain external or internal signals. In some cases, a transcription factor that binds to a *cis*-acting site and regulates a certain gene may be present in a cell and may even bind to its appropriate *cis*-acting site but will only become active when modified structurally (for example, by phosphorylation) or by binding to another molecule such as a hormone. These modifications to transcription factors may also be regulated in tissue- or temporal-specific ways. In addition, different transcription factors may compete for binding to the same DNA sequence or to one of two overlapping sequences. Transcription factor concentrations in the cell and the strength with which each factor binds to the DNA will dictate which factor binds and hence will determine the level of transcription initiation dictated by the factor that binds. Finally, multiple transcription factors that bind to several different enhancers and promoter elements within a gene-regulatory region can interact with each other to fine-tune the levels and timing of transcription initiation.

The Human Metallothionein IIA Gene: Multiple *Cis*-Acting Elements and Transcription Factors

The human **metallothionein IIA gene (*hMTIIA*)** provides an example of how the transcription of one gene can be regulated by the interplay of multiple promoter and enhancer elements and the transcription factors that bind to them. The product of the *hMTIIA* gene is a protein that binds to heavy metals such as zinc and cadmium, thereby protecting cells from the toxic effects of high levels of these metals. It is also implicated in protecting cells from the effects of oxidative stress. The gene is expressed at low levels in all cells but is transcribed at high levels when cells are exposed to heavy metals or to steroid hormones such as glucocorticoids.

The *cis*-acting regulatory elements controlling transcription of the *hMTIIA* gene include promoter, enhancer, and silencer elements (Figure 17–8). Each *cis*-acting element is a short DNA sequence that has specificity for binding to one or more transcription factors.

The *hMTIIA* gene contains the core promoter elements TATA box and Inr, which specify the start of transcription. The factors that bind to core promoters will be described in the next section of this chapter. The proximal promoter element, GC, binds the SP1 factor, which is present at all times in most eukaryotic cells and stimulates transcription at low levels in most cells. Basal levels of expression are also regulated by the BLE (basal element) and ARE (AP factor response element) regions, which can be considered enhancer elements. These *cis*-elements bind the activator proteins 1, 2, and 4 (AP1, AP2, and AP4), which are present at various levels in different cell types and can be activated in response to extracellular growth signals. The BLE element contains overlapping binding sites for the AP1 and AP4

factors, providing some degree of selectivity in how these factors stimulate transcription of *hMTIIA* when bound to the BLE in different cell types. High levels of transcription are stimulated by the presence of the enhancers MRE (metal response element) and GRE (glucocorticoid response element). The metal-inducible transcription factor (MTF1) binds to the MRE element in response to the presence of heavy metals. The glucocorticoid receptor protein binds to the GRE, but only when the receptor protein is also bound to the glucocorticoid steroid hormone. The glucocorticoid receptor is normally located in the cytoplasm of the cell; however, when glucocorticoid hormone enters the cytoplasm, it binds to the receptor and causes a conformational change that allows the receptor to enter the nucleus, bind to the GRE, and stimulate *hMTIIA* gene transcription. In addition to activation, transcription of the *hMTIIA* gene can be repressed by the actions of the repressor protein PZ120, which binds over the transcription start region.

The presence of multiple regulatory elements and transcription factors that bind to them allows the *hMTIIA* gene to be transcriptionally activated or repressed in response to subtle changes in both extracellular and intracellular conditions.

Functional Domains of Eukaryotic Transcription Factors

We have described transcription factors as proteins that bind to DNA and activate or repress transcription initiation. These actions are achieved through the presence of two functional domains (clusters of amino acids that carry out a specific function) within each of these proteins. One domain, the **DNA-binding domain**, binds to specific DNA sequences present in the *cis*-acting regulatory site. The other domain, the ***trans*-activating** or ***trans*-repression domain**, activates or represses transcription. The *trans*-activating and *trans*-repression domains bring about their effects by interacting with other transcription factors or RNA polymerase, as we will discuss in the next section.

The DNA-binding domains of eukaryotic transcription factors have various characteristic three-dimensional

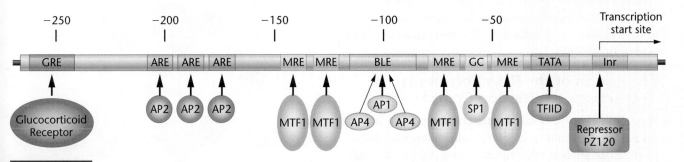

FIGURE 17–8   The human metallothionein IIA gene promoter and enhancer regions, containing multiple *cis*-acting regulatory sites. The transcription factors controlling both basal and induced levels of MTIIA transcription are indicated below the gene, with arrows showing their binding sites.

structural motifs. Examples include the helix–turn–helix (HTH), zinc finger, and basic leucine zipper (bZIP) motifs.

The **helix–turn–helix (HTH)** motif, present in both prokaryotic and eukaryotic transcription factors, is characterized by a certain geometric conformation rather than a distinctive amino acid sequence. The presence of two adjacent α-helices separated by a "turn" of several amino acids (hence the name of the motif) enables the protein to bind to DNA.

The **zinc-finger** motif is found in a wide range of transcription factors that regulate gene expression related to cell growth, development, and differentiation. A typical zinc-finger protein contains clusters of two cysteines and two histidines at repeating intervals. These clusters bind zinc atoms, fold into loops, and interact with specific DNA sequences.

The **basic leucine zipper (bZIP)** motif contains a region called a leucine zipper that allows protein–protein dimerization. When two bZIP-containing molecules dimerize, the leucine residues "zip" together. The resulting dimer contains two basic α-helical regions adjacent to the zipper that bind to the phosphate residues and specific bases in their DNA-binding site.

Transcription factors also contain *trans*-activating or *trans*-repressing domains that are distinct from their DNA-binding domains. These domains can occupy from 30 to 100 amino acids.

17.6
Activators and Repressors Interact with General Transcription Factors at the Promoter

We have now discussed the first steps in eukaryotic transcription regulation: first, chromatin must be remodeled and modified to allow transcription regulatory proteins to bind to their specific *cis*-acting sites; second, transcription factors bind to *cis*-acting sites and bring about positive and negative effects on the transcription initiation rate—often in response to extracellular signals or in tissue- or time-specific ways. The next question for us to consider is, how do these *cis*-acting regulatory elements and their DNA-binding factors act to influence transcription initiation? To answer this question, we must first discuss how eukaryotic RNA polymerase II and its general transcription factors assemble at promoters.

Formation of the RNA Polymerase II Transcription Initiation Complex

A number of proteins called **general transcription factors** are needed at a promoter in order to initiate either basal-level or enhanced levels of transcription. These proteins assemble in a specific order, forming a transcription **pre-initiation complex (PIC)** that provides a platform for RNA polymerase II to bind to a promoter and initiate transcription. In this

discussion, we will restrict our example of PIC formation to focused promoters with TATA boxes—the type of promoter for which the most information is available.

The general transcription factors that assist RNA polymerase II at a core promoter are called **TFIID** (*Transcription Factor for RNA polymerase IID*), **TFIIB, TFIIA, TFIIE, TFIIF, TFIIH**, and **Mediator**. The general transcription factors and their interactions with the core promoter and RNA polymerase II are outlined in Figure 17–9.

The first step in the formation of a PIC is the binding of TFIID to one or more core promoter elements. TFIID is a multi-subunit complex that contains **TBP** (*TATA Binding Protein*) and approximately 13 proteins called **TAFs** (*TBP Associated Factors*). As its name implies, TBP binds to the TATA box. In addition, a subset of TAFs binds to Inr elements, as well as DPEs and MTEs. TFIIA interacts with

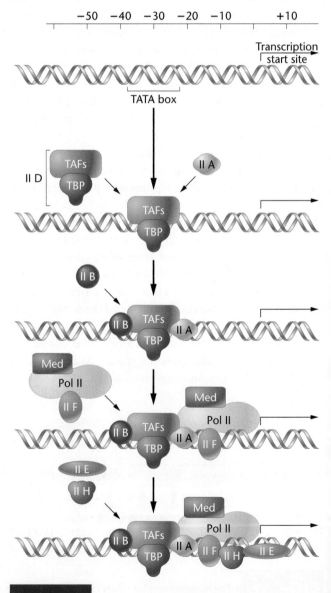

FIGURE 17–9 The assembly of general transcription factors required for the initiation of transcription by RNA polymerase II.

TFIID and assists the binding of TFIID to the core promoter. Once TFIID has made contact with the core promoter, TFIIB binds to BRE elements on one or both sides of the TATA box. Once TFIID and TFIIB have bound the core promoter, the other general transcription factors interact with RNA polymerase II and help recruit it to the promoter. The fully formed PIC mediates the unwinding of promoter DNA at the start site and the transition of RNA polymerase II from transcription initiation to elongation. RNA polymerase then clears the promoter as it proceeds down the DNA template in an **elongation complex**. Several of the general transcription factors, specifically TFIID, TFIIE, TFIIH, and Mediator, remain on the core promoter to help set up the next PIC.

Interactions of General Transcription Factors with Activators and Repressors

Transcription activators and repressors may increase or decrease the rate of transcription initiation in several ways. One way in which activators and repressors may influence transcription is to bind to chromatin near a gene's promoter. Once bound, these proteins may recruit chromatin remodeling complexes. The remodeling complexes may open regions of the promoter for further interactions with the transcription machinery, such as the PIC, or they may modify chromatin into repressive structures that inhibit transcription initiation.

A second way that activators and repressors may affect transcription is to make direct contacts with general transcription factors and either enhance or inhibit the ability of general transcription factors to associate with a promoter. Some activators, when bound to enhancers, interact with other proteins called **coactivators** in a complex known as an **enhanceosome**. The enhanceosome then makes contacts with one or more general transcription factors in the PIC, bending or looping out the intervening DNA (Figure 17–10).

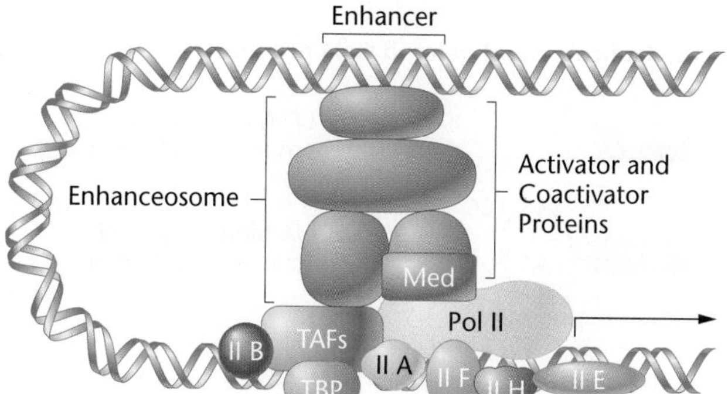

By enhancing the rate of PIC assembly, or by influencing its stability, transcription activators stimulate the rate of transcription initiation. In contrast, transcription can be repressed by the actions of repressor proteins bound at silencer DNA elements. Repressors can inhibit the formation of a PIC, stimulate or recruit chromatin-remodeling proteins that create repressive chromatin structures, or block the association of a gene's regulatory elements with activators, or with the PIC.

In addition to chromatin remodeling and PIC formation, transcription activators may increase the rate of DNA unwinding within the gene and accelerate the release of RNA polymerase from the promoter into the transcribed region of the gene.

NOW SOLVE THIS

17–2 The hormone estrogen converts the estrogen receptor (ER) protein from an inactive molecule to an active transcription factor. The ER binds to *cis*-acting sites that act as enhancers, located near the promoters of a number of genes. In some tissues, the presence of estrogen appears to activate transcription of ER-target genes, whereas in other tissues, it appears to repress transcription of those same genes. Offer an explanation as to how this may occur.

■ HINT: *This problem involves an understanding of how transcription enhancers and repressors work. The key to its solution is to consider the many ways that trans-acting factors can interact at enhancers to bring about changes in transcription initiation.*

FIGURE 17–10 Formation of a DNA loop allows factors that bind to an enhancer or silencer at a distance from a promoter to interact with general transcription factors in the pre-initiation complex and to regulate the level of transcription.

17.7

Gene Regulation in a Model Organism: Transcription of the *GAL* Genes of Yeast

One of the first model systems used to study eukaryotic gene regulation was the **GAL gene system** in yeast. The *GAL* system comprises four structural genes (*GAL1, GAL10, GAL2,* and *GAL7*) and three regulatory genes (*GAL4, GAL80,* and *GAL3*). The products of the structural genes transport galactose into the cell and metabolize the sugar. The products of the regulatory genes positively and negatively control the transcription of the structural genes.

Transcription of the *GAL* structural genes is **inducible**— that is, transcription is regulated by the presence or absence of the substrate, galactose. In the absence of galactose, the *GAL* structural genes are not transcribed. If galactose is added to the growth medium, transcription begins immediately, and the mRNA concentration increases a thousand-fold. Null mutations in the regulatory gene *GAL4* prevent activation, indicating that transcription is under **positive control**—that is, the activator protein must be present to turn on gene transcription. In this section, we will examine how transcription of two *GAL* genes, *GAL1* and *GAL10*, is positively regulated (Figure 17–11(a)).

Transcription of these two genes is controlled by a central control region called **UAS$_G$** (*u*pstream *a*ctivating *se*quence of *GAL* genes), of approximately 170 bp. In yeast, UAS elements are functionally similar to the enhancers found in higher eukaryotes. The chromatin structure of the UAS$_G$ is constitutively open, or **DNase hypersensitive**, meaning that it is free of nucleosomes. Within the UAS$_G$ are four binding sites for the Gal4 protein (**Gal4p**), which is encoded by the *GAL4* gene. These sites are permanently occupied by Gal4p, whether or not the *GAL1* and *GAL10* genes are transcribed. Gal4p is, in turn, negatively regulated

by the Gal80 protein (**Gal80p**), which is the product of the *GAL80* gene.

In the absence of galactose, Gal80p is always bound to Gal4p, covering Gal4p's transcription activation domain, which is shown in Figure 17–11(b) as a dark patch. Transcription activation occurs when galactose interacts with the Gal3 protein (**Gal3p**), encoded by the *GAL3* gene. When bound to galactose, the Gal3p molecule undergoes a conformational change that allows it to interact with the UAS$_G$-bound Gal4p/Gal80p complex. This interaction disrupts the association of Gal4p with Gal80p, exposing the Gal4p activation domain.

Gal4p is a protein consisting of 881 amino acids. It includes a DNA-binding domain that recognizes and binds to DNA sequences in the UAS$_G$ and a *trans*-activating domain that activates transcription [Figure 17–12(a)]. Researchers have identified these functional domains by making deletions in the *GAL4* gene and assaying the gene products for their ability to bind DNA and to activate transcription. [Figure 17–12(b)].

Data from many studies suggest that transcription activation results from contacts between the activating domain of Gal4p and other proteins. Research shows that the first factor to interact with the Gal promoter is a coactivator known as SAGA, which associates directly with Gal4p. Following the recruitment of SAGA, the general transcription factor Mediator interacts with Gal4p and enters the

(a) *GAL* gene cluster

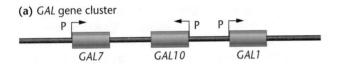

(b)

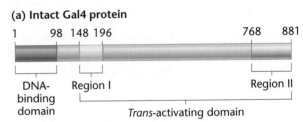

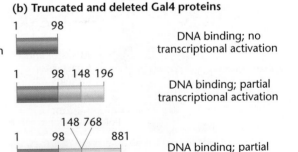

FIGURE 17–11   Model of *GAL1* and *GAL10* transcriptional activation. (a) The structure of the yeast genome region containing three of the *GAL* genes—*GAL7*, *GAL10*, and *GAL1*. Arrows indicate transcription start sites and the directions of transcription. Promoter regions are indicated as "P"s. (b) The *GAL10* and *GAL1* gene UAS$_G$ region. The UAS$_G$ region has four binding sites for Gal4p, shown as purple boxes. Gal4p molecules are shown as blue circles, and Gal80p molecules are shown as red triangles or squares. Induction is indicated by a change in the conformation of Gal80p, and its association with Gal4p, which exposes the transcription activation domain of Gal4p.

(a) Intact Gal4 protein

DNA-binding domain / Region I / Region II / *Trans*-activating domain

(b) Truncated and deleted Gal4 proteins

1   98   DNA binding; no transcriptional activation

1   98   148   196   DNA binding; partial transcriptional activation

1   98   148   768   881   DNA binding; partial transcriptional activation

FIGURE 17–12   Structure and function of the Gal4p activator. (a) Gal4p contains a DNA-binding domain, shown in dark blue, and two transcriptional activation regions, shown in light blue. The Gal80p binding region resides in the amino acid region 851–881. (b) Effects of various deletions on the activity of Gal4p.

complex. Finally, TFIID, TFIIH, TFIIE, and TFIIF, as well as RNA polymerase II are recruited into a PIC on *GAL* promoters. Another step that is stimulated by Gal4p is nucleosome remodeling. Gal4p recruits the nucleosome-remodeling complex SWI/SNF to *GAL* promoters, followed by the removal of promoter nucleosomes.

17.8

Posttranscriptional Gene Regulation Occurs at All the Steps from RNA Processing to Protein Modification

We have noted that regulation of gene expression occurs at many points along the pathway from DNA to protein. Although transcriptional control exerts the primary effect on gene regulation in eukaryotes, **posttranscriptional regulation** plays an equal, if not more significant, role. Prior to translation, eukaryotic mRNA transcripts can be processed by the removal of noncoding introns, the precise splicing together of the remaining exons, the addition of a cap at the mRNA 5′ end, and the synthesis of a poly-A tail at its 3′ end. The mature messenger RNA is then exported to the cytoplasm, where it is translated and degraded. Each of these can be regulated to control the quantity of a protein product. In addition, the stability and activity of the protein product can be regulated.

In this section, we examine several mechanisms of posttranscriptional gene regulation that are especially important in eukaryotes—control of alternative splicing, mRNA stability, and translation. In a subsequent section, we will discuss a newly discovered method of posttranscriptional gene regulation—RNA silencing.

Alternative Splicing of mRNA

Alternative splicing can generate different forms of mRNA from identical pre-mRNA molecules, so that expression of one gene can give rise to a number of proteins, with similar or different functions.

Changes in splicing patterns can have many different effects on the translated protein. Even small changes can alter the protein's enzymatic activity, receptor-binding capacity, or protein localization in the cell.

Figure 17–13 presents the example of alternative splicing of the pre-mRNA transcribed from the **calcitonin/calcitonin gene-related peptide gene (*CT/CGRP* gene)**. In thyroid cells, the *CT/CGRP* primary transcript is spliced in such a way that the mature mRNA contains the first four exons only. In these cells, the exon 4 polyadenylation signal is used to process the mRNA and add the poly-A tail. This mRNA is translated into the calcitonin peptide, a 32-amino

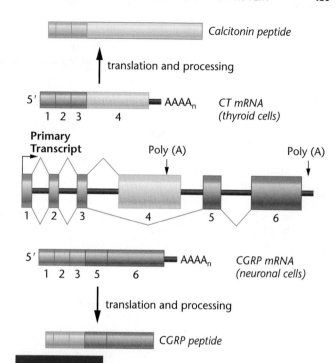

FIGURE 17–13   Alternative splicing of the *CT/CGRP* gene transcript. The primary transcript, which is shown in the middle of the diagram, contains six exons. The primary transcript can be spliced into two different mRNAs, both containing the first three exons but differing in their final exons. The *CT* mRNA contains exon 4, with polyadenylation occurring at the end of the fourth exon. The *CGRP* mRNA contains exons 5 and 6, with polyadenylation occurring at the end of exon 6. The *CT* mRNA is produced in thyroid cells. After translation, the resulting protein is processed into the calcitonin peptide. In contrast, the *CGRP* mRNA is produced in neuronal cells, and after translation, its protein product is processed into the CGRP peptide.

acid peptide hormone that is involved in regulating calcium. In the brain and peripheral nervous system, the *CT/CGRP* primary transcript is spliced to include exons 5 and 6, but not exon 4. In these cells, the exon 6 polyadenylation site is recognized. The *CGRP* mRNA encodes a 37-amino acid peptide with hormonal activities in a wide range of tissues. Through alternative splicing, two peptide hormones with different structures, locations, and functions are synthesized from the same gene.

Alternative splicing increases the number of proteins that can be made from each gene. As a result, the number of proteins that an organism can make—its **proteome**—is not the same as the number of genes in the genome, and protein diversity can exceed gene number by an order of magnitude. Alternative splicing is found in all metazoans but is especially common in vertebrates, including humans. Scientists estimate that at least two-thirds of the genes in the human genome can undergo alternative splicing. Thus, humans can produce several hundred thousand different proteins

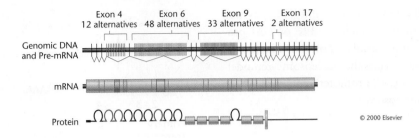

FIGURE 17–14   Alternative splicing of the *Dscam* gene mRNA. (Top) Organization of the *Dscam* gene in *Drosophila melanogaster* and the transcribed pre-mRNA. The *Dscam* gene encodes a protein that guides axon growth during development. Each mRNA will contain one of the 12 possible exons for exon 4 (red), one of the 48 possible exons for exon 6 (blue), one of the 33 possible exons for exon 9 (green), and one of the 2 possible exons for exon 17 (yellow). Counting all possible combinations of these exons, the *Dscam* gene could encode 38,016 different versions of the DSCAM protein.

(or perhaps more) from the approximately 25,000 genes in the haploid genome.

Given the existence of alternative splicing, how many different polypeptides can be derived from the same pre-mRNA? One answer to that question comes from research on the *Dscam* gene in *Drosophila*. During development, cells of the nervous system must accurately connect with each other. Even in *Drosophila*, with only about 250,000 neurons, this is a formidable task. Neurons have cellular processes called axons that form connections with other nerve cells. The *Drosophila* **Dscam gene** encodes a protein that guides axon growth, ensuring that neurons are correctly wired together. In *Dscam* pre-mRNA, exons 4, 6, 9, and 17 each consist of an array of possible alternatives (**Figure 17–14**). These are spliced into the mature mRNA in an exclusive fashion, so that each exon is represented by no more than one of its possible alternatives. There are 12 alternatives for exon 4; 48 alternatives for exon 6; 33 alternatives for exon 9; and 2 alternatives for exon 17. The number of possible combinations that could be formed in this way suggests that, theoretically, the *Dscam* gene can produce 38,016 different proteins. Although this is an impressive number of possible different mRNAs and protein isoforms, does the *Drosophila* nervous system require all these alternatives? Recent research suggests that it does.

Scientists have shown that each neuron expresses a different subset of Dscam protein isoforms. In addition, *in vitro* studies show that each Dscam protein isoform can bind to the same Dscam protein isoform but not to others. Even a small change in amino acid sequence reduces or eliminates the binding between two Dscam molecules. *In vivo* studies show that cells expressing the same isoforms of Dscam interact with each other. Therefore, it appears that the diversity of Dscam protein isoforms in neurons provides a kind of molecular identity tag for each neuron, helping to guide it to the correct target and preventing the tangling of extensions from different neurons.

A more extreme example is provided by **para**, another gene expressed in the nervous system of *Drosophila*. *Para* not only has at least 6 sites of alternative splicing, leading to 48 possible different mRNA variants, but also undergoes another posttranscriptional modification called RNA editing, at 11 positions. **RNA editing** involves base substitutions made after transcription and splicing. With both alternative splicing and editing, the *para* gene can theoretically produce more than 1 million different transcripts.

The *Drosophila* genome contains about 13,000 genes, but the *Dscam* gene alone can produce 2.5 times that many proteins. Because alternative splicing is far more common in vertebrates, the combinations of proteins that can be produced from the human genome may be astronomical.

Alternative Splicing and Human Diseases

Mutations that affect regulation of splicing contribute to several genetic disorders. One of these disorders, **myotonic dystrophy**, provides an example of how defects in alternative RNA splicing can lead to a wide range of symptoms. Myotonic dystrophy (DM) is the most common form of adult muscular dystrophy, affecting 1 in 8000 individuals. It is an autosomal dominant disorder that occurs in two forms—DM1 and DM2. Both of these diseases show a wide range of symptoms, including muscle wasting, **myotonia** (difficulty relaxing muscles), insulin resistance, cataracts, testicular atrophy, behavior and cognitive defects, cardiac muscle problems, and hair follicle tumors.

DM1 is caused by the expansion of the trinucleotide repeat CTG in the 3′-untranslated region of the **DMPK gene**. In unaffected individuals, the *DMPK* gene contains between 5 and 35 copies of the CTG repeat sequence, whereas in DM1 patients, the gene contains between 150 and 2000 copies. The severity of the symptoms are directly related to the number of copies of the repeat sequence. DM2 is caused by an expansion of the repeat sequence CCTG within the first intron of the **ZNF9 gene**. Affected individuals may have up to 11,000 copies of the repeat sequence in the *ZNF9* intron. In DM2, the severity of symptoms is not related to the number of repeats.

Recently, scientists have discovered that DM1 and DM2 are caused not by changes in the protein products of the *DMPK* or *ZNF9* genes, but by the toxic effects of their repeat-containing RNAs. These RNAs accumulate and form inclusions within the nucleus. In the case of *ZNF9*, only the CCUG sequence repeat itself accumulates in the nucleus, as the remainder of the intron is degraded after splicing of the gene. It appears that the accumulated RNAs bind to, and sequester, proteins that would normally be involved in regulating the alternative splicing patterns of a large number of other RNAs. These RNAs include those whose products are required for the proper functioning of muscle and neural tissue. So far, scientists have discovered over 20 genes that are inappropriately spliced in DM1 muscle, heart, and brain. Often, the fetal splicing patterns occur in DM1 and DM2 patients, and the normal transitions to adult splicing patterns are lacking. Such defects in the regulation of RNA splicing are known as **spliceopathies**.

Other human diseases are caused by expansions of trinucleotide repeats. Two of these, fragile-X syndrome and Huntington disease, are discussed in Chapter 8 and Chapter 24, respectively.

Sex Determination in *Drosophila*: A Model for Regulation of Alternative Splicing

As outlined in Chapter 7, sex in *Drosophila* is determined by the ratio of X chromosomes to sets of autosomes (X:A). When the ratio is 0.5 (1X:2A), males are produced, even when no Y chromosome is present; when the ratio is 1.0 (2X:2A), females are produced. Intermediate ratios (2X:3A) produce intersexes. Chromosomal ratios are interpreted by a small number of genes that initiate a cascade of splicing events, resulting in the production of male or female somatic cells and the corresponding male or female phenotypes. Three major genes in this pathway are *Sex lethal* (*Sxl*), *transformer* (*tra*), and *doublesex* (*dsx*). We will review some of the key steps in this process.

The regulatory gene at the beginning of this cascade (Figure 17–15) is the gene **Sex lethal** (*Sxl*), which encodes an RNA-binding protein. In females, transcription factors encoded by genes on the X chromosome are thought to activate transcription of the *Sxl* gene. In males, the lower concentration of these transcription factors is not sufficient to activate transcription of *Sxl*. As a result of this differential regulation of transcription, the SXL protein is expressed only in female embryos. The presence (in females) or absence (in males) of SXL protein begins cascades of

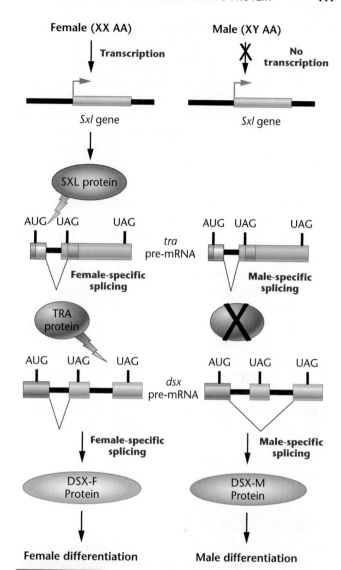

FIGURE 17–15 Regulation of pre-mRNA splicing that determines male and female sexual development in *Drosophila*. The ratio of X chromosomes to autosomes (AA) affects transcription of the *Sxl* gene. The presence of SXL protein begins a cascade of pre-mRNA splicing events that culminate in female-specific gene expression and production of the DSX-F transcription factor. In the absence of SXL protein, a male-specific pattern of pre-mRNA splicing results in male-specific patterns of gene expression induced by the DSX-M protein.

pre-mRNA splicing events that are specific for females or males. In the presence of SXL, female splicing patterns are expressed, which override the default male splicing patterns.

One of the targets of SXL protein is the pre-mRNA encoded by the **transformer** (*tra*) gene. This pre-mRNA is transcribed in both male and female cells. When SXL protein is present, *tra* pre-mRNAs are spliced to produce mature mRNAs that are translated into functional TRA protein. If SXL is absent, *tra* pre-mRNAs are spliced in such a way that a translation stop codon remains in the mature mRNA. Translation of this mRNA results in a truncated, nonfunctional protein. SXL alters the splicing pattern of *tra* pre-mRNA by binding to the pre-mRNA and altering recognition of splicing signals by the splicing machinery.

The next gene in the cascade, **doublesex (dsx)**, is a critical control point in the development of sexual phenotype. It produces a functional mRNA and protein in both females and males. However, the pre-mRNA is processed in a sex-specific manner to produce different transcripts and hence different DSX proteins. In females, the functional TRA protein acts as a splicing factor that binds to the *dsx* pre-mRNA and directs splicing in a female-specific pattern. In males, no TRA protein is present, and splicing of the *dsx* pre-mRNA results in a male-specific mRNA and protein. The female DSX protein (DSX-F) and the male protein (DSX-M) are both transcription factors, but act in different ways. DSX-F represses the transcription of genes whose products control male sexual development, whereas DSX-M activates the transcription of genes whose products control male sexual development. In addition, DSX-M represses the transcription of genes that control female sexual development.

In sum, the product of the *Sxl* gene acts as a switch that selects the pathway of sexual development by controlling splicing of the *dsx* transcript. The SXL protein is produced only in embryos with an X:A ratio of 1. Failure to control the splicing of the *dsx* transcript in a female mode results in the default splicing of the transcript in a male mode, leading to the production of a male phenotype.

NOW SOLVE THIS

17–3 Some mutations in the *tra* gene of *Drosophila* cause XX females to appear as males. In contrast, other mutations in *tra* cause XY male flies to appear as normal females. Which of these phenotypes would you expect if a mutation in *tra* resulted in a null allele? Which would you expect if the mutation produced a constitutively active *tra* gene product?

■ HINT: *This problem involves an understanding of the SXL pathway in* Drosophila. *The key to its solution is to review the functions of the normal TRA protein in* Drosophila *males and females.*

Control of mRNA Stability

The **steady-state level** of an mRNA is its amount in the cell as determined by a combination of the rate at which the gene is transcribed and the rate at which the mRNA is degraded. The steady-state level determines the amount of mRNA that is available for translation. All mRNA molecules are degraded at some point after their synthesis, but the lifetime of an mRNA, defined in terms of its **half-life** or $t_{1/2}$, can vary widely between different mRNAs and can be regulated in response to the needs of the cell.

An mRNA may be degraded along three general pathways. First, an mRNA may be targeted for degradation by enzymes that shorten the length of the poly-A tail. In newly synthesized mRNAs, the poly-A tail is about 200 nucleotides long and binds a protein known as the **poly-A binding protein**. The binding of this protein to the poly-A tail helps to stabilize the mRNA. If the poly-A tail is shortened to less than about 30 nucleotides, the mRNA becomes unstable and acts as a substrate for exonucleases that degrade the RNA in either a 5′ to 3′ or 3′ to 5′ direction. Second, **decapping enzymes** can remove the 7-methylguanosine cap, which also renders the mRNA unstable. Third, an mRNA may be cleaved internally by an endonuclease, providing unprotected ends at which exonuclease degradation may proceed. Examples of endonucleolytic cleavages are those that occur during **nonsense-mediated decay** and those triggered by RNA interference (which will be discussed in Section 17.9). Nonsense-mediated mRNA decay occurs when translation terminates at premature stop codons. Endonucleases attack the mRNA near the stop codon, leaving unprotected ends that are then degraded by exonucleases. Nonsense-mediated decay is an important mechanism for removing mutated mRNA molecules that could, if efficiently translated, result in accumulations of aberrant protein products.

What are the mechanisms by which mRNA stability can be regulated? One way that an mRNA's half-life can be altered is through specific RNA-sequence elements that recruit degrading or stabilizing complexes. One well-studied mRNA stability element is the **adenosine-uracil rich element (ARE)**—a stretch of ribonucleotides that consist of A and U ribonucleotides. These AU-rich elements are usually located in the 3′ untranslated regions of mRNAs that have short, regulated half-lives. These ARE-containing mRNAs encode proteins that are involved in cell growth or transcription control and need to be rapidly modulated in abundance. In cells that are not growing or require low levels of gene expression, specific complexes bind to the ARE elements of these mRNA molecules, bringing about shortening of the poly-A tail and rapid mRNA degradation. It is estimated that approximately 10 percent of mammalian mRNAs contain these instability elements.

Translational and Posttranslational Regulation

The ultimate end-point of gene expression is the presence of an active gene product. There are many ways by which the quantity and activity of a gene product can be regulated. In some cases, the translation of an mRNA can be regulated to produce the correct quantity of protein product. In other cases, the stability of a protein can be modulated or the protein can be modified after translation to change its structure and affect its activity.

An important example of posttranslational regulation is that of the **p53 protein.** The p53 protein is essential to protect normal cells from the effects of DNA damage and other stresses. It is a transcription factor that increases the transcription of a number of genes whose products are involved in cell-cycle arrest, DNA repair, and programmed cell death. Under normal conditions, the levels of p53 are extremely low in cells, and the p53 that is present is inactive. When cells suffer DNA damage or metabolic stress, the level of p53 protein increases dramatically. In addition, p53 becomes an active transcription factor.

The changes in the levels and activity of p53 are due to a combination of increased protein stability and modifications to the protein. In unstressed cells, p53 is bound by another protein called **Mdm2.** The Mdm2 protein binds to the transcriptional activation domain of the p53 molecule, blocking its ability to induce transcription. In addition, Mdm2 acts as a ubiquitin ligase, adding ubiquitin residues onto the p53 protein. **Ubiquitin** (a "ubiquitous" molecule) is a small protein that tags other proteins for degradation by proteolytic enzymes. The presence of ubiquitin on p53 results in p53 degradation. When cells are stressed, Mdm2 and p53 are modified by phosphorylation and acetylation, resulting in the release of Mdm2 from p53. As a consequence, p53 proteins are stabilized, the levels of p53 increase, and the protein is able to act as a transcription factor. An added level of control is that p53 is a transcription factor that induces the transcription of the *Mdm2* gene. Hence, the presence of active p53 triggers a negative feedback loop that creates more Mdm2 protein, which rapidly returns p53 to its rare and inactive state.

17.9

RNA Silencing Controls Gene Expression in Several Ways

In the last decade, the discovery that small RNA molecules control gene expression has given rise to a new field of research. First discovered in plants, short RNA molecules, ~21 nucleotides long, are now known to regulate gene expression in the cytoplasm of plants, animals, and fungi by repressing translation and triggering the degradation of mRNAs. This form of sequence-specific postranscriptional regulation is known as **RNA interference (RNAi).** More recently, researchers have discovered that short RNAs act in the nucleus to alter chromatin structure and bring about repression of transcription. Together, these phenomena are known as **RNA-induced gene silencing.** We will outline some basic features of RNA-induced gene silencing, while keeping in mind that this field is rapidly expanding and advancing. In addition, we will discuss some of the ways in which RNAi is being used in biotechnology and medicine.

RNAi was first discovered during laboratory research, in studies of plant and animal gene expression. In one research project, Andrew Fire and Craig Mello injected roundworm (*Caenorhabditis elegans*) cells with either single-stranded or double-stranded RNA molecules—both containing sequences complementary to the mRNA of the *unc-22* gene. Although they expected that the single-stranded antisense RNA molecules would suppress *unc-22* gene expression by binding to the endogenous sense mRNA, they were surprised to discover that the injection of double-stranded *unc-22* RNA was 10- to 100-fold more powerful in repressing expression of the *unc-22* mRNA. They studied the phenomenon further and published their results in the journal *Nature* in 1998. They reported that the presence of double-stranded RNA acts to degrade the mRNA if the mRNA is complementary in sequence to one strand of the double-stranded RNA. Only a few molecules of double-stranded RNA are needed to bring about the degradation of large amounts of mRNA. Fire and Mello's research opened up an entirely new and surprising branch of molecular biology, with far-reaching implications for practical applications. For their insights into RNAi, they were awarded the Nobel Prize for Physiology or Medicine in 2006.

The Molecular Mechanisms of RNA-induced Gene Silencing

Two types of short RNA molecules are involved in RNA-induced gene silencing: The **small interfering RNAs (siRNAs)** and the **microRNAs (miRNAs).** Although they arise from different sources, their mechanisms of action are similar. Both types of RNA are short, double-stranded molecules, between 21 and 24 ribonucleotides long. The siRNAs are derived from longer RNA molecules that are linear, double-stranded, and located in the cell cytoplasm. In nature, these siRNA precursors arise within cells as a result of virus infection or the expression of transposons—both of which synthesize double-stranded RNA molecules as part of their life cycles. RNAi may be a method by which cells recognize these double-stranded RNAs and inactivate them, protecting the organism from external or internal assaults. Another source of siRNA molecules is in the research lab. Scientists are now able to introduce double-stranded RNAs into cells for research or therapeutic purposes. In the cytoplasm, double-stranded RNA molecules are recognized by an enzyme complex known as **Dicer** and are cleaved by Dicer into siRNAs.

The miRNAs are derived from single-stranded RNAs that are transcribed within the nucleus from the cell's own genome and that contain a double-stranded stem-loop structure. Nuclease enzymes within the nucleus recognize these stem-loop structures and cleave them from the longer single-stranded RNA. The stem-loop RNA fragments are exported from the nucleus into the cytoplasm where they are further processed by the Dicer complex into short, linear, double-stranded miRNAs. Over the last few years, scientists have discovered that significant amounts of eukaryotic genomes are transcribed by RNA polymerase II into RNA products that contain no open reading frames and are not translated into protein products. These RNAs are transcribed either from sequences within the introns of other protein-coding genes or from their own promoters. So far, more than 700 of these noncoding RNA genes have been discovered in the human genome. *Arabidopsis* has more than 130, and *C. elegans* has more than 100. This is probably an underestimate, and scientists speculate that eukaryotic genomes may contain thousands of genes that are transcribed into short noncoding RNAs, which may regulate the expression of more than half of all protein-coding genes. It is estimated that 30 percent of human genes are at least in part regulated by miRNA-related mechanisms.

How do these noncoding RNAs work to negatively regulate gene expression? The several different pathways involved in RNA-induced gene silencing are outlined in Figure 17–16.

The RNAi pathway takes several steps. First, siRNA or miRNA molecules associate with an enzyme complex called the **RNA-induced silencing complex (RISC)**. Second, within the RISC, the short double-stranded RNA is denatured and the sense strand is degraded. Third, the RNA/RISC complex becomes a functional and highly specific agent of RNAi, seeking out mRNA molecules that are complementary to the antisense RNA contained in the RISC. At this point, RNAi can take one of two different pathways. If the antisense RNA in the RISC is perfectly complementary to the mRNA, the RISC will cleave the mRNA. The cleaved mRNA is then degraded by ribonucleases. If the antisense RNA within the RISC is not exactly complementary to the mRNA, the RISC complex stays bound to the mRNA, interfering with the ability of ribosomes to translate the mRNA. Hence, RNAi can silence gene expression by affecting either mRNA stability or translation.

In addition to repressing mRNA translation and triggering mRNA degradation, siRNAs and miRNAs can also repress the transcription of specific genes and larger regions of the genome. They do this by associating with a different complex—the **RNA-induced initiation of transcription silencing complex (RITS)**. The antisense RNA strand within the RITS targets the RITS complex to specific gene promoters or larger regions of chromatin. RITS then recruits chromatin remodeling enzymes to these regions. These enzymes methylate histones and DNA, resulting in heterochromatin formation

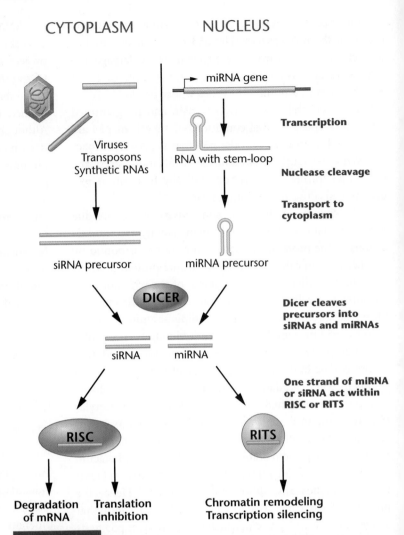

FIGURE 17–16 Mechanisms of gene regulation by RNA-induced gene silencing. The siRNA or miRNA precursors are processed into short double-stranded RNA molecules by the Dicer complex in the cytoplasm. They are then recognized by either the RISC complex or the RITS complex, and one strand is degraded. In the RNAi pathway, the RISC complex, guided by the antisense single-stranded RNA, recognizes target mRNA substrates, marking them for degradation or translation inhibition. In the transcription silencing pathway, the RITS complex acts in the nucleus, by recognizing genomic DNA that is complementary to the single strands of the miRNAs or siRNAs. The RITS complex recruits chromatin remodeling proteins that modify chromatin and repress transcription.

and subsequent transcriptional silencing. As a result of their effects on chromatin-mediated gene silencing, miRNA molecules are thought to be involved in epigenetic phenomena such as gene imprinting and X-chromosome inactivation.

RNAi pathways are also able to repress transcription in indirect ways. Transcription factor mRNAs are frequent targets for RNAi-mediated silencing. When the levels of specific transcription factors are reduced in a cell, transcription of genes whose expression depends on these factors is also repressed.

Recent studies are demonstrating that RNA-induced gene silencing mechanisms operate during normal development and control the expression of batteries of genes involved in tissue-specific cellular differentiation. In addition, scientists have discovered that abnormal activities of miRNAs contribute to the occurrence of cancers, diabetes, and heart disease.

RNA-Induced Gene Silencing in Biotechnology and Medicine

Recently, geneticists have applied RNAi as a powerful research tool. RNAi technology allows investigators to specifically create single-gene defects without having to induce inherited gene mutations. RNAi-mediated gene silencing is relatively specific and inexpensive, and it allows scientists to rapidly analyze gene function. Several dozen scientific supply companies now manufacture synthetic siRNA molecules of specific ribonucleotide sequence for use in research. These molecules can be introduced into cultured cells to knock out specific gene products.

In addition to its use in laboratory research, RNAi is being developed as a potential pharmaceutical agent. In theory, any disease caused by overexpression of a specific gene, or even normal expression of an abnormal gene product, could be attacked by therapeutic RNAi. Viral infections are an obvious target, and scientists have had promising results using RNAi in tissue cultures to reduce the severity of infection by several types of viruses such as HIV, influenza, and polio. In animal models, siRNA molecules have successfully treated virus infections, eye diseases, cancers, and inflammatory bowel disease. In these studies, the synthetic siRNA molecules were applied to the surfaces of mucous membranes, or into eyes, brain tissues, or the lower intestine. Research into RNAi pharmaceuticals has made rapid progress, and the Phase I (toxicity) stages of the first human clinical trials have been completed. Phase II and III (efficacy) clinical trials have begun to test the use of siRNAs directed against the product of the vascular endothelial growth factor gene, as a drug to treat age-related macular degeneration. Clinical trials have also been started to test RNAi in the treatment of Hepatitis B and respiratory syncytial virus infections. Other clinical trials are planned for siRNA treatment of influenza and hepatitis C virus infections, as well as some solid tumors.

Scientists are particularly enthusiastic about the potential uses of RNAi in the diagnosis and treatment of cancers. Recent studies show that the expression profiles of miRNA genes are characteristic of each tumor type. This observation may lead to more precise methods of diagnosing tumors, predicting their course, and planning treatments. In addition, certain cancers appear to have defects in miRNA gene expression. Treatment of these tumors with synthetic siRNAs may be able to correct these defects and reverse the cancer phenotype. Finally, some cancers are characterized by overexpression or abnormal expression of one or several key proto-oncogenes. If RNAi methods can target these specific gene products, they might help treat cancers that have become resistant to other methods such as radiation or chemotherapy.

New as it is, the science of RNAi holds powerful promise for molecular medicine. Analysts expect the first RNAi drugs to be available within the next decade.

EXPLORING GENOMICS

Tissue-Specific Gene Expression

MG™ *Study Area: Exploring Genomics*

In this chapter, we discussed how gene expression can be regulated in many complex ways. Recall that one aspect of gene expression regulation we considered is the way promoter, enhancer, and silencer sequences can govern transcriptional initiation of genes to allow for tissue-specific gene expression. All cells and tissues of an organism possess the same genome, and many genes are expressed in all cell and tissue types. However, muscle cells, blood cells, and all other tissue types express genes that are largely tissue-specific (i.e., they have limited or no expression in other tissue types). In this exercise, we use BLAST to learn more about tissue-specific gene-expression patterns.

■ **Exercise – Tissue-Specific Gene Expression**

In this exercise, we return to the National Center for Biotechnology Information site (NCBI) and use the search tool **BLAST, Basic Local Alignment Search Tool**, which you were introduced to in the "Exploring Genomics" exercise for Chapter 10.

1. Access BLAST from the NCBI Web site at http://www.ncbi.nlm.nih.gov/BLAST.

2. The following are GenBank accession numbers for four different genes that show tissue-specific expression patterns.

(continued)

Exploring Genomics—continued

You will perform your searches on these genes.

NM_021588.1
NM_00739.1
AY260853.1
NM_004917

3. For each gene, carry out a nucleotide BLAST search using the accession numbers for your sequence query. Refer to the "Exploring Genomics" feature in Chapter 10 to refresh your memory on BLAST searches. Because the accession numbers are for nucleotide sequences, be sure to use the "nucleotide blast" (blastn) program when running your searches. Once you enter "blastn" under the "Choose Search Set" category, you should set the database to "Others (nr etc.)," so that you are not searching an organism-specific database.

4. For the top alignments for each gene, the "Links" column (far right) contains colored boxes labeled U (for UniGene expression data), E (Gene Expression Profiles), and G (Gene Information). Each of these boxes will link you to information about the gene. The UniGene link will show you a UniGene report. For some genes, upon entering UniGene you may need to click a link above the gene name before retrieving a UniGene report. Be sure to explore the "Expression Profile" link under the "Gene Expression" category in each UniGene report. Expression profiles will show a table of gene expression patterns in different tissues.

Also explore the "GEO profiles" link under the "Gene Expression" category of the UniGene reports, when avail-

able. These links will take you to a number of gene-expression studies related to each gene of interest. Explore these resources for each gene, and then answer the following questions:

a. What is the identity of each sequence, based on sequence alignment? How do you know this?

b. What species was each gene cloned from?

c. Which tissue(s) are known to express each gene?

d. Does this gene show regulated expression during different times of development?

e. Which gene shows the most restricted pattern of expression by being expressed in the fewest tissues?

CASE STUDY | A mysterious muscular dystrophy

A man in his early 30s suddenly developed weakness in his hands and neck, followed a few weeks later by burning muscle pain—all symptoms of late-onset muscular dystrophy. His internist ordered genetic tests to determine whether he had one of the inherited muscular dystrophies, focusing on Becker muscular dystrophy, myotonic dystrophy Type I, and myotonic dystrophy Type II. These tests were designed to detect mutations in the related *dystrophin*, *DMPK*, and *ZNF9* genes. The testing ruled out Becker muscular dystrophy. While awaiting the results of the *DMPK* and *ZNF9* gene tests, the internist explained that the possible mutations were due to expanded tri- and tetranucleotide repeats, but not in the protein-coding portions of the genes. She went on to say that the resulting disorders were the result not of changes in the encoded proteins, which appear to

be normal, but instead altered RNA splicing patterns, whereby the RNA splicing remnants containing the nucleotide repeats disrupt normal splicing of the transcripts of other genes. This discussion raises several interesting questions about the diagnosis and genetic basis of the disorders.

1. What is alternative splicing, where does it occur, and how could disrupting it affect the expression of the affected gene(s)?

2. What role might the expanded tri- and tetranucleotide repeats play in the altered splicing?

3. How does this contrast with other types of muscular dystrophy, such as Becker muscular dystrophy and Duchenne muscular dystrophy?

Summary Points

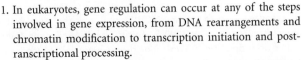

 For activities, animations, and review quizzes, go to the study area at www.masteringgenetics.com

1. In eukaryotes, gene regulation can occur at any of the steps involved in gene expression, from DNA rearrangements and chromatin modification to transcription initiation and post-transcriptional processing.

2. Programmed DNA rearrangements occur in some genes in specific cell types. These rearrangements are necessary for normal regulated expression of these genes.

3. Eukaryotic gene regulation at the chromatin level may involve gene-specific chromatin remodeling, histone modifications, or DNA modifications.

4. Eukaryotic transcription is regulated at gene-specific promoter, enhancer, and silencer elements.

5. Transcription factors influence transcription rates by binding to *cis*-acting regulatory sites within or adjacent to a gene promoter.

6. Transcription factors are thought to act by enhancing or repressing the association of general transcription factors with the core promoter. They may also assist in chromatin remodeling.

7. Posttranscriptional gene regulation includes alternative splicing of nascent RNA, RNA transport, or changes in mRNA stability.

8. Alternative splicing increases the number of gene products encoded by a single gene.

9. Postranscriptional gene regulation at the levels of translation and protein stability also affect the levels of active gene product.

10. RNA-induced gene silencing is a postranscriptional mechanism of gene regulation that affects the translatability or stability of mRNA as well as transcription.

INSIGHTS AND SOLUTIONS

1. As a research scientist, you have decided to study transcription regulation of a gene whose DNA has been cloned and sequenced. To begin your study, you obtain the DNA clone of the gene, which includes the gene and at least 1 kb of upstream DNA. You then create a number of subclones of this DNA, containing various deletions in the gene's upstream region. These deletion templates are shown in the figure below. To test these DNA templates for their ability to direct transcription of the gene, you prepare two different types of *in vitro* transcription systems. The first is a defined system containing purified RNA polymerase II and the purified general transcription factors TFIID, TFIIB, TFIIE, TFIIF, and TFIIH. The second system consists of a crude nuclear extract, which is made by extracting most of the proteins from the nuclei of cultured cells. When you test your two transcription systems using each of your templates, you obtain the following results:

Undeleted template

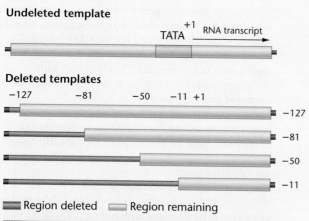

Deleted templates

DNA added	Nuclear extract	Purified system
undeleted	++++	+
−127 deletion	++++	+
−81 deletion	++++	+
−50 deletion	+	+
−11 deletion	o	o

+ Low-efficiency transcription
++++ High-efficiency transcription
o No transcription

(a) Why is there no transcription from the −11 deletion template in both the crude extract and the purified system?

(b) How do the results for the nuclear extract and the purified system differ, for the *undeleted* template? How would you interpret this result?

(c) For each of the various deletion templates, compare the results obtained from both the nuclear extract and the purified systems.

(d) What do these data tell you about the transcription regulation of this gene?

Solution:
(a) The lack of transcription from the −11 template suggests that some essential DNA sequences are missing from this deletion template. As the −50 template does show some transcription in both the crude extract and purified system, it is likely that the essential missing sequences, at least for basal levels of transcription, lie between −50 and −11. As the TATA box is located in this region, its absence in the −11 template may be the reason for the lack of transcription.

(b) The undeleted template containing large amounts of upstream DNA is sufficient to promote high levels of transcription in a nuclear extract, but only low levels in a purified system. These data suggest that something is missing in the purified system, compared with the nuclear extract, and this component is important for high levels of transcription from this promoter. As crude nuclear extracts are not defined in content, it would not be clear from these data what factors in the extract are the essential ones.

(c) Both the −127 and −81 templates act the same way as the undeleted template in both the nuclear extract and the purified system—high levels of transcription in nuclear extracts but low levels in a purified system. In contrast, the −50 template shows only low levels of transcription in both systems. These results indicate that all of the sequences necessary for high levels of transcription in a crude system are located between −81 and −50.

(d) First, these data tell you that general transcription factors alone are not sufficient to specify high efficiencies of transcription from this promoter. The DNA sequence elements through which the general transcription factors work are located within 50 bp of the transcription start site. Second, the data tell you that the promoter for this gene is likely a member of the "focused" class of promoters, with one defined transcription start site and an essential TATA box. Third, high levels of transcription require sequences between −81 and −50 relative to the transcription start site. These sequences interact with some component(s) of crude nuclear extracts.

2. Scientists estimate that more than 15 percent of disease-causing mutations involve errors in alternative splicing. However, there is an interesting case in which an exon deletion appears to enhance dystrophin production in muscle cells of Duchenne muscular dystrophy (DMD) patients. A deletion of exon 45 of the *dystrophin* gene is the most frequent DMD-causing mutation. But some individuals with Becker muscular dystrophy (BMD), a milder form of muscular dystrophy, have deletions of both exons 45 and 46 (van Deutekom and van Ommen, 2003. *Nat. Rev. Genetics* 4: 774–783). Provide a possible explanation for why BMD patients, with a deletion of both exon 45 and 46, produce more dystrophin than DMD patients do.

Solution: Having a deletion of one exon has several possible effects on a gene product. One possibility is that the mRNA transcribed from the exon-deleted *dystrophin* gene is unstable, leading to a lack of dystrophin protein production. Even if the mRNA is stable, the resulting mutated dystrophin protein could be targeted for rapid degradation, leading to the absence of stable active protein. Another possibility is that the deletion of one exon creates a frameshift mutation, which creates a premature translation stop codon. As the dystrophin gene has 79 exons

(continued)

Insights and Solutions—continued

spanning over 2.6 million base pairs of DNA, a frameshift mutation in exon 45 could create a stop codon near the middle of the gene. The mRNA transcribed from this gene could be targeted for nonsense-mediated mRNA degradation, due to the presence of the premature stop codon. Any mRNA escaping degradation would encode a shorter than normal dystrophin protein, which likely would be rapidly degraded.

It is possible that a deletion encompassing both exon 45 and 46 could restore the reading frame of the dystrophin protein in exon 47. The protein product of this gene would be missing amino acid sequences encoded by the two missing exons; however, the protein itself could still have some activity, partially restoring the wild-type phenotype.

As described in the van Deutekom and van Ommen paper, about 60 percent of all DMD mutations involve deletions of an exon. One strategy being considered for DMD gene therapy is to restore the normal reading frame by causing other exons to be skipped during pre-mRNA splicing. If successful, this type of therapy could convert a severe DMD phenotype into a milder form of the disease, similar to BMD.

Problems and Discussion Questions

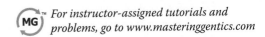 *For instructor-assigned tutorials and problems, go to www.masteringgentics.com*

HOW DO WE KNOW?

1. In this chapter, we focused on how eukaryotic genes are regulated at different steps in their expression, from chromatin modifications to control of protein stability. At the same time, we found many opportunities to consider the methods and reasoning by which much of this information was acquired. From the explanations given in the chapter,

    (a) How do we know that promoter and enhancer sequences control the initiation of transcription in eukaryotes?

    (b) How do we know that eukaryotic transcription factors bind to DNA sequences at or near promoter regions?

    (c) How do we know that double-stranded RNA molecules can control gene expression?

2. Why is gene regulation more complex in a multicellular eukaryote than in a prokaryote? Why is the study of this phenomenon in eukaryotes more difficult?

3. List and define the levels of gene regulation discussed in this chapter.

4. Provide a definition of chromatin remodeling, and give two examples of this phenomenon.

5. Describe the organization of the interphase nucleus. Include in your presentation a description of chromosome territories, interchromosomal compartments, and transcription factories. Explain how chromosome-painting techniques have helped reveal the organization of the interphase nucleus.

6. A number of experiments have demonstrated that areas of the genome that are relatively inert transcriptionally are resistant to DNase I digestion; however, those areas that are transcriptionally active are DNase I sensitive. Describe how DNase I resistance or sensitivity might indicate transcriptional activity.

7. Provide a brief description of two different types of histone modification.

8. Present an overview of the manner in which chromatin can be remodeled. Describe the manner in which these remodeling processes influence transcription.

9. Distinguish between the *cis*-acting regulatory elements referred to as promoters and enhancers.

10. Is the binding of a transcription factor to its DNA recognition sequence necessary and sufficient for an initiation of transcription at a regulated gene? What else plays a role in this process?

11. Compare the control of gene regulation in eukaryotes and prokaryotes at the level of initiation of transcription. How do the regulatory mechanisms work? What are the similarities and differences in these two types of organisms in terms of the specific components of the regulatory mechanisms? Address how the differences or similarities relate to the biological context of the control of gene expression.

12. Many promoter regions contain CAAT boxes containing consensus sequences CAAT or CCAAT approximately 70 to 80 bases upstream from the transcription start site. How might one determine the influence of CAAT boxes on the transcription rate of a given gene?

13. Recent research indicates that promoters may fall into two classes: *focused* or *dispersed*. How do these differ, and which genes tend to be associated with each?

14. Present an overview of RNA-induced gene silencing achieved through RNA interference (RNAi) and microRNAs (miRNAs). How do the silencing processes begin, and what major components participate?

15. How does the inherent imprecision of recombination events during immunoglobulin gene rearrangement contribute to the diversity of the immune response?

16. Explain the structural features of the Initiator (Inr), BRE, DPE, and MTE elements of focused promoters.

17. DNA supercoiling, which occurs when coiling tension is generated ahead of the replication fork, is relieved by DNA gyrase. Supercoiling may also be involved in transcription regulation. Researchers discovered that transcriptional enhancers operating over a long distance (2500 base pairs) are dependent on DNA supercoiling, while enhancers operating over shorter distances (110 base pairs) are not so dependent (Liu et al., 2001. *Proc. Natl. Acad. Sci. [USA]* 98: 14,883–14,888). Using a diagram, suggest a way in which supercoiling may positively influence enhancer activity over long distances.

18. In some organisms, including mammals, there is an inverse relationship between the presence of 5-methylcytosine (m^5C) in CpG sequences and gene activity. In addition, m^5C may be involved in the recruitment of proteins that convert chromatin regions from transcriptionally active to inactive states. Overall, genomic DNA is relatively poor in CpG sequences due to the conversion of m^5C to thymine; however, unmethylated CpG

islands are often associated with active genes. Researchers have determined that patterns of DNA methylation in spermatozoa vary with age in rats and suggest that such age-related alterations in DNA methylation may be one mechanism underlying age-related abnormalities in mammals (Oakes et al., 2003. *Proc. Natl. Acad. Sci.* 100: 1775–1780). In light of the above information, provide an explanation that relates paternal age-related alterations in DNA methylation to birth abnormalities.

19. In the hypothetical extraterrestrial organism, the Quagarre, the immunoglobulin genes undergo rearrangements similar to those in humans, with two main differences. One difference is that the κ light-chain V regions are located on chromosome 4 and the other κ light-chain regions are located on chromosome 10. Another difference is that the Quagarre immunoglobulin genes do not rearrange until the creature is 13 years old. By examining a mitotic spread of chromosomes from a Quagarre B cell, how could you estimate the creature's age?

Extra-Spicy Problems

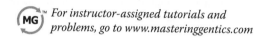 *For instructor-assigned tutorials and problems, go to www.masteringgentics.com*

20. Because the degree of DNA methylation appears to be a relatively reliable genetic marker for some forms of cancer, researchers have explored the possibility of altering DNA methylation as a form of cancer therapy. Initial studies indicate that while hypomethylation suppresses the formation of some tumors, other tumors thrive. Why would one expect different cancers to respond differently to either hypomethylation or hypermethylation therapies?

21. Explain how the following mutations would affect the transcription of the yeast *GAL1* gene.
    (a) A deletion within the *GAL4* gene that removes the region encoding amino acids 1 to 100.
    (b) A deletion of the entire *GAL3* gene.
    (c) A mutation within the *GAL80* gene that blocks the ability of Gal80 protein to interact with Gal3p.
    (d) A deletion of one of the four UAS$_G$ elements upstream from the *GAL1* gene.
    (e) A point mutation in the *GAL1* core promoter that alters the sequence of the TATA box.

22. The interphase nucleus appears to be a highly structured organelle with chromosome territories, interchromosomal compartments, and transcription factories. In cultured human cells, researchers have identified approximately 8000 transcription factories per cell, each containing an average of eight tightly associated RNA polymerase II molecules actively transcribing RNA. If each RNA polymerase II molecule is transcribing a different gene, how might such a transcription factory appear? Provide a simple diagram that shows eight different genes being transcribed in a transcription factory and include the promoters, structural genes, and nascent transcripts in your presentation.

23. A particular type of anemia in humans, called β-thalassemia, results from a severe reduction or absence of a normal β chain of hemoglobin. A variety of studies have explored the use of 5-azacytidine for the treatment of such patients. How might administration of 5-azacytidine be an effective treatment for β-thalassemia? Would you consider adverse side-effects likely? Why?

24. The addition of a "5'-cap" and a "3' poly-A" tail to many nuclear RNA species occurs prior to export of mature mRNAs to the cytoplasm. Such modifications appear to influence mRNA stability. Many assays for gene regulation involve the use of reporter genes such as luciferase from the North American firefly (*Photinus pyralis*). When the luciferase gene is transcribed and translated, the protein product can be easily assayed in a test tube. In the presence of luciferin, ATP, and oxygen, the luciferase enzyme produces light that can be easily quantified.

Assuming that luciferase mRNA can be obtained and differentially modified (5'-capped, poly-A tailed) in a test tube, suggest an assay system that would allow you to determine the influence of 5'-capping and poly-A tail addition on mRNA stability.

25. DNA methylation is commonly associated with a reduction of transcription. The following data come from a study of the impact of the location and extent of DNA methylation on gene activity in human cells. A bacterial gene, luciferase, was cloned next to eukaryotic promoter fragments that were methylated to various degrees, *in vitro*. The chimeric plasmids were then introduced into tissue culture cells, and the luciferase activity was assayed. These data compare the degree of expression of luciferase with differences in the location of DNA methylation (Irvine et al., 2002. *Mol. and Cell. Biol.* 22: 6689–6696). What general conclusions can be drawn from these data?

DNA Segment	Patch Size of Methylation (kb)	Number of Methylated CpGs	Relative Luciferase Expression
Outside transcription unit (0–7.6 kb away)	0.0	0	490X
	2.0	100	290X
	3.1	102	250X
	12.1	593	2X
Inside transcription unit	0.0	0	490X
	1.9	108	80X
	2.4	134	5X
	12.1	593	2X

26. Many scientists regard selective nuclear transport of RNAs as a form of genetic regulation in eukaryotes. The discovery of what appears to be an array of nuclear fibrous elements adds to that appeal. In a recent literature review (Pederson, 2000. *Molecular Biology of the Cell* 11: 799–805), the author reported data arising from various labeling experiments designed to determine whether gene regulation in eukaryotes is related to nuclear sorting of pre-mRNAs. The vast majority of the results indicated that nuclear poly-A RNA moves by diffusion. What influence would these results have on models that relate gene regulation to nuclear RNA transport?

27. Mouse quaking is caused by a recessive dysmyelination mutation (*qk*), which appears to involve defects in alternative splicing of an evolutionarily conserved signal transduction protein.
Homozygotes (*qk/qk*) suffer tremors during exertion. Mature
mice may experience seizures and remain motionless for many
seconds. Below is a figure that describes the splicing patterns
in various genotypes of quaking and normal mice. Portions
of brain MAG (myelin-associated glycoprotein) RNA (including exons 11, 12, and 13) from young (14-day) and adult
(2-month) mice are represented. The relative concentration of
each RNA is indicated in the gel diagram below. (Figure and
data are modified from Wu, J., et al. 2002. *Proc. Natl. Acad. Sci.*
99: 4233–4238.)

   Given this information, describe the alternative splicing patterns with regards to exons included/excluded as a function of
age and genotype.

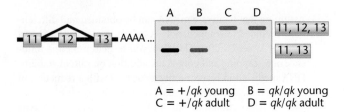

28. Incorrectly spliced RNAs often lead to human pathologies. Scientists have examined human cancer cells for splice-specific changes
and found that many of the changes disrupt tumor-suppressor
gene function (Xu and Lee, 2003. *Nucl. Acids Res.* 31: 5635–5643).
In general, what would be the effects of splicing changes on these
RNAs and the function of tumor-suppressor gene function? How
might loss of splicing specificity be associated with cancer?

29. In a classical genetic analysis, scientists create mutations within
or surrounding cloned gene sequences. They then introduce
these mutated gene sequences into cell cultures or model organisms and examine the phenotypes. Geneticists are now adding RNA-induced gene silencing to their gene-analysis toolbox.
Describe how you would use RNA-induced gene silencing to
study the role of the *Drosophila tra* gene in sex determination.
Next, describe how you would study *tra* function, using a classical genetic analysis. What are the advantages and limitations
of each research approach?

30. During an examination of the genomic sequences surrounding
the human β-globin gene, you discover a region of DNA that
bears sequence resemblance to the GRE of the human metallothionein IIA (*hMTIIA*) gene. Describe experiments that you
would design to test (1) whether this sequence was necessary
for accurate β-globin gene expression and (2) whether this sequence acted in the same way as the *hMTIIA* gene's GRE.

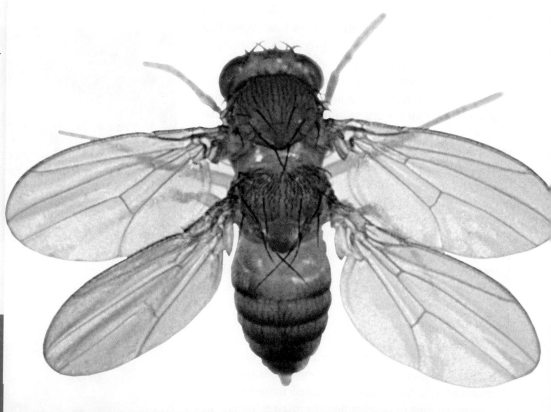

This unusual four-winged *Drosophila* has developed an extra set of wings as a result of a mutation in a homeotic selector gene.

18

Developmental Genetics

Over the last two decades molecular biology and genomic analysis have shown that in spite of wide diversity in the size and shape of adults, all multicellular organisms share many genes, genetic pathways, and molecular signaling mechanisms in the developmental events leading from the zygote to the adult. At the cellular level, development is marked by three important events: **specification**, when genetic and positional cues confer a spatially discrete identity on cells, **determination**, the time when a specific developmental fate for a cell becomes fixed, and **differentiation**, the process by which a cell achieves its final form and function. In addition, comparative genomics has shown that higher organisms share many evolutionary and developmental relationships. Genetic analysis has identified many regulatory genes that control networks of structural genes involved in developmental processes, and we are beginning to understand how the action and interaction of these genes control basic developmental processes in eukaryotes.

In this chapter, the primary emphasis will be on how genetics has been used to study development. This area, called developmental genetics, has contributed tremendously to our understanding of developmental processes because genetic information is required for the molecular and cellular functions mediating developmental events and contributes to the continually changing phenotype of the newly formed organism.

(a)

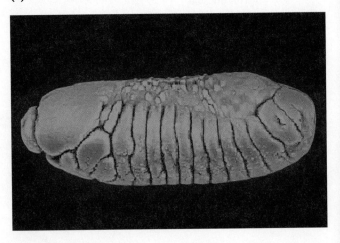

(b)

FIGURE 18–1   (a) A *Drosophila* embryo and (b) the adult fly that develops from it.

18.1

Differentiated States Develop from Coordinated Programs of Gene Expression

Animal genomes contain tens of thousands of genes, but only a small subset of these control the events that shape the adult body (Figure 18–1).

Developmental geneticists study mutant alleles of these genes to ask important questions about development:

- What genes are expressed?

- When are they expressed?

- In what parts of the developing embryo are they expressed?

- How is the expression of these genes regulated?

- What happens when these genes are defective?

These questions provide a foundation for exploring the molecular basis of developmental processes such as determination, induction, cell–cell communication, and cellular

differentiation. Genetic analysis of mutant alleles is used to establish a causal relationship between the presence or absence of inducers, receptors, transcriptional events, and cell and tissue interactions, and the observable morphological events that accompany development.

A useful way to define **development** is to say that it is the *attainment of a differentiated state* by all the cells of an organism (except for stem cells). For example, a cell in a blastula-stage embryo (when the embryo is just a ball of uniform-looking cells) is undifferentiated, while a red blood cell synthesizing hemoglobin in the adult body is differentiated. How do cells get from the undifferentiated to the differentiated state? The process involves progressive activation of different gene sets in different cells of the embryo. From a genetics perspective, one way of defining the different cell types that form during development in multicellular organisms is to identify and catalog the genes that are active in each cell type. In other words, development depends on patterns of differential gene expression.

The idea that differentiation is accomplished by activating and inactivating genes at different times and in different cell types is called the **variable gene activity hypothesis**. Its underlying assumptions are, first, that each cell contains an entire genome, and, second, that differential transcription of selected genes controls the development and differentiation of each cell. In multicellular organisms, evolution has conserved the genes involved in development, the patterns of differential transcription, and the ensuing developmental mechanisms. As a result, scientists are able to learn about development in multicellular organisms in general by dissecting these mechanisms in a small number of genetically well-characterized model organisms.

18.2 Evolutionary Conservation of Developmental Mechanisms Can Be Studied Using Model Organisms

Genetic analysis of development in a wide range of organisms has demonstrated that all animals use a common set of developmental mechanisms and signaling systems. For example, most of the differences in shape between zebras and zebrafish are controlled by different patterns of expression in a single cluster of genes called the homeotic genes, and not by expression of many different genes scattered across the genome. Sequencing projects have confirmed that genes from a wide range of organisms are highly conserved at the level of DNA sequence. This homology in genes and regulatory mechanisms means that many aspects of normal human embryonic development and associated genetic disorders can be studied in model organisms such as *Drosophila,* which, unlike humans, can be genetically manipulated (see Chapter 1 for a discussion of model organisms in genetics).

Although many developmental mechanisms are similar among all animals, selection has generated new and alternative pathways for transforming zygotes into adult organisms. Several genetic mechanisms underlie these evolutionary changes including mutation, gene duplication and divergence, the assignment of new functions to old genes, and the recruitment of genes to new developmental pathways. The emphasis in this chapter, however, will be on the similarities among species.

Analysis of Developmental Mechanisms

In this chapter, we cannot survey all aspects of development, nor can we explore the genetic analysis of all developmental mechanisms triggered by the fusion of sperm and egg. Rather, we will focus on several general processes in development:

- how the adult body plan of animals is laid down in the embryo

- the program of gene expression that turns undifferentiated cells into differentiated cells

- the role of cell–cell communication in development

We will use three model systems—*Drosophila melanogaster, Arabidopsis thaliana,* and *Caenorhabditis elegans*—to illustrate these developmental processes and related topics. We will examine the role of differential gene expression in the progressive restriction of developmental options leading to the formation of the adult body plan in the model organisms *Drosophila* and *Arabidopsis*. We will then expand the discussion to include the selection of pathways that result in differentiated cells in plants and animals, and consider the role of cell–cell communication in development in *C. elegans*.

18.3 Genetic Analysis of Embryonic Development in *Drosophila* Reveals How the Animal Body Axis Is Specified

How specific cells in an embryo turn gene sets on or off at precisely timed stages of development is a central question in developmental biology. The study of model organisms gives us an insight into the identity of regulatory genes and the networks they control during development. In particular, the genetic and molecular analysis of embryonic development in *Drosophila* highlights the key role of molecular components in the oocyte cytoplasm in controlling gene expression.

Overview of *Drosophila* Development

The development of *Drosophila* from zygote to adult takes about 10 days and includes several distinct phases: the embryo, three larval stages, the pupal stage, and the adult stage (Figure 18–2). Internally, the cytoplasm of the fertilized egg is organized into a series of maternally constructed molecular gradients that play key roles in determining the developmental fates of nuclei located in specific regions of the embryo.

Immediately after fertilization, the zygote nucleus undergoes a series of nuclear divisions without cytokinesis [Figure 18–3(a) and (b)], forming a syncytial blastoderm (a syncytium is any cell with more than one nucleus). At about the tenth division, nuclei migrate to the periphery of the egg into cytoplasm containing localized gradients of maternally derived mRNA transcripts and proteins [Figure 18–3(c)]. After several more divisions, the nuclei become enclosed in plasma membranes [Figure 18–3(d)] and form cells.

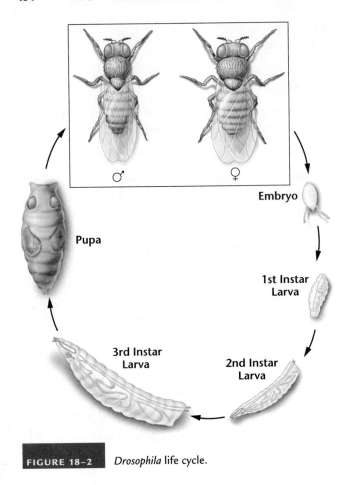

Drosophila life cycle.

Cells that form at the posterior pole of the embryo [Figure 18–3(c) and (d)] become germ cells. If nuclei from other regions of the embryo are transplanted into the posterior cytoplasm, they will form germ cells, confirming that the cytoplasm in this region contains maternal components that direct nuclei to form germ cells.

Transcriptional programs activated in the non–germ-cell nuclei form the embryo's anterior–posterior (head to tail) and dorsal–ventral (back to front) axes of symmetry, leading to the formation of a segmented embryo [Figure 18–3(e)]. At a later stage of development, under control of the *Hox* gene set (discussed in a later section), these segments give rise to the differentiated structures of the adult fly [Figure 18–3(f)].

Genetic Analysis of Embryogenesis

Two different gene sets control embryonic development in *Drosophila*: maternal-effect genes and zygotic genes (Figure 18–4). During development of the egg, products of maternal-effect genes (mRNA and proteins) are placed in the egg cytoplasm. Many of these products are distributed in a gradient or concentrated in specific regions of the egg. Female flies homozygous for certain recessive mutations of maternal-effect genes are sterile: none of their embryos receive wild-type gene products from their mother, so all the embryos develop abnormally. Maternal-effect genes encode transcription factors and proteins that regulate gene expression. At specific stages of embryonic development, these gene products activate or repress expression of the zygotic genome in a temporal and spatial sequence.

Zygotic genes are transcribed in the nuclei of the developing embryo. Flies with certain homozygous mutations in zygotic genes exhibit embryonic lethality. In a cross between two flies heterozygous for a recessive zygotic mutation, one-fourth of the embryos (the recessive homozygotes) therefore fail to develop normally and die. In *Drosophila*, many zygotic genes are transcribed in specific regions of the embryo in response to the distribution of maternal-effect proteins.

Much of our knowledge about the genes that regulate *Drosophila* development is based on the work of Christiane Nüsslein-Volhard, Eric Wieschaus, and Ed Lewis, who were awarded the 1995 Nobel Prize for Physiology or Medicine. Ed Lewis initially identified and studied one of these regulatory genes in the 1970s. In the late 1970s, Nüsslein-Volhard and Wieschaus devised a screening strategy to identify all the genes that control development in *Drosophila*. Their method required examining thousands of offspring of mutagenized flies, looking for recessive embryonic lethal mutations with phenotypic defects in body segments and other external structures. The parents were thus identified as heterozygous carriers of these mutations, which the researchers grouped into three classes: *gap, pair-rule,* and *segment polarity* genes. In 1980, on the basis of their observations, Nüsslein-Volhard and Wieschaus proposed a model in which embryonic development is initiated by gradients of maternal-effect gene products. The positional information laid down by these molecular gradients is interpreted by expression of two sets of zygotic genes: (1) **segmentation genes** (gap, pair-rule, and segment polarity genes) and (2) **homeotic selector genes**. Action of the segmentation genes divides the embryo into a series of stripes or segments and defines the number, size, and polarity of each segment. The homeotic genes specify the identity of each segment and the adult structures formed from the segments (Figure 18–4).

The model developed by Nüsslein-Volhard and Wieschaus is shown in Figure 18–5. Most maternal-effect gene products placed in the egg during oogenesis are activated immediately after fertilization and help establish the anterior–posterior axis of the embryo [Figure 18–5(a)]. Many maternal gene products encode transcription factors that activate transcription of the gap genes, whose expression divides the embryo into a series of regions corresponding to the head, thorax, and abdomen of the adult [Figure 18–5(b)]. Gap proteins, in turn, act as other sets of transcription factors that activate pair-rule genes, whose products divide the embryo into smaller regions about two segments wide [Figure 18–5(c)]. The pair-rule genes

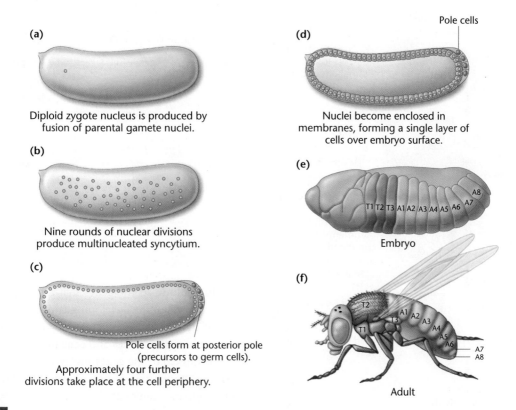

(a) Diploid zygote nucleus is produced by fusion of parental gamete nuclei.

(b) Nine rounds of nuclear divisions produce multinucleated syncytium.

(c) Pole cells form at posterior pole (precursors to germ cells). Approximately four further divisions take place at the cell periphery.

(d) Pole cells
Nuclei become enclosed in membranes, forming a single layer of cells over embryo surface.

(e) T1 T2 T3 A1 A2 A3 A4 A5 A6 A7 A8
Embryo

(f) T2 T1 T3 A1 A2 A3 A4 A5 A6 A7 A8
Adult

FIGURE 18–3 Early stages of embryonic development in *Drosophila*. (a) Fertilized egg with zygotic nucleus (2*n*), shortly after fertilization. (b) Nuclear divisions occur about every 10 minutes. Nine rounds of division produce a multinucleate cell, the syncytial blastoderm. (c) At the tenth division, the nuclei migrate to the periphery or cortex of the egg, and four additional rounds of nuclear division occur. A small cluster of cells, the pole cells, form at the posterior pole about 2.5 hours after fertilization. These cells will form the germ cells of the adult. (d) About 3 hours after fertilization, the nuclei become enclosed in membranes, forming a single layer of cells over the embryo surface, creating the cellular blastoderm. (e) The embryo at about 10 hours after fertilization. At this stage, the segmentation pattern of the body is clearly established. Behind the segments that will form the head, T1–T3 are thoracic segments, and A1–A8 are abdominal segments. (f) The adult fly showing the structures formed from each segment of the embryo.

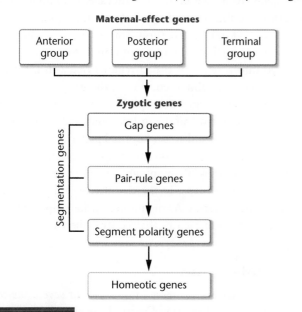

Maternal-effect genes

| Anterior group | Posterior group | Terminal group |

↓
Zygotic genes

Segmentation genes {
Gap genes
↓
Pair-rule genes
↓
Segment polarity genes
}
↓
Homeotic genes

FIGURE 18–4 The hierarchy of genes involved in establishing the segmented body plan in *Drosophila*. Gene products from the maternal genes regulate the expression of the first three groups of zygotic genes (gap, pair-rule, and segment polarity, collectively called the segmentation genes), which in turn control expression of the homeotic genes.

in turn activate the segment polarity genes, which divide each segment into anterior and posterior regions [Figure 18–5(d)]. The collective action of the maternal genes that form the anterior–posterior axis and the segmentation genes define the field of action for the homeotic (*Hox*) genes [Figure 18–5(e)].

NOW SOLVE THIS

18–1 Suppose you perform a screen for maternal-effect mutations in *Drosophila* affecting external structures of the embryo and your screen identifies more than 100 mutations that affect external structures. From their screening, other researchers concluded that there are about 40 maternal-effect genes. How do you reconcile your results with those of the other researchers?

■ HINT: *This problem involves an understanding of how mutant screens work. Once mutants are identified, they must be screened by complementation analysis (Chapter 4). The key to its solution lies in remembering the differences between genes and alleles.*

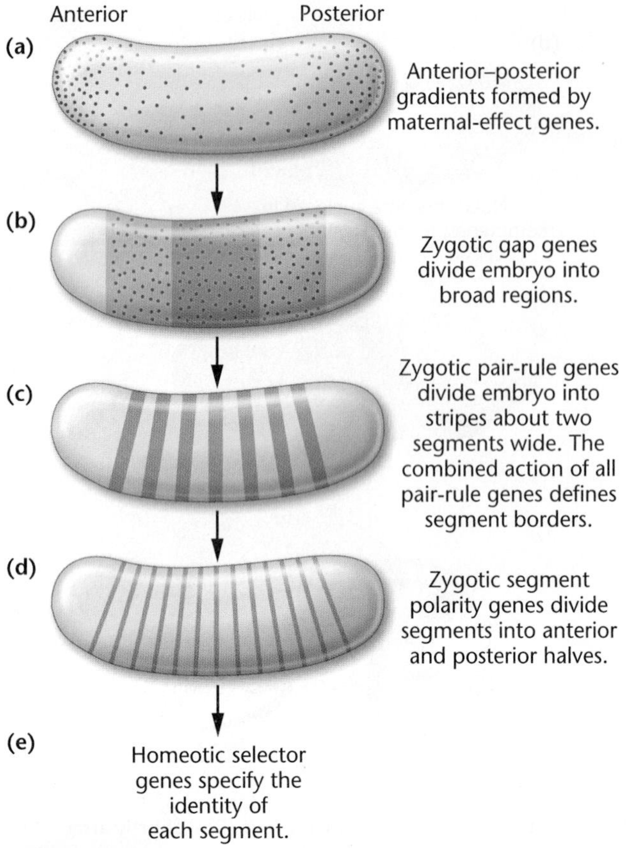

FIGURE 18–5   (a) Progressive restriction of cell fate during development in *Drosophila*. Gradients of maternal proteins are established along the anterior–posterior axis of the embryo. (b), (c), and (d) Three groups of segmentation genes progressively define the body segments. (e) Individual segments are given identity by the homeotic genes.

TABLE 18.1

Segmentation Genes in *Drosophila*

Gap Genes	Pair-Rule Genes	Segment Polarity Genes
Krüppel	*hairy*	*engrailed*
knirps	*even-skipped*	*wingless*
hunchback	*runt*	*cubitis*
giant	*fushi-tarazu*	*hedgehog*
tailless	*paired*	*fused*
huckebein	*odd-paired*	*armadillo*
caudal	*odd-skipped*	*patched*
	sloppy-paired	*gooseberry*
		paired
		naked
		disheveled

In addition to the genes that determine the anterior–posterior axis of the developing embryo, the dorsal–ventral axis of the embryo is organized by a combination of maternal and zygotic genes and gene products. Our discussion will be limited to the genes involved in establishing the anterior–posterior axis. Let us now examine each member of this group in greater detail.

Gap Genes

The embryonic **gap genes** are activated or inactivated by gene products previously expressed along the anterior–posterior axis and by other genes of the maternal gradient system. When mutated, these genes produce large gaps in the embryo's segmentation pattern. Mutants of the *hunchback* gene lose head and thorax structures, *Krüppel* mutants lose thoracic and abdominal structures, and *knirps* mutants lose most abdominal structures. Transcription of wild-type gap genes (which encode transcription factors) divides the embryo into a series of broad regions that will form the head, thorax, and abdomen. Within these regions, specific patterns of gene expression specify both the type of segment that will form and the order of segments in the body of the larva, pupa, and adult. Regional patterns of gap genes expression in different parts of the embryo correlate roughly with the location of their mutant phenotypes: *hunchback* at the anterior, *Krüppel* in the middle (Figure 18–6), and *knirps* at the posterior. As mentioned earlier, gap genes encode transcription factors that control the expression of pair-rule genes.

Pair-Rule Genes

Pair-rule genes are expressed in a series of seven narrow bands or stripes that extend around the circumference of the embryo. Expression of this gene set first establishes the boundaries of segments and then establishes the developmental fate of the cells within each segment by controlling expression of the segment polarity genes. Mutations

18.4

Zygotic Genes Program Segment Formation in *Drosophila*

The expression or repression of zygotic genes during embryonic development occurs in response to the positional gradient of maternal-effect gene products in the cytoplasm. The sequential expression of three subsets of segmentation genes divides the embryo into a series of segments along its anterior–posterior axis. These segmentation genes are normally transcribed in the developing embryo, and mutations of these genes have embryo-lethal phenotypes.

Over 20 segmentation genes (Table 18.1) have been identified. They are classified on the basis of their mutant phenotypes: (1) mutations in gap genes delete a group of adjacent segments, (2) mutations in pair-rule genes affect every other segment and eliminate a specific part of each affected segment, and (3) mutations in segment polarity genes cause defects in homologous portions of each segment.

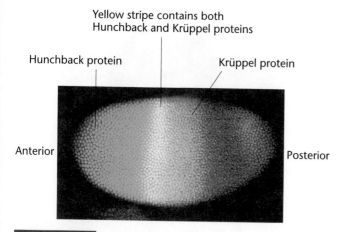

Yellow stripe contains both
Hunchback and Krüppel proteins

Hunchback protein

Krüppel protein

Anterior

Posterior

FIGURE 18–6   Expression of gap genes in a *Drosophila* embryo. The Hunchback protein is shown in orange, and Krüppel is indicated in green. The yellow stripe is created when cells contain both Hunchback and Krüppel proteins. Each dot in the embryo is a nucleus.

in pair-rule genes eliminate segment-size sections at every other segment. At least eight pair-rule genes act to divide the embryo into a series of stripes. However, the boundaries of these stripes overlap, so that within each area of overlap, cells express a different combination of pair-rule genes (Figure 18–7). The transcription of the pair-rule genes is mediated by the action of gap gene products and maternal gene products, but the resolution of this segmentation pattern into highly delineated stripes results from

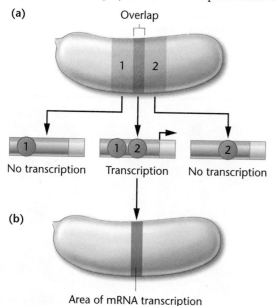

(a)

Overlap

1   2

1 | No transcription

1 2 | Transcription

2 | No transcription

(b)

Area of mRNA transcription

FIGURE 18–7   New patterns of gene expression can be generated by overlapping regions containing two different gene products. (a) Transcription factors 1 and 2 are present in an overlapping region of expression. If both transcription factors must bind to the promoter of a target gene to trigger expression, the gene will be active only in cells containing both factors (most likely in the zone of overlap). (b) The expression of the target gene in the restricted region of the embryo.

(a)

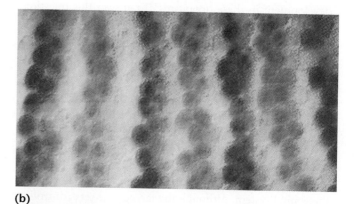

(b)

FIGURE 18–8   Stripe pattern of pair-rule gene expression in *Drosophila* embryo. This embryo is stained to show patterns of expression of the genes *even-skipped* and *fushi-tarazu;* (a) low-power view and (b) high-power view of the same embryo.

the interaction among the gene products of the pair-rule genes themselves (Figure 18–8).

Segment Polarity Genes

Expression of **segment polarity genes** is controlled by transcription factors encoded by pair-rule genes. Within each segment created by pair-rule genes, segment polarity genes become active in a single band of cells that extends around the embryo's circumference (Figure 18–9). This divides the

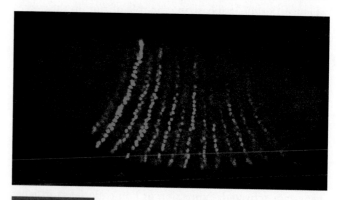

FIGURE 18–9   The 14 stripes of expression of the segment polarity gene *engrailed* in a *Drosophila* embryo.

embryo into 14 segments. The products of the segment polarity genes control the cellular identity within each of them and establish the anterior–posterior pattern (the polarity) within each segment.

Segmentation Genes in Mice and Humans

We have seen that segment formation in *Drosophila* depends on the action of three subsets of segmentation genes. If *Drosophila* is to be a useful model for understanding general principles of animal development, it is logical to ask whether these gene families are found in humans and other mammals, and if so, do they control aspects of embryonic development in these organisms? To answer this question, let's examine *runt*, one of the pair-rule genes in *Drosophila*. In late stages of development, it controls aspects of sex determination and formation of the nervous system. The gene encodes a protein that regulates transcription of its target genes, and contains a 128-amino-acid DNA-binding region (called the runt domain) that is highly conserved in mouse and human proteins. In fact, *in vitro* experiments show that the *Drosophila* and mouse runt proteins are functionally interchangeable. In mice, *runt* is expressed early in development and controls formation of blood cells, bone, and the genital system. Although the target gene sets controlled by *runt* are different in *Drosophila* and the mouse, in both organisms, expression of *runt* specifies the fate of uncommitted embryonic cells by regulating transcription of target genes.

In humans, mutation in *RUNX2*, a human homolog of *runt*, causes cleidocranial dysplasia (CCD), an autosomal dominantly inherited trait. Those affected with CCD have a hole in the top of their skull because their fontanel does not close. Their collar bones (clavicles) do not develop, enabling them to fold their shoulders across their chest (Figure 18–10). Mice with one mutant copy of the *runt* homolog have a phenotype similar to that seen in humans; mice with two mutant copies of the gene have no bones at all. Their skeletons contain only cartilage, much like sharks (Figure 18–11), emphasizing the role of *runt* in these species as an important gene controlling the initiation of bone formation.

18.5
Homeotic Selector Genes Specify Parts of the Adult Body

As boundaries are established by expression of segmentation genes, the homeotic (from the Greek word for "same") genes are activated. Expression of homeotic selector genes determines which adult structures will be formed by each body segment. In *Drosophila*, this includes the antennae,

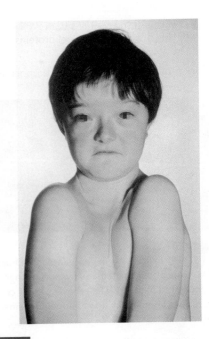

FIGURE 18–10   A boy affected with cleidocranial dysplasia (CCD). This disorder, inherited as an autosomal dominant trait, is caused by mutation in a human *runt* gene, *RUNX2*. Affected heterozygotes have a number of skeletal defects, including a hole in the top of the skull where the infant fontanel fails to close, and collar bones that do not develop or form only small stumps. Because the collar bones do not form, CCD individuals can fold their shoulders across their chests. *Reprinted by permission from Macmillan Publishers Ltd.: Fig. 1 on p. 244 from:* British Dental Journal *195: 243–248 2003. Greenwood, M. and Meechan, J. G. "General medicine and surgery for dental practitioners." Copyright © Macmillan Magazines Limited.*

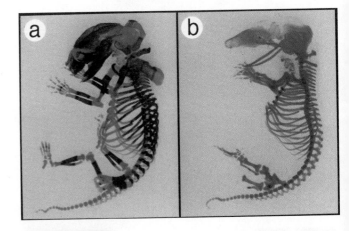

FIGURE 18–11   Bone formation in normal mice and mutants for the *runt* gene *Runx2*. (a) Normal mouse embryos at day 17.5 show cartilage (blue) and bone (brown). (b) The skeleton of a 17.5-day homozygous mutant embryo. Only cartilage has formed in the skeleton. There is complete absence of bone formation in the mutant mouse. Expression of a normal copy of the *Runx2* gene is essential for specifying the developmental fate of bone-forming osteoblasts.

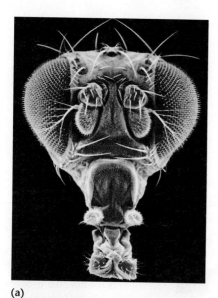

FIGURE 18–12   *Antennapedia (Antp)* mutation in *Drosophila.* (a) Head from wild-type *Drosophila,* showing the antenna and other head parts. (b) Head from an *Antp* mutant, showing the replacement of normal antenna structures with legs. This is caused by activation of the *Antp* gene in the head region.

mouth parts, legs, wings, thorax, and abdomen. Mutants of these genes are called **homeotic mutants** because the structure formed by one segment is transformed into that formed by another segment. For example, the wild-type allele of *Antennapedia (Antp)* specifies formation of a leg on the second segment of the thorax. Dominant gain-of-function *Antp* mutations cause this gene to be expressed in the head as well, and in mutant flies the antenna is transformed into a leg (Figure 18–12).

Hox Genes in *Drosophila*

The *Drosophila* genome contains two clusters of homeotic selector genes (called *Hox* genes) on chromosome 3 that encode transcription factors (Table 18.2). The *Antennapedia (ANT-C)* cluster contains five genes that specify structures in the head and the first two segments of the thorax [Figure 18–13(a)]. The second cluster, the *bithorax (BX-C)*

TABLE 18.2	
Hox* Genes of *Drosophila	
Antennapedia Complex	**Bithorax Complex**
labial	*Ultrabithorax*
Antennapedia	*abdominal A*
Sex combs reduced	*Abdominal B*
Deformed	
proboscipedia	

(a)

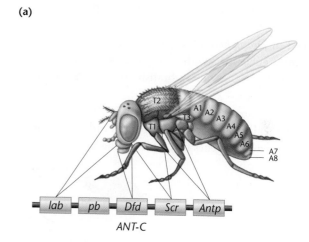

(b)

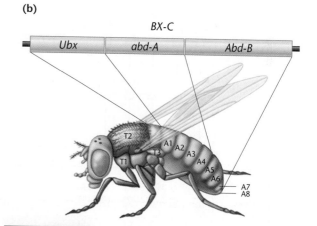

FIGURE 18–13   Genes of the *Antennapedia* complex and the adult structures they specify. (a) In the *ANT-C* complex, the *labial* (*lab*) and *Deformed* (*Dfd*) genes control the formation of head segments. The *Sex comb reduced* (*Scr*) and *Antennapedia* (*Antp*) genes specify the identity of the first two thoracic segments, T1 and T2. The remaining gene in the complex, *proboscipedia* (*pb*), may not act during embryogenesis but may be required to maintain the differentiated state in adults. In mutants, the labial palps are transformed into legs. (b) In the *BX-C* complex, *Ultrabithorax* (*Ubx*) controls formation of structures in the posterior compartment of T2 and structures in T3. The two other genes, *abdominal A* (*abdA*) and *Abdominal B* (*AbdB*), specify the segmental identities of the eight abdominal segments (A1–A8).

complex, contains three genes that specify structures in the posterior portion of the second thoracic segment, the entire third thoracic segment, and the abdominal segments [Figure 18–13(b)].

Hox genes (listed in Table 18.2) have two properties in common. First, each contains a 180-bp domain known as a **homeobox**. (*Hox* is a contraction of homeobox.) The homeobox encodes a DNA-binding sequence of 60 amino acids known as a **homeodomain**. Second, in most species, expression of the genes is colinear with the anterior to posterior organization of the body. Genes at the 3'end of a cluster are expressed at the anterior end of the embryo, those in the middle are expressed in the middle of the embryo, and genes at the 5' end of a cluster are expressed at the embryo's posterior region (**Figure 18–14**). Although first identified in *Drosophila*, *Hox* genes are found in the genomes of most eukaryotes with segmented body plans, including nematodes, sea urchins, zebrafish, frogs, mice, and humans (**Figure 18–15**).

To summarize, genes that control development in *Drosophila* act in a temporally and spatially ordered cascade, beginning with the genes that establish the anterior–posterior

(a) Expression domains of homeotic genes

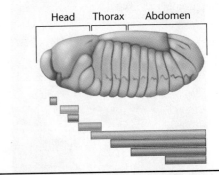

(b) Chromosomal locations of homeotic genes

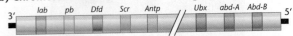

FIGURE 18–14 The colinear relationship between the spatial pattern of expression and chromosomal locations of homeotic genes in *Drosophila*. (a) *Drosophila* embryo and the domains of homeotic gene expression in the embryonic epidermis and central nervous system. (b) Chromosomal location of homeotic selector genes. Note that the order of genes on the chromosome correlates with the sequential anterior borders of their expression domains.

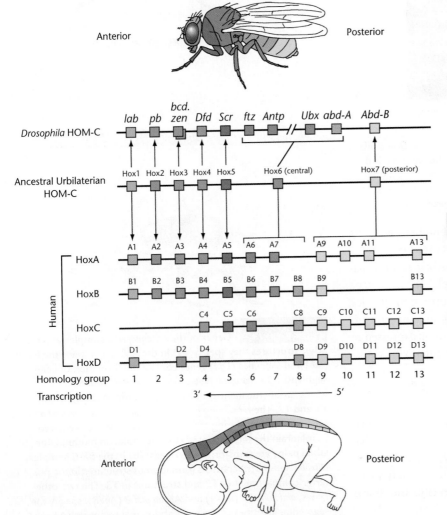

FIGURE 18–15 Conservation of organization and patterns of expression in *Hox* genes. (Top) The structures formed in adult *Drosophila* are shown, with the colors corresponding to members of the *Hox* cluster that control their formation. (Middle) The reconstructed *Hox* cluster of the common ancestor to all bilateral organisms contains seven genes. (Bottom) The arrangement and expression patterns of the four clusters of *Hox* genes in an early human embryo. Some of the posterior genes are expressed in the limbs. The expression pattern is inferred from that observed in mice. As in *Drosophila*, genes at the 3' end of the cluster form anterior structures, and genes at the 5' end of the cluster form posterior structures. Genes homologous to the same ancestral sequence (because of duplications) are indicated by brackets.

(and dorsal–ventral) axis of the egg and early embryo. Gradients of maternal mRNAs and proteins along the anterior–posterior axis activate gap genes, which subdivide the embryo into broad bands. Gap genes in turn activate pair-rule genes, which divide the embryo into segments. The final group of segmentation genes, the segment polarity genes, divides each segment into anterior and posterior regions arranged linearly along the anterior–posterior axis. The segments are then given identity by the *Hox* genes. Therefore, this progressive restriction of developmental potential of the *Drosophila* embryo's cells (all of which occurs during the first third of embryogenesis) involves a cascade of gene action, with regulatory proteins acting on transcription, translation, and signal transduction.

Hox Genes and Human Genetic Disorders

Although first described in *Drosophila*, *Hox* genes with a high degree of homology are found in the genomes of all animals where they play a fundamental role in shaping the body and its appendages. In vertebrates, the conservation of sequence, the order of genes in the *Hox* clusters, and their pattern of expression suggest that, as in *Drosophila*, these genes control development along the anterior–posterior axis and the formation of appendages (Figure 18–16). However, in vertebrates, including mice and humans, there are four clusters of *Hox* genes: *HOXA, HOXB, HOXC,* and *HOXD* instead of a single cluster as in *Drosophila*. This means that in vertebrates, not just one, but a combination of 2–4 *Hox* genes is involved in forming specific structures. As a result, homeotic mutations in individual vertebrate *Hox* genes do

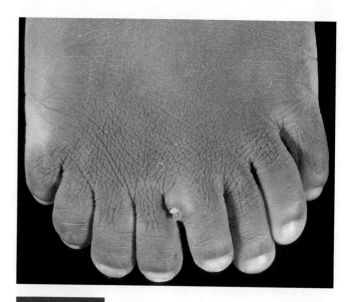

FIGURE 18–17   Mutations in posterior *Hox* genes (*HOXD13* in this case) in humans result in malformations of the limbs, shown here as extra toes. This condition is known as synpolydactyly. Mutations in *HOXD13* are also associated with abnormalities of the bones in the hands and feet.

not produce complete transformations as in *Drosophila*, where mutation of a single *Hox* gene can transform a haltere into a wing (see the photo at the beginning of this chapter). In spite of these differences, the role for *HOXD* genes in human development was confirmed by the discovery that several inherited limb malformations are caused by mutations in *HOXD* genes. For example, mutations in *HOXD13* cause synpolydactyly (SPD), a malformation characterized by extra fingers and toes, and abnormalities in bones of the hands and feet (Figure 18–17).

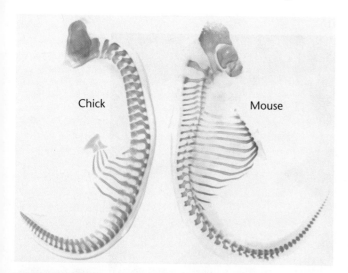

FIGURE 18–16   Patterns of *Hox* gene expression control the formation of structures along the anterior–posterior axis of bilaterally symmetrical animals in a species-specific manner. In the chick (left) and the mouse (right), expression of the same set of *Hox* genes is differentially programmed in time and space to produce different body forms.

NOW SOLVE THIS

18–2 In *Drosophila*, both *fushi tarazu* (*ftz*) and *engrailed* (*eng*) genes encode homeobox transcription factors and are capable of eliciting the expression of other genes. Both genes work at about the same time during development and in the same region to specify cell fate in body segments. To discover if *ftz* regulates the expression of *engrailed* genes, if *engrailed* regulates *ftz*, or if both are regulated by another gene, you perform a mutant analysis. In *ftz⁻* embryos (*ftz/ftz*) engrailed protein is absent; in *engrailed* embryos (*eng/eng*) *ftz* expression is normal. What does this tell you about the regulation of these two genes—does the *engrailed* gene regulate *ftz*, or does the *ftz* gene regulate *engrailed*?

■ HINT: *This problem involves an understanding of regulation of gene expression by trans-acting factors (see Chapter 17). The key to its solution is analysis of the genetic background of each mutant strain.*

18.6

Plants Have Evolved Developmental Systems That Parallel Those of Animals

Plants and animals diverged from a common unicellular ancestor about 1.6 billion years ago, after the origin of eukaryotes and probably before the rise of multicellular organisms. Genomic analysis of mutants in plants and animals indicates that basic mechanisms of developmental pattern formation evolved independently in animals and plants. We have already examined genetic systems that control development and pattern formation in animals, using *Drosophila* as a model organism.

In plants, pattern formation has been studied using flower development in *Arabidopsis thaliana* (Figure 18–18), a small plant in the mustard family, as a model organism. A cluster of undifferentiated cells, called the *floral meristem*, gives rise to flowers (Figure 18–19). Each flower consists of four organs—sepals, petals, stamens, and carpels—that develop from concentric rings of cells within the meristem (Figure 18–20). Each organ develops from a different concentric ring, or whorl of cells.

Homeotic Genes in *Arabidopsis*

Three classes of floral homeotic genes control the development of these organs (Table 18.3). Class A genes acting alone specify sepals, class A and class B genes expressed together

FIGURE 18–18   The flowering plant *Arabidopsis thaliana,* used as a model organism in plant genetics.

specify petals, and together, class B and class C genes control stamen formation. Class C genes acting alone specify carpels. During flower development [Figure 18–21(a)], Class A genes are active in whorls 1 and 2 (sepals and petals), class B genes are expressed in whorls 2 and 3 (petals and stamens), and class C genes are expressed in whorls 3 and 4 (stamens and carpels). The organ formed depends on the expression

(a)

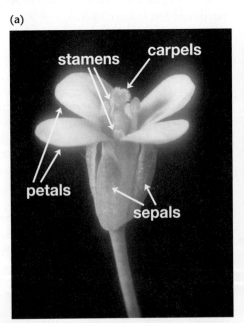

(b)

FIGURE 18–19   (a) Parts of the *Arabidopsis* flower. The floral organs are arranged concentrically. The sepals form the outermost ring, followed by petals and stamens, with carpels on the inside. (b) View of the flower from above.

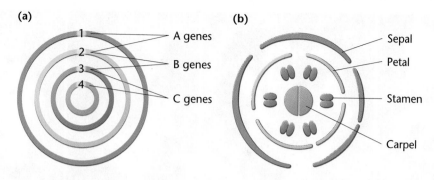

FIGURE 18–20   Cell arrangement in the floral meristem. (a) The four concentric rings, or whorls, labeled 1–4, give rise to (b) arrangement of the sepals, petals, stamens, and carpels, respectively, in the mature flower.

pattern of the three gene classes. Expression of class A genes in whorl 1 causes sepals to form. Expression of class A *and* class B genes in whorl 2 leads to petal formation. Expression of class B and class C genes in whorl 3 leads to stamen formation. In whorl 4, expression of class C genes causes carpel formation.

As in *Drosophila,* mutations in homeotic genes cause organs to form in abnormal locations. For example, in *APETALA2* mutants (*ap2*), the order of organs is carpel, stamen, stamen, and carpel instead of the normal order, sepal, petal, stamen, and carpel [Figure 18–21(b)]. In class B loss-of-function mutants (*ap3, pi*) petals become sepals, and stamens are transformed into carpels [Figure 18–21(c)] and the order of organs becomes sepal, sepal, carpel, carpel. Plants carrying a mutation for the class C gene *AGAMOUS* will have petals in whorl 3 (instead of stamens) and sepals in whorl 4 (instead of carpels), and the order of organs will be sepal, petal, petal, and sepal [Figure 18–21(d)].

Evolutionary Divergence in Homeotic Genes

Drosophila and *Arabidopsis* use different sets of nonhomologous master regulatory genes to establish the body axis and specify the identity of structures along the axis. In *Drosophila,* this task is accomplished in part by the *Hox* genes, which encode a set of transcription factors sharing a homeobox domain. In *Arabidopsis,* the floral homeotic genes belong to a different family of transcription factors, called the **MADS-box proteins**, characterized by a common sequence of 58 amino acids with no similarity in amino acid sequence or protein structure with the *Hox* genes. Both gene sets encode transcription factors, both sets are master regulators of development expressed in a pattern of overlapping domains, and both specify identity of structures.

Reflecting their evolutionary origin from a common ancestor, the genomes of both *Drosophila* and *Arabidopsis* contain members of the homeobox and MADS-box genes, but these genes have been adapted for different uses in the plant and animal kingdoms, indicating that developmental mechanisms evolved independently in each group.

In both plants and animals, the action of transcription factors depends on changes in chromatin structure that make genes available for expression. Mechanisms of transcription initiation are conserved in plants and animals, as is reflected in the homology of genes in *Drosophila* and *Arabidopsis* that maintain patterns of expression initiated by regulatory gene sets. Action of the floral homeotic genes is controlled by a gene called *CURLY LEAF.* This gene shares significant homology with members of a *Drosophila* gene family called *Polycomb.* This family of regulatory genes controls expression of homeobox genes during development. Both *CURLY LEAF* and *Polycomb* encode proteins that alter chromatin conformation and shut off gene expression. Thus, although different genes are used to control development, both plants and animals use an evolutionarily conserved mechanism to regulate expression of these gene sets.

TABLE 18.3

Homeotic Selector Genes in *Arabidopsis**	
Class A	APETALA1 (AP1)
	APETALA2 (AP2)
Class B	APETALA3 (AP3)
	PISTILLATA (PI)
Class C	AGAMOUS (AG)

*By convention, wild-type genes in *Arabidopsis* use capital letters.

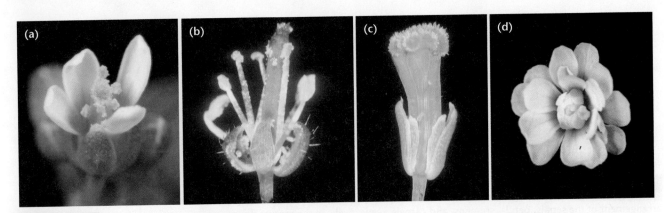

FIGURE 18–21 (a) The individual and combined action of Class A, B, and C genes form the sepals, petals, stamens, and carpels of wild-type flowers of *Arabidopsis*. (b) Homeotic *APETALA2* (*ap2*) mutant flower (a Class A mutant), has carpels, stamens, stamens, and carpels. (c) Class B mutants (*ap3* and *pi*) have sepals, sepals, carpels, and carpels. (d) Class C mutants (*ag*) have petals and sepals at places where stamens and carpels should form.

18.7

Cell–Cell Interactions in Development Are Modeled in *C. elegans*

During development in multicellular organisms, cell–cell interactions influence the transcriptional programs and developmental fate of surrounding cells. Cell–cell interaction is an important process in the embryonic development of most eukaryotic organisms, including *Drosophila*, as well as vertebrates, including mice and humans.

Signaling Pathways in Development

In early development, animals use a number of signaling pathways to regulate development; after organ formation begins, other signal pathways are added to those already in use. These newly activated pathways act both independently and in coordinated networks to elicit specific transcriptional responses. The signal networks establish anterior–posterior polarity and body axes, coordinate pattern formation, and direct the differentiation of tissues and organs. The signaling pathways used in early development and some of the developmental processes they control are listed in Table 18.4. After an introduction to the components and interactions of one of these systems—the **Notch signaling pathway**—we will briefly examine its role in the development of the vulva in the nematode, *Caenorhabditis elegans*.

The Notch Signaling Pathway

The genes in the Notch pathway are named after the *Drosophila* mutants that were used to identify components of this signal transduction system. Notch works through direct cell–cell contact to control the developmental fate of the interacting cells. The *Notch* gene (and the equivalent gene in other organisms) encodes a signal receptor embedded in the plasma membrane (Figure 18–22). The signal is another membrane protein encoded by the *Delta* gene (and its equivalents). Because both the signal and receptor are membrane proteins, the Notch signal system works between adjacent cells. When the Delta signal protein binds to the Notch receptor protein, the cytoplasmic tail of the Notch protein is cut off and binds to a cytoplasmic protein encoded by the *Su(H)* (suppressor of *Hairless*) gene. This protein complex moves into the nucleus and binds to transcriptional cofactors, activating transcription of a

TABLE 18.4

Signaling Pathways Used in Early Embryonic Development

Wnt Pathway
Dorsalization of body
Female reproductive development
Dorsal–ventral differences

TGF-β Pathway
Mesoderm induction
Left–right asymmetry
Bone development

Hedgehog Pathway
Notochord induction
Somitogenesis
Gut/visceral mesoderm

Receptor Tyrosine Kinase Pathway
Mesoderm maintenance

Notch Signaling Pathway
Blood cell development
Neurogenesis
Retina development

Source: Taken from Gerhart, J. 1999. 1998 Warkany lecture: Signaling pathways in development. *Teratology* 60: 226–239.

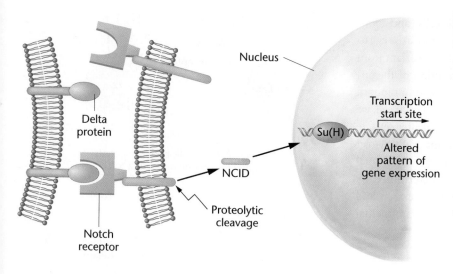

FIGURE 18–22 Components of the Notch signaling pathway in *Drosophila*. The cell carrying the Delta transmembrane protein is the sending cell; the cell carrying the transmembrane Notch protein receives the signal. Binding of Delta to Notch triggers a proteolytic-mediated activation of transcription. The fragment cleaved from the cytoplasmic side of the Notch protein, called the Notch intracellular domain (NCID), combines with the Su(H) protein and moves to the nucleus where it activates a program of gene transcription.

gene set that controls a specific developmental pathway (Figure 18–22).

One of the main roles of the Notch signal system is specifying the fate of equivalent cells in a population. In its simplest form, this interaction involves two neighboring cells that are developmentally equivalent. We will explore the role of the Notch signaling system in development of the vulva in *C. elegans*, after a brief introduction to nematode embryogenesis.

Overview of *C. elegans* Development

The nematode *C. elegans* is widely used to study the genetic control of development. This organism has several advantages for such studies: (1) the genetics of the organism are well known, (2) its genome has been sequenced, and (3) adults contain a small number of cells that follow a highly deterministic developmental program. Adults are about 1 mm long and develop from a fertilized egg in about two days (Figure 18–23). The life cycle includes an embryonic stage (about 16 hours), four larval stages (L1 through L4), and the adult stage. Adults are either XX self-fertilizing hermaphrodites that can make both eggs and sperm, or XO males. Self-fertilization of mutagen-treated hermaphrodites is used to develop homozygous stocks of mutant strains, and hundreds of such mutants have been generated, catalogued, and mapped.

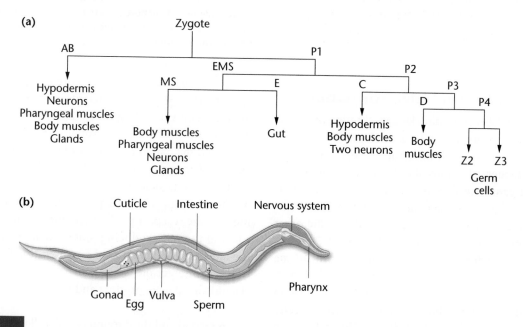

FIGURE 18–23 (a) A truncated cell lineage chart for *C. elegans*, showing early divisions and the tissues and organs formed from these lineages. Each vertical line represents a cell division, and horizontal lines connect the two cells produced. For example, the first division of the zygote creates two new cells, AB and P1. During embryogenesis, cell divisions will produce the 959 somatic cells of the adult hermaphrodite worm. (b) An adult *C. elegans* hermaphrodite. This nematode, about 1 mm in length, consists of 959 cells and is widely used as a model organism to study the genetic control of development.

Adult hermaphrodites have 959 somatic cells (and about 2000 germ cells). The lineage of each cell, from fertilized egg to adult, has been mapped (Figure 18–23) and, is invariant from individual to individual, with the single exception of the anchor cell/ventral uterine cell specification, which is a random event (see Figure 18–24). Knowing the lineage of each cell, we can easily follow events caused by mutations that alter cell fate or by killing specific cells with laser microbeams or ultraviolet irradiation. In *C. elegans* hermaphrodites, the developmental fate of cells in the reproductive system is determined by cell–cell interaction, illustrating how gene expression and cell–cell interaction work together to specify developmental outcomes.

NOW SOLVE THIS

18–3 Vulval development in *C. elegans* begins when two neighboring cells (Z1.ppp and Z4.aaa) interact with each other by cell–cell signaling involving two components: a membrane-bound signal molecule and a membrane-bound receptor. By chance, one cell produces more signal, which in turn, causes its neighbor to produce more receptor. The cell producing more signal becomes the anchor cell, and the cell producing more receptor becomes the ventral uterine cell. This form of cell–cell interaction is called the Notch/Delta signaling system. Although it is a widely used signaling mechanism in metazoans, this pathway works only in adjacent cells. Why is this so, and what are the advantages and disadvantages of such a system?

■ HINT: *This problem involves an understanding of how adjacent cell–cell signaling systems work. The key to its solution is recognizing the dynamic interactions between signal and receptor systems in this form of cell–cell signaling.*

Genetic Analysis of Vulva Formation

Adult *C. elegans* hermaphrodites lay eggs through the vulva, an opening near the middle of the body (Figure 18–23). The vulva is formed in stages during larval development and involves three sequential rounds of cell–cell interactions.

In *C. elegans*, two developmentally equivalent adjacent cells, Z1.ppp and Z4.aaa, interact with each other so that one becomes the gonadal anchor cell and the other becomes a precursor to the uterus (Figure 18–24). The determination of which cell becomes which occurs during the second larval stage (L2) and is controlled by the Notch receptor gene, *lin-12*. In recessive *lin-12(0)* mutants (a loss-of-function mutant), both cells become anchor cells. The dominant mutation *lin-12(d)* (a gain-of-function mutation) causes both to become uterine precursors. Thus, expression of the *lin-12* gene causes the selection of the uterine pathway, since in the absence of the LIN-12 (Notch) receptor, both cells become anchor cells.

(a)

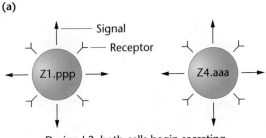

During L2, both cells begin secreting signal for uterine differentiation

(b)

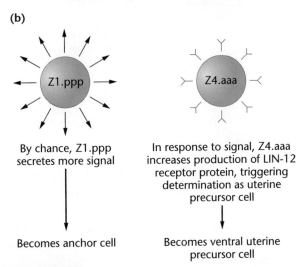

By chance, Z1.ppp secretes more signal

In response to signal, Z4.aaa increases production of LIN-12 receptor protein, triggering determination as uterine precursor cell

Becomes anchor cell

Becomes ventral uterine precursor cell

FIGURE 18–24   Cell–cell interaction in anchor cell determination. (a) During L2, two neighboring cells begin the synthesis of signal and receptor molecules for the induction of uterine differentiation. (b) By chance, cell Z1.ppp produces more of these signals, causing cell Z4.aaa to increase production of the receptor for signals. The action of increased signals causes Z4.aaa to become the ventral uterine precursor cell and allows Z1.ppp to become the anchor cell.

However, the situation is more complex than it first appears. Initially, the two neighboring cells are developmentally equivalent (Figure 18–24). Each synthesizes low levels of the Notch signal protein (encoded by the *lag-2* gene) *and* the Notch receptor protein. By chance, one cell ends up producing more of the signal (LAG-2 protein). This causes its neighboring cell to increase production of the receptor (LIN-12 protein). The cell producing more of the receptor protein becomes the uterine precursor, and the cell producing more signal protein becomes the anchor cell. The critical factor in this first round of cell–cell interaction is the balance between the LAG-2 (Delta) signal gene product and the LIN-12 (Notch) gene product.

A second and third round of cell–cell interactions leads to formation of the vulva itself (Figure 18–25). The second round involves the anchor cell (located in the gonad) and six of its neighboring cells (called precursor cells) located in the skin. The precursor cells, named P3.p to P8.p, are called

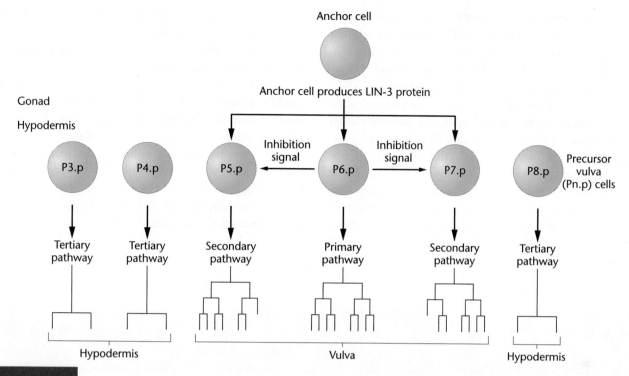

FIGURE 18–25   Cell lineage determination in *C. elegans* vulva formation. A signal from the anchor cell in the form of LIN-3 protein is received by three precursor vulval cells (Pn.p cells). The cell closest to the anchor cell becomes the primary vulval precursor cell, and adjacent cells become secondary precursor cells. The primary cell produces a signal that activates the *lin-12* gene in secondary cells, preventing them from becoming primary cells. Flanking precursor cells, which receive no signal from the anchor cell, become skin (hypodermis) cells, instead of vulval cells.

Pn.p cells. The fate of each Pn.p cell is specified by its position relative to the anchor cell.

The anchor cell synthesizes the LIN-3 signal protein which is received and processed by three adjacent Pn.p precursor cells (Pn.p 5–7). The cell closest to the anchor cell (usually Pn.p 6) becomes the primary vulval precursor cell, and the adjacent cells (Pn.p 5 and 7) become secondary precursor cells. Once the primary vulval cell has been established, in a third round of cell–cell interaction a signal protein made by the primary vulval cell activates the *lin-12* gene in the secondary cells and prevents them from becoming primary precursor cells. The other precursor cells (Pn.p 3, 4, and 8) receive no signal from the anchor cell and become skin cells.

18.8

Programmed Cell Death Is Required for Normal Development

During normal development, programmed cell death is a genetically controlled event that helps shape tissues and organs. One well-known example of programmed cell death is the formation of digits in the vertebrate limb. This process requires the death of the cells between the digits

(Figure 18–26), a process called **apoptosis.** The genes that control apoptosis were first identified in *C. elegans.* In *C. elegans,* as in many other organisms, normal development relies on programmed cell death. The number of cells that die during the worm's development is always the same: 131 of 1090 in hermaphrodites and 147 of 1178 in males. In addition, the time in development at which a given cell dies and the identity of the cells that die are always the same.

Mutational analysis indicates that, although programmed cell death occurs in cells in different lineages, all

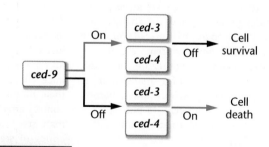

FIGURE 18–26   In *C. elegans,* the genetic pathway controlling cell death, the gene *ced-9* acts as a binary switch. If *ced-9* is active, it represses the action of *ced-3* and *ced-4,* and the cell remains alive. If *ced-9* is inactive, *ced-3* and *ced-4* are expressed, initiating a cascade of events that results in cell death, a process known as apoptosis.

cells use the same genetic pathway. In *C. elegans,* at least 15 genes are involved in cell death. These genes control four processes: (1) decisions about cell death, (2) implementation of decisions, (3) engulfment of dying cells, and (4) degradation of cell debris within the engulfing cells.

Expression of *ced-3* and *ced-4* is necessary for execution of the cell death program; mutations that inactivate either of these genes result in survival of cells that normally die. Expression of *ced-3* and *ced-4* is controlled by *ced-9*. Gain-of-function mutations that cause constitutive expression or overexpression of *ced-9* prevent cell death. Conversely, loss-of-function mutants that inactivate *ced-9* cause embryonic lethality, meaning that *ced-9* works by inactivating *ced-3* and *ced-4* in surviving cells (Figure 18–26). In other words, *ced-9*

is a *binary switch gene* for apoptosis. Cells that express *ced-9* survive and those that do not, die.

The *ced-9* gene of *C. elegans* has a human homolog, *bcl-2,* a proto-oncogene that controls apoptosis in humans. Mutations that overexpress *bcl-2* prevent death in cells that would normally die. In humans, overexpression of *bcl-2* is found in follicular lymphoma, a form of cancer. Transfer of a cloned human *bcl-2* gene into *ced-9* null mutant *C. elegans* embryos prevents apoptosis, indicating that nematodes and mammals share a common pathway for this process. Thus, the homology in molecules and mechanisms among species across the phylogenetic tree, illustrated throughout this chapter, extends to the molecules and mechanisms required for cells to die.

GENETICS, TECHNOLOGY, AND SOCIETY

Stem Cell Wars

Stem cell research may be the most controversial research area since the beginning of recombinant DNA technology in the 1970s. Although stem cell research is the focus of presidential proclamations, media campaigns, and ethical debates, few people understand it sufficiently to evaluate its pros and cons.

Stem cells are primitive cells that replicate indefinitely and have the capacity to differentiate into cells with specialized functions, such as the cells of heart, brain, liver, and muscle tissue. All the cells that make up the approximately 200 distinct types of tissues in our bodies are descended from stem cells. Some types of stem cells are defined as *totipotent,* meaning that they have the ability to differentiate into any mature cell type in the body. Other types of stem cells are *pluripotent,* meaning that they are able to differentiate into any of a smaller number of mature cell types. In contrast, mature, fully differentiated cells do not replicate or undergo transformations into different cell types.

In the last few years, several research teams have isolated and cultured human pluripotent stem cells. These cells remain undifferentiated and grow indefinitely in culture dishes. When treated with growth factors or hormones, these pluripotent

stem cells differentiate into cells that have the characteristics of neural, bone, kidney, liver, heart, or pancreatic cells.

The fact that pluripotent stem cells grow prolifically in culture and differentiate into more specialized cells has created great excitement. Some foresee a day when stem cells may be a cornucopia from which to harvest unlimited numbers of specialized cells to replace cells in damaged and diseased tissues. Hence, stem cells could be used to treat Parkinson disease, type 1 diabetes, chronic heart disease, Alzheimer disease, and spinal cord injuries. Some predict that stem cells will be genetically modified to eliminate transplant rejection or to deliver specific gene products, thereby correcting genetic defects or treating cancers. The excitement about stem cell therapies has been fueled by reports of dramatically successful experiments in animals. For example, mice with spinal cord injuries regained their mobility and bowel and bladder control after they were injected with human stem cells. Both proponents and critics of stem cell research agree that stem cell therapies could be revolutionary. Why, then, should stem cell research be so contentious?

The answer to that question lies in the source of the pluripotent stem cells. Until recently, all pluripotent stem cell

lines were derived from five-day-old embryonic blastocysts. Blastocysts at this stage consist of 50–150 cells, most of which will develop into placental and supporting tissues for the early embryo. The inner cell mass of the blastocyst consists of about 30 to 40 pluripotent stem cells that can develop into all the embryo's tissues. *In vitro* fertilization clinics grow fertilized eggs to the five-day blastocyst stage prior to uterine transfer. Embryonic stem cell (ESC) lines are created by taking the inner cell mass out of five-day blastocysts and growing the cells in culture dishes.

The fact that early embryos are destroyed in the process of establishing human ESC lines disturbs people who believe that preimplantation embryos are persons with rights; however, it does not disturb people who believe that these embryos are too primitive to have the status of a human being. Both sides in the debate invoke fundamental questions of what constitutes a human being.

Recently, scientists have developed several types of pluripotent stem cells without using embryos. One of the most promising types—known as *induced pluripotent stem (iPS) cells*—uses adult somatic cells as the source of pluripotent stem cell lines. To prepare iPS cells, scientists isolate somatic cells (such as cells from skin) and infect them with engineered retroviruses that integrate into the cells' DNA. These retroviruses contain several cloned human genes that encode products responsible for converting the somatic cells into immortal, pluripotent stem cells.

The development of iPS cell lines has generated renewed enthusiasm for stem cell research, as these cells bypass the ethical problems associated with the use of human embryos. In addition, they may become sources of patient-specific pluripotent stem cell lines that can be used for transplantation, without immune system rejection.

At the present time, it is unknown whether stem cells of any type will be as miraculous as predicted; however, if stem cell research progresses at its current rapid pace, we won't have long to wait.

Your Turn

Take time, individually or in groups, to answer the following questions. Investigate the references and links to help you understand the technologies and controversies surrounding stem cell research.

1. What, in your opinion, are the scientific and ethical problems that still surround stem cell research? Are these problems solved by the new methods of creating pluripotent stem cells?

You can find descriptions of some new methods of generating pluripotent stem cell lines, and the ethical issues that accompany these methods, in Kastenberg, Z. J. and Odorico, J. S. 2008. Alternative sources of pluripotency: science, ethics, and stem cells. Transplantation Rev.22: 215–222.

2. What are the current stem cell research laws in your region, and how do these laws compare with national regulations?

A starting point for information about stem cell research regulations can be found on the Stem Cell Information Web site of the National Institutes of Health (http://stemcells.nih.gov).

3. Do you oppose or support stem cell research? Why, or why not?

An interesting online poll, along with arguments for and against stem cell research, is offered by the Public Broadcasting Corporation, at http://www.pbs.org/wgbh/nova/body/stem-cell-poll.html.

4. What, in your opinion, is the most significant development in stem cell research in the last year?

Some ideas to start your search are: the PubMed Web site (http://www.ncbi.nlm.nih.gov/sites/entrez? db=PubMed) and the New York Times online Stem Cell page (http://topics.nytimes.com/top/news/health/diseasesconditionsandhealthtopics/stemcells).

CASE STUDY | One foot or another

In humans the *HOXD* homeotic gene cluster plays a critical role in limb development. In one large family, 16 of 36 members expressed one of two dominantly inherited malformations of the feet known as rocker bottom foot (CVT) or claw foot (CMT). One individual had one foot with CVT and the other with CMT. Genomic analysis identified a single missense mutation in the *HOXD10* gene, resulting in a single amino acid substitution in the homeodomain of the encoded transcription factor. This region is crucial for making contact and binding to the target genes controlled by this protein. All family members with the foot malformations were heterozygotes; all unaffected members were homozygous for the normal allele.

1. Given that affected heterozygotes carry one normal allele of the *HOXD10* gene, how might a dominant mutation in a gene encoding a transcription factor lead to a developmental malformation?
2. How can two clinically different disorders result from the same mutation?
3. What might we learn about the control of developmental processes from an understanding of how this mutation works?

Summary Points

 For activities, animations, and review quizzes, go to the study area at www.masteringgenetics.com

1. Developmental genetics, which explores the mechanisms by which genetic information controls development and differentiation, is one of the major areas of study in biology. Geneticists are investigating this topic by isolating developmental mutations and identifying the genes involved in developmental processes.

2. During embryogenesis, the activity of specific genes is controlled by the internal environment of the cell, including localized cytoplasmic components. In flies, the regulation of early events is mediated by the maternal cytoplasm, which then influences zygotic gene expression. As development proceeds, both the cell's internal environment and its external environment become further altered by the presence of early gene products and communication with other cells.

3. In *Drosophila*, both genetic and molecular studies have confirmed that the egg contains information specifying the body plan of the larva and adult.

4. Extensive genetic analysis of embryonic development in *Drosophila* has led to the identification of maternal-effect genes whose products establish the anterior–posterior axis of the embryo. In addition, these maternal-effect genes activate sets of zygotic segmentation genes, initiating a cascade of gene regulation that ends with the determination of segment identity by the homeotic selector genes. These same gene sets control aspects of embryonic development in all bilateral animals, including humans.

5. Flower formation in *Arabidopsis* is controlled by homeotic genes, but these gene sets are from a different gene family than the homeotic selector genes of *Drosophila* and other animals.

6. In *C. elegans,* the strictly determined lineage of each cell allows developmental biologists to study the cell–cell signaling required for organogenesis and to determine which genes are required for the normal process of programmed cell death.

INSIGHTS AND SOLUTIONS

1. In the slime mold *Dictyostelium,* experimental evidence suggests that cyclic AMP (cAMP) plays a central role in the developmental program leading to spore formation. The genes encoding the cAMP cell-surface receptor have been cloned, and the amino acid sequence of the protein components is known. To form reproductive structures, free-living individual cells aggregate together and then differentiate into one of two cell types, prespore cells or prestalk cells. Aggregating cells secrete waves or oscillations of cAMP to foster the aggregation of cells and then continuously secrete cAMP to activate genes in the aggregated cells at later stages of development. It has been proposed that cAMP controls cell–cell interaction and gene expression. It is important to test this hypothesis by using several experimental techniques. What different approaches can you devise to test this hypothesis, and what specific experimental systems would you employ to test them?

   Solution: Two of the most powerful forms of analysis in biology involve the use of biochemical analogs (or inhibitors) to block gene transcription or the action of gene products in a predictable way, and the use of mutations to alter genes and their products. These two approaches can be used to study the role of cAMP in the developmental program of *Dictyostelium.* First, compounds chemically related to cAMP, such as GTP and GDP, can be used to test whether they have any effect on the processes controlled by cAMP. In fact, both GTP and GDP lower the affinity of cell-surface receptors for cAMP, effectively blocking the action of cAMP.

   Mutational analysis can be used to dissect components of the cAMP receptor system. One approach is to use transformation with wild-type genes to restore mutant function. Similarly, because the genes for the receptor proteins have been cloned, it is possible to construct mutants with known alterations in the component proteins and transform them into cells to assess their effects.

2. In the sea urchin, early development may occur even in the presence of actinomycin D, which inhibits RNA synthesis. However, if actinomycin D is present early in development but is removed a few hours later, all development stops. In fact, if actinomycin D is present only between the sixth and eleventh hours of development, events that normally occur at the fifteenth hour are arrested. What conclusions can be drawn concerning the role of gene transcription between hours 6 and 15?

   Solution: Maternal mRNAs are present in the fertilized sea urchin egg. Thus, a considerable amount of development can take place without transcription of the embryo's genome. Because development past 15 hours is inhibited by prior treatment with actinomycin D, it appears that transcripts from the embryo's genome are required to initiate or maintain these events. This transcription must take place between the sixth and fifteenth hours of development.

3. If it were possible to introduce one of the homeotic genes from *Drosophila* into an *Arabidopsis* embryo homozygous for a homeotic flowering gene, would you expect any of the *Drosophila* genes to negate (rescue) the *Arabidopsis* mutant phenotype? Why or why not?

   Solution: The *Drosophila* homeotic genes belong to the *Hox* gene family, whereas *Arabidopsis* homeotic genes belong to the MADS-box protein family. Both gene families are present in *Drosophila* and *Arabidopsis,* but they have evolved different functions in the animal and the plant kingdoms. As a result, it is unlikely that a transferred *Drosophila Hox* gene would rescue the phenotype of a MADS-box mutant, but only an actual experiment would confirm this.

Problems and Discussion Questions

 For instructor-assigned tutorials and problems, go to www.masteringgenetics.com

HOW DO WE KNOW?

1. In this chapter we focused on how differential gene expression guides the processes that lead from the fertilized egg to the adult. At the same time, we found many opportunities to consider the methods and reasoning by which much of this information was acquired. From the explanations given in the chapter, what answers would you propose to the following fundamental questions?

   (a) How do we know how many genes control development in an organism like *Drosophila?*

   (b) What experimental evidence is available to show that molecular gradients in the egg control development?

   (c) How do we know that a genetic program specifying a body part can be changed?

   (d) What genetic evidence shows that chemical signals between cells control developmental events?

   (e) How do we know whether a signaling system in vulva development works only on adjacent cells or uses signals that can affect more distant cells?

2. Carefully distinguish between the terms *differentiation* and *determination.* Which phenomenon occurs initially during development?

3. Nuclei from almost any source may be injected into *Xenopus* oocytes. Studies have shown that these nuclei remain active in transcription and translation. How can such an experimental system be useful in developmental genetic studies?

4. Distinguish between the syncytial blastoderm stage and the cellular blastoderm stage in *Drosophila* embryogenesis.

5. (a) What are maternal-effect genes? (b) When are gene products from these genes made, and where are they located? (c) What aspects of development do maternal-effect genes control? (d) What is the phenotype of maternal-effect mutations?

6. (a) What are zygotic genes, and when are their gene products made? (b) What is the phenotype associated with zygotic gene mutations? (c) Does the maternal genotype contain zygotic genes?

7. List the main classes of zygotic genes. What is the function of each class of these genes?

8. Experiments have shown that any nuclei placed in the polar cytoplasm at the posterior pole of the *Drosophila* egg will differentiate into germ cells. If polar cytoplasm is transplanted into the anterior end of the egg just after fertilization, what will happen to nuclei that migrate into this cytoplasm at the anterior pole?

9. How can you determine whether a particular gene is being transcribed in different cell types?

10. You observe that a particular gene is being transcribed during development. How can you tell whether the expression of this gene is under transcriptional or translational control?

11. What are *Hox* genes? What properties do they have in common? Are all homeotic genes *Hox* genes?

12. The homeotic mutation *Antennapedia* causes mutant *Drosophila* to have legs in place of antennae and is a dominant gain-of-function mutation. What are the properties of such mutations? How does the *Antennapedia* gene change antennae into legs?

13. The *Drosophila* homeotic mutation *spineless aristapedia* (ss^a) results in the formation of a miniature tarsal structure (normally part of the leg) on the end of the antenna. What insight does (ss^a) provide concerning the role of genes during determination?

14. Embryogenesis and oncogenesis (generation of cancer) share a number of features including cell proliferation, apoptosis, cell migration and invasion, formation of new blood vessels, and differential gene activity. Embryonic cells are relatively undifferentiated, and cancer cells appear to be undifferentiated or dedifferentiated. Homeotic gene expression directs early development, and mutant expression leads to loss of the differentiated state or an alternative cell identity. M. T. Lewis (2000. *Breast Can. Res.* 2: 158–169) suggested that breast cancer may be caused by the altered expression of homeotic genes. When he examined 11 such genes in cancers, 8 were underexpressed while 3 were overexpressed compared with controls. Given what you know about homeotic genes, could they be involved in oncogenesis?

15. Early development depends on the temporal and spatial interplay between maternally supplied material and mRNA and the onset of zygotic gene expression. Maternally encoded mRNAs must be produced, positioned, and degraded (Surdej and Jacobs-Lorena, 1998. *Mol. Cell Biol.* 18: 2892–2900). For example, transcription of the *bicoid* gene that determines anterior–posterior polarity in *Drosophila* is maternal. The mRNA is synthesized in the ovary by nurse cells and then transported to the oocyte, where it localizes to the anterior ends of oocytes. After egg deposition, *bicoid* mRNA is translated, and unstable bicoid protein forms a decreasing concentration gradient from the anterior end of the embryo. At the start of gastrulation, *bicoid* mRNA has been degraded. Consider two models to explain the degradation of *bicoid* mRNA: (1) degradation may result from signals within the mRNA (intrinsic model), or (2) degradation may result from the mRNA's position within the egg (extrinsic model). Experimentally, how could one distinguish between these two models?

16. In *Arabidopsis*, flower development is controlled by sets of homeotic genes. How many classes of these genes are there, and what structures are formed by their individual and combined expression?

17. Formation of germ cells in *Drosophila* and many other embryos is dependent on their position in the embryo and their exposure to localized cytoplasmic determinants. Nuclei exposed to cytoplasm in the posterior end of *Drosophila* eggs (the pole plasm) form cells that develop into germ cells under the direction of maternally derived components. R. Amikura et al. (2001. *Proc. Nat. Acad. Sci.* (*USA*) 98: 9133–9138) consistently found mitochondria-type ribosomes outside mitochondria in the germ plasm of *Drosophila* embryos and postulated that they are intimately related to germ-cell specification. If you were studying this phenomenon, what would you want to know about the activity of these ribosomes?

18. One of the most interesting aspects of early development is the remodeling of the cell cycle from rapid cell divisions, apparently lacking G1 and G2 phases, to slower cell cycles with measurable G1 and G2 phases and checkpoints. During this remodeling, maternal mRNAs that specify cyclins are deadenylated, and zygotic genes are activated to produce cyclins. Audic et al. (2001. *Mol. and Cell. Biol.* 21: 1662–1671) suggest that deadenylation requires transcription of zygotic genes. Present a diagram that captures the significant features of these findings.

Extra-Spicy Problems

 ™ *For instructor-assigned tutorials and problems, go to www.masteringgenetics.com*

19. A number of genes that control expression of *Hox* genes in *Drosophila* have been identified. One of these homozygous mutants is *extra sex combs*, where some of the head and all of the thorax and abdominal segments develop as the last abdominal segment. In other words, all affected segments develop as posterior segments. What does this phenotype tell you about which set of *Hox* genes is controlled by the *extra sex combs* gene?

20. The *apterous* gene in *Drosophila* encodes a protein required for wing patterning and growth. It is also known to function in nerve development, fertility, and viability. When human and mouse genes whose protein products closely resemble *apterous* were used to generate transgenic *Drosophila* (Rincon-Limas

et al., 1999. *Proc. Nat. Acad. Sci.* [*USA*] 96: 2165–2170), the *apterous* mutant phenotype was *rescued*. In addition, the whole-body expression patterns in the transgenic *Drosophila* were similar to normal *apterous*. (a) What is meant by the term *rescued* in this context? (b) What do these results indicate about the molecular nature of development?

21. The floral homeotic genes of *Arabidopsis* belong to the MADS-box gene family, while in *Drosophila*, homeotic genes belong to the homeobox gene family. In both *Arabidopsis* and *Drosophila*, members of the *Polycomb* gene family control expression of these divergent homeotic genes. How do *Polycomb* genes control expression of two very different sets of homeotic genes?

22. The identification and characterization of genes that control sex determination has been another focus of investigators working with *C. elegans*. As with *Drosophila*, sex in this organism is determined by the ratio of X chromosomes to sets of autosomes. A diploid wild-type male has one X chromosome, and a diploid wild-type hermaphrodite has two X chromosomes. Many different mutations have been identified that affect sex determination. Loss-of-function mutations in a gene called *her-1* cause an XO nematode to develop into a hermaphrodite and have no effect on XX development. (That is, XX nematodes are normal hermaphrodites.) In contrast, loss-of-function mutations in a gene called *tra-1* cause an XX nematode to develop into a male. Deduce the roles of these genes in wild-type sex determination from this information.

23. Based on the information in Problem 22 and the analysis of the phenotypes of single- and double-mutant strains, a model for sex determination in *C. elegans* has been generated. This model proposes that the *her-1* gene controls sex determination by establishing the level of activity of the *tra-1* gene, which in turn, controls the expression of genes involved in generating the various sexually dimorphic tissues. Given this information, (a) does the *her-1* gene product have a negative or a positive effect on the activity of the *tra-1* gene? (b) What would be the phenotype of a *tra-1, her-1* double mutant?

24. Below is a microarray assessment of gene expression among developmentally significant categories (tissues and organs) in the model organism *Arabidopsis thaliana* (Schmid et al., 2005). Relative gene activities are presented on the ordinate, while several tissue types are presented on the abscissa.

(a) Are gene-expression patterns reasonably compatible with expectations?

(b) The general developmental program of plants contrasts with that of animals in that plants develop continuously, with new organs being added throughout their life span. How might such a developmental program help explain the expression of photosynthetic genes in flowers and seeds?

(c) Typically, single or small numbers of genes are studied to determine their impact on development. Here, genome-wide analysis reveals output from general classes containing many genes. How might such a global approach further our understanding of the role of genes in development?

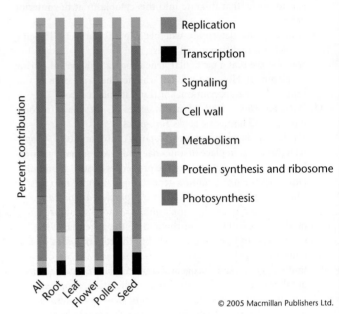

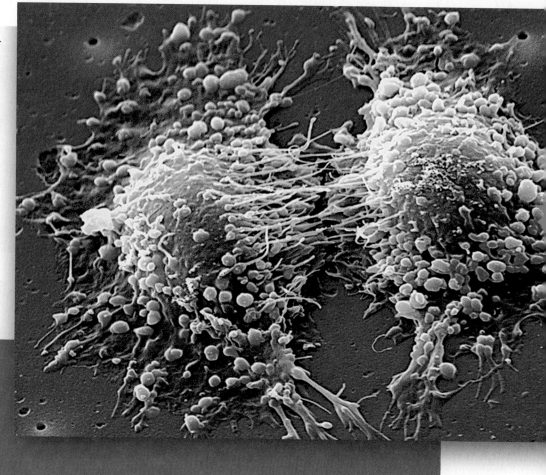

Colored scanning electron micrograph of two prostate cancer cells in the final stages of cell division (cytokinesis). The cells are still joined by strands of cytoplasm.

19

Cancer and Regulation of the Cell Cycle

CHAPTER CONCEPTS

- Cancer is a group of genetic diseases affecting fundamental aspects of cellular function, including DNA repair, cell-cycle regulation, apoptosis, and signal transduction.

- Most cancer-causing mutations occur in somatic cells; only about 1 percent of cancers have a hereditary component.

- Mutations in cancer-related genes lead to abnormal proliferation and loss of control over how cells spread and invade surrounding tissues.

- The development of cancer is a multistep process requiring mutations in genes controlling many aspects of cell proliferation and metastasis.

- Cancer cells show high levels of genomic instability, leading to the accumulation of multiple mutations in cancer-related genes.

- Epigenetic effects such as DNA methylation and histone modifications may play significant roles in the development of cancers.

- Mutations in proto-oncogenes and tumor-suppressor genes contribute to the development of cancers.

- Oncogenic viruses introduce oncogenes into infected cells and stimulate cell proliferation.

- Environmental agents contribute to cancer by damaging DNA.

ancer is the leading cause of death in Western countries. It strikes people of all ages, and one out of three people will experience a cancer diagnosis sometime in his or her lifetime. Each year, more than 1 million cases of cancer are diagnosed in the United States, and more than 500,000 people die from the disease.

Over the last 30 years, scientists have discovered that cancer is a genetic disease at the somatic cell level, characterized by the presence of gene products derived from mutated or abnormally expressed genes. The combined effects of numerous abnormal gene products lead to the uncontrolled growth and spread of cancer cells. Although some mutated cancer genes may be inherited, most are created within somatic cells that then divide and form tumors. Completion of the Human Genome Project and numerous large-scale rapid DNA sequencing studies have opened the door to a wealth of new information about the mutations that trigger a cell to become cancerous. This new understanding of cancer genetics is also leading to new gene-specific treatments, some of which are now entering clinical trials. Some scientists predict that gene-targeted therapies will replace chemotherapies within the next 25 years.

The goal of this chapter is to highlight our current understanding of the nature and causes of cancer. As we will see, cancer is a genetic disease that arises from the accumulation of mutations in genes controlling many basic aspects of cellular function. We will examine the relationship between genes and cancer, and consider how mutations, chromosomal changes, epigenetics, and environmental agents play roles in the development of cancer.

19.1

Cancer Is a Genetic Disease at the Level of Somatic Cells

Perhaps the most significant development in understanding the causes of cancer is the realization that cancer is a genetic disease. Genomic alterations that are associated with cancer range from single-nucleotide substitutions to large-scale chromosome rearrangements, amplifications, and deletions (Figure 19–1). However, unlike other genetic diseases, cancer is caused by mutations that occur predominantly in somatic cells. Only about 1 percent of cancers are associated with germ-line mutations that increase a person's susceptibility to certain types of cancer. Another important difference between cancers and other genetic diseases is that cancers rarely arise from a single mutation in a single gene, but from the accumulation of mutations in many genes—as many as six to twelve. The mutations that lead to cancer affect multiple cellular functions, including repair of DNA

(a)

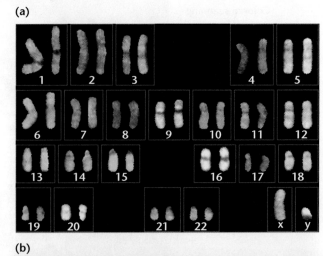

(b)

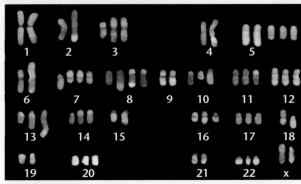

FIGURE 19–1   (a) Spectral karyotype of a normal cell. (b) Karyotype of a cancer cell showing translocations, deletions, and aneuploidy—characteristic features of cancer cells.

damage, cell division, apoptosis, cellular differentiation, migratory behavior, and cell–cell contact.

What Is Cancer?

Clinically, cancer is defined as a large number of complex diseases, up to a hundred, that behave differently depending on the cell types from which they originate. Cancers vary in their ages of onset, growth rates, invasiveness, prognoses, and responsiveness to treatments. However, at the molecular level, all cancers exhibit common characteristics that unite them as a family.

All cancer cells share two fundamental properties: (1) abnormal cell growth and division (**proliferation**), and (2) defects in the normal restraints that keep cells from spreading and colonizing other parts of the body (**metastasis**). In normal cells, these functions are tightly controlled by genes that are expressed appropriately in time and place. In cancer cells, these genes are either mutated or are expressed inappropriately.

It is this combination of uncontrolled cell proliferation and metastatic spread that makes cancer cells dangerous. When a cell simply loses genetic control over cell growth, it may grow into a multicellular mass, a **benign tumor**. Such a tumor can often be removed by surgery and may cause no serious harm. However, if cells in the tumor also have the ability to break loose, enter the bloodstream, invade other tissues, and form secondary tumors (**metastases**), they become malignant. **Malignant tumors** are often difficult to treat and may become life threatening. As we will see later in the chapter, there are multiple steps and genetic mutations that convert a benign tumor into a dangerous malignant tumor.

The Clonal Origin of Cancer Cells

Although malignant tumors may contain billions of cells, and may invade and grow in numerous parts of the body, all cancer cells in the primary and secondary tumors are clonal, meaning that they originated from a common ancestral cell that accumulated specific mutations. This is an important concept in understanding the molecular causes of cancer and has implications for its diagnosis.

Numerous data support the concept of cancer clonality. For example, reciprocal chromosomal translocations are characteristic of many cancers, including leukemias and lymphomas (two cancers involving white blood cells). Cancer cells from patients with **Burkitt's lymphoma** show reciprocal translocations between chromosome 8 (with translocation breakpoints at or near the *c-myc* gene) and chromosomes 2, 14, or 22 (with translocation breakpoints at or near one of the immunoglobulin genes). Each Burkitt's lymphoma patient exhibits unique breakpoints in his or her *c-myc* and immunoglobulin gene DNA sequences; however, all lymphoma cells within that patient contain identical translocation breakpoints. This demonstrates that all cancer cells in each case of Burkitt's lymphoma arise from a single cell, and this cell passes on its genetic aberrations to its progeny.

Another demonstration that cancer cells are clonal is their pattern of X-chromosome inactivation. As explained in Chapter 7, female humans are mosaic, with some cells containing an inactivated paternal X chromosome and other cells containing an inactivated maternal X chromosome. X-chromosome inactivation occurs early in development and takes place at random. All cancer cells within a tumor, both primary and metastatic, within one female individual, contain the same inactivated X chromosome. This supports the concept that all the cancer cells in that patient arose from a common ancestral cell.

The Cancer Stem Cell Hypothesis

A concept that is related to the clonal origin of cancer cells is that of the cancer stem cell. Many scientists now believe that tumors are comprised of a mixture of cells, many of which do not proliferate. Those that do proliferate and give rise to all the cells within the tumor are known as **cancer stem cells**. Stem cells are cells that have the capacity for self-renewal—a process in which the stem cell divides unevenly, creating one daughter cell that goes on to differentiate into a mature cell type and one that remains a stem cell.

(Stem cells are also discussed in Chapters 18 and 27.) The cancer stem cell hypothesis contrasts the random or stochastic model. This model predicts that every cell within a tumor has the potential to form a new tumor.

Although scientists still actively debate the existence of cancer stem cells, evidence is accumulating that cancer stem cells do exist, at least in some tumors. Cancer stem cells have been identified in leukemias as well as in solid tumors of the brain, breast, colon, ovary, pancreas, and prostate. It is still not clear what fraction of any tumor is comprised of cancer stem cells. For example, human acute myeloid leukemias contain less than 1 cancer stem cell in 10,000. In contrast, some solid tumors may contain as many as 40 percent cancer stem cells.

Scientists are also not sure about the origins of cancer stem cells. It is possible that they may arise from normal adult stem cells within a tissue, or they may be created from more differentiated cells that acquire properties similar to stem cells after accumulating numerous mutations.

Cancer As a Multistep Process, Requiring Multiple Mutations

Although we know that cancer is a genetic disease initiated by mutations that lead to uncontrolled cell proliferation and metastasis, a single mutation is not sufficient to transform a normal cell into a tumor-forming (tumorigenic), malignant cell. If it were sufficient, then cancer would be far more prevalent than it is. In humans, mutations occur spontaneously at a rate of about 10^{-6} mutations per gene, per cell division, mainly due to the intrinsic error rates of DNA replication. Because there are approximately 10^{16} cell divisions in a human body during a lifetime, a person might suffer up to 10^{10} mutations per gene somewhere in the body, during his or her lifetime. However, only about one person in three will suffer from cancer.

The phenomenon of age-related cancer is another indication that cancer develops from the accumulation of several mutagenic events in a single cell. The incidence of most cancers rises exponentially with age (Figure 19–2). If a single mutation were sufficient to convert a normal cell to a malignant one, then cancer incidence would appear to be independent of age. The age-related incidence of cancer suggests that many independent mutations, occurring randomly, and with a low probability, are necessary before a cell is transformed into a malignant cancer cell. Another indication that cancer is a multistep process is the delay that occurs between exposure to **carcinogens** (cancer-causing agents) and the appearance of the cancer. For example, an incubation period of five to eight years separated exposure of people to the radiation of the atomic explosions at Hiroshima and Nagasaki and the onset of leukemias.

The multistep nature of cancer development is supported by the observation that cancers often develop in progressive steps, beginning with mildly aberrant cells and progressing to cells that are increasingly tumorigenic and malignant. This

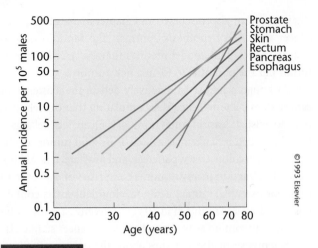

FIGURE 19–2   The incidence of most cancers rises exponentially with age. The graph shows the logarithmic plot of the incidence rate with the logarithmic plot of the patient's age. (*Adapted from Vogelstein, B. and Kinzler, K.W. 1993. The multistep nature of cancer.* Trends in Genetics *9: 138–141*).

progressive nature of cancer is illustrated by the development of colon cancer, as discussed in Section 19.6.

Each step in **tumorigenesis** (the development of a malignant tumor) appears to be the result of two or more genetic alterations that release the cells progressively from the controls that normally operate on proliferation and malignancy. This observation suggests that the progressive genetic alterations that create a cancer cell confer selective advantages to the cell and are propagated through cell divisions during the creation of tumors.

Scientists are now applying some of the recent advances in DNA sequencing in order to identify all of the somatic mutations that occur during the development of a cancer cell. These studies compare the DNA sequences of genomes from cancer cells and normal cells derived from the same patient. Data from these studies are revealing that tens of thousands of somatic mutations are present in cancer cells. Researchers believe that only a handful of these mutations—called **driver mutations**—give a growth advantage to a tumor cell. The remainder of the mutations may be acquired over time, perhaps as a result of the increased levels of DNA damage that accumulate in cancer cells, but these mutations have no direct contribution to the cancer phenotype. These are known as **passenger mutations**. The total number of driver mutations that occur in any particular cancer is still unclear; however, scientists expect that the presence of fewer than a dozen mutated genes may be sufficient to create a cancer cell.

As we will discover in subsequent sections of this chapter, the genes that undergo mutations leading to cancer (called oncogenes and tumor-suppressor genes) are those that control DNA damage repair, the cell cycle, cell–cell contact, and programmed cell death. We will now investigate each of these fundamental processes, the genes that control them, and how mutations in these genes may lead to cancer.

19.2

Cancer Cells Contain Genetic Defects Affecting Genomic Stability, DNA Repair, and Chromatin Modifications

Cancer cells show higher than normal rates of mutation, chromosomal abnormalities, and genomic instability. Many researchers believe that the fundamental defect in cancer cells is a derangement of the cells' normal ability to repair DNA damage. This loss of genomic integrity leads to a general increase in the mutation rate for every gene in the genome, including those whose products control aspects of cell proliferation, programmed cell death, and metastasis. The high level of genomic instability seen in cancer cells is known as the **mutator phenotype**. In addition, recent research has revealed that cancer cells contain aberrations in the types and locations of chromatin modifications, particularly DNA and histone methylation patterns.

Genomic Instability and Defective DNA Repair

Genomic instability in cancer cells is characterized by the presence of gross defects such as translocations, aneuploidy, chromosome loss, DNA amplification, and chromosome deletions (Figures 19–1 and 19–3). Cancer cells that are

(a) Double minutes

(b) Heterogeneous staining region

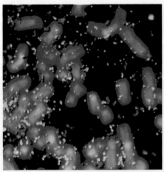 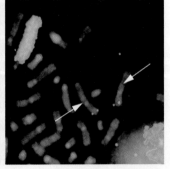

FIGURE 19–3   DNA amplifications in neuroblastoma cells. (a) Two cancer genes (*MYCN* in red and *MDM2* in green) are amplified as small DNA fragments that remain separate from chromosomal DNA within the nucleus. These units of amplified DNA are known as double minute chromosomes. Normal chromosomes are stained blue. (b) Multiple copies of the *MYCN* gene are amplified within one large region called a heterogeneous staining region (green). Single copies of the *MYCN* gene are visible as green dots at the ends of the normal parental chromosomes (white arrows). Normal chromosomes are stained red.

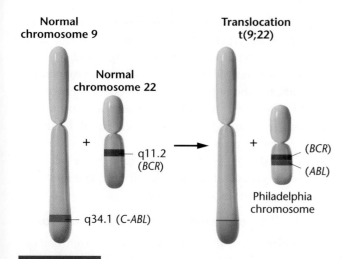

Normal chromosome 9

Normal chromosome 22

+ q11.2 (*BCR*)

q34.1 (*C-ABL*)

Translocation t(9;22)

+ (*BCR*) (*ABL*)

Philadelphia chromosome

FIGURE 19-4 A reciprocal translocation involving the long arms of chromosomes 9 and 22 results in the formation of a characteristic chromosome, the Philadelphia chromosome, which is associated with chronic myelogenous leukemia (CML). The t(9;22) translocation results in the fusion of the *C-ABL* proto-oncogene on chromosome 9 with the *BCR* gene on chromosome 22. The fusion protein is a powerful hybrid molecule that allows cells to escape control of the cell cycle, contributing to the development of CML.

grown in cultures in the lab also show a great deal of genomic instability—duplicating, losing, and translocating chromosomes or parts of chromosomes. Often cancer cells show specific chromosomal defects that are used to diagnose the type and stage of the cancer. For example, leukemic white blood cells from patients with **chronic myelogenous leukemia (CML)** bear a specific translocation, in which the *C-ABL* gene on chromosome 9 is translocated into the *BCR* gene on chromosome 22. This translocation creates a structure known as the **Philadelphia chromosome** (Figure 19–4). The *BCR-ABL* fusion gene codes for a chimeric BCR-ABL protein. The normal ABL protein is a **protein kinase** that acts within signal transduction pathways, transferring growth factor signals from the external environment to the nucleus. The BCR-ABL protein is an abnormal signal transduction molecule in CML cells, which stimulates these cells to proliferate even in the absence of external growth signals.

In keeping with the concept of the cancer mutator phenotype, a number of inherited cancers are caused by defects in genes that control DNA repair. For example, xeroderma pigmentosum (XP) is a rare hereditary disorder that is characterized by extreme sensitivity to ultraviolet light and other carcinogens. Patients with XP often develop skin cancer. Cells from patients with XP are defective in nucleotide excision repair, with mutations appearing in any one of seven genes whose products are necessary to carry out DNA repair. XP cells are impaired in their ability to repair DNA lesions such as thymine dimers induced by UV light.

The relationship between XP and genes controlling nucleotide excision repair is also described in Chapter 17.

Another hereditary cancer, **hereditary nonpolyposis colorectal cancer (HNPCC)**, is also caused by mutations in genes controlling DNA repair. HNPCC is an autosomal dominant syndrome, affecting about one in every 200 to 1000 people. Patients affected by HNPCC have an increased risk of developing colon, ovary, uterine, and kidney cancers. Cells from patients with HNPCC show higher than normal mutation rates and genomic instability. At least eight genes are associated with HNPCC, and four of these genes control aspects of DNA mismatch repair. Inactivation of any of these four genes—*MSH2, MSH6, MLH1,* and *MLH3*—causes a rapid accumulation of genome-wide mutations and the subsequent development of colorectal and other cancers.

The observation that hereditary defects in genes controlling nucleotide excision repair and DNA mismatch repair lead to high rates of cancer lends support to the idea that the mutator phenotype is a significant contributor to the development of cancer.

Chromatin Modifications and Cancer Epigenetics

The field of cancer epigenetics is providing new perspectives on the genetics of cancer. **Epigenetics** is the study of factors that affect gene expression but that do not alter the nucleotide sequence of DNA. Epigenetic effects can be inherited from one cell to its progeny cells and may be present in either somatic or germ-line cells. DNA methylation and histone modifications such as acetylation and phosphorylation are examples of epigenetic modifications. The genomic patterns and locations of these modifications can affect gene expression. For example, DNA methylation is thought to be responsible for the gene silencing associated with parental imprinting, heterochromatin gene repression, and X-chromosome inactivation. The effects of epigenetic factors on gene expression and hereditary disease are discussed further in Chapter 17 and Special Topics, p. 517.

Cancer cells contain altered DNA methylation patterns. Overall, there is much less DNA methylation in cancer cells than in normal cells. At the same time, the promoters of some genes are hypermethylated in cancer cells. These changes are thought to result in the release of transcription repression over the bulk of genes that would be silent in normal cells—including cancer-causing genes—while at the same time repressing transcription of genes that would regulate normal cellular functions such as DNA repair and cell-cycle control. Methylation profiles in cancer cells are now being used to help diagnose tumors and predict their course.

Histone modifications are also disrupted in cancer cells. Genes that encode histone acetylases, deacetylases, methyltransferases, and demethylases are often mutated or aberrantly expressed in cancer cells. The large numbers of epigenetic abnormalities in tumors have prompted some scientists to speculate that there may be more epigenetic defects in cancer cells than there are gene mutations. In addition, because epigenetic modifications are reversible, it may be possible to treat cancers using epigenetic-based therapies. Although the field of cancer epigenetics is still in its infancy, it has already provided major insights into tumorigenesis as well as new clinical applications.

19.3

Cancer Cells Contain Genetic Defects Affecting Cell-Cycle Regulation

One of the fundamental aberrations in all cancer cells is a loss of control over cell proliferation. Cell proliferation is the process of cell growth and division that is essential for all development and tissue repair in multicellular organisms. Although some cells, such as epidermal cells of the skin or blood-forming cells in the bone marrow, continue to grow and divide throughout an organism's lifetime, most cells in adult multicellular organisms remain in a nondividing, quiescent, and differentiated state. **Differentiated cells** are those that are specialized for specific functions, such as photoreceptor cells of the retina or muscle cells of the heart. The most extreme examples of nonproliferating cells are nerve cells, which divide little, if at all, even to replace damaged tissue. In contrast, many differentiated cells, such as those in the liver and kidney, are able to grow and divide when stimulated by extracellular signals and growth factors. In this way, multicellular organisms are able to replace dead and damaged tissue. However, the growth and differentiation of cells must be strictly regulated; otherwise, the integrity of

organs and tissues would be compromised by the presence of inappropriate types and quantities of cells. Normal regulation over cell proliferation involves a large number of gene products that control steps in the cell cycle, programmed cell death, and the response of cells to external growth signals. In cancer cells, many of the genes that control these functions are mutated or aberrantly expressed, leading to uncontrolled cell proliferation.

In this section, we will review steps in the cell cycle, some of the genes that control the cell cycle, and how these genes, when mutated, lead to cancer.

The Cell Cycle and Signal Transduction

The cellular events that occur in sequence from one cell division to the next comprise the **cell cycle** (Figure 19–5). The **interphase** stage of the cell cycle is the interval between mitotic divisions. During this time, the cell grows and replicates its DNA. During **G1**, the cell prepares for DNA synthesis by accumulating the enzymes and molecules required for DNA replication. G1 is followed by **S phase**, during which the cell's chromosomal DNA is replicated. During **G2**, the cell continues to grow and prepare for division. During **M phase**, the duplicated chromosomes condense, sister chromosomes separate to opposite poles, and the cell divides in two. These phases of the cell cycle are also discussed in more detail in Chapter 2.

In early to mid-G1, the cell makes a decision either to enter the next cell cycle or to withdraw from the cell cycle into quiescence. Continuously dividing cells do not exit the cell cycle but proceed through G1, S, G2, and M phases; however, if the cell receives signals to stop growing, it enters the **G0** phase of the cell cycle. During G0, the cell remains metabolically active but does not grow or divide.

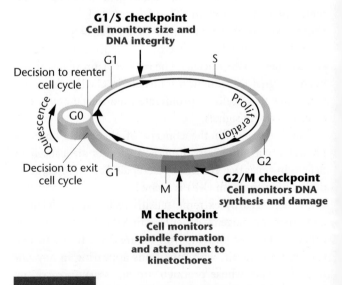

FIGURE 19–5   Checkpoints and proliferation decision points monitor the progress of the cell through the cell cycle.

Most differentiated cells in multicellular organisms can remain in this G0 phase indefinitely. Some, such as neurons, never reenter the cell cycle. In contrast, cancer cells are unable to enter G0, and instead, they continuously cycle. Their *rate* of proliferation is not necessarily any greater than that of normal proliferating cells; however, they are not able to become quiescent at the appropriate time or place.

Cells in G0 can often be stimulated to reenter the cell cycle by external growth signals. These signals are delivered to the cell by molecules such as growth factors and hormones that bind to cell-surface receptors, which then relay the signal from the plasma membrane to the cytoplasm. The process of transmitting growth signals from the external environment to the cell nucleus is known as **signal transduction**. Ultimately, signal transduction initiates a program of gene expression that propels the cell out of G0 back into the cell cycle. Cancer cells often have defects in signal transduction pathways. Sometimes, abnormal signal transduction molecules send continuous growth signals to the nucleus even in the absence of external growth signals. An example of abnormal signal transduction due to mutations in the *ras* gene is described in Section 19.4. In addition, malignant cells may not respond to external signals from surrounding cells—signals that would normally inhibit cell proliferation within a mature tissue.

Cell-Cycle Control and Checkpoints

In normal cells, progress through the cell cycle is tightly regulated, and each step must be completed before the next step can begin. There are at least three distinct points in the cell cycle at which the cell monitors external signals and internal equilibrium before proceeding to the next stage. These are the **G1/S**, the **G2/M**, and **M checkpoints** (Figure 19–5). At the G1/S checkpoint, the cell monitors its size and determines whether its DNA has been damaged. If the cell has not achieved an adequate size, or if the DNA has been damaged, further progress through the cell cycle is halted until these conditions are corrected. If cell size and DNA integrity are normal, the G1/S checkpoint is traversed, and the cell proceeds to S phase. The second important checkpoint is the G2/M checkpoint, where physiological conditions in the cell are monitored prior to mitosis. If DNA replication or repair of any DNA damage has not been completed, the cell cycle arrests until these processes are complete. The third major checkpoint occurs during mitosis and is called the M checkpoint. At this checkpoint, both the successful formation of the spindle-fiber system and the attachment of spindle fibers to the kinetochores

associated with the centromeres are monitored. If spindle fibers are not properly formed or attachment is inadequate, mitosis is arrested.

In addition to regulating the cell cycle at checkpoints, the cell controls progress through the cell cycle by means of two classes of proteins: **cyclins** and **cyclin-dependent kinases (CDKs)**. The cell synthesizes and destroys cyclin proteins in a precise pattern during the cell cycle (Figure 19–6). When a cyclin is present, it binds to a specific CDK, triggering activity of the CDK/cyclin complex. The CDK/cyclin complex then selectively phosphorylates and activates other proteins that in turn bring about the changes necessary to advance the cell through the cell cycle. For example, in G1 phase, CDK4/cyclin D complexes activate proteins that stimulate transcription of genes whose products (such as DNA polymerase δ and DNA ligase) are required for DNA replication during S phase. Another CDK/cyclin complex, CDK1/cyclin B, phosphorylates a number of proteins that bring about the events of early mitosis, such as nuclear membrane breakdown, chromosome condensation, and cytoskeletal reorganization (Figure 19–7). Mitosis can only be completed, however, when cyclin B is degraded and the protein phosphorylations characteristic of M phase are reversed. Although a large number of different protein kinases exist in cells, only a few are involved in cell-cycle regulation.

The cell cycle is regulated by an interplay of genes whose products either promote or suppress cell division. Mutation or misexpression of any of the genes controlling the cell cycle can contribute to the development of cancer. For example, if genes that control the G1/S or G2/M checkpoints are

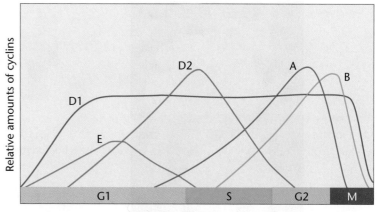

FIGURE 19–6   Relative expression times and amounts of cyclins during the cell cycle. Cyclin D1 accumulates early in G1 and is expressed at a constant level through most of the cycle. Cyclin E accumulates in G1, reaches a peak, and declines by mid-S phase. Cyclin D2 begins accumulating in the last half of G1, reaches a peak just after the beginning of S, and then declines by early G2. Cyclin A appears in late G1, accumulates through S phase, peaks at the G2/M transition, and is rapidly degraded. Cyclin B peaks at the G2/M transition and declines rapidly in M phase.

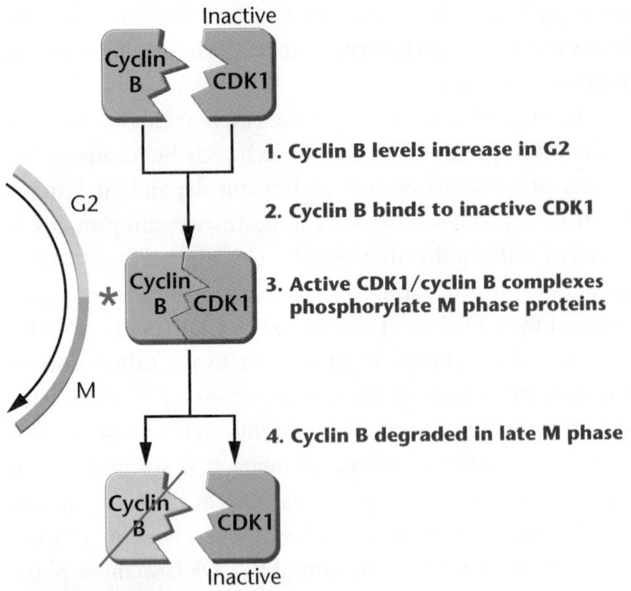

Inactive

Cyclin B — CDK1

G2

1. **Cyclin B levels increase in G2**

2. **Cyclin B binds to inactive CDK1**

Cyclin B — CDK1
*

3. **Active CDK1/cyclin B complexes phosphorylate M phase proteins**

M

4. **Cyclin B degraded in late M phase**

Cyclin B — CDK1

Inactive

FIGURE 19–7 CDK1 and cyclin B control the transition from G2 to M phase. In late G2 phase, cyclin B accumulates and forms complexes with inactive CDK1 molecules. CDK1 is activated within the complexes and adds phosphate groups to cellular components. These phosphorylated molecules bring about the structural and biochemical changes that are necessary for M phase. In late M phase, cyclin B is degraded, CDK1 becomes inactive, and M phase phosphorylations are reversed.

mutated, the cell may continue to grow and divide without repairing DNA damage. As these cells continue to divide, they accumulate mutations in genes whose products control cell proliferation or metastasis. Similarly, if genes that control progress through the cell cycle, such as those that encode the cyclins, are expressed at the wrong time or at incorrect levels, the cell may grow and divide continuously and may be unable to exit the cell cycle into G0. The result in both cases is that the cell loses control over proliferation and is on its way to becoming cancerous.

Control of Apoptosis

As already described, if DNA replication, repair, or chromosome assembly is defective, normal cells halt their progress through the cell cycle until the condition is corrected. This reduces the number of mutations and chromosomal abnormalities that accumulate in normal proliferating cells. However, if DNA or chromosomal damage is so severe that repair is impossible, the cell may initiate a second line of defense—a process called **apoptosis**, or **programmed cell death**. Apoptosis is a genetically controlled process whereby the cell commits suicide. Besides its role in preventing cancer, apoptosis is also initiated during normal multicellular development in order to eliminate certain cells that do not contribute to the final adult organism. The steps in apoptosis are the same for damaged cells and for cells being eliminated during development: nuclear DNA becomes fragmented, internal cellular structures are disrupted, and the cell dissolves into small spherical structures known as apoptotic bodies Figure 19–8 (a). In the final step, the apoptotic bodies are engulfed by the immune system's phagocytic cells. A series of proteases called **caspases** are responsible for initiating apoptosis and for digesting intracellular components.

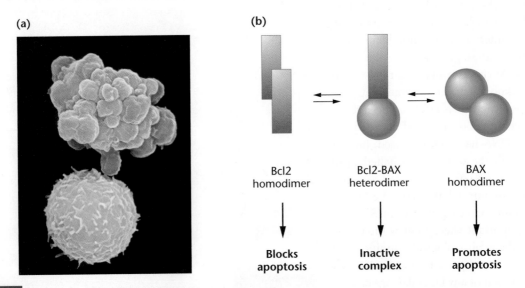

(a)

(b)

Bcl2 homodimer

Bcl2-BAX heterodimer

BAX homodimer

Blocks apoptosis

Inactive complex

Promotes apoptosis

FIGURE 19–8 (a) Normal white blood cell (bottom) and a white blood cell undergoing apoptosis (top). Apoptotic bodies appear as grape-like clusters on the cell surface. (b) The relative concentrations of the Bcl2 and BAX proteins regulate apoptosis. A normal cell contains a balance of Bcl2 and BAX, which form inactive heterodimers. A relative excess of Bcl2 results in the formation of Bcl2 homodimers, which prevent apoptosis. Cancer cells with Bcl2 overexpression are resistant to chemotherapies and radiation therapies. A relative excess of BAX results in the formation of BAX homodimers, which induce apoptosis. In normal cells, activated p53 protein induces transcription of the *BAX* gene and inhibits transcription of the *Bcl2* gene, leading to cell death. In many cancer cells, p53 is defective, preventing the apoptotic pathway from removing the cancer cells.

Apoptosis is genetically controlled in that regulation of the levels of specific gene products such as Bcl2 and BAX (Figure 19–8b) can trigger or prevent apoptosis. By removing damaged cells, programmed cell death reduces the number of mutations that are passed to the next generation, including those in cancer-causing genes. Some of the same genes that control cell-cycle checkpoints can trigger apoptosis. These genes are mutated in many cancers. As a result of the mutation or inactivation of these checkpoint genes, the cell is unable to repair its DNA or undergo apoptosis. This inability leads to the accumulation of even more mutations in genes that control growth, division, and metastasis.

19.4

Proto-oncogenes and Tumor-suppressor Genes Are Altered in Cancer Cells

Two general categories of cancer-causing genes are mutated or misexpressed in cancer cells—the proto-oncogenes and the tumor-suppressor genes (Table 19.1). **Proto-oncogenes** encode transcription factors that stimulate expression of other genes, signal transduction molecules that stimulate cell division, and cell-cycle regulators that move the cell through the cell cycle. Their products are important for normal cell functions, especially cell growth and division. When normal cells become quiescent and cease division, they repress the expression of most proto-oncogenes or modify the activities of their products. In cancer cells, one or more proto-oncogenes are altered in such a way that the

activities of their products cannot be regulated in a normal fashion. This is sometimes due to mutations that result in an abnormal protein product. In other cases, proto-oncogenes may be overexpressed or expressed at an incorrect time. If a proto-oncogene is continually in an "on" state, its product may constantly stimulate the cell to divide. When a proto-oncogene is mutated or abnormally expressed and contributes to the development of cancer, it is known as an **oncogene**—a cancer-causing gene. Oncogenes are proto-oncogenes that have experienced a gain-of-function alteration. As a result, only one allele of a proto-oncogene needs to be mutated or misexpressed in order to contribute to cancer. Hence, oncogenes confer a dominant cancer phenotype.

Tumor-suppressor genes are genes whose products normally regulate cell-cycle checkpoints or initiate the process of apoptosis. In normal cells, proteins encoded by tumor-suppressor genes halt progress through the cell cycle in response to DNA damage or growth-suppression signals from the extracellular environment. When tumor-suppressor genes are mutated or inactivated, cells are unable to respond normally to cell-cycle checkpoints, or are unable to undergo programmed cell death if DNA damage is extensive. This leads to the accumulation of more mutations and the development of cancer. When both alleles of a tumor-suppressor gene are inactivated, and other changes in the cell keep it growing and dividing, cells may become tumorigenic.

The following are examples of proto-oncogenes and tumor-suppressor genes that contribute to cancer when mutated. Approximately 400 oncogenes and tumor-suppressor genes are now known, and more will likely be discovered as cancer research continues.

TABLE 19.1

Some Proto-oncogenes and Tumor-suppressor Genes

Proto-oncogene	Normal Function	Alteration in Cancer	Associated Cancers
c-myc	Transcription factor, regulates cell cycle, differentiation, apoptosis	Translocation, amplification, point mutations	Lymphomas, leukemias, lung cancer, many types
c-kit	Tyrosine kinase, signal transduction	Mutation	Sarcomas
RARα	Hormone-dependent transcription factor, differentiation	Chromosomal translocations with PML gene, fusion product	Acute promyelocytic leukemia
E6	Human papillomavirus encoded oncogene, inactivates p53	HPV infection	Cervical cancer
Cyclins	Bind to CDKs, regulate cell cycle	Gene amplification, overexpression	Lung, esophagus, many types

Tumor Suppressor	Normal Function	Alteration in Cancer	Associated Cancers
RB1	Cell-cycle checkpoints, binds E2F	Mutation, deletion, inactivation by viral oncogene products	Retinoblastoma, osteosarcoma, many types
APC	Cell–cell interaction	Mutation	Colorectal cancers, brain, thyroid
p53	Transcription regulation	Mutation, deletion, viruses	Many types
BRCA1, BRCA2	DNA repair	Point mutations	Breast, ovarian, prostate cancers

The *ras* Proto-oncogenes

Some of the most frequently mutated genes in human tumors are those in the ***ras* gene family**. These genes are mutated in more than 30 percent of human tumors. The *ras* gene family encodes signal transduction molecules that are associated with the cell membrane and regulate cell growth and division. Ras proteins normally transmit signals from the cell membrane to the nucleus, stimulating the cell to divide in response to external growth factors (Figure 19–9). Ras proteins alternate between an inactive (switched off) and an active (switched on) state by binding either guanosine diphosphate (GDP) or guanosine triphosphate (GTP). When a cell encounters a growth factor (such as platelet-derived growth factor or epidermal growth factor), growth factor receptors on the cell membrane bind to the growth factor, resulting in autophosphorylation of the cytoplasmic portion of the growth factor receptor. This causes recruitment of proteins known as nucleotide exchange factors to the plasma membrane.

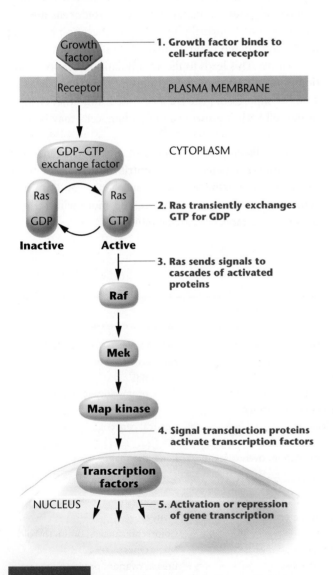

FIGURE 19–9 A signal transduction pathway mediated by Ras.

These nucleotide exchange factors cause Ras to release GDP and bind GTP, thereby activating Ras. The active, GTP-bound form of Ras then sends its signals through cascades of protein phosphorylations in the cytoplasm. The end-point of these cascades is activation of nuclear transcription factors that stimulate expression of genes whose products drive the cell from quiescence into the cell cycle. Once Ras has sent its signals to the nucleus, it hydrolyzes GTP to GDP and becomes inactive. Mutations that convert the *ras* proto-oncogene to an oncogene prevent the Ras protein from hydrolyzing GTP to GDP and hence freeze the Ras protein into its "on" conformation, constantly stimulating the cell to divide.

The *p53* Tumor-suppressor Gene

The most frequently mutated gene in human cancers—mutated in more than 50 percent of all cancers—is the ***p53* gene**. This gene encodes a nuclear protein that acts as a transcription factor, repressing or stimulating transcription of more than 50 different genes.

Normally, the p53 protein is continuously synthesized but is rapidly degraded and therefore is present in cells at low levels. In addition, the p53 protein is normally bound to another protein called **MDM2**, which has several effects on p53. The presence of MDM2 on the p53 protein tags p53 for degradation and sequesters the transcriptional activation domain of p53. It also prevents the phosphorylations and acetylations that convert the p53 protein from an inactive to an active form. Several types of cellular stress events bring about rapid increases in the nuclear levels of activated p53 protein. These include chemical damage to DNA, double-stranded breaks in DNA induced by ionizing radiation, and the presence of DNA-repair intermediates generated by exposure of cells to ultraviolet light. In response to these signals, MDM2 dissociates from p53, making p53 more stable and unmasking its transcription activation domain. Increases in the levels of activated p53 protein also result from increases in protein phosphorylation, acetylation, and other post-translational modifications (Figure 19–10). Activated p53 protein acts as a transcription factor that stimulates expression of the *MDM2* gene. As the levels of MDM2 increase, p53 protein is again bound by MDM2, returned to an inactive state, and targeted for degradation, in a negative feedback loop.

The p53 protein initiates several different responses to DNA damage including cell-cycle arrest followed by DNA repair and apoptosis if DNA cannot be repaired. These responses are accomplished by p53 acting as a transcription factor that stimulates or represses the expression of genes involved in each response.

In normal cells, p53 can arrest the cell cycle at the G1/S and G2/M checkpoints, as well as retarding the progression of the cell through S phase. To arrest the cell cycle at the G1/S checkpoint, activated p53 protein stimulates transcription of a gene encoding the p21 protein. The p21 protein inhibits

(a) p53 in unstressed cells

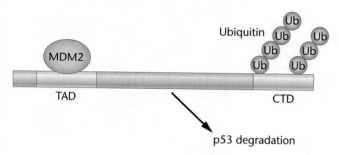

(b) After DNA damage and cell stress

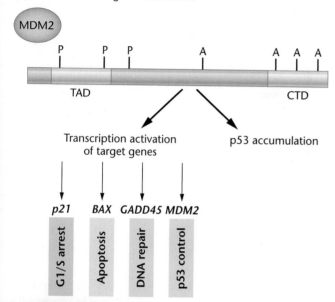

FIGURE 19–10 Steps in the regulation of p53 levels and activity. (a) In normal unstressed cells, p53 is kept inactive and at low abundance by MDM2, which binds to the transactivation domain (TAD) and stimulates the addition of ubiquitin onto lysine residues in the carboxy-terminal domain (CTD). The presence of ubiquitin promotes p53 degradation. (b) After various types of cellular stress including DNA damage, cellular kinases add phosphates (P's) to serines and threonines in the TAD, leading to dissociation of MDM2 and subsequent loss of ubiquitin. As the levels of p53 increase in the nucleus, histone acetyl transferases add acetyl groups (A's) to lysines in the CTD, which increases p53 stability and affinity for specific DNA sequences within the promoter regions of target genes. Examples of genes that are transcriptionally stimulated by p53 are *p21* (leading to G1/S cell cycle arrest), *BAX* (stimulating apoptosis), *GADD45* (contributing to DNA repair), and *MDM2* (returning p53 to an inactive and low abundance state).

the CDK4/cyclin D1 complex, hence preventing the cell from moving from G1 phase into S phase. Activated p53 protein also regulates expression of genes that retard the progress of DNA replication, thus allowing time for DNA damage to be repaired during S phase. By regulating expression of other genes, activated p53 can block cells at the G2/M checkpoint, if DNA damage occurs during S phase.

Activated p53 can also instruct a damaged cell to commit suicide by apoptosis. It does so by activating the transcription of the *Bax* gene and repressing transcription of the *Bcl2* gene. In normal cells, the BAX protein is present in a heterodimer with the Bcl2 protein, and the cell remains viable (Figure 19–8). But when the levels of BAX protein increase in response to p53 stimulation of *Bax* gene transcription, BAX homodimers are formed, and these homodimers activate the cellular changes that lead to apoptosis. In cancer cells that lack functional p53, BAX protein levels do not increase in response to cell damage, and apoptosis may not occur.

Cells lacking functional p53 are unable to arrest at cell-cycle checkpoints or to enter apoptosis in response to DNA damage. As a result, they move unchecked through the cell cycle, regardless of the condition of the cell's DNA. Cells lacking p53 have high mutation rates and accumulate the types of mutations that lead to cancer. Because of the importance of the *p53* gene to genomic integrity, it is often referred to as the "guardian of the genome."

The *RB1* Tumor-suppressor Gene

The loss or mutation of the ***RB1*** (**retinoblastoma 1**) tumor-suppressor gene contributes to the development of many cancers, including those of the breast, bone, lung, and bladder. The *RB1* gene was originally identified as a result of studies on **retinoblastoma**, an inherited disorder in which tumors develop in the eyes of young children. Retinoblastoma occurs with a frequency of about 1 in 15,000 individuals. In the familial form of the disease, individuals inherit one mutated allele of the *RB1* gene and have an 85 percent chance of developing retinoblastomas as well as an increased chance of developing other cancers. All somatic cells of patients with hereditary retinoblastoma contain one mutated allele of the *RB1* gene. However, it is only when the second normal allele of the *RB1* gene is lost or mutated in certain retinal cells that retinoblastoma develops. In individuals who do not have this hereditary condition, retinoblastoma is extremely rare, as it requires at least two separate somatic mutations in a retinal cell in order to inactivate both copies of the *RB1* gene (Figure 19–11).

The **retinoblastoma protein (pRB)** is a tumor-suppressor protein that controls the G1/S cell-cycle checkpoint. The pRB protein is found in the nuclei of all cell types at all stages of the cell cycle. However, its activity varies throughout the cell cycle, depending on its phosphorylation state. When cells are in the G0 phase of the cell cycle, the pRB protein is nonphosphorylated and binds to transcription factors such as E2F, inactivating them (Figure 19–12). When the cell is stimulated by growth factors, it enters G1 and approaches S phase. Throughout the G1 phase, the pRB protein becomes phosphorylated by the CDK4/cyclin D1 complex. Phosphorylated pRB releases its bound regulatory proteins. When E2F and other regulators are released by pRB, they are free to induce the expression of over

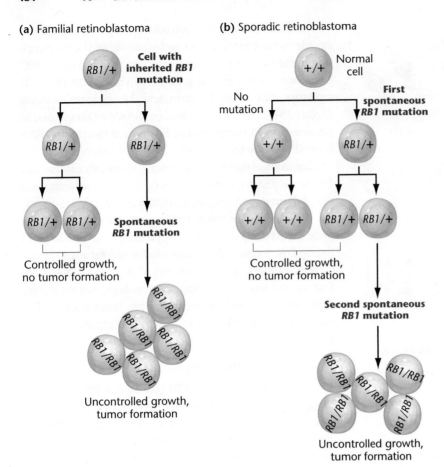

(a) Familial retinoblastoma

(b) Sporadic retinoblastoma

FIGURE 19–11   (a) In familial retinoblastoma, one mutation (designated as
RB1) is inherited and present in all cells.
A second mutation at the retinoblastoma locus in any retinal cell contributes
to uncontrolled cell growth and tumor
formation. (b) In sporadic retinoblastoma,
independent mutations in both alleles of
the retinoblastoma gene within a single cell
are acquired sequentially, also leading to
tumor formation.

30 genes whose products are required for the transition from
G1 into S phase. After cells traverse S, G2, and M phases, pRB
reverts to a nonphosphorylated state, binds to regulatory proteins such as E2F, and keeps them sequestered until required
for the next cell cycle. In normal quiescent cells, the presence
of the pRB protein prevents passage into S phase. In many
cancer cells, including retinoblastoma cells, both copies of the
RB1 gene are defective, inactive, or absent, and progression
through the cell cycle is not regulated.

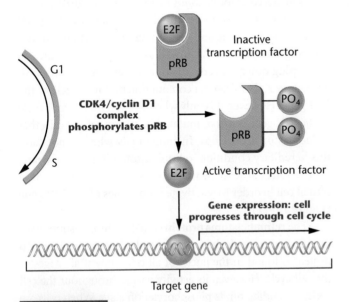

FIGURE 19–12   During G0 and early G1, pRB interacts with
and inactivates transcription factor E2F. As the cell moves from
G1 to S phase, a CDK4/cyclinD1 complex forms and adds phosphate groups to pRB. As pRB becomes phosphorylated, E2F is
released and becomes transcriptionally active, allowing the cell
to pass through S phase. Phosphorylation of pRB is transitory;
as CDK/cyclin complexes are degraded and the cell moves
through the cell cycle to early G1, pRB phosphorylation declines,
allowing pRB to reassociate with E2F.

NOW SOLVE THIS

19–2 People with a genetic condition known as Li–Fraumeni
syndrome inherit one mutant copy of the *p53* gene. These
people have a high risk of developing a number of different cancers, such as breast cancer, leukemia, bone cancer,
adrenocortical tumors, and brain tumors. Explain how mutations in one cancer-related gene can give rise to such a
diverse range of tumors.

■ HINT: *This problem involves an understanding of how tumor-
suppressor genes regulate cell growth and behavior. The key to its
solution is to consider which cellular functions are regulated by the
p53 protein and how the absence of p53 could affect each of these
functions. Also, read about loss of heterozygosity in Section 19.6.*

19.5
Cancer Cells Metastasize and Invade Other Tissues

As discussed at the beginning of this chapter, uncontrolled growth alone is insufficient to create a malignant and life-threatening cancer. Cancer cells must also become malignant, acquiring the ability to disengage from the original tumor site, to enter the blood or lymphatic system, to invade surrounding tissues, and to develop into secondary tumors. In order to leave the site of the primary tumor and invade other tissues, tumor cells must dissociate from the primary tumor and secrete proteases that digest components of the **extracellular matrix** and **basal lamina**, which normally surround and separate the body's tissues. The extracellular matrix and basal lamina are composed of proteins and carbohydrates. They surround and separate body tissues, form the scaffold for tissue growth, and inhibit the migration of cells. The ability to invade the extracellular matrix is also a property of some normal cell types. For example, implantation of the embryo in the uterine wall during pregnancy requires cell migration across the extracellular matrix. In addition, white blood cells reach sites of infection by penetrating capillary walls. The mechanisms of invasion are probably similar in these normal cells and in cancer cells. The difference is that, in normal cells, the invasive ability is tightly regulated, whereas in tumor cells, this regulation has been lost.

Once cancer cells have disengaged from the primary tumor and traversed tissue barriers, they enter the blood or lymphatic system and may become lodged in microvessels of other tissues. At this point the cells may undergo a second round of invasion to enter the new tissue and grow into new (metastatic) tumors. Only a small percentage of circulating cancer cells—about 0.01 percent—survive to establish metastatic tumors. Other important features of metastatic cells are increased cell motility, the capacity to stimulate new blood vessel formation, and the ability to escape detection by the host's immune system.

Metastasis is controlled by a large number of genes, including those that encode cell-adhesion molecules, cytoskeleton regulators, and proteolytic enzymes. For example, epithelial tumors have a lower than normal level of the **E-cadherin glycoprotein**, which is responsible for cell–cell adhesion in normal tissues. Also, proteolytic enzymes such as **metalloproteinases** are present at higher than normal levels in many highly malignant tumors. For example, breast cancer cells that metastasize to bone abnormally express the metalloproteinase gene *MMP1*. Those that spread to the lungs overexpress the *MMP1* and *MMP2* genes. It has been shown that the level of aggressiveness of a tumor correlates positively with the levels of proteolytic enzymes expressed by

the tumor. In addition, malignant cells are not susceptible to the normal controls conferred by regulatory molecules such as **tissue inhibitors of metalloproteinases (TIMPs)**.

Like the tumor-suppressor genes that are mutated in primary cancers, **metastasis-suppressor genes** are mutated or disrupted in metastatic tumors. Less than a dozen of these metastasis-suppressor genes have been identified so far, but all appear to affect the growth and development of metastatic tumors and not the primary tumor. One example is the CD82 protein, encoded by the *CD82* gene. This protein normally inhibits functions related to metastasis such as invasiveness and cell motility. It does this by directly interacting with proteins involved in these functions. In metastatic tumors, the expression of *CD82* is reduced or lost. The expression of *CD82* and other metastasis-suppressor genes is reduced by epigenetic mechanisms rather than by mutation. This observation provides hope that researchers can develop antimetastasis therapies that target the epigenetic silencing of metastasis-suppressor genes.

19.6
Predisposition to Some Cancers Can Be Inherited

Although the vast majority of human cancers are sporadic, a small fraction (1 to 2 percent) have a hereditary or familial component. At present, about 50 forms of hereditary cancer are known (Table 19.2).

TABLE 19.2

Some Inherited Predispositions to Cancer

Tumor Predisposition Syndromes	Chromosome	Gene Affected
Early-onset familial breast cancer	17q	BRCA1
Familial adenomatous polyposis	5q	APC
Familial melanoma	9p	CDKN2
Gorlin syndrome	9q	PTCH1
Hereditary nonpolyposis colon cancer	2p	MSH2, 6
Li-Fraumeni syndrome	17p	p53
Multiple endocrine neoplasia, type 1	11q	MEN1
Multiple endocrine neoplasia, type 2	10q	RET
Neurofibromatosis, type 1	17q	NF1
Neurofibromatosis, type 2	22q	NF2
Retinoblastoma	13q	pRb
Von Hippel–Lindau syndrome	3p	VHL
Wilms tumor	11p	WT1

Most inherited cancer-susceptibility alleles, though transmitted in a Mendelian dominant fashion, are not sufficient in themselves to trigger development of a cancer. At least one other somatic mutation in the other copy of the gene must occur in order to drive a cell toward tumorigenesis. In addition, mutations in still other genes are usually necessary to fully express the cancer phenotype. As mentioned earlier, inherited mutations in the *RB1* gene predispose individuals to developing various cancers. Although the normal somatic cells of these patients are heterozygous for the *RB1* mutation, cells within their tumors contain mutations in both copies of the gene. The phenomenon whereby the second, wild-type, allele is mutated in a tumor is known as **loss of heterozygosity**. Although loss of heterozygosity is an essential first step in expression of these inherited cancers, further mutations in other proto-oncogenes and tumor-suppressor genes are necessary for the tumor cells to become fully malignant.

The development of hereditary colon cancer illustrates how inherited mutations in one allele of a gene contribute only one step in the multistep pathway leading to malignancy. About 1 percent of colon cancer cases result from a genetic predisposition to cancer known as **familial adenomatous polyposis (FAP)**. In FAP, individuals inherit one mutant copy of the *APC* **(adenomatous polyposis) gene** located on the long arm of chromosome 5. Mutations include deletions, frameshift, and point mutations. The normal function of the *APC* gene product is to act as a tumor suppressor controlling cell–cell contact and growth inhibition by interacting with the β-catenin protein. The presence of a heterozygous *APC* mutation causes the epithelial cells of the colon to partially escape cell-cycle control, and the cells divide to form small clusters of cells called **polyps** or adenomas. People who are heterozygous for this condition develop hundreds to thousands of colon and rectal polyps early in life. Although it is not necessary for the second allele of the *APC* gene to be mutated in polyps at this stage, in the majority of cases, the second *APC* allele becomes mutant in a later stage of cancer development. The relative order of mutations in the development of colon cancer is shown in Figure 19–13.

The second mutation in polyp cells that contain an *APC* gene mutation occurs in the *ras* proto-oncogene. The combined *APC* and *ras* gene mutations bring about the development of intermediate adenomas. Cells within these adenomas have defects in normal cell differentiation. In addition, these cells will grow in culture and are not growth-inhibited by contact with other cells—a process known as **transformation**. The third step toward malignancy requires loss of function of both alleles of the *DCC* (*deleted in colon cancer*) gene. The *DCC* gene product is thought to be involved with cell adhesion and differentiation. Mutations in both *DCC* alleles result in the formation of late-stage adenomas with a number of finger-like outgrowths (villi). When late adenomas progress to cancerous adenomas, they usually suffer loss of functional *p53* genes. The final steps toward malignancy involve mutations in an unknown number of genes associated with metastasis.

19–3 Although tobacco smoking is responsible for a large number of human cancers, not all smokers develop cancer. Similarly, some people who inherit mutations in the tumor-suppressor genes *p53* or *RB1* never develop cancer. Explain these observations.

■ HINT: *This problem asks you to consider the reasons why only some people develop cancer as a result of environmental factors or mutations in tumor-suppressor genes. The key to its solution is to consider the steps involved in the development of cancer and the number of abnormal functions in cancer cells. Also, consider how genetics may affect DNA repair functions.*

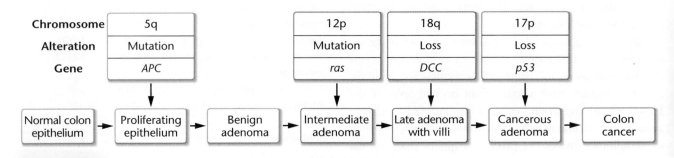

FIGURE 19–13 A model for the multistep development of colon cancer. The first step is the loss or inactivation of one allele of the *APC* gene on chromosome 5. In FAP cases, one mutant *APC* allele is inherited. Subsequent mutations involving genes on chromosomes 12, 17, and 18 in cells of benign adenomas can lead to a malignant transformation that results in colon cancer. Although the mutations on chromosomes 12, 17, and 18 usually occur at a later stage than those involving chromosome 5, the sum of changes is more important than the order in which they occur.

Viruses Contribute to Cancer in Both Humans and Animals

Viruses that cause cancer in animals have played a significant role in the search for knowledge about the genetics of human cancer. Most cancer-causing animal viruses are RNA viruses known as **retroviruses**. In humans, most of the known cancer viruses are DNA viruses.

To understand how retroviruses cause cancer in animals, it is necessary to know how these viruses replicate in cells. When a retrovirus infects a cell, its RNA genome is copied into DNA by the **reverse transcriptase** enzyme, which is brought into the cell with the infecting virus. The DNA copy then enters the nucleus of the infected cell, where it integrates at random into the host cell's genome. The integrated DNA copy of the retroviral RNA is called a **provirus**. The proviral DNA contains powerful enhancer and promoter elements in its U5 and U3 sequences at the ends of the provirus (Figure 19–14). The proviral promoter uses the host cell's transcription proteins, directing transcription of the viral genes (*gag, pol,* and *env*). The products of these genes are the proteins and RNA genomes that make up the new retroviral particles. Because the provirus is integrated into the host genome, it is replicated along with the host's DNA during the cell's normal cell cycle. A retrovirus may not kill a cell, but it may continue to use the cell as a factory to replicate more viruses that will then infect surrounding cells.

A retrovirus may cause cancer in three different ways. First, the proviral DNA may integrate by chance near one of the cell's normal proto-oncogenes. The strong promoters and enhancers in the provirus then stimulate high levels or inappropriate timing of transcription of the proto-oncogene, leading to stimulation of host-cell proliferation. Second, a retrovirus may pick up a copy of a host proto-oncogene and integrate it into its genome (Figure 19–14). The cellular proto-oncogene may be mutated during the process of transfer into the virus, or it may be expressed at abnormal levels because it is now under the control of viral promoters. Retroviruses that carry these cell-derived oncogenes can infect and transform normal cells into tumor cells, and are known as **acute transforming retroviruses**. Through the study of many acute transforming viruses of animals, scientists have identified dozens of proto-oncogenes. Third, a retrovirus may contain a normal viral gene whose product can either stimulate the cell cycle or act as a gene-expression regulator for both cellular and viral genes. As a result, expression of such a viral gene may lead to inappropriate cell growth or to abnormal expression of cancer-related cellular genes.

So far, no acute transforming retroviruses have been identified in humans; however, several human retroviruses, such as **human immunodeficiency virus (HIV)** and the **human T-cell leukemia virus (HTLV-1)** are associated with human cancers. These retroviruses are thought to stimulate cancer development through the third mechanism, described in the previous paragraph.

DNA viruses also contribute to the development of human cancers in a variety of ways. Because viruses are comprised solely of a nucleic acid genome surrounded by a protein coat, they must utilize the host cell's biosynthetic machinery in order to reproduce themselves. To access the host's DNA-synthesizing enzymes, viruses require the host cell to be in an actively growing state. Thus, many DNA viruses contain genes encoding products that stimulate the cell cycle. These products often interact with tumor-suppressor proteins, inactivating them. If the host cell survives the infection, it may lose control of the cell cycle and begin its journey to carcinogenesis.

It is thought that, worldwide, about 15 percent of human cancers are associated with viruses, making virus infection the second greatest risk factor for cancer, next to tobacco smoking. The most significant contributors to virus-induced cancers are the **papillomaviruses (HPV 16 and 18)**, **human T-cell leukemia virus (HTLV-1)**, **hepatitis B virus, human herpesvirus 8**, and **Epstein-Barr virus**. Like other risk factors for cancer, including hereditary predisposition

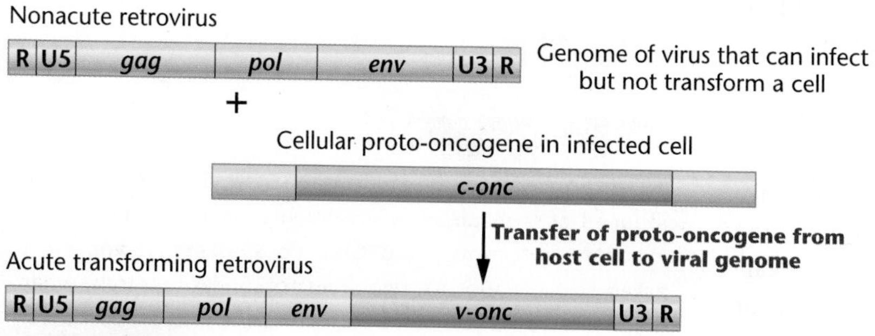

FIGURE 19–14 The genome of a typical retrovirus is shown at the top of the diagram. The genome contains repeats at the termini (R), the U5 and U3 regions that contain promoter and enhancer elements, and the three major genes that encode viral structural proteins (*gag* and *env*) and the viral reverse transcriptase (*pol*). RNA transcripts of the entire viral genome comprise the new viral genomes. If the retrovirus acquires all or part of a host-cell proto-oncogene (*c-onc*), this gene (now known as a *v-onc*) is expressed along with the viral genes, leading to overexpression or inappropriate expression of the *v-onc* gene. The *v-onc* gene may also acquire mutations that enhance its transforming ability.

to certain cancers, virus infection alone is not sufficient to trigger human cancers. Other factors, including DNA damage or the accumulation of mutations in one or more of a cell's oncogenes and tumor-suppressor genes, are required to move a cell down the multistep pathway to cancer.

19.8

Environmental Agents Contribute to Human Cancers

Any substance or event that damages DNA has the potential to be carcinogenic. Unrepaired or inaccurately repaired DNA introduces mutations, which, if they occur in proto-oncogenes or tumor-suppressor genes, can lead to abnormal regulation of the cell cycle or disruption of controls over apoptosis or metastasis.

Our environment, both natural and human-made, contains abundant carcinogens. These include chemicals, radiation, some viruses, and chronic infections. Perhaps the most significant carcinogen in our environment is tobacco smoke, which contains at least 60 chemicals that interact with DNA and cause mutations. Epidemiologists estimate that about 30 percent of human cancer deaths are associated with cigarette smoking. Smokers have a 20-fold increased risk of developing lung cancer, which kills more than one million people, worldwide, each year.

Diet is often implicated in the development of cancer. Consumption of red meat and animal fat is associated with some cancers, such as colon, prostate, and breast cancer. The mechanisms by which these substances may contribute to carcinogenesis may involve stimulation of cell division through hormones or creation of carcinogenic chemicals during cooking. Alcohol may cause inflammation of the liver and contribute to liver cancer.

Although most people perceive the human-made, industrial environment to be a highly significant contributor to cancer, it may account for only a small percentage of total cancers, and only in special situations. Some of the most mutagenic agents, and hence potentially the most carcinogenic, are natural substances and natural processes. For example, **aflatoxin**, a component of a mold that grows on peanuts and corn, is one of the most carcinogenic chemicals known. Most chemical carcinogens, such as **nitrosamines**, are components of synthetic substances and are found in some preserved meats; however, many are naturally occurring. For example, natural pesticides and antibiotics found in plants may be carcinogenic, and the human body itself creates alkylating agents in the acidic environment of the gut. Nevertheless, these observations do not diminish the serious cancer risks to specific populations who are exposed to human-made carcinogens such as synthetic pesticides or asbestos.

DNA lesions brought about by natural radiation (X rays, ultraviolet light), natural dietary substances, and substances in the external environment contribute the majority of environmentally caused mutations that lead to cancer. In addition, normal metabolism creates oxidative end products that can damage DNA, proteins, and lipids. It is estimated that the human body suffers about 10,000 damaging DNA lesions per day due to the actions of oxygen free radicals. DNA repair enzymes deal successfully with most of this damage; however, some damage may lead to permanent mutations. The process of DNA replication itself is mutagenic. Hence, substances such as growth factors or hormones that stimulate cell division are ultimately mutagenic and perhaps carcinogenic. Chronic inflammation due to infection also stimulates tissue repair and cell division, resulting in DNA lesions accumulating during replication. These mutations may persist, particularly if cell-cycle checkpoints are compromised due to mutations or inactivation of tumor-suppressor genes such as *p53* or *RB1*.

Both ultraviolet (UV) light and ionizing radiation (such as X rays and gamma rays) induce DNA damage. UV in sunlight is well accepted as an inducer of skin cancers. Ionizing radiation has clearly shown itself to be a carcinogen in studies of populations exposed to neutron and gamma radiation from atomic blasts such as those in Hiroshima and Nagasaki. Another significant environmental component, radon gas, may be responsible for up to 50 percent of the ionizing radiation exposure of the U.S. population and could contribute to lung cancers in some populations.

NOW SOLVE THIS

19-4 Cancer can arise spontaneously, but can also be induced as a result of environmental factors such as sun exposure, infections, and tobacco smoking. If you were asked to help allocate resources to cancer research, what emphasis would you place on research to find cancer cures, compared to that placed on education about cancer prevention?

■ HINT: *This problem asks you to consider the outcomes of two different approaches to cancer research. The key to its solution is to think about the relative rates of environmentally induced and spontaneous cancers. (An interesting source of information on this topic is Ames, B. N. et al. 1995. The causes and prevention of cancer. Proc. Natl. Acad. Sci. USA 92: 5258–5265.)*

The Cancer Genome Anatomy Project (CGAP)

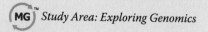 *Study Area: Exploring Genomics*

A research group headed by Dr. Victor Velculescu of Johns Hopkins University reported that breast and colon cancers contain about 11 gene mutations that may contribute to the cancer phenotype. The research group analyzed 13,023 of the 21,000 known genes in the human genome, comparing the DNA sequences from normal cells and cancer cells. Most of the mutations that were specific to cancer cells were not previously known to be associated with cancer.

Dr. Velculescu's study was one of the first in *The Cancer Genome Atlas (TCGA)* project, a $1.5 billion federal project designed to systematically scan the human genome to find genes that are mutated in many different cancers. In this exercise, we will explore aspects of Dr. Velculescu's research by mining information available in the online database, **The Cancer Genome Anatomy Project (CGAP)**. The purpose of the CGAP is to understand the expression profiles of genes from normal, precancer, and cancer cells. Data within CGAP is made available to all cancer researchers via its Web site, at http://cgap.nci.nih.gov/cgap.html.

■ **Exercise – Colon Cancer and the TBX22 Gene**

One gene that Dr. Velculescu's research group discovered to be mutated in colon cancers—*TBX22*—was not previously suspected to contribute to this cancer. What is *TBX22*, and how do you think a mutated *TBX22* gene would contribute to the development of colon cancer?

1. To begin your search for the answers, go to CGAP at http://cgap.nci.nih.gov/cgap.html.

2. Click the "Genes" button near the top of the page.

3. From the list of "Gene Tools" in the left-hand margin, select "Gene Finder."

4. Select "Homo sapiens" in the "Select organism" box, and type TBX22 in the "Enter a unique identifier" box. Submit the query.

5. Select "Gene Info" in the right-hand column of the table.

6. Explore the many sources of information about *TBX22* from various database links listed on this page.

Prepare a brief written or verbal report on what you learned during your explorations and which sources you used to reach your conclusions about *TBX22*.

CASE STUDY | I thought it was safe

A middle-aged woman taking the breast cancer drug Tamoxifen for ten years became concerned when she saw a news report with disturbing information. In some women, the drug made their cancer more aggressive and more likely to spread. Other women with breast cancer, the report stated, do not respond to Tamoxifen at all, and 30 to 40 percent of women who take the drug eventually become resistant to chemotherapy. The woman contacted her oncologist to ask some questions:

1. How can some people react one way to a cancer treatment and others react a different way?
2. Why do most cancers eventually become resistant to a specific chemotherapeutic drug?
3. Why does it seem that some drugs are thought to be safe one day and declared unsafe the next day?

Summary Points

 For activities, animations, and review quizzes, go to the study area at www.masteringgenetics.com

1. Cancer cells show two fundamental properties: abnormal cell proliferation and a propensity to spread and invade other parts of the body.
2. Cancers are clonal, meaning that all cells within a tumor originate from a single cell that contained a number of mutations.
3. The development of cancer is a multistep process, requiring mutations in several cancer-related genes.

4. Cancer cells show high rates of mutation, chromosomal abnormalities, genomic instability, and abnormal patterns of chromatin modifications.
5. Cancer cells have defects in cell-cycle progression, checkpoint controls, and programmed cell death.
6. Proto-oncogenes are normal genes that promote cell growth and division. When proto-oncogenes are mutated or misexpressed in cancer cells, they are known as oncogenes.

7. Tumor-suppressor genes normally regulate cell-cycle checkpoints and apoptosis. When tumor-suppressor genes are mutated or inactivated, cells cannot correct DNA damage. This leads to accumulations of mutations that may cause cancer.

8. The ability of cancer cells to metastasize requires defects in gene products that control a number of functions such as cell adhesion, proteolysis, and tissue invasion.

9. Inherited mutations in cancer-susceptibility genes are not sufficient to trigger cancer. Other somatic mutations in proto-oncogenes or tumor-suppressor genes are necessary for the development of hereditary cancers.

10. Tumor viruses contribute to cancers by introducing viral oncogenes, interfering with tumor-suppressor proteins, or altering expression of a cell's proto-oncogenes.

11. Environmental agents such as chemicals, radiation, viruses, and chronic infections contribute to the development of cancer. The most significant environmental factors that affect human cancers are tobacco smoke, diet, and natural radiation.

INSIGHTS AND SOLUTIONS

1. In disorders such as retinoblastoma, a mutation in one allele of the *RB1* gene can be inherited from the germ line, causing an autosomal dominant predisposition to the development of eye tumors. To develop tumors, a somatic mutation in the second copy of the *RB1* gene is necessary, indicating that the mutation itself acts as a recessive trait. Given that the first mutation can be inherited, in what ways can a second mutational event occur?

Solution: In considering how this second mutation arises, we must look at several types of mutational events, including changes in nucleotide sequence and events that involve whole chromosomes or chromosome parts. Retinoblastoma results when both copies of the *RB1* locus are lost or inactivated. With this in mind, you must first list the phenomena that can result in a mutational loss or the inactivation of a gene.

One way the second *RB1* mutation can occur is by a nucleotide alteration that converts the remaining normal *RB1* allele to a mutant form. This alteration can occur through a nucleotide substitution or through a frameshift mutation caused by the insertion or deletion of nucleotides during replication. A second mechanism involves the loss of the chromosome carrying the normal allele. This event would take place during mitosis, resulting in chromosome 13 monosomy and leaving the mutant copy of the gene as the only *RB1* allele. This mechanism does not necessarily involve loss of the entire chromosome; deletion of the long arm (*RB1* is on 13q) or an interstitial deletion involving the *RB1* locus and some surrounding material would have the same result. Alternatively, a chromosome aberration involving loss of the normal copy of the *RB1* gene might be followed by duplication of the chromosome carrying the mutant allele. Two copies of chromosome 13 would be restored to the cell, but the normal *RB1* allele would not be present. Finally, a recombination event followed by chromosome segregation could produce a homozygous combination of mutant *RB1* alleles.

2. Proto-oncogenes can be converted to oncogenes in a number of different ways. In some cases, the proto-oncogene itself becomes amplified up to hundreds of times in a cancer cell. An example is the *cyclin D1* gene, which is amplified in some cancers. In other cases, the proto-oncogene may be mutated in a limited number of specific ways, leading to alterations in the gene product's structure. The *ras* gene is an example of a proto-oncogene that becomes oncogenic after suffering point mutations in specific regions of the gene. Explain why these two proto-oncogenes (*cyclin D1* and *ras*) undergo such different alterations in order to convert them into oncogenes.

Solution: The first step in solving this question is to understand the normal functions of these proto-oncogenes and to think about how either amplification or mutation would affect each of these functions.

The cyclin D1 protein regulates progression of the cell cycle from G1 into S phase, by binding to CDK4 and activating this kinase. The cyclin D1/CDK4 complex phosphorylates a number of proteins including pRB, which in turn activate other proteins in a cascade that results in transcription of genes whose products are necessary for DNA replication in S phase. The simplest way to increase the activity of cyclin D1 would be to increase the number of cyclin D1 molecules available for binding to the cell's endogenous CDK4 molecules. This can be accomplished by several mechanisms, including amplification of the *cyclin D1* gene. In contrast, a point mutation in the *cyclin D1* gene would most likely interfere with the ability of the cyclin D1 protein to bind to CDK4; hence, mutations within the gene would probably repress cell-cycle progression rather than stimulate it.

The *ras* gene product is a signal transduction protein that operates as an on/off switch in response to external stimulation by growth factors. It does so by binding either GTP (the "on" state) or GDP (the "off" state). Oncogenic mutations in the *ras* gene occur in specific regions that alter the ability of the Ras protein to exchange GDP for GTP. Oncogenic Ras proteins are locked in the "on" conformation, bound to GTP. In this way, they constantly stimulate the cell to divide. An amplification of the *ras* gene would simply provide more molecules of normal Ras protein, which would still be capable of on/off regulation. Hence, simple amplification of *ras* would less likely be oncogenic.

Problems and Discussion Questions

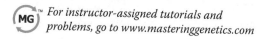

For instructor-assigned tutorials and problems, go to www.masteringgenetics.com

HOW DO WE KNOW?

1. In this chapter, we focused on cancer as a genetic disease, with an emphasis on the relationship between cancer, the cell cycle, and DNA damage, as well as on the multiple steps that lead to cancer. At the same time, we found many opportunities to consider the methods and reasoning by which much of this information was acquired. From the explanations given in the chapter,

   (a) How do we know that malignant tumors arise from a single cell that contains mutations?

   (b) How do we know that cancer development requires more than one mutation?

   (c) How do we know that cancer cells contain defects in DNA repair?

2. What events occur in each phase of the cell cycle? Which phase is most variable in length?

3. Where are the major regulatory points in the cell cycle?

4. List the functions of kinases and cyclins, and describe how they interact to cause cells to move through the cell cycle.

5. (a) How does pRB function to keep cells at the G1 checkpoint?

   (b) How do cells get past the G1 checkpoint to move into S phase?

6. What is the difference between saying that cancer is inherited and saying that the predisposition to cancer is inherited?

7. As a genetic counselor, you are asked to assess the risk for a couple with a family history of retinoblastoma who are thinking about having children. Both the husband and wife are phenotypically normal, but the husband has a sister with familial retinoblastoma in both eyes. What is the probability that this couple will have a child with retinoblastoma? Are there any tests that you could recommend to help in this assessment?

8. What is apoptosis, and under what circumstances do cells undergo this process?

9. Define tumor-suppressor genes. Why is a mutation in a single copy of a tumor-suppressor gene expected to behave as a recessive gene?

10. A genetic variant of the retinoblastoma protein, called PSM-RB (phosphorylation site mutated RB), is not able to be phosphorylated by the action of CDK4/cyclinD1 complex. Explain why PSM-RB is said to have a constitutive growth-suppressing action on the cell cycle.

11. Part of the Ras protein is associated with the plasma membrane, and part extends into the cytoplasm. How does the Ras protein transmit a signal from outside the cell into the cytoplasm? What happens in cases where the *ras* gene is mutated?

12. If a cell suffers damage to its DNA while in S phase, how can this damage be repaired before the cell enters mitosis?

13. Distinguish between oncogenes and proto-oncogenes. In what ways can proto-oncogenes be converted to oncogenes?

14. Of the two classes of genes associated with cancer, tumor-suppressor genes and oncogenes, mutations in which group can be considered gain-of-function mutations? In which group are the loss-of-function mutations? Explain.

15. How do translocations such as the Philadelphia chromosome contribute to cancer?

16. Explain why many oncogenic viruses contain genes whose products interact with tumor-suppressor proteins.

17. DNA sequencing has provided data to indicate that cancer cells may contain tens of thousands of somatic mutations, only some of which confer a growth advantage to a cancer cell. How do scientists describe and categorize these recently discovered populations of mutations in cancer cells?

18. How do normal cells protect themselves from accumulating mutations in genes that could lead to cancer? How do cancer cells differ from normal cells in these processes?

19. Describe the difference between an acute transforming virus and a virus that does not cause tumors.

20. Epigenetics is a relatively new area of genetics with a focus on phenomena that affect gene expression but do not affect DNA sequence. Epigenetic effects are quasi-stable and may be passed to progeny somatic or germ-line cells. What are known causes of epigenetic effects, and how do they relate to cancer?

21. Radiotherapy (treatment with ionizing radiation) is one of the most effective current cancer treatments. It works by damaging DNA and other cellular components. In which ways could radiotherapy control or cure cancer, and why does radiotherapy often have significant side effects?

22. Genetic tests that detect mutations in the *BRCA1* and *BRCA2* oncogenes are widely available. These tests reveal a number of mutations in these genes—mutations that have been linked to familial breast cancer. Assume that a young woman in a suspected breast cancer family takes the *BRCA1* and *BRCA2* genetic tests and receives negative results. That is, she does not test positive for the mutant alleles of *BRCA1* or *BRCA2*. Can she consider herself free of risk for breast cancer?

23. Explain the apparent paradox that both hypermethylation and hypomethylation of DNA are often found in the same cancer cell.

24. As part of a cancer research project, you have discovered a gene that is mutated in many metastatic tumors. After determining the DNA sequence of this gene, you compare the sequence with those of other genes in the human genome sequence database. Your gene appears to code for an amino acid sequence that resembles sequences found in some serine proteases. Conjecture how your new gene might contribute to the development of highly invasive cancers.

Extra-Spicy Problems

For instructor-assigned tutorials and problems, go to www.masteringgenetics.com

25. A study by Bose and colleagues (1998. *Blood* 92: 3362–3367) and a previous study by Biernaux and others (1996. *Bone Marrow Transplant* 17: (Suppl. 3) S45–S47) showed that *BCR-ABL* fusion gene transcripts can be detected in 25 to 30 percent of healthy adults who do not develop chronic myelogenous leukemia (CML). Explain how these individuals can carry a fusion gene that is transcriptionally active and yet do not develop CML.

26. Those who inherit a mutant allele of the *RB1* gene are at risk for developing a bone cancer called osteosarcoma. You suspect that in these cases, osteosarcoma requires a mutation in the second *RB1* allele, and you have cultured some osteosarcoma cells and obtained a cDNA clone of a normal human *RB1* gene. A colleague sends you a research paper revealing that a strain of cancer-prone mice develop malignant tumors when injected with osteosarcoma cells, and you obtain these mice. Using these three resources, what experiments would you perform to determine (a) whether osteosarcoma cells carry two *RB1* mutations, (b) whether osteosarcoma cells produce any pRB protein, and (c) if the addition of a normal *RB1* gene will change the cancer-causing potential of osteosarcoma cells?

27. The table in this problem summarizes some of the data that have been collected on *BRCA1* mutations in families with a high incidence of both early-onset breast cancer and ovarian cancer.
    (a) Note the coding effect of the mutation found in kindred group 2082. This results from a single base-pair substitution. Draw the normal double-stranded DNA sequence for this codon (with the 5' and 3' ends labeled), and show the sequence of events that generated this mutation, assuming that it resulted from an uncorrected mismatch event during DNA replication.
    (b) Examine the types of mutations that are listed in the table and determine if the *BRCA1* gene is likely to be a tumor-suppressor gene or an oncogene.
    (c) Although the mutations listed in the table are clearly deleterious and cause breast cancer in women at very young ages, each of the kindred groups had at least one woman who carried the mutation but lived until age 80 without developing cancer. Name at least two different mechanisms (or variables) that could underlie variation in the expression of a mutant phenotype and propose an explanation for the incomplete penetrance of this mutation. How do these mechanisms or variables relate to this explanation?

Predisposing Mutations in *BRCA1*

Kindred	Codon	Nucleotide Change	Coding Effect	Frequency in Control Chromosomes
1	24	−11 bp	Frameshift or splice	0/180
2	1313	C → T	Gln → Stop	0/170
3	1756	Extra C	Frameshift	0/162
4	1775	T → G	Met → Arg	0/120
5	NA*	?	Loss of transcript	NA*

Source: 1994. Science 266: 66–71. © AAAS.

*NA indicates not applicable, as the regulatory mutation is inferred, and the position has not been identified.

28. The following table shows neutral polymorphisms found in control families (those with no increased frequency of breast and ovarian cancer).

    Examine the data in the table and answer the following questions:
    (a) What is meant by a neutral polymorphism?
    (b) What is the significance of this table in the context of examining a family or population for *BRCA1* mutations that predispose an individual to cancer?
    (c) Is the PM2 polymorphism likely to result in a neutral missense mutation or a silent mutation?
    (d) Answer part (c) for the PM3 polymorphism.

Neutral Polymorphisms in *BRCA1*

Name	Codon Location	Base in Codon[†]	Frequency in Control Chromosomes*			
			A	C	G	T
PM1	317	2	152	0	10	0
PM6	878	2	0	55	0	100
PM7	1190	2	109	0	53	0
PM2	1443	3	0	115	0	58
PM3	1619	1	116	0	52	0

*The number of chromosomes with a particular base at the indicated polymorphic site (A, C, G, or T) is shown.

[†]Position 1, 2, or 3 of the codon.

29. Prostate cancer is a major cause of cancer-related deaths among men. Epigenetic changes that regulate gene expression are involved in both the initiation and progression of such cancers. Following is a table that lists the number of genes known to be hypermethylated in prostate cancer cells (modified from Long-Cheng, L. et al., 2005. *J. Natl. Cancer Inst.* 97: 103–115). For each category of genes, speculate on the mechanism(s) by which cancer initiation or progression might be influenced by hypermethylation.

DNA Hypermethylation of	Number of Known Genes
Hormonal response genes	5
Cell-cycle control genes	2
Tumor cell invasion genes	8
DNA damage repair genes	2
Signal transduction genes	4

SPECIAL TOPICS IN MODERN GENETICS

DNA Forensics

Genetics is arguably the most influential science today—dramatically affecting technologies in fields as diverse as agriculture, archaeology, medical diagnosis, and disease treatment.

One of the areas that has been the most profoundly altered by modern genetics is forensic science. **Forensic science** (or *forensics*) uses technological and scientific approaches to answer questions about the facts of criminal or civil cases. Prior to 1986, forensic scientists had a limited array of tools with which to link evidence to specific individuals or suspects. These included some reliable methods such as blood typing and fingerprint analysis, but also many unreliable methods such as bite mark comparisons and hair microscopy.

Since the first forensic use of **DNA profiling** in 1986 (Box 1), DNA forensics (also called **forensic DNA fingerprinting** or **DNA typing**) has become an important method for police to identify sources of biological materials. DNA profiles can now be obtained from saliva left on cigarette butts or postage stamps, pet hairs found at crime scenes, or bloodspots the size of pinheads. Even biological samples that are degraded by fire or time are yielding DNA profiles that help the legal system determine identity, innocence, or guilt. Investigators now scan large databases of stored DNA profiles in order to match profiles generated from crime scene evidence. DNA profiling has proven the innocence of hundreds of people who were convicted of serious crimes and even sentenced to death. Forensic scientists have used DNA profiling to identify victims of mass disasters such as the Asian Tsunami of 2004 and the September 11, 2001 terrorist attacks in New York. They have also used forensic DNA analysis to identify endangered species and animals trafficked in the illegal wildlife trade. The power of DNA forensic analysis has captured the public imagination, and DNA forensics is featured in several popular television series.

Even biological samples degraded by fire or time are yielding DNA profiles that help determine identity, innocence, or guilt.

The applications of DNA profiling extend beyond forensic investigations. These include paternity and family relationship testing, identification of plant materials, verification of military casualties, and evolutionary studies.

BOX 1
The Pitchfork Case: The First Criminal Conviction Using DNA Profiling

In the mid-1980s, the bodies of two schoolgirls, Lynda Mann and Dawn Ashworth, were found in Leicestershire, England. Both girls had been raped, strangled, and their bodies left in the bushes. In the absence of useful clues, the police questioned a local mentally retarded porter named Richard Buckland who had a history of previous sexual offenses. During interrogation, Buckland confessed to the murder of Dawn Ashworth; however, police did not know whether he was also responsible for Lynda Mann's death. In 1986, in order to identify the second killer, the

police asked Dr. Alec Jeffreys of the University of Leicester to try a new method of DNA analysis called DNA fingerprinting. Dr. Jeffreys had developed a method of analyzing DNA regions called *variable number of tandem repeats* (VNTRs), which vary in length between members of a population. Dr. Jeffreys's VNTR analysis revealed a match between the DNA profiles from semen samples obtained from both crime scenes, suggesting that the same person was responsible for both rapes. However, neither of the DNA profiles matched those from a blood sample taken from Richard Buckland. Having eliminated their only suspect, the police embarked on the first mass DNA dragnet in history, by requesting blood samples from every adult male in the region. Although 4000

men offered samples, one did not. Colin Pitchfork, a bakery worker, paid a friend to give a blood sample in his place, using forged identity documents. Their plan was detected when their conversation was overheard at a local pub. The conversation was reported to police, who then arrested Pitchfork, obtained his blood sample, and sent it for analysis. His DNA profile matched the profiles from the semen samples left at both crime scenes. Pitchfork confessed to the murders, pleaded guilty, and was sentenced to life in prison. The Pitchfork Case was not only the first criminal case resolved by forensic DNA profiling, but also the first case in which DNA profiling led to the exoneration of an innocent person.

It is important for all of us to understand the basics of forensic DNA analysis. As informed citizens, we need to monitor its uses and potential abuses. Although DNA profiling is well validated as a technique and is considered the gold standard of forensic identification, it is not without controversy and the need for legislative oversight.

In this Special Topics chapter, we will explore how DNA profiling works and how the results of profiles are interpreted. We will learn about DNA databases, the potential problems associated with DNA profiling, and the future of this powerful technology.

DNA Profiling Methods

VNTR-Based DNA Fingerprinting

The era of DNA-based human identification began in 1984, with Dr. Alec Jeffreys's publication on DNA loci known as **minisatellites**, or **variable number of tandem repeats** (**VNTRs**). As described in Chapter 12, VNTRs are located in noncoding regions of the genome and are made up of DNA sequences of between 15 and 100 bp long, with each unit repeated a number of times. The number of repeats found at each VNTR locus varies from person to person, and hence VNTRs can be from 1 to 20 kilobases (kb) in length, depending on the person. For example, the VNTR

5′- GACTGCCTGCTAAGAT**GACTGCCTGCTAAGAT**GA
CTGCCTGCTAAGAT-3′

is comprised of three tandem repeats of a 16-nucleotide sequence (highlighted in bold).

VNTRs are useful for DNA profiling because there are as many as 30 different possible alleles (repeat lengths) at any VNTR in a population. This creates a large number of possible genotypes. For example, if one examined four different VNTR loci within a population, and each locus had 20 possible alleles, there would be more than 2 billion (4^{20}) possible genotypes in this four-locus profile.

To create a VNTR profile (also known as a DNA fingerprint), scientists extract DNA from a tissue sample and digest it with a restriction enzyme that cleaves on either side of the VNTR repeat region (ST Figure 1–1). The digested DNA is separated by gel electrophoresis and subjected to Southern blot analysis (which is described in detail in Chapter 20). Briefly, separated DNA is transferred from the gel to a membrane and hybridized with a radioactive probe that recognizes DNA sequences within the VNTR region. After exposing the membrane to X-ray film, the pattern of bands is measured, with larger VNTR repeat alleles remaining near the top of the gel and smaller VNTRs, which migrate more rapidly through the gel, being closer to the bottom. The pattern of bands is the same for a given individual, no matter what tissue is used as the source of the DNA. If enough VNTRs are analyzed, each person's DNA profile will be unique (except, of course, for identical twins) because of the huge number of possible VNTRs and alleles. In practice, scientists analyze about five or six loci to create a DNA profile.

A significant limitation of VNTR profiling is that it requires a relatively large sample of DNA (10,000 cells or

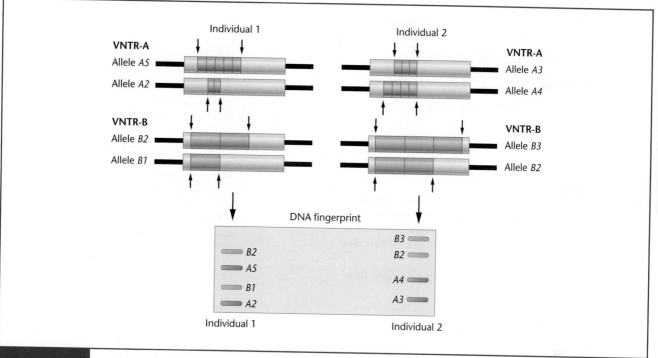

ST FIGURE 1–1   **DNA fingerprint at two VNTR loci for two individuals**. VNTR alleles at two loci (*A* and *B*) are shown for two different individuals. Arrows mark restriction-enzyme cutting sites that flank the VNTRs. Restriction-enzyme digestion produces a series of fragments that can be separated by gel electrophoresis and detected as bands on a Southern blot (bottom). The number of repeats at each locus is variable, so the overall pattern of bands is distinct for each individual. The DNA fingerprint profile shows that these individuals share one allele (B2).

about 50 μg of DNA)—more than is usually found at a typi-cal crime scene. In addition, the DNA must be relatively in-tact (nondegraded). As a result, VNTR profiling has been used most frequently when large tissue samples are avail-able—such as in paternity testing. Although VNTR profil-ing is still used in some cases, it has mostly been replaced by more sensitive methods, as described next.

Autosomal STR DNA Profiling

The development of the polymerase chain reaction (PCR) revolutionized DNA profiling. PCR methods are described in detail in Chapter 20. Using PCR-amplified DNA samples, scientists are able to generate DNA profiles from trace sam-ples (e.g., the bulb of single hairs or a few cells from a blood-stain) and from samples that are old or degraded (such as a bone found in a field or an ancient Egyptian mummy).

The majority of human forensic DNA profiling is now done using commercial kits that amplify and analyze regions of the genome known as **microsatellites**, or **short tandem repeats (STRs)**. STRs are similar to VNTRs, but the repeated motif is shorter—between two and nine base pairs, repeated from 7 to 40 times. For example, one locus known as D8S1179 is made up of the four base-pair sequence TCTA, repeated 7 to 20 times, depending on the allele. There are 19 possible alleles of the locus that are found within a population.

Although hundreds of STR loci are present in the human ge-nome, only a subset is used for DNA profiling. The FBI and other law enforcement agencies have selected 13 STR loci to be used as a core set for forensic analysis (ST Figure 1–2).

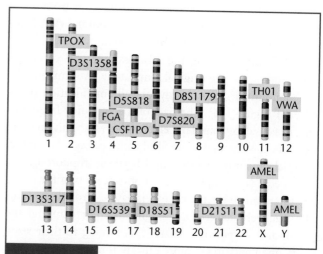

ST FIGURE 1–2   **Chromosomal positions of the 13 core STR loci used for forensic DNA profiling**. The *AMEL* (*Amelo-genin*) locus is included with the 13 core loci and is used to determine the gender of the person providing the DNA sample. The *AMEL* locus on the X chromosome contains a 6-nucleotide deletion compared to that on the Y chromosome.

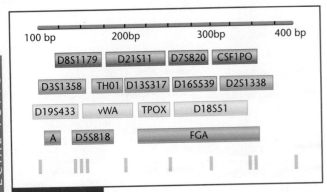

ST FIGURE 1–3   Relative size ranges and fluorescent dye labeling colors of 16 STR products generated by a commercially available DNA profiling kit. The DNA fragments shown in orange at the bottom of the diagram are DNA size markers. The *AMEL* locus is indicated as an *A*.

(From Applied Biosystems AmpFLSTR Identifiler kit, http://www.appliedbiosystems.com)

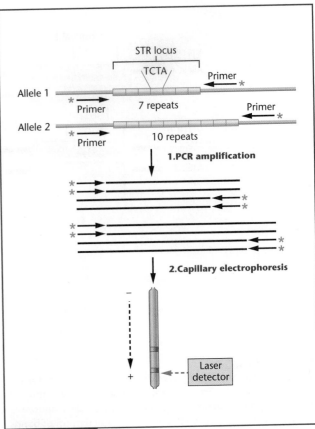

ST FIGURE 1–4   Steps in the PCR amplification and analysis of one STR locus (*D8S1179*). In this example, the person is heterozygous at the *D8S1179* locus: One allele has 7 repeats and one has 10 repeats. Primers are specific for sequences flanking the STR locus and are labeled with a blue fluorescent dye. The double-stranded DNA is denatured, the primers are annealed, and each allele is amplified by PCR in the presence of all four dNTPs and Taq DNA polymerase. After amplification, the labeled products are separated according to size by capillary electrophoresis, followed by fluorescence detection.

Several commercially available kits are currently used for forensic DNA analysis of STR loci. The methods vary slightly, but generally involve the following steps. As shown in ST Figure 1–3, each primer set is tagged by one of four fluorescent dyes—blue, green, yellow, or red. Each primer set is designed to amplify DNA fragments, the sizes of which vary depending on the number of repeats within the region amplified. For example, the primer sets that amplify the D19S433, vWA, TPOX, and D18S51 STR loci are all labeled with a yellow fluorescent tag. The sizes of the amplified DNA fragments produced allow scientists to differentiate between the yellow-labeled products. For example, the amplified products from the D19S433 locus range from about 100 to 150 bp in length, whereas those from the vWA locus range from about 150 to 200 bp, and so on.

After amplification, the DNA sample will contain a small amount of the original template DNA sample and a large amount of fluorescently labeled amplification products (ST Figure 1–4). The sizes of the amplified fragments are measured by **capillary electrophoresis**. This method uses thin glass tubes that are filled with a polyacrylamide gel material similar to that used in slab gel electrophoresis. The amplified DNA sample is loaded onto the top of the capillary tube, and an electric current is passed through the tube. The negatively charged DNA fragments migrate through the gel toward the positive electrode, according to their sizes. Short fragments move more quickly through the gel, and larger ones more slowly. At the bottom of the tube, a laser detects each fluorescent fragment as it migrates through the tube. The data are analyzed by software that calculates both the sizes of the fragments and their quantities, and these are represented as peaks on a graph (ST Figure 1–5).

Typically, automated capillary electrophoresis systems analyze as many as 16 samples at a time, and the analysis takes approximately 30 minutes.

After DNA profiling, the profile can be directly compared to a profile from another person, from crime scene evidence, or from other profiles stored in DNA profile databases (ST Figure 1–6). The STR profile genotype of an individual is expressed as the number of times the STR sequence is repeated. For example, in the profile shown in ST Figure 1–6, the person's profile would be expressed as shown in ST Table 1.1.

Scientists interpret STR profiles using statistics, probability, and population genetics, and these methods will be discussed in the section Interpreting DNA Profiles.

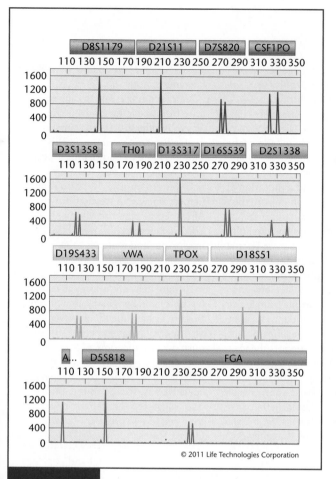

© 2011 Life Technologies Corporation

ST FIGURE 1–5 An electropherogram showing the results of a DNA profile analysis using the 16-locus STR profile kit shown in ST Figure 1–3. Heterozygous loci show up as double peaks and homozygous loci as single, higher peaks. The sizes of each allele can be calculated from the peak locations relative to the size axis shown at the top of each panel. The single peak for the *AMEL* (*A*) locus indicates that this DNA profile is that of a female individual, as described in the text.

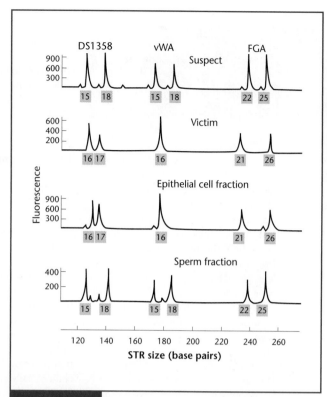

ST FIGURE 1–6 Electropherogram showing the STR profiles of four samples from a rape case. Three STR loci were examined from samples taken from a suspect, a victim, and two fractions from a vaginal swab taken from the victim. The *x*-axis shows the DNA size ladder, and the *y*-axis indicates relative fluorescence intensity. The number below each allele indicates the number of repeats in each allele, as measured against the DNA size ladder. Notice that the STR profile of the sperm sample taken from the victim matches that of the suspect.

Y-Chromosome STR Profiling

In many forensic applications, it is important to differentiate the DNA profiles of two or more people in a mixed sample. For example, vaginal swabs from rape cases usually contain a mixture of female cells and male sperm cells. In addition, some crime samples contain evidence material from a number of male suspects. In these types of cases, STR profiling of Y-chromosome DNA is useful. There are more than 200 STR loci on the Y chromosome that are useful for DNA profiling; however, fewer than 20 of these are used routinely for forensic analysis. PCR amplification of Y-chromosome STRs uses specific primers that do not amplify DNA on the X chromosome.

One limitation of Y-chromosome DNA profiling is that it cannot differentiate between the DNA from fathers and

sons, or from male siblings. This is because the Y chromosome is directly inherited from the father to his sons, as a single unit. The Y chromosome does not undergo recombination, meaning that less genetic variability exists on the Y chromosome than on autosomal chromosomes. Therefore, all patrilineal relatives share the same Y chromosome profile. Even two apparently unrelated males may share the same Y profile, if they also shared a distant male ancestor.

ST TABLE 1.1

STR Profile Genotypes from the Four Profiles Shown in ST Figure 1–6

STR Locus	Profile Genotype from			
	Suspect	Victim	Epithelial Cells	Sperm Fraction
DS1358	15, 18	16, 17	16, 17	15, 18
vWA	15, 18	16, 16	16, 16	15, 18
FGA	22, 25	21, 26	21, 26	22, 25

BOX 2
Thomas Jefferson's DNA: Paternity and Beyond

For more than two centuries, historians debated whether U.S. President Thomas Jefferson had fathered one or more children with Sally Hemings, one of his slaves. In 1997, in an attempt to resolve the controversy, scientists analyzed the Y-chromosome DNA profiles from the Jefferson and Hemings male lineages. Because Jefferson did not have any surviving male-line descendants, the researchers tested the Y chromosome from five patrilineal descendants of Jefferson's paternal uncle. They also analyzed the DNA from a patrilineal descendant of Eston Hemings Jefferson, Sally Hemings's youngest son, as well as DNA from three patrilineal descendants of Jefferson's sister's sons, and five patrilineal descendants of Thomas Woodson,

who claimed to be the first child of Sally Hemings. DNA profiles were generated from 11 Y-chromosome STR loci, one larger minisatellite (MSY1), and seven SNP-like loci that differ by one nucleotide. All five of the Jefferson-line males shared identical Y-chromosome profiles, except for one individual who had a mutation at one STR locus. The profiles from the Woodson family line did not match those from the Jefferson line, excluding Thomas Woodson as a son of Thomas Jefferson. Similarly, and not unexpectedly, Jefferson's nephews did not share his Y-chromosome profile. However, the Y-chromosome profile from Eston Hemings's male descendant was identical to profiles from the Jefferson line. Although these data support the idea that Jefferson fathered at least one child of Sally Hemings, it does not exclude the possibility that another male Jefferson (such as Thomas Jefferson's brother, his brother's sons, or another male individual sharing the Jefferson

Y chromosome) could have been Eston Hemings's father.

Another interesting finding to arise from the Jefferson DNA study was that Jefferson's Y-chromosome profile pattern (haplotype K2) is extremely rare in Europeans. It occurs predominantly in men from Africa and western Eurasia. It is also found in men of Jewish ancestry from North Africa, Egypt, and the Middle East. In 2007, geneticists discovered the K2 haplotype (and matches to Jefferson's Y-chromosome profile) in 2 out of 85 men with the surname Jefferson in Britain; hence, it is likely that Jefferson's family originally came from Britain, as he claimed. However, it also suggests that sometime in the past, a Y chromosome of African or Middle Eastern origin entered the Jefferson line. To add to the Jefferson speculations, some commentators have suggested that Jefferson was the "first Jewish president" of the United States.

Although these features of Y-chromosome profiles present limitations for some forensic applications, they are useful for identifying missing persons when a male relative's DNA is available for comparison. They also allow researchers to trace paternal lineages in genetic genealogy studies (Box 2).

Mitochondrial DNA Profiling

Another important addition to DNA profiling methods is **mitochondrial DNA (mtDNA)** analysis. Between 200 and 1700 mitochondria are present in each human somatic cell. Each mitochondrion contains one or more 16-kb circular DNA chromosomes. Mitochondria divide within cells and are distributed to daughter cells after cell division. Mitochrondria are passed from the human egg cell to the zygote during fertilization; however, as sperm cells contribute few if any mitochondria to the zygote, they do not contribute these organelles to the next generation. Therefore, all cells in an individual contain multiple copies of identical mitochondria derived from the mother. Like Y-chromosome DNA, mtDNA undergoes little if any recombination and is inherited as a single unit.

Scientists create mtDNA profiles by amplifying regions of mtDNA that show variability between unrelated individuals and populations. Two commonly used regions are known as **hypervariable segment I and II (HVSI and HVSII)** (ST Figure 1–7). After PCR amplification, the DNA

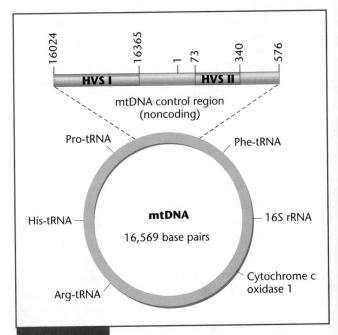

ST FIGURE 1–7 **Human mtDNA molecule, showing the locations of HVS I and II regions relative to other mtDNA genes.** The mtDNA control region is comprised of 1122 base pairs. Forensic DNA analysis involves sequencing the two HVS regions (610 base pairs total) and comparing sequences between individuals or samples. Sequence differences include short dinucleotide duplications as well as single base-pair changes.

sequence within these regions is determined by automated DNA sequencing. Scientists then compare the sequence with sequences from other individuals or crime samples, to determine whether or not they match.

The fact that mtDNA is present in high copy numbers in cells makes its analysis useful in cases where crime samples are small, old, or degraded. mtDNA profiling is particularly useful for identifying victims of mass murders or disasters, such as the Srebrenica massacre of 1995 and the World Trade Center attacks of 2001, where reference samples from relatives are available (Box 3). The main disadvantage of mtDNA profiling is that it is not possible to differentiate between the mtDNA from maternal relatives or from siblings. Like Y-chromosome profiles, mtDNA profiles may be shared by two apparently unrelated individuals who also share a distant ancestor—in this case a maternal ancestor. Researchers use mtDNA profiles in scientific studies of genealogy, evolution, and human population migrations.

Mitochondrial DNA analyses have also been useful in wildlife forensics cases. Billions of dollars are generated from the illegal wildlife trade, throughout the world. Often, the identification of the species or origin of plant and animal material is the key to successful prosecution of wildlife trafficking cases. A case of illegal smuggling of bird eggs in Australia, solved by mitochondrial sequence analysis, is presented in Box 4.

Single-Nucleotide Polymorphism Profiling

Single-nucleotide polymorphisms (SNPs) are single-nucleotide differences between two DNA molecules. They may be base-pair changes or small insertions or deletions (ST Figure 1–8). SNPs occur randomly throughout the genome and on mtDNA, every 500 to 1000 nucleotides.

This means that there are potentially millions of loci in the human genome that can be used for profiling. However, as SNPs usually have only two alleles, many SNPs (50 or more) must be used to create a DNA profile that can distinguish between two individuals as efficiently as STRs.

Scientists analyze SNPs by using specific primers to amplify the regions of interest. The amplified DNA regions are then analyzed by a number of different methods such as automated DNA sequencing or hybridization to immobilized probes on DNA microarrays that distinguish between DNA molecules with single-nucleotide differences.

Forensic SNP profiling has one major advantage over STR profiling. Because a SNP involves only one nucleotide

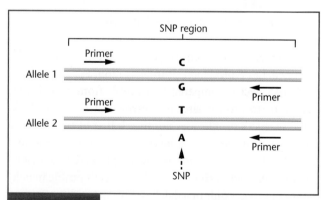

ST FIGURE 1–8   Example of a single-nucleotide polymorphism (SNP) from an individual who is heterozygous at the SNP locus. The arrows indicate the locations of PCR primers used to amplify the SNP region, prior to DNA sequence analysis. If this SNP locus only had two known alleles—the C and T alleles—there would be three possible genotypes in the population: CC, TT, and CT. The individual in this example has the CT genotype.

BOX 4

The Pascal Della Zuana Case: DNA Barcodes and Wildlife Forensics

On August 2, 2006, a freelance photographer named Pascal Della Zuana was stopped by customs officers at Australia's Sydney International Airport. While questioning him about his flight from Thailand to Australia, officers noticed that he was wearing an unusual white vest under his outer clothing. Inside the vest, they discovered 23 concealed bird eggs.

Due to Australia's strict quarantine regulations, the eggs had to be treated with radiation in order to sterilize them. Unable to hatch the eggs, authorities turned to DNA typing in an attempt to identify the origin and species of the eggs.

The eggs were sent to Dr. Rebecca Johnson at the DNA Laboratory at the Australian Museum for forensic identification. Dr. Johnson took a small sample from each egg and extracted the DNA. She used PCR methods to amplify an approximately 650-bp region of the mitochondrial genome, within the cytochrome c oxidase 1 gene. She then organized these sequences into a format known as a DNA barcode. In order to identify the species, Dr. Johnson compared each DNA barcode to barcode entries in a large DNA barcode database compiled at the University of Guelph in Canada. The database contains mitochondrial DNA barcode sequences from hundreds of universities and museums throughout the world, cataloging more than 70,000 different species.

The results of Dr. Johnson's barcode sequence comparisons were dramatic. Della Zuana's vest had concealed eggs of exotic bird species such as macaws, African grey and Eclectus parrots, as well as a rare threatened species, the Moluccan cockatoo.

On January 20, 2007, Pascal Della Zuana was found guilty of contravening the Convention on International Trade in Endangered Species (CITES), as well as three Australian Customs and Quarantine Acts. He was fined $10,000 and sentenced to two years in prison.

During the court case, it was learned that, if hatched, the birds would have fetched about $250,000 on the black market. The worldwide smuggling of wildlife and wildlife parts is thought to be worth as much as US$150 billion each year— surpassed only by drugs and arms in terms of illegal profit.

---

of a DNA molecule, the theoretical size of DNA required for a PCR reaction is the size of the two primers and one more nucleotide (i.e., about 50 nucleotides). This feature makes SNP analysis suitable for analyzing DNA samples that are severely degraded. Despite this advantage, SNP profiling has not yet become routine in forensic applications. More frequently, researchers use SNP profiling of Y-chromosome and mtDNA loci for lineage and evolution studies.

Interpreting DNA Profiles

After a DNA profile is generated, its significance must be determined. In a typical forensic investigation, a profile derived from a suspect is compared to a profile from an evidence sample or to profiles already present in a DNA database. If the suspect's profile does not match that of the evidence profile or database entries, investigators can conclude that the suspect is not the source of the sample(s) that generated the other profile(s). However, if the suspect's profile matches the evidence profile or a database entry, the interpretation becomes more complicated. In this case, one could conclude that the two profiles either came from the same person—or they came from two different people who share the same DNA profile by chance. To determine the significance of any DNA profile match, it is necessary to estimate the probability that the two profiles are a random match.

The **profile probability**, or **random match probability** method gives a numerical probability that a person chosen at random from a population would share the same DNA profile as the evidence or suspect profiles. The following example demonstrates how to arrive at a profile probability (ST Table 1.2).

The first locus examined in this DNA profile (D5S818) has two alleles: 11 and 13. Population studies show that the 11 allele of this locus appears at a frequency of 0.361 in this population and the 13 allele appears at a frequency of 0.141. In population genetics, the frequencies of two different alleles at a locus are given the designation p and q, following the Hardy–Weinberg Law described in Chapter 25. We assume that the person having this DNA profile received the 11 and 13 alleles at random from each parent. Therefore, the probability that this person received allele 11 from the mother and allele 13 from the father is expressed as $p \times q = pq$. In addition, the probability that the person received allele 11 from the father and allele 13 from the mother is also pq. Hence, the total probability that this person would have the 11, 13 genotype at this locus, by chance, is $2pq$. As we see from ST Table 1.2, $2pq$ is 0.102 or approximately 10 percent. It is obvious from this sample that using a DNA profile of only one locus would not be very informative, as about 10 percent of the population would also have the D5S818 11, 13 genotype.

The discrimination power of the DNA profile increases when we add more loci to the analysis. The next locus of this

ST TABLE 1.2

A Profile Probability Calculation Based on Analysis of Five STR Loci

STR Locus	Alleles from Profile	Allele Frequency from Population Database*	Genotype Frequency Calculation
D5S818	11	0.361	$2pq = 2 \times 0.361 \times 0.141 = 0.102$
	13	0.141	
TPOX	11	0.243	$p^2 = 0.243 \times 0.243 = 0.059$
	11	0.243	
D8S1179	13	0.305	$2pq = 2 \times 0.305 \times 0.031 = 0.019$
	16	0.031	
CSF1PO	10	0.217	$p^2 = 0.217 \times 0.217 \times 0.047$
	10	0.217	
D19S433	13	0.253	$2pq = 2 \times 0.253 \times 0.369 = 0.187$
	14	0.369	

Genotype frequency from this 5-locus profile $= 0.102 \times 0.059 \times 0.019 \times 0.047 \times 0.187 = 0.0000009 = 9 \times 10^{-7}$

*A U.S. Caucasian population database (Butler, J.M., et al. 2003. *J. Forensic Sci.* 48: 908–911.)

person's DNA profile (TPOX) has two identical alleles—the 11 allele. Allele 11 appears at a frequency of 0.243 in this population. The probability of inheriting the 11 allele from each parent is $p \times p = p^2$. As we see in the table, the genotype frequency at this locus would be 0.059, which is about 6 percent of the population. If this DNA profile contained only the first two loci, we could calculate how frequently a person chosen at random from this population would have the genotype shown in the table, by multiplying the two genotype probabilities together. This would be $0.102 \times 0.059 = 0.006$. This analysis would mean that about 6 persons in 1000 (or 1 person in 166) would have this genotype. The method of multiplying all frequencies of genotypes at each locus is known as the **product rule**. It is the most frequently used method of DNA profile interpretation and is widely accepted in U.S. courts.

By multiplying all the genotype probabilities at the five loci, we arrive at the genotype frequency for this DNA profile: 9×10^{-7}. This means that approximately 9 people in every 10 million (or about 1 person in a million), chosen at random from this population, would share this 5-locus DNA profile.

The Uniqueness of DNA Profiles

As we increase the number of loci analyzed in a DNA profile, we obtain smaller probabilities of a random match. Theoretically, if a sufficient number of loci were analyzed, we could be *almost* certain that the DNA profile was unique. At the present time, law enforcement agencies in North America use a core set of 13 STR loci to generate DNA profiles. A hypothetical genotype comprised of the most common alleles of each STR locus in the core STR profile would be expected to occur only once in a population of 10 billion people. Hence, the frequency of this profile would be 1 in 10 billion.

Although this would suggest that most DNA profiles generated by analysis of the 13 core STR loci would be

unique on the planet, several situations can alter this interpretation. For example, identical twins share the same DNA, and their DNA profiles will be identical. Identical twins occur at a frequency of about 1 in 250 births. In addition, siblings can share one allele at any DNA locus in about 50 percent of cases and can share both alleles at a locus in about 25 percent of cases. Parents and children also share alleles, but are less likely than siblings to share both alleles at a locus. When DNA profiles come from two people who are closely related, the profile probabilities must be adjusted to take this into account. The allele frequencies and calculations that we describe here are based on assumptions that the population is large and has little relatedness or inbreeding. If a DNA profile is analyzed from a person in a small interrelated group, allele frequency tables and calculations may not apply.

The Prosecutor's Fallacy

It is sometimes stated, by both the legal profession and the public, that "the suspect must be guilty given that the chance of a random match to the crime scene sample is 1 in 10 billion—greater than the population of the planet." This type of statement is known as the **prosecutor's fallacy** because it equates guilt with a numerical probability derived from one piece of evidence, in the absence of other evidence. A match between a suspect's DNA profile and crime scene evidence does not necessarily prove guilt, for many reasons such as human error or contamination of samples, or even deliberate tampering. In addition, a DNA profile that does not match the evidence does not necessarily mean that the suspect is innocent. For example, a suspect's profile may not match that from a semen sample at a rape scene, but the suspect could still have been involved in the crime, perhaps by restraining the victim. For these and other reasons, DNA profiles must be interpreted in the context of all the

BOX 5

The Kennedy Brewer Case:
Two Bite-Mark Errors
and One Hit

I n 1992 in Mississippi, Kennedy Brewer was arrested and charged with the rape and murder of his girlfriend's 3-year-old daughter, Christine Jackson. Although a semen sample had been obtained from Christine's body, there was not sufficient DNA for profiling. Forensic scientists were also unable to identify the ABO blood group from the bloodstains left at the crime scene. The prosecution's only evidence came from a forensic bite-mark specialist who testified that the 19 "bite marks" found on Christine's body matched imprints made by Brewer's two top teeth. Even though the specialist had recently been discredited by the American Board of Forensic Odontology, and

the defense's expert dentistry witness testified that the marks on Christine's body were actually postmortem insect bites, the court convicted Brewer of capital murder and sexual battery and sentenced him to death.

In 2001, more sensitive DNA profiling was conducted on the 1992 semen sample. The profile excluded Brewer as the donor of the semen sample. It also excluded two of Brewer's friends, and Y-chromosome profiles excluded Brewer's male relatives. Despite these test results, Brewer remained in prison for another five years, awaiting a new trial. In 2007, the Innocence Project took on Brewer's case and retested the DNA samples. The profiles matched those of another man, Justin Albert Johnson, a man with a history of sexual assaults who had been one of the original suspects in the case. Johnson subsequently confessed to

Christine Jackson's murder, as well as to another rape and murder—that of a 3-year-old girl named Courtney Smith. Levon Brooks, the ex-boyfriend of Courtney's mother, had been convicted of murder in the Smith case, also based on bite-mark testimony by the same discredited expert witness.

On February 15, 2008, all charges against Kennedy Brewer were dropped, and he was exonerated of the crimes. Levon Brooks was subsequently exonerated of the Smith murder in March of 2008.

Since 1989, more than 250 people in the United States have been exonerated of serious crimes, based on DNA profile evidence. Seventeen of these people had served time on death row. In more than 100 of these exoneration cases, the true perpetrator has been identified, often through searches of DNA databases.

evidence in a case. A more detailed description of problems with DNA profiles is given in the next section.

DNA Profile Databases

Many countries throughout the world maintain national DNA profile databases. The first of these databases was established in the UK in 1995 and now contains more than 5 million profiles—representing almost 10 percent of the population. In the UK, DNA samples can be taken from anyone arrested for an offense that could lead to a prison sentence. Although highly controversial, DNA profiles from suspects can be stored permanently in the national DNA database, even if the person is not convicted.

In the United States, both state and federal governments have DNA profile databases. The entire system of databases along with tools to analyze the data is known as the **Combined DNA Index System (CODIS)** and is maintained by the FBI. At the beginning of 2010, there were more than 8 million DNA profiles stored within the CODIS system. The two main databases in CODIS are the **convicted offender database**, which contains DNA profiles from individuals convicted of certain crimes, and the **forensic database**, which contains profiles generated from crime scene evidence. In addition, some states have DNA profile databases containing profiles from suspects and from unidentified human remains and missing persons. Suspects who are not convicted can request that their profiles be removed from the databases.

DNA profile databases have proven their value in many different situations. As of January 2010, use of CODIS databases resulted in more than 100,000 profile matches that assisted criminal investigations and missing persons searches. (Box 5). Despite the value of DNA profile databases, they remain a concern for many people who question the privacy and civil liberties of individuals versus the needs of the state.

Technical and Ethical Issues Surrounding DNA Profiling

Although DNA profiling is sensitive, accurate, and powerful, it is important to be aware of its limitations. One limitation is that most criminal cases have either no DNA evidence for analysis, or DNA evidence that would not be informative to the case. In some cases, potentially valuable DNA evidence exists but remains unprocessed and backlogged. Another serious problem is that of human error. There are cases in which innocent people have been convicted of violent crimes based on DNA samples that had been inadvertently switched during processing. DNA evidence samples from crime scenes are often mixtures derived from any number of people present at the crime scene or even from people who were not present, but whose biological material (such as hair or saliva) was indirectly introduced to the site. Crime scene evidence is often degraded, yielding partial DNA profiles that are difficult to interpret.

One of the most disturbing problems with DNA profiling is its potential for deliberate tampering. DNA profile technologies are so sensitive that profiles can be generated from only a few cells—or even from fragments of synthetic DNA. There have been cases in which criminals have introduced biological material to crime scenes, in an attempt to affect forensic DNA profiles. It is also possible to manufacture artificial DNA fragments that match STR loci of a person's DNA profile. In 2010, a research paper[1] reported methods for synthesizing DNA of a known STR profile, mixing the DNA with body fluids, and depositing the sample on crime scene items. When subjected to routine forensic analysis, these artificial samples generated perfect STR profiles. In the future, it may be necessary to develop methods to detect the presence of synthetic or cloned DNA in crime scene samples. It has been suggested that such detections could be done, based on the fact that natural DNA contains epigenetic markers such as methylation.

Many of the ethical questions related to DNA profiling involve the collection and storage of biological samples and DNA profiles. Should police be able to collect DNA samples without a suspect's knowledge or consent? Who should have their DNA profiles stored on a database? Should law enforcement agencies reveal the identities of people whose DNA profiles partially match those of a suspect, on the chance that the two individuals are related? Should researchers have access to DNA databases for research purposes? Could DNA profiles be associated with regions of the genome that might reveal information about a person's health, racial background, or appearance—and if so, should that be admissible evidence?

As DNA profiling becomes more sophisticated and prevalent, we should carefully consider both the technical and ethical questions that surround this powerful new technology.

Selected Readings and Resources

Journal Articles

Brettell, T.A., et al., 2009. Forensic science. *Anal. Chem.* 81: 4695–4711.

Butler, J.M., et al., 2007. STRs vs. SNPs: Thoughts on the future of forensic DNA testing. *Forensic Sci Med Pathol* 3: 200–205.

Frumkin, D., et al., 2010. Authentication of forensic DNA samples. *Forensic Sci International* 4: 95–103.

Gill, P., Jeffreys, A.J., and Werrett, D.J. 2005. Forensic applications of DNA "fingerprints." *Nature* 318: 577–579.

Houck, M.M. 2006. CSI: Reality. *Scientific American* 295(1): 84–89.

Jobling, M.A., and Gill, P. 2004. Encoded evidence: DNA in forensic analysis. *Nature Reviews Genetics* 5: 739–750.

King, T.E., et al. 2007. Thomas Jefferson's Y chromosome belongs to a rare European lineage. *Am J Phys Anthropology* 132: 584–589.

Roewer, L. 2009. Y chromosome STR typing in crime casework. *Forensic Sci Med Pathol.* 5(2): 77–84.

Whittall, H. 2008. The forensic use of DNA: Scientific success story, ethical minefield. *Biotechnol J* 3: 303–305.

Web Sites

Brenner, C.H., Forensic Mathematics of DNA Matching. http://dna-view.com/profile.htm

Butler, J.M. and Reeder, D.J. Short Tandem Repeat DNA Internet DataBase. http://www.cstl.nist.gov/div831/strbase/

Wikipedia: DNA Profiling. http://en.wikipedia.org/wiki/DNA_profiling

DNA initiative: advancing criminal justice through DNA technology. http://www.dna.gov

The Innocence Project. http://www.innocenceproject.org

Berson, S.B. Debating DNA collection, from Office of Justice Programs, National Institute of Justice Journal, November 2009. http://www.nij.gov/journals/264/debating-DNA.htm

CODIS-NDIS Statistics. Federal Bureau of Investigation Web site, http://www.fbi.gov/hq/lab/codis/clickmap.htm

---

[1]Frumkin, D., et al. 2010. Authentication of forensic DNA samples. *Forensic Sci Int Genetics* 4: 95–103.

SPECIAL TOPICS IN MODERN GENETICS

Genomics and Personalized Medicine

In the near future, personalized medicine will allow physicians to predict which diseases you will develop, which therapeutics will work for you, and which drug dosages are appropriate.

Physicians have always practiced personalized medicine in order to make effective treatment decisions for their patients. Doctors take into account a patient's symptoms, family history, lifestyle, and data derived from many types of medical tests. However, within the last 20 years, personalized medicine has taken a new and potentially powerful direction based on genetics and genomics. Today, the phrase *personalized medicine* is used to describe the application of information from a patient's unique genetic profile in order to select effective treatments that have minimal side-effects and to detect disease susceptibility prior to development of the disease.

Despite the immense quantities of medical information and pharmaceuticals that are available, the diagnosis and treatment of human disease remains an imperfect process. It is sometimes difficult or impossible to accurately diagnose some conditions. In addition, some patients do not respond to treatments, while others may develop side-effects that can be annoying or even life-threatening. As much of the basis for disease susceptibility and the variation that patients exhibit toward drug treatments are genetically determined, progress in genetics, genomics, and molecular biology has the potential to significantly advance medical diagnosis and treatment.

The sequencing of the human genome, the cataloging of genetic sequence variants, and the linking of sequence variants with disease susceptibility form the basis of the newly emerging field of personalized medicine. By 2009, the number of diseases detectable by genetic tests rose to approximately 2000. In addition, a rapidly growing list of genetic tests help physicians determine whether a patient will have an adverse drug reaction and whether a particular pharmaceutical will be effective for that patient.

Although much of the promise of personalized medicine remains in the future, significant progress is underway. As genome technologies advance and the cost of sequencing personal genomes declines, it is becoming easier to examine a patient's unique genomic profile in order to diagnose diseases and prescribe treatments. Proponents of personalized medicine foresee a future in which each person will have his or her genome sequence determined at birth and will have the sequence stored in a digital form within a personal computerized medical file. Medical practitioners will use automated methods to scan the sequence information within these files for clues to disease susceptibility and reactions to drugs. In the near future, genomic profiling and personalized medicine will allow physicians to predict which diseases you will develop, which therapeutics will work for you, and which drug dosages are appropriate.

In this Special Topics chapter, we will outline the current uses of genetic and genomic-based personalized medicine in disease diagnosis and drug selection. In addition, we will outline the future directions for personalized medicine, as well as some ethical and technical challenges associated with it.

Drug type		
Anti-depressants (SSRIs)	38%	
Asthma drugs	40%	
Diabetes drugs	43%	
Arthritis drugs	50%	
Alzheimer's drugs	70%	
Cancer drugs	75%	

© 2009 Personalized Medical Coalition

ST FIGURE 2–1 **Variations in patient response to drugs.** This figure gives a general summary of the percentages of patients for which a particular class of drugs is effective.

Personalized Medicine and Pharmacogenomics

At present, the most developed aspect of the new personalized medicine is in the field of pharmacogenomics. **Pharmacogenomics** is the study of how an individual's entire genetic makeup determines the body's response to drugs. The term *pharmacogenomics* is used interchangeably with *pharmacogenetics*, which refers to the study of how sequence variation within specific candidate genes affects an individual's drug responses.

In pharmacogenomics, scientists take into account many aspects of drug metabolism and how genetic traits affect these aspects. When a drug enters the body, it interacts with various proteins including carriers, cell-surface receptors, transporters, and metabolizing enzymes. These proteins affect a drug's target site of action, absorption, pharmacological response, breakdown, and excretion. Because there are so many interactions that occur between a drug and proteins within the patient, many genes and many different genetic polymorphisms can affect a person's response to a drug.

In this subsection, we examine two ways in which genomics and personalized medicine are changing the field of pharmacogenomics: by optimizing drug therapies and by reducing adverse drug reactions.

Optimizing Drug Therapies

When it comes to drug therapy, it is clear that "one size does not fit all." On average, a drug will be effective in only about 50 percent of patients who take it (ST Figure 2–1). This situation means that physicians often must switch their patients from one drug to another until they find one that is effective. Not only does this waste time and resources, but also it may be dangerous to the patient who is exposed to a variety

of different pharmaceuticals and who may not receive appropriate treatment in time to combat a progressive illness.

Pharmacogenomics increases the efficacy of drugs by targeting those drugs to subpopulations of patients who will benefit. One of the most common current applications of personalized pharmacogenomics is in the diagnosis and treatment of cancers. Large-scale sequencing studies show that each tumor is genetically unique, even though it may fall into a broad category based on cytological analysis or knowledge of its tissue origin. Given this genomic variability, it is important to understand each patient's mutation profile to select an appropriate treatment—particularly those newer treatments based on the molecular characteristics of tumors (Box 1).

One of the first success stories in personalized medicine was that of the **HER-2** gene and the use of the drug **Herceptin®** in breast cancer. The human epidermal growth factor receptor 2 (*HER-2*) gene is located on chromosome 17 and codes for a transmembrane tyrosine kinase receptor protein called HER-2. These receptors are located within the cell membranes of normal breast epithelial cells and are responsible for sending signals to the cell nucleus that result in the transcription of genes involved in cell growth and division. In a normal situation, an extracellular growth factor (ligand) binds to a HER-2 protein, which then dimerizes with another HER family receptor and phosphorylates tyrosine residues on specific target proteins within the cell. These phosphorylated proteins are then activated and phosphorylate a series of other proteins, in a process known as **signal transduction**. The end result of this cascade of signal transduction is that transcription factors in the nucleus stimulate the transcription of genes whose products regulate cell proliferation.

BOX 1
The Story of Pfizer's Crizotinib

In 2007, Beverly Sotir was diagnosed with advanced non-small-cell lung cancer (NSCLC). Beverly, a 68-year-old grandmother and nonsmoker, received standard chemotherapy, but her cancer continued to proliferate. She was given six months to live. At this same time, an apparently unrelated scientific study was underway by the pharmaceutical company, Pfizer. Pfizer had developed a compound called crizotinib, which was designed to inhibit the activity of MET, a tyrosine kinase that is abnormal in a number of tumors. Although crizotinib also inhibited another kinase called ALK (anaplastic lymphoma kinase), scientists did not consider it significant. After clinical trials for crizotinib began, an article was published* describing a chromosomal translocation found in a small number of NSCLCs. This translocation fused the *ALK* gene to another gene called *EML4,* leading to production of a fusion protein that stimulated cancer cell growth. Pfizer immediately changed its clinical trial to include NSCLC patients. Beverly's doctors at the Dana-Farber Cancer Institute in Boston tested her tumors, discovered that they contained the *ALK/EML4* fusion gene, and enrolled Beverly in the trials. The results were dramatic. Within six months, Beverly's tumors shrunk by more than 50 percent and some disappeared entirely. As of September 2010, Beverly continued to do well.

Results of the Phase I and II clinical trials for crizotinib showed that tumors shrunk or stabilized in 90 percent of the 82 patients whose tumors contained the *ALK* fusion gene. Those patients who responded well to treatment had positive responses for up to 15 months. Scientists report that the *ALK* fusion gene tends to occur most frequently in young NSCLC patients who have never smoked. Approximately 4 percent of patients with NSCLC have this translocation in their tumor cells. Although this appears a small percentage of people who might benefit from crizotinib, this means that about 45,000 people a year, worldwide, may be eligible for this treatment. Crizotinib is now in Phase III clinical trials.

*Choi, S.M., et al. 2007. Identification of the transforming EML4-ALK fusion gene in non-small-cell lung cancer. *Nature* 448: 561–566.

In about 25 percent of invasive breast cancers, the *HER-2* gene is amplified and the protein is overexpressed on the cell surface. In some breast cancers, the *HER-2* gene is present in as many as 100 copies per cell. The presence of *HER-2* gene amplification is associated with increased tumor invasiveness, metastasis, and cell proliferation, as well as a poorer patient prognosis.

Using recombinant DNA technology, Genentech Corporation in California developed a monoclonal antibody known as trastuzumab (or Herceptin®) that is designed to bind specifically to the extracellular region of the HER-2 receptor. When bound to the receptor, Herceptin® appears to inhibit the signaling capability of HER-2 and may also flag the HER-2-expressing cell for destruction by the patient's immune system. In cancer cells that overexpress HER-2, Herceptin treatment causes cell-cycle arrest, and in some cases, death of the cancer cells.

Because Herceptin will only act on breast cancer cells that have amplified *HER-2* genes, it is important to know the HER-2 phenotype of each cancer. In addition, Herceptin has potentially serious side-effects. Hence, its use must be limited to those who could benefit from the treatment. A number of molecular assays have been developed to determine the gene and protein status of breast cancer cells. Two of the most commonly used tests are based on **immunohistochemistry (IHC)** and **fluorescence *in situ* hybridization (FISH)**. In IHC assays, an antibody that binds to HER-2 protein molecules is added to fixed tissue on a slide. The antibody is bound to another molecule that reacts to produce a visual stain. After washing and staining, the tissues are observed under a microscope. The level of HER-2 staining is assessed from "0" (fewer than 20,000 HER-2 molecules per cell) to "+3" (approximately 2 million molecules per cell) (ST Figure 2–2(a)). In FISH, DNA or RNA molecules with sequence complementarity to the *HER-2* gene sequence are added to the fixed tissue on the slide. These DNA or RNA probes are labeled with a fluorescent tag molecule. After hybridizing the probes to the tissue and washing off excess probe, the location and intensity of the probe are determined by observing the tissue under a fluorescence microscope. The number of *HER-2* genes is assessed by comparing the fluorescence signal of the *HER-2* probe with a control signal from another gene that is not amplified in the cells (ST Figure 2–2(b)).

Herceptin has had a major effect on the treatment of HER-2 positive breast cancers. When Herceptin is used in combination with chemotherapy, there is a 25 to 50 percent increase in survival, compared with the use of chemotherapy alone. Herceptin is now one of the biggest selling biotechnology products in the world, generating more than $5 billion in annual sales.

There are now dozens of drugs whose prescription and use depend on the genetic status of the target cells. Approximately 10 percent of FDA-approved drugs have labels that include pharmacogenomic information (ST Table 2.1). For example, about 40 percent of colon cancer patients respond

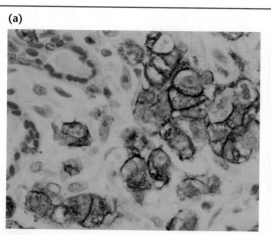
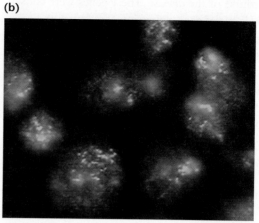

ST FIGURE 2–2 *HER-2* **gene and protein assays.** (a) Normal and breast cancer cells within a biopsy sample, stained by HER-2 immunohistochemistry. Cell nuclei are stained blue. Cancer cells that overexpress HER-2 protein stain brown. (b) Cancer cells from the same tumor assayed in (a), assayed for *HER-2* gene copy number by fluorescence *in situ* hybridization. Cancer cell nuclei appear green under the fluorescence microscope and the *HER-2* gene DNA appears bright yellow. Large clumps of yellow stain indicate *HER-2* gene amplification (more than 20 copies per nucleus).

Source: The Case for Personalized Medicine, Figure 1, Page 5. © 2009 Personalized Medicine coalition. Used by permission.

SPECIAL TOPIC II

to the drugs **Erbitux®** (cetuximab) and **Vectibix®** (panitumab). These two drugs are monoclonal antibodies that bind to **epidermal growth factor receptors (EGFRs)** on the surface of cells and inhibit the EGFR signal transduction pathway. In order to work, cancer cells must express EGFR on their surfaces and must also have a wild-type *K-RAS* gene. The **K-RAS** gene encodes a small signaling protein that is part of the EGFR pathway. Mutations in *K-RAS* codons 12, 13, and 61 are common in colorectal cancers, and the presence of these mutations makes treatment with these drugs ineffective. Prior to treatment with Erbitux or Vectibix, tumor cells from patients are screened for the presence of EGFR and a wild-type *K-RAS* gene. The EGFR is detected with an immunohistochemistry test, as described for HER-2. *K-RAS* mutations are detected using an assay based

on the polymerase chain reaction (PCR). The PCR method was described previously in Chapter 20. The *K-RAS* assay is based on the observation that the *Taq* DNA polymerase can distinguish between a perfect primer match and a one-nucleotide mismatch at the 3′ end of the primer. By including primers that hybridize perfectly with only the mutations in codons 12, 13, or 61, only the mutated *K-RAS* genes will be amplified by PCR. The primers are labeled with a fluorescing molecule, and the presence of a PCR product is visualized by automated equipment.

Another example of treatment decisions being informed by genetic tests is that of the **Oncotype DX®** Assay (Genomic Health Inc.). This assay analyzes the expression (amount of mRNA) from 21 genes in breast cancer samples, in order to help physicians select appropriate treatments and

ST TABLE 2.1

Examples of Personalized Medicine Drugs and Diagnostics

Therapy	Gene Test	Description
Herceptin® (trastuzumab)	*HER-2* amplification	Breast cancer test to accompany herceptin use
Erbitux® (cetuximab)	*EGFR* expression, *K-RAS* mutations	Protein and mutation analysis prior to treatment
Gleevec® (imatinib)	*BCR/ABL* fusion	Gleevec used in treatment of Philadelphia chromosome[+] chronic myelogenous leukemia
Gleevec® (imatinib)	*C-KIT*	Gleevec used in stomach cancers expressing mutated *C-KIT*
Tarceva® (erlotinib)	*EGFR* expression	Lung cancer for EGFR[+] tumors
Drugs/surgery	MLH1, MSH2, MSH6	Gene mutations related to colon cancers
Hormone/chemotherapies	**Oncotype DX® test**	Selection of breast cancer patients for chemotherapy
Chemotherapies	**Aviara Cancer TYPE ID®**	Classifies 39 tumor types using gene-expression assays
Rituximab	**PGx Predict®** (*FcRIIIa*)	Detects CD-20 variants that predict response to rituximab in Non-Hodgkin's lymphoma

predict the course of the disease. These genes were chosen because their levels of gene expression correlate with breast cancer recurrence after initial treatment. To perform the assay, scientists extract nucleic acids from tumor tissue and remove the DNA by treating the sample with the enzyme DNase I. They then convert the mRNA molecules to single-stranded DNA molecules (cDNA) using reverse transcriptase. The cDNA is amplified in PCR reactions using primers that are specific for each of the 21 genes. This technique, known as reverse transcription PCR (RT-PCR), is described in detail in Chapter 20. The level of amplification of each cDNA during the PCR reaction reflects its concentration. Based on the mRNA expression levels revealed in the assay results, scientists calculate a "Recurrence Score," estimating the likelihood that the cancer will recur within a ten-year period. Those patients with a low-risk rating would likely not benefit by adding chemotherapy to their treatment regimens and so can be treated with hormones alone. Those with higher risk scores would likely benefit from more aggressive therapies.

Reducing Adverse Drug Reactions

Every year, about 2 million people in the United States have serious side-effects from pharmaceutical drugs, and approximately 100,000 people die. The costs associated with these **adverse drug reactions (ADRs)** are estimated to be $136 billion annually. Although some ADRs result from drug misuse, others result from a patient's inherent physiological reactions to a drug.

Sequence variations in a large number of genes can affect drug responsiveness (ST Table 2.2). Of particular significance are the genes that encode the cytochrome P450 families of enzymes. These family members are encoded by

57 different genes. The products of the *CYP2A6, CYP2B6, CYP2C9, CYP2C19, CYP2D6, CYP2E1,* and *CYP3A4* genes are responsible for metabolizing most clinically important pharmaceutical drugs. People with some gene variants metabolize and eliminate drugs slowly, which can lead to accumulations of the drug and overdose side-effects. In contrast, other people have variants that cause drugs to be eliminated quickly, leading to reduced effectiveness. An example of gene variants that affect drug responses is that of *CYP2D6*. This member of the cytochrome P450 family encodes the debrisoquine hydroxylase enzyme, which is involved in the metabolism of approximately 25 percent of all pharmaceutical drugs, including diazepam, acetaminophen, clozapine, beta blockers, tamoxifen, and codeine. There are more than 70 variant alleles of this gene. Some mutations reduce the activity of the enzyme, and others can increase it. Approximately 80 percent of people are homozygous or heterozygous for the wild-type *CYP2D6* gene and are known as extensive metabolizers (ST Figure 2–3). Approximately 10 to 15 percent of people are homozygous for alleles that decrease activity (poor metabolizers), and the remainder of the population have duplicated genes (ultra-rapid metabolizers). Poor metabolizers are at increased risk for ADRs, whereas ultra-rapid metabolizers may not receive sufficient dosages to have an effect on their conditions.

Another example of gene variants affecting drug metabolism is the enzyme **thiopurine S-methyltransferase (TPMT)**, encoded by the *TPMT* gene. This enzyme metabolizes a large number of drugs, including many psychoactive drugs and thiopurine drugs used to treat cancers. There are more than 28 variant alleles of this gene, most of which are **single-nucleotide polymorphisms (SNPs)** that create amino acid substitutions and reduced enzyme activity.

ST TABLE 2.2

Examples of Variant Gene Products that Affect Drug Responses

Gene Product	Variant Phenotype	Drugs Affected	Response
Acetyl transferase NAT2	Slow, rapid acetylators	Isoniazid, sulfamethazine, dapsone, paraminosalicylic acid, heterocyclic amines	Slow: toxic neuritis, lupus erythematosus, bladder cancer; Rapid: colorectal cancer
Thiopurine methyltransferase	Poor TPMT Methylators	6-mercaptopurine, 6-thioguanine, azathioprin	Bone marrow toxicity, liver damage
Catechol O-methyl transferase	High, low methylators	Levodopa, methyl dopa	Low or increased response
CYP2C19	Poor, extensive hydroxylators	Mephenytoin, hexobarbital, proguanil, etc.	Poor or increased toxicity, poor efficacy (proguanil)
β_2 Adrenoceptor	Enhanced receptor downregulation	Albuterol, ventolin	Poor asthma control
5-HT2A serotonergic receptor	Multiple polymorphisms	Clozapine	Variable drug efficiencies
Multiple drug resistance transporter	Overexpression in cancer	Vinblastin, doxorubicin, paclitaxel, etc.	Drug resistance

Source: Adapted from Table 1 of Mancinelli, L., et al. 2000. Pharmacogenomics: The promise of personalized medicine. *AAPS PharmSci* 2(1): Article 4.

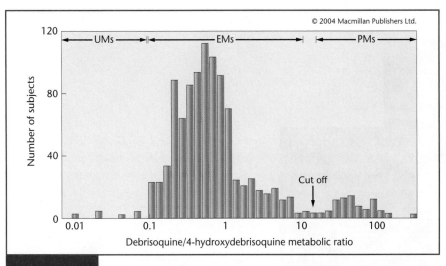

ST FIGURE 2–3   *CYP2D6* **pharmacogenetic profile in a Swedish population.** Individuals were tested for their ability to metabolize debrisoquine to 4-hydroxydebrisoquine, as an indication of the efficiency of debrisoquine hydroxylase enzyme activity. The population sample was divided into the categories UMs (ultra-rapid metabolizers), EMs (extensive metabolizers), and PMs (poor metabolizers). The "cut-off" label indicates the cut-off between extensive and poor metabolizers.

Genetic and enzymatic tests are now able to detect variations in TPMT activity, allowing tailoring of therapeutic doses to individual patients. TPMT variants and their effects are also discussed in Chapter 22.

In 2005, the FDA approved a microarray gene test called the **AmpliChip® CYP450** assay (Roche Diagnostics) that detects 29 genetic variants of two genes—*CYP2D6* and *CYP2C19*. This test detects SNPs as well as gene duplications and deletions. The AmpliChip CYP450 assay is an example of a genotyping microarray, such as those described in Chapter 22. To perform the assay, DNA is extracted from a patient's blood sample. The *CYP2D6* and *CYP2C19* genes are amplified from the DNA sample, using mixtures of gene-specific primers and PCR amplification. After PCR amplification, the amplification products are cleaved into short fragments (50–200 nucleotides) using limited digestion with the enzyme **DNase I**. The short fragments of amplified DNA are labeled at their 3′ ends with biotin. In the next step, the biotin-labeled fragments are hybridized onto the microarray. The microarray is a 20 × 20 μm glass surface that holds approximately 15,000 different synthetic oligonucleotides (called probes) in a specific grid pattern (ST Figure 2–4). Each probe location contains up to 10^7 copies of a specific probe. The sequences of the probes are complementary to known sequence variations in the two genes. The PCR amplified DNA fragments are hybridized to the microarray at temperatures that allow only perfect matches of probe and DNA fragment to anneal. After washing off the excess amplified DNA, the presence of hybridized amplification products is visualized by staining the microarray with a fluorescent stain that binds to biotin molecules. An automated scanner detects the location of stained DNA, and the stain intensity reflects the amount of hybridized DNA on each probe. After scanning, the data are analyzed by computer software, and the *CYP2D6*/*CYP2C19* genotype of the individual is generated.

Another example of pharmacogenomics in personalized medicine is that of the *CYP2C9* and *VKORC1* genes and the drug **warfarin**. Warfarin (also known as Coumadin) is an anticoagulant drug that is prescribed to prevent blood clots after surgery and to aid people with cardiovascular conditions who are prone to clots. Warfarin inhibits the vitamin K-dependent synthesis of several clotting factors. There is an approximately ten-fold variability between patients in the doses of warfarin that have a therapeutic response. In the past, physicians attempted to adjust the doses of warfarin through a trial-and-error process during the first year of treatment. If the dosage of warfarin is too high, the patient may experience serious hemorrhaging; if it is too low, the patient may develop life-threatening blood clots. It is estimated that 20 percent of patients are hospitalized during their first six months of treatment due to warfarin side-effects.

Variations in warfarin activity are affected by polymorphisms in several genes, particularly *CYP2C9* and *VKORC1*. Two single-nucleotide polymorphisms in *CYP2C9* lead to reduced elimination of warfarin and increased risk of hemorrhage. The first mutation is a C to T transition at codon 430, which leads to an arginine to cysteine amino acid substitution at amino acid 144 and a 30 percent lower activity of the *CYP2C9* gene product. The second mutation creates an isoleucine to leucine substitution at codon 359 and a 90 percent lower activity. About 25 percent of Caucasians are heterozygous for one of these polymorphisms, and 5 percent appear to be homozygous. About 5 percent of patients of Asian and African descent carry these variants. Patients who are heterozygous or homozygous for some alleles of *CYP2C9* require a 10 to 90 percent lower dose of warfarin. The *VKORC1* gene encodes **vitamin K epoxide reductase complex subunit 1,** a vitamin K-activating enzyme that is required for the formation of clotting factors. The activity of this enzyme is inhibited by warfarin. A common variant of *VKORC1* is a G to A transversion in the promoter region of the gene, leading to lower levels of VKORC1 and clotting factors. This mutation

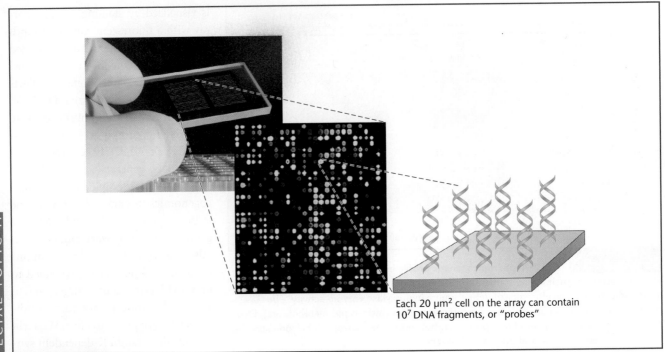

Each 20 μm² cell on the array can contain 10^7 DNA fragments, or "probes"

ST FIGURE 2–4   **Views of a DNA microarray.** The microarray cartridge is shown on the left, and a magnified version of the microarray is shown in the center. The 20-μm² microarray contains synthetic DNA oligonucletoide probes complementary to variants within genes such as *CYP2D6* and *CYP2C19*. Each probe type is located in a specific region of the microarray, called a probe cell. Within each probe cell, there are approximately 10^7 copies of the probe. After hybridization, washing, and staining, the microarray is scanned with a laser. The amount of light emitted by each fluorescent PCR amplification product is proportional to the amount of bound product on that probe cell.

leads to an increased sensitivity to warfarin, requiring lower doses of the drug. About 50 percent of Caucasians are heterozygous for this mutation and 14 percent are homozygous. Among Asians, about 18 percent are heterozygous and 80 percent are homozygous. Among African Americans, about 20 percent are heterozygous and fewer than 1 percent are homozygous.

Recently, the FDA recommended the use of *CYP2C9* and *VKORC1* genetic tests to predict the likelihood that a patient may have an adverse reaction to warfarin. Several companies now offer tests to detect the polymorphisms described earlier, using methods based on PCR amplification and allele-specific primers. It is estimated that the use of warfarin genetic tests could prevent 17,000 strokes and 85,000 serious hemorrhages per year. The savings in health care could reach $1.1 billion per year.

Pharmacogenomic tests and treatments, and the genetic information on which they are based, are rapidly advancing. A source of updated information on all aspects of pharmacogenomics can be found on the Pharmacogenomics Knowledge Base, which is described in Box 2.

Personalized Medicine and Disease Diagnosis

As of 2009, there were genetic tests for approximately 2000 different diseases (ST Figure 2–6). These tests are categorized according to their uses and can fall into one or more of the following groups: **Diagnostic tests** detect the presence or absence of gene variants linked to a suspected genetic disorder in a symptomatic patient; **predictive tests** detect a gene mutation in patients with a family history of having a known genetic disorder (for example, Huntington disease or BRCA-linked breast cancer); **carrier tests** help physicians identify patients who carry a gene mutation linked to a disorder that might be passed on to their offspring (for example Tay–Sachs or cystic fibrosis); **prenatal tests** detect potential genetic diseases in a fetus (for example, Down syndrome); and **preimplantation tests** are performed on early embryos in order to select embryos for implantation that do not carry a suspected disease. Most of these genetic tests

BOX 2

The Pharmacogenomics Knowledge Base (PharmGKB): Genes, Drugs, and Diseases on the Web

The Pharmacogenomics Knowledge Base (PharmGKB) is a publicly available Internet database and information source developed by Stanford University. It is funded by the National Institutes of Health (NIH) and forms part of the NIH Pharmacogenomics Research Network, a U.S. research consortium. The goal of PharmGKB is to provide researchers and the general public with information that will increase the understanding of how genetic variation contributes to an individual's reaction to drugs. On the PharmGKB Web site (see ST Figure 2–5), you may search for genes and more than 650 variants that affect drug reactions, information on a large number of drugs, diseases and their genetic links, pharmacogenomic pathways, gene tests, and relevant publications. Visit the PharmGKB Web site at http://www.pharmgkb.org

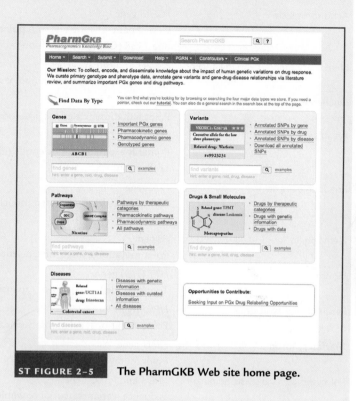

ST FIGURE 2–5   **The PharmGKB Web site home page.**

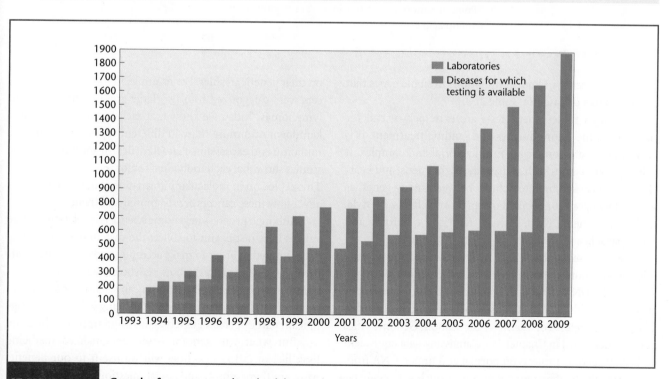

ST FIGURE 2–6   **Growth of gene tests and testing laboratories from 1993 to 2009.** From the GeneTests Web site at http://www.ncbi.nlm.nih.gov/projects/GeneTests/static/whatsnew/labdirgrowth.shtml.

ST TABLE 2.3

Some Single-Gene Defects for Which Genetic Tests Are Available

Disease	Gene Mutation	Description
Achondroplasia	FGFR3 gene. 99% of patients have a G to A point mutation at nucleotide 1138 (G380R substitution)	Abnormal bone growth
Hereditary breast/ovarian cancer	BRCA1 and BRCA2 genes. Deletions, duplications, and point mutations	Predisposition to breast, ovarian, prostate, and other cancer
Duchenne muscular dystrophy	DMD gene. Point mutations, deletions, insertions, splicing mutations	Early-onset progressive muscular weakness, heart disease
Fragile-X syndrome	FMR1 gene. Primarily expanded trinucleotide (CGG) repeats and loss of function	Mental retardation, developmental disorders
Friedrich's ataxia	FXN gene. 98% of cases have expanded trinucleotide (GAA) repeats in intron 1	Ataxia, muscle weakness, spasticity, heart and other organ dysfunctions
Hemophilia A	F8 gene. Point mutations, insertions, deletions, inversions	Factor VIII blood-clotting defects, bleeding
Huntington disease	HTT (HD) gene. Trinucleotide (CAG) repeat expansions	Midlife onset of progressive motor and cognitive disorders
Lesch-Nyhan syndrome	HPRT1 gene. Point mutations, deletions, duplications	Developmental, motor, and cognitive disorders
Marfan syndrome	FBN1 gene. Point mutations, splicing mutations, deletions	Connective tissue disorders affecting numerous organs
Polycystic kidney disease, dominant	PKD1 and PKD2 genes. Sequence variants, partial or whole-gene deletions and duplications	Cysts in kidney, liver, and other organs, vascular abnormalities
Sickle cell disease	HBB gene. Point mutation leading to Glu to Val substitution at amino acid 6	Early-onset anemia

Source: GeneReviews at (http://www.ncbi.nlm.nih.gov/bookshelf/br.fcgi?book=gene).

detect the presence of known mutations in single genes that are linked to a disease (ST Table 2.3).

Although these genetic tests are extremely useful for detecting some future diseases and guiding treatment, it is clear that most disorders are multifactorial and complex. It is likely that diseases such as diabetes, Alzheimer's, and heart disease are caused by interactions between many genes, as well as by factors contributed by epigenetic effects, lifestyle, and environment. These diseases tend to be chronic and have a significant burden on health-care systems.

Genome sequencing, SNP identification, and genome-wide association studies (GWAS) are beginning to reveal some of the DNA variants that may contribute to the risk of developing multifactorial diseases such as cancer, heart disease, and diabetes. For example, the Cancer Genome Atlas project (described in Chapter 19) is amassing data equivalent to 20,000 genome projects on normal and tumor DNA from patients with 20 different types of cancer. Such studies are revealing that cancers that were once classified broadly (such as "prostate cancer") are in fact many different diseases based

on their genetic profiles. For example, in the past, blood cancers were categorized into two large groups: leukemias and lymphomas. Today, we know that each category can be broken down into more than 40 different types, based on gene mutation and expression characteristics. In addition, genome studies show that each individual tumor is unique genetically. This explosion in molecular information is now beginning to affect how these cancers are diagnosed and treated.

Although progress in genome science is accelerating, these studies have just begun to collect the vast amount of genetic data required in order to make accurate predictions and treatment choices for many diseases. As whole-genome sequencing becomes faster and more economical, scientists predict that genomics and personal genome sequencing will become a significant part of personalized diagnosis and treatment by 2015.

But what will genome-based personalized medicine look like in 2015? And how will we use it to our benefit? Also, are there ethical and social questions that we need to address as we enter the age of the new personalized medicine? The next sections will address these questions.

Analyzing One Personal Genome

In 2010, the journal *Lancet* published a report illustrating the type of information that we can currently obtain from a personal genome sequence.[1] The patient was a healthy 40-year-old male who had a family history of arthritis, aortic aneurysm, coronary artery disease, and sudden cardiac death. The researchers obtained his genomic DNA sequence using a rapid single-molecule sequencing method (Box 3). By comparing the patient's sequence with other human genome sequences in databases, a total of 2.6 million SNPs and 752 copy number variations were discovered. The researchers then sorted through the genome sequence data to determine which of these variants might have an effect on phenotype. This was accomplished by searching known SNPs in several large databases, manually creating their own disease-associated SNP database, and calculating likelihood ratios for various disease risks. The analysis required the combined efforts of more than two dozen scientists and clinicians over a period of about a year, and information gleaned from more than a dozen sequence databases, new and existing sequence analysis tools, and hundreds of individually accessed research papers.

To determine how this patient may respond to pharmaceutical drugs, the researchers searched the **PharmGKB database** (see Box 2) for the presence of known variants within pharmacogenomically important genes. The patient was found to have 63 clinically relevant SNPs within genes associated with drug reactions. In addition, his genome contained six previously unknown SNPs that could alter amino acid sequences in drug-response genes. For example, the genome sequence revealed that this patient was heterozygous for a null mutation in the *CYP2C19* gene. This mutation could make him sensitive to a range of drugs including those used to treat aspects of heart disease. He would also be more sensitive than normal to warfarin, based on SNPs within his *VKORC1* and *CYP4F2* genes. In contrast, the patient's sequence contained gene variants associated with good responses to statins; however, other gene variants suggested that he might require higher-than-normal statin dosages.

The search for mutations within genes that directly affect disease conditions revealed several potentially damaging variants. The patient was heterozygous for a SNP within the *CFTR* gene that would change a glycine to arginine at position 458. This mutation could lead to cystic fibrosis if it was passed on to a son or daughter who also inherited a defective *CFTR* gene from the other parent. Similarly, the patient was heterozygous for a recessive mutation in the hereditary haemochromatosis protein precursor gene (*HFE*), which is associated with the development of haemochromatosis, a serious condition leading to toxic accumulations of iron. Also, the patient was heterozygous for a recessive mutation in the solute carrier family 3 (*SLC3A1*) gene. This mutation is linked to cystinuria, an inherited disorder characterized by inadequate excretion of cysteine and development of kidney stones. The scientists discovered a heterozygous SNP within the parafibromin (*CDC73*) gene that would create a prematurely terminated protein. This gene is a tumor-suppressor gene linked to the development of hyperparathyroidism and parathyroid tumors. The presence of this SNP increased the risk that the patient might develop these types of tumors, if any of the patient's cells experienced a loss-of-heterozygosity mutation in the other copy of the gene.

The analysis of this patient's genome sequence for the purpose of predicting future development of multifactorial disease was more challenging. Genome-wide association studies have revealed large numbers of sequence variants that are associated with complex diseases; however, each of these variants most often contributes only a small part of the susceptibility to disease. Because not all variants have been discovered or characterized, it is difficult to establish a numerical risk score for each of these diseases based on the presence of one or more SNPs. As an example, the researchers discovered SNPs within three genes (*TMEM43*, *DSP*, and *MYBPC3*) that may be associated with sudden cardiac death. However, the exact effects of two of these SNPs are still unclear, and the other SNP had not previously been described. The patient had five SNPs in genes associated with an increased risk of developing myocardial infarction and two SNPs associated with a lower risk. Among the SNPs associated with increased risk, a variant in the apolipoprotein A precursor (*LPA*) gene is associated with a five-fold increased plasma lipoprotein(a) concentration and a two-fold increased risk of coronary artery disease. By taking into consideration the simultaneous potential effects of many different SNPs, as well as the patient's own environmental and personal lifestyle factors, the researchers concluded that the patient's genetics contributed to a significantly increased risk for eight conditions (such as type 2 diabetes, obesity, and coronary artery disease) and a decreased risk for seven conditions (such as Alzheimer's disease). The patient was offered the services of clinical geneticists, counselors, and clinical lab directors in order to help interpret the information generated from the genome sequence. Genetic counseling covered areas such as psychological and reproductive implications of genetic disease risk, the possibilities of discrimination based on genetic test results, and the uncertainties in risk assessments.

[1]Ashley, E.A., et al. 2010. Clinical assessment incorporating a personal genome. *Lancet* 375: 1525–1535.

BOX 3

How to Sequence a Human Genome

The personal genome sequence described in this chapter was the first human genome to be sequenced using a method known as true single-molecule sequencing (tSMS™). Other methods such as the Sanger method and the second-generation high-throughput methods described in Chapter 21 require cloning or PCR amplification of template DNA prior to sequencing. In contrast, the tSMS method directly sequences individual DNA strands with minimum processing. The sequencing of this genome took about a week, was performed with one machine and the services of three people, and cost $48,000. The sequence was that of Dr. Stephen Quake, a Stanford University professor who developed the technology and headed the research group.

The sequencing machine used in this study is called a HeliScope™, a commercially available refrigerator-sized instrument that processes, visualizes and analyzes the sequence data. The HeliScope analyzes about 1 billion individual DNA molecules simultaneously per day as they are undergoing copying synthesis. The following is a brief summary of the tSMS method.

1. DNA is extracted from the patient's sample, fragmented into short pieces about 100 to 200 nucleotides long, and denatured into single strands.

2. PolyA tails are added to the 3′ ends of each single strand, and each strand is tagged with a fluorescent molecule. These are the template strands for DNA synthesis.

3. The template strands are hybridized to billions of oligo-T molecules immobilized on the surface of a flow cell.

4. The positions of each of these template-oligo-T molecules is visualized and located on the flow cell using the HeliScope's laser illumination and a high magnification camera.

5. The fluorescent tag molecules are cleaved from the templates and washed off.

6. Fluorescently labeled nucleotides (A, T, C, or G) along with a DNA polymerase are added, one nucleotide per reaction cycle, to the flow cell. If a template strand contains a complementary nucleotide next to the oligo-T primer sequence (let's say a G), then a fluorescent C will be incorporated during the C cycle (ST Figure 2–7).

7. After washing away the DNA polymerase and excess fluorescent C, those templates that contained a G

at that position will fluoresce in the HeliScope and their location will be established (ST Figure 2–8).

8. After imaging the incorporation of the G, the fluorescent tag on the incorporated G nucleotides is cleaved off and washed away along with the DNA polymerase.

9. In the next cycle, DNA polymerase and another fluorescently tagged nucleotide (let's say T) are added to the flow cell. The T nucleotides will be incorporated into the growing complementary DNA strands if the template contains an A nucleotide next to the growing 3′ end.

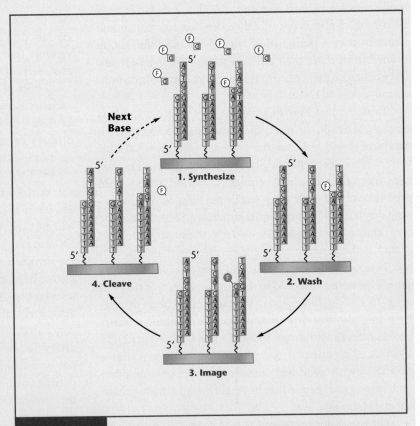

ST FIGURE 2–7 **Steps in true single-molecule sequencing (tSMS) technique.** Oligo-T molecules, attached to a solid substrate, anneal to genomic DNA fragments. Step 1: A fluorescently labeled nucleotide (F-C) and DNA polymerase are added. F-C molecules are synthesized onto the oligo-T molecules, if a G residue is present in the genomic DNA. Steps 2 and 3: Excess F-C and DNA polymerase are washed off, and the presence of a labeled oligo-T molecule is imaged and the location of the label registered. Step 4: Fluorescent label (F) is cleaved from the incorporated F-C molecules. The steps are repeated with other fluorescently labeled nucleotides until synthesis reaches approximately 32 nucleotides.

(Continued)

BOX 3
How to Sequence a Human Genome (*Continued*)

10. The presence of a fluorescent T will be detected and mapped by the HeliScope, and the process is repeated through enough cycles to synthesize about 32 nucleotides of complementary DNA on each template.

11. The HeliScope determines the sequence of each of the template strands based on these cycles of synthesis and fluorescent imaging. It then performs sequence alignments of all the individual sequences in order to compile the entire genomic sequence.

The methods and analysis of tSMS are illustrated in a video available on the Helicos BioSciences Corporation Web site at: http://www.helicosbio .com/Technology/TrueSingleMolecule Sequencing/tabid/64/Default.aspx

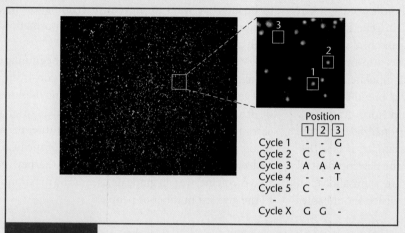

| | Position | | |
	1	2	3
Cycle 1	-	-	G
Cycle 2	C	C	-
Cycle 3	A	A	A
Cycle 4	-	-	T
Cycle 5	C	-	-
Cycle X	G	G	-

ST FIGURE 2–8 **Images taken by a HeliScope single-molecule sequencer.** A magnified view of a flow cell is shown on the left. On the right is a higher magnification of part of the flow cell, showing the position of individual molecules that have incorporated a fluorescent G (F-G) nucleotide during "Cycle X" of the most recent synthesis cycle. In this case, molecules 1 and 2 have incorporated an F-G, but molecule 3 has not.

Technical, Social, and Ethical Challenges

There are still many technical hurdles to overcome before personalized medicine will become a standard part of medical care. The technologies of genome sequencing and alignment, microarray analysis, and SNP detection need to be faster, more accurate, and cheaper. Scientists expect that these challenges will be overcome in the near future; however, personalized genome analysis needs to be used with caution until the technology becomes highly accurate and reliable. Even a low rate of error in genetic sequences or test results could lead to erroneous diagnoses and inappropriate treatments. Perhaps an even greater challenge lies in the ability of scientists to store and interpret the vast amount of emerging sequence data. Each personal genome generates the letter-equivalent of 200 large phone books, which must be stored in databases, mined for relevant sequence variants and meaning assigned to each SNP. To undertake these kinds of analyses, scientists need to analyze the results from large-scale population genotyping studies that will link sequence variants to phenotype, disease, or drug responses. Experts suggest that such studies will take the coordinated efforts of public and private research teams, and more than a decade to complete. Scientists will also need to develop efficient automated systems and algorithms to deal with this massive amount of information. Moreover, these data analyses will need to take into account the fact that genetic variants contribute only partially to personal phenotype. Personalized medicine will also need to integrate information about environmental, personal lifestyle, and epigenetic factors.

Another technical challenge for personalized medicine is the development of automated health information technologies. Health-care providers will need to use electronic health records to store, retrieve, and analyze each patient's genomic profile, as well as to compare this information with constantly advancing knowledge about genes and disease. Currently, fewer than 10 percent of hospitals and physicians in the United States have access to these types of information technologies.

Personalized medicine has a number of societal implications. To make personalized medicine available to everyone, the costs of genetic tests, as well as the genetic counseling that accompanies them, must be reimbursed by insurance companies, even in cases where there is no prior disease or symptoms. Regulatory changes are required to ensure that genetic tests and genomic sequencing are accurate and that the data generated are reliably stored in databases that guarantee the patient's privacy. At the present time, less than 1 percent of genetic tests are regulated by agencies such as the FDA.

Personalized medicine also requires changes to medical education. In the future, physicians will be expected to use genomics information as part of their patient management. For this to be possible, medical schools will need to train future physicians to interpret and explain genetic data. In addition, more genetic counselors and genomics specialists will be required. These specialists will need to understand genomics and disease, as well as to manipulate bioinformatic data. As of 2010, there were only about 2500 genetic counselors and 1100 clinical geneticists in North America.

The ethical aspects of the new personalized medicine are also diverse and challenging. For example, it is sometimes argued that the costs involved in the development of genomics and personalized medicine are a misallocation of limited resources. Some argue that science should solve larger problems facing humanity, such as the distribution of food and clean water, before embarking on personalized medicine. Similarly, some critics argue that such highly specialized and expensive medical care will not be available to everyone and represents a worsening of economic inequality. There are also concerns about how we will protect the privacy of genome information that is contained in databases and private health-care records. In addition, there needs to be effective ways to prevent discrimination in employment or insurance coverage, based on information derived from genomic analysis.

Most experts agree that we are at the beginning of a personalized medicine revolution. Information from genetics and genomics research is already increasing the effectiveness of drugs and enabling health-care providers to predict diseases prior to their occurrence. In the future, personalized medicine will touch almost every aspect of medical care. By addressing the upcoming challenges of the new personalized medicine, we can guide its use for the maximum benefit to the greatest number of people.

Selected Readings and Resources

Journal Articles

Ashley, E.A., et al. 2010. Clinical assessment incorporating a personal genome. *Lancet* 375: 1525–1535.

Collins, F. 2010. Has the revolution arrived? *Nature* 464: 674–675.

Hamburg, M.A., and Collins, F.S. 2010. The path to personalized medicine. *New England Journal of Medicine* 363: 301–304.

Manolio, T.A., et al. 2009. Finding the missing heritability of complex diseases. *Nature* 461: 747–753.

Ormond, K.E., et al. 2010. Challenges in the clinical application of whole-genome sequencing. *Lancet* 375: 1749–1751.

Pushkarev, D., et al. 2009. Single-molecule sequencing of an individual human genome. *Nature Biotech* 27: 847–850.

Ross, J.S. 2009. The HER-2 receptor and breast cancer: ten years of targeted anti-HER-2 therapy and personalized medicine. *The Oncologist* 14: 320–368.

Venter, J.C. 2010. Multiple personal genomes await. *Nature* 464: 676–677.

Weinshilboum, R., and Wang, L. 2004. Pharmacogenomics: Bench to bedside. *Nature Rev Drug Disc* 3: 739–748.

Web Sites

Personalized Medicine Coalition, 2009. The Case for Personalized Medicine. http://www.personalizedmedicinecoalition.org/sites/default/files/TheCaseforPersonalizedMedicine_5_5_09.pdf

University of Washington, 2010. GeneTests Web site: http://www.ncbi.nlm.nih.gov/sites/GeneTests/?db=GeneTests

U.S. Department of Energy Genome Program's Human Genome Project Information, 2008. Pharmacogenomics. http://www.ornl.gov/sci/techresources/Human_Genome/medicine/pharma.shtml

U.S. Food and Drug Administration, 2010. Table of Valid Genomic Biomarkers in the Context of Approved Drug Labels. http://www.fda.gov/Drugs/ScienceResearch/ResearchAreas/Pharmacogenetics/ucm083378.htm

SPECIAL TOPICS IN MODERN GENETICS

Epigenetics

The newly emerging field of epigenetics is providing us with a basis for understanding how heritable changes other than those in DNA sequence can influence phenotypic variation.

The somatic cells of the human body contain 20,000 to 25,000 genes. Yet in any of these cells, only a relatively small percentage of all genes are active. In the more than 200 different cell types present in the body, different cell-specific gene sets are transcribed, while the rest of the genome is transcriptionally inactive. In addition, programs of gene expression become more and more restricted during development and differentiation as embryonic cells gradually become specialized adult cells with distinct phenotypes. The prevailing view has been that regulation of gene expression is coordinated by promoter, promoter–proximal, enhancer, and other *cis*-regulatory elements as well as DNA-binding proteins and transcription factors. This regulation can occur at any of the steps in gene expression. For example, steps in transcriptional regulation, as well as mRNA processing and other stages of posttranscriptional regulation, control the amount of gene product synthesized from a DNA template. However, as we learn more about genome organization and the regulation of gene expression, it is clear that classical genetic mechanisms cannot explain how some phenotypes arise. Monozygotic twins, for example, have identical genotypes but are not always phenotypically identical. Although for each gene, one allele is inherited maternally and one paternally, in some cases, only the maternal or paternal allele is expressed, while the other is transcriptionally silent.

The newly emerging field of epigenetics is providing us with a basis for understanding how heritable changes other than those in DNA sequence can influence phenotypic variation. These advances greatly extend our understanding of the molecular basis of gene regulation and apply to wide-ranging areas including genetic disorders, cancer, and behavior.

An **epigenetic trait** is a stable, mitotically and meiotically heritable phenotype that results from changes in gene expression without alterations in the DNA sequence. **Epigenetics** is the study of the ways in which these changes alter cell- and tissue-specific patterns of gene expression. Epigenetic regulation of gene expression uses reversible modifications of DNA and chromatin structure to mediate the interaction of the genome with a variety of environmental factors and to generate changes in the patterns of gene expression in response to these factors. The **epigenome** refers to the epigenetic state of a cell. During its life span, an organism has one genome, but this genome can be modified in diverse cell types at different times to produce many epigenomes.

Current research efforts are focused on several aspects of epigenetics: how an epigenome arises in developing and differentiated cells, what mechanisms maintain these states, and how they are transmitted via mitosis and meiosis, making them heritable traits. In addition, because epigenetically controlled alterations to the genome are associated with cancer and some common diseases such as diabetes and asthma, efforts are also being directed at the development of drugs that can modify or reverse disease-associated epigenetic changes in cells.

Several systems and pathways that result in the establishment, maintenance, and inheritance of the epigenetic state are recognized. These pathways are organized into three categories, but other mechanisms and categories undoubtedly remain to be elucidated. The first category includes environmental signals called *epigenators* that are received by the cell and that stimulate a response via an intracellular pathway. Responses to epigenator signals are called *epigenetic initiators*. Components of this second category produce epigenetic changes. These initiators include protein–protein signal transduction pathways, DNA-binding proteins, and noncoding RNAs. The actions of initiators define the location at which epigenetic changes in chromatin will take place. DNA sequence recognition is a necessary part of this response. Once the epigenetic modifications have occurred, they are maintained by molecular elements called *epigenetic maintainers*. Components of this third category conserve and sustain the epigenetic changes in the present and future generations. Epigenetic maintainers are not sequence-specific, they operate anywhere in the genome, and they depend on initiators to specify the loci at which chromatin modifications will take place. Epigenetic maintainers ensure that epigenetic modifications are transmitted by mitosis to daughter cells, or by meiosis to gametes and to subsequent generations. Maintainers include DNA methylation and histone modifications. ST Figure 3–1 shows these components of the epigenetic pathway along with examples. It should be noted that histone modifications are part of the general process of transcriptional

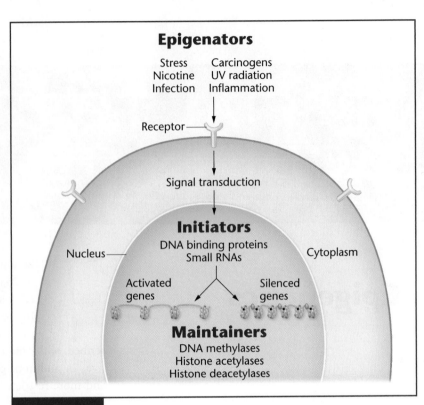

ST FIGURE 3–1   An epigenetic pathway involves the reception and processing of external signals resulting in alterations that change and maintain patterns of gene expression.

activation; this means that some, but not all, histone modifications are epigenetic events.

Epigenetic modification of the genome provides a clear-cut response to environmental signals by mediating changes in gene expression that persist through several generations. Epigenetics has been implicated in many aspects of biology, including the progressive restriction of gene expression during development, allele-specific expression seen in gene imprinting, and environment–genome interactions during prenatal development that affect adult phenotypes. Abnormal regulation of the epigenome leads to human genetic disorders including Prader–Willi syndrome, Angelman syndrome, and Beckwith–Weidemann syndrome. The loss or alteration of other epigenetic states results in cancer.

Knowledge of the mechanisms of epigenetic modifications to the genome, how these modifications are maintained and transmitted, and their relationship to basic biological processes will be important for our understanding of development, disease processes, reproduction, and the evolution of adaptation to the environment.

Here we will focus on how epigenetic alterations impact some genetic diseases, cancer, and environment–genome interactions. Because epigenetic changes are potentially reversible, we will also examine how knowledge of molecular mechanisms of epigenetics is being used to develop drugs and treatments for human diseases.

SPECIAL TOPIC III

BOX 1
The Beginning of Epigenetics

C.H. Waddington is credited with coining the term *epigenetics* in the 1940s to describe how environmental influences on developmental events can affect the phenotype of the adult. He showed that environmental alterations during development induced alternative phenotypes in organisms with identical genotypes. Using *Drosophila melanogaster*, Waddington found that wing vein patterns could be altered by administering heat shocks during pupal development. Adults with these environmentally induced changes were used to establish strains that showed the alternative phenotype without the need for continued environmental stimulus. He called this phenomenon "genetic assimilation." In other words, interactions between the environment and the genome during certain stages of development produce phenotypic effects that are heritable.

In the 1970s, Holliday and Pugh proposed that changes in programs of gene expression during development may depend on the methylation of specific bases in DNA, and that altering these patterns of methylation might affect the resulting phenotype. In addition, the progressive restriction of developmental pathways during differentiation may be explained by a program of methylation that silences gene sets at specific stages of embryogenesis. Several factors, including Waddington's work, the methylation model of Holliday and Pugh, and the discovery that expression of genes from both the maternal and paternal genomes is required for normal development, all helped set the stage for the birth of epigenetics and epigenomics as fields of scientific research.

Epigenetic Alterations to the Genome

Unlike the genome, which is identical in all cell types of an organism, the epigenome is cell-type specific and heritable. Like the genome, the epigenome can be transmitted to daughter cells by mitosis and to future generations by meiosis. Unlike genomic mutations, epigenetic modifications are reversible, opening the possibility for the development of drugs for treatment of conditions linked to dysfunction of epigenetic processes. In the following sections, we will examine mechanisms of epigenetics and their role in imprinting, cancer, and environment–genome interactions, providing a snapshot of the many roles played by this recently discovered mechanism of gene regulation.

There are three major epigenetic mechanisms: (1) reversible modification of DNA by the addition or removal of methyl groups; (2) modification of histones by the addition or removal of chemical groups; and (3) regulation of gene expression by small, noncoding RNA molecules.

Methylation

In mammals, methylation of DNA takes place after replication and involves the addition of a methyl group ($-CH_3$) to cytosine, a reaction catalyzed by methyltransferase enzymes. DNA methylation also occurs during the differentiation of adult cells. In both instances, methylation takes place almost exclusively on cytosine bases adjacent to a guanine, a combination called a CpG dinucleotide. Many of these dinucleotides are clustered in regions, called CpG islands, located in and near promoter sequences adjacent to genes (ST Figure 3–2). Islands adjacent to essential genes (housekeeping genes) and cell-specific genes are unmethylated, making these genes available for transcription. Other genes with adjacent methylated CpG islands are transcriptionally silenced. The methyl groups in CpG islands occupy the major groove of DNA, and block the binding of transcription factors necessary to form transcription complexes.

The bulk of methylated CpG dinucleotides are not adjacent to genes, but are found in repetitive DNA sequences located in heterochromatic regions of the genome, including the centromere. Methylation of these sequences contributes to silencing the transcription and replication of transposable elements such as LINE and SINE sequences (see Chapter 12), which constitute a major fraction of the human genome. Heterochromatic methylation also maintains chromosome stability by preventing translocation and other chromosomal abnormalities.

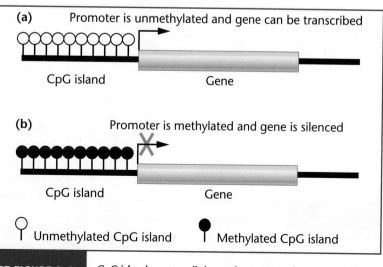

(a) Promoter is unmethylated and gene can be transcribed

CpG island   Gene

(b) Promoter is methylated and gene is silenced

CpG island   Gene

○ Unmethylated CpG island   ● Methylated CpG island

ST FIGURE 3–2   CpG islands are usually located upstream of promoter regions.

As part of dosage compensation, X chromosomes in mammalian females are inactivated by converting them to heterochromatin. These inactivated chromosomes have altered patterns of DNA methylation (see Chapter 7 for a detailed explanation of X inactivation). As mentioned above, CpG methylation in euchromatic regions causes a parent-specific pattern of gene transcription.

Histone Modification

In addition to DNA methylation, histone modification is an important epigenetic mechanism of gene regulation. Recall that chromatin is composed of DNA wound around an octamer core of histone proteins to form nucleosomes. Amino acids in the N-terminal region of these histones can be covalently modified in several ways, including acetylation, methylation, and phosphorylation (ST Figure 3–3). These modifications occur at conserved amino acid sequences in the N-terminal histone tails which protrude from the nucleosome. Chemical modification of histones alters the structure of chromatin, making genes accessible or inaccessible for transcription. Normally, when histones are modified by acetylation, a reaction catalyzed by the enzyme histone acetyltransferase (HAT), chromatin structure becomes "open," making genes on these modified nucleosomes available for transcription (ST Figure 3–4(a)). This modification is reversible, and acetyl groups can be removed by another enzyme, histone deacetylase (HDAC), changing the chromatin to a "closed" configuration, and silencing genes by making them unavailable for transcription (ST Figure 3–4(b)). However, the program of histone modification depends on other events, including the presence or absence of methyl groups on histones.

The program of amino acid modification is complex, and the same type of modification can often have opposite transcriptional outcomes. For example, methylation of lysine 4 or 27 on histone H3 causes transcriptional activation, while methylation of lysine 9 causes transcriptional silencing. We are learning that specific combinations of histone modifications control the transcriptional status of a chromatin region. For example, whether or not lysine 9 on histone H3 will be methylated is controlled by modifications made elsewhere on this protein. On one hand, if serine 10 is phosphorylated, methylation of lysine 9 is inhibited. On the other hand, if lysine 14 is deacetylated, methylation of lysine 9 is facilitated. The sum of the complex patterns and interactions of histone modifications that change chromatin organization and gene expression is called the **histone code**. The details of how this level of regulation works are beginning to emerge but are not yet fully understood.

ST FIGURE 3–3   Clusters of histones, nucleosomes are the focus of epigenetic modifications. Ac = acetyl groups, Me = methyl groups, P = phosphate groups.

RNA Interference

In addition to DNA methylation and histone modification, small, noncoding RNA molecules (discussed in Chapter 17) also participate in epigenetic regulation of gene expression. After transcription, these small interfering RNA (siRNA) molecules associate with protein complexes to form RNA-Induced Silencing Complexes (RISCs). RISCs bind to mRNA molecules that carry sequences complementary to siRNA in the RISC. If the siRNA is not perfectly complementary to the mRNA, the binding interferes with translation, resulting in downregulation of gene expression. If, however, the siRNA in the RISC *is* perfectly complementary to sequences in the mRNA, the mRNA is cleaved and destroyed, effectively silencing the gene.

Recently, it has been discovered that siRNAs can silence genes by directly interfering with transcription initiation. This does not involve any changes in existing epigenetic

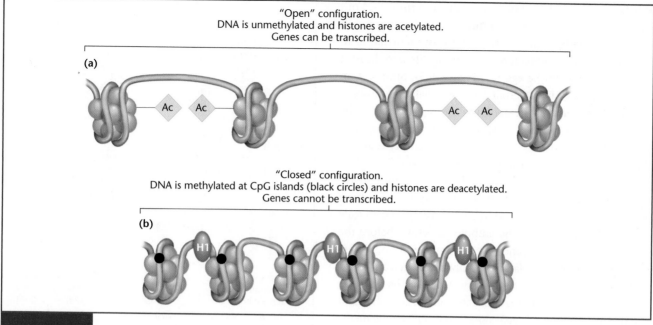

ST FIGURE 3–4   Epigenetic modifications to the genome.

promoter modifications, nor does it require new modifications. Instead, siRNAs complementary to promoter regions bind to a promoter. Binding blocks the assembly of the pre-initiation complex by preventing binding of transcription factor TFIIB and RNA polymerase.

Short RNA molecules can also associate with protein complexes to form RNA-Induced Transcriptional Silencing (RITS) complexes. RITS complexes initiate formation of facultative heterochromatin that silences genes located within these newly created heterochromatic regions. Unlike the heterochromatin at telomeres and centromeres, which is constitutive, the heterochromatic state in facultative heterochromatin is reversible and can be converted to euchromatin, with genes in this region once again accessible for transcription.

In sum, epigenetic modifications alter chromatin structure by several mechanisms including DNA methylation, histone acetylation, and RNA interference, without changing the sequence of DNA. These epigenetic changes create an epigenome that in turn, can regulate normal development or generate responses to environmental signals.

Epigenetics and Imprinting

Mammals inherit a maternal and a paternal copy of each autosomal gene, and either or both copies of these genes can be expressed in the offspring. Imprinted genes show expression of only the maternal allele or the paternal allele. This parent-specific pattern of allele expression is laid down during gamete formation. Differential methylation of CpG-rich regions produce allele-specific imprinting and subsequent gene silencing.

In mice, fewer than 100 genes were thought to be imprinted, but recent work has identified more than 1300 imprinted genes. In humans, more than 150 candidate genes are thought to be imprinted, but the findings in mice suggest that many more imprinted genes remain to be identified in humans.

Once a gene has been methylated and imprinted, it remains transcriptionally silent during embryogenesis and development. Most imprinted genes direct aspects of growth during prenatal development. For example, in mice, genes on the maternal X chromosome are expressed in the placenta, while genes on the paternal X chromosome are silenced. At the level of individual genes, having only one functional allele makes these genes highly susceptible to the deleterious effects of mutations. Because imprinted genes are clustered, mutation in one gene can have an impact on the function of adjacent imprinted genes, amplifying its impact on the phenotype. Mutations in imprinted genes can arise by changes in the DNA sequence or by epigenetic changes, called **epimutations**, both of which are heritable changes in the activity of a gene.

The pattern of imprinting in mammals is reprogrammed every generation. For example, females receive a maternal and a paternal set of chromosomes. In somatic cells and in germ cells, the maternal chromosome set has female imprints, and the paternal set contains male imprints.

When gamete formation begins in female germ cells, both chromosome sets have their imprints erased and are each reprogrammed by changing the pattern of methylation to carry a female imprint pattern that is transmitted to the next generation through the egg (ST Figure 3–5). Similarly, in male germ cells, the paternal and maternal chromosome sets have their imprints erased and are reprogrammed by methylation to become a male imprinted set. Reprogramming occurs at two stages: in the parental germ cells and in the developing embryo just before implantation. In the first stage, erasure by demethylation and reprogramming by re-methylation lay down a parent-specific imprinting pattern in germ cells of the parent. In the second stage, large-scale demethylation occurs in the embryo sometime before the 16-cell stage of development. After implantation, differential genomic remethylation recalibrates which maternal alleles and which paternal alleles will be inactivated. It is important to remember that once imprinted, alleles remain inactive in all cells, while genes silenced by epigenetic methylation can be reactivated by external signals during or after differentiation.

Most human disorders associated with imprinting have their origins during fetal growth and development. Imprinting defects cause Prader–Willi syndrome, Angelman syndrome, Beckwith–Wiedemann syndrome, and several other diseases (ST Table 3.1). However, given the number of candidate genes and the possibility that additional imprinted genes remain to be discovered, the overall number of imprinting-related genetic disorders may be much higher.

In humans, most known imprinted genes encode growth factors or other growth-regulating genes. Generally, maternally expressed alleles of imprinted genes suppress growth, and paternal alleles enhance growth. One autosomal dominant disorder of imprinting, Beckwith–Wiedemann syndrome (BWS), offers insight into how disruptions of epigenetically imprinted genes lead to an abnormal phenotype. BWS is a prenatal overgrowth disorder with abdominal wall defects, enlarged organs, large birth weight, and predisposition to cancer. BWS is not caused by mutation, nor is it associated with any chromosomal aberration. Instead it is a disorder of imprinting and is caused by abnormal methylation patterns.

Genes linked to BWS are located in a cluster of imprinted genes on the short arm of chromosome 11. This cluster contains more than a dozen imprinted genes, some of which are paternally expressed, while others are maternally expressed and all genes in this cluster regulate growth during prenatal development. The imprinted region is subdivided into two separately regulated domains, one of which contains the closely linked genes *IGF2* (insulin growth factor 2) and *H19*. Normally, the paternal allele of *IGF2* is expressed, and the allele on the maternal homolog is imprinted and silenced. In the case of *H19,* the situation

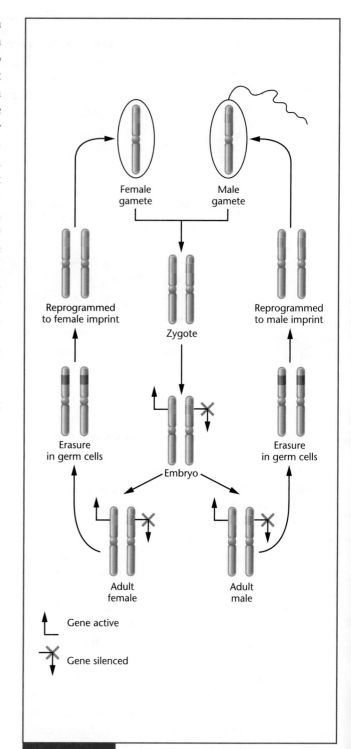

ST FIGURE 3–5   Imprinting patterns are reprogrammed each generation.

is usually the reverse: The maternal allele is expressed, and the paternal allele is imprinted and silenced. The protein encoded by *IGF2* is a growth factor, and the product of the *H19* gene is a long, noncoding RNA that is a growth repressor. Expression of these genes is normally controlled by an imprinting control region (ICR) located within its chromosomal domain.

Some Imprinting Disorders in Humans

Disorder	Locus
Albright hereditary osteodystrophy	20q13
Angelman syndrome	15q11-q15
Beckwith–Wiedemann syndrome	11p15
Prader–Willi syndrome	15q11-q15
Silver–Russell syndrome	Chromosome 7
Uniparental disomy 14	Chromosome 14

Abnormal maternal imprinting in one or both domains can lead to BWS, but many affected individuals have a loss of imprinting of the maternal *IGF2* allele. This causes both the maternal and paternal alleles to be transcriptionally active, resulting in the overgrowth of tissues characteristic of this disease.

The known number of imprinted genes represent only a small fraction of the genome, but they play major roles in controlling growth during embryonic and prenatal development. Because they act so early in life, external or internal factors that disturb the epigenetic pattern of imprinting or the expression of imprinted genes can have serious phenotypic consequences.

Following the introduction of *in vitro* fertilization (IVF) and embryo cloning in the production of farm animals, a high incidence of birth defects and neonatal deaths was noted. Studies showed that many of these problems were linked to abnormal epigenetic alterations during embryonic growth and development. These findings in animals raised the possibility that the use of IVF and other assisted reproductive technologies (ART) in humans may cause problems with imprinted genes.

Several studies have shown that children born after IVF are at risk for low or very low birth weight, a condition that may result from abnormal imprinting. The use of IVF and other ART procedures has also been associated with a three- to six-fold increased risk of Beckwith–Wiedemann syndrome (BWS). One study examined the epigenome of children with BWS who were born after a procedure called intracytoplasmic sperm injection (ICSI), where a single sperm is injected into the oocyte. The results showed that a majority of these children had specific changes in both imprinting and epigenetic patterns of methylation. Other studies have shown that more than 90 percent of children born with BWS after ART had imprinting defects. Because imprinting disorders are uncommon (BWS occurs with a frequency of about 1 in 15,000 births), large-scale and longitudinal studies will be needed to resolve a causal relationship among imprinting abnormalities, growth disorders, and ART.

Epigenetics and Cancer

Epidemiological studies investigate the role of environmental factors in normal phenotypic variation and as risk factors for disease. In many cases it is difficult to obtain quantitative measures of the effects of multidimensional environmental influences on an organism. For some complex diseases, there are strong links with environmental factors such as the association between smoking and lung cancer. The discovery that epigenetics mediates the process of changing patterns of gene expression in response to environmental signals offers a new and potentially more direct approach to understanding the interactions between the genome and the environment in diseases such as cancer.

Following the discovery of cancer-associated genes, including tumor-suppressor genes and proto-oncogenes, research into the genetic basis of cancer focused mainly on mutant alleles of genes involved in several cellular functions, including the cell cycle, differentiation, adhesion, and apoptosis. Until recently, the conventional view has been that cancer is clonal in origin and begins in a cell that has accumulated a suite of dominant and recessive mutations that allow it to escape control of the cell cycle. Subsequent mutations allow cells of the tumor to acquire metastatic properties, and the cancer spreads to other locations in the body where it establishes new malignant tumors. Converging lines of evidence are clarifying the role that epigenetic changes play in the initiation and maintenance of malignancy. These findings help explain properties of cancer cells that are difficult to explain by the action of mutant alleles alone. Evidence for the role of epigenetic changes in cancer now challenges the conventional paradigm for the origin of cancer and establishes epigenomic changes as a major pathway for the formation and spread of malignant cells.

The relationship between epigenetics and cancer was first noted in the 1980s by Feinberg and Vogelstein who observed that colon cancer cells had much lower levels of methylation than normal cells derived from the same tissue. Subsequent research by many investigators showed that global hypomethylation is a property of all cancers examined to date. In the ensuing years, it has become clear that the epigenetic states of normal cells are greatly altered in cancer cells and that other epigenetic changes, including selective hypermethylation and gene silencing, are also present in cancer cells. Cancer is now being viewed as a disease that involves both epigenetic *and* genetic changes that lead to alterations in gene expression (ST Figure 3–6).

DNA hypomethylation reverses the inactivation of genes, leading to unrestricted transcription of many gene sets including oncogenes. It also relaxes control over imprinted genes, causing cells to acquire new growth properties. Hypomethylation of repetitive DNA sequences in

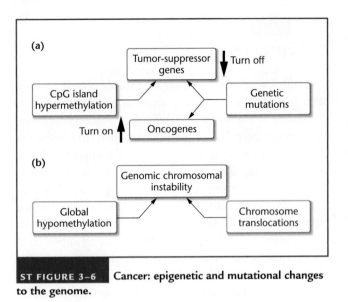

ST FIGURE 3–6 **Cancer: epigenetic and mutational changes to the genome.**

heterochromatic regions is associated with an increase in chromosome rearrangements and changes in chromosome number, both of which are characteristic of cancer cells. In addition, hypomethylation of repetitive sequences leads to transcriptional activation of transposable DNA sequences such as LINEs and SINEs, further increasing genomic instability.

While widespread hypomethylation is a hallmark of cancer cells, hypermethylation at CpG islands and inactivation of certain genes, including tumor-suppressor genes (ST Table 3.2), are also found in many cancers, often in a

ST TABLE 3.2

Some Cancer-Related Genes Inactivated by Hypermethylation in Human Cancers

Gene	Locus	Function	Related Cancers
BRCA1	17q21	DNA repair	Breast, ovarian
APC	5q21	Nucleo-cytoplasmic signaling	Colorectal, duodenal
MLH1	3p21	DNA repair	Colon, stomach
RB1	13q14	Cell-cycle control point	Retinoblastoma, osteosarcoma
AR	Xq11-12	Nuclear receptor for androgen; transcriptional activator	Prostate
ESR1	6q25	Nuclear receptor for estrogen; transcriptional activator	Breast, colorectal

tumor-specific pattern. For example *BRCA1* is hypermethylated and inactivated in breast and ovarian cancer, and *MLH1* is hypermethylated in some forms of colon cancer. Inactivation of tumor-suppressor genes by hypermethylation is thought to play an important complementary role to mutational changes that accompany the transformation of normal cells into malignant cells. For example, in a bladder cancer cell line, one allele of the cell cycle control gene *CDKN2A* is mutated, and the other, normal allele is inactivated by hypermethylation of its CpG island. The inactivation of both alleles allows these cells to escape control of the cell cycle and divide continuously. In many clinical cases, the combination of mutation and hypermethylation occurs in familial forms of cancer.

However, genes other than tumor-suppressor genes are also hypermethylated in some cancer cells; these include genes that control or participate in DNA repair, differentiation, apoptosis, and drug resistance. In fact, the majority of hypermethylated genes in cancer cells are not tumor-suppressor genes, suggesting that the pattern of hypermethylation may result from a widespread deregulation of the methylation process rather than a targeted event.

In addition to altered patterns of methylation, many cancer cells also have disrupted histone modification profiles. In some cases, mutations in the genes encoding members of the histone-modifying proteins histone acetyltransferase (HAT) and histone deacetylase (HDAC) are linked to the development of cancer. For example, those affected with Rubenstein–Taybi syndrome inherit a germ-line mutation that produces a dysfunctional HAT and have a greater than 300-fold increased risk of cancer. In other cases, HDAC complexes are selectively recruited to tumor-suppressor genes by mutated, oncogenic DNA binding proteins. Action of the HDAC complexes at these genes converts the chromatin to a closed configuration and inhibits transcription, causing the cell to lose control of the cell cycle.

The mechanisms that cause epigenetic changes in cancer cells are not known, partly because they take place very early in the conversion of a normal cell to a cancerous one, and partly because by the time the cancer is detected, alterations in the methylation pattern have already occurred. The fact that such changes occur very early in the transformation process has led to the proposal that initiating epigenetic changes leading to cancer may occur in stem cells residing in normal tissue. Three lines of evidence support this idea. First, epigenetic mechanisms can replace mutations as a way of silencing individual tumor-suppressor genes or activating oncogenes. Second, global hypomethylation may cause genomic instability and the large-scale chromosomal changes that are a characteristic feature of cancer. Third, because epigenetic modifications can silence multiple genes, they are more effective than serial mutations of single genes in transforming normal cells into malignant cells. A model of cancer based on epigenetic changes in colon stem cells as

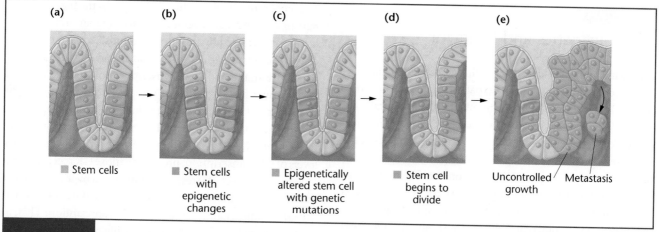

(a) ■ Stem cells

(b) ■ Stem cells with epigenetic changes

(c) ■ Epigenetically altered stem cell with genetic mutations

(d) ■ Stem cell begins to divide

(e) Uncontrolled growth / Metastasis

ST FIGURE 3–7 Epigenetic stem cell model proposes that both epigenetic changes and mutations are involved in the origins of cancer.

SPECIAL TOPIC III

initiating events in carcinogenesis followed by mutational events is shown in ST Figure 3–7

In addition to changing ideas about the origins of cancer, the fact that epigenetic changes are potentially reversible makes the development of therapeutic drugs for cancer treatment a possible new approach to chemotherapy. The focus of epigenetic therapy is the reactivation of genes that have been silenced by methylation or histone modification, essentially reprogramming the pattern of gene expression in cancer cells. Several drugs for altering epigenetic genome modifications are in clinical trials, and one (decitabine, marketed as Vidaza) has been approved by the U.S. Food and Drug Administration for treatment of acute myeloid leukemia and myelodysplastic syndrome, a precursor to leukemia. This drug is an analog of cytidine

and is incorporated into DNA during replication during the S phase of the cell cycle. Methylation enzymes (methyltransferases) bind irreversibly to decitabine, preventing methylation of DNA at many other sites, effectively reducing the amount of methylation in cancer cells. Other drugs that inhibit histone deacetylases (HDAC) are also being investigated for use in epigenetic therapy. Experiments with cancer cell lines indicate that inhibiting HDAC activity results in the reexpression of tumor-suppressor genes. One HDAC inhibitor now in clinical trials may be approved for the treatment of some forms of lymphoma. Further research into the mechanisms and locations of epigenetic genome modification in cancer cells will allow the design of more potent specific drugs to target epigenetic events as a form of cancer therapy.

The discovery that epigenetic changes may be as important as genetic changes in the origin, maintenance, and metastasis of cancers has opened new avenues of cancer research. Key discoveries about epigenetic mechanisms include the finding of tumor-specific deregulation of genes by altered DNA methylation profiles and histone modifications, the discovery that epigenetic changes

in histones or DNA methylation are interconnected, and the recognition that epigenetic changes can affect hundreds of genes in a single cancer cell. These advances were made in the span of a few years, and while it is clear that epigenetics plays a key role in cancer, many questions remain to be answered before we can draw conclusions about the relative contributions of genetics and epigenetics to the development of cancer. Some of these questions are as follows:

- Is global hypomethylation in cancer cells a cause or effect of the malignant condition?

- Do these changes arise primarily in stem cells or in differentiated cells?
- Once methylation alterations begin, what triggers hypermethylation in cancer cells?
- Is hypermethylation a process that targets certain gene classes, or is it a random event?
- Can we develop drugs that target cancer cells and reverse tumor-specific epigenetic changes?
- Can we target specific genes for reactivation, while leaving others inactive?

Epigenetics and Behavior

A growing body of evidence shows that epigenetic changes, including alterations in methylation patterns, and histone modification may be important components of behavioral phenotypes. In mice, two regions of the brain show preferential expression of parental genes. In one of these (the hypothalamus), sex-specific imprinting was more prominent in female offspring, suggesting parental effects on the function of this brain region in daughters. Altogether, parent-of-origin effects have been shown in more than 800 genes, supporting the idea that imprinting in different regions of the brain may represent a major form of epigenetic regulation.

In humans, epigenetic changes have been documented in the progression of neurodegenerative disorders and neuropsychiatric diseases with altered behavioral phenotypes. Epigenetic changes in the nervous system have been documented in cases of Alzheimer disease, Parkinson's disease, Huntington disease, and neuropsychiatric diseases including schizophrenia and bipolar disorder. Because the appearance of these disorders involves genetic predispositions, events in development, environmental effects, and the existence of transgenerational epigenetic effects, it has been difficult to define the precise role of epigenomic changes in neural disorders. In addition, different brain regions in unaffected individuals show different epigenetic profiles, adding a layer of complexity to the interpretation of results. Perhaps the most relevant problem is the difficulty of obtaining brain tissue for analysis. Most of the data on epigenetic alterations have been derived from samples obtained at autopsy, a situation that makes it impossible to directly correlate these alterations with the origin or progression of these disorders. However, in spite of these difficulties, there is enough evidence to suggest that in the future, epigenetics should be considered as part of the diagnosis and treatment of complex neural disorders.

One of the fastest-growing and perhaps most controversial theories related to epigenomics concerns the idea that epigenetic alterations linked to environmental signals during development or early in life influence behavior (and physical health) later in adult life. One of the first studies in behavioral epigenomics, published by Szyf and Meaney and their colleagues in 2004, showed that variations in rat maternal behavior act as epigenators, changing the way the way the pup's brain responds to stress during adulthood. Mice with nurturing mothers adapted better to stressful situations than mice that received little or no nurturing. In rats and humans, stress activates the hypothalamic-pituitary-adrenal (HPA) axis, which dampens the reaction to stress by increasing glucocorticoid levels in the blood.

In adult rats that experienced nurturing care early in life, expression of glucocorticoid receptors (GR) on cells in the hypothalamic region of the brain is increased. These receptors sense levels of glucocorticoids in the blood and are part of the feedback loop that reduces the amount of stress hormones released by the adrenal glands. Adult rats that received less nurturing care as newborns have a higher level of methylation and reduced transcription of the GR receptor gene. Lower levels of the GR receptor do not dampen the stress response in these rats. Rats that experienced high levels of care had 90 percent less methylation at a transcription factor binding site in the GR promoter region, allowing higher levels of GR transcription, producing more GR receptors and lowering the effect of stress. In adults raised by less nurturing mothers, drugs that reverse the high levels of methylation at the GR promoter allow enhanced GR transcription. This, in turn, reversed the effect of poor early-life nurturing and lowered the response to stress. Later work showed that female rats raised by more nurturing mothers are more attentive to their own pups, showing the role of epigenetic maintainers in transmitting these changes to offspring.

Following this research, other studies in rats and mice showed that environments experienced by pregnant animals or newborns affected the behavior and health of the offspring as adults. These experiments also showed that these behavioral changes were mediated by epigenetic methylation of DNA and modification of histones that alter chromatin configuration leading to altered levels of gene expression.

The question of whether these findings can be extrapolated to humans has become a source of some controversy. It has been established that adversity early in life, such as child abuse, increases the risk of depression and suicide. Can epigenetics be involved? In humans, there is little or no direct evidence that epigenetic changes in the brain early in life affect behavior later in life. For example, some studies show a correlation between low economic status (childhood poverty) and heart disease in adulthood. The question is whether epigenetic changes are involved in these and similar situations. Because it is not possible to obtain and analyze brain samples from individuals at intervals from infancy through adulthood, researchers have used other sources for material, including cord blood and white blood cells. Results using these cell types have been inconclusive at best. Whether epigenetic changes can be invoked as an explanation for the role of environmental conditions in adult behavior and disease will depend on results from work using the target human tissues, not by extrapolating findings from experimental animals or studies using surrogate cell types.

Epigenetics and the Environment

Environmental agents including nutrition, chemicals, and physical factors such as temperature can alter gene expression by affecting the epigenetic state of the genome. In humans it is difficult to determine the relative contributions of environmental or learned behavior as factors in changing the epigenome, but there is evidence that changes in nutrition and exposure to agents that affect the endocrine system can have effects on individuals in subsequent generations.

Women who were pregnant during the 1944–1945 famine in the Netherlands had children with increased risk of obesity, diabetes, and coronary heart disease, and members of the F2 generation also had abnormal patterns of weight gain and growth. In the United States, the drug diethylstilbestrol (DES) was prescribed for several decades to women in the belief that it would prevent complications of pregnancy and miscarriages. After it was discovered that girls and young women who were exposed to DES during prenatal development have a 40-fold increased risk of vaginal cancer, and are also at higher risk for breast cancer, use of the drug was discontinued. Males exposed to DES *in utero* have no identifiable physical problems or risk factors, but may have neurological and behavioral alterations. In addition, children of women who took DES have an increased risk of having children with birth defects, irregular menstrual periods, and a higher than normal frequency of infertility. Epigenetic modification of patterns of gene expression in embryogenesis can be invoked to explain the transgenerational transmission of these developmental problems. For the present, however, this is a questionable explanation because there is no direct evidence linking DES to epigenetic changes.

The clearest evidence for the role of environmental factors in epigenetic modifications comes from studies in experimental animals. A reduced protein diet fed to rats during pregnancy results in permanent changes in the expression of several genes in the F1 and F2 offspring. Increased expression of liver genes is associated with hypomethylation of their respective promoter regions, and other evidence indicates that epigenetic changes triggered by this diet modification were gene-specific.

In mice, coat color is controlled by the dominant allele *Agouti* (*A*). In homozygous *AA* mice, the allele is active only during a specific time during hair development, producing a yellow band on an otherwise black hair shaft, resulting in the agouti phenotype. A nonlethal mutant allele (*A^{vy}*) causes yellow pigment formation along the entire hair shaft, producing a yellow fur color. This allele is the result of the insertion of a transposable element near the transcription start site of the *Agouti* gene. A promoter element within the transposon is responsible for this change in gene expression. The degree of methylation in the transposon's promoter is related to the amount of yellow pigment deposited in the hair shaft and varies from individual to individual. The result is a wide variation in coat color in genetically identical mice (ST Figure 3–8), ranging from yellow (unmethylated) to pseudoagouti (highly methylated). In addition to a gradation in coat color, there is also a gradation in body weight. Yellow mice are more obese than the brown, pseudoagouti

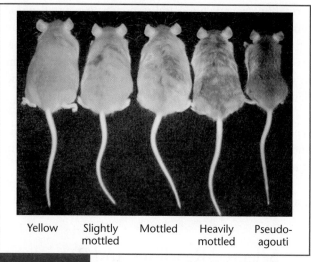

| Yellow | Slightly mottled | Mottled | Heavily mottled | Pseudo-agouti |

ST FIGURE 3–8 Variable expression of yellow phenotype in mice caused by diet-related epigenetic changes in the genome.

mice. Alleles such as *A^{vy}* that show variable expression from individual to individual in genetically identical strains caused by epigenetic modifications are called *metastable epialleles*. Metastable refers to the changeable nature of the epigenetic modifications, and epiallele refers to the heritability of the epigenetic status of the allele.

To evaluate the role of environmental factors in modifying the epigenome, the diet of pregnant mice was supplemented with methylation precursors, including folic acid, vitamin B_{12}, and choline. In the offspring, variation in coat color was reduced and shifted toward the pseudoagouti phenotype. The shift in coat color was accompanied by increased methylation of the transposon's promoter. Methylation profiles in the *A^{vy}* allele from tissues representing the three major cell types in the body (ectoderm, mesoderm, and endoderm) were also altered, showing that methylation at the *A^{vy}* allele occurred very early in embryonic development, before differentiation of the embryonic stem cells. These findings have applications to epigenetic diseases in humans. For example, the risk of colorectal cancer is linked directly to folate dietary deficiency and activity differences in enzymes leading to the synthesis of methyl donors.

Epigenome Projects

We conclude by discussing several projects that are underway to map the human epigenome, including the NIH Roadmap Epigenomics Project. This undertaking is based on the idea that many aspects of health and susceptibility to disease are related to epigenetic regulation or misregulation of gene activity. The program is focused on how epigenetic mechanisms controlling stem cell differentiation and organ formation generate biological responses to external and internal stimuli that result in disease. The Human Epigenome Atlas is one program of the NIH Roadmap Project. The atlas collects and catalogs data on a set of human epigenomes to serve as reference standards; it is being used to provide detailed information about epigenomic modifications at specific loci, cell types, physiological states, as well as genotypes. This atlas allows researchers to perform integrative and comparative analysis of epigenomic data across genomic regions or entire genomes.

Another project, called the Human Epigenome Project, is a multinational, public/private consortium established to identify, map, and establish the functional significance of all DNA methylation patterns in the human genome. Analysis of these methylation patterns may show that genetic responses to environmental cues mediated by epigenetic changes are a pathway to disease. Even as these projects are in their early stages of growth, the information already available strongly suggests that we are on the threshold of a new era in genetics, one in which we can study the impact of environmental factors on the genome at the molecular level. The results of these projects may help explain how environmental settings in early life can affect predisposition to adulthood diseases.

Selected Readings and Resources

Journal Articles

Burdge, G.C., and Lillicrop, K.A. 2010. Nutrition, epigenetics, and developmental plasticity: implications for understanding human disease. *Ann. Rev. Nutr.* 30: 315–339.

Butler, M.G. 2009. Genomic imprinting disorders in humans: a mini-review. *J. Assist. Reprod. Genet.* 20: 477–486.

Dulac, C. 2010. Brain function and chromatin plasticity. *Nature* 465: 728–735.

Iacobuzio-Donahue, C.A. 2009. Epigenetic changes in cancer. *Ann. Rev.* Pathol. Mech. Dis. 4: 229–249.

Petronis, A. 2010. Epigenetics as a unifying principle in the aetiology of complex traits and diseases. *Nature* 465: 721–727.

Relton, C.L., and Smith, G.D. 2010. Epigenetic epidemiology of common complex disease: prospects for prediction, prevention, and treatment. *PLoS Medicine* 7: (10) e1000356.

Web Sites

National Institutes of Health Roadmap for Epigenomics. http://www.nihroadmap.nih.gov/epigenomics/initiatives.asp

Human Epigenome Project. www.epigenome.org

Human Epigenome Atlas. http://www.genboree.org/epigenomeatlas/index.rhtml

Computational Epigenetics Group. http://www.computational-epigenetics.de

SPECIAL TOPIC III

SPECIAL TOPICS IN MODERN GENETICS

Stem Cells

The tiny cluster of cells shown in ST Figure 4–1(a) makes for an interesting photo, and such images have been widely shown around the world in scientific journals and in popular media. To the nonscientist this little clump of cells may not look like much, yet contained within it are cells that without question are the most controversial and intensely scrutinized cells in all of biology. This cluster of cells is an early-stage human embryo that is being used to isolate **stem cells** (ST Figure 4–1(b)). Stem cells invoke emotional and controversial responses from scientists, clergy, politicians, and the general public. For some people the isolation and use of these cells engender excitement, fear, anger, and a range of other emotions.

> **Stem cells have great potential and promise in regenerative applications for the treatment of human diseases.**

Much of the excitement about stem cells in the scientific and medical communities comes from their largely untapped and unproven *potential* for treating human conditions (see ST Table 4.1) through **regenerative medicine**—creating cells, tissues, and organs for tissue or organ repair and replacement. In this mini-chapter we will explore what stem cells are and what properties make them unique, examine different sources of stem cells and how they can be isolated or produced, and consider potential applications involving stem cells. We will also discuss regulations on stem cell use and ethical issues regarding stem cells.

What Are Stem Cells?

Developmental biologists have known about stem cells for decades, yet only relatively recently have stem cells generated tremendous attention from the scientific community worldwide. So what are stem cells, and what makes them such attractive candidates for repairing failing tissues and organs? As you will soon learn, there are many different types of stem cells, but in general they all share two basic characteristics that make them distinctive from other cell types (ST Figure 4–2): **self-renewal** and **differentiation** into specialized cell types:

1. Self-renewal: Stem cells grow and divide (proliferate) indefinitely by mitosis to create populations of identical stem cells.

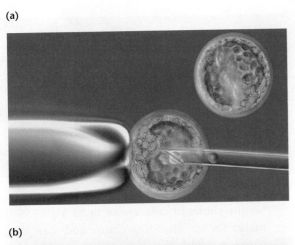

(a)

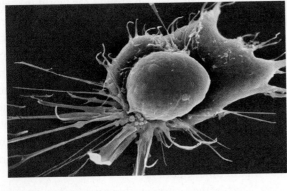

(b)

ST FIGURE 4–1   **Human Embryonic Stem Cells.** (a) Isolating human embryonic stem cells (hESCs). An early-stage human embryo secured by a holding pipette (left) is being used to isolate hESCs using a suction pipette (right). (b) An isolated hESC growing in culture.

2. Differentiation into all cell and tissue types of the body. Under control of key signals, stem cells can differentiate into specialized cell types such as skin, muscle, bone, or blood cells.

As we discussed in Chapter 18, differentiation is a complex process involving many genes that must be activated and silenced in carefully coordinated temporal patterns of expression, and differentiating cells rely on chemical signals such as growth factors and hormones from other cells to help them change. Some stem cells possess greater differentiation ability than others, and scientists refer to this ability as the *potency* of the cell type. A **totipotent** cell, such as the zygote, can form not only all adult body cell types, but also the specialized tissues needed for development of the embryo, such as the placenta. Many types of stem cells are called **pluripotent** because they have the potential to eventually differentiate into a variety of different cell types to form all of the 220 cell types in the human body.

ST TABLE 4.1

Potential U.S. Patient Populations for Stem Cell-Based Therapies

The conditions listed below occur in many forms and thus not every person with these diseases could potentially benefit from stem cell-based therapies. Nonetheless, the widespread incidence of these conditions suggests that stem cell research could help millions of Americans.

Disease condition	Number of patients in the United States
Cardiovascular disease	58 million
Autoimmune diseases	30 million
Diabetes	16 million
Osteoporosis	10 million
Cancers	8.2 million
Alzheimer's disease	5.5 million
Parkinson's disease	5.5 million
Burns (severe)	0.3 million
Spinal-cord injuries	0.25 million
Birth defects	0.15 million/year

Source: Derived from Perry (2000).

Sources and Types of Stem Cells

Where do stem cells come from? At one time, it was commonly thought that stem cells were only present in an embryo. We now know that there are several sources and different types of stem cells. To understand what stem cells are, we need to briefly consider the development of a human embryo (ST Figure 4–3). As you know, following fertilization of a sperm and an egg cell, the fertilized egg is called a **zygote**. The zygote divides rapidly and after three to five days first forms a compact ball of about 12 cells called a **morula**, meaning "little mulberry." Around five to seven days after fertilization, the dividing cells create an embryo consisting of a small hollow cluster of approximately 100 cells called a

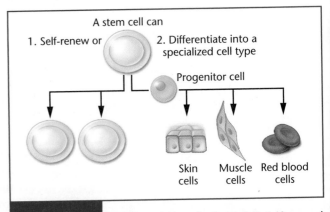

ST FIGURE 4–2   **Characteristics of a Stem Cell.** Self-renewal and differentiation into different cell types are two key characteristics of stem cells.

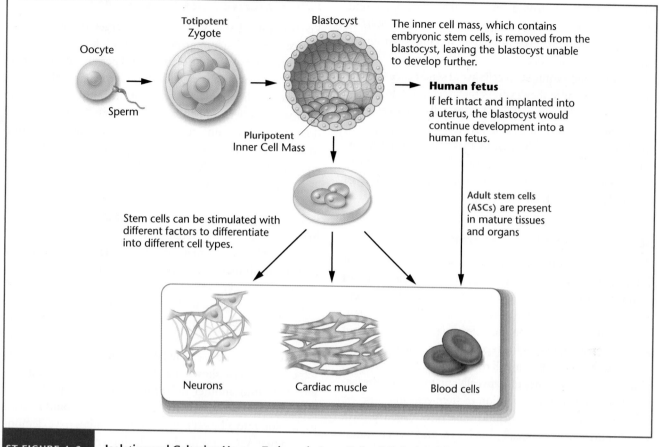

Isolating and Culturing Human Embryonic Stem Cells. Cells isolated from the inner cell mass of human embryos can be grown in culture as a source of hESCs. Under the proper growth conditions, hESCs can be stimulated to differentiate into virtually all cell types in the body.

blastocyst. The blastocyst is approximately one-seventh of a millimeter in diameter and contains an outer row of single cells called the *trophoblast*; this layer develops to form the fetal portion of the placenta that nourishes the developing embryo (ST Figure 4–3).

Within the blastocyst is a small cluster of around 30 cells that form a structure known as the **inner cell mass.** The inner cell mass is the source of **human embryonic stem cells (hESCs).** During embryonic development, cells of the inner cell mass develop to form the embryo itself, and hESCs can differentiate to form all cell types in the body. hESCs are pluripotent, as are other types of cells we will discuss later in this Special Topics section.

Successful isolation and culturing of the first hESCs from a human blastocyst was reported in 1998 by James Thomson of the University of Wisconsin at Madison who had cultured ESCs from rhesus monkeys two years earlier. Also in 1998, John Gearhart and colleagues at Johns Hopkins University isolated embryonic germ cells, primitive cells that form the gametes—sperm and egg cells—from human fetal tissue and demonstrated that these cells can develop into different cell types. When hESCs are isolated, scientists use a holding pipette that applies a brief suction to "hold" the blastocyst in place (see ST Figure 4–1(a)), and a glass micropipette is then inserted into the blastocyst to gently remove cells from the inner cell mass, which are then cultured in dishes and flasks in the lab.

These discoveries followed the work of other scientists who had isolated stem cells in species such as mice, pigs, cows, rabbits, and sheep. Stem cell researchers credit much of what is now known about isolating hESCs from pioneering work done with mice in the 1980s.

Initially, the main source of hESCs was leftover embryos produced by assisted reproductive technologies such as **in vitro fertilization (IVF).** In IVF, multiple eggs are removed from a woman and fertilized *in vitro*. Resulting embryos are then implanted into a woman's uterus, but typically only a few of the embryos produced this way are implanted. Excess embryos are typically frozen at ultra-low temperatures for use by the couple in the future if desired. Alternatively, these embryos may be destroyed or, with the couple's consent, they may be donated for research. An estimated 400,000 of IVF-generated embryos are stored in clinics within the United States alone.

Human embryonic stem cells avoid senescence (cell aging) in part because they express high levels of telomerase. Several groups have maintained stem cells for over three years and over 600 rounds of division without apparent problems. Cultured cells such as these that can be maintained and grown successively are called **cell lines**. Stem cells also grow rapidly and can be frozen for long periods of time and still retain their properties.

Stimulating ESCs to Differentiate

Under the right conditions, when stimulated with different factors, ESC lines and other types of cells that will be discussed in this chapter, can be coaxed to differentiate into different types of cells *in vitro* (ST Figure 4–3). This *directed differentiation* of stem cells into specific differentiated cells of interest is key for creating tissues for regenerative medicine applications. A major focus of stem cell research is to determine what controls the pluripotency of stem cells and to identify the factors that stimulate differentiation of stem cells into discrete cell types. These signals include substances called growth factors, hormones, and small proteins (peptides) that stimulate differentiation in tissue-specific ways. Many of these signals activate developmental stage-specific and tissue-specific transcription factors, which in turn direct differentiation of pluripotent stem cells into differentiated cell types.

For example, signaling systems that involve transforming growth factor-β (TGF-β), bone morphogenetic proteins (BMPs), and other growth differentiation factors act on a gene for a transcription factor called *Nanog*. Nanog is one key protein that maintains hESCs in an undifferentiated, pluripotent state. In the laboratory under the proper culturing conditions, ESCs from humans, mice, rats, and other species have been shown to differentiate into a myriad of cells including skin cells; brain cells (both neurons and glial cells such as oligodendrocytes, which support, nourish, and protect neurons); cardiomyocytes (cardiac muscle cell precursors); cartilage (chondrocytes); insulin-secreting pancreatic beta cells, spermatozoa, and osteoblasts (bone-forming cells); liver cells (hepatocytes); muscle cells including smooth muscle, which forms the walls of blood vessels; skeletal muscle cells, which form the muscles that attach to and move the skeleton; and cardiac muscle cells (myocytes), which form the muscular walls of the heart. Once differentiated, scientists frequently examine cells for the expression of cell-type specific genes, one way to confirm the cell type produced. For instance, when chondrocytes are differentiated from hESCs, they express a chondrogenic transcription factor called *SOX9*.

Adult-Derived Stem Cells (ASCs)

Research on hESCs is controversial largely because of the source of ESCs—a human embryo. It was long thought that the early embryo was the primary and perhaps the only major source of stem cells, but **adult-derived stem cells (ASCs)** do reside in differentiated tissues of the body. ASCs appear in small numbers, but they can be isolated; this has been done from the brain, intestine, hair, skin, pancreas, bone marrow, fat, mammary glands, teeth, muscle, and blood and almost every adult tissue (ST Figure 4–3).

Opponents of hESC research have often claimed that ASCs are a more acceptable alternative than hESCs because isolating ASCs does not require the destruction of an embryo. ASCs can be harvested from people by fine-needle biopsy, where a thin-diameter needle is inserted into muscle or bone tissue. It may even be possible to isolate ASCs from cadavers. We also know that ASCs are present in fat (adipose) tissue, which could potentially be an outstanding source of stem cells, especially if you consider that over 500,000 liters of fat tissue collected by liposuction and other cosmetic surgery techniques are discarded in the United States each year! Experiments have shown that ASCs from one tissue can differentiate into another different specialized cell type. For instance, an ASC isolated from muscle tissue could be used to develop into a blood cell. But other studies have demonstrated that ASCs may not be as pluripotent as hESCs. Much more research is required to determine whether ASCs will be as valuable as hESCs are expected to be.

Amniotic Fluid-Derived Stem Cells

Stem cells can be isolated from human amniotic fluid, the protective fluid that surrounds a developing fetus. In the lab, these **amniotic fluid-derived stem cells** have been coaxed to become neurons, muscle cells, adipocytes, bone, blood vessels, and liver cells. It is not entirely clear if these cells are truly different from hESCs or ASCs, but if they are, they may produce a key breakthrough in stem cell technologies.

Cancer Stem Cells

Cancer stem cells (CSCs), also called tumor-initiating cells, have been identified and implicated in the development of cancers, tumor progression, tumor metastasis, and the recurrence of cancers. Similar to normal stem cells, CSCs can self-renew and can differentiate to form tissues from which they were derived. Certain CSCs grow slowly in clusters or *niches* within a tissue. It is not clear what properties CSCs may have besides the ability to form a tumor. Researchers are also not sure if CSCs are derived from normal cells that have undergone mutations or if they are involved in cancer tumor resistance to chemotherapies, but these cells are a focus of intense research and potential therapeutic treatments for the treatment of cancers.

Tests for Pluripotency

When scientists work with stem cells—either isolated cells such as hESCs or ASCs, or cells created by nuclear reprogramming as we will discuss in the next section—one important aspect of stem cell research is to test for the pluripotency of the isolated cells. This is important for determining whether cells are able to form other cell types, and this property is indicative of a stem cell's potential to do so *in vivo* or *in vitro*. It turns out that embryonic cells are pluripotent for a relatively brief period of time during development. When properly isolated and cultured, ESCs can be maintained in a pluripotent state *in vitro* indefinitely. Frequently, heterogeneic populations of ESCs are often produced in which cells have different capacities to differentiate *in vivo* or *in vitro*. The concept of pluripotency is not difficult, but actually determining the potency of a stem cell line is not so easy.

One key test for pluripotency is to determine whether stem cells can form the three *primary germ layers:* ectoderm, mesoderm, and endoderm. During embryological development, the inner cell mass forms three specialized layers of tissue—the primary germ layers, which then form the specific tissues of the body. The outer layer, ectoderm, gives rise to skin, brain, and nervous tissue. The middle layer, or mesoderm, gives rise to blood cells, the heart, bone, kidneys, muscles, and cartilage. The innermost layer, the endoderm, develops into lungs, the liver, and the digestive system. *In vitro* it has been demonstrated that ESCs can differentiate into cell types representing cells that would originate from the primary germ layers. *In vitro* tests are also often used to determine whether stem cells can aggregate into a cluster of cells called an embryoid body.

Stem cell biologists use *in vivo* tests to demonstrate pluripotency as a way to validate the viability of stem cells, especially ESCs. In these assays, ESCs are injected into mice to see if the cells respond to signals *in vivo* to stimulate tissue development. Immunodeficient mice—those lacking an immune system—are used so that the host mice do not reject the injected stem cells. When stem cells are placed into these mice, tumors form including **teratomas** (ST Figure 4–4). Within these tumors differentiated tissues typically form, and these tumors can be analyzed to determine which tissue types have formed. These types of tumors appear as a disorganized mass of cells but also often possess many differentiated cells and tissues including bits of bone, hair, and teeth. Incidentally, these tumors are also the source of embryonal carcinoma (EC) cells, which have properties of stem cells and were frequently used in the past for a variety of experimental applications, including making transgenic animals. Making a teratoma demonstrates one aspect of the potency of stem cell lines.

ST FIGURE 4–4   **Teratoma.** A teratoma is a mass of disorganized tissue that can occur naturally in individuals but that can also occur when stem cells are tested for pluripotency. Teratomas typically consist of a disorganized mass of tissue, including several differentiated tissue types such as hair, skin, and bone. Shown here is a tooth growing within a teratoma.

For many scientists, the most significant demonstration of pluripotency *in vivo* is to inject ESCs into a developing embryo at the blastocyst stage; the resulting fetus can then be examined to determine whether injected ESCs contributed to development of all three primary germ layers (ST Figure 4–5). For example, as shown in ST Figure 4–5, if ESCs from a black mouse were put into a blastocyst-stage embryo for a white mouse, then the coat color of a mouse born from this chimera blastocyst would produce mottled color fur with both black and white fur cells. A variety of other genetic markers can also be followed to determine which cells arose from the injected ESCs, including expression of human tissue genes and proteins if hESCs were injected into a mouse blastocyst. Of course, when working with hESCs, mice are used for this pluripotency demonstration by injecting hESCs into mouse embryos because ethically these types of experiments cannot be done with human embryos.

In vivo and *in vitro,* a range of molecular techniques are commonly utilized to examine the potency of all kinds of stem cells. This includes the DNA methylation status of pluripotent cell-specific genes, assaying for a range of stem cell-specific protein markers, which are commonly examined in stem cells but not other cell types, and microarray studies to analyze stem cell-specific patterns of gene expression.

We have yet to discuss another main source of stem cells which have scientists particularly optimistic about their applications. But these cells are not isolated from an embryo or an adult tissue as a source of stem cells; rather, scientists are using nuclear reprogramming to turn differentiated cells into stem cells. Because of the complexities involved in this approach, we consider these cells separately in the next section.

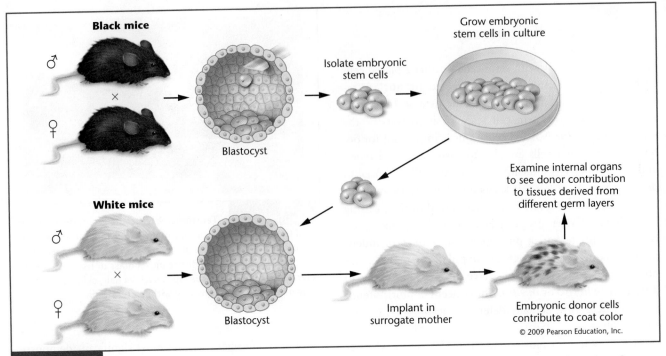

Black mice

♂   ×   ♀

Blastocyst

Isolate embryonic stem cells

Grow embryonic stem cells in culture

Examine internal organs to see donor contribution to tissues derived from different germ layers

White mice

♂   ×   ♀

Blastocyst

Implant in surrogate mother

Embryonic donor cells contribute to coat color

© 2009 Pearson Education, Inc.

ST FIGURE 4–5   **Tests for Pluripotency.** Evidence that ESCs are pluripotent and can contribute to all tissues is important for evaluating the viability of stem cells. In this example, mouse ESCs from an embryo derived from mice with black fur are transferred to a blastocyst from mice with white fur. When this hybrid blastocyst is implanted into a surrogate female mouse, the resulting embryos will be chimeric mice. Various tissues, representing development of the three primary germ layers, will have formed from both the white and black stem cells if the black stem cells are truly pluripotent.

Nuclear Reprogramming Approaches for Producing Pluripotent Stem Cells

Research on stem cell biology is an extremely active field. One primary area of focus continues to be alternative approaches for producing pluripotent stem cells without destroying an embryo. One of the most promising new approaches for creating and isolating stem cells without an embryo involves **nuclear reprogramming of somatic cells**. It was previously thought that once cells differentiated to become a specific, specialized cell type, for example, a skin cell, their differentiation fate was irreversible. But we now know this is not the case. Scientists have been working on reprogramming cell fates for about 50 years, and in recent years, stem cell biologists have worked intensely in this area in efforts designed to produce pluripotent stem cells without destroying an embryo. The basic idea is to take a differentiated, adult cell and to alter its patterns of gene expression in order to reprogram the cell to an early stage in its differentiation pathway—that is, to push the cells back

to an undifferentiated, pluripotent state. The mechanisms by which cells are reprogrammed back to a pluripotent state are largely unknown. So reprogramming is not as easy as it may sound. For example, epigenetic changes that occur in a genome as a cell differentiates can be difficult to erase or reverse. Some scientists have raised significant questions about whether cells that are reprogrammed maintain stable epigenetic states or whether shifts in epigenetic changes such as DNA methylation or modifications in histones will occur in reprogrammed cells.

Currently, there are several approaches to nuclear reprogramming (ST Figure 4–6). For example, **somatic cell nuclear transfer** involves transplanting a somatic cell nucleus into an enucleated egg cell (recall that this approach was used to clone Dolly the sheep). Nuclear reprogramming of the somatic cell genome is initiated, and an embryo forms based on genes expressed by the somatic cell genome. The resulting embryo can be grown to the blastocyst stage and used to isolate ESCs (ST Figure 4–6(a), although this approach has not been particularly effective for generating hESCs. Somatic cells can also be fused to stem cells in an attempt to reprogram the somatic cells. One of the first successful reprogramming techniques involved fusing hESCs

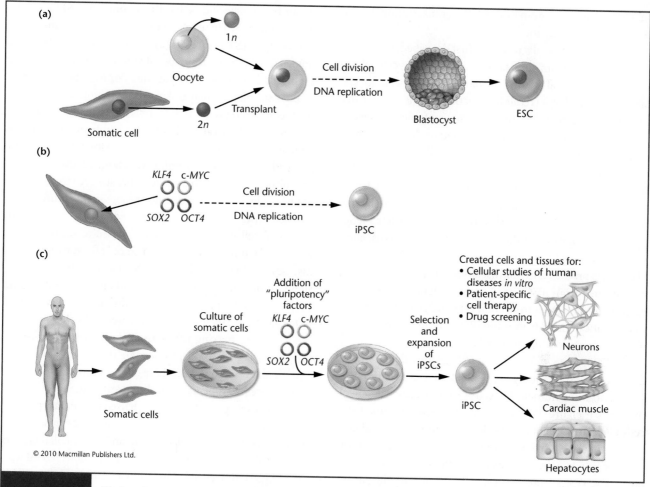

ST FIGURE 4–6   **Nuclear Reprogramming to Produce Pluripotent Stem Cells.** (a) Nuclear transfer to create embryonic stem cells from a somatic cell. (b) Transcription factor transduction to create iPSCs including patient-specific iPSCs (c).

SPECIAL TOPIC IV

with skin cells called *fibroblasts*. The hybrid cells generated displayed several properties of hESCs *in vitro* and *in vivo*.

A third approach to nuclear reprogramming has involved the use of transcription-factor genes to induce reprogramming to create **induced pluripotent stem cells (iPSCs)**.

Induced Pluripotent Stem Cells

Several laboratories have achieved success with a nuclear reprogramming of somatic cells from adult mice, humans, and pigs to create iPSCs. This approach has been heralded as a revolution in stem cell biology research which has demonstrated outstanding potential as an alternative way to generate stem cells. In 2006, Shinya Yamanaka of Kyoto University in Japan created the first iPSCs. Yamanaka used

retroviruses to deliver four transgenes *Oct4*, *Sox2*, *c-myc*, and *Klf4* into mouse fibroblasts (ST Figure 4–6(b)). Expression of these four genes, which encode transcription factors involved in cell development, "reprogrammed" the fibroblasts back to an earlier stage of differentiation producing iPSCs. iPSCs demonstrate many properties of hESCs, such as self-renewal and pluripotency, and they appear to be indistinguishable from hESCs. These iPSCs also express genes such as *Nanog* and *Oct*, which are characteristic markers known to be expressed in undifferentiated hESCs. Other experiments demonstrated that iPSCs could differentiate into other cell types, including neural cells and cardiac muscle cells. This strategy is also being used by scientists to convert cell types, both stem cells and non-stem cells, from one type to another. For example, mouse fibroblasts have been converted into functional neurons *in vitro* using a combination of the transcription factors *Ascl1*, *Brn2*, *Pou3f2*, and *Myt1l*.

Since the Yamanaka work was first described, iPSCs have been produced from human, rat, and monkey cells. iPSCs researchers have also used Yamanaka's approach to create mice from iPSCs (recall from earlier in this chapter that this approach is a key test of pluripotency) to determine that reprogrammed cells are highly pluripotent. In these experiments, researchers fused together a two-cell embryo to create a single-cell tetraploid blastocyst that develops a placenta but not embryonic cells to become a body. Mouse iPSCs were then injected into tetraploid blastocysts, and these blastocysts were transferred to a surrogate mother. These injected iPSCs pushed development of the blastocysts into embryos, and a small percentage (22 births from 624 transferred embryos) of blastocysts produced live births, although clearly some of the mice created from iPSCs had a high death rate. Nonetheless, reprogramming DNA from adult cells and generating genetically identical individuals using iPSCs could potentially be a new way to clone adult mammals.

In 2007, two different research groups demonstrated that they had successfully derived **patient-specific iPSCs** by reprogramming human skin cells taken directly from a volunteer (ST Figure 4–6(c). One group reprogrammed skin cells taken from the face of a 36-year-old woman and connective tissue cells from the joints of a 69-year-old man, and another group used skin cells from the foreskin of a newborn boy. iPSCs were also recently produced without using the *c-MYC* gene. This is a potentially important advance because *c-MYC* is a known oncogene (cancer-causing gene).

Researchers are also trying to avoid problems associated with random integration into the genome which occurs when retroviral vectors are used to deliver genes for nuclear *reprogramming*. One approach involved introducing OCT4, SOX2, *c-MYC*, and KLF4 proteins into adult somatic cells to reprogram them into iPSCs, thus eliminating the need to use viral vectors for reprogramming. Human neural stem cells have also been coaxed to produce iPSCs by introducing OCT4 protein into these cells—without the addition of genes or the other three transcriptional factors. Producing viral vector-free iPSCs will continue to be a focus of iPSC researchers.

In 2010, researchers at the Mount Sinai School of Medicine demonstrated that fetal skin cells in amniotic fluid could be readily reprogrammed into iPSCs with greater efficiency than other somatic cells. In this work, skin cells derived from amniotic fluid were cultured and transfected with plasmids expressing *Oct4*, *Sox2*, *c-myc*, and *Klf4*. The iPSCs developed by this approach showed gene-expression patterns characteristic of stem cells and telomerase activity, and were capable of differentiation both *in vivo* and *in vitro*.

These promising results with iPSCs demonstrate that nuclear reprogramming is a potentially viable way to generate donor patient-specific stem cells without the need for an embryo and hESCs. These cells could be used for patient-specific cell therapies (ST Figure 4–6(c). In addition, with iPSC technologies, it is theoretically possible to create disease-specific stem cells from individual patients. For example, one could take a tissue biopsy, such as skin, from a person with a particular disease and reprogram those cells into stem cells that could then be used to create cell types for combating the disease. Patient-specific stem iPSCs could be used for cell-based therapies without the risk of immune rejection, but even the most optimistic iPSC researchers believe that such cell therapies will not be ready for at least a decade or more in the future.

Scientists are also very excited about how reprogrammed cells from patients can be used to study "disease in a dish." Cultured reprogrammed cells from patients with diseases are being studied to help scientists better understand human disease progression and disease processes and use drug screening tests to determine the effectiveness of potential drug treatments for diseased cells. Labs around the world are making iPSCs from patients with a disease, differentiating these cells to become the tissues affected by a particular disease so that these diseases can be modeled *in vitro* and cells can be used for drug treatments.

As an example of the approach used to generate patient-specific iPSCs from an individual with a disease, researchers at the Harvard Stem Cell Institute and other collaborators used fibroblasts from patients with amyotrophic lateral sclerosis (ALS; Lou Gehrig's Disease) to create patient-specific neural cells, motor neurons, and supporting cells (glial cells) called astrocytes. ALS is a fatal neurodegenerative disease with no cure. It causes deterioration of motor neurons in the brain and spinal cord, leading to progressive loss of motor skills including the ability to walk, talk, and breathe. The Harvard work was a major development because, although significant progress has been made in identifying genetic defects associated with neurodegenerative diseases such as ALS, Parkinson's disease, and Alzheimer's disease, there is a lack of good animal or tissue model systems for *in vivo* or *in vitro* studies to better understand these diseases and to develop treatment approaches.

As shown in ST Figure 4–7, fibroblasts were taken from two elderly sisters with ALS-associated mutations (one of these sisters had developed ALS) in the gene for superoxide dismutase (*SOD1*) and reprogrammed to produce iPSCs. These cells were then treated with retinoic acid and the protein sonic hedgehog to stimulate differentiation of iPSCs into motor neurons and astrocytes. This work demonstrated that it is possible to generate patient-specific iPSCs from elderly individuals with a disease; it should prove to be very valuable for researchers studying motor neuron function in ALS patients and developing therapies for ALS and other neurodegenerative diseases.

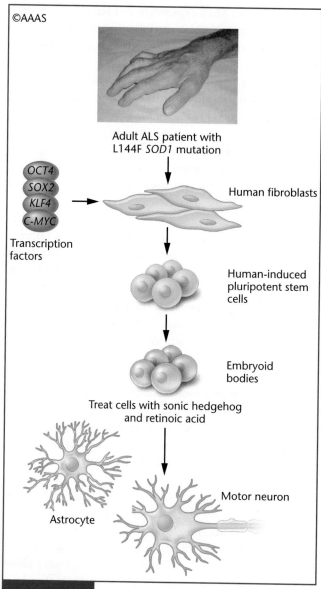

Adult ALS patient with
L144F *SOD1* mutation

Transcription
factors

OCT4
SOX2
KLF4
C-MYC

Human fibroblasts

Human-induced
pluripotent stem
cells

Embryoid
bodies

Treat cells with sonic hedgehog
and retinoic acid

Astrocyte

Motor neuron

ST FIGURE 4–7 **Generating Patient-Specific iPSCs from Patients with Amyotrophic Lateral Sclerosis.** Fibroblasts from skin biopsies of two elderly sisters with ALS-associated mutations in the *SOD1* gene were used to create patient-specific iPSCs that were differentiated into motor neurons and astrocytes.

On the surface, iPSCs circumvent some of the legal and ethical controversies associated with ESCs. But as promising as iPSCs are, iPSCs present a number of challenges that will need to be addressed. For example, scientists still do not fully understand how pluripotent iPSCs may be and how to best control the potency of these cells. In addition, iPSCs:

- are relatively inefficient to produce (only about one in 1000 somatic cells exposed to most reprogramming approaches becomes an iPSC);
- require constant feeding to maintain viable cell lines;

- show low viability compared to other cell types once they have been stored frozen;
- can be prone to forming tumors;
- occasionally show spontaneous differentiation into mature cell types when in culture; and
- can sometimes be difficult for directing differentiation into particular cell types.

Research with iPSCs is progressing at an astonishing and exciting pace as scientists work to better understand the properties and capabilities of these cells.

RNA-Induced Pluripotent Stem Cells

In late 2010, a paper describing a new approach for coaxing mature cells into iPSCs generated a great deal of attention for a new technique that could be used to quickly and safely reprogram adult cells to create stem cells called **RNA-induced pluripotent stem cells (RiPS)**. Use of a cocktail of *RNA molecules* corresponding to transcripts for the same genes utilized in traditional reprogramming (*OCT4, SOX2, c-MYC,* and *KLF4*) generated RiPS from somatic cells. This approach produced iPSCs in about half the time (two weeks compared to more than a month for iPSCs) as other approaches and reprogrammed approximately 4 percent of the cells treated, which makes this method approximately 100 times more efficient than the previously described gene-delivery approaches for making iPSCs. Watch for exciting new developments in the next few years involving nuclear reprogramming, iPSCs, and RiPS.

Potential Applications of Stem Cells

The Centers for Disease Control's National Center for Human Statistics indicates that approximately 3000 Americans die every day from diseases that may one day potentially be treated by stem cell technologies. In the future, stem cell research may affect the lives of millions of people throughout the world. The tremendous promise and controversy surrounding stem cells has made stem cell research and related topics regular front-page news items and TV headlines. There are many potential applications for stem cells—from using stem cells to grow healthy tissues to studying stem cells to understand and treat birth defects and genetic diseases, to genetically manipulating stem cells for delivering genes in gene therapy approaches, to creating whole tissues in the laboratory using tissue engineering. Many scientists believe that stem cell technologies will play key roles in developing treatments for diseases such as stroke, heart disease, Parkinson's disease,

Alzheimer's disease, ALS, chronic spinal cord injuries, diabetes, and other conditions (ST Table 4.1).

The words *potential* and *promise* are frequently used when discussing stem cell applications, but use of these cells for treating disease is still largely unproven. There are an alarming number of unregulated and unproved stem cell clinics around the world (do a Google search!) that promote the effectiveness of stem cells in treating and curing diseases. Here we consider representative examples of stem cell applications to date. Although we will focus on modern applications of stem cell technologies, some applications have been around for decades. For example, bone marrow transplantation is a form of stem cell therapy. Bone marrow contains ASCs. During a bone marrow transplant, stem cells are transferred from a healthy donor to a needy recipient, wherein the cells regenerate various blood cell types as needed.

Patients with leukemia, a cancer that causes white blood cells to divide abnormally, producing immature cells, frequently require chemotherapy or radiation treatment to destroy the defective white blood cells; as a result chemotherapy and radiation greatly weaken the patient's immune system. Leukemia treatments may also involve blood transfusions to replace white blood cells and red blood cells damaged by chemotherapy. Using stem cells to make white blood cells is becoming an effective way to treat leukemia. In addition, stem cells from umbilical cord blood have also been used to provide red blood cells for sickle-cell patients and individuals with other blood deficiencies. Isolating stem cells from cord blood is becoming so popular that, in many U.S. states, parents can opt to pay to have cord blood stem cells frozen indefinitely should their child need them in future.

It is important to recognize that the most effective and the safest stem cell treatments in the future are likely to involve differentiating stem cells *in vitro* into desired cell and tissue types and then introducing those cells into a patient as appropriate, instead of directly injecting stem cells into a person. Currently studies that have used multi- or pluripotent stem cells of any source and introduced those into animals or patients have largely been plagued by problems. What do you think is the main issue here? The main problem is how does one control differentiation of pluripotent stem cells into desired tissues of interest if the stem cells are injected into a tissue, say skeletal muscle? When stem cells have been introduced this way, it is difficult to control how they will respond to differentiation cues *in vivo*; as a result, they often develop into many different and undesirable cell and tissue types contrary to their intended application.

So far there have been a number of promising results in animal models and in human clinical trials using stem cells for tissue repair. Repair of heart tissue has shown strong potential. Heart attacks are a leading cause of death in the United States, resulting in the death of nearly half a million people each year. Stem cells might be used to replace dead and dying cardiac muscle cells following trauma such as a heart attack. Adult cardiac muscle cells do not heal well on their own.

Death of cardiac muscle cells weakens the heart and can prevent it from beating with the proper strength to maintain normal blood flow. A number of different research groups have reported success using adult stem cells in animal models to improve cardiac function following heart attacks. For example, researchers have successfully injected ASCs from mouse bone marrow as well as human ASCs into damaged areas of the mouse heart (ST Figure 4–8). These stem cells can develop into cardiac muscle cells, form electrical connections with healthy muscle cells, and improve heart function by over 35 percent. A similar approach was used to transplant hESCs into the ventricular wall of damaged hearts in pigs and in humans involved in clinical trials. In both examples, transplanted stem cells differentiated to form cardiac muscle cells that restored a significant percentage of electrical activity and contractility to the damaged areas. One study used stem cells from human umbilical cord blood to improve cardiac function in rats.

It is unclear how long stem cells used in these experiments might remain viable and how long this type of approach might be effective. Scientists are actively working on devising noninvasive monitoring methods (using molecular-genetic approaches) and diagnostic imaging techniques to track the fate of transplanted stem cells. Scientists are optimistic that this approach may someday work in humans. Might it be possible in the future for a physician to order a few grams of cardiac muscle cells from a regenerative medicine lab to transplant into a heart attack patient in much the same way that surgeons routinely order blood from a blood bank for a transfusion during a surgical procedure?

Advanced Cell Technology in Massachusetts has an ongoing clinical trial using hESCs to treat a rare form of retinal degeneration that causes blindness. This work was the second-ever clinical trial involving hESCs approved by the FDA. Researchers at Emory University in Atlanta, Georgia are working on a Phase I trial, approved by the FDA in 2009, to study the safety of stem cells directly injected into the spinal cord of ALS patients. This is the first such clinical trial involving stem cells for the treatment of ALS in the United States. Scientists do not expect the injected stem cells to create new motor neurons, but they believe the stem cells may protect functional motor neurons from the effects of ALS and therefore slow the progression of the disease.

Scientists working in regenerative medicine are also particularly optimistic about their ability to produce viable organs from stem cells. Imagine the ability to make kidneys, bladders, spleens, and other organs from stem cells and to have a repository of organs available for transplantation.

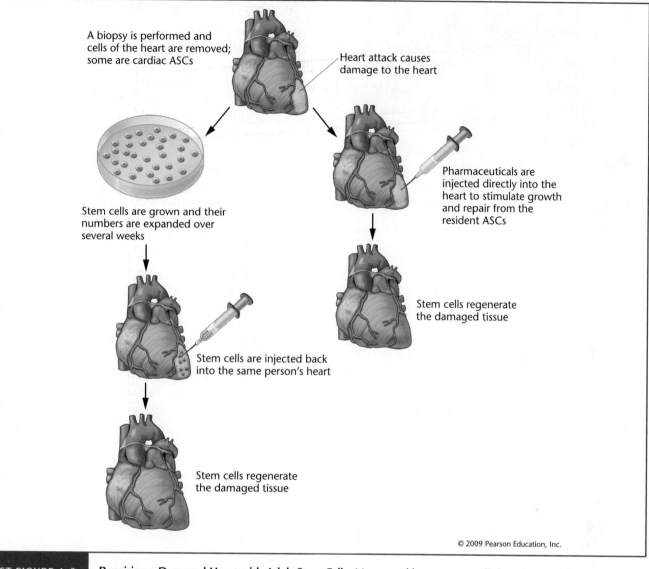

A biopsy is performed and cells of the heart are removed; some are cardiac ASCs

Heart attack causes damage to the heart

Stem cells are grown and their numbers are expanded over several weeks

Pharmaceuticals are injected directly into the heart to stimulate growth and repair from the resident ASCs

Stem cells are injected back into the same person's heart

Stem cells regenerate the damaged tissue

Stem cells regenerate the damaged tissue

© 2009 Pearson Education, Inc.

ST FIGURE 4–8 **Repairing a Damaged Heart with Adult Stem Cells.** Mouse and human stem cells have been used to repair areas of the mouse heart damaged by a heart attack.

These organs could be designed to lack key surface proteins that would prevent them from being recognized and rejected by the recipient's immune system; in effect then they could be transplanted with little concern for immune rejection. These possibilities are still largely speculative. Sheets of skin and other organs such as the urinary bladder have been created *in vitro* (ST Figure 4–9). In this example, biodegradable scaffolding materials were used to form the shape of the bladder, and stem cells were seeded onto the scaffolding and then differentiated to form bladder tissues. Making a urinary bladder is an impressive accomplishment indeed, but creating an organ that is effectively a holding tank for urine is still quite different from making an organ with a multitude of different tissue types and more complex functional

capacities—such as a spleen, pancreas, or lung. Yet these results are encouraging examples of what we can expect from regenerative medicine in the future.

In the last few years, researchers have disproved a long-standing belief that the human brain and spinal cord cannot grow new neurons. Adult stem cells have been isolated from the brain and used to make neurons in culture, and scientists have already demonstrated that ESCs can be differentiated to form neurons that can be injected into mice and rats to improve neural function in animals with spinal cord injuries. Researchers at Johns Hopkins University have demonstrated that human stem cell transplants can enable rats with paralyzed hind limbs to walk. Many companies have announced preclinical data indicating that human-derived neural stem

iPSCs for Treating Sickle-Cell Disease

Rudolph Jaenisch and colleagues at Massachusetts Institute for Technology generated a great deal of excitement in 2007 with reports of the first therapeutic application of iPSCs. Jaenisch and colleagues used iPSCs to correct sickle-cell anemia in a mouse homozygous for the human β^S-globin (sickle) allele. In this work, researchers used a humanized mouse model of sickle-cell anemia, so named because this is a mouse strain in which mouse globin genes were replaced by human genes (using a "knock-in" approach), including the human β^S-globin allele (ST Figure 4–10). These humanized mice display many of the key symptoms, such as anemia, that occur in humans with sickle-cell disease. Skin cells were isolated from the globin humanized adult mice, and the cells were reprogrammed using retroviruses encoding the four genes we described earlier in the chapter when describing iPSCs: (*Oct4*, *Sox2*, *c-myc*, and *Klf4*).

Scientists then used homologous recombination to correct the iPSCs by replacing the β^S allele of the β-globin gene with the β^A allele (the allele that does not cause disease; ST Figure 4–10). The corrected iPSCs were differentiated *in vitro* to produce hematopoietic progenitor cells that were subsequently transplanted into anemic mice with the β^S alleles, which had been irradiated to destroy blood stem cells. Transplanted iPSCs restored hematopoietic

cells can be used to restore lost motor function in mice with spinal cord injuries. The FDA approved the California-based biotechnology company Geron Corporation for the first U.S. clinical trial to use hESCs to treat individuals with spinal cord injuries. Although much work remains to be done in the spinal cord repair and regeneration field, researchers are optimistic that neural stem cell transplants may be ready for human clinical trials in the next three to five years, offering hope to the many individuals who are affected by spinal cord injuries.

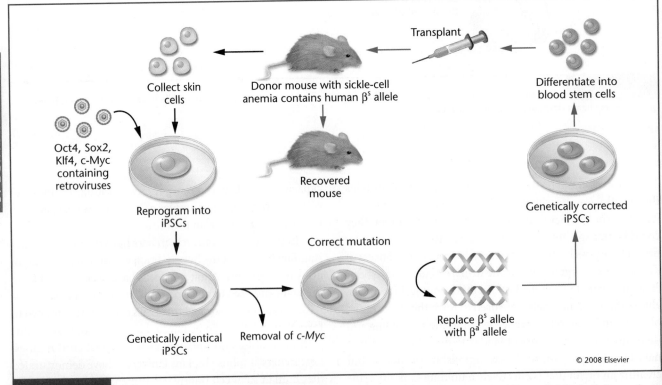

© 2008 Elsevier

ST FIGURE 4–10   **iPSCs for Treating Sickle-Cell Disease.** iPSCs have been used to ameliorate the effects of sickle-cell disease in a humanized mouse model. Blue arrows show the pathway for using corrected iPSCs to successfully treat the donor mouse.

stem cells in the diseased mice and dramatically reduced symptoms of sickle-cell disease. Although many barriers need to be addressed before such approaches are viable options for treating human diseases, this early-stage work has created great optimism about the potential of treating human diseases with iPSCs.

Therapeutic Cloning

When reproductive cloning of mammals by somatic cell nuclear transfer became possible, it became apparent that similar techniques might be used for **therapeutic cloning**. The intent of reproductive cloning is to create a baby. Dolly was the first of many other mammals to be produced by reproductive cloning. Unlike reproductive cloning, therapeutic cloning could provide stem cells that are a genetic match to a patient who requires a transplant. In therapeutic cloning, the basic concept is that the genome from a patient's somatic cell (for instance, skin cells) could be injected into an enucleated egg—an egg that has had its nucleus removed—that is then stimulated to divide in culture to create an embryo (ST Figure 4–11). The embryo produced will not be used to produce a child; instead, the embryo will be grown for several days until it reaches the blastocyst stage so that it can be used to harvest ESCs. Stem cells isolated from this embryo could then be grown in culture and then introduced into the patient who donated the somatic cells used to make the ESC (ST Figure 4–11).

Prior to the development of iPSCs, therapeutic cloning was seen as a possible way of providing patient-specific stem cells that could be used to treat disease without fear of immune rejection by the recipient because he or she was the original source of the cells. In theory, stem cells from a patient can also be used to create cell lines from humans with genetic diseases to provide scientists with unprecedented potential to study and learn more about human disease conditions. We say in theory because it has not been proven that therapeutic cloning in humans can work.

Many scientists do not like the term *therapeutic cloning* because it implies creating a human clone. Creating stem cells to treat human diseases is not the same as cloning a human being. With the development of iPSCs, many scientists who supported therapeutic cloning are instead turning their attention to

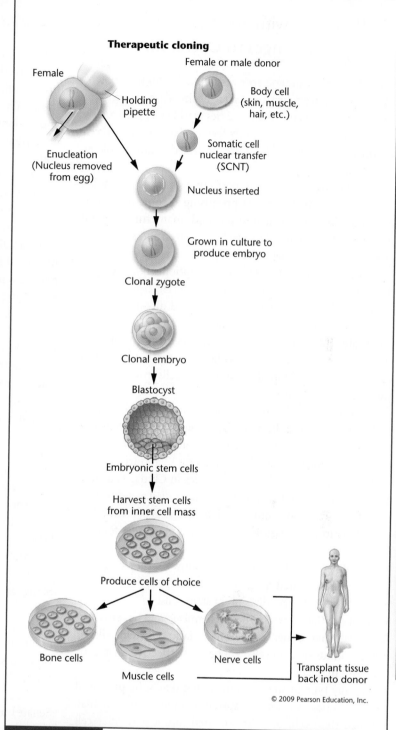

ST FIGURE 4–11 **Therapeutic Cloning.** In reproductive cloning, the goal is to produce a cloned baby. In therapeutic cloning, stem cells that are genetically identical to the cells taken from a patient are produced to provide patient-specific stem cell therapy

iPSCs to avoid ethical concerns associated with creating and destroying a human embryo for the purposes of isolating hESCs.

Problems with Stem Cells and Challenges to Overcome

Often stem cell treatments are over-publicized for their potential to cure disease, which can lead to a false sense of optimism about such treatments. Many studies worldwide, particularly those in countries with relatively few regulations on stem cell treatments for humans, have demonstrated major problems with stem cell applications.

A long road lies ahead in perfecting stem cell technologies for patient treatment. As promising as some of the potential applications of stem cell sound, many fundamental questions about stem cells must be answered and significant challenges must be met before stem cell technologies can be considered viable and safe treatment options. For instance, how can directed differentiation of stem cells be controlled *in vitro* and *in vivo* if stem cells are placed directly into the body? One problem with injecting stem cells instead of differentiated adult tissues is that scientists cannot fully control the spread of stem cells to other places in the body, nor can they control differentiation of stem cells into other tissues than their intended use.

Injected ESCs in animal models have formed tumors, including teratomas. In other experiments *in vivo*, cells differentiated from stem cells generated troubling abnormalities in chromosome number such as trisomy 12 and trisomy 17. Patients in unregulated stem cell clinics in China, Thailand, Korea, Romania, and other countries have died as a direct result of complications from receiving injections of stem cells.

Other important questions to be answered are as follows:

- How do stem cells self-renew, and how do they maintain an undifferentiated state?
- What factors trigger division of stem cells?
- What are the growth signals (chemical, genetic, environmental) that influence the differentiation of stem cells?
- What kinds of stem cells will be most effective for regenerative medicine?
- Can any type of stem cell, including iPSCs, be produced in a practical way for widespread therapeutic treatments?
- When stem cells are differentiated into a particular cell type, can it be assured that they will not dedifferentiate to an earlier developmental state once they are introduced into patients?
- Which diseases can be most effectively treated by stem cell technologies?
- What are the best ways to deliver stem cells effectively and safely?

Answers to these and many other questions will help scientists and physicians in their quest to develop stem cell–based applications for treating human disease.

Stem Cell Regulations

Although international guidelines on ethical use of stem cells have been established, there is currently no international policy governing stem cell use. In the United States, the FDA has rules regarding the purity of stem cells and their applications in clinical trials and medical products. Shortly after hESCs were first isolated, in the United States the National Bioethics Advisory Committee and the National Institutes of Health (NIH) began working on guidelines for hESC research. On August 9, 2001, President George W. Bush announced a ban on using federal funds to create an embryo for the purpose of isolating embryonic stem cells. This ban did provide for the use of federal funds for research on 78 cell lines that had already been established and available through the NIH prior to this date. The logic behind this ruling was that leftover embryos from IVF had already been used to create these cell lines.

However, only 21 of these lines became available for research, and many of the lines turned out to be far less valuable than initially believed. Some failed to grow without differentiating. Others showed genetic instability, with abnormalities in chromosome number. Some lines were contaminated by mouse feeder cells (which were commonly used to provide key nutrients to growing stem cells in the early days of ESC work). Many scientists and stem cell advocacy groups rallied hard for lifting the ban on federal funding for creating new hESC lines.

The scientific community was generally concerned that banning the use of federal funding for hESC work would lead to a "brain drain" in which the best stem cell researchers would leave the United States for countries with less restrictive policies on stem cell research and put the United States at a competitive disadvantage in developing new technologies involving stem cells. Many private companies took the lead in stem cell research, using private funding, and in many cases these companies moved research operations to states and countries that were most supportive of stem cell research.

In 2006 a bill to lift the ban on federal funding passed the Senate but was subsequently vetoed by President Bush. Congress tried again with a bill in 2007, which was also quickly vetoed by President Bush. In March 2009 President Barack Obama issued an executive order to have the ban lifted and charged the NIH to develop guidelines for governing federal funding of ESC work within 4 months. By April 2009 the NIH had released draft guidelines to allow federal funding for research on hESCs (but not for the use of federal funds to derive hESCs). In August 2010, a District of Columbia federal judge issued a preliminary injunction that temporarily blocked President Obama's 2009 order to expand embryonic stem cell funding. This ruling created major turmoil for stem cell researchers relying on federal funding for their work. This injunction prohibited the use of federal funds for hESC work, although it did allow work from funds previously awarded to continue.

In September 2010 the U.S. Court of Appeals temporarily stayed the injunction and by December a federal appellate panel began its review of the ruling. To demonstrate the impact of this ruling on U.S.-based companies in stem cell research, stock share prices for such companies dropped over 8 percent when it was first announced. The NIH was forced to order an immediate shutdown of hESC research by its investigators. By late April 2011, the U.S. Court of Appeals vacated the injunction but it is possible this case will eventually wind up in the Supreme Court. Visit the NIH stem cells website at http://stemcells.nih.gov/ for updates on current regulations and other resources regarding stem cells and their uses.

Many private foundations are providing hundreds of millions of dollars for stem cell researchers, and several states have enacted legislation to create stem cell research institutes as well as to provide state-funded support for ES research. Some of the more active states include California (which in 2004 passed Proposition 71 approving a budget of nearly $300 million in bonds and over $3 billion overall), New Jersey, Connecticut, Illinois, Maryland, and Wisconsin. Around the world, regulations vary on the production of new hESC cell lines and policies on therapeutic cloning. For instance, production of new lines and therapeutic cloning is legal in the United Kingdom, Israel, South Korea, China, and Singapore. Therapeutic cloning is banned in Brazil, Australia, and the European Union (although hESCs can be derived from unused IVF embryos where legal in member nations).

Legitimate stem cell research centers have been established around the world in countries considered major powers as well as many relatively small countries (Belgium, Sweden, Turkey, Israel, Switzerland, etc.). But as mentioned earlier, many stem cell treatment clinics have also sprouted up around the world, and patients desperate for cures have taken to traveling to other countries seeking stem cell–based cures, which are often overhyped and promoted as successful despite baseless information and adverse side effects. Often called stem cell tourists, patients desperate for a cure have been known to travel around the world for unapproved stem cell treatments, and the number of individuals involved in this "tourism" continues to grow at an alarming rate. The European Union has laws similar to FDA regulations in the United States governing stem cells. But it is clear that international regulations need to be established to govern safe applications of stem cells and to avoid the fraudulent and tragic cases of inappropriate, unethical applications of stem cells. Such inappropriate use of stem cells and their resulting tragedies will only impede progress and erode confidence in the legitimate treatments being developed.

Ethical Issues Involving Stem Cells

Clergy, politicians, researchers, and the general public continue to passionately debate the merits of stem cells. One issue at the root of these debates is the source of human stem cells, in particular hESCs, and their potential uses and abuses. In large part, hESCs are controversial because of their source—the early human embryo. Knowing that hESCs may have enormous potential for treating and curing many devastating diseases and providing people with an opportunity for healthier, longer lives, what do you think about their use?

The range of ethical questions surrounding stem cells and their applications is seemingly endless.

- Is it acceptable to produce a human embryo for the sole purpose of destroying it for other uses?
- Some fear that stem cells and cloning technologies will cause a great need for human eggs to support research. Is it acceptable to pay women to collect their eggs surgically?
- What is the moral status of early embryos created by therapeutic cloning?
- What rights does a cell donor have to stem cell lines or technologies created from cells they donated?
- Should tissue donors share in the commercial potential and monetary awards of stem cell lines created from cells they donated?

Some people believe that a person is formed at the moment an egg is fertilized, and so they consider therapeutic cloning equivalent to deliberately killing a child for the benefit of another person. Others believe that the early embryo is a cluster of living cells with the *potential* for forming a person, but that the early embryo itself is not a human being. Also, how can we justify destroying embryos that are developed through *in vitro* fertilization approaches? Why not use these embryos in an attempt to reduce pain and suffering in other humans?

Scientists define life in many ways. Biologists agree that the cluster of cells called the blastocyst is alive at the cellular level. Although all life forms warrant respect, the blastocyst is not a person because it does not have limbs, a nervous system, organs, or other physical features of a human individual. So a major source of debate continues about whether we should assign moral status to human embryos, and if so at what stage. Does the moral value of an embryo increase as it develops? Or is the embryo's moral value equal to that of a baby or adult? If an early embryo is deemed a living person, then it has all rights of other living persons. Consequently, intentionally destroying an embryo is immoral.

Significant ethical issues need to be addressed regarding individuals who donate cells used to derive stem cells of any type. Couples using IVF to have a child sign informed consent forms, which gives them legal authority over the embryos created from their gametes after infertility treatments are over. Recently, the NIH rejected dozens of hESC lines carrying mutations for different diseases because the agency determined that overly broad language in informed consent forms did not meet ethical requirements to inform donating couples of what their embryos might be used for. Of these rejected cell lines, many were created from embryos evaluated by preimplantation genetic diagnosis to have genetic defects for particular diseases.

Tax-paying citizens must decide how their money will be spent and what they believe is ethical, responsible, and safe research. Various public opinion polls have reported mixed feedback from Americans about the use of hESCs.

Recent polls indicate that the percentage of Americans who consider the use of hESCs morally acceptable has gradually increased from about 50 to 65 percent over the past eight years. Generally, Americans expect the highest level of health care in the world. If scientists in another country use stem cells to produce treatments for Parkinson's disease, Alzheimer's disease, and others, how will the American public feel about not having access to such technologies in the United States?

How about the ethical implications of using any type of stem cell for regenerative medicine? Some have voiced opposition to the potential use of iPSCs because they do not believe scientists should be creating replacement tissues and organs.

Selected Resources

Journal Articles

Cowan, C. A., Atienza, J., Melton, D. A., et al. 2005. Nuclear Reprogramming of Somatic Cells After Fusion with Human Embryonic Stem Cells. *Science*, 309: 1369–1373.

De Coppi, P., Bartsch, G., Siddiqui, M. M, et al. 2007. Isolation of Amniotic Stem Cell Lines with Potential for Therapy. *Nature Biotechnology*, 25: 100–106.

Hall, S. S. (2011). Diseases in a Dish. Scientific American, 304: 41–45.

Hanna, J. H., Saha, K., and Jaenisch, R. 2010. Pluripotency and Cellular Reprogramming: Facts, Hypotheses, Unresolved Issues. *Cell*, 143: 508–525.

Hanna, J., Wernig, M., Markoulaki, S., et al. 2007. Treatment of Sickle Cell Anemia Mouse Model with iPS Cells Generated from Autologous Skin. *Science*, 318: 1920–1923.

Hochedlinger, K. Your Inner Healers. 2010. *Scientific American*, 302: 46–53, May.

Meissner, A., Wernig, M., and Jaenisch, R. (2007). Direct Reprogramming of Genetically Unmodified Fibroblasts into Pluripotent Stem Cells. *Nature Biotechnology*, 25: 1177–1181.

Pera, M. F., and Tam, P. P. L. 2010. Extrinsic Regulation of Pluripotent Stem Cells. *Nature*, 465: 713–720.

Shamblott, M. J., Axelman, J., Wang, S., et al. 1998. Derivation of Pluripotent Stem Cells from Cultured Human Primordial Germ Cells. *Proceedings of the National Academy of Sciences U.S.A.*, 95: 13726–13731.

Takahashi, K., Tanabe, K., Ohnuki, M., et al. 2007. Induction of Pluripotent Stem Cells from Adult Human Fibroblasts by Defined Factors. *Cell*, 131: 861–872.

Thomson, J. A., Itskovitz-Eldor, J., Shapiro, S. S., et al. 1998. Embryonic Stem Cell Lines Derived from Human Blastocysts. *Science*, 282: 1145–1147.

Yamanaka, S., and Blau, H. M. 2010. Nuclear Reprogramming to a Pluripotent State by Three Approaches. *Nature*, 465: 704–712.

Yu, J., Vodyanik, M. A., Smuga-Otto, K., et al. 2007. Induced Pluripotent Stem Cell Lines from Human Somatic Cells. *Science*, 318: 1917–1920.

Web Sites

Human Embryonic Stem Cell Animation. Sumanas, Inc. 2007. http://www.sumanasinc.com/webcontent/animations/content/stemcells_scnt.html

Stem Book. http://www.stembook.org/

Stem Cell Information: The National Institutes of Health resource for stem cell research. http://stemcells.nih.gov/

Stem Cell Resources.org: The Science of Education. http://www.stemcellresources.org/

University of Michigan: Stem Cell Research. http://www.umich.edu/stemcell/

An agarose gel containing separated DNA fragments stained with the DNA-binding dye ethidium bromide and visualized under ultraviolet light.

20

Recombinant DNA Technology

CHAPTER CONCEPTS

- Recombinant DNA technology creates combinations of DNA sequences from different sources.

- A common application of recombinant DNA technology is to clone a DNA segment of interest.

- For some cloning applications, specific DNA segments are inserted into vectors to create recombinant DNA molecules that are transferred into eukaryotic or prokaryotic host cells such as bacteria, where the recombinant DNA replicates as the host cells divide.

- DNA libraries are collections of cloned DNA and were historically used to isolate specific genes.

- DNA segments can be quickly amplified and cloned millions of times using the polymerase chain reaction (PCR).

- DNA, RNA, and proteins can be analyzed using a range of molecular techniques.

- DNA sequencing reveals the nucleotide composition of cloned DNA, and rapid advances in sequencing technologies have advanced many areas of modern genetics research particularly genomics.

- Recombinant DNA technology has revolutionized our ability to investigate the genomes of diverse species and led to the modern revolution in genomics.

n 1971, a paper published by Kathleen Danna and Daniel Nathans marked the beginning of the recombinant DNA era. The paper described the isolation of an enzyme from a bacterial strain and the use of the enzyme to cleave viral DNA at specific nucleotide sequences. It contained the first published photograph of DNA cut with such an enzyme, now called a restriction enzyme.

Using restriction enzymes and a number of other resources, researchers of the mid- to late 1970s developed various techniques to create, replicate, and analyze **recombinant DNA** molecules—DNA created by joining together pieces of DNA from different sources. The methods used to copy or **clone** DNA, called **recombinant DNA technology** and often known as "gene splicing" in the early days, were a major advance in research in molecular biology and genetics, allowing scientists to isolate and study specific DNA sequences. For their contributions to the development of this technology, Nathans, Hamilton Smith, and Werner Arber were awarded the 1978 Nobel Prize in Physiology or Medicine.

The power of recombinant DNA technology is astonishing, enabling geneticists to identify and isolate a single gene or DNA segment of interest from a genome. Subsequently, through cloning, large quantities of identical copies of this specific DNA molecule can be produced. These identical copies, or clones, can then be manipulated for numerous purposes, including research into the structure and organization of the DNA, studying gene expression, and producing important commercial products from the protein encoded by a gene. The fundamental techniques involved in recombinant DNA technology subsequently led to the field of genomics, enabling scientists to sequence and analyze entire genomes. We will consider these topics in detail in Chapters 21 and 22. In this chapter, we review basic methods of recombinant DNA technology used to isolate, replicate, and analyze DNA.

20.1

Recombinant DNA Technology Began with Two Key Tools: Restriction Enzymes and DNA Cloning Vectors

Although natural genetic processes such as crossing over produce recombined DNA molecules, the term *recombinant DNA* is generally reserved for molecules produced by artificially joining DNA obtained from different sources. The methods used to create these molecules were largely derived from nucleic acid biochemistry, coupled with genetic techniques developed for the study of bacteria and viruses.

We will begin our discussion of recombinant DNA technology by considering two important tools used to construct and amplify recombinant DNA molecules: DNA-cutting enzymes called **restriction enzymes** and **DNA cloning vectors**. The use of restriction enzymes and cloning vectors was largely responsible for advancing the field of molecular biology because of a wide range of techniques that are based on recombinant DNA technology.

Restriction Enzymes Cut DNA at Specific Recognition Sequences

Restriction enzymes are produced by bacteria as a defense mechanism against infection by viruses. They restrict or prevent viral infection by degrading the DNA of invading viruses. More than 3500 restriction enzymes have been identified, and over 250 are commercially produced and available for use by researchers. A restriction enzyme recognizes and binds to DNA at a specific nucleotide sequence called a **restriction site** (Figure 20–1). The enzyme then cuts both strands of the DNA within that sequence by cleaving the phosphodiester backbone of DNA. Scientists commonly refer to this as "digestion" of DNA. The usefulness of restriction enzymes in cloning derives from their ability to accurately and reproducibly cut genomic DNA into fragments. Restriction enzymes represent sophisticated molecular scissors for cutting DNA into fragments of desired sizes. The size of DNA restriction fragments produced by digesting DNA with a particular enzyme can be estimated from the number of times a given restriction enzyme cuts the DNA. Restriction sites are present randomly in the genome. Enzymes with a four-base recognition sequence—such as the enzyme *Alu*I, which recognizes the sequence AGCT—will cut, on average, every 256 base pairs ($4^n = 4^4 = 256$) if all four nucleotides are present in equal proportions, producing many small fragments. The actual fragment sizes produced by DNA digestion with a given restriction enzyme vary because the number and location of recognition sequences are not always distributed randomly in DNA.

Most recognition sequences exhibit a form of symmetry described as a **palindrome**: the nucleotide sequence reads the same on both strands of the DNA when read in the 5′ to 3′ direction. Each restriction enzyme recognizes its particular recognition sequence and cuts the DNA in a characteristic cleavage pattern. The most common recognition sequences are four or six nucleotides long, but some contain eight or more nucleotides. Enzymes such as *Eco*RI and *Hind*III make offset cuts in the DNA strands, thus producing fragments with single-stranded overhanging ends called **cohesive ends**, while others such as *Alu*I and *Bal*I cut both strands at the same nucleotide pair, producing DNA fragments with double-stranded ends called **blunt-end** fragments. Four common restriction enzymes, their restriction sites, and source microbes are indicated in Figure 20–1.

One of the first restriction enzymes to be identified was isolated from *Escherichia coli* strain R and was designated *Eco*RI. DNA fragments produced by *Eco*RI digestion

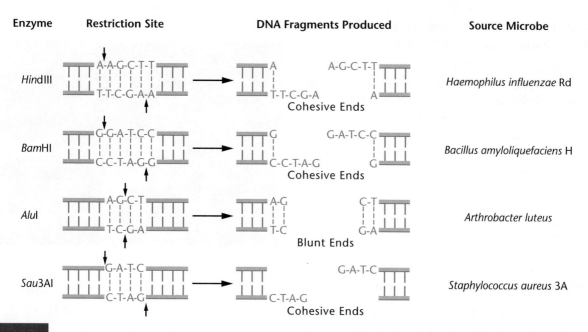

FIGURE 20–1 Common restriction enzymes, with their restriction sites, DNA cutting patterns, and sources. Arrows indicate the location in the DNA cut by each enzyme.

(Figure 20–2) have cohesive ends or so-called sticky ends because they can base-pair with complementary single-stranded ends on other DNA fragments cut using *Eco*RI. When mixed together, single-stranded ends of DNA fragments from different sources cut with the same restriction enzyme can **anneal**, or stick together, by hydrogen bonding of complementary base pairs in single-stranded ends. Addition of the enzyme **DNA ligase** (Do you remember the role of DNA ligase in DNA replication as discussed in Chapter 11?) to DNA fragments will seal the phosphodiester backbone of DNA to covalently join the fragments together to form recombinant DNA molecules (Figure 20–2).

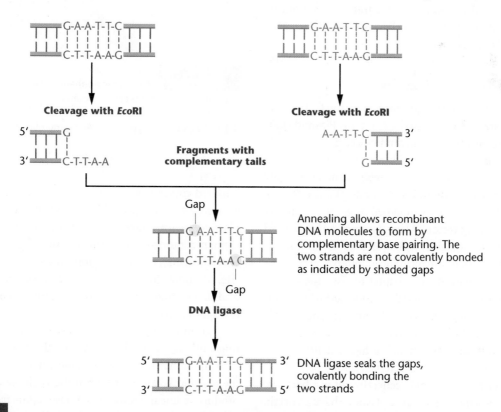

FIGURE 20–2 DNA from different sources is cleaved with *Eco*RI and mixed to allow annealing. The enzyme DNA ligase forms phosphodiester bonds between these fragments to create an intact recombinant DNA molecule.

DNA Vectors Accept and Replicate DNA Molecules to Be Cloned

Scientists recognized that DNA fragments produced by restriction enzyme digestion could be copied or cloned if they had a technique for replicating the fragments. The second key tool that allowed DNA cloning was the development of DNA **cloning vectors**. **Vectors** are DNA molecules that accept DNA fragments and replicate inserted DNA fragments when vectors are placed into host cells.

Many different vectors are available for cloning. Vectors differ in terms of the host cells they are able to enter and in the size of inserts they can carry, but most DNA vectors have several key properties.

- A vector contains several restriction sites that allow insertion of the DNA fragments to be cloned.

- Vectors must be introduced into host cells to allow for independent replication of the vector DNA and any DNA fragment it carries.

- To distinguish host cells that have taken up vectors from host cells that have not, the vector should carry a **selectable marker gene** (usually an antibiotic resistance gene or the gene for an enzyme absent from the host cell).

- Many vectors incorporate specific sequences that allow for sequencing inserted DNA.

- The vector and its inserted DNA fragment should be easy to isolate from the host cell to recover cloned DNA for different applications.

Bacterial Plasmid Vectors

Genetically modified bacterial **plasmids** were the first vectors developed and are still widely used for cloning. Plasmid cloning vectors were derived from naturally occurring plasmids. Recall from Chapter 6 that plasmids are extrachromosomal, double-stranded DNA molecules that replicate independently from the chromosomes within bacterial cells (**Figure 20–3a**). Plasmids have been extensively modified by genetic engineering to serve as cloning vectors. Many commercially prepared plasmids are readily available with a range of useful features (Figure 20–3b). Plasmids are introduced into bacteria by the process of **transformation** (see Chapter 6). Two main techniques are widely used for bacterial transformation. One approach involves treating cells with calcium ions and using a brief heat shock to pulse DNA into cells. The other technique, called **electroporation**, uses a brief, but high-intensity, pulse of electricity to move DNA into bacterial cells.

Only one or a few plasmids generally enter a bacterial host cell by transformation. Because plasmids have an origin of replication (*ori*) that allows for plasmid replication, many plasmids can increase their copy number to produce several

(a)

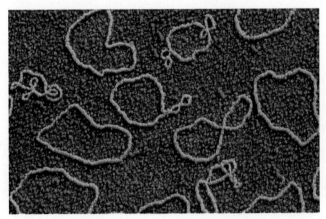

(b)

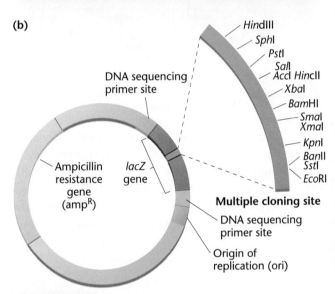

FIGURE 20–3　(a) A color-enhanced electron micrograph of circular plasmid molecules isolated from *E. coli*. Genetically engineered plasmids are used as vectors for cloning DNA. (b) A diagram of a typical DNA cloning plasmid.

hundred copies in a single host cell. These plasmids greatly enhance the number of DNA clones that can be produced. Plasmid vectors have also been genetically engineered to contain a number of restriction sites for commonly used restriction enzymes in a region called the **multiple cloning site**. Multiple cloning sites allow scientists to clone a range of different fragments generated by many commonly used restriction enzymes.

Cloning DNA with a plasmid generally begins by cutting both the plasmid DNA and the DNA to be cloned with the same restriction enzyme (**Figure 20–4**). Typically the plasmid is cut once within the multiple cloning site to produce a linear vector. DNA restriction fragments from the DNA to be cloned are added to the linearized vector in the presence of DNA ligase. Sticky ends of DNA fragments

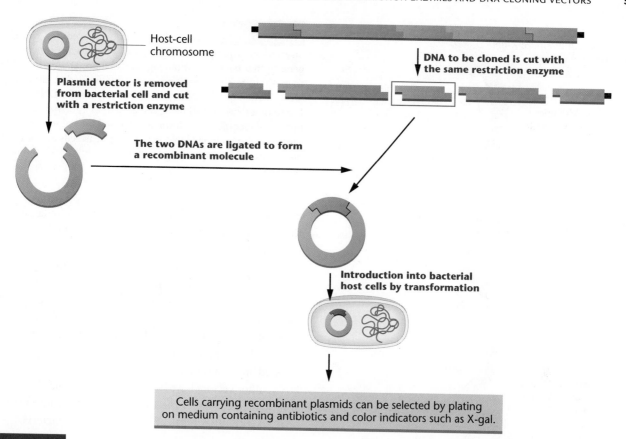

FIGURE 20–4   Cloning with a plasmid vector involves cutting both plasmid and the DNA to be cloned with the same restriction enzyme. The DNA to be cloned is spliced into the vector and transferred to a bacterial host for replication. Bacterial cells carrying plasmids with DNA inserts can be identified by selection and then isolated. The cloned DNA is then recovered from the bacterial host for further analysis.

anneal, joining the DNA to be cloned and the plasmid. DNA ligase is then used to create phosphodiester bonds to seal nicks in the DNA backbone, thus producing recombinant DNA, which is then introduced into bacterial host cells by transformation. Once inside the cell, plasmids replicate quickly to produce multiple copies.

However, when cloning DNA using plasmids, not all plasmids will incorporate DNA to be cloned. For example, a plasmid cut with a particular restriction enzyme can close back on itself (self-ligation) if cut ends of the plasmid rejoin. Obviously then, such nonrecombinant plasmids are not desired. Also, during transformation, not all host cells will take up plasmids. Therefore it is important that bacterial cells containing recombinant DNA can be readily identified in a cloning experiment. One way this is accomplished is through the use of selectable marker genes described earlier. Genes that provide resistance to antibiotics such as ampicillin and genes such as the *lacZ* gene are very effective selectable marker genes. Figure 20–5 provides an example of how these genes can be used to identify bacteria containing recombinant plasmids. This process is often referred to as **"blue-white" selection** for a reason that will soon become obvious. In blue-white selection a plasmid is used that contains the *lacZ* gene incorporated into the multiple

cloning site. The *lacZ* gene encodes the enzyme β-galactosidase which, as you learned in Chapter 16, is used to cleave the disaccharide lactose into its component monosaccharides glucose and galactose. Blue-white selection takes advantage of the enzymatic activity of β-galactosidase.

Using this approach, one can easily identify transformed bacterial cells containing recombinant or nonrecombinant plasmids. If a DNA fragment is inserted anywhere in the multiple cloning site, the *lacZ* gene is disrupted and will not produce functional copies of β-galactosidase. Transformed bacteria in this experiment are plated on agar plates that contain an antibiotic—ampicillin in this case. Nontransformed bacteria cannot grow well on these plates because they do not have the *amp*R gene and will be prevented from growing by the antibiotic. But these agar plates also contain a substance called X-gal (technically 5-bromo-4-chloro-3-indolyl-β-D-galactopyranoside). X-gal is similar to lactose in structure. It can be cleaved by β-galactosidase, and when this happens it turns blue. As a result, bacterial cells carrying nonrecombinant plasmids (those that have closed up on themselves and do not contain inserted DNA) have a functional *lacZ* gene and produce β-galactosidase, which cleaves X-gal in the medium, and these cells turn blue. Recombinant bacteria with plasmids containing an inserted DNA fragment

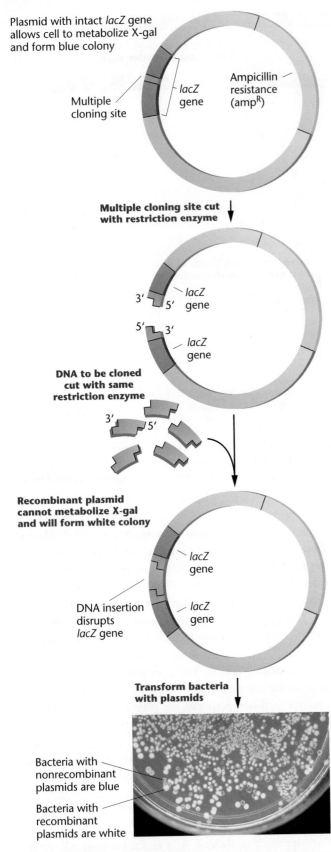

Plasmid with intact *lacZ* gene allows cell to metabolize X-gal and form blue colony

Multiple cloning site

lacZ gene

Ampicillin resistance (amp^R)

Multiple cloning site cut with restriction enzyme

3′ 5′ *lacZ* gene

5′ 3′ *lacZ* gene

DNA to be cloned cut with same restriction enzyme

3′ 5′

Recombinant plasmid cannot metabolize X-gal and will form white colony

lacZ gene

lacZ gene

DNA insertion disrupts *lacZ* gene

Transform bacteria with plasmids

Bacteria with nonrecombinant plasmids are blue

Bacteria with recombinant plasmids are white

FIGURE 20–5   In blue-white selection procedures, DNA inserted into multiple cloning site of a plasmid disrupts the *lacZ* gene so that bacteria containing recombinant DNA are unable to metabolize X-gal, resulting in white colonies that allow direct identification of bacterial colonies carrying cloned DNA inserts. Photo of a Petri dish showing the growth of bacterial cells after uptake of recombinant plasmids. Cells in blue colonies contain vectors without cloned DNA inserts, whereas cells in white colonies contain vectors carrying DNA inserts.

to flasks of bacterial culture broth and grown in large quantities, after which it is relatively easy to isolate recombinant plasmids from these cells.

Plasmids are still the workhorses for many applications of recombinant DNA technology, but they have one limitation: because they are small, they can only accept inserted pieces of DNA up to about 25 kilobases (kb) in size and most plasmids can often only accept substantially smaller pieces. Therefore as recombinant DNA technology has developed and it has become desirable to clone large pieces of DNA, other vectors have been developed primarily for their ability to accept larger pieces of DNA and because they can be used with other types of host cells beside bacteria.

Other Types of Cloning Vectors

Phage vector systems were among the earliest vectors used in addition to plasmids. These included genetically modified strains of λ **phage**. The genome of λ phage, a virus that infects *E. coli*, has been completely mapped and sequenced, and the λ phage genome has been modified to incorporate many of the important features of cloning vectors described earlier in this chapter, including a multiple cloning site. Phage vectors were popular for quite some time and are still in use today because they can carry inserts up to 45 kb, more than twice as long as DNA inserts in most plasmid vectors. DNA fragments are ligated into the phage vector to produce recombinant λ vectors that are subsequently packaged into phage protein heads *in vitro* and introduced into bacterial host cells growing on Petri plates. Inside the bacteria, the vectors replicate and form many copies of infective phage, each of which carries a DNA insert. As they reproduce, they lyse their bacterial host cells, forming the clear spots known as plaques (described in Chapter 6), from which phage can be isolated and the cloned DNA can be recovered.

Bacterial artificial chromosomes (BACs) and **yeast artificial chromosomes (YACs)** are two other examples of vectors that can be used to clone large fragments of DNA. For example, the mapping and analysis of large eukaryotic genomes such as the human genome required cloning vectors that could carry very large DNA fragments such as segments of an entire chromosome. BACs are essentially very large but low copy number (typically one or two copies/bacterial cell)

form white colonies on X-gal medium because the plasmids in these cells are not producing functional β-galactosidase (Figure 20–5). Bacteria in these white colonies are clones of each other—genetically identical cells with copies of recombinant plasmids. White colonies can be transferred

plasmids that can accept DNA inserts in the 100- to 300-kb range. Like natural chromosomes, a YAC has telomeres at each end, origins of replication, and a centromere. These components are joined to selectable marker genes and to a cluster of restriction-enzyme recognition sequences for insertion of foreign DNA. Yeast chromosomes range in size from 230 kb to over 1900 kb, making it possible to clone DNA inserts from 100 to 1000 kb in YACs. The ability to clone large pieces of DNA in these vectors makes them an important tool in genome sequencing projects, including the Human Genome Project (see Chapter 21).

Expression vectors are designed to ensure mRNA expression of cloned gene with the purpose of producing many copies of the gene's encoded protein in a host cell. Expression vectors are available for both prokaryotic and eukaryotic host cells. For many research applications that involve studies of protein structure and function, producing a recombinant protein in bacteria (or other host cells) and purifying the protein is a routine approach, although it is not always easy to properly express a protein that maintains its biological function. The biotechnology industry also relies heavily on expression vectors to produce commercially valuable protein products from cloned genes, a topic we will discuss in Chapter 22.

Ti Vectors for Plant Cells

Introducing genes into plants is a common application that can be done in many ways, and we will discuss some aspects of genetic engineering of plants in Chapter 22. One widely used approach to insert genes into plant cells involves the soil bacterium *Rhizobium radiobacter,* which infects plant cells and produces tumors (called crown galls) in many species of plants. Formerly *Agrobacterium tumefaciens* this bacterium was renamed based on genomic analysis. *Rhizobium* contains a plasmid called the **Ti plasmid** (tumor-inducing), and tumor formation is associated with the presence of particular genes in the Ti plasmid. In the wild, when Ti plasmid-carrying bacteria infect plant cells, a segment of the Ti plasmid, known as *T-DNA,* is transferred into the genome of the host plant cell. Genes in the T-DNA segment control tumor formation and the synthesis of compounds required for growth of the infecting bacteria.

Restriction sites in Ti plasmids can be used to insert foreign DNA, and recombinant vectors are introduced into *Agrobacterium* by transformation. Tumor-inducing genes from Ti plasmids are removed from the vector so that the recombinant vector does not result in tumor production. *Agrobacterium* containing recombinant DNA is mixed with plant cells (not all types of plant cells can be infected by *Agrobacterium*). Once inside the cell, the foreign DNA is inserted into the plant genome when the T-DNA integrates into a host-cell chromosome. Plant cells carrying a recombinant Ti plasmid can be grown in tissue culture. The presence of certain compounds in the culture

medium plant cells stimulates the formation of roots and shoots, and eventually a mature plant carrying a foreign gene.

Host Cells for Cloning Vectors

Besides deciding which DNA cloning vector to use, another cloning consideration worthy of discussion is which host cells are used to accept recombinant DNA for cloning. There are many different reasons why particular host cells are chosen for a recombinant DNA experiment depending on the purpose of the work. Whereas *E. coli* is widely used as a prokaryotic host cell of choice when working with plasmids, the yeast *Saccharomyces cerevisiae* is extensively used as a host cell for the cloning and expression of eukaryotic genes. There are several reasons: (1) Although yeast is a eukaryotic organism, it can be grown and manipulated in much the same way as bacterial cells. (2) The genetics of yeast has been intensively studied, providing a large catalog of mutants and a highly developed genetic map (prior to completion of the yeast genome). (3) The entire yeast genome has been sequenced, and genes in the organism have been identified. (4) To study the

NOW SOLVE THIS

20–1 An ampicillin-resistant, tetracycline-resistant plasmid, pBR322, is cleaved with *Pst*I, which cleaves within the ampicillin resistance gene. The cut plasmid is ligated with *Pst*I-digested *Drosophila* DNA to prepare a genomic library, and the mixture is used to transform *E. coli* K12.

(a) Which antibiotic should be added to the medium to select cells that have incorporated a plasmid?

(b) If recombinant cells were plated on medium containing ampicillin or tetracycline and medium with both antibiotics, on which plates would you expect to see growth of bacteria containing plasmids with *Drosophila* DNA inserts?

(c) How can you explain the presence of colonies that are resistant to both antibiotics?

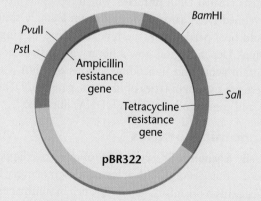

■ HINT: *This problem involves an understanding of antibiotic selectable marker genes in plasmids and antibiotic DNA selection for identifying bacteria transformed with recombinant plasmid DNA. The key to its solution is to recognize that inserting foreign DNA into the plasmid vector disrupts one of the antibiotic resistance genes in the plasmid.*

function of some eukaryotic proteins, it is necessary to use a host cell that can modify the protein, by adding carbohydrates, for example, after it has been synthesized, to convert it to a functional form (bacteria cannot carry out some of these modifications). (5) Yeast have been used for centuries in the baking and brewing industries and is considered to be a safe organism for producing proteins for vaccines and therapeutic agents.

The choice of host cells also extends to a number of other cell types, including insect cells. A variety of different human cell types can be grown in culture and used to express genes and proteins. Such cell lines can also then be subjected to various approaches for gene or protein functional analysis, including drug testing for effectiveness at blocking or influencing a particular recombinant protein being expressed, particularly if the cell lines are of a human disease condition such as cancer.

20.2
DNA Libraries Are Collections of Cloned Sequences

Only relatively small DNA segments—representing only a single gene or even a portion of a gene—are produced by cloning DNA into vectors, particularly plasmids. Even when several hundred genes are introduced into larger vectors such as BACs or YACs, one still needs a method for identifying the DNA pieces that were cloned. Consider this: in the cloning discussions we have had so far, we have described *how* DNA can be inserted into vectors and cloned—a relatively straightforward process—but we have not discussed how one knows what particular DNA sequence they have cloned. Simply cutting DNA and inserting into vectors does not tell you what gene or sequences have been cloned.

During the first several decades of DNA cloning, scientists created **DNA libraries**, which represent a collection of cloned DNA samples derived from a single source that could be a particular tissue type, cell type, or single individual. Depending on how a library is constructed, it may contain genes and noncoding regions of DNA. Generally, there are two main types of libraries, genomic DNA libraries and complementary DNA (cDNA) libraries.

Genomic Libraries

Ideally, a **genomic library** consists of many overlapping fragments of the genome, with at least one copy of every DNA sequence in an organism's genome, which in summary span the entire genome. In making a genomic library, DNA is extracted from cells or tissues and cut with restriction enzymes, and the resulting fragments are inserted into vectors using techniques that we discussed in the previous section. Since some vectors (such as plasmids) can carry only a few thousand base pairs of inserted DNA, selecting the vector so that the library

contains the whole genome in the smallest number of clones is an important consideration. Because genomic DNA is the foreign DNA introduced into vectors, genomic libraries contain coding and noncoding segments of DNA such as introns.

When working with large genomes such as the Human Genome, the choice of a vector is a primary consideration when making such libraries. As a result, YACs were commonly used to accommodate large sizes of DNA necessary to span the approximately 3 billion bp of DNA in the human genome. If, for example, a human genome library was constructed using plasmid vectors with an average insert size of 5 kb, then more than 2.4 million clones would be required for a 99 percent probability of recovering any given sequence from the genome. Because of its size, this library would be difficult to use efficiently. However, if the library was constructed in a YAC vector with an average insert size of 1 Mb, then the library would only need to contain about 14,000 YACs, making it relatively easy to use. Vectors with large cloning capacities such as YACs were essential tools for the Human Genome Project.

As you will learn in Chapter 21, **whole-genome shotgun cloning** approaches (see Figure 21–1) and new sequencing methodologies (called next-generation sequencing) are readily replacing traditional genomic DNA libraries because they effectively allow one to sequence all of the DNA fragments in a genomic DNA sample without the need for inserting DNA fragments into vectors and cloning them in host cells. In Chapter 21 we will also consider how DNA sequence analysis using bioinformatics allows one to identify protein-coding and noncoding sequences in cloned DNA.

Complementary DNA (cDNA) Libraries

Complementary DNA (cDNA) libraries offer certain advantages over genomic libraries and continue to be a particularly useful methodology for gene cloning. This is primarily because a cDNA library contains DNA copies—called cDNA—which are made from the mRNA molecules of a cell population and therefore represent the genes being expressed in the cells at the time the library was made. cDNA is complementary to the nucleotide sequence of the mRNA, and so unlike a genomic library, which contains all of the DNA in a genome—gene coding and noncoding sequences—a cDNA library contains only expressed genes. As a result, cDNA libraries have been particularly useful for identifying and studying genes expressed in certain cells or tissues under certain conditions—for example, during development, cell death, cancer, and other biological processes. One can also use these libraries to compare expressed genes from normal tissues and diseased tissues. For instance, this approach has been particularly valuable for identifying genes involved in cancer formation, such as those genes that contribute to progression from a normal cell to a cancer cell and genes involved in cancer cell metastasis (spreading).

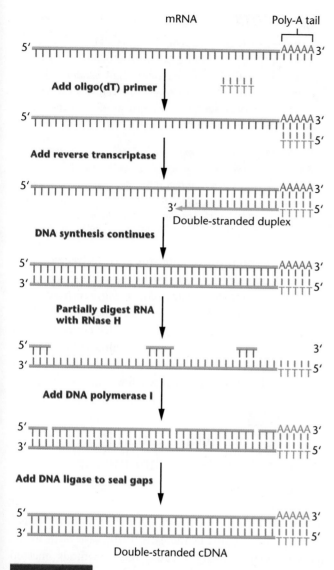

FIGURE 20-6 Producing cDNA from mRNA. Because many eukaryotic mRNAs have a poly-A tail of variable length at one end, a short oligo-dT molecule annealed to this tail serves as a primer for the enzyme reverse transcriptase. Reverse transcriptase uses the mRNA as a template to synthesize a complementary DNA strand (cDNA) and forms an mRNA/cDNA double-stranded duplex. The mRNA is digested with the enzyme RNAse H, producing gaps in the RNA strand. The 3′ ends of the remaining RNA serve as primers for DNA polymerase I, which synthesizes a second DNA strand. The result is a double-stranded cDNA molecule that can be cloned into a suitable vector or used directly as a probe for library screening.

A cDNA library is prepared by isolating mRNA from a population of cells of interest. Typically one picks cells that express an abundance of mRNA for the genes to be cloned. Nearly all mRNA isolated from eukaryotic cells has a string of adenines on the 3′ end called the poly-A tail (recall this from Chapter 13), and we can take advantage of this structural feature of mRNA. The next step in making a cDNA library is to mix mRNAs with oligo(dT) primers—short, single-stranded sequences of T nucleotides that anneal to

the poly-A tail (Figure 20–6). The enzyme **reverse transcriptase** extends the oligo (dT) primer and synthesizes a complementary DNA copy of the mRNA sequence. The product of this reaction is an mRNA–DNA double-stranded hybrid molecule. The RNA in the hybrid molecule can be enzymatically digested or chemically degraded and a second, opposing strand of DNA synthesized by DNA polymerase. Over time all of the RNA in hybrid molecules is replaced with DNA creating double-stranded cDNA molecules that are complementary to the mRNA. Alternatively, cDNA can also be synthesized using primers that bind randomly to mRNA and the first strand of cDNA to eventually create double-stranded cDNA from the original mRNA strand.

The cDNA molecules are subsequently inserted into vectors, usually plasmids. Because one typically wouldn't know what restriction enzymes could be used to cut this DNA, to insert cDNA into a plasmid one usually needs to attach linker sequences to the ends of the cDNA. Linkers are short double-stranded oligonucleotides containing a restriction enzyme recognition sequence (e.g., *Eco*RI). After attachment to the cDNAs, the linkers are cut with *Eco*RI and ligated to vectors treated with the same enzyme. Transfer of vectors carrying cDNA molecules to host cells and cloning is used to make a cDNA library.

These libraries provide a snapshot of the genes that were transcriptionally active in a tissue at a particular time because the relative amount of cDNA in a particular library is equivalent to the amount of starting mRNA isolated from the tissue and used to make the library. Many different cDNA libraries are available from cells and tissues in specific stages of development, different organs such as brain, muscle, and kidney, and tissues from different disease states such as cancer. These libraries provide an instant catalog of all the genes active in a cell at a specific time and have been very valuable tools for scientists isolating and studying genes in particular tissues.

Specific Genes Can Be Recovered from a Library by Screening

Genomic and cDNA libraries often consist of several hundred thousand different DNA clones, much like a large book library may have many books but only a few of interest to your studies in genetics. So how can libraries be used to locate a specific gene of interest in a library? To find a specific gene, we need to identify and isolate only the clone or clones containing that gene. We must also determine whether a given clone contains all or only part of the gene we are studying. Several methods allow us to sort through a library and isolate specific genes of interest, and this approach is called **library screening**. The choice of method often depends on the circumstances and available information about the gene being sought.

Often, probes are used to screen a library to recover clones of a specific gene. A **probe** is any DNA or RNA sequence that is complementary to some part of a cloned sequence present in the library—the target gene or sequence to be identified. A probe is also labeled or tagged in different ways so that it can be identified. When used in a hybridization reaction, the probe binds to any complementary DNA sequences present in one or more clones. Probes can be labeled with radioactive isotopes, or increasingly these days probes are labeled with nonradioactive compounds that undergo chemical or color reactions to indicate the location of a specific clone in a library.

Probes are derived from a variety of sources—often related genes isolated from another species can be used if enough of the DNA sequence is conserved. For example, genes cloned in rats, mice, or even *Drosophila* that have conserved sequence similarity to human genes have been used as probes to identify human genes during library screening.

To screen a library with a probe, bacterial clones from the library are grown on nutrient agar plates, where they form hundreds or thousands of colonies (Figure 20–7). A replica of the colonies on each plate is made by gently pressing a nylon or nitrocellulose membrane onto the plate's surface; this transfers the pattern of bacterial colonies from the plate to the membrane. The membrane is processed to lyse the bacterial cells, denature the double-stranded DNA released from the cells into single strands, and bind these strands to the membrane.

The DNA on the membrane is screened by incubation with a labeled nucleic acid probe. First, the probe is heated and quickly cooled to form single-stranded molecules, and then it is added to a solution containing the membrane. If the nucleotide sequence of any of the DNA on the membrane is complementary to the probe, a double-stranded DNA–DNA hybrid molecule will form (one strand from the probe and the other from the cloned DNA on the membrane). This step is called **hybridization**.

After incubation of the probe and the membrane, unbound probe molecules are washed away, and the membrane is assayed to detect the hybrid molecules that remain. If a radioactive probe has been used, the membrane is overlaid with a piece of X-ray film. Radioactivity from the probe molecules bound to DNA on the membrane will expose the film, producing dark spots on the film. These spots represent colonies on the plate containing the cloned gene of interest (Figure 20–7). The positions of spots on the film are used as a guide to identify and recover the corresponding colonies on the plate. The cloned DNA they contain can be used in further experiments. With some nonradioactive probes, a chemical reaction emits photons of light (chemiluminescence) to expose the photographic film and reveal the location of colonies carrying the gene of interest.

As we have discussed here, libraries enable scientists to clone DNA and then identify individual genes in the library. Cloning DNA from libraries is still a technique with valuable applications. However, as you will learn in Chapter 21,

NOW SOLVE THIS

20–2 For question 20-16, you are given a cDNA library of human genes prepared in a bacterial plasmid vector. You are also given the cloned genomic DNA sequence yeast gene that encodes EF-1a, a protein that is highly conserved among eukaryotes. Outline how you would use these resources to identify the human cDNA clone encoding EF-1a.

■ HINT: *This problem asks you to use a cDNA library to isolate the human EF-1a gene when you have access to the yeast EF-1a gene. The key to its solution is to remember that cDNA clones do not have all the sequences of a genomic library, but do contain coding sequences (exons) and they can be identified with the proper probe.*

the basic methods of recombinant DNA technology were the foundation for the development of powerful techniques for whole-genome cloning and sequencing, which led to the **genomics** era of modern genetics and molecular biology. Increasingly, genomic techniques, in which entire genomes are being sequenced without creating libraries, are replacing many traditional recombinant DNA approaches that cloned or identified individual or a few genes at a time.

20.3
The Polymerase Chain Reaction Is a Powerful Technique for Copying DNA

As we will discuss in Chapter 22, the recombinant DNA techniques developed in the early 1970s gave birth to the biotechnology industry because these methods enabled scientists to clone human genes, such as the insulin gene, whose protein product could be used for therapeutic purposes. However, cloning DNA using vectors and host cells is labor intensive and time consuming. In 1986, another technique, called the **polymerase chain reaction (PCR)**, was developed. This advance revolutionized recombinant DNA methodology and further accelerated the pace of biological research. The significance of this method was underscored by the awarding of the 1993 Nobel Prize in Chemistry to Kary Mullis, who developed the technique.

PCR is a rapid method of DNA cloning that extends the power of recombinant DNA research and in many cases eliminates the need to use host cells for cloning. Although library cloning techniques still have specific applications of value, PCR is a method of choice for many applications, whether in molecular biology, human genetics, evolution, development, conservation, or forensics.

By copying a specific DNA sequence through a series of *in vitro* reactions, PCR can amplify target DNA sequences that are initially present in very small quantities in a population of other DNA molecules. When using PCR to clone

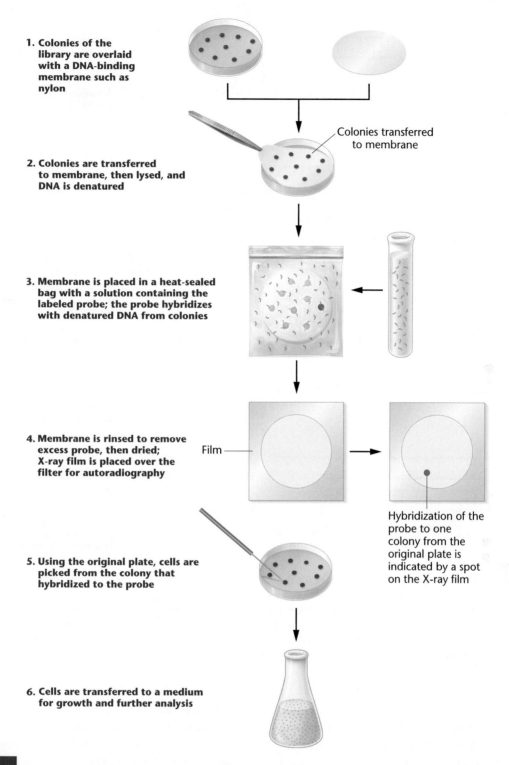

1. **Colonies of the library are overlaid with a DNA-binding membrane such as nylon**

Colonies transferred to membrane

2. **Colonies are transferred to membrane, then lysed, and DNA is denatured**

3. **Membrane is placed in a heat-sealed bag with a solution containing the labeled probe; the probe hybridizes with denatured DNA from colonies**

4. **Membrane is rinsed to remove excess probe, then dried; X-ray film is placed over the filter for autoradiography**

Film

Hybridization of the probe to one colony from the original plate is indicated by a spot on the X-ray film

5. **Using the original plate, cells are picked from the colony that hybridized to the probe**

6. **Cells are transferred to a medium for growth and further analysis**

FIGURE 20–7 Screening a library constructed using a plasmid vector to recover a specific gene. The library, present in bacteria on Petri plates, is overlaid with a DNA-binding membrane, and colonies are transferred to the membrane. Colonies on the membrane are lysed, and the DNA is denatured to single strands. The membrane is placed in a hybridization bag along with buffer and a labeled single-stranded DNA probe. During incubation, the probe forms a double-stranded hybrid with any complementary sequences on the membrane. The membrane is removed from the bag and washed to remove excess probe. Hybrids are detected by placing a piece of X-ray film over the membrane and exposing it for a short time. The film is developed, and hybridization events are visualized as spots on the film. Colonies containing the insert that hybridized to the probe are identified from the orientation of the spots. Cells are picked from this colony for growth and further analysis.

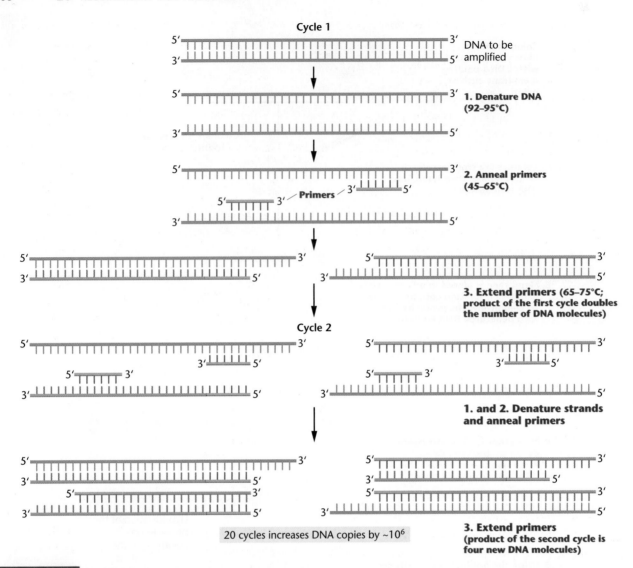

FIGURE 20-8 In the polymerase chain reaction (PCR), the target DNA is denatured into single strands; each strand is then annealed to short, complementary primers. DNA polymerase extends the primers in the 5' to 3' direction, using the single-stranded DNA as a template. The result after one round of replication is a doubling of DNA molecules to create two newly synthesized double-stranded DNA molecules. Repeated cycles of PCR can quickly amplify the original DNA sequence more than a millionfold. Note: shown here is a relatively short sequence of DNA being amplified. Typically much longer segments of DNA are used for PCR and the primers bind somewhere within the DNA molecule and not so close to the end of the actual molecule.

DNA, double-stranded target DNA to be cloned is placed in a tube with DNA polymerase, Mg^{+2} (as an important cofactor for DNA polymerase), and the four deoxyribonucleoside triphosphates. In addition, as a prerequisite for PCR, some information about the nucleotide sequence of the target DNA is required. This sequence information is used to synthesize two oligonucleotide **primers**: short (typically about 20 nt long) single-stranded DNA sequences, one complementary to the 5' end of one strand of target DNA to be amplified and another primer complementary to the opposing strand of target DNA at its 3' end. When added to a sample of double-stranded DNA that has been denatured into single strands, the primers bind to complementary

nucleotides flanking the sequence to be cloned. DNA polymerase can then extend the 3' end of each primer to synthesize second strands of the target DNA. Therefore, one complete reaction process, called a **cycle**, doubles the number of DNA molecules in the reaction (**Figure 20–8**). Repetition of the process produces large numbers of copies of the DNA very quickly. If desired, the products of PCR can be cloned into plasmid vectors for further use.

The amount of amplified DNA produced is theoretically limited only by the number of times these cycles are repeated, although several factors prevent PCR reactions from amplifying very long stretches of DNA or amplifying DNA indefinitely. Most routine PCR applications involve a series of three reaction steps in a cycle. These three steps are as follows:

1. **Denaturation:** The double-stranded DNA to be cloned is *denatured* into single strands. The DNA can come from many sources, including genomic DNA, mummified remains, fossils, or forensic samples such as dried blood or semen, single hairs, or dried samples from medical records. Heating to 92–95°C for about 1 minute denatures the double-stranded DNA into single strands.

2. **Hybridization/Annealing:** The temperature of the reaction is lowered to a temperature between 45°C and 65°C, which causes the primer binding, also called hybridization or annealing, to the denatured, single-stranded DNA. As described earlier, the primers are short oligonucleotides complementary to sequences flanking the target DNA. The primers serve as starting points for DNA polymerase to synthesize new DNA strands complementary to the target DNA. Factors such as primer length, base composition of primers (GC-rich primers are more thermally stable than AT-rich primers), and whether or not all bases in a primer are complementary to bases in the target sequence are among primary considerations when selecting a hybridization temperature for an experiment.

3. **Extension:** The reaction temperature is adjusted to between 65°C and 75°C, and DNA polymerase uses the primers as a starting point to synthesize new DNA strands by adding nucleotides to the ends of the primers in a 5′ to 3′ direction.

PCR is a chain reaction because the number of new DNA strands is doubled in each cycle, and the new strands, along with the old strands, serve as templates in the next cycle. Each cycle takes 2 to 5 minutes and can be repeated immediately, so that in less than 3 hours, 25 to 30 cycles result in over a million-fold increase in the amount of DNA (Figure 20–8). This process is automated by instruments called *thermocyclers,* or simply PCR machines, that can be programmed to carry out a predetermined number of cycles. As a result, large amounts of a specific DNA sequence are produced that can be used for many purposes, including cloning into plasmid vectors, DNA sequencing, clinical diagnosis, and genetic screening.

A key requirement for PCR is the type of DNA polymerase used in PCR reactions. Multiple PCR cycles involve repetitive heating and cooling of samples, which eventually lead to heat denaturation and loss of activity of most proteins. PCR reactions rely on thermostable forms of DNA polymerase capable of withstanding multiple heating and cooling cycles without significant loss of activity. PCR became a major tool when DNA polymerase was isolated from *Thermus aquaticus,* a bacterium living in

the hot springs of Yellowstone National Park. Called *Taq Polymerase,* this enzyme is capable of tolerating extreme temperature changes and was the first thermostable polymerase used for PCR.

PCR-based DNA cloning has several advantages over library cloning approaches. PCR is rapid and can be carried out in a few hours, rather than the days required for making and screening DNA libraries. PCR is also very sensitive and amplifies specific DNA sequences from vanishingly small DNA samples, including the DNA in a single cell. This feature of PCR is invaluable in several kinds of applications, including genetic testing, forensics, and molecular paleontology. With carefully designed primers, DNA samples that have been partially degraded, contaminated with other materials, or embedded in a matrix (such as amber) can be recovered and amplified, when conventional cloning would be difficult or impossible. A wide variety of PCR-based techniques involve different variations of the basic technique described here. Several commonly used variations are discussed below.

Limitations of PCR

Although PCR is a valuable technique, it does have limitations: some information about the nucleotide sequence of the target DNA must be known in order to synthesize primers. In addition, even minor contamination of the sample with DNA from other sources can cause problems. For example, cells shed from the skin of a researcher can contaminate samples gathered from a crime scene or taken from fossils, making it difficult to obtain accurate results. PCR reactions must always be performed in parallel with carefully designed and appropriate controls. Also, PCR typically cannot amplify particularly long segments of DNA. Normally, DNA polymerase in a PCR reaction only extends primers for relatively short distances and does not continue processively until it reaches the other end of long template strands of DNA. Because of this characteristic, scientists often use PCR to amplify pieces of DNA that are several thousand nucleotides in length, which is fine for most routine applications.

> **NOW SOLVE THIS**
>
> **20–3** Question 20-28 asks you to calculate the annealing temperature for a primer in a PCR experiment. To do this you need to consider the %GC in a primer. Based on what you know about the complementary base pairs, why is annealing temperature dependent on %AT and %GC in a primer?
>
> ■ HINT: *This problem involves an understanding of the relationship between hybridization temperature in a PCR reaction and the nucleotide composition of PCR primers. The key to its solution is to consider what you know about hydrogen bonding arrangements of complementary base pairs in DNA.*

Applications of PCR

Cloning DNA by PCR has been one of the most widely used techniques in genetics and molecular biology for over 20 years. PCR and its variations have many other applications as well. In short, PCR is one of the most versatile techniques in modern genetics. As you will learn in Chapter 22, gene-specific primers provide a way of using PCR for screening mutations involved in genetic disorders, allowing the location and nature of a mutation to be determined quickly. Primers can be designed to distinguish between target sequences that differ by only a single nucleotide. This makes it possible to synthesize allele-specific probes for genetic testing; thus PCR is important for diagnosing genetic disorders. PCR is also a key diagnostic methodology for the detection of bacteria and viruses (such as hepatitis or HIV) in humans, and pathogenic bacteria such as *E. coli* and *Staphylococcus aureus* in contaminated food.

PCR techniques are particularly advantageous when studying samples from single cells, fossils, or a crime scene, where a single hair or even a saliva-moistened postage stamp is the source of the DNA. In Special Topics on DNA Forensics (p. 493) we will discuss how PCR is used in human identification, including remains identification, and in forensic applications. Using PCR, researchers can also explore uncharacterized DNA regions adjacent to known regions and even sequence DNA. PCR has been used to enforce the worldwide ban on the sale of certain whale products and to settle arguments about the pedigree background of purebred dogs.

One commonly used PCR method is called **reverse transcription PCR (RT-PCR)**. RT-PCR is a powerful methodology for studying gene expression, that is, mRNA production by cells or tissues. In RT-PCR, RNA is isolated from cells or tissues to be studied, and reverse transcriptase is used to generate double-stranded cDNA molecules, as described earlier when we discussed preparation of cDNA libraries. This reaction is followed by PCR to amplify cDNA with a set of primers specific for the gene of interest. Amplified cDNA fragments are then separated on an agarose gel. Because the amount of amplified cDNA in an RT-PCR reaction is based on the relative number of mRNA molecules in the starting reaction, RT-PCR can be used to evaluate relative levels of gene expression in different samples. The amplified cDNA can be inserted into plasmid vectors, which are replicated to produce a cDNA library. RT-PCR is more sensitive than conventional cDNA preparation and is a powerful tool for identifying mRNAs that may be present in only one or two copies per cell.

Finally, in discussing PCR approaches, one of the most valuable modern PCR techniques involves a method called **quantitative real-time PCR (qPCR)** or simply real-time PCR. This approach makes it possible to determine the amount of PCR product made during an experiment, which enables researchers to quantify amplification reactions as they occur in "real time" (Figure 20–9). There are several

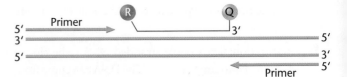

1. Hybridization. Forward and reverse PCR primers bind to denatured target DNA. TaqMan probe with reporter (R) and quencher (Q) dye binds to target DNA between the primers. When probe is intact, emission by the reporter dye is quenched.

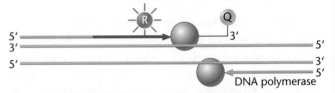

2. Extension. As DNA polymerase extends the forward primer, it reaches the TaqMan probe and cleaves the reporter dye from the probe. Released from the quencher, the reporter can now emit light excited by a laser.

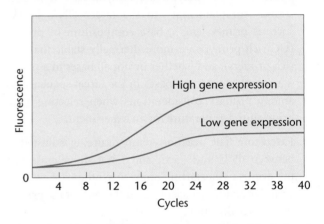

3. Detection. Emitted light from the reporter is detected and interpreted to produce a plot that quantitates the amount of PCR product produced with each cycle.

FIGURE 20–9   The TaqMan approach to quantitative real-time PCR (qPCR) involves a pair of PCR primers along with a probe sequence complementary to the target gene. The probe contains a reporter dye (R) at one end and a quencher dye (Q) at the other end. When the quencher dye is close to the reporter dye, it interferes with fluorescence released by the reporter dye. When *Taq* DNA polymerase extends a primer to synthesize a strand of DNA, it cleaves the reporter dye off of the probe allowing the reporter to give off energy. (b) Each subsequent PCR cycle removes more reporter dyes so that increased light emitted from the dye can be captured by a computer to produce a readout of fluorescence intensity with each cycle.

ways to run qPCR experiments, but the basic procedure involves the use of specialized (and expensive) thermal cyclers that use a laser to scan a beam of light through the top or bottom of each PCR tube. Each reaction tube contains either a dye-containing probe or DNA-binding dye that emits fluorescent light when illuminated by the laser. The light emitted by these dyes correlates to the amount of PCR product amplified. Light from each tube is captured by a detector that relays information to a computer to provide a readout on the amount of fluorescence produced after each cycle and the precise number of molecules in the original sample.

Two commonly used approaches for qPCR involve the use of a dye called SYBR Green and TaqMan probes. SYBR Green is a dye that binds double-stranded DNA. As more double-stranded DNA is copied with each round of real-time PCR, there are more DNA copies to bind SYBR Green, which increases the amount of fluorescent light emitted. TaqMan probes are complementary to specific regions of the target DNA between where the forward and reverse primers for PCR bind (Figure 20–9). TaqMan probes contain two dyes. One dye, the reporter, is located at the 5′ end of the probe and can release fluorescent light when excited by laser light from the thermal cycler. The other dye, called a quencher, is attached to the 3′ end of the probe. When these two dyes are close to each other, the quencher dye interferes with the fluorescent light released from the reporter dye. However, as *Taq* DNA polymerase extends each primer, it removes the reporter dye from the end of the probe (and eventually removes the entire probe). Now that the reporter dye is separated from the quencher, the fluorescent light released by the reporter can be detected by the thermal cycler. A computer analyzes the detected light to produce a plot displaying the amount of fluorescence emitted with each cycle. Because real-time PCR does not involve running gels, it is a powerful and rapid technique for measuring and quantitating changes in gene expression, particularly when multiple samples and different genes are being analyzed.

NOW SOLVE THIS

20–4 Question 20-14 refers to creating a genomic DNA library from the African okapi. If you were the first person in the world attempting to use PCR to amplify particular genes from this library, what strategies might you use to design PCR primers for your experiments if the okapi genome has yet to be sequenced?

■ HINT: *This problem asks you to design PCR primers to amplify the β-globin gene from a species whose genome you just sequenced. The key to its solution is to remember that you have at your disposal sequence data for the human β-globin gene and consider that PCR experiments require the use of primers that bind to complementary bases in the DNA to be amplified.*

20.4

Molecular Techniques for Analyzing DNA

The identification of genes and other DNA sequences by cloning or by PCR plays a very powerful role in analyzing genomic structure and function. In addition to cloning and PCR methods, a wide range of molecular techniques is available to geneticists, molecular biologists, and almost anyone who does research involving DNA and RNA, particularly those who study the structure, expression, and regulation of genes. In the following sections, we consider some of the most commonly used molecular methods that provide information about the organization and function of cloned sequences. Throughout later sections of the text you will see these and other techniques discussed in the context of certain applications in modern genetics.

Restriction Mapping

Historically, one of the first steps in characterizing a DNA clone was the construction of a **restriction map**. A restriction map establishes the number of, order of, and distances between restriction-enzyme cleavage sites along a cloned segment of DNA, thus providing information about the length of the cloned insert and the location of restriction-enzyme cleavage sites within the clone. The data the maps provide can be used to reclone fragments of a gene or compare its internal organization with that of other cloned sequences.

Before DNA sequencing and bioinformatics became popular, restriction maps were created experimentally by cutting DNA with different restriction enzymes and separating DNA fragments by gel electrophoresis, a method that separates fragments by size, with the smallest pieces moving farthest through the gel (see Chapter 10, refer to Figure 10–20). The fragments form a series of bands that can be visualized by staining the DNA with ethidium bromide and illuminating it with ultraviolet light (**Figure 20–10**). The cutting pattern of fragments generated can then be interpreted

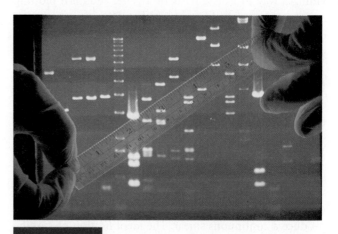

FIGURE 20–10 An agarose gel containing separated DNA fragments stained with the DNA-binding dye (ethidium bromide) and visualized under ultraviolet light. Smaller fragments migrate faster and farther than do larger fragments, resulting in the distribution shown. Molecular techniques involving agarose gel electrophoresis are routinely used in a wide range of applications.

to determine the location of restriction sites for different enzymes.

Because of recent advances in DNA sequencing and the use of bioinformatics, most restriction maps are now created by simply using software to identify restriction enzyme cutting sites in sequenced DNA. The Exploring Genomics exercise in this chapter involves a Web site, Webcutter, which is commonly used for generating restriction maps. Restriction maps were an important way of characterizing cloned DNA and could be constructed in the absence of any other information about the DNA, including whether or not it encodes a gene or has other functions.

Restriction digestion of clones (that is, cutting the clones with restriction enzymes) can still play an important role in mapping genes to specific human chromosomes and to defined regions of individual chromosomes. In addition, if a restriction site maps close to a mutant allele, this site can be used as a marker in genetic testing to identify carriers of recessively inherited disorders or to prenatally diagnose a fetal genotype. This topic will be discussed in Chapter 22.

Nucleic Acid Blotting

Several of the techniques described in this chapter rely on hybridization between complementary nucleic acid (DNA or RNA) molecules. One of the most widely used methods for detecting such hybrids is called Southern blotting (after Edwin Southern, who devised it). The **Southern blot** method can be used to identify which clones in a library contain a given DNA sequence (such as ribosomal DNA, a β-globin gene, etc.) and to characterize the size of the fragments. Southern blots can also be used to identify fragments carrying specific genes in genomic DNA digested with a restriction enzyme. Fragments of genomic clones isolated by Southern blots can in turn be isolated and recloned, providing a way to isolate parts of a gene. Southern blotting has also been a valuable tool for identifying the number of copies of a particular sequence or gene that are present in a genome.

Southern blotting has two components: separation of DNA fragments by gel electrophoresis and hybridization of the fragments using labeled probes. Gel electrophoresis can be used, as shown above, to characterize the number of fragments produced by restriction digestion and to estimate their molecular weights. However, restriction-enzyme digestion of large genomes—such as the human genome, with more than 3 billion nucleotides—will produce so many different fragments that they will run together on a gel to produce a continuous smear. The identification of specific fragments in these cases is accomplished in the next step: hybridization characterizes the DNA sequences present in the fragments. The DNA to be characterized by Southern blot hybridization can come from several sources, including clones selected from a library or genomic DNA.

To make a Southern blot, DNA is often cut into fragments with one or more restriction enzymes, and the fragments are separated by gel electrophoresis (Figure 20–11). Sometimes if smaller pieces of DNA are being analyzed (for example plasmid DNA), undigested DNA may be used for Southern blot. In preparation for hybridization, the DNA in the gel is denatured with alkaline treatment to form single-stranded fragments. The gel is then overlaid with a DNA-binding membrane, usually nylon. Transfer of the DNA fragments to the membrane is accomplished by placing the membrane and gel on a wick (often a sponge) sitting in a buffer solution. Layers of paper towels or blotting paper are placed on top of the membrane and held in place with a weight. Capillary action draws buffer up through the gel, transferring the DNA fragments from the gel to the membrane.

The membrane is placed in a heat-sealed bag with a labeled, single-stranded DNA probe for hybridization. For many years Southern blots were primarily carried out with radioactively labeled probes, but as was mentioned when describing library screening, probes that use fluorescent and chemiluminescent labels are now commonly used. DNA fragments on the membrane that are complementary to the probe's nucleotide sequence bind to the probe to form double-stranded hybrids. Excess probe is then washed away, and the hybridized fragments are visualized on a piece of film (Figures 20–11 and 20–12).

To produce Figure 20–12, researchers cut samples of genomic DNA with several restriction enzymes. The pattern of fragments obtained for each restriction enzyme is shown in Figure 20–12(a). A Southern blot of this gel is illustrated in Figure 20–12(b). The probe hybridized to complementary sequences, identifying fragments of interest.

Southern blotting led to the development of other blotting approaches. RNA blotting was subsequently called **Northern blot analysis** or simply **Northern blotting**, and following a naming scheme that correlates with the directionality of a compass, a related blotting technique involving proteins is known as **Western blotting**. Western blotting is a widely used technique for analyzing proteins. Thus part of the historical significance of Southern blotting is that it led to the development of other very important blotting methods that are key tools for studying nucleic acids and proteins.

To determine whether a gene is actively being expressed in a given cell or tissue type, Northern blotting probes for the presence of mRNA complementary to a cloned gene (Figure 20–13). To do this, mRNA is extracted from a specific cell or tissue type and separated by gel electrophoresis. The resulting pattern of RNA bands is transferred to a membrane, as in Southern blotting. The membrane is then exposed to a labeled single-stranded DNA probe derived from a cloned copy of the gene. If mRNA complementary to the DNA probe is present, the complementary sequences

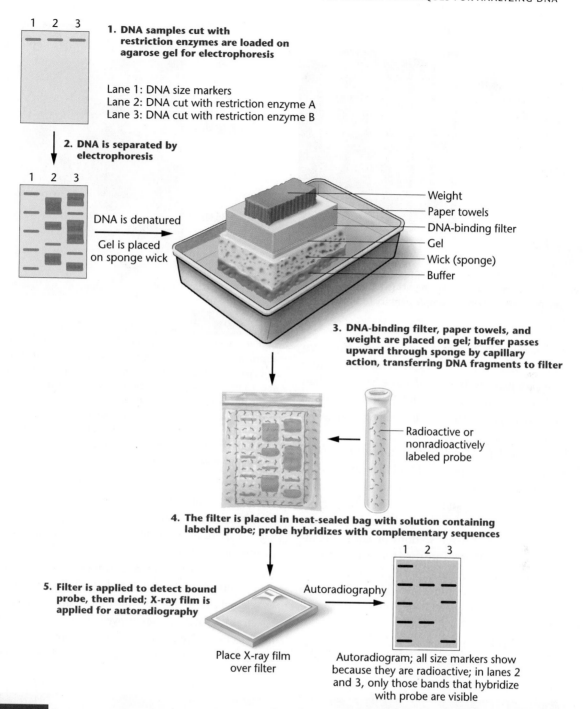

1. **DNA samples cut with restriction enzymes are loaded on agarose gel for electrophoresis**

Lane 1: DNA size markers
Lane 2: DNA cut with restriction enzyme A
Lane 3: DNA cut with restriction enzyme B

2. **DNA is separated by electrophoresis**

DNA is denatured

Gel is placed on sponge wick

Weight
Paper towels
DNA-binding filter
Gel
Wick (sponge)
Buffer

3. **DNA-binding filter, paper towels, and weight are placed on gel; buffer passes upward through sponge by capillary action, transferring DNA fragments to filter**

Radioactive or nonradioactively labeled probe

4. **The filter is placed in heat-sealed bag with solution containing labeled probe; probe hybridizes with complementary sequences**

5. **Filter is applied to detect bound probe, then dried; X-ray film is applied for autoradiography**

Autoradiography

Place X-ray film over filter

Autoradiogram; all size markers show because they are radioactive; in lanes 2 and 3, only those bands that hybridize with probe are visible

FIGURE 20–11 In the Southern blotting technique, samples of the DNA to be probed are cut with restriction enzymes and the fragments are separated by gel electrophoresis. The pattern of fragments is visualized and photographed under ultraviolet illumination. Then the gel is placed on a sponge wick that is in contact with a buffer solution and covered with a DNA-binding membrane. Layers of paper towels or blotting paper are placed on top of the membrane and held in place with a weight. Capillary action draws the buffer through the gel, transferring the pattern of DNA fragments from the gel to the membrane. The DNA fragments on the membrane are then denatured into single strands and hybridized with a labeled DNA probe. The membrane is washed to remove excess probe and overlaid with a piece of X-ray film for autoradiography. The hybridized fragments show up as bands on the X-ray film.

will hybridize and be detected as a band on the film. Northern blots provide information about the expression of specific genes and are used to study patterns of gene expression in embryonic tissues, cancer, and genetic disorders. Northern blots also detect alternatively spliced mRNAs (multiple types of transcripts derived from a single gene) and are used to derive other information about transcribed mRNAs. If marker RNAs of known size are run as controls, Northern blots can be used to measure the size of a gene's mRNA transcripts. Measuring band density gives an estimate of the relative transcriptional activity of the gene. Thus, Northern blots characterize and quantify the transcriptional activity

(a)   (b)

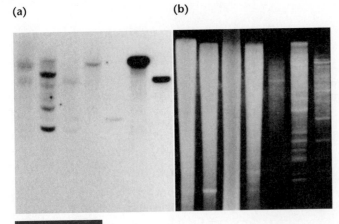

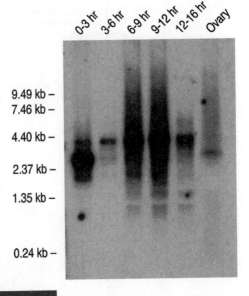

FIGURE 20–12   (a) Agarose gel stained with ethidium bromide to show DNA fragments. (b) Exposed X-ray film of a Southern blot prepared from the gel in part (a). Only those bands containing DNA sequences complementary to the probe show hybridization.

FIGURE 20–13   Northern blot analysis of *dfmr1* gene expression in *Drosophila* ovaries and embryos. A *dfmr1* transcript of approximately 2.8 kb is present in ovaries and 0- to 3-hr-old embryos. The *dfmr1* transcript peaks in abundance between 9 and 12 hr of embryonic development, when it measures 4.0 kb. These data suggest that *dfmr1* gene expression may be regulated at the levels of transcription or transcript processing during embryogenesis. The *dfmr1* gene is a homolog of the human *FMR1* gene. Loss-of-function mutations in *FMR1* result in human fragile-X mental retardation.

of genes in different cells, tissues, and organisms. Northern blots are still used to study RNA expression, but because PCR-based techniques are faster and more sensitive than blotting methods, techniques such as RT-PCR are often the preferred approach, particularly for measuring changes in gene expression.

Finally, as noted in Chapter 10, **fluorescent *in situ* hybridization** or **FISH**, is a powerful tool that involves hybridizing a probe directly to a chromosome or RNA without blotting (see Figure 10–19 and Figures 18–8 and 18–9). FISH can be carried out with isolated chromosomes on a slide or directly *in situ* in tissue sections or entire organisms, particularly when embryos are used for various studies in developmental genetics (Figure 20–14). For example, in developmental studies one can identify which cell types in an embryo express different genes during different stages of development.

For *in situ* hybridization, a probe for a particular sequence is labeled with nucleotides tagged with a particular dye that will fluoresce under fluorescent light. When hybridized to a chromosome, for example, the probe can reveal the specific location of a gene on a particular chromosome. Variations of the FISH technique are also being used to produce karyotypes (sometimes called **spectral karyotypes**) in which individual chromosomes can be detected using probes labeled with dyes that will fluoresce at different wavelengths (see Chapter 8 opening figure and Figure 19–1). Spectral karyotyping has proven to be extremely valuable for detecting deletions, translocations, duplications, and other anomalies in chromosome structure, such as the chromosomal

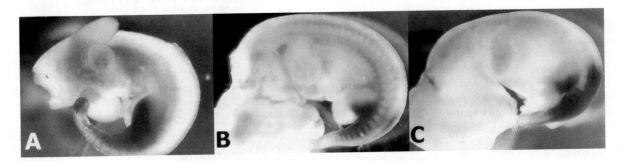

FIGURE 20–14   *In situ* hybridization of whole mouse embryos, showing distribution of the *Hoxc11* mRNA during development. RNA is detectable as dark blue staining, near the posterior end of the embryo. Head of the embryo is at the left; tail at the right. At 10.5 (A), 11.5 (B), and 12.5 (C) days of gestation, *Hoxc11* mRNA is concentrated in hindlimbs, vertebrae, and cells that will later form kidney and reproductive organs. These data suggest that the *Hoxc11* gene product is involved in the early development of these structures.

rearrangements we discussed in Chapter 8 and for detecting chromosomal abnormalities in cancer cells as was discussed in Chapter 19.

20.5

DNA Sequencing Is the Ultimate Way to Characterize DNA Structure at the Molecular Level

In a sense, a cloned DNA molecule or any DNA, from a single gene to an entire genome, is completely characterized at the molecular level only when its nucleotide sequence is known. The ability to sequence DNA has greatly enhanced our understanding of genome organization and increased our knowledge of gene structure, function, and mechanisms of regulation.

Historically the most commonly used method of DNA sequencing was developed by Fred Sanger and his colleagues and is known as **dideoxynucleotide chain-termination sequencing** or simply **Sanger sequencing**. In this technique, a double-stranded DNA molecule whose sequence is to be determined is converted to single strands that are used as a template for synthesizing a series of complementary strands. The DNA to be sequenced is mixed with a primer that is complementary to the target DNA or vector along with DNA polymerase, and the four deoxyribonucleotide triphosphates (dATP, dCTP, dGTP, and dTTP) are added to each tube.

The key to the Sanger technique is the addition of a small amount of one modified deoxyribonucleotide (Figure 20–15),

called a **dideoxynucleotide** (abbreviated ddNTP). Notice that dideoxynucleotides have a 3′ hydrogen instead of a 3′ hydroxyl group. Dideoxynucleotides are called chain-termination nucleotides because they lack the 3′ oxygen required to form a phosphodiester bond with another nucleotide. Thus when ddNTPs are included in a reaction as DNA synthesis takes place, the polymerase occasionally inserts a dideoxynucleotide instead of a deoxyribonucleotide into a growing DNA strand. Since the dideoxynucleotide has no 3′-OH group, it cannot form a 3′ bond with another nucleotide, and DNA synthesis terminates because DNA polymerase cannot add new nucleotides to a ddNTP. The Sanger reaction takes advantage of this key modification.

For example, in Figure 20–16, notice that the shortest fragment generated is a sequence that has added ddCTP to the 3′ end of the primer and the chain has terminated. Over time as the reaction proceeds eventually there will be a ddNTP inserted at every location in the newly synthesized DNA so that each strand synthesized differs in length by one nucleotide and is terminated by a ddNTP. This allows for separation of these DNA fragments by gel electrophoresis, which can then be used to determine the sequence.

When the Sanger technique was first developed, four separate reaction tubes, each with a different single ddNTP (e.g., ddATP, ddCTP, ddGTP, and ddTTP), were used. These reactions typically used either a radioactively labeled primer or radioactively labeled ddNTP for analysis of the sequence following polyacrylamide gel electrophoresis and autoradiography. This approach involved large polyacrylamide gels in which each reaction was loaded on a separate lane of the gel and ladder-like banding patterns revealed by autoradiography were read to determine the sequence. This original approach could typically read about 800 bases of 100 DNA molecules simultaneously. *Read length,* that is, the amount of sequence that can be generated in a single individual reaction and the total amount of DNA sequence generated in a sequence *run,* which is effectively read length times the number of reactions an instrument can run during a given period of time, have become hot areas for innovation in sequencing technology.

In the past 20 years, modifications of the Sanger technique have led to technologies that now allow sequencing reactions to occur in a single tube in which each of the four ddNTPs is labeled with a different-colored fluorescent dye (Figure 20–16). These reactions were often carried out in PCR-like fashion using cycling reactions that permit greater read and run capabilities. The reaction products are separated through a single, ultrathin-diameter polyacrylamide tube gel called a capillary gel (capillary gel electrophoresis). As DNA fragments move through the gel, they are scanned with a laser. The laser stimulates fluorescent dyes on each DNA fragment, which then emit different wavelengths of

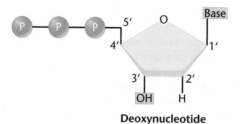

Deoxynucleotide

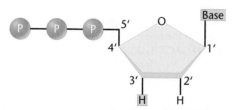

Dideoxynucleotide (ddNTP)

FIGURE 20–15 Deoxynucleotides (top) have an OH group at the 3′ position in the deoxyribose molecule. Dideoxynucleotides (bottom) lack an OH group and have only hydrogen (H) at this position. Dideoxynucleotides can be incorporated into a growing DNA strand, but the lack of a 3′-OH group prevents formation of a phosphodiester bond with another nucleotide, terminating further elongation of the template strand.

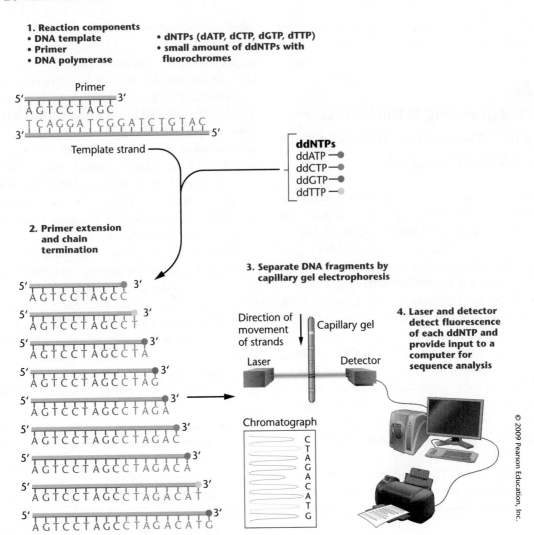

FIGURE 20–16 Computer-automated DNA sequencing using the chain-termination (Sanger) method. (1) A primer is annealed to a sequence adjacent to the DNA being sequenced (usually near the multiple cloning site of a cloning vector). (2) A reaction mixture is added to the primer–template combination. This includes DNA polymerase, the four dNTPs, and small molar amounts of dideoxy-nucleotides (ddNTPs) labeled with fluorescent dyes. All four ddNTPs are added to the same tube, and during primer extension, all possible lengths of chains are produced. During primer extension, the polymerase occasionally (randomly) inserts a ddNTP instead of a dNTP, terminating the synthesis of the chain because the ddNTP does not have the OH group needed to attach the next nucleotide. Over the course of the reaction, all possible termination sites will have a ddNTP inserted. The products of the reaction are added to a single lane on a capillary gel, and the bands are read by a detector and imaging system. This process is now automated, and robotic machines, such as those used in the Human Genome Project, sequence several hundred thousand nucleotides in a 24-hour period and then store and analyze the data automatically. The sequence is obtained by extension of the primer and is read from the newly synthe-sized strand, not the template strand. Thus, the sequence obtained begins with 5'-CTAGACATG-3'.

light for each ddNTP. Emitted light is captured by a detector that amplifies and feeds this information into a computer to convert the light patterns into a DNA sequence that is technically called an electropherogram (Figure 20–17). The data are represented as a series of colored peaks, each corresponding to one nucleotide in the sequence.

Since the early 1990s DNA sequencing has largely been performed through computer-automated Sanger-reaction-based technology and is referred to as **computer-automated**

high-throughput DNA sequencing. Such systems generate relatively large amounts of sequence DNA. Computer-automated sequences can achieve read lengths of approximately 1000 bp with about 99.999 percent accuracy for about $0.50 per kb. Automated DNA sequencers often contain multiple capillary gels (as many as 96) that are several feet long and can process several thousand bases of sequences so that many of these instruments made it possible to generate over 2 million bp of sequences in a day! These systems became essential for

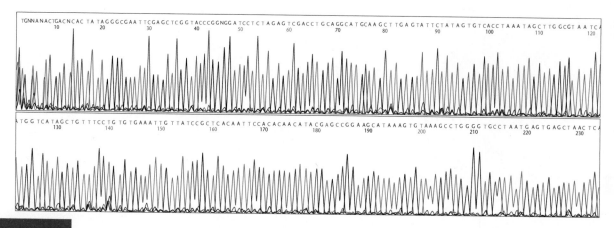

FIGURE 20–17   Output of a computer-automated DNA sequencing chromatograph or electropherogram. Each peak represents the correct nucleotide in the sequence. The sequence extending from the primer (which is not shown here) starts at the upper left of the diagram and extends to the right. The bases labeled as N are ambiguous and cannot be identified with certainty. These ambiguous base readings are more likely to occur near the primer because the quality of sequence determination deteriorates the closer the sequence is to the primer. The separated bases are read in order along the axis from left to right. Thus, this sequence begins as 5′-TGNNANACTGACNCAC. Numbers below the bases indicate length of the sequence in base pairs.

enabling the rapidly accelerating progress of the Human Genome Project.

Sequencing Technologies Have Progressed Rapidly

Nearly three decades since Fred Sanger was awarded part of the 1980 Nobel Prize in Chemistry (which he shared with Walter Gilbert and Paul Berg) for sequencing technology, DNA sequencing technologies have undergone an incredible evolution to dramatically improve sequencing capabilities. New innovations in sequencing technology are developing quickly. Sanger sequencing approaches (particularly those involving computer-automated instruments such as capillary electrophoresis) still have their place in everyday routine applications that require sequencing, such as sequencing a relatively short piece of DNA amplified by PCR.

When it comes to sequencing entire genomes, however, Sanger sequencing technologies are becoming outdated. The costs of Sanger sequencing are relatively high compared to newer technologies, and Sanger sequencing output, even with computer-automated DNA sequencing, is simply not high enough to support the growing demand for genomic data. This demand is being driven in large part by personalized genomics (see Chapter 21) and the desire to reveal the genetic basis of human diseases, which will require tens of thousands of individual genome sequences. As we will discuss in Chapter 21, the race is on to develop sequencing technologies that will allow for the complete sequencing of an individual human genome for $1000. There is every indication that technology will allow this to happen relatively soon thanks to the development of **next-generation sequencing (NGS) technologies**.

Next-Generation Sequencing Technologies

The development of genomics has spurred a demand for sequencers that are faster and capable of generating millions of bases of DNA sequences in a relatively short time, leading to the development of NGS approaches. Such sequencing technologies dispense with the Sanger technique and capillary electrophoresis methods in favor of sophisticated, parallel formats (simultaneous reaction formats) that use state-of-the art fluorescence imaging techniques. NGS technologies are providing an unprecedented capacity for generating massive amounts of DNA sequence data rapidly (up to 200 times faster than Sanger approaches in some cases!) and at dramatically reduced costs per base.

The desire for next-generation sequencing within the research community and challenges such as the $1000 genome have led to an intense race among many companies eager to produce NGS methods. Next-generation sequencing started around 2005. Some of the first instruments were capable of producing as much data as 50 capillary electrophoresis systems and are up to 200 times faster and cheaper than conventional Sanger approaches. Major companies such as Applied Biosystems (ABI), Roche 454 Life Sciences, Helicos Biosciences (which has developed a single-molecule sequencing approach), and Illumina are leading this effort.

In 2005, Roche 454 Life Sciences was the first company to commercialize a next-generation sequencing technology (Figure 20–18). This approach can be used to sequence genomes using a so-called solid-phase method in which beads are attached to fragmented genomic DNA, which

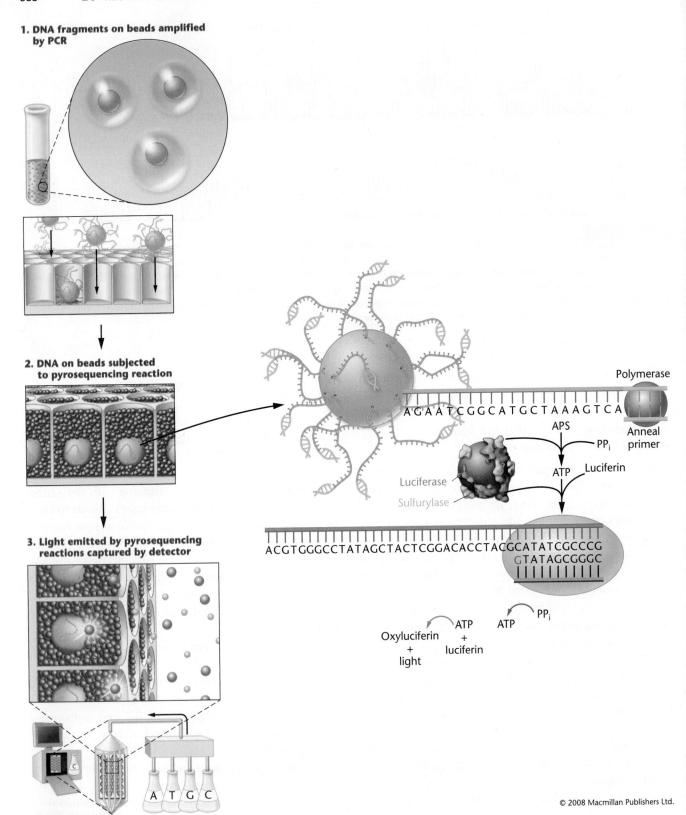

1. DNA fragments on beads amplified by PCR

2. DNA on beads subjected to pyrosequencing reaction

3. Light emitted by pyrosequencing reactions captured by detector

Polymerase

AGAATCGGCATGCTAAAGTCA

APS

PP$_i$

Anneal primer

ATP   Luciferin

Luciferase

Sulfurylase

ACGTGGGCCTATAGCTACTCGGACACCTACGCATATCGCCCG
GTATAGCGGGC

ATP   ATP   PP$_i$

Oxyluciferin
+
light

ATP
+
luciferin

A T G C

FIGURE 20–18   Roche 454: One example of next-generation sequencing technology. Roche 454 sequencing technology binds DNA fragments to beads. DNA fragments are amplified by PCR and then added to wells (Step 1) and subjected to pyrosequencing as described in the text. (Step 2). During pyrosequencing an individual fluorescent nucleotide, the blue G shown in this case, is flowed over the well. If it is added to a primer by DNA polymerase it generates inorganic pyrophosphate (PPi), which is converted to ATP by sulfurylase (blue arrow). The firefly enzyme luciferase uses ATP to convert luciferin to oxyluciferin-producing light. (Step 3) Light captured by a detection system is analyzed to track the pattern of nucleotides added to each well. The flow cycle is subsequently repeated with each of the other three nucleotides. Continual cycling of this process generates sequencing reads of approximately 400 bases.

is then PCR amplified in separate water droplets in oil for each bead and loaded into multiwell plates and mixed with DNA polymerase. Multiwell plates often contain more than one million wells with one bead per well—each serving as a reaction tube for sequencing (see Figure 20–18). Next a sequencing technology called **pyrosequencing** is used to sequence DNA on the beads in each well. In pyrosequencing, a single, labeled nucleotide (e.g., dATP) is flowed over the wells. Each well contains a single bead along with primers annealed to the DNA on the bead. When a complementary nucleotide crosses a template strand adjacent to the primer, it is added to the 3′ end of the primer by DNA polymerase. Incorporation of a nucleotide results in the release of pyrophosphate, which initiates a series of chemiluminescent (light-releasing) reactions that ultimately produce light using the firefly enzyme luciferase. Emitted light from the reaction is captured and recorded to determine when a single nucleotide has been incorporated into a strand. By rapidly repeating the nucleotide flow step with each of the four nucleotides to determine which base is next in the sequence, this approach can generate read lengths of about 400 bases and on the order of 400 million bases (Mbp) of data per 10-hour run. NGS sequencing technologies are also creating major data management challenges for saving and storing such large image data files.

Methods developed by Illumina use a slightly different approach, attaching DNA fragments to a solid support (similar to the microarrays we discuss in Chapter 21) and then using sequencing reactions similar to the dideoxy chain termination reactions pioneered by Sanger. This approach generates shorter read lengths (~100 bp total length), but the number of sequencing reads per run is much higher—about 300 million reads per flow. Similarly, the company Applied Biosystems (ABI) has developed an approach called **SOLiD** (**supported oligonucleotide ligation and detection**) that can produce 6 *gigabases* of sequence data per run! The SOLiD method combines a variety of approaches to sequence DNA fragments that are linked to beads and amplified, similar to the 454 approach. However, different sequencing technologies are used that can provide greater output of sequencing data per instrument run. The instrumentation needed to run these platforms is expensive, but given the massive amounts of sequence data the NGS methods can generate, the average cost per base is much lower than Sanger sequencing. Through 2006, new sequencing technologies were cutting sequencing costs in half about every two years (Figure 20–19).

Work on third-generation sequencers is actively underway, and there is every reason to believe these instruments will be available by around 2015. For example, one promising approach on the immediate horizon involves the use of nanotechnology by pushing single-stranded

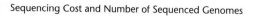

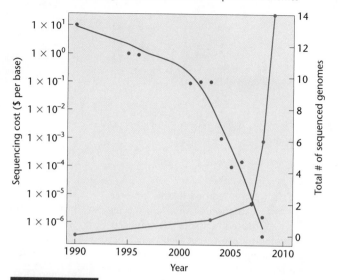

Sequencing Cost and Number of Sequenced Genomes

FIGURE 20–19 Rapid advances in sequencing technologies have greatly increased sequencing productivity and decreased sequencing costs.

DNA fragments into nanopores and then cleaving off individual bases to produce a signal that can be captured. This method does not involve DNA amplification or fluorescent tags and thus provides direct sequencing of the DNA in a single strand.

The genomics research community has embraced NGS technologies. Which approaches will eventually emerge as the sequencing methods of choice for the short-term future is unclear, but what is clear is that the landscape of sequencing capabilities has dramatically changed for the better and never before have scientists had the ability to generate so much sequence data so quickly. And keep an eye on the $1000 genome: there is every reason to believe sequencing technology will get us there soon.

DNA Sequencing and Genomics

Finishing this chapter with DNA sequencing is a great introduction to Chapter 21 where we present a detailed discussion of genomics and many related topics. The principal techniques of recombinant DNA technology, particularly DNA cloning, were essential for making genome projects possible, but no technology has had a greater influence on our ability to study genomes than DNA sequencing. Rapid advances in sequencing technology that pushed the capabilities of computer-automated sequencing were driven by the demands of genome scientists (particularly those working on the Human Genome Project) to rapidly generate more sequence with greater accuracy and at lower cost.

Manipulating Recombinant DNA: Restriction Mapping and Designing PCR Primers

(MG) *Study Area: Exploring Genomics*

As you learned in this chapter, restriction enzymes are sophisticated "scissors" that molecular biologists use to cut DNA, and they are routinely used in genetics and molecular biology laboratories for recombinant DNA experiments. Yet another advantage of the genomics revolution has been the development of a wide variety of online tools to assist scientists working with restriction enzymes and manipulating recombinant DNA for different applications, such as restriction mapping and designing primers for PCR experiments. Here we explore **Webcutter** and **Primer3**, two sites that make recombinant DNA experiments much easier.

■ **Exercise I – Creating a Restriction Map in Webcutter**

Suppose you had cloned and sequenced a gene and you wanted to design a probe approximately 600 bp long that could be used to analyze expression of this gene in different human tissues by Northern blot analysis. Not too long ago, you had primarily two ways to approach this task. You could digest the cloned DNA with whatever restriction enzymes were in your freezer, then run agarose gels and develop restriction maps in the hope of identifying cutting sites that would give you the size fragment you wanted. Or you could scan the sequence with your eyes, looking for restriction sites of interest—a very time-consuming and eye-straining effort! Internet sites such as **Webcutter** take the guesswork out of developing restriction maps and make it relatively easy to design experiments for manipulating recombinant DNA. In this exercise, you will use Webcutter to create a restriction map of human DNA with the enzymes *Eco*RI, *Bam*HI, and *Pst*I.

1. Access **Webcutter** at http://rna.lundberg.gu.se/cutter2. Copy the sequence of cloned human DNA

shown below and paste it into the text box in Webcutter. (*Hint*: Access this sequence from the Companion Web site so that you can copy and paste the sequence into Webcutter.)

Human DNA sequence

CCCCAGGAGACCTGGTTGTG
GAATTCTGTGTGTGAGTGGTT
GACCTTCCTCCATCCCCTG
GTCCTTCCCTTCCCTTCCCGAGGCA
CAGAGAGACAGGGCAGGATCCAC
GTGCCCATTGTGGAGGCAGAGA
AAAGAGAAAGTGTTTTATATACG
GTACTTATTTAATATCCCTTTTTA
ATTAGAAATTAAAACAGTTAATTTA
ATTAAAGAGTAGGGTTTTTTTTCAG
TATTCTTGGTTAATATTTAATTTCAAC
TATTTATGAGATGTATCTTTT
GCTCTCTCTTGCTCTCTTATTTG
TACCGGTTTTTGTATATAAAATTCAT
GTTTCCAATCTCTCTCTCCCT
GATCGGTGACAGTCACTAGCT
TATCTTGAACAGATATTTAATTTTGC
TAACACTCAGCTCTGCCCTCCCC
GATCCCCTGGCTCCCCAG
CACACATTCCTTTGAAATA
AGGTTTCAATATACATCTACATAC
TATATATATATTTGGCAACTTG
TATTTGTGTGTATATATATATATATAT
GTTTATGTATATATGTGATTCT
GATAAAATAGACATTGCTATTCT
GTTTTTTATATGTAAAAACAAAA
CAAGAAAAAATAGAGAATTTCA
CATACTAAATCTCTCTCCTTTTTTA
ATTTTAATATTTGTTATCATTTATT
TATTGGTGCTACTGTTTATCCG
TAATAATTGTGGGGAAAAGATAT
TAACATCACGTCTTTGTCTCTAGT
GCAGTTTTTCGAGATATTCCGTAG
TACATATTTATTTTTAAACAACGA
CAAAGAAATACAGATATATCTTA
AAAAAAAAAAAGCATTTTGTATTA
AAGAATTTAATTCTGATCTGCAGCT
CAAAAAAA AAAAAA

2. Scroll down to "Please indicate which enzymes to include in the analysis."

Click the button indicating "Use only the following enzymes." Select the restriction enzymes *Eco*RI, *Bam*HI, and *Pst*I from the list provided, then click "Analyze sequence." (*Note*: Use the command, control, or shift key to select multiple restriction enzymes.)

3. After examining the results provided by Webcutter, create a table showing the number of cutting sites for each enzyme and the fragment sizes that would be generated by digesting with each enzyme. Draw a restriction map indicating cutting sites for each enzyme with distances between each site and the total size of this piece of human DNA.

■ **Exercise II – Designing a Recombinant DNA Experiment**

Now that you have created a restriction map of your piece of human DNA, you need to ligate the DNA into a plasmid DNA vector that you can use to make your probe (molecular biologists often refer to this as subcloning). To do this, you will need to determine which restriction enzymes would best be suited for cutting both the plasmid and the human DNA.

1. Below is a plasmid DNA sequence. Copy this sequence into the text box in Webcutter and identify cutting sites for the same enzymes you used in Exercise I. Then answer the following questions:

a. What is the total size of the plasmid DNA analyzed in Webcutter?

b. Which enzyme(s) could be used in a recombinant DNA experiment to ligate the plasmid to the *largest* DNA fragment from the human gene? Briefly explain your answer.

c. What size recombinant DNA molecule will be created by ligating these fragments?

d. Draw a simple diagram showing the cloned DNA inserted into the plasmid and indicate the restriction-enzyme cutting site(s) used to create this recombinant plasmid.

Plasmid DNA sequence

```
TATAAATATAGAATAATGAAT
CATATAAACATATCATTATTCATT
TATTTACATTTAAAATTATT
GTTTCAGTATCTTTAATTTATTATG
TATATATAAAAATAACTTACAATTT
TATTAATAAACAATATATGTTTAT
TAATTCATGTTTTGTAATTTAT
GGGATAGCGATTTTTTTTACT
GTCTGTATTTTTCTTTTTAATTAT
GTTTTAATTGTATTTTATTTTTAT
TATTGTTCTTTTTATAGTATTATTT
TAAAACAAAATGTATTTTCTA
AGAACTTATAATAATAATAAATATA
AATTTTAATAAAAATTATATT
TATCTTTTACAATATGAACATA
AAGTACAACATTAATATATAGCTTT
TAATATTTTTATTCCTAATCATG
TAAATCTTAAATTTTTCTTTTTA
AACATATGTTAAATATTTATTTCT
CATTATATATAAGAACATATT
TATTAAATCTAGAATTCTATAGT
GAGTCGTATTACAATTCACTG
GCCGTCGTTTTACAACGTCGT
GACTGGGAAAACCCTGGCGT
TACCCAACTTAATCGCCTTGCAG
CACATCCCCCTTTCGCCAGCTG
GCGTAATAGCGAAGAGGCCC
GCACCGATCGCCCTTCCCAA
CAGTTGCGCAGCCTGAATG
GCGAATGGCGCCTGATGCGG
TATTTTCTCCTTACGCATCTGT
GCGGTATTTCACACCGCATAT
GGTGCACTCTCAGTACAATCT
GCTCTGATGCCGCATAGTTA
AGCCAGCCCCGACACCCGC
CAACACCCGCTGACGCGCCCT
GACGGGCTTGTCTGCTCCCG
GCATCCGCTTACAGACAAGCT
GTGACCGTCTCCGGGAGCTG
CATGTGTCAGAGGTTTTCACC
GTCATCACCGAAACGCGCGAGA
CGAAAGGGCCTCGTGATAC
GCCTATTTTTATAGGTTAATGTCAT
GATAATAATGGTTTCTTAGACGT
CAGGTGGCACTTTTCGGGGAAAT
GTGCGCGGAACCCCTATTTGTT
TATTTTTCTAAATACATTCAAATAT
GTATCCGCTCATGAGACAATA
ACCCTGATAAATGCTTCAATA
ATATTGAAAAAGGAAGAGTAT
GAGTATTCAACATTTCCGTGTC
GCCCTTATTCCCTTTTTTGCG
```

```
GCATTTTGCCTTCCTGTTTTTGCT
CACCCAGAAACGCTGGTGAAAG
TAAAAGATGCTGAAGATCAGTT
GGGTGCACGAGTGGGTTACATC
GAACTGGATCTCAACAGCGGTAA
GATCCTTGAGAGTTTTCGCCCC
GAAGAACGTTTTCCAATGATGAG
CACTTTTAAAGTTCTGCTATGT
GGCGCGGTATTATCCCGTATT
GACGCCGGGCAAGAGCAACTC
GGTCGCCGCATACACTATTCT
CAGAATGACTTGGTTGAGTACT
CACCAGTCACAGAAAAGCATCT
TACGGATGGCATGACAGTAAGAGA
ATTATGCAGTGCTGCCATAAC
CATGAGTGATAACACTGCG
GCCAACTTACTTCTGACAAC
GATCGGAGGACCGAAGGAGC
TAACCGCTTTTTTGCACAACAT
GGGGGATCATGTAACTCGCCTT
GATCGTTGGGAACCGGAGCT
GAATGAAGCCATACCAAACGAC
GAGCGTGACACCACGATGCCTG
TAGCAATGCCAACAACGTTGCG
CAAACTATTAACTGGCGAACTACT
TACTCTAGCTTCCCGGCAACAAT
TAATAGACTGGATGGAGGCG
GATAAAGTTGCAGGACCACTTCT
GCGCTCGGCCCTTCCGGCTG
GCTGGTTTATTGCTGATAAATCTG
GAGCCGGTGAGCGTGGGTCTC
GCGGTATCATTGCAGCACT
GGGGCCAGATGGTAAGCCCTCCC
GTATCGTAGTTATCTACACGAC
GGGGAGTCAGGCAACTATGGAT
GAACGAAATAGACAGATCGCT
GAGATAGGTGCCTCACTGAT
TAAGCATTGGTAACTGTCAGAC
CAAGTTTACTCATATATACTTTA
GATTGATTTAAAACTTCATTTT
TAATTTAAAAGGATCTAG
GTGAAGATCCTTTTTGATAA
TCTCATGACCAAAATCCCTTA
ACGTGAGTTTTCGTTCCACT
GAGCGTCAGACCCCGTAGAAA
GATCAAAGGATCTTCTTGAGA
TCCTTTTTTTCTGCGCGTAATCT
GCTGCTTGCAAACAAAAAAA
CCACCGCTACCAGCGGTGGTTT
GTTTGCCGGATCAAGAGCTAC
```

2. As you prepare to carry out this sub-cloning experiment, you find that the expiration dates on most of your restriction enzymes have long since passed. Rather than run an experiment with old enzymes, you decide to purchase new enzymes. Fortunately, a site called **REBASE®: The Restriction**

Enzyme Database can help you. Over 300 restriction enzymes are commercially available rather inexpensively, but scientists are always looking for ways to stretch their research budgets as far as possible. REBASE is excellent for locating enzyme suppliers and enzyme specifics, particularly if you need to work with an enzyme that you are unfamiliar with. Visit **REBASE®** at http://rebase.neb.com/rebase/rebase .html to identify companies that sell the restriction enzyme(s) you need for this experiment.

■ **Exercise III – Designing PCR Primers**

Giving this experiment more thought, you decide to try reverse transcriptase PCR (RT-PCR) first instead of Northern blotting because RT-PCR is a faster and more sensitive way to detect gene expression. Picking correct primers for a PCR experiment is not a trivial process. You have to be sure the primers can amplify the gene of interest, and you need to avoid primer self-annealing—or having primers bind to each other—among many other considerations. Fortunately, primer design is another task made much easier by the Internet. In this exercise, you will use **Primer3**, a PCR primer design site from the Whitehead Institute for Biomedical Research.

1. Access **Primer3** at http://frodo.wi.mit .edu/primer3. Copy the human DNA sequence from Exercise I into the text box, then click "Pick Primers."

2. On the next page, the sequences for the best recommended primers will appear at the top of the screen. Answer the following:

   a. What is the length, in base pairs, of the left (forward) primer and right (reverse) primer? Where does each of these primers bind in the gene sequence?

   b. The hybridization temperature for a PCR reaction is often set around 5 degrees below the melting temperature, or T_m (refer to Chapter 10 for a discussion of melting temperature). Based on the T_m for these primers, what might be the optimal hybridization temperature for this experiment?

   c. What size PCR product would you expect these primers to generate if you ran the DNA amplified by this PCR reaction on an agarose gel?

CASE STUDY | Should we worry about recombinant DNA technology?

Early in the 1970s, when recombinant DNA research was first developed, scientists realized that there may be unforeseen dangers, and after a self-imposed moratorium on all such research, they developed and implemented a detailed set of safety protocols for the construction, storage, and use of genetically modified organisms. These guidelines then formed the basis of regulations adopted by the federal government. Over time, safer methods were developed, and these stringent guidelines were gradually relaxed or in many cases, eliminated altogether. Now, however, the specter of bioterrorism has refocused attention on the potential misuses of recombinant DNA technology. For example, individuals or small groups might use the information in genome databases coupled with recombinant DNA technology to construct or reconstruct agents of disease, such as the smallpox virus or the deadly influenza virus.

1. Do you think that the question of recombinant DNA research regulation by university and corporations should be revisited to monitor possible bioterrorist activity?

2. Should freely available access to genetic databases, including genomes, and gene or protein sequences be continued, or should it be restricted to individuals who have been screened and approved for such access?

3. Forty years after its development, the use of recombinant DNA technology is widespread and is found even in many middle school and high school biology courses. Are there some aspects of gene splicing that might be dangerous in the hands of an amateur?

Summary Points

 For activities, animations, and review quizzes, go to the study area at www.masteringgenetics.com

1. Recombinant DNA technology was made possible by the discovery of specific proteins called restriction enzymes, which cut DNA at specific recognition sequences, producing fragments that can be joined together with other DNA fragments to form recombinant DNA molecules.

2. Recombinant DNA molecules can be transferred into any of several types of host cells where cloned copies of the DNA are produced during host-cell replication. Many kinds of host cells may be used for replication, including bacteria, yeast, and mammalian cells.

3. The polymerase chain reaction (PCR) allows DNA to be amplified without host cells and is a rapid, sensitive method with wide-ranging applications.

4. Historically, DNA libraries have been important for producing collections of cloned genes to identify genes and gene-regulatory regions of interest.

5. Once cloned, DNA sequences are analyzed through a variety of molecular techniques that allow scientists to study gene structure, expression, and function.

6. By determining the nucleotide sequence of a DNA segment, DNA sequencing is the ultimate way to characterize DNA at the molecular level.

7. Rapid advances in next-generation sequencing technologies have led to greatly increased sequencing capacities at reduced costs over historically used sequencing methods, providing scientists with unprecedented access to sequence data.

INSIGHTS AND SOLUTIONS

1. The recognition sequence for the restriction enzyme *Sau* 3AI is GATC (see Figure 20–1); in the recognition sequence for the enzyme *Bam*HI—GGATCC—the four internal bases are identical to the *Sau*3AI sequence. The single-stranded ends produced by the two enzymes are identical. Suppose you have a cloning vector that contains a *Bam*HI recognition sequence and you also have foreign DNA that was cut with *Sau*3AI.

(a) Can this DNA be ligated into the *Bam*HI site of the vector, and if so, why? (b) Can the DNA segment cloned into this sequence be cut from the vector with *Sau*3AI? With *Bam*HI? What potential problems do you see with the use of *Bam*HI?

Solution: (a) DNA cut with *Sau*3AI can be ligated into the vector's *Bam*HI cutting site because the single-stranded ends generated by the two enzymes are identical. (b) The DNA can be cut from the vector with *Sau*3AI because the recognition sequence for this enzyme (GATC) is maintained on each side of

the insert. Recovering the cloned insert with *Bam*HI is more problematic. In the ligated vector, the conserved sequences are GGATC (left) and GATCC (right). The correct base for recognition by *Bam*HI will *follow* the conserved sequence (to produce GGATCC on the left) only about 25 percent of the time, and the correct base will *precede* the conserved sequence (and produce GGATCC on the right) about 25 percent of the time as well. Thus, *Bam*HI will be able to cut the insert from the vector (0.25 × 0.25 = 0.0625), or only about 6 percent, of the time.

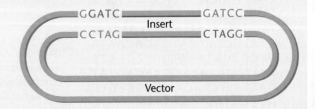

Problems and Discussion Questions

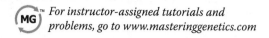

For instructor-assigned tutorials and problems, go to www.masteringgenetics.com

HOW DO WE KNOW?

1. In this chapter we focused on how specific DNA sequences can be copied, identified, characterized, and sequenced. At the same time, we found many opportunities to consider the methods and reasoning underlying these techniques. From the explanations given in the chapter, what answers would you propose to the following fundamental questions?

   (a) In a recombinant DNA cloning experiment, how can we determine whether DNA fragments of interest have been incorporated into plasmids and, once host cells are transformed, which cells contain recombinant DNA?

   (b) When using DNA libraries to clone genes, what combination of techniques are used to identify a particular gene of interest?

   (c) What steps make PCR a chain reaction that can produce millions of copies of a specific DNA molecule in a matter of hours without using host cells?

   (d) How has DNA sequencing technology evolved in response to the emerging needs of genome scientists?

2. What roles do restriction enzymes, vectors, and host cells play in recombinant DNA studies?

3. The human insulin gene contains a number of sequences that are removed in the processing of the mRNA transcript. In spite of the fact that bacterial cells cannot excise these sequences from mRNA transcripts, explain how a gene like this can be cloned into a bacterial cell and produce insulin.

4. What role does DNA ligase perform in a DNA cloning experiment? How does the action of DNA ligase differ from the function of restriction enzymes?

5. Although many cloning applications involve introducing recombinant DNA into bacterial host cells, many other cell types are also used as hosts for recombinant DNA. Why?

6. Using DNA sequencing on a cloned DNA segment, you recover the nucleotide sequence shown below. Does this segment contain a palindromic recognition sequence for a restriction enzyme? If so, what is the double-stranded sequence of the palindrome, and what enzyme would cut at this sequence? (Consult Figure 20–1 for a list of restriction sites.)

   CAGTATCCTAGGCAT

7. Restriction sites are palindromic; that is, they read the same in the 5′ to 3′ direction on each strand of DNA. What is the advantage of having restriction sites organized this way? Restriction sites are palindromic organized in this way?

8. List the advantages and disadvantages of using plasmids as cloning vectors. What advantages do BACs and YACs provide over plasmids as cloning vectors?

9. What are the advantages of using a restriction enzyme whose recognition site is relatively rare? When would you use such enzymes?

10. The introduction of genes into plants is a common practice that has generated not only a host of genetically modified foodstuffs, but also significant worldwide controversy. Interestingly, a tumor-inducing plasmid is often used to produce genetically modified plants. Is the use of a tumor-inducing plasmid the source of such controversy?

11. In the context of recombinant DNA technology, of what use is a probe?

12. If you performed a PCR experiment starting with only one copy of double-stranded DNA, approximately how many DNA molecules would be present in the reaction tube after 15 cycles of amplification?

13. In a control experiment, a plasmid containing a *Hin*dIII recognition sequence within a kanamycin resistance gene is cut with *Hin*dIII, re-ligated, and used to transform *E. coli* K12 cells. Kanamycin-resistant colonies are selected, and plasmid DNA from these colonies is subjected to electrophoresis. Most of the colonies contain plasmids that produce single bands that migrate at the same rate as the original intact plasmid. A few colonies, however, produce two bands, one of original size and one that migrates much higher in the gel. Diagram the origin of this slow band as a product of ligation.

14. You have just created the world's first genomic library from the African okapi, a relative of the giraffe. No genes from this genome have been previously isolated or described. You wish to isolate the gene encoding the oxygen-transporting protein β-globin from the okapi library. This gene has been isolated from humans, and its nucleotide sequence and amino acid sequence are available in databases. Using the information available about the human β-globin gene, what two strategies can you use to isolate this gene from the okapi library?

15. What advantages do cDNA libraries provide over genomic DNA libraries? Describe cloning applications where the use of a genomic library is necessary to provide information that a cDNA library cannot.

16. A colleague provided you with a cDNA library of human genes cloned into bacterial plasmid vectors. Armed with the cloned yeast gene that encodes EF-1a, a protein that is highly conserved among eukaryotes, you use the yeast gene as a probe to isolate a cDNA clone containing the human EF-1a gene. You sequence the human gene and find that it is 1384 nucleotide pairs long. Using this cDNA clone as a probe, you isolate the DNA encoding EF-1a from a human genomic library. The genomic clone is sequenced and found to be 5282 nucleotides long. What accounts for the difference in length observed between the cDNA clone and the genomic clone?

17. You have recovered a cloned DNA segment from a vector and determine that the insert is 1300 bp in length. To characterize this cloned segment, you isolate the insert and decide to construct a restriction map. Using enzyme I and enzyme II, followed by gel electrophoresis, you determine the number and size of the fragments produced by enzymes I and II alone and in combination, as recorded in the following table. Construct a restriction map from these data, showing the positions of the restriction-enzyme cutting sites relative to one another and the distance between them in units of base pairs.

Enzymes	Restriction Fragment Sizes (bp)
I	350, 950
II	200, 1100
I and II	150, 200, 950

18. To create a cDNA library, cDNA can be inserted into vectors and cloned. In the analysis of cDNA clones, it is often difficult to find clones that are full length—that is, many clones are shorter than the mature mRNA molecules from which they are derived. Why is this so?

19. Although the capture and trading of great apes has been banned in 112 countries since 1973, it is estimated that about 1000 chimpanzees are removed annually from Africa and smuggled into Europe, the United States, and Japan. This illegal trade is often disguised

by simulating births in captivity. Until recently, genetic identity tests to uncover these illegal activities were not used because of the lack of highly polymorphic markers (markers that vary from one individual to the next) and the difficulties of obtaining chimpanzee blood samples. A study was reported in which DNA samples were extracted from freshly plucked chimpanzee hair roots and used as templates for PCR. The primers used in these studies flank highly polymorphic sites in human DNA that result from variable numbers of tandem nucleotide repeats. Several offspring and their putative parents were tested to determine whether the offspring were "legitimate" or the product of illegal trading. The data are shown in the following Southern blot.

Examine the data carefully and choose the best conclusion.

(a) None of the offspring is legitimate.

(b) Offspring B and C are not the products of these parents and were probably purchased on the illegal market. The data are consistent with offspring A being legitimate.

(c) Offspring A and B are products of the parents shown, but C is not and was therefore probably purchased on the illegal market.

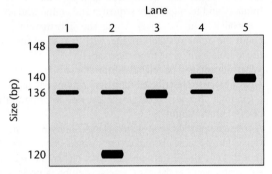

Lane 1: father chimpanzee
Lane 2: mother chimpanzee
Lanes 3–5: putative offspring A, B, C

(d) There are not enough data to draw any conclusions. Additional polymorphic sites should be examined.

(e) No conclusion can be drawn because "human" primers were used.

20. List the steps involved in screening a genomic library. What must be known before starting such a procedure? What are the potential problems with such a procedure, and how can they be overcome or minimized?

21. To estimate the number of cleavage sites in a particular piece of DNA with a known size, you can apply the formula, $N/4^n$ where N is the number of base pairs in the target DNA and n is the number of bases in the recognition sequence of the restriction enzyme. If the recognition sequence for *Bam*HI is GGATCC and the λ phage DNA contains approximately 48,500 bp, how many cleavage sites would you expect?

22. In a typical PCR reaction, describe what is happening in stages occurring at temperature ranges (a) 90–95°C, (b) 50–70°C, and (c) 70–75°C?

23. We usually think of enzymes as being most active at around 37°C, yet in PCR the DNA polymerase is subjected to multiple exposures of relatively high temperatures and seems to function appropriately at 70–75°C. What is special about the DNA polymerizing enzymes typically used in PCR?

24. How are dideoxynucleotides (ddNTPs) structurally different from deoxynucleotides (dNTPs), and how does this structural difference make ddNTPs valuable in chain-termination methods of DNA sequencing?

25. Assume you have conducted a DNA sequencing reaction using the chain-termination (Sanger) method. You performed all the steps correctly and electrophoresced the resulting DNA fragments correctly, but when you looked at the sequencing gel, many of the bands were duplicated (in terms of length) in other lanes. What might have happened?

26. How is fluorescent *in situ* hybridization (FISH) used to produce a spectral karyotype?

Extra-Spicy Problems

MG™ *For instructor-assigned tutorials and problems, go to www.masteringgenetics.com*

27. The gel presented here shows the pattern of bands of fragments produced with several restriction enzymes. The enzymes used are identified above and below the gel and six possible restriction maps shown in the column to the right.

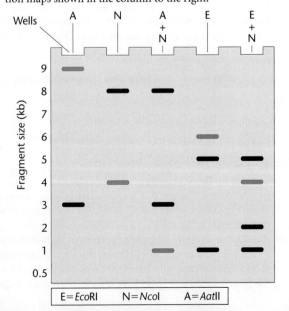

E = *Eco*RI   N = *Nco*I   A = *Aat*II

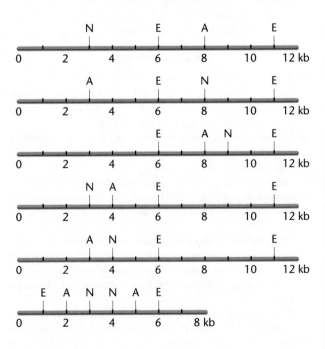

One of the six restriction maps shown is consistent with the pattern of bands shown in the gel.

(a) From your analysis of the pattern of bands on the gel, select the correct map and explain your reasoning.

(b) In a Southern blot prepared from this gel, the highlighted bands (pink) hybridized with the gene *pep*. Where is the *pep* gene located?

28. A widely used method for calculating the annealing temperature for a primer used in PCR is 5 degrees below the T_m (°C), which is computed by the equation $81.5 + 0.41$ (%GC)– $(675/N)$, where %GC is the percentage of GC nucleotides in the oligonucleotide and N is the length of the oligonucleotide. Notice from the formula that both the GC content *and* the length of the oligonucleotide are variables. Assuming you have the following oligonucleotide as a primer, compute the annealing temperature for PCR. What is the relationship between T_m (°C) and %GC? Why? (*Note:* In reality, this computation provides only a starting point for empirical determination of the most useful annealing temperature.)

5′-TTGAAAATATTTCCCATTGCC-3′

29. Most of the techniques described in this chapter (blotting, cloning, PCR, etc.) are dependent on intermolecular attractions (annealing) between different populations of nucleic acids. Length of the strands, temperature, and percentage of GC nucleotides weigh considerably on intermolecular associations. Two other components commonly used in hybridization protocols are monovalent ions and formamide. A formula that takes monovalent ion (Na^+) and formamide concentrations into consideration to compute a T_m (temperature of melting) is as follows:

$$T_m = 81.5 + 16.6(\log M[Na^+]) + 0.41\%GC - 0.72(\%formamide)$$

(a) For the following concentrations of Na^+ and formamide, calculate the T_m. Assume 45% GC content.

Na$^+$	% Formamide
0.825	20
0.825	40
0.165	20
0.165	40

(b) Given that formamide competes for hydrogen bond locations of nucleic acid bases and monovalent cations are attracted to the negative charges of nucleic acids, explain why the T_m varies as described in part (a).

30. The U.S. Department of Justice has established a database that catalogs PCR amplification products from short tandem repeats of the Y (Y-STRs) chromosome in humans. The database contains polymorphisms of five U.S. ethnic groups (African Americans, European Americans, Hispanics, Native Americans, and Asian Americans) as well as worldwide population.

(a) Given that STRs are repeats of varying lengths, for example $(TCTG)_{9-17}$ or $(TAT)_{6-14}$, explain how PCR could reveal differences (polymorphisms) among individuals. How could the Department of Justice make use of those differences?

(b) Y-STRs from the nonrecombining region of the Y chromosome (NRY) have special relevance for forensic purposes. Why?

(c) What would be the value of knowing the ethnic population differences for Y-STR polymorphisms?

(d) For forensic applications, the probability of a "match" for a crime scene DNA sample and a suspect's DNA often culminates in a guilty or innocent verdict. How is a "match" determined, and what are the uses and limitations of such probabilities?

31. There are a variety of circumstances under which rapid results using multiple markers in PCR amplifications is highly desired, such as in forensics, pathogen analysis, or detection of genetically modified organisms. In multiplex PCR, multiple sets of primers are used, often with less success than when applied to PCR as individual sets. Numerous studies have been conducted to optimize procedures, but each has described the process as time consuming and often unsuccessful. Considering the information given in question 29, why should multiplex PCR be any different than single primer set PCR in terms of dependability and ease of optimization?

Alignment comparing DNA sequence for the leptin gene from dogs (top) and humans (bottom). Vertical lines and shaded boxes indicate identical bases. *LEP* encodes a hormone that functions to suppress appetite. This type of analysis is a common application of bioinformatics and a good demonstration of comparative genomics.

```
AGGCCCAACAAGCACAGCCGGGGAAGGAAAATGCGTTGTGGACCTCTGTGCCGATTCCTG
|||||||| ||||| || || |||||||||||| ||| ||| | ||||| ||||| ||
AGGCCCAAGGAAGC-CATCCTGGGAAGGAAAATGCATTGGGGAACCCTGTGCGGATTCTTG

TGGCTTTGGCCCTATCTGTCCTGTGTTGAAGCTGTGCCAATCCGAAAAGTCCAGGATGAC
|||||||||||||||||||||| | || ||| ||||||||||| |||| ||||||| ||||||
TGGCTTTGGCCCTATCTTTTCTATGTCCAAGCTGTGCCCATCCAAAAAGTCCAAGATGAC

ACCAAAACCCTCATCAAGACGATTGTCGCCAGGATCAATGACATTTCACACACGCAGTCT
||||||||||||||||||||||| |||||| |||||||||||||||||||||||||||||||
ACCAAAACCCTCATCAAGACAATTGTCACCAGGATCAATGACATTTCACACACGCAGTCA

GTCTCCTCCAAACAGAGGGTCGCTGGTCTGGACTTCATTCCTGGGCTCCAACCAGTCCTG
|||||||||||||||| ||| | ||| ||||||||||||||||||||||||||| || |||||
GTCTCCTCCAAACAGAAAGTCACCGGTTTGGACTTCATTCCTGGGCTCCACCCCATCCTG

AGTTTGTCCAGGATGGACCAGACGTTGGCCATCTACCAACAGATCCTCAACAGTCTGCAT
| || |||| |||||||||||| |||| |||||||||||||||||||| |||| ||| |
ACCTTATCCAAGATGGACCAGACACTGGCAGTCTACCAACAGATCCTCACCAGTATGCCT

TCCAGAAATGTGGTCCAAATATCTAATGACCTGGAGAACCTCCGGGACCTTCTCCACCTG
|||||||| ||| |||||||||||| || ||||||||||||||||||||| ||||| ||| ||
TCCAGAAACGTGATCCAAATATCCAACGACCTGGAGAACCTCCGGGATCTTCTTCACGTG

CTGGCCTCCTCCAAGAGCTGCCCCTTGCCCCGGGCCAGGGGCCTGGAGACCTTTGAGAGC
||||||| |||| |||||||||| ||||||| ||||||| ||||||||||||| || |||
CTGGCCTTCTCTAAGAGCTGCCACTTGCCCTGGGCCAGTGGCCTGGAGACCTTGGACAGC

CTGGGCGGCGTCCTGGAAGCCTCACTCTACTCCACAGAGGTGGTGGCTCTGAACAGACTG
|||||| || ||||||||||| ||| |||||||||||||||||||||||| |||| ||| |||
CTGGGGGGTGTCCTGGAAGCTTCAGGCTACTCCACAGAGGTGGTGGCCCTGAGCAGGCTG
```

Genomics, Bioinformatics, and Proteomics

CHAPTER CONCEPTS

- Genomics applies recombinant DNA, DNA sequencing methods, and bioinformatics to sequence, assemble, and analyze genomes.

- Disciplines in genomics encompass several areas of study, including structural and functional genomics, comparative genomics, and metagenomics, and have led to an "omics" revolution in modern biology.

- Bioinformatics merges information technology with biology and mathematics to store, share, compare, and analyze nucleic acid and protein sequence data.

- The Human Genome Project has greatly advanced our understanding of the organization, size, and function of the human genome.

- Ten years after completion of the Human Genome Project, a new era of genomics studies is providing deeper insights into the human genome.

- Comparative genomics analysis of model prokaryotes and eukaryotes has revealed similarities and differences in genome size and organization.

- Metagenomics is the study of genomes from environmental samples and is valuable for identifying microbial genomes.

- Transcriptome analysis provides insight into patterns of gene expression and gene-regulatory activity of a genome.

- Proteomics focuses on the protein content of cells and on the structures, functions, and interactions of proteins.

- Systems biology approaches attempt to uncover complex interactions among genes, proteins, and other cellular components.

The term **genome**, meaning the complete set of DNA in a single cell of an organism, was coined in 1920, at a time when geneticists began to turn from the study of individual genes to a focus on the larger picture. To begin to characterize all of the genes in an organism's DNA, geneticists typically followed a two-part approach: (1) identify spontaneous mutations or collect mutants produced by chemical or physical agents, and (2) generate linkage maps using mutant strains as discussed in Chapter 5.

These effective but extremely time-consuming strategies were used to identify genes in many of the classic model organisms discussed in this book, such as *Drosophila*, maize, mice, bacteria, and yeast, as well as in viruses, such as bacteriophages. These approaches formed the technical backbone of genetic analysis and still have their applications today; however, they have several major limitations. For instance, conventional mutational analysis and linkage requires that at least one mutation for each gene was available before all the genes in a genome could be identified. Obtaining mutants and carrying out linkage studies is very time consuming, and when mutations are lethal or have no clear phenotype, they can be difficult or impossible to map. In addition, although researchers can generate mutations in animal models in a laboratory, they cannot do the same with humans; thus identifying human genes by mutational analysis is largely limited to linkage mapping of inherited or spontaneously acquired mutant genes with clear phenotypes. Another fundamental limitation of these approaches is that, although they can be used for identifying and characterizing gene loci, they do not lead to a determination of DNA sequence. Nor are they particularly useful for studying noncoding areas of the genome such as DNA regulatory sequences.

In 1977, as recombinant DNA-based techniques were developed, Fred Sanger and colleagues began the field of **genomics**, the study of genomes, by using a newly developed method of DNA sequencing to sequence the 5400-nucleotide genome of the virus ϕX174. Other viral genomes were sequenced in short order, but even this technology was slow and labor-intensive, limiting its use to small genomes. In the 1980s, geneticists interested in mapping human genes began using recombinant DNA technology to map DNA sequences to specific chromosomes. Initially, most of these sequences were not actually full-length genes but marker sequences such as restriction fragment length polymorphisms (RFLPs). Once assigned to chromosomes, these markers were used in pedigree analysis to establish linkage between the markers and disease phenotypes for genetic disorders. This approach, called **positional cloning**, was used to map, isolate, clone, and sequence the genes for cystic fibrosis, neurofibromatosis, and dozens of other disorders. Positional cloning identified one gene at a time, and yet by the mid-1980s, it

had been used to assign more than 3500 genes and markers to human chromosomes.

At this time it was estimated that there were approximately 100,000 genes in the human genome, and it was readily apparent that mapping by using existing methods would be a laborious, time-consuming, and nearly insurmountable task. As you will soon learn, this estimate for gene number turned out to be fairly inaccurate. During the next three decades, the development of computer-automated DNA sequencing methods made it possible to consider sequencing the larger and more complex genomes of eukaryotes, including the 3.1 billion nucleotides that comprise the human genome. The development of recombinant DNA technologies coupled with the advent of computer-automated DNA sequencing methods, bioinformatics, and now next-generation sequencing technologies is responsible for rapidly accelerating the field of genomics. Genomic technologies have developed so quickly that modern biological research is currently experiencing a genomics revolution. In this chapter, we will examine basic technologies used in genomics and then discuss examples of genome data and different disciplines of genomics. We will also discuss *transcriptome analysis,* the study of genes expressed in a cell or tissue (the "transcriptome"), and *proteomics,* the study of proteins present in a cell or tissue. The chapter concludes with a brief look at *systems biology,* a new area of contemporary biology that incorporates and integrates genomics, transcriptome analysis, and proteomics data. In Chapter 22, we will continue our discussion of genomics by discussing many modern applications of recombinant DNA and genomic technologies.

21.1

Whole-Genome Shotgun Sequencing Is a Widely Used Method for Sequencing and Assembling Entire Genomes

As discussed in Chapter 20, recombinant DNA technology made it possible to generate DNA libraries that could be used to identify, clone, and sequence specific genes of interest. But a primary limitation of library screening and even of most polymerase chain reaction (PCR) approaches is that they typically can identify only relatively small numbers of genes at a time. Genomics allows the sequencing of entire genomes. **Structural genomics** focuses on sequencing genomes and analyzing nucleotide sequences to identify genes and other important sequences such as gene-regulatory regions.

The most widely used strategy for sequencing and assembling an entire genome involves variations of a method called **whole-genome shotgun sequencing** also known as

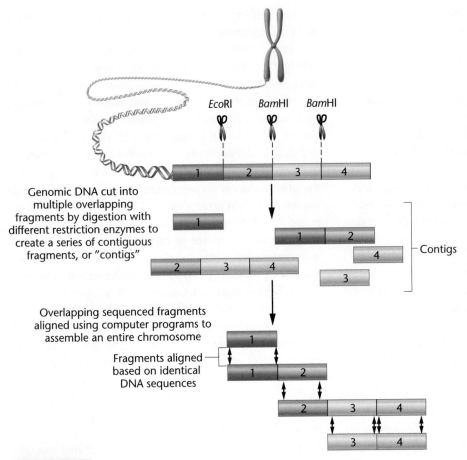

Genomic DNA cut into multiple overlapping fragments by digestion with different restriction enzymes to create a series of contiguous fragments, or "contigs"

Overlapping sequenced fragments aligned using computer programs to assemble an entire chromosome

Fragments aligned based on identical DNA sequences

FIGURE 21–1   An overview of whole-genome shotgun sequencing and assembly. This approach shows one strategy that involves using restriction enzymes to digest genomic DNA into contigs, which are then sequenced and aligned using bioinformatics to identify overlapping fragments based on sequence identity. Notice that *Eco*RI digestion of the portion of DNA depicted here produces two fragments (contigs 1, 2–4), whereas digestion with *Bam*HI produces produces three fragments (contigs 1–2, 3, 4).

a basic example of DNA shearing using restriction enzymes. Increasingly, nonenzymatic approaches for shearing DNA are being used, especially for some of the next-generation sequencing approaches. Recall from Chapter 20 that **restriction enzymes** are DNA-digesting enzymes that cut the phosphodiester backbone of DNA at specific sequences. Different restriction enzymes can be used so that chromosomes are cut at different sites; or sometimes, **partial digests** of DNA using the same restriction enzyme are used. With partial digests, DNA is incubated with restriction enzymes for only a short period of time, so that not every site in a particular sequence is cut to completion by an individual enzyme. Either way, restriction digests of whole chromosomes generate thousands to millions of overlapping DNA fragments. For example, a 6-bp cutter such as *Eco*-RI creates about 700,000 fragments when used to digest the human genome! Because these overlapping fragments are adjoining segments that collectively form one continuous DNA molecule within a chromosome, they are called **contiguous fragments**, or "**contigs.**"

In the next section, we will discuss the importance of bioinformatics to genomics. One of the earliest bioinformatics applications to be developed for genomic purposes was the use of algorithm-based software programs for creating a DNA-sequence **alignment**, in which similar sequences of bases, such as contigs, are lined up for comparison. Alignment identifies overlapping sequences, allowing scientists to reconstruct their order in a chromosome. Figure 21–2 shows an example of contig alignment and assembly for a portion of human chromosome 2. For simplicity, this figure shows relatively short sequences for each contig, which in actuality would be much longer. The figure is also simplified in that, in actual alignments, assembled sequences do not always overlap only at their ends.

The whole-genome shotgun sequencing method was developed by J. Craig Venter and colleagues at The Institute for Genome Research (TIGR). In 1995, TIGR scientists used this approach to sequence the 1.83-million-bp genome of the bacterium *Haemophilus influenzae*.

shotgun cloning. In simple terms, this technique is analogous to you and a friend taking your respective copies of this genetics textbook and randomly ripping the pages into strips about 5 to 7 inches long. Each chapter represents a chromosome, and all of the letters in the entire book are the "genome." Then you and your friend would go through the painstaking task of comparing the pieces of paper to find places that match, overlapping sentences—areas where there are similar sentences on different pieces of paper. Eventually, in theory, many of the strips containing matching sentences would overlap in ways that you could use to reconstruct the pages and assemble the order of the entire text.

Figure 21–1 shows a basic overview of whole-genome shotgun sequencing. First, an entire chromosome is cut into short, overlapping fragments, either by mechanically shearing the DNA in various ways (such as excessive heat treatment or sonication in which sonic energy is used to break DNA) or by using restriction enzymes to cleave the DNA at different locations. For simplicity, here we present

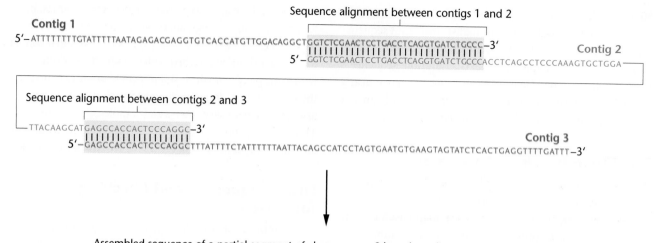

FIGURE 21-2 DNA-sequence alignment of contigs on human chromosome 2. Single-stranded DNA for three different contigs from human chromosome 2 is shown in blue, red, or green. The actual sequence from chromosome 2 is shown, but in reality, contig alignment involves fragments that are several thousand bases in length. Alignment of the three contigs allows a portion of chromosome 2 to be assembled. Alignment of all contigs for a particular chromosome would result in assembly of a completely sequenced chromosome.

This was the first completed genome sequence from a free-living organism, and it demonstrated "proof of concept" that shotgun sequencing could be used to sequence an entire genome. Even after the genome for *H. influenzae* was sequenced, many scientists were skeptical that a shotgun approach would work on the larger genomes of eukaryotes. But variations of shotgun approaches are now the predominant methods for sequencing genomes, including those of *Drosophila*, dog, several hundred species of bacteria, humans, and many other organisms, as you will read about later in this chapter.

Cutting a genome into contigs is not particularly difficult; however, a primary hurdle that had to be overcome to advance whole-genome sequencing was the question of how to sequence millions or billions of base pairs in a timely and cost-effective way. This was a major challenge for scientists working on the Human Genome Project (Section 21.4). The Sanger sequencing method discussed in Chapter 20 was the predominant sequencing technique for a long time; however, a major limitation of this technique was that even the best sequencing gels would typically yield only several hundred base pairs in each run and relatively few runs could be completed in a day so the overall production of sequence data was quite slow compared with modern techniques. Obviously, it would be very time consuming to manually sequence an entire genome by the Sanger method. The major technological breakthrough that made genomics possible was the development of computer-automated sequencers.

High-Throughput Sequencing and Its Impact on Genomics

As we discussed in Chapter 20, the first **computer-automated DNA sequencing instruments** utilized dideoxynucleotides (ddNTPs) labeled with fluorescent dyes (refer to Figure 20–16). A single reaction tube is used, and sequencing reaction mixtures are separated on an ultra-thin-diameter polyacrylamide tube gel called a **capillary gel**. As DNA fragments move through the gel, they are scanned with a laser beam. The laser stimulates the fluorescent dye on each DNA fragment, causing each ddNTP that was incorporated into newly synthesized DNA strands to emit different wavelengths of light. The emitted light is collected by a detector, which amplifies and then feeds this information to a computer to process and convert into the DNA sequence.

Many of the early computer-automated sequencers, designed for so-called **high-throughput sequencing**, could process millions of base pairs in a day. These sequencers contained multiple capillary gels that are several feet long. Some run as many as 96 capillary gels at a time, each producing around 900 bases of sequence. Because these sequencers are computer automated, they can work around the clock, generating over 2 million bases of sequence in a day. In the past 10 years, high-throughput sequencing has increased the productivity of DNA-sequencing technology over 500-fold. The total number of bases that could be sequenced in a single reaction was doubling about every 24 months. At the same time, this increase in efficiency brought about a dramatic decrease in cost, from about $1.00 to less than

$0.001 per base pair. As we will discuss in Section 21.4, without question the development of high-throughput sequencing was essential for the Human Genome Project. And as you know from Chapter 20, next-generation sequencers now enable genome scientists to produce sequence nearly 50,000 times faster than sequencers in 2000 with greater output, improved accuracy, and reduced cost.

The Clone-by-Clone Approach

Prior to the widespread use of whole-genome sequencing approaches, genomes were being assembled using a **clone-by-clone** approach, also called **map-based cloning** (Figure 21–3). Initial progress on the Human Genome Project was based on this methodology, in which individual DNA fragments from restriction digests of chromosomes are aligned to create the restriction maps of a chromosome. These restriction fragments are then ligated into vectors such as bacterial artificial chromosomes (BACs) or yeast artificial chromosomes (YACs) to create libraries of contigs. Recall from Chapter 20 that BACs and YACs are good cloning vectors for replicating large fragments of DNA. Often, libraries from individual chromosomes were prepared.

Prior to the development of high-throughput approaches capable of sequencing several thousand bases in a single run, DNA fragments in BACs and YACs would often be further digested into smaller, more easily manipulated pieces that were then subcloned into cosmids or plasmids so that they could be sequenced in their entirety (Figure 21–3). After each fragment was sequenced and then analyzed for alignment overlaps, a chromosome could be assembled. The bioinformatics approaches we will discuss in the next section would then be used to identify possible protein-coding genes and assign them a location on the chromosome. For example, Figure 21–3 shows the use of map-based cloning to sequence part of chromosome 11, including part of the human β-globin gene.

Compared to whole-genome sequencing, the clone-by-clone approach is cumbersome and time consuming because of the time required to clone DNA fragments into different vectors, transform bacteria or yeast, select individual clones from the library for sequencing, and then carry out sequence analysis and assembly on relatively short sequences. Essentially, the clone-by-clone approach is the organized sequencing of contigs from a restriction map instead of random sequencing and assembly. As whole-genome sequencing approaches have become the most common method for assembling genomes, map-based cloning approaches are now primarily used to resolve the problems often encountered during whole-genome sequencing. For example, highly repetitive sequences in a chromosome can be difficult to align correctly in order to identify overlaps because with such sequences one cannot know for sure whether portions that are nearly identical are overlapping fragments or belong to different parts of a highly repetitive chromosome.

Frequently, there are also gaps between aligned contigs. In the textbook-ripping analogy, after you compare all your pieces of paper, you may be left with some very small ones that contain too few words to be matched with certainty to any others and with some pieces for which you just could not find matches. In these instances, a clone-by-clone approach may enable you to assemble the necessary contigs and complete the chromosome. Thus, whole-genome sequencing and map-based sequencing are commonly combined in the task of sequencing a genome.

Draft Sequences and Checking for Errors

It is common for a draft sequence of a genome to be announced several years before a final sequence is released. Draft sequences often contain gaps in areas that, for any number of reasons, may have been difficult to analyze. The decision to designate a sequence as "final" is dictated by the amount of error genome scientists are willing to accept as a cutoff.

NOW SOLVE THIS

21–1 To deal with the problems of correctly annotating microbial genomes, Marie Skovgaard and her colleagues (2001. *Trends Genet.* 17: 425–428) compared the annotated number of genes in genomes derived from sequence analysis to the number of known proteins in each organism as reported in a protein database. The results of their study are summarized in the accompanying graph. The errors range from a few percent for *M. genitalium* to almost 100 percent for *A. pernix*. The general trend shown in the graph is that the error rate increases as the GC content of the genome increases. What might account for this? What precautions should be taken in annotating the genomes with high GC content?

■ HINT: *This problem asks you to analyze the relationship between GC content of a genome and error rates in annotation. The key to its solution is to consider possible challenges of aligning contigs from GC-rich genomes or from any sequences that have an abundance of repeated bases.*

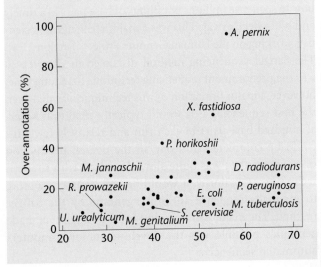

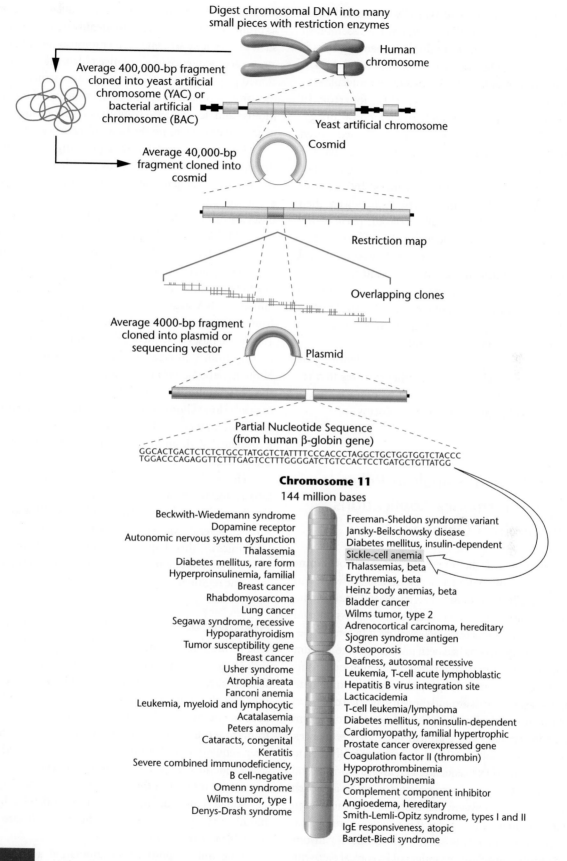

Digest chromosomal DNA into many
small pieces with restriction enzymes

Human chromosome

Average 400,000-bp fragment
cloned into yeast artificial
chromosome (YAC) or
bacterial artificial
chromosome (BAC)

Yeast artificial chromosome

Cosmid

Average 40,000-bp
fragment cloned into
cosmid

Restriction map

Overlapping clones

Average 4000-bp fragment
cloned into plasmid or
sequencing vector

Plasmid

Partial Nucleotide Sequence
(from human β-globin gene)

GGCACTGACTCTCTCTGCCTATGGTCTATTTTCCCACCCTAGGCTGCTGGTGGTCTACCC
TGGACCCAGAGGTTCTTTGAGTCCTTTGGGGATCTGTCCACTCCTGATGCTGTTATGG

Chromosome 11
144 million bases

Beckwith-Wiedemann syndrome
Dopamine receptor
Autonomic nervous system dysfunction
Thalassemia
Diabetes mellitus, rare form
Hyperproinsulinemia, familial
Breast cancer
Rhabdomyosarcoma
Lung cancer
Segawa syndrome, recessive
Hypoparathyroidism
Tumor susceptibility gene
Breast cancer
Usher syndrome
Atrophia areata
Fanconi anemia
Leukemia, myeloid and lymphocytic
Acatalasemia
Peters anomaly
Cataracts, congenital
Keratitis
Severe combined immunodeficiency
B cell-negative
Omenn syndrome
Wilms tumor, type I
Denys-Drash syndrome

Freeman-Sheldon syndrome variant
Jansky-Beilschowsky disease
Diabetes mellitus, insulin-dependent
Sickle-cell anemia
Thalassemias, beta
Erythremias, beta
Heinz body anemias, beta
Bladder cancer
Wilms tumor, type 2
Adrenocortical carcinoma, hereditary
Sjogren syndrome antigen
Osteoporosis
Deafness, autosomal recessive
Leukemia, T-cell acute lymphoblastic
Hepatitis B virus integration site
Lacticacidemia
T-cell leukemia/lymphoma
Diabetes mellitus, noninsulin-dependent
Cardiomyopathy, familial hypertrophic
Prostate cancer overexpressed gene
Coagulation factor II (thrombin)
Hypoprothrombinemia
Dysprothrombinemia
Complement component inhibitor
Angioedema, hereditary
Smith-Lemli-Opitz syndrome, types I and II
IgE responsiveness, atopic
Bardet-Biedi syndrome

FIGURE 21–3   A clone-by-clone, or map-based, approach to genome sequencing involves cloning overlapping DNA fragments (contigs) into vectors. Different vectors, such as BACs, YACs, cosmids, and plasmids, are used depending on the size of each DNA fragment being analyzed. Overlapping clones are then sequenced and aligned to assemble an entire chromosome.

Chromosome segments are typically sequenced more than once to ensure a high level of accuracy. The assembly of a final genomic sequence from multiple sequencing runs is known as **compiling**. One way to compile and error check is to sequence complementary strands of a DNA molecule separately and then use base-pairing rules to check for errors. In one case, researchers using the shotgun method on the genome of the bacterium *Pseudomonas aeruginosa* sequenced the 6.3 million nucleotides seven times to ensure that the final sequence would be accurate. Yet even with this level of redundancy, the assembler software recognized 1604 regions that required further clarification. These regions were then reanalyzed and re-sequenced. Finally, relevant parts of the shotgun sequence were compared with the sequences of two widely separated genomic regions obtained by conventional cloning. The 81,843 nucleotides cloned and sequenced by the clone-by-clone method were in perfect agreement with the sequence obtained by the shotgun method. This level of care in checking for accuracy is not unusual; similar precautions are taken in almost every genome project.

Once compiled, a genome is analyzed to identify gene sequences, regulatory elements, and other features that reveal important information. In the next section we discuss the central role of bioinformatics in this process.

21.2

DNA Sequence Analysis Relies on Bioinformatics Applications and Genome Databases

Genomics necessitated the rapid development of **bioinformatics**, the use of computer hardware and software and mathematics applications to organize, share, and analyze data related to gene structure, gene sequence and expression, and protein structure and function. However, even before whole-genome sequencing projects had been initiated, a large amount of sequence information from a range of different organisms was accumulating as a result of gene cloning by recombinant DNA techniques. Scientists around the world needed databases that could be used to store, share, and obtain the maximum amount of information from protein and DNA sequences. Thus, bioinformatics software was already being widely used to compare and analyze DNA sequences and to create private and public databases. Once genomics emerged as a new approach for analyzing DNA, however, bioinformatics became even more important than before. Today, it is a dynamic area of biological research, providing new career opportunities for anyone interested in merging an understanding of biological data with information technology, mathematics, and statistical analysis.

Among the most important applications of bioinformatics are to compare DNA sequences, as in contig alignment, discussed in the previous section; to identify genes in a genomic DNA sequence; to find gene-regulatory regions, such as promoters and enhancers; to identify structural sequences, such as telomeric sequences, in chromosomes; to predict the amino acid sequence of a putative polypeptide encoded by a cloned gene sequence; to analyze protein structure and predict protein functions on the basis of identified domains and motifs; and to deduce evolutionary relationships between genes and organisms on the basis of sequence information.

High-throughput DNA sequencing techniques were developed nearly simultaneously with the expansion of the Internet. As genome data accumulated, many DNA-sequence databases became freely available online. Databases are essential for archiving and sharing data with other researchers and with the public. One of the most important genomic databases, called **GenBank**, is maintained by the National Center for Biotechnology Information (NCBI) in Washington, D.C., and is the largest publicly available database of DNA sequences. GenBank shares and acquires data from databases in Japan and Europe; it contains more than 100 billion bases of sequence data from over 100,000 species; and it doubles in size roughly every 14–18 months! The Human Genome Nomenclature Committee, supported by the NIH, establishes rules for assigning names and symbols to newly cloned human genes. As sequences are identified and genes are named, each sequence deposited into GenBank is provided with an **accession number** that scientists can use to access and retrieve that sequence for analysis.

The NCBI is an invaluable source of public access databases and bioinformatics tools for analyzing genome data. You have already been introduced to NCBI and GenBank through several Exploring Genomics exercises. In Exploring Genomics for this chapter, you will use NCBI and GenBank to compare and align contigs in order to assemble a chromosome segment.

Annotation to Identify Gene Sequences

One of the fundamental challenges of genomics is that, although genome projects generate tremendous amounts of DNA sequence information, these data are of little use until they have been analyzed and interpreted. Genome projects accumulate nucleotide sequences, and then scientists have to make sense of those sequences. Thus, after a genome has been sequenced and compiled, scientists are faced with the task of identifying gene-regulatory sequences and other sequences of interest in the genome so that gene maps can be developed. This process, called **annotation**, relies heavily on bioinformatics, and a wealth of different software tools are available to carry it out.

One initial approach to annotating a sequence is to compare the newly sequenced genomic DNA to the known sequences already stored in various databases. The NCBI provides access to **BLAST (Basic Local Alignment Search Tool)**, a very popular software application for searching through

ref | NT_039455.6 | Mm8_39495_36
Mus musculus chromosome 8 genomic contig, strain C57BL/6J
Features in this part of subject sequence: insulin receptor
Score = 418 bits (226), Expect = 2e-114
Identities = 262/280 (93%), Gaps = 0/280 (0%)

```
Query   1        CAGGCCATCCCGAAAGCGAAGATCCCTTGAAGAGGTGGGCAATGTGACAGCCACTACACC   60
                 |||||||||||||||||||||||||||||||||||||||| |||||||||||||| ||||
Sbjct   174891   CAGGCCATCCCGAAAGCGAAGATCCCTTGAAGAGGTGGGGAATGTGACAGCCACCACACT   174832

Query   61       CACACTTCCAGATTTTCCCAACATCTCCTCCACCATCGCGCCCACAAGCCACGAAGAGCA   120
                 ||||||||||||| ||||| | |||||||| |||||| | ||||||||| || || |||||
Sbjct   174831   CACACTTCCAGATTTCCCCAACGTCTCCTCTACCATTGTGCCCACAAGTCAGGAGGAGCA   174772

Query   121      CAGACCATTTGAGAAAGTAGTAAACAAGGAGTCACTTGTCATCTCTGGCCTGAGACACTT   180
                 |||| ||||||||||||| || ||||||||||||||||||||||||||||||||||||||
Sbjct   174771   CAGGCCATTTGAGAAAGTGGTGAACAAGGAGTCACTTGTCATCTCTGGCCTGAGACACTT   174712

Query   181      CACTGGGTACCGCATTGAGCTGCAGGCATGCAATCAGGACTCCCCAGAAGAGAGGTGCAG   240
                 |||||||||||||||||||||||||||||||||||| || |||||||| ||||||||||||
Sbjct   174711   CACTGGGTACCGCATTGAGCTGCAGGCATGCAATCAAGATTCCCCAGATGAGAGGTGCAG   174652

Query   241      CGTGGCTGCCTACGTCAGTGCCCGGACCATGCCTGAAGGT   280
                 ||||||||||||||||||||||||||||||||||||| |||
Sbjct   174651   TGTGGCTGCCTACGTCAGTGCCCGGACCATGCCTGAAGGT   174612
```

FIGURE 21–4 BLAST results showing a 280-base sequence of a chromosome 12 contig from rats (*Rattus norvegicus*, the "query") aligned with a portion of chromosome 8 from mice (*Mus musculus*, the "subject") that contains a partial sequence for the insulin receptor gene. Vertical lines indicate exact matches. The rat contig sequence was used as a query sequence to search a mouse database in GenBank. Notice that the two sequences show 93 percent identity, strong evidence that this rat contig sequence contains a gene for the insulin receptor.

banks of DNA and protein sequence data. Using BLAST, we can compare a segment of genomic DNA to sequences throughout major databases such as GenBank to identify portions that align with or are the same as existing sequences. Figure 21–4 shows a representative example of a sequence alignment based on a BLAST search. Here a 280-bp chromosome 12 contig from the rat was used to search a mouse database to determine whether a sequence in the rat contig matched a known gene in mice. Notice that the rat contig (the query sequence in the BLAST search) aligned with base pairs 174,612 to 174,891 of mouse chromosome 8. The accession number for the mouse chromosome sequence, NT_039455.6, is indicated at the top of the figure. BLAST searches calculate a **similarity score**—also called the **identity** value—determined by the sum of identical matches between aligned sequences divided by the total number of bases aligned. Gaps, indicating missing bases in the two sequences, are usually ignored in calculating similarity scores. The aligned rat and mouse sequences were 93 percent similar and showed no gaps in the alignment. Notice that the BLAST report also provides an "Expect" value, or **E value**, based on the number of matches that would be expected by chance in the aligned sequences. Significant alignments, indicating that DNA sequences are significantly similar, have E values less than 1.0.

Because this mouse sequence on chromosome 8 is known to contain an insulin receptor gene (encoding a pro-

tein that binds the hormone insulin), it is highly likely that the rat contig sequence also contains an insulin receptor gene. We will return to the topic of similarity in Sections 21.3 and 21.6, where we consider how similarity between gene sequences can be used to infer function and to identify evolutionarily related genes through comparative genomics.

Hallmark Characteristics of a Gene Sequence Can Be Recognized during Annotation

A major limitation of this approach to annotation is that it only works if similar gene sequences are already in a database. Fortunately, it is not the only way to identify genes. Whether the genome under study is from a eukaryote or a prokaryote, several hallmark characteristics of genes can be searched for using bioinformatics software (Figure 21–5). We discussed many of these characteristics of a "typical" gene in Chapters 13 and 17. For instance, gene-regulatory sequences found upstream of genes are marked by identifiable sequences such as promoters, enhancers, and silencers. Recall from Chapter 17 that TATA box, GC box, and CAAT box sequences are often present in the promoter region of eukaryotic genes. Recall also that splice sites between **exons** and **introns** contain a predictable sequence (most introns begin with CT and end with AG) and such splice site sequences are important for determining intron and exon boundaries. Interestingly, current

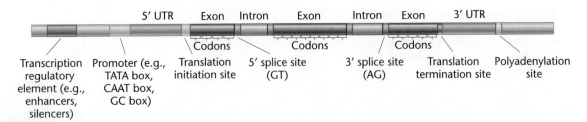

FIGURE 21–5   Characteristics of a protein-coding gene that can be used during annotation to identify a gene in an unknown sequence of genomic DNA. Most eukaryotic genes are organized into coding segments (exons) and noncoding segments (introns). When annotating a genome sequence to determine whether it contains a gene, it is necessary to distinguish between introns and exons, gene-regulatory sequences, such as promoters and enhancers, untranslated regions (UTRs), and gene termination sequences.

estimates indicate that only 6 percent of human genes are transcribed from a single, linear stretch of DNA that does not contain any introns.

Downstream elements, such as termination sequences and well-defined sequences at the end of a gene, where a polyadenylation sequence signals the addition of a poly-A tail to the 3′ end of a mRNA transcript are also important for annotation (Figure 21–5). Annotation can sometimes be a little bit easier for prokaryotic genes than for eukaryotic genes because there are no introns in prokaryotic genes. Gene-prediction programs are used to annotate sequences. These programs incorporate search elements for many of the criteria mentioned above and have become invaluable applications of bioinformatics.

Yet even with bioinformatics, identifying a gene in a particular sequence of DNA is not always straightforward, particularly when one is studying genes that do not code for proteins. In fact, a reasonable question whenever one sequences a genome is "where are the genes?" In other words, how does one know what sequences of a genome are genes and which sequences are not genes or parts of a gene? Consider the sequence presented in Figure 21–6(a),

FIGURE 21–6   Annotation of a DNA sequence containing part of the human β-globin gene. By convention, the sequence is presented in groups of ten nucleotides, although in reality the sequence is continuous. (a) The location of genes, if any, in this sequence is not readily apparent from a cursory glance. (b) The analyzed sequence, showing the location of an upstream regulatory sequence (green). The red box indicates a start triplet representing a start codon in mRNA. Open reading frames for three exons of the human β-globin gene are shown in blue. (c) Diagrammatic representation of three exons (Exons 1, 2, and 3) for the human β-globin gene encoded by the sequence shown in (a).

(a)

```
gagccacacc  ctagggttgg  ccaatctact  cccaggagca  gggagggcag  gagccagggc
tgggcataaa  agtcagggca  gagccatcta  ttgcttacat  ttgcttctga  cacaactgtg
ttcactagca  acctcaaaca  gacaccatgg  tgcacctgac  tcctgaggag  aagtctgccg
ttactgccct  gtggggcaag  gtgaacgtgg  atgaagttgg  tggtgaggcc  ctgggcaggt
tggtatcaag  gttacaagac  aggtttaagg  agaccaatag  aaactgggca  tgtggagaca
gagaagactc  ttgggtttct  gataggcact  gactctctct  gcctattggt  ctattttccc
acccttaggc  tgctggtggt  ctaccttgg  acccagaggt  tctttgagtc  ctttgggggat
ctgtccactc  ctgatgctgt  tatgggcaac  cctaaggtga  aggctcatgg  caagaaagtg
ctcggtgcct  ttagtgatgg  cctggctcac  ctggacaacc  tcaagggcac  ctttgccaca
ctgagtgagc  tgcactgtga  caagctgcac  gtggatcctg  agaacttcag  ggtgagtcta
tgggaccctt  gatgttttct  ttcccctct  tttctatggt  taagttcatg  tcataggaag
gggagaagta  acagggtaca  gtttagaatg  ggaaacagac  gaatgattgc  atcagtgtgg
aagtctcagg  atcgttttag  tttcttttat  ttgctgttca  taacaattgt  tttcttttgt
ttaattcttg  ctttcttttt  ttttcttctc  cgcaatttt   actattatac  ttaatgcctt
aacattgtgt  ataacaaaag  gaaatatctc  tgagatacat  taagtaactt  aaaaaaaac
tttacacagt  ctgcctagta  cattactatt  tggaatatat  gtgtgcttat  ttgcatattc
ataatctccc  tactttattt  tcttttattt  ttaattgata  cataatcatt  atacatattt
atgggttaaa  gtgtaatgtt  ttaatatgtg  tacacatatt  gaccaaatca  gggtaatttt
gcatttgtaa  ttttaaaaaa  tgctttcttc  ttttaatata  cttttttgtt  tatcttattt
ctaatactttt ccctaatctc  tttctttcag  ggcaataatg  atacaatgta  tcatgcctct
ttgcaccatt  ctaaagaata  acagtgataa  tttctgggtt  aaggcaatag  caatatttct
gcatataaat  atttctgcat  ataaattgta  actgatgtaa  gaggtttcat  attgctaata
gcagctacaa  tccagctacc  attctgcttt  tattttatgg  ttgggataag  gctggattat
tctgagtcca  agctaggccc  ttttgctaat  catgttcata  cctcttatct  tcctcccaca
gctcctgggc  aacgtgctgg  tctgtgtgct  ggcccatcac  tttggcaaag  aattcacccc
accagtgcag  gctgcctatc  agaaagtggt  ggctggtgtg  gctaatgccc  tggcccacaa
gtatcactaa  gctcgctttc  ttgctgtcca  atttctatta  aaggttcctt  tgttccctaa
gtccaactac  taaactgggg  gatattatga  agggccttga  gcatctggat  tctgcctaat
aaaaaacatt  tattttcatt  gcaatgatgt  atttaaatta  tttctgaata  ttttactaaa
```

(b)

```
gagccacacc  ctagggttgg  ccaatctact  cccaggagca  gggagggcag  gagccagggc
tgggcataaa  agtcagggca  gagccatcta  ttgcttacat  ttgcttctga  cacaactgtg
ttcactagca  acctcaaaca  gacaccatgg  tgcacctgac  tcctgaggag  aagtctgccg   ┐
ttactgccct  gtggggcaag  gtgaacgtgg  atgaagttgg  tggtgaggcc  ctgggcaggt   ├ Exon 1
tggtatcaag  gttacaagac  aggtttaagg  agaccaatag  aaactgggca  tgtggagaca   
gagaagactc  ttgggtttct  gataggcact  gactctctct  gcctattggt  ctattttccc
acccttaggc  tgctggtggt  ctaccttgg  acccagaggt  tctttgagtc  ctttgggggat  ┐
ctgtccactc  ctgatgctgt  tatgggcaac  cctaaggtga  aggctcatgg  caagaaagtg   ├ Exon 2
ctcggtgcct  ttagtgatgg  cctggctcac  ctggacaacc  tcaagggcac  ctttgccaca   
ctgagtgagc  tgcactgtga  caagctgcac  gtggatcctg  agaacttcag  ggtgagtcta   
tgggaccctt  gatgttttct  ttcccctct  tttctatggt  taagttcatg  tcataggaag
gggagaagta  acagggtaca  gtttagaatg  ggaaacagac  gaatgattgc  atcagtgtgg
aagtctcagg  atcgttttag  tttcttttat  ttgctgttca  taacaattgt  tttcttttgt
ttaattcttg  ctttcttttt  ttttcttctc  cgcaatttt   actattatac  ttaatgcctt
aacattgtgt  ataacaaaag  gaaatatctc  tgagatacat  taagtaactt  aaaaaaaac
tttacacagt  ctgcctagta  cattactatt  tggaatatat  gtgtgcttat  ttgcatattc
ataatctccc  tactttattt  tcttttattt  ttaattgata  cataatcatt  atacatattt
atgggttaaa  gtgtaatgtt  ttaatatgtg  tacacatatt  gaccaaatca  gggtaatttt
gcatttgtaa  ttttaaaaaa  tgctttcttc  ttttaatata  cttttttgtt  tatcttattt
ctaatactttt ccctaatctc  tttctttcag  ggcaataatg  atacaatgta  tcatgcctct
ttgcaccatt  ctaaagaata  acagtgataa  tttctgggtt  aaggcaatag  caatatttct
gcatataaat  atttctgcat  ataaattgta  actgatgtaa  gaggtttcat  attgctaata
gcagctacaa  tccagctacc  attctgcttt  tattttatgg  ttgggataag  gctggattat
tctgagtcca  agctaggccc  ttttgctaat  catgttcata  cctcttatct  tcctcccaca
gctcctgggc  aacgtgctgg  tctgtgtgct  ggcccatcac  tttggcaaag  aattcacccc  ┐
accagtgcag  gctgcctatc  agaaagtggt  ggctggtgtg  gctaatgccc  tggcccacaa  ├ Exon 3
gtatcactaa  gctcgctttc  ttgctgtcca  atttctatta  aaggttcctt  tgttccctaa  
gtccaactac  taaactgggg  gatattatga  agggccttga  gcatctggat  tctgcctaat
aaaaaacatt  tattttcatt  gcaatgatgt  atttaaatta  tttctgaata  ttttactaaa
```

(c)

```
[EXON 1]  [EXON 2]                                    [EXON 3]
|___|___|___|___|___|___|___|___|___|___|___|___|___|___|___|___| kb
0.0          0.5              1.0              1.5
```

which shows a portion of the human genome. From a casual inspection, it is not clear whether this sequence contains any genes and, if so, how many. Analysis of the sequence, however, reveals identifiable features that provide clues to the presence of a protein-coding gene. In addition, protein-coding genes contain one or more **open reading frames (ORFs)**, sequences of triplet nucleotides that, after transcription and mRNA splicing, are translated into the amino acid sequence of a protein. ORFs typically begin with an initiation sequence, usually ATG, which transcribes into the AUG start codon of an mRNA molecule, and end with a termination sequence, TAA, TAG, or TGA, which correspond to the stop codons of UAA, UAG, and UGA in mRNA. Genetic information is encoded in groups of three nucleotides (triplets), but it is not always clear whether to begin the analysis of a sequence at the first nucleotide, the second, or the third. Typically, the sequence adjacent to a promoter is examined for a start (initiation) triplet; however, ORFs can be used to identify a gene even when a promoter sequence is not apparent. Software programs can then analyze the ORFs three nucleotides at a time. The discovery of an ORF starting with an ATG followed at some distance by a termination sequence is usually a good indication that the coding region of a gene has been identified.

The way genes are organized in eukaryotic genomes (including the human genome) makes direct searching for ORFs more difficult in them than in prokaryotic genomes. First, many eukaryotic genes have introns. As a result, many, if not most, eukaryotic genes are not organized as continuous ORFs; instead, the gene sequences consist of ORFs (exons) interspersed with introns. Second, genes in humans and other eukaryotes are often widely spaced, increasing the chances of finding false ORFs in the regions between gene clusters.

Annotation of the sequence shown in Figure 21–6(a) reveals several identifiable indicators that the sequence contains a protein-coding gene: it includes a promoter sequence, an initation codon, and three *exons* [Figure 21–6(b)]. The two unshaded regions between the exons represent introns that would be spliced out following transcription when the mRNA is processed [Figure 21–6(b, c)]. Using this sequence as the query in a search of genomic databases would reveal that it is the sequence of a single gene, the human β-globin gene.

Software designed for ORF analysis of eukaryotic genomes is highly valuable. Often such programs are used to make computational predictions of all ORFs (the ORFeome!) in a sequenced genome as a way to estimate the number of potential protein-coding genes in a genome. In addition to the features already mentioned, such software can be used to "translate" ORFs into possible polypeptide sequences as a way to predict the polypeptide encoded by a gene. Shown in **Figure 21–7** is a partial sequence for the first exon of the human

(a) *Homo sapiens TUBA3C* (bp 1-300)

```
  1 ggttgaggtcaagtagtagcgttgggctgcggcagcggaggagctcaacatgcgtgagtg
 61 tatctctatccacgtgggcaggcaggagtccagatcggcaatgcctgctgggaactgta
121 ctgcctggaacatggaattcagcccgatggtcagatgccaagtgataaaaccattggtgg
181 tggggacgactccttcaacacgttcttcagtgagactggagctggcaagcacgtgcccag
241 agcagtgtttgtggacctggagcccactgtggtcgatgaagtgcgcacaggaacctatag (300)
```

(b) Predicted polypeptides

5' to 3' Frame 1
G **Stop** G Q V V A L G C G S G G A Q H A **Stop** V Y L Y P R G A G R S P D R Q C L L G T V L P G T W N S A R W S D A K **Stop** **Stop** N H W W W G R L L Q H V L Q **Stop** D W S W Q A R A Q S S V C G P G A H C G R **Stop** S A H R N L **Stop**

5' to 3' Frame 2
V E V K **Stop** **Stop** R W A A A A E E L N **Met** R E C I S I H V G Q A G V Q I G N A C W E L Y C L E H G I Q P D G Q **Met** P S D K T I G G G D D S F N T F F S E T G A G K H V P R A V F V D L E P T V V D E V R T G T Y

5' to 3' Frame 3
L R S S S S V G L R Q R R S S T C V S V S L S T W G R Q E S R S A **Met** P A G N C T A W N **Met** E F S P **Met** V R C Q V I K P L V V G T T P S T R S S V R L E L A S T C P E Q C L W T W S P L W S **Met** K C A Q E P I

3' to 5' Frame 1
L **Stop** V P V R T S S T T V G S R S T N T A L G T C L P A P V S L K N V L K E S S P P P **Met** V L S L G I **Stop** P S G **Stop** I P C S R Q Y S S Q Q A L P I W T P A C P T W I E I H S R **Met** L S S S A A A A Q R Y Y L T S T

3' to 5' Frame 2
Y R F L C A L H R P Q W A P G P Q T L L W A R A C Q L Q S H **Stop** R T C **Stop** R S R P H H Q W F Y H L A S D H R A E F H V P G S T V P S R H C R S G L L P A P R G **Stop** R Y T H A C **Stop** A P P L P Q P N A T T **Stop** P Q

3' to 5' Frame 3
I G S C A H F I D H S G L Q V H K H C S G H V L A S S S L T E E R V E G V V P T T N G F I T W H L T I G L N S **Met** F Q A V Q F P A G I A D L D S C L P H V D R D T L T H V E L L R C R S P T L L L D L N

FIGURE 21–7 Predicted polypeptide sequences translated from potential ORFs in the human *TUBA3C* gene. (a) Nucleotides 1–300 of the first exon in the human *TUBA3C* gene. (b) A translation program predicts six possible polypeptide sequences from this exon. Which predicted sequence is correct?

Note: Methionine is highlighted using the three letter amino acid code (Met).

tubulin alpha 3c gene (*TUBA3C*). Prediction programs scan potential ORFs in the 5' to 3' direction on both strands of a section of genomic DNA to predict possible reading frames in each direction. Figure 21–7 shows the results for the six possible reading frames in the sequence of interest. Amino acids are shown using the single-letter code for each residue. Notice the very different results obtained for each of the six frames. For instance the 5' to 3' ORF 1 contains several stop codons interspersed among amino acids but no methionine residues that are evidence of a start codon. Other ORFs would contain too many methionines to produce a functional polypeptide. For this exon of *TUBA3C*, the 5' to 3' ORF 2 is correct.

Prediction programs can also search for **codon bias**, the more frequent use of one or two codons to encode an amino acid that can be specified by a number of different codons. For example, alanine can be encoded by GCA, GCT, GCC, and GCG. If the codons were used randomly, each would be used about 25 percent of the time. Yet in the human genome, GCC is used 41 percent of the time, and GCG only 11 percent of the time. Codon bias is present in exons but should not be present in introns or intergenic spacers.

NOW SOLVE THIS

21–2 The process of annotating a sequenced genome is continual. When the first annotated sequence of the *Drosophila* genome was released, it predicted 13,601 protein-coding genes within the euchromatic region of the genome. Shown here are selected data from the next two annotated versions that were released (modified from Misra et al., 2002. *genomebiology3(12)@genomebiology.com/2002/3/12/RESEARCH/ 0083*).

(a) Assuming a uniform distribution in Release 3, approximately how many base pairs of DNA lie between protein-coding genes in *Drosophila*?

(b) On average, approximately how many exons are reported per gene in Release 3?

(c) Approximately how many introns are there per gene?

(d) What appears to be the most significant difference between Release 2 and Release 3?

Criteria	Release 2	Release 3
Total length of euchromatin	116.2 Mb	116.8 Mb
Total protein-coding genes	13,474	13,379
Protein-coding exons	50,667	54,934
Introns	48,381	48,257
Genes with alternative transcripts	689	2729

■ HINT: *This problem asks you specific questions about two releases of the annotated genome of Drosophila. The key to its solution is to consider potential relationships between protein-coding genes, the number of protein-coding exons, and alternative splicing.*

21.3

Functional Genomics Attempts to Identify Potential Functions of Genes and Other Elements in a Genome

Reading a genome sequence is a surefire cure for insomnia. What is exciting is not the sequence of the nucleotides but the information that the sequence contains. After a genome has been annotated and ORFs have been identified, the next analytical task is to assign putative functions to all possible genes in the sequence. As the term suggests, **functional genomics** is the study of gene functions, based on the resulting RNAs or possible proteins they encode, and the functions of other components of the genome, such as gene-regulatory elements. Functional genomics can involve experimental approaches to confirm or refute computational predictions about genome functions (such as the number of protein-coding genes), and it also considers how genes are expressed and the regulation of gene expression.

Predicting Gene and Protein Functions by Sequence Analysis

Some newly identified genomic sequences may already have had functions assigned to their genes by classic methods such as mutagenesis and linkage mapping, but many other genes that have been sequenced have not yet been correlated with a function. One approach to assigning functions to genes is to use sequence similarity searches, as described in the previous section. Programs such as BLAST are used to search through databases to find alignments between the newly sequenced genome and genes that have already been identified, either in the same or in different species. You were introduced to this approach for predicting gene function in Figure 21–4, when we demonstrated how sequence similarity to the mouse gene was used to identify a gene in a rat contig as the insulin receptor gene. Inferring gene function from similarity searches is based on a relatively simple idea. If a genome sequence shows statistically significant similarity to the sequence of a gene whose function is known, then it is likely that the genome sequence encodes a protein with a similar or related function.

For example, Figure 21–8 indicates functional categories that have been assigned to genes in the small flowering plant *Arabidopsis thaliana* genome on the basis of similarity searches. To geneticists, *Arabidopsis* is the fruit fly of the plant world. *Arabidopsis* is small, has a short generation time, and a relatively small genome distributed on five chromosomes. More importantly, the plant kingdom evolved independently from the animal kingdom, and analysis of flowering plants can provide insight into the ways in which evolutionary and coevolutionary adaptations have shaped their genomes. At least half the genes in the *Arabidopsis* genome are identical to or closely related to genes found in bacteria and humans, but it also contains genes encoding dozens of protein families that may be unique to plants.

Another major benefit of similarity searches is that they are often able to identify **homologous genes**, genes that are evolutionarily related. After the human genome was sequenced, many ORFs in it were identified as protein-coding genes based on their alignment with related genes of known function in other species. As an example, Figure 21–9 compares portions of the human leptin gene (*LEP*) with its homolog in mice (*ob/ Lep*). These two genes are over 85 percent identical in sequence. The leptin gene was first discovered in mice. The match between the *LEP*-containing DNA sequence in humans and the mouse homolog sequence confirms the identity and leptin-coding function of this gene in human genomic DNA.

As an interesting aside, the leptin gene (also called *ob*, for obesity, in mice) is highly expressed in fat cells (adipocytes). This gene produces the protein hormone leptin, which targets cells in the brain to suppress appetite. Knockout mice lacking a functional *ob* gene grow dramatically overweight. A similar phenotype has been observed in small numbers

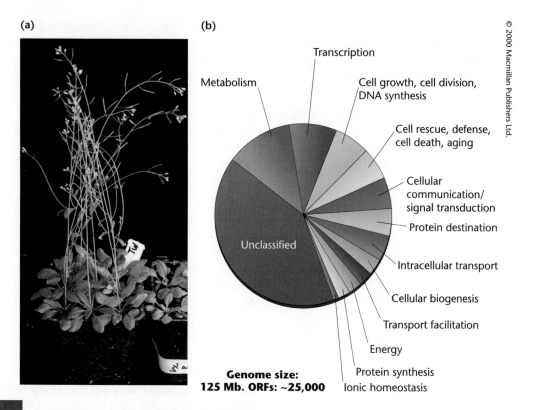

(a)   (b)

Transcription

Metabolism

Cell growth, cell division, DNA synthesis

Cell rescue, defense, cell death, aging

Cellular communication/ signal transduction

Protein destination

Unclassified

Intracellular transport

Cellular biogenesis

Transport facilitation

Energy

Protein synthesis

Ionic homeostasis

Genome size: 125 Mb. ORFs: ~25,000

FIGURE 21–8   *Arabidopsis thaliana* (a), a small plant used as a genetic model for studying genome organization and development of flowering plants. (b) Assignment of *Arabidopsis* genes to functional categories based on homology searches.

of humans with particular mutations in *LEP*. Although it is important to note that weight control is not regulated by a single gene, the discovery of leptin has provided significant insight into lipid metabolism and weight disorders in humans. Further studies on leptin will be important for understanding more about the genetics of weight disorders.

If homologous genes in different species are thought to have descended from a gene in a common ancestor, the genes are known as **orthologs**. In Section 21.6 we will consider the globin gene family. Mouse and human α-globin genes are orthologs evolved from a common ancestor. Homologous genes in the same species are called **paralogs**. The α- and β-globin subunits in humans are paralogs resulting

from a gene-duplication event. Paralogs often have similar or identical functions.

Predicting Function from Structural Analysis of Protein Domains and Motifs

When a gene sequence is used to predict a polypeptide sequence, the polypeptide can be analyzed for specific structural domains and motifs. Identification of **protein domains**, such as ion channels, membrane-spanning regions, DNA-binding regions, secretion and export signals, and other structural aspects of a polypeptide that are encoded by a DNA sequence, can in turn be used to predict protein function. Recall from Chapter 17, for example, that the structures of many DNA-binding proteins have characteristic patterns, or **motifs**, such as the helix-turn-helix, leucine zipper, or zinc-finger motifs. These motifs can often easily be searched for using bioinformatics software, and their identification in a sequence is a common strategy for inferring the possible functions of a protein.

Human *LEP* gene

```
GTCACCAGGATCAATGACATTTCACACACG- - -TCAGTCTCCTCCAAACAGAAAGTCACC
|||||||||||||||||||||||||||||    || || ||| |||| |||| ||||||
GTCACCAGGATCAATGACATTTCACACACGCAGTCGGTATCCGCCAAGCAGAGGGTCACT
```
Mouse *ob* gene

```
GGTTTGGACTTCATTCCTGGGCTCCACCCCATCCTGACCTTATCCAAGATGGACCAGACA
|| |||||||||||||||||||| ||||||| |||| || ||||||||||||||||||
GGCTTGGACTTCATTCCTGGGCTTCACCCCATTCTGAGTTTGTCCAAGATGGACCAGACT
```

```
CTGGCAGTCTACCAACAGATCCTCACCAGTATGCCTTCCAGAAACGTGATCCAAATATCC
||||||||||| |||||| |||||||||  ||||||||  ||| ||| | || ||| ||
CTGGCAGTCTATCAACAGGTCCTCACCAGCCTGCCTTCCCAAAATGTGCTGCAGATAGCC
```

FIGURE 21–9   Comparison of the human *LEP* and mouse *ob/ Lep* genes. Partial sequences for these homologs are shown with the human *LEP* gene on top and the mouse *ob/Lep* gene sequence below it. Notice from the number of identical nucleotides, indicated by vertical lines, that the nucleotide sequence for these two genes is very similar. Gaps are indicated by horizontal dashes.

Investigators Are Using Genomics Techniques Such as Chromatin Immunoprecipitation to Investigate Aspects of Genome Function and Regulation

In this chapter and in Chapter 22 we will consider a range of different genomic techniques that investigators are using that are valuable for functional genomics studies. One example is a technique called **chromatin immunoprecipitation (ChIP),** and another is a related technique that couples ChIP to microarrays or gene chips, called **ChIP-on-chip.** These techniques are designed to map protein–DNA interactions and are useful for identifying genes that are regulated by DNA-binding transcription factors.

To perform these techniques, researchers treat tissues or single cells such as yeast with formaldehyde (Figure 21–10).

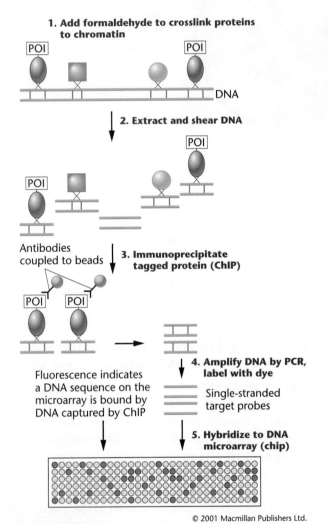

© 2001 Macmillan Publishers Ltd.

FIGURE 21–10   Genome-wide screens for transcription factor binding sites can be carried out using the ChIP-on-chip technique. In this method, formaldehyde is added to tissues or cultured cells to crosslink DNA-binding proteins currently attached to chromatin when formaldehyde was added. Then the DNA is extracted from cells and sheared into small fragments. An antibody or antibodies that recognize specific DNA binding proteins of interest (POI), such as a transcription factor, are added to the mixture and the antibodies attach to the POI. Then the antibody, together with its protein-DNA fragment, is pulled out of the mixture (immunoprecipitated). The immunoprecipitated DNA fragments are purified and often amplified by PCR, then labeled with a fluorescent dye, and hybridized to a DNA microarray. A positive hybridization signal indicates a DNA sequence that was bound to the POI.

Any proteins that are tightly bound to DNA will be crosslinked to the DNA by the formaldehyde. The researchers then extract DNA from the cells and shear it into small fragments. To isolate those DNA fragments that are bound to a specific protein, they add an antibody that recognizes (attaches to) the protein. These antibodies are usually attached to a bead or resin that enables the antibody to then be precipitated (by immunoprecipitation) from the mixture, along with the protein to which it binds and any DNA fragments that are crosslinked to the protein. Immunoprecipitation is usually accomplished by centrifugation of the antibody/protein/DNA complexes. Because the antibody has a bead attached to it, the mass of the bead is used to centrifuge the antibody/protein/DNA complexes to the bottom of centrifuge tube. The investigators can then purify the immunoprecipitated DNA fragments by removing them from the antibody/protein complex, amplify these DNA fragments by PCR, and label the amplified fragments with a fluorescent tag.

The labeled DNA fragments are hybridized to a DNA microarray (described in Section 21.9) containing synthetic oligonucleotides or cloned DNAs representing the organism's entire genome. Any spot on the microarray that hybridizes to the labeled DNA represents a DNA sequence that bound the protein. Because each spot on the DNA microarray is known, the identity of each positive signal can be determined. DNA isolated by ChIP can also be sequenced to identify DNA sequences that are bound by DNA binding proteins. In one application of this genome-wide method, more than 200 previously unknown targets of a transcription activator protein that functions at the G1/S cell-cycle interface were identified.

21.4

The Human Genome Project Revealed Many Important Aspects of Genome Organization in Humans

Now that you have a general idea of the strategies used for analyzing a genome, let's look at the largest genomics project completed to date. The **Human Genome Project (HGP)** was a coordinated international effort to determine the sequence of the human genome and to identify all the genes it contains. It has produced a plethora of information, much of which is still being analyzed and interpreted. What is clear so far, from all the different kinds of genomes sequenced, is that humans and all other species share a common set of genes essential for cellular function and reproduction, confirming that all living organisms arose from a common ancestor.

Origins of the Project

The publicly funded Human Genome Project began in 1990 under the direction of James Watson, the co-discoverer of the double-helix structure of DNA. Eventually the public project was led by Dr. Francis Collins, who had previously led a research team involved in identifying the *CFTR* gene as the cause of cystic fibrosis. In the United States, the Collins-led HGP was coordinated by the Department of Energy and the National Center of Human Genome Research, a division of the National Institutes of Health. It established a 15-year plan with a proposed budget of $3 billion to identify all human genes, originally thought to number between 80,000 and 100,000, to sequence and map them all, and to sequence the approximately 3 billion base pairs thought to comprise the 24 chromosomes (22 autosomes, plus X and Y) in humans. Other primary goals of the HGP included the following:

- To establish functional categories for all human genes

- To analyze genetic variations between humans, including the identification of single-nucleotide polymorphisms (SNPs)

- To map and sequence the genomes of several model organisms used in experimental genetics, including *E. coli, S. cerevisiae, C. elegans, D. melanogaster,* and *M. musculus* (mouse)

- To develop new sequencing technologies, such as high-throughput computer-automated sequencers, in order to facilitate genome analysis

- To disseminate genome information among both scientists and the general public

Lastly, to deal with the impact that genetic information would have on society, the HGP set up the **ELSI program** (standing for Ethical, Legal, and Social Implications) to consider ethical, legal, and social issues arising from the HGP and to ensure that personal genetic information would be safeguarded and not used in discriminatory ways.

As the HGP grew into an international effort, scientists in 18 countries were involved in the project. Much of the work was carried out by the International Human Genome Sequence Consortium, involving nearly 3000 scientists working at 20 centers in six countries (China, France, Germany, Great Britain, Japan, and the United States).

In 1999, a privately funded human genome project led by J. Craig Venter at **Celera Genomics** (aptly named from a word meaning "swiftness") was announced. Celera's goal was to use whole-genome shotgun sequencing and computer-automated high-throughput DNA sequencers to sequence the human genome more rapidly than HGP. The public project had proposed using a clone-by-clone approach to sequence the genome. Recall that Venter and

colleagues had proven the potential of shotgun sequencing in 1995 when they completed the genome for *H. influenzae*. Celera's announcement set off an intense competition between the two teams, which both aspired to be first with the human genome sequence. This contest eventually led to the HGP finishing ahead of schedule and under budget after scientists from the public project began to use high-throughput sequencers and whole-genome sequencing strategies as well.

Major Features of the Human Genome

In June 2000, the leaders of the public and private genome projects met at the White House with President Clinton and jointly announced the completion of a draft sequence of the human genome. In February 2001, they each published an analysis covering about 96 percent of the euchromatic region of the genome. The public project sequenced euchromatic portions of the genome 12 times and set a quality control standard of a 0.01 percent error rate for their sequence. Although this error rate may seem very low, it still allows about 600,000 errors in the human genome sequence. Celera sequenced certain areas of the genome more than 35 times when compiling the genome.

The remaining work of completing the sequence by filling in gaps clustered around centromeres, telomeres, and repetitive sequences, correcting misaligned segments, and re-sequencing portions of the genome to ensure accuracy. In 2003 genome sequencing and error-fixing were deemed sufficient to pass the international project's definition of completion—that it contained fewer than 1 error per 10,000 nucleotides and that it covered 95 percent of the gene-containing portions of the genome. Yet even at the time of "completion" there were still some 350 gaps in the sequence that continue to be worked on.

And of course the HGP did not sequence the genome of every person on Earth. The assembled genomes largely consist of haploid genomes pooled from different individuals so that they provide a *reference genome* representative of major, common elements of a human genome widely shared among populations of humans. Examples of major features of the human genome are summarized in Table 21.1. As you can see in this table, many unexpected observations have provided us with major new insights. The genome is not static! Genome variations, including the abundance of repetitive sequences scattered throughout the genome, verify that the genome is indeed dynamic, revealing many evolutionary examples of sequences that have changed in structure and location. In many ways, the HGP has revealed just how little we know about our genome.

Two of the biggest surprises discovered by the HGP were that less than 2 percent of the genome codes for proteins and that there are only around 20,000 protein-coding

TABLE 21.1

Major Features of the Human Genome

- The human genome contains 3.1 billion nucleotides, but protein-coding sequences make up only about 2 percent of the genome.

- The genome sequence is ~99.9 percent similar in individuals of all nationalities. Single-nucleotide polymorphisms (SNPs) and copy number variations (CNVs) account for genome diversity from person to person.

- The genome is dynamic. At least 50 percent of the genome is derived from transposable elements, such as LINE and *Alu* sequences, and other repetitive DNA sequences.

- The human genome contains approximately 20,000 protein-coding genes, far fewer than the predicted number of 80,000–100,000 genes.

- The average size of a human gene is ~25 kb including gene regulatory regions, introns, and exons. On average, mRNAs produced by human genes are ~3000 nt long.

- Many human genes produce more than one protein through alternative splicing, thus enabling human cells to produce a much larger number of proteins (perhaps as many as 200,000) from only ~20,000 genes.

- More than 50 percent of human genes show a high degree of sequence similarity to genes in other organisms; however, more than 40 percent of the genes identified have no known molecular function.

- Genes are not uniformly distributed on the 24 human chromosomes. Gene-rich clusters are separated by gene-poor "deserts" that account for 20 percent of the genome. These deserts correlate with G bands seen in stained chromosomes. Chromosome 19 has the highest gene density, and chromosome 13 and the Y chromosome have the lowest gene densities.

- Chromosome 1 contains the largest number of genes, and the Y chromosome contains the smallest number.

- Human genes are larger and contain more and larger introns than genes in the genomes of invertebrates, such as *Drosophila*. The largest known human gene encodes dystrophin, a muscle protein. This gene, associated in mutant form with muscular dystrophy, is 2.5 Mb in length (Chapter 14), larger than many bacterial chromosomes. Most of this gene is composed of introns.

- The number of introns in human genes ranges from 0 (in histone genes) to 234 (in the gene for *titin*, which encodes a muscle protein).

genes. Recall that the number of genes had originally been estimated to be about 100,000, based in part on a prediction that human cells produce about 100,000 proteins. At least half of the genes show sequence similarity to genes shared by many other organisms, and as you will learn in Section 21.7, a majority of human genes are similar in sequence to genes from closely related species such as chimpanzees. The exact number of human genes is still not certain. One reason

is that it is unclear whether or not many of the presumed genes produce functional proteins. Genome scientists continue to annotate the genome, and as mentioned earlier, functional genomics studies have important roles in determining whether or not computational predictions about the number of protein-coding and non–protein-coding genes are accurate.

The number of genes is much lower than the number of predicted proteins in part because many genes code for multiple proteins through **alternative splicing**. Recall from Chapter 13 that alternative splicing patterns can generate multiple mRNA molecules, and thus multiple proteins, from a single gene, through different combinations of intron–exon splicing arrangements. Initial estimates suggested that over 50 percent of human genes undergo alternative splicing to produce multiple transcripts and multiple proteins. Recent studies suggest that ~94–95 percent of human pre-mRNAs contain multiple exons that are processed to produce multiple transcripts and potentially multiple different protein products. Clearly alternative splicing produces an incredible diversity of proteins beyond simple predictions based on the number of genes in the human genome.

Functional categories have been assigned for human genes, primarily on the basis of (1) functions determined previously (for example, from recombinant DNA cloning of human genes and known mutations involved in human diseases), (2) comparison to known genes and predicted protein sequences from other species, and (3) predictions based on annotation and analysis of protein functional domains and motifs (Figure 21–11). Although functional categories and assignments continue to be revised, the functions of over 40 percent of human genes remain unknown. Determining human gene functions, deciphering complexities of gene-expression regulation and gene interaction, and uncovering the relationships between human genes and phenotypes are among the many challenges for genome scientists.

The HGP has also shown us that in all humans, regardless of racial and ethnic origins, the genomic sequence is approximately 99.9 percent the same. As we discuss in other chapters most genetic differences between humans result from **single-nucleotide polymorphisms (SNPs)** and **copy number variations (CNVs)**. Recall that SNPs are single base changes in the genome and variations of many SNPs are associated with disease conditions. For example, SNPs cause sickle-cell anemia and cystic fibrosis. In Chapters 22 and 29 we will examine how SNPs can be detected and used for diagnosis and treatment of disease.

After the draft sequence of the human genome was completed, it initially appeared that most genetic variations between individuals (the 0.1 percent differences) were due to SNPs. While SNPs are important contributing factors to genome variation, structural differences that we discussed in Chapter 12 such as deletions, duplications, inversions,

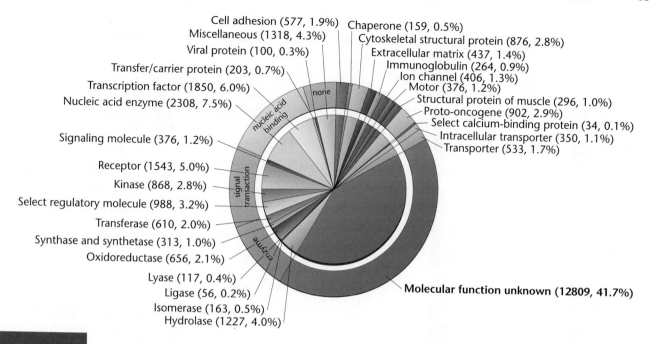

Cell adhesion (577, 1.9%)
Miscellaneous (1318, 4.3%)
Viral protein (100, 0.3%)
Transfer/carrier protein (203, 0.7%)
Transcription factor (1850, 6.0%)
Nucleic acid enzyme (2308, 7.5%)
Signaling molecule (376, 1.2%)
Receptor (1543, 5.0%)
Kinase (868, 2.8%)
Select regulatory molecule (988, 3.2%)
Transferase (610, 2.0%)
Synthase and synthetase (313, 1.0%)
Oxidoreductase (656, 2.1%)
Lyase (117, 0.4%)
Ligase (56, 0.2%)
Isomerase (163, 0.5%)
Hydrolase (1227, 4.0%)

Chaperone (159, 0.5%)
Cytoskeletal structural protein (876, 2.8%)
Extracellular matrix (437, 1.4%)
Immunoglobulin (264, 0.9%)
Ion channel (406, 1.3%)
Motor (376, 1.2%)
Structural protein of muscle (296, 1.0%)
Proto-oncogene (902, 2.9%)
Select calcium-binding protein (34, 0.1%)
Intracellular transporter (350, 1.1%)
Transporter (533, 1.7%)

none
nucleic acid binding
signal transduction
enzyme

Molecular function unknown (12809, 41.7%)

FIGURE 21–11 A preliminary list of the functional categories to which genes in the human genome have been assigned on the basis of similarity to proteins of known function. Among the most common genes are those involved in nucleic acid metabolism (7.5 percent of all genes identified), transcription factors (6.0 percent), receptors (5 percent), hydrolases (4 percent), protein kinases (2.8 percent), and cytoskeletal structural proteins (2.8 percent). A total of 12,809 predicted proteins (41 percent) have unknown functions, indicative of the work that is still needed to fully decipher our genome.

and CNVs which can span millions of bp of DNA, play much more important roles in genome variation than previously thought. As we discussed in Chapters 8 and 12, recall that CNVs are duplications or deletions of relatively large sections of DNA on the order of several hundred or several thousand base pairs. Many of the CNVs that vary the most among genomes appear to be at least 1 kilobase.

Although most human DNA is present in two copies per cell, one from each parent, CNVs are segments of DNA that are duplicated or deleted, resulting in variations in the number of copies of a DNA segment inherited by individuals. In some cases CNVs are major deletions removing entire genes, other deletions affect gene function by frameshifts in the reading code. CNV sequences that are duplicated can result in overexpression of a particular gene, yet many deleted and duplicated CNVs do not present clearly identifiable phenotypes.

Current estimates of the number of CNVs in an individual genome range from about 12 CNVs to perhaps 4–5 dozen per person. Some studies estimate that there may be as many as 1500 CNVs greater than 1 kb among the human genome. Other studies claim there are more than 1.5 million deletions of less than 100 bp that contribute to genome variation between individuals.

It is now possible to access databases and other sites on the Internet that display maps for all human chromosomes. You will visit a number of these databases in Exploring Genomics exercises. Figure 21–12(a) displays a partial gene map for chromosome 12 that was taken from an NCBI database called Map Viewer. You may already have used Map Viewer for the Exploring Genomics exercises in Chapters 5 and 12. This image shows an ideogram, or cytogenetic map, of chromosome 12. To the right of the ideogram is a column showing the contigs (arranged lying vertically) that were aligned to sequence this chromosome. The Hs UniG column displays a histogram representation of gene density on chromosome 12. Notice that relatively few genes are located near the centromere. Gene symbols, loci, and gene names (by description) are provided for selected genes; in this figure only 20 genes are shown. When accessing these maps on the Internet, one can magnify, or zoom in on, each region of the chromosome, revealing all genes mapped to a particular area.

You can see that most of the genes listed here have been assigned descriptions based on the functions of their products, some of which are transmembrane proteins, some enzymes such as kinases, some receptors, including several involved in olfaction, and so on. Other genes are described in terms of hypothetical products; they are presumed to be genes based on the presence of ORFs, but their function remains unknown [Figure 21–12(a)].

The HGP's most valuable contribution will perhaps be the identification of disease genes and the development of new treatment strategies as a result. Thus, extensive maps have been developed for genes implicated in human disease conditions. The disease gene map of chromosome 21 shown

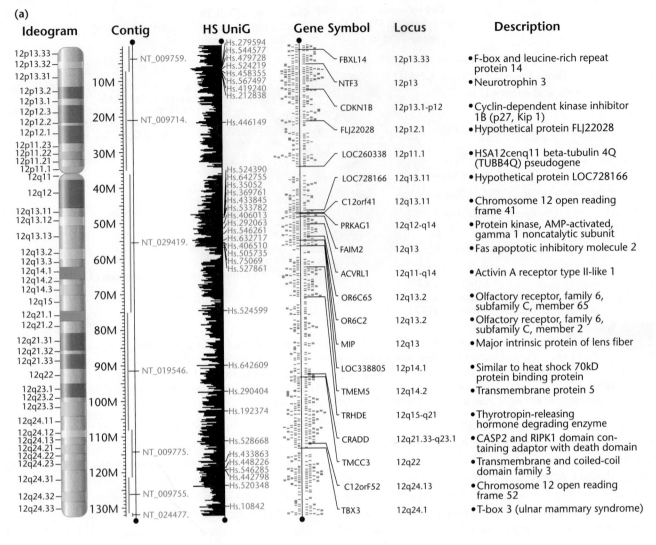

FIGURE 21–12   (a) A gene map for chromosome 12 from the NCBI database Map Viewer. (b) Partial map of disease genes on human chromosome 21. Maps such as this depict genes thought to be involved in human genetic disease conditions.

in Figure 21–12(b) indicates genes involved in amyotrophic lateral sclerosis (ALS), Alzheimer disease, cataracts, deafness, and several different cancers. In Chapter 22 we discuss implications of the HGP for the identification of genes involved in human genetic diseases, and for disease diagnosis, detection, and gene therapy applications.

21.5

The "Omics" Revolution Has Created a New Era of Biological Research

The Human Genome Project and the development of genomics techniques have been largely responsible for launching a new era of biological research—the era of "omics." It seems that every year, more areas of biological research are being described as having an omics connection. Some examples of "omics" are

- proteomics—the analysis of all the proteins in a cell or tissue

- metabolomics—the analysis of proteins and enzymatic pathways involved in cell metabolism

- glycomics—the analysis of the carbohydrates of a cell or tissue

- toxicogenomics—the analysis of the effects of toxic chemicals on genes, including mutations created by toxins and changes in gene expression caused by toxins

- metagenomics—the analysis of genomes of organisms collected from the environment

- pharmacogenomics—the development of customized medicine based on a person's genetic profile for a particular condition

- transcriptomics—the analysis of all expressed genes in a cell or tissue

We will consider several of these genomics disciplines in other parts of this chapter.

As evidence of the impact of genomics, a new field of nutritional science called nutritional genomics, or **nutrigenomics**, has emerged. Nutrigenomics focuses on understanding the interactions between diet and genes. We have all had routine medical tests for blood pressure, blood sugar levels, and heart rate. Based on these tests, your physician may recommend that you change your diet and exercise more to lose weight, or that you reduce your intake of sodium to help lower your blood pressure. Now several companies claim to provide nutrigenomics tests that analyze your genomes for genes thought to be associated with

different medical conditions or aspects of nutrient metabolism. The companies then provide a customized nutrition report, recommending diet changes for improving your health and preventing illness, based on your genes! It remains to be seen whether this approach as currently practiced is of valid scientific or nutritional value.

Stone-Age Genomics

In yet another example of how genomics has taken over areas of DNA analysis, a number of labs around the world are involved in analyzing "ancient" DNA. These so-called **stone-age genomics** studies are generating fascinating data from miniscule amounts of ancient DNA obtained from bone and other tissues such as hair that are tens of thousands to about 100,000 years old, and often involve samples from extinct species. Analysis of DNA from a 2400-year-old Egyptian mummy, bison, mosses, platypus, mammoths, Pleistocene-age cave bears and polar bears, and Neanderthals are some of the most prominent examples of stone-age genomics.

In 2005, researchers from McMaster University in Canada and Pennsylvania State University published about 13 million bp from a 27,000-year-old woolly mammoth. This study revealed a ~98.5 percent sequence identity between mammoths and African elephants. Subsequent studies by other scientists have used whole-genome shotgun sequencing of mitochondrial and nuclear DNA from Siberian mammoths to provide data on the mammoth genome. These studies suggest that the mammoth genome differs from the African elephant by as little as 0.6 percent. These studies are also great demonstrations of how stable DNA can be under the right conditions, particularly when frozen.

Perhaps even more intriguing are similarities that have been revealed between the mammoth and human genomes. For example, as shown in Figure 21–13, when the gene sequences from human chromosomes were aligned with sequences from the mammoth genome, approximately 50 percent of mammoth genes showed sequence alignment with human genes on autosomes. Incidentally, notice that this figure also shows the relative number of genes from James Watson's genome (which we will discuss in the next section) compared to the human genome reference sequence.

In Section 21.6 we will discuss recent work on the Neanderthal genome. Obtaining the genome of a human ancestor this old was previously unimaginable, and this work is providing new insights into our understanding of human evolution.

10 Years after the HGP: What Is Next?

In the ten years since completion of a draft sequence of the human genome, studies on the human genome continue at a rapid pace. For example, as the HGP was being completed, a group of about three dozen research teams around the world

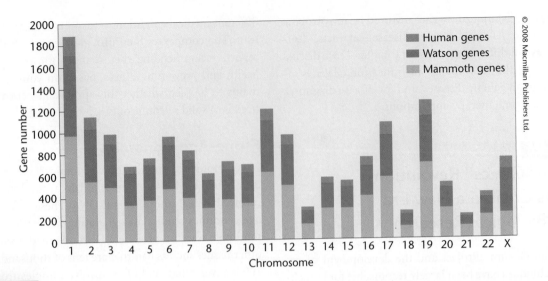

FIGURE 21–13 Plot showing the number of genes on each human chromosome (blue), the average fraction of protein-coding bases that align to Roche 454 reads from James D. Watson's genome (green), and the fraction of coding bases that align to one or more mammoth reads (orange), using predicted elephant genes that map to the human chromosome based on sequence similarity—approximately 50 percent for each autosome, but only 31 percent for the X chromosome because the mammoth used for this study was male.

began the **Encyclopedia of DNA Elements (ENCODE) Project**. The main goal of ENCODE is to use both experimental approaches and bioinformatics to identify and analyze functional elements (such as transcriptional start sites, promoters, and enhancers) that regulate expression of human genes. Prior to ENCODE approximately 532 promoters had been identified, but now in excess of 775 promoters have been identified in the human genome, with many other potential promoter sequences being analyzed.

As a result of the HGP, many other major theme areas for human genome research have emerged including a cancer genome project, analysis of the epigenome (including a Human Epigenome Project that is creating hundreds of maps of epigenetic changes in different cell and tissue types and evaluating potential roles of epigenetics in complex diseases), characterization of SNPs (the International HapMap Project) and CNVs for their role in genome variation, disease, and pharmacogenomics applications. We have discussed aspects of a cancer genome project (Cancer Genome Atlas Project) in Chapter 19. The epigenome is covered in depth in Chapter 26. SNPs and pharmacogenomics are discussed in Chapter 22 and Chapter 29. Here we consider two areas of human genome research that are extensions of the HGP: (1) the analysis of personal genomes, including haploid genomes, and (2) the Human Microbiome Project.

Personalized Genome Projects and Personal Genomics

As we discussed earlier in this chapter and in Chapter 20, high-throughput sequencing and most recently next-generation sequencing technologies, capable of generating longer sequence reads at higher speeds with greater accuracy, have greatly reduced the cost of DNA sequencing, and expectations for continued cost reductions along with continued technological advances are high (see Figure 21–14). These expectations have led several companies to propose personalized genome sequencing for individual people. In 2006, the X Prize Foundation announced the Archon X Prize for Genomics, an award of $10 million to the first private group that develops technology capable of sequencing 100 human genomes with a high degree of accuracy in 10 days for under $10,000 per genome. Other groups are working on sequencing a personalized genome for a mere $1000! Two programs funded by the National Institutes of Health are challenging scientists to develop sequencing technologies to complete a human genome for $1000 by 2014 (see the "Genetics, Technology, and Society" essay in Chapter 22).

Pursuit of the $1000 genome has become an indicator that DNA sequencing may eventually be affordable enough for individuals to consider acquiring a readout of their own genetic blueprint. The genome of James D. Watson, who together with Francis Crick discovered the structure of DNA, was the focus of "Project Jim" by the Connecticut company 454 Life Sciences, which wanted to sequence the genome of a high-profile person and decided that the co-discoverer of DNA structure and the first director of the U.S. Human Genome Project should be that person. This company used their next-generation pyrosequencing approach (see Figure 21–18) for Project Jim, and within two years it was announced that six-fold coverage of Watson's genome was complete at a rough cost of just under $1 million. James Watson was then presented with two DVDs containing his genome sequence.

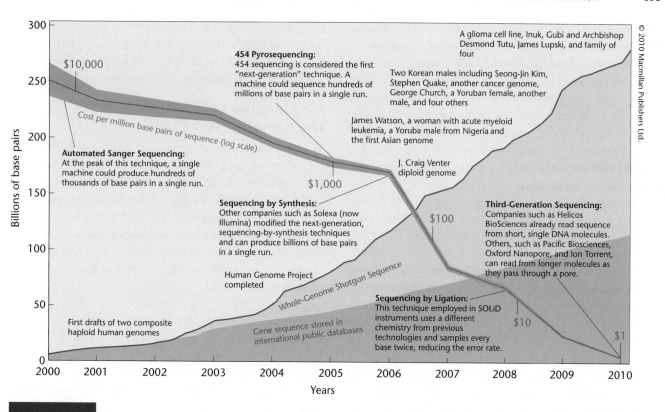

FIGURE 21–14 Human genome sequence explosion. Sequencing costs have steadily declined since 2000 due to innovations in sequencing technology. As a result, notice that the amount of whole-genome shotgun sequencing data, gene sequencing data stored in public databases, traces archive data (raw sequence data), and now Sequence Read Archive (SRA) data from next-generation sequencing—which includes data on several individual genomes—has dramatically increased.

Human genome pioneer J. Craig Venter, whose accomplishments we have discussed in several chapters, had his genome completed by the J. Craig Venter Institute and deposited into GenBank in May 2007. George Church of Harvard and his colleagues have started a **Personal Genome Project (PGP)** and have recruited volunteers to provide DNA for individual genome sequencing on the understanding that the genome data will be made publically available. Church's genome has been completed and been made available online. The concept of a personalized genome project raises the obvious question: would you have your genome sequenced for $10,000, $1000, or even for free?

Since the Watson and Venter genomes were completed, in 2008 the first complete genome sequence was provided for an individual "ancient" human, a Palaeo-Eskimo, obtained from ~4000-year-old permafrost-preserved hair. This work recovered about 78 percent of the diploid genome and revealed many interesting SNPs (of which about 7 percent have not been previously reported). As of early 2010, thirteen individual human genome sequences have been reported, including sequences for a Yoruba African, two individuals of northwest European origin, a Han Chinese individual, two persons from Korea, African Archbishop Desmond Tutu, and a family of four among several others.

One especially intriguing example of personal genomics involves a genetics researcher who examined his own genome for insight about a medical condition. Dr. Richard Gibbs of Baylor College of Medicine in Texas led a group that sequenced the whole genome of his colleague Dr. James Lupski, a medical geneticist who has **Charcot-Marie-Tooth (CMT) disease**. This disease is a neurological condition that causes muscle weakness. Interestingly, mutations in over 30 genes, many of which were identified by Dr. Lupski, are involved in CMT, although Lupski did not carry any of these mutations. A comparison of Lupski's genome to the HGP reference sequence revealed many SNPs and other variations, but it was unclear how many of these were simply sequencing errors or variations that were not involved in CMT.

Focusing on genes previously linked to CMT and other neurological conditions, researchers found that Lupski's genome had two different mutations in the gene *SH3TC2*, which is expressed in Schwann cells that wrap around certain neurons to form the myelin sheath essential for impulse conductions in nerves. In one *SH3TC2* allele a nonsense mutation was revealed, and in Lupski's second allele for *SH3TC2* a new missense mutation was found. When genetic tests for these alleles were carried out on Lupski's parents and seven siblings, the nonsense mutation was found in one parent and two siblings who did not have the disease. The missense mutation was found in another parent and one grandparent, neither of whom had the disorder. Only siblings who inherited both mutated alleles had CMT disease. Many consider this the first clinically relevant success for personal genome sequencing—at least for identifying disease genes.

Another particularly beneficial aspect of personal genome projects is the insight they are providing regarding genome variation. The HGP combined samples from different individuals to create a reference genome for a *haploid genome*. Personal genome projects sequence a diploid genome, and because of this such projects indicate that haploid genome comparisons may underestimate the extent of genome variation between individuals by five-fold or more. For example, when Venter's genome was analyzed, over 4 million variations were found between his maternal and paternal chromosomes alone. From what we are learning about personal genomes, genome variation between individuals may be closer to 0.5 percent than 0.1 percent, and in a 3 billion bp genome this is a significant difference in sequence variation. Integrating genome data from several complete individual genomes of individuals from different ethnic groups will also be of great value in evolutionary genetics to address fundamental questions about human diversity, ancestry, and migration patterns.

The Human Microbiome Project

In 2008 the National Institutes of Health announced plans for the **Human Microbiome Project**, a $115 million, five-year project to complete the genomes of an estimated 600–1000 microorganisms, bacteria, viruses, and yeast that live on and inside humans. Microorganisms comprise ~1–2 percent of the human body, outnumbering human cells by about 10 to 1. Many microbes, such as *E. coli* in the digestive tract, have important roles in human health, and of course other microbes make us ill. The Human Microbiome Project has several major goals, including:

- Determining if individuals share a core human microbiome.

- Understanding whether changes in the microbiome can be correlated with changes in human health.

- Developing new methods, including bioinformatics tools, to support analysis of the microbiome.

- Addressing ethical, legal, and social implications raised by human microbiome research. Does this sound familiar? Recall that addressing ethical, legal, and social issues was a goal of the HGP.

The Human Microbiome Project is still very much in its infancy as a project, but it has already revealed that over 3.3 million human gut microbe genes characterized to date appear to be very similar among over 100 individuals. The saliva microbiome is also highly similar from individual to individual, and a Human Oral Microbiome database has been established from these studies, which seek to develop linkages between the oral microbiome and oral health.

No Genome Left Behind and the Genome 10K Plan

Without question new sequencing technologies that have been developed as a result of the HGP are an important part of the transformational effect the HGP has had on modern biology. About ten years ago, a room full of sequencers and several million dollars were required to sequence the 97-Mb genome of *C. elegans*. As a sign of modern times in the world of genomics, in 2009 two sequencers and $500,000 produced a reasonably complete draft of the 750-Mb cod genome—in a month!

Modern sequencing technologies are asking some to consider the question, "What would you do if you could sequence everything?" Partners around the world including genome scientists and museum curators have proposed to sequence 10,000 vertebrate genomes, the **Genome 10K** plan. Shortly after the HGP finished, the National Human Genome Research Institute (NHGRI) assembled a list of mammals and other vertebrates as priorities for genome sequencing in part because of their potential benefit for learning about the human genome through comparative genomics. Genome 10K will also provide insight into genome evolution and speciation. This ambitious plan proposes to assemble 10,000 genomes in five years—about one genome a day!

21.6

Comparative Genomics Analyzes and Compares Genomes from Different Organisms

As of 2010, the genomes of over 3800 prokaryotic and eukaryotic organisms—including many model organisms and a number of viruses—have been sequenced. This is quite extraordinary progress in a relatively short time span! Among these organisms are yeast (*Saccharomyces cerevisiae*)—the first eukaryotic genome to be sequenced to bacteria such as *E. coli*, the nematode roundworm (*Caenorhabditis elegans*), the thale cress plant (*Arabidopsis thaliana*), mice (*Mus musculus*), zebrafish (*Danio rerio*), and of course *Drosophila*. In the past few years, genomes for chimpanzees, dogs, chickens, sea urchins, honey bees, pufferfish, rice, and wheat have all been sequenced.

These studies have demonstrated not only significant differences in genome organization between prokaryotes and eukaryotes but also many similarities between genomes of nearly all species. In this section we provide a basic overview of genome organization in prokaryotes and eukaryotes and discuss interesting aspects of genomes in selected organisms. Analysis of the growing number of genome sequences confirms that all living organisms are related and descended from a common ancestor. Similar gene sets are used in all organisms for basic cellular functions, such as DNA replication, transcription, and translation. These genetic relationships are the rationale for the use of model organisms to study inherited human disorders, the effects of the environment on genes, and interactions of genes in complex diseases, such as cardiovascular disease, diabetes, neurodegenerative conditions, and behavioral disorders.

Comparative genomics compares the genomes of different organisms to answer questions about genetics and other aspects of biology. It is a field with many research and practical applications, including gene discovery and the development of model organisms to study human diseases. It also incorporates the study of gene and genome evolution and the relationship between organisms and their environment. Comparative genomics uses a wide range of techniques and resources, such as the construction and use of nucleotide and protein databases containing nucleic acid and amino acid sequences, fluorescent *in situ* hybridization (FISH), and the creation of gene knockout animals. Comparative genomics can reveal genetic differences and similarities between organisms to provide insight into how those differences contribute to differences in phenotype, life cycle, or other attributes, and to ascertain the evolutionary history of those genetic differences.

Prokaryotic and Eukaryotic Genomes Display Common Structural and Functional Features and Important Differences

Since most prokaryotes have small genomes amenable to shotgun cloning and sequencing, many genome projects have focused on prokaryotes, and more than 900 additional projects to sequence prokaryotic genomes are now under way. Many of the prokaryotic genomes already sequenced are from organisms that cause human diseases, such as cholera, tuberculosis, and leprosy. Traditionally, the bacterial genome has been thought of as relatively small (less than 5 Mb) and contained within a single circular DNA molecule. *E. coli,* used as the prototypical bacterial model organism in genetics, has a genome with these characteristics. However, the flood of genomic information now available has challenged the validity of this viewpoint for bacteria in general (Table 21.2). Although most prokaryotic genomes are small, their sizes vary across a surprisingly wide range. In fact, there is some overlap in size between larger bacterial genomes (30 Mb in *Bacillus megaterium*) and smaller eukaryotic genomes (12.1 Mb in yeast). Gene number in

TABLE 21.2

Genome Size and Gene Number in Selected Prokaryotes

	Genome Size (Mb)	Number of Genes
Archaea		
Methanosarcina berkeri	4.84	3680
Archaeoglobus fulgidis	2.17	2437
Methanococcus jannaschii	1.66	1783
Nanoarchaeum equitans	0.49	552
Thermoplasma acidophilium	1.56	1509
Eubacteria		
Pseudomonas aeruginosa	6.30	5570
Rhizobium radiobacter	4.67	5419
Escherichia coli	4.64	4289
Bacillus subtilis	4.21	4779
Haemophilus influenzae	1.83	1738
Aquifex aeolicus	1.55	1749
Rickettsia prowazekii	1.11	834
Mycoplasma pneumonia	0.82	680
Mycoplasma genitalium	0.58	483

bacterial genomes also demonstrates a wide range, from less than 500 to more than 5000 genes, a ten-fold difference.

In addition, although many bacteria have a single, circular chromosome, there is substantial variation in chromosome organization and number among bacterial species. An increasing number of genomes composed of linear DNA molecules are being identified, including the genome of *Borrelia burgdorferi,* the organism that causes Lyme disease. Sequencing of the *Vibrio cholerae* genome (the organism responsible for cholera) revealed the presence of two circular chromosomes. Other bacteria that have genomes with two or more chromosomes include *Rhizobium radiobacter* (formerly *Agrobacterium tumefaciens*), *Deinococcus radiodurans,* and *Rhodobacter sphaeroides.* The finding that some bacterial species have multiple chromosomes raises questions both about how replication and segregation of their chromosomes are coordinated during cell division and about what undiscovered mechanisms of gene regulation may exist in bacteria. The answers may provide clues about the evolution of multichromosome eukaryotic genomes.

We can make two generalizations about the organization of protein-coding genes in bacteria. First, gene density is very high, averaging about one gene per kilobase of DNA. For example, the genome of *E. coli* strain K12, which was sequenced in 1997 as the second prokaryotic genome to be sequenced, is 4.6 Mb in size and it contains 4289 protein-coding genes in its single, circular chromosome. This close packing of genes in prokaryotic genomes means that a very high proportion of the DNA (approximately 85 to 90 percent) serves as coding DNA. Typically, only a small amount of a bacterial genome is noncoding DNA, often in the form of regulatory sequences or of transposable elements that can move from one place to another in the genome.

TABLE 21.3

Comparison of Selected Genomes

Organism (Scientific Name)	Approximate Size of Genome (in million [megabase, Mb] or billion [gigabase, Gb] bases) (Date Completed)	Number of Genes	Approximate Percentage of Genes Shared with Humans
Bacterium (*Escherichia coli*)	4.1 Mb (1997)	4403	not determined
Chicken (*Gallus gallus*)	1 Gb (2004)	~20,000–23,000	60%
Dog (*Canis familiaris*)	2.5 Gb (2003)	~18,400	75%
Chimpanzee (*Pan troglodytes*)	~3 Gb (2005)	~20,000–24,000	98%
Fruit fly (*Drosophila melanogaster*)	165 Mb (2000)	~13,600	50%
Human (*Homo sapiens*)	~2.9 Gb (2004)	~20,000	100%
Mouse (*Mus musculus*)	~2.5 Gb (2002)	~30,000	80%
Rat (*Rattus norvegicus*)	~2.75 Gb (2004)	~22,000	80%
Rhesus macaque (*Macaca mulatta*)	2.87 Gb (2007)	~20,000	93%
Rice (*Oryza sativa*)	389 Mb (2005)	~41,000	not determined
Roundworm (*Caenorhabditis elegans*)	97 Mb (1998)	19,099	40%
Sea urchin (*Strongylocentrotus purpuratus*)	814 Mb (2006)	~23,500	60%
Thale cress (plant) (*Arabidopsis thaliana*)	140 Mb (2000)	~27,500	not determined
Yeast (*Saccharomyces cerevisiae*)	12 Mb (1996)	~5700	30%

Adapted from Palladino, M. A. *Understanding the Human Genome Project*, 2nd ed. Benjamin Cummings, 2006.

Note: Billion bp (gigabase, Gb).

The second generalization we can make is that bacterial genomes contain operons (recall from Chapter 16 that operons contain multiple genes functioning as a transcriptional unit whose protein products are part of a common biochemical pathway). In *E. coli,* 27 percent of all genes are contained in operons (almost 600 operons). In other bacterial genomes, the organization of genes into transcriptional units is challenging our ideas about the nature of operons. For example, in *Aquifex aeolicus,* one polygenic transcription unit contains six genes involved in several different cellular processes with no apparent common relationships: two genes for DNA recombination, one for lipid synthesis, one for nucleic acid synthesis, one for protein synthesis, and one that encodes a protein for cell motility. Other polygenic transcription units in this species also contain genes with widely different functions. This finding, combined with similar results from other genome projects, raises interesting questions about the consensus that operons encode products that control a single metabolic pathway in bacterial cells.

The basic features of eukaryotic genomes are similar in different species, although genome size in eukaryotes is highly variable (Table 21.3). Genome sizes range from about 10 Mb in fungi to over 100,000 Mb in some flowering plants (a ten thousand-fold range); the number of chromosomes per genome ranges from two into the hundreds (about a hundred-fold range), but the number of genes varies much less dramatically than either genome size or chromosome number.

Eukaryotic genomes have several features not found in prokaryotes:

- **Gene density**. In prokaryotes, gene density is close to 1 gene per kilobase. In eukaryotic genomes, there is a wide range of gene density. In yeast, there is about 1 gene/2 kb,

in *Drosophila,* about 1 gene/13 kb, and in humans, gene density varies greatly from chromosome to chromosome. Human chromosome 22 has about 1 gene/64 kb, while chromosome 13 has 1 gene/155 kb of DNA.

- **Introns**. Most eukaryotic genes contain introns. There is wide variation among genomes in the number of introns they contain and also wide variation from gene to gene. The entire yeast genome has only 239 introns, whereas just a single gene in the human genome can contain more than 100 introns. Regarding intron size, generally the size in eukaryotes is correlated with genome size. Smaller genomes have smaller average introns, and larger genomes have larger average intron sizes. But there are exceptions, For example, the genome of the pufferfish (*Fugu rubripes*) has relatively few introns.

- **Repetitive sequences**. The presence of introns and the existence of repetitive sequences are two major reasons for the wide range of genome sizes in eukaryotes. In some plants, such as maize, repetitive sequences are the dominant feature of the genome. The maize genome has about 2500 Mb of DNA, and more than two-thirds of that genome is composed of repetitive DNA. In the human, as discussed previously, about half of the genome is repetitive DNA.

Comparative Genomics Provides Novel Information about the Genomes of Model Organisms and the Human Genome

As mentioned earlier, the Human Genome Project sequenced genomes from a number of model nonhuman organisms too, including *E. coli, Arabidopsis thaliana, Saccharomyces cerevisiae, Drosophila melanogaster,* the nematode roundworm *Caenorhabditis elegans,* and the mouse *Mus musculus.* Complete genome sequences of such organisms have been invaluable for comparative genomics studies of gene function in these organisms and in humans. As shown in Table 21.3, the number of genes humans share with other species is very high, ranging from about 30 percent of the genes in yeast to ~80 percent in mice and ~98 percent in chimpanzees. The human genome even contains around 100 genes that are also present in many bacteria. Comparative genomics has shown us that many mutated genes involved in human disease are also present in model organisms. For instance, approximately 60 percent of genes mutated in nearly 300 human diseases are also found in *Drosophila.* These include genes involved in prostate, colon, and pancreatic cancers; cardiovascular disease; cystic fibrosis; and several other conditions. Here we consider how comparative genomics studies of several model organisms (dogs, chimpanzees, Rhesus monkeys, and sea urchins) and the Neanderthal genome have revealed interesting elements of the human genome.

The Dog Genome

In 2005 the genome for "man's best friend" was completed, and it revealed that we share about 75 percent of our genes with dogs (*Canis familiaris*), providing a useful model with which to study our own genome [Figure 21–15]. Dogs have a genome that is similar in size to the human genome: about 2.5 billion base pairs with an estimated 18,400 genes [Figure 21–15(a)]. The dog offers several advantages for studying heritable human diseases. Dogs share many genetic disorders with humans, including over 400 single-gene disorders, sex-chromosome aneuploidies, multifactorial diseases (such as epilepsy), behavioral conditions (such as obsessive-compulsive disorder), and genetic predispositions to cancer, blindness,

heart disease, and deafness. The molecular causes of at least 60 percent of inherited diseases in dogs, such as point mutations and deletions, are similar or identical to those found in humans. In addition, at least 50 percent of the genetic diseases in dogs are breed-specific, so that the mutant allele segregates in relatively homogeneous genetic backgrounds. Dog breeds resemble isolated human populations in having a small number of founders and a long period of relative genetic isolation. These properties make individual dog breeds useful as models of human genetic disorders.

In addition, differences in biology and behavior among dog breeds are well documented. Domestic dogs show greater variation in body size than all other living terrestrial vertebrates. The size difference between a Chihuahua and a Great Dane is an excellent example [Figure 21–15(b)]. Mapping and comparing DNA sequence differences (polymorphisms) among breeds may help identify genes that contribute to both physiological and behavioral differences. For instance, in 2007, genome-wide analysis of different large and small dog breeds revealed a locus on chromosome 15 where a single-nucleotide polymorphism in the insulin-like growth factor 1 gene (*Igf1*) is common in all small breeds of dogs but virtually absent from large dogs. It is well known that *Igf1* plays important roles in growth-hormone-regulated increases in muscle mass and bone growth during adolescence in humans. This study provides very strong evidence that mutation of *Igf1* is a primary determinant of body size in small dogs.

Dog breeders are now using genetic tests to screen dogs for inherited disease conditions, for coat color in Labrador retrievers and poodles, and for fur length in Mastiffs. Undoubtedly, we can expect many more genetic tests for dogs in the near future, including DNA analysis for size, type of tail, speed, sense of smell, and other traits deemed important by breeders and owners.

The Chimpanzee Genome

Although the chimpanzee (*Pan troglodytes*) genome was not part of the HGP, its nucleotide sequence was completed in 2004. Overall, the chimp and human genome sequences differ by less than 2 percent, and 98 percent of the genes are the same. Comparisons between these genomes offer some interesting insights into what makes some primates humans and others chimpanzees.

The speciation events that separated humans and chimpanzees occurred less than 6.3 million years ago (mya). Genomic analysis indicates that these species initially diverged but then exchanged genes again before separating completely. Their separate evolution after this point is exhibited in such differences as are seen between the sequence of chimpanzee chromosome 22 and its human ortholog, chromosome 21 (Table 21.4; chimps have 48 chromosomes and humans have 46, so the numbering is different). These chromosomes have accumulated nucleotide substitutions

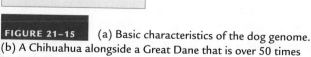

(a)

Dog Genome

Size: 2.47 Gb

Chromosomes: 39

ORFs: 19,300

Repetitive DNA: 31%

(b)

FIGURE 21–15 (a) Basic characteristics of the dog genome. (b) A Chihuahua alongside a Great Dane that is over 50 times its mass. The genetic basis for varied phenotypes in dogs, such as body size, is being revealed by analyzing the dog genome.

TABLE 21.4

Comparisons between Human Chromosome 21 and Chimpanzee Chromosome 22

	Human 21	Chimpanzee 22
Size (bp)	33,127,944	32,799,845
%G + C Content	40.94	41.01
CpG Islands	950	885
SINEs (*Alu* elements)	15,137	15,048
Genes	284	272
Pseudogenes	98	89

that total 1.44 percent of the sequence. The most surprising difference is the discovery of 68,000 nucleotide insertions or deletions, collectively called **indels**, in the chimp and human chromosomes, a frequency of 1 indel every 470 bases. Many of these are *Alu* insertions in human chromosome 21. Although the overall difference in the nucleotide sequence is small, there are significant differences in the encoded genes. Only 17 percent of the genes analyzed encode identical proteins in both chromosomes; the other 83 percent encode genes with one or more amino acid differences.

Differences in the time and place of gene expression also play a major role in differentiating the two primates. Using DNA microarrays (discussed in Section 21.8), researchers compared expression patterns of 202 genes in human and chimp cells from brain and liver. They found more species-specific differences in expression of brain genes than liver genes. To further examine these differences, Svante Pääbo and colleagues compared expression of 10,000 genes in human and chimpanzee brains and found that 10 percent of genes examined differ in expression in one or more regions of the brain. More importantly, these differences are associated with genes in regions of the human genome that have been duplicated subsequent to the divergence of chimps and humans. This finding indicates that genome evolution, speciation, and gene expression are interconnected. Further work on these segmental duplications and the genes they contain may identify genes that help make us human.

The Rhesus Monkey Genome

The Rhesus macaque monkey (*Macaca mulatta*), another primate, has served as one of the most important model organisms in biomedical research. Macaques have played central roles in our understanding of cardiovascular disease, aging, diabetes, cancer, depression, osteoporosis, and many other aspects of human health. They have been essential for research on AIDS vaccines and for the development of polio vaccines. The macaque's genome is the first monkey genome to have been sequenced. A main reason geneticists are so excited about the completion of this sequencing project is that macaques provide a more distant evolutionary window

that is ideally suited for comparing and analyzing human and chimpanzee genomes. As we discussed in the preceding section, humans and chimpanzees shared a common ancestor approximately 6 mya. But macaques split from the ape lineage that led to chimpanzees and humans about 25 mya. The macaque and human genome have thus diverged farther from one another, as evidenced by the ~93 percent sequence identity between humans and macaques compared to the ~98 percent sequence identity shared by humans and chimpanzees.

The macaque genome was published in 2007, and it was no surprise to learn that it consists of 2.87 billion bp (similar to the size of the human genome) contained in 22 chromosomes (20 autosomes, an X, and a Y) with ~20,000 protein-coding genes. Although comparative analyses of this genome are ongoing, a number of interesting features have been revealed so far. As in humans, about 50 percent of the genome consists of repeat elements (transposons, LINEs, SINEs). Gene duplications and gene families are abundant, including cancer gene families found in humans.

A number of interesting surprises have also been observed. For instance, recall from Chapter 4 and elsewhere our discussion about the genetic disorder phenylketonuria (PKU), an autosomal recessive inherited condition in which individuals cannot metabolize the amino acid phenylalanine due to mutation of the phenylalanine hydroxylase (*PAH*) gene. The histidine substitution encoded by a mutation in the *PAH* gene of humans with PKU appears as the wild-type amino acid in the protein from healthy macaques. Further analysis of the macaque genome and comparison to the human and chimpanzee genome will be invaluable for geneticists studying genetic variations that played a role in primate evolution.

The Sea Urchin Genome

In 2006, researchers from the Sea Urchin Genome Sequencing Consortium completed the 814 million bp genome of the sea urchin *Strongylocentrotus purpuratus* [pictured in Figure 21–23(a)]. Sea urchins are shallow-water marine invertebrates that have served as important model organisms, particularly for developmental biologists. One reason is that the sea urchin is a nonchordate deuterostome, and humans, with their spinal cord, are chordate deuterostomes. Fossil records indicate that sea urchins appeared during the Early Cambrian period, around 520 mya.

A combination of whole-genome shotgun sequencing and map-based cloning in BACs was used to complete the genome. Sea urchins have an estimated 23,500 genes, including representative genes for just about all major vertebrate gene families. Sequence alignment and homology searches demonstrate that the sea urchin contains many genes with important functions in humans, yet interestingly, important genes in flies and worms, such as certain cytochrome P-450 genes that play a role in the breakdown of toxic compounds, are missing from sea urchins. The sea urchin genome also has an abundance (~25 to 30 percent) of **pseudogenes**—nonfunctional relatives of protein-coding genes (we meet pseudogenes again in the next subsection). Sea urchins have a smaller average intron size than humans, supporting the general trend revealed by comparative genomics that intron size is correlated with overall genome size.

Another genome trend that urchins share with other eukaryotes is the presence of genes involved in innate immunity, the inborn defense mechanisms that provide broad-spectrum protection against many pathogens. Sea urchins have an extraordinarily rich number of genes providing innate immunity. For example, one very important category of innate immunity genes, the Toll-like receptors (TLRs), produce transmembrane proteins that are essential for pathogen recognition in nearly every cell type of vertebrates. Sea urchins have over 200 *Tlr* genes compared to 11 in humans. The abundance of these and other important innate immunity genes in sea urchins has led to categorizing these genes as the urchin "defensome." This characteristic of sea urchins may help explain how these organisms have adapted so well to the pathogen-loaded environments of seabeds.

Urchins have nearly 1000 genes for sensing light and odor, indicative of great sensory abilities. In this respect, their genome is more typical of vertebrates than invertebrates. A number of orthologs of human genes involved in hearing and balance are present in the sea urchin, as are many human-disease-associated orthologs, including protein kinases, GTPases, transcription factors, TLRs, transporters, and low-density lipoprotein receptors. Sea urchins and humans share approximately 7000 orthologs.

Another interesting aspect of the sea urchin genome project is that it has identified genes previously thought to be vertebrate-specific. One example, the *WntA* gene, important for patterning during embryonic development, as discussed in Chapter 18, was thought to be absent from nonvertebrate deuterostomes. The sea urchin genome also contains genes that are not present in chordates. Further analysis of the urchin genome is expected to make important contributions to our understanding of evolutionary transitions between invertebrates and vertebrates.

The Neanderthal Genome and Modern Humans

In early 2009, a team of scientists led by Svante Pääbo at the Max Planck Institute for Evolutionary Anthropology in Germany and 454 Life Sciences reported completion of a rough draft of the Neanderthal (*Homo neanderthalensis*) genome encompassing more than 3 billion bp of Neanderthal DNA and about two-thirds of the genome. Previously, in 1997, Pääbo's lab sequenced portions of Neanderthal mitochon-

drial DNA from a fossil. In late 2006, Pääbo's group along with a number of scientists in the United States reported the first sequence of ~65,000 bp of nuclear DNA isolated from Neanderthal bone samples from Croatia. Bones from three females who lived in Vindija Cave in Croatia about 38,000 to 44,000 years ago were used to produce the draft sequence of the Neanderthal nuclear genome.

Because Neanderthals are members of the human family, and closer relatives to humans than chimpanzees, the Neanderthal genome is expected to provide an unprecedented opportunity to use comparative genomics to advance our understanding of evolutionary relationships between modern humans and our predecessors. In particular, scientists are interested in identifying areas in the genome where humans have undergone rapid evolution since splitting (diverging) from Neanderthals. Much of this analysis involves a comparative genomics approach to compare the Neanderthal genome to the human and chimpanzee genomes.

The human and Neanderthal genomes are 99 percent identical. Comparative genomics has identified 78 protein-coding sequences in humans that seem to have arisen since the divergence from Neanderthals and that may have helped modern humans adapt. Some of these sequences are involved in cognitive development and sperm motility. Of the many genes shared by these species, *FOXP2* is a gene that has been linked to speech and language ability. There are many genes that influence speech, so this finding does not mean that Neanderthals spoke as we do. But because Neanderthals had the same modern human *FOXP2* gene scientists have speculated that Neanderthals possessed linguistic abilities.

A 2008 study of the Neanderthal mitochondrial genome by Pääbo's team obtained 39 million sequence reads averaging 69 bp in length. Of 16,568 nucleotides present in mtDNA of modern humans, on average 206 differed from those of Neanderthals. These differences fall outside the variation among modern humans and led the research team to estimate that the most common point of ancestry, the divergence date between the two mtDNA lineages, was 660,000 years ago (±140,000). This study also provided interesting observations about rates of amino acid substitutions in key enzymes in the mitochondrial electron transport chain.

The realization that modern humans and Neanderthals lived in overlapping ranges as recently as 30,000 years ago has led to speculation about the interactions between modern humans and Neanderthals. Genome studies suggest that interbreeding took place between Neanderthals and modern humans an estimated 45,000 to 80,000 years ago in the eastern Mediterranean. In fact, the genome of non-African *H. sapiens* contains approximately 1–4 percent of sequence inherited from Neanderthals. These exciting studies, previously thought to be impossible, are having ramifications in many areas of human evolution, and it will be interesting indeed to follow the progress of this work.

Comparative Genomics Is Useful for Studying the Evolution and Function of Multigene Families

Comparative genomics has also proven to be valuable for identifying members of **multigene families**, groups of genes that share similar but not identical DNA sequences through duplication and descent from a single ancestral gene. Their gene products frequently have similar functions, and the genes are often, but not always, found at a single chromosomal locus. A group of related multigene families is called a **superfamily**. Sequence data from genome projects is providing evidence that multigene families are present in many, if not all, genomes. To demonstrate how analysis of multigene families provides insight into eukaryotic genome evolution and function, we will now examine the globin gene superfamily, whose members encode very similar but not identical polypeptide chains with closely related functions. Other well-characterized gene superfamilies include the histone, tubulin, actin, and immunoglobulin (antibody) gene superfamilies.

Recall that paralogs, which we defined in Section 21.3, are homologous genes present in the same single organism, believed to have evolved by gene duplication. The globin genes that encode the polypeptides in hemoglobin molecules are a paralogous multigene superfamily that arose by duplication and dispersal to occupy different chromosomal sites. One of the best-studied examples of gene family evolution is the **globin gene superfamily** (Figure 21–16).

In this family, an ancestral gene encoding an oxygen transport protein was duplicated about 800 mya, producing two sister genes, one of which evolved into the modern-day myoglobin gene. **Myoglobin** is an oxygen-carrying protein found in muscle. The other gene underwent further duplication and divergence about 500 mya and formed prototypes of the α-globin and β-globin genes. These genes encode proteins found in **hemoglobin**, the oxygen-carrying molecule in red blood cells. Additional duplications within these genes occurred within the last 200 million years. Events subsequent to each duplication dispersed these gene subfamilies to different chromosomes, and in the human genome, each now resides on a separate chromosome. Similar patterns of evolution are observed in other gene families, including the trypsin–chymotrypsin family of proteases, the homeotic selector genes of animals, and the rhodopsin family of visual pigments.

Adult hemoglobin is a tetramer containing two α- and two β-polypeptides (refer to Figure 14–20 for the structure of hemoglobin). Each polypeptide incorporates a heme group that reversibly binds oxygen. The α-globin gene cluster on chromosome 16 and the β-globin gene cluster on chromosome 11 share nucleotide- and amino acid–sequence similarity (Figure 21–17), but the highest degree of sequence similarity is found within subfamilies.

The **α-globin** gene subfamily [Figure 21–18(a)] contains three genes: the ζ (zeta) gene, expressed only in early embryogenesis, and two copies of the α gene, expressed during the fetal (α_1) and adult stages (α_2). In addition, the cluster contains two pseudogenes (similar to ζ and α_1), which in this family are designated by the prefix ψ (psi) followed by the symbol of the gene they most resemble. Thus, the designation $\psi\alpha_1$ indicates a pseudogene of the fetal α_1 gene.

The organization of the α-globin subfamily members and the locations of their introns and exons demonstrate several characteristic features [Figure 21–18(a)]. First, as is common in eukaryotes, the DNA encoding the three functional α genes occupies only a small portion of the 30-kb region containing the subfamily. Most of the DNA in this region is intergenic spacer DNA. Second, each functional gene in this subfamily contains two introns at precisely the same positions. Third, the nucleotide sequences

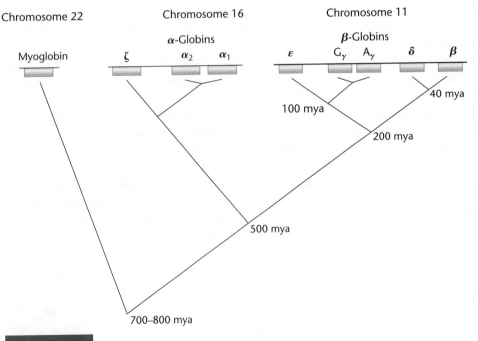

FIGURE 21–16 The evolutionary history of the globin gene superfamily. A duplication event in an ancestral gene gave rise to two lineages about 700 to 800 million years ago (mya). One line led to the myoglobin gene, which is located on chromosome 22 in humans; the other underwent a second duplication event about 500 mya, giving rise to the ancestors of the α-globin and β-globin gene subfamilies. Duplications beginning about 200 mya formed the β-globin gene subfamily. In humans, the α-globin genes are located on chromosome 16, and the β-globin genes are on chromosome 11.

```
α-globin  V – L S P A D K T N V K A A W G K V G A H A G E Y G A E A L E R M F L S F P T T K T Y F P H F – D L S H
β-globin  V H L T P E E K S A V T A L W G K V – – N V D E V G G E A L G R L L V V Y P W T Q R F F E S F G D L S T


α-globin  – – – G S A Q V K G H G K K V A D A L T N A V A H V D D M P N A L S A L S D L H A H K L R V D P V N
β-globin  A V M G N P K V K A H G K K V L G A F S D G L A H L D N L K G T F A T L S E L H C D K L H V D P E N


α-globin  L L S H C L L V T L A A H L P A E F T P A V H A S L D K F L A S V S T V L T S K Y R 141 amino acids
β-globin  L L G N V L V C V L A H H F G K E F T P P V Q A A Y Q K V V A G V A N A L A H K Y H 146 amino acids
```

FIGURE 21–17   The amino acid sequences of the α- and β-globin proteins, depicted using the single-letter abbreviations for the amino acids (see Figure 15–16). Shaded areas indicate identical amino acids. The two proteins are slightly different in length. α-globin contains 141 amino acids, while β-globin is 146 amino acids long. Gaps in the two sequences, representing areas that do not align, are indicated by horizontal dashes (–).

within corresponding exons are nearly identical in the ζ and α genes. Each of these genes encodes a polypeptide chain of 141 amino acids. However, their intron sequences are highly divergent, even though the introns are about the same size. Note that much of the nucleotide sequence of each gene is contained in these noncoding introns.

The human **β-globin** gene cluster contains five genes spaced over 60 kb of DNA [Figure 21–18(b)]. In this and the α-globin gene subfamily, the order of genes on the chromosome parallels their order of expression during development. Three of the five genes are expressed before birth. The ε (epsilon) gene is expressed only during embryogenesis,

while the two nearly identical γ genes (G_γ and A_γ) are expressed only during fetal development. The polypeptide products of the two γ genes differ only by a single amino acid. The two remaining genes, δ and β, are expressed after birth and throughout life. A single pseudogene, $\psi\beta_1$, is present in this subfamily. All five functional genes in this cluster encode proteins with 146 amino acids and have two similar-sized introns at exactly the same positions. The second intron in the β-globin subfamily is significantly larger than its counterpart in the functional α-globin subfamily. These features reflect the evolutionary history of each subfamily and the events such as gene duplication, nucleotide substitution,

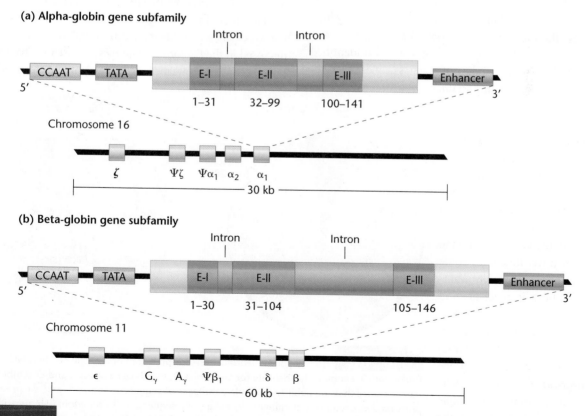

FIGURE 21–18   Organization of (a) the α-globin gene subfamily on chromosome 16 and (b) the β-globin gene subfamily on chromosome 11. Also shown in each case is the internal organization of the α_1 gene and the β gene, respectively. Both genes contain three exons (E-I, E-II, E-III) and two introns. The numbers below the exons indicate the location of amino acids in the gene product encoded by each exon.

and chromosome translocations that produced the present-day globin superfamily.

21.7

Metagenomics Applies Genomics Techniques to Environmental Samples

Metagenomics, also called **environmental genomics**, is the use of whole-genome shotgun approaches to sequence genomes from entire communities of microbes in environmental samples of water, air, and soil. Oceans, glaciers, deserts, and virtually every other environment on Earth are being sampled for metagenomics projects. Human genome pioneer J. Craig Venter left Celera in 2003 to form the J. Craig Venter Institute, and his group has played a central role in developing metagenomics as an emerging area of genomics research.

One of the institute's major initiatives has been a global expedition to sample marine and terrestrial microorganisms from around the world and to sequence their genomes. Through this project, called the *Sorcerer II* Global Ocean Sampling (GOS) Expedition, Venter and his researchers traveled the globe by yacht, in a sailing voyage described as a modern-day version of Charles Darwin's famous voyage on the *H.M.S. Beagle*.

A key benefit of metagenomics is its potential for teaching us more about millions of species of bacteria, of which only a few thousand have been well characterized. Many new viruses, particularly bacteriophages, are also identified through metagenomics studies of water samples. Metagenomics is providing important new information about genetic diversity in microbes that is key to understanding complex interactions between microbial communities and their environment, as well as allowing phylogenetic classification of newly identified microbes. Metagenomics also has great potential for identifying genes with novel functions, some of which have potentially valuable applications in medicine and biotechnology.

The general method used in metagenomics to sequence genomes for all microbes in a given environment involves isolating DNA directly from an environmental sample without requiring cultures of the microbes or viruses. Such an approach is necessary because often it

is difficult to replicate the complex array of growth conditions the microbes need to survive in culture. For the *Sorcerer II* GOS project, samples of water from different layers in the water column were passed through high-density filters of various sizes to capture the microbes. DNA was then isolated from the microbes and subjected to shotgun sequencing and genome assembly. High-throughput sequencers on board the yacht operated nearly around the clock. One of the earliest expeditions by this group sequenced bacterial genomes from the Sargasso Sea off Bermuda. This project yielded over 1.2 million novel DNA sequences from 1800 microbial species, including 148 previously unknown bacterial species, and identified hundreds of photoreceptor genes. Many aquatic microorganisms rely on photoreceptors for capturing light energy to power photosynthesis. Scientists are interested in learning more about photoreceptors to help develop ways in which photosynthesis may be used to produce hydrogen as a fuel source. Medical researchers are also very interested in photoreceptors because in humans and many other species, photoreceptors in the retina of the eye are key proteins that detect light energy and transduce electrical signals that the brain eventually interprets to create visual images.

By early 2007, the GOS database contained approximately 6 billion bp of DNA from more than 400 uncharacterized microbial species! These sequences included 7.7 million previously uncharacterized sequences, encoding more than 6 million different potential proteins. This is almost twice the total number of previously characterized proteins in all other known databases worldwide (such as the Swiss-Prot database discussed in the Exploring Genomics exercise for Chapter 14). Figure 21–19(a) shows the kingdom assignments for predicted

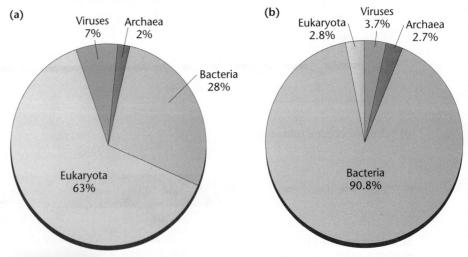

FIGURE 21–19 (a) Kingdom identifications for predicted proteins in NCBInr, NCBI Prokaryotic Genomes, the Institute for Genomics Research Gene Indices, and Ensembl databases. Notice that the publicly available databases of sequenced genomes and the predicted proteins they encode are dominated by eukaryotic sequences. (b) Kingdom identifications for novel predicted proteins in the Global Ocean Sampling (GOS) database. Bacterial sequences dominate this database, demonstrating the value of metagenomics for revealing new information about microbial genomes and microbial communities.

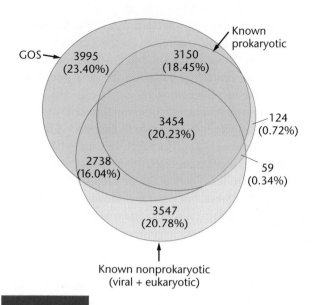

Venn diagram representation of 17,067 medium and large clusters of protein families grouped according to three categories: GOS, known prokaryotic sequences, and known nonprokaryotic sequences. Notice that the GOS project identified 3995 medium and large clusters of gene families that are unique to the GOS database and do not show significant homology to known protein families in prokaryotes or nonprokaryotes (viral and eukaryotes).

protein sequences in publicly available databases worldwide, such as the NCBI-nonredundant protein database (NCBInr), which accesses GenBank, Ensembl, and other well-known databases. Eukaryotic sequences comprise the majority (63 percent) of predicted proteins in these databases. Reviewing the kingdom assignments of approximately 6 million predicted proteins in the Global Ocean Sampling (GOS) dataset shows that, in contrast, the largest majority (90.8 percent) of sequences in this database are from the bacterial kingdom [Figure 21–19(b)].

The GOS Expedition also examined protein families corresponding to the predicted proteins encoded by the genome sequences in the GOS database: 17,067 families were medium (between 20 and 200 proteins) and large-sized (>200 proteins) clusters. A **Venn diagram**, like the image shown in Figure 21–20, is a common way to represent overlapping data in genomics datasets. In this figure, overlapping ovals indicate numbers of protein families belonging to certain sets of categories. Each area of the diagram is labeled with the number and percentage of families out of the 17,067 GOS medium- and large-sized clusters in each category or area of overlap. Of the 17,067 clusters in the GOS database, 3995 did not show significant homology to known protein families in prokaryotes, viruses, or eukaryotes (Figure 21–20). The results summarized in Figures 21–19 and 21–20 demonstrate the value of the GOS Expedition and of metagenomics for identifying novel microbial genes and potential proteins.

Many other high-profile metagenomics applications have emerged recently, including the use of metagenomics to identify viruses and fungi thought to be involved in colony collapse disorder, a malady that has resulted in the loss of 50–90 percent of the honey bee population in beekeeping operations throughout the United States.

Transcriptome Analysis Reveals Profiles of Expressed Genes in Cells and Tissues

Sequencing a genome is a major endeavor, and even once any genome has been sequenced and annotated, a formidable challenge still remains: that of understanding genome function by analyzing the genes it contains and the ways the genes expressed by the genome are regulated. **Transcriptome analysis**, also called **transcriptomics** or **global analysis of gene expression**, studies the expression of genes by a genome both qualitatively—by identifying which genes are expressed and which genes are not expressed—and quantitatively—by measuring varying levels of expression for different genes.

As we know, all cells of an organism possess the same genome, but in any cell or tissue type, certain genes will be highly expressed, others expressed at low levels, and some not expressed at all. Transcriptome analysis provides gene-expression profiles that for the same genome may vary from cell to cell or from tissue type to tissue type. Identifying genes expressed by a genome is essential for understanding how the genome functions. Transcriptome analysis provides insights into (1) normal patterns of gene expression that are important for understanding how a cell or tissue type differentiates during development, (2) how gene expression dictates and controls the physiology of differentiated cells, and (3) mechanisms of disease development that result from or cause gene-expression changes in cells. In Chapter 22, we will consider why gene-expression analysis is gradually becoming an important diagnostic tool in certain areas of medicine. For example, examining gene-expression profiles in a cancerous tumor can help diagnose tumor type, determine the likelihood of tumor metastasis (spreading), and develop the most effective treatment strategy.

A number of different techniques can be used for transcriptome analysis. PCR-based methods are useful because of their ability to detect genes that are expressed at low levels. **DNA microarray analysis** is widely used because it enables researchers to analyze all of a sample's expressed genes simultaneously.

Most microarrays, also known as **gene chips**, consist of a glass microscope slide onto which single-stranded DNA molecules are attached, or "spotted," using a computer-controlled high-speed robotic arm called an arrayer. Arrayers are fitted with a number of tiny pins. Each pin is immersed in a small amount of solution containing millions of copies of a different single-stranded DNA molecule. For example, many microarrays are made with single-stranded sequences of complementary DNA (cDNA) or expressed sequenced tags (ESTs)—short fragments of cloned DNA from expressed genes. The arrayer fixes the DNA onto the slide at specific locations (points, or spots) that are recorded by a computer. A single microarray can have over 20,000 different spots of DNA, each containing a unique sequence for a different gene. Entire genomes are available on microarrays, including the

human genome. As you will learn in Chapter 22, researchers are also using microarrays to compare patterns of gene expression in tissues in response to different conditions, to compare gene-expression patterns in normal and diseased tissues, and to identify pathogens.

To prepare a microarray for use in transcriptome analysis, scientists typically begin by extracting mRNA from cells or tissues (**Figure 21–21**). The mRNA is usually then reverse transcribed to synthesize cDNA tagged with fluorescently labeled nucleotides. The mRNA or cDNA can be labeled in a number of ways, but most methods involve the use of fluorescent dyes. Typically, microarray studies often involve

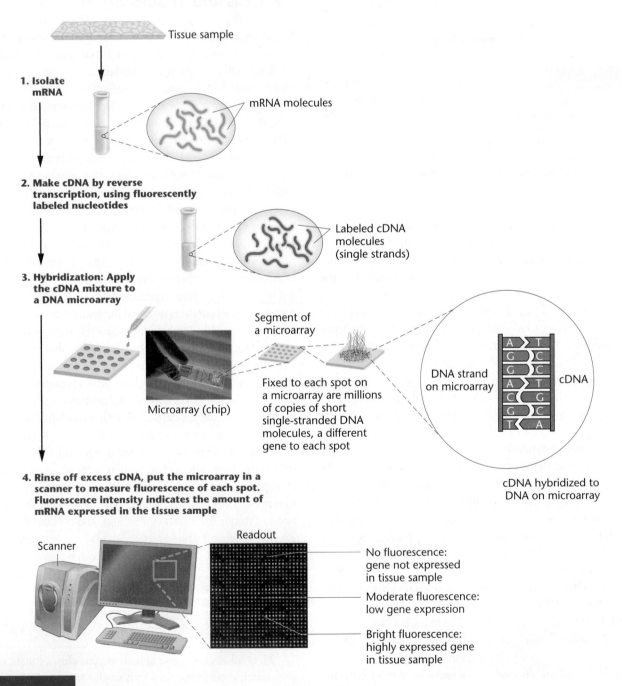

FIGURE 21–21   Microarray analysis for analyzing gene-expression patterns in a tissue.

comparing gene expression in different cell or tissue samples. cDNA prepared from one tissue is usually labeled with one color dye, red for example, and cDNA from another tissue labeled with a different-colored dye, such as green. Labeled cDNAs are then denatured and incubated overnight with the microarray so that they will hybridize to spots on the microarray that contain complementary DNA sequences. Next, the microarray is washed, and then it is scanned by a laser that causes the cDNA hybridized to the microarray to fluoresce. The patterns of fluorescent spots reveal which genes are expressed in the tissue of interest, and the intensity of spot fluorescence indicates the relative level of expression. The brighter the spot, the more the particular mRNA is expressed in that tissue.

Microarrays are dramatically changing the way gene expression patterns are analyzed. As discussed in Chapter 20, Northern blot analysis was one of the earliest methods used for analyzing gene expression. Then PCR techniques proved to be rapid and more sensitive approaches. The biggest advantage of microarrays is that they enable thousands of genes to be studied simultaneously. As a result, however, they can generate an overwhelming amount of gene-expression data. In addition, even when properly controlled, microarrays often yield variable results. For example, one experiment under certain conditions may not always yield similar patterns of gene expression as another identical experiment. Some of these differences can be due to real differences in gene expression, but others can be the result of variability in chip preparation, cDNA synthesis, probe hybridization, or washing conditions, all of which must be carefully controlled to limit such variability. Commercially available microarrays can reduce the variability that can result when individual researchers make their own arrays.

Computerized microarray data analysis programs are essential for organizing gene-expression profile data from microarrays. For instance, **cluster algorithm** programs can be used to retrieve spot-intensity data from different locations on a microarray and to group gene-expression data from one or multiple microarrays into cluster images incorporating results from many experiments. Cluster analysis groups genes according to whether they show increased (upregulated) or decreased (downregulated) expression under the experimental conditions examined. Figure 21–22(a) shows hierarchical clusters of upregulated and downregulated gene-expression patterns for the yeast *Saccharomyces cerevisiae* grown under different experimental culture conditions. Notice that different culture times reveal different patterns of downregulation (yellow oval) and upregulation (white oval). The identities of the genes in these regulated clusters indicate that many of the genes affected by the growth conditions of this experiment are genes involved in cell division.

(a)

Experiments 1–15

| 10min | 30min | 50min | 70min | 90min | 110min | 130min | 150min | 170min | 190min | 210min | 230min | 250min | 270min | 290min |

Genes

	unknown
SCW11	cell wall biogenesis glucanase (putative)
CLB6	cell cycle B-type cyclin; S phase
KCC4	bud growth protein kinase
MCD1	mitosis, sister chromatid cohesion unknown
POL30	DNA replication DNA polymerase processivity factor
	unknown
RFA1	DNA replication replication factor A, 69 kD subunit
	unknown
STP3	tRNA splicing
TOS6	unknown; similar to Mid2p
ISR1	staurosporine resistance protein kinase
SRO4	bud site selection, plasma membrane protein
CSI2	cell wall biogenesis chitin synthase 3 subunit
SVS1	vanadate resistance
CLN2	cell cycle G1/S cyclin

Cluster of down-regulated genes

Cluster of up-regulated genes

(b)

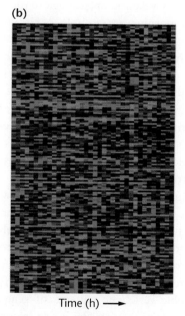

Time (h) ⟶

FIGURE 21–22 (a) Hierarchical clusters in a microarray experiment using RNA samples from *Saccharomyces cerevisiae* grown for varying times in culture (experiments 1–15). Gene identities are labeled to the right of the array image. Red color in the array indicates upregulation in the experimental sample, and green indicates downregulation in the sample, as compared to a control culture. The intensity of the color indicates the magnitude of up- or downregulation. Brighter spots represent higher levels of expression than dimmer or black spots. The yellow oval highlights a cluster of downregulated genes, and the white oval highlights a cluster of upregulated genes. (b) High (red), intermediate (black), and low (green) levels of gene expression for *Drosophila* genes that exhibit a circadian rhythm. RNA samples from *Drosophila* were collected every four hours for six days and then used for microarray analysis. Each row shows all 36 responses of a single gene. Each column shows the gene-expression pattern for each time point sampled. Genes were arranged from top to bottom according to the time of peak activity.

(a)

(b)

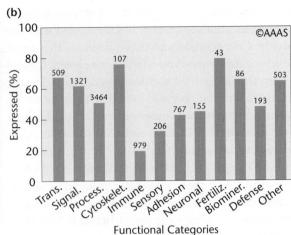

FIGURE 21–23 (a) The sea urchin *Strongylocentrotus purpuratus*. (b) Transcriptome analysis of genes expressed in the sea urchin embryo. The *y*-axis of the histogram displays the percentage of annotated genes in different functional categories expressed in the embryo. The number at the top of each bar represents the total number of annotated genes in the corresponding functional category. Trans., transcription factors; Signal, signaling genes; Process, basic cellular processes, such as metabolism; Cytoskelet., cytoskeletal genes; Fertiliz., fertilization; and Biominer., biomineralization.

Figure 21–22(b) reveals an interesting gene-expression profile for genes in *Drosophila* in which they display repeating patterns of expression as part of a **circadian rhythm** response. Circadian rhythms are oscillations in biological activity that occur on a regular cycle of time, such as 24 hours. Reductions in brain wave electrical activity that occur when you sleep, followed by the increased brain wave activity that occurs to wake you up, just before your alarm clock sounds in the morning, are examples of circadian responses. In the *Drosophila* microarray, each horizontal row shows expression results for a different gene. Notice how every several hours the expression for most genes is upregulated (red) and then downregulated (green) several hours later in a rhythmic pattern that repeats over the time course of this experiment (6 days). Results such as these are representative of the types of gene-expression profiles that can be revealed by microarrays. You will see similar representations of microarray data in Chapter 24.

In Section 21.6 we briefly discussed the sea urchin genome and the importance of the sea urchin as a key model organism. Sea urchins have a relatively simple body plan [Figure 21–23(a)]. They contain approximately 1500 cells with only a dozen cell types. Yet sea urchin development progresses through complex changes in gene expression that resemble vertebrate patterns of gene-expression changes during development. This is one reason urchins are a valuable model organism for developmental biology.

Scientists at NASA's Ames Genome Research Facility in Moffett, California, used microarrays to carry out a transcriptome study on sea urchin gene expression during the first two days of development (to the mid-late gastrula stage, about 48 hours postfertilization). This work revealed that approximately 52 percent of all genes in the sea urchin are active during this period of development: 11,500 of the sea urchin's 23,500 known genes were expressed in the embryo. The functional categories of genes expressed in the embryo were diverse, including genes for about 70 percent of the nearly 300 transcription factors in the sea urchin genome, along with genes involved in cell signaling, immunity, fertilization, and metabolism [Figure 21–23(b)]. Incredibly, 51,000 RNAs of unknown function were also expressed. Studies are underway to explain the differences between gene number and transcripts expressed, although it is already known that many sea urchin genes are extensively processed through alternative splicing. Further analysis of the sea urchin genome will undoubtedly reveal interesting aspects of gene function during sea urchin development and advance our understanding of the genetics of embryonic development in both invertebrates and vertebrates.

Now that we have considered genomes and transcriptomes, we turn our attention to the ultimate end products of most genes, the proteins encoded by a genome.

21.9

Proteomics Identifies and Analyzes the Protein Composition of Cells

As more genomes have been sequenced and studied, biologists in many different disciplines have focused increasingly on understanding the complex structures and functions of the proteins the genomes encode. This interest is not surprising given that in most of the genomes sequenced to date, many newly discovered genes and their putative proteins have no known function. Keep in mind, in the ensuing discussion, that although every cell in the body contains an equivalent set of genes, not all cells express the same genes and proteins. **Proteome** is a term that represents the complete set of proteins encoded by a genome, but it is also often used to mean the entire complement of proteins in a cell. This definition would then include proteins that a cell acquired from another cell type.

Proteomics—the complete identification, characterization, and quantitative analysis of the proteome of a cell, tissue, or organism—can be used to reconcile differences between the number of genes in a genome and the number of different proteins produced. But equally important, proteomics also provides information about a protein's structure and function; posttranslational modifications; protein–protein, protein–nucleic acid, and protein–metabolite interactions; cellular localization of proteins; protein stability and aspects of translational and posttranslational levels of gene-expression regulation; and relationships (shared domains, evolutionary history) to other proteins. Proteomics projects have been used to characterize major families of proteins for some species. For example, about two-thirds of the *Drosophila* proteome has been well catalogued using proteomics.

Proteomics is also of clinical interest because it allows comparison of proteins in normal and diseased tissues, which can lead to the identification of proteins as biomarkers for disease conditions. Proteomic analysis of mitochondrial proteins during aging, proteomic maps of atherosclerotic plaques from human coronary arteries, and protein profiles in saliva as a way to detect and diagnose diseases are examples of such work.

Reconciling the Number of Genes and the Number of Proteins Expressed by a Cell or Tissue

Recall from Chapter 14 the one gene:one polypeptide hypothesis of George Beadle and Edward Tatum. As we have discussed in that chapter and elsewhere, genomics has revealed that the link between gene and gene product is often much more complex. Genes can have multiple transcription start sites that produce several different types of transcripts. Alternative splicing and editing of pre-mRNA molecules can generate dozens of different proteins from a single gene. Remember the current estimate that over 50 percent of human genes produce more than one protein by alternative splicing. As a result, proteomes are substantially larger than genomes. For instance, the ~20,000 genes in the human genome encode ~100,000 proteins, although some estimates suggest that the human proteome may be as large as 150,000–200,000 proteins.

Proteomes undergo dynamic changes that are coordinated in part by regulation of gene-expression patterns—the transcriptome. However, a number of other factors affect the proteome profile of a cell, further complicating the analysis of protein function. For instance, many proteins are modified by co-translational or posttranslational events, such as cleavage of signal sequences that target a protein for an organelle pathway, propeptides, or initiator methionine residues; by linkage to carbohydrates and lipids; or by the addition of chemical groups through methylation, acetylation, and phosphorylation and other modifications. Over a hundred different mechanisms of posttranslational modification are known. In addition, many proteins work via elaborate protein–protein interactions or as part of a large molecular complex.

Well before a draft sequence of the human genome was available, scientists were already discussing the possibility of a "Human Proteome Project." One reason such a project never came to pass is that there is no single human proteome: different tissues produce different sets of proteins. But the idea of such a project led to the **Protein Structure Initiative (PSI)** by the National Institute of General Medical Sciences (NIGMS), a division of the National Institutes of Health, involving over a dozen research centers. Initiated in 2000, PSI is a multiphase project designed to analyze the three-dimensional structures of more than 4000 protein families. Proteins with interesting potential therapeutic properties are a top priority for the PSI, and to date the structures of over 1000 proteins have been determined. Developing computation protein structural prediction methods, solving unique protein structures, disseminating PSI information, and focusing on the biological relevance of the work are major goals. There also are a number of other ongoing projects dedicated to identifying proteome profiles that correlate with diseases such as cancer and diabetes.

Proteomics Technologies: Two-Dimensional Gel Electrophoresis for Separating Proteins

In Chapter 20, we explained how in the early days of DNA cloning and recombinant DNA technology, before genomics, scientists often cloned, sequenced, and studied one or a few genes at a time over a span of several years. Until relatively recently, a similar approach (involving different techniques, however) typically applied to studying individual proteins or a relatively small number of interrelated proteins simultaneously. But now, with proteomics technologies, scientists have the ability to study thousands of proteins simultaneously, generating enormous amounts of data quickly and dramatically changing ways of analyzing the protein content of a cell.

The early history of proteomics dates back to 1975 and the development of **two-dimensional gel electrophoresis (2DGE)** as a technique for separating hundreds to thousands of proteins with high resolution. In this technique, proteins isolated from cells or tissues of interest are loaded onto a polyacrylamide tube gel and first separated by **isoelectric focusing**,

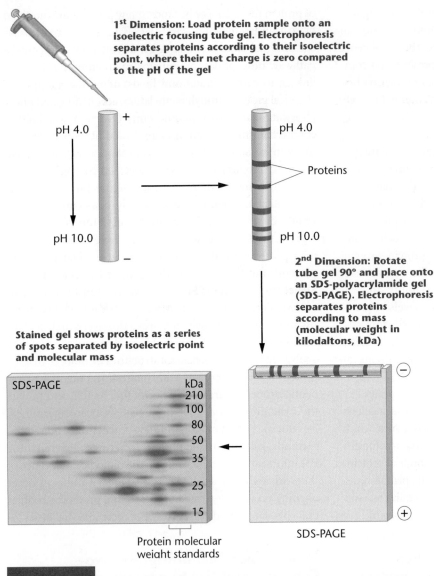

1st Dimension: Load protein sample onto an isoelectric focusing tube gel. Electrophoresis separates proteins according to their isoelectric point, where their net charge is zero compared to the pH of the gel

pH 4.0

pH 10.0

pH 4.0

Proteins

pH 10.0

2nd Dimension: Rotate tube gel 90° and place onto an SDS-polyacrylamide gel (SDS-PAGE). Electrophoresis separates proteins according to mass (molecular weight in kilodaltons, kDa)

Stained gel shows proteins as a series of spots separated by isoelectric point and molecular mass

SDS-PAGE

kDa
210
100
80
50
35
25
15

Protein molecular weight standards

SDS-PAGE

FIGURE 21–24   Two-dimensional gel electrophoresis (2DGE) is a useful method for separating proteins in a protein extract from cells or tissues that contains a complex mixture of proteins with different biochemical properties.

in human platelets (thrombocytes). Particularly abundant protein spots in this gel have been labeled with the names of identified proteins. With thousands of different spots on the gel, how are the identities of the proteins ascertained?

In some cases, 2D gel patterns from experimental samples can be compared to gels run with reference standards containing known proteins with well-characterized migration patterns. Many reference gels for different biological samples such as human plasma are available, and computer software programs can be used to align and compare the spots from different gels. In the early days of 2DGE, proteins were often identified by cutting spots out of a gel and sequencing the amino acids the spots contained. Only relatively small sequences of amino acids can typically be generated this way; rarely can an entire polypeptide be sequenced using this technique. BLAST and similar programs can be used to search protein databases containing amino acid sequences of known proteins. However, because of alternative splicing or post-translational modifications, peptide sequences may not always match easily with the final product, and the identity of the protein may have to be confirmed by another approach. As you will learn in the next section, proteomics has incorporated other techniques to aid in protein identification, and one of these techniques is mass spectrometry.

which causes proteins to migrate according to their electrical charge in a pH gradient. During isoelectric focusing, proteins migrate until they reach the location in the gel where their net charge is zero compared to the pH of the gel (Figure 21–24). Then in a second migration, perpendicular to the first, the proteins are separated by their molecular mass using **sodium dodecyl sulfate polyacrylamide gel electrophoresis (SDS-PAGE)**. In this step, the tube gel is rotated 90° and placed on top of an SDS polyacrylamide gel; an electrical current is applied to the gel to separate the proteins by mass.

Proteins in the 2D gel are visualized by staining with Coomassie blue, silver stain, or other dyes that reveal the separated proteins as a series of spots in the gel (Figure 21–25). It is not uncommon for a 2D gel loaded with a complex mixture of proteins to show several thousand spots in the gel, as in Figure 21–25, which displays the complex mixture of proteins

Proteomics Technologies: Mass Spectrometry for Protein Identification

As important as 2DGE has been for protein analysis, **mass spectrometry (MS)** has been instrumental to the development of proteomics. Mass spectrometry techniques analyze ionized samples in gaseous form and measure the **mass-to-charge (m/z) ratio** of the different ions in a sample. Proteins analyzed by mass spectra generate m/z spectra that can be correlated with an m/z database containing known protein sequences to discover the protein's identity. Certain MS applications can provide peptide sequences directly from spectra. Some of the most valuable proteomics applications of this technology are to identify an unknown protein or proteins in a complex mix of proteins, to sequence peptides, to

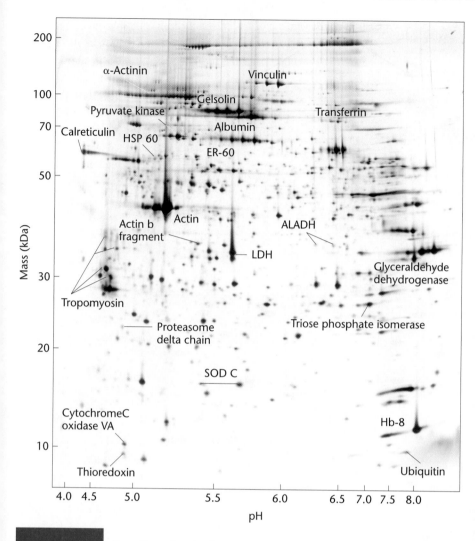

FIGURE 21–25 Two-dimensional gel separations of human platelet proteins. Each spot represents a different polypeptide separated by molecular weight (*y*-axis) and isoelectric point, pH (*x*-axis). Known protein spots are labeled by name based on identification by comparison to a reference gel or by determination of protein sequence using mass spectrometry techniques. Notice that many spots on the gel are unlabeled, indicating proteins of unknown identity.

identify posttranslational modifications of proteins, and to characterize multiprotein complexes.

One commonly used mass spectrometry approach is **matrix-assisted laser desorption ionization (MALDI)**. MALDI is ideally suited for identifying proteins and is widely used for proteomic analysis of tissue samples treated under different conditions. The proteins are first extracted from cells or tissues of interest and separated by 2DGE, after which MALDI (described below) is used to identify the proteins in the different spots. Figure 21–26 shows an example in which two different sets of cells grown in culture are analyzed for protein differences. Just about any source providing a sufficient number of cells can be used: blood, whole tissues, and organs; tumor samples; microbes; and many other substances. Many proteins involved in cancer have been identified by the use of MALDI to compare protein profiles in normal tissue and tumor samples.

Protein spots are cut out of the 2D gel, and proteins are purified out of each gel spot. Computer-automated high-throughput instruments are available that can pick all of the spots out of a 2D gel. Isolated proteins are then enzymatically digested with a protease (a protein-digesting enzyme) such as trypsin to create a series of peptides. This proteolysis produces a complex mixture of peptides determined by the cleavage sites for the protease in the original protein. Each type of protein produces a characteristic set of peptide fragments, and these are identified by MALDI as follows.

In MALDI, peptides are mixed with a low molecular weight, and ultraviolet (UV) light-absorbing acidic matrix material (such as dihydroxybenzoic acid) is then applied to a metal plate. A UV laser, often a nitrogen laser at a wavelength of 337 nm, is then fired at the sample. As the matrix absorbs energy from the laser, heat accumulating on the matrix vaporizes and ionizes the peptide fragments. Released ions are then analyzed for mass; MALDI displays the m/z ratio of each ionized peptide as a series of peaks representative of the molecular masses of peptides in the mixture and their relative abundance (Figure 21–27). Because different proteins produce different sets of peptide fragments, MALDI produces a peptide "fingerprint" that is characteristic of the protein being analyzed.

Databases of MALDI-generated m/z spectra for different peptides can be analyzed to look for matches between m/z spectra of unknown samples and those of known proteins. One limitation of this approach is database quality. An unknown protein from a 2D gel can only be identified by MALDI if proteomics databases have a MALDI spectrum for that protein. But as is occurring with genomics databases, proteomics databases with thousands of well-characterized proteins from different organisms are rapidly developing.

MALDI is often coupled with a protein biochemistry technique for mass analysis called **time of flight (TOF)**. TOF moves ionized peptide fragments through an electrical field in a vacuum. Each ion has a speed that varies with its mass. The speed with which each ion crosses the vacuum chamber can be

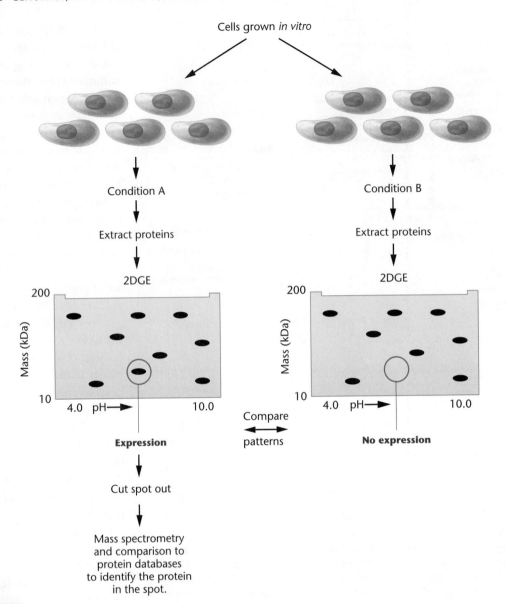

Cells grown *in vitro*

Condition A

Extract proteins

2DGE

Condition B

Extract proteins

2DGE

Expression

Compare patterns

No expression

Cut spot out

Mass spectrometry and comparison to protein databases to identify the protein in the spot.

FIGURE 21–26 In a typical proteomic analysis, cells are exposed to different conditions (such as different growth conditions, drugs, or hormones). Then proteins are extracted from these cells and separated by 2DGE, and the resulting patterns of spots are compared for evidence of differential protein expression. Spots of interest are cut out from the gel, digested into peptide fragments, and analyzed by mass spectrometry to identify the protein they contain.

measured, and differences in the ions' kinetic energy can be used to develop a mass-dependent velocity profile—a MALDI-TOF spectrum—that shows each ion's "time of flight." MALDI-TOF spectra can then be compared to databases of spectra for known proteins, as described above for MALDI spectra.

Many other methods involve mass spectrometry. Some incorporate liquid chromatography (LC) to separate proteins by mass and then employ **tandem mass spectrometry (MS/MS)** approaches to generate m/z spectra. Also emerging are new mass spectrometry techniques that do not involve the running of gels. As we mentioned when discussing genomics, high-throughput 2DGE instruments and mass spectrometers can process thousands of samples in a

single day. Instruments with faster sample-processing times and increased sensitivity are under development. These instruments may soon make "shotgun proteomics" a viable approach for characterizing entire proteomes.

Protein microarrays are also becoming valuable tools for proteomics research. These are designed around the same basic concept as microarrays (gene chips) and are often constructed with antibodies that specifically recognize and bind to different proteins. These microarrays are used, among other applications, for examining protein–protein interactions, for detecting protein markers for disease diagnosis, and for studying in biosensors designed to detect pathogenic microbes and potentially infectious bioweapons.

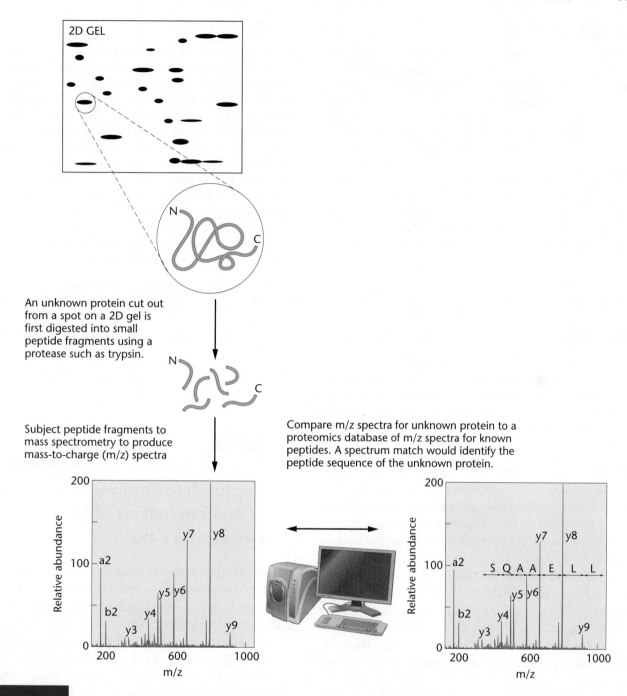

FIGURE 21–27 Mass spectrometry for identifying an unknown protein isolated from a 2D gel. The mass-to-charge spectrum (m/z) (determined, for example, by MALDI) for trypsin-digested peptides from the unknown protein can be compared to a proteomics database for a spectrum match to identify the unknown protein. The peptide in this example was revealed to have the amino acid sequence serine (S)-glutamine (Q)-alanine (A)-alanine (A)-glutamic acid (E)-leucine (L)-leucine (L), shown in single-letter amino acid code.

Identification of Collagen in *Tyrannosaurus rex* and *Mammut americanum* Fossils

Recently, a team of scientists reported results of mass spectrometry analysis of bone tissue from a *Tyrannosaurus rex* skeleton excavated from the Hell Creek Formation in eastern Montana and estimated to be 68 million years old. As mentioned earlier,

DNA has been recovered from fossils, but the general assumption has been that proteins degrade in fossilized materials and cannot be recovered. This study demonstrated that fossilization does not fully destroy all proteins in well-preserved fossils under certain conditions. This research also demonstrates the power and sensitivity of mass spectrometry as a proteomics tool.

In this work, medullary tissue was removed from the inside of the left and right femoral bones. Medullary tissue is

21–3 Annotation of a proteome attempts to relate each protein to a function in time and space. Traditionally, protein annotation depended on an amino acid sequence comparison between a query protein and a protein with known function. If the two proteins shared a considerable portion of their sequence, the query would be assumed to share the function of the annotated protein. Following is a representation of this method of protein annotation involving a query sequence and three different human proteins. Note that the query sequence aligns to common domains within the three other proteins. What argument might you present to suggest that the function of the query is not related to the function of the other three proteins?

—— Query amino acid sequence

Region of amino acid sequence match to query

■ HINT: *This problem asks you to think about sequence similarities between four proteins and predict functional relationships. The key to its solution is to remember that although protein domains may have related functions, proteins can contain several different interacting domains that determine protein function.*

porous, spongy bone that contains bone marrow cells, blood vessels, and nerves. *T. rex* proteins extracted from the tissue showed cross-reactivity with antibodies to chicken collagen and were digested by the collagen-specific protease collagenase. These results suggested that the *T. rex* protein samples contained collagen, a major matrix component of bone, ligaments, tendons, and skin. To definitively identify the presence of collagen, tryptic peptides from the *T. rex* samples were analyzed by liquid chromatography and mass spectrometry (LC/MS; Figure 21–27). The m/z spectra for one of the *T. rex* peptides was identified from a database of m/z spectra as corresponding to collagen. Compare the spectrum for a collagen peptide in Figure 21–28(a) to that of a synthetic version of a collagen peptide [Figure 21–28(b)], and you will notice that the m/z ratios for all major ions align almost identically, confirming that the *T. rex* sequence is collagen. The *T. rex* peptide also contained a hydroxyl group attached to a proline residue. Proline hydroxylation is a characteristic feature of collagen. Furthermore, the amino acid sequence of *T. rex* collagen peptide aligned with an isoform of chicken collagen, demonstrating sequence similarity. Such work has provided excellent experimental evidence to support the widely accepted theory that birds and dinosaurs are close relatives.

Similar results were obtained for 160,000- to 600,000-year-old mastodon (*Mammut americanum*) peptides that showed matches to collagen from extant species, including collagen isoforms from humans, chimps, dogs, cows, chickens, elephants, and mice.

21–4 Because of its accessibility and biological significance, the proteome of human plasma has been intensively studied and used to provide biomarkers for such conditions as myocardial infarction (troponin) and congestive heart failure (B-type natriuretic peptide). Polanski and Anderson (Polanski, M., and Anderson, N. L., *Biomarker Insights*, 2: 1–48, 2006) have compiled a list of 1261 proteins, some occurring in plasma, that appear to be differentially expressed in human cancers. Of these 1261 proteins, only 9 have been recognized by the FDA as tumor-associated proteins. First, what advantage should there be in using plasma as a diagnostic screen for cancer? Second, what criteria should be used to validate that a cancerous state can be assessed through the plasma proteome?

■ HINT: *This problem asks you to consider criteria that are valuable for using plasma proteomics as a diagnostic screen for cancer. The key to its solution is to consider proteomics data that you would want to evaluate to determine whether a particular protein is involved in cancer.*

21.10

Systems Biology Is an Integrated Approach to Studying Interactions of All Components of an Organism's Cells

We conclude this chapter by discussing **systems biology**, an emerging discipline that incorporates data from genomics, transcriptomics, proteomics, and other areas of biology, as well as engineering applications and problem-solving approaches. Identifying genes and proteins by mutational analysis of genomes has been a very important and successful approach for characterizing genes when mutants showing visible phenotypes are found to be part of similar biochemical pathways. However, even extensive mutational analysis and screening will not provide a full understanding of complex cellular processes such as signal transduction pathways, metabolic pathways, and regulation of cell division, DNA replication, and gene expression. A more comprehensive, more integrated approach is needed.

As we mentioned earlier in this chapter, until relatively recently, much of what has been learned about gene and protein functions at the cellular, molecular, and biochemical levels has been acquired primarily through decades of work by scientists studying the functions of individual genes or relatively small numbers of genes and proteins. Many researchers have spent entire careers studying one gene or protein. However, just when it seems as if we know all there is to

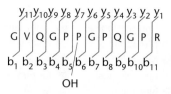

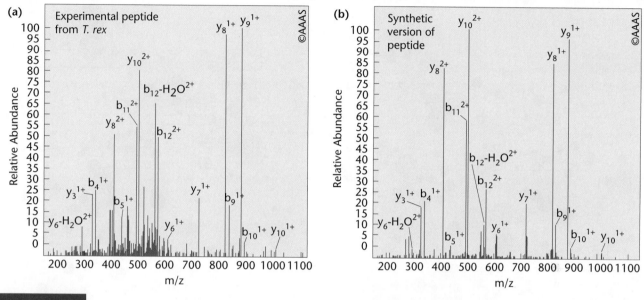

FIGURE 21–28 (a) Mass spectrometry (MS) patterns for a trypsin-digested peptide sequence—GVQPP(OH)GPQGPR—from *T. rex*. The peptide sequence, shown here in single-letter amino acid code, contains a charged hydroxyl group characteristic of collagen. (b) Mass spectrometry of a synthetic version of collagen peptide shows good alignment with the m/z spectra for fragmented ions from the *T. rex* peptide, thus confirming the *T. rex* sequence as collagen and demonstrating the value of MS techniques.

know about even the most well-characterized protein, another study reveals that it possesses novel functions. Such revelations demonstrate the incompleteness of our understanding of the extreme complexity of genes and proteins in a cell. As a simple analogy, you could study the individual components of your cell phone, but until you focused on how the many components interact, you would not truly understand how a cell phone works.

Linking genomic studies to gene function and the physiology of different disease states is essential. For example, genome-wide association studies have uncovered many loci associated with diseases but translating such information to make valuable and reliable diagnostic predictions and to develop drug targets of key proteins involved in disease is necessary. A major challenge in modern biology is to unravel the interactions of genes and proteins that enable complex processes such as cell division; this is what systems biology is all about. Proteins occasionally function alone, but more typically they work in complex interconnected networks under the regulation and control of other proteins or metabolites. Networks of interacting proteins form the regulatory framework controlling how cells respond to environmental signals, metabolize nutrients, move organelles, divide, and carry out many other processes. As genomics

and proteomics have advanced, the discipline of systems biology has emerged as a more holistic approach to studying cell function by analyzing interactions among all of the molecular components of a biological system. Systems biology considers genes, proteins, metabolites, and other interacting molecules of a cell in order to understand molecular interactions and to integrate such information into models that can be used to better understand the biological functions of an organism.

In many ways, systems biology is interpreting genomic information in the context of the structure, function, and regulation of biological pathways. As is well known, biological systems are very complex. By studying relationships between all components in an organism, biologists are trying to build a "systems"-level understanding of how organisms function. Systems biologists typically combine recently acquired genomics and proteomics data with years of more traditional studies of gene and protein structure and function. Much of this data is retrieved from databases such as PubMed, GenBank, and other newly emerging genomics, transcriptomics, and proteomics resources. Systems models are used to diagram interactions within a cell or an entire organism, such as protein–protein interactions, protein–nucleic acid interactions, and protein–metabolite interactions

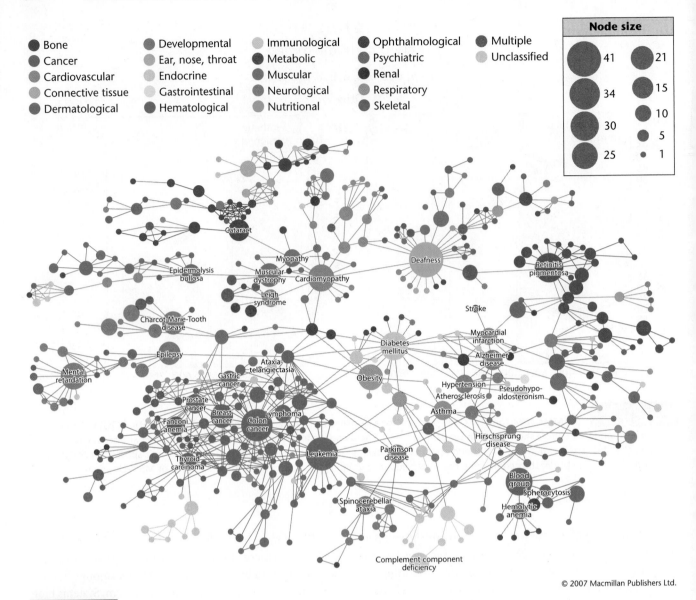

Node size

41	21
34	15
30	10
25	5
	1

- Bone
- Cancer
- Cardiovascular
- Connective tissue
- Dermatological
- Developmental
- Ear, nose, throat
- Endocrine
- Gastrointestinal
- Hematological
- Immunological
- Metabolic
- Muscular
- Neurological
- Nutritional
- Ophthalmological
- Psychiatric
- Renal
- Respiratory
- Skeletal
- Multiple
- Unclassified

© 2007 Macmillan Publishers Ltd.

FIGURE 21–29 A systems biology model of human disease gene interactions. The model shows nodes corresponding to 22 specific disorders colored by class. Node size is proportional to the number of genes contributing to the disorder.

(e.g., enzyme-substrate binding). These models help systems biologists understand the components of interacting pathways and the interrelationships of molecules in an interacting pathway. In recent years, the term **interactome** has arisen to describe the interacting components of a cell. Systems biologists use several different types of models to diagram protein interaction pathways. One of the most common model types is a **network map**—a sketch showing interacting proteins, genes, and other molecules. These diagrams are essentially the equivalent of an electrical wiring diagram. One disadvantage of network maps is that they are static diagrams that typically lack information about when and where each interaction occurs. Even so, they are a useful foundation for generating computational models that al-

low the running of simulations to determine how signaling events occur. For example, major groups of kinases, enzymes that phosphorylate other proteins to affect their activity, have been network mapped to show their interactions with each other. Because kinases play such important roles in the regulation of most critical cellular processes, such information about the "kinome" has been very valuable for companies developing drug treatments targeted to certain metabolic pathways.

Network maps are helping scientists model intricate potential interactions of molecules involved in normal and disease processes. **Figure 21–29** shows an example of a network map. This map depicts a human disease network model illustrating the complexity of interactions between

EXPLORING GENOMICS

Contigs, Shotgun Sequencing, and Comparative Genomics

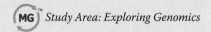

Study Area: Exploring Genomics

In this chapter, we discussed how whole-genome shotgun sequencing methods can be used to assemble chromosome maps. Recall that in the technique of shotgun cloning, chromosomal DNA is first digested with different restriction enzymes to create a series of overlapping DNA fragments called contiguous sequences, or "contigs." The contigs are then subjected to DNA sequencing, after which bioinformatics-based programs are used to arrange the contigs in their correct order on the basis of short overlapping sequences of nucleotides.

In this Exploring Genomics exercise you will carry out a simulation of contig alignment to help you to understand the underlying logic of this approach to creating sequence maps of a chromosome. For this purpose, you will return to the **National Center for Biotechnology Information BLAST** site that was used in other Exploring Genomics exercises and apply a DNA alignment program called bl2seq.

■ Exercise I – Arranging Contigs to Create a Chromosome Map

1. Access BLAST from the NCBI Web site at http://www.ncbi.nlm.nih.gov/ BLAST. Locate and select the "Align two sequences using BLAST (bl2seq)" category at the bottom of the BLAST homepage. The bl2seq feature allows you to compare two DNA sequences at a time to check for sequence similarity alignments.

2. Below are eight contig sequences taken from an actual human chromosome sequence deposited in GenBank. For this exercise we have used short fragments; however, in reality contigs are usually several thousand base pairs long. To complete this exercise, copy and paste two sequences into the Align feature of BLAST and then run an alignment (by clicking on "Align"). (*Hint*: Access these sequences from the Companion Web site so that you can copy and paste the sequences easily.) Repeat these steps with other combinations of two sequences to determine which sequences overlap, and then use your findings to create a sequence map that places overlapping contigs in their proper order. Here are a few tips to consider:

   ■ Develop a strategy to be sure that you analyze alignments for all pairs of contigs.

   ■ Only consider alignment overlaps that show 100 percent sequence similarity.

Sequence A

```
CTAATTTTTTTTGTATTTTTA
ATAGAGACGAGGTGTCAC
CATGTTGGACAGGCTGGTCTC
GAACTCCTGACCTCAGGTGATCT
GCCCACCTCAGCCTCCCAAAGT
GCTGGGATTACAAGCAT
GAGCCACCACTCCCAGGCTT
```

```
TATTTTCTATTTTTTAATTA
CAGCCATCCTAGTGAATGT
GAAGTAGTATCTCACTGAG
GTTTTGATTTGCATTTTTCTAT
GACAATGAACAATGTTTCAT
GTGCTTGTTGGCTGTTT
GTATATCCTTTTTGGAGA
AATACCAATTCATGTCCTTT
GCCCATTTTTAAAGTGGATTG
CATGTCTTTTTGTTGTTTAGTTG
TAAAGATGTGGGTTTTTCTTTT
GAGACGGAGTCTCGCTGTC
GCCTAGGCTGGAGTGCAGAG
GCATGATCTCGGCTGACT
GCAATCCCCACCTCCTG
GCATCAAGAAGTTCTCCT
GCCTCAGCCTTCCAAG
TAGCTGGGTTTACAGATGC
```

Sequence B

```
CTTTATCTCAGGACAATGAACCC
GCAAGGAGAGGAAGAGC
CAGTAATTCTATAGAGACTC
GGAGGCGCAGGGGGCAC
GCTTAGTTAGAGTGGT
GGTGGTATTTTCAGT
GTTTTCTGGTTTTATGATA
AACACAAGCATCAATGTCT
CAAGACTTTCATCTTTATCTT
TTTTTTTTTTTTTTTTTTTTTTCTT
GAGACAGGGTTTCCCTCTGT
CACCCAGGCTGGAGTGCATTG
GTGGTGTGATCTTGGCTTTCT
GTAACCTCGGGGCTTCT
GGGCTCAAGCCGTTCTAC
TACCTCAGCCTCCCAAATAGC
```

(continued)

genes involved in 22 different human diseases. Look at the cluster of turquoise-colored nodes corresponding to genes involved in several different cancers. One aspect of the map that should be immediately obvious is that a number of cancers share interacting genes even though the cancers affect different organs. Knowing the genes involved and the

protein interaction networks for different cancers is a major breakthrough for informing scientists about target genes and proteins to consider for therapeutic purposes.

Systems biology is becoming increasingly important in the drug discovery and development process, where its approaches can help scientists and physicians develop a

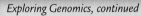

Exploring Genomics, continued

TAGAACTACAAGCGTGT
GCTGCCACACCTGGCTAATT
TGTTGTATTTTATTTAT
TCATTTATTTTTGTGAAGAC
AAGGTCTTGCCATGTTGCC
CAGGCTGGTCTCAGACT
CCTGGGCTCAAGCAATCCACCC
ACCTTAGTCTCCCAAAGT
GCTGGGATTACAGGC
GTGAGCCACCACACCCA

Sequence C

GGAATTTCACTCTTGTT
GCCCAAGTTGGAGTGCAAT
GGCGCGATCTCAGCTCACT
GCAACCTCCGCCTCCCAG
GTTCAAACGATTCTCCT
GCTTCACTCTCCCCAGTAGCT
GGGATTACAGGCTGCACCAC
CACACCTGGCTAATTTTTTTT
GTATTTTAATAGAGACGAG
GTGTCACCATGTTGGACAG
GCTGGTCTCGAACTCCT
GACCTCAGGTGATCT
GCCCACCTCAGCCTCCCAAAGT
GCTGGGATTACAAGCAT
GAGCCACCACTCCCAGGC

Sequence D

GCTTCATCTTTCTCTTCACC
GTAAAACAGGAAAGTGT
GTGGTGACCAGTATTTTA
AGGGAAAGGCACTTACAGAGA
ATTAAGCATTTGACAAAATT
TATTTACAGATATTTGTCTGTG
GACCACTTCCGCACCAGCT
GTGCATGAGAGGGCTCATT
GCTCTGAATTTGCCTCCTT
GTCTGCACCCAGGAGA
CCGTTTCCCAGATCACG
CAAACGCTGCCTTCTCCCCA
CACCAGGGCCCTCAGCAT
GGGAATGACCTTCCAGC
GCTGCACGTTTCCAATCCAT
GCTCTGTTTTTCAGTTCTG
GCTCACAGAGGACTGCT
GGTTGCAAGCAAACTT
GTATCTGGGTCTTCA

Sequence E

CCTTAAGTGATCTACCTGTCTCT
GCCTCTCAAAGTGCTGGGATTG
CAGGCATAAGCCGCCATGCCC
GGCCCAAAGTTTCTTTATAT
GTGCTGGATACTAGGCCCG
TAACAGATATACAATTTGTA
AATATTTTCTCTCATTTTGAA
GATTTTCTTTTCACTTTCTTGATA
ATGTCCTTTGTGTATTTTTT
GATAATGTCCTTTGATA
CACAAAAGTTTTTAAGTTT
GATGAAGTTCAATTTACCTAT
TATTTTCTTTTGTTGTTCAT

Sequence F

TGTTTGTTTGTTTGTTTT
GTTTCATTTTGTTTTTGAGA
CAGAGTCTTGTTCTGTC
GCCCAGTGTAGAGTGCAGT
GGCATAATCTCGGCTCACC
GTAACCTCCGCCTCCC
GGGTTCAAGCAACTCTGCCT
GCCTCAGCCTCCCAAGGAGCT
GGGATTATAGACGCCCAC
CACCATGCCTGGTTAATTTTT
GTAGTTTTTTTTAGTAGAGAT
GGGGTTTTGCCATTTTGGCCAG
GCTGGTCTTGAACTCCTGACCT
CAGGTGATCTGCCCACCCTG
GCCTCTCAAAGTGCTGGGAT
TACAGGTGTGAGCTGCCA
CACTCGGCCACAACAAATTTTT
GCACCAGTTGCTCACA

Sequence G

TCCTTTGATACACAAAAGTTTT
TAAGTTTGATGAAGTTCAATT
TACCTATTATTTTCTTTT
GTTGTTCATTCATTTTGT
GTCCTATGTAGGAATC
TATTGCCAAATTCAAGGT
GATAAAGATTTACCCCTAT
GTTTCCTTCTAAGAGTTT
TATTGTTTTAGCCCTGATATT
TAGCTAAACTTAATTGATT
TATTAAGTTTAATTTTCCTAT

GTGGTATGAAGTCATT
TATCTTCTTTAGTTCAGGATC
CAAGTGAAAGGGGCATCTTC
TATCTGGGACATGCCATTCT
CATGACAGAGGAAAAAGA
CAAAAAACTGACACATACAAT
GACTTTAAAACTTCACTCA

Sequence H

GGGTTTTTCTTTTGAGACG
GAGTCTCGCTGTCGCCTAG
GCTGGGAGTGCAGAGGCAT
GATCTCGGCTGACTG
CAATCCCCACCTCCTGGCAT
CAAGAAGTTCTCCTGCCT
CAGCCTTCCAAGTAGCTGGGTT
TACAGATGCCCACCACCAT
GCCTGGCTGGTTTTTGTATTT
TAGTAGACACGGGGTTTTAC
CATGTTGGCCGGGCTGGTCT
GGAACTCCTAACCTTAAGT
GATCTACCTGTCTCTGCCTCT
CAAAGTGCTGGGATTGCAG
GCATAAGCCGCCATGCCCG
GCCCAAAGTTTCTTTATAT
GTGCTGGATACTAGGCCCG
TAACAGATATACAATTTGTAAA

3. On the basis of your alignment results, answer the following questions, referring to the sequences by their letter codes (A through H):

   (a) What is the correct order of overlapping contigs?

   (b) What is the length, measured in number of nucleotides, of each sequence overlap between contigs?

   (c) What is the total size of the chromosome segment that you assembled?

   (d) Did you find any contigs that do not overlap with any of the others? Explain.

4. Run a nucleotide-nucleotide BLAST search (BLASTn) on any of the overlapping contigs to determine which chromosome these contigs were taken from, and report your answer.

conceptual framework of gene and protein interactions in human disease that can then serve as the rationale for effective drug design. Understanding disease development and progression by defining interaction networks of molecules in normal and diseased tissue will be important for detecting and treating complex diseases such as cancer. Many databases are now being developed to model interactomes for human diseases, including breast and prostate cancer, diabetes, asthma, and cardiovascular disease. Systems biology is also being used to create biofuels and to design genetically modified organisms for cleaning up the environment, among a range of other exciting applications.

CASE STUDY | Bioprospecting in Darwin's wake

The Global Ocean Sampling (GOS) expedition followed the route of Charles Darwin's voyage to chart the genetic diversity of microbes in the marine environment. Metagenomics was used to catalog DNA sequences and their encoded proteins from thousands of previously undescribed organisms present in samples collected in diverse oceanic regions. Although many samples remain to be analyzed, the project has already identified more than 1.2 million new genes and thousands of species previously unknown to scientists, generated data on more than 1700 previously unknown families of proteins, and assembled information about more than 6 million specific proteins. Some, like those forming light-driven proton pumps and those that function in nitrogen fixation, may have immediate applications in technology and agriculture. For now, however, the emphasis is on better understanding the distribution and diversity of microbes in the oceans. This massive project has raised several questions about bioprospecting in the waters of coastal nations.

1. Although sampling sites are selected in consultation with scientists in the host countries, who owns the organisms collected along the coast of these countries?

2. Who will own any processes or products developed from these genetic resources?

3. How can the findings from this metagenomics survey of the ocean be applied?

4. Is it surprising that so many previously undiscovered organisms are represented in the samples collected on this expedition?

Summary Points

 For activities, animations, and review quizzes, go to the study area at www.masteringgenetics.com

1. High-throughput computer-automated DNA sequencing methods coupled with bioinformatics enable scientists to assemble sequence maps of entire genomes.

2. Bioinformatics is essential for the analysis of genomes and proteomes. Bioinformatics applies computer hardware and software together with statistical approaches to analyze biological sequence data.

3. Annotation is used to identify protein-coding DNA sequencing and noncoding sequences such as regulatory elements, while bioinformatics programs are used to identify open reading frames that predict possible polypeptides coded for by a particular sequence.

4. Functional genomics predicts gene function based on sequence analysis.

5. The Human Genome Project revealed many surprises about human genetics, including gene number, the high degree of DNA sequence similarity between individuals and between humans and other species, and showed that many genes encode multiple proteins.

6. Genomics has led to other related "omics" disciplines that are rapidly changing how modern biologists study DNA, RNA, and proteins and many aspects of cell function.

7. The genomes for many important model organisms have been completed. Genomic analysis of model prokaryotes and eukaryotes has revealed similarities and important fundamental differences in genome size, gene number, and genome organization.

8. Studies in comparative genomics are revealing fascinating similarities and differences in genomes from different organisms, including the identification and analysis of gene families.

9. Metagenomics, or environmental genomics, sequences genomes of microorganisms from environmental samples, often identifying new sequences that encode proteins with novel functions.

10. DNA chips or microarrays are valuable for transcriptome analysis in studying expression patterns for thousands of genes simultaneously.

11. Methods such as two-dimensional gel electrophoresis and mass spectrometry are valuable for analyzing proteomes—the protein content of a cell.

12. Systems biology approaches are designed to provide an integrated understanding of the interactions between genes, proteins, and other molecules that govern complex biological processes.

INSIGHTS AND SOLUTIONS

1. One of the main problems in annotation is deciding how long a putative ORF must be before it is accepted as a gene. Shown here are three different ORF scans of the same *E. coli* genome region—the region containing the *lacY* gene. Regions shaded in brown indicate ORFs. The top scan was set to accept ORFs of 50 nucleotides as genes. The middle and bottom scans accepted ORFs of 100 and 300 nucleotides as genes, respectively. How many putative genes are detected in each scan? The longest ORF covers 1254 bp; the next longest, 234 bp; and the shortest, 54 bp. How can we decide the actual number of genes in this region? In this type of ORF scan, is it more likely that the number of genes in the genome will be overestimated or underestimated? Why?

   Solution: Generally one can examine conserved sequences in other organisms to indicate that an ORF is likely a coding region. One can also match a sequence to previously described sequences that are known to code for proteins. The problem is not easily solved—that is, deciding which ORF is actually a gene. The shorter the ORFs scan, the more likely the overestimate of genes because ORFs longer than 200 are less likely to occur by chance. For these scans, notice that the 50 bp scans produce the highest number of possible genes, whereas the 300 bp scan produces the lowest number (1) of possible genes.

2. Sequencing of the heterochromatic regions (repeat-rich sequences concentrated in centromeric and telomeric areas) of the *Drosophila* genome indicates that within 20.7 Mb, there are 297 protein-coding genes (Bergman et al. 2002. *genomebiology3 (12)@genomebiology.com/2002/3/12/RESEARCH/0086*). Given that the euchromatic regions of the genome contain 13,379 protein-coding genes in 116.8 Mb, what general conclusion is apparent?

   Solution: Gene density in euchromatic regions of the *Drosophila* genome is about one gene per 8730 base pairs,

50

Sequenced strand

Complementary strand

100

Sequenced strand

Complementary strand

300

Sequenced strand

Complementary strand

while gene density in heterochromatic regions is one gene per 70,000 bases (20.7 Mb/297). Clearly, a given region of heterochromatin is much less likely to contain a gene than the same-sized region in euchromatin.

Problems and Discussion Questions

 For instructor-assigned tutorials and problems, go to www.masteringgenetics.com

HOW DO WE KNOW?

1. In this chapter, we focused on the analysis of genomes, transcriptomes, and proteomes and considered important applications and findings from these endeavors. At the same time, we found many opportunities to consider the methods and reasoning by which much of this information was acquired. From the explanations given in the chapter, what answers would you propose to the following fundamental questions:

   (a) How do we know which contigs are part of the same chromosome?

   (b) How do we know if a genomic DNA sequence contains a protein-coding gene?

   (c) What evidence supports the concept that humans share substantial sequence similarities and gene functional similarities with model organisms?

   (d) How can proteomics identify differences between the number of protein-coding genes predicted for a genome and the number of proteins expressed by a genome?

   (e) What evidence indicates that gene families result from gene duplication events?

   (f) How have microarrays demonstrated that, although all cells of an organism have the same genome, some genes are expressed in almost all cells, whereas other genes show cell- and tissue-specific expression?

2. What is functional genomics? How does it differ from comparative genomics?

3. Compare and contrast whole-genome shotgun sequencing to a map-based cloning approach.

4. What, if any, features do bacterial genomes share with eukaryotic genomes?

5. What is bioinformatics, and why is this discipline essential for studying genomes? Provide two examples of bioinformatics applications.

6. List and describe three major goals of the Human Genome Project.

7. How do high-throughput techniques such as computer-automated and next-generation sequencing and mass spectrometry facilitate research in genomics and proteomics? Explain.

8. BLAST searches and related applications are essential for analyzing gene and protein sequences. Define BLAST, describe basic features of this bioinformatics tool, and provide an example of information provided by a BLAST search.

9. What are pseudogenes, and how are they produced?

10. Describe the human genome in terms of genome size, the percentage of the genome that codes for proteins, how much is composed of repetitive sequences, and how many genes it contains. Describe two other features of the human genome.

11. Compare the organization of bacterial genes to that of eukaryotic genes. What are some of the major differences?

12. The Human Genome Project has demonstrated that in humans of all races and nationalities approximately 99.9 percent of the sequence is the same, yet different individuals can be identified by DNA fingerprinting techniques. What is one primary variation in the human genome that can be used to distinguish different individuals? Briefly explain your answer.

13. Annotation involves identifying genes and gene-regulatory sequences in a genome. List and describe characteristics of a genome that are hallmarks for identifying genes in an unknown sequence. What characteristics would you look for in a prokaryotic genome? A eukaryotic genome?

14. Through the Human Genome Project (HGP) a relatively accurate human genome sequence was published in 2003 from combined samples from different individuals. It serves as a reference for a haploid genome. Recently, genomes of a number of individuals have been sequenced under the auspices of the Personal Genome Project (PGP). How do results from the PGP differ from those of the HGP?

15. Describe the significance of the Genome 10K plan.

16. It can be said that modern biology is experiencing an "omics" revolution. What does this mean? Explain your answer.

17. Metagenomics studies generate very large amounts of sequence data. Provide examples of genetic insight that can be learned from metagenomics.

18. What are gene microarrays? How are microarrays used?

19. In a draft annotation and overview of the human genome sequence, F.A. Wright et al. (*Genome Biol.* 2001: 2(7): Research0025) presented a graph similar to the one shown here. The graph details the approximate number of genes from each chromosome that are expressed only in embryos. Review earlier information in the text on human chromosomal aneuploids and correlate that information with the graph. Does this graph provide insight as to why some aneuploids occur and others do not?

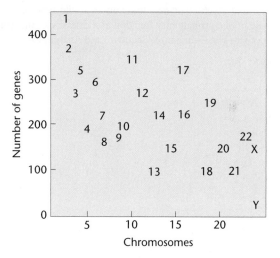

20. Annotation of the human genome sequence reveals a discrepancy between the number of protein-coding genes and the number of predicted proteins actually expressed by the genome. Proteomic analysis indicates that human cells are capable of synthesizing more than 100,000 different proteins and perhaps three times this number. What is the discrepancy, and how can it be reconciled?

Extra-Spicy Problems

 For instructor-assigned tutorials and problems, go to www.masteringgenetics.com

21. Genomic sequencing has opened the door to numerous studies that help us understand the evolutionary forces shaping the genetic makeup of organisms. Using databases containing the sequences of 25 genomes, scientists (Kreil, D.P. and Ouzounis, C.A., *Nucl. Acids Res.* 29: 1608–1615, 2001) examined the relationship between GC content and global amino acid composition. They found that it is possible to identify thermophilic species on the basis of their amino acid composition alone, which suggests that evolution in a hot environment selects for a certain whole organism amino acid composition. In what way might evolution in extreme environments influence

genome and amino acid composition? How might evolution in extreme environments influence the interpretation of genome sequence data?

22. The β-globin gene family consists of 60 kb of DNA, yet only 5 percent of the DNA encodes gene products. Account for as much of the remaining 95 percent of the DNA as you can.

23. In a sequence of 99.4 percent of the euchromatic regions of human chromosome 1, Gregory et al. (Gregory, S.G. et al., *Nature*, 441: 315–321, 2006) have identified 3141 gene structures.

(a) How does one identify a gene within a raw sequence of bases in DNA?

(b) What procedures are often used to verify likely gene assignments?

(c) Given that chromosome 1 contains approximately 8 percent of the human genome, and assuming that there are approximately 20,000–25,000 genes, would you consider chromosome 1 to be "gene rich"?

24. M. Stoll and colleagues have compared candidate loci in humans and rats in search of loci in the human genome that are likely to contribute to the constellation of factors leading to hypertension. Through this research, they identified 26 chromosomal regions that they consider likely to contain hypertension genes. How can comparative genomics aid in the identification of genes responsible for such a complex human disease? The researchers state that comparisons of rat and human candidate loci to those in the mouse may help validate their studies. Why might this be so?

25. Comparisons between human and chimpanzee genomes indicate that a gene that may function as a wild type or normal gene in one primate may function as a disease-causing gene in another (The Chimpanzee Sequence and Analysis Consortium, *Nature*, 437: 69–87, 2005). For instance, the *PPARG* locus (regulator of adipocyte differentiation) is associated with type 2 diabetes in humans but functions as a wild-type gene in chimps. What factors might cause this apparent contradiction? Would you consider such apparent contradictions to be rare or common? What impact might such findings have on the use of comparative genomics to identify and design therapies for disease-causing genes in humans?

26. The discovery that *M. genitalium* has a genome of 0.58 Mb and only 470 protein-coding genes has sparked interest in determining the minimum number of genes needed for a living cell. In the search for organisms with smaller and smaller genomes, a new species of Archaea, *Nanoarchaeum equitans*, was discovered in a high-temperature vent on the ocean floor. This prokaryote has one of the smallest cell sizes ever discovered, and its genome is only about 0.5 Mb. However, organisms such as *M. genitalium*, *N. equitans*, and other microbes with very small genomes are either parasites or symbionts. How does this affect the search for a minimum genome? Should the definition of the minimum genome size for a living cell be redefined?

Transgenic pigs generated by incorporating a viral vector carrying the jellyfish gene encoding green fluorescent protein into the pig genome. A non-transgenic pig is in the center of the photograph.

22

Applications and Ethics of Genetic Engineering and Biotechnology

CHAPTER CONCEPTS

- Recombinant DNA technology, genetic engineering, and biotechnology have revolutionized medicine and agriculture.

- Genetically modified plants and animals can serve as bioreactors to produce therapeutic proteins and other valuable protein products.

- Genetic modifications of plants have resulted in herbicide- and pest-resistant crops, and crops with improved nutritional value; similarly, transgenic animals are being created to produce therapeutic proteins and to protect animals from disease.

- A synthetic genome has been assembled and transplanted into a donor bacterial strain elevating interest in potential applications of synthetic biology.

- Applications of recombinant DNA technology and genomics have become essential for diagnosing genetic disorders, determining genotypes, and scanning the human genome to detect diseases.

- Genome-wide association studies (GWAS) scan for hundreds or thousands of genetic differences in an attempt to link genome variations to particular traits and diseases.

- Pharmacogenomics and rational drug design have led to customized medicines based primarily on a person's genotype.

- Gene therapy by transfer of cloned copies of functional alleles into target tissues is used to treat genetic disorders.

- Almost all applications of genetic engineering and biotechnology present unresolved ethical dilemmas that involve important moral, social, and legal issues.

Since the dawn of recombinant DNA technology in the 1970s, scientists have harnessed **genetic engineering** not only for biological research, but also for applications in medicine, agriculture, and biotechnology. Genetic engineering refers to the alteration of an organism's genome and typically involves the use of recombinant DNA technologies to add a gene or genes to a genome, but it can also involve gene removal. The ability to manipulate DNA *in vitro* and to introduce genes into living cells has allowed scientists to generate new varieties of plants, animals, and other organisms with specific gene traits, and to manufacture cheaper and more effective therapeutic products. These new varieties of organisms are called **genetically modified organisms**, or **GMOs**. Industry analysts estimate that genetic engineering will lead to U.S. commercial products worth over $55 billion by 2011. Many of these commercial products will be developed by the biotechnology industry.

Biotechnology is the use of living organisms to create a product or a process that helps improve the quality of life for humans or other organisms. As you will soon learn, biotechnology as a modern industry began in earnest shortly after recombinant DNA technology developed. But biotechnology is actually a science dating back to ancient civilization and the use of microbes to make many important products, including beverages such as wine and beer, vinegar, breads, and cheeses. Modern biotechnology relies heavily on recombinant DNA technology, genetic engineering, and genomics applications, and these areas will be the focus of this chapter. Existing products and new developments that occur seemingly every day make the biotechnology industry one of the most rapidly developing branches of the workforce worldwide, encompassing nearly 5000 companies in 54 countries.

The development of the biotechnology industry and the rapid growth in the number of applications for DNA technologies have raised serious concerns about using our power to manipulate genes and to apply gene technologies. Genetic engineering and biotechnology have the potential to provide solutions to major problems globally and to significantly alter how humans deal with the natural world; hence, they raise ethical, social, and economic questions that are unprecedented in human experience. These complex issues cannot be fully explored in the context of an introductory genetics textbook.

This chapter will therefore present only a selection of applications that illustrate the power of genetic engineering and biotechnology and the complexity of the dilemmas they engender. We will begin by explaining how genetic engineering has modified agriculturally important plants and animals. We briefly describe how genetic engineering has affected the production of pharmaceutical products, and we examine the impact of genetic technologies on the diagnosis

and treatment of human diseases, including gene therapy approaches. Finally, we explore some of the social, ethical, and legal implications of genetic engineering and biotechnology.

DNA forensics and applications involving technologies that use DNA for identification will be discussed in Special Topics on Forensics, p. 493. Similarly, an expanded discussion of personalized medicine based on genetics will be presented in Special Topics on Genomics and Personalized Medicine, p. 504.

22.1

Genetically Engineered Organisms Synthesize a Wide Range of Biological and Pharmaceutical Products

The most successful and widespread application of recombinant DNA technology has been production by the biotechnology industry of recombinant proteins as **biopharmaceutical** products—particularly, therapeutic proteins to treat diseases. Prior to the recombinant DNA era, biopharmaceutical proteins such as insulin, clotting factors, or growth hormones were purified from tissues such as the pancreas, blood, or pituitary glands. Clearly, these sources were in limited supply, and the purification processes were expensive. In addition, products derived from these natural sources could be contaminated by disease agents such as viruses. Now that human genes encoding important therapeutic proteins can be cloned and expressed in a number of nonhuman host-cell types, we have more abundant, safer, and less expensive sources of biopharmaceuticals. **Biopharming** is a commonly used term to describe the production of valuable proteins in genetically modified (GM) animals and plants.

In this section, we outline several examples of therapeutic products that are produced by expression of cloned genes in transgenic host cells and organisms. It should not surprise you that cancers, arthritis, diabetes, heart disease, and infectious diseases such as AIDS are among the major diseases that biotechnology companies are targeting for treatment by recombinant therapeutic products. Table 22.1 provides a short list of important recombinant products currently synthesized in transgenic bacteria, plants, yeast, and animals.

Insulin Production in Bacteria

Many therapeutic proteins have been produced by introducing human genes into bacteria. In most cases, the human gene is cloned into a plasmid, and the recombinant vector is introduced into the bacterial host. Large quantities of the transformed bacteria are grown, and the recombinant human protein is recovered and purified from bacterial extracts.

TABLE 22.1

Examples of Genetically Engineered Biopharmaceutical Products Available or under Development

Gene Product	Condition Treated	Host Type
Erythropoitin	Anemia	*E. coli*; cultured mammalian cells
Interferons	Multiple sclerosis, cancer	*E. coli*; cultured mammalian cells
Tissue plasminogen activator tPA	Heart attack, stroke	Cultured mammalian cells
Human growth hormone	Dwarfism	Cultured mammalian cells
Monoclonal antibodies against vascular endothelial growth factor (VEGF)	Cancers	Cultured mammalian cells
Human clotting factor VIII	Hemophilia A	Transgenic sheep, pigs
C1 inhibitor	Hereditary angioedema	Transgenic rabbits
Recombinant human antithrombin	Hereditary antithrombin deficiency	Transgenic goats
Hepatitis B surface protein vaccine	Hepatitis B infections	Cultured yeast cells, bananas
Immunoglobulin IgG1 to HSV-2	Herpesvirus infections	Transgenic soybeans glycoprotein B
Recombinant monoclonal antibodies	Passive immunization against rabies (also used in diagnosing rabies), cancer, rheumatoid arthritis	Transgenic tobacco, soybeans, cultured mammalian cells
Norwalk virus capsid protein	Norwalk virus infections	Potato (edible vaccine)
E. coli heat-labile enterotoxin	*E. coli* infections	Potato (edible vaccine)

The first human gene product manufactured by recombinant DNA technology was human insulin, called Humulin, which was licensed for therapeutic use in 1982 by the **U.S. Food and Drug Administration (FDA)**, the government agency responsible for regulating the safety of food and drug products and medical devices. In 1977, scientists at Genentech, the San Francisco biotechnology company cofounded in 1976 by Herbert Boyer (one of the pioneers of using plasmids for recombinant DNA technology) and Robert Swanson isolated and cloned the gene for insulin and expressed it in bacterial cells. Genentech, short for "genetic engineering technology," is also generally regarded as the world's first biotechnology company.

Previously, insulin was chemically extracted from the pancreas of cows and pigs obtained from slaughterhouses. **Insulin** is a protein hormone that regulates glucose metabolism. Individuals who cannot produce insulin have diabetes, a disease that, in its more severe form (type I), affects more than 2 million individuals in the United States. Although synthetic human insulin can now be produced by another process, a look at the original genetic engineering method is instructive, as it shows both the promise and the difficulty of applying recombinant DNA technology.

Clusters of cells embedded in the pancreas synthesize a precursor polypeptide known as preproinsulin. As this polypeptide is secreted from the cell, amino acids are cleaved from the end and the middle of the chain. These cleavages produce the mature insulin molecule, which contains two polypeptide chains (the *A* and *B* chains) joined by disulfide bonds. The *A* subunit contains 21 amino acids, and the *B* subunit contains 30.

In the original bioengineering process, synthetic genes that encode the *A* and *B* subunits were constructed by oligonucleotide synthesis (63 nucleotides for the *A* polypeptide and 90 nucleotides for the *B* polypeptide). Each synthetic oligonucleotide was inserted into a separate vector, adjacent to the *lacZ* gene encoding the bacterial form of the enzyme β-galactosidase. When transferred to a bacterial host, the *lacZ* gene and the adjacent synthetic oligonucleotide were transcribed and translated as a unit. The product is a **fusion protein**—that is, a hybrid protein consisting of the amino acid sequence for β-galactosidase attached to the amino acid sequence for one of the insulin subunits (**Figure 22–1**). The fusion proteins were purified from bacterial extracts and treated with cyanogen bromide, a chemical that cleaves the fusion protein from the β-galactosidase. When the fusion products were mixed, the two insulin subunits spontaneously united, forming an intact, active insulin molecule. The purified injectable insulin was then packaged for use by diabetics.

Shortly after insulin became available, growth hormone—used to treat children who suffer from a form of dwarfism—was cloned. Soon, recombinant DNA technology made that product readily available too, as well as a wide variety of other medically important proteins that were once difficult to obtain in adequate amounts. Since recombinant insulin ushered in the biotechnology era, well over 200 recombinant products have entered the market worldwide.

Transgenic Animal Hosts and Pharmaceutical Products

Although bacteria have been widely used to produce therapeutic proteins, there are some disadvantages in using prokaryotic hosts to synthesize eukaryotic proteins. One problem is that bacterial cells often cannot process and modify eukaryotic proteins. As a result, they frequently cannot add the carbohydrates and phosphate groups to proteins that are

(a)

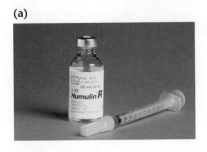

(b)

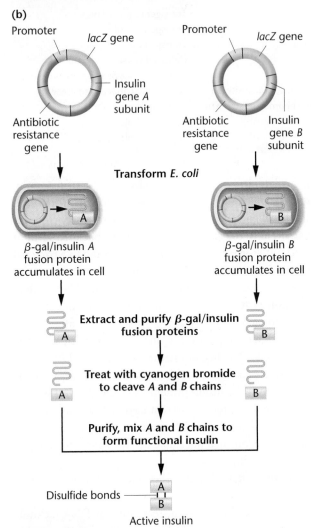

FIGURE 22–1   (a) Humulin, a recombinant form of human insulin, was the first therapeutic protein produced by recombinant DNA technology to be approved for use in humans. (b) To synthesize recombinant human insulin, synthetic oligonucleotides encoding the insulin A and B chains were inserted (in separate vectors) at the tail end of a cloned *E. coli lacZ* gene. The recombinant plasmids were transformed into *E. coli* host cells, where the β-gal/insulin fusion protein was synthesized and accumulated in the cells. Fusion proteins were then extracted from the host cells and purified. Insulin chains were released from β-galactosidase by treatment with cyanogen bromide. The insulin subunits were purified and mixed to produce a functional insulin molecule.

needed for full biological activity. In addition, eukaryotic proteins produced in prokaryotic cells often do not fold into the proper three-dimensional configuration and are therefore inactive. To overcome these difficulties and increase yields, many biopharmaceuticals are now produced in eukaryotic hosts. As seen in Table 22.1, eukaryotic hosts may include cultured eukaryotic cells (plant or animal) or transgenic farm animals. A herd of goats or cows serve as very effective **bioreactors** or **biofactories**—living factories—that will continuously make milk containing the desired therapeutic protein that can then be isolated in a noninvasive way.

Yeast are also valuable hosts for expressing recombinant proteins. Even insect cells are valuable for this purpose, through the use of a gene delivery system (virus) called **baculovirus**. Recombinant baculovirus containing a gene of interest is used to infect insect cell lines, which then express the protein at high levels. Baculovirus-insect cell expression is particularly useful for producing human recombinant proteins that are heavily glycosylated. Regardless of the host, therapeutic proteins may then be purified from the host cells—or when transgenic farm animals are used, isolated from animal products such as milk.

An example of a biopharmaceutical product synthesized in transgenic animals is the human protein α1-**antitrypsin**. A deficiency of the enzyme α1-antitrypsin is associated with the heritable form of emphysema, a progressive and fatal respiratory disorder common among people of European ancestry. To produce α1-antitrypsin for use in treating this disease, the human gene was cloned into a vector at a site adjacent to a sheep promoter sequence that specifically activates transcription in milk-producing cells of the sheep mammary glands. Genes placed next to this promoter are expressed only in mammary tissue. This fusion gene was microinjected into sheep zygotes fertilized *in vitro*. The fertilized zygotes were transferred to surrogate mothers. The resulting transgenic sheep developed normally and produced milk containing high concentrations of functional human α1-antitrypsin. This human protein is present in concentrations of up to 35 grams per liter of milk and can be easily extracted and purified. A small herd of lactating transgenic sheep can provide an abundant supply of this protein.

In 2006, recombinant human **antithrombin**, an anticlotting protein, became the world's first drug extracted from the milk of farm animals to be approved for use in humans. Scientists at GTC Biotherapeutics of Framingham, Massachusetts, introduced the human antithrombin gene into goats. By placing the gene adjacent to a promoter for beta casein, a common protein in milk, GTC scientists were able to target antithrombin expression in the mammary gland. As a result, antithrombin protein is highly expressed in the milk. In one year, a single goat will produce the equivalent amount of antithrombin that in the past would have been isolated from ~90,000 blood collections.

In a similar example involving a nonbiopharmaceutical application, transgenic "silk-milk" goats have been generated that express spider-silk proteins in their milk. These goats are a rich source of silk proteins used for various commercial applications such as manufacturing bulletproof vests.

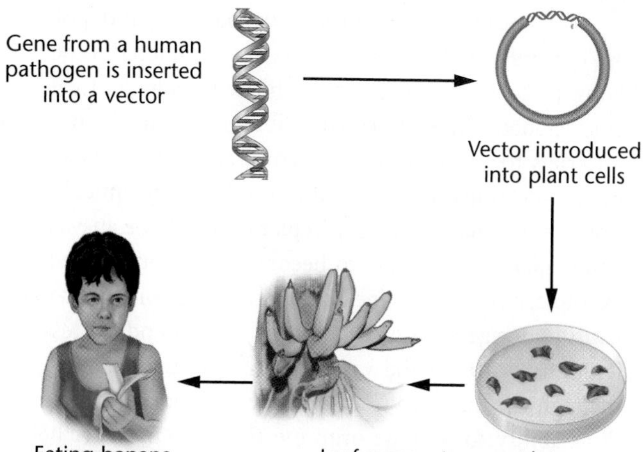

Recombinant DNA Approaches for Vaccine Production and Transgenic Plants with Edible Vaccines

One of the most promising applications of recombinant DNA technology for therapeutic purposes may be the production of vaccines. Vaccines stimulate the immune system to produce antibodies against disease-causing organisms and thereby confer immunity against specific diseases. Traditionally, two types of vaccines have been used: **inactivated vaccines**, which are prepared from killed samples of the infectious virus or bacteria; and **attenuated vaccines**, which are live viruses or bacteria that can no longer reproduce but can cause a mild form of the disease. Inactivated vaccines include the vaccines for rabies and influenza; vaccines for tuberculosis, cholera, and chickenpox are examples of attenuated vaccines.

Genetic engineering is being used to produce a relatively new type of vaccine called a **subunit vaccine**, which consists of one or more surface proteins from the virus or bacterium but not the entire virus or bacterium. This surface protein acts as an antigen that stimulates the immune system to make antibodies that act against the organism from which it was derived. One of the first subunit vaccines was made against the **hepatitis B virus**, which causes liver damage and cancer. The gene that encodes the hepatitis B surface protein was cloned into a yeast expression vector, and the cloned gene was expressed in yeast host cells. The protein was then extracted and purified from the host cells and packaged for use as a vaccine.

In 2005, the FDA approved **Gardasil**, a subunit vaccine produced by the pharmaceutical company Merck and the first cancer vaccine to receive FDA approval. Gardasil targets four strains of **human papillomavirus (HPV)** that cause ~70 percent of cervical cancers. Approximately 70 percent of sexually active women will be infected by an HPV strain during their lifetime. Gardasil is designed to provide immune protection against HPV prior to infection but is not effective against existing infections. You may have heard of Gardasil through media coverage of the legislation pending in several states that would require all adolescent school girls to receive a Gardasil vaccination regardless of whether or not they are sexually active.

Developing countries face serious difficulties in manufacturing, transporting, and storing vaccines. Most vaccines

Gene from a human pathogen is inserted into a vector

Vector introduced into plant cells

Eating banana triggers immune response to pathogen

Leaf segments sprout into whole plants carrying gene from human pathogen

FIGURE 22–2 To make an edible vaccine, a gene from a pathogen (a disease-causing agent, such as a virus or bacterium) is transferred into a vector, which is then introduced into plant cells. In this example, infection of banana plant leaf segments transfers the vector and the pathogen's gene into the nuclei of banana leaf cells. The leaf segments are grown into mature banana trees that express the pathogenic gene. Eating the raw banana produced by these plants triggers an immune response to the protein encoded by the pathogen's gene, conferring immunity to infection by this pathogen.

need refrigeration and must be injected under sterile conditions. In many rural areas, refrigeration and sterilization facilities are not available. In addition, in many cultures people are fearful of being injected with needles. To overcome these problems, scientists are attempting to develop vaccines that can be synthesized in edible food plants (Figure 22–2). These vaccines would be inexpensive to produce, would not require refrigeration, and would not have to be given under sterile conditions by trained medical personnel.

Plants offer several other advantages for expressing recombinant proteins. For instance, once a transgenic plant is made, it can easily be grown and replicated in a greenhouse or field, and it will provide a constant source of recombinant protein. In addition, the cost of expressing a recombinant protein in a transgenic plant is typically much lower than making the same protein in bacteria, yeast, or mammalian cells.

No recombinant proteins expressed in transgenic plants have yet been approved for use by the FDA as therapeutic proteins for humans, although about a dozen products are close to making it through final clinical trials. Some edible vaccines are now in clinical trials. For example, a vaccine against a bacterium that causes cholera has been produced in genetically engineered potatoes and used to successfully vaccinate human volunteers.

Tests using vaccine-producing bananas and potatoes are currently under way. Bananas are considered to be perhaps the best edible vaccine candidate for a hepatitis B vaccine. Genetically engineered edible plants are being used for trials to vaccinate infants, children, and adults against many infectious diseases. But a number of technical questions about vaccine delivery in plants need to be answered if edible plant vaccines are to become more widely used. For example, how can vaccine dose be carefully controlled when fruits and vegetables grow to different sizes and express different amounts of the vaccine? Will vaccine proteins pass through the digestive tract unaltered so that they maintain their ability to provide immune protection? Nevertheless, these are exciting prospects.

Lastly, it warrants mentioning that **DNA-based vaccines** have been attempted for many years, and recently there has been renewed interest in using these vaccines to protect against viral pathogens. In this approach, DNA encoding proteins from a particular pathogen are inserted into a plasmid vector, which is then injected directly into an individual or delivered via a viral vector similar to the way certain viruses are used for gene therapy. The idea here is that pathogen proteins encoded by the delivered DNA would be produced and trigger an immune response that could provide protection should an immunized person be exposed to the pathogen in the future. For example, trials are underway using plasmid DNA encoding protein antigens from HIV as an attempt to vaccinate individuals against HIV. Thus far a major limitation of DNA-based vectors has been that they typically result in very low production of protein encoded by delivered genes and thus the immune response in vaccinated persons is insufficient to provide the desired protection. Work on DNA-based vaccines continues to be an active area of exploration but whether they will ever have significant roles in the vaccine market remains to be seen.

22.2
Genetic Engineering of Plants Has Revolutionized Agriculture

For millennia, farmers have manipulated the genetic makeup of plants and animals to enhance food production. Until the advent of genetic engineering 30 years ago, these genetic manipulations were primarily restricted to **selective breeding**—the selection and breeding of naturally occurring or mutagen-induced variants. In the last 50 to 100 years, genetic improvement of crop plants through the traditional methods of artificial selection and genetic crosses has resulted in dramatic increases in productivity and nutritional enhancement. For example, maize yields have increased fourfold over the last 60 years, and more than half of this increase is due to genetic improvement by artificial selection and selective breeding (Figure 22–3). Modern maize has substantially larger ears and kernels than the predecessor crops, including hybrids from which it was bred.

Recombinant DNA technology provides powerful new tools for altering the genetic constitution of agriculturally important organisms. Scientists can now identify, isolate, and clone genes that confer desired traits, then specifically and efficiently introduce these into organisms. As a result, it is possible to quickly introduce insect resistance, herbicide resistance, or nutritional characteristics into farm plants and animals, a primary purpose of **agricultural biotechnology**.

Zea canina   Hybrid   *Zea mays*

FIGURE 22–3   Selective breeding is one of the oldest methods of genetic alteration of plants. Shown here is teosinte (*Zea canina*, left), a selectively bred hybrid (center), and modern corn (*Zea mays*).

(a)

Total number of trials = 798

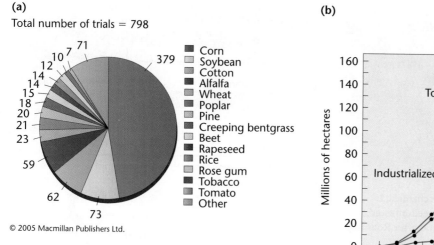

© 2005 Macmillan Publishers Ltd.

(b)

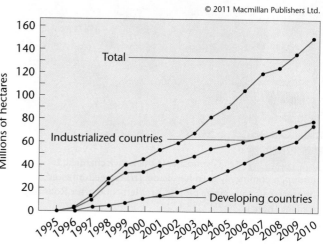

© 2011 Macmillan Publishers Ltd.

FIGURE 22–4 (a) Analysis of nearly 800 transgenic crop trials worldwide shows that GM varieties of corn, soybean, cotton, alfalfa, and wheat are among the most commonly manipulated crops (*Nature Biotechnology* 23(3), P. 281, March 2005). (b) During the past 10 years, transgenic crops have been rapidly adopted in both industrialized and developing countries (Ernst & Young, *Beyond Borders: Global Biotechnology Report* 2006, www.ey.com/beyondborders. Adapted by permission from Macmillan Publishers Ltd: March 2, 2011, *Nature* 471, 10–11, Copyright 2011, www.nature.com.).

In this section we primarily consider genetic manipulations to produce transgenic crop plants of agricultural value. In Section 22.3, we will discuss examples of genetic manipulations of agriculturally important animals.

Worldwide, over 150 million hectares have been planted with genetically engineered crops as of 2010, (the most recent date for which such data are available), particularly herbicide- and pest-resistant soybeans, corn, cotton, and canola; over 50 different transgenic crop varieties are available, including alfalfa, corn, rice, potatoes, tomatoes, tobacco, wheat, and cranberries [Figure 22–4(a)]. These crops are planted by over 8 million farmers in 17 countries. As evident in Figure 22–4(b), both industrialized and developing countries are taking advantage of transgenic crops. Since 1996, there has been a 4000 percent increase in GM crop acreage worldwide.

In 2005, the 10-year anniversary of commercial biotech crops, the one-billionth biotech acre was planted. American farmers planted 111 million acres of GM corn, soybeans, and cotton in 2004, a 17 percent increase from the year before. In the United States 86 percent of the soybeans, 78 percent of the cotton, and 46 percent of the corn are genetically engineered to resist pest or herbicides.

Several of the main reasons for generating transgenic crops include:

- Improving the growth characteristics and yield of agriculturally valuable crops

- Increasing the nutritional value of crops

- Providing crop resistance against insect and viral pests, drought, and herbicides

In addition, many new GM crops that will soon be on the market will be designed for ethanol production and for making biodiesel fuel—that is, for providing sustainable sources of energy.

The first commercially available GM food was called the Flavr Savr tomato. Designed by researchers at Calgene (now a division of Monsanto, Inc.), the Flavr Savr was designed to increase the shelf life of tomatoes by allowing them to stay ripe for several weeks without softening—a common problem for tomatoes. Calgene scientists used antisense RNA technology to inhibit an enzyme called polygalacturonase, which digests pectin in the cell wall of tomatoes. Pectin digestion occurs naturally once tomatoes are picked from the plant, and this process is a main reason why tomatoes soften as they age. This GM approach was generally effective, but the attempts to remedy shipping problems that continued to cause bruising increased the cost of these tomatoes. This and public skepticism about the safety of the first GM food are two of the reasons Flavr Savr was eventually taken off the market.

Insights from plant genome sequencing projects will undoubtedly be the catalyst for analysis of genetic diversity in crop plants, identification of genes involved in crop domestication and breeding traits, and subsequent enhancement of a variety of desirable traits through genetic engineering. In the past several years genome projects have been completed for many major food and industrial crops, including the three crops that account for most of the world's caloric intake: maize, rice, wheat. We will now examine other, more successful examples of genetically engineered plants used in agriculture. Some of the ethical and social concerns associated with these practices will be examined in Section 22.7.

(a)   (b)

FIGURE 22-5 (a) Glyphosate is the active chemical in Roundup, a commonly used herbicide. (b) A weed-infested glyphosate-resistant maize plot before (left) and after Roundup treatment (right).

Transgenic Crops for Herbicide and Pest Resistance

Damage from weed infestation destroys about 10 percent of crops worldwide. In an attempt to combat this problem, farmers often apply herbicides to the soil to kill weeds prior to seeding a field crop. As the most efficient herbicides also kill crop plants, herbicide uses are limited. The creation of herbicide-resistant crops has opened the way to efficient weed control and increased yields of some major agricultural crops. At present, over 75 percent of soybeans and cotton in the United States are resistant to the herbicide **glyphosate**. You may be familiar with glyphosate because it is the active herbicide in Roundup, which is commonly sold through home improvement stores for keeping sidewalks and patios weed-free (see Figure 22–5).

Glyphosate is effective at very low concentrations, is not toxic to humans, and is rapidly degraded by soil microorganisms. It kills plants by inhibiting the action of a chloroplast enzyme called **EPSP synthase**. This enzyme is important in amino acid biosynthesis in both bacteria and plants. Without the ability to synthesize vital amino acids, plants wither and die.

Recall from Chapter 20 that *Rhizobium radiobacter* (formerly called *Agrobacterium tumefaciens*) is a soil microbe that can infect wounded plants and create crown gall tumors. *R. radiobacter* contains **Ti plasmids**, so named because they contain tumor-inducing genes. Modified versions of Ti plasmids that lack tumor-inducing genes and contain other features, such as antibiotic resistance, have been widely used as vectors for introducing genes into plants. To produce a glyphosate-resistant crop plant, researchers began by isolating and cloning an EPSP synthase gene from a glyphosate-resistant strain of *E. coli*. Next, they cloned the *EPSP* gene into a Ti plasmid between promoter sequences derived from a plant virus and transcription termination sequences derived from a plant gene (Figure 22–6). This recombinant vector was then transformed into *R. radiobacter*.

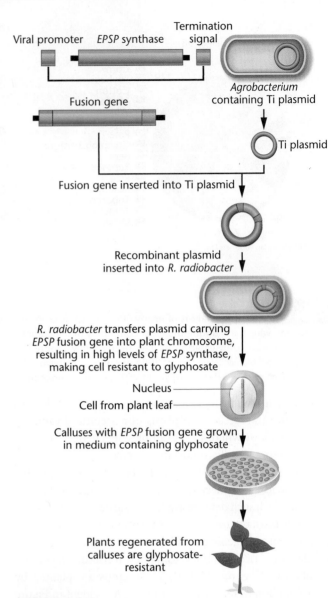

FIGURE 22-6 To create glyphosate-resistant transgenic plants, the EPSP synthase gene from bacteria is fused to a promoter such as the promoter from the cauliflower mosaic virus. This fusion gene is then ligated into a Ti-plasmid vector, and the recombinant vector is transformed into *R. radiobactier* host cells. *R. radiobacter* infection of cultured plant cells transfers the EPSP synthase fusion gene into a plant-cell chromosome. Cells that acquire the gene are able to synthesize large quantities of EPSP synthase, making them resistant to the herbicide glyphosate. Resistant cells are selected by growth in herbicide-containing medium. Plants regenerated from these cells are herbicide-resistant.

The Ti plasmid-carrying bacteria were then used to infect plant cells derived from plant leaves. The clumps of cells (calluses) that formed after infection with *R. radiobacter* were tested for their ability to grow in the presence of glyphosate. Glyphosate-resistant calluses were grown into transgenic plants and sprayed with glyphosate at concentrations four times higher than that needed to kill wild-type plants. Transgenic plants that expressed EPSP synthase grew and developed, while the control plants withered and died. Figure 22–5(b) demonstrates the effectiveness of glyphosate as a herbicide and the resistance of glyphosate-resistant soybeans.

Similar transgenic techniques have been used to make plants resistant to several other herbicides, to pathogens such as viruses, and also to insect pests. Some of the most well-described and controversial GM crops are the so-called **Bt crops**, designed to be resistant to insects. The bacterium *Bacillus thuringiensis* (Bt) produces a protein that when ingested by insects and larvae will crystallize in the gut, killing pests such as corn-borer larvae that are responsible for millions of dollars of crop damage worldwide. Initially, applications of Bt involved spraying these bacteria on crops. But recombinant DNA technology has enabled scientists to produce Bt transgenic crops with built-in insecticide protection. The *cry* genes that encode the Bt crystalline protein have been effectively introduced into a number of different crops, including corn, cotton, tomatoes, and tobacco.

Bt crops have been hailed as one of the greatest success stories of agricultural biotechnology, but they have also been one of the most controversial. Some studies had suggested a correlation between decreases in Monarch butterfly populations and ingestion of pollen from Bt corn (Monarchs do not feed on the corn itself). More recently, several long-term studies have demonstrated that exposure to Bt crops has no apparent effects on the Monarch; however, the possibility of danger to nontarget insect species must be considered whenever pest-resistant crops are used in the wild. In addition, concerns have been raised about the possibility that insect pests might develop resistance to Bt. Nonetheless, based on the success of Bt crops, many other transgenic crops are under development, including plants with increased tolerance to viral pests, drought, and salty soils.

Nutritional Enhancement of Crop Plants

Because crop plants can become deficient in some of the nutrients required in the human diet, biotechnology is being used to produce crops with enhanced nutritional value. One example is the production of **"golden rice,"** with enhanced levels of β-carotene, a precursor to vitamin A

(a)

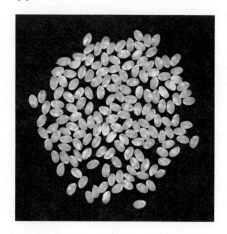

(b)

Geranylgeranyl diphosphate (C_{20}) Geranylgeranyl diphosphate (C_{20})

Block exists here in white rice grains   Phytoene synthase

Phytoene (C_{40})

Phytoene desaturase

ζ-carotene

ζ-carotene desaturase

Lycopene

Lycopene isomerase
α, β-lycopene cyclase

β-carotene

© 2005 Macmillan Publishers Ltd.

FIGURE 22–7 (a) Golden rice, a strain genetically modified to produce β-carotene, a precursor to vitamin A. Many children in countries where rice is a dietary staple lose their eyesight because of diets deficient in vitamin A. (b) White rice lacks the enzyme phytoene synthase, which is responsible for converting C_{20} into phytoene, a rate-limiting step in the production of β-carotene. Introducing the phytoene synthase gene into rice is one way to overcome this block and produce golden rice enriched in β-carotene.

(Figure 22–7). Vitamin A deficiency is prevalent in many areas of Asia and Africa, and more than 500,000 children a year become permanently blind as a result of this deficiency. Rice is a major staple food in these regions but does not contain vitamin A.

To create golden rice, scientists transferred into the rice genome, by recombinant DNA technology, three genes encoding enzymes required for the biosynthetic pathway leading to β-carotenoid synthesis. Two of these genes came from the daffodil and one from a bacterium. Although golden rice is currently available for planting, it produced only moderate levels of β-carotene. New varieties with higher levels of β-carotene production are in development. One of these strains, Golden Rice 2, incorporates the phytoene synthase gene from maize instead of daffodils and produces about 20 times more β-carotene than the original golden rice. Acceptance of genetically engineered varieties of rice, opposition and regulations on transgenic crops in some countries, and efficient distribution of Golden Rice remain challenges to its wider use. Nonetheless, encouraged by the effectiveness of Golden Rice 2, researchers are working on developing rice with enhanced iron and protein content.

Many other varieties of nutritionally enhanced food crops have been, and are being, developed. These include plants with augmented levels of key fatty acids, antioxidants, and other vitamins and minerals. These efforts are directed at addressing nutrient deficiencies affecting more than 40 percent of the world's population. Other expected developments include decaffeinated teas and coffees, as well as crops enhanced for traits affecting taste, growth rates, yields, color, storage, ripening, and similar characteristics.

(a)

(b)

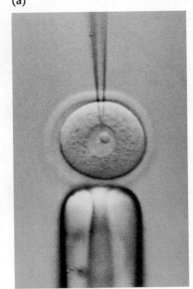

FIGURE 22–8   (a) Scientist microinjecting cloned DNA into a fertilized egg. The injections are performed by manipulating the egg and microinjection needle under a light microscope, seen in the background. The injection procedure is displayed on the screen in the foreground. The egg is held by a suction pipette (seen below the egg). (b) A transgenic mouse with its nontransgenic sibling. The mouse on the left is transgenic for a rat growth hormone gene, cloned downstream from a mouse metallothionein promoter. When the transgenic mouse was fed zinc, the metallothionein promoter induced the transcription of the growth hormone gene, stimulating the growth of the transgenic mouse.

22.3
Transgenic Animals with Genetically Enhanced Characteristics Have the Potential to Serve Important Roles in Biotechnology

Although genetically engineered plants are major players in modern agriculture, commercial applications of transgenic animals are less widespread. Nonetheless, some high-profile examples of genetically engineered farm animals have aroused public interest and controversy.

Making a Transgenic Animal: The Basics

The method of creating transgenic animals is conceptually relatively simple, although there are species-specific challenges associated with creating transgenics. Many of the prevailing techniques used to make transgenics were developed in mice. One method is to isolate newly fertilized eggs from a female mouse (or female of the desired animal species) and to inject purified cloned DNA containing a vector and the transgene of interest into the nucleus of the egg (Figure 22–8). In a relatively small percentage of transgenic eggs, the transgenic DNA becomes inserted into the egg cell genome by recombination due to the action of naturally occurring DNA recombination enzymes. Newer approaches involving stem cells are also popular for creating transgenic animals.

Injected eggs are then placed into the oviduct of a so-called pseudopregnant mouse, a mouse previously impregnated by mating with a male mouse. The pseudopregnant mouse offers a uterus that is receptive to implantation of the egg containing transgenic DNA. Once baby mice are born, researchers screen the transgenic mice by obtaining DNA from a sample of tail tissue, purifying the DNA, and performing PCR to verify that the transgene is present in the animal's genome. As long as the integrated DNA is present in germ cells, the transgene will be inherited in all of the offspring generated with the transgenic mouse but typically most F1 generation transgenic mice produced this way are not homozygous for the transgene. Sibling matings of F1 animals can then be used to generate homozygous transgenic animals.

Transgenic animals overexpressing certain genes, expressing human genes, and expressing mutant genes are among examples of transgenics that are valuable models for basic and applied research to understand gene function. Here we highlight several interesting examples of transgenic animals created with the purpose of producing a potentially commercially valuable biotechnology product.

Examples of Transgenic Animals

Oversize mice containing a human growth hormone transgene were some of the first transgenic animals created. Attempts to create farm animals containing transgenic growth hormone genes have not been particularly successful, probably because growth is a complex, multigene trait. One notable exception is the transgenic Atlantic salmon, bearing copies of a Chinook salmon growth hormone gene adjacent to a constitutive promoter. These salmon mature quickly, grow 400 to 600 percent faster than nontransgenic salmon, and appear to have no adverse health effects from the added gene (Figure 22–9).

As discussed in Section 22.1, currently, the major uses for transgenic farm animals are as bioreactors to produce useful pharmaceutical products, but a number of other interesting transgenic applications are under development. Several of these applications are designed to increase milk production or increase nutritional value of milk. Significant research efforts are also being made to protect farm animals against common pathogens that cause disease and animal loss (including potential bioweapons that could be used in a terrorist attack on food animals) and put the food supply at risk. For instance, controlling mastitis in cattle by creating transgenic cows has shown promise (Figure 22–10). **Mastitis** is an infection of the mammary glands. It is the

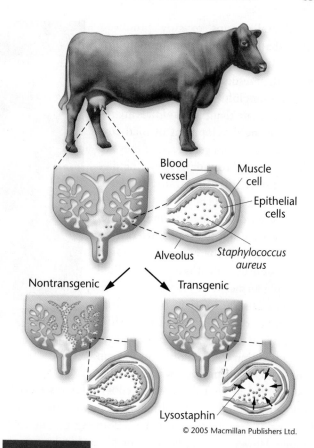

© 2005 Macmillan Publishers Ltd.

FIGURE 22–10 Transgenic cows for battling mastitis. The mammary glands of nontransgenic cows are highly susceptible to infection by the skin microbe *Staphylococcus aureus*. Transgenic cows express the lysostaphin transgene in milk, where it can kill *S. aureus* before they can multiply in sufficient numbers to cause inflammation and damage mammary tissue.

most costly disease affecting the dairy industry, leading to over $2 billion in losses in the United States. Mastitis can block milk ducts, reducing milk output, and can also contaminate the milk with pathogenic microbes. Infection by the bacterium *Staphylococcus aureus* is the most common cause of mastitis, and most cattle with mastitis typically do not respond well to conventional treatments with antibiotics. As a result, mastitis is a significant cause of herd reduction.

In an attempt to create cattle resistant to mastitis, transgenic cows were generated that possessed the lysostaphin gene from *Staphylococus simulana*. Lysostaphin is an enzyme that specifically cleaves components of the *S. aureus* cell wall. Transgenic cows expressing this protein in milk produce a natural antibiotic that wards off *S. aureus* infections. These transgenic cows do not completely solve the mastitis problem because lysostaphin is not effective against other microbes such as *E. coli* and *S. uberis* that occasionally cause mastitis; moreover, there is also the potential that *S. aureus* may develop resistance to lysostaphin. Nonetheless, scientists are cautiously optimistic that transgenic approaches have a strong future for providing farm animals with a level of protection against major pathogens.

FIGURE 22–9 Transgenic Atlantic salmon (bottom) overexpressing a growth hormone (GH) gene display rapidly accelerated rates of growth compared to wild strains and nontransgenic domestic strains (top). GH salmon weigh an average of nearly 10 times more than nontransgenic strains.

Several groups recently produced cattle that lack the prion protein *PrP* gene. Misfolded configurations of the PrP protein result in **mad cow disease**, or bovine spongioform encephalitis. Early results indicate that these animals are resistant to the development of mad cow disease, but it is not yet known if they are fully protected from developing the disease. Another successful transgenic farm animal is **EnviroPig**, a pig that expresses the gene encoding the enzyme phytase. These pigs are able to break down dietary phosphorus, thereby reducing excretion of phosphorus, which is a major pollutant in pig farms.

Scientists at Yorktown Industries of Austin, Texas, created the **GloFish**, a transgenic strain of zebrafish (*Danio rerio*) containing a red fluorescent protein gene from sea anemones. Marketed as the first GM pet in the United States, GloFish fluoresce bright pink when illuminated by ultraviolet light (**Figure 22–11**). GM critics describe these fish as an abuse of genetic technology. However, GloFish may not be as frivolous a use of genetic engineering as some believe. A variation of this transgenic model, incorporating a heavy-metal-inducible promoter adjacent to the red fluorescent protein gene, has shown promise in a bioassay for heavy metal contamination of water. When these transgenic zebrafish are in water contaminated by mercury and other heavy metals, the promoter becomes activated, inducing transcription of the red fluorescent protein gene. In this way, zebrafish fluorescence can be used as a bioassay to measure heavy metal contamination and uptake by living organisms.

FIGURE 22–11   GloFish, marketed as the world's first GM-pet, are a controversial product of genetic engineering.

22.4

Synthetic Genomes, Genome Transplantation, and the Emergence of Synthetic Biology

Studying genomes has led to the fundamental question, "what is the minimum number of genes necessary to support life?" Scientists are interested in this question in part because the answers are expected to provide ways to create artificial cells or designer organisms based on genes encoded by a **synthetic genome** constructed artificially. We can use the small genomes of obligate parasites to speculate on the minimum number of genes *required to maintain life.* For example, the bacterium *Mycoplasma genitalium,* a human parasitic pathogen, is among the simplest self-replicating prokaryotes known and has served as a model for understanding the minimal elements of a genome necessary for a self-replicating cell. *M. genitalium* has a genome of 580 kb. *M. genitalium* can cause diseases in a wide range of hosts, including insects, plants, and humans. In humans, it causes genital infections.

In 1995, a team involving J. Craig Venter, Hamilton Smith, and Clyde Hutchison III sequenced the *M. genitalium* genome. With 485 protein-coding genes, it is one of the smallest bacterial genomes to be sequenced. In contrast, the 1.8-Mb genome of *Haemophilus influenzae* (the first bacterial genome sequenced) has 1783 genes. The availability of genome sequences allows us to ask which of the 485 genes carried by *M. genitalium* are close to the minimum gene set needed for life.

Can we define life in terms of a number of specific genes? A combination of comparative and experimental methods can be used to answer this question including comparative genomics approaches to compare the *M. genitalium* genome to the genomes of other relatives such as *M. pneumoniae* and *H. influenzae.* By comparing the nucleotide sequences of the *M. genitalium* genes with the *H. influenzae* genes, researchers identified 240 genes that are orthologous for these species. In addition to the 240 shared genes, 16 genes with different sequences but identical functions were identified. These represent essential functions performed by nonorthologous genes. Thus, comparative genomics estimates that 256 genes may represent the minimum gene set needed for life.

J. Craig Venter and his colleagues used an experimental approach to determine how many of the 480 *M. genitalium* genes are essential for life. They used transposon-based methods to selectively mutate genes in *M. genitalium.* Mutations in essential genes would produce a lethal phenotype, but mutation of nonessential genes would not affect viability or show any obvious negative impacts. They found that about 100 of the 480 genes were nonessential and that the minimum gene set for *M. genitalium* is about 381 genes.

A Synthetic Genome and Genome Transplantation Creates a Bacterial Strain

In 2008 scientists from the J. Craig Venter Institute (JCVI) reported a complete chemical synthesis of the 580-kb *M. genitalium* genome, but they could not demonstrate the functionality of the synthetic genome they produced because it could not be transplanted into another bacterium. **Genome transplantation** is effectively the true test of the functionality of a synthetic genome. From Chapter 20, recall that if a plasmid containing a gene of interest were introduced into bacteria, the expression of genes in the plasmid whether they were antibiotic resistance genes or a human gene such as insulin would demonstrate successful transformation of bacteria. Transplanting an entire synthetic genome with the expectation that genes in the synthetic genome would completely transform the phenotype of the cell is essential.

In 2010 JCVI scientists published the first report of a functional synthetic genome. In this approach they designed and had chemically synthesized more than 1000, 1080-bp segments called cassettes covering the entire 1.08-Mb *M. mycoides* genome (Figure 22–12[a]). To assemble these segments correctly, the sequences had 80-bp sequences at each end which overlapped with their neighbor sequences. These sequences were cloned in *E. coli*. Then, using the yeast *Saccharomyces cerevisiae*, a homologous recombination approach was used to assemble the sequences into 11 separate 10-kb assemblies that were eventually combined to completely span the entire 1.08-Mb *M. mycoides* genome (Figure 22–12[b]).

The entire assembled genome, called JCVI-syn1.0, was then transplanted into a close relative *M. capricolum* as recipient cells resulting in a new cell with the JCVI-syn1.0 genotype and phenotype of a new strain of *M. mycoides*. JDVI-syn1.0 was forced into the existing natural genome of *M. capricolum* (although many of the mechanistic details about genome transplantation are unclear). As shown in Figure 22–12(c), JCVI determined that the recipient cells were taken over to become JCVI-syn.10 *M. mycoides* in part because they were shown to express the *lacZ* gene which was incorporated into the synthetic genome (Figure 22–12[c]). Selection for tetracycline resistance and a determination that recipient cells also made proteins characteristic of *M. mycoides* and not *M. capricolum* were also used to verify strain conversion.

One particularly impressive accomplishment of these experiments was that the synthetic DNA was "naked" DNA, because it did not contain any proteins from *M. mycoides*. Therefore it was capable of transcribing all of the appropriate genes and translating all of the protein products necessary for life as *M. mycoides*! This is not a trivial accomplishment. The synthetic genome effectively rebooted the *M. capricolum* recipient cells to change them from one form to another. When this work was announced, J. Craig Venter claimed: "This is equivalent to changing a Macintosh computer into a PC by inserting a new piece of PC software." The JCVI team is currently working on stripping out 100-kb segments and recombining them again to determine which segments are needed to form the minimal genome to create *M. mycoides*.

Venter and others have used recombinant DNA technology to construct synthetic copies of viral genomes. This approach has demonstrated that a minimal genome can be created and assembled for nonliving entities such as viruses. For example, the genome for polio virus and the 1918 influenza strain responsible for the pandemic flu have been assembled this way. But Venter's recent work with *M. mycoides* JCVIsyn1.0 9, a decade long-project that cost about $40 million and involved about 20 people, is being hailed as a defining moment in the emerging field of **synthetic biology**.

This work did not create life from an inanimate object since it was based on converting one living strain into another. Also the *Mycoplasma* strains used lack a cell wall typically found in other bacteria, which could be a barrier to genome transplantation in other bacterial species. There are many fundamental questions about synthetic genomes and genome transplantation that need to be answered. But clearly these studies provided key "proof of concept" that synthetic genomes could be produced, assembled, and successfully transplanted to create a microbial strain encoded by a synthetic genome and bring scientists closer to producing novel synthetic genomes incorporating genes for specific traits of interest.

What are some of the potential applications of synthetic genomes and synthetic biology? JCVI claims that their ultimate goal is to create microorganisms that can be used to synthesize biofuels. Other possibilities exist such as creating synthetic microbes with genomes engineered to express gene products to degrade pollutants (bioremediation), the synthesis of new biopharmaceutical products, genetically programmed bacteria to help us heal, and the ability to make "prosthetic genomes." Work on synthetic genomes and synthetic biology has led to speculation of a future world in which new bacteria, and perhaps new animal and plant cells, can be designed and even programmed to be controlled as we want them to! In the future, could synthetic genomes be used to create life from inanimate components? Stay tuned!

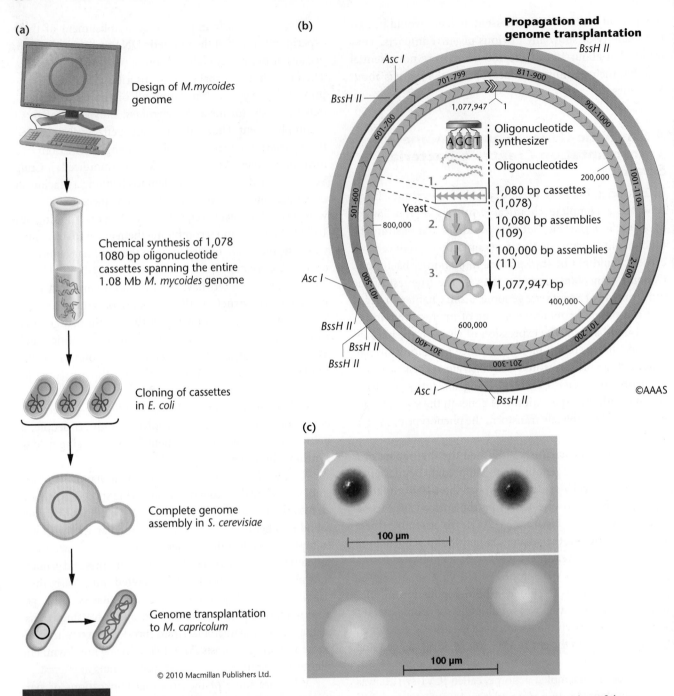

(a) Design of *M.mycoides* genome

Chemical synthesis of 1,078 1080 bp oligonucleotide cassettes spanning the entire 1.08 Mb *M. mycoides* genome

Cloning of cassettes in *E. coli*

Complete genome assembly in *S. cerevisiae*

Genome transplantation to *M. capricolum*

© 2010 Macmillan Publishers Ltd.

(b) **Propagation and genome transplantation**

Oligonucleotide synthesizer
Oligonucleotides
1. 1,080 bp cassettes (1,078)
Yeast
2. 10,080 bp assemblies (109)
100,000 bp assemblies (11)
3. 1,077,947 bp

©AAAS

(c)

100 μm

100 μm

FIGURE 22–12 Building a synthetic version of the 1.08 Mb *Mycoplasma mycoides* genome JCVI-syn1.0. (a) Overview of the approach used to produce *M. mycoides* JCVI-syn1.0. (b) Assembly of the synthetic *M. mycoides* genome in yeast occurred in three steps. (1) 1080-bp segments (cassettes) were produced from overlapping synthetic oligonucleotides (orange arrows), (2) segments were recombined in sets of 10 to produce 109 ~10 kb assemblies (blue arrows), then (3) these were combined in sets of 10 to produce 11 ~100 kb assemblies (green arrows). Assemblies carried out in yeast to create the entire synthetic genome called JCVI-syn1.0. (c) Images of *M. mycoides* JCVI-syn1.0 (top) and wild-type *M. mycoides* (bottom). Cells with the synthetic genome express the *lacZ* gene and are thus blue in color, wild-type cells do not express the *lacZ* gene and are white in color.

22.5

Genetic Engineering and Genomics Are Transforming Medical Diagnosis

Geneticists are now applying knowledge about the human genome and the genetic basis of many diseases to a wide range of medical applications. Gene-based technologies have already had a major impact on the diagnosis of disease and are revolutionizing medical treatments and the development of specific and effective pharmaceuticals. In large part as a result of the Human Genome Project, researchers are identifying genes involved in both single-gene diseases and complex genetic traits. Methods based on recombinant DNA and genomics technologies are making this possible.

In this section, we provide an overview of representative examples that demonstrate how gene-based technologies are being used to diagnose genetic diseases.

Using DNA-based tests, scientists can directly examine a patient's DNA for mutations associated with disease. Gene testing was one of the first successful applications of recombinant DNA technology, and currently more than 900 gene tests are in use. These tests usually detect DNA mutations associated with single-gene disorders that are inherited in a Mendelian fashion. Examples of such genetic tests are those that detect sickle-cell anemia, cystic fibrosis, Huntington disease, hemophilias, and muscular dystrophies. Other genetic tests have been developed for complex disorders such as breast and colon cancers. Gene tests are used to perform prenatal diagnosis of genetic diseases, to identify carriers, to predict the future development of disease in adults, to confirm the diagnosis of a disease detected by other methods, and to identify genetic diseases in embryos created by *in vitro* fertilization.

For genetic testing of adults, DNA from white blood cells is commonly used. Alternatively, many genetic tests can be carried out on cheek cells collected by swabbing the inside of the mouth, or hair cells. Some genetic testing can be carried out on gametes. For prenatal diagnosis, fetal cells are obtained by **amniocentesis** or **chorionic villus sampling**. Figure 22–13 shows the procedure for amniocentesis, in which a small volume of the amniotic fluid surrounding the fetus is removed. Amniotic fluid contains fetal cells that can be used for karyotyping, genetic testing, and other

procedures. For chorionic villus sampling, cells from the fetal portion of the placental wall (the chorionic villi) are sampled through a vacuum tube, and analyses can be carried out on this tissue.

Another approach called **fetal cell sorting** may eventually replace amniocentesis and chorionic villus sampling because it is noninvasive for the fetus. In pregnant women a small number of fetal cells are present in the maternal bloodstream. An instrument known as a fluorescence-activated cell sorter can be used to separate fetal cells from a maternal blood sample based on proteins expressed on the fetal cells but not the maternal cells. Captured fetal cells can then be subjected to genetic analysis, usually involving techniques that involve PCR (such as allele-specific oligonucleotide testing described later in this section).

Genetic Tests Based on Restriction Enzyme Analysis

A classic method of genetic testing is **restriction fragment length polymorphism (RFLP) analysis**. As we will discuss in the next section, PCR-based methods have largely replaced RFLP analysis; however, applications of this approach are still used occasionally, and for historical purposes it is also helpful to compare RFLP analysis to new approaches, which were largely not widespread prior to completion of the HGP. To illustrate this method, we examine the prenatal diagnosis of **sickle-cell anemia**. As we have discussed before, this disease is an autosomal recessive condition common in people with family origins in areas of West Africa, the Mediterranean

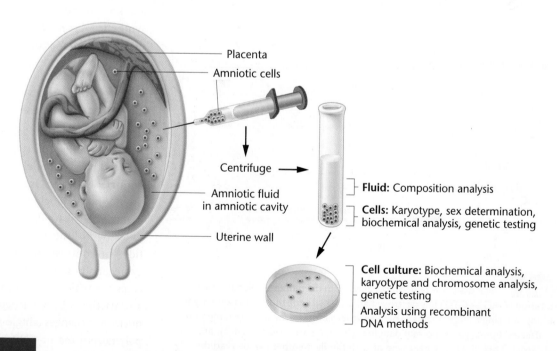

FIGURE 22–13 For amniocentesis, the position of the fetus is first determined by ultrasound, and then a needle is inserted through the abdominal and uterine walls to recover amniotic fluid and fetal cells for genetic or biochemical analysis.

basin, and parts of the Middle East and India. It is caused by a single amino acid substitution in the β-globin protein, as a consequence of a single-nucleotide substitution in the β-globin gene. The single-nucleotide substitution also eliminates a cutting site in the β-globin gene for the restriction enzymes *Mst*II and *Cvn*I. As a result, the mutation alters the pattern of restriction fragments seen on Southern blots. These differences in restriction cutting sites are used to prenatally diagnose sickle-cell anemia and to establish the parental genotypes and the genotypes of other family members who may be heterozygous carriers of this condition.

DNA is extracted from tissue samples and digested with *Mst*II. This enzyme cuts three times within a region of the normal β-globin gene, producing two small DNA fragments. In the mutant sickle-cell allele, the middle *Mst*II site is destroyed by the mutation, and one large restriction fragment is produced by *Mst*II digestion (**Figure 22–14**). The restriction-enzyme-digested DNA fragments are separated by gel electrophoresis, transferred to a nylon membrane, and visualized by Southern blot hybridization, using a probe from this region. Figure 22–14 shows the results of RFLP analysis for sickle-cell anemia in one family. Both parents (I-1 and I-2) are heterozygous carriers of the mutation. *Mst*II digestion of the parents' DNA produces a large band (because of the mutant allele) and two smaller bands (from the normal allele) in each case. The parents' first child (II-1) is homozygous normal because she has only the two smaller bands. The second child (II-2) has sickle-cell anemia; he has only one large band and is homozygous for the mutant allele. The fetus (II-3) has a large band and two small bands and is therefore heterozygous for sickle-cell anemia. He or she will be unaffected but will be a carrier.

NOW SOLVE THIS

22–3 You are asked to assist with a prenatal genetic test for a couple, each of whom is found to be a carrier for a deletion in the β-globin gene that produces β-thalassemia when homozygous. The couple already has one child who is unaffected and is not a carrier. The woman is pregnant, and the couple wants to know the status of the fetus. You receive DNA samples obtained from the fetus by amniocentesis and from the rest of the family by extraction from white blood cells. Using a probe that binds to the mutant allele, you obtain the following blot. Is the fetus affected? What is its genotype for the β-globin gene?

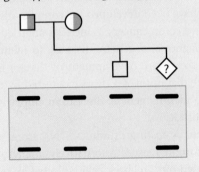

■ HINT: *This problem is concerned with interpreting results of RFLP analysis of a fetus. The key to its solution is to remember that differences in the number and location of restriction sites create RFLPs that can be used to determine genotype.*

Only about 5 to 10 percent of all point mutations can be detected by restriction enzyme analysis because most mutations occur in regions of the genome that do not contain restriction enzyme cutting sites. However, now that the HGP has been completed and many disease-associated mutations are known, geneticists can employ synthetic oligonucleotides to detect these mutations, as described next.

Genetic Tests Using Allele-Specific Oligonucleotides

Another method of genetic testing involves the use of synthetic DNA probes known as **allele-specific oligonucleotides (ASOs)**. Scientists use these short, single-stranded fragments of DNA to identify alleles that differ by as little as a single nucleotide. In contrast to restriction enzyme analysis, which is limited to cases for which a mutation changes a restriction site, ASOs detect single-nucleotide changes (**single-nucleotide polymorphisms** or **SNPs**), including those that do not affect restriction

Region recognized by probe

β^S-globin gene
GTG

5' ───■━━▮▮━━──── 3' →

↑   ↑
*Mst*II   *Mst*II

Normal β^A-globin gene
GAG

5' ──■━▮▮━━──── 3' →

↑   ↑   ↑
*Mst*II   *Mst*II   *Mst*II

Genotypes: β^A/β^S   β^A/β^S   β^A/β^A   β^S/β^S   β^A/β^S

FIGURE 22–14 RFLP diagnosis of sickle-cell anemia. In the mutant β-globin allele (β^S), a point mutation (GAG → GTG) has destroyed a cutting site for the restriction enzyme *Mst*II, resulting in a single large fragment on a Southern blot. In the pedigree, the family has one unaffected homozygous normal daughter (II-1), an affected son (II-2), and an unaffected carrier fetus (II-3). The genotype of each family member can be read directly from the blot and is shown below each lane.

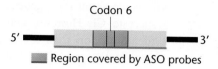

Region of β-globin gene amplified by PCR

Codon 6

5′ ▬▬▬ 3′

▨ Region covered by ASO probes

DNA is spotted onto binding filters and
hybridized with ASO probe

(a) Genotypes   *AA*   *AS*   *SS*

⬤   ◔   ◯

Normal (*β*^A) ASO: 5′ – CTCCTGAGGAGAAGTCTGC – 3′

(b) Genotypes   *AA*   *AS*   *SS*

◯   ◔   ⬤

Mutant (*β*^S) ASO: 5′ – CTCCTGTGGAGAAGTCTGC – 3′

FIGURE 22–15 Allele-specific oligonucleotide (ASO) testing for the β-globin gene and sickle-cell anemia. The β-globin gene is amplified by PCR, using DNA extracted from white blood cells or cells obtained by amniocentesis. The amplified DNA is then denatured and spotted onto strips of DNA-binding membranes. Each strip is hybridized to a specific ASO and visualized on X-ray film after hybridization and exposure. (a) Results observed when the three possible genotypes are hybridized to an ASO from the normal β-globin allele: *AA*-homozygous individuals have normal hemoglobin that has two copies of the normal β-globin gene and will show heavy hybridization; *AS*-heterozygous individuals carry one normal β-globin allele and one mutant allele and will show weaker hybridization; *SS*-homozygous sickle-cell individuals carry no normal copy of the β-globin gene and will show no hybridization to the ASO probe for the normal β-globin allele. (b) Results observed when DNA for the three genotypes are hybridized to the probe for the sickle-cell β-globin allele: no hybridization by the *AA* genotype, weak hybridization by the heterozygote (*AS*), and strong hybridization by the homozygous sickle-cell genotype (*SS*).

enzyme cutting sites. As a result, this method offers increased resolution and wider application. Under proper conditions, an ASO will hybridize only with its complementary DNA sequence and not with other sequences, even those that vary by as little as a single nucleotide.

Genetic testing using ASOs and PCR analysis are now available to screen for many disorders, such as sickle-cell anemia. In the case of sickle-cell screening, DNA is extracted, and a region of the β-globin gene is amplified by PCR. A small amount of the amplified DNA is spotted onto strips of a DNA-binding membrane, and each strip is hybridized to an ASO synthesized to resemble the relevant sequence from either a normal or mutant β-globin gene (**Figure 22–15**). The ASO is tagged with a molecule that is either radioactive or fluorescent, to allow for visualization of the ASO hybridized to DNA on the membrane. This rapid, inexpensive, and highly accurate technique is used to diagnose a wide range of genetic disorders caused by point mutations. Although

highly effective, SNPs can affect probe binding leading to false positive or false negative results that may not reflect a genetic disorder, particularly if precise hybridization conditions are not used. Sometimes DNA sequencing is carried out on amplified gene segments to confirm identification of a mutation.

Because ASO testing makes use of PCR, small amounts of DNA can be analyzed. As a result, ASO testing is ideal for **preimplantation genetic diagnosis (PGD)**. PGD is the genetic analysis of single cells from embryos created by *in vitro* fertilization (**Figure 22–16**). When sperm and eggs are mixed to create zygotes, the early-stage embryos are grown in culture. A single cell can be removed from an early-stage embryo using a vacuum pipette to gently aspirate one cell away from the embryo (Figure 22–16). This could possibly kill the embryo, but if it is done correctly the embryo will often continue to divide normally. DNA from the removed cell is then typically analyzed by FISH (for chromosome analysis) or by ASO testing [Figure 22–16]. The genotypes for each cell can then be used to decide which embryos will be implanted into the uterus. Any alleles that can be detected by ASO testing can be used for PGD. Sickle-cell anemia, cystic fibrosis, and dwarfism are often tested for by PGD, but alleles for many other conditions are often analyzed.

ASOs can also be used to screen for disorders that involve deletions instead of single-nucleotide mutations. An example is the use of ASOs to diagnose **cystic fibrosis (CF)**. CF is an autosomal recessive disorder associated with a

NOW SOLVE THIS

22–4 The DNA sequence surrounding the site of the sickle-cell mutation in the β-globin gene, for normal and mutant genes, is as follows.

Each type of DNA is denatured into single strands and applied to a DNA-binding membrane. The membrane containing the two spots is hybridized to an ASO of the sequence

5′-GACTCCTGAGGAGAAGT-3′

Which spot, if either, will hybridize to this probe?

5′-GACTCCTGAGGAGAAGT-3′

3′-CTGAGGACTCCTCTTCA-5′

Normal DNA

5′-GACTCCTGTGGAGAAGT-3′

3′-CTGAGGACACCTCTTCA-5′

Sickle-cell DNA

■ HINT: *This problem asks you to analyze results of an ASO test. The key to its solution is to understand that ASO analysis is done under conditions that allow only identical nucleotide sequences to hybridize to the ASO on the membrane.*

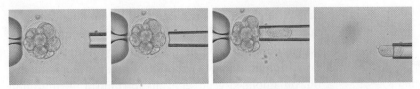

At the 8–16 cell stage, one cell from an embryo is gently removed with a suction pipette. The remaining cells continue to grow in culture.

DNA from an isolated cell is amplified by PCR with primers for the β-globin gene. Small volumes of denatured PCR products are spotted onto two separate DNA binding membranes.

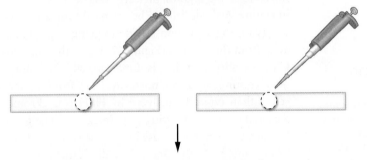

One membrane is hybridized to a probe for the normal β-globin allele (β^A) and the other membrane is hybridized to a probe for the mutant β-globin allele (β^S).

Membrane hybridized to a probe for the normal β-globin allele (β^A)

Membrane hybridized to a probe for the mutant β-globin allele (β^S)

In this example, hybridization of the PCR products to the probes for both the β^A and β^S alleles reveals that the cell analyzed by PGD has a carrier genotype ($\beta^A\beta^S$) for sickle-cell anemia.

FIGURE 22–16   A single cell from an early-stage human embryo created by *in vitro* fertilization can be removed and subjected to preimplantation genetic diagnosis (PGD) by ASO testing. DNA from each cell is isolated, amplified by PCR with primers specific for the gene of interest, then subjected to ASO analysis as shown in Figure 22–15. In this example, a region of the *β-globin* gene was amplified and analyzed by ASO testing to determine the sickle-cell genotype for this cell.

defect in a protein called the **cystic fibrosis transmembrane conductance regulator (CFTR)**, which regulates chloride ion transport across the plasma membrane. A small deletion called *Δ508* is found in 70 percent of all mutant copies of the *CFTR* gene. To detect carriers of the *Δ508* mutation, ASOs are made by PCR from cloned samples of the normal and mutant alleles. DNA extracted from white blood cells of the individuals to be tested is amplified by PCR and then hybridized to each ASO (**Figure 22–17**). In affected individuals, only the ASO from the mutant allele hybridizes. In heterozygotes, both ASOs hybridize, and in normal homozygotes, only the ASO from the normal allele hybridizes.

CF affects approximately 1 in 2000 individuals of northern European descent. Screening for CF can be used in these populations to detect carriers and to counsel people about their genetic status with respect to CF. However, geneticists have evidence of more than 1000 different mutations for this gene. Not all of these alleles can be screened for because ASO tests have not been generated for all of these variants. Keep in mind also that an ASO test can only screen for specific alleles targeted by the test. Thus, a negative result does not necessarily eliminate someone from having other mutant alleles for a particular condition. Furthermore, it is likely that still more CF mutations remain to be identified. Consequently, CF screening is not widespread but will no doubt become commonplace when tests can detect a majority of all possible mutations.

Genetic Testing Using DNA Microarrays and Genome Scans

Both RFLP and ASO analyses are efficient methods of screening for gene mutations; however, they can only detect the presence of one or a few specific mutations whose identity and locations in the gene are known. There is also a need for genetic tests that detect complex mutation patterns or previously unknown mutations in genes associated with genetic diseases and cancers. For example, the gene that is responsible for cystic fibrosis (the *CFTR* gene) contains 27 exons and encompasses 250 kilobases of genomic DNA. As mentioned earlier, of the 1000 known mutations of the *CFTR* gene, about half of these are point mutations, insertions, and deletions—and these are widely distributed throughout the gene. Moreover, additional *CFTR* mutations may yet be discovered. Similarly, over 500 different mutations are known to occur within the tumor suppressor *p53* gene, and any of these mutations may be associated with, or predispose a patient to, a variety of cancers. In order to screen for mutations in these genes, comprehensive, high-throughput methods are required.

Recall from Chapter 21 that one emerging high-throughput screening technique is based on the use of **DNA microarrays**. DNA microarrays (also called DNA chips or gene chips) are small, solid supports, usually glass or polished quartz-based, on which known fragments of DNA are deposited in a precise pattern. Each spot on a DNA microarray

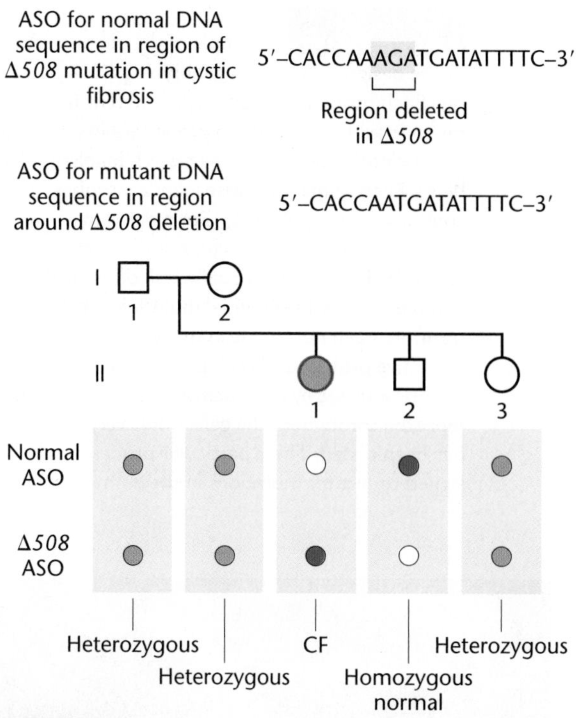

ASO for normal DNA sequence in region of Δ508 mutation in cystic fibrosis

5′–CACCAAAGATGATATTTTC–3′

Region deleted in Δ508

ASO for mutant DNA sequence in region around Δ508 deletion

5′–CACCAATGATATTTTC–3′

Normal ASO

Δ508 ASO

Heterozygous

Heterozygous

CF

Homozygous normal

Heterozygous

FIGURE 22–17 Detecting a deletion in the *CFTR* gene by ASO testing. ASOs for the region spanning the most common mutation in CF (the Δ508 allele) are prepared from cloned copies of the normal allele and the Δ508 allele, which contains a small deletion, and spotted onto DNA-binding membranes. In screening, the CF alleles carried by an individual are amplified by PCR, labeled, and hybridized to the ASOs on the membranes. The genotype of each family member can then be read directly from the membranes. DNA from the parents (I-1 and I-2) hybridizes to both ASOs, indicating that they each carry one normal allele and one mutant allele and are therefore heterozygous. DNA from II-1 hybridizes only to the Δ508 ASO, indicating that this family member is homozygous for the mutation and has cystic fibrosis. DNA from II-2 hybridizes only to the ASO from the normal CF allele, indicating that this individual carries two normal alleles. DNA from II-3 shows two hybridization spots, so the person is heterozygous and a carrier for CF.

is called a **field** (sometimes also called a feature). The DNA fragments that are deposited in a DNA microarray field—called probes—are single-stranded and may be oligonucleotides synthesized *in vitro* or longer fragments of DNA created from cloning or PCR amplification. There are typically over a million identical molecules of DNA in each field. The numbers and types of DNA sequences on a microarray are dictated by the type of analysis that is required. For example, each field on a microarray might contain a DNA sequence derived from each member of a gene family, or sequence variants from one or several genes of interest, or a sequence derived from each gene in an organism's genome. Some microarrays use identical sequences as probes in each field for a particular gene; other microarrays use many different probes for the same gene. What makes DNA microarrays so amazing

FIGURE 22–18 A commercially available DNA microarray, called a GeneChip, marketed by Affymetrix, Inc. This microarray can be used to analyze expression for approximately 50,000 RNA transcripts. It contains 22 different probes for each transcript and allows scientists to simultaneously assess the expression levels of most of the genes in the human genome.

is the immense amount of information that can be simultaneously generated from a single array. DNA microarrays the size of postage stamps (just over 1 cm square) can contain up to 500,000 different fields, each representing a different DNA sequence. In Chapter 21, you learned about the use of microarrays for transcriptome analysis. Scientists are now using DNA microarrays in a wide range of applications, including the detection of mutations in genomic DNA and the detection of gene expression patterns in diseased tissues.

Most human genes are available on a human genome microarray (**Figure 22–18**). Geneticists often use a type of DNA microarray known as a **genotyping microarray** to detect mutations in specific genes. Probes on a genotyping microarray consist of short oligonucleotides, about 20 nucleotides long. These probes are designed to methodically scan through the gene of interest, one nucleotide at a time, checking for the presence of a mutation at each position in the gene. Each position in the gene is tested by a set of five oligonucleotides (and hence five fields, arranged in a column) that are identical in sequence except for one nucleotide that differs in each of them, being either A, C, G, T, or a deletion (**Figure 22–19**).

The genomic DNA sample to be tested on such a microarray is cleaved into small fragments, and the DNA regions of interest are amplified by PCR, labeled with a fluorescent dye, and then hybridized to the microarray. DNA molecules with a sequence that exactly matches the sequence present in a microarray field will hybridize to that field. DNA molecules that differ by one or more nucleotides will hybridize less efficiently or not at all, depending on the hybridization conditions. After washing off material that does not hybridize to the microarray, scientists analyze the microarray with a fluorescence scanner to determine which fields hybridized to the test DNA sample and which fields did not.

Region of wild-type
p53 gene to be tested: 5′ – acttgtcatg gcgactgtcc acctttgtgc – 3′

```
Set 1   Probe 1            3′ – tgaacagtaa cgctgacagg – 5′
        Probe 2            3′ – tgaacagtac cgctgacagg – 5′
        Probe 3            3′ – tgaacagtat cgctgacagg – 5′
        Probe 4            3′ – tgaacagtag cgctgacagg – 5′
        Probe 5            3′ – tgaacagta– cgctgacagg – 5′

Set 2   Probe 1            3′ – gaacagtac agctgacagg t – 5′
        Probe 2            3′ – gaacagtac cgctgacagg t – 5′
        Probe 3            3′ – gaacagtac tgctgacagg t – 5′
        Probe 4            3′ – gaacagtac ggctgacagg t – 5′
        Probe 5            3′ – gaacagtac –gctgacagg t – 5′

Set 3   Probe 1            3′ – aacagtac cactgacagg tg – 5′
        Probe 2            3′ – aacagtac ccctgacagg tg – 5′
        Probe 3            3′ – aacagtac ctctgacagg tg – 5′
        Probe 4            3′ – aacagtac cgctgacagg tg – 5′
        Probe 5            3′ – aacagtac c–ctgacagg tg – 5′
```

FIGURE 22–19 Example of the oligonucleotide probe design for use on a genotyping DNA microarray. Each probe would occupy a different field on the microarray. Each set of five probes is aligned under the nucleotide position to be tested (highlighted in blue). Each of the otherwise identical probes in a set contains either an A, C, T, G or a deletion at the test position. One probe in each set is complementary to the wild-type DNA sequence; all other probes are complementary to potential mutations at the same position. Each nucleotide in the gene is tested with a set of five probes, and each probe set is offset from the previous set by one nucleotide. The pattern is repeated throughout the gene until every nucleotide has been tested. These particular probes are designed to test for mutations in the human *P53* gene.

Under stringent hybridization conditions, the PCR-amplified DNA fragment will hybridize most efficiently to its exact complement but less efficiently to the other possible sequence variants. Hence, the strongest hybridization among the five fields will be to the oligonucleotide probe that contains the correct nucleotide at the tested position. The sequence of the test DNA can therefore be determined by reading through the columns and noting the fields to which the hybridization is most intense (Figure 22–20). DNA microarrays have been designed to scan for mutations in many disease-related genes, including the *p53* gene, which is mutated in a majority of human cancers, and the *BRCA1* gene, which, when mutated, predisposes women to breast cancer. Figure 22–20 shows results for a genotypic microarray for the *p53* gene. In this example, hybridization of PCR-amplified DNA fragments from a patient appears as green or blue-green spots on the microarray [Figure 22–20(b)]. Interpreting the hybridization pattern reveals the *p53* gene sequence from the patient ("target").

In addition to testing for mutations in single genes, DNA microarrays can contain probes that detect SNPs. SNPs occur randomly about every 100 to 300 nucleotides throughout the human genome, both inside and outside of genes. SNPs crop up in an estimated 15 million positions in the genome where these single-based changes reveal differences from one person to the next. Certain SNP sequences at a specific locus are shared by certain segments of the population. In addition, certain SNPs cosegregate with genes associated with some disease conditions. By correlating the presence or absence of a particular SNP with a genetic disease, scientists are able to use the SNP as a genetic testing marker. The presence of SNPs as probes on a DNA microarray allows scientists to simultaneously screen thousands of genes that might be involved in single-gene diseases as well as those involved in disorders exhibiting multifactorial inheritance. This technique, known as **genome scanning**, makes it possible to analyze a person's DNA for dozens or hundreds of disease alleles, including

(a)

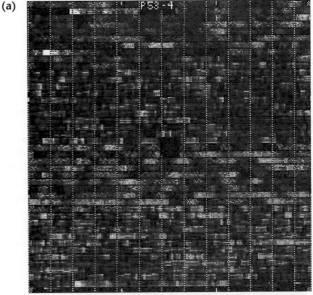

(b)

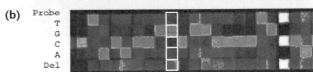

FIGURE 22–20 A *p53* GeneChip (Affymetrix, Inc.) after hybridization to *p53* PCR products. (a) This DNA microarray contains over 65,000 fields, each about 50 micrometers square. The microarray tests for mutations within the entire 1.2-kb coding sequence of the human *p53* gene. The dotted vertical lines are generated by control fields and allow for alignment of the DNA chip during scanner analysis. (b) An enlarged portion of the *p53* microarray. Each square is a field to which the test DNA has or has not hybridized. Each row represents one of the four nucleotides or a deletion in the test position of the probe. Each column represents the nucleotide position being tested. The column outlined in white shows that the test DNA sample has a C at this position in the *p53* sequence based on its hybridization to the G in this field. The column marked with an asterisk (*) is the alignment control. The sequence of the *p53* gene is shown under the photo.

those that might predispose the person to heart attacks, asthma, diabetes, Alzheimer disease, and other genetically defined disease subtypes. Although genome scanning approaches are not widely used yet, perhaps in decades to come, all newborns will undergo genome scanning to determine their lifetime risks for suffering from genetic disorders.

Finally, **array comparative genomic hybridization (CGH)** is a microarray-based technique that is used to identify copy number variations (CNVs) throughout the genome which may be involved in genetic disease conditions. Recall from Chapter 21 and elsewhere in the book that CNVs are relatively large (~500 bp to several Mb) insertions or deletions, which together with SNPs comprise a majority of genome variations between individuals. Originally developed to identify chromosomal changes in tumor sample, a CGH array uses oligonucleotides that span the entire genome of a species based on the reference sequence for that species.

Human genome CGH arrays are made using bacterial artificial chromosomes (BACs) that contain reference sequences spanning the entire genome. Test genomic DNA from an individual with a particular disorder and control DNA from a healthy individual are labeled differentially with fluorescent tags and then hybridized to the CGH array. Once scanned, fluorescent ratios on the array are analyzed and used to indicate the relative presence or absence of hybridization in the samples. Low or no binding of a DNA sample to a particular field on the array can be one indication of the presence of a CNV. Array CGH is a rapid way of examining an individual genome for CNVs, but these do not reveal very small variations in the genome; nor do they recognize translocations or inversions that can contribute to genetic diseases.

Genetic Analysis Using Gene-Expression Microarrays

Gene-expression microarrays are effective for analyzing gene-expression patterns in genetic diseases because the progression of a tissue from a healthy to a diseased state is almost always accompanied by changes in expression of hundreds to thousands of genes. These arrays provide a powerful tool for diagnosing genetic disorders and gene-expression changes. Expression microarrays may contain probes for only a few specific genes thought to be expressed differently in cell types or may contain probes representing each gene in the genome. Although microarray techniques provide novel information about gene expression, keep in mind that DNA microarrays do not directly provide us with information about protein levels in a cell or tissue. We often infer what predicted protein levels may be based on mRNA expression patterns, but this may not always be accurate.

In a typical expression microarray analysis, mRNA is isolated from two different cell or tissue types—for example, normal cells and cancer cells arising from the same cell type [Figure 22–21(a)]. The mRNA samples contain transcripts

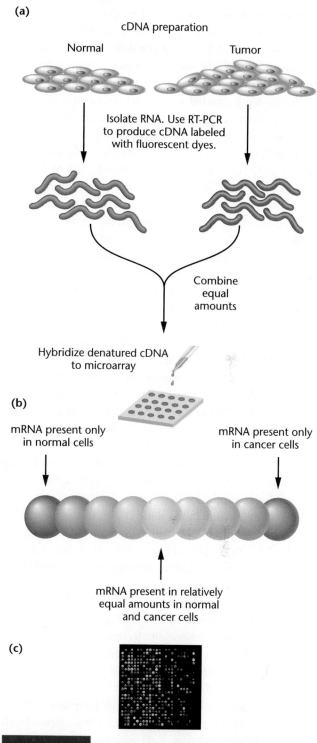

FIGURE 22–21 (a) Microarray procedure for analyzing gene expression in normal and cancer cells. (b) The method shown here is based on a two-channel microarray in which cDNA samples from the two different tissues are competing for binding to the same probe sets. Colors of dots on an expression microarray represent levels of gene expression. In this example, green dots represent genes expressed only in one cell type (e.g., normal cells), and red dots represent genes expressed only in another cell type (e.g., cancer cells). Intermediate colors represent different levels of expression of the same gene in the two cell types. (c) A small portion of a DNA microarray, showing different levels of hybridization to each field.

from each gene that is expressed in that cell type. Some genes are expressed more efficiently than others; therefore, each type of mRNA is present at a different level. The level of each mRNA can be used to develop a gene-expression profile that is characteristic of the cell type. Isolated mRNA molecules are converted into cDNA molecules, using reverse transcriptase. The cDNAs from the normal cells are tagged with fluorescent dye-labeled nucleotides (for example, green), and the cDNAs from the cancer cells are tagged with a different fluorescent dye-labeled nucleotide (for example, red). The labeled cDNAs are mixed together and applied to a DNA microarray. The cDNA molecules bind to complementary single-stranded probes on the microarray but not to other probes. Keep in mind that each field or feature does not consist of just one probe, but rather they contain thousands of copies of the probe. After washing off the nonbinding cDNAs, scientists scan the microarray with a laser, and a computer captures the fluorescent image pattern for analysis. The pattern of hybridization appears as a series of colored dots, with each dot corresponding to one field of the microarray [Figure 22–21(b)].

The color patterns revealed on the microarray fields [Figure 22–21(c)] provide a sensitive measure of the relative levels of each cDNA in the mixture. In the example shown here, if an mRNA is present only in normal cells, the probe representing the gene encoding that mRNA will appear as a green dot because only "green" cDNAs have hybridized to it. Similarly, if an mRNA is present only in the cancer cells, the microarray probe for that gene will appear as a red dot. If both samples contain the same cDNA, in the same relative amounts, both cDNAs will hybridize to the same field, which will appear yellow [Figure 22–21(b)]. Intermediate colors indicate that the cDNAs are present at different levels in the two samples.

Expression microarray profiling has revealed that certain cancers have distinct patterns of gene expression and that these patterns correlate with factors such as the cancer's stage, clinical course, or response to treatment. In one such experiment, scientists examined gene expression in both normal white blood cells and in cells from a white blood cell cancer known as **diffuse large B-cell lymphoma (DLBCL)**. About 40 percent of patients with DLBCL respond well to chemotherapy and have long survival times. The other 60 percent respond poorly to therapy and have short survival. The investigators assayed the expression profiles of 18,000 genes and discovered that there were two types of DLBCL, with almost inverse patterns of gene expression (Figure 22–22). One type of DLBCL, called *GC B-like*, had an expression pattern dramatically different from that of a second type, called *activated B-like*. Patients with the activated B-like pattern of gene expression had much lower survival rates than patients with the GC B-like pattern. The researchers concluded that DLBCL is actually two different

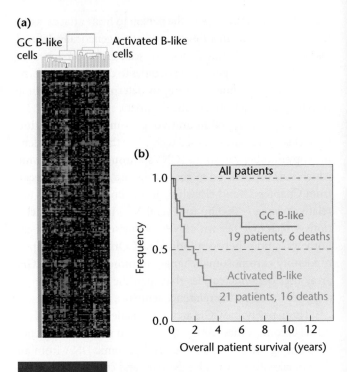

FIGURE 22–22   (a) Gene-expression analysis generated from expression DNA microarrays that analyzed 18,000 genes expressed in normal and cancerous lymphocytes. Each row represents a summary of the gene expression from one particular gene; each column represents data from one cancer patient's sample. The colors represent ratios of relative gene expression compared to normal control cells. Red represents expression greater than the mean level in controls, green represents expression lower than in the controls, and the intensity of the color represents the magnitude of difference from the mean. In this summary analysis, the cancer patients' samples are grouped by how closely their gene-expression profiles resemble each other. The cluster of cancer patients' samples marked with orange at the top of the figure are GC B-like DLBCL cells. The blue cluster contains samples from cancer patients within the activated B-like DLBCL group. (b) Gene-expression profiling and survival probability. Patients with activated B-like profiles have a much higher rate of death (16 in 21) than those with GC B-like profiles (6 in 19). Data such as these demonstrate the value of microarray analysis for diagnosing disease conditions.

diseases with different outcomes. Once this type of analysis is introduced into routine clinical use, it may be possible to adjust therapies for each group of cancer patients and to identify new specific treatments based on gene-expression profiles. Similar gene-expression profiles have been generated for many other cancers, including breast, prostate, ovarian, and colon cancer. Gene-expression microarrays are providing tremendous insight into both substantial and subtle variations in genetic diseases.

Several companies are now promoting "nutrigenomics" services in which they claim to use genotyping and gene-expression microarrays to identify allele polymorphisms and gene-expression patterns for genes involved in nutrient metabolism. For example, polymorphisms in genes such as that for apolipoprotein A (*APOA1*), involved in lipid metabolism, and that for *MTHFR* (methylenetetrahydrofolate reductase), involved in metabolism of folic acid, have been implicated in cardiovascular disease. Nutrigenomics companies claim that microarray analysis of a patient's DNA sample for genes such as these and others enables them to judge whether a patient's allele variations or gene-expression profiles warrant dietary changes to potentially improve health and reduce the risk of diet-related diseases.

Application of Microarrays for Gene Expression and Genotype Analysis of Pathogens

Microarrays are also providing infectious disease researchers with powerful new tools for studying pathogens. Genotyping microarrays are being used to identify strains of emergent viruses, such as the virus that causes the highly contagious condition called Severe Acute Respiratory Syndrome (SARS) as well as the H5N1 avian influenza virus, the cause of bird flu, which has killed about two dozen people in Asia, leading to the slaughter of over 80 million chickens and causing concern about possible pandemic outbreaks.

Whole-genome transcriptome analysis of pathogens is being used to inform researchers about genes that are important for pathogen infection and replication (**Figure 22–23**). In this approach, bacteria, yeast, protists, or viral pathogens are used to infect host cells *in vitro*, and then expression microarrays are used to analyze pathogen gene-expression profiles. Patterns of gene activity during pathogen infection of host cells and replication are useful for identifying pathogens and understanding mechanisms of infection. But

of course a primary goal of infectious disease research is to prevent infection. Gene-expression profiling is also a valuable approach for identifying important pathogen genes and the proteins they encode that may prove to be useful targets for subunit vaccine development or for drug treatment strategies to prevent or control infectious disease. This strategy primarily informs researchers about how a pathogen responds to its host.

Similarly, researchers are evaluating host responses to pathogens (**Figure 22–24**). This type of detection has been accelerated in part by the need to develop pathogen-detection strategies for military and civilian use both for detecting outbreaks of naturally emerging pathogens such as SARS and avian influenza and for potential detection of outbreaks such as anthrax (caused by the bacterium *Bacillus anthracis*) that could be the result of a bioterrorism event. Host-response gene-expression profiles are developed by exposing a host to a pathogen and then using expression microarrays to analyze host gene-expression patterns.

Figure 22–24 shows the different gene-expression profiles for mice following exposure to *Neisseria meningitidis*, the SARS virus, or *E. coli*. In this example, although there are several genes that are upregulated or downregulated by each pathogen, notice how each pathogen strongly induces different prominent clusters of genes that reveal a host gene-expression response to the pathogen and provide a signature of pathogen infection. Comparing such host gene-expression profiles following exposure to different pathogens provides researchers with a way to quickly diagnose and classify infectious diseases. In the future, scientists expect to develop databases of both pathogen and host response expression profile data that can be used to identify pathogens efficiently.

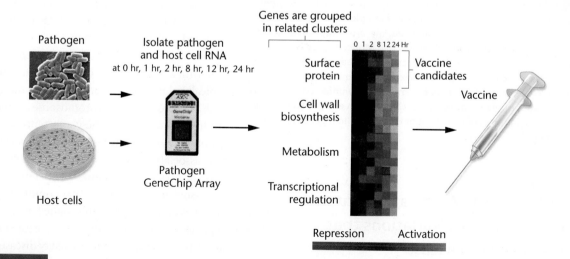

FIGURE 22–23 Whole-genome expression profiling enables researchers to identify genes that are actively expressed when pathogens infect host cells and replicate. Knowing which genes are critical for pathogen infection and replication can help scientists target proteins as vaccine and drug-treatment candidates for preventing or treating infectious diseases.

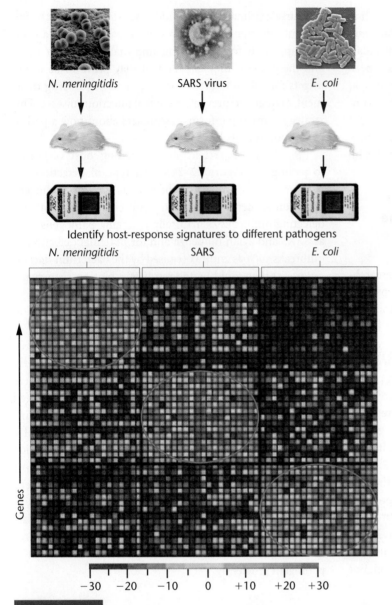

N. meningitidis   SARS virus   *E. coli*

Identify host-response signatures to different pathogens

N. meningitidis   SARS   *E. coli*

Genes

−30 −20 −10 0 +10 +20 +30

FIGURE 22–24 Gene-expression microarrays can reveal host-response signatures for pathogen identification. In this example, mice were infected with different pathogens: *Neisseria meningitidis,* the virus that causes Severe Acute Respiratory Syndrome (SARS), and *E. coli.* Mouse tissues were then used as the source of mRNA for gene-expression microarray analysis. Increased expression compared to uninfected control mice is shown in shades of yellow. Decreased expression compared to uninfected controls is indicated in shades of blue. Notice that each pathogen elicits a somewhat different response in terms of which major clusters of host genes are activated by pathogen infection (circles).

22.6

Genome-Wide Association Studies Identify Genome Variations that Contribute to Disease

Microarray-based genomic analysis has led geneticists to employ powerful new strategies called **genome-wide association**

studies **(GWAS)** in their quest to identify genes that may influence disease risk. During the past 5 years there has been a dramatic expansion in the number of GWAS being reported. For example, GWAS for height differences, autism, obesity, diabetes, macular degeneration, myocardial infarction, arthritis, hypertension, several cancers, bipolar disease, autoimmune diseases, Crohn's disease, schizophrenia, amyotrophic lateral sclerosis, and multiple sclerosis are among the many GWAS that have been widely publicized in the scientific literature and popular press. As of fall 2010, GWAS have resulted in approximately 700 publications linking nearly 3000 genetic variations to about 150 traits.

How are GWAS carried out? In a GWAS, the genomes of thousands of unrelated individuals with a particular disease are analyzed, typically by microarray analysis, and results are compared with genomes of individuals without the disease as an attempt to identify genetic variations that may confer risk of developing the disease. Many GWAS involve large-scale use of SNP microarrays that can probe on the order of 500,000 SNPs to evaluate results from different individuals. Other GWAS approaches can look for specific gene differences or evaluate CNVs or changes in the epigenome, such as methylation patterns in particular regions of a chromosome. By determining which SNPs, CNVs, or epigenome changes co-occur in individuals with the disease, scientists can calculate the disease risk associated with each variation. Analysis of GWAS results requires statistical analysis to predict the relative potential impact (association or risk) of a particular genetic variation on development of a disease phenotype.

Figure 22–25 shows a typical representation of one way that results from GWAS are commonly reported. Called a Manhattan plot, such representations are "scatterplots" that are used to display data with a large number of data points. The *x*-axis typically plots a particular position in the genome; in this case loci on each chromosome are plotted in a different color code. The *y*-axis plots results of a genotypic association test. There are several ways that association can be calculated. Shown here is a negative log of *p*-values that shows loci determined to be significantly associated with a particular condition. The top line of this plot establishes a threshold value for significance. Marker sequences with significance levels exceeding 10^{-5}, corresponding to 5.0 on the *y*-axis, are likely disease-related sequences (Figure 22–25).

One prominent study that brought the potential of GWAS to light involved research in the United States and Iceland

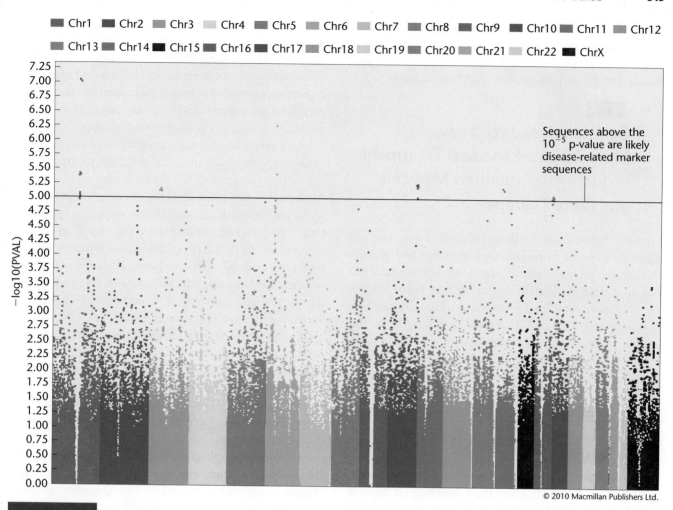

■ Chr1 ■ Chr2 ■ Chr3 ■ Chr4 ■ Chr5 ■ Chr6 ■ Chr7 ■ Chr8 ■ Chr9 ■ Chr10 ■ Chr11 ■ Chr12
■ Chr13 ■ Chr14 ■ Chr15 ■ Chr16 ■ Chr17 ■ Chr18 ■ Chr19 ■ Chr20 ■ Chr21 ■ Chr22 ■ ChrX

Sequences above the
10^{-5} p-value are likely
disease-related marker
sequences

© 2010 Macmillan Publishers Ltd.

FIGURE 22–25 A GWAS study for type 2 diabetes revealed 386,371 genetic markers, clustered here by chromosome number. Markers above the red line appeared to be significantly associated with the disease.

on 4500 patients with a history of heart attack, technically a myocardial infarction (MI), and 12,767 control patients. This work was done with microarrays containing 305,953 SNPs. Among the most notable results, the study revealed variations in tumor-suppressor genes on chromosome 9 (*CDKN2A* and *CDKN2B*). Twenty-one percent of MI individuals were homozygous for deleterious mutations in the *CDKN2A* and *CDKN2B* genes, and these individuals showed a 1.6× enhanced risk of MI compared with noncarriers, including individuals homozygous for wild-type alleles. These variations were correlated with those in people of European descent, but so far these same marker genes have not been revealed in African Americans. Does this mean that these genes are not MI risk factor genes among African Americans?

This question reflects one of many questions and ethical concerns about patients involved in GWAS and their emotional responses to knowing about genetic risk data. For example,

■ What does it mean if an individual has 3, 5, 9, or 30 risk alleles for a particular condition?

■ How do we categorize rare, common, and low-frequency risk alleles to determine the overall risk for developing a disease?

■ GWAS often reveal dozens of DNA variations, but many variations have only a modest effect on risk. How does one explain to a person that he or she has a gene variation that changes a risk difference for a particular disease from 12 to 16 percent over an individual's lifetime? What does this information mean?

■ If the sum total of GWAS for a particular condition reveals about 50 percent of the risk alleles, what are the other missing elements of heritability that may contribute to developing a complex disease?

In some cases, risk data revealed by GWAS may help patients and physicians develop diet and exercise plans designed to minimize the potential for developing a particular disease. But the number of risk genes identified by most GWAS is showing us that, unlike single-gene disorders, complex genetic disease conditions involve a multitude of genetic factors contributing to the total risk for developing a condition. We need such information to make meaningful progress in disease diagnosis and treatment, which is ultimately a major purpose of what GWAS are all about.

22.7

Genomics Leads to New, More Targeted Medical Treatment Including Personalized Medicine and Gene Therapy

Genomic technologies are changing medical diagnosis and allowing scientists to manufacture abundant and effective therapeutic proteins. The examples already available today are a strong indication that in the near future, we will see even more transformative medical treatments based on genomics and advanced DNA-based technologies. In this section, we provide a brief introduction to pharmacogenomics and rational drug design, topics that will be considered in greater detail in Special Topics on Genomics and Personalized Medicine, p. 504, and we examine gene therapy approaches to treat genetic diseases.

Pharmacogenomics and Rational Drug Design

Every year, more than 2 million Americans experience serious side effects of medications, and more than 100,000 die from adverse drug reactions. In addition, most drugs are effective in only about 50 percent of the population. Until now, the selection of effective medications for each individual has been a random, trial-and-error process. The new field of **pharmacogenomics** promises to lead to more specific, effective, and personally customized drugs that are designed to complement each person's individual genetic makeup.

In some ways, pharmacogenomics began in the 1950s, when scientists discovered that reactions to drugs had a hereditary component. We now know that many genes affect how different individuals react to drugs. Some of these genes encode products such as cell-surface receptors that bind a drug and allow it to enter a cell, as well as enzymes that metabolize drugs. For example, liver enzymes encoded by the cytochrome *P450* gene family affect the metabolism of many modern drugs, including those used to treat cardiovascular and neurological conditions. DNA sequence variations in these genes result in enzymes with different abilities to metabolize and utilize these drugs. Thus, gene variants that encode inactive forms of the cytochrome P450 enzymes are associated with a patient's inability to break down drugs in the body, leading to drug overdoses. A genetic test that recognizes some of these variants is currently being used to screen patients who are recruited into clinical trials for new drugs.

Another example is the reaction of certain people to the thiopurine drugs used to treat childhood leukemias (**Figure 22–26**). Some individuals have sequence variations

Individuals respond differently to the anti-leukemia drug 6-mercaptopurine.	The diversity in responses is due to mutations in a gene called thiopurine methyltransferase (*TPMT*).	After a simple blood test, individuals can be given doses of medication that are tailored to their genetic profile.

Most people metabolize the drug quickly. Doses need to be high enough to treat leukemia and prevent relapses.

Normal TPMT enzyme

High dose for TPMT homozygote

Others metabolize the drug slowly and need lower doses to avoid toxic side effects of the drug.

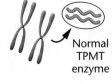

Normal and mutant TPMT (✹) enzyme

Moderate dose for TPMT heterozygote

A small portion of people metabolize the drug so poorly that its effects can be fatal

Mutant TPMT enzyme

Low dose for an extra slow metabolizer (TPMT-deficient homozygote)

FIGURE 22–26 Pharmacogenomics approaches to drug development are saving lives. Different individuals with the same disease, in this case childhood leukemia, often respond differently to a drug treatment because of subtle differences in gene expression. The dose of an anticancer drug such as 6-MP that works for one person may be toxic for another person. A simple gene or enzyme test to identify genetic variations can enable physicians to prescribe a drug treatment and dosage based on a person's genetic profile.

in the gene encoding the enzyme thiopurine methyltransferase (TPMT), which breaks down thiopurines. Anticancer drugs such as 6-mercaptopurine (6-MP) that are commonly used to treat leukemia have a thiopurine structure. In individuals with mutations in the *TPMT* gene, thiopurine cancer drugs can build up to toxic levels. As a result, although some patients, such as those who are homozygous for the wild-type *TPMT* gene, respond well to 6-MP treatment, others who are heterozygotes or homozygous for mutations in *TPMT* can have severe or even fatal reactions to 6-MP. At first a genetic test was developed to detect *TPMT* gene variants, but now a simple blood test is enough to enable clinicians to tailor the drug dosage to the individual. As a result of this new technology, toxic effects of 6-MP have decreased, and survival rates for childhood leukemia patients treated with 6-MP have increased—a great example of pharmacogenomics in action.

Several methods are being developed for expanding the uses of pharmacogenomics. One promising method involves the detection of SNPs. Perhaps researchers will be able to identify a shared SNP sequence in the DNA of people who also share a heritable reaction to a drug. If the SNP segregates with a part of the genome containing the gene responsible for the drug reaction, it may be possible to devise gene tests based on the SNP, without even knowing the identity of the gene responsible for the drug reaction. In the future, DNA microarrays may be used to screen a patient's genome for multiple drug reactions.

Knowledge from genetics and molecular biology is also contributing to the development of new drugs targeted at specific disease-associated molecules. Most drug development is currently based on trial-and-error testing of chemicals in lab animals, in the hope of finding a chemical that has a useful effect. In contrast, **rational drug design** involves the synthesis of specific chemical substances that affect specific gene products. An example of a rational drug design product is the new drug imatinib, trade name **Gleevec**, used to treat chronic myelogenous leukemia (CML). Geneticists had discovered that CML cells contain the Philadelphia chromosome, which results from a reciprocal translocation between chromosomes 9 and 22. Gene cloning revealed that the t(9;22) translocation creates a fusion of the *C-ABL* proto-oncogene with the *BCR* gene. This *BCR-ABL* fusion gene encodes a powerful fusion protein that causes cells to escape cell-cycle control. The fusion protein, which acts as a tyrosine kinase, is not present in noncancer cells from CML patients.

To develop Gleevec, chemists used high-throughput screens of chemical libraries to find a molecule that bound to the BCR-ABL enzyme. After chemical modifications to make the inhibitory molecule bind more tightly, tests showed that it specifically inhibited BCR-ABL activity. Clinical trials revealed that Gleevec was effective against CML,

with minimal side effects and a higher remission rate than that seen with conventional therapies. Gleevec is now used to treat CML and several other cancers. With scientists discovering more genes and gene products associated with diseases, rational drug design promises to become a powerful technology within the next decade.

Gene Therapy

Although drug treatments are often effective in controlling symptoms of genetic disorders, the ideal outcome of medical treatment is to cure these diseases. In an effort to cure genetic diseases, scientists are actively investigating **gene therapy**—a therapeutic technique that aims to transfer normal genes into a patient's cells. In theory, the normal genes will be transcribed and translated into functional gene products, which, in turn, will bring about a normal phenotype.

One key to gene therapy is having a delivery system to transfer genes into a patient. In many gene therapy trials, scientists often used genetically modified retroviruses as vectors. An example is a vector based on a mouse virus called **Moloney murine leukemia virus (MLV)**. Disabled forms of adeno-associated virus (AAV), which in its native form infects ~80–90 percent of the population during childhood, and nonviral methods are being used to transfer genes into cells include chemically assisted transfer of genes across cell membranes, and fusion of cells with artificial vesicles containing cloned DNA sequences. Retroviral vectors are created by removing a cluster of three genes from the virus and inserting a cloned human gene. After being packaged in a viral protein coat, the recombinant vector is used to infect cells. Once inside a cell, the virus cannot replicate itself because of the missing viral genes. In the cell, the recombinant virus with the inserted human gene moves to the nucleus, integrates into a site on a chromosome, and becomes part of the genome. If the inserted gene is expressed, it produces a normal gene product that may be able to correct the mutation carried by the affected individual. In initial attempts at gene therapy, several heritable disorders, including severe combined immunodeficiency (SCID), familial hypercholesterolemia, and cystic fibrosis were treated.

Human gene therapy began in 1990 with the treatment of a young girl named Ashanti DeSilva [**Figure 22–27**(a)], who has a heritable disorder called **severe combined immunodeficiency (SCID)**. Individuals with SCID have no functional immune system and usually die from what would normally be minor infections. Ashanti has an autosomal form of SCID caused by a mutation in the gene encoding the enzyme **adenosine deaminase (ADA)**. Her gene therapy began when clinicians isolated some of her white blood cells, called T cells [Figure 22–27(b)]. These cells, which are key

(a)   **(b)**

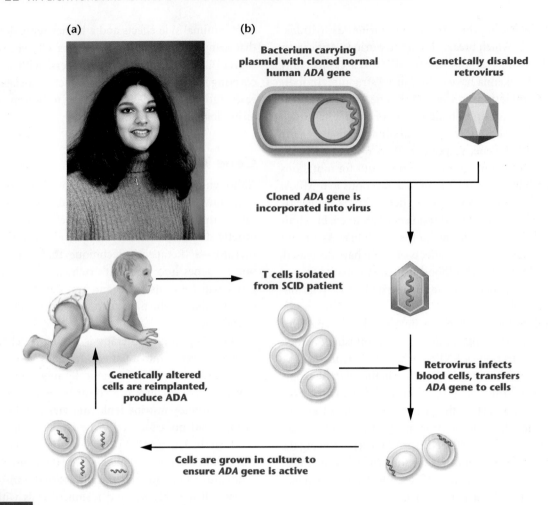

FIGURE 22–27   (a) Ashanti DeSilva, the first person to be treated by gene therapy. (b) To treat SCID using gene therapy, a cloned human *ADA* gene is transferred into a viral vector, which is then used to infect white blood cells removed from the patient. The transferred *ADA* gene is incorporated into a chromosome and becomes active. After growth to enhance their numbers, the cells are inserted back into the patient, where they produce ADA, allowing the development of an immune response.

components of the immune system, were mixed with a retroviral vector carrying an inserted copy of the normal *ADA* gene. The virus infected many of the T cells, and a normal copy of the *ADA* gene was inserted into the genome of some T cells. After being mixed with the vector, the T cells were grown in the laboratory and analyzed to make sure that the transferred *ADA* gene was expressed (Figure 22–27). Then a billion or so genetically altered T cells were injected into Ashanti's bloodstream. Some of these T cells migrated to her bone marrow and began dividing and producing daughter cells that also produce ADA. She now has ADA protein expression in 25 to 30 percent of her T cells, which is enough to allow her to lead a normal life.

To date, gene therapy has successfully restored the health of about 20 children affected by SCID. Although gene therapy was originally developed as a treatment for single-gene (monogenic) inherited diseases, the technique was quickly adapted for the treatment of acquired diseases

such as cancer, neurodegenerative diseases, cardiovascular disease, and infectious diseases, such as HIV. In the case of HIV, scientists are exploring ways to deliver immune system-stimulating genes that could make individuals resistant to HIV infection or cripple the virus in HIV-positive persons. There are nearly 1000 gene therapy trials actively underway in the United States alone. Over a 10-year period, from 1990 to 1999, more than 4000 people underwent gene therapy for a variety of genetic disorders. These trials often failed and thus led to a loss of confidence in gene therapy.

Hopes for gene therapy plummeted even further in September 1999 when teenager Jesse Gelsinger died while undergoing gene therapy to treat a liver disease condition. His death was triggered by a massive inflammatory response to the vector, a modified **adenovirus**, one of the viruses that cause colds and respiratory infections. Large numbers of adenovirus vectors bearing the *ornithine* *transcarbamylase (OTC)* gene were injected into his hepatic artery. The virus vectors were expected to lodge in his liver, enter the liver cells, and trigger the production of OTC protein. In turn, the OTC protein might correct his genetic defect and perhaps cure him of his liver disease. However, within hours of his first treatment, a massive immune reaction surged through Jesse's body. He developed a high fever, his lungs filled with fluid, multiple organs shut down, and he died four days later of acute respiratory failure.

In the aftermath of the tragedy, several government and scientific inquiries were conducted. Investigators learned that clinical trial scientists had not reported other adverse reactions to gene therapy and that some of the scientists were affiliated with private companies that could benefit financially from the trials. They found that serious side effects seen in animal studies were not explained to patients during informed-consent discussions, and that some clinical trials were proceeding too quickly in the face of data suggesting a need for caution. The U.S. Food and Drug Administration (FDA) scrutinized gene therapy trials across the country, halted a number of them, and shut down several gene therapy programs. Other research groups voluntarily suspended their gene therapy studies. Tighter restrictions on clinical trial protocols were imposed to correct some of the procedural problems that emerged from the Gelsinger case. Jesse's death had dealt a severe blow to the struggling field of gene therapy—a blow from which it was still reeling when a second tragedy hit.

The outlook for gene therapy brightened in 2000, when a group of French researchers reported the first large-scale success in gene therapy. Nine children with a fatal X-linked form of SCID developed functional immune systems after being treated with a retroviral vector carrying a normal gene. Published reports of the study were greeted with enthusiasm by the gene therapy community. But elation turned to despair in 2003, when it became clear that 2 of the 10 children who had been cured of X-SCID had developed leukemia as a direct result of their therapy, and one died as a result of the treatment. In two of the children, their cancer cells contained the retroviral vector, inserted near or into a gene called *LMO2*. This insertion activated the *LMO2* gene, causing uncontrolled white blood cell proliferation and development of leukemia. FDA immediately halted 27 similar gene therapy clinical trials, and once again gene therapy underwent a profound reassessment. In 2005, a third child in the French X-SCID study developed leukemia, likely as a result of gene therapy.

Up until the apparent success of the French X-SCID clinical trials, gene therapy had suffered not only from the scandals and scrutiny that emerged from Jesse Gelsinger's death but also from the skepticism of many scientists and the general public about the feasibility of this much-ballyhooed therapeutic technique. To date, no human gene therapy product has been approved for sale. Critics of gene therapy continue to berate research groups for undue haste, conflicts of interest, and sloppy clinical trial management, and for promising much but delivering little. Most problems associated with gene therapy have been traced to the vectors used to transfer therapeutic genes into cells.

These vectors, including MLV and adenovirus, have several serious drawbacks. First, integration of retroviral genomes (including the human therapeutic gene) into the host cell's genome occurs only if the host cells are replicating their DNA. In the body, only a small number of cells in any tissue are dividing and replicating their DNA. Second, most viral vectors are capable of causing an immune response in the patient, as happened in Jesse Gelsinger's case. Third, insertion of viral genomes into host chromosomes can activate or mutate an essential gene, as in the case of the three French patients. Viral integrase, the enzyme that allows for viral genome integration into the host genome, interacts with chromatin-associated proteins, often steering integration toward transcriptionally active genes. Unfortunately, it is not yet possible to reliably target insertion of therapeutic genes into specific locations in the genome, but targeted gene delivery is a major area of active research.

Fourth, retroviruses cannot carry DNA sequences much larger than 8 kb. Many human genes exceed this size. Finally, there is a possibility that a fully infectious virus could be created if the vector were to recombine with another viral genome already present in the host cell.

To overcome these problems, new viral vectors and strategies for transferring genes into cells are being developed in an attempt to improve the action and safety of vectors. Researchers hope that the use of new gene delivery systems will circumvent the problems inherent in earlier vectors, as well as allow regulation of both insertion sites and the levels of gene product produced from the therapeutic genes.

In addition to the vector delivery issues addressed above, a number of other barriers must be overcome if gene therapy is to become a viable approach for reliably treating many genetic disorders. Issues include:

- What is the proper route for gene delivery in different kinds of disorders? For example, what is the best way to treat brain or muscle tissues? Tissue-specific gene delivery approaches are key.

- What percentage of cells in an organ or tissue need to express a therapeutic gene to alleviate the effects of a genetic disorder?

- What amount of a therapeutic gene product must be produced to provide lasting improvement of the condition, and how can sufficient production be ensured? Currently many gene therapy approaches provide only short-lived delivery of the therapeutic gene and its protein.

■ Will it be possible to use gene therapy to treat diseases that involve multiple genes?

■ Can expression of therapeutic genes be controlled in a patient?

Several well-publicized studies carried out from 2008 to 2010 have been encouraging. For example, researchers from the University of Pennsylvania used gene therapy to restore retinal cone cell function and day vision in dogs with a condition called congenital achromatopsia. Achromatopsia is a rare, autosomal recessive condition (1 in 30,000 to 50,000 humans) and affects cone cells in the retina that are essential for color vision and aspects of visual acuity. The therapy cured both young and older canines and appears to be permanent.

University of Pennsylvania and Children's Hospital of Philadelphia researchers also reported beneficial results of treatments for Leber's congenital amaurosis (LCA), a degenerative disease of the retina that affects 1 in 50,000 to 1 in 100,000 infants each year and causes severe blindness. Young adult patients with defects in the *RPE65* gene were given injections of the normal gene. Similar human trial results were reported in studies in Italy and the United Kingdom, where human trials for LCA first began. Complete vision was not restored to these patients. Several months after a single treatment of the gene the patients are still legally blind, but they can see more light, some of them can read lines of an eye chart, and two who had stumbled through an obstacle course were able to navigate through it.

Researchers at the University of Paris and Harvard Medical School have reported that 2 years after gene therapy treatment for β-thalassemia, a blood disorder that involves a defect in the β-globin chain of hemoglobin which reduces the production of hemoglobin, a young man no longer needs transfusions and appears to be healthy. A modified, disabled HIV was used to carry a copy of the normal gene, although there have been reports of therapeutic gene integration near a growth factor gene called *HMGA2* resulting in activation of this gene, reminiscent of what occurred in the French X-SCID trials.

Scientists are also working on gene replacement approaches that involve removing a defective gene from the genome. Recent work with enzymes called **zinc-finger nucleases** have shown promise in animal models and cultured cells. These enzymes can create site-specific cleavage in the genome and when coupled with certain integrases may lead to gene editing by cutting out defective sequences and introducing normal homologous sequences into the genome. Encouraging breakthroughs have taken place in this area using model organisms such as mice; however, this technology has not advanced sufficiently for use in humans. Attempts

have been made to use **antisense oligonucleotides** in order to inhibit translation of mRNAs from defective genes, but this approach to gene therapy has generally not yet proven to be reliable.

The recent emergence of RNA interference as a powerful gene-silencing tool has reinvigorated gene therapy approaches by gene silencing. As you learned in Chapter 17, **RNA interference (RNAi)** is a form of gene-expression regulation (see Figure 17–16). In animals short, double-stranded RNA molecules are delivered into cells where the enzyme Dicer chops them into 21-nt long pieces called **small interfering RNAs (siRNAs)**. siRNAs then join with an enyzme complex called the **RNA inducing silencing complex (RISC)**, which shuttles the siRNAs to their target mRNA, where they bind by complementary base pairing. The RISC complex can block siRNA-bound mRNAs from being translated into protein or can lead to degradation of siRNA-bound mRNAs so they cannot be translated into protein.

A main challenge to RNAi-based therapeutics so far has been *in vivo* delivery of double-stranded RNA or siRNA. RNAs degrade quickly in the body. It is also hard to get them to penetrate cells and to target the right tissue. For example, how does one deliver RNA-based therapies to cancer cells but not to noncancerous, healthy cells? Two common delivery approaches are to inject the siRNA directly or to deliver them via a plasmid vector that is taken in by cells and transcribed to make double-stranded RNA that can be cleaved by Dicer into siRNAs.

Several RNAi clinical trials to treat blindness are underway in the United States. One RNAi strategy to treat a form of blindness called macular degeneration targets a gene called *VEGF*. The VEGF protein promotes blood vessel growth. Overexpression of this gene, causing excessive production of blood vessels in the retina, leads to impaired vision and eventually blindness. Many expect that this disease will soon become the first condition to be treated by RNAi therapy. Other disease candidates for treatment by RNAi include several different cancers, diabetes, multiple sclerosis, and arthritis.

The question remains whether gene therapy can ever recover from past setbacks and fulfill its promise as a cure for genetic diseases. Many scientists feel that we should continue gene therapy research and clinical trials despite the setbacks. However, those working today have a more sober view of its progress. Clinical trials for any new therapy are potentially dangerous, and often animal studies will not accurately reflect the reaction of individual humans to a new drug or procedure. Inevitably, more adverse reactions to gene therapy will emerge in the clinical trials, even as the methods become more effective. Perhaps we should view gene therapy as we have antibiotics, organ transplants, and manned space travel. There will be setbacks and even

NOW SOLVE THIS

22–5 Gene therapy for human genetic disorders involves transferring a copy of the normal human gene into a vector and using the vector to transfer the cloned human gene into target tissues. Presumably, the gene enters the target tissue and becomes active, and the gene product relieves the symptoms.

(a) Why are disorders such as muscular dystrophy difficult to treat by gene therapy?

(b) What are the potential problems of using retroviruses as vectors?

(c) Should gene therapy involve germ-line tissue instead of somatic tissue? What are some of the potential ethical problems associated with the former approach?

■ HINT: *This problem asks you to think about potential challenges associated with gene therapy for muscular dystrophy. The key to its solution is to consider the type of tissues affected in muscular dystrophy patients.*

tragedies, but step by small step, we will move toward a technology that could—someday—provide cures for many severe genetic diseases.

22.8

Genetic Engineering, Genomics, and Biotechnology Create Ethical, Social, and Legal Questions

Geneticists use recombinant DNA and genomic technologies to identify genes, diagnose and treat genetic disorders, produce commercial and pharmaceutical products, and solve crimes. However, the applications that arise from these technologies raise important ethical, social, and legal issues that must be identified, debated, and resolved. Here we present a brief overview of some current ethical debates concerning the uses of genetic technologies.

Concerns about Genetically Modified Organisms and GM Foods

Most GM food products contain an introduced gene encoding a protein that confers a desired trait (for example, herbicide resistance or insect resistance). Much of the concern over GM plants centers on issues of consumer safety and environmental consequences. Are GM plants safe to eat? In general, if the proteins are not found to be toxic or allergenic and do not have other negative physiological effects, they are not considered to be a significant hazard to health. In Europe and Asia, labeling of food containing genetically modified ingredients is mandatory. But in the United States such labeling is not required at the present time, and foods

with less than 5 percent of their content from genetically modified organisms (GMOs) can be labeled as GMO-free. Certified organic foods, as designed by the USDA, must be GMO-free.

Environmental concerns generally have to do with any risks posed by releasing genetically modified organisms into the environment. Environmental risks include possible gene transfer by cross breeding with wild plants, toxicity, and invasiveness of the modified plant, resulting in loss of natural species (loss of biodiversity). Although laboratory and field studies suggest that cross-pollination and gene transfer can occur between some genetically engineered plants and wild relatives, there is little evidence that this has occurred in nature. If, for example, glyphosate resistance was transferred from cultivated plants such as canola into wild relatives, the herbicide-resistant weeds could make herbicide treatment ineffective. Biotechnology companies have engineered transgenic plants into sterile forms that are unable to transfer their genes into other plants. Built-in sterility was also designed to ensure that farmers could not produce their own seed from genetically modified crops, guaranteeing that biotechnology companies would have exclusive distribution of each year's crop. This, in itself, is an ethical issue, particularly in underdeveloped countries with limited resources to purchase genetically modified seeds.

Genetic Testing and Ethical Dilemmas

When the Human Genome Project was first discussed, scientists and the general public raised concerns about how genome information would be used and how the interests of both individuals and society can be protected. To address these concerns, the **Ethical, Legal, and Social Implications (ELSI) Program** was established as an adjunct to the Human Genome Project. The ELSI Program considers a range of issues, including the impact of genetic information on individuals, the privacy and confidentiality of genetic information, and implications for medical practice, genetic counseling, and reproductive decision making. Through research grants, workshops, and public forums, ELSI is formulating policy options to address these issues.

ELSI focuses on four areas in its deliberations concerning these various issues: (1) privacy and fairness in the use and interpretation of genetic information, (2) ways to transfer genetic knowledge from the research laboratory to clinical practice, (3) ways to ensure that participants in genetic research know and understand the potential risks and benefits of their participation and give informed consent, and (4) public and professional education. It is hoped that, as the Human Genome Project moves from generating information about the genetic basis of disease to improving treatments, promoting prevention, and developing cures, these and other ethical concerns will have been thoroughly studied and an international consensus developed on appropriate policies and laws.

Many of the potential benefits and consequences of genetic testing are not always clear. For example,

- We have the technologies to test for genetic diseases for which there are no effective treatments. But *should* we test people for these disorders?

- With present genetic testing technologies, a negative result does not necessarily rule out future development of a disease; nor does a positive result always mean that an individual will get the disease. How can we effectively communicate the results of testing and the actual risks to those being tested?

- What information should people have before deciding to have a genome scan or a genetic test for a single disorder?

- How can we protect the information revealed by such tests?

- Since sharing of patient data through electronic medical records is a significant concern, what issues of consent need to be considered?

- How can we define and prevent genetic discrimination?

Earlier in this chapter we discussed preimplantation genetic diagnosis (PGD), which provides couples with the ability to screen embryos created by *in vitro* fertilization for genetic disorders. As we learn more about genes involved in human traits, will other, nondisease-related genes be screened for by PGD? Will couples be able to select embryos with certain genes encoding desirable traits for height, weight, intellect, and other physical or mental characteristics? What do you think of using genetic testing to purposely select for an embryo with a genetic disorder? Recently there have been several well-publicized cases of couples seeking to use prenatal diagnosis or PGD to select for embryos with dwarfism and deafness.

As identification of genetic traits becomes more routine in clinical settings, physicians will need to ensure genetic privacy for their patients. There are significant concerns about how genetic information could be used in negative ways by employers, insurance companies, governmental agencies, or the general public. Genetic privacy and prevention of genetic discrimination will be increasingly important in the coming years. Currently, no federal laws regarding genetic privacy and genetic discrimination exist, so individuals must rely on the trustworthiness of the people with whom they are dealing. In 2008, the **Genetic Information Nondiscrimination Act** was signed into law in the United States. This legislation is designed to prohibit the improper use of genetic information in health insurance and employment.

Direct-to-Consumer Genetic Testing and Regulating the Genetic Test Providers

The past decade has seen dramatic developments in **direct-to-consumer (DTC) genetic tests**. A simple Web search will reveal many companies offering DTC genetic tests. There are approximately 1900 diseases for which such tests are now available (in 1993 there were about 100 such tests). Most DTC tests require that a person mail a saliva sample, hair sample, or cheek cell swab to the company. For a range of pricing options, DTC companies largely use SNP-based tests such as ASO tests to screen for different mutations. For example, in 2007 Myriad Genetics, Inc. began a major DTC marketing campaign of its tests for *BRCA1* and *BRCA2*. Mutations in these genes increase risk of developing breast and ovarian cancer. DTC testing companies report absolute risk, the probability that an individual will develop a disease, but how such risks results are calculated is highly variable and subject to certain assumptions.

Such tests are controversial for many reasons. For example, the test is purchased online by individual consumers and requires no involvement of a physician or other health-care professionals such as a nurse or genetic counselor to administer or to interpret results. There are significant questions about the quality, effectiveness, and accuracy of such products because currently the DTC industry is largely self-regulated. The FDA does not regulate DTC genetic tests. There is at present no comprehensive way for patients to make comparisons and evaluations about the range of tests available and their relative quality.

Most companies make it clear that they are not trying to diagnose or prevent disease, nor that they are offering health advice, so what is the purpose of the information that test results provide to the consumer?

Web sites and online programs from DTC companies provide information on what advice a person should pursue if positive results are obtained. But is this enough? If results are not understood, might negative tests not provide a false sense of security? Just because a woman is negative for *BRCA1* and *BRCA2* mutations *does not* mean that one cannot develop breast or ovarian cancer.

In June 2010, the FDA announced that five genetic test manufacturers (Illumina, Pathway Genomics, NaviGenics, 23andMe, and deCODE Genetics) would need FDA approval before their tests could be sold to consumers. This action was prompted when Pathway Genomics announced plans to market a DTC kit for "comprehensive genotyping" in the pharmacy chain Walgreens. Pathway Genomics and the other companies have been selling their DTCs through company Web sites for several years. Pathway and others claim that because their DTC kits are Clinical Laboratory Improvement Amendments (CLIA) approved that no further regulation was required. CLIA regulates certain laboratory tests but is not part of the FDA. This scenario in particular prompted discussion on how the FDA will oversee DTC genetic tests. However, at the time of publication of this edition, the FDA has not revealed any definitive plans to regulate or oversee DTC genetic tests. There are varying opinions on the regulatory issue. Some believe that the FDA has no business regulating DTC tests and that consumers should be free to purchase products according to their own needs or interests. Others insist that the FDA must regulate DTCs in the interest of protecting consumers.

In 2010, the National Institutes of Health announced it will create the **Genetic Testing Registry (GTR)** designed to increase transparency by publicly sharing information about the utility of their tests, research for the general public, patients, health-care workers, genetic counselors, insurance companies, and others. The GTR is intended to allow individuals and families access to key resources to make more well-informed decisions about their health and genetic tests. But participation in the GTR by DTC companies has not been made mandatory yet, so will companies involved in genetic testing participate?

Ethical Concerns Surrounding Gene Therapy

Gene therapy raises several ethical concerns, and many forms of gene therapy are sources of intense debate. At present, all gene therapy trials are restricted to using somatic cells as targets for gene transfer. This form of gene therapy is called **somatic gene therapy**; only one individual is affected, and the therapy is done with the permission and informed consent of the patient or family.

Two other forms of gene therapy have not been approved, primarily because of the unresolved ethical issues surrounding them. The first is called **germ-line therapy**, whereby germ cells (the cells that give rise to the gametes—i.e., sperm and eggs) or mature gametes are used as targets for gene transfer. In this approach, the transferred gene is incorporated into all the future cells of the body, including the germ cells. This means that individuals in future generations will also be affected, without their consent. Is this kind of procedure ethical? Do we have the right to make this decision for future generations? Thus far, the concerns have outweighed the potential benefits, and such research is prohibited.

The second unapproved form of gene therapy—which raises an even greater ethical dilemma—is termed **enhancement gene therapy**, whereby people may be "enhanced" for some desired trait. This use of gene therapy is extremely controversial and is strongly opposed by many people. Should genetic technology be used to enhance human potential? For example, should it be permissible to use gene therapy to increase height, enhance athletic ability, or extend intellectual potential? Presently, the consensus is that enhancement therapy, like germ-line therapy, is an unacceptable use of gene therapy. However, there is an ongoing debate, and many issues are still unresolved. For example, the FDA now permits growth hormone produced by recombinant DNA technology to be used as a growth enhancer, in addition to its medical use for the treatment of growth-associated genetic disorders. Critics charge that the use of a gene product for enhancement will lead to the use of transferred genes for the same purpose. The outcome of these debates may affect not only the fate of individuals but the direction of our society as well.

DNA and Gene Patents

Intellectual property (IP) rights are also being debated as an aspect of the ethical implications of genetic engineering, genomics, and biotechnology. Patents on intellectual property (isolated genes, new gene constructs, recombinant cell types, GMOs) can be potentially lucrative for the patent-holders but may also pose ethical and scientific problems. Why is protecting IP important for companies? Consider this issue. If a company is willing to spend millions or billions of dollars and several years doing research and development (R&D) to produce a valuable product, then shouldn't it be afforded a period of time to protect its discovery so that it can recover R&D costs and made a profit on its product?

Genes in their natural state as products of nature cannot be patented. Consider the possibilities for a human gene that has been cloned and then patented by the scientists who did the cloning. The person or company holding the patent could require that anyone attempting to do research with the patented gene pay a licensing fee for its use. Should a diagnostic test or therapy result from the research, more fees and royalties may be demanded, and as a result the costs of a genetic test may be too high for many patients to afford. But limiting or preventing the holding of patents for genes or genetic tools could reduce the incentive for pursuing the research that produces such genes and tools, especially for companies that need to profit from their research. Should scientists and companies be allowed to patent DNA sequences from naturally living organisms? And should there be a lower or an upper limit to the size of those sequences? For example, should patents be awarded for small pieces of genes, such as expressed sequence tags (ESTs), just because some individual or company wants to claim a stake in having cloned a piece of DNA first, even if no one knows whether the DNA sequence has a use? Can or should investigators be allowed to patent the entire genome of any organism they have sequenced?

To date the U.S. Patent and Trademark Office has granted patents for more than 35,000 genes or gene sequences,

GENETICS, TECHNOLOGY, AND SOCIETY

Personal Genome Projects and the Race for the $1000 Genome

t took $3 billion and close to 15 years to sequence the entire human genome, as part of the Human Genome Project. This ambitious international project yielded the nucleotide sequence of the 3 billion base pairs of DNA that comprise the human genome. The reference sequence created by the Human Genome Project was not derived from one person's DNA, but from a composite of DNA samples from numerous anonymous donors.

When the Human Genome Project was completed, the goal of routinely sequencing the genomes of individual humans seemed remote. However, the development of high-throughput sequencing technologies, capable of generating long sequence reads at high speeds with great accuracy, has reduced the cost of DNA sequencing dramatically. As mentioned in Chapter 21, the race is on to sequence a complete human genome for $1000!

The $1000 genome is coming closer to reality. In 2009, Stanford University professor Stephen Quake announced that he had sequenced his entire genome in a few weeks for under $50,000, using a new sequencing technology and a single sequencing machine. By 2010, two California companies (Illumina and Life Technologies) offered sequencing instruments for sale that were capable of sequencing a human genome in one day for less than $6000.

As the $1000 genome approaches, scientists are making plans for even more ambitious projects. An international research consortium has initiated the "1000 Genomes Project," which aims to sequence the genomes of 1000 volunteers from various backgrounds including African, Asian, and European. The goal is to produce an extensive catalog of human DNA sequence variation. George Church of Harvard University and his colleagues have started the Personal Genome Project—a project that aims to sequence the genomes of 100,000 individuals. Volunteers for the Personal Genome Project must provide their DNA samples on the understanding that their genome data will be made publicly available. Church's genome has already been made available online. The goal of the project is to correlate genome sequences with phenotypic characteristics, from height and hair color to disease predisposition.

The race continues to lower the cost and time to sequence an individual genome. Ready or not, we are entering the age of personal genomics.

Your Turn

1. Would you have your genome sequenced, if the price was affordable? Why, or why not?

You can find a discussion of the pros and cons of knowing your own genome sequence in

a series of articles under the heading of "My Genome. So What?" (Nature **456**: 1, 2008).

2. Would you make your genome sequence publicly available? How might such information be misused?

The issue of genetic privacy is particularly important when considering making genomic sequences available to all. Read about the potential uses and misuses of genome sequences in: Taylor, P. 2008. When consent gets in the way. *Nature* **456**:32–33.

3. What sorts of information do you think you could obtain by examining your own genome sequence?

The relationship between DNA sequence and predictable phenotype is still a murky one. Read about this in: Maher, B. 2008. Personal genomes: The case of the missing heritability. *Nature* **456**: 18–21.

4. Private companies are now offering personal DNA sequencing along with interpretation. What services do they offer? Do you think that these services should be regulated, and if so, in what way?

Investigate one such company, 23andMe, at http://www.23andMe.com. *You can read about the pros and cons of direct-to-consumer sequencing companies in* Ng, P.C. et al. 2010. An agenda for personalized medicine. *Nature* 461:724–726.

including an estimated 20 percent of human genes. Incidentally the patenting of human genes has led some to use the term *patentome*! Some scientists are concerned that to award a patent for simply cloning a piece of DNA is awarding a patent for too little work. Given that computers do most of the routine work of genome sequencing, who should get the patent? What about individuals who figure out *what* to do with the gene? What if a gene sequence has a role in a disease for which a genetic therapy may be developed? Many scientists believe that it is more appropri-

ate to patent novel technology and applications that make use of gene sequences than to patent the gene sequences themselves.

Congress is considering legislation that would ban the patenting of human genes and any sequences, functions, or correlations to naturally occurring products from a gene. The patenting of genetic tests is also under increased scrutiny in part because of concerns that a patented test can create monopolies in which patients cannot get a second opinion if only one company holds the rights to conduct a particular genetic test. Recent analysis has estimated that as many as 64 percent of patented tests for disease genes make it very difficult or impossible for other groups to propose a different way to test for the same disease.

In 2010 a landmark case brought by the American Civil Liberties Union against Myriad Genetics contended that Myriad could not patent the *BRCA1* and *BRCA2* sequences used to diagnose breast cancer. A U.S. District Court judge ruled Myriad's patents invalid on the basis that DNA in an isolated form is not fundamentally different from how it exists in the body. Myriad lost and has been essentially accused of having a monopoly on its tests, which have existed for a little over a decade based on its exclusive licenses in the United States.

Patents and Synthetic Biology

The J. Craig Venter Institute (JCVI) has filed two patent applications for what is being called "the world's first-ever human-made life form." The patents are intended to cover the minimal genome of *M. genitalium,* which JCVI believes are the genes essential for self-replication. One of these patent applications is designed to claim the rights to synthetically constructed organisms. Another U.S. patent recently issued to another group of researchers covers application of a minimal genome for *E. coli,* which has generated even more concern given its relative importance compared with *M. genitalium.* What do you think? Should it be possible to patent a minimal genome or a synthetic organism?

CASE STUDY | A first for gene therapy

In 1990, a young girl who was born without a functional immune system became the first person to undergo gene therapy. In an attempt to treat her autosomal recessive disorder, known as severe combined immunodeficiency (SCID), cloned copies of the gene encoding the missing enzyme (adenosine deaminase, or ADA) were inserted into some of her white blood cells, which were injected back into her bloodstream. Expression of the normal ADA allele led to the development of a functional immune system, allowing the girl to lead a normal life. An understanding of this spectacular success depends on knowing several details of this process:

1. Is it important that the cloned gene becomes part of a chromosome when inserted into a cell?

2. Does the cloned gene replace the defective copy of the ADA gene?

3. Why were white blood cells chosen as targets for the transferred genes?

4. Would you expect that production of 50 percent of the normal levels of ADA would be enough to restore immune function?

Summary Points

 For activities, animations, and review quizzes, go to the study area at www.masteringgenetics.com

1. Recombinant DNA technology can be used to produce valuable biopharmaceutical protein products such as therapeutic proteins for treating disease.

2. Genetically modified (GM) plants, designed to improve crop yield and nutritional value, and to increase resistance to herbicides, pests, and severe weather, are becoming prevalent worldwide.

3. Transgenic animals with improved growth characteristics or desirable phenotypes are being genetically engineered for a number of different applications.

4. Interest in synthetic biology and its potential applications has been spurred by the creation of a synthetic genome successfully used in a genome transplantation experiment in bacteria.

5. A variety of different molecular techniques, including restriction fragment length polymorphism analysis, allele-specific

oligonucleotides tests, and DNA microarrays, can be used to identify genotypes associated with both normal and disease phenotypes.

6. Genome-wide association studies can reveal genetic variations linked with disease conditions within populations.

7. Pharmacogenomics provides the basis for customized medical intervention based on an individual's genotype.

8. Gene therapy involves the delivery of therapeutic genes or the inhibition of expression of defective genes to treat genetic disorders.

9. Applications of genetic engineering and biotechnology involve a range of ethical, social, and legal dilemmas with important scientific and societal implications.

INSIGHTS AND SOLUTIONS

1. Recently reported work by Petukhova et al. (Genome-wide association study in alopecia areata implicates both innate and adaptive immunity, *Nature* 466:113–117, 2010) involved a GWAS study to analyze 1054 cases of patients with alopecia areata (AA) and 3278 controls. Alopecia areata is a condition that leads to major hair loss and affects approximately 5.3 million people in the United States alone.

   (a) A Manhattan plot from this work is shown below:

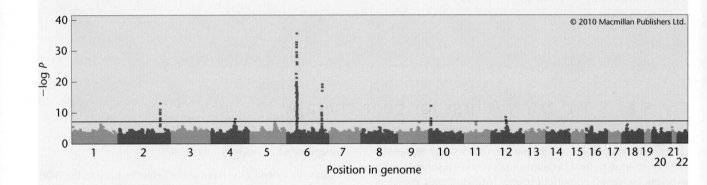

   Based on your interpretation of this plot, which chromosomes were associated with loci that may contribute to AA?

   (b) Of the 139 SNPs significantly associated with AA, several genes are involved in controlling the activation and proliferation of regulatory T lymphocytes (Treg cells) and cytotoxic T lymphocytes, genes involved in antigen presentation to immune cells, immune regulatory molecules such as the interleukins, and genes expressed in the hair follicle itself. Speculate how these candidate genes may help scientists understand how AA progresses as a disease.

   Solution:

   (a) Investigators identified eight genomic regions with SNPs that exceed the genome-wide significance value of 5×10^{-7} (red line). These regions were clustered on chromosomes 2, 4, 6, 9, 10, 11, and 12.

   (b) AA is an autoimmune disease in which the immune system attacks hair follicles, resulting in hair loss that can permeate across the entire scalp and even the whole body. AA hair follicles are attacked by T cells. The identification of candidate genes involved in T-cell proliferation, immune system regulation, and follicular development may potentially help investigators develop cures for AA.

2. Infection by HIV-1 (human immunodeficiency virus) weakens the immune system and results in the symptoms of AIDS (acquired immunodeficiency syndrome). Specifically, HIV infects and kills cells of the immune system that carry a cell-surface receptor known as CD4. An HIV surface protein known as gp120 binds to the CD4 receptor and allows the virus to enter the cell. The gene encoding the CD4 protein has been cloned. How might this clone be used along with recombinant DNA techniques to combat HIV infection?

   Solution: Researchers hope that clones of the *CD4* gene can be used in the design of systems for the targeted delivery of drugs and toxins to combat the infection. For example, because infection depends on an interaction between the viral gp120 protein and the CD4 protein, the cloned *CD4* gene has been modified to produce a soluble form of the protein (sCD4) that, because of its solubility, would circulate freely in the body. The idea is that HIV might be prevented from infecting cells if the gp120 protein of the virus first encounters and binds to extra molecules of the soluble form of the CD4 protein. Once bound to the extra molecules, the virus would be unable to bind to CD4 proteins on the surface of immune system cells. Studies in cell-culture systems indicate that the presence of sCD4 effectively prevents HIV infection of tissue culture cells. However, studies in HIV-positive humans have been somewhat disappointing, mainly because the strains of HIV used in the laboratory are different from those found in infected individuals. In another strategy, the *CD4* gene has been fused with genes encoding bacterial toxins. The resulting fusion protein contains CD4 regions that should bind to gp120 on the surface of HIV-infected cells and toxin regions that should then kill the infected cell. In tissue culture experiments, cells infected with HIV are killed by this fusion protein, whereas uninfected cells survive.

Problems and Discussion Questions

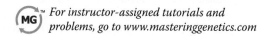 *For instructor-assigned tutorials and problems, go to www.masteringgenetics.com*

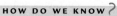

1. In this chapter, we focused on a number of interesting applications of genetic engineering, genomics, and biotechnology. At the same time, we found many opportunities to consider the methods and reasoning by which much of this information was acquired. From the explanations given in the chapter, what answers would you propose to the following fundamental questions:

   (a) How do we determine whether genetically modified plants are safe for human consumption?

   (b) What experimental evidence confirms that we have introduced a useful gene into a transgenic organism and that it performs as we anticipate?

   (c) How can we use DNA analysis to determine that a human fetus has sickle-cell anemia?

   (d) How can DNA microarray analysis be used to identify specific genes that are being expressed in a specific tissue?

   (e) How are GWAS carried out, and what information do they provide?

   (f) What are some of the technical reasons why gene therapy is difficult to carry out effectively?

2. What are some of the reasons why GM crops are controversial? Describe some of the primary concerns that have been raised about GM foods.

3. Should the United States require mandatory labeling of all foods that contain GMOs? Explain your answer.

4. The human insulin gene contains introns. Since bacterial cells will not excise introns from mRNA, how can a gene like this be cloned into a bacterial cell that will produce insulin?

5. One of the main safety issues associated with genetically modified crops is the potential for allergenicity caused by introducing an allergen or by changing the level of expression of a host allergen. Based on the observation that common allergenic proteins often contain identical stretches of a few (six or seven) amino acids, researchers developed a method for screening transgenic crops to evaluate potential allergenic properties (Kleter & Peijnenburg, 2002. *BMC Struct. Biol.* 2: 8). How do you think they accomplished this?

6. Why are most recombinant human proteins produced in animal or plant hosts instead of bacterial host cells?

7. There are more than 1000 cloned farm animals in the United States. In the near future, milk from cloned cows and their offspring (born naturally) may be available in supermarkets. These cloned animals have not been transgenically modified, and they are no different than identical twins. Should milk from such animals and their natural-born offspring be labeled as coming from cloned cows or their descendants? Why?

8. One of the major causes of sickness, death, and economic loss in the cattle industry is *Mannheimia haemolytica*, which causes bovine pasteurellosis, or shipping fever. Noninvasive delivery of a vaccine using transgenic plants expressing immunogens would reduce labor costs and trauma to livestock. An early step toward developing an edible vaccine is to determine whether an injected version of an antigen (usually a derivative of the pathogen) is capable of stimulating the development of antibodies in a test organism. The following table assesses the ability of a transgenic portion of a toxin (Lkt) of *M. haemolytica* to stimulate development of specific antibodies in rabbits.

   (a) What general conclusion can you draw from the data?

   (b) With regards to development of a usable edible vaccine, what work remains to be done?

Immunogen Injected	Antibody Production in Serum
Lkt50*—saline extract	+
Lkt50—column extract	+
Mock injection	−
Pre-injection	−

*Lkt50 is a smaller derivative of Lkt that lacks all hydrophobic regions. + indicates at least 50 percent neutralization of toxicity of Lkt; − indicates no neutralization activity.
Source: Modified from Lee et al. 2001. *Infect. and Immunity* 69: 5786–5793.

9. Describe how the team from the J. Craig Venter Institute created a synthetic genome. How did they demonstrate that the genome converted the recipient strain of bacteria into a different strain?

10. Suppose you develop a screening method for cystic fibrosis that allows you to identify the predominant mutation $\Delta 508$ and the next six most prevalent mutations. What must you consider before using this method to screen a population for this disorder?

11. Sequencing the human genome and the development of microarray technology promises to improve our understanding of normal and abnormal cell behavior. How are microarrays dramatically changing our understanding of complex diseases such as cancer?

12. A couple with European ancestry seeks genetic counseling before having children because of a history of cystic fibrosis (CF) in the husband's family. ASO testing for CF reveals that the husband is heterozygous for the $\Delta 508$ mutation and that the wife is heterozygous for the *R117* mutation. You are the couple's genetic counselor. When consulting with you, they express their conviction that they are not at risk for having an affected child because they each carry different mutations and cannot have a child who is homozygous for either mutation. What would you say to them?

13. When genome scanning technologies become widespread, medical records will contain the results of such testing. Who should have access to this information? Should employers, potential employers, or insurance companies be allowed to have this information? Would you favor or oppose having the government establish and maintain a central database containing the results of individuals' genome scans?

14. What limits the use of differences in restriction enzyme sites as a way of detecting point mutations in human genes?

15. What is the main purpose of genome-wide association studies (GWAS)? How can information from GWAS be used to inform scientists and physicians about genetic diseases?

16. Recombinant adenoviruses have been used in a number of preclinical studies to determine the efficacy of gene therapy for rheumatoid arthritis and osteoarthritis. In the viruses, genes can be delivered by injection to the tissues that need them. Christopher Evans and colleagues (2001. *Arthritis Res.* 3: 142–146)

estimated that approximately 20 percent of all human gene therapy trials have used adenoviruses for gene delivery. The death of a patient in 1999 after infusion of adenoviral vectors has caused concern. As you consider the use of viral vectors as therapy-delivery vehicles for human pathologies, what factors seem of paramount concern?

17. Define somatic gene therapy, germ-line therapy, and enhancement gene therapy. Which of these is currently in use?

18. Provide examples of major barriers that need to be addressed if gene therapy is to become a safe and reliable treatment for genetic diseases.

19. *Transductional targeting* is a preferred route for the delivery of therapeutics for human diseases. It involves the development of tissue-specific interactions between the viral vector and a specific tissue. A genetic approach that has been used involves engineering the capsid of an adeno-associated virus (type 2) vector to target specific human cell types. Nongenetic approaches are also possible. Speculate on problems associated with the genetic approach of capsid alteration and problems that might be associated with nongenetic approaches to transductional targeting.

20. The development of safe vectors for human gene therapy has been a goal since 1990. Among the problems associated with viral-based vectors is that many of the viruses (i.e., SV40) have transformation properties thought to be mediated by binding and inactivating gene products such as p53, retinoblastoma protein (pRB), and others. SV40-based vectors that are deficient in binding p53, pRB, and other proteins have been developed. Why would you specifically want to avoid inactivating p53, pRB, and related proteins?

21. Dominant mutations can be categorized according to whether they increase or decrease the overall activity of a gene or gene product. Although a loss-of-function mutation (a mutation that inactivates the gene product) is usually recessive, for some genes, one dose of the normal gene product, encoded by the normal allele, is not sufficient to produce a normal phenotype. In this case, a loss-of-function mutation in the gene will be dominant, and the gene is said to be *haploinsufficient*. A second category of dominant mutation is the gain-of-function mutation, which results in a new activity or increased activity or expression of a gene or gene product. The gene therapy technique currently used in clinical trials involves the "addition" to somatic cells of a normal copy of a gene. In other words, a normal copy of the gene is inserted into the genome of the mutant somatic cell, but the mutated copy of the gene is not removed or replaced. Will this strategy work for either of the two aforementioned types of dominant mutations?

22. In mice transfected with the rabbit β-globin gene, the rabbit gene is active in several tissues, including the spleen, brain, and kidney. In addition, some transfected mice suffer from thalassemia (a form of anemia) caused by an imbalance in the coordinate production of α- and β-globins. Which problems associated with gene therapy are illustrated by these findings?

23. The Genetic Testing Registry is intended to provide better information to patients, but companies involved in genetic testing are not required to participate. Should company participation be mandatory? Why or why not? Explain your answers?

24. Should the FDA regulate direct-to-consumer genetic tests, or should these tests be available as a "buyer beware" product?

Extra-Spicy Problems

 For instructor-assigned tutorials and problems, go to www.masteringgenetics.com

25. Host transgenic mammals are often made by injecting transgenic DNA into a pronucleus, a process that is laborious, technically demanding, and typified by yields of less than 1 percent. Lately, the engineered lentivirus, a retrovirus, has been used to generate transgenic pigs, rats, mice, and cattle with greater than 10 percent efficiency (Whitelaw, 2004). Lentivectors have reverse transcriptase activity, can infect both dividing and nondividing cells, and are replication-defective. A lentivector carrying the reporter gene, which codes a green fluorescent protein, has produced a remarkable set of pigs (refer to the chapter-opening figure). Present, in a labeled diagram, a strategy for producing a transgenic mammal carrying a pharmaceutically important gene using a lentivector.

26. Yeager, M., et al. (*Nature Genetics* 39: 645–649, 2007) and Sladek, R. et al. (*Nature* 445: 881–885, 2007) have used single-nucleotide polymorphisms (SNPs) in genome-wide association studies (GWAS) to identify novel risk loci for prostate cancer and type 2 diabetes mellitus, respectively. Each study suggests that disease-risk genes can be identified that significantly contribute to the disease state. Given your understanding of such complex diseases, what would you consider as reasonable factors to consider when interpreting the results of GWAS studies?

27. In March, 2010 Judge R. Sweet ruled to invalidate Myriad Genetics' patents on the *BRCA1* and *BRCA2* genes. Sweet

wrote that since the genes are part of the natural world, they are not patentable. Myriad Genetics also holds patents on the development of a direct-to-consumer test for the *BRCA1* and *BRCA2* genes.

(a) Would you agree with Judge Sweet's ruling to invalidate the patenting of the *BRCA1* and *BRCA2* genes? If you were asked to judge the patenting of the direct-to-consumer test for the *BRCA1* and *BRCA2* genes, how would you rule?

(b) J. Craig Venter has filed a patent application for his "first-ever human made life form." This patent is designed to cover the genome of *M. genitalium*. Would your ruling for Venter's "organism" be different from Judge Sweet's ruling on patenting of the *BRCA1* and *BRCA2* genes?

28. A number of mouse models for human cystic fibrosis (CF) exist. Each of these mouse strains is transgenic and bears a different specific *CFTR* gene mutation. The mutations are the same as those seen in the varieties of human CF. These transgenic CF mice are being used to study the range of different phenotypes that characterize CF in humans. They are also used as models to test potential CF drugs. Unfortunately, most transgenic mouse CF strains do not show one of the most characteristic symptoms of human CF, that of lung congestion. Can you think of a reason why mouse CF strains do not display this symptom of human CF?

23

Quantitative Genetics and Multifactorial Traits

CHAPTER CONCEPTS

- Quantitative inheritance results in a range of measurable phenotypes for a polygenic trait.

- With some exceptions, polygenic traits tend to demonstrate continuous variation.

- Quantitative traits can be explained in Mendelian terms whereby certain alleles have an additive effect on the traits under study.

- The study of polygenic traits relies on statistical analysis.

- Heritability values estimate the genetic contribution to phenotypic variability under specific environmental conditions.

- Twin studies allow an estimation of heritability in humans.

- Quantitative trait loci (QTLs) can be mapped and identified.

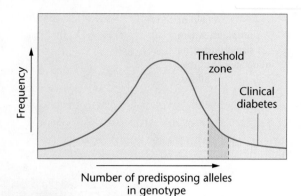

p to this point in the text, most of our examples of phenotypic variation have been those that have been assigned to distinct and separate categories; for example, human blood type was A, B, AB, or O; squash fruit shape was spherical, disc shaped, or elongated; and fruit fly eye color was red or white (see Chapter 4). Typically in these traits, a genotype will produce a single identifiable phenotype, although phenomena such as variable penetrance and expressivity, pleiotropy, and epistasis can obscure the relationship between genotype and phenotype.

In this chapter, we will look at traits that are not as clear cut, including many that are of medical or agricultural importance. The traits on which we focus show much more variation, often falling into a continuous range of phenotypes that are more difficult to classify into distinct categories. Most of these traits show *continuous variation*, including, for example, height in humans, milk and meat production in cattle, yield and seed protein content in various crops. Continuous variation across a range of phenotypes is measured and described in quantitative terms, so this genetic phenomenon is known as **quantitative inheritance**. And because the varying phenotypes result from the input of genes at more than one, and often many, loci, they are also said to be **polygenic** (literally "of many genes").

To further complicate the link between the genotype and phenotype, the genotype generated at fertilization establishes a quantitative range within which a particular individual can fall. However, the final phenotype is often also influenced by environmental factors to which that individual is exposed. Human height, for example, is genetically influenced, but is also affected by environmental factors such as nutrition. Quantitative (polygenic) traits whose phenotypes result from both gene action and environmental influences are often termed **multifactorial,** or **complex traits**. Often these terms are used interchangeably. For consistency throughout the chapter, we will utilize the term *multifactorial* in our discussions.

In this chapter, we will examine examples of quantitative inheritance, multifactorial traits, and some of the statistical techniques used to study them. We will also consider how geneticists assess the relative importance of genetic versus environmental factors contributing to continuous phenotypic variation, and we will discuss approaches to identifying and mapping genes that influence quantitative traits.

FIGURE 23–1 A graphic depiction of predisposing alleles characteristic of a threshold trait within a population, illustrated by Type II diabetes.

numbers. Examples of meristic traits include the number of seeds in a pod or the number of eggs laid by a chicken in a year. These are quantitative traits, but they do not have an infinite range of phenotypes: for example, a pod may contain 2, 4, or 6 seeds, but not 5.75. **Threshold traits** are polygenic (and frequently multifactorial), but they are distinguished from continuous and meristic traits by having a small number of discrete phenotypic classes. Threshold traits are currently of heightened interest to human geneticists because an increasing number of diseases are now thought to show this pattern of polygenic inheritance. One example is **Type II diabetes**, also known as adult-onset diabetes because it typically affects individuals who are middle aged or older. A population can be divided into just two phenotypic classes for this trait—individuals who have Type II diabetes and those who do not—so at first glance, it may appear to more closely resemble a simple monogenic trait. However, no single adult-onset diabetes gene has been identified. Instead, the combination of alleles present at multiple contributing loci gives an individual a greater or lesser likelihood of developing the disease. These varying levels of liability form a continuous range: at one extreme are those at very low risk for Type II diabetes, while at the other end of the distribution are those whose genotypes make it highly likely they will develop the disease (Figure 23–1). As with many threshold traits, environmental factors also play a role in determining the final phenotype, with diet and lifestyle having significant impact on whether an individual with moderate to high genetic liability will actually develop Type II diabetes.

23.1

Not All Polygenic Traits Show Continuous Variation

In addition to quantitative traits that display continuous variation, there are two other classes of polygenic traits. **Meristic traits** are those in which the phenotypes are described by whole

23.2

Quantitative Traits Can Be Explained in Mendelian Terms

The question of whether continuous phenotypic variation could be explained in Mendelian terms caused considerable controversy in the early 1900s. Some scientists argued that, although Mendel's unit factors, or genes, explained patterns of discontinuous segregation with discrete phenotypic classes, they could not also account for the range of phenotypes seen in quantitative patterns of inheritance. However, geneticists William Bateson and G. Udny Yule, adhering to a Mendelian explanation, proposed the **multiple-factor** or **multiple-gene hypothesis**, in which many genes, each individually behaving in a Mendelian fashion, contribute to the phenotype in a *cumulative* or *quantitative* way.

The Multiple-Gene Hypothesis for Quantitative Inheritance

The **multiple-gene hypothesis** was initially based on a key set of experimental results published by Hermann Nilsson-Ehle in 1909. Nilsson-Ehle used grain color in wheat to test the concept that the cumulative effects of alleles at multiple loci produce the range of phenotypes seen in quantitative traits. In one set of experiments, wheat with red grain was crossed to wheat with white grain (**Figure 23–2**). The F_1 generation demonstrated an intermediate pink color, which at first sight suggested incomplete dominance of two alleles at a single locus. However, in the F_2 generation, Nilsson-Ehle did not observe the typical segregation of a monohybrid cross. Instead, approximately 15/16 of the plants showed some degree of red grain color, while 1/16 of the plants showed white grain color. Careful examination of the F_2 revealed that grain with color could be classified into four different shades of red. Because the F_2 ratio occurred in sixteenths, it appears that two genes, each with two alleles, control the phenotype and that they segregate independently from one another in a Mendelian fashion.

If each gene has one potential **additive allele** that contributes approximately equally to the red grain color and one potential **nonadditive allele** that fails to produce any red pigment, we can see how the multiple-factor hypothesis could account for the various grain color phenotypes. In the P_1 both parents are homozygous; the red parent contains only additive alleles (*AABB* in Figure 23–2), while the white parent contains only nonadditive alleles (*aabb*). The F_1 plants are heterozygous (AaBb), contain two additive (*A* and *B*) and two nonadditive (*a* and *b*) alleles, and express the intermediate pink phenotype. Each of the F_2 plants has 4, 3, 2, 1, or 0 additive alleles. F_2 plants with no additive alleles are white (*aabb*) like one of the P_1 parents, while F_2 plants with 4 additive alleles are red (*AABB*) like the other P_1 parent. Plants with 3, 2, or 1 additive alleles constitute the other three categories of red color observed in the F_2 generation. The greater the number of additive alleles in the genotype, the more intense the red color expressed in the phenotype,

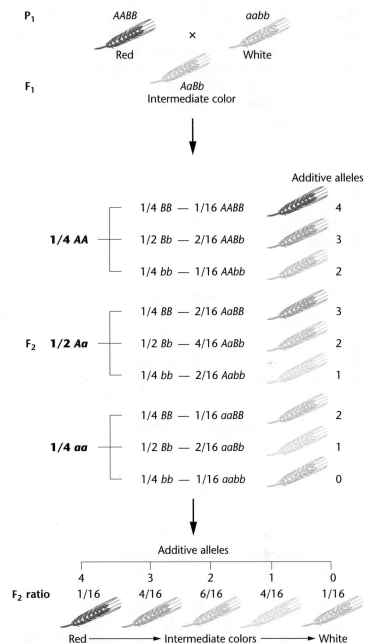

FIGURE 23–2 How the multiple-factor hypothesis accounts for the 1:4:6:4:1 phenotypic ratio of grain color when all alleles designated by an uppercase letter are additive and contribute an equal amount of pigment to the phenotype.

as each additive allele present contributes equally to the cumulative amount of pigment produced in the grain.

Nilsson-Ehle's results showed how continuous variation could still be explained in a Mendelian fashion, with additive alleles at multiple loci influencing the phenotype in a quantitative manner, but each individual allele segregating according to Mendelian rules. As we saw in Nilsson-Ehle's initial cross, if two loci, each with two alleles, were involved, then five F_2 phenotypic categories in a 1:4:6:4:1 ratio would be expected. However, there is no reason why three, four, or more loci cannot function in a similar fashion in controlling various quantitative phenotypes. As more quantitative loci become involved, greater and greater numbers of classes appear in the F_2 generation in more complex ratios. The number of phenotypes and the expected F_2 ratios for crosses involving up to four gene pairs are illustrated in Figure 23–3.

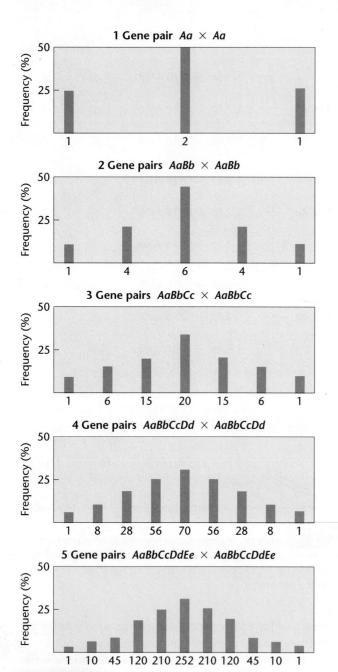

FIGURE 23–3 The genetic ratios (on the X-axis) resulting from crossing two heterozygotes when polygenic inheritance is in operation with 1–5 gene pairs. The histogram bars indicate the distinct F_2 phenotypic classes, ranging from one extreme (left end) to the other extreme (right end). Each phenotype results from a different number of additive alleles.

Additive Alleles: The Basis of Continuous Variation

The multiple-gene hypothesis consists of the following major points:

1. Phenotypic traits showing continuous variation can be quantified by measuring, weighing, counting, and so on.

2. Two or more gene loci, often scattered throughout the genome, account for the hereditary influence on the phenotype in an *additive way*. Because many genes may be involved, inheritance of this type is called *polygenic*.

3. Each gene locus may be occupied by either an *additive* allele, which contributes a constant amount to the phenotype, or a *nonadditive* allele, which does not contribute quantitatively to the phenotype.

4. The contribution to the phenotype of each additive allele, though often small, is approximately equal. While we now know this is not always true, we have made this assumption in the above discussion.

5. Together, the additive alleles contributing to a single quantitative character produce substantial phenotypic variation.

Calculating the Number of Polygenes

Various formulas have been developed for estimating the number of **polygenes**, the genes contributing to a quantitative trait. For example, if the ratio of F_2 individuals resembling *either* of the two extreme P_1 phenotypes can be determined, the number of polygenes involved (n) may be calculated as follows:

$$1/4^n = \text{ratio of } F_2 \text{ individuals expressing either extreme phenotype}$$

In the example of the red and white wheat grain color summarized in Figure 23–2, 1/16 of the progeny are either red *or* white like the P_1 phenotypes. This ratio can be substituted on the right side of the equation to solve for n:

$$\frac{1}{4^n} = \frac{1}{16}$$

$$\frac{1}{4^2} = \frac{1}{16}$$

$$n = 2$$

Table 23.1 lists the ratio and the number of F_2 phenotypic classes produced in crosses involving up to five gene pairs.

For low numbers of polygenes (n), it is sometimes easier to use the equation

$$(2n + 1) = \text{the number of distinct phenotypic categories observed}$$

TABLE 23.1

Determination of the Number of Polygenes (n) Involved in a Quantitative Trait

n	Individuals Expressing Either Extreme Phenotype	Distinct Phenotypic Classes
1	1/4	3
2	1/16	5
3	1/64	7
4	1/256	9

For example, when there are two polygenes involved ($n = 2$), then $(2n + 1) = 5$ and each phenotype is the result of 4, 3, 2, 1, or 0 additive alleles. If $n = 3$, $2n + 1 = 7$ and each phenotype is the result of 6, 5, 4, 3, 2, 1, or 0 additive alleles. Thus, working backwards with this rule and knowing the number of phenotypes, we can calculate the number of polygenes controlling them.

It should be noted, however, that both of these simple methods for estimating the number of polygenes involved in a quantitative trait assume not only that all the relevant alleles contribute equally and additively, but also that phenotypic expression in the F_2 is not affected significantly by environmental factors. As we will see later, for many quantitative traits, these assumptions may not be true.

NOW SOLVE THIS

23–1 A homozygous plant with 20-cm diameter flowers is crossed with a homozygous plant of the same species that has 40-cm diameter flowers. The F_1 plants all have flowers 30 cm in diameter. In the F_2 generation of 512 plants, 2 plants have flowers 20 cm in diameter, 2 plants have flowers 40 cm in diameter, and the remaining 508 plants have flowers of a range of sizes in between.

(a) Assuming that all alleles involved act additively, how many genes control flower size in this plant?

(b) What frequency distribution of flower diameter would you expect to see in the progeny of a backcross between an F_1 plant and the large-flowered parent?

■ HINT: *This problem provides F_1 and F_2 data for a cross involving a quantitative trait and asks you to calculate the number of genes controlling the trait. The key to its solution is to remember that unless you know the total number of distinct F_2 phenotypes involved, then the ratio (not the number) of parental phenotypes reappearing in the F_2 must be used in your determination of the number of genes involved.*

23.3

The Study of Polygenic Traits Relies on Statistical Analysis

Before considering the approaches that geneticists use to dissect how much of the phenotypic variation observed in a population is due to genotypic differences among individuals and how much is due to environmental factors, we need to consider the basic statistical tools they use for the task. It is not usually feasible to measure expression of a polygenic trait in every individual in a population, so a random subset of individuals is usually selected for measurement to provide a *sample*. It is important to remember that the accuracy of the final results of the measurements depends on whether the sample is truly random and representative of the population from which it was drawn. Suppose, for example, that a student wants to determine the average height of the 100 students in his genetics class, and for his sample he measures the two students sitting next to him, both of whom happen to be centers on the college basketball team. It is unlikely that this sample will provide a good estimate of the average height of the class, for two reasons: First, it is too small; second, it is not a representative subset of the class (unless all 100 students are centers on the basketball team).

If the sample measured for expression of a quantitative trait is sufficiently large and also representative of the population from which it is drawn, we often find that the data form a **normal distribution**; that is, they produce a characteristic bell-shaped curve when plotted as a frequency histogram (**Figure 23–4**). Several statistical concepts are useful in the analysis of traits that exhibit a normal distribution, including the mean, variance, standard deviation, standard error of the mean, and covariance.

The Mean

The mean provides information about where the central point lies along a range of measurements for a quantitative trait.

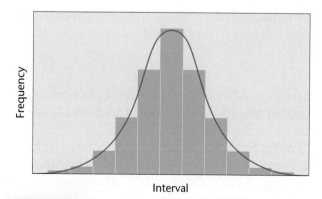

FIGURE 23–4 Normal frequency distribution, characterized by a bell-shaped curve.

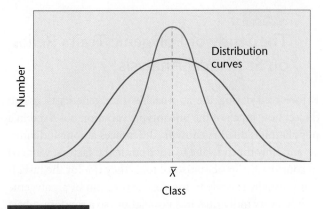

FIGURE 23–5 Two normal frequency distributions with the same mean but different amounts of variation.

Figure 23–5 shows the distribution curves for two different sets of phenotypic measurements. Each of these sets of measurements clusters around a central value (as it happens, they both cluster around the same value). This clustering is called a central tendency, and the central point is the mean.

Specifically, the **mean** ($\overline{X}$) is the arithmetic average of a set of measurements and is calculated as

$$\overline{X} = \frac{\Sigma X_i}{n}$$

where $\overline{X}$ is the mean, ΣX_i represents the sum of all individual values in the sample, and n is the number of individual values.

The mean provides a useful descriptive summary of the sample, but it tells us nothing about the range or spread of the data. As illustrated in Figure 23–5, a symmetrical distribution of values in the sample may, in one case, be clustered near the mean. Or a set of measurements may have the same mean but be distributed more widely around it. A second statistic, the variance, provides information about the spread of data around the mean.

Variance

The **variance** (s^2) for a sample is the average squared distance of all measurements from the mean. It is calculated as

$$s^2 = \frac{\Sigma(X_i - \overline{X})^2}{n - 1}$$

where the sum (Σ) of the squared differences between each measured value (X_i) and the mean ($\overline{X}$) is divided by one less than the total sample size $n - 1$.

As Figure 23–5 shows, it is possible for two sets of sample measurements for a quantitative trait to have the same mean but a different distribution of values around it. This range will be reflected in different variances. Estimation of variance can be useful in determining the degree of genetic control of traits when the immediate environment also influences the phenotype.

TABLE 23.2	
Sample Inclusion for Various s Values	
Multiples of s	**Sample Included (%)**
$\overline{X} \pm 1s$	68.3
$\overline{X} \pm 1.96s$	95.0
$\overline{X} \pm 2s$	95.5
$\overline{X} \pm 3s$	99.7

Standard Deviation

Because the variance is a squared value, its unit of measurement is also squared (m^2, g^2, etc.). To express variation around the mean in the original units of measurement, we can use the square root of the variance, a term called the **standard deviation** (s):

$$s = \sqrt{s^2}$$

Table 23.2 shows the percentage of individual values within a normal distribution that fall within different multiples of the standard deviation. The values that fall within one standard deviation to either side of the mean represent 68 percent of all values in the sample. More than 95 percent of all values are found within two standard deviations to either side of the mean. This means that the standard deviation s can also be interpreted in the form of a probability. For example, a sample measurement picked at random has a 68 percent probability of falling within the range of one standard deviation.

Standard Error of the Mean

If multiple samples are taken from a population and measured for the same quantitative trait, we might find that their means vary. Theoretically, larger, truly random samples will represent the population more accurately, and their means will be closer to each other. To measure the accuracy of the sample mean we use the **standard error of the mean** ($S_{\overline{X}}$), calculated as

$$S_{\overline{X}} = \frac{s}{\sqrt{n}}$$

where s is the standard deviation and $\sqrt{n}$ is the square root of the sample size. Because the standard error of the mean is computed by dividing s by $\sqrt{n}$, it is always a smaller value than the standard deviation.

Covariance

Often geneticists working with quantitative traits find they have to consider two phenotypic characters simultaneously. For example, a poultry breeder might investigate the correlation between body weight and egg production in hens: Do heavier birds tend to lay more eggs? The **covariance** statistic measures how much variation is common to both quantitative traits. It is calculated by taking the deviations from the mean for each trait (just as we did for estimating variance) for each individual in the sample. This gives a pair of values for each individual. The two values are multiplied together, and the sum of all these individual products is then divided by one fewer than the number in the sample. Thus the covariance cov_{XY} of two sets of trait measurements, X and Y, is calculated as

$$cov_{XY} = \frac{\Sigma\left[(X_i - \overline{X})(Y_i - \overline{Y})\right]}{n - 1}$$

The covariance can then be standardized as yet another statistic, the **correlation coefficient (r)**. The calculation of r is

$$r = cov_{XY}/S_X S_Y$$

where S_X is the standard deviation of the first set of quantitative measurements X, and S_Y is the standard deviation of the second set of quantitative measurements Y. Values for the correlation coefficient r can range from -1 to $+1$. Positive r values mean that an increase in measurement for one trait tends to be associated with an increase in measurement for the other, while negative r values mean that increases in one trait are associated with decreases in the other. Therefore, if heavier hens do tend to lay more eggs, a positive r value can be expected. A negative r value, on the other hand, suggests that greater egg production is more likely from less heavy birds. One important point to note about correlation coefficients is that even significant r values—close to $+1$ or -1—do not prove that a cause-and-effect relationship exists between two traits. Correlation analysis simply tells us the extent to which variation in one quantitative trait is associated with variation in another, not what causes that variation.

Analysis of a Quantitative Character

To apply these statistical concepts, let's consider a genetic experiment that crossed two different homozygous varieties of tomato. One of the tomato varieties produces fruit averaging 18 oz in weight, whereas fruit from the other averages 6 oz. The F_1 obtained by crossing these two varieties has fruit weights ranging from 10 to 14 oz. The F_2 population contains individuals that produce fruit ranging from 6 to 18 oz. The results characterizing both generations are shown in **Table 23.3**.

NOW SOLVE THIS

23–2 The following table shows measurements for fiber lengths and fleece weight in a small flock of eight sheep.

	Sheep Fiber Length (cm)	Fleece Weight (kg)
1	9.7	7.9
2	5.6	4.5
3	10.7	8.3
4	6.8	5.4
5	11.0	9.1
6	4.5	4.9
7	7.4	6.0
8	5.9	5.1

(a) What are the mean, variance, and standard deviation for each trait in this flock?
(b) What is the covariance of the two traits?
(c) What is the correlation coefficient for fiber length and fleece weight?
(d) Do you think greater fleece weight is correlated with an increase in fiber length? Why or why not?

This problem provides data for two quantitative traits in a flock of sheep. After making numerous statistical calculations, you are asked in part (d) to determine if the traits are correlated.

■ HINT: *This problem provides data for two quantitative traits and asks you to make numerous statistical calculations, ultimately determining if the traits are correlated. The key to its solution is that once the calculation of the correlation coefficient (r) is completed, you must interpret that value—whether it is positive or negative, and how close to zero it is.*

The mean value for the fruit weight in the F_1 generation can be calculated as

$$\overline{X} = \frac{\Sigma X_i}{n} = \frac{626}{52} = 12.04$$

The mean value for fruit weight in the F_2 generation is calculated as

$$\overline{X} = \frac{\Sigma X_i}{n} = \frac{872}{72} = 12.11$$

Although these mean values are similar, the frequency distributions in Table 23.3 show more variation in the F_2

TABLE 23.3

Distribution of F_1 and F_2 Progeny Derived from a Theoretical Cross Involving Tomatoes

						Weight (oz.)								
		6	7	8	9	10	11	12	13	14	15	16	17	18
Number of	F_1					4	14	16	12	6				
Individuals	F_2	1	1	2	0	9	13	17	14	7	4	3	0	1

generation. The range of variation can be quantified as the sample variance s^2, calculated, as we saw above, as the sum of the squared differences between each value and the mean, divided by one less than the total number of observations.

$$s^2 = \frac{\Sigma(X_i - \bar{X})^2}{n - 1}$$

When the above calculation is made, the variance is found to be 1.29 for the F_1 generation and 4.27 for the F_2 generation. When converted to the standard deviation ($s = \sqrt{s^2}$), the values become 1.13 and 2.06, respectively. Therefore, the distribution of tomato weight in the F_1 generation can be described as 12.04 $\pm$ 1.13, and in the F_2 generation it can be described as 12.11 $\pm$ 2.06.

Assuming that both tomato varieties are homozygous at the loci of interest and that the alleles controlling fruit weight act additively, we can estimate the number of polygenes involved in this trait. Since 1/72 of the F_2 offspring have a phenotype that overlaps one of the parental strains (72 total F_2 offspring; one weighs 6 oz, one weighs 18 oz; see Table 23.3), the use of the formula $1/4^n = 1/72$ indicates that n is between 3 and 4, providing evidence of the number of genes that control fruit weight in these tomato strains.

23.4

Heritability Values Estimate the Genetic Contribution to Phenotypic Variability

The question most often asked by geneticists working with multifactorial traits and diseases is how much of the observed phenotypic variation in a population is due to genotypic differences among individuals and how much is due to environment. The term **heritability** is used to describe *what proportion of total phenotypic variation in a population is due to genetic factors*. For a multifactorial trait in a given population, a high heritability estimate indicates that much of the variation can be attributed to genetic factors, with the environment having less impact on expression of the trait. With a low heritability estimate, environmental factors are likely to have a greater impact on phenotypic variation within the population.

The concept of heritability is frequently misunderstood and misused. It should be emphasized that heritability indicates neither how much of a trait is genetically determined nor the extent to which an individual's phenotype is due to genotype. In recent years, such misinterpretations of heritability for human quantitative traits have led to controversy, notably in relation to measurements such as intelligence quotients, or IQs. Variation in heritability estimates for IQ among different racial groups tested led to incorrect suggestions that unalterable genetic factors control differences in

intelligence levels among humans of different ancestries. Such suggestions misrepresented the meaning of heritability and ignored the contribution of genotype-by-environment interaction variance (see p. 667) to phenotypic variation in a population. Moreover, heritability is not fixed for a trait. For example, a heritability estimate for egg production in a flock of chickens kept in individual cages might be high, indicating that differences in egg output among individual birds are largely due to genetic differences, as they all have very similar environments. For a different flock kept outdoors, heritability for egg production might be much lower, as variation among different birds may also reflect differences in their individual environments. Such differences could include how much food each bird manages to find and whether it competes successfully for a good roosting spot at night. Thus a heritability estimate tells us the proportion of phenotypic variation that can be attributed to genetic variation *within a certain population in a particular environment*. If we measure heritability for the same trait among different populations in a range of environments, we frequently find that the calculated heritability values have large standard errors. This is an important point to remember when considering heritability estimates for traits in human populations. A mean heritability estimate of 0.65 for human height does not mean that your height is 65 percent due to your genes, but rather that in the populations sampled, on average, *65 percent of the overall variation in height could be explained by genotypic differences among individuals.*

With this subtle but important distinction in mind, we will now consider how geneticists divide the phenotypic variation observed in a population into genetic and environmental components. As we saw in the previous section, variation can be quantified as a sample variance: taking measurements of the trait in question from a representative sample of the population and determining the extent of the spread of those measurements around the sample mean. This gives us an estimate of the total **phenotypic variance** in the population (V_P.) Heritability estimates are obtained by using different experimental and statistical techniques to partition V_P into **genotypic variance** (V_G) and **environmental variance** (V_E) components.

An important factor contributing to overall levels of phenotypic variation is the extent to which individual genotypes affect the phenotype differently depending on the environment. For example, wheat variety A may yield an average of 20 bushels an acre on poor soil, while variety B yields an average of 17 bushels. On good soil, variety A yields 22 bushels, while variety B averages 25 bushels an acre. There are differences in yield between the two genotypically distinct varieties, so variation in wheat yield has a genetic component. Both varieties yield more on good soil, so yield is also affected by environment. However, we also see that the two varieties do not respond to better soil conditions

(a)

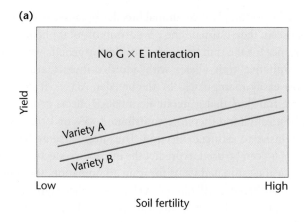

(b)

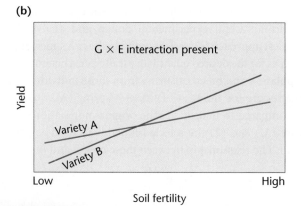

FIGURE 23–6 Differences in yield between two wheat varieties at different soil fertility levels. (a) No genotype-by-environment, or G × E, interaction: The varieties show genetic differences in yield but respond equally to increasing soil fertility. (b) G × E interaction present: Variety A outyields B at low soil fertility, but B yields more than A at high-fertility levels.

equally: The genotype of wheat variety B achieves a greater increase in yield on good soil than does variety A. Thus, we have differences in the interaction of genotype, with environment contributing to variation for yield in populations of wheat plants. This third component of phenotypic variation is **genotype-by-environment interaction variance ($V_{G \times E}$)** (Figure 23–6).

We can now summarize all the components of total phenotypic variance V_P using the following equation:

$$V_P = V_G + V_E + V_{G \times E}$$

In other words, total phenotypic variance can be subdivided into genotypic variance, environmental variance, and genotype-by-environment interaction variance. When obtaining heritability estimates for a multifactorial trait, researchers often assume that the genotype-by-environment interaction variance is small enough that it can be ignored or combined with the environmental variance. However, it is worth remembering that this kind of approximation is another reason heritability values are *estimates* for a given population in a particular context, not a *fixed attribute* for a trait.

Animal and plant breeders use a range of experimental techniques to estimate heritabilities by partitioning measurements of phenotypic variance into genotypic and environmental components. One approach uses inbred strains containing genetically homogeneous individuals with highly homozygous genotypes. Experiments are then designed to test the effects of a range of environmental conditions on phenotypic variability. Variation *between* different inbred strains reared in a constant environment is due predominantly to genetic factors. Variation *among* members of the same inbred strain reared under different conditions is more likely to be due to environmental factors. Other approaches involve analysis of variance for a quantitative trait among offspring from different crosses, or comparing expression of a trait among offspring and parents reared in the same environment.

Broad-Sense Heritability

Broad-sense heritability (represented by the term H^2) measures the contribution of the genotypic variance to the total phenotypic variance. It is estimated as a proportion:

$$H^2 = \frac{V_G}{V_P}$$

Heritability values for a trait in a population range from 0.0 to 1.0. A value approaching 1.0 indicates that the environmental conditions have little impact on phenotypic variance, which is therefore largely due to genotypic differences among individuals in the population. Low values close to 0.0 indicate that environmental factors, not genotypic differences, are largely responsible for the observed phenotypic variation within the population studied. Few quantitative traits have very high or very low heritability estimates, suggesting that both genetics and environment play a part in the expression of most phenotypes for the trait.

The genotypic variance component V_G used in broad-sense heritability estimates includes all types of genetic variation in the population. It does not distinguish between quantitative trait loci with alleles acting additively as opposed to those with epistatic or dominance effects. Broad-sense heritability estimates also assume that the genotype-by-environment variance component is negligible. While broad-sense heritability estimates for a trait are of general genetic interest, these limitations mean this kind of heritability is not very useful in breeding programs. Animal or plant breeders wishing to develop improved strains of livestock or higher-yielding crop varieties need more precise heritability estimates for the traits they wish to manipulate in a population. Therefore, another type of estimate, narrow-sense heritability, has been devised that is of more practical use.

Narrow-Sense Heritability

Narrow-sense heritability (h^2) is the proportion of phenotypic variance due to additive genotypic variance alone.

Genotypic variance can be divided into subcomponents representing the different modes of action of alleles at quantitative trait loci. As not all the genes involved in a quantitative trait affect the phenotype in the same way, this partitioning distinguishes among three different kinds of gene action contributing to genotypic variance. **Additive variance, V_A**, is the genotypic variance due to the additive action of alleles at quantitative trait loci. **Dominance variance, V_D**, is the deviation from the additive components that results when phenotypic expression in heterozygotes is not precisely intermediate between the two homozygotes. **Interactive variance, V_I**, is the deviation from the additive components that occurs when two or more loci behave epistatically. The amount of interactive variance is often negligible, and so this component is often excluded from calculations of total genotypic variance.

The partitioning of the total genotypic variance V_G is summarized in the equation

$$V_G = V_A + V_D + V_I$$

and a narrow-sense heritability estimate based only on that portion of the genotypic variance due to additive gene action becomes

$$h^2 = \frac{V_A}{V_P}$$

Omitting V_I and separating V_P into genotypic and environmental variance components, we obtain

$$h^2 = \frac{V_A}{V_E + V_A + V_D}$$

Heritability estimates are used in animal and plant breeding to indicate the potential response of a population to artificial selection for a quantitative trait. Narrow-sense heritability, h^2, provides a more accurate prediction of selection response than broad-sense heritability, H^2, and therefore h^2 is more widely used by breeders.

Artificial Selection

Artificial selection is the process of choosing specific individuals with preferred phenotypes from an initially heterogeneous population for future breeding purposes. Theoretically, if artificial selection based on the same trait preferences is repeated over multiple generations, a population can be developed containing a high frequency of individuals with the desired characteristics. If selection is for a simple trait controlled by just one or two genes subject to little environmental influence, generating the desired population of plants or animals is relatively fast and easy. However, many traits of economic importance in crops and livestock, such as grain yield in plants, weight gain or milk yield in cattle, and speed or stamina in horses, are polygenic and frequently multifactorial. Artificial selection for such traits is slower and more complex. Narrow-sense heritability estimates are

valuable to the plant or animal breeder because, as we have just seen, they estimate the proportion of total phenotypic variance for the trait that is due to additive genetic variance. Quantitative trait alleles with additive impact are those most easily manipulated by the breeder. Alleles at quantitative trait loci that generate dominance effects or interact epistatically (and therefore contribute to V_D or V_I) are less responsive to artificial selection. Thus narrow-sense heritability, h^2, can be used to predict the impact of selection. The higher the estimated value for h^2 in a population, the more likely the breeder will observe a change in phenotypic range for the trait in the next generation after artificial selection.

Partitioning the genetic variance components to calculate h^2 and predict response to selection is a complex task requiring careful experimental design and analysis. The simplest approach is to select individuals with superior phenotypes for the desired quantitative trait from a heterogeneous population and breed offspring from those individuals. The mean score for the trait of those offspring ($M2$) can then be compared to that of: (1) the original population's mean score (M) and (2) the selected individuals used as parents ($M1$). The relationship between these means and h^2 is

$$h^2 = \frac{M2 - M}{M1 - M}$$

This equation can be further simplified by defining $M2 - M$ as the **selection response (R)**—the degree of response to mating the selected parents—and $M1 - M$ as the **selection differential (S)**—the difference between the mean for the whole population and the mean for the selected population—so h^2 reflects the ratio of the response observed to the total response possible. Thus,

$$h^2 = \frac{R}{S}$$

A narrow-sense heritability value obtained in this way by selective breeding and measuring the response in the offspring is referred to as an estimate of **realized heritability**.

As an example of a realized heritability estimate, suppose that we measure the diameter of corn kernels in a population where the mean diameter M is 20 mm. From this population, we select a group with the smallest diameters, for which the mean $M1$ equals 10 mm. The selected plants are interbred, and the mean diameter $M2$ of the progeny kernels is 13 mm. We can calculate the realized heritability h^2 to estimate the potential for artificial selection on kernel size:

$$h^2 = \frac{M2 - M}{M1 - M}$$

$$h^2 = \frac{13 - 20}{10 - 20}$$

$$= \frac{-7}{-10}$$

$$= 0.70$$

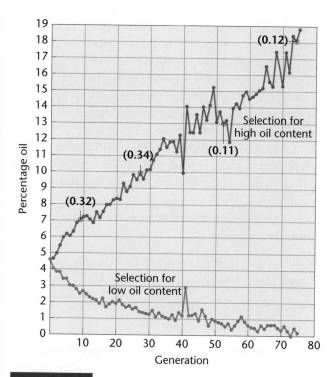

FIGURE 23–7 Response of corn selected for high and low oil content over 76 generations. The numbers in parentheses at generations 9, 25, 52, and 76 for the "high oil" line indicate the calculation of heritability at these points in the continuing experiment.

TABLE 23.4

Estimates of Heritability for Traits in Different Organisms

Trait	Heritability (h^2)
Mice	
Tail length	60%
Body weight	37
Litter size	15
Chickens	
Body weight	50
Egg production	20
Egg hatchability	15
Cattle	
Birth weight	45
Milk yield	44
Conception rate	3

This value for narrow-sense heritability indicates that the selection potential for kernel size is relatively high.

The longest running artificial selection experiment known is still being conducted at the State Agricultural Laboratory in Illinois. Since 1896, corn has been selected for both high and low oil content. After 76 generations, selection continues to result in increased oil content (**Figure 23–7**). With each cycle of successful selection, more of the corn plants accumulate a higher percentage of additive alleles involved in oil production. Consequently, the narrow-sense heritability h^2 of increased oil content in succeeding generations has declined (see parenthetical values at generations 9, 25, 52, and 76 in Figure 23–7) as artificial selection comes closer and closer to optimizing the genetic potential for oil production. Theoretically, the process will continue until all individuals in the population possess a uniform genotype that includes all the additive alleles responsible for high oil content. At that point, h^2 will be reduced to zero, and response to artificial selection will cease. The decrease in response to selection for low oil content shows that heritability for low oil content is approaching this point.

Table 23.4 lists narrow-sense heritability estimates expressed as percentage values for a variety of quantitative traits in different organisms. As you can see, these h^2 values vary, but heritability tends to be low for quantitative traits

that are essential to an organism's survival. Remember, this does not indicate the absence of a genetic contribution to the observed phenotypes for such traits. Instead, the low h^2 values show that natural selection has already largely optimized the genetic component of these traits during evolution. Egg production, litter size, and conception rate are examples of how such physiological limitations on selection have already been reached. Traits that are less critical to survival, such as body weight, tail length, and wing length, have higher heritabilities because more genotypic variation for such traits is still present in the population. Remember also that any single heritability estimate can only provide information about one population in a specific environment. Therefore, narrow-sense heritability is a more valuable predictor of response to selection when estimates are calculated for many populations and environments and show the presence of a clear trend.

23.5

Twin Studies Allow an Estimation of Heritability in Humans

Human twins are useful subjects for examining how much phenotypic variance for a multifactorial trait is due to the genotype as opposed to the environment. In these studies, the underlying principle has been that **monozygotic (MZ),** or **identical, twins** are derived from a single zygote that divides mitotically and then spontaneously splits into two separate cells. Both cells give rise to a genotypically identical embryo. **Dizygotic (DZ),** or **fraternal, twins,** on the other hand, originate from two separate fertilization events and are only as genetically similar as any two siblings, with an average of 50 percent of their alleles in common. For a given trait, therefore, phenotypic differences between pairs of MZ twins will be equivalent to the environmental variance (V_E) (because the genotypic variance is zero). Phenotypic differences between DZ twins, however, display both environmental variance (V_E) and approximately half the genotypic variance (V_G). Comparing the extent of phenotypic variance for the same trait in MZ and DZ sets of twins provides an estimate of broad-sense heritability for the trait.

Twins are said to be **concordant** for a given trait if both express it or neither expresses it. If one expresses the trait and the other does not, the pair is said to be **discordant**. Comparison of concordance values of MZ versus DZ twins reared together illustrates the potential value for heritability assessment. (See the Now Solve This feature on p. 671, for example.)

Before any conclusions can be drawn from twin studies, the data must be examined carefully. For example, if concordance values approach 90 to 100 percent in MZ twins, we might be inclined to interpret that as a large genetic contribution to the phenotype of the trait. In some cases—for example, blood types and eye color—we know that this is indeed true. In the case of contracting measles, however, a high concordance value merely indicates that the trait is almost always induced by a factor in the environment—in this case, a virus.

It is more meaningful to compare the *difference* between the concordance values of MZ and DZ twins. If concordance values are significantly higher in MZ twins, we suspect a strong genetic component in the determination of the trait. In the case of measles, where concordance is high in both types of twins, the environment is assumed to be the major contributing factor. Such an analysis is useful because phenotypic characteristics that remain similar in different environments are likely to have a strong genetic component.

Twin Studies Have Several Limitations

Interesting as they are, human twin studies contain some unavoidable sources of error. For example, MZ twins are often treated more similarly by parents and teachers than are DZ twins, especially when the DZ siblings are of different sex. This circumstance may inflate the environmental variance for DZ twins. Another possible error source is interactions between the genotype and the environment that produce variability in the phenotype. These interactions can increase the total phenotypic variance for DZ twins compared to MZ twins raised in the same environment, influencing heritability calculations. Overall, heritability estimates for human traits based on twin studies should therefore be considered approximations and examined very carefully before any conclusions are drawn.

Although they must often be viewed with caution, classical twin studies, based on the assumption that MZ twins share the same genome, have been valuable for estimating heritability over a wide range of traits including multifactorial disorders such as cardiovascular disease, diabetes, and mental illness, for example. These disorders clearly have genetic components, and twin studies provide a foundation for studying interactions between genes and environmental factors. However, results from genomics research have challenged the view that MZ twins are truly identical and have forced a reevaluation of both the methodology and the results of twin studies. Such research has also opened the way to new approaches to the study of interactions between the genotype and environmental factors.

The most relevant genomic discoveries about twins include the following:

- By the time they are born, MZ twins do not necessarily have identical genomes.

- Gene-expression patterns in MZ twins change with age, leading to phenotypic differences.

We will address these points in order. First, MZ twins develop from a single fertilized egg, where sometime early in development the resulting cell mass separates into two distinct populations creating two independent embryos. Until that time, MZ twins have identical genotypes. Subsequently, however, the genotypes can diverge slightly. For example, differences in *copy number variation* (*CNV*)—variation in the number of copies of numerous large DNA sequences (usually 1000 bp or more)—may arise, differentially producing genetically distinct populations of cells in each embryo (see Chapter 8, p. 210, for a discussion of CNV). This creates a condition called *somatic mosaicism,* which may result in a milder disease phenotype in some disorders and may play a similar role in phenotypic discordance observed in some pairs of MZ twins.

At this point, it is difficult to know for certain how often CNV arises after MZ twinning, but one estimate suggests that such differences are believed to occur in 10 percent of all twin pairs. In those pairs where it does occur, one estimate is that such divergence takes place in 15 to 70 percent of the somatic cells. In one case, a CNV difference between MZ twins has been associated with chronic lymphocytic leukemia in one twin, but not the other.

The second genomic difference between MZ twins involves **epigenetics**—the chemical modification of their DNA and associated histones. An international study of epigenetic modifications in adult European MZ twins showed that MZ twin pairs are epigenetically identical at birth, but adult MZ twins show significant differences in the *methylation patterns* of both DNA and histones. Such epigenetic changes in turn affect patterns of gene expression. The accumulation of epigenetic changes and gene-expression profiles may explain some of the observed phenotypic

23–3 The following table gives the percentage of twin pairs studied in which both twins expressed the same phenotype for a trait (concordance). Percentages listed are for concordance for each trait in monozygotic (MZ) and dizygotic (DZ) twins. Assuming that both twins in each pair were raised together in the same environment, what do you conclude about the relative importance of genetic versus environmental factors for each trait?

Trait	MZ %	DZ %
Blood types	100	66
Eye color	99	28
Mental retardation	97	37
Measles	95	87
Hair color	89	22
Handedness	79	77
Idiopathic epilepsy	72	15
Schizophrenia	69	10
Diabetes	65	18
Identical allergy	59	5
Cleft lip	42	5
Club foot	32	3
Mammary cancer	6	3

■ HINT: *This problem asks you to evaluate the relative importance of genetic versus environmental contributions to specific traits by examining concordance values in MZ versus DZ twins. The key to its solution is to examine the difference in concordance values and to factor in what you have learned about the genetic differences between MZ and DZ twins.*

discordance and susceptibility to diseases in adult MZ twins. A difference in DNA methylation patterns is observed in MZ twins discordant for Beckwith-Wiedemann syndrome, a genetic disorder associated with developmental overgrowth of certain tissues and organs and an increased risk of cancer.

Progressive, age-related genomic modifications may be the result of MZ twins being exposed to different environmental factors, or from failure of epigenetic marking following DNA replication. These findings also indicate that concordance studies in DZ twins must take into account genetic as well as *epigenetic differences* that contribute to discordance in these twin pairs.

The realization that epigenetics may play an important role in the development of phenotypes promises to make twin studies an especially valuable tool in dissecting the interactions among genes and the role of environmental factors in the production of phenotypes. Once the degree of epigenetic differences between MZ and DZ twin pairs has been defined, molecular studies on DNA and histone modification can link changes in gene expression with differences in the concordance rates between MZ and DZ twins.

23.6

Quantitative Trait Loci Are Useful in Studying Multifactorial Phenotypes

Environmental effects, interaction among segregating alleles, and the large number of genes that may contribute to a polygenic phenotype make it difficult to: (1) identify all genes that are involved; and (2) determine the effect of each gene on the phenotype. However, because many quantitative traits are of economic or medical relevance, it is often desirable to obtain this information. In such studies, a chromosome region is identified as containing one or more genes contributing to a quantitative trait is known as a **quantitative trait locus (QTL)**.* When possible, the relevant gene or genes contained within a QTL are isolated and studied.

The modern approach used to find and map QTLs involves looking for associations between DNA markers and phenotypes. One way to do this is to begin with individuals from two lines created by artificial selection that are highly divergent for a phenotype (fruit weight, oil content, bristle number, etc.). For example, **Figure 23–8** illustrates a generic case of QTL mapping. Over many generations of artificial selection, two divergent lines become highly homozygous, which facilitates their use in QTL mapping. Individuals from each of the lines with divergent phenotypes [generation 25 in Figure 23–8(a)] are used as parents to create an F_1 generation whose members will be heterozygous at most of the loci contributing to the trait. Additional crosses, either among F_1 individuals or between the F_1 and the inbred parent lines, result in F_2 generations that carry different portions of the parental genomes [Figure 23–8(b)] with different QTL genotypes and associated phenotypes. This segregating F_2 is known as the **QTL mapping population**.

Researchers then measure phenotypic expression of the trait among individuals in the mapping population and identify genomic differences among individuals by using chromosome-specific DNA markers such as restriction *fragment length polymorphisms (RFLPs), microsatellites,* and *single-nucleotide polymorphisms (SNPs)* (see Chapter 21).

---

*We utilize QTLs to designate the plural form, quantitative trait loci.

(a)

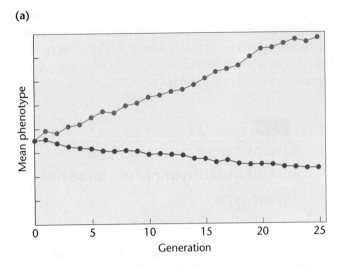

(b)

$P_1 \times P_1$

$F_1 \times F_1$

F_2

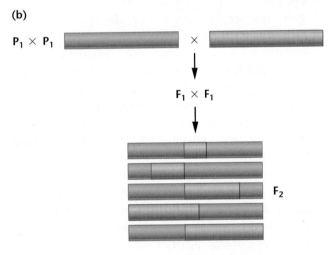

(c)

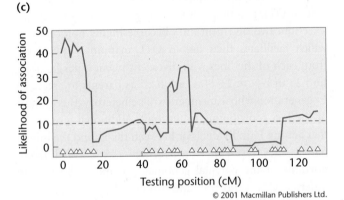

© 2001 Macmillan Publishers Ltd.

FIGURE 23–8   (a) Individuals from highly divergent lines created by artificial selection are chosen from generation 25 as parents. (b) The thick bars represent the genomes of individuals selected from the divergent lines as parents. These individuals are crossed to produce an F_1 generation (not shown). An F_2 generation is produced by crossing members of the F_1. As a result of crossing over individual members of the F_2 generation carry different portions of the P_1 genome, as shown by the colored segments of the thick bars. DNA markers and phenotypes in individuals of the F_2 generation are analyzed. (c) Statistical methods are used to determine the probability that a DNA marker is associated with a QTL that affects the phenotype. The results are plotted as the likelihood of association against chromosomal location. Units on genetic maps are measured in centimorgans (cM), determined by crossover frequencies. Peaks above the horizontal line represent significant results. The data shows five possible QTLs, with the most significant findings at about 10 cM and 60 cM.

phenotypic expression of the trait. When this occurs, the marker locus and the QTL are said to *cosegregate*. Consistent cosegregation establishes the presence of a QTL at or near the DNA marker along the chromosome—in other words, the marker and QTL are linked. When numerous QTLs for a given trait have been located, a genetic map is created, showing the probability that specific chromosomal regions are associated with the phenotype of interest [Figure 23–8(c)]. Further research using genomic techniques identify genes in these regions that contribute to the phenotype.

QTL mapping has been extensively used in agriculture, including plants such as corn, rice, wheat, and tomatoes (**Table 23.5**), and livestock such as cattle, pigs, sheep, and chickens. For example, hundreds of QTLs have been located in the tomato, and its genome has been sequenced. Many chromosome regions responsible for quantitative traits such as fruit size, shape, soluble solid content, and acidity have been identified. These QTLs are distributed on all 12 chromosomes representing the haploid genome of this plant.

TABLE 23.5

QTLs for Quantitative Phenotypes

Organism	Quantitative Phenotype	QTLs Identified
Tomato	Soluble solids	7
	Fruit mass	13
	Fruit pH	9
	Growth	5
	Leaflet shape	9
	Height	9
Maize	Height	11
	Leaf length	7
	Grain yield	18
	Number of ears	10

Source: Used with permission of Annual Reviews of Genetics, from "Mapping Polygenes" by S.D. Tanksley, *Annual Review of Genetics*, Vol. 27:205–233, Table 1, December 1993. Permission conveyed through Copyright Clearance Center, Inc.

Computer-based statistical analysis is used to search for linkage between the markers and a component of phenotypic variation associated with the trait. If a DNA marker (such as those markers described above) *is not* linked to a QTL, then the phenotypic mean score for the trait will not vary among individuals with different genotypes at that marker locus. However, if a DNA marker *is* linked to a QTL, then different genotypes at that marker locus will also differ in their

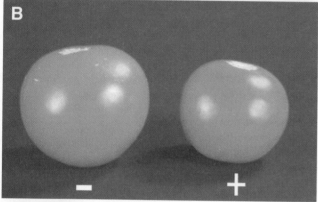

FIGURE 23–9   (a) The ancestral and modern-day tomato. (b) Phenotypic effect of the *fw2.2* transgene in the tomato. When the allele causing small fruit is transferred to a plant that normally produces large fruit, the fruit is reduced in size (+). A control fruit (−) that has not been transformed is shown for comparison.

A research program conducted by Steven Tanksley and his colleagues at Cornell University has focused on mapping and characterizing quantitative traits in the tomato, including fruit shape and weight. We will describe one aspect of this research, which represents a model approach in the study of QTLs. While the cultivated tomato can weigh up to 1000 grams, fruit from the related wild species thought to be the ancestor of the modern tomato weighs only a few grams [**Figure 23–9(a)**]. QTL mapping has identified more than 28 QTLs related to this thousand-fold variation in fruit weight. More than ten years of work was required to localize, identify, and clone one of these QTLs, called *fw2.2* (on chromosome 2). Within this QTL, a specific gene, *ORFX*, has been identified, and alleles at this locus are responsible for about 30 percent of the variation in fruit weight.

The *ORFX* gene has been isolated, cloned, and transferred between plants, with interesting results. One allele of

ORFX is present in all wild small-fruited varieties of tomatoes investigated, while another allele is present in all domesticated large-fruited varieties. When a cloned *ORFX* gene from small-fruited varieties is transferred to a plant that normally produces large tomatoes, the transformed plant produces fruits that are greatly reduced in weight [(Figure 23–8(b)]. In the varieties studied by Tanksley's group, the reduction averaged 17 grams, a statistically significant phenotypic change caused by the action of a gene found within a single QTL.

Further analysis of *ORFX* revealed that this gene encodes a protein that negatively regulates cell division during fruit development. Differences in the time of gene expression and differences in the amount of transcript produced lead to small or large fruit. Higher levels of expression mediated by transferred *ORFX* alleles exert a negative control over cell division, resulting in smaller tomatoes.

Yet *ORFX* and other related genes cannot account for all the observed variation in tomato size. Analysis of another QTL, *fas* (located on chromosome 11), indicates that the development of extreme differences in fruit size resulting from artificial selection also involves an increase in the number of compartments in the mature fruit. The small, ancestral stocks produce fruit with two to four seed compartments, but the large-fruited present-day strains have eight or more compartments. Thus, the QTLs that affect fruit size in tomatoes work by controlling at least two developmental processes: cell division early in development and the determination of the number of ovarian compartments.

The discovery that QTLs can control levels of gene expression has led to new, molecular definitions of phenotypes associated with quantitative traits. For example, the phenotype investigated may be the amount of an RNA transcript produced by a gene (**expression QTLs, or eQTLs**), or the amount of protein produced (**protein QTLs, or pQTLs**). These molecular phenotypes are polygenically controlled in the same way as more conventional phenotypes, such as fruit weight. Gene expression, for example, is controlled by *cis* factors, including promoters, and by *trans*-acting transcription factors (see Chapter 17 for a discussion of gene regulation in eukaryotes).

The use of these new methods of QTL analysis moves the field in a new direction by focusing on regulatory networks and protein–protein interactions, and systems biology. In plant biology, eQTLs are being used to study flowering time, pathogen resistance, and the influence of the environment on developmental events. These new methods will be useful not only in agriculture, but in other fields as well, including the dissection of multifactorial traits and diseases in humans such as *cleft lip*, *spina bifida*, *Type II diabetes*, and *coronary artery disease*.

The Green Revolution Revisited: Genetic Research with Rice

Of the 6.7 billion people now living on Earth, over 800 million do not have enough to eat. That number is expected to grow by an additional 1 million people each year for the next several decades. How will we be able to feed the estimated 8 billion people on Earth by 2025?

The past gives us some reasons to be optimistic. In the 1950s and 1960s, in the face of looming population increases, plant scientists around the world set about to increase the production of crop plants, including the three most important grains—rice, wheat, and maize. These efforts became known as the *Green Revolution*. The approach was three-pronged: (1) to increase the use of fertilizers, pesticides, and irrigation; (2) to bring more land under cultivation; and (3) to develop improved varieties of crop plants by intensive plant breeding.

The results were dramatic. Developing nations more than doubled their production of rice, wheat, and maize between 1961 and 1985. Nations that were facing widespread famine in the 1960s were able to feed themselves and became major exporters of grain.

The Green Revolution saved millions of people from starvation and improved the quality of life for millions more; however, its effects may be diminishing. The rate of increase in grain yields has slowed since the 1980s, due to slower growth in irrigation development and fertilizer use. If food production is to keep pace with the projected increase in the world's population, we will have to depend more and more on the genetic improvement of crop plants to provide higher yields. Is this possible, or are we approaching the theoretical limits of yield in important crop plants? Recent work with rice suggests that the answer to this question is a resounding no.

About half of the Earth's population depends on rice for basic nourishment. The Green Revolution for rice began in 1960, aided by the establishment of the International Rice Research Institute (IRRI). One of their major developments was the breeding of a rice variety with improved disease resistance and higher yield. The IRRI research team crossed a Chinese rice variety (*Dee-geo-woo-gen*) and an Indonesian variety (*Peta*) to create a new cultivar known as IR8. IR8 produced a greater number of rice kernels per plant, when grown in the presence of fertilizers and irrigation. Under these cultural practices, IR8 plants were so top-heavy with grain that they tended to fall over—a trait called "lodging." To reduce lodging, IRRI breeders crossed IR8 with a dwarf native variety to create semi-dwarf lines. Due in part to the adoption of the semi-dwarf IR8 lines, the world production of rice doubled in 25 years.

Predictions suggest that a 40 percent increase in the annual rice harvest may be necessary to keep pace with anticipated population growth during the next 30 years. As land and water resources become more scarce and the prices of fertilizers and pesticides grow, more emphasis will be placed on creating new rice varieties that have even higher yields and greater disease resistance.

Geneticists are now using several strategies to develop more productive rice. Conventional hybridization and selection techniques, such as those used to create the IR8 strain, continue to produce varieties that contribute to a productivity increase of between 1 and 10 percent a year. Several quantitative trait loci (QTLs) from wild rice appear to contribute to increased yields, and scientists are attempting to introduce these traits into current dwarf varieties of domestic rice. Genomics and genetic engineering are also contributing to the new Green Revolution for rice. In 2002, the rice genome was the first cereal crop genome to be sequenced. Research is now concentrated on assigning a function to all the genes in the rice genome. Once identified and characterized, these genes may be transferred into crop plants, speeding the creation of rice varieties with desirable traits, such as disease resistance, tolerance to drought and salinity, and improved nutritional content.

Your Turn

Take time, individually or in groups, to answer the following questions. Investigate the references and links, to help you understand some of the technologies and issues surrounding the new Green Revolution.

1. How do you think that genomics and genetic engineering will contribute to the development of more productive rice varieties?

   One source of information on this topic is: Khush, G.S. 2005. What it will take to feed 5.0 billion rice consumers in 2030. *Plant Mol Biol.* 59: 1–6.

2. A serious complication in our efforts to feed the Earth's population is global climate change. Given your assessment of current climate change predictions, what traits should we introduce into rice plants, and what methods can we use to do this?

   To learn more about the effects of global climate change on agriculture, read Battisti, D.S. and Naylor, R.L. 2009. Historical warnings of future food insecurity with unprecedented seasonal heat. *Science* 323: 240–244. *Also, search for "climate change" publications on the Web site of the International Rice Research Institute* (http://www.irri.org).

3. Despite its benefits, the Green Revolution has been the subject of controversy. What are the main criticisms of the Green Revolution, and how can we mitigate some of the Green Revolution's negative aspects?

   A summary of criticisms of the Green Revolution can be found on the Wikipedia Web site (http://en.wikipedia.org/wiki/Green_revolution).

CASE STUDY | A genetic flip of the coin

On July 11, 2008, twin sons were born to Stephan Gerth from Germany and Addo Gerth from Ghana. Stephan is very fair-skinned with blue eyes and straight hair; Addo is dark-skinned, with brown eyes and curly hair. The first born of the twins, Ryan, is fair-skinned, with blue eyes and straight hair; his brother, Leo, has light brown skin, brown eyes, and curly hair. Although the twins' hair texture and eye color were the same as those of one or the other parent, the twins had different skin colors, intermediate to that of their parents. Experts explained that the blending effect of skin color in the twins resulted from quantitative inheritance involving at least three different gene pairs, whereas hair texture and eye color are not quantitatively inherited. Using this as an example of quantitative genetics, we can ask the following questions:

1. What approach is used in estimating how many gene pairs are involved in a quantitative trait? Why would this be extremely difficult in the case of skin color in humans?

2. Would either parent need to have mixed-race ancestry for the twins to be so different?

3. Would twins showing some parental traits (hair texture, eye color) but a blending of other traits (skin color in this case) seem to be a commonplace event, or are we looking at a "one in a million" event?

Summary Points

For activities, animations, and review quizzes, go to the study area at www.masteringgenetics.com

1. Quantitative inheritance results in a range of phenotypes due to the action of additive alleles from two or more genes, as influenced by environmental factors.
2. Numerous statistical methods are essential during the analysis of quantitative traits, including the mean, variance, standard deviation, standard error, covariance, and the correlation coefficient.
3. Heritability is an estimate of the relative contribution of genetic versus environmental factors to the range of phenotypic variation seen in a quantitative trait in a particular population and environment.
4. Twin studies are used to estimate the heritability of multifactorial traits in humans.
5. Quantitative trait loci, or QTLs, may be identified and mapped using DNA markers.

INSIGHTS AND SOLUTIONS

1. In a certain plant, height varies from 6 to 36 cm. When 6-cm and 36-cm plants were crossed, all F_1 plants were 21 cm. In the F_2 generation, a continuous range of heights was observed. Most were around 21 cm, and 3 of 200 were as short as the 6-cm P_1 parent.

   (a) What mode of inheritance does this illustrate, and how many gene pairs are involved?

   (b) How much does each additive allele contribute to height?

   (c) List all genotypes that give rise to plants that are 31 cm.

   Solution

   (a) Polygenic inheritance is illustrated when a trait is continuous and when alleles contribute additively to the phenotype. The 3/200 ratio of F_2 plants is the key to determining the number of gene pairs. This reduces to a ratio of 1/66.7, very close to 1/64. Using the formula $1/4^n = 1/64$ (where 1/64 is equal to the proportion of F_2 phenotypes as extreme as either P_1 parent), $n = 3$. Therefore, three gene pairs are involved.

   (b) The variation between the two extreme phenotypes is

   $$36 - 6 = 30 \text{ cm}$$

   Because there are six potential additive alleles ($AABBCC$), each contributes

   $$30/6 = 5 \text{ cm}$$

   to the base height of 6 cm, which results when no additive alleles ($aabbcc$) are part of the genotype.

   (c) All genotypes that include five additive alleles will be 31 cm (5 alleles × 5 cm/allele + 6 cm base height = 31 cm). Therefore, $AABBCc$, $AABbCC$, and $AaBBCC$ are the genotypes that will result in plants that are 31 cm.

2. In a cross separate from the above-mentioned F_1 crosses, a plant of unknown phenotype and genotype was testcrossed, with the following results:

   $$1/4 \ 11 \text{ cm}$$
   $$2/4 \ 16 \text{ cm}$$
   $$1/4 \ 21 \text{ cm}$$

   An astute genetics student realized that the unknown plant could be only one phenotype but could be any of three genotypes. What were they?

Solution: When testcrossed (with *aabbcc*), the unknown plant must be able to contribute either one, two, or three additive alleles in its gametes in order to yield the three phenotypes in the offspring. Since no 6-cm offspring are observed, the unknown plant never contributes all nonadditive alleles (*abc*). Only plants that are homozygous at one locus and heterozygous at the other two loci will meet these criteria. Therefore, the unknown parent can be any of three genotypes, all of which have a phenotype of 26 cm:

AABbCc

AaBbCC

AaBBCc

For example, in the first genotype (*AABbCc*),

AABbCc × *aabbcc*

yields

1/4 *AaBbCc* 21 cm

1/4 *AaBbcc* 16 cm

1/4 *AabbCc* 16 cm

1/4 *Aabbcc* 11 cm

which is the ratio of phenotypes observed.

3. The mean and variance of corolla length in two highly inbred strains of *Nicotiana* and their progeny are shown in the following table. One parent (P₁) has a short corolla, and the other parent (P₂) has a long corolla. Calculate the broad-sense heritability (H^2) of corolla length in this plant.

Strain	Mean (mm)	Variance (mm)
P_1 short	40.47	3.12
P_2 long	93.75	3.87
F_1 ($P_1 \times P_2$)	63.90	4.74
F_2 ($F_1 \times F_1$)	68.72	47.70

Solution: The formula for estimating heritability is $H^2 = V_G/V_P$, where V_G and V_P are the genetic and phenotypic components of variation, respectively. The main issue in this problem is obtaining some estimate of two components of phenotypic variation: genetic and environmental factors. V_P is the combination of genetic and environmental variance. Because the two parental strains are true breeding, they are assumed to be homozygous, and the variance of 3.12 and 3.87 is considered to be the result of environmental influences. The average of these two values is 3.50. The F_1 is also genetically homogeneous and gives us an additional estimate of the impact of environmental factors. By averaging this value along with that of the parents,

$$\frac{4.74 + 3.50}{2} = 4.12$$

we obtain a relatively good idea of environmental impact on the phenotype. The phenotypic variance in the F_2 is the sum of the genetic (V_G) and environmental (V_E) components. We have estimated the environmental input as 4.12, so 47.70 minus 4.12 gives us an estimate of V_G of 43.58. Heritability then becomes 43.58/47.70, or 0.91. This value, when interpreted as a percentage, indicates that about 91 percent of the variation in corolla length is due to genetic influences.

Problems and Discussion Questions

 For instructor-assigned tutorials and problems, go to www.masteringgenetics.com

HOW DO WE KNOW?

1. In this chapter, we focused on a mode of inheritance referred to as quantitative genetics, as well as many of the statistical parameters utilized to study quantitative traits. Along the way, we found opportunities to consider the methods and reasoning by which geneticists acquired much of their understanding of quantitative genetics. From the explanations given in the chapter, what answers would you propose to the following fundamental questions:
   (a) How do we know that threshold traits are actually polygenic even though they may have as few as two discrete phenotypic classes?
   (b) How can we ascertain the number of polygenes involved in the inheritance of a quantitative trait?
   (c) What findings led geneticists to postulate the multiple-factor hypothesis that invoked the idea of additive alleles to explain inheritance patterns?
   (d) How do we assess environmental factors to determine if they impact the phenotype of a quantitatively inherited trait?
   (e) How do we know that monozygotic twins are not identical genotypically as adults?

2. What is the difference between continuous and discontinuous variation? Which of the two is most likely to be the result of polygenic inheritance?

3. Define the following: (a) polygenic, (b) additive alleles, (c) correlation, (d) monozygotic and dizygotic twins, (e) heritability, and (f) QTL.

4. A dark-red strain and a white strain of wheat are crossed and produce an intermediate, medium-red F_1. When the F_1 plants are interbred, an F_2 generation is produced in a ratio of 1 dark-red: 4 medium-dark-red: 6 medium-red: 4 light-red: 1 white. Further crosses reveal that the dark-red and white F_2 plants are true breeding.
   (a) Based on the ratios in the F_2 population, how many genes are involved in the production of color?
   (b) How many additive alleles are needed to produce each possible phenotype?
   (c) Assign symbols to these alleles and list possible genotypes that give rise to the medium-red and light-red phenotypes.
   (d) Predict the outcome of the F_1 and F_2 generations in a cross between a true-breeding medium-red plant and a white plant.

5. Height in humans depends on the additive action of genes. Assume that this trait is controlled by the four loci R, S, T, and U and that environmental effects are negligible. Instead of additive versus nonadditive alleles, assume that additive and partially additive alleles exist. Additive alleles contribute two units, and partially additive alleles contribute one unit to height.
(a) Can two individuals of moderate height produce offspring that are much taller or shorter than either parent? If so, how?
(b) If an individual with the minimum height specified by these genes marries an individual of intermediate or moderate height, will any of their children be taller than the tall parent? Why or why not?

6. An inbred strain of plants has a mean height of 24 cm. A second strain of the same species from a different geographical region also has a mean height of 24 cm. When plants from the two strains are crossed together, the F_1 plants are the same height as the parent plants. However, the F_2 generation shows a wide range of heights; the majority are like the P_1 and F_1 plants, but approximately 4 of 1000 are only 12 cm high, and about 4 of 1000 are 36 cm high.
(a) What mode of inheritance is occurring here?
(b) How many gene pairs are involved?
(c) How much does each gene contribute to plant height?
(d) Indicate one possible set of genotypes for the original P_1 parents and the F_1 plants that could account for these results.
(e) Indicate three possible genotypes that could account for F_2 plants that are 18 cm high and three that account for F_2 plants that are 33 cm high.

7. Erma and Harvey were a compatible barnyard pair, but a curious sight. Harvey's tail was only 6 cm long, while Erma's was 30 cm. Their F_1 piglet offspring all grew tails that were 18 cm. When inbred, an F_2 generation resulted in many piglets (Erma and Harvey's grandpigs), whose tails ranged in 4-cm intervals from 6 to 30 cm (6, 10, 14, 18, 22, 26, and 30). Most had 18-cm tails, while 1/64 had 6-cm tails and 1/64 had 30-cm tails.
(a) Explain how these tail lengths were inherited by describing the mode of inheritance, indicating how many gene pairs were at work, and designating the genotypes of Harvey, Erma, and their 18-cm-tail offspring.
(b) If one of the 18-cm F_1 pigs is mated with one of the 6-cm F_2 pigs, what phenotypic ratio would be predicted if many offspring resulted? Diagram the cross.

8. In the following table, average differences of height, weight, and fingerprint ridge count between monozygotic twins (reared together and apart), dizygotic twins, and nontwin siblings are compared:

Trait	MZ Reared Together	MZ Reared Apart	DZ Reared Together	Sibs Reared Together
Height (cm)	1.7	1.8	4.4	4.5
Weight (kg)	1.9	4.5	4.5	4.7
Ridge count	0.7	0.6	2.4	2.7

Based on the data in this table, which of these quantitative traits has the highest heritability values?

9. What kind of heritability estimates (broad sense or narrow sense) are obtained from human twin studies?

10. List as many human traits as you can that are likely to be under the control of a polygenic mode of inheritance.

11. Corn plants from a test plot are measured, and the distribution of heights at 10-cm intervals is recorded in the following table:

Height (cm)	Plants (no.)
100	20
110	60
120	90
130	130
140	180
150	120
160	70
170	50
180	40

Calculate (a) the mean height, (b) the variance, (c) the standard deviation, and (d) the standard error of the mean. Plot a rough graph of plant height against frequency. Do the values represent a normal distribution? Based on your calculations, how would you assess the variation within this population?

12. The following variances were calculated for two traits in a herd of hogs.

Trait	V_P	V_G	V_A
Back fat	30.6	12.2	8.44
Body length	52.4	26.4	11.70

(a) Calculate broad-sense (H^2) and narrow-sense (h^2) heritabilities for each trait in this herd.
(b) Which of the two traits will respond best to selection by a breeder? Why?

13. The mean and variance of plant height of two highly inbred strains (P_1 and P_2) and their progeny (F_1 and F_2) are shown here.

Strain	Mean (cm)	Variance
P_1	34.2	4.2
P_2	55.3	3.8
F_1	44.2	5.6
F_2	46.3	10.3

Calculate the broad-sense heritability (H^2) of plant height in this species.

14. A hypothetical study investigated the vitamin A content and the cholesterol content of eggs from a large population of chickens. The variances (V) were calculated, as shown here:

Variance	Vitamin A	Trait Cholesterol
V_P	123.5	862.0
V_E	96.2	484.6
V_A	12.0	192.1
V_D	15.3	185.3

(a) Calculate the narrow-sense heritability (h^2) for both traits.
(b) Which trait, if either, is likely to respond to selection?

15. In a herd of dairy cows the narrow-sense heritability for milk protein content is 0.76, and for milk butterfat it is 0.82. The correlation coefficient between milk protein content and butterfat is 0.91. If the farmer selects for cows producing more butterfat in their milk, what will be the most likely effect on milk protein content in the next generation?

16. In an assessment of learning in *Drosophila*, flies were trained to avoid certain olfactory cues. In one population, a mean of 8.5 trials was required. A subgroup of this parental population that was trained most quickly (mean = 6.0) was interbred, and their progeny were examined. These flies demonstrated a mean training value of 7.5. Calculate realized heritability for olfactory learning in *Drosophila*.

17. Suppose you want to develop a population of *Drosophila* that would rapidly learn to avoid certain substances the flies could detect by smell. Based on the heritability estimate you obtained in Problem 16, do you think it would be worth doing this by artificial selection? Why or why not?

18. In a population of tomato plants, mean fruit weight is 60 g and (h^2) is 0.3. Predict the mean weight of the progeny if tomato plants whose fruit averaged 80 g were selected from the original population and interbred.

19. In a population of 100 inbred, genotypically identical rice plants, variance for grain yield is 4.67. What is the heritability for yield? Would you advise a rice breeder to improve yield in this strain of rice plants by selection?

20. Many traits of economic or medical significance are determined by quantitative trait loci (QTLs) in which many genes, usually scattered throughout the genome, contribute to expression.
(a) What general procedures are used to identify such loci?
(b) What is meant by the term *cosegregate* in the context of QTL mapping? Why are markers such as RFLPs, SNPs, and microsatellites often used in QTL mapping?

Extra-Spicy Problems

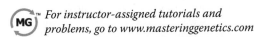

For instructor-assigned tutorials and problems, go to www.masteringgenetics.com

21. In 1988, Horst Wilkens investigated blind cavefish, comparing them with members of a sibling species with normal vision that are found in a lake [Wilkens, H. (1988). *Ecol. Biol.* 23: 271–367]. We will call them cavefish and lakefish. Wilkens found that cavefish eyes are about seven times smaller than lakefish eyes. F_1 hybrids have eyes of intermediate size. These data, as well as the $F_1 \times F_1$ cross and those from backcrosses ($F_1 \times$ cavefish and $F_1 \times$ lakefish), are depicted on the right:
Examine Wilkens's results and respond to the following questions:
(a) Based strictly on the F_1 and F_2 results of Wilkens's initial crosses, what possible explanation concerning the inheritance of eye size seems most feasible?
(b) Based on the results of the F_1 backcross with cavefish, is your explanation supported? Explain.
(c) Based on the results of the F_1 backcross with lakefish, is your explanation supported? Explain.
(d) Wilkens examined about 1000 F_2 progeny and estimated that 6–7 genes are involved in determining eye size. Is the sample size adequate to justify this conclusion? Propose an experimental protocol to test the hypothesis.
(e) A comparison of the embryonic eye in cavefish and lakefish revealed that both reach approximately 4 mm in diameter. However, lakefish eyes continue to grow, while cavefish eye size is greatly reduced. Speculate on the role of the genes involved in this problem.

22. A 3-inch plant was crossed with a 15-inch plant, and all F_1 plants were 9 inches. The F_2 plants exhibited a "normal distribution," with heights of 3, 4, 5, 6, 7, 8, 9, 10, 11, 12, 13, 14, and 15 inches.
(a) What ratio will constitute the "normal distribution" in the F_2?
(b) What will be the outcome if the F_1 plants are testcrossed with plants that are homozygous for all nonadditive alleles?

23. In a cross between a strain of large guinea pigs and a strain of small guinea pigs, the F_1 are phenotypically uniform, with an average size about intermediate between that of the two parental strains. Among 1014 F_2 individuals, 3 are about the same size as the small parental strain and 5 are about the same size as

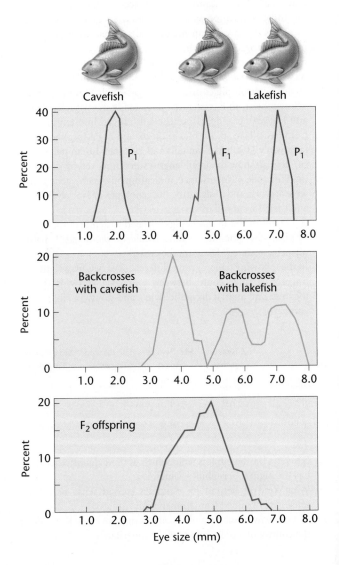

the large parental strain. How many gene pairs are involved in the inheritance of size in these strains of guinea pigs?

24. Type A1B brachydactyly (short middle phalanges) is a genetically determined trait that maps to the short arm of chromosome 5 in humans. If you classify individuals as either having or not having brachydactyly, the trait appears to follow a single-locus, incompletely dominant pattern of inheritance. However, if one examines the fingers and toes of affected individuals, one sees a range of expression from extremely short to only slightly short. What might cause such variation in the expression of brachydactyly?

25. In a series of crosses between two true-breeding strains of peaches, the F_1 generation was uniform, producing 30-g peaches. The F_2 fruit mass ranges from 38 to 22 g at intervals of 2 g.
(a) Using these data, determine the number of polygenic loci involved in the inheritance of peach mass.
(b) Using gene symbols of your choice, give the genotypes of the parents and the F_1.

26. Students in a genetics laboratory began an experiment in an attempt to increase heat tolerance in two strains of *Drosophila melanogaster*. One strain was trapped from the wild six weeks before the experiment was to begin; the other was obtained from a *Drosophila* repository at a university laboratory. In which strain would you expect to see the most rapid and extensive response to heat-tolerance selection, and why?

27. Consider a true-breeding plant, *AABBCC*, crossed with another true-breeding plant, *aabbcc*, whose resulting offspring are *AaBbCc*. If you cross the F_1 generation, and independent assortment is operational, the expected fraction of offspring in each phenotypic class is given by the expression $N![M!(N - M!)]$ where N is the total number of alleles (six in this example) and M is the number of uppercase alleles. In a cross of *AaBbCc* $\times$ *AaBbCc*, what proportion of the offspring would be expected to contain two uppercase alleles?

28. Canine hip dysplasia is a quantitative trait that continues to affect most large breeds of dogs in spite of approximately 40 years of effort to reduce the impact of this condition. Breeders and veterinarians rely on radiographic and universal registries to facilitate the development of breeding schemes for reducing its incidence. Data (Wood and Lakhani. 2003. *Vet. Rec.* 152: 69–72) indicate that there is a "month-of-birth" effect on hip dysplasia in Labrador retrievers and Gordon setters, whereby the frequency and extent of expression of this disorder vary depending on the time of year dogs are born. Speculate on how breeders attempt to "select" out this disorder and what the month-of-birth phenomenon indicates about the expression of polygenic traits?

29. Floral traits in plants often play key roles in diversification, in that slight modifications of those traits, if genetically determined, may quickly lead to reproductive restrictions and evolution. Insight into genetic involvement in flower formation is often acquired through selection experiments that expose realized heritability. Lendvai and Levin (2003) conducted a series of artificial selection experiments on flower size (diameter) in *Phlox drummondii*. Data from their selection experiments are presented in the following table in modified form and content.

Year	Treatment	Mean (mm)
1997	Control	30.04
	Selected parents	34.13
	Offspring	32.21
1998	Control	28.11
	Selected parents	31.98
	Offspring	31.90
1999	Control	29.68
	Selected parents	31.81
	Offspring	33.74

(a) Considering that differences in control values represent year-to-year differences in greenhouse conditions, calculate (in mm) the average response to selection over the three-year period.

(b) Calculate the realized heritability for each year and the overall realized heritability.
(c) Assuming that the realized heritability in phlox is relatively high, what factors might account for such a high response?
(d) In terms of evolutionary potential, is a population with high heritability likely to be favored compared to one with a low realized heritability?

Genetically engineered *Caenorhabditis elegans* roundworms that have turned blue in response to an environmental stress, such as toxins or heat.

24

Genetics of Behavior

CHAPTER CONCEPTS

- Behavior is a complex response to stimuli that is mediated both by genes and by the environment.

- The behavior-first approach is used to establish that behavioral traits are heritable. If so, genetic crosses are used to identify the number and chromosomal location of genes contributing to the trait.

- The gene-first approach utilizes mutagenesis to establish strains carrying single-gene mutations that contribute to a behavioral response. Subsequent comparisons of the normal and mutant alleles provide insight into the underlying mechanism controlling a behavioral response.

- The behavior-first and gene-first approaches have been successfully used to dissect behavioral responses in *Drosophila*, making it a useful model organism for the study of nervous system function and the mechanisms that underlie human behavioral disorders.

- Many human behavioral disorders are complex responses to stimuli. These responses are mediated by genes, environmental factors, and interactions among genes and the environment. More recently, newly developed genomic techniques are being used to study the role of genes in human behavior.

Behavior is generally defined as a reaction to stimuli, whether internal or external, that alter an organism's response to its environment. Animals run, remain still, or counterattack in the presence of a predator; birds build complex and distinctive nests in response to a combination of internal and external signals; plants bend toward light; and humans behave in both simple and complex ways guided by their intellect, emotions, and culture.

Efforts to study the role of heredity in behavior began in the nineteenth century with Francis Galton, who systematically studied behavior and heredity. In one of his early studies, he examined his own family (he was a cousin of Charles Darwin) trying to find a pattern of inheritance associated with intelligence. By the early 1900s, clear-cut cases of genetic influence on behavior had been identified, but at the time, behavior was primarily of interest to psychologists, who were concerned with learned or conditioned behavior. Such behaviors were thought to reflect the influence of the environment to the exclusion of the genotype. The difference between these two approaches to studying behavior was the starting point for what has been called the nature versus nurture debate. In this simplistic (and false) dichotomy, behavior is controlled entirely either by genes or by the environment.

This controversy presented a distorted view of the nature of behavioral responses. After all, any behavior must rely on the expression of an individual's genotype, which takes place within a hierarchy of environmental settings (that is, gene expression depends on interactions within the cell, the tissue, the organ, the organism, and finally the population and surrounding environment). Without genes and their environments, there could be no behavior. Nevertheless, the nature–nurture controversy flourished well into the 1950s. By then, it became clear that behavioral patterns were the result of the interaction of genetic and environmental factors. Because the nervous system senses the environment, processes the information, and initiates the response that we perceive as behavior, much of the study of behavior in genetics and molecular biology focuses on the development, structure, and function of the nervous system. Since about 1950, research into genetic components of behavior has intensified, and appreciation for the importance of genetics in understanding behavior has increased.

With this brief history in mind, here are some of the basic questions we will answer in this chapter.

- How do we define behavior?

- What evidence is there that behavior has a biological or a genetic basis?

- How is behavior genetics studied?

- How does a gene influence or control a specific behavior?

- How do genes interact with the environment to produce a behavioral response?

- What is the pathway from a gene to a behavioral phenotype?

The definition of behavior presented at the beginning of this chapter, based on observable changes in the relationship of a cell or an organism to its environment as the result of some stimulus, broadly identifies the phenotype that geneticists use as a starting point for investigating the biological basis of behavior. Evidence that some behaviors have a genetic basis comes from several sources: observable species-specific behaviors, such as courtship rituals, and artificial selection, which draws upon individual variations in behavior among organisms in populations that can be used to establish strains with heritable differences in behavior. In humans, twin studies and adoption studies have provided evidence for the role of heredity in behavioral responses.

Two different approaches have been used to study the genetic control of behavior, to define the interactions between genotype and environmental factors, and to dissect the pathways leading from genes to a behavioral phenotype. One of these approaches is a top-down, or *behavior-first,* method in which a specific behavior is identified in an organism, and then, genetic crosses are used to produce strains that bred true for either a high level or a low level of this behavioral response. Once these strains are established, further crosses identify and analyze the genetic components of the behavior. The second approach is a bottom-up, or *gene-first,* approach in which mutagenesis followed by screening is used to identify single-gene mutations associated with variant or abnormal behaviors. Analysis of the molecular mechanism of gene action in these mutant strains often provided a direct explanation of the behavior. Each approach has its advantages and shortcomings, but in spite of their differences, both share the same goals: to establish the inherited nature of a specific behavior, to identify and enumerate the genes or gene systems involved in the behavior being studied, to map these genes or gene systems to specific chromosomes, and to elucidate the molecular mechanisms by which these genes influence a behavioral response. A recently developed third approach to studying the genetics of behavior is based on methods that allow entire genomes to be scanned to identify genes and markers for genes and genome variations related to specific behaviors.

The prevailing view today is that most behaviors are complex traits involving a number of genes, as well as interactions between and among these genes and environmental influences. In this scenario, the genotype provides the biological underpinnings or mental ability (or both) required to execute a behavior and further determines the limitations of environmental influences. In this chapter, we will examine the three approaches to studying the genetic basis of behavior, evaluate the strengths and weaknesses of each, and consider what we have learned about human behavior from these methods and the use of model organisms.

24.1

The Behavior-First Approach Can Establish Genetic Strains with Behavioral Differences

The behavior-first approach begins by selecting organisms exhibiting a specific behavior from members of a genetically heterogeneous population. If genetic strains can be established that uniformly express this behavior, and if the trait can be transferred by genetic crosses to another strain that initially does not exhibit the behavior under study, genetic involvement in the behavior is confirmed. Often, selection is used to establish two strains: one showing high levels and one showing low levels of the behavior under investigation. This bidirectional selection creates two lines with progressively greater differences in behavior. If the genetic background of the strains remains constant, this method maximizes differences in genes that control the trait of interest, and can minimize differences in all other genes. The study of emotional behavior in mice and geotaxis in *Drosophila* will be used to illustrate this approach.

Mapping Genes for Anxiety in Mice

To identify and map genes controlling emotional behavior in mice, Jonathon Flint and colleagues have used quantitative trait loci (QTL) mapping (see Chapter 23 for a discussion of QTLs). In one experiment, heterozygotes from crosses between C57BL/6J (a nonemotional, low-anxiety strain) and BALB/cJ (an emotional, high-anxiety strain) were intercrossed, and the progeny were selected over multiple generations for extremes of high (H1a) or low (L1a) anxiety. These inbred strains were subjected to several different behavioral tests of anxiety, including the open field, the elevated plus maze (Figure 24–1), and a test called the light–dark box. In this last-named test, mice are placed in an environment in which there is a dark, enclosed box, from which they can emerge into a brightly lit area. Fearful animals prefer to remain in the box and do not explore the lighted areas of the environment. The phenotypically most extreme progeny recovered from these anxiety tests were analyzed for 84 microsatellite markers and four different measures of emotionality. A QTL on chromosome 1 was identified and localized to a small region of the chromosome spanning 66 Mb. Analysis of over 1600 mice in replicated

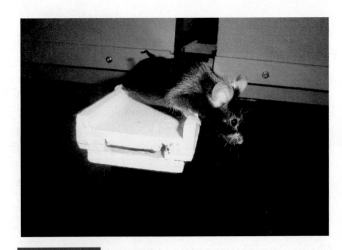

FIGURE 24–1 A mouse explores a new environment from the open arm of an elevated plus maze, a standard device for measuring fear in mice. This experiment was one of a series designed to identify genes involved in fear and anxiety in mice.

populations consistently gave the same results. Similar experiments using microsatellite markers have shown that QTLs controlling other aspects of behavior are distributed on all mouse chromosomes (Figure 24–2).

To advance beyond the mapping of a QTL region, which may contain dozens of genes, and to identify the specific loci involved in emotional behavior requires the combined use of genetics and genomics. Genes in the chromosome 1 QTL region identified as having a behavioral phenotype are being tested in genetic complementation assays to detect interactions among these loci. DNA sequencing of these regions is being used to identify and clone these genes. Once identified, the mechanisms of action of these behavioral loci and their relevance to human behavior remain to be assessed.

Selection for Geotaxis in *Drosophila*

A **taxis** (pronounced *tak sis*) is a movement toward or away from an external stimulus. The response may be positive (moving toward the stimulus) or negative (away from it). Many types of

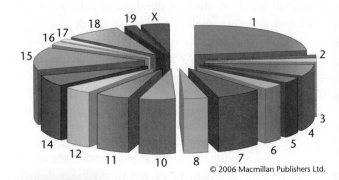

© 2006 Macmillan Publishers Ltd.

FIGURE 24–2 Loci for genes controlling fear and other forms of emotional behavior are distributed on 17 of 19 autosomes in the mouse genome. Significant numbers of these genes are located on chromosomes 1, 15, and 18.

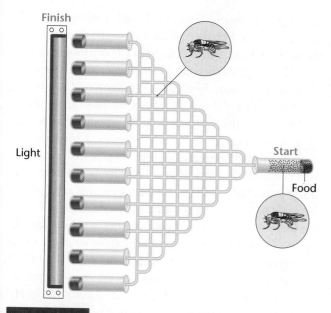

FIGURE 24–3 Schematic drawing of a maze used to study geotaxis in *Drosophila*.

stimuli elicit this form of behavior, including chemicals (chemotaxis), gravity (geotaxis), and light (phototaxis).

To investigate geotaxis in *Drosophila*, Jerry Hirsch and his colleagues designed a maze that tests about 200 flies per trial, as shown in **Figure 24–3**. Once in the maze, flies are repeatedly required to choose between moving up or down. Flies that turn up at each intersection will collect in tubes at the top of the maze; those that always turn down will collect in tubes at the bottom; and those making both "up" and "down" decisions will end up somewhere in between. Flies can be selected both for positive (going up in the maze) and for negative geotaxis (going down in the maze), establishing the existence of a genetic influence on this behavioral response. As shown in **Figure 24–4**, mean

scores of flies in this maze vary from about $+4$ to -6, indicating that selection is stronger for negative geotaxis than for positive geotaxis.

The two lines have undergone selection for almost 40 years, encompassing over 1000 generations and the testing of more than 80,000 flies. Throughout the experiment, clear-cut, but fluctuating, differences have been observed. These results indicate that this complex behavior in *Drosophila* is genetically controlled and is a polygenic trait. Using the lines selected for positive and negative geotaxis, Hirsch and his colleagues analyzed the relative contribution of genes on different chromosomes to this behavior. The effects of chromosomes X, 2, and 3 on geotaxis were identified in an ingenious way. **Figure 24–5** shows how this

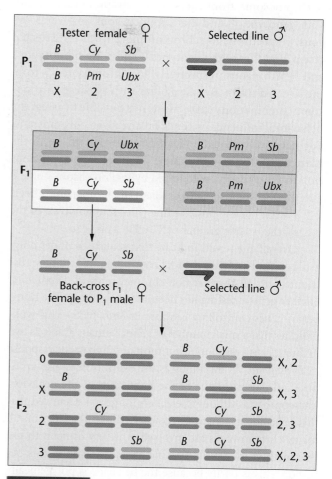

FIGURE 24–5 In this mating scheme in *Drosophila*, the effect of genes located on specific chromosomes that contribute to geotaxis can be assessed. The progeny produced by backcrossing the female contain all combinations of chromosomes. Examining the presence or absence of dominant marker phenotypes in these flies makes it possible to determine which chromosomes from the selected strain are present. Subsequent testing for geotaxis is then performed. (B = *Bar eyes*; Cy = *Curly wing*; Sb = *Stubble bristles*; Pm = *Plum eye color*; Ubx = *ultrabithorax*) The designations alongside each genotype in the F_2 generation (e.g., X, 2, and 3) indicate which chromosomes are heterozygous.

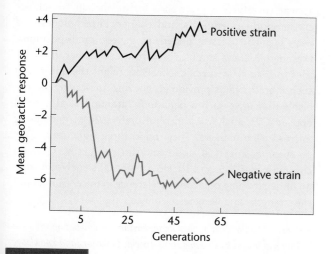

FIGURE 24–4 Selection for positive and negative geotaxis in *Drosophila* over many generations. The scores represent the sum of upward ($+$) or downward ($-$) turns in the maze.

is accomplished. A male from a selected line is crossed to a "tester" female, whose chromosomes are each marked with a dominant mutation (these chromosomes are called **balancer chromosomes**). Each balancer chromosome carries one or more inversions to suppress the recovery of any crossover products. An F_1 female will be heterozygous for each chromosome and is backcrossed to a male from the original line. In the F_2, female offspring contain all combinations of the marker chromosomes and the chromosomes from the selected strain. The dominant mutations carried by the marker chromosomes make it possible to recognize which chromosomes from the selected lines are present in homozygous or heterozygous configurations. Flies with genotypes 0 and X (Figure 24–5) are homozygous and heterozygous, respectively, for the X chromosome from the selected strain. Similarly, flies with genotypes 0 and 2 are homozygous and heterozygous, respectively, for chromosome 2 from the selected strain, and flies with genotypes 0 and 3 are homozygous and heterozygous, respectively, for chromosome 3 from the selected strain. By testing flies with all these chromosome combinations in the maze, it is possible to assess the behavioral influence of genes on any given chromosome. For negatively geotaxic flies (the ones that go up in the maze), genes on chromosome 2 make the largest contribution to the phenotype, followed by loci on chromosome 3 and the X chromosome (a gradient of chromosome effects $2 > 3 > X$). For the positive line (flies that go down in the maze), the reverse gradient ($X > 3 > 2$) is seen.

Overall, the results indicate that geotaxis is under polygenic control and that the loci controlling this trait are distributed on all three major chromosomes of *Drosophila*. Further genetic testing has been used to estimate the number of genes controlling geotaxis in *Drosophila*. This work indicates that a small number of genes, perhaps as few as two to four loci, are responsible for most of the geotactic response in *Drosophila*. As ingenious as this work is, it is also arduous, and illustrates one of the limitations of the behavior-first approach. Although selection for geotaxis showed that this behavior is genetically controlled and that loci on all major chromosomes are involved, it is very difficult to use this method to identify specific genes involved in this behavior. This is partly because the behavior is polygenically controlled, and many of these genes may make only a small contribution to the phenotype.

Genomic technology including microarray analysis (described in Chapters 21 and 22) offers a way to identify specific genes that influence complex polygenic traits, including behavior. Microarrays are one example of the high-throughput experimental methods that are revolutionizing genetics and molecular biology. Instead of measuring changes in expression one gene at a time, microarrays make it possible to measure changes in hundreds or thousands of genes in a single experiment. Genes to be used in microarray analysis can be selected on the basis of their expression patterns in specific tissues, in specific cells, or in whole genome scans.

To study geotaxis genes in *Drosophila*, researchers used a two-step process. First, cDNA microarrays containing about one-third of the genes in the *Drosophila* genome were used to analyze mRNA levels from the strains with the highest (*Hi5*) and lowest (*Lo*) geotaxis behavior. A small number of genes showed reproducible differences in expression and were chosen for further analysis. To narrow the list of candidate genes further, the researchers used mutant alleles of these genes that were associated with neurological defects. Ten lines, including several control lines, were selected for the second step in this gene-identification strategy: analysis of geotaxis behavior. The geotactic behavior and mRNA levels of three genes in the mutant flies showed a correlation between levels of mRNA expression and behavior in *Hi5* and *Lo* flies. Individually, each of these genes made small, incremental contributions to geotaxis, as would be expected for a polygenically controlled trait. The mechanism by which these three genes, *cryptochrome* (*cry*), *Pendulin* (*Pen*), and *Pigment-dispersing-factor* (*Pdf*), are involved in geotaxis is unknown. However, all three are involved in the development and function of the brain and nervous system, and their role in geotaxis is currently being investigated, but whether they are the primary genes involved or are downstream targets of other, unidentified genes is not known. However, the use of this two-step strategy is significant because it illustrates the successful integration of classical genetics with genomics to identify genes involved in a polygenic trait.

NOW SOLVE THIS

24–1 Certain inbred strains of mice (i.e., C57BL and C3H/2) appear to exhibit a preference for alcohol in free-choice consumption experiments, and a variety of techniques have been applied to assess underlying genetic predispositions. Presumably, these strains differ from each other in the alleles they carry at specific loci. How might the analysis of such inbred strains provide insight into molecular mechanisms that drive such a behavior? Tarantino et al. (1998) applied QTL analysis to alcohol preference in mice using microsatellite markers. They found three significant QTLs on chromosomes 1, 4, and 9 and three suggestive QTLs on chromosomes 2, 3, and 10. How might QTL analysis lead to an understanding of alcohol preference in mice? Summarize the strengths and weaknesses of each approach.

■ HINT: *This problem involves an understanding of what information can be obtained by studying strain-specific behaviors and knowledge of what QTLs are, and the key to its solution depends on knowing how QTLs are mapped using chromosome-specific markers.*

24.2
The Gene-First Approach Analyzes Mutant Alleles to Study Mechanisms That Underlie Behavior

Although we discussed geotaxis in *Drosophila* as an example of the behavior-first approach, much has also been learned using this organism for gene-first studies of behavior. Although *Drosophila* and humans separated from a common ancestor millions of years ago, molecular aspects of cellular function in the nervous system overlap considerably in these two organisms. As a result, work with mutant strains of *Drosophila* has provided insight into the fundamental mechanisms that govern nerve impulse transmission. As the name implies, the gene-first approach to studying behavior first isolates mutant strains with a behavioral abnormality and then investigates the structural and/or functional abnormalities that underlie this altered behavior.

The gene-first approach has several advantages over the behavior-first approach. Using this approach allows researchers to quantify the impact of single genes on a complex behavioral response. Use of this method must take into account the polygenic nature of behavior and the interaction of contributing genes. In this context, the gene-first approach can in theory, be used to identify, map, clone, and sequence all genes contributing to a behavior. This approach is simplified in *Drosophila*, because researchers have developed an extensive and sophisticated set of genetic tools,

especially marker chromosomes and collections of strains with deleted and duplicated chromosome regions, as well as mutants with cytological markers for mosaic analysis.

In the following sections, we will explore the use of mutant analysis in dissecting mechanisms of nerve impulses and their direct application to human disorders, and then we will discuss one form of complex behavior, learning, to further illustrate how *Drosophila* serves as a model organism for studying the molecular aspects of behavior.

Genes Involved in Transmission of Nerve Impulses

Figure 24–6 shows a neuron (nerve cell) in which impulses are generated at one end (on a dendrite) and move along the cell to the other end (the axon), which transmits the impulse to adjacent neurons. During this process, the impulse is propagated by the transport of sodium and potassium ions across the plasma membrane of the neuron. The movement of these charged ions can be monitored by measuring changes in the electrical potential of the membrane.

To screen for genes that control the generation and transmission of nerve impulses, Barry Ganetzky and his colleagues screened behavioral mutants of *Drosophila* to identify those with electrophysiological abnormalities in the generation and propagation of nerve impulses. Two general classes of such mutants have been isolated: those with defects in the movement of sodium and those with defects in potassium transport (Table 24.1). One mutant identified in the screening procedure is a temperature-sensitive allele of the *paralytic* gene. Flies homozygous for this mutant

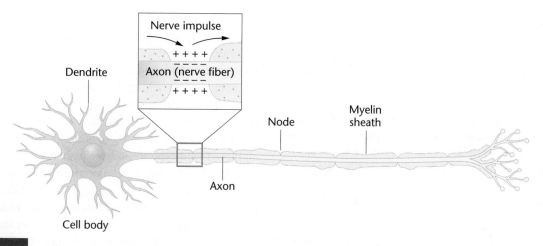

FIGURE 24–6 A nerve cell (neuron) has extensions called dendrites that carry impulses toward the cell body and also has one or more axons that carry impulses away from the cell body. The axon is encased in a sheath of myelin, except at nodes spaced at intervals along the axon. Nerve impulses are accelerated by jumping from node to node. Electrodes placed on either side of the neuron's plasma membrane can record the electric potential across the membrane. When the neuron is at rest, there is more sodium outside the cell and more potassium inside the cell. When a nerve impulse is generated, sodium moves into the cell, and potassium moves out, altering the electric potential. As the impulse moves away, ions are pumped across the membrane to restore the original electric potential.

TABLE 24.1

Behavioral Mutants of *Drosophila* in Which Nerve Impulse Transmission Is Affected

Mutation	Map Location	Ion Channel Affected	Phenotype
nap^{ts}	2-56.2	Sodium	Adults and larvae paralyzed at 37.5°C; reversible at 25°C
$para^{ts}$	1-53.9	Sodium	Adults paralyzed at 29°C, larvae at 37°C; reversible at 25°C
Tip-E	3-13.5	Sodium	Adults and larvae paralyzed at 39–40°C; reversible at 25°C
sei^{ts}	2-10.6	Sodium	Adults paralyzed at 38°C, larvae unaffected; adults recover at 25°C
Sh	1-57.7	Potassium	Aberrant leg shaking in adults exposed to ether
eag	1-50.0	Potassium	Aberrant leg shaking in adults exposed to ether
Hk	1-30.0	Potassium	Ether-induced leg shaking
sio	3-85.0	Potassium	At 22°C, adults are weak fliers; at 38°C, adults are weak, uncoordinated

allele become paralyzed when exposed to temperatures at or above 29°C, but recover rapidly when the temperature is lowered to 25°C. Electrophysiological studies showed that mutant flies have defective sodium transport associated with the conduction of nerve impulses.

Subsequently, Ganetzky and his colleagues mapped, isolated, and cloned the *paralytic* gene. This locus encodes a sodium channel protein, which controls the movement of sodium across the membrane of nerve cells in many different organisms, from flies to humans. These same researchers found that another mutant allele, called *Shaker* (originally isolated over 40 years ago as a behavioral mutant), encodes a potassium channel protein. This gene has also been cloned and characterized. Because the mechanism of nerve impulse conduction has been highly conserved during animal evolution, the cloned *Drosophila* genes were used as probes to isolate the equivalent human ion channel genes. Study of the human genes is providing new insights into the molecular basis of neuronal activity. The sodium channel gene is defective in a heritable form of cardiac arrhythmia. The

identification and cloning of the human gene now make it possible to screen for family members at risk for this potentially fatal condition.

Drosophila Can Learn and Remember

Researchers gain a great advantage when they can study the genetics of a complex behavior, such as learning, in a model organism that is as convenient to manipulate as *Drosophila*. However, this is possible only if the model organism can perform the complex behavior the researchers are interested in studying. Thus, before using *Drosophila* to study learning, researchers needed to know whether *Drosophila* could learn. Seymour Benzer's lab was one of the first to identify genes that control learning and memory in *Drosophila*. To do so, Benzer's group used an olfactory-based shock-avoidance learning system in which flies are presented with a pair of odors, one of which is associated with an electric shock. Flies quickly learn to avoid the odor associated with the shock. For several reasons, this response is thought to be learned. First, performance is associated with the pairing of a stimulus or response with reinforcement. Second, the response is reversible: Flies can be trained to select an odor they previously avoided. Third, flies exhibit short-term memory for the training they have received.

The demonstration that *Drosophila* can learn opened the way to selecting mutants that are defective in learning and memory (**Table 24.2**). To accomplish this, males from an inbred wild-type strain were treated with mutagens and mated to females from the same strain. Their progeny were recovered and mated to produce strains that carried a mutagenized X chromosome. Mutations that affected learning were selected by testing for responses in the olfactory shock apparatus. A number of learning-deficient mutants, including *dunce, turnip, rutabaga,* and *cabbage*, were isolated in this way. In addition, a memory-deficient mutant, *amnesia*, which learns normally, but forgets four times faster than normal, was recovered. Each of these mutations represents a single gene defect that affects a specific form of behavior. Because of the method used to recover them, all

NOW SOLVE THIS

24–2 Assume that you discovered a fruit fly that walked with a limp and was continually off balance as it moved. Describe how you would determine whether this behavior was due to an injury (induced in the environment) or to an inherited trait. Assuming that it was inherited, what are the various ways a gene might lead to this trait? Describe how you would determine the mechanism of gene expression experimentally if the gene were X-linked.

■ HINT: *This problem involves an understanding of how traits are inherited and transmitted to offspring in Mendelian crosses and the expected outcome of crosses involving recessive, dominant, and X-linked genes. The key to its solution is to first determine whether your analysis will use the behavior-first or the gene-first method.*

TABLE 24.2

Some *Drosophila* Mutations That Affect Learning and Memory

Behavior Genes	Molecular Function	Behavior Phenotype, Function
dunce (dnc)	cAMP-specific phosphodiesterase	Locomotor rhythms, ethanol tolerance
rutabaga (rut)	Adenylate cyclase	Courtship learning, ethanol tolerance, grooming
amnesiac (amn); cheapdate (chpd)	Neuropeptide	Ethanol tolerance
latheo (lat)	DNA-replication factor	Larval feeding
Shaker (Sh)	Voltage-sensitive potassium channel	Courtship suppression, gestation defect, ether sensitivity
G proteins 60 A	Heterotrimeric G protein	Visual behavior, cocaine sensitivity
DCO; cAMP-dependent protein kinase I (Pka-C1)	Protein Ser/Thr kinase	Locomotor rhythms, ethanol tolerance
cAMP-response-element-binding protein B at 17A	Transcription factor	Locomotor rhythms *(CrebB17A); dCREB*
Calcium/calmodulin-dependent protein kinase II (CaMKII)	Protein Ser/Thr kinase	Courtship suppression
Neurofibromatosis 1 (Nf1)	Ras GTPase activator	Embryonic, nervous system defects

Source: Adapted by permission from Macmillan Publishers Ltd: *Nature Reviews Genetics*, "Drosophila: Genetics Meets Behavior," by M.B. Sokolowski. *Nature Reviews Genetics* 2:879–890, table 1, p. 882. Copyright 2001, www.nature.com/nrg.

the mutants found so far are X-linked genes. Using other methods, autosomal genes that control learning and memory have been identified.

Dissecting the Mechanisms and Neural Pathways in Learning

Screening for learning and memory-defective *Drosophila* was successful in identifying a number of single genes involved in these behaviors. Some of these mutants acted in different stages of the same biochemical pathway, leading to the identification of neural mechanisms and pathways involved in learning and memory. For example, one group of *Drosophila* learning mutations is defective in a signal transduction system that plays an important role in learning.

In cells of the nervous system and many other cell types, cyclic adenosine monophosphate (cAMP) is produced from adenosine triphosphate (ATP) in the cytoplasm in response to signals received at the cell surface (see Figure 24–7). Once produced, cyclic AMP (cAMP) activates protein kinases, which, in turn, phosphorylate proteins, transfer the signal to the nucleus, and initiate a cascade of changes in gene expression.

Behavioral mutants of *Drosophila* were among the first to show the link between cAMP metabolism and learning. The *rutabaga (rut)* locus encodes adenylyl cyclase, the enzyme that synthesizes cAMP from ATP. In a mutant allele of *rutabaga*, a missense mutation destroys the catalytic activity of adenylyl cyclase, leading to learning deficiency in homozygous flies. The *dunce (dnc)* locus encodes the structural gene for the enzyme cAMP phosphodiesterase, which degrades adenylyl cyclase. Mutant alleles of *dunce* also lead to learning deficiencies. The *turnip (tur)* mutation occurs in a gene encoding a G protein, a class of molecules that bind guanosine-triphosphate (GTP) and, in turn, activate adenylyl

cyclase. Analysis of these mutations also identified parts of the nervous system where sensory information is processed and where both short-term and long-term memory are formed.

Cells in the *Drosophila* brain, called mushroom body neurons (Figure 24–8), integrate and process sensory input from olfactory stimuli and from electric shocks. When both stimuli arrive at a mushroom body simultaneously, adenylyl

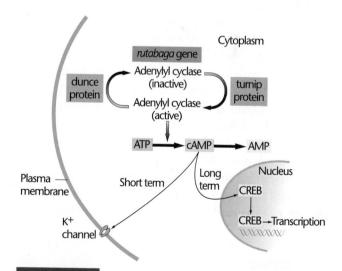

FIGURE 24–7 The pathway for cyclic AMP (cAMP) synthesis and degradation. ATP is converted into cAMP by the enzyme adenylyl cyclase; cAMP affects short-term memory via the potassium channel in the plasma membrane and affects long-term memory by regulating the transcription factor CREB. In *Drosophila*, the *rutabaga* gene encodes adenylyl cyclase. When produced, the enzyme is inactive; it is converted to its active form by a protein encoded by the *turnip* gene. The enzyme is inactivated by a protein encoded by the *dunce* gene. Mutations in any of these genes affect learning and memory in *Drosophila*.

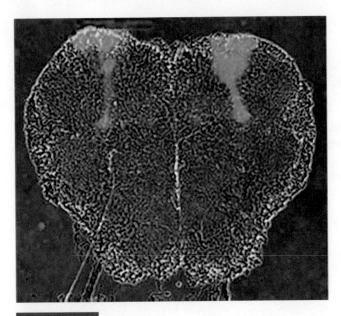

FIGURE 24–8   Green fluorescent protein (GFP) labeling of the neurons in the mushroom body of the *Drosophila* brain. Cells in this region are important in learning and memory.

cyclase is stimulated and activates a G-protein-linked receptor, which leads to elevated cAMP levels. The increase in cAMP leads to changes in the excitability of the mushroom body neuron. Elevated levels of cAMP upregulate expression of a protein kinase, which enhances short-term and long-term memory. Short-term memory is mediated by changes in potassium channels. Long-term memory is associated with changes in gene expression within mushroom body neurons, which is controlled by a transcriptional activator, CREB (cAMP response element binding protein). CREB binds to elements upstream of cAMP inducible genes and changes the pattern of gene expression, encoding memory.

The biochemical pathways that control learning and memory in *Drosophila* are similar to those in other organisms, including mice and humans. The mammalian homologs of the *Drosophila* genes are active in parts of the mouse brain involved in learning and memory. New genetic screening methods in both mice and flies are being used to identify new genes and other biochemical pathways involved in learning.

24.3
Human Behavior Has Genetic Components

The genetic control of behavior has proven more difficult to characterize in humans than in other organisms, partly because the types of responses considered to be the most interesting forms of human behavior, including aspects of *intelligence, language, personality,* and *emotion,* are difficult

to study. Two problems arise in studying such behaviors. First, these traits are difficult to define objectively and to measure quantitatively. Second, they are affected by environmental factors and there is a wide range of individual variation in the responses to these factors. In each case, the environment is extremely important in shaping, limiting, or facilitating the final phenotype.

Historically, the study of human behavior genetics has been hampered for other reasons as well. Many early studies of human behavior were conducted by psychologists with limited training in genetics. Second, traits involving intelligence, personality, and emotion have great social and political significance. Consequently, research findings concerning these traits are likely to be distorted by sensationalism when reported to the public. In fact, because the study of these traits often comes close to infringing upon individual liberties, such as the right to privacy, the studies themselves, much less their conclusions, very frequently stir up controversy.

In lamenting the gulf between psychology and genetics in the study of human behavior, C. C. Darlington wrote in 1963, "Human behavior has thus become a happy hunting ground for literary amateurs. And the reason is that psychology and genetics, whose business it is to explain behavior, have failed to face the task together." Since 1963, some progress has been made in bridging this gap, but the genetics of human behavior remains somewhat controversial.

Here we will consider three approaches to the study of human behavior genetics: analysis of single genes with behavioral components, the use of animal models, and genomic methods. Like other areas of behavioral genetics, the study of human behavior depends on the analysis of single genes, as well as on the investigation of complex, multigenic traits with environmental components.

Single Genes and Behavior: Huntington Disease

Neurodegenerative disorders affect millions of people in the United States and across the world. These diseases are associated with the progressive accumulation of misfolded proteins in intra- and extracellular spaces, leading to the death of brain cells, with behavioral consequences. Some of these, such as familial Alzheimer disease (AD), Parkinson's disease, amyotrophic lateral sclerosis (ALS), and **Huntington disease (HD),** can be caused by mutations in single genes.

HD is inherited as an autosomal dominant disorder that affects about 1 in 10,000 people, causing cell death in selected brain cells. Symptoms usually appear in the fifth decade of life as a gradual loss of motor function and coordination. As structural degeneration of the brain progresses, personality changes occur. The symptoms are caused by brain-cell death in specific brain regions, including the cerebral cortex and the striatum. Most victims die within 10 to 15 years after the onset of the disease.

The symptoms of HD usually appear between the ages of 35 and 45, generally after someone has had children, who then must live with the knowledge that they face a 50 percent probability of developing the disorder (affected individuals are usually heterozygotes). The *HD* gene, on chromosome 4, was one of the first genes to be located by using restriction fragment length polymorphism (RFLP) analysis. The gene encodes a large protein (350 kDa) called huntingtin (Htt) that is essential for the survival of certain neurons in adult brains. Mutant HD alleles have an expanded number of cytosine-adenine-guanine (CAG) trinucleotide repeats in exon 1 (see Chapter 15 for a discussion of trinucleotide repeats). Normal alleles of the *HD* gene carry 7 to 34 CAG repeats, which specify insertion of multiple copies of the amino acid glutamine near the amino terminus of the huntingtin protein. In mutant alleles, expansion of the CAG repeats places additional glutamine residues in the encoded protein. The addition of more than 40 CAG repeats and the additional glutamine residues cause Htt to become toxic and cause cell death in certain regions of the brain, leading to behavioral abnormalities.

Transgenic animal models of neurodegenerative diseases caused by single-gene mutations, including AD, ALS, and HD, have been developed. These model systems are used to study the molecular events of synthesis, processing, and aggregation of mutant proteins, and the links between protein accumulation, cell death, and behavioral changes.

A Transgenic Mouse Model of Huntington Disease

Several animal models of HD have been developed to study the normal function of huntingtin, to model the disease process and study its effects at the molecular level, and to develop drugs that slow or stop the degeneration and death of brain cells. To study the relation between CAG repeat length and disease progression, Danilo Tagle and his colleagues constructed transgenic mice carrying a human *HD* gene with 16, 48, or 89 copies of the CAG repeat. In the vector used for gene transfer, the *HD* gene was placed adjacent to a promoter and an enhancer to ensure high levels of expression (see Chapter 22 for a discussion of transgenic organisms). Mice carrying the transgene were monitored from birth to death to determine the age of onset and progression of abnormal behavioral phenotypes. Animals carrying 48- or 89-repeat *HD* genes showed behavioral abnormalities as early as 8 weeks, and by 20 weeks, these mice showed both behavioral and motor-coordination abnormalities compared with the control animals and mice carrying human *HD* genes with 16 CAG repeats.

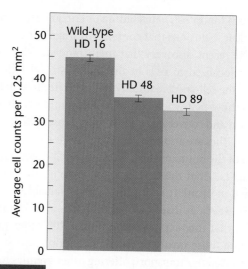

FIGURE 24–9 Relative levels of neuronal loss in HD transgenic mice. Cell counts show a significant reduction of certain neurons (small-medium neurons) in the corpus striatum of the brains of HD48 mutants (middle column) and HD89 mutants (right column) compared to wild-type (left column) mice. Cell loss in this brain tissue is also found in humans with HD, making these transgenic mice valuable models to study the course of this disease.

At various ages, brains of wild-type mice and of transgenic animals carrying mutant alleles with 16, 48, and 89 copies of the CAG repeat were examined for changes in structure. Degenerating neurons and cell loss were evident in mice carrying 48 and 89 repeats, but no changes were seen in brains of wild-type mice or of those carrying a 16-repeat transgene (Figure 24–9).

The behavioral changes and the degeneration in specific brain regions in the transgenic mice parallel the progression of HD in humans. These mice are now used to study early changes in brain structure that occur before the onset of early symptoms, and in the development of experimental treatments to slow or reverse cell loss. Having a mouse model for the disease allows researchers to administer treatment at specific times in disease progression and to evaluate the outcome of treatments in the presymptomatic stages of HD.

Mechanisms of Huntington Disease

Although the *HD* gene was mapped, isolated, and characterized over 25 years ago, the mechanism by which mutant forms of the Htt protein (mHtt) cause HD is still unknown. Htt is expressed in all cells of the body, with highest levels in the brain and testes. The mutant form of Htt causes cell death initially in striatal cells in the brain. Several mechanisms have been proposed to explain the region-specific pattern of cell death, including abnormal mitochondrial metabolism, increased activity of proteases associated with cell death (apoptosis), misfolding of Htt, and the formation of nuclear inclusions that contain Htt fragments.

The recent discovery that mHtt binds to Rhes, a small protein localized to cells in the brain's striatal region may explain the cell-specific toxicity of the mHtt protein. Experiments show that when bound to Rhes, mHtt is chemically modified and converted to a form that causes cell death. Low levels of Rhes are also present in the cerebral cortex, another region of the brain in which Htt causes cell death. This finding suggests that the development of drugs that block the interaction of mHtt with Rhes may be valuable in treating HD. However, as discussed below, Htt is a multifunctional protein that interacts with more than 100 other proteins, including those involved with transcription, cell signaling, and intracellular transport, offering other avenues for treatment. Recent work has focused on the role of mHtt in the disruption of transcriptional regulation.

In affected individuals, onset of HD is associated with decreases in mRNA levels for genes encoding certain neurotransmitter receptors. In a transgenic HD mouse model, researchers used cDNA arrays to screen nearly 6000 mRNAs from striatal cells. They found lower levels of mRNA from a small set of genes involved in signaling pathways that are critical to striatal cell function. Similar results were found in another transgenic HD mouse model and are consistent with findings from HD patients, implicating aberrant gene expression as one of the underlying causes of HD. The mutant form of Htt interacts with and binds to several transcription factors and may alter patterns of gene expression by making these proteins unavailable. Some proteins that interact with Htt have histone-modifying activity and are found in Htt protein aggregates in brains of transgenic HD mice and patients with HD.

Using transgenic HD mice, several groups showed that drugs which alter histone modification increased survival and reduced cell death. Similar drugs are now being used in human clinical trials on HD patients, and other drugs are being screened in mouse models for therapeutic effects.

Complex Behavioral Traits: Schizophrenia

Although genomic techniques have been used successfully to identify mutant alleles associated with single-gene behavioral disorders (such as *HD*), other behaviors are polygenic with strong environmental components, making their genetics more difficult to dissect. For example, schizophrenia is a brain disorder affecting about 1 percent of the population and is an example of a complex genetic disease. **Schizophrenia** is a collection of mental disorders characterized by avoidance of social contact and by bizarre and sometimes delusional behavior. Those affected by the

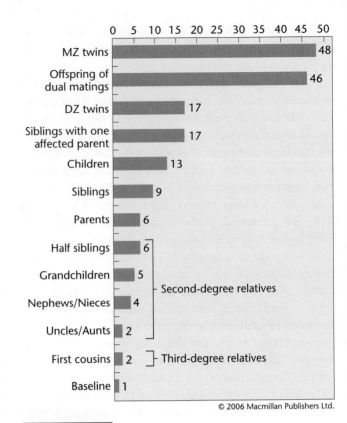

FIGURE 24–10   Relative lifetime risk of schizophrenia for nonrelated individuals and for relatives of a schizophrenic proband.

condition are unable to lead normal lives and are periodically disabled by its symptoms. It is clearly a familial disorder, with relatives of schizophrenics having a much higher incidence of the condition than the general population (Figure 24–10). Furthermore, the closer the genetic or biological relationship to an affected individual, the greater is a person's probability of developing the disorder. Twin studies have helped establish a genetic link to schizophrenia. If both twins express a trait, they are said to be concordant for that trait; if one expresses the trait, but the other does not, they are discordant for that trait. In studies of monozygotic and dizygotic twins for schizophrenia, concordance is higher in monozygotic twins than in dizygotic twins. Although these results suggest that a genetic component exists, they do not reveal the precise genetic basis of schizophrenia. Other studies emphasize the importance of the effects of pleiotropy, epistasis, as well as shared environmental factors, in the development of this disorder. The current view is that schizophrenia is a genetically mediated but not genetically determined neurodevelopmental disorder. The genetic contributions to schizophrenia have been difficult to identify. Simple monohybrid inheritance, dihybrid inheritance, and multiple-gene control have been discounted as causes of schizophrenia. It does not seem likely that only one or two loci are involved, nor is it likely that the disorder is strictly quantitative, as in polygenic inheritance. Schizophrenia is currently regarded as a multifactorial and quantitative trait, influenced by genes, by the environment, and by interactions between the genotype and environmental factors.

The search for the role of genes in schizophrenia parallels the development of technologies that allow new ways of searching for genes associated with complex traits. Beginning in the late 1980s, linkage studies with RFLP markers identified chromosome regions that segregated in families with affected members. By 2000, a small number of candidate genes had been identified on chromosomes 6, 8, and 13. The role of some of the encoded proteins in nerve impulse transmission supports the idea that these genes may play a role in schizophrenia, although there is some controversy about their importance in the disease process.

New experimental methods developed during and after the Human Genome Project have used haplotype mapping and high-throughput sequencing approaches to study schizophrenia. Foremost among these are genome-wide association studies (GWAS) that utilize some of the more than 10 million single-nucleotide polymorphisms (SNPs) present in the human genome (SNPs are discussed in Chapter 21). Subsets of 300,000 to 500,000 SNPs organized in closely linked clusters called **haplotypes** spread across the genome are used with sequencing platforms to survey thousands of genomes, looking for an association between haplotypes and a trait, in this case, schizophrenia (Figure 24–11). The haplotypes are markers that identify

chromosome regions that may contain genes associated with schizophrenia.

The evidence from GWAS on schizophrenia indicates that no single gene or allele makes a significant contribution to this disorder. Instead, the results point to the involvement of hundreds of genes that each contributes only a small amount to schizophrenia. GWAS are based on the assumption that common gene variants (present in more than 5 percent of the population) contribute to disease risk (the common disease/common variant hypothesis, abbreviated as CDCV). In this case, common variants identified by GWAS contribute only about 4 to 30 percent of the risk for schizophrenia.

The failure of GWAS to identify the majority of genes that are risk factors for schizophrenia and other disorders has generated a debate among human geneticists. Some argue that by greatly increasing the number of genomes scanned, GWAS will be able to identify many low-effect genes for a specific disorder. Others point out that there is no way to know the sample size needed to identify all genes that contribute to any disorder and in fact, GWAS may be the wrong strategy. In this view, it is thought that searching for rare allele variants that may underlie relative common disorders (the common disease/rare variant (CDRV) hypothesis) may be the best strategy to identify alleles responsible for many complex disorders. Large-scale whole-genome resequencing to identify allelic variants associated with gene–phenotype associations are proposed as a means of identifying these variants. Still others argue that other mechanisms, such as gene–gene interactions, protein–protein interactions, epigenetic alterations, or copy number variations (CNVs), may be important in schizophrenia and other psychiatric disorders.

A recent study showed that *de novo* CNVs representing both deletions and duplications were three to four times more frequent in cases of schizophrenia than in normal individuals. Many genes in these CNVs are involved in brain development, providing clues to the cellular basis of schizophrenia. Following this study, other work showed that CNVs distributed in other regions of the genome are disproportionately represented in schizophrenia; this finding emphasizes the need for the use of several techniques in dissecting this and other complex behavioral disorders.

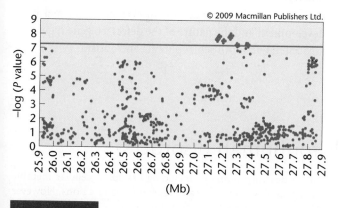

© 2009 Macmillan Publishers Ltd.

(Mb)

FIGURE 24–11 The results of a genome-wide association study for the association of SNPs and schizophrenia on chromosome 6p22.1. SNPs above the statistical threshold for association (the red line) are shown as red diamonds. Seven SNPs spanning 209 kb of DNA show significance. Red circles represent SNPs at the threshold of statistical significance, and green circles indicate SNPs with no association. The X axis marks the location of the SNPs in million-base pairs (Mb) within the 6p22.1 region.

Schizophrenia and Autism Are Related Neurodevelopmental Disorders

Like schizophrenia, family studies and twin studies show that autism spectrum disorders (ASD) are complex, multifactorial diseases with a genetic component. It is estimated that ASD affects about 3.4 per 1000 children between the ages of 3 and 10 years. These disorders are characterized by severe and pervasive deficits in (1) thinking, (2) language skills, (3) feelings, and (4) social interaction. These disorders range from a severe form called autism through several intermediates to a mild form called Asperger's syndrome.

Some rare mutations and chromosomal rearrangements are associated with ASD, but taken together, account for only about 5 to 15 percent of all cases. GWAS results have identified SNPs associated with susceptibility to ASD. Some of these SNPs are also risk factors for schizophrenia, indicating that these disorders may be related. Other studies reveal the existence of rare copy number variants (CNVs) associated with the development of both disorders.

CNVs covering seven genomic regions have been linked to both schizophrenia and ASD (**Table 24.3**). In two of these regions, deletions are associated with an increased risk of autism, and duplications are associated with an increased risk of schizophrenia (16p11.2, 22q13.3). In two others (1q21.1, 22.q11.21), deletions increase the risk of schizophrenia, and duplications are associated with ASD. The region at 16p13.1 supports a model of overlap between the two conditions.

Further work has shown that genes in these chromosomal regions play important roles in the development of brain cells, the maturation of synapses, and the interconnection of neurons. These genes have been mapped into functional pathways using methods of systems biology, and reveal interconnected nodes of gene interaction that are important in the formation and function of parts of the nervous system.

Expression profiles show that schizophrenia and ASD are disorders that represent opposite sides of the same coin. ASD is associated with increased activity in these developmental pathways leading to cellular overgrowth and increased brain and head size, while schizophrenia is characterized by a reduced activity in these pathways, resulting in developmental undergrowth and reduced brain and head size. Mapping of other genes in these chromosomal regions into functional networks should advance our knowledge of normal brain development and the nature of the defects in gene regulation associated with both schizophrenia and autism.

TABLE 24.3

Frequencies of CNV deletions or duplications associated with autism or schizophrenia

CNV Locus	Condition	Deletion Cases	Duplication Cases
1q21.1	Autism	2	**10**
	Schizophrenia	**15**	4
15q13.3	Autism	3	2
	Schizophrenia	**10**	4
16p11.2	Autism	**14**	5
	Schizophrenia	5	**24**
16p13.1	Autism	0	**3**
	Schizophrenia	8	**23**
17p12	Autism	4	1
	Schizophrenia	**8**	0
22q11.21	Autism	1	**8**
	Schizophrenia	**16**	1
22q13.3	Autism	**5**	0
	Schizophrenia	0	**4**

Entries in bold indicate that the CNV is statistically documented as a risk factor for the condition specified.
From: Crespi, B, et al., 2010. Comparative genomics of autism and schizophrenia. Proc. Nat. Acad. Sci. 107: 1736–1741; their Table 1, page 1737.
Contact: crespi@sfu.ca

CASE STUDY | Primate models for human disorders

Huntington disease (HD) is a dominantly inherited, adult-onset neurodegenerative disorder that affects about 1 in 10,000 people. Onset is around 50 years of age, there is no treatment, and the disorder inevitably causes dementia and loss of motor control, and is fatal. To study the cellular and molecular mechanisms associated with HD, animal models, including mice and *Drosophila*, have been created by transferring mutant human *HD* genes into these organisms. However, some of the behavioral changes seen in HD mice and fruit flies are difficult to correlate with those seen in affected humans. To more closely replicate the cognitive and motor changes in HD, researchers have created a primate model using rhesus macaques, which are one of our closest relatives and share many similarities with humans, including life span, cellular metabolism, as well as endocrine and reproductive functions. However, their use raises several important questions.

1. What are the ethical concerns related to research with primate models of human disease?

2. Does it seem likely that results from using rhesus macaques will translate directly to an understanding of the disease process in humans and lead to treatment?

3. If the primate HD model is successful, should it be used for other human behavioral disorders such as Alzheimer or Parkinson disease?

HomoloGene: Searching for Behavioral Genes

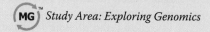 *Study Area: Exploring Genomics*

This chapter discussed some of the complexities of understanding behavior—in humans and other species—as a product of both genes and environment. In this exercise, we will explore the NCBI database **HomoloGene** to search for genes implicated in behavioral conditions. HomoloGene is a database of homologs from several eukaryotic species, and it contains genes that have been implicated in behavioral conditions both in animal models and in humans. Some of these are mutant genes, which, on the basis of single-nucleotide polymorphisms or linkage mapping, are thought to contribute to a behavioral phenotype. Keep in mind that correlation of a gene with a particular behavioral condition does not necessarily mean that the gene is the cause of the condition. Nor does it mean that the gene is the only one involved in the behavioral condition or that the condition is entirely

due to genetics. Nonetheless, exploring HomoloGene is an interesting way to search for putative behavioral genes and learn more about them.

■ Exercise I – HomoloGene

1. Access **HomoloGene** from the ENTREZ site at http://www.ncbi.nlm.nih.gov/gquery/ gquery.fcgi.

2. Search HomoloGene using the following list of terms related to behavioral conditions in humans. What human genes are listed in the top two categories of gene homologs that result from the HomoloGene search for each condition? Name these genes and identify their chromosomal loci. Notice the list of other species thought to share homologs of each gene.

   a. alcoholism

   b. depression

   c. nicotine addiction

   d. seasonal affective disorder

   e. intelligence

3. For each of the top two categories of homologs, click on the category title to open a report on those genes. Then use the links for individual genes to learn more about the functions of those genes. In addition to other good resources, the "Additional Links" category at the bottom of each gene report page will provide you with access to the Online Mendelian Inheritance in Man (OMIM) site that we have used for other exercises.

4. Search HomoloGene for any behavioral conditions you are interested in to see if the database contains genes that may be implicated in those conditions.

Summary Points

 For activities, animations, and review quizzes, go to the study area at www.masteringgenetics.com

1. Behavior is a multifactorial trait controlled by genes, environmental factors, and interactions among these genes and the environment.

2. The behavior-first approach can generate strains that exhibit specific behaviors but cannot be used to identify genes associated with those behaviors.

3. The gene-first approach can be used to analyze the effect of single genes on complex behaviors that are controlled by multiple

genes, but is difficult to use in the study of naturally-occurring variation.

4. Because of the large number of genetic resources available, *Drosophila* and other organisms are useful or the study of behavioral disorders in humans.

5. Genomic technology has opened new avenues of investigation into the development of the human nervous system and the mechanisms of behavioral disorders.

INSIGHTS AND SOLUTIONS

1. Bipolar disorder is a mental illness associated with pervasive and wide mood swings. It is estimated that 1 in 100 individuals has or will suffer from this disorder at least once in his or her lifetime. Genetic studies indicate that bipolar disorder is familial and that mutations in several genes are involved. In an attempt to identify genes associated with bipolar disorder (this condition was formerly called manic depression), researchers began by using conventional genetic analysis. In 1987, two separate studies using pedigree analysis, RFLP analysis, and a variety of other genetic markers reported a linkage between manic depression and, in one report, markers on the X chromosome and, in the other report, markers on the short arm of chromosome 11. At the time, these reports were hailed as landmark discoveries. It was thought that genes controlling this behavioral disorder could be identified, mapped, and sequenced in much the same way as genes for cystic fibrosis, neurofibromatosis, and other disorders. It was hoped that this discovery would open the way to the development of therapeutic strategies based on knowledge of the nature of the gene product and its mechanism of action. One of the challenges of human genetics results from the fact that some disorders show variable times of onset, and the bipolar studies provide an example of this. Following the original analysis identifying markers, some of the controls (unaffected individuals) who did not carry the markers subsequently developed bipolar disorder. This outcome weakened the evidence for linkage. These findings do not exclude the role of major genes on the X chromosome and chromosome 11 in bipolar disorder, but they do exclude linkage to the markers used in the original reports. This setback not only caused embarrassment and confusion, but also forced a reexamination of the validity of the newly developed recombinant DNA-based methods used in mapping other human genes and of the analysis of data from linkage studies involving complex behavioral traits. Several factors have been proposed to explain the flawed conclusions reported in the original studies. What do you suppose some of these factors were?

Solution: Although some of the criticisms were directed at the choice of markers, most of the factors at work in this situation appear to be related to the phenotype of manic depression. At least three confounding factors have been identified. One factor relates to the diagnosis of bipolar disorder itself. The phenotype is complex and not as easily quantified as height or weight. In addition, mood swings are a universal part of everyday life, and it is not always easy to distinguish transient mood alterations and the role of environmental factors from disorders having a biological or genetic basis (or both). A second confounding factor is that, in the populations studied, there may in fact be more than one major X-linked gene and more than one major autosomal gene that trigger bipolar disorder. A third factor is age of onset. There is a positive correlation between age and the appearance of the symptoms of bipolar disorder. Therefore, at the time of a pedigree study, younger individuals who will be affected later in life may not show any signs of bipolar disorder. Taken together, these factors illustrate the difficulty in researching the genetic basis of complex behavioral traits. Individually or in combinations, the factors described might skew the results enough so that the guidelines for proof that are adequate for other traits are not stringent enough for behavioral traits with complex phenotypes and complex underlying causes.

Problems and Discussion Questions

 For instructor-assigned tutorials and problems, go to www.masteringgenetics.com

HOW DO WE KNOW?

1. In this chapter we focused on how genes that control the development, structure, and function of the nervous system and interactions with environmental factors produce behavior. At the same time, we found many opportunities to consider the methods and reasoning by which much of this information was acquired. From the explanations given in the chapter, what answers would you propose to the following fundamental questions:
   (a) How do we know there is genetic variation for a specific behavior that is present in strains of a species?
   (b) How do we know that geotaxis in *Drosophila* is under genetic control?
   (c) How do we know that *Drosophila* can learn and remember?
   (d) How do we know how the mutant gene product in Huntington disease functions?
   (e) How do we know that schizophrenia has genetic components?

2. Contrast the two principal approaches used in studying behavioral genetics. What are the advantages and disadvantages of each method with respect to the type of information gained?

3. In humans, the chemical phenylthiocarbamide (PTC) is either tasted or not. When the offspring of various combinations of taster and nontaster parents are examined, the following data are obtained:

Parents	Offspring
Both tasters	All tasters
Both tasters	1/2 tasters
	1/2 nontasters
Both tasters	3/4 tasters
	1/4 nontasters
One taster and One nontaster	All tasters
One taster One nontaster	1/2 tasters 1/2 nontasters
Both nontasters	All nontasters

Based on these data, how is PTC tasting behavior inherited?

4. Jerry Hirsch and colleagues have used the behavior-first approach to study the genes responsible for geotaxis in *Drosophila*. Their elegant genetic analysis has shown that this behavior is under genetic control, and they estimate the number of genes responsible for geotactic behavior. Although the approach has been successful, what are the limitations of the behavior-first approach as used here? What steps did Hirsch and his colleagues take to overcome those limitations? Has this approach been successful?

5. Various approaches have been applied to study the genetics of problem and pathological gambling (PG), and within-family vulnerability has been well documented. However, family studies, while showing clusters within blood relatives, cannot separate genetic from environmental influences. Eisen (2001) applied "twin studies" using 3359 twin pairs from the Vietnam-era Twin Registry and found that a substantial portion of the variance associated with PG can be attributed to inherited factors. How might twin studies be used to distinguish environmental from genetic factors in complex behavioral traits such as PG?

6. *Caenorhabditis elegans* has become a valuable model organism for the study of development and genetics for a variety of reasons. The developmental fate of each cell (1031 in males and 959 in hermaphrodites) has been mapped. *C. elegans* has only 302 neurons whose pattern of connectivity is known, and it displays a variety of interesting behaviors including chemotaxis, thermotaxis, and mating behavior, to name a few. In addition, isogenic lines have been established. What advantage would the use of *C. elegans* have over other model organisms in the study of animal behavior? What are likely disadvantages?

7. Using the behavioral phenotypes of a series of 18 mutants falling into five phenotypic classes, Thomas (1990) described the genetic program that controls the cyclic defecation motor program in *Caenorhabditis elegans*. The first step in each cycle is contraction of the posterior body muscles, followed by contraction of the anterior body muscles. Finally, anal muscles open and expel the intestinal contents. Below is a list of selected mutants (simplified) that cause defects in defecation.

Gene	Phenotype
exp	failure to open the anus and expel intestinal contents
unc	failure to contract anterior body muscles
pbo	failure to contract posterior body muscles
aex	failure to open the anus and expel intestinal contents and failure to contract anterior body muscles
cha	alteration of cycle periodicity, all other components being completely functional

Present a simple schematic that illustrates genetic involvement in *C. elegans* defecation. Account for the action of the *aex* gene and speculate on the involvement and placement of the *cha* gene.

8. Discuss why the study of human behavior genetics has lagged behind that of other organisms.

9. J. P. Scott and J. L. Fuller studied 50 traits in five pure breeds of dogs. Almost all the traits varied significantly in the five breeds, but very few bred true in crosses. What can you conclude with respect to the genetic control of these behavioral traits?

Extra-Spicy Problems

 For instructor-assigned tutorials and problems, go to www.masteringgenetics.com

10. Although not discussed in this chapter, *C. elegans* is a model system whose life cycle makes it an excellent choice for the genetic dissection of many biological processes. *C. elegans* has two natural sexes: hermaphrodite and male. The hermaphrodite is essentially a female that can generate sperm as well as oocytes, so reproduction can occur by hermaphrodite self-fertilization or hermaphrodite–male mating. In the context of studying mutations in the nervous system, what is the advantage of hermaphrodite self-fertilization with respect to the identification of recessive mutations and the propagation of mutant strains?

11. Hypothetical data concerning the genetic effect of *Drosophila* chromosomes on geotaxis are shown here:

| | Chromosome | | |
	X	2	3
Positive geotaxis	+0.2	+0.1	+3.0
Unselected	+0.1	−0.2	+1.0
Negative geotaxis	−0.1	−2.6	+0.1

What conclusions can you draw?

12. In July 2006, a population of flies, *Drosophila melanogaster,* rode the space shuttle *Discovery* to the International Space Station (ISS) where a number of graviperception experiments and observations were conducted over a nine-generation period. Frozen specimens were collected by astronauts and returned to Earth. Researchers correlated behavioral and physiological responses to microgravity with changes in gene activity by analyzing RNA and protein profiles. The title of the project is "*Drosophila* Behavior and Gene Expression in Microgravity." If you were in a position to conduct three experiments on the behavioral aspects of these flies, what would they be? How would you go about assaying changes in gene expression in response to microgravity? Given that humans share over half of the genome and proteins of *Drosophila*, how would you justify the expense of such a project in terms of improving human health?

13. Of a variety of investigational approaches that have been applied to the study of schizophrenia, two separate approaches, twin studies and genomic analysis, have provided strong support for a genetic component.
(a) Provide a summary of the contribution that each has played in understanding schizophrenia.
(b) Speculate on the value that genomic analysis using samples from monozygotic and dizygotic twins might have for enhancing research on schizophrenia.

14. Describe the use of single-nucleotide polymorphisms (SNPs) in the study of genetic causes of schizophrenia.

15. Autism, a relatively common complex of human disorders, can range from severe to mild. What evidence indicates a possible link between genetics and autism?

16. An interesting and controversial finding by Wedekind and colleagues (1995) suggested that a sense of smell, flavored with various HLA (Human Leukocyte Antigen) haplotypes, plays a role in mate selection in humans. Women preferred T-shirts from men with HLA haplotypes unlike their own. The HLA haplotypes are components of the Major Histocompatibility Complex (MHC) and are known to have significant immunological functions. In addition to humans, mice and fish also condition mate preference on MHC constitution. The three-spined stickleback (*Gasterosteus aculeatus*) has recently been studied to determine whether females use an MHC odor-based system to select males as mates. Examine the following data (modified from Milinski et al., 2005) and provide a summary conclusion. Why might organisms evolve a mate selection scheme for assessing and optimizing MHC diversity?

	Male with optimal MHC	Male with nonoptimal MHC
Gravid female time spent (seconds)	350	250
Number spawning females	11	3

The table compares odor-based choices made by gravid females for males with optimal MHC alleles and males with nonoptimal MHC alleles. Gravid females chose different amounts of time (seconds) for exposure to males with optimal MHC alleles versus those with nonoptimal MHC alleles. The number of females spawning when exposed to different males with different MHC complements is also presented in the table.

These lady-bird beetles, from the Chiricahua Mountains in Arizona, show considerable phenotypic variation.

25

Population and Evolutionary Genetics

n the mid-nineteenth century, Alfred Russel Wallace and Charles Darwin identified natural selection as the mechanism of evolution. In his 1859 book, *On the Origin of Species,* Darwin presented comprehensive evidence that populations and species are not fixed, but change, or evolve, over time as a result of natural selection. However, Wallace and Darwin could not explain either the origin of the variations that provide the raw material for evolution or the mechanisms by which such variations are passed from parents to offspring. Gregor Mendel published his work on the inheritance of traits in 1866, but it received little notice at the time. The rediscovery of Mendel's work in 1900 began a 30-year effort to reconcile the concept of genes and alleles with the theory of evolution by natural selection. As twentieth-century biologists applied the principles of Mendelian genetics to populations, both the source of variation (mutation) and the mechanism of inheritance (segregation of alleles) were explained. We now view evolution as a consequence of changes in alleles and allele frequencies in populations over time. This union of population genetics with the theory of natural selection generated a new view of the evolutionary process, called *neo-Darwinism.*

In addition to natural selection, other forces including mutation, migration, and drift, individually and collectively, alter allele frequencies and bring about evolutionary divergence that eventually may result in **speciation**, the formation of new species. Speciation is facilitated by environmental diversity. If a population is spread over a geographic range encompassing a number of ecologically distinct subenvironments with different selection pressures, isolated populations occupying these areas may gradually adapt and become genetically distinct from one another. Genetically differentiated populations may remain in existence, become extinct, reunite with each other, or continue to diverge until they become reproductively isolated. Populations that are reproductively isolated are regarded as separate species. Genetic changes within populations can modify a species over time, transform it into another species, or cause it to split into two or more species.

Population geneticists investigate patterns of genetic variation within and among groups of interbreeding individuals. Because mutations change the genetic structure of populations and form the basis for evolutionary change, population genetics has become an important subdiscipline of evolutionary biology. In this chapter, we examine the processes of **microevolution**—defined as evolutionary change within populations of a species—and then consider how molecular aspects of these processes can be extended to **macroevolution**—defined as evolutionary events leading to the emergence of new species and other taxonomic groups.

25.1

Genetic Variation Is Present in Most Populations and Species

A **population** is a group of individuals belonging to the same species that live in a defined geographic area and actually or potentially interbreed. In thinking about the human population, we can define it as everyone who lives in the United States, or in Sri Lanka, or we can specify a population as all the residents of a particular small town or village.

The genetic information carried by members of a population constitutes that population's **gene pool**. At first glance, it might seem that a population that is well adapted to its environment must be highly homozygous because you assume that the most favorable allele at each locus is present at a high frequency. In addition, a look at most populations of plants and animals reveals many phenotypic similarities among individuals. However, a large body of evidence indicates that, in reality, most populations contain a high degree of heterozygosity. This built-in genetic diversity is often concealed because it is not necessarily apparent phenotypically; hence, detecting it is not a simple task. Nevertheless, the diversity within a population can be revealed by several methods.

Detecting Genetic Variation by Artificial Selection

One way to determine whether genetic variation affects a phenotypic character is to use artificial selection. A phenotype that is not associated with variation in the genes that control the phenotype will not respond to selection; if genetic variation does exist, the phenotype can change over a few generations. A dramatic example of this test is the domestic dog. The broad array of sizes, shapes, colors, and behaviors seen in different breeds of dogs all arose from the effects of selection on the genetic variation present in wild wolves, from which all domestic dogs are descended. Genetic and archaeological evidence indicates that the domestication of dogs took place at least 15,000 years ago and possibly much earlier. On a shorter time scale, laboratory selection experiments on the fruit fly *Drosophila melanogaster* have generated significant changes over a few generations in almost every phenotype imaginable, including size, shape, developmental rate, fecundity, and behavior.

Variations in Nucleotide Sequence

The most direct way to estimate genetic variation is to compare the nucleotide sequences of genes carried by individuals in a population. To do this, Martin Kreitman analyzed variation in the *alcohol dehydrogenase* gene (*Adh*) in *Dro-*

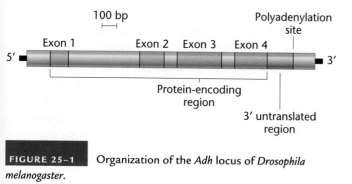

FIGURE 25-1 Organization of the *Adh* locus of *Drosophila melanogaster*.

Consensus *Adh* sequence:	Exon 3 C C C C	Intron 3 G G A A T	Exon 4 C T C C A*C T A G
Strain			
Wa-S	T T • A	C A • T A	A C • • • • • • •
Fl1-S	T T • A	C A • T A	A C • • • • • • •
Ja-S	• • • •	• • • • •	• • • T • T • C A
Fl-F	• • • •	• • • • •	• • G T C T C C •
Ja-F	• • A •	• • G • •	• • G T C T C C •

© 1983 Macmillan Publishers Ltd.

FIGURE 25-2 DNA sequence variation in parts of the *Drosophila Adh* gene in a sample of the 11 laboratory strains derived from the five natural populations. The dots represent nucleotides that are the same as the consensus sequence; letters represent nucleotide polymorphisms. An A/C polymorphism (A*) in codon 192 creates the two *Adh* alleles (F and S). All other polymorphisms are silent or noncoding.

sophila melanogaster (Figure 25–1). At the protein level, this gene has two alleles, *Adh-f* and *Adh-s*. The encoded proteins differ by only a single amino acid (thr versus lys at codon 192). To determine whether the amount of genetic variation detectable at the protein level (one amino acid difference) corresponds to the variation at the nucleotide level, Kreitman cloned and sequenced *Adh* genes from five natural populations of *Drosophila* (Figure 25–2).

The 11 cloned genes isolated from these five populations contained a total of 43 nucleotide variations in the *Adh* sequence of 2721 base pairs. These variations are distributed throughout the gene: 14 in exon-coding regions, 18 in introns, and 11 in the untranslated flanking regions. Of the 14 variations in coding regions, only one leads to an amino acid replacement—the one in codon 192, producing the two alleles. The other 13 coding-region nucleotide substitutions do not change the encoded amino acid, and as such, are silent variations in this gene.

As another example, consider the *CF* gene, which encodes the cystic fibrosis transmembrane conductance regulator (CFTR), one of the most intensively studied human genes. Recessive loss-of-function mutations in the *CFTR* gene cause **cystic fibrosis**, a disease that affects secretory glands

and lungs, leading to susceptibility to bacterial infections. Almost 1800 different mutations in the *CFTR* gene have been identified. Among these are missense mutations, amino acid deletions, nonsense mutations, frameshifts, and splice defects.

Figure 25–3 shows a map of the 27 exons in the *CFTR* gene, with many exons identified by function. The histogram above the map shows the locations of some of the disease-causing mutations and the number of copies of each mutation that have been identified. One mutation, a 3-bp deletion in exon 10 called Δ*F508* accounts for 67 percent of all mutant cystic fibrosis alleles, but several other mutations were present in at least 100 of the chromosomes surveyed. In populations of European ancestry, between 1 in 44 and 1 in 20 individuals are heterozygous carriers of mutant alleles. Note that Figure 25–3 includes only the sequence variants that alter the function of the CFTR protein. There are

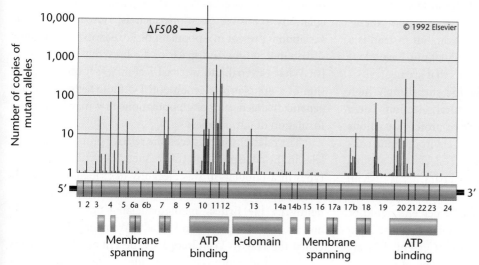

FIGURE 25-3 The locations of disease-causing mutations in the cystic fibrosis gene. The histogram shows the number of copies of each mutation geneticists have found. (The vertical axis is on a logarithmic scale.) The genetic map below the histogram shows the locations and relative sizes of the 27 exons of the *CFTR* locus. The boxes at the bottom indicate the functions of different domains of the CFTR protein.

undoubtedly many more *CFTR* alleles with silent sequence variants that do not change the amino acid sequence of the protein and do not affect its function.

Studies of other organisms, including the rat, the mouse, and the mustard plant *Arabidopsis thaliana,* have produced similar estimates of nucleotide diversity in various genes. The results point to the presence of an enormous reservoir of genetic variability within most populations and show that, at the DNA level most, and perhaps all, genes exhibit diversity from individual to individual. Alleles representing these variations are distributed among members of a population.

Explaining the High Level of Genetic Variation in Populations

The finding that populations harbor considerable genetic variation at the amino acid and nucleotide levels came as a surprise to many evolutionary biologists. The early consensus had been that selection would favor a single optimal (wild-type) allele at each locus and that, as a result, populations would have high levels of homozygosity. This expectation was shown conclusively to be wrong, and considerable research and argument has ensued concerning the forces that maintain such high levels of genetic variation.

The **neutral theory** of molecular evolution, proposed by Motoo Kimura in 1968, proposes that mutations leading to amino acid substitutions are usually detrimental, with only a very small fraction being favorable. Some mutations are neutral, that is, they are functionally equivalent to the allele they replace. Mutations that are favorable or detrimental are preserved or removed from the population, respectively, by natural selection. However, the frequency of the neutral alleles in a population will be determined by mutation rates and random genetic drift, and not by selection. Some neutral mutations will drift to fixation in the population; other neutral mutations will be lost. At any given time, a population may contain several neutral alleles at any particular locus. The diversity of alleles at most loci does not, therefore, reflect the action of natural selection, but instead is a function of population size (larger populations have more variation) and the fraction of mutations that are neutral.

The alternative explanation for the surprisingly high genetic variation in populations is natural selection. There are several examples in which enzyme or protein variations are maintained by adaptation to certain environmental conditions. The well-known advantage of sickle-cell anemia heterozygotes when infected by malarial parasites is such an example.

Fitness differences of a fraction of a percent would be sufficient to maintain such a variation, but at that level their presence would be difficult to measure. Current data are therefore insufficient to determine what fraction of molecular genetic variation is neutral and what fraction is subject to selection. The neutral theory nonetheless serves a crucial function: It points out that some genetic variation is expected simply as a result of mutation and drift. In addition, the neutral theory provides a working hypothesis for studies of molecular evolution. In other words, biologists must find positive evidence that selection is acting on allele frequencies at a particular locus before they can reject the simpler assumption that only mutation and drift are at work.

25.2

The Hardy–Weinberg Law Describes Allele Frequencies and Genotype Frequencies in Populations

Populations are dynamic; they expand and contract through changes in birth and death rates, migration, or contact with other populations. Often some individuals within a population will produce more offspring than others, contributing a disproportionate fraction of their alleles to the next generation. Thus, differential reproduction in a population can, over time, lead to changes in the allele and genotype frequencies in subsequent generations. Changes in allele frequencies in a population that do not result in reproductive isolation are examples of microevolution. In the following sections, we will discuss microevolutionary changes in population gene pools, and later in this chapter we will consider macroevolution and the process of speciation.

Often when we examine a single genetic locus in a population, we find that the distribution of alleles at this locus produces individuals with different genotypes. The calculation of allele frequencies and genotype frequencies in the population, and the determination of how these frequencies change from one generation to the next are key elements of population genetics. Population geneticists use these calculations to answer questions such as: How much genetic variation is present in a population? Are genotypes randomly distributed in time and space, or do discernible patterns exist? What external and internal factors affect the composition of a population's gene pool? Do these factors produce genetic divergence among populations that may lead to the formation of new species?

The relationship between the relative proportions of alleles in the gene pool and the frequencies of different genotypes in a population was elegantly described in the early 1900s in a simple mathematical model developed independently by the British mathematician Godfrey H. Hardy and the German physician Wilhelm Weinberg. This model, called the **Hardy–Weinberg Law**, describes what happens to alleles and genotypes in an "ideal" population that is infinitely large with random mating and is not subject to any evolutionary forces such as mutation, migration, or selection. Under these conditions, the Hardy–Weinberg model makes two predictions:

1. The frequencies of alleles in the gene pool do not change over time.

2. If two alleles at a locus, *A* and *a*, are considered, then as we will show later in this chapter, after one generation of random mating, the frequencies of the genotypes *AA:Aa:aa* in the population can be calculated as

$$p^2 + 2pq + q^2 = 1$$

where p = frequency of allele *A* and q = frequency of allele *a*.

A population that meets these criteria, and in which the frequencies of p and q, two alleles at a given locus result in the predicted genotypic frequencies, is said to be in Hardy–Weinberg equilibrium. As we will see later in this chapter, it is rare for a real population to conform totally to the Hardy–Weinberg model and for all allele and genotype frequencies to remain unchanged for generation after generation.

The Hardy–Weinberg model uses the Mendelian principles of segregation along with simple probability to explain the relationship between allele and genotype frequencies in a population. To demonstrate how this works, we will consider a single autosomal locus with two alleles, *A* and *a*, in a population where the frequency of *A* is 0.7 and the frequency of *a* is 0.3. Note that $0.7 + 0.3 = 1$, indicating that all the alleles for gene *A* present in the gene pool are accounted for. We assume that individuals mate randomly, following Hardy–Weinberg requirements, so for any one zygote, the probability that the female gamete will contain *A* is 0.7, and the probability that the male gamete will contain *A* is also 0.7. The probability that *both* gametes will contain *A* is $0.7 \times 0.7 = 0.49$. Thus we predict that genotype *AA* will occur 49 percent of the time. The probability that a zygote will be formed from a female gamete carrying *A* and a male gamete carrying *a* is $0.7 \times 0.3 = 0.21$, and the probability of a female gamete carrying *a* being fertilized by a male gamete carrying A is $0.3 \times 0.7 = 0.21$, so the frequency of genotype *Aa* is $0.21 + 0.21 = 0.42 = 42$ percent. Finally, the probability that a zygote will be formed from two gametes carrying *a* is $0.3 \times 0.3 = 0.09$, so the frequency of genotype *aa* is 9 percent. As a check on our calculations, note that $0.49 + 0.42 + 0.09 = 1.0$, confirming that we have accounted for all of the zygotes. These calculations are summarized in **Figure 25–4**.

We started with the frequency of a particular allele in a specific gene pool, and we calculated the probability that certain genotypes would be produced from this pool. When the zygotes develop into adults and reproduce, what will be

FIGURE 25–4 Calculating genotype frequencies from allele frequencies. Gametes represent samples drawn from the gene pool to form the genotypes of the next generation. In this population, the frequency of the *A* allele is 0.7, and the frequency of the *a* allele is 0.3. The frequencies of the genotypes in the next generation are calculated as 0.49 for *AA*, 0.42 for *Aa*, and 0.09 for *aa*. Under the Hardy–Weinberg Law, the frequencies of *A* and *a* remain constant from generation to generation.

the frequency distribution of alleles in the new gene pool? Under the assumptions of the Hardy–Weinberg Law, we assume that all genotypes have equal rates of survival and reproduction. This means that in the next generation, all genotypes contribute equally to the new gene pool. The *AA* individuals constitute 49 percent of the population, and we can predict that the gametes they produce will constitute 49 percent of the gene pool. These gametes all carry allele *A*. Similarly, *Aa* individuals constitute 42 percent of the population, so we predict that their gametes will constitute 42 percent of the new gene pool. Half (0.5) of these gametes will carry allele *A*. Thus, the frequency of allele *A* in the gene pool is $0.49 + (0.5)0.42 = 0.7$. The other half of the gametes produced by *Aa* individuals will carry allele *a*. The *aa* individuals constitute 9 percent of the population, so their gametes will constitute 9 percent of the new gene pool. All these gametes carry allele *a*. Thus, we can predict that the allele *a* in the new gene pool is $(0.5)0.42 + 0.09 = 0.3$. As a check on our calculation, note that $0.7 + 0.3 = 1.0$, accounting for all of the gametes in the gene pool of the new generation.

We have arrived where we began: with a gene pool where the frequency of allele *A* is 0.7 and the frequency of allele *a* is 0.3. For the general case of the Hardy–Weinberg model, we use variables instead of numerical values for the allele frequencies. Imagine a gene pool in which the frequency of allele *A* is p and the frequency of allele *a* is q, such that $p + q = 1$. If we randomly draw male and female gametes from the gene pool and pair them to make a zygote, the probability that both will carry allele *A* is $p \times p$. Thus, the frequency of genotype *AA* among the zygotes is p^2. The probability that the female gamete carries *A* and the male gamete carries *a* is $p \times q$, and the probability that the female gamete carries *a* and the male gamete carries *A* is $q \times p$.

Sperm

FIGURE 25–5   The general description of allele and genotype frequencies under Hardy–Weinberg assumptions. The frequency of allele A is p, and the frequency of allele a is q. After mating, the three genotypes AA, Aa, and aa have the frequencies p^2, $2pq$, and q^2, respectively.

Thus, the frequency of genotype Aa among the zygotes is $2pq$. Finally, the probability that both gametes carry a is $q \times q$, making the frequency of genotype aa in the zygotes q^2. Therefore, the distribution of genotypes among the zygotes is

$$p^2 + 2pq + q^2 = 1$$

This is summarized in Figure 25–5.

These calculations demonstrate the two main predictions of the Hardy–Weinberg Law: (1) allele frequencies in our population do not change from one generation to the next, and (2) genotype frequencies after one generation of random mating can be predicted from the allele frequencies. In other words, this population does not change or evolve with respect to the locus we have examined. Remember, however, the assumptions about the theoretical population described by the Hardy–Weinberg Law:

1. Individuals of all genotypes have equal rates of survival and equal reproductive success—that is, there is no selection.

2. No new alleles are created or converted from one allele into another by mutation.

3. Individuals do not migrate into or out of the population.

4. The population is infinitely large, which in practical terms means that the population is large enough that sampling errors and other random effects are negligible.

5. Individuals in the population mate randomly.

These assumptions are what make the Hardy–Weinberg Law so useful in population genetics research. By specifying the conditions under which the population cannot evolve, the Hardy–Weinberg Law can be used to identify the real-world forces that cause allele frequencies to change. In other words, by holding certain conditions constant, use of Hardy–Weinberg isolates the forces of evolution and allows them to be

quantified. It also can be used to identify "neutral genes" in a population gene pool—those not being operated on by the forces of evolution.

The Hardy–Weinberg Law has three additional important consequences:

1. It shows that dominant traits do not necessarily increase from one generation to the next.

2. It demonstrates that **genetic variability** can be maintained in a population since, once established in an ideal population, allele frequencies remain unchanged.

3. Using Hardy–Weinberg assumptions, if we know the frequency of just one genotype we can calculate the population frequencies of all other genotypes at that locus. This is particularly useful in human genetics because we can calculate the frequency of heterozygous carriers for recessive genetic disorders even when all we know is the frequency of affected individuals.

NOW SOLVE THIS

25–1 The ability to taste the compound PTC is controlled by a dominant allele T, while individuals homozygous for the recessive allele t are unable to taste PTC. In a genetics class of 125 students, 88 can taste PTC and 37 cannot. Calculate the frequency of the T and t alleles in this population and the frequency of the genotypes.

■ HINT: *This problem involves an understanding of how to determine allele and genotype frequencies using the Hardy–Weinberg Law. The key to its solution lies in determining which allele frequency (p or q) you must estimate first when homozygous dominant and heterozygous genotypes have the same phenotype.*

25.3

The Hardy–Weinberg Law Can Be Applied to Human Populations

To show how allele frequencies are measured in a real population, let's consider a gene that influences susceptibility to infection by HIV-1, the virus responsible for AIDS (acquired immunodeficiency syndrome). A small number of individuals who make high-risk choices (such as unprotected sex with HIV-positive partners) remain uninfected. Some of these individuals are homozygous for a mutant allele of a gene called *CCR5*.

Calculating Allele Frequency

The *CCR5* gene (Figure 25–6) encodes a protein called the C-C chemokine receptor-5, often abbreviated CCR5.

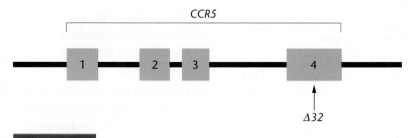

FIGURE 25–6 Organization of the *CCR5* gene in region 3p21.3. The gene contains 4 exons and 2 introns. The arrow shows the location of the 32-bp deletion in exon 4 that confers resistance to HIV-1 infection.

Chemokines are signaling molecules associated with the immune system. The CCR5 protein is also a receptor for strains of HIV-1, allowing it to gain entry to cells. The mutant allele of the *CCR5* gene contains a 32-bp deletion in a coding region, making the encoded protein shorter and nonfunctional. In individuals homozygous for this mutation, HIV-1 cannot enter their cells. The normal allele is called *CCR51* (also called *1*), and the mutant allele is called *CCR5-Δ32* (also called *Δ32*).

Homozygous *Δ32/Δ32* individuals are resistant to HIV-1 infection; heterozygotes (*1/Δ32*) are susceptible to HIV-1 infection but progress more slowly to AIDS. Table 25.1 summarizes the genotypes possible at the *CCR5* locus and the phenotypes associated with each.

The discovery of the *CCR5-Δ32* allele generates two important questions: Which human populations harbor the *Δ32* allele, and how common is it? To address these questions, teams of researchers surveyed people from a variety of populations. Genotypes were determined by direct DNA analysis (Figure 25–7). In one population, 79 individuals had genotype *1/1*, 20 were *1/Δ32*, and 1 was *Δ32/Δ32*. This population has 158 *1* alleles carried by the *1/1* individuals plus 20 *1* alleles carried by *1/Δ32* individuals, for a total of 178. The frequency of the *CCR51* allele in the sample population is thus 178/200 = 0.89 = 89 percent. Copies of the *CCR5-Δ32* allele were carried by 20 *1/Δ32* individuals, plus 2 carried by the *Δ32/Δ32* individual, for a total of 22. The frequency of the *CCR5-Δ32* allele is thus 22/200 = 0.11 = 11%. Notice that $p + q = 1$, confirming that we have accounted for the entire gene pool. Table 25.2 shows two methods for computing the frequencies of the *1* and *Δ32* alleles in the population surveyed.

Can we expect the *CCR5-Δ32* allele to increase in human populations in which it is currently rare? If these populations meet the five assumptions of the Hardy–Weinberg Law, then the frequency of the *Δ32* allele will not change. However, as we shall see in later sections of this chapter, when the assumptions of the Hardy–Weinberg Law are not met—because of natural selection, mutation, migration, or genetic drift—the allele frequencies in a population may change from one generation to the next. The Hardy–Weinberg Law tells geneticists where to look to find the causes of evolution in populations.

Testing for Hardy–Weinberg Equilibrium

One way to establish whether one (or more) of the Hardy–Weinberg assumptions does not hold in a given population is to determine whether the population's genotypes are in equilibrium. To do this, we first determine the frequencies of the genotypes, either directly from the phenotypes (if heterozygotes are recognizable) or by analyzing proteins or DNA sequences. We then calculate the allele frequencies from the genotype frequencies, as demonstrated earlier. Finally, we use the allele frequencies in the parental generation to predict the offspring's genotype frequencies. According to the Hardy–Weinberg Law, the genotype frequencies are predicted to fit the $p^2 + 2pq + q^2 = 1$ relationship. If they do not, then one or more of the assumptions are invalid for the population in question.

To demonstrate this, let's start with a population that includes 283 individuals, of which 223 have genotype *1/1*; 57 have genotype *1/Δ32*; and 3 have genotype

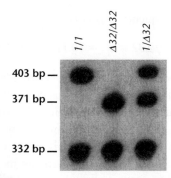

FIGURE 25–7 Allelic variation in the *CCR5* gene. Michel Samson and colleagues used PCR to amplify a part of the *CCR5* gene containing the site of the 32-bp deletion, cut the resulting DNA fragments with a restriction enzyme, and ran the fragments on an electrophoresis gel. Each lane reveals the genotype of a single individual. The *1* allele produces a 332-bp fragment and a 403-bp fragment; the *Δ32* allele produces a 332-bp fragment and a 371-bp fragment. Heterozygotes produce three bands.

TABLE 25.1

CCR5 Genotypes and Phenotypes

Genotype	Phenotype
1/1	Susceptible to sexually transmitted strains of HIV-1
1/Δ32	Susceptible but may progress to AIDS slowly
Δ32/Δ32	Resistant to most sexually transmitted strains of HIV-1

Methods of Determining Allele Frequencies from Data on Genotypes

(a) Counting Alleles

	Genotype			
	1/1	*1/Δ32*	*Δ32/Δ32*	Total
Number of individuals	79	20	1	100
Number of *1* alleles	158	20	0	178
Number of *Δ32* alleles	0	20	2	22
Total number of alleles	158	40	2	200

Frequency of *CCR51* in sample: 178/200 = 0.89 = 89%

Frequency of *CCR5-Δ32* in sample: 22/200 = 0.11 = 11%

(b) From Genotype Frequencies

	Genotype			
	1/1	*1/Δ32*	*Δ32/Δ32*	Total
Number of individuals	79	20	1	100
Genotype frequency	79/100 = 0.79	20/100 = 0.20	1/100 = 0.01	1.00

Frequency of *CCR51* in sample: 0.79 + (0.5)0.20 = 0.89 = 89%

Frequency of *CCR5-Δ32* in sample: (0.5)0.20 + 0.01 = 0.11 = 11%

Δ32/Δ32. These numbers represent genotype frequencies of $223/283 = 0.788, 57/283 = 0.201$, and $3/283 = 0.011$, respectively. From the genotype frequencies, we compute the *CCR5-1* allele frequency as 0.89 and the frequency of the *CCR5-Δ32* allele as 0.11. From these allele frequencies, we can use the Hardy–Weinberg Law to determine whether this population is in equilibrium. The allele frequencies predict the genotype frequencies as follows:

Expected frequency of genotype

$$1/1 = p^2 = (0.89)^2 = 0.792$$

Expected frequency of genotype

$$1/\Delta 32 = 2pq = 2(0.89)(0.11) = 0.196$$

Expected frequency of genotype

$$\Delta 32/\Delta 32 = q^2 = (0.11)^2 = 0.012$$

These expected frequencies are nearly identical to the observed frequencies. Our test of this population has failed to provide evidence that Hardy–Weinberg assumptions are being violated. The conclusion is confirmed by a χ^2 analysis (see Chapter 3). The χ^2 value in this case is tiny: 0.00023. To reject the null hypothesis at even the most generous, accepted level, $p = 0.05$, the χ^2 value would have to be 3.84. (In a test for Hardy–Weinberg equilibrium, the degrees of freedom are given by $k - 1 - m$, where k is the number of genotypes and m is the number of independent allele frequencies estimated from the data. Here, $k = 3$ and $m = 1$ since calculating only one allele frequency allows us to determine the other by subtraction. Thus, we have $3 - 1 - 1 = 1$ degree of freedom.)

Calculating Frequencies for Multiple Alleles in Hardy–Weinberg Populations

We commonly find several alleles of a single gene in a population. The ABO blood group in humans (discussed in Chapter 4) is such an example. Recall that the locus *I* (isoagglutinin) has three alleles I^A, I^B, and i, yielding six possible genotypic combinations ($I^A I^A$, $I^B i$, ii, $I^A I^B$, $I^A i$, $I^B i$). Remember that in this case I^A and I^B are codominant alleles and that both of these are dominant to i. The result is that homozygous $I^A I^A$ and heterozygous $I^A i$ individuals are phenotypically identical, as are $I^B I^B$ and $I^B i$ individuals, so we can distinguish only four phenotypic combinations.

By adding another variable to the Hardy–Weinberg equation, we can calculate both the genotype and allele frequencies for the situation involving three alleles. Let p, q, and r represent the frequencies of alleles I^A, I^B, and i, respectively. Note that because there are three alleles

$$p + q + r = 1$$

Under Hardy–Weinberg assumptions, the frequencies of the genotypes are given by

$$(p + q + r)^2 = p^2 + q^2 + r^2 + 2pq + 2pr + 2pq = 1$$

If we know the frequencies of blood types for a population, we can then estimate the frequencies for the three alleles of the ABO system. For example, in one population sampled, the following blood-type frequencies are observed: A = 0.53, B = 0.13, O = 0.26. Because the i allele is recessive, the population's frequency of type O blood equals the proportion of the recessive genotype r^2. Thus,

$$r^2 = 0.26$$
$$r = \sqrt{0.26}$$
$$r = 0.51$$

Using r, we can calculate the allele frequencies for the I^A and I^B alleles. The I^A allele is present in two genotypes, $I^A I^A$ and $I^A i$. The frequency of the $I^A I^A$ genotype is represented by p^2 and the $I^A i$ genotype by $2pr$. Therefore, the combined frequency of type A blood and type O blood is given by

$$p^2 + 2pr + r^2 = 0.53 + 0.26$$

TABLE 25.3

Calculating Genotype Frequencies for Multiple Alleles in a Hardy–Weinberg Population Where the Frequency of Allele $I^A = 0.38$, Allele $I^B = 0.11$, and Allele $i = 0.51$

Genotype	Genotype Frequency	Phenotype	Phenotype Frequency
$I^A I^A$	$p^2 = (0.38)^2 = 0.14$	A	0.53
$I^A i$	$2pr = 2(0.38)(0.51) = 0.39$		
$I^B I^B$	$q^2 = (0.11)^2 = 0.01$	B	0.12
$I^B i$	$2qr = 2(0.11)(0.51) = 0.11$		
$I^A I^B$	$2pr = 2(0.38)(0.11) = 0.084$	AB	0.08
ii	$r^2 = (0.51)^2 = 0.26$	O	0.26

If we factor the left side of the equation and take the sum of the terms on the right

$$(p + r)^2 = 0.79$$
$$p + r = \sqrt{0.79}$$
$$p = 0.89 - r$$
$$p = 0.89 - 0.51 = 0.38$$

Having calculated p and r, the frequencies of allele I^A and allele i, we can now calculate the frequency for the I^B allele:

$$p + q + r = 1$$
$$q = 1 - p - r$$
$$= 1 - 0.38 - 0.51$$
$$= 0.11$$

The phenotypic and genotypic frequencies for this population are summarized in Table 25.3.

Calculating Allele Frequencies for X-linked Traits

The Hardy–Weinberg Law can be used to calculate allele and genotype frequencies for X-linked traits, as long as we remember that in an XY sex-determination system, the homogametic (XX) sex will have two copies of an X-linked allele, whereas the heterogametic sex (XY) only has one copy. Thus, for mammals (including humans) where the female is XX and the male is XY, the frequency of the X-linked allele in the gene pool and the frequency of males expressing the X-linked trait will be the same. This is because each male only has one X chromosome, and the probability of any individual male receiving an X chromosome with the allele in question must be equal to the frequency of the allele. The probability of any individual female having the allele in question on both X chromosomes will be q^2, where q is the frequency of the allele.

To illustrate this for a recessive X-linked trait, let's consider the example of red–green color blindness, which affects 8 percent of human males. The frequency of the color blindness allele is therefore 0.08; in other words, 8 percent of X chromosomes carry it. The other 92 percent of X chromosomes carry the dominant allele for normal red–green color vision. If we define p as the frequency of the normal allele and q as the frequency of the color-blindness allele, then $p = 0.92$ and $q = 0.08$. The frequency of color-blind females (with two affected X chromosomes) is $q^2 = (0.08)^2 = 0.0064$, and the frequency of carrier females (having one normal and one affected X chromosome) is $2pq = 2(0.08)(0.92) = 0.147$. In other words, 14.7 percent of females carry the allele for red–green color blindness and can pass it to their children, although they themselves have normal color vision.

An important consequence of the difference in allele frequency for X-linked genes between male and female gametes is that for a rare recessive allele, the trait will be expressed at a much higher frequency among XY individuals than among those who are XX. So, for example, diseases such as hemophilia and Duchenne muscular dystrophy (DMD) in humans, both of which are caused by recessive mutations on the X chromosome, are much more common in boys, who need only inherit a single copy of the mutated allele to suffer from the disease. Girls who inherit two affected X chromosomes will also have the disease; but with a rare allele, the probability of this occurrence is small.

Calculating Heterozygote Frequency

In another application, the Hardy–Weinberg Law allows us to estimate the frequency of heterozygotes in a population. The frequency of a recessive trait can usually be determined by counting such individuals in a sample of the population. With this information and the Hardy–Weinberg Law, we can then calculate the allele and genotype frequencies.

Cystic fibrosis, an autosomal recessive trait, has an incidence of about $1/2500 = 0.0004$ in people of northern European ancestry. Individuals with cystic fibrosis are easily distinguished from the population at large by such symptoms as extra-salty sweat, excess amounts of thick mucus in the lungs, and susceptibility to bacterial infections. Because this is a recessive trait, individuals with cystic fibrosis must be homozygous. Their frequency in a population is represented by q^2, provided that mating has been random in the previous generation. The frequency of the recessive allele is therefore

$$q = \sqrt{q^2} = \sqrt{0.0004} = 0.02$$

Since $p + q = 1$, then the frequency of p is

$$p = 1 - q = 1 - 0.02 = 0.98$$

In the Hardy–Weinberg equation, the frequency of heterozygotes is $2pq$. Thus,

$$2pq = 2(0.98)(0.02)$$
$$= 0.04 \text{ or 4 percent, or } 1/25$$

Heterozygotes for cystic fibrosis are rather common in the population (about 1/25, or 4 percent), even though the incidence of homozygous recessives is only 1/2500, or 0.04 percent. Calculations such as these are estimates because the population may not meet all Hardy–Weinberg assumptions.

NOW SOLVE THIS

25–2 If the albino phenotype occurs in 1/10,000 individuals in a population at equilibrium and albinism is caused by an autosomal recessive allele *a*, calculate the frequency of (a) the recessive mutant allele; (b) the normal dominant allele; (c) heterozygotes in the population; and (d) matings between heterozygotes.

■ HINT: *This problem involves an understanding of how to use the Hardy–Weinberg Law to calculate allele frequencies and heterozygote frequencies. The key to its solution lies in solving the problem in the order presented.*

25.4

Natural Selection Is a Major Force Driving Allele Frequency

To understand evolution, we must understand the forces that transform the gene pools of populations and can lead to the formation of new species. Chief among the mechanisms transforming populations is **natural selection**, discovered independently by Alfred Russel Wallace and Charles Darwin. The Wallace–Darwin concept of natural selection can be summarized as follows:

1. Individuals of a species exhibit variations in phenotype— for example, differences in size, agility, coloration, defenses against enemies, ability to obtain food, courtship behaviors, and flowering times.

2. Many of these variations, even small and seemingly insignificant ones, are heritable and passed on to offspring.

3. Organisms tend to reproduce in an exponential fashion. More offspring are produced than can survive. This causes members of a species to engage in a struggle for

survival, competing with other members of the species for scarce resources. Offspring also must avoid predators, and in sexually reproducing species, adults must compete for mates.

4. In the struggle for survival, individuals with particular phenotypes will be more successful than others, allowing the former to survive and reproduce at higher rates.

As a consequence of natural selection, populations and species change. The phenotypes that confer improved ability to survive and reproduce become more common, and the phenotypes that confer poor prospects for survival and reproduction may eventually disappear. Under certain conditions, populations that at one time could interbreed may lose that capability, thus segregating their adaptations into particular niches. If selection continues, it may result in the appearance of new species.

Detecting Natural Selection in Populations

Recall that measuring allele frequencies and genotype frequencies using the Hardy–Weinberg Law is based on certain assumptions about an ideal population: large population size, lack of migration, presence of random mating, absence of selection and mutation, and equal survival rates of offspring.

However, if all genotypes do not have equal rates of survival or do not leave equal numbers of offspring, then allele frequencies may change from one generation to the next. To see why, let's imagine a population of 100 individuals in which the frequency of allele *A* is 0.5 and that of allele *a* is 0.5. Assuming the previous generation mated randomly, we find that the genotype frequencies in the present generation are $(0.5)^2 = 0.25$ for *AA*, $2(0.5)(0.5) = 0.5$ for *Aa*, and $(0.5)^2 = 0.25$ for *aa*. Because our population contains 100 individuals, we have 25 *AA* individuals, 50 *Aa* individuals, and 25 *aa* individuals. Now suppose that individuals with different genotypes have different rates of survival: All 25 *AA* individuals survive to reproduce, 90 percent or 45 of the *Aa* individuals survive to reproduce, and 80 percent or 20 of the *aa* individuals survive to reproduce. When the survivors reproduce, each contributes two gametes to the new gene pool, giving us $2(25) + 2(45) + 2(20) = 180$ gametes. What are the frequencies of the two alleles in the surviving population? We have 50 *A* gametes from *AA* individuals, plus 45 *A* gametes from *Aa* individuals, so the frequency of allele *A* is $(50 + 45)/180 = 0.53$. We have 45 *a* gametes from *Aa* individuals, plus 40 *a* gametes from *aa* individuals, so the frequency of allele *a* is $(45 + 40)/180 = 0.47$.

These differ from the frequencies we started with. The frequency of allele *A* has increased, and the frequency of allele *a* has decreased. A difference among individuals in survival or reproduction rate (or both) is an example of **natural selection**. Natural selection is the principal force that shifts allele frequencies within large populations and is one of the most important agents of evolutionary change.

Fitness and Selection

Selection occurs whenever individuals with a particular genotype enjoy an advantage in survival and reproduction over other genotypes. However, selection may vary from less than 1 to 100 percent. In the previous hypothetical example, selection was strong. Weak selection might involve just a fraction of a percent difference in the survival rates of different genotypes. Advantages in survival and reproduction ultimately translate into increased genetic contribution to future generations. An individual's genetic contribution to future generations is called its **fitness**. Genotypes associated with high rates of reproductive success are said to have high fitness, whereas genotypes associated with low reproductive success are said to have low fitness.

Hardy–Weinberg analysis also allows us to examine fitness. By convention, population geneticists use the letter w to represent fitness. Thus, w_{AA} represents the relative fitness of genotype AA, w_{Aa} the relative fitness of genotype Aa, and w_{aa} the relative fitness of genotype aa. Assigning the values $w_{AA} = 1$, $w_{Aa} = 0.9$, and $w_{aa} = 0.8$ would mean, for example, that all AA individuals survive, 90 percent of the Aa individuals survive, and 80 percent of the aa individuals survive, as in the previous hypothetical case.

Let's consider selection against deleterious alleles. Fitness values $w_{AA} = 1$, $w_{Aa} = 1$, and $w_{aa} = 0$ describe a situation in which a is a homozygous lethal allele. Because homozygous recessive individuals die without leaving offspring, the frequency of allele a will decrease. The change in the frequency of allele a is described by the equation

$$q_g = \frac{q_0}{1 + gq_0}$$

where q_g is the frequency of allele a in generation g, q_0 is the frequency of a in generation zero, and g is the number of generations that have passed.

Figure 25–8 shows what happens to a lethal recessive allele with an initial frequency of 0.5. At first, because of the high percentage of aa genotypes, the frequency of allele a declines rapidly. The frequency of a is halved in only two generations. By the sixth generation, the frequency is halved again. By now, however, the majority of a alleles are carried by heterozygotes. Because a is recessive, these heterozygotes are not selected against. Consequently, as more time passes, the frequency of allele a declines ever more slowly. As long as heterozygotes continue to mate, it is difficult for selection to completely eliminate a recessive allele from a population.

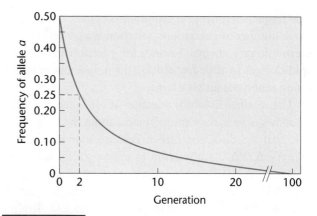

FIGURE 25-8   Change in the frequency of a lethal recessive allele, a. The frequency of a is halved in two generations and halved again by the sixth generation. Subsequent reductions occur slowly because the majority of a alleles are carried by heterozygotes.

Figure 25–9 shows the outcome of different degrees of selection against a nonlethal recessive allele, a. In this case, the intensity of selection varies from strong (red curve) to weak (blue curve), as well as intermediate values (yellow, purple, and green curves). In each example, the frequency of the deleterious allele, a, starts at 0.99 and declines over time. However, the rate of decline depends heavily on the strength of selection. When selection is strong and only 90 percent of the heterozygotes and 80 percent of the aa homozygotes survive (red curve), the frequency of allele a drops from 0.99 to less than 0.01 in about 85 generations. However, when selection is weak, and 99.8 percent of the heterozygotes and 99.6 percent of the aa homozygotes survive (blue curve), it takes 1000 generations for the frequency of allele a to drop from 0.99 to 0.93. Two important conclusions can be drawn from this example. First, over thousands of generations, even weak selection can cause substantial changes in

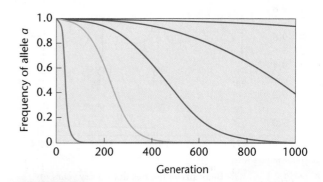

FIGURE 25-9   The effect of selection on allele frequency. The rate at which a deleterious allele is removed from a population depends heavily on the strength of selection.

allele frequencies; because evolution generally occurs over a large number of generations, selection is a powerful force in evolutionary change. Second, for selection to produce rapid changes in allele frequencies, the differences in fitness among genotypes must be large.

The manner in which selection affects allele frequencies allows us to make some inferences about the *CCR5-Δ32* allele that we discussed earlier. Because individuals with genotype *Δ32/Δ32* are resistant to most sexually transmitted strains of HIV-1, while individuals with genotypes *1/1* and *1/Δ32* are susceptible, we might expect AIDS to act as a selective force causing the frequency of the *Δ32* allele to increase over time. Indeed, it probably will, but the increase in frequency is likely to be slow in human terms. In fact, it will take about 100 generations (about 2000 years) for the frequency of the *Δ32* allele to reach just 0.11. In other words, the frequency of the *Δ32* allele will probably not change much over the next few generations in most populations that currently harbor it.

There Are Several Types of Selection

The phenotype is the result of the combined influence of the individual's genotype at many different loci and the effects of the environment. Selection for these complex traits can be classified as (1) directional, (2) stabilizing, or (3) disruptive.

In **directional selection** (Figure 25–10) phenotypes at one end of the spectrum become selected for or against, usually as a result of changes in the environment. A carefully documented example comes from research by Peter and Rosemary Grant and their colleagues, who study the medium ground finches (*Geospiza fortis*) of Daphne Major Island in the Galapagos Islands. The beak size of these birds varies enormously. For example, in 1976, some birds in the

population had beaks less than 7 mm deep, while others had beaks more than 12 mm deep. In 1977, a severe drought killed some 80 percent of the finches. Big-beaked birds survived at higher rates than small-beaked birds because when food became scarce, the big-beaked birds were able to eat a greater variety of seeds. When the drought ended in 1978, the offspring of the survivors inherited their parents' big beaks. Between 1976 and 1978, the beak depth of the average finch in the Daphne Major population increased by just over 0.5 mm, shifting the average beak size in the direction of one phenotypic extreme.

Stabilizing selection tends to favor intermediate phenotypes, with those at both extremes being selected against. Over time, this will reduce the phenotypic variance in the population but without a significant shift in the mean. One of the clearest demonstrations of stabilizing selection is shown by a study of human birth weight and survival for 13,730 children born over an 11-year period. Figure 25–11 shows the distribution of birth weight and the percentage of mortality at 4 weeks of age. Infant mortality increases on either side of the optimal birth weight of 7.5 pounds. Stabilizing selection acts to keep a population well adapted to its present environment.

Disruptive selection is a case where selection acts against intermediate phenotypes and in favor of *both* phenotypic extremes. It can be viewed as the opposite of stabilizing selection because the intermediate types are selected against. This will result in a population with an increasingly bimodal distribution for the trait, as we can see in Figure 25–12. In one set of experiments using *Drosophila*, after several generations of disruptive artificial selection for bristle number, in which only flies with high- or low-bristle numbers were allowed to breed, most flies could be easily placed in a low- or high-bristle category. In natural populations, such a situation might exist for a population in a heterogeneous environment.

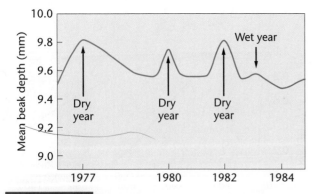

FIGURE 25–10 Beak size in finches during dry years increases because of strong selection. Between droughts, selection for large beak size is not as strong, and birds with smaller beak sizes survive and reproduce, increasing the number of birds with smaller beaks.

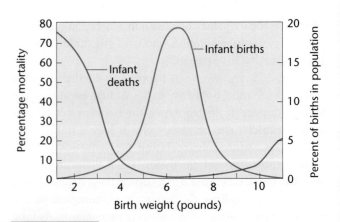

FIGURE 25–11 Relationship between birth weight and mortality in humans.

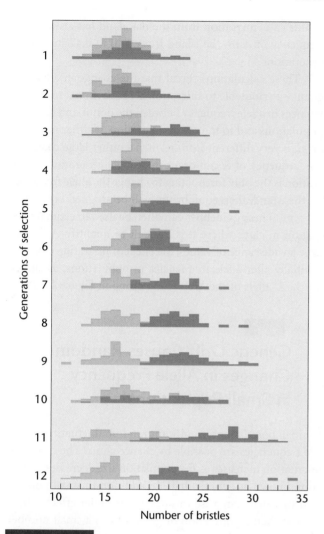

FIGURE 25–12 The effect of disruptive selection on bristle number in *Drosophila*. When individuals with the highest and lowest bristle number were selected, the population showed a nonoverlapping divergence in only 12 generations.

25.5

Mutation Creates New Alleles in a Gene Pool

Each generation, a population's gene pool is reshuffled to produce new genotypes in the offspring. The enormous genetic variation present in the gene pool allows assortment and recombination to produce new genotypic combinations continuously. But assortment and recombination do not produce new alleles. **Mutation** alone acts to create new alleles. It is important to keep in mind that mutational events occur at random—that is, without regard for any possible benefit or disadvantage to the organism. In this section, we consider whether mutation, by itself, is a significant factor in changing allele frequencies.

To determine whether mutation is a significant force in changing allele frequencies, we must measure the rate at which mutations are produced. As most mutations are recessive, it is difficult to observe mutation rates directly in diploid organisms. Indirect methods use probability and statistics or large-scale screening programs to estimate mutation rates. For certain dominant mutations, however, a direct method of measurement can be used. To ensure accuracy, several conditions must be met:

1. The allele must produce a distinctive phenotype that can be distinguished from similar phenotypes produced by recessive alleles.

2. The trait must be fully expressed or completely penetrant so that mutant individuals can be identified.

3. An identical phenotype must never be produced by nongenetic agents such as drugs or chemicals.

Mutation rates can be defined as the number of new mutant alleles per given number of gametes. Suppose that for a certain recessively inherited gene that undergoes mutation to a dominant allele, 2 out of 100,000 births exhibit a mutant phenotype. In these cases, both sets of parents are phenotypically normal. Because the zygotes that produced these births each carry two copies of the gene, we have actually surveyed 200,000 copies of the gene (or 200,000 gametes). If we assume that the affected births are each heterozygous, we have uncovered 2 new mutant alleles out of 200,000 alleles surveyed. Thus, the mutation rate is 2/200,000 or 1/100,000, which in scientific notation is written as 1×10^{-5}. In humans, a dominantly inherited form of dwarfism known as **achondroplasia** fulfills the requirements for measuring mutation rates. Individuals with this skeletal disorder have an enlarged skull, short arms and legs, and can be diagnosed by X-ray examination at birth. In a survey of almost 250,000 births, the mutation rate (μ) for achondroplasia has been calculated as

$$\mu = 1.4 \times 10^{-5} \pm 0.5 \times 10^{-5}$$

Knowing the rate of mutation, we can estimate the extent to which mutation can cause allele frequencies to change from one generation to the next. We represent the normal allele as *d* and the allele for achondroplasia as *D*.

Imagine a population of 500,000 individuals in which everyone has genotype *dd*. The initial frequency of *d* is 1.0, and the initial frequency of *D* is 0. If each individual contributes two gametes to the gene pool, the gene pool will contain 1,000,000 gametes, all carrying allele *d*. In this collection of alleles, 1.4 of every 100,000 *d* alleles mutate into a *D* allele. The frequency of allele *d* is now $(1,000,000 - 14)/1,000,000 = 0.999986$, and the frequency of allele *D* is $14/1,000,000 = 0.000014$. From these

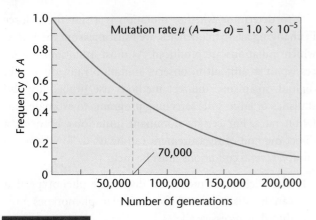

FIGURE 25-13 Replacement rate of an allele by mutation alone, assuming an average mutation rate of 1.03×10^{-5}.

numbers, it will clearly be a long time before mutation, by itself, causes any appreciable change in the allele frequencies in this population (**Figure 25–13**). In other words, mutation generates new alleles but by itself does not alter allele frequencies at an appreciable rate.

25.6

Migration and Gene Flow Can Alter Allele Frequencies

Occasionally, a species becomes divided into smaller, geographically separated populations. In such populations, various evolutionary forces, including selection, migration, and drift, can alter allele frequencies. **Migration** occurs when individuals move between the populations. Imagine a species in which a given locus has two alleles, A and a. There are two populations of this species, one on a mainland and one on an island. The frequency of A on the mainland is represented by p_m, and the frequency of A on the island is p_i. If there is migration from the mainland to the island, the frequency of A in the next generation on the island $(p_{i'})$ is given by

$$p_{i'} = (1 - m)p_i + mp_m$$

where m represents migrants from the mainland to the island.

As an example of how migration might affect the frequency of A in the next generation on the island $(p_{i'})$, assume that $p_i = 0.4$ and $p_m = 0.6$ and that 10 percent of the parents of the next generation are migrants from the mainland $(m = 0.1)$. In the next generation, the frequency of allele A on the island will therefore be

$$p_{i'} = \left[(1 - 0.1) \times 0.4\right] + (0.1 \times 0.6)$$
$$= 0.36 + 0.06$$
$$= 0.42$$

In this case, migration from the mainland has changed the frequency of A on the island from 0.40 to 0.42 in a single generation.

These calculations reveal that the change in allele frequency attributable to migration is proportional to the differences in allele frequency between the donor and recipient populations and to the rate of migration. If either m is large or p_m is very different from p_i, then a rather large change in the frequency of A can occur in a single generation. If migration is the only force acting to change the allele frequency on the island, then equilibrium will be attained only when $p_i = p_m$. These guidelines can often be used to estimate migration in cases where it is difficult to quantify. As m can have a wide range of values, the effect of migration can substantially alter allele frequencies in populations, as shown for the I^B allele of the ABO blood group in **Figure 25–14**.

25.7

Genetic Drift Causes Random Changes in Allele Frequency in Small Populations

In small populations, significant random fluctuations in allele frequencies are possible by chance alone. The degree of fluctuation increases as the population size decreases, a situation known as **genetic drift**. In addition to small population size, drift can arise through the **founder effect**, which occurs when a population originates from a small number of individuals, whose gene pool may not reflect that of the larger population from which the founders are drawn. Although the population may later increase to a large size, the genes carried by all members are derived only from those of the founders (assuming no mutation, migration, or selection, and the presence of random mating). Drift can also arise via a **genetic bottleneck**. Bottlenecks develop when a large population undergoes a drastic but temporary reduction in numbers. Even though the population recovers, its genetic diversity has been greatly reduced.

Founder Effects in Human Populations

Allele frequencies in certain human populations demonstrate the role of genetic drift in natural populations. Native Americans living in the southwestern United States have a high frequency of oculocutaneous albinism (OCA). In the Navajo, who live primarily in northeast Arizona, albinism occurs with a frequency of 1 in 1500–2000, compared with whites (1 in 36,000) and African-Americans (1 in 10,000). There are four different forms of OCA (OCA1–4), all with varying degrees of melanin deficiency in the skin, eyes, and hair. OCA2 is caused by mutations in the P gene, which encodes a plasma membrane protein. To investigate the

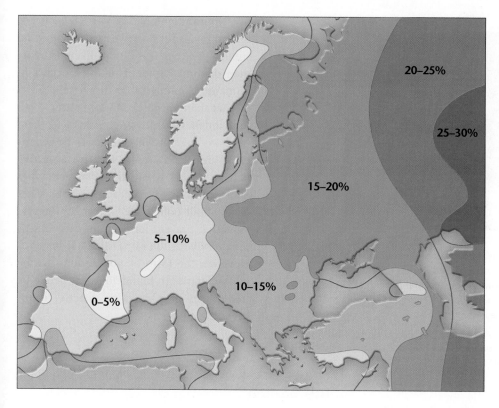

20–25%

25–30%

15–20%

5–10%

10–15%

0–5%

FIGURE 25–14 Affected migration as a force in evolution. The I^B allele of the *ABO* locus is present in a gradient from east to west. This allele shows the highest frequency in central Asia and the lowest in northeast Spain. The gradient parallels the waves of Mongol migration into Europe following the fall of the Roman Empire and is a genetic relic of human history.

genetic basis of albinism in the Navajo, researchers screened for mutations in the *P* gene. In their study, all Navajo with albinism were homozygous for a 122.5-kb deletion in the *P* gene, spanning exons 10–20 (Figure 25–15). This deletion allele was not present in 34 individuals belonging to other Native American populations.

Using a set of PCR primers, researchers were able to identify homozygous affected individuals and heterozygous carriers (Figure 25–16) and surveyed 134 normally

pigmented Navajo and 42 members of the Apache, a tribe closely related to the Navajo. Based on this sample, the heterozygote frequency in the Navajo is estimated to be 4.5 percent. No carriers were found in the Apache population that was studied.

The 122.5-kb deletion allele causing OCA2 was found only in the Navajo population and not in members of other Native American tribes in the southwestern United States, suggesting that the mutant allele is specific to the Navajo

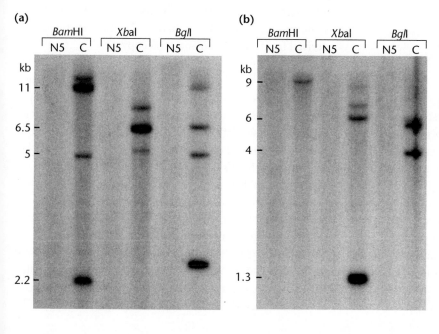

(a)

BamHI		XbaI		BglI	
N5	C	N5	C	N5	C

kb
11 —
6.5 —
5 —

2.2 —

(b)

BamHI		XbaI		BglI	
N5	C	N5	C	N5	C

kb
9 —
6 —
4 —

1.3 —

FIGURE 25–15 Genomic DNA digests from a Navajo affected with albinism (N5) and a normally pigmented individual (C). (a) Hybridization with a probe covering exons 11–15 of the *P* gene; there are no hybridizing fragments detected in N5. (b) Hybridization with a probe covering exons 15–20 of the *P* gene; there are no hybridizing fragments detected in N5. This confirms the presence of a deletion in affected individuals.

Courtesy of Murray Brilliant, "A 122.5 kilobase deletion of P gene underlies the high prevalence of oculocutaneous albinism type 2 in the Navajo population." From: American Journal Human Genetics *72:62–72, Figure 1, p. 65. Published by University of Chicago Press.*

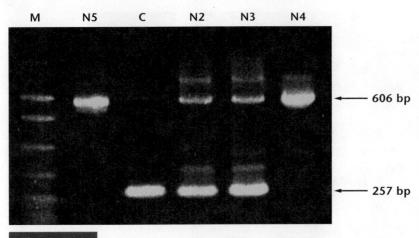

and may have arisen in a single individual who was one of a small number of founders of the Navajo population. Using other methods, workers estimated the age of the mutation to be between 400 and 11,000 years. To narrow this range, they relied on tribal history. Navajo oral tradition indicates that the Navajo and Apache became separate populations between 600 and 1000 years ago. Because the deletion is not found in the Apaches, it probably arose in the Navajo population after the tribes split. On this basis, the deletion is estimated to be 400 to 1000 years old and probably arose as a founder mutation.

25.8

Nonrandom Mating Changes Genotype Frequency but Not Allele Frequency

We have explored how violations of the first four assumptions of the Hardy–Weinberg Law, in the form of selection, mutation, migration, and genetic drift can cause allele frequencies to change. The fifth assumption is that the members of a population mate at random; in other words, any one genotype has an equal probability of mating with any other genotype in the population. Nonrandom mating can change the frequencies of genotypes in a population. Subsequent selection for or against certain genotypes has the potential to affect the overall frequencies of the alleles they contain, but it is important to note that nonrandom mating *does not itself directly change allele frequencies.*

Nonrandom mating can take one of several forms. In **positive assortive mating** similar genotypes are more likely to mate than dissimilar ones. This often occurs in humans: A number of studies have indicated that many people are more attracted to individuals who physically resemble them (and are therefore more likely to be genetically similar as well). **Negative assortive mating** occurs when dissimilar genotypes are more likely to mate; some plant species have inbuilt pollen/stigma recognition systems that prevent fertilization between individuals with the same alleles at key loci. However, the form of nonrandom mating most commonly found to affect genotype frequencies in population genetics is **inbreeding**.

Inbreeding

Inbreeding occurs when mating individuals are more closely related than any two individuals drawn from the population at random; loosely defined, inbreeding is mating among relatives. For a given allele, inbreeding increases the proportion of homozygotes in the population. A completely inbred population will theoretically consist only of homozygous genotypes.

To describe the amount of inbreeding in a population, geneticist Sewall Wright devised the **coefficient of inbreeding (F)**. F quantifies the probability that the two alleles of a given gene in an individual are identical *because they are descended from the same single copy of the allele in an ancestor.* If $F = 1$, all individuals in the population are homozygous, and both alleles in every individual are derived from the same ancestral copy. If $F = 0$, no individual has two alleles derived from a common ancestral copy.

One method of estimating F for an individual is shown in Figure 25–17. In this pedigree, the fourth-generation female (shaded pink) is the daughter of first cousins (yellow). Suppose her great-grandmother (green) was a carrier of a recessive lethal allele, *a*. What is the probability that this fourth-generation female will inherit two copies of her great-grandmother's lethal allele? For this to happen, (1) the great-grandmother had to pass a copy of the allele to her son, (2) her son had to pass it to his daughter, and (3) his daughter had to pass it to her daughter (the pink female). Also, (4) the great-grandmother had to pass a copy of the allele to her daughter, (5) her daughter had to pass it to her son, and (6) her son had to pass it to his daughter (the pink female). Each of the six necessary events has an individual probability of 1/2, and they *all* have to happen, so the probability that the

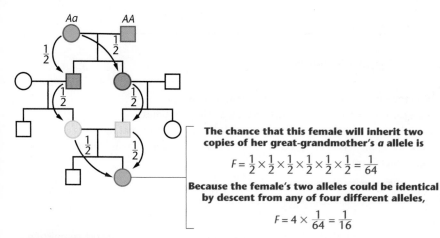

The chance that this female will inherit two copies of her great-grandmother's *a* allele is

$$F = \frac{1}{2} \times \frac{1}{2} \times \frac{1}{2} \times \frac{1}{2} \times \frac{1}{2} \times \frac{1}{2} = \frac{1}{64}$$

Because the female's two alleles could be identical by descent from any of four different alleles,

$$F = 4 \times \frac{1}{64} = \frac{1}{16}$$

FIGURE 25–17 Calculating the coefficient of inbreeding (*F*) for the offspring of a first-cousin marriage.

pink female will inherit two copies of her great-grandmother's lethal allele is $(1/2)^6 = 1/64$. To calculate an overall value of *F* for the pink female as a child of a first-cousin marriage, remember that she could also inherit two copies of any of the other three alleles present in her great-grandparents. Because any of four possibilities would give the pink female two alleles identical by descent from an ancestral copy,

$$F = 4 \times (1/64) = 1/16$$

NOW SOLVE THIS

25–3 A prospective groom, who is healthy, has a sister with cystic fibrosis (CF), an autosomal recessive disease. Their parents are normal. The brother plans to marry a woman who has no history of CF in her family. What is the probability that they will produce a CF child? They are both Caucasian, and the overall frequency of CF in the Caucasian population is 1/2500—that is, 1 affected child per 2500. (Assume the population meets the Hardy–Weinberg assumptions.)

■ HINT: *This problem involves an understanding of how recessive traits are inherited and how to use family history to calculate the probability that an individual is heterozygous. The key to its solution lies in knowing the risk factors for the man and the woman.*

25.9

Reduced Gene Flow, Selection, and Genetic Drift Can Lead to Speciation

A **species** can be defined as a group of actually or potentially interbreeding organisms that is reproductively isolated in nature from all other such groups. In sexually reproducing

organisms, speciation transforms the gene pool of the parental species or divides a single gene pool into two or more separate and distinct gene pools. Changes in morphology or physiology and adaptation to an ecological niche may also occur but are not necessary components of the speciation event. Speciation can take place gradually or within a few generations.

As we saw earlier, most populations contain considerable genetic variation, and different populations within a species may carry different alleles or allele frequencies at a variety of loci. The genetic divergence of these populations can be caused by natural selection, genetic drift, or both. In an earlier section, we saw that the migration of individuals between populations, together with the gene flow that accompanies that migration, tends to homogenize allele frequencies among populations. In other words, migration counteracts the tendency of populations to diverge.

When gene flow between populations is reduced or absent, the populations may diverge to the point that members of one population are no longer able to interbreed successfully with members of the other. When populations reach the point where they are reproductively isolated from one another, they have become different species, according to the biological species concept. The genetic changes that result in reproductive isolation between or among populations and that lead to the formation of new species (or higher taxonomic groups) are an example of macroevolution.

The biological barriers that prevent or reduce interbreeding between populations are called **reproductive isolating mechanisms**, classified in **Table 25.4**. These mechanisms may be ecological, behavioral, seasonal, mechanical, or physiological.

Prezygotic isolating mechanisms prevent individuals from mating in the first place. Individuals from different populations may not find each other at the right time, may not recognize each other as suitable mates, or may try to mate but find that they are unable to do so.

Postzygotic isolating mechanisms create reproductive isolation even when the members of two populations are willing and able to mate with each other. For example, genetic divergence may have reached the stage where the viability or fertility of hybrids is reduced. Hybrid zygotes may be formed, but all or most may be inviable. Alternatively, the hybrids may be viable, but may be sterile or suffer from reduced fertility. Yet again, the hybrids themselves may be fertile, but their progeny may have lowered viability or fertility. In all these situations, hybrids will not reproduce and are genetic dead-ends. These postzygotic mechanisms act at

Reproductive Isolating Mechanisms

Prezygotic Mechanisms
Prevent fertilization and zygote formation

1. **Geographic or ecological:** The populations live in the same regions but occupy different habitats.
2. **Seasonal or temporal:** The populations live in the same regions but are sexually mature at different times.
3. **Behavioral (only in animals):** The populations are isolated by different and incompatible behavior before mating.
4. **Mechanical:** Cross-fertilization is prevented or restricted by differences in reproductive structures (genitalia in animals, flowers in plants).
5. **Physiological:** Gametes fail to survive in alien reproductive tracts.

Postzygotic Mechanisms
Fertilization takes place and hybrid zygotes are formed, but these are nonviable or give rise to weak or sterile hybrids.

1. **Hybrid nonviability or weakness.**
2. **Developmental hybrid sterility:** Hybrids are sterile because gonads develop abnormally or meiosis breaks down before completion.
3. **Segregational hybrid sterility:** Hybrids are sterile because of abnormal segregation into gametes of whole chromosomes, chromosome segments, or combinations of genes.
4. **F_2 breakdown:** F_1 hybrids are normal, vigorous, and fertile, but the F_2 contains many weak or sterile individuals.

Source: From *Processes of Organic Evolution*, 3rd Edition by G. Ledyard Stebbins, © 1977, p. 143. Reprinted by permission of Pearson Education, Inc.

or beyond the level of the zygote and are generated by genetic divergence.

Postzygotic isolating mechanisms waste gametes and zygotes and lower the reproductive fitness of hybrid survivors. Selection will therefore favor the spread of alleles that reduce the formation of hybrids, leading to the development of prezygotic isolating mechanisms, which in turn prevent interbreeding and the formation of hybrid zygotes and offspring. In animal evolution, one of the most effective prezygotic mechanisms is behavioral isolation, involving courtship behavior.

Changes Leading to Speciation

The Isthmus of Panama, which created a land bridge connecting North and South America and simultaneously separated the Caribbean Sea from the Pacific Ocean, formed roughly 3 million years ago. To study populations separated by the formation of the isthmus, researchers matched seven Caribbean species of snapping shrimp (Figure 25–18) with

a similar Pacific species to form a pair. Members of each pair were closer to each other in structure and appearance than either was to any other species in its own ocean. Analysis of allele frequencies and mitochondrial DNA sequences confirmed that the members of each pair were one another's closest genetic relatives.

The interpretation of these data is that, prior to the formation of the isthmus, the ancestors of each pair were members of a single species. When the isthmus closed, each of the seven ancestral species was divided into two separate populations, one in the Caribbean and the other in the Pacific.

Meeting in a dish in a lab for the first time in 3 million years, would Caribbean and Pacific members of a species pair recognize each other as suitable mates? Males and females were paired together, and the relative inclination of Caribbean–Pacific couples to mate versus that of Caribbean–Caribbean or Pacific–Pacific couples was calculated. For three of the seven species pairs, transoceanic couples refused to mate altogether. For the other four species pairs, transoceanic couples were 33, 45, 67, and 86 percent as likely to mate with each other as were same-ocean pairs. Of the same-ocean couples that mated, 60 percent produced viable clutches of eggs. Of the transoceanic couples that mated, only 1 percent produced viable clutches. We can conclude from these results that 3 million years of separation has resulted in complete or nearly complete speciation, involving strong pre- and postzygotic isolating mechanisms for all seven species pairs.

The Rate of Macroevolution and Speciation

How much time is required for speciation? In many cases, divergence of populations and the appearance of new species occur over a long period of time. In other cases, however,

FIGURE 25–18 A snapping shrimp (genus *Alpheus*).

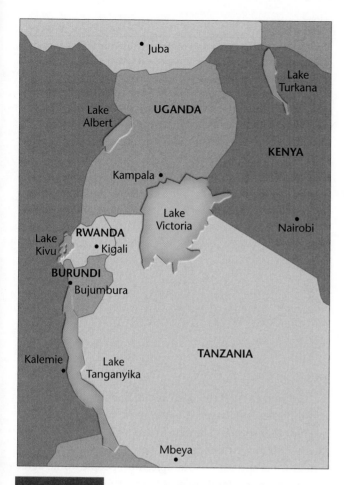

The lakes in the Rift Valley of east Africa are home to most species of cichlids, a fresh water fish.

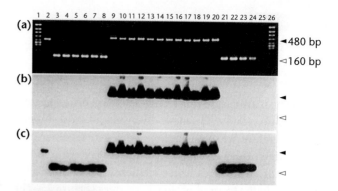

FIGURE 25–20 Use of repetitive DNA elements to trace species relationships in cichlids from Lake Tanganyika. (a) Photograph of an agarose gel showing PCR fragments generated from primers that flank the AFC family of SINES. Large fragments containing this family are present in all samples of genomic DNA from members of the Lamprologini tribe of cichlids (lanes 9–20). DNA from the species in lane 2 has a similar but shorter fragment, which may represent another repetitive sequence. DNA from other species (lanes 3–8 and 21–24) produce short, non-SINE-containing fragments. (b) A Southern blot of the gel from (a) probed with DNA from the AFC family of SINES. AFC is present in DNA from all species in the Lamprologini tribe (lanes 9–20), but not in DNA from other species (lanes 2–8 and 21–24). (c) A second Southern blot of the gel from (a) probed with the genomic sequence at which the AFC SINE inserts. All species examined (lanes 2–24) contain the insertion sequence. The larger fragments in Lamprologini DNA (lanes 9–20) correspond to those in the previous blot, showing that the SINE is inserted at the same site in all tribal species. In sum, these data show that the AFC SINE is present only in members of this tribe, is present in all member species, and is inserted at the same site in all cases. These results are interpreted as showing a common origin for the species of the tribe in question.

genetic divergence, reproductive isolation, and speciation can be surprisingly rapid.

The Rift Valley lakes of east Africa (Figure 25–19) support hundreds of species of cichlid fish. Cichlids are highly specialized for different niches. Some eat algae floating on the water's surface, whereas others are bottom feeders, insect feeders, mollusk eaters, and predators on other fish species. Lake Tanganyika is an old lake, formed about 20 million years ago and more than 200 species of cichlids have been identified in this lake, with many more remaining to be discovered. Genetic analysis indicates that a majority of the species in this lake are descended from a single common ancestor and that they diverged within their home lake.

The cichlid species in Lake Tanganyika are organized into 16 groups called tribes; all species in a tribe are thought to be descended from a common ancestral species. In a study of cichlid origins in Lake Tanganyika, Norihiro Okada and colleagues examined the insertion of a novel family of repetitive DNA sequences called short interspersed elements (SINES) (discussed in Chapter 11) into the genomes of cichlid species in Lake Tanganyika. SINES are a type of retroposon, and the random integration of a SINE at a locus is most likely an

irreversible event. If a SINE is present at the same locus in the genome of all species examined, this is strong evidence that all of those species descend from a common ancestor. Using a SINE called AFC, Okada's team screened 33 species of cichlids belonging to four tribes. In each tribe, the SINE was present at all the sites tested, indicating that the species in each tribe are descended from a single ancestral species (Figure 25–20). Cichlid speciation in Lake Tankanyika has taken place over several million years and produced the most diverse collection of species of this fresh water fish.

In contrast, Lake Victoria, a neighboring Rift Valley lake, is only about 400,000 years old and has dried out several times in its history, most recently, about 15,000 years ago. The 500 or so cichlid species in this lake today are thought to have diversified from a very small number of founding species. If true, this means that new species were formed at a very rapid rate, averaging 1 species about every 30 years. If further research using SINES and other molecular markers can confirm the number of founding species and establish their tribal organization, it would provide evidence for the fastest evolutionary divergence of species ever documented in vertebrates.

Phylogeny Can Be Used to Analyze Evolutionary History

Speciation is associated with changes in the genetic structure of populations and with genetic divergence of those populations. Therefore, we should be able to use genetic differences among present-day species to reconstruct their evolutionary histories, or phylogenies. These relationships are most often presented in the form of phylogenetic trees (**Figure 25–21**), which show the ancestral relationships among a group of organisms. These groups can be species, or larger groups. In a phylogenetic tree, branches represent lineages over time. The length of a branch can be arbitrary, or it can be derived from a time scale, showing the length of time between speciation events. Points at which lines diverge, called nodes, show when a species split into two or more species. Each node represents a common ancestor of the species diverging at that node. The tips of the branches represent species (or a larger group) alive today (or those that ended in extinction). Groups that consist of an ancestral species and all its descendants are called a **monophyletic** group. The root of a phylogenetic tree represents the oldest common ancestor to all the groups shown in the tree.

Constructing Phylogenetic Trees from Amino Acid Sequences

In an important early example of phylogenetic reconstruction, W. M. Fitch and E. Margoliash assembled data on the amino acid sequence for cytochrome c in a variety of organisms. **Cytochrome c** is a component of the electron transport chain in mitochondria, and its amino acid sequence has evolved very slowly. For example, its amino acid sequence in humans and chimpanzees is identical, and humans and rhesus monkeys show only one amino acid difference. This similarity is remarkable considering that the fossil record indicates that the lines leading to humans and monkeys diverged from a common ancestor approximately 20 million years ago.

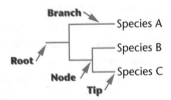

Elements of a phylogenetic tree showing the relationships among species. The root represents a common ancestor to all species on the tree. Branches represent lineages through time. The points at which branches separate are called nodes, and at the tips of branches are living species (or those that have gone extinct.

TABLE 25.5

Amino Acid Differences and Minimal Mutational Distances between Cytochrome c in Humans and Other Organisms

Organism	(a) Amino Acid Differences	(b) Minimal Mutational Distance
Human	0	0
Chimpanzee	0	0
Rhesus monkey	1	1
Rabbit	9	12
Pig	10	13
Dog	10	13
Horse	12	17
Penguin	11	18
Moth	24	36
Yeast	38	56

©AAAS

Source: From W.M. Fitch and E. Margoliash, Construction of phylogenetic trees, *Science* 155: 279–284, January 20, 1967. Reprinted with permission from AAAS.

Column (a) of **Table 25.5** shows the number of amino acid differences between cytochrome c in humans and in various other species. The data in this table are broadly consistent with our intuitions about how closely related we are to these other species. For example, most people would agree that we are more closely related to other mammals than we are to insects, and we are more closely related to insects than we are to yeast. Accordingly, our cytochrome c differs in 10 amino acids from that of dogs, in 24 amino acids from that of moths, and in 38 amino acids from that of yeast.

Amino acid changes are the product of nucleotide changes, and more than one nucleotide change may be required to change a given amino acid. When the necessary nucleotide changes that account for all amino acid differences in a protein are totaled, the **minimal mutational distance** between the genes of any two species is established. Column (b) in Table 25.5 shows such an analysis of the genes encoding cytochrome c. As expected, these values are larger than the corresponding number of amino acids separating humans from the other nine organisms listed.

Data on the minimal mutational distances between the cytochrome c genes of 19 organisms was used to reconstruct their evolutionary history. The result is a phylogenetic tree showing the relationships among the species studied (**Figure 25–22**). The black dots on the tips of the branches represent existing species, whose inferred common ancestors are linked to them by green lines that diverge at nodes (red dots). The common ancestors are connected to still earlier common ancestors, culminating in a single common ancestor for all the species on the tree, represented by the red dot on the extreme left.

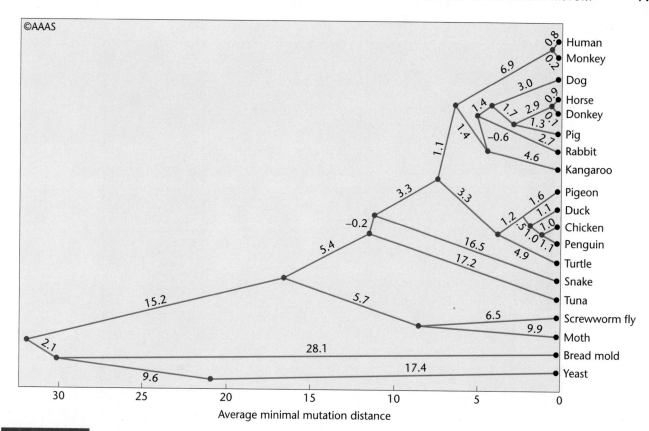

Average minimal mutation distance

FIGURE 25-22 Phylogenetic tree constructed by comparing homologies in cytochrome c amino acid sequences. *Reprinted with permission from Fitch, W.M. and Margoliash, E. 1967. Construction of phylogenetic trees.* Science *279: 279–284, Figure 2.*

Molecular Clocks Measure the Rate of Evolutionary Change

In many cases, we would like to estimate not only which members of a set of species are most closely related, but also when their common ancestors lived. Sometimes we can do so, thanks to **molecular clocks**—amino acid sequences or nucleotide sequences in which evolutionary changes accumulate at a constant rate over time.

Research on the influenza A virus shows how molecular clocks are constructed and used. Parts of the hemagglutinin gene from viral strains isolated at various times over a 20-year period were sequenced. A phylogenetic tree was constructed using the number of nucleotide differences among these strains [Figure 25–23(b)]. Most of the strains are now extinct and have no descendants. Plotting the number of nucleotide substitutions between strains against the year in which they were isolated shows that the points fall very close to a straight line [**Figure 25–23(a)**]. This means that the nucleotide substitutions in the hemagglutinin gene have accumulated at a steady rate, and thus serve as a molecular clock to track the evolution of the influenza A virus.

This molecular clock is also used to compare the sequence of the hemagglutinin gene of new flu viruses as they appear each year and to estimate the time that has passed since each diverged from a common ancestor.

Molecular clocks must be carefully calibrated and used with caution. For example, the data indicate that strains of influenza A that jump from birds to humans have evolved much more rapidly than strains that have remained in birds. Hence, a molecular clock calibrated from human strains of the virus would be highly misleading if applied to bird strains.

Analysis of Genetic Divergence between Neanderthals and Modern Humans

Fossil evidence indicates that the Neanderthals, *Homo neanderthalensis,* lived in Europe and western Asia from some 300,000 years ago until they disappeared about 30,000 years ago. For at least 30,000 years, Neanderthals coexisted with anatomically modern humans (*Homo sapiens*) in several regions. Genetic analysis using DNA sequencing has helped answer several questions about Neanderthals and modern humans: (1) Were Neanderthals direct ancestors of modern humans? (2) Did Neanderthals and our species, *H. sapiens,* interbreed, so that our genomes carry Neanderthal genes? Or did the Neanderthals die off and become extinct leaving no genetic heritage? (3) What can we say about the similarities and differences between our genome and that of the Neanderthals?

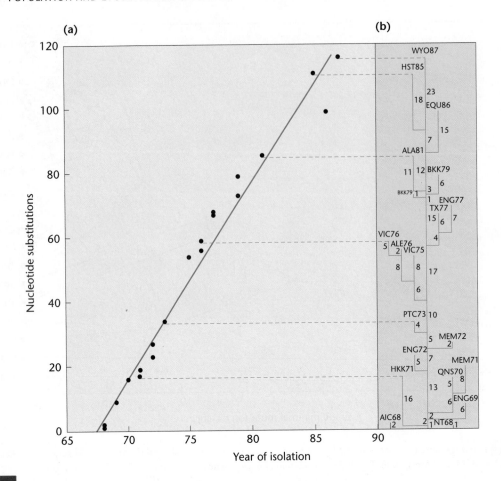

(a)

(b)

FIGURE 25–23   Phylogenetic molecular clock in the influenza A hemagglutinin gene. (a) Number of nucleotide differences between the first isolate as a function of year of isolation. (b) Estimate of the phylogeny of the isolates.

To answer these and other questions, researchers used two approaches, both of which involved sequencing DNA recovered from Neanderthal bones. One approach analyzed mitochondrial DNA recovered from skeletons found in Feldhofer cave in Germany and Mezmaiskaya cave in the Caucasus Mountains east of the Black Sea. The Feldhofer DNA and mitochondrial DNA (mtDNA) from more than 2000 present-day humans were used to construct a phylogenetic tree [**Figure 25–24**(a)]. From the structure of the tree, researchers concluded that Neanderthals are a distant relative of modern humans. The mtDNA recovered from the Mezmaiskaya skeleton was sequenced and compared to the sequence from the Feldhofer skeleton. Although the two Neanderthal mtDNA sequences are from locations more than 1000 miles apart, they vary by only about 3.5 percent. This indicates that the Neanderthal samples derive from a single gene pool. Phylogenetic analysis comparing the two Neanderthal sequences with those of modern humans and chimpanzees places the Neanderthals in a group that is clearly distinct from modern humans [Figure 25–24(b)]. The conclusion from the study of mitochondrial DNA is that while Neanderthals and humans have a common ancestor, the

Neanderthals were a separate hominid line and did not contribute mitochondrial genes to *H. sapiens*.

The Neanderthal Genome Project

The second approach to studying Neanderthal DNA was an ambitious project to isolate and sequence the Neanderthal genome. The international research team began by isolating DNA from three female Neanderthals whose skeletons were recovered from a cave in Croatia. Using newly developed DNA sequencing techniques, they assembled a draft sequence of the Neanderthal genome. As part of the project, they also sequenced the genomes of five individuals living in different parts of the world.

The Neanderthal genome contains about 3.2 billion base pairs, the same as the genome of our species, and is 99.7 percent identical to our genome. Both the Neanderthal genome and our genome are about 98.8 percent identical to that of the chimpanzee.

By comparing the Neanderthal genome sequence with the genomes of five present-day humans and with the chimpanzee genome, researchers were able to identify amino acid-coding differences in 78 genes that developed after the

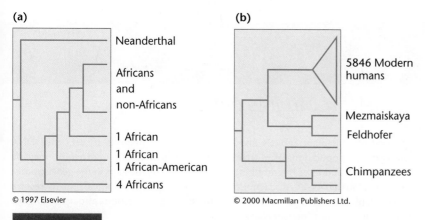

(a)

Neanderthal

Africans and non-Africans

1 African

1 African
1 African-American
4 Africans

© 1997 Elsevier

(b)

5846 Modern humans

Mezmaiskaya

Feldhofer

Chimpanzees

© 2000 Macmillan Publishers Ltd.

FIGURE 25–24 (a) A phylogenetic tree estimated from mitochondrial DNA sequences of one Neanderthal and over 2000 modern humans. *From Krings, M. et al., 1997. Cell 90: 19–30. Figure 7A, p. 26, copyright 1997 with permission from Elsevier;* (b) A phylogenetic tree estimated from analysis of over 5000 modern humans, the Neanderthal samples from the Feldhofer cave and the Mezmaiskaya cave, and from chimpanzees.

From Ovchinnikov, I.V. et al., 2000. Molecular analysis of Neanderthal DNA from the northern Caucasus. Nature 404: 490–493 copyright 2000 Macmillan Publishers Ltd.

split between the Neanderthal and human lineages. These genes are involved in cognitive development, skin morphology, and skeletal development. In addition, there is evidence that several regions of the modern human genome have undergone positive selection in the interval since Neanderthals and modern humans last shared a common ancestor.

Perhaps the most unexpected finding in this analysis is that the genomes of present-day humans in Europe and Asia, but not Africa, are composed of 1–4 percent Neanderthal sequences. Interbreeding with Neanderthals may have occurred somewhere in the Middle East, before humans migrated into Europe and Asia. No Neanderthal contributions

to African genomes were detected, but it is possible that a larger sampling of African populations will determine whether some Africans carry Neanderthal genes.

Comparative genomics combined with the fossil record has allowed scientists to construct a phylogenetic tree showing the pattern and times of divergence of Neanderthals and our species (**Figure 25–25**). Assuming that chimpanzees and humans last shared a common ancestor about 6.5 million years ago, the tree shows that Neanderthals and humans last shared a common ancestor about 706,000 years ago and that the isolating split between Neanderthals and human populations occurred about 370,000 years ago. From these studies, several conclusions can be drawn. First, Neanderthals are not direct ancestors of our species. Second, Neanderthals and members of our species may have interbred, but from the results available, it appears that Neanderthals did not make major contributions to our genome. As a species, Neanderthals are extinct, but some of their genes survive as part of our genome. Third, from what we know about the Neanderthal genome, we share most genes and other sequences with them.

More exciting answers to the question about the similarities and differences between our genome and that of Neanderthals will be derived from further analysis and will allow us to identify key differences that define our species and, in the process, revolutionize the field of human evolution.

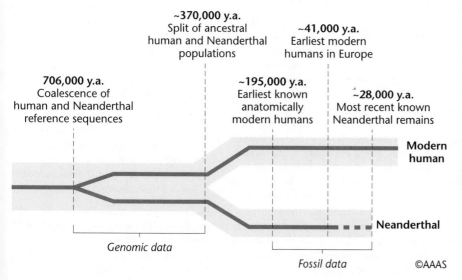

~370,000 y.a.
Split of ancestral human and Neanderthal populations

~41,000 y.a.
Earliest modern humans in Europe

706,000 y.a.
Coalescence of human and Neanderthal reference sequences

~195,000 y.a.
Earliest known anatomically modern humans

~28,000 y.a.
Most recent known Neanderthal remains

Modern human

Neanderthal

Genomic data

Fossil data

©AAAS

— Evolutionary lineage of human and Neanderthal reference sequences

  Evolutionary lineage of ancestral human and Neanderthal populations

FIGURE 25–25 Estimated times of divergence of human and Neanderthal genomic sequences and, subsequently, of their populations, relative to landmark events in both human and Neanderthal evolution. These estimates are based on sequencing about 65,000 base pairs of Neanderthal DNA (y.a. = years ago).

From Noonan, J.P. et al. 2006. Sequencing and analysis of Neanderthal genomic DNA. Science 314: 1113–1118.

Tracking Our Genetic Footprints out of Africa

Over the last century, paleoanthropologists have applied a variety of sophisticated scientific tools to explore our origins and human kinships.

Based on the physical traits and distribution of hominid fossils, most paleoanthropologists agree that a large-brained, tool-using hominid they call *Homo erectus* appeared in east Africa about 2 million years ago. This species used simple stone tools and hunted, but did not fish, build houses or fireplaces, or follow ritual burial practices. About 1.7 million years ago, *H. erectus* spread into Eurasia and south Asia. Most scientists also agree that *H. erectus* likely developed into several hominid types, including Neanderthals (in Europe) and Peking man or Java man (in Asia). These hominids were anatomically robust, with large, heavy skeletons and skulls. Neanderthals and other *H. erectus* groups disappeared 50,000 to 30,000 years ago—around the same time that anatomically modern humans (*H. sapiens*) appeared all over the world. It is at this point in our history—when ancient hominids gave way to anatomically modern humans—that controversy arises.

At present, two main hypotheses explain the origins of modern humans: the multiregional hypothesis and the out-of-Africa hypothesis. The multiregional hypothesis is based primarily on archaeological and fossil evidence. It proposes that *H. sapiens* developed gradually and simultaneously all over the world from existing *H. erectus* groups, including Neanderthals. Interbreeding between these groups eventually made *H. sapiens* a genetically homogeneous species. Natural selection over 1.5 million years then created the regional variants that we see today. In the multiregional view, our genetic makeup should include contributions from many *H. erectus* groups, including Neanderthals. In contrast, the out-of-Africa hypothesis, based primarily on genetic analyses of modern human populations, contends that *H. sapiens* evolved from the descendants of *H. erectus* in sub-Saharan Africa about 200,000 years ago. A small band of *H. sapiens* (fewer than 1000) then left Africa, expanded, and migrated into Europe and Asia around 60,000 years ago. By about 40,000 years ago, populations of *H. sapiens* reached Australia and later migrated into North America. In the out-of-Africa model, *H. sapiens* replaced all the preexisting *H. erectus* types, without interbreeding. In this way, *H. sapiens* became the only species in the genus by about 30,000 years ago.

Although the out-of-Africa hypothesis is still debated, most genetic evidence appears to support it. Humans all over the globe are remarkably similar genetically. DNA sequences from any two people chosen at random are 99.9 percent identical. More genetic identity exists between two persons chosen at random from a human population than between two chimpanzees chosen at random from a chimpanzee population. Interestingly, about 90 percent of the genetic differences that do exist occur between individuals rather than between populations. This unusually high degree of genetic relatedness in all humans around the world supports the idea that our species arose recently from a small founding group of humans.

Studies of mitochondrial DNA sequences from current human populations reveal that the highest levels of genetic variation occur within African populations. Africans show twice the mitochondrial DNA-sequence diversity of non-Africans. This implies that the earliest branches of *H. sapiens* diverged in Africa and had a longer time to accumulate mitochondrial DNA mutations, which are thought to accumulate at a constant rate over time.

DNA sequences from mitochondrial, Y-chromosome, and chromosome-21 markers support the idea that human roots are in east Africa and that the migration out of Africa occurred through Ethiopia, along the coast of the Arabian Peninsula, and outward to Eurasia and Southeast Asia. Recent data based on nuclear microsatellite variants and whole-genome single nucleotide polymorphism (SNP) analysis further support the notion that humans migrated out of Africa and dispersed throughout the world from a small founding population.

As with any explanation of human origins, the out-of-Africa hypothesis is actively debated. As methods to sequence DNA from ancient fossils improve, it may be possible to fill the gaps in the genetic pathway leading out of Africa and to resolve those age-old questions about our origins.

Your Turn

1. The sequencing of Neanderthal DNA shows that it is so different from ours that Neanderthals may have been a separate species and that Neanderthals and *H. sapiens* diverged about 600,000 years ago. However, the debate continues about whether *H. sapiens* and Neanderthals interbred, and if so, to what extent. Discuss evidence that supports, and refutes, the idea that Neanderthals and *H. sapiens* interbred.

Start your investigations by reading Green, R.E. et al. 2010. A draft sequence of the Neanderthal genome. *Science* 328: 710–722.

2. If all people on Earth are very similar genetically, how did we come to have such a range of physical differences, which some describe as racial differences? How has modern genomics contributed to the debate about the validity and definition of the term *race*?

An interesting discussion of race and how genomic studies are changing our concepts of human ancestry can be found on the Wikipedia site: http://en.wikipedia.org/wiki/Race_ (classification_of_humans). *A study of human population structure, based on microsatellite analysis, can be found at* Rosenberg, N.A. et al. 2002. Genetic structure of human populations. *Science 298:* 2381–2385.

3. Geneticists study mitochondrial and Y-chromosome DNA to determine the ancestry of modern humans. Why are these two types of DNA used in lineage studies? What is meant by the terms *mitochondrial Eve* and *Y-chromosome Adam*?

To read the original paper hypothesizing a mitochondrial Eve, see Cann, R.L. et al. 1987. Mitochondrial DNA and human evolution. *Nature* 325: 31–36. *For a discussion of Y-chromosome Adam, see* Gibbons, A. 1997. Y Chromosome shows that Adam was an African. *Science* 278: 804–805.

CASE STUDY | An unexpected outcome

A newborn screening program identified a baby with a rare autosomal recessive disorder called arginosuccinate aciduria (AGA), which causes high levels of ammonia to accumulate in the blood. Symptoms usually appear in the first week after birth and can progress to include severe liver damage, developmental delay, and mental retardation. AGA occurs with a frequency of about 1 in 70,000 births. There is no history of this disorder in either the father's or mother's family. This case raises several questions:

1. Since it appears that the unaffected parents are heterozygotes, would it be considered unusual that there would be no family history of the disorder? How would they be counseled about risks to future children?

2. If the disorder is so rare, what is the frequency of heterozygous carriers in the population?

3. What are the chances that two heterozygotes will meet and have an affected child?

Summary Points

 For activities, animations, and review quizzes, go to the study area at www.masteringgenetics.com

1. Genetic variation is a characteristic of most populations. In some cases, this can be observed at the phenotypic level, but analysis at the amino acid and nucleotide levels provides a more direct way to estimate genetic variation.

2. Using the assumptions of the Hardy–Weinberg Law, it is possible to estimate allele and gene frequencies in populations.

3. The Hardy-Weinberg Law can be applied to determining allele and genotype frequencies for multiple alleles and X-linked alleles, as well as calculating the frequency of heterozygotes for a given gene.

4. Natural selection changes allele frequency in populations leading to evolutionary change. Selection for quantitative traits can involve directional selection, stabilizing selection, or disruptive selection.

5. In addition to natural selection, other forces act on allele frequencies in populations. These include mutation, migration,

and genetic drift. Nonrandom mating changes genotype frequencies but does not change allele frequencies.

6. The formation of new species depends on the formation of subpopulations and the accumulation of enough genetic differences that, when reunited, members of the separated populations cannot interbreed.

7. Phylogenetic analysis using morphology, amino acid sequences, or nucleotide sequences can be used to construct phylogenetic trees showing the evolutionary relationships among a group of organisms. When calibrated with molecular clocks, the evolutionary changes on a phylogenetic tree can be calibrated with a time scale.

8. Phylogenetic analysis combined with genome sequence data from humans, Neanderthals, and other hominids has helped scientists reconstruct the relationship between Neanderthals and our species.

INSIGHTS AND SOLUTIONS

1. Tay–Sachs disease is caused by loss-of-function mutations in a gene on chromosome 15 that encodes a lysosomal enzyme. Tay–Sachs is inherited as an autosomal recessive condition. Among Ashkenazi Jews of Central European ancestry, about 1 in 3600 children is born with the disease. What fraction of the individuals in this population are carriers?

Solution: If we let p represent the frequency of the wild-type enzyme allele and q the total frequency of recessive loss-of-function alleles, and if we assume that the population is in Hardy–Weinberg equilibrium, then the frequencies of the genotypes are given by p^2 for homozygous normal, $2pq$ for carriers, and q^2 for individuals with Tay–Sachs. The frequency of Tay–Sachs alleles is thus

$$q = \sqrt{q^2} = \sqrt{\frac{1}{3600}} = 0.017$$

Since $p + q = 1$, we have

$$p = 1 - q = 1 - 0.017 = 0.983$$

Therefore, we can estimate that the frequency of carriers is

$$2pq = 2(0.983)(0.017) = 0.033 \text{ or about 1 in 30}$$

2. A single plant twice the size of others in the same population suddenly appears. Normally, plants of that species reproduce by self-fertilization and by cross-fertilization. Is this new giant plant simply a variant, or could it be a new species? How would you determine which it is?

Solution: One of the most widespread mechanisms of speciation in higher plants is polyploidy, the multiplication of entire sets of chromosomes. The result of polyploidy is usually a larger plant with larger flowers and seeds. There are two ways of testing the new variant to determine whether it is a new species. First, the giant plant should be crossed with a normal-sized plant to see whether the giant plant produces viable, fertile offspring. If it does not, then the two different types of plants would appear to be reproductively isolated. Second, the giant plant should be cytogenetically screened to examine its chromosome complement. If it has twice the number of its normal-sized neighbors, it is a tetraploid that may have arisen spontaneously. If the chromosome number differs by a factor of two and the new plant is reproductively isolated from its normal-sized neighbors, it is a new species.

Problems and Discussion Questions

 For instructor-assigned tutorials and problems, go to www.masteringgenetics.com

HOW DO WE KNOW?

1. In this chapter we focused on how changes in gene frequency are related to the process of species formation. At the same time, we found many opportunities to consider the methods and reasoning by which much of this information was acquired. From the explanations given in the chapter, what answers would you propose to the following fundamental questions?
   (a) How do we determine how much genetic variation exists in a population?
   (b) What evidence has been obtained from laboratory studies to indicate that natural selection is the cause of genetic differences among natural populations of a species?
   (c) How can the minimum genetic divergence between two species be determined?
   (d) How can we determine the last common ancestor shared by two divergent species?

2. What types of nucleotide substitutions will not be detected by amino acid sequencing of a gene's protein product?

3. The genetic difference between two *Drosophila* species, *D. heteroneura* and *D. sylvestris*, as measured by nucleotide diversity, is about 1.8 percent. The difference between chimpanzees (*P. troglodytes*) and humans (*H. sapiens*) is about the same, yet the latter species are classified in different genera. In your opinion, is this valid? Explain why.

4. The use of nucleotide sequence data to measure genetic variability is complicated by the fact that the genes of higher eukaryotes are complex in organization and contain 5′ and 3′ flanking regions as well as introns. Researchers have compared the nucleotide sequence of two cloned alleles of the γ-globin gene from a single individual and found a variation of 1 percent. Those differences include 13 substitutions of one nucleotide for another and 3 short DNA segments that have been inserted in one allele or deleted in the other. None of the changes takes place in the gene's exons (coding regions). Why do you think this is so, and should it change our concept of genetic variation?

5. Calculate the frequencies of the *AA*, *Aa*, and *aa* genotypes after one generation if the initial population consists of 0.2 *AA*, 0.6 *Aa*, and 0.2 *aa* genotypes and meets the requirements of the Hardy–Weinberg relationship. What genotype frequencies will occur after a second generation?

6. Consider a rare disorder in a population caused by an autosomal recessive mutation. From the frequencies of the disorder in the population given, calculate the percentage of heterozygous carriers:
   (a) 0.0064
   (b) 0.000081
   (c) 0.09
   (d) 0.01
   (e) 0.10

7. What must be assumed in order to validate the answers in Problem 6?

8. In a population where only the total number of individuals with the dominant phenotype is known, how can you calculate the percentage of carriers and homozygous recessives?

9. Determine whether the following two sets of data represent populations that are in Hardy–Weinberg equilibrium (use χ^2 analysis if necessary):
   (a) *CCR5* genotypes: *1/1*, 60 percent; *1/Δ32*, 35.1 percent; *Δ32/Δ32*, 4.9 percent
   (b) Sickle-cell hemoglobin: *AA*, 75.6 percent; *AS*, 24.2 percent; *SS*, 0.2 percent

10. If 4 percent of a population in equilibrium expresses a recessive trait, what is the probability that the offspring of two individuals who do not express the trait will express it?

11. Consider a population in which the frequency of allele *A* is $p = 0.7$ and the frequency of allele *a* is $q = 0.3$, and in which the alleles are codominant. What will be the allele frequencies after one generation if the following occurs?
    (a) $w_{AA} = 1, w_{Aa} = 0.9$, and $w_{aa} = 0.8$
    (b) $w_{AA} = 1, w_{Aa} = 0.95$, and $w_{aa} = 0.9$
    (c) $w_{AA} = 1, w_{Aa} = 0.99, w_{aa} = 0.98$
    (d) $w_{AA} = 0.8, w_{Aa} = 1, w_{aa} = 0.8$

12. If the initial allele frequencies are $p = 0.5$ and $q = 0.5$ and allele *a* is a lethal recessive, what will be the frequencies after 1, 5, 10, 25, 100, and 1000 generations?

13. Under what circumstances might a lethal dominant allele persist in a population?

14. Assume that a recessive autosomal disorder occurs in 1 of 10,000 individuals (0.0001) in the general population and that in this population about 2 percent (0.02) of the individuals are carriers for the disorder. Estimate the probability of this disorder occurring in the offspring of a marriage between first cousins. Compare this probability to the population at large.

15. One of the first Mendelian traits identified in humans was a dominant condition known as *brachydactyly*. This gene causes an abnormal shortening of the fingers or toes (or both). At the time, some researchers thought that the dominant trait would spread until 75 percent of the population would be affected (because the phenotypic ratio of dominant to recessive is 3:1). Show that the reasoning was incorrect.

16. Achondroplasia is a dominant trait that causes a characteristic form of dwarfism. In a survey of 50,000 births, five infants with achondroplasia were identified. Three of the affected infants had affected parents, while two had normal parents. Calculate the mutation rate for achondroplasia and express the rate as the number of mutant genes per given number of gametes.

17. A recent study examining the mutation rates of 5669 mammalian genes (17,208 sequences) indicates that, contrary to popular belief, mutation rates among lineages with vastly different generation lengths and physiological attributes are remarkably constant (Kumar, S., and Subramanian, S. 2002. *Proc. Natl. Acad. Sci. [USA]* 99: 803–808). The average rate is estimated at 12.2×10^{-9} per bp per year. What is the significance of this finding in terms of mammalian evolution?

18. List the barriers that prevent interbreeding and give an example of each.

19. What are the two groups of reproductive isolating mechanisms? Which of these is regarded as more efficient, and why?

20. What is the neutral theory of molecular evolution as proposed by Kimura (1968)? What determines the frequency of neutral alleles in a population?

21. What are considered significant factors in maintaining the surprisingly high level of genetic variation in natural populations?

22. A botanist studying waterlilies in an isolated pond observed three leaf shapes in the population: round, arrowhead, and scalloped. Marker analysis of DNA from 125 individuals showed the round-leafed plants to be homozygous for allele *r1*, while the plants with arrowhead leaves were homozygous for a different allele at the same locus, *r2*. Plants with scalloped leaves showed DNA profiles with both the *r1* and *r2* markers. Frequency of the *r1* marker was estimated at 0.81. If the botanist counted 20 plants with scalloped leaves in the pond, what is the inbreeding coefficient *F* for this population?

23. A farmer plants transgenic Bt corn that is genetically modified to produce its own insecticide. Of the corn borer larvae feeding on these Bt corn plants, only 10 percent survive unless they have at least one copy of the dominant resistance allele *B* that confers resistance to the Bt insecticide. When the farmer first plants Bt corn, the frequency of the *B* resistance allele in the corn borer population is 0.02. What will be the frequency of the resistance allele after one generation of corn borers fed on Bt corn?

24. In an isolated population of 50 desert bighorn sheep, a mutant recessive allele *c* has been found to cause curled coats in both males and females. The normal dominant allele *C* produces straight coats. A biologist studying these sheep counts four with curled coats and takes blood samples for DNA marker analysis, which reveals that 17 of the straight-coated sheep are carriers of the *c* allele. What is the inbreeding coefficient *F* for this population?

25. To increase genetic diversity in the bighorn sheep population described in Problem 24, 10 sheep are introduced from a population in a different region where the *c* mutation is absent. Assuming that random mating occurs between the original and the introduced sheep, and that the *c* allele is selectively neutral, what will be the frequency of *c* in the next generation?

Extra-Spicy Problems

 For instructor-assigned tutorials and problems, go to www.masteringgenetics.com

26. A form of dwarfism known as Ellis–van Creveld syndrome was first discovered in the late 1930s, when Richard Ellis and Simon van Creveld shared a train compartment on the way to a pediatrics meeting. In the course of conversation, they discovered that they each had a patient with this syndrome. They published a description of the syndrome in 1940. Affected individuals have a short-limbed form of dwarfism and often have defects of the lips and teeth, and polydactyly (extra fingers). The largest pedigree for the condition was reported in an Old Order Amish population in eastern Pennsylvania by Victor McKusick and his colleagues (1964). In that community, about 5 per 1000 births are affected, and in the population of 8000, the observed frequency is 2 per 1000. All affected individuals have unaffected parents, and all affected cases can trace their ancestry to Samuel King and his wife, who arrived in the area in 1774. It is known that neither King nor his wife was affected with the disorder. There are no cases of the disorder in other Amish communities, such as those in Ohio or Indiana.

(a) From the information provided, derive the most likely mode of inheritance of this disorder. Using the Hardy–Weinberg Law, calculate the frequency of the mutant allele in the population and the frequency of heterozygotes, assuming Hardy–Weinberg conditions.

(b) What is the most likely explanation for the high frequency of the disorder in the Pennsylvania Amish community and its absence in other Amish communities?

27. The original source of new alleles, upon which selection operates, is mutation, a random event that occurs without regard to selectional value to the organism. Although many model organisms have been used to study mutational events in populations, some investigators have developed abiotic, molecular models. Soll (2006) examined one such model to study the relationship between both deleterious and advantageous mutations and population size in a ligase molecule composed of RNA (ribozyme). Soll found that the smaller the population of molecules, the more likely it was that not only deleterious mutations but also advantageous mutations would disappear. Why would population size influence the survival of both types of mutations (deleterious and advantageous) in populations?

28. A number of comparisons of nucleotide sequences among hominids and rodents indicate that inbreeding may have occurred more in hominid than in rodent ancestry. When a population bottleneck of approximately 10,000 individuals occurred, Knight (2005) and Bakewell (2007) suggest that this may have left early humans with a greater chance for genetic disease. Why would a population bottleneck influence the frequency of genetic disease?

29. Shown below are two homologous lengths of the α and β chains of human hemoglobin. Consult the genetic code dictionary (Figure 13.7) and determine how many amino acid substitutions may have occurred as a result of a single-nucleotide substitution. For any that cannot occur as the result of a single change, determine the minimal mutational distance.

α:	Ala	Val	Ala	His	Val	Asp	Asp	Met	Pro
β:	Gly	Leu	Ala	His	Leu	Asp	Asn	Leu	Lys

30. Determine the minimal mutational distances between these amino acid sequences of cytochrome c from various organisms. Compare the distance between humans and each organism.

Human:	Lys	Glu	Glu	Arg	Ala	Asp
Horse:	Lys	Thr	Glu	Arg	Glu	Asp
Pig:	Lys	Gly	Glu	Arg	Glu	Asp
Dog:	Thr	Gly	Glu	Arg	Glu	Asp
Chicken:	Lys	Ser	Glu	Arg	Val	Asp
Bullfrog:	Lys	Gly	Glu	Arg	Glu	Asp
Fungus:	Ala	Lys	Asp	Arg	Asn	Asp

31. Recent reconstructions of evolutionary history are often dependent on assigning divergence in terms of changes in amino acid or nucleotide sequences. For example, a comparison of cytochrome c shows 10 amino acid differences between humans and dogs, 24 differences between humans and moths, and 38 differences between humans and yeast. Such data provide no information as to the absolute times of divergence for humans, dogs, moths, and yeast. How might one calibrate the molecular clock to an absolute time clock? What problems might one encounter in such a calibration?

32. In a recent study of cichlid fish inhabiting Lake Victoria in Africa, Nagl et al. (1998. *Proc. Natl. Acad. Sci. [USA]* 95: 14,238–14,243) examined suspected neutral sequence polymorphisms in noncoding genomic loci in 12 species and their putative river-living ancestors. At all loci, the same polymorphism was found in nearly all of the tested species from Lake Victoria, both lake-dwelling and river-dwelling. Different polymorphisms at these loci were found in cichlids at other African lakes.

    (a) Why would you suspect neutral sequences to be located in noncoding genomic regions?

    (b) What conclusions can be drawn from these polymorphism data in terms of cichlid ancestry in these lakes?

33. Given that there are approximately 400 cichlid species in Lake Victoria and that it dried up almost completely about 14,000 years ago, what evidence indicates that extremely rapid evolutionary adaptation rather than extensive immigration occurred?

34. What genetic changes take place during speciation?

35. Some critics have warned that the use of gene therapy to correct genetic disorders will affect the course of human evolution. Evaluate this criticism in light of what you know about population genetics and evolution, distinguishing between somatic gene therapy and germ-line gene therapy.

36. Comparisons of Neanderthal mitochondrial DNA with that of modern humans indicate that they are not related to modern humans and did not contribute to our mitochondrial heritage. However, this does not necessarily rule out some forms of interbreeding causing the modern European gene pool to be derived from both Neanderthals and early humans (called Cro-Magnons). Before the Neanderthal genome was sequenced, Caramelli et al. (2003. *Proc. Natl. Acad. Sci. [USA]* 100: 6593–6597) analyzed mitochondrial DNA sequences from 25,000-year-old Cro-Magnon (early modern humans belonging to our species) remains and compared them to four Neanderthal specimens and a large dataset derived from present-day modern humans. The results are shown in the graph.

    The x-axis represents the age of the specimens in thousands of years; the y-axis represents the average genetic distance. Modern humans are indicated by filled squares; Cro-Magnons, open squares; and Neanderthals, diamonds.

    (a) What can you conclude about the relationship between Cro-Magnons and modern Europeans? What about the relationship between Cro-Magnons and Neanderthals?

    (b) From these data, does it seem likely that Neanderthals made any contributions to the Cro-Magnon gene pool or the modern European gene pool?

    (c) How can you reconcile these results with those obtained from comparing the nuclear genome sequence of Neanderthals and modern humans?

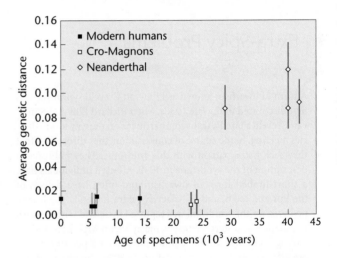

26

Conservation Genetics

CHAPTER CONCEPTS

- Species require genetic diversity for long-term survival and adaptation.

- Small, isolated populations are particularly vulnerable to genetic effects.

- Endangered populations that recover in size may not redevelop genetic diversity.

- Conservation and breeding efforts focus on retaining genetic diversity for long-term species survival.

A s the twenty-first century progresses, the diversity of life on Earth is under increasing pressure from the direct and indirect effects of explosive human population growth. Approximately 10 million *Homo sapiens* lived on the planet 10,000 years ago. This number grew to 100 million 2000 years ago and to 2.5 billion by 1950. Within the span of a single lifetime, the world's human population more than doubled to 5.5 billion in 1993 and is projected to reach as high as 19 billion by 2100 (Figure 26–1).

The effect of accelerating human population growth on other species has been dramatic. Data from the 2010 World Conservation Union (IUCN) Red List of Threatened Species show that globally, 25% of mammalian, 13% of avian, 41% of amphibian, 33% of reef-forming coral, and 30% of conifer species are threatened. The 2008 Red List data show that 8400 species of vascular plant species are also threatened worldwide. Not only wild species are at risk. Genetic diversity in domesticated plants and animals is also being lost as many traditional crop varieties and livestock breeds disappear. The Food and Agriculture Organization (FAO) estimates that since 1900, 75 percent of the genetic diversity in agricultural crops has been lost. Out of approximately 5000 different breeds of domesticated farm animals worldwide, one-third are at risk.

Why should we be concerned about losing **biodiversity**— that is, *the biological variation represented by these different plants and animals*? After all, the fossil record shows that many different plants and animals have become extinct in the course of evolution, even before humans existed, and other species have taken their places. Biologists are concerned, however, at the accelerated rate of species extinctions we are witnessing today, all of which can be ascribed directly or indirectly to human impact. Deliberate hunting

FIGURE 26–2   The brown tree snake (*Boiga irregularis*).

or harvesting of plants and animals by humans, and habitat destruction through human development activities have greatly reduced the populations of many species. An additional problem in many parts of the world is the deliberate or accidental introduction by humans of invasive nonnative plants or animals that prey on or compete with native species, further jeopardizing their survival. For example, the brown tree snake, *Boiga irregularis* (Figure 26–2), was accidentally transported by ship in the late 1940s to the island of Guam. An aggressive predator, the brown tree snake has so far caused the extinction of 12 bird and 4 lizard species that were previously native to Guam. This introduced species continues to threaten native biodiversity on the island.

One of the greatest threats to biodiversity, however, may be climate change due to global warming. Scientists are increasingly concerned that this phenomenon will lead to the extinction of many species, especially those adapted to live in colder environments. For example, the IUCN 2008 Red List includes the polar bear as an endangered species and classifies the 22,000 remaining polar bears (Figure 26–3) as

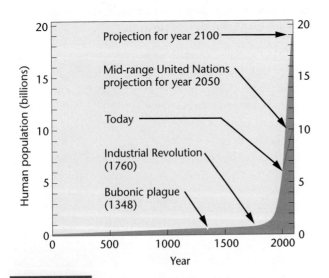

FIGURE 26–1   Growth in human population over the past 2000 years and projected through 2100.

FIGURE 26–3   The polar bear (*Ursus maritimus*).

vulnerable due to fears that their habitat will be lost from global warming, and a recent study by scientists from the Woods Hole Oceanographic Institute predicts that loss of sea ice will result in the disappearance of 95 percent of emperor penguin populations by 2100.

As a result of species extinction, biologists fear that **ecosystems**—*the complex webs of interdependent but diverse plants and animals found together in the same environment*—may collapse if key sustaining species are lost. This portends possible consequences for our own long-term survival. Other scientists have pointed out that we lose the unknown economic potential of unexploited plants and animals if we allow them to become extinct. Rare species sometimes turn out to be immensely valuable. For example, an obscure bacterium, *Thermus aquaticus*, was first discovered in a thermal pool in Yellowstone National Park and is known to exist in only a few hot springs throughout the world. DNA polymerase isolated from this microorganism functions at very high temperatures and is critical to the *polymerase chain reaction (PCR)*, a process for *in vitro* replication of DNA that we have alluded to throughout this book. This research technique is essential to the multibillion dollar global biotechnology industry. Quite apart from the practical benefits of biodiversity, scientists and nonscientists have made the case that human life will be diminished both spiritually and aesthetically if we do not strive to maintain the fascinating and often beautiful variety of living organisms that share the planet with us.

Conservation biologists work to understand and maintain biodiversity, studying the factors that lead to species decline and the ways species can be preserved. The new field of **conservation genetics** has emerged in the last 25 years as scientists have begun to recognize that genetics will be an important tool in maintaining and restoring population viability. The applications of genetics to conservation biology are multifaceted; in this chapter, we will explore just a few of them. Underlying the increasingly important role of genetics in conservation biology is the recognition that biodiversity depends on genetic diversity and that maintaining biodiversity in the long term is unlikely if genetic diversity is lost.

26.1

Genetic Diversity Is the Goal of Conservation Genetics

While *biodiversity* refers to the variation represented by all existing species of plants and animals on this planet at any given time, **genetic diversity** is not as easy to define or to study. Genetic diversity can be considered on two levels: *interspecific diversity* and *intraspecific diversity*.

Diversity between species, or **interspecific diversity**, is the diversity reflected in the number of different plant and

(a)  (b)

FIGURE 26–4 (a) A tropical rainforest is an ecosystem with high interspecific diversity. (b) A desert plain in Namibia, a country in southern Africa.

animal species present in an ecosystem. Some ecosystems have a very high level of interspecific diversity, such as a tropical rainforest in which hundreds of different plant and animal species may be found within a few square meters [**Figure 26–4**(a)]. Other ecosystems, especially where plants and animals must adapt to a harsh environment, may have much lower numbers of species [Figure 26–4(b)]. Species inventories—lists of the different plant and animal species found in a particular environment—are useful for identifying diversity hot spots: geographic areas with especially high interspecific diversity, where conservation efforts can be focused. Conservation biologists working at the ecosystem level are interested in preventing species from being lost and in restoring species that were once part of the system but are no longer present. The reestablishment of the gray wolf (*Canis lupus*) in Yellowstone National Park and the release of captive-bred California condors (*Gymnogyps californianus*) into the mountain ranges they previously occupied in the southwestern United States are examples of current attempts by conservation biologists to restore missing species to their ecosystems.

Intraspecific diversity—the diversity within a species—itself has two components: *Intrapopulation diversity* is the genetic variation occurring between individuals within a single population of a given species; and *interpopulation diversity* is the variation occurring between different populations of the same species. Genetic variation within populations can be measured as the frequency of individuals in the population that are heterozygous at a given locus or as the number of different alleles at a locus that are present in the population gene pool. When DNA-profiling techniques are used, the percentage of polymorphic loci—represented by bands on the DNA profile that are different in different individuals—can be calculated to indicate the extent of genetic diversity in a population. In outbreeding species, most intraspecific genetic diversity is found at the intrapopulation level.

Significant interpopulation diversity can occur if populations are separated geographically and there is no migration or exchange of gametes between them. Similarly, predominantly inbreeding species (such as self-fertilizing plants) tend to have greater levels of interpopulation than intrapopulation diversity. In these species, a limited number of genotypes dominate an individual population, but there is greater differentiation between populations. Understanding the breeding system and the distribution of genetic variation in an endangered species is important to its conservation, not only to ensure continued production of offspring but also to determine the best strategy for maintaining intraspecific diversity. For example, would it be more effective to preserve a few large populations or many small, distinct ones? This information can then be used to guide conservation or restoration efforts.

Loss of Genetic Diversity

Loss of genetic diversity in nondomesticated species is often associated with a reduction in population size owing in part to excessive hunting or harvesting. For example, biologists blame commercial overfishing for the collapse in the early 1990s of the deep-sea cod populations off the Newfoundland coast. The cod fishery in that area was once one of the most productive in the world, but despite being protected from fishing since 1992, these fish populations have not recovered. Habitat loss is also a major cause of population decline. As the global human population increases, more land is developed for housing and transport systems or is put into agricultural production, reducing or eliminating areas that were once home to wild plants and animals. The shrinking of available habitat reduces populations of wild species and often also isolates them from each other, as individual populations become trapped in pockets of undeveloped land surrounded by areas taken over for agriculture, urban development, or other human uses. This process is known as **population fragmentation**. When populations are no longer in contact with each other, **gene flow**, *the gradual exchange of alleles between two populations, brought about through migration or gamete exchange between them*, ceases, and an important mechanism for maintaining genetic variation is lost.

In domesticated species, loss of genetic diversity is not usually the result of habitat loss or collapsing population numbers; there is little risk that cows or corn as species will become extinct any time soon. Reduction in diversity within domesticated species can instead be traced to changes in agricultural practice and consumer demand. Modern farming techniques have greatly increased production levels, but they have also led to greater genetic uniformity. As farmers switch to new crop varieties or improved livestock strains on a large scale, they abandon cultivation of many older local types, which may then disappear if efforts are not made to preserve them.

For example, Seed Savers Exchange, an organization working to preserve traditional fruit and vegetable varieties, reports that in 1904 a total of 7098 apple varieties were recorded in North America by U.S. Department of Agriculture scientist W. H. Ragan. Fewer than 14 percent of those varieties can still be found today, with just 15 varieties accounting for over 90 percent of the apples sold in U.S. grocery stores. Modern varieties may be better adapted for modern agricultural production, but many older types contain useful genes that can still play a vital role in survival functions, such as resistance to disease, cold, or drought.

Identifying Genetic Diversity

For many years, population geneticists based estimates of intraspecific diversity on phenotypic differences between individuals, such as different colors of seeds or flowers or variation in markings (Figure 26–5). DNA analysis is a more modern and direct molecular approach for detecting and quantifying genetic differences between individuals. This technique has become an important tool for assessing intra- and interpopulation genetic variation. Nuclear, mitochondrial, and chloroplast DNA can all be analyzed to determine levels of genetic diversity. We have already described applications of DNA analysis using either restriction enzymes or PCR in Chapter 20. Conservation biologists have used similar techniques to examine the levels and distribution of genetic variation both within and between populations and to subsequently guide conservation efforts.

One technique that is widely used to analyze genetic diversity in plant and animal populations takes advantage of the presence within the genome of **STRs** (**short tandem repeats**),

FIGURE 26–5   Phenotypic variation in seed color and markings in the common bean (*Phaseolus vulgaris*) reveals high levels of intraspecific diversity.

also known as **microsatellites**, which were introduced in Chapter 12. STRs consist of short DNA sequences (two to nine bases) repeated a variable number of times and are found throughout the genome. The number of times a repeat is present at a given STR locus often varies between individuals. This variation is detected by sequencing the DNA regions on either side of the STR locus and then creating single-stranded DNA primers complementary to these flanking sequences that will selectively anneal to them during the polymerase chain reaction (PCR), which is explained in more detail in Chapter 20. This allows targeted amplification of the microsatellite regions so that they can be analyzed and the number of repeats determined. Although the STR regions themselves do not contain expressed genes, conservation biologists can use microsatellite variation among individuals within populations as an indirect estimate of the amount of overall genetic variation present.

For example, researchers from the University of Arizona have used microsatellite analysis to compare genetic variation in the three remaining populations of the Mauna Loa silversword, an endangered plant found only in Hawaii. Biologists working to conserve this plant needed to know whether they should try to increase genetic diversity by redistributing seed among the populations. However, STR analysis not only revealed unexpected amounts of diversity still present within populations but also showed that the three populations were genetically quite different from one another. The seed redistribution plan was abandoned in favor of preserving the unique identity of each population.

Another application of DNA markers such as microsatellites in conservation involves *PCR-based DNA fingerprinting*. The use of DNA-profiling techniques in forensic science was discussed in Chapter 22. Similar techniques can be used to uncover illegal trade in endangered plant and animal species that are protected by law. For example, although international trade in ivory is illegal, poaching of African elephants and smuggling of ivory worth millions of dollars continue. To determine the origin of ivory found in a secret compartment in a shipping container in Hong Kong in 2008, an international group of researchers used microsatellite markers to analyze DNA extracted from the seized ivory. They compared this microsatellite profile with more than 600 reference samples of DNA collected from the droppings of wild elephant populations across Africa and found that it matched those of forest elephants from southern Gabon. This finding provided a vital clue about poaching rings that can be used to combat this illicit trade.

Unlawful trade in animal products also occurs in the United States. In a recent study by scientists at the University of Florida, amplification of mtDNA sequences from turtle meat sold in Louisiana and Florida revealed that one quarter of the samples were actually alligator, not turtle. Most of the rest were from small fresh water turtles, indicating that populations of fresh water turtle species were declining through overharvesting and thus were in need of protection. Further refinements to DNA extraction and profiling techniques allowed accurate identification of illegally harvested sea turtle eggs cooked and served in restaurants, and led to prosecution of the poachers.

New methods for analyzing DNA marker data also permit estimation of past changes in populations, based on the rate at which mtDNA is known to accumulate spontaneous point mutations. In a controversial recent study, scientists at Stanford University led by Dr. Stephen Palumbi measured sequence diversity at key points in mtDNA from humpback whales and compared their findings with the amount of diversity expected for a given population size. The Stanford researchers found much more mtDNA sequence diversity than expected and concluded that before being hunted almost to extinction in the nineteenth and twentieth centuries, the global humpback whale population must have numbered over one million, ten times higher than historical estimates based on records from whaling ships. Because current international agreements could permit hunting of whales to resume when populations have recovered to 54 percent of what is thought to be their original carrying capacity, Palumbi's findings are important for the future management of this species.

26.2

Population Size Has a Major Impact on Species Survival

Some species have never been numerous, especially those that are adapted to survive in unusual habitats. Biologists refer to such species as *naturally rare. Newly rare* species, on the other hand, are those whose numbers are in decline because of pressures such as habitat loss. Populations of such species may not only be small in number but also fragmented and isolated from other populations. Both decreased population size and increased isolation have important genetic consequences, leading to an increased risk of loss of diversity.

How small must a population be before it is considered endangered? The answer varies somewhat with species, but small populations can quickly become vulnerable to genetic phenomena that increase the risk of extinction. In general, a population of fewer than 100 individuals is considered extremely sensitive to these problems, which include genetic drift, inbreeding, and reduction in gene flow. The effects of such problems on species survival are substantial. Studies of bighorn sheep, for example, have shown that populations of fewer than 50 are highly likely to become extinct within 50 years. Projections based on computer models show that for all species, populations of fewer than 10,000 are likely to be limited in adaptive genetic variation and that at least 100,000 individuals must be present if a population is to show long-term sustainability.

Determining the number of individuals a population must contain in order to have long-term sustainability is complicated by the fact that not all members of a population are equally likely to produce offspring: some will be infertile, too young, or too old. The *effective population size* (N_e) is defined as the number of individuals in a population having an equal probability of contributing gametes to the next generation. N_e is almost always smaller than the *absolute population size* (N). The effective population size can be calculated in different ways, depending on the factors that are preventing individuals in a population from contributing equally to the next generation. In a sexually reproducing population that contains different numbers of males and females, for example, the effective population size is calculated as

$$N_e = \frac{4(N_m N_f)}{N_m + N_f}$$

where N_m is the number of males and N_f the number of females in the population. Hence a population of 100 males and 100 females would have an effective size of $4(100 \times 100)/(100 + 100) = 200$. In contrast, if there were 180 males and only 20 females, the effective population size would be $4(180 \times 20)/(180 + 20) = 72$.

Effective population size is also influenced by fluctuations in absolute population size from one generation to the next. Here, the effective population size is the harmonic mean of the numbers in each generation, so

$$N_e = 1 \left/ \frac{1}{t}\left(\frac{1}{N_1} + \frac{1}{N_2} + \cdots + \frac{1}{N_t}\right)\right.$$

where t is the total number of generations being considered. For example, if a population went through a temporary reduction in size in generation 2, so that $N_1 = 100$, $N_2 = 10$, and $N_3 = 100$, then

$$N_e = 1 \left/ \frac{1}{3}\left(\frac{1}{100} + \frac{1}{10} + \cdots + \frac{1}{100}\right)\right. = \frac{1}{0.04} = 25$$

In this case, although the actual mean number of individuals in the population over three generations was 70, the effective population size during that time was only 25. A severe temporary reduction in size such as this is known as a **population bottleneck**. Bottlenecks occur when a population or species is reduced to a few reproducing individuals whose offspring then increase in numbers over subsequent generations to reestablish the population. Although the number of individuals may be restored to healthier levels, genetic diversity in the newly expanded population is often severely reduced because gametes from the handful of surviving individuals functioning as parents only represent a subset of the original gene pool.

26–1 A population of endangered lowland gorillas is studied by conservation biologists in the wild. The biologists count 15 gorillas but observe that the population consists of two harems, each dominated by a different single male. One harem contains eight females and the other has five. What is the effective population size of the gorilla population?

■ HINT: *This question asks you to calculate effective population size N_e for an endangered gorilla population. The key to its solution is to consider what unusual feature of the social structure of the gorilla population will affect N_e.*

Captive-breeding programs, in which a few surviving individuals from an endangered species are removed from the wild to rebuild the population in a protected environment, inevitably create population bottlenecks. Bottlenecks also occur naturally when a small number of individuals from one population migrate to establish a new population elsewhere. When a new population derived from a small subset of individuals has significantly less genetic diversity than the original population, it exhibits the **founder effect** (Chapter 25). Reduced levels of genetic diversity due to a founder effect can persist for many generations, as shown by studies of two species of Antarctic fur seal, *Arctocephalus gazella* and *Arctocephalus tropicalis*. Seal hunters in the eighteenth and nineteenth centuries severely reduced the populations of these species, eliminating them from parts of their natural range in the Southern Ocean.* Although the number of Antarctic fur seals has now rebounded and the two species have recolonized much of their original habitat, mtDNA fingerprinting reveals that a founder effect can still be detected in *A. gazella*, with reduced genetic variation in current populations descended from a handful of surviving individuals.

The cheetah (*Acinonyx jubatus*), shown in **Figure 26–6**, is another well-studied example of a species with reduced genetic variation resulting from at least one severe population bottleneck that occurred in its recent history. Allozyme studies of South African cheetah populations have shown levels of genetic variation that are less than 10 percent of those found in other mammals. The abnormal spermatozoa and poor reproductive rates commonly observed in cheetahs are thought to be linked to the lack of genetic diversity, although when and how the population bottleneck occurred in this species is still unclear.

---

*In 2000, this fifth world ocean was delimited by oceanographers, incorporating southern portions of the Atlantic, Indian, and Pacific oceans.

FIGURE 26–6 The cheetah (*Acinonyx jubatus*).

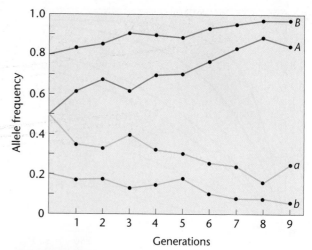

FIGURE 26–7 Change in frequencies over ten generations for two sets of alleles, *A/a* and *B/b*, in a theoretical population subject to genetic drift.

26.3

Genetic Effects Are More Pronounced in Small, Isolated Populations

Small isolated populations, such as those found in threatened and endangered species or produced by population fragmentation, are especially vulnerable to genetic drift, inbreeding, and reduction in gene flow. These phenomena act on the gene pool in different ways, but ultimately have similar effects in that they all can further reduce genetic diversity and long-term species viability.

Genetic Drift

If the number of breeding individuals in a population is reduced in size, fewer gametes will form the next generation. The alleles carried by these gametes may not be a representative sample of all those present in the population; purely by chance, some alleles may be underrepresented or not present at all, which will cause changes in allele frequency over time, resulting in **genetic drift**. This phenomenon has already been described in Chapter 25. A serious result of genetic drift in populations with a small effective population size is the loss of genetic variation. Genetic drift is a random process, so either deleterious or advantageous alleles can become fixed within a small population. In other words, one allele becomes the only version of that gene present in the gene pool of the population. This means a useful allele can be lost even if it has the potential to increase fitness or long-term adaptability. The probability that an allele will be fixed through drift is the same as its initial frequency. Thus, if a locus has two alleles, *A* and *a,* and if the initial frequency of *A* is 0.8, then the probability of *A* becoming fixed is 0.8, or 80 percent, and the probability of *A* being lost through drift

is 0.2 or 20 percent. Figure 26–7 is a graph of the effect of genetic drift on a theoretical population.

Inbreeding

In small populations, the chance of **inbreeding**—meaning mating between closely related individuals—is greater. As described in Chapter 25, inbreeding increases the proportion of homozygotes in a population, thus expanding the possibility that an individual may be homozygous for a deleterious allele. We have already seen in Chapter 25 how the extent of inbreeding in a population can be quantified as the **inbreeding coefficient** (*F*), which measures the probability that two alleles of a given gene are derived from a common ancestral allele. Remember that *the inbreeding coefficient is inversely related to the frequency of heterozygotes in the population* and can be calculated as

$$F = \frac{2pq - H}{2pq}$$

where $2pq$ is the expected frequency of heterozygotes based on the Hardy–Weinberg Law and *H* is the actual frequency of heterozygotes in the population.

In a declining population that has become small enough for drift to occur, heterozygosity (*H*) will decrease with each generation. The smaller the effective population size, the more rapid the decrease in *H* and the resulting increase in *F,* as can be seen from the equation

$$H_t/H_0 = \left(1 - \frac{1}{2N_e}\right)^t$$

where H_0 is the initial frequency of heterozygotes, H_t is the frequency of heterozygotes after t generations, and N_e is the effective population size. Figure 26–8 compares rates of increase in *F* for different effective population sizes.

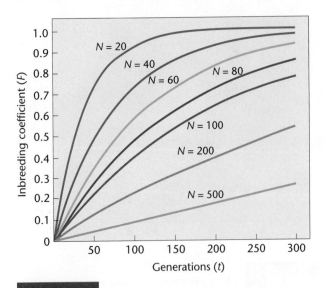

FIGURE 26–8 Increase in inbreeding coefficient (F) in theoretical populations as the population size (N) decreases.

What effect does inbreeding have on the long-term survival of a species? If numbers of individuals in a species remain high, inbreeding in a single population may not immediately reduce the amount of genetic variation present in

NOW SOLVE THIS

26–2 A wildlife biologist studied four generations of a population of rare Ethiopian jackals. When the study began, there were 47 jackals in the population, and analysis of microsatellite loci from these animals showed a heterozygote frequency of 0.55. In the second generation, an outbreak of distemper occurred in the population, and only 17 animals survived to adulthood. These jackals produced 20 surviving offspring, which in turn gave rise to 35 progeny in the fourth generation.

(a) What was the effective population size for the four generations of this study?

(b) Based on its effective population size, what is the heterozygote frequency of the jackal population in generation 4?

(c) What is the inbreeding coefficient in generation 4, assuming an inbreeding coefficient of $F = 0$ at the beginning of the study, no change in microsatellite allele frequencies in the gene pool, and random mating in all generations?

■ HINT: *This problem asks you to calculate effective population sizes, a heterozygote frequency, and an inbreeding coefficient for a jackal population that crashed and then recovered. The key to its solution is to first calculate N_e (the effective population size) for each generation, and remember that heterozygote frequency and inbreeding coefficient are inversely related.*

the overall gene pool. Self-pollinating plants, for example, often show high levels of homozygosity and relatively little genetic variation within single populations. However, they tend to have considerable variation between *different populations,* each of which has adapted to slightly different local environmental conditions. On the other hand, in most outbreeding species (those where gene flow occurs freely between populations), including all mammals, inbreeding is associated with reduced fitness and lower survival rates among offspring. This **inbreeding depression** can result from increased homozygosity for deleterious alleles. The number of deleterious alleles present in the gene pool of a population is called the **genetic load** (or *genetic burden*).

In some species, inbreeding accompanied by selection against less fit individuals homozygous for deleterious alleles has resulted in the elimination of these alleles from the gene pool, a process known as *purging the genetic load.* Species that have successfully purged their genetic load do not show continued reduction in fitness, even after many generations of inbreeding. This is true of numerous domesticated species, especially self-pollinating plants such as wheat. However, computer simulation experiments have shown that it may take 50 generations or more to complete the purging process, during which time inbreeding depression will still occur.

Alternatively, inbreeding depression can result from heterozygous individuals having a higher level of fitness than either of the corresponding homozygotes. In this case, the long-term survival of the population requires that inbreeding be avoided and that the levels of all alleles in the gene pool be maintained. Both of these requirements can be difficult to satisfy in a species that has already suffered a significant reduction in population size.

The effects of inbreeding depression and loss of genetic variation in a small, isolated population have been documented in the case of the Isle Royale gray wolves (Figure 26–9).

FIGURE 26–9 The Isle Royale gray wolf (*Canis lupus*).

Around 1950, a pair of gray wolves apparently crossed an ice bridge from the Canadian mainland to Isle Royale in Lake Superior. The island had no other wolves and had an abundance of moose, which became the wolves' main food source. By 1980, the Isle Royale wolf population had increased to over 50 individuals. Over the next decade, however, wolf numbers declined to fewer than a dozen with no new litters being born, despite plentiful food and no apparent sign of disease. The genetic variation of the remaining wolves was examined by mtDNA analysis and nuclear DNA fingerprinting. It was found that the Isle Royale wolves had levels of homozygosity that were twice as high as those of wolves in an adjacent mainland population. Furthermore, all the wolves possessed the same mtDNA genotype, consistent with descent from the same female. Hence, the degree of relatedness between individual Isle Royale wolves was equivalent to that of full siblings. This finding suggests that the wolves' reproductive failure was due to inbreeding depression, a phenomenon that has also been seen in captive-wolf populations.

Reduction in Gene Flow

Gene flow, described earlier as the gradual exchange of alleles between populations, is brought about by the dispersal of gametes or the migration of individuals. It is an important mechanism for introducing new alleles into a gene pool and increasing genetic variation. Migration is the main route for gene flow in animals; we examined the effect on population allele frequencies of migration and gene flow in Chapter 25. In plants, gene flow occurs not through movement of individuals but as a result of cross-pollination between different populations and through seed dispersal. Isolation and fragmentation of populations in rare and declining species significantly reduce gene flow and the potential for maintaining genetic diversity. As we have already discussed, habitat loss is a major threat to species survival. It is not unusual for a threatened or an endangered species to be restricted to small separate pockets of the remaining habitat. This isolates and fragments the surviving populations so that movement of individuals can no longer occur between them, thus preventing gene flow.

The term **metapopulation** describes *a population consisting of spatially separated subpopulations with limited gene flow*, especially if local extinctions and replacements of some of the subpopulations occur over time. One well-studied metapopulation is that of the endangered red-cockaded woodpecker (*Picoides borealis*) (Figure 26–10), which was once common in pinewoods throughout the southeastern United States. Habitat loss because of logging has reduced the remaining populations of this bird to

FIGURE 26–10   The red-cockaded woodpecker (*Picoides borealis*).

small scattered sites isolated from each other, with virtually no migration between them. Studies using allozymes and DNA profiling show that the smallest surviving woodpecker populations (where $N = <100$) have suffered the greatest loss of genetic diversity and are at most risk of inbreeding depression, compared with larger populations where $N = >100$. Management of the red-cockaded woodpecker now includes efforts to increase the genetic diversity of the smallest populations by introducing birds from different larger populations to artificially re-create the gene flow by migration that would have occurred in the original unfragmented distribution of the species.

Another species in which the effects of gene-flow reduction have been studied is the North American brown bear, *Ursus arctos*. A team of Canadian and U.S. researchers measured heterozygosity in different brown bear populations using amplified microsatellite markers to create DNA profiles for individual bears. The researchers found that levels of heterozygosity in the brown bear population living in and around Yellowstone Park were only two-thirds as high as those in brown bear populations from Canada and mainland Alaska. They concluded that this was due to the isolation and reduced migration of the Yellowstone bears. The habitat of the Canadian and Alaskan bear populations was much less fragmented, allowing migration of individuals and consequent gene flow between populations. The researchers also examined an island population of brown bears on the Kodiak archipelago off the Alaskan coast. Here, heterozygosity was even lower—less than one-half that of the mainland populations—providing further evidence for the effect of small population size, restricted migration, and restricted gene flow on the reduction of genetic diversity.

26.4
Genetic Erosion Threatens Species' Survival

The loss of previously existing genetic diversity from a population or species is referred to as **genetic erosion**. Why does it matter if a population loses genetic diversity, especially if the numbers of individuals remain high? Genetic erosion has two important effects on a population. First, it can result in the loss of potentially useful alleles from the gene pool, thus reducing the ability of the population to adapt to changing environmental conditions and increasing its risk of extinction. Several decades before the dawn of modern genetics, Charles Darwin recognized the importance of diversity to long-term species survival and evolutionary success. In *On the Origin of Species* (1859), Darwin wrote:

> *"The more diversified the descendants of any one species . . . by so much will they be better enabled to seize on many widely diversified places in the polity of nature, and so enabled to increase in numbers."*

Although the importance of allelic diversity for population survival under changing environmental conditions can be presumed for an endangered species, it has actually been demonstrated in several weed species whose gene pools contain alleles conferring resistance to chemical herbicide. If the weeds are not sprayed, the plants with these resistance alleles enjoy no particular selective advantage, and the resistance allele often persists at a relatively low frequency. However, if herbicide is applied, the individuals with the resistance allele are much more likely to survive and generate a new resistant population, while populations whose gene pools lack the resistance allele will be eliminated.

The second important effect of genetic erosion is a reduction in levels of heterozygosity. At the population level, reduced heterozygosity will be seen as an increase in the number of individuals homozygous at a given locus. At the individual level, a decrease in the number of heterozygous loci within the genotype of a particular plant or animal will occur. As we have seen, loss of heterozygosity is a common consequence of reduced population size. Alleles may be lost through genetic drift or because individuals carrying them die without reproducing. Smaller populations also increase

the likelihood of inbreeding, which inevitably increases homozygosity.

Obviously, once an allele is lost from a gene pool, the potential for heterozygosity is greatly reduced. If there were originally only two alleles at a given locus, loss of one means that heterozygosity at the locus is completely eliminated and the other allele is now fixed. The level of homozygosity that can be tolerated varies with species. Studies of populations showing higher-than-normal levels of homozygosity have documented a range of deleterious effects, including reduced sperm viability and reproductive abnormalities in African lions, increased offspring mortality in elephant seals, and reduced nesting success in woodpeckers.

As yet, no evidence has emerged from field studies that conclusively links the extinction of a wild population to genetic erosion. However, laboratory studies using *Drosophila melanogaster* to model evolutionary events show that loss of genetic variation does reduce the ability of a population to adapt to changing environmental conditions. Fruit flies, with their small size and rapid generation time of only 10 to 14 days, are a useful model organism for studying evolutionary events, especially because large populations can be readily developed and maintained through multiple generations.

In one set of experiments carried out by scientists at Macquarie University in Sydney, Australia, *Drosophila* populations were reduced to a single pair for up to three generations to simulate a population bottleneck and were then allowed to increase in number. The capacity of the bottlenecked populations to tolerate increasing levels of sodium chloride was compared with that of normal outbred populations. Researchers found that bottlenecked populations with low levels of heterozygosity were less tolerant of high salt concentrations (Figure 26–11).

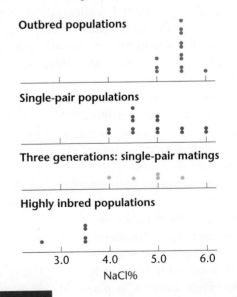

FIGURE 26–11 Effects of bottlenecks in various populations on evolutionary potential in *Drosophila*, as shown by distributions of NaCl concentrations at extinction.

Other experiments comparing inbred and outbred *Drosophila* populations exposed to environmental stresses such as high temperatures and ethanol presence have shown that populations with even a low level of inbreeding have a much greater probability of extinction at lower levels of environmental stress than outbred populations. Experimental results such as these indicate that genetic erosion does indeed reduce the long-term viability of a population by reducing its capacity to adapt to changing environmental conditions.

26.5

Conservation of Genetic Diversity Is Essential to Species Survival

Scientists working to maintain biological diversity face several dilemmas. Should they focus on preserving individual populations, or should they take a broader approach by trying to conserve not just one species but all the interdependent plants and animals in an ecosystem? How can genetic diversity be maintained in a species whose numbers are declining? Can genetic diversity lost from a population be restored? Early conservation efforts often focused only on the population size of an endangered species. Biologists now recognize that a complex interplay of factors must be considered during conservation efforts, including the habitat and role of a species within an ecosystem, as well as the importance of genetic variation for long-term survival.

Ex Situ Conservation: Captive Breeding

Ex situ (Latin for off-site) **conservation** involves removing plants or animals from their original habitat to an artificially maintained location such as a zoo or botanic garden. This living collection can then form the basis of a captive-breeding program.

These programs, where a few surviving individuals from an endangered species are removed from the wild to breed, raise their offspring, and rebuild the population in a protected environment, have been instrumental in bringing a number of species back from the brink of extinction. However, such programs can have undesirable genetic consequences that potentially jeopardize the long-term survival of the species even after population numbers have been restored. A captive-breeding program is rarely initiated until very few individuals are left in the wild, when the original genetic diversity of the species is already depleted. The breeding program is frequently based on a small number of captured individuals, so genetic diversity is further reduced by the founder effect. Since captive-breeding facilities often accommodate only a small number of individuals in the breeding group, inbreeding is difficult to avoid, especially for animal species that form harems (breeding groups dominated by a single male), so that is considerably smaller than

N. Finally, unintended selection for genotypes more suited to captive-breeding conditions over time can reduce the overall capacity of the recovered population to adapt and survive in the wild.

How can this type of program be managed to minimize these genetic effects? As we have already seen, loss of genetic diversity measured as heterozygosity over *t* generations can be expressed as

$$H_t/H_0 = \left(1 - \frac{1}{2N_e}\right)^t$$

This equation indicates that the loss of genetic diversity H_t/H_0 will be greater with a smaller effective population size and a larger number of generations *t*. We can expand this equation for a captive-breeding population as follows:

$$H_t/H_0 = \left[1 - (1/2N_{fo})\right]\left\{1 - 1/\left[2N(N_e/N)\right]\right\}^{t-1}$$

where N_{fo} is the effective size of the founding population, N is the mean population size, and N_e is the mean effective population size of the group over *t* generations. Examination of this equation shows that maximum genetic diversity in a captive-breeding group will be maintained by (1) using the largest possible number of founding individuals to maximize N_{fo}, (2) maximizing N_e/N so that as many individuals as possible produce offspring each generation, and (3) minimizing the number of generations in captivity to reduce *t*. The importance of maximizing N_{fo} to increase allelic diversity in the founding population can also be seen in Figure 26–12. Successful programs of this sort for endangered species are managed with these three goals in mind. In addition, keeping good pedigree records and exchanging individuals between different breeding programs when possible will reduce inbreeding.

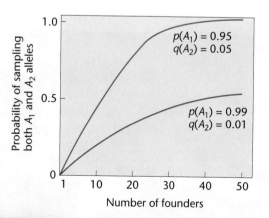

FIGURE 26–12 Effect of captive-population founder number on the probability of maintaining both A_1 and A_2 alleles at a locus.

26–3 Twenty endangered red pandas are taken into captivity to found a breeding group.

(a) What is the probability that at least one of the captured pandas has the genotype B_1/B_2 if $p(B_1) = 0.99$?

(b) Careful management keeps the N_e/N ratio for the red panda breeding population at 0.42. If the overall size of the captive population is maintained at 50 individuals, what proportion of the heterozygosity present in the founding population will still be present after five generations in captivity?

■ HINT: *This problem asks you to estimate the potential for genetic erosion in a captive population of rare red pandas. The key to its solution is to consider that if N_e/N is maintained at 0.42 and actual population size at 50, you can calculate N_e. Then apply the equation given in the discussion above.*

Rescue of the Black-Footed Ferret through Captive Breeding

The black-footed ferret, *Mustela nigripes* (Figure 26–13), is an excellent example of a species that has been successfully subjected to captive breeding. This ferret was once widespread throughout the plains of the western United States, but by the 1970s it was considered to be extinct. Decades of trapping and poisoning of both the ferret and its main prey, the prairie dog, had decimated their populations. However, in 1981 a small surviving colony of black-footed ferrets was discovered on a ranch near Meeteetse, Wyoming. Conservation biologists first tried to conserve this ferret population *in situ*, but in 1985 the colony was infected with canine distemper, which nearly wiped it out. Of the 18 ferrets that were saved and transferred to a captive facility, only 8 were considered to be sufficiently unrelated to found a new popula-

tion. The breeding program has produced more than 6500 ferrets, and the species is now being reintroduced to parts of its original range. The goal was to establish populations of 1500 ferrets in sustainable wild colonies by 2010. By the end of 2008, more than 2300 captive-bred ferrets had been released at 18 different locations, and reintroduced populations in Arizona, South Dakota, and Wyoming are now actively breeding and considered self-sustaining.

Although the recovery of the black-footed ferret appears to be a success, the small founder group of eight individuals caused a severe bottleneck for this species. Additional loss of genetic diversity occurred in the captive-breeding program because of drift, limited reproduction from several of the founder females, and a high rate of breeding by one founder male. In short, the ferrets still face the risks of inbreeding and further genetic erosion that are characteristic of this type of program.

Genetic management strategies, such as using DNA markers to identify the most genetically varied individuals, maintaining careful pedigree records to avoid mating closely related animals, and developing techniques for artificial insemination and sperm cryopreservation, have helped to conserve the genetic diversity that remained in the species. Conservation geneticists working on the recovery project estimate that all existing black-footed ferrets share about 12 percent of their genome. This is roughly the equivalent of being full cousins. The effects of this degree of genetic similarity in the expanding ferret population are unclear. Occasional abnormalities such as webbed feet and kinked or short tails have been observed in the captive population, but without a noninbred population available for comparison, it is unclear whether this is evidence that inbreeding is increasing the homozygosity for deleterious alleles.

Scientists disagree as to whether the long-term future of the black-footed ferret is jeopardized by the severe bottleneck and subsequent loss of genetic diversity the species has experienced. Some geneticists suggest that because the one surviving Meeteetse population was so isolated, it may have already become sufficiently inbred to purge any deleterious alleles. Other researchers point out that studies of black-footed ferret DNA extracted from museum specimens show that the ferrets alive today have lost significant genetic diversity compared with earlier prebottleneck populations, and they suggest that loss of fitness because of inbreeding will inevitably be seen over time. This view is supported by a 2008 study using microsatellite analysis to assess levels of genetic diversity in reintroduced ferret populations; the study found that animals in smaller populations with less diversity tended to have shorter limbs and decreased body size. The authors of the study recommended repeated annual introductions of additional ferrets to newly reintroduced colonies to build numbers rapidly and minimize inbreeding effects.

FIGURE 26–13   The black-footed ferret (*Mustela nigripes*).

Ex Situ Conservation and Gene Banks

Another form of *ex situ* conservation is provided by establishing **gene banks**. In contrast to housing entire animals or plants, these collections instead provide long-term storage and preservation for reproductive components, such as sperm, ova, and frozen embryos in the case of animals, and seeds, pollen, and cultured tissue in the case of plants. Many more individual genotypes can be preserved for longer periods in a gene bank than in a living collection. Cryopreserved gametes or seeds can be used to reconstitute lost or endangered animals or plants after many years in storage. Because they are expensive to construct and maintain, most gene banks are used to conserve the genetic legacy of domesticated species having economic value.

Gene banks have been established in many countries to help preserve genetic material of agricultural importance, such as traditional crop varieties that are no longer grown or old livestock breeds that are becoming rare. One of the most important *ex situ* collections in the United States is the National Center for Genetic Resources Preservation, a U.S. Department of Agriculture (USDA) facility in Fort Collins, Colorado, which maintains more than 300,000 different crop varieties and related wild species. Some of the accessions are stored as seeds and others as cryogenically preserved tissue from which whole plants can be regenerated. (See this chapter's opening photograph.) Animal genetic resources, including frozen semen and embryos from endangered livestock breeds, are also preserved at this facility.

Ex situ conservation using gene banks, though often vital, has several disadvantages compared to other methods of conservation. A major problem with gene banks is that even large collections cannot contain all the genetic variation that is present in a species. Conservation geneticists attempt to address this problem by identifying a **core collection** for a species. The core collection is *a subset of individual genotypes, carefully chosen to contain as much as possible of the species' genetic variation*; preserving the core collection takes priority over randomly collecting and preserving large numbers of genotypes. Another disadvantage of *ex situ* conservation is that the artificial conditions under which a species is preserved in a living collection or gene bank often create their own selection pressures. When seeds of a rare plant species are maintained in cold storage, for example, selection may occur for those genotypes better adapted to withstand the low temperatures, and genotypes may be lost that would actually have greater fitness in the plant's natural environment. Yet another problem posed by *ex situ* conservation is that while the greatest biological diversity in both domesticated and nondomesticated species is frequently found in underdeveloped countries, most *ex situ* collections are situated in developed countries that have the resources to establish and maintain them. This leads to conflict over who owns and has access to the potentially valuable genetic resources maintained in such collections.

In Situ Conservation

In situ (Latin for on-site) **conservation** attempts to preserve the population size and biological diversity of a species while it remains in its original habitat. The use of species inventories to identify diversity hot spots is an important tool for determining the best places to establish parks and reserves where plants and animals can be protected from hunting or collecting and where their habitat can be preserved.

For domesticated species, there is increasing interest in "on-farm" preservation in which additional resources and financial incentives encourage farmers to maintain traditional crop varieties and livestock breeds. Nondomesticated species with economic potential have also been targeted for *in situ* preservation. In 1998, the USDA established its first *in situ* conservation sites for a wild plant in order to protect populations of the native rock grape (*Vitis rupestris*) in several eastern states. The rock grape is prized by winegrowers not for its fruit but for its roots; grape vines grafted onto wild rock grape rootstock are resistant to phylloxera, a serious pest of wine grapes.

The advantage of *in situ* conservation for the rock grape, as with other species, is that larger populations with greater genetic diversity can be maintained. Another advantage is that species conserved *in situ* continue to live and reproduce in the environments to which they are adapted, which reduces the likelihood that novel selection pressures will produce undesirable changes in allele frequency. However, the continuing increase in global human population makes setting aside suitable areas for *in situ* conservation an ever greater challenge. For example, even in a large preserve such as Yellowstone National Park, the migration and gene flow of species such as the North American brown bear may eventually be eliminated, with a consequent loss of genetic diversity. This problem is even more acute in smaller, more fragmented areas of protected habitat.

Population Augmentation

What genetic considerations should accompany efforts to restore populations or species that are in decline? As we have already seen, populations that experience bottlenecks continue to suffer from low levels of genetic diversity, even after their numbers have recovered. We have also seen that inbreeding and drift contribute to genetic erosion in small populations, and fragmentation interrupts migration and gene flow, further reducing diversity. Captive-breeding programs designed to restore a critically endangered species beginning with only a few surviving individuals risk genetic erosion in the renewed population from founder effect and inbreeding, as in the case of the black-footed ferret.

FIGURE 26–14   The Florida panther (*Puma concolor coryi*).

An alternative strategy used by conservation biologists is **population augmentation**—*boosting the numbers of a declining population by transplanting and releasing individuals of the same species captured or collected from more numerous populations elsewhere.* Attempts to reestablish gene flow in severely fragmented populations of the endangered red-cockaded woodpecker by augmenting the smallest populations were described earlier. Other population augmentation projects in the United States have involved bighorn sheep and grizzly bears in the Rocky Mountains.

This method has also been employed with the Florida panther (Figure 26–14), using an isolated population of less than 30 animals confined to the area around the Big Cypress Swamp and Everglades National Park in south Florida. As we will discuss in the Genetics, Technology, and Society essay at the end of this chapter, genetic studies revealed high levels of inbreeding in the Florida panther, with reduced fitness because of severe reproductive abnormalities and increased susceptibility to parasite infections. In an effort to combat these defects, eight unrelated female panthers from Texas were released into Florida in 1995 and allowed to interbreed with the Florida population. The augmentation program is now considered a success, as it produced more

genetically diverse family groups, reduced the rates of genetic defects, and improved population survival. The case of the Florida panther shows how augmentation can increase population numbers, as well as genetic diversity, if the transplanted individuals are unrelated to those in the population to which they are introduced. However, the Florida panther also demonstrates a potential problem with population augmentation, namely, **genetic swamping**. This occurs *when the gene pool of the original population is overwhelmed by different genotypes from the transplanted individuals and loses its identity.* Further augmentation of the Florida panther population remains controversial, as some biologists argue that the unique features that allow the Florida panther to be classified as a separate subspecies will be lost if breeding takes place with other panthers.

Although not apparent in the Florida panther augmentation program, another difficulty that can be caused by population augmentation is **outbreeding depression**, *where reduced fitness occurs in the progeny of matings between genetically diverse individuals.* Outbreeding depression occurring in the F_1 generation is thought to be due to the offspring being less well adapted to local environmental conditions than either parent. This phenomenon has been documented in some plant species in which seeds of the same species, but from a different location, were used to revegetate a damaged area by cross-pollination with the remaining local plants. Outbreeding depression that occurs in the F_2 and later generations is due to the disruption of **coadapted gene complexes**—*groups of alleles that have evolved to work together to produce the best level of fitness in an individual.* This type of outbreeding depression has been documented in hybrid offspring from matings between fish from different salmon populations in Alaska. These studies suggest that restoring the most beneficial type and amount of genetic diversity in a population is more complicated than previously thought, reinforcing the argument that the best long-term strategy for species survival is to prevent the loss of diversity in the first place.

CASE STUDY | The flip side of the green revolution

To increase food production, the Green Revolution (see Chapter 23) fostered the development of inbred strains of high-yield crop plants. This approach was very successful, and it is estimated that over one billion people worldwide now derive all or part of their food supply from these new crops. However, this program has a dark side. In many countries, as they switched to the high-yield strains, farmers abandoned the traditional varieties of crop plants, which contained a high degree of genetic diversity. For example, 50 years ago, farmers in India used to plant more than 30,000 varieties of rice. Now it is estimated that 50 to 75 percent of all rice fields are planted with just 10 varieties. In the United States some 90 percent of

all varieties of crop plants grown a century ago have disappeared from seed catalogs and are not preserved in seed banks. This dramatic loss of genetic variation has generated some controversy and raised several questions.

1. Might the Green Revolution represent short-term gains but long-term losses?

2. Why is it important to preserve the genetic diversity of crop plants?

3. Will climate change adversely affect the new and old strains of crop plants and perhaps reduce the food supply?

GENETICS, TECHNOLOGY, AND SOCIETY

Gene Pools and Endangered Species: The Plight of the Florida Panther

The story of the Florida panther provides a dramatic example of an animal brought to the verge of extinction and the challenges that must be overcome to restore it to a healthy place in the ecosystem. The Florida panther is an endangered population of North American cougars (also known as pumas or mountain lions). These panthers once roamed the southeastern corner of the United States, from South Carolina and Arkansas to the southern tip of Florida; however, they now occupy less than 5 percent of their original habitat, in southern Florida. After human settlement began, panthers were killed by hunting, poisoning, highway collisions, and loss of habitat that supports their prey—primarily wild deer and hogs. By 1967, Florida panthers were listed as endangered, and only about 30 of the animals were left. Population estimates predicted that the Florida panther would be extinct by the year 2055.

Because of geographical isolation and inbreeding, Florida panthers had the lowest levels of genetic heterozygosity of any population of cougar. The loss of genetic diversity manifested itself in the appearance of some severe genetic defects. For example, almost 80 percent of panther males born after 1989 in the Big Cypress region showed an autosomal dominant or X-linked recessive condition known as *cryptorchidism*—failure of one or both testicles to descend. This defect is associated with low testosterone levels and reduced sperm count. Life-threatening congenital heart defects also appeared in the Florida panther population, possibly because of an autosomal dominant gene defect. In addition, some immune deficiencies emerged, making the animals more susceptible to diseases and further contributing to the population's decline. Other less serious genetic features arose, such as a kink in the tail and a whorl of fur on the back.

Over the last two decades, federal and state agencies, as well as private individuals, have implemented a *Florida Panther Recovery Program*. The program's goal is to exceed 500 breeding animals and give the panther a 95 percent probability of survival while retaining up to 90 percent of its genetic diversity. The plan includes strict protection, enlargement and improvement of the panther's habitat, and education of landowners. Wildlife underpasses have been constructed along highways in panther territory, and these have significantly reduced panther highway fatalities (which account for about half of panther deaths).

To retard the detrimental effects of inbreeding, eight wild female Texas panthers were released into Florida panther territory in 1995. Two of the females survived and gave birth to litters of healthy kittens. Since then, the panther population has increased to about 130 individuals, and kittens with Texas panther ancestry show a more than two-fold increase in survival rates over purebred Florida kittens. Genetic defects, such as heart disease, have declined by one-half, and hybrid panthers are noticeably larger and stronger than their purebred counterparts. Although the population augmentation program is considered successful, it is of concern to some biologists, who worry that the Texas cougar genes may genetically swamp the distinct Florida panther population.

Paradoxically, the success of the restoration program has become a problem. Now that the population has reached about 130 animals, the Florida panther has almost exceeded the capacity of its existing habitat. Biologists estimate that the population must reach at least 250 individuals in order to be self-sustaining. To reach this number, the panthers will need to expand their territory, putting them in direct competition with human expansion in Florida. Biologists are now evaluating potential new territories for the growing population of Florida panthers—including parts of Louisiana, Arkansas, and South Carolina.

Despite the success of the restoration program, the survival of the Florida panther is far from certain. The panther's comeback will require years of monitoring and frequent intervention. In addition, people must be willing to share their land with wild creatures that are dangerous and do not directly further human interests. Public support for the return of the Florida panther has been strong, however, so there may be hope for this unique, impressive animal.

Your Turn

Take time, individually or in groups, to answer the following questions. Investigate the references and links to help you understand some of the challenges surrounding the Florida Panther Recovery Program.

1. Recent genetic studies suggest that the Florida panther may be too closely related to other North American cougars to merit classification as a subspecies. How were these genetic studies done, and how would you interpret the results of the studies, as they relate to the Florida panther?

To learn about genetic relatedness among American cougars, see Culver, M., et al. 2000. Genetic Ancestry of the American Puma (Puma concolor). J. Heredity 91(3): 186–197.

2. How do you think the data in the Culver et al. study could both support and complicate efforts to restore the endangered Florida panther?

A discussion of the relative merits of genetic and traditional classification methods can be found in the Florida Panther Recovery Plan, 3rd edition 2008, U.S. Fish and Wildlife Service, pages 7–12. You can access this work through the Florida Panther Net (http://www.panther.state.fl.us).

3. Do you think that genetics or habitat is the most important factor involved in the Florida Panther Recovery Program?

An interesting discussion of this question can be found in the New York Times article, "A Rare Predator Bounces Back (Now Get It Out Of Here)" at http://www.nytimes.com/2006/03/14/science/14pant.html?scp=1sq=florida+panther&st=nyt.

Summary Points

 For activities, animations, and review quizzes, go to the study area at www.masteringgenetics.com

1. Biodiversity is diminished as increasing numbers of plants and animal species are threatened with extinction. Conservation genetics applies principles of population genetics to the preservation and restoration of threatened species. A major concern of conservation geneticists is the maintenance of genetic diversity.

2. Genetic diversity includes interspecific diversity, which is reflected by the number of different species present in an ecosystem, and intraspecific diversity, which is reflected by genetic variation within a population or between different populations of the same species. Genetic diversity can be measured by examining different phenotypes in the population or, at the molecular level, by using allozyme analysis or DNA-profiling techniques.

3. Major declines in a species' population numbers reduce genetic diversity and contribute to the risk of extinction as a result of genetic drift, inbreeding, or loss of gene flow. Populations that suffer severe reductions in effective size and then recover have passed through a population bottleneck and often show reduced genetic diversity.

4. Loss of genetic diversity reduces the capacity of a population to adapt to changing environmental conditions, because useful alleles may disappear from the gene pool. Reduced genetic diversity also results in greater levels of homozygosity in a population, often leading to the manifestation of deleterious alleles and inbreeding depression.

5. Conservation of genetic diversity depends on *ex situ* methods, such as living collections, captive-breeding programs, and gene banks, as well as *in situ* approaches, such as the establishment of parks and preserves.

6. Population augmentation, in which individuals are transplanted into a declining population from a more numerous population of the same species located elsewhere, can be used to increase numbers and genetic diversity. However, the risk of outbreeding depression and genetic swamping accompanies this process.

INSIGHTS AND SOLUTIONS

1. Is a rare species found as several fragmented subpopulations more vulnerable to extinction than an equally rare species found as one larger population? What factors should be considered when managing fragmented populations of a rare species?

Solution: A rare species in which the remaining individuals are divided among smaller, isolated subpopulations can appear to be less vulnerable. If one subpopulation becomes extinct through local causes, such as disease or habitat loss, then the remaining subpopulations may still survive. However, genetic drift will cause smaller populations to experience more rapidly increasing homozygosity over time compared with larger populations. Even with random mating, the change in heterozygosity from one generation to the next because of drift can be calculated as

$$H_1 = H_0(1 - 1/2N)$$

where H_0 is the frequency of heterozygotes in the present generation, H_1 is the frequency of heterozygotes in the next generation, and N is the number of individuals in the population. Thus, in a small population of 50 individuals with an initial heterozygote frequency of 0.5, heterozygosity will decline to $0.5(1 - 1/100) = 0.4995$, a loss of 0.5 percent in just one generation. In a larger population of 500 individuals and the same initial heterozygote frequency, heterozygosity after one generation will be $0.5(1 - 1/100) = 0.4995$, a loss of only 0.05 percent.

Therefore, smaller populations are likely to show the effects of homozygosity for deleterious alleles sooner than larger populations, even with random mating. If populations are fragmented so that movement of individuals or gametes between them is prevented, management options could include transplanting individuals from one subpopulation to another to enable gene flow to occur. Establishment of "wildlife corridors" of undisturbed habitat that connect fragmented populations could be considered. In captive populations, exchange of breeding adults (or their gametes through shipment of preserved semen or pollen) can be undertaken. Promotion of increased population numbers, however, is vital to prevent further genetic erosion through drift.

Problems and Discussion Questions

 For instructor-assigned tutorials and problems, go to www.masteringgenetics.com

HOW DO WE KNOW?

1. In this chapter, we have focused on conservation genetics, emphasizing how geneticists assess genetic diversity and work to maintain species survival. Along the way, we found many opportunities to consider the methods and reasoning by which several of these practices were developed. From the explanations given in the chapter, what answers would you propose to the following fundamental questions?
(a) How do we know the extent of genetic diversity in a species?
(b) How do we know that diminished genetic diversity occurs and that it is detrimental to species survival?

(c) How do we know that genetic drift occurs and that it is detrimental to species survival?

(d) How do we know that population augmentation works in enhancing a species' likelihood of surviving?

(e) How do we attempt to "conserve" existing genetic diversity?

2. Chondrodystrophy, a lethal form of dwarfism, has recently been reported in captive populations of the California condor and has killed embryos in 5 out of 169 fertile eggs. Chondrodystrophy in condors appears to be caused by an autosomal recessive allele with an estimated frequency of 0.09 in the gene pool of this species.

(a) How do you think California condor populations should be managed in the future to minimize the effect of this lethal allele?

(b) What are the advantages and disadvantages of attempting to eliminate it from the gene pool?

3. A geneticist is studying three loci, each with one dominant and one recessive allele, in a small population of rare plants. She estimates the frequencies of the alleles at each of these loci as $A = 0.75, a = 0.25; B = 0.80, b = 0.20; C = 0.95, c = 0.05$. What is the probability that all the recessive alleles will be lost from the population through genetic drift?

4. How are genetic drift and inbreeding similar in their effects on a population? How are they different?

5. You are the manager of a game park in Africa with a native herd of just 16 black rhinos, an endangered species worldwide. Describe how you would manage this herd to establish a viable population of black rhinos in the park. What genetic factors would you take into consideration in your management plan?

6. Compare the causes and effects of inbreeding depression and outbreeding depression.

7. Cloning, using the techniques similar to those pioneered by the Scottish scientists who produced Dolly the sheep, has been proposed as a way to increase the numbers of some highly endangered mammalian species. Discuss the advantages and disadvantages of using such an approach to aid long-term species survival.

8. In a population of wild poppies found in a remote region of the mountains of eastern Mexico, almost all of the members have pale yellow flowers, but breeding experiments show that pale yellow is recessive to deep orange. Using the tools of the conservation geneticists described in this chapter, how could you experimentally determine whether the prevalence of the recessive phenotype among the eastern Mexican poppy population is due to natural selection or simply due to the effects of genetic drift and/or inbreeding?

9. Contrast *ex situ* conservation techniques with *in situ* conservation techniques.

10. Describe how the captive-breeding *ex situ* conservation approach is applied to a severely endangered species.

11. Explain why a low level of genetic diversity in a species is a detriment to the survival of that species.

12. Contrast allozyme analysis with RFLP analysis as measures of genetic diversity.

13. Use your analysis of the red panda in the Now Solve It (26–3) to make suggestions for managing the captive-breeding population of red pandas to maintain as much genetic diversity as possible.

Extra-Spicy Problems

 For instructor-assigned tutorials and problems, go to www.masteringgenetics.com

14. *Antechinus agilis*, a small mouse-like marsupial found in Australia, mates for a 10- to 15-day period in August during which females ovulate and after which males die of stress. Individual females rely on stored sperm from multiple male sources to fertilize their eggs. Using DNA profiling, researchers determined that females are capable of releasing a mix of sperm for fertilization—regardless of the time of male access to the female or her time of ovulation—that maintains high genetic heterogeneity in the face of male absenteeism (Shimmin et al., 2000). Does the reproductive strategy of *Antechinus* provide any clues as to the lengths organisms will go to maintain genetic variation?

15. As revealed by microsatellite analysis, habitat loss and fragmentation have led to a significant loss of heterozygosity among nine populations of Florida black bear (*Ursus americanus floridanus*). Researchers determined that although each population was in Hardy–Weinberg equilibrium, mean heterozygosity was lower in smaller, less interconnected populations, and each population was genetically distinct (Dixon et al., 2007). No relationship was found between the degree of genetic differences and the distance from neighboring populations.

(a) Is their finding of a lower mean heterozygosity in smaller, less interconnected populations expected?

(b) Given that the Florida black bear inhabits relatively populated areas, what would explain the lack of a relationship between genetic differences and interpopulational distances?

(c) Considering that low heterozygosity often contributes to reduced species survival, what steps might be taken to enhance the survival of the Florida black bear?

16. Once nearly extinct, black-footed ferrets, which prey on prairie dogs, have increased in numbers from 18 in 1985 to approximately 700 presently occupying a recovery area of Buffalo Gap National Grassland (O'Neill, 2007). However, because prairie dogs eat grass that feeds cattle, some ranchers want to reduce prairie dog populations in that area, where approximately 30,000 of their cattle graze. A plan for balancing the welfare of the ranchers, the prairie dogs, and the black-footed ferret is under consideration by the U.S. Forest Service.

(a) What impact would a reduction of prairie dog populations have on black-footed ferret populations?

(b) Conservationists argue that exposing the black-footed ferret population to a second bottleneck would irreversibly harm ferret survival. Some might argue that since the ferret population has in the past recovered from one bottleneck, it should be able to recover from a second. Do you agree?

(c) Given that the black-footed ferret is one of America's most endangered mammals, what kind of compromise would you negotiate among the farmer, the prairie dog, and the ferret?

17. Przewalski's horse (*Equus przewalskii*), thought by some biologists to represent the ancestral species from which modern horses were domesticated, is classified as a separate species from the domestic horse, although members of the two species

can mate and produce fertile offspring. Przewalski's horse was hunted to extinction in its native habitat in the Asian steppes by the 1920s. The few hundred Przewalski's horses alive today are descended from 12 surviving individuals that had been taken into captivity. The founder breeding group also included a domestic mare. Currently, there is interest in reintroducing Przewalski's horse to its original range. What advice would you give to the conservation biologists managing the reintroduction project?

18. DNA profiles based on different kinds of molecular markers are increasingly used to measure levels of genetic diversity in populations of threatened and endangered species. Do you think these marker-based estimates of genetic diversity are reliable indicators of a population's potential for survival and adaptation in a natural environment? Why, or why not?

19. Seed banks provide managed protection for the conservation of species of economic and noneconomic importance. One study quantified considerable fitness decay in seeds maintained in long-term storage (Schoen et al., 1998). When germination falls below 65–85 percent, regeneration (planting and seed collection) of a finite sample of the stored seed type is recommended. What genetic consequences might you expect to accompany the conservation practice of long-term seed storage and regeneration?

20. According to one writer, "One of the first questions a resource manager asks about threatened and endangered species is: How bad is it?" (Holmes, 2001). In other words, the manager seeks a population viability analysis that includes estimates of extinction risk. What factors would you consider significant in providing an estimate of extinction risk for a species?

21. It is believed that most microsatellite DNA sequences are selectively neutral and highly heterozygous in natural populations. Considering that cheetahs underwent a population bottleneck approximately 12,000 years ago, North American pumas 10,000 years ago, and Gir Forest lions 1000 years ago, in which of these species would you expect to see the highest degree of microsatellite polymorphism? the lowest?

22. For many conservation efforts, scientists lack sufficient data to make definitive conservation decisions. Yet the allocation of resources cannot be delayed until such data are available. In these cases, conservation efforts are often directed toward three classes of species: flagships (high-profile species), umbrellas (species requiring large areas for habitat), and biodiversity indicators (species representing diverse, especially productive habitats) (Andelman and Fagan, 2000). In terms of protecting threatened species, what advantages and disadvantages might accompany investing scarce talent and resources in each of these classes?

Appendix A

SELECTED READINGS

Chapter 1 Introduction to Genetics

Amman, N. H. 2008. In defense of GM crops. *Science* 322: 1465–1466.

Bilen, J., and Bonini, N. M. 2005. *Drosophila* as a model for human neurodegenerative disease. *Annu. Rev. Genet.* 39: 153–171.

Campbell, A. M., and Heyer, L. J. (2007). *Discovering genomics, proteomics, and bioinformatics,* 2nd ed. San Francisco, CA: Benjamin Cummings.

Chen, M., Shelton, A., and Ye, G. Y. 2011. Insect-resistant genetically-modified rice in China: From research to commercialization. *Ann. Rev. Entomol.* 56: 81–101.

Dale, P. J., Clarke, B., and Fontes, E. M. G. 2002. Potential for the environmental impact of transgenic crops. *Nature Biotech.* 20: 567–574.

Daya, S., and Berns, K. I. 2008. Gene therapy using adeno-associated virus vectors. *Clin. Microbiol. Rev.* 21: 83–93.

Ehrnhoefer, D. E., Butland, S. L., Pouladi, M. A., and Hayden, M. R. 2009. Mouse models of Huntington disease: Variations on a theme. *Dis. Models Mech.* 2: 123–129.

Lonberg, N. 2005. Human antibodies from transgenic animals. *Nature Biotech.* 23: 1117–1125.

Müller, B., and Grossnicklaus, U. 2010. Model organisms—A historical perspective. *J. Proteomics* 73: 2054–2063.

Pearson, H. 2006. What is a gene? *Nature* 441: 399–401.

Pray, C. E., Huang, J., Hu, R., and Rozelle, S. 2002. Five years of Bt cotton in China—The benefits continue. *The Plant Journal* 31: 423–430.

Primrose, S. B., and Twyman, R. M. 2004. *Genomics: Applications in human biology.* Oxford: Blackwell Publishing.

Wisniewski, J-P., Frange, N., Massonneau, A., and Dumas, C. 2002. Between myth and reality: Genetically modified maize, an example of a sizeable scientific controversy. *Biochimie* 84: 1095–1103.

Chapter 2 Mitosis and Meiosis

Alberts, B., et al. 2007. *Molecular biology of the cell,* 5th ed. New York: Garland Publ.

Brachet, J., and Mirsky, A. E. 1961. *The cell: Meiosis and mitosis,* Vol. 3. Orlando, FL: Academic Press.

DuPraw, E. J. 1970. *DNA and chromosomes.* New York: Holt, Rinehart & Winston.

Glotzer, M. 2005. The molecular requirements for cytokinesis. *Science* 307: 1735–1739.

Glover, D. M., Gonzalez, C., and Raff, J. W. 1993. The centrosome. *Sci. Am.* (June) 268: 62–68.

Golomb, H. M., and Bahr, G. F. 1971. Scanning electron microscopic observations of surface structures of isolated human chromosomes. *Science* 171: 1024–1026.

Hartwell, L. H., and Karstan, M. B. 1994. Cell cycle control and cancer. *Science* 266: 1821–1828.

Hartwell, L. H., and Weinert, T. A. 1989. Checkpoint controls that ensure the order of cell cycle events. *Science* 246: 629–634.

Ishiguro, K., and Watanabe Y. 2007. Chromosome cohesion in mitosis and meiosis. *J. Cell Sci.* 120: 367–369.

Mazia, D. 1961. How cells divide. *Sci. Am.* (Jan.) 205: 101–120.

———. 1974. The cell cycle. *Sci. Am.* (Jan.) 235: 54–64.

McIntosh, J. R., and McDonald, K. L. 1989. The mitotic spindle. *Sci. Am.* (Oct.) 261: 48–56.

Nature Milestones. 2001. *Cell Division.* London: Nature Publishing Group.

Watanabe Y. 2005. Shugoshin: Guardian spirit at the centromere. *Curr. Opin. Cell Biol.* 17: 590–595.

Westergaard, M., and von Wettstein, D. 1972. The synaptinemal complex. *Annu. Rev. Genet.* 6: 71–110.

Chapter 3 Mendelian Genetics

Bennett, R. L., et al. 1995. Recommendations for standardized human pedigree nomenclature. *Am. J. Hum. Genet.* 56: 745–752.

Carlson, E. A. 1987. *The gene: A critical history,* 2nd ed. Philadelphia: Saunders.

Dunn, L. C. 1965. *A short history of genetics.* New York: McGraw-Hill.

Henig, R. M. 2001. *The monk in the garden: The lost and found genius of Gregor Mendel, the father of genetics.* New York: Houghton-Mifflin.

Klein, J. 2000. Johann Mendel's field of dreams. *Genetics* 156: 1–6.

Miller, J. A. 1984. Mendel's peas: A matter of genius or of guile? *Sci. News* 125: 108–109.

Olby, R. C. 1985. *Origins of Mendelism,* 2nd ed. London: Constable.

Orel, V. 1996. *Gregor Mendel: The first geneticist.* Oxford: Oxford University Press.

Peters, J., ed. 1959. *Classic papers in genetics.* Englewood Cliffs, NJ: Prentice-Hall.

Schwartz, J. 2008. *In pursuit of the gene: From Darwin to DNA.* Cambridge, MA: Harvard University Press.

Sokal, R. R., and Rohlf, F. J. 1995. *Biometry,* 3rd ed. New York: W. H. Freeman.

Stern, C., and Sherwood, E. 1966. *The origins of genetics: A Mendel source book.* San Francisco: W. H. Freeman.

Stubbe, H. 1972. *History of genetics: From prehistoric times to the rediscovery of Mendel's laws.* Cambridge, MA: MIT Press.

Sturtevant, A. H. 1965. *A history of genetics.* New York: Harper & Row.

Tschermak-Seysenegg, E. 1951. The rediscovery of Mendel's work. *J. Hered.* 42: 163–172.

Welling, F. 1991. Historical study: Johann Gregor Mendel 1822–1884. *Am. J. Med. Genet.* 40: 1–25.

Chapter 4 Modification of Mendelian Ratios

Bartolomei, M. S., and Tilghman, S. M. 1997. Genomic imprinting in mammals. *Annu. Rev. Genet.* 31: 493–525.

Brink, R. A., ed. 1967. *Heritage from Mendel.* Madison: University of Wisconsin Press.

Bultman, S. J., Michaud, E. J., and Woychik, R. P. 1992. Molecular characterization of the mouse *agouti* locus. *Cell* 71: 1195–1204.

Carlson, E. A. 1987. *The gene: A critical history*, 2nd ed. Philadelphia: Saunders.

Cattanach, B. M., and Jones, J. 1994. Genetic imprinting in the mouse: Implications for gene regulation. *J. Inherit. Metab. Dis.* 17: 403–420.

Dunn, L. C. 1966. *A short history of genetics.* New York: McGraw-Hill.

Feil, R., and Khosla, S. 1999. Genomic imprinting in mammals: An interplay between chromatin and DNA methylation. *Trends in Genet.* 15: 431.

Foster, M. 1965. Mammalian pigment genetics. *Adv. Genet.* 13: 311–339.

Grant, V. 1975. *Genetics of flowering plants.* New York: Columbia University Press.

Harper, P. S., et al. 1992. Anticipation in myotonic dystrophy: New light on an old problem. *Am. J. Hum. Genet.* 51: 10–16.

Howeler, C. J., et al. 1989. Anticipation in myotonic dystrophy: Fact or fiction? *Brain* 112: 779–797.

Morgan, T. H. 1910. Sex-limited inheritance in *Drosophila. Science* 32: 120–122.

Peters, J. A., ed. 1959. *Classic papers in genetics.* Englewood Cliffs, NJ: Prentice-Hall.

Phillips, P. C. 1998. The language of gene interaction. *Genetics* 149: 1167–1171.

Race, R. R., and Sanger, R. 1975. *Blood groups in man*, 6th ed. Oxford: Blackwell.

Sapienza, C. 1990. Parental imprinting of genes. *Sci. Am.* (Oct.) 363: 52–60.

Siracusa, L. D. 1994. The *agouti* gene: Turned on to yellow. *Cell* 10: 423–428.

Waters, D. J., and Wildasin, K. 2006. Cancer clues from pet dogs. *Sci. Am.* (Dec.) 295: 96–101.

Yoshida, A. 1982. Biochemical genetics of the human blood group ABO system. *Am. J. Hum. Genet.* 34: 1–14.

Chapter 5 Chromosome Mapping in Eukaryotes

Allen, G. E. 1978. *Thomas Hunt Morgan: The man and his science.* Princeton, NJ: Princeton University Press.

Chaganti, R., Schonberg, S., and German, J. 1974. A manyfold increase in sister chromatid exchange in Bloom syndrome lymphocytes. *Proc. Natl. Acad. Sci.* 71: 4508–4512.

Creighton, H. S., and McClintock, B. 1931. A correlation of cytological and genetical crossing over in *Zea mays. Proc. Natl. Acad. Sci.* 17: 492–497.

Douglas, L., and Novitski, E. 1977. What chance did Mendel's experiments give him of noticing linkage? *Heredity* 38: 253–257.

Ellis, N. A., et al. 1995. The Bloom syndrome gene product is homologous to RecQ helicases. *Cell* 83: 655–666.

Ephrussi, B., and Weiss, M. C. 1969. Hybrid somatic cells. *Sci. Am.* (Apr.) 220: 26–35.

Latt, S. A. 1981. Sister chromatid exchange formation. *Annu. Rev. Genet.* 15: 11–56.

Lindsley, D. L., and Grell, E. H. 1972. *Genetic variations of Drosophila melanogaster.* Washington, DC: Carnegie Institute of Washington.

Morgan, T. H. 1911. An attempt to analyze the constitution of the chromosomes on the basis of sex-linked inheritance in *Drosophila. J. Exp. Zool.* 11: 365–414.

Morton, N. E. 1955. Sequential test for the detection of linkage. *Am. J. Hum. Genet.* 7: 277–318.

———. 1995. LODs—Past and present. *Genetics* 1 40: 7–12.

Neuffer, M. G., Jones, L., and Zober, M. 1968. *The mutants of maize.* Madison, WI: Crop Sci. Soc. of America.

Perkins, D. 1962. Crossing over and interference in a multiply marked chromosome arm of *Neurospora. Genetics* 47: 1253–1274.

Ruddle, F. H., and Kucherlapati, R. S. 1974. Hybrid cells and human genes. *Sci. Am.* (July) 231: 36–49.

Stahl, F. W. 1979. *Genetic recombination.* New York: W. H. Freeman.

Stern, C. 1936. Somatic crossing over and segregation in *Drosophila melanogaster. Genetics* 21: 625–631.

Sturtevant, A. H. 1913. The linear arrangement of six sex-linked factors in *Drosophila,* as shown by their mode of association. *J. Exp. Zool.* 14: 43–59.

———. 1965. *A history of genetics.* New York: Harper & Row.

Voeller, B. R., ed. 1968. *The chromosome theory of inheritance: Classical papers in development and heredity.* New York: Appleton-Century-Croft.

Wellcome Trust Case Control Consortium, 2007. Genome association study of 14,000 cases of seven common diseases and 3,000 shared controls. *Nature* 447: 661–676.

Wolff, S., ed. 1982. *Sister chromatid exchange.* New York: Wiley-Interscience.

Chapter 6 Genetic Analysis and Mapping in Bacteria and Bacteriophages

Adelberg, E. A. 1960. *Papers on bacterial genetics.* Boston: Little, Brown.

Benzer, S. 1962. The fine structure of the gene. *Sci. Am.* (Jan.) 206: 70–86.

Birge, E. A. 1988. *Bacterial and bacteriophage genetics—An introduction.* New York: Springer-Verlag.

Brock, T. 1990. *The emergence of bacterial genetics.* Cold Spring Harbor, NY: Cold Spring Harbor Press.

Bukhari, A. I., Shapiro, J. A., and Adhya, S. L., eds. 1977. *DNA insertion elements, plasmids, and episomes.* Cold Spring Harbor, NY: Cold Spring Harbor Press.

Cairns, J., Stent, G. S., and Watson, J. D., eds. 1966. *Phage and the origins of molecular biology.* Cold Spring Harbor, NY: Cold Spring Harbor Press.

Campbell, A. M. 1976. How viruses insert their DNA into the DNA of the host cell. *Sci. Am.* (Dec.) 235: 102–113.

Hayes, W. 1968. *The genetics of bacteria and their viruses,* 2nd ed. New York: Wiley.

Hershey, A. D., and Rotman, R. 1949. Genetic recombination between host range and plaque-type mutants of bacteriophage in single cells. *Genetics* 34: 44–71.

Hotchkiss, R. D., and Marmur, J. 1954. Double marker transformations as evidence of linked factors in deoxyribonucleate transforming agents. *Proc. Natl. Acad. Sci. (USA)* 40: 55–60.

Jacob, F., and Wollman, E. L. 1961. Viruses and genes. *Sci. Am.* (June) 204: 92–106.

Kohiyama, M., et al. 2003. Bacterial sex: Playing voyeurs 50 years later. *Science* 301: 802–803.

Kruse, H., and Sorum, H. 1994. Transfer of multiple drug resistance plasmids between bacteria of diverse origins in natural microenvironments. *Appl. Environ. Microbiol.* 60: 4015–4021.

Lederberg, J. 1986. Forty years of genetic recombination in bacteria: A fortieth anniversary reminiscence. *Nature* 324: 627–628.

Luria, S. E., and Delbruck, M. 1943. Mutations of bacteria from virus sensitivity to virus resistance. *Genetics* 28: 491–511.

Lwoff, A. 1953. Lysogeny. *Bacteriol. Rev.* 17: 269–337.

Miller, J. H. 1992. *A short course in bacterial genetics.* Cold Spring Harbor, NY: Cold Spring Harbor Press.

Miller, R. V. 1998. Bacterial gene swapping in nature. *Sci. Am.* (Jan.) 278: 66–71.

Morse, M. L., Lederberg, E. M., and Lederberg, J. 1956. Transduction in *Escherichia coli* K12. *Genetics* 41: 141–156.

Novick, R. P. 1980. Plasmids. *Sci. Am.* (Dec.) 243: 102–127.

Smith-Keary, P. F. 1989. *Molecular genetics of Escherichia coli.* New York: Guilford Press.

Stahl, F. W. 1987. Genetic recombination. *Sci. Am.* (Nov.) 256: 91–101.

Stent, G. S. 1966. *Papers on bacterial viruses,* 2nd ed. Boston: Little, Brown.

Wollman, E. L., Jacob, F., and Hayes, W. 1956. Conjugation and genetic recombination in *Escherichia coli* K12. *Cold Spring Harb. Symp. Quant. Biol.* 21: 141–162.

Zinder, N. D. 1958. Transduction in bacteria. *Sci. Am.* (Nov.) 199: 38–46.

Chapter 7 Sex Determination and Sex Chromosomes

Amory, J. K., et al. 2000. Klinefelter's syndrome. *Lancet* 356: 333–335.

Arnold, A. P., Itoh, Y., and Melamed, E. 2008. A bird's-eye view of sex chromosome dosage compensation. *Annu. Rev. Genomics Hum. Genet.* 9: 109–127.

Court-Brown, W. M. 1968. Males with an XYY sex chromosome complement. *J. Med. Genet.* 5: 341–359.

Davidson, R., Nitowski, H., and Childs, B. 1963. Demonstration of two populations of cells in human females heterozygous for glucose-6-phosphate dehydrogenase variants. *Proc. Natl. Acad. Sci. (USA)* 50: 481–485.

Hodgkin, J. 1990. Sex determination compared in *Drosophila* and *Caenorhabditis. Nature* 344: 721–728.

Hook, E. B. 1973. Behavioral implications of the humans XYY genotype. *Science* 179: 139–150.

Hughes, J. F., et al. 2010 Chimpanzee and human Y chromosomes are remarkably divergent in structure and gene content. *Nature* 463: 536–539.

Irish, E. E. 1996. Regulation of sex determination in maize. *BioEssays* 18: 363–369.

Jacobs, P. A., et al. 1974. A cytogenetic survey of 11,680 newborn infants. *Ann. Hum. Genet.* 37: 359–376.

Jegalian, K., and Lahn, B. T. 2001. Why the Y is so weird. *Sci. Am.* (Feb.) 284: 56–61.

Koopman, P., et al. 1991. Male development of chromosomally female mice transgenic for *Sry. Nature* 351: 117–121.

Lahn, B. T., and Page, D. C. 1997. Functional coherence of the human Y chromosome. *Science* 278: 675–680.

Lucchesi, J. 1983. The relationship between gene dosage, gene expression, and sex in *Drosophila. Dev. Genet.* 3: 275–282.

Lyon, M. F. 1972. X-chromosome inactivation and developmental patterns in mammals. *Biol. Rev.* 47: 1–35.

Marshall Graves, J. A. 2009. Birds do it with a Z gene. *Nature* 461: 177–178.

McMillen, M. M. 1979. Differential mortality by sex in fetal and neonatal deaths. *Science* 204: 89–91.

Ng, K., Pullirsch, D., Leeb, M., and Wutz, A. 2007. Xist and the order of silencing. *EMBO Reports* 8: 34–39.

Nora, E. P., and Heard, E. 2009. X chromosome inactivation: When dosage counts. *Cell* 139: 865–868.

Penny, G. D., et al. 1996. Requirement for Xist in X chromosome inactivation. *Nature* 379: 131–137.

Pieau, C. 1996. Temperature variation and sex determination in reptiles. *BioEssays* 18: 19–26.

Quinn, A. E., et al. 2007. Temperature sex reversal implies sex gene dosage in a reptile. *Science* 316: 411.

Smith, C. A., et al. 2009. The avian Z-linked gene DMRT1 is required for male sex determination in the chicken. *Nature* 461: 267–270.

Straub, T., and Becker, P. B. 2007. Dosage compensation: The beginning and end of a generalization. *Nat. Rev. Genet.* 8: 47–57.

Warner, D. A., and Shine, R. 2008. The adaptive significance of temperature-dependent sex determination in a reptile. *Nature* 451: 566–567.

Westergaard, M. 1958. The mechanism of sex determination in dioecious flowering plants. *Adv. Genet.* 9: 217–281.

Whitelaw, E. 2006. Unravelling the X in sex. *Dev. Cell* 11: 759–762.

Witkin, H. A., et al. 1996. Criminality in XYY and XXY men. *Science* 193: 547–555.

Xu, N., Tsai, C-L., and Lee, J. T. 2006. Transient homologous chromosome pairing marks the onset of X inactivation. *Science* 311: 1149–1152.

Chapter 8 Chromosome Mutations: Variation in Number and Arrangement

Antonarakis, S. E. 1998. Ten years of genomics, chromosome 21, and Down syndrome. *Genomics* 51: 1–16.

Ashley-Koch, A. E., et al. 1997. Examination of factors associated with instability of the *FMR1* CGG repeat. *Am. J. Hum. Genet.* 63: 776–785.

Beasley, J. O. 1942. Meiotic chromosome behavior in species, species hybrids, haploids, and induced polyploids of *Gossypium. Genetics* 27: 25–54.

Blakeslee, A. F. 1934. New jimson weeds from old chromosomes. *J. Hered.* 25: 80–108.

Boue, A. 1985. Cytogenetics of pregnancy wastage. *Adv. Hum. Genet.* 14: 1–58.

Carr, D. H. 1971. Genetic basis of abortion. *Ann. Rev. Genet.* 5: 65–80.

Conrad, D. F., et al. 2010. Origins and functional impact of copy number variation in the human genome. *Nature* 464: 704–10.

Croce, C. M. 1996. The *FHIT* gene at 3p14.2 is abnormal in lung cancer. *Cell* 85: 17–26.

DeArce, M. A., and Kearns, A. 1984. The fragile X syndrome: The patients and their chromosomes. *J. Med. Genet.* 21: 84–91.

Epstein, C. J. 2006. Down's syndrome: Critical genes in a critical region. *Nature* 441: 582–583.

Feldman, M., and Sears, E. R. 1981. The wild gene resources of wheat. *Sci. Am.* (Jan.) 244: 102–112.

Galitski, T., et al. 1999. Ploidy regulation of gene expression. *Science* 285: 251–254.

Gersh, M., et al. 1995. Evidence for a distinct region causing a catlike cry in patients with 5p deletions. *Am. J. Hum. Genet.* 56: 1404–1410.

Hassold, T. J., et al. 1980. Effect of maternal age on autosomal trisomies. *Ann. Hum. Genet.* (London) 44: 29–36.

Hassold, T. J., and Hunt, P. 2001. To err (meiotically) is human: The genesis of human aneuploidy. *Nat. Rev. Gen.* 2: 280–291.

Hassold, T., and Jacobs, P. A. 1984. Trisomy in man. *Annu. Rev. Genet.* 18: 69–98.

Hecht, F. 1988. Enigmatic fragile sites on human chromosomes. *Trends Genet.* 4: 121–122.

Hulse, J. H., and Spurgeon, D. 1974. *Triticale. Sci. Am.* (Aug.) 231: 72–81.

Kaiser, P. 1984. Pericentric inversions: Problems and significance for clinical genetics. *Hum. Genet.* 68: 1–47.

Lewis, E. B. 1950. The phenomenon of position effect. *Adv. Genet.* 3: 73–115.

Lewis, W. H., ed. 1980. *Polyploidy: Biological relevance.* New York: Plenum Press.

Lynch, M., and Conery, J. S. 2000. The evolutionary fate and consequences of duplicate genes. *Science* 290: 1151–1154.

Madan, K. 1995. Paracentric inversions: A review. *Hum. Genet.* 96: 503–515.

Nelson, D. L., and Gibbs, R. A. 2004. The critical region in trisomy 21. *Science* 306: 619–621.

Ohno, S. 1970. *Evolution by gene duplication.* New York: Springer-Verlag.

Oostra, B. A., and Verkerk, A. J. 1992. The fragile X syndrome: Isolation of the *FMR-1* gene and characterization of the fragile X mutation. *Chromosoma* 101: 381–387.

Patterson, D. 1987. The causes of Down syndrome. *Sci. Am.* (Aug.) 257: 52–61.

Patterson, D., and Costa, A. 2005. Down syndrome and genetics—A case of linked histories. *Nature Reviews Genetics* 6: 137–145.

Shepard, J., et al. 1983. Genetic transfer in plants through interspecific protoplast fusion. *Science* 21: 683–688.

Shepard, J. F. 1982. The regeneration of potato plants from protoplasts. *Sci. Am.* (May) 246: 154–166.

Stranger, B. E., et al., 2007. Relative impact of nucleotide and copy number variation on gene expression phenotypes. *Science* 315: 848–854.

Taylor, A. I. 1968. Autosomal trisomy syndromes: A detailed study of 27 cases of Edwards syndrome and 27 cases of Patau syndrome. *J. Med. Genet.* 5: 227–252.

Tjio, J. H., and Levan, A. 1956. The chromosome number of man. *Hereditas* 42: 1–6.

Wilkins, L. E., Brown, J. X., and Wolf, B. 1980. Psychomotor development in 65 home-reared children with cri-du-chat syndrome. *J. Pediatr.* 97: 401–405.

Chapter 9   Extranuclear Inheritance

Adams, K. L., et al. 2000. Repeated, recent and diverse transfers of a mitochondrial gene to the nucleus in flowering plants. *Nature* 408: 354–357.

Choman, A. 1998. The myoclonic epilepsy and ragged-red fiber mutation provides new insights into human mitochondrial function and genetics. *Am. J. Hum. Genet.* 62: 745–751.

Freeman, G., and Lundelius, J. W. 1982. The developmental genetics of dextrality and sinistrality in the gastropod *Lymnaea peregra. Wilhelm Roux Arch.* 191: 69–83.

Gray, M. W., Burger, G., and Lang, B. 1999. Mitochondrial evolution. *Science* 283: 1476–1481.

Green, B. R., and Burton, H. 1970. *Acetabularia* chloroplast DNA: Electron microscopic visualization. *Science* 168: 981–982.

Grivell, L. A. 1983. Mitochondrial DNA. *Sci. Am.* (Mar.) 248: 78–89.

He, Y., et al. 2010. Heteroplasmic mitochondrial DNA mutations in normal and tumour cells. *Nature* 464: 610–614.

Hiendleder, S. 2007. Mitochondrial DNA inheritance after SCNT. *Adv. Exp. Med Biol.* 591: 103–116.

Ishikawa, K. 2008. ROS-generating mitochondrial DNA mutations can regulate tumor cell metastasis. *Science* 320: 661–664.

Larson, N. G., and Clayton, D. A. 1995. Molecular genetic aspects of human mitochondrial disorders. *Ann. Rev. Genet.* 29: 151–178.

Margulis, L. 1970. *Origin of eukaryotic cells.* New Haven, CT: Yale University Press.

Mitchell, M. B., and Mitchell, H. K. 1952. A case of maternal inheritance in *Neurospora crassa. Proc. Natl. Acad. Sci. (USA)* 38: 442–449.

Nüsslein-Volhard. C. 1996. Gradients that organize embryo development. *Sci. Am.* (Aug.) 275: 54–61.

Preer, J. R. 1971. Extrachromosomal inheritance: Hereditary symbionts, mitochondria, chloroplasts. *Annu. Rev. Genet.* 5: 361–406.

Sager, R. 1965. Genes outside the chromosomes. *Sci. Am.* (Jan.) 212: 70–79.

_____. 1985. Chloroplast genetics. *BioEssays* 3: 180–184.

Schwartz, R. M., and Dayhoff, M. O. 1978. Origins of prokaryotes, eukaryotes, mitochondria and chloroplasts. *Science* 199: 395–403.

Sonneborn, T. M. 1959. Kappa and related particles in *Paramecium. Adv. Virus Res.* 6: 229–256.

Sturtevant, A. H. 1923. Inheritance of the direction of coiling in *Limnaea. Science* 58: 269–270.

Taylor, R. W., and Turnbull, D. M., 2005. Mitochondrial DNA mutations in human disease. *Nature Reviews Genetics* 6: 389–402.

Wallace, D. C. 1997. Mitochondrial DNA in aging and disease. *Sci. Am.* (Aug.) 277: 40–59.

_____. 1999. Mitochondrial diseases in man and mouse. *Science* 283: 1482–1488.

Chapter 10   DNA Structure and Analysis

Adleman, L. M. 1998. Computing with DNA. *Sci. Am.* (Aug.) 279: 54–61.

Alloway, J. L. 1933. Further observations on the use of pneumococcus extracts in effecting transformation of type *in vitro. J. Exp. Med.* 57: 265–278.

Avery, O. T., MacLeod, C. M., and McCarty, M. 1944. Studies on the chemical nature of the substance inducing transformation of pneumococcal types: Induction of transformation by a desoxyribonucleic acid fraction isolated from pneumococcus type III. *J. Exp. Med.* 79: 137–158. (Reprinted in Taylor, J. H. 1965. *Selected papers in molecular genetics.* Orlando, FL: Academic Press.)

Britten, R. J., and Kohne, D. E. 1968. Repeated sequences in DNA. *Science* 161: 529–540.

Chargaff, E. 1950. Chemical specificity of nucleic acids and mechanism for their enzymatic degradation. *Experientia* 6: 201–209.

Darnell, J. E. 1985. RNA. *Sci. Am.* (Oct.) 253: 68–87.

Dawson, M. H. 1930. The transformation of pneumococcal types: I. The interconvertibility of type-specific *S. pneumococci. J. Exp. Med.* 51: 123–147.

Dickerson, R. E., et al. 1982. The anatomy of A-, B-, and Z-DNA. *Science* 216: 475–485.

Dubos, R. J. 1976. *The professor, the Institute and DNA: Oswald T. Avery, his life and scientific achievements.* New York: Rockefeller University Press.

Felsenfeld, G. 1985. DNA. *Sci. Am.* (Oct.) 253: 58–78.

Franklin, R. E., and Gosling, R. G. 1953. Molecular configuration in sodium thymonucleate. *Nature* 171: 740–741.

Griffith, F. 1928. The significance of pneumococcal types. *J. Hyg.* 27: 113–159.

Guthrie, G. D., and Sinsheimer, R. L. 1960. Infection of protoplasts of *Escherichia coli* by subviral particles. *J. Mol. Biol.* 2: 297–305.

Hershey, A. D., and Chase, M. 1952. Independent functions of viral protein and nucleic acid in growth of bacteriophage. *J. Gen. Phys.* 36: 39–56. (Reprinted in Taylor, J. H. 1965. *Selected papers in molecular genetics.* Orlando, FL: Academic Press.)

Levene, P. A., and Simms, H. S. 1926. Nucleic acid structure as determined by electrometric titration data. *J. Biol. Chem.* 70: 327–341.

McCarty, M. 1985. *The transforming principle: Discovering that genes are made of DNA.* New York: W. W. Norton.

Olby, R. 1974. *The path to the double helix.* Seattle: University of Washington Press.

Pauling, L., and Corey, R. B. 1953. A proposed structure for the nucleic acids. *Proc. Natl. Acad. Sci. (USA)* 39: 84–97.

Rich, A., Nordheim, A., and Wang, A. H.-J. 1984. The chemistry and biology of left-handed Z-DNA. *Annu. Rev. Biochem.* 53: 791–846.

Spizizen, J. 1957. Infection of protoplasts by disrupted T2 viruses. *Proc. Natl. Acad. Sci. (USA)* 43: 694–701.

Varmus, H. 1988. Retroviruses. *Science* 240: 1427–1435.

Watson, J. D. 1968. *The double helix.* New York: Atheneum.

Watson, J. D., and Crick, F. C. 1953a. Molecular structure of nucleic acids: A structure for deoxyribose nucleic acids. *Nature* 171: 737–738.

———. 1953b. Genetic implications of the structure of deoxyribose nucleic acid. *Nature* 171: 964.

Wilkins, M. H. F., Stokes, A. R., and Wilson, H. R. 1953. Molecular structure of desoxypentose nucleic acids. *Nature* 171: 738–740.

Chapter 11   DNA Replication and Recombination

Blackburn, E. H. 1991. Structure and function of telomeres. *Nature* 350: 569–572.

DeLucia, P., and Cairns, J. 1969. Isolation of an *E. coli* strain with a mutation affecting DNA polymerase. *Nature* 224: 1164–1166.

Gilbert, D. M. 2001. Making sense of eukaryotic DNA replication origins. *Science* 294: 96–100.

Greider, C. W. 1996. Telomeres, telomerase, and cancer. *Sci. Am.* (Feb.) 274: 92–97.

———. 1998. Telomerase activity, cell proliferation, and cancer. *Proc. Natl. Acad. Sci. (USA)* 95: 90–92.

Holliday, R. 1964. A mechanism for gene conversion in fungi. *Genet. Res.* 5: 282–304.

Holmes, F. L. 2001. *Meselson, Stahl, and replication of DNA: A history of the "most beautiful experiment in biology."* New Haven, CT: Yale University Press.

Kim, J., Kaminker, P., and Campisi, J. 2002. Telomeres, aging and cancer: In search of a happy ending. *Oncogene* 21: 503–511.

Kornberg, A. 1960. Biological synthesis of DNA. *Science* 131: 1503–1508.

Kornberg, A., and Baker, T. 1992. *DNA replication,* 2nd ed. New York: W. H. Freeman.

Krogh, B. O., and Symington, L. S. 2004. Recombination proteins in yeast. *Annu. Rev. Genet.* 38: 233–271.

Luke, B., and Linguer, J. 2009. TERRA: Telomere repeat-containing RNA. *EMBO J.* 28: 2503–2510.

Meselson, M., and Stahl, F. W. 1958. The replication of DNA in *Escherichia coli. Proc. Natl. Acad. Sci. (USA)* 44: 671–682.

Mitchell, M. B. 1955. Aberrant recombination of pyridoxine mutants of *Neurospora. Proc. Natl. Acad. Sci. (USA)* 41: 215–220.

Okazaki, T., et al. 1979. Structure and metabolism of the RNA primer in the discontinuous replication of prokaryotic DNA. *Cold Spring Harbor Symp. Quant. Biol.* 43: 203–222.

Radman, M., and Wagner, R. 1988. The high fidelity of DNA duplication. *Sci. Am.* (Aug.) 259: 40–46.

———. 1987. Genetic recombination. *Sci. Am.* (Feb.) 256: 90–101.

Redon, S., et. al. 2010. The noncoding RNA TERRA is a natural ligand of human telomerase. *Nucleic Acid Res.* 38: 5797–5806.

Takeda, D. Y., and Dutta, A. 2005. DNA replication and progression through S phase. *Oncogene* 24: 2827–2843.

Taylor, J. H., Woods, P. S., and Hughes, W. C. 1957. The organization and duplication of chromosomes revealed by autoradiographic studies using tritium-labeled thymidine. *Proc. Natl. Acad. Sci. (USA)* 48: 122–128.

Wang, J. C. 1987. Recent studies of DNA topoisomerases. *Biochim. Biophys. Acta* 909: 1–9.

Whitehouse, H. L. K. 1982. *Genetic recombination: Understanding the mechanisms.* New York: Wiley.

Chapter 12   Chromosome Structure and DNA Sequence Organization

Angelier, N., et al. 1984. Scanning electron microscopy of amphibian lampbrush chromosomes. *Chromosoma* 89: 243–253.

Beerman, W., and Clever, U. 1964. Chromosome puffs. *Sci. Am.* (Apr.) 210: 50–58.

Carbon, J. 1984. Yeast centromeres: Structure and function. *Cell* 37: 352–353.

Chen, T. R., and Ruddle, F. H. 1971. Karyotype analysis utilizing differential stained constitutive heterochromatin of human and murine chromosomes. *Chromosoma* 34: 51–72.

DuPraw, E. J. 1970. *DNA and chromosomes.* New York: Holt, Rinehart & Winston.

Gall, J. G. 1981. Chromosome structure and the *C*-value paradox. *J. Cell Biol.* 91: 3s–14s.

Hewish, D. R., and Burgoyne, L. 1973. Chromatin substructure. The digestion of chromatin DNA at regularly spaced sites by a nuclear deoxyribonuclease. *Biochem. Biophys. Res. Comm.* 52: 504–510.

Korenberg, J. R., and Rykowski, M. C. 1988. Human genome organization: *Alu,* LINES, and the molecular organization of metaphase chromosome bands. *Cell* 53: 391–400.

Kornberg, R. D. 1975. Chromatin structure: A repeating unit of histones and DNA. *Science* 184: 868–871.

Kornberg, R. D., and Klug, A. 1981. The nucleosome. *Sci. Am.* (Feb.) 244: 52–64.

Lorch, Y., et al. 2006. Chromatin remodeling by nucleosome disassembly *in vitro*. *Proc. Nat. Acad. Sci.* 103: 3090–3093.

Luger, K., et al. 1997. Crystal structure of the nucleosome core particle at 2.8 resolution. *Nature* 389: 251–256.

Moyzis, R. K. 1991. The human telomere. *Sci. Am.* (Aug.) 265: 48–55.

Olins, A. L., and Olins, D. E. 1974. Spheroid chromatin units (ν bodies). *Science* 183: 330–332.

———. 1978. Nucleosomes: The structural quantum in chromosomes. *Am. Sci.* 66: 704–711.

Singer, M. F. 1982. SINES and LINES: Highly repeated short and long interspersed sequences in mammalian genomes. *Cell* 28: 433–434.

Chapter 13 The Genetic Code and Transcription

Barrell, B. G., Air, G., and Hutchinson, C. 1976. Overlapping genes in bacteriophage φX174 *Nature* 264: 34–40.

Barrell, B. G., Banker, A. T., and Drouin, J. 1979. A different genetic code in human mitochondria. *Nature* 282: 189–194.

Bass, B. L., ed. 2000. *RNA editing*. Oxford: Oxford University Press.

Brenner, S., Jacob, F., and Meselson, M. 1961. An unstable intermediate carrying information from genes to ribosomes for protein synthesis. *Nature* 190: 575–580.

Brenner, S., Stretton, A. O. W., and Kaplan, D. 1965. Genetic code: The nonsense triplets for chain termination and their suppression. *Nature* 206: 994–998.

Cattaneo, R. 1991. Different types of messenger RNA editing. *Annu. Rev. Genet.* 25: 71–88.

Cech, T. R. 1986. RNA as an enzyme. *Sci. Am.* (Nov.) 255(5): 64–75.

———. 1987. The chemistry of self-splicing RNA and RNA enzymes. *Science* 236: 1532–1539.

Chambon, P. 1981. Split genes. *Sci. Am.* (May) 244: 60–71.

Cramer, P., et al. 2000. Architecture of RNA polymerase II and implications for the transcription mechanism. *Science* 288: 640–649.

Crick, F. H. C. 1962. The genetic code. *Sci. Am.* (Oct.) 207: 66–77.

———. 1966a. The genetic code: III. *Sci. Am.* (Oct.) 215: 55–63.

———. 1966b. Codon–anticodon pairing: The wobble hypothesis. *J. Mol. Biol.* 19: 548–555.

Crick, F. H. C., Barnett, L., Brenner, S., and Watts-Tobin, R. J. 1961. General nature of the genetic code for proteins. *Nature* 192: 1227–1232.

Darnell, J. E. 1983. The processing of RNA. *Sci. Am.* (Oct.) 249: 90–100.

Dickerson, R. E. 1983. The DNA helix and how it is read. *Sci. Am.* (Dec.) 249: 94–111.

Dugaiczk, A., et al. 1978. The natural ovalbumin gene contains seven intervening sequences. *Nature* 274: 328–333.

Fiers, W., et al. 1976. Complete nucleotide sequence of bacteriophage MS2 RNA: Primary and secondary structure of the replicase gene. *Nature* 260: 500–507.

Gamow, G. 1954. Possible relation between DNA and protein structures. *Nature* 173: 318.

Hall, B. D., and Spiegelman, S. 1961. Sequence complementarity of T2-DNA and T2-specific RNA. *Proc. Natl. Acad. Sci. (USA)* 47: 137–146.

Hamkalo, B. 1985. Visualizing transcription in chromosomes. *Trends Genet.* 1: 255–260.

Khorana, H. G. 1967. Polynucleotide synthesis and the genetic code. *Harvey Lectures* 62: 79–105.

Makalowska, I., et al. 2005. Overlapping genes in vertebrate genomes. *Comput Biol Chem.* 29: 1–12.

Miller, O. L., Hamkalo, B., and Thomas, C. 1970. Visualization of bacterial genes in action. *Science* 169: 392–395.

Nirenberg, M. W. 1963. The genetic code: II. *Sci. Am.* (Mar.) 190: 80–94.

O'Malley, B., et al. 1979. A comparison of the sequence organization of the chicken ovalbumin and ovomucoid genes. In *Eucaryotic gene regulation*, R. Axel et al., eds., pp. 281–299. Orlando, FL: Academic Press.

Reed, R., and Maniatis, T. 1985. Intron sequences involved in lariat formation during pre-mRNA splicing. *Cell* 41: 95–105.

Ridley, M., 2006. *Francis Crick: Discoverer of the genetic code*. New York: HarperCollins.

Roberts, J. W. 2006. RNA polymerase: A scrunching machine. *Science* 314: 1097–1098.

Sharp, P. A. 1994. Nobel lecture: Split genes and RNA splicing. *Cell* 77: 805–815.

Steitz, J. A. 1988. Snurps. *Sci. Am.* (June) 258(6): 56–63.

Volkin, E., and Astrachan, L. 1956. Phosphorus incorporation in *E. coli* ribonucleic acids after infection with bacteriophage T2. *Virology* 2: 149–161.

Watson, J. D. 1963. Involvement of RNA in the synthesis of proteins. *Science* 140: 17–26.

Woychik, N. A., and Jampsey, M. 2002. The RNA polymerase II machinery: Structure illuminates function. *Cell* 108: 453–464.

Chapter 14 Translation and Proteins

Anfinsen, C. B. 1973. Principles that govern the folding of protein chains. *Science* 181: 223–230.

Atkins, J. F., and Baranov, P. V. 2007. Duality in the genetic code. *Nature* 448: 1004–1005.

Bartholome, K. 1979. Genetics and biochemistry of phenylketonuria—Present state. *Hum. Genet.* 51: 241–245.

Beadle, G. W., and Tatum, E. L. 1941. Genetic control of biochemical reactions in *Neurospora*. *Proc. Natl. Acad. Sci. (USA)* 27: 499–506.

Beet, E. A. 1949. The genetics of the sickle-cell trait in a Bantu tribe. *Ann. Eugenics* 14: 279–284.

Ben-Sham, A. 2010. Crystal structure of the eukaryotic ribosome. *Science* 330: 1203–1208.

Brenner, S. 1955. Tryptophan biosynthesis in *Salmonella typhimurium*. *Proc. Natl. Acad. Sci. (USA)* 41: 862–863.

Doolittle, R. F. 1985. Proteins. *Sci. Am.* (Oct.) 253: 88–99.

Fischer, N., et al. 2010. Ribosome dynamics and tRNA movement by time-resolved electron cryomicroscopy. *Nature* 466: 329–336.

Frank, J. 1998. How the ribosome works. *Amer. Scient.* 86: 428–439.

Garrod, A. E. 1902. The incidence of alkaptonuria: A study in chemical individuality. *Lancet* 2: 1616–1620.

———. 1909. *Inborn errors of metabolism*. London: Oxford University Press. (Reprinted 1963, Oxford University Press, London.)

Garrod, S. C. 1989. Family influences on A. E. Garrod's thinking. *J. Inher. Metab. Dis.* 12: 2–8.

Ingram, V. M. 1957. Gene mutations in human hemoglobin: The chemical difference between normal and sickle-cell hemoglobin. *Nature* 180: 326–328.

Koshland, D. E. 1973. Protein shape and control. *Sci. Am.* (Oct.) 229: 52–64.

Lake, J. A. 1981. The ribosome. *Sci. Am.* (Aug.) 245: 84–97.

Maniatis, T., et al. 1980. The molecular genetics of human hemoglobins. *Annu. Rev. Genet.* 14: 145–178.

Neel, J. V. 1949. The inheritance of sickle-cell anemia. *Science* 110: 64–66.

Nirenberg, M. W., and Leder, P. 1964. RNA codewords and protein synthesis. *Science* 145: 1399–1407.

Nomura, M. 1984. The control of ribosome synthesis. *Sci. Am.* (Jan.) 250: 102–114.

Pauling, L., Itano, H. A., Singer, S. J., and Wells, I. C. 1949. Sickle-cell anemia: A molecular disease. *Science* 110: 543–548.

Ramakrishnan, V. 2002. Ribosome structure and the mechanism of translation. *Cell* 108: 557–572.

Rich, A., and Houkim, S. 1978. The three-dimensional structure of transfer RNA. *Sci. Am.* (Jan.) 238: 52–62.

Rich, A., Warner, J. R., and Goodman, H. M. 1963. The structure and function of polyribosomes. *Cold Spring Harbor Symp. Quant. Biol.* 28: 269–285.

Richards, F. M. 1991. The protein-folding problem. *Sci. Am.* (Jan.) 264: 54–63.

Rould, M. A., et al. 1989. Structure of *E. coli* glutaminyl-tRNA synthetase complexed with tRNA^Gln and ATP at 2.8 resolution. *Science* 246: 1135–1142.

Srb, A. M., and Horowitz, N. H. 1944. The ornithine cycle in *Neurospora* and its genetic control. *J. Biol. Chem.* 154: 129–139.

Warner, J., and Rich, A. 1964. The number of soluble RNA molecules on reticulocyte polyribosomes. *Proc. Natl. Acad. Sci. (USA)* 51: 1134–1141.

Wimberly, B. T., et al. 2000. Structure of the 30S ribosomal subunit. *Nature* 407: 327–333.

Yuan, J., et al. 2010. Distinct genetic code expansion strategies for selenocysteine and pyrrolysine are reflected in different aminoacyl-tRNA formation systems. *FEBS Lett.* 584: 342–349.

Yusupov, M. M., et al. 2001. Crystal structure of the ribosome at 5.5A resolution. *Science* 292: 883–896.

Chapter 15 Gene Mutation and DNA Repair

Arnheim, N., and Calabrese, P. 2009. Understanding what determines the frequency and pattern of human germline mutations. *Nature Rev. Genetics* 10: 478–488.

Cairns, J., Overbaugh, J., and Miller, S. 1988. The origin of mutants. *Nature* 335: 142–145.

Cleaver, J. E. 2005. Cancer in xeroderma pigmentosum and related disorders of DNA repair. *Nature Reviews Cancer* 5: 564–571.

Comfort, N. C. 2001. *The tangled field: Barbara McClintock's search for the patterns of genetic control.* Cambridge, MA: Harvard University Press.

Ellegren, H., Smith, N. G. C, and Webster, M. T. 2003. Mutation rate variation in the mammalian genome. *Curr. Opin. Genet. Dev.* 13: 562–568.

Jiricny, J. 1998. Eukaryotic mismatch repair: An update. *Mutation Research* 409: 107–121.

Luria, S. E., and Delbrück, M. (1943). Mutations of bacteria from virus sensitivity to virus resistance. *Genetics* 28(6): 491–511.

Mirkin, S. M. 2007. Expandable DNA repeats and human disease. *Nature* 447: 932–940.

Mortelmans, K., and Zeigler, E. 2000. The Ames *Salmonella/microsome* mutagenicity assay. *Mutation Research* 455: 29–60.

O'Driscoll, M., and Jeggo, P. A. 2006. The role of double-strand break repair— insights from human genetics. *Nature Reviews Genetics* 7: 45–51.

O'Hare, K. 1985. The mechanism and control of *P* element transposition in *Drosophila*. *Trends Genet.* 1: 250–254.

Rosenberg, S. M. 2001. Evolving responsively: adaptive mutation. *Nature Rev. Genetics* 2: 504–515.

Chapter 16 Regulation of Gene Expression in Prokaryotes

Antson, A. A., et al. 1999. Structure of the trp RNA-binding attenuation protein, TRAP, bound to RNA. *Nature* 401: 235–242.

Beckwith, J. R., and Zipser, D., eds. 1970. *The lactose operon.* Cold Spring Harbor, NY: Cold Spring Harbor Laboratory Press.

Bertrand, K., et al. 1975. New features of the regulation of the tryptophan operon. *Science* 189: 22–26.

Breaker, Ronald R. 2008. Complex riboswitches. *Science* 319: 1795–1797.

Gilbert, W., and Müller-Hill, B. 1966. Isolation of the *lac* repressor. *Proc. Natl. Acad. Sci. (USA)* 56: 1891–1898.

————. 1967. The *lac* operator is DNA. *Proc. Natl. Acad. Sci. (USA)* 58: 2415–2421.

Jacob, F., and Monod, J. 1961. Genetic regulatory mechanisms in the synthesis of proteins. *J. Mol. Biol.* 3: 318–356.

Lewis, M., et al. 1996. Crystal structure of the lactose operon repressor and its complexes with DNA and inducer. *Science* 271: 1247–1254.

Maumita, M., et al. 2003. Riboswitches control fundamental biochemical pathways in *Bacillus subtilis* and other bacteria. *Cell* 113: 577–586.

Serganov, A. 2009. The long and the short of riboswitches. *Curr Opin Struct Biol.* 19: 251–259.

Stroynowski, I., and Yanofsky, C. 1982. Transcript secondary structures regulate transcription termination at the attenuator of *S. marcescens* tryptophan operon. *Nature* 298: 34–38.

Valbuzzi, A., and Yanofsky, C. 2001. Inhibition of the *B. subtilis* regulatory protein TRAP by the TRAP-inhibitory protein, AT. *Science* 293: 2057–2061.

Yanofsky, C. 1981. Attenuation in the control of expression of bacterial operons. *Nature* 289: 751–758.

Chapter 17 Regulation of Gene Expression in Eukaryotes

Alt, F. W., et al. 1987. Development of the primary antibody repertoire. *Science* 238: 1079–1087.

Becker, P. B., and Hörz, W. 2002. ATP-dependent nucleosome remodeling. *Annu. Rev. Biochem.* 71: 247–273.

Black, D. L. 2000. Protein diversity from alternative splicing: A challenge for bioinformatics and postgenomic biology. *Cell* 103: 367–370.

Black, D. L. 2003. Mechanisms of alternative pre-mRNA splicing. *Annu. Rev. Biochem.* 72: 291–336.

Bumcrot, D., et al. 2006. RNAi therapeutics: a potential new class of pharmaceutical drugs. *Nature Chemical Biology* 2: 711–718.

Henikoff, S. 2008. Nucleosome destabilization in the epigenetic regulation of gene expression. *Nature Reviews Genetics* 9: 15–20.

Jackson, D. A. 2003. The anatomy of transcription sites. *Curr. Opin. Cell Biol.* 15: 311–317.

Juven-Gershon, T., and Kadonaga, J. T. 2009. Regulation of gene expression via the core promoter and the basal transcriptional machinery. *Dev. Biol.* 339: 225–229.

Licatalosi, D. D., and Darnell, R. B. 2010. RNA processing and its regulation: global insights into biological networks. *Nature Rev. Genetics* 11: 75–87.

Mello, C. C., and Conte, D., Jr. 2004. Revealing the world of RNA interference. *Nature* 431: 338–342.

Newbury, S. F. 2006. Control of mRNA stability in eukaryotes. *Biochem. Soc. Trans.* 34: 30–34.

Rana, T. M. 2007. Illuminating the silence: understanding the structure and function of small RNAs. *Nature Mol. Cell Biol.* 8: 23–34.

Turner, B. M. 2000. Histone acetylation and an epigenetic code. *Bioessays* 22: 836–845.

Wheeler, T. M. and Thornton, C. A. 2007. Myotonic dystrophy: RNA-mediated muscle disease. *Curr. Opin. Neurol.* 20: 572–576.

Chapter 18 Developmental Genetics

Davidson, E. H., and Levine, M. S. 2008. Properties of developmental gene regulatory networks. *Proc. Nat. Acad. Sci.* 115: 20063–20066.

De Leon, S. B-T., and Davidson, E. H. 2007. Gene regulation: Gene control networks in development. *Annu. Rev. Biophys. Biomol. Struct.* 36: 191–212.

Feng, S., Jacobsen, S. E., and Reik, W. 2010. Epigenetic reprogramming in plant and animal development. *Science* 330: 622–627.

Goodman, F. 2002. Limb malformations and the human *HOX* genes. *Am. J. Med. Genet.* 112: 256–265.

Gridley, T. 2003. Notch signaling and inherited human diseases. *Hum. Mol. Genet.* 12: R9–R13.

Inoue, T., *et al.* 2005. Gene regulatory special feature: Transcriptional network underlying *Caenorhabditis elegans* vulval development. *Proc. Nat. Acad. Sci.* 102: 4972–4977.

Kestler, H. A., Wawra, C., Kracher, B., and Kuhl, M. 2008. Network modeling of signal transduction: Establishing the global view. *Bioessays* 30: 1110–1125.

Krizek, B. A., and Fletcher, J. C. 2006. Molecular mechanisms of flower development: An armchair guide. *Nat. Rev. Genet.* 6: 688–698.

Lynch, J. A., and Roth, S. 2011. The evolution of dorsal-ventral patterning mechanism in insects. *Genes Dev.* 25: 107–118.

Maeda, R. K., and Karch, F. 2006. The ABC of the BX-C: The bithorax complex explained. *Development* 133: 1413–1422.

Nüsslein-Volhard, C., and Weischaus, E. 1980. Mutations affecting segment number and polarity in *Drosophila*. *Nature* 287: 795–801.

Reyes, J. C. 2006. Chromatin modifiers that control plant development. *Curr. Opin. Plant Biol.* 9: 21–27.

Schroeder, M. D., et al., 2004. Transcriptional control in the segmentation gene network of *Drosophila*. *PLOS Biol.* 2: 1396–1410.

Schvartsman, S. Y., Coppey, M., and Berezhkovskii, A. 2008. Dynamics of maternal morphogen gradients in *Drosophila*. *Curr. Opin. Genet. Dev.* 18: 342–347.

Verakasa, A., Del Campo, M., and McGinnis, W. 2000. Developmental patterning genes and their conserved functions: From model organisms to humans. *Mol. Genet. Metabol.* 69: 85–100.

Wang, M., and Sternberg, P. W. 2001. Pattern formation during *C. elegans* vulval induction. *Curr. Top. Dev. Biol.* 51: 189–220.

Yuan, J., and Kroemer, G. 2010. Alternative cell death mechanisms in development and beyond. *Genes Dev.* 24: 2592–2602.

Chapter 19 Cancer and Regulation of the Cell Cycle

Alison, M. R., et al. 2010. Stem cells in cancer: instigators and propagators? *J. Cell Sci.* 123: 2357–2368.

Bernards, R., and Weinberg, R. A. 2002. A progression puzzle. *Nature* 418: 823.

Brown, M. A. 1997. Tumor suppressor genes and human cancer. *Adv. Genet.* 36: 45–135.

Compagni, A., and Christofori, G. 2000. Recent advances in research on multistage tumorigenesis. *Brit. J. Cancer* 83: 1–5.

Esteller, M. 2007. Cancer epigenomics: DNA methylomes and histone-modification maps. *Nature Reviews Cancer* 8: 286–297.

Futreal, P. A., et al. 2004. A census of human cancer genes. *Nature Reviews Cancer* 4: 177–183.

Gray, J. 2010. Genomics of metastasis. *Nature* 464: 989–990.

Hartwell, L. H., and Kastan, M. B. 1994. Cell cycle control and cancer. *Science* 266: 1821–1827.

Ledford, H. 2010. The cancer genome challenge. *Nature* 464: 972–974.

Lengauer, C., Kinzler, K. W., and Vogelstein, B. 1997. Genetic instability in colorectal cancer. *Nature* 386: 623–627.

Nurse, P. 1997. Checkpoint pathways come of age. *Cell* 91: 865–867.

Sherr, C. J. 1996. Cancer cell cycles. *Science* 274: 1672–1677.

Varmus, H. 2006. The new era in cancer research. *Science* 312: 1162–1164.

Vazquez, J., et al. 2008. The genetics of the p53 pathway, apoptosis and cancer therapy. *Nature Reviews Drug Discov.* 7: 979–987.

Vogelstein, B., and Kinzler, K. W. 1993. The multistep nature of cancer. *Trends in Genetics* 9: 138–141.

Yokota, J. 2000. Tumor progression and metastasis. *Carcinogenesis* 21: 497–503.

Chapter 20 Recombinant DNA Technology

Brownlee, C. 2005. Danna and Nathans: Restriction enzymes and the boon to modern molecular biology. *Proc. Nat. Acad. Sci.* 103: 5909.

Cohen, S. N. 1975. The manipulation of genes. *Sci. Am.* (Jan.) 233: 24–33.

Mardis, E. R. 2008. Next-generation DNA sequencing methods. *Annu. Rev. Genomics Hum. Genet.* 9: 387–402.

Mardis, E. R. 2011. A decade's perspective on DNA sequencing technology. *Nature*, 470: 198–203.

Mullis, K. B. 1990. The unusual origin of the polymerase chain reaction. *Sci. Am.* (Apr.) 262: 56–65.

Rothberg, J. M., and Leamon, J. H. 2008. The development and impact of 454 sequencing. *Nat. Biotech.* 26: 1117–1124.

Sanger, F., et al. 1977. DNA sequencing with chain-terminating inhibitors. *Proc. Natl. Acad. Sci. (USA)* 74: 5463–5467.

Shendure, J., and Ji, H. 2008. Next-generation DNA sequencing. *Nature Biotechnology.* 26: 1135–1145.

Snyder, M., Du, J., and Gerstein, M. 2010. Personal genome sequencing: Current approaches and challenges. *Genes & Development.* 24: 423-431.

Southern, E. 1975. Detection of specific sequences among DNA fragments separated by gel electrophoresis. *J. Mol. Biol.* 98: 503–507.

Chapter 21 Genomics, Bioinformatics, and Proteomics

Asara, J. M., et al. 2007. Protein sequences from Mastodon and *Tyrannosaurus Rex* revealed by mass spectrometry. *Science* 316: 280–285.

Chandonia, J. M. 2006. The impact of structural genomics: expectations and outcomes. *Science* 311: 347–351.

Church, G. M. 2006. Genomes for all. *Sci. Am.* (Jan.) 294: 47–54.

Conrad, D. F., et al. 2010. Origins and functional impact of copy number variations in the human genome. *Nature* 464: 704–712.

Domon, B., and Aebersold, R. 2006. Mass spectrometry and protein analysis. *Science* 312: 212–217.

Fleming, K., et al. 2006. The proteome: structure, function and evolution. *Philos. Trans. R. Soc. Lond. B Biol. Sci.* 361: 441–451.

Green, R. E., et al. 2008. A complete Neandertal mitochondrial genome sequence determined by high-throughput sequencing. *Cell* 134: 416–426.

Green, R. E., et al. 2010. A draft sequence of the Neandertal genome. *Science* 328: 710–722.

Hood, L., Heath, J. R., Phelps, M. E., and Lin, B. 2004. Systems biology and new technologies enable predictive and preventative medicine. *Science* 306: 640–643.

Hoskins, R. A., et al., 2007. Sequence finishing and mapping of *Drosophila melanogaster* heterochromatin. *Science* 316: 1625–1628.

Hugenholtz, P., and Tyson, G. W. 2008. Metagenomics. *Nature* 455: 481–483.

International Human Genome Sequencing Consortium. 2001. Initial sequencing and analysis of the human genome. *Nature* 409: 860–921.

International Human Genome Sequencing Consortium. 2004. Finishing the euchromatic sequence of the human genome. *Nature* 431: 931–945.

Joyce, A. R., and Palsson, B. O. 2006. The model organism as a system integrating "omics" data sets. *Nat. Rev. Mol. Cell Biol.* 7: 901–909.

Lander, E. S. 2011. Initial impact of sequencing the human genome. *Nature* 470: 187–197.

Loscalzo, J., Kohane, I., and Barabasi, A-L. 2007. Human disease classification in the postgenomic era: A complex systems approach to human pathobiology. *Molecular Systems Biology* 124: 32–42.

Nei, M., and Rooney, A. P. 2005. Concerted and birth-and-death evolution of multigene families. *Annu. Rev. Genet.* 39: 197–218.

Nezvizhskii, A. I., Vitek, O., and Aebersold, R. 2007. Analysis and validation of proteomic data generated by tandem mass spectrometry. *Nature Methods* 4: 787–797.

Noonan, J. P., et al. 2006. Sequencing and analysis of Neanderthal DNA. *Science* 314: 1113–1118.

Pennisi, E. 2006. The dawn of Stone Age genomics. *Science* 314: 1068–1071.

Qin, J. et al. 2010. A human gun microbial gene catalogue established by metagenomic sequencing. *Nature*, 464: 59–64.

Rhesus Macaque Genome Sequencing and Analysis Consortium. 2007. Evolutionary and biomedical insights from the Rhesus Macaque genome. *Science* 316: 222–234.

Schweitzer, M. H., et al., 2007. Analyses of soft tissue from *Tyrannosaurus rex* suggest the presence of protein. *Science* 316: 277–280.

Sea Urchin Genome Sequencing Consortium. 2006. The genome of the sea urchin *Stronglyocentrotus purpuratus.* *Science* 314: 941–956.

Switnoski, M., Szczerbal, I., and Nowacka, J. 2004. The dog genome map and its use in mammalian comparative genomics. *J. Appl. Physiol.* 45: 195–214.

The 1000 Genomes Project Consortium. 2010. A map of human genome variation from population-scale sequencing. *Nature* 467: 1061–1073.

Thieman, W. J., and Palladino, M. A. 2009. *Introduction to biotechnology,* 2nd ed. San Francisco, CA: Benjamin Cummings.

Venter, J. C., et al. 2001. The sequence of the human genome. *Science* 291: 1304–1351.

Venter, J. C., et al. 2004. Environmental genome shotgun sequencing of the Sargasso Sea. *Science* 304: 6–74.

Wong, D. T. 2008, Salivary diagnostics. *American Scientist*, 96: 37–43.

Yooseph, S., Sutton, G., Rusch, D. B., Halpern, A. L., Williamson, S. J., et al. 2007. *The Sorcerer II* global ocean sampling expedition: Expanding the universe of protein families. *PLoS Biology* 5(3): 0432–0466.

Chapter 22 Applications and Ethics of Genetic Engineering and Biotechnology

Altshuler, D., Daly, M. J., and Lander, E. S. 2008. Genetic Mapping in Human Disease. *Science* 322: 881–888.

Cavazzanna-Calvo M., et al. 2000. Gene therapy of severe combined immunodeficiency (SCID)-XI disease. *Science* 288: 669–672.

Cavazzanna-Calvo, M., et al. 2010. Transfusion independence and *HMGA2* activation after gene therapy of human β-thalassemia. *Nature* 467: 318-322.

Chrispeels, M. J., and Sadava, D. E. 2003. *Plants, genes, and crop biotechnology,* 2nd ed. Sudbury, MA: Jones and Barlett Publishers.

Colavito, M. C. 2007. *Gene therapy.* M. A. Palladino, ed. San Francisco, CA: Benjamin Cummings.

Collins, F. S., and Barker, A. D. 2007. Mapping the cancer genome. *Scientific American* 296: 50–57.

Dykxhoorn, D. M., and Liberman, J. 2006. Knocking down disease with siRNAs. *Cell* 126: 231–235.

Engler, O. B., et al. 2001. Peptide vaccines against hepatitis B virus: From animal model to human studies. *Mol. Immunol.* 38: 457–465.

Friend, S. H., and Stoughton, R. B. 2002. The magic of microarrays. *Sci. Am.* (Feb.) 286: 44–50.

Gibson, D. G., et al. 2010. Creation of a bacterial cell controlled by a chemically synthesized genome. *Science* 329: 52–56.

Green, E. D., and Guyer, M. S. 2011. Charting a course for genomic medicine from base pairs to bedside. *Nature*, 470: 204–213.

Hacein-Bey-Abina, S., Von Kalle, C., Schmidt, M., et al. 2003. *LM02*-Associated clonal T cell proliferation in two patients after gene therapy for SCID-X1. *Science*, 302: 415–419.

Knoppers, B. M., Bordet, S., and Isasi, R. M. 2006. Preimplantation genetic diagnosis: An overview of socio-ethical and legal considerations. *Annu. Rev. Genom. Hum. Genet.* 7: 201–221.

Manolio, T. A., et al. 2009. Finding the missing heritability of complex diseases. *Nature* 461: 747–753.

Schillberg, S., Fischer, T., and Emans, N. 2003. Molecular farming of recombinant antibodies in plants. *Cell Mol. Life Sci.* 60: 433–445.

Shastry, B. S. 2006. Pharmacogenetics and the concept of individualized medicine. *Pharmacogenetics Journal* 6: 16–21.

Stix, G. 2006. Owning the stuff of life. *Sci. Am.* (Feb.) 294: 76–83.

Wang, D. G., et al. 1998. Large-scale identification, mapping, and genotyping of single-nucleotide polymorphisms in the human genome. *Science* 280: 1077–1082.

Whitelaw, C. B. A. 2004. Transgenic livestock made easy. *Trends in Biotechnology* 22(4): 257–259.

Chapter 23 Quantitative Genetics and Multifactorial Traits

Browman, K. W. 2001. Review of statistical methods of QTL mapping in experimental crosses. *Lab Animal* 30: 44–52.

Cong, B., Barrero, L. S., and Tanksley, S. D. 2008. Regulatory changes in a YABBY-like transcription factor led to evolution of extreme fruit size during tomato domestication. *Nat. Genet.* 40: 800–804.

Cong, B., Liu, J., and Tanksley, S. 2002. Natural alleles at a tomato fruit size quantitative trait locus differ by heterochronic regulatory mutations. *Proc. Nat. Acad. Sci.* 99: 13606–13611.

Crow, J. F. 1993. Francis Galton: Count and measure, measure and count. *Genetics* 135: 1.

Druka, A., et al. 2010. Expression of quantitative trait loci analysis in plants. *Plant Biotech.* 8: 10–27.

Falconer, D. S., and Mackay, F. C. 1996. *Introduction to quantitative genetics,* 4th ed. Essex, England: Longman.

Farber, S. 1980. *Identical twins reared apart.* New York: Basic Books.

Feldman, M. W., and Lewontin, R. C. 1975. The heritability hangup. *Science* 190: 1163–1166.

Frary, A., et al. 2000. *fw2.2:* A quantitative trait locus key to the evolution of tomato fruit size. *Science* 289: 85–88.

Gupta, V., et al. 2009. Genome analysis and genetic enhancement of tomato. *Crit. Rev. Biotech.* 29: 152–181.

Haley, C. 1991. Use of DNA fingerprints for the detection of major genes for quantitative traits in domestic species. *Anim. Genet.* 22: 259–277.

———. 1996. Livestock QTLs: Bringing home the bacon. *Trends Genet.* 11: 488–490.

Johnson, W. E., et al. 2010. Genetic restoration of the Florida Panther. *Science* 329: 1641–1644.

Kaminsky, A. A., et al. 2009. DNA methylation profiles in monozygotic and dizygotic twins. *Nat. Genet.* 41: 240–245.

Lander, E., and Botstein, D. 1989. Mapping Mendelian factors underlying quantitative traits using RFLP linkage maps. *Genetics* 121: 185–199.

Lander, E., and Schork, N. 1994. Genetic dissection of complex traits. *Science* 265: 2037–2048.

Lynch, M., and Walsh, B. 1998. *Genetics and analysis of quantitative traits.* Sunderland, MA: Sinauer Associates.

Macy, T. F. C. 2001. Quantitative trait loci in *Drosophila. Nature Reviews Genetics* 2: 11–19.

Newman, H. H., Freeman, F. N., and Holzinger, K. T. 1937. *Twins: A study of heredity and environment.* Chicago: University of Chicago Press.

Paterson, A., Deverna, J., Lanini, B., and Tanksley, S. 1990. Fine mapping of quantitative traits loci using selected overlapping recombinant chromosomes in an interspecific cross of tomato. *Genetics* 124: 735–742.

Tanksley, S. D. 2004. The genetic, developmental and molecular bases of fruit size and shape variation in tomato. *Plant Cell* 16: S181–S189.

Zar, J. H. 1999. *Biostatistical analysis,* 4th ed. Upper Saddle River, NJ: Prentice Hall.

Chapter 24 Genetics and Behavior

Conrad, D. F., Pinto, D., Redon, R., Feuk, L., Gokcumen, O., Zhang, Y., et al. 2010. Origin and functional impact of copy number variation in the human genome. *Nature* 464: 704–712.

Davis, R. L. 2005. Olfactory memory formation in *Drosophila*: From molecular to systems neuroscience. *Annu. Rev. Neurosci.* 28: 275–302.

Desbonnet, L., Waddington, J. L., and O'Tuathaigh, C. M. 2009. Mutant models for genes associated with schizophrenia. *Biochem. Soc. Trans.* 37: 308–312.

Dickson, B. J. 2008. Wired for sex: the neurobiology of *Drosophila* mating decisions. *Science* 322: 904–909.

Flint, J., and Shifman, S. 2008. Animal models of psychiatric diseases. *Curr. Opin. Genet. Dev.* 18: 235–240.

Ganetzky, B. 1994. Cysteine strings, calcium channels and synaptic transmission. *BioEssays* 16: 461–463.

Hakak, Y., et al. 2001. Genome-wide expression analysis reveals dysregulation of myelination-related genes in chronic schizophrenia. *Proc. Nat. Acad. Sci.* 98: 4746–4751.

Hovatta, I., and Barlow, C. 2008. Molecular genetics of anxiety in mice and men. *Ann. Med.* 40: 92–109.

Mackay, T. F. C. and Anholt, R. H. 2006. Of flies and man: *Drosophila* as a model for human complex traits. *Annu. Rev. Hum. Genet.* 7: 39–367.

Reddy, P. H., et al. 1999. Transgenic mice expressing mutated full-length *HD* cDNA: A paradigm for locomotor changes and selective neuronal loss in Huntington's disease. *Phil. Trans. R. Soc. Lond.* B: 354: 1035–1045.

Ricker, J. P., and Hirsch, J. 1988. Genetic changes occurring over 500 generations in lines of *Drosophila melanogaster* selected divergently for geotaxis. *Behav. Genet.* 18: 13–24.

Rockenstein, E., Crews, L., and Masliah, E. 2007. Transgenic animal models of neurodegenerative diseases and their application to treatment development. *Advanced Drug Del. Rev.* 59: 1093–1102.

Siegel, R. W., Hall, J. C., Gailey, D. A., and Kyriacou, C. P. 1984. Genetic elements of courtship in *Drosophila*: Mosaics and learning mutants. *Behav. Genet.* 14: 383–410.

Stoltenberg, S. F., Hirsch, J., and Berlocher, S. H. 1995. Analyzing correlations of three types in selected lines of *Drosophila melanogaster* that have evolved stable extreme geotactic performance. *J. Comp. Psychol.* 105: 85–94.

Toma, D. P., White, K. P., Hirsch, J., and Greenspan, R. J. 2002. Identification of genes involved in *Drosophila melanogaster* geotaxis, a complex behavioral trait. *Nat. Genet.* 31: 349–353.

Tully, T. 1996. Discovery of genes involved with learning and memory: An experimental synthesis of Hirschian and Benzerian perspectives. *Proc. Nat. Acad. Sci.* 93: 13460–13467.

Turri, M. G., et al. 2001. QTL analysis identifies multiple behavioral dimensions in ethological tests of anxiety in laboratory mice. *Curr. Biol.* 11: 725–734.

Wellcome Trust Case Control Consortium. 2010. Genome-wide association study of CNVs in 16,000 cases of eight common diseases and 3,000 shared controls. *Nature* 464: 713–722.

Willis-Owen, S. A. G., and Flint, J. 2006. The genetic basis of emotional behavior in mice. *Eur. J. Hum. Genet.* 14: 721–728.

Chapter 25 Population and Evolutionary Genetics

Abzanhov, A., et al., 2004. Bmp4 and morphological variation of beaks in Darwin's finches. *Science* 305: 1462–1465.

Ansari-Lari, M. A., et al. 1997. The extent of genetic variation in the *CCR5* gene. *Nature Genetics* 16: 221–222.

Carrington, M., Kissner, T., et al. 1997. Novel alleles of the chemokine-receptor gene *CCR5*. *Am. J. Hum. Genet.* 61: 1261–1267.

Green, R. E., et al. 2010. A draft sequence of the Neandertal genome. *Science* 328: 710–722.

Karn, M. N., and Penrose, L. S. 1951. Birth weight and gestation time in relation to maternal age, parity and infant survival. *Ann. Eugen.* 16: 147–164.

Kerr, W. E., and Wright, S. 1954. Experimental studies of the distribution of gene frequencies in very small populations of *Drosophila melanogaster*. *Evolution* 8: 172–177.

Knowlton, N., et al., 1993. Divergence in proteins, mitochondrial DNA, and reproductive compatibility across the Isthmus of Panama. *Science* 260: 1629–1632.

Kreitman, M. 1983. Nucleotide polymorphism at the alcohol dehydrogenase locus of *Drosophila melanogaster*. *Nature* 304: 412–417.

Lamb, R. S., and Irish, V. F. 2003. Functional divergence within the *APETALA3/PISTILLATA* floral homeotic gene lineages. *Proc. Natl. Acad. Sci.* 100: 6558–6563.

Leibert, F., et al. 1998. The *DCCR5* mutation conferring protection against HIV-1 in Caucasian populations has a single and recent origin in northeastern Europe. *Hum. Molec. Genet.* 7: 399–406.

Markow, T., et al. 1993. HLA polymorphism in the Havasupai: Evidence for balancing selection. *Am. J. Hum. Genet.* 53: 943–952.

Noonan, J. P., et al. 2006. Sequencing and analysis of Neanderthal genomic DNA. *Science* 314: 1113–1118.

Presgraves, D. C. 2010. The molecular evolutionary basis for species formation. *Nat. Rev. Genet.* 11: 175–180.

Rasmussen, M., et al. 2010. Ancient human genome sequence of an extinct Paleo-Eskimo. *Nature* 463: 757–762.

Reznick, D. N., and Ricklefs, R. E. 2009. Darwin's bridge between microevolution and macroevolution. *Nature* 457: 837–842.

Salzburger, W. 2009. The interaction of sexually and naturally selected traits in the adaptive radiations of cichlid fishes. *Mol. Evol.* 18: 169–185.

Schulter, D. 2009. Evidence for ecological speciation and its alternative. *Science* 323: 737–741.

Spinney, L. 2010. Dreampond revisited. *Nature* 466: 174–175.

Stiassny, M. L. J., and Meyer, A. 1999. Cichlids of the Rift Lakes. *Sci. Am.* (Feb.) 280: 64–69.

Takahashi, K., et al., 1998. A novel family of short interspersed repetitive elements (SINES) from cichlids: The pattern of insertion of SINES at orthologous loci support the proposed monophyly of four major groups of cichlid fishes in Lake Tanganyika. *Mol. Biol. Evol.* 15: 391–407.

Yi, Z., et al. 2003. A 122.5 kilobase deletion of the *P* gene underlies the high prevalence of oculocutaneous albinism type 2 in the Navajo population. *Am. J. Hum. Genet.* 72: 62–72.

Chapter 26 Conservation Genetics

Baker, C. S., and Palumbi, S. R. 1994. Which whales are hunted? A molecular genetic approach to monitoring whaling. *Science* 265: 1538–1539.

Daniels, S. J., and Walters, J. R. 2000. Inbreeding depression and its effects on natal dispersal in red-cockaded woodpeckers. *Condor* 102: 482–491.

Dobson, A., and Lyles, A. 2000. Black-footed ferret recovery. *Science* 288: 985.

Frankham, R. 1995. Conservation genetics. *Ann. Rev. Genet.* 29: 305–327.

Friar, E. A., et al. 2001. Population structure in the endangered Mauna Loa silversword, *Argyroxiphium kauense*, and its bearing on reintroduction. *Molecular Ecology* 10: 1657–1663.

Gharrett, A. J., and Smoker, W. W. 1991. Two generations of hybrids between even-year and odd-year pink salmon (*Oncorhynchus gorbuscha*): A test for outbreeding depression? *Canadian J. Fish. Aquat. Sci.* 48: 426–438.

Hedrick, P. W. 2001. Conservation genetics: Where are we now? *Trends Ecol. Evol.* 16: 629–636.

Johnson, W. E., et al. 2010. Genetic restoration of the Florida panther. *Science* 329: 1641–1644.

Lacy, R. C. 1997. Importance of genetic variation to the viability of mammalian populations. *J. Mammalogy* 78: 320–335.

Moore, M. K., et al. 2003. Use of restriction fragment length polymorphisms to identify sea turtle eggs and cooked meats to species. *Conserv. Genet.* 4: 95–103.

Paetkau, D., et al. 1998. Variation in genetic diversity across the range of North American brown bears. *Conserv. Biol.* 12: 418–429.

Ralls, K., et al. 2000. Genetic management of chondrodystrophy in California condors. *Animal Conserv.* 3: 145–153.

Roman, J., and Bowen, B. W. 2000. The mock turtle syndrome: Genetic identification of turtle meat purchased in the southeastern United States of America. *Animal Conserv.* 3: 61–65.

Roman, J., and Palumbi, S. R. 2003. Whales before whaling in the North Atlantic. *Science* 301: 508–510.

Wayne, R. K., et al. 1991. Conservation genetics of the endangered Isle Royale gray wolf. *Conserv. Biol.* 5: 41–51.

Wilson, E. O., ed. 1988. *Biodiversity*. Washington, DC: National Academy of Sciences.

Wynen, L. P., et al. 2000. Postsealing genetic variation and population structure of two species of fur seal. *Mol. Ecol.* 9: 299–314.

Appendix B

ANSWERS TO SELECTED PROBLEMS

Chapter 1

2. Based on the parallels between Mendel's model of heredity and the behavior of chromosomes, the chromosome theory of inheritance emerged. It states that inherited traits are controlled by genes residing on chromosomes that are transmitted by gametes.

4. A gene variant is called an allele. There can be many such variants in a population.

6. Genetic information is encoded in DNA by the sequence of bases.

8. 20^5

10. Recently, biotechnology has allowed genes to be moved in a variety of ways to generate transgenic plants. Such plants can be engineered to increase their ecological breadth, disease resistance, and/or nutrient value. Wheat, rice, corn, beans, and cassava are being modified to enhance nutritional value by increasing vitamin and mineral content.

12. Both genomics and proteomics examine cellular components; genomics examines DNA, while proteomics examines proteins. Both disciplines examine the structure and function of such macromolecules in time and space within cells. To manage the volume of information produced by the fields of genomics and proteomics, a relatively new field, bioinformatics, has evolved. Bioinformatics involves the development of hardware and software needed to process, store, and analyze large amounts of DNA and protein data.

14. Some mechanism should be in place to protect the investments of individuals and institutions that develop needed and useful products, such as selected stretches of human DNA. However, safeguards, both ethical and economic, need to be developed to ensure that relatively free and fair access exists when vital issues are in question. Any mechanism needs to protect investors as well as consumers.

16. This question is open to many "answers" depending on the individual. Although it may be difficult to put yourself in this position, consider not only what your decision would be but also why you would make such a decision.

   Often, as a person ages, their perspective changes; for instance, how would the possibility of having children influence your decision?

18. Such groups seek to reclaim community involvement and decision-making in governmental and industrial applications of biotechnology. They consider that profit-driven motives may compromise benefits that such technology may provide. They question the safety of genetically modified organisms to human health, ecological harmony, not to mention biotechnical applications to weapons development.

Chapter 2

ANSWERS TO NOW SOLVE THIS

2-1. (a) 32 (b) 16
2-2. (a) 8 (b) 8 (c) 8

2-3. Not necessarily; if crossing over occurs in meiosis I, then the chromatids in the secondary oocyte are not identical. Once they separate during meiosis II, unlike chromatids reside in the ootid and the second polar body.

SOLUTIONS TO PROBLEMS AND DISCUSSION QUESTIONS

2. (a) Chromatin contains the genetic material that is responsible for maintaining hereditary information (from one cell to daughter cells and from one generation to the next) and production of the phenotype.

   (b) The *nucleolus (pl. nucleoli)* is a structure that is produced by activity of the nucleolar organizer region in eukaryotes. Composed of ribosomal DNA and protein, it is the structure for the production of ribosomes.

   (c) The *ribosome* is the structure where various RNAs, enzymes, and other molecular species assemble the primary sequence of a protein. That is, amino acids are placed in order as specified by messenger RNA.

   (d) The *mitochondrion (pl. mitochondria)* is a membrane-bound structure located in the cytoplasm of eukaryotic cells. It is the site of oxidative phosphorylation and production of relatively large amounts of ATP.

   (e) The *centriole* is a cytoplasmic structure involved (through the formation of spindle fibers) in the migration of chromosomes during mitosis and meiosis primarily in animal cells.

   (f) The *centromere* serves as an attachment point for sister chromatids and a region where spindle fibers attach to chromosomes (kinetochore). The centromere divides during mitosis and meiosis II, thus aiding in the partitioning of chromosomal material to daughter cells.

4. Overall length and centromere position are but two factors required for homology. Most importantly, genetic content in nonhomologous chromosomes is expected to be quite different. Other factors including banding pattern and time of replication during S phase would also be expected to vary among nonhomologous chromosomes.

6. Metacentric (a), submetacentric (b), acrocentric (c), telocentric (d).

8. Major divisions of the cell cycle include interphase and mitosis. Interphase is composed of four phases: G1, G0, S, and G2. During the S phase, chromosomal DNA doubles. Refer to the text figures for a diagram of mitosis. Notice that, in contrast to meiosis, there is no pairing of homologous chromosomes in mitosis and the chromosome number does not change.

10. (a) *Synapsis* is the point-by-point pairing of homologous chromosomes during prophase of meiosis I.

(b) *Bivalents* are those structures formed by the synapsis of homologous chromosomes. In other words, there are two chromosomes (and four chromatids) that make up a bivalent.

(c) *Chiasmata* is the plural form of chiasma and refers to the structure, when viewed microscopically, of crossed chromatids.

(d) *Crossing over* is the exchange of genetic material between chromatids. It is a method of providing genetic variation through the breaking and rejoining of chromatids.

(e) *Chromomeres* are bands of chromatin that look different from neighboring patches along the length of a chromosome.

(f) *Sister chromatids* are "post-S phase" structures of replicated chromosomes. Sister chromatids are genetically identical (except where mutations have occurred) and are originally attached to the same centromere.

(g) *Tetrads* are synapsed homologous chromosomes thereby composed of four chromatids.

(h) *Dyads* are composed of two chromatids joined by a centromere.

(i) At anaphase II of meiosis, the centromeres divide and sister chromatids (*monads*) go to opposite poles.

12. During meiosis I chromosome number is reduced to haploid complements. This is achieved by synapsis of homologous chromosomes and their subsequent separation. It would seem to be more mechanically difficult for genetically identical daughters to form from mitosis if homologous chromosomes paired. By having chromosomes unpaired at metaphase of mitosis, only centromere division is required for daughter cells to eventually receive identical chromosomal complements.

14. First, through independent assortment of chromosomes at anaphase I of meiosis, daughter cells (secondary spermatocytes and secondary oocytes) may contain different sets of maternally and paternally derived chromosomes. Second, crossing over, which happens at a much higher frequency in meiotic cells as compared with mitotic cells, allows maternally and paternally derived chromosomes to exchange segments, thereby increasing the likelihood that daughter cells (that is, secondary spermatocytes and secondary oocytes) are genetically unique.

16. There would be sixteen combinations with the addition of another chromosome pair.

18. One half of each tetrad will have a maternal homolog: $(1/2)^{10}$.

20. In angiosperms, meiosis results in the formation of microspores (male) and megaspores (female), which give rise to the haploid male and female gametophyte stage. Micro- and megagametophytes produce the pollen and the ovules, respectively. Following fertilization, the sporophyte is formed.

22. The folded-fiber model is based on each chromatid consisting of a single fiber wound like a skein of yarn. Each fiber consists of DNA and protein. A coiling process occurs during the transition of interphase chromatin into more condensed chromosomes during prophase of mitosis or meiosis. Such condensation leads to a 5000-fold contraction in the length of the DNA within each chromatid.

24. 50, 50, 50, 100, 200

26. Duplicated chromosomes A^m, A^p, B^m, B^p, C^m, and C^p will align at metaphase, with the centromeres dividing and sister chromatids going to opposite poles at anaphase.

28. As long as you have accounted for eight possible combinations in the previous problem, there would be no new ones added in this problem.

30. See the products of nondisjunction of chromosome *C* at the end of meiosis I as follows:

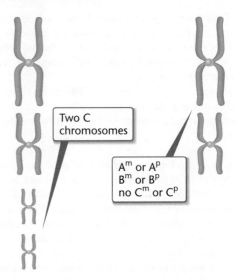

Two C chromosomes

A^m or A^p
B^m or B^p
no C^m or C^p

At the end of meiosis II, assuming that, as the problem states, the *C* chromosomes separate as dyads instead of monads during meiosis II, you would have monads for the A and B chromosomes, and dyads (from the cell on the left) for both *C* chromosomes as one possibility.

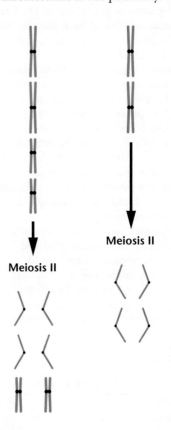

Meiosis II

Meiosis II

32. (a) The chromosome number in the somatic tissues of the hybrid would be the summation of the haploid numbers of each parental species: $7 + 14 = 21$.

(b) Given that no homologous chromosome pairing occurs at metaphase I, one would expect 21 univalents and random "1×0" separation of chromosomes at anaphase I.

That is, there would be a random separation of univalents to either pole at anaphase I usually leading to inviable haploid cells and thus sterility.

Chapter 3

ANSWERS TO NOW SOLVE THIS

3-1. P = checkered; p = plain.

$$\text{Cross (a): } PP \times PP \quad \text{or} \quad PP \times Pp$$

Notice in cross (d) that the checkered offspring, when crossed to plain, produce only checkered F_2 progeny and in cross (g) when crossed to checkered still produce only checkered progeny. From this additional information, one can conclude that in the progeny of cross (a) there are no heterozygotes and the original cross must have been $PP \times PP$.

$$\text{Cross (b): } PP \times pp$$

Cross (c): Because all the offspring from this cross are plain, there is no doubt that the genotype of both parents is pp.

Genotypes of all individuals:

		Progeny	
	P_1 Cross	Checkered	Plain
(a)	$PP \times PP$	PP	
(b)	$PP \times pp$	Pp	
(c)	$pp \times pp$		pp
(d)	$PP \times pp$	Pp	
(e)	$Pp \times pp$	Pp	pp
(f)	$Pp \times Pp$	PP, Pp	pp
(g)	$PP \times Pp$	PP, Pp	

3-2. Suggested symbolism:

w = wrinkled seeds g = green cotyledons
W = round seeds G = yellow cotyledons

(a) Notice a 3:1 ratio for seed shape, therefore $Ww \times Ww$; and no green cotyledons, therefore $GG \times GG$ or $GG \times Gg$. Putting the two characteristics together gives

$$WwGG \times WwGG$$
$$\text{or}$$
$$WwGG \times WwGg$$

(b) $wwGg \times WwGg$
(c) $WwGg \times WwGg$
(d) $WwGg \times wwgg$

3-3. (a)

Offspring:

Genotypes	Ratio	Phenotypes
AABBCC	(1/16)	
AABBCc	(1/16)	
AABbCC	(1/16)	
AABbCc	(1/16)	
AaBBCC	(2/16)	$A\_B\_C\_ = 12/16$
AaBBCc	(2/16)	
AaBbCC	(2/16)	
AaBbCc	(2/16)	
aaBBCC	(1/16)	
aaBBCc	(1/16)	$aaB\_C\_ = 4/16$
aaBbCC	(1/16)	
aaBbCc	(1/16)	

(b)

Offspring:

Genotypes	Ratio	Phenotypes
AaBBCC	1/8	$A\_BBC\_ = 3/8$
AaBBCc	2/8	
AaBBcc	1/8	$A\_BBcc = 1/8$
aaBBCC	1/8	$aaBBC\_ = 3/8$
aaBBCc	2/8	
aaBBcc	1/8	$aaBBcc = 1/8$

(c) There will be eight (2^n) different kinds of gametes from each of the parents, therefore a 64-box Punnett square. Doing this problem by the forked-line method helps considerably.

Simply multiply through each component to arrive at the final genotypic frequencies.

For the phenotypic frequencies, set up the problem in the following manner:

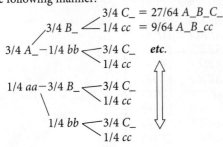

$$3/4\ C\_ = 27/64\ A\_B\_C\_$$
$$3/4\ B\_ \ \diagdown 1/4\ cc = 9/64\ A\_B\_cc$$

$$3/4\ A\_ -1/4\ bb \diagup\begin{array}{l}3/4\ C\_ \quad \textbf{\textit{etc.}}\\ 1/4\ cc\end{array}$$

$$1/4\ aa -3/4\ B\_ \diagdown\begin{array}{l}3/4\ C\_\\ 1/4\ cc\end{array}$$

$$1/4\ bb \diagdown\begin{array}{l}3/4\ C\_\\ 1/4\ cc\end{array}$$

SOLUTIONS TO PROBLEMS AND DISCUSSION QUESTIONS

2. (a) P_1:

Phenotypes:	Black	×	White
Genotypes:	WW		ww
F_1:	Ww (Black)		

$F_1 \times F_1$:

Phenotypes:	Black	×	Black
Genotypes:	Ww		Ww

F_2:

Phenotypes:	Black	Black	Black	White
Genotypes:	WW	Ww	Ww	ww

(b) white × white

   all white

(c) Ww × Ww

4. Unit Factors in Pairs, Dominance and Recessiveness, Segregation

6. *Pisum sativum* is easy to cultivate. It is naturally self-fertilizing, but it can be crossbred. It has several visible features (e.g., tall or short, red flowers or white flowers) that are consistent under a variety of environmental conditions, yet contrast due to genetic circumstances. Seeds could be obtained from local merchants.

8. $WWgg = 1/16$

10. 1. Factors occur in pairs.
  2. Some genes have dominant and recessive alleles.
  3. Alleles segregate from each other during gamete formation. When homologous chromosomes separate from each other at anaphase I, alleles will go to opposite poles of the meiotic apparatus.
  4. One gene pair separates independently from other gene pairs. Different gene pairs on the same homologous pair of chromosomes (if far apart) or on nonhomologous chromosomes will separate independently from each other during meiosis.

12. There are two characteristics presented here: body color and wing length. First, assign meaningful gene symbols.

Body color	*Wing length*
E = gray body color	V = long wings
e = ebony body color	v = vestigial wings

(a) P_1:

EEVV × eevv

F_1: *EeVv* (gray, long)

F_2: This will be the result of a Punnett square with 16 boxes.

Phenotypes	Ratio	Genotypes	Ratio
gray, long	9/16	EEVV	1/16
		EEVv	2/16
		EeVV	2/16
		EeVv	4/16
gray, vestigial	3/16	EEvv	1/16
		Eevv	2/16
ebony, long	3/16	eeVV	1/16
		eeVv	2/16
ebony, vestigial	1/16	eevv	1/16

(b) P_1: *EEvv × eeVV*

F_1: It is important to see that the results from this cross will be exactly the same as those in part (a) above. The only difference is that the recessive genes are coming from both parents, rather than from one parent only as in (a). The F_2 ratio will be the same as (a) also. When you have genes on the autosomes (not X-linked), independent assortment, complete dominance, and no gene interaction (see later) in a cross involving double heterozygotes, the offspring ratio will be 9:3:3:1.

(c) P_1: *EEVV × EEvv*

F_1: *EEVv* (gray, long)

F_2: Notice that all the offspring will have gray bodies and you will get a 3:1 ratio of long to vestigial wings. You should see this before you even begin working through the problem. Even though this cross involves two gene pairs it will give a "monohybrid" type of ratio because one of the gene pairs is homozygous (body color) and **one** gene pair is heterozygous (wing length).

Phenotypes	Ratio	Genotypes	Ratio
gray, long	3/4	EEVV	1/4
		EEVv	2/4
gray, vestigial	1/4	EEvv	1/4

14.

Phenotypes	*Genotypes*
P_1: Yellow × green	GG × gg
F_1: all yellow	Gg
F_2: 6022 yellow	1/4 GG; 2/4 Gg
2001 green	1/4 gg

$$GG \times GG = \text{all } GG$$

$$Gg \times Gg = 1/4\ GG;\ 2/4\ Gg;\ 1/4\ gg$$

16.

Seed shape	*Seed color*
W = round	G = yellow
w = wrinkled	g = green

P_1: *WWgg × wwGG*

F_1:   *WwGg* cross to *wwgg*

(which is a typical testcross)

The offspring will occur in a typical 1:1:1:1 as

1/4 *WwGg* (round, yellow)

1/4 *Wwgg* (round, green)

1/4 *wwGg* (wrinkled, yellow)

1/4 *wwgg* (wrinkled, green)

18. (a)

Expected ratio	Observed (o)	Expected (e)
3/4	882	885.75
1/4	299	295.25

$$\chi^2 = \Sigma\, (o - e)^2/e = .064$$

By looking at the χ^2 table with 1 degree of freedom (because there were two classes, therefore $n - 1$ or 1 degree of freedom), we find a probability (p) value between 0.9 and 0.5.

We would therefore say that there is a "good fit" between the observed and expected values.

(b)

Expected ratio	Observed (o)	Expected (e)
3/4	705	696.75
1/4	224	232.25

$$\chi^2 = 0.39$$

The p value in the table for 1 degree of freedom is still between 0.9 and 0.5; however, because the χ^2 value is larger in (b) we should say that the deviations from expectation are greater. The deviation in each case can be attributed to chance.

20. Use of the $p = 0.10$ as the "critical" value for rejecting or failing to reject the null hypothesis instead of $p = 0.05$ would allow more null hypotheses to be rejected. As the critical p value is increased, it takes a smaller χ^2 value to cause rejection of the null hypothesis. It would take less difference between the expected and observed values to reject the null hypothesis; therefore the stringency of failing to reject the null hypothesis is increased.

22. (a) There are two possibilities. Either the trait is dominant, in which case I-1 is heterozygous as are II-2 and II-3, or the trait is recessive and I-1 is homozygous and I-2 is heterozygous. Under the condition of recessiveness, both II-1 and II-4 would be heterozygous; II-2 and II-3 are homozygous.

(b) recessive: parents *Aa*, *Aa*

(c) recessive: parents *Aa*, *Aa*

(d) recessive or dominant, not sex-linked; if recessive, parents *Aa*, *aa*

24. $p = [8!\,(3/4)^6(1/4)^2]/6!2!$

26. Most likely, the attending physician misdiagnosed the syndrome of the first child by telling the parents that the birth defects were not genetic. Given the birth of the second child with Smith-Lemli-Opitz syndrome, it is highly likely that both parents were carriers for a recessive mutant gene causing Smith-Lemli-Opitz syndrome. Under that circumstance, there is a 25 percent chance that each of their children would be affected. The probability that two children of heterozygous parents would be affected would be 0.25 × 0.25 = 0.0625 or a little over 6 percent.

28. (a) The first task is to draw out an accurate pedigree (one of several possibilities):

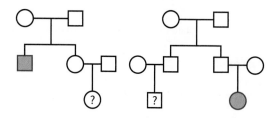

(b) $1/3 \times 1/4 = 1/12$

(c) The probability that neither is heterozygous is: $2/3 \times 3/4 = 6/12$

(d) The probability that one is heterozygous is:

$$(1/3 \times 3/4) + (2/3 \times 1/4) = 5/12$$

Also, there are only three possibilities: both are heterozygous, neither is heterozygous, and at least one is heterozygous. You have already calculated the first two probabilities; the last is simply $1 - (1/12 + 6/12) = 5/12$.

30. (a) straight wings = w^+ curled wings = w
short bristles = b^+ long bristles = b

(b) Cross #1: $w^+/w;\ b^+/b$ × $w^+/w;\ b^+/b$
Cross #2: $w^+/w;\ b/b$ × $w^+/w;\ b/b$
Cross #3: $w/w;\ b/b$ × $w^+/w;\ b^+/b$
Cross #4: $w^+/w^+;\ b^+/b$ × $w^+/w^+;\ b^+/b$
(one parent could be w^+/w)
Cross #5: $w/w;\ b^+/b$ × $w^+/w;\ b^+/b$

32. (a) Set I Expected Numbers:
Tall = 26.25 Short = 8.75
Set II Expected Numbers:
Tall = 262.5 Short = 87.5
For Set I the χ^2 value would be:
$(30 - 26.25)^2/26.25 + (5 - 8.75)^2/8.75$
   $= 2.15$ with p being between 0.2 and 0.05

so one would accept the null hypothesis of no significant difference between the expected and observed values.

For Set II, the χ^2 value would be:

21.43 and p <0.001 and one would reject the null hypothesis and assume a significant difference between the observed and expected values.

(b) In most cases, more confidence is gained as the sample size increases; however, depending on the organism or experiment, there are practical limits on sample size.

34. Given that dentinogenesis imperfecta is inherited as a dominant allele and the man's mother had normal teeth, the man with six children must be heterozygous. Therefore the probability that their first child will be a male with dentinogenesis imperfecta would be 1/2 (passage of the allele) × 1/2 (probability of child being male) = 1/4. The probability that three of their six children will have the disease is best determined by application of the following formula.

$$p = \frac{n!}{s!t!}\,a^s b^t$$

n = total number of events (six in this case)
s = number of times outcome a happens (three in this case)
t = number of times outcome b happens (three in this case)

a = probability of being normal in this family (1/2)
b = probability of having the disease in this family (1/2)

$$p = \frac{6!}{3!\,3!}\,1/2^3\,1/2^3$$

The overall probability (p) would be 5/16 or about .31 (31%).

Chapter 4

ANSWERS TO NOW SOLVE THIS

4-1. (a) Parents: sepia × cream

Cross: $c^k c^a \times c^d c^a$

2/4 sepia; 1/4 cream; 1/4 albino

(b) Parents: sepia × cream

($c^k c^d \times c^k c^d$ or $c^k c^d \times c^k c^a$) the cream parent could be $c^d c^d$ or $c^d c^a$.

Crosses: $c^k c^a \times c^d c^d$

1/2 sepia; 1/2 cream*

*(if parents are assumed to be homozygous)

or $c^k c^a$ × $c^d c^a$

1/2 sepia; 1/4 cream; 1/4 albino

(c) Parents: sepia × cream

Because the sepia guinea pig had two full color parents, which could be

Cc^k, Cc^d, or Cc^a

(not CC because sepia could not be produced), its genotype could be

$c^k c^k$, $c^k c^d$, or $c^k c^a$

Because the cream guinea pig had two sepia parents

($c^k c^d \times c^k c^d$ or $c^k c^d \times c^k c^a$) the cream parent could be $c^d c^d$ or $c^d c^a$.

Crosses:

$c^k c^k \times c^d c^d \Longrightarrow$ all sepia

$c^k c^k \times c^d c^a \Longrightarrow$ all sepia

$c^k c^d \times c^d c^d \Longrightarrow$ 1/2 sepia; 1/2 cream

$c^k c^d \times c^d c^a \Longrightarrow$ 1/2 sepia; 1/2 cream

$c^k c^a \times c^d c^d \Longrightarrow$ 1/2 sepia; 1/2 cream

$c^k c^a \times c^d c^a \Longrightarrow$ 1/2 sepia; 1/4 cream; 1/4 albino

(d) Parents: sepia × cream

Because the sepia parent had a full color parent and an albino parent

($Cc^k \times c^a c^a$), it must be $c^k c^a$. The cream parent had two full color parents that could be Cc^d or Cc^a; therefore it could be $c^d c^d$ or $c^d c^a$.

Crosses:

$c^k c^a \times c^d c^d \Longrightarrow$ 1/2 sepia; 1/2 cream

$c^k c^a \,3\, c^d c^a \Longrightarrow$ 1/2 sepia; 1/4 cream; 1/4 albino

4-2. A = pigment; a = pigmentless (colorless) B = purple; b = red

$AaBb \times AaBb$

$A\_B\_$ = purple
A_bb = red
$aaB\_$ = colorless
$aabb$ = colorless

4-3. For all three pedigrees, let a represent the mutant gene and A represent its normal allele.

(a) This pedigree is consistent with an X-linked recessive trait because the male would contribute an X chromosome carrying the a mutation to the aa daughter. The mother would have to be heterozygous Aa.

(b) This pedigree is consistent with an X-linked recessive trait because the mother could be Aa and transmit her a allele to her one son (a/Y) and her A allele to her other son.

(c) This pedigree is not consistent with an X-linked recessive mode of inheritance because the aa mother has an A/Y son.

SOLUTIONS TO PROBLEMS AND DISCUSSION QUESTIONS

2. Seeing the 1:2:1 ratio in the offspring of

roan × roan

confirms the hypothesis of incomplete dominance as the mode of inheritance.

Symbolism:

AA = red

aa = white

Aa = roan

$AA \times AA \Longrightarrow AA$

$aa \times aa \Longrightarrow aa$

$AA \times aa \Longrightarrow Aa$

$Aa \times Aa$

1/4 AA; 2/4 Aa; 1/4 aa

4. $Pp \times Pp$

1/4 PP (**lethal**)

2/4 Pp (platinum)

1/4 pp (silver)

Therefore, the ratio of surviving foxes is 2/3 platinum, 1/3 silver. The P allele behaves as a recessive in terms of lethality (seen only in the homozygote) but as a dominant in terms of coat color (seen in the homozygote).

6.

Blood Group (phenotype)	Genotype(s)
A	$I^A I^A$, $I^A i$
B	$I^B I^B$, $I^B i$
AB	$I^A I^B$
O	ii

I^A and I^B are codominant (notice the AB blood group), while being dominant to i.

8. The only *blood type* that would exclude a male from being the father would be AB, because no i allele is present. Because many individuals in a population could have genotypes with the i allele, one could not prove that a particular male was the father by this method.

10. The simplest explanation is that the homozygous creepers combination is lethal.

12. Three independently assorting characteristics are being dealt with: flower color (incomplete dominance), flower shape (dominant/recessive), and plant height (dominant/recessive).

   Establish appropriate gene symbols:

   Flower color:
   $$RR = \text{red}; Rr = \text{pink}; rr = \text{white}$$

   Flower shape:
   $$P = \text{personate}; p = \text{peloric}$$

   Plant height:
   $$D = \text{tall}; d = \text{dwarf}$$

   $RRPPDD \times rrppdd$ ⬇

   $RrPpDd$ (pink, personate, tall)

   Use *components* of the forked-line method as follows:

   $$2/4 \text{ pink} \times 3/4 \text{ personate} \times 3/4 \text{ tall} = 18/64$$

14. (a) This is a case of incomplete dominance in which, as shown in the third cross, the heterozygote (palomino) produces a typical 1:2:1 ratio.

   $C^{ch}C^{ch}$ = chestnut

   $C^{c}C^{c}$ = cremello

   $C^{ch}C^{c}$ = palomino

   (b) The F_1 resulting from matings between cremello and chestnut horses would be expected to be all palomino. The F_2 would be expected to fall in a 1:2:1 ratio as in the third cross in part (a) above.

16. (a) In a cross of

   $$AACC \times aacc$$

   the offspring are all $AaCc$ (agouti) because the C allele allows pigment to be deposited in the hair and when it is, it will be agouti.

   F_2 offspring would have the following "simplified" genotypes with the corresponding phenotypes:

   $A\_C\_$ = 9/16 (agouti)

   A_cc = 3/16 (colorless because cc is epistatic to A)

   $aaC\_$ = 3/16 (black)

   $aacc$ = 1/16 (colorless because cc is epistatic to aa)

   The two colorless classes are phenotypically indistinguishable; therefore the final ratio is 9:3:4.

   (b) Results of crosses of female agouti

   $$(A\_C\_) \times aacc \text{ (males)}$$

   are given in three groups:

   (1) To produce an even number of agouti and colorless offspring, the female parent must have been $AACc$ so that half of the offspring are able to deposit pigment because of C; when they do, they are all agouti (having received only A from the female parent).

   (2) To produce an even number of agouti and black offspring, the mother must have been Aa, and so that no colorless offspring were produced, the female must have been CC. Her genotype must have been $AaCC$.

   (3) Notice that half of the offspring are colorless; therefore the female must have been Cc. Half of the pigmented offspring are black and half are agouti; therefore the female must have been Aa. Overall, the $AaCc$ genotype seems appropriate.

   The final ratio would be

   3/8 (gray)
   1/8 (yellow)
   4/8 (albino)

18. (a) Since this is a 9:3:3:1 ratio with no albino phenotypes, the parents must each have been double heterozygotes and incapable of producing the cc genotype.

   Genotypes:

   $$AaBbCC \times AaBbCC$$
   $$\text{or}$$
   $$AaBbCC \times AaBbCc$$

   Phenotypes:

   $$\text{gray} \times \text{gray}$$

   (b) Since there are no black offspring, there are no combinations in the parents that can produce aa. The 4/16 proportion indicates that the C locus is heterozygous in both parents.

   If the parents are

   $$AABbCc \times AaBbCc$$
   $$\text{or}$$
   $$AABbCc \times AABbCc$$

   then the results would follow the pattern given.

   $$\text{Phenotypes:} \quad \text{gray} \times \text{gray}$$

   (c) Notice that 16/64 or 1/4 of the offspring are albino; therefore the parents are both heterozygous at the C locus. Second, notice that without considering the C locus, there is a 27:9:9:3 ratio that reduces to a 9:3:3:1 ratio.

   Given this information, the genotypes must be

   $$AaBbCc \times AaBbCc$$

   Phenotypes: gray × gray

   (d) Genotypes:

   $$aaBbCc \times aabbCc$$

   Phenotypes: black × cream

(e)

Genotypes:

$$aaBbCc \times aaBbcc$$

Phenotypes: black × albino

20. The initial cross must have been

$$AABB \times aabb$$

There are two gene pairs involved.

(a) $A\_B\_ = 9/16$ (tall)

$A\_bb = 3/16$ (dwarf)

$aaB\_ = 3/16$ (dwarf)

$aabb = 1/16$ (dwarf)

(b) There are three different classes of dwarf plants. Within each of the 3/16 classes there are two types:

$A\_bb = 3/16$ (dwarf)

$= 1/3$ *AAbb* and $2/3$ *Aabb*

and

$aaB\_ = 3/16$ (dwarf)

$= 1/3$ *aaBB* and $2/3$ *aaBb*

Therefore, the true breeding dwarf plants would be the following:

AAbb, aaBB, and *aabb*

and they would constitute 3/7 of the dwarf group.

22.

Cross #1 = (c)

Cross #2 = (d)

Cross #3 = (b)

Cross #4 = (e)

Cross #5 = (a)

24. The mating is $X^{RG}X^{rg}; I^A i \times X^{RG}Y; I^A i$

The final product of the independent probabilities is

$$1/2 \times 1 \times 1/4 = 1/8$$

26. Assuming that the parents are homozygous, the crosses would be as follows. Notice that the X symbol may remain to remind us that the *sd* gene is on the X chromosome. It is extremely important that one account for both the mutant genes and each of their wild-type alleles.

$P_1: X^{sd}X^{sd}; e^+/e^+ \times X^+/Y; e/e$

$F_1:$

$1/2\ X^+X^{sd}; e^+/e$ (female, normal)

$1/2\ X^{sd}/Y; e^+/e$ (male, scalloped)

$F_2:$

Phenotypes:

3/16 normal females
3/16 normal males
1/16 ebony females
1/16 ebony males
3/16 scalloped females

3/16 scalloped males
1/16 scalloped, ebony females
1/16 scalloped, ebony males

Forked-line method:

$P_1:$ $X^{sd}X^{sd}; e^+/e^+$ × $X^+/Y; e/e$
⇓

$F_1:$ $1/2\ X^+X^{sd}; e^+/e$ (female, normal)
$1/2\ X^{sd}/Y; e^+/e$ (male, scalloped)

$F_2:$

	Wings	Color
1/4	females, ——3/4 normal —— 3/16	
	normal ——1/4 ebony —— 1/16	
1/4	females, ——3/4 normal —— 3/16	
	scalloped 1/4 ebony —— 1/16	
1/4	males, ——3/4 normal —— 3/16	
	normal ——1/4 ebony —— 1/16	
1/4	males, —— 3/4 normal—— 3/16	
	scalloped 1/4 ebony —— 1/16	

28. (a) $P_1: X^vX^v; +/+ \times X^+/Y; b^r/b^r$ ⬎

$F_1:$
$1/2\ X^+X^v; +/b^r$ (female, normal)
$1/2\ X^v/Y; +/b^r$ (male, vermilion)

$F_2:$

3/16 = females, normal
1/16 = females, brown eyes
3/16 = females, vermilion eyes
1/16 = females, white eyes
3/16 = males, normal
1/16 = males, brown eyes
3/16 = males, vermilion eyes
1/16 = males, white eyes

(b) $P_1: X^+X^+; b^r/b^r \times X\ X^v/Y; +/+$ ⬎

$F_1:$
$1/2\ X^+X^v; +/b^r$ (female, normal)
$1/2\ X^+/Y; +/b^r$ (male, normal)

$F_2:$

6/16 = females, normal
2/16 = females, brown eyes
3/16 = males, normal
1/16 = males, brown eyes
3/16 = males, vermilion eyes
1/16 = males, white eyes

(c) $P_1: X^vX^v; b^r/b^r \times X\ X^+/Y; +/+$ ⬎

$F_1: 1/2\ X^+X^v; +/b^r$ (female, normal)
$1/2\ X^v/Y; +/b^r$ (male, vermilion)

$F_2:$
3/16 = females, normal
1/16 = females, brown eyes
3/16 = females, vermilion eyes
1/16 = females, white eyes
3/16 = males, normal
1/16 = males, brown eyes
3/16 = males, vermilion eyes
1/16 = males, white eyes

30. (a,b) The condition is therefore *recessive*. In the second cross, note that the father is not shaded yet the daughter (II-4) is. If the condition is recessive, then it must also be *autosomal*.

(c) II-1 = *AA* or *Aa*
II-6 = *AA* or *Aa*
II-9 = *Aa*

32. P$_1$: female: *HH* × male: *hh*
F$_1$:
all hen-feathering
F$_2$:

1/2 females	1/2 males
1/4 *HH* hen-feathering	hen-feathering
2/4 *Hh* hen-feathering	hen-feathering
1/4 *hh* hen-feathering	cock-feathering

All of the offspring would be hen-feathered except for 1/8 of the males, which are cock-feathered.

34. Phenotypic expression is dependent on the genome of the organism, the immediate molecular and cellular environment of the genome, and numerous interactions between a genome, the organism, and the environment.

36. (a) *AAB_* × *aaBB* (other configurations possible, but each must give all offspring with *A* and *B* dominant alleles)

(b) *AaB_* × *aaBB* (other configurations are possible, but no *bb* genotypes can be produced)

(c) *AABb* × *aaBb*

(d) *AABB* × *aabb*

(e) *AaBb* × *Aabb*

(f) *AaBb* × *aabb*

(g) *aaBb* × *aaBb*

(h) *AaBb* × *AaBb*

Those genotypes that will breed true will be as follows:

black = *AABB*

golden = all genotypes that are *bb*

brown = *aaBB*

38. The homozygous dominant type is lethal. Polled is caused by an independently assorting dominant allele, while horned is caused by the recessive allele to polled.

40. Given the degree of outcrossing, that the gene is probably quite rare and therefore heterozygotes are uncommon, and that the frequency of transmission is high, it is likely that this form of male precocious puberty is caused by an autosomal dominant, sex-limited gene.

42. Given the following genotypes of the parents:

aabb = crimson
AABB = white

the F$_1$ consist of *AaBb* genotypes with a rose phenotype.

In the F$_2$ the following genotypes correspond to the given phenotypes:

*AAB_* = white 4/16
AaBB = magenta 2/16
AaBb = rose 4/16
Aabb = orange 2/16
aaBB = yellow 1/16

aaBb = pale yellow 2/16
aabb = crimson 1/16

Gene interaction is occurring along with the absence of complete dominance.

44. Beatrice, Alice of Hesse, and Alice of Athlone are carriers. There is a 1/2 chance that Princess Irene is a carrier.

Chapter 5

ANSWERS TO NOW SOLVE THIS

5-1. (a)

1/4 *AaBb*
1/4 *Aabb*
1/4 *aaBb*
1/4 *aabb*

(b)

1/4 *AaBb*
1/4 *Aabb*
1/4 *aaBb*
1/4 *aabb*

(c)

1/2 *AB/ab*
1/2 *ab/ab*

Under this condition *A* and *B* are *coupled*. If, however, *A* and *B* are not coupled, then the symbolism would be *Ab/aB* × *ab/ab*.

The offspring would occur as follows:

1/2 *Ab/ab*
1/2 *aB/ab*

5-2. Adding the crossover percentages together (6.9 + 7.1) gives 14 percent, which would be the map distance between the two genes.

5-3. (a,b) + *b c*/ *a* + +

$$a - b = \frac{32 + 38 + 0 + 0}{1000} \times 100 = 7 \text{ map units}$$

$$b - c = \frac{11 + 9 + 0 + 0}{1000} \times 100 = 2 \text{ map units}$$

(c) The progeny phenotypes that are missing are + + *c* and *a b* +, which, of 1000 offspring, 1.4 (.07 × .02 × 1000) would be expected. Perhaps by chance or some other unknown selective factor, they were not observed.

SOLUTIONS TO PROBLEMS AND DISCUSSION QUESTIONS

2. The biological significance of genetic exchange and recombination appears to be to generate genetic variation in gametes, thereby leading to genetic variation in organisms. By reshuffling genes, new combinations are generated, which may then be of evolutionary advantage. In addition, because chromosomal position can influence gene function, variation is created by *position effect*.

4. With some qualification, especially around the centromeres and telomeres, one can say that crossing over is somewhat randomly distributed over the length of the chromosome. Two loci that are far apart are more likely to have a crossover between them than two loci that are close together.

6. If the probability of one event is

then the probability of two events occurring at the same time will be

$$1/X^2$$

8. Each cross must be set up in such a way as to reveal crossovers because it is on the basis of crossover frequency that genetic maps are developed. It is necessary that genetic heterogeneity exist so that different arrangements of genes, generated by crossing over, can be distinguished. The organism that is heterozygous must be the sex in which crossing over occurs. In other words, it would be useless to map genes in *Drosophila* if the male parent is the heterozygote since crossing over is not typical in *Drosophila* males. Lastly, the cross must be set up so that the phenotypes of the offspring readily reveal their genotypes. The best arrangement is one where a fully heterozygous organism is crossed with an organism that is fully recessive for the genes being mapped.

10. Notice that the most frequent phenotypes in the offspring, the parentals, are colored, green (88) and colorless, yellow (92). This indicates that the heterozygous parent in the testcross is coupled

$$RY/ry \times ry/ry$$

with the two dominant genes on one chromosome and the two recessives on the homolog. Seeing that there are 20 crossover progeny among the 200, or 20/200, the map distance would be 10 map units (20/200 × 100 to convert to percentages) between the *R* and *Y* loci.

Notice that the distribution of phenotypes is the same, regardless of the contribution of the crossover classes.

12.

d	*b*	*pr*	*vg*	*c*	*adp*
31	48	54	67	75	83

Map Units

The expected map units between *d* and *c* would be 44, *d* and *vg* would be 36, and *d* and *adp* 52. However, because there is a theoretical maximum of 50 map units possible between two loci in any one cross, that distance would be below the 52 determined by simple subtraction.

14. (a) P₁:

$$sc\ s\ v\ /sc\ s\ v \times +\ +\ +/Y$$

F₁:

$$+\ +\ +/sc\ s\ v \times sc\ s\ v/Y$$

(b) Using method I or II for determining the sequence of genes, examine the parental classes and compare the arrangement with the double-crossover (least frequent) classes.

$$\underbrace{sc\ v\ s}$$
$$+\ +\ +$$

$$sc - v = \frac{150 + 156 + 10 + 14}{1000} \times 100$$

$$= 33\% \text{ (map units)}$$

$$v - s = \frac{46 + 30 + 10 + 14}{1000} \times 100$$

$$= 10\% \text{ (map units)}$$

$$\underbrace{sc\text{-------}v\text{-----}s}_{}$$
$$\quad 33 \qquad 10$$

(c,d) The coefficient of coincidence = .727 which indicates that there were fewer double crossovers than expected; therefore, positive chromosomal interference is present.

16. (a) P₁: $D + +/ + + + \times + e\ p/ + e\ p$
    F₁: $D + +/ + e\ p \times + e\ p/ + e\ p$

F₂:	
$D + +/ + e\ p$	Dichete
$+ e\ p/ + e\ p$	ebony, pink
$D\ e +/ + e\ p$	Dichete, ebony
$+ + p/ + e\ p$	pink
$D + p/ + e\ p$	Dichete, pink
$+ e +/ + e\ p$	ebony
$D\ e\ p/ + e\ p$	Dichete, ebony, pink
$+ + +/ + e\ p$	wild type

(b) Determine which gene is in the middle by comparing the parental classes with the double-crossover classes.

F₁:

$$D + +/ + p\ e \times + p\ e/ + p\ e$$

$$D - p = \frac{12 + 13 + 2 + 3}{1000} \times 100$$

$$= 3.0 \text{ map units}$$

$$p - e = \frac{84 + 96 + 2 + 3}{1000} \times 100$$

$$= 18.5 \text{ map units}$$

18. $+ cu/ + cu$

20. Because sister chromatids are genetically identical (with the exception of rare new mutations), crossing over between sisters provides no increase in genetic variability. Somatic crossing over would have no influence on the offspring produced.

22. (a) There would be $2^n = 8$ genotypic and phenotypic classes, and they would occur in a 1:1:1:1:1:1:1:1 ratio.

(b) There would be two classes, and they would occur in a 1:1 ratio.

(c) There are 20 map units between the *A* and *B* loci, and locus *C* assorts independently from both *A* and *B* loci.

24. In contrast to the other organisms mentioned, a single human mating pair produces relatively few offspring and the haploid number of chromosomes is relatively high (23), so there are rather small numbers of identifiable genes per chromosome. In addition, accurate medical records are often difficult to obtain, and the life cycle is relatively long.

26. Assign the following symbols for example:
    R = Red   r = yellow
    O = Oval   o = long
    Progeny A: *Ro/rO × rroo* = 10 map units
    Progeny B: *RO/ro × rroo* = 10 map units

28. 10 map units

30. For Cross 1:

$$\frac{36 + 14}{100} \times 100 = 50 \text{ map units}$$

Because there are 50 map units between genes a and b, they are not linked.

For Cross 2:

$$\frac{3 + 9}{100} \times 100 = 12 \text{ map units}$$

Because genes a and b are not linked, they could be on nonhomologous chromosomes or far apart (50 map units or more) on the same chromosome. Because genes c and b are linked and therefore on the same chromosome, it is also possible that genes a and c are on different chromosome pairs. Under that condition, the NP and P (parental ditypes) would be equal; however, there is a possibility that the following arrangement occurs and that genes a and c are linked.

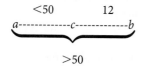

32.(a)

Problem	Tetrad in Class
1	NP
2	T
3	P
4	NP
5	T
6	P
7	T

(b) If P = NP they are independently assorting. In the problem given here, P = 44 and NP = 2. Therefore, the genes are linked.

(c)

For the centromere to d distance:

$$\frac{17 + 1 + 3 + 1}{69} \times 100 = 31.9$$

now divide by 2 = 15.9 map units

The map would be, according to these figures:

C-----------c---------------d

or

c-------------C----------------------d

If the two genes were on opposite sides of the centromere, the highest number of tetrad arrangements (5 in #5) would be associated with three crossover events. Since the likelihood of multiple crossovers decreases as the number of crossovers increases, it would seem reasonable that the genes are on the same side of the centromere.

(d) $= \dfrac{2 + 12}{69} = 20$ map units

(e)

The discrepancy between the two mapping systems is caused by the manner in which first and second division segregation products are scored. For instance, in tetrad arrangement #1 there are actually two crossovers between the d gene and the centromere, but it is still scored as a first division segregation. In tetrad arrangement #4, three crossovers occur between the d gene and the centromere, but they are scored as one. If one were to draw out all the crossovers needed to produce the tetrad arrangements in this problem, it would become clear that there are many crossovers between the d gene and the centromere that go undetected in the scoring of the arrangements of the d gene itself. This will cause one to underestimate the distance and give the discrepancy noted. One could account for these additional crossover classes to make the map more accurate.

(f) Tetrad class #6

34. Look for overlap between chromosome number in given clones and genes expressed. Note that $ENO1$ is expressed in clones B, D, and E; chromosomes 1 and 5 are common to these clones. However, since $ENO1$ is not expressed in clone C, which is missing chromosome 1 (and has chromosome 5), $ENO1$ must be on chromosome 1.

 $MDH1$: chromosome 2
 $PEPS$: chromosome 4
 $PMG1$: chromosome 1

36. There is no crossing over in males, and if two genes are on the same chromosome, there will be complete linkage of the genes in the male gametes. In females, crossing over will produce parental and crossover gametes. What you will have is the following gametes from the females (left) and males (right):

$bw^+ \; st^+$	1/4		$bw^+ \; st^+$	1/2
$bw^+ \; st$	1/4		$bw \;\; st$	1/2
$bw \;\; st^+$	1/4			
$bw \;\; st$	1/4			

Combining these gametes will give the ratio presented in the table of results.

38. Mapping the distance between B and m would be as follows:

$$(57 + 64)/ (226 + 218 + 57 + 64) \times 100$$

$$= 121/565 \times 100 = 21.4 \text{ map units}$$

We would conclude that the *ebony* locus is either far away from B and m (50 map units or more) or that it is on a different chromosome. In fact, *ebony* is on a different chromosome.

40. (a) Data presented in this table show an inverse correlation between recombination frequency and live-born children having various trisomies. As the frequency of crossing over decreases, the frequency of trisomy increases. While these data indicate a correlation, other factors such as intrauterine survival are also likely to play a role in determining trisomy live-born frequencies.

(b) If positive interference does spread out crossovers among and within chromosomes, then ensuring that crossovers are distributed among all the chromosomes (and all portions of chromosomes) may reduce non-disjunction and therefore be of selective advantage. This model assumes that the total number of crossovers per oocyte is limiting.

Chapter 6

ANSWERS TO NOW SOLVE THIS

6-1. T C H R O M B A K

All of the genes can be linked together to give a consistent map, and the ends overlap, indicating that the map is circular. The order is reversed in two of the crosses indicating the orientation of transfer is reversed.

6-2. Multiplying .031 × .012 gives .00037 or approximately 0.04%. From this information, one would consider no linkage between these two loci. Notice that this frequency is approximately the same as the frequency in the second experiment, where the loci are transformed independently.

6-3. For Group A, d and f are in the same complementation group (gene), while e is in a different one. Therefore, $e \times f$ = lysis.

For Group B, all three mutations are in the same gene, hence $b \times i$ = no lysis.

In Group C, j and k are in different complementation groups as are j and l. It would be impossible to determine whether l and k are in the same or different complementation group if the rII region had more than two cistrons. However, because only two complementation regions exist, and both are not in the same one as j, k and l must both be in the other.

SOLUTIONS TO PROBLEMS AND DISCUSSION QUESTIONS

2. Conjugation is dependent on the F factor, which, by a variety of mechanisms, can direct genetic exchange between two bacterial cells. Transformation is the uptake of exogenous DNA by cells. Transduction is the exchange of genetic material using a bacteriophage.

4. (a) In an $F^+ \times F^-$ cross, the transfer of the F factor produces a recipient bacterium, which is F^+. Any gene may be transferred on an F', and the frequency of transfer is relatively low. Crosses that are Hfr × F^- produce recombinants at a higher frequency than the $F^+ \times F^-$ cross. The transfer is oriented (nonrandom), and the recipient cell remains F^-.

(b) Bacteria that are F^+ possess the F factor, while those that are F^- lack the F factor. In Hfr cells the F factor is integrated into the bacterial chromosome, and in F' bacteria, the F factor is free of the bacterial chromosome, yet it possesses a piece of the bacterial chromosome.

6. Participating bacteria typically consist of two types, prototrophs and auxotrophs. If a miminal medium is used first, the auxotrophs would be unable to grow. The transfer to minimal medium allows the detection of recombinant bacteria. Mutant bacteria would not be identifiable if participating bacteria were transferred to a complete medium.

8. The F^+ element can enter the host bacterial chromosome and upon returning to its independent state, may pick up a piece of a bacterial chromosome. When combined with a bacterium with a complete chromosome, a partial diploid, or merozygote, is formed.

10. During transformation, incoming DNA forms a complex, with the host chromosome leading to a double helix that contains one host DNA strand and one incoming DNA strand. Since these two single strands are different in base sequence (otherwise genetic recombination would not occur), the term *heteroduplex* seems appropriate. After one round of replication, the heteroduplex DNA is resolved into a double-stranded structure identical to the host's original DNA, and one double-stranded mutant DNA.

12. A *plaque* results when bacteria in a "lawn" are infected by a phage and the progeny of the phage destroy (lyse) the bacteria. A somewhat clear region is produced that is called a plaque.

Lysogeny is a complex process whereby certain temperate phages can enter a bacterial cell and, instead of following a lytic developmental path, integrate their DNA into the bacterial chromosome. In doing so, the bacterial cell becomes lysogenic. The latent, integrated phage chromosome is called a *prophage*.

14. The c gene is in the middle. The map distances are as follows:

$$a \text{ to } c = (740 + 670 + 90 + 110)/10,000$$

$$= 16.1 \text{ map units}$$

$$c \text{ to } b = (160 + 140 + 90 + 110)/10,000$$

$$= 5 \text{ map units}$$

Negative interference is occurring.

16. Starting with a single bacteriophage, one lytic cycle produces 200 progeny phages; three more lytic cycles would produce $(200)^4$ or 1,600,000,000 phages.

18. Both mutant (rII) and wild-type strains of phage T4 can grow on *E. coli* B; however, only wild type T4 can grow on K12. Therefore, the degree of recombination between various rII strains can be determined by examining the number of plaques on K12 because only recombinant, wild-type, strains can grow. A greater number of T4 phage can grow on *E. coli* B.

20. $2(5 \times 10^1)/(2 \times 10^5) = 5 \times 10^{-4}$

22. (a)

Combination	Complementation
1, 2	−
1, 3	+
2, 4	+
4, 5	−

24. Because the frequency of double transformants is quite high (compare the trp^+tyr^+ transformants in A and B experiments), one may conclude that the genes are quite closely linked together. Part B in the experiment gives one the frequencies of transformations of the individual genes and the frequency of transformants receiving two pieces of DNA (2 in the data table). One must know these numbers in order to estimate the actual number of trp^+tyr^+ cotransformations.

26. The concentration of the original phage suspension would be:

$$10 \times 10^6 \times 17 = 1.7 \times 10^8$$

plaque-forming units per ml or pfu/ml.

28. Since g cotransforms with f, it is likely to be in the c b f "linkage group" and would be expected to cotransform with each. One would not expect transformation with a, d, or e.

30. (a) No, all functional groups do not impact similarly on conjugative transfer of R27. Regions 1, 2, and 4 appear to be least influenced by mutation because transfer is at 100 percent.

(b) Regions 3, 5, 6, 8, 9, 10, 12, 13, and 14 appear to have the most impact on conjugation because when mutant, conjugation is abolished.

(c) Regions 7 and 11, when mutant, only partially abolish conjugation; therefore they probably have less impact on conjugation than those listed in part (b).

(d) The data in this problem provide some insight into the complexity of the genetic processes involved in bacterial conjugation. The regions that have the most impact on conjugation fall into three different functional groups. In addition, notice that regions 1, 2, 4, 7, and 11, those that appear to have little, if any, impact on conjugation, are functionally related as indicated by their shading.

Chapter 7

ANSWERS TO NOW SOLVE THIS

7-1. In mammals, the scheme of sex determination is dependent on the presence of a piece of the Y chromosome. If present, a male is produced. In *Bonellia viridis*, the female proboscis produces some substance that triggers a morphological, physiological, and behavioral developmental pattern that produces males. To elucidate the mechanism, one could attempt to isolate and characterize the active substance by testing different chemical fractions of the proboscis. Second, mutant analysis usually provides critical approaches into developmental processes. Depending on characteristics of the organism, one could attempt to isolate mutants that lead to changes in male or female development. Third, by using micro-tissue transplantations, one could attempt to determine which anatomical "centers" of the embryo respond to the chemical cues of the female.

7-2. (a) Something is missing from the male-determining system of sex determination either at the level of the genes, gene products, or receptors, etc. (b) The *SOX9* gene, or its product, is probably involved in male development. Perhaps it is activated by *SRY*. (c) There is probably some evolutionary relationship between the *SOX9* gene and *SRY*. There is considerable evidence that many other genes and pseudogenes are also homologous to *SRY*. (d) Normal female sexual development does not require the *SOX9* gene or gene product(s).

7-3. If one assumes that the ovarian cell was engaging in X chromosome inactivation, the ovarian cell that Rainbow donated to create CC contained an activated *black* gene and an inactivated *orange* gene (from X-inactivation). This would mean that as CC developed, her cells did not change that inactivation pattern. Therefore, unlike Rainbow, CC developed without any cells that specified orange coat color. The result is CC's black and white tiger-tabby coat. Had the inactivation pattern been reversed such that both the *black* and *orange* alleles were originally active and subsequently inactivated randomly, CC would still not have appeared identical to Rainbow because the pattern of inactivation is random and the distribution of white patches is variable.

SOLUTIONS TO PROBLEMS AND DISCUSSION QUESTIONS

2. (a) The term *homomorphic* refers to the situation where both the sex chromosomes have the same form. The term *heteromorphic* refers to the condition in many organisms where there are two different forms (morphs) of chromosomes such as X and Y.

(b) In *isogamous* species, there is little visible difference between the haploid vegetative cells that reproduce asexually and the haploid gametes that are involved in sexual reproduction. The two gametes that fuse during mating are morphologically indistinguishable and are called *isogametes*. An organism that is *heterogamous* is one in which there are two morphologically distinct gametes.

6. The *Protenor* form of sex determination involves the XX/XO condition, while the *Lygaeus* mode involves the XX/XY condition.

8. Mammals possess a system of X chromosome inactivation whereby one of the two X chromosomes in females becomes a chromatin body or Barr body. If one of the two X chromosomes is randomly inactivated, the dosage of genetic information is more or less equivalent in males (XY) and females (XX).

10. The Y chromosome is male determining in humans, and it is a particular region of the Y chromosome that causes maleness, the sex-determining region (SRY). SRY encodes a product called the testis-determining factor (TDF), which causes the undifferentiated gonadal tissue to form testes. Individuals with the 47, XXY complement are males, while those with 45, XO are females. In *Drosophila* it is the balance between the number of X chromosomes and the number of haploid sets of autosomes that determines sex. In contrast to humans, XO *Drosophila* are males and the XXY complement is female.

12. (a) female $X^{rw}Y$ $\times$ male X^+X^+

F₁: females:	X^+Y (normal)
males:	$X^{rw}X^+$ (normal)
F₂: females:	X^+Y (normal)
	$X^{rw}Y$ (reduced wing)
males:	$X^{rw}X^+$ (normal)
	X^+X^+ (normal)

(b) female $X^{rw}X^{rw} \times$ male X^+Y

F_1: females: $X^{rw}X^+$ (normal)
    males: $X^{rw}Y$ (reduced wing)

F_2: females: $X^{rw}X^+$ (normal)
             $X^{rw}X^{rw}$ (reduced wing)
    males: X^+Y (normal)
             $X^{rw}Y$ (reduced wing)

14. Because attached-X chromosomes have a mother-to-daughter inheritance and the father's X is transferred to the son, one would see daughters with the white-eye phenotype and sons with the miniature wing phenotype.

16. Because synapsis of chromosomes in meiotic tissue is often accompanied by crossing over, it would be detrimental to sex-determining mechanisms to have sex-determining loci on the Y chromosome transferred, through crossing over, to the X chromosome.

18.

Klinefelter syndrome (XXY)	= 1
Turner syndrome (XO)	= 0
47,XYY	= 0
47,XXX	= 2
48,XXXX	= 3

20. Unless other markers, cytological or molecular, are available, one cannot test the Lyon hypothesis with homozygous X-linked genes.

22. Dosage compensation and the formation of Barr bodies occur only when there are two or more X chromosomes. Males normally have only one X chromosome; therefore such mosaicism cannot occur.

24. Instead of X chromosome inactivation as seen in humans (mammals in general), male X-linked genes in *Drosophila* are transcribed at twice the rate as comparable genes on the X chromosome in *Drosophila* females.

26. Several possibilities are discussed in the text. One could account for the significant departures from a 1:1 ratio of males to females by suggesting that at anaphase I of meiosis, the Y chromosome more often goes to the pole that produces the more viable sperm cells. One could also speculate that the Y-bearing sperm has a higher likelihood of surviving in the female reproductive tract, or that the egg surface is more receptive to Y-bearing sperm. At this time the mechanism is unclear.

28. Because of the homology between the *red* and *green* genes, there exists the possibility for an irregular synapsis, which, following crossing over, would give a chromosome with only one (*green*) of the duplicated genes.

30. The presence of the Y chromosome provides a factor (or factors) that leads to the initial specification of maleness and the formation of testes. Subsequent expression of secondary sex characteristics must be dependent on the interaction of the normal X-linked *Tfm* allele with testosterone. Without such interaction, differentiation takes the female path. To test the dominant nature of the *Tfm* allele, one could degenerate an XXY male that is heterozygous for the *Tfm* allele. Should the same testicular feminization phenotype occur, the dominant nature of the *Tfm* allele would be supported. If the *Tfm* phenotype was eliminated in the heterozygote, one would have support for the model that the normal *Tfm* allele is needed to interact with testosterone.

32. In snapping turtles, sex determination is strongly influenced by temperature such that males are favored in the 26–34°C range. Lizards, on the other hand, appear to have their sex determined by factors other than temperature in the 20–40°C range.

34. (a) The figures below depict only those chromosomes relevant to the *Mm* and *Dd* genotypes. With the *MmDd* genotype, both dyads separate intact to the secondary spermatocytes and are potentially capable of fertilizing the egg. (Only one configuration of the *MmDd* genotype is presented here.)

Primary Spermatocyte

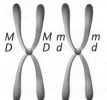

Secondary Spermatocyte

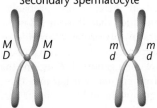

Below is the *Mm dd* genotype that leads to fragmentation of the *m*-bearing chromosome. Thus, only the *M*-bearing chromosome is available for fertilization.

Primary Spermatocyte

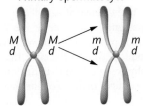

Secondary Spermatocyte

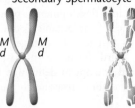

(b) There have been many attempts to control pest species by hampering the production or fertility of one sex. In this case, if a sex ratio distorter could be successfully integrated into a large enough pool of males, then female numbers may drop. Whether one could successfully manage a pest population with such a method remains to be tested.

36. Do sex cells respond to signals carried in body fluids (nonautonomous), or are they cell specific (autonomous)? Reciprocal transplantation tests can be used to determine the influence of the anatomical environment for cellular differentiation. Since male/female gynandromorphs could be produced in the first place, it argues for an autonomous form of sex differentiation in contrast with that seen in mammals.

Chapter 8

ANSWERS TO NOW SOLVE THIS

8-1. If the father had hemophilia, it is likely that the Turner syndrome individual inherited the X chromosome from the father and no sex chromosome from the mother. If nondisjunction occurred in the mother, either during meiosis I or meiosis II, an egg with no X chromosome can be the result.

8-2. The sterility of interspecific hybrids is often caused from a high proportion of univalents in meiosis I. As such, viable gametes are rare and the likelihood of two such gametes "meeting" is remote. Even if partial homology of chromosomes allows some pairing, sterility is usually the rule. The horticulturist may attempt to reverse the sterility by treating the sterile hybrid with colchicine. Such a treatment, if successful, may double the chromosome number, and each chromosome would then have a homolog with which to pair during meiosis.

8-3. The rare double crossovers within the boundaries of a paracentric or pericentric inversion produce only minor departures from the standard chromosomal arrangement as long as the crossovers involve the same two chromatids. With two-strand double crossovers, the second crossover negates the first. However, three-strand and four-strand double crossovers have consequences that lead to anaphase bridges as well as a high degree of genetically unbalanced gametes.

SOLUTIONS TO PROBLEMS AND DISCUSSION QUESTIONS

2. With a diploid chromosome number of 18 ($2n$), a haploid (n) would have 9 chromosomes, a triploid ($3n$) would have 27 chromosomes, and a tetraploid ($4n$) would have 36 chromosomes. A trisomic would have one extra chromosome (19), and a monosomic would have one less than the diploid (17).

4. Individuals with Down syndrome, while suffering congenital defects with tendencies toward respiratory disease and leukemia, can live well into adulthood. Individuals with Patau or Edwards syndrome live less than four months on the average. Comparing the different sizes of the involved chromosomes (21, 13, and 18, respectively) in the text for example, suggests that the larger the chromosome, the lower the likelihood of lengthy survival. In addition, it would be expected that certain chromosomes, because of their genetic content, may have different influences on development.

6. Karyotype analysis of spontaneously aborted fetuses has shown that a significant percentage of abortuses are trisomic and every chromosome can be involved. Other forms of aneuploidy (monosomy, nullisomy) are less represented.

8. American cultivated cotton has 26 pairs of chromosomes: 13 large and 13 small. Old World cotton has 13 pairs of large chromosomes, and American wild cotton has 13 pairs of small chromosomes. It is likely that an interspecific hybridization occurred followed by chromosome doubling. These events probably produced a fertile amphidiploid (allotetraploid). Experiments have been conducted to reconstruct the origin of American cultivated cotton.

10. While there is the appearance that crossing over is suppressed in inversion "heterozygotes," the phenomenon extends from the fact that the crossover chromatids end up being abnormal in genetic content. As such, they fail to produce viable (or competitive) gametes, or they lead to zygotic or embryonic death.

12. Modern globin genes resulted from a duplication event in an ancestral gene about 500 million years ago. Mutations occurred over time and a chromosomal aberration separated the duplicated genes, leaving the eventual α cluster on chromosome 16 and the eventual β cluster on chromosome 11.

14. By having the genes in an inversion, crossover chromatids are not recovered and therefore, are not passed on to future generations. Translocations offer an opportunity for new gene combinations by associations of genes from nonhomologous chromosomes. Under certain conditions, such new combinations may be of selective advantage. Meiotic conditions have evolved so that segregation of translocated chromosomes yields a relatively uniform set of gametes.

16. Given the basic chromosome set of 9 unique chromosomes (a haploid complement), other forms with the "n multiples" are forms of autopolyploidy. Individual organisms with 27 chromosomes ($3n$) are more likely to be sterile because there are trivalents at meiosis I, which cause a relatively high number of unbalanced gametes to be formed.

18. The cross would be as follows:

$$WWWW \times wwww$$
(assuming that chromosomes pair as bivalents at meiosis)

F_1: $WWww$
F_2: $1WW$ $1WW$ $4Ww$ $1ww$

$1WW$
$4WW$ fill in Punnett square
$1ww$ 35 W phenotypes and 1w phenotype

20. While a number of mechanisms for *bobbed* reversion have been documented, one based on meiotic recombination occurs through "unequal crossing over." When redundant chromosomal regions synapse, homologs can misalign. If crossing over occurs in the misaligned segments, one chromatid can gain chromosomal material at the expense of the other chromatid. As a chromosome gains rRNA genes, it harbors a selective advantage and produces flies that outcompete those with nonreverted *bobbed* mutations. Eventually, a stock that originally contained *bobbed Drosophila* contains what appear to be wild-type flies over time.

22. (a) In all probability, crossing over in the inversion loop of an inversion (in the heterozygous state) had produced defective, unbalanced chromatids, thus leading to stillbirths and/or malformed children. (b) It is probable that a significant proportion (perhaps 50 percent) of the children of the man will be similarly influenced by the inversion. (c) Since the karyotypic abnormality is observable, it may be possible to detect some of the abnormal chromosomes of the fetus by amniocentesis or CVS. However, depending on the type of inversion and the ability to detect minor changes in banding patterns, not all abnormal chromosomes may be detected.

24. (a) Reciprocal translocation

(b)

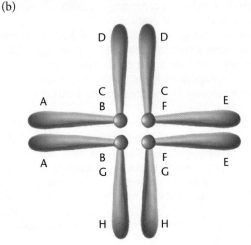

(c) Notice that all chromosomal segments are present and there is no apparent loss of chromosomal material. However, if the breakpoints for the translocation occurred within genes, then an abnormal phenotype may be the result. In addition, a gene's function is sometimes influenced by its position (its neighbors, in other words). If such "position effects" occur, then a different phenotype may result.

26. The symbolism t(14;21) indicates that a translocation (t) has occurred between chromosomes 14 and 21. Generally, a Down syndrome individual with a t(14;21) karyotype has 46 chromosomes.

28. Notice that a chromosome in this question is defined as having two sisters joined at the centromere. This is the expected chromosome structure at the end of meiosis I.

(a) In light of this information, meiosis I must have produced the abnormal oocytes with more or less than 24 chromosomes, indicating multiple conditions of nondisjunction. More likely, the oocytes consisted of "22 1/2" chromosomes, those 22 normal dyads and a single monad.

(b) The result will be a monosomic and a normal zygote, assuming that the half chromosome (monad) migrates, intact, to one pole or the other.

(c) In all likelihood, premature division of the centromere (at meiosis I) probably causes the single (nonduplicated) chromosome at meiosis II.

(d) We generally consider nondisjunction occurring at meiosis I to consist of intact chromosomes, two sister chromatids, failing to separate appropriately. These data indicate that some forms of aneuploidy result from premature division of the centromere at meiosis I as in the figure below.

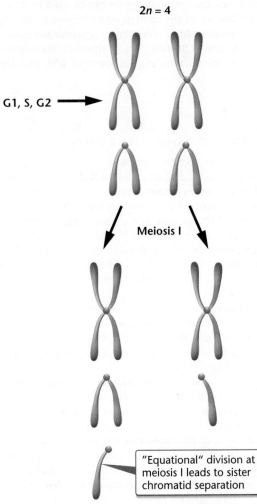

30. It is likely that mitotic nondisjunction contributed to the mosaic condition. If one of the X chromosomes failed to be included in a daughter mitotic cleavage cell, then a substantial proportion of the child's cells would be XO. Expression of Turner syndrome characteristics would depend on the percentage and location of the XO cell population.

32. *Trisomic rescue* is a condition in which an original trisomic zygote occurs but a particular chromosome is lost. In the

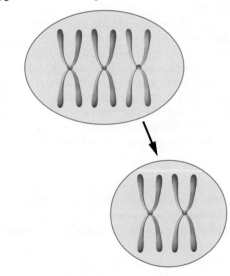

diagram above, two chromosomes originally came from the father (blue), and one came from the mother (gold). The one from the mother was eliminated during mitosis, leaving the two chromosomes from the father.

Monosomic rescue occurs when a chromosome is originally present in a zygote, and usually through mitotic nondisjunction, a duplicate of that chromosome occupies the cell.

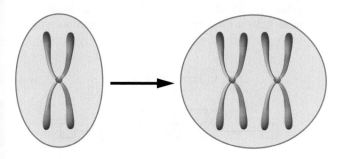

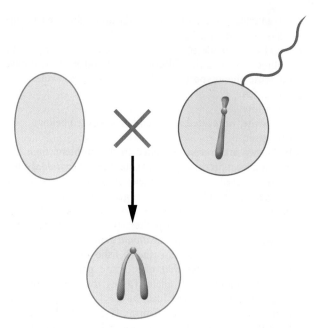

Gamete complementation occurs when a cell without either homolog of a given chromosome is fertilized by a gamete with two copies of that homolog. The resulting zygote contains two homologous chromosomes from one parent.

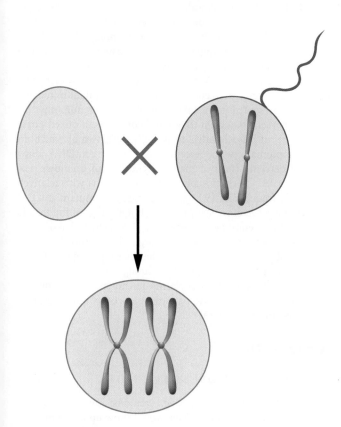

Isochromosome formation occurs when a chromosome contains two copies of one arm and has lost the homolog. Such a homolog loss can originate in mitosis or meiosis. The isochromosome compensates for the nullisomic condition. Effectively, the chromosome is "uniparental" because most the chromosomal material of a given homolog is from one parent.

Chapter 9

ANSWERS TO NOW SOLVE THIS

9-1. The *mt*⁺ strain is the donor of the *cp*DNA since the inheritance of resistance or sensitivity is dependent on the status of the *mt*⁺ gene. In this organism, chloroplasts obtain their characteristics from the *mt*⁺ strain, while mitochondria obtain their characteristics from the *mt*⁻ strain.

9-2. (a) neutral
(b) segregational (nuclear mutations)
(c) suppressive

9-3. In many cases, molecular components of mitochondria are recruited from the cytoplasm, having been synthesized from nuclear genes.

9-4. From an organismic standpoint, individuals with the most severe mitochondrial defects tend to be less reproductively successful. At the cellular level, mitochondria with mutations in protein-coding genes tend to be selected against. Such purifying selection tends to favor nonmutant mitochondria.

SOLUTIONS TO PROBLEMS AND DISCUSSION QUESTIONS

2. The pattern of inheritance is more often from one parent to the offspring. One does not see both parents contributing to the characteristics of the offspring as is the case with Mendelian (chromosomal) forms of inheritance. Standard Mendelian ratios (3:1) are usually not present. In general, the results of reciprocal crosses differ.

> *Female mutant × male wild*
>
> all offspring mutant
>
> *Female wild × male mutant*
>
> all offspring wild

In sex-linked inheritance, the pattern is often from grandfather through carrier mother to son. Patterns of extranuclear inheritance are often not influenced by the sex of the individual.

4. Because the ovule source furnishes the cytoplasm to the embryo and thus the chloroplasts, the offspring will have the same phenotype as the plant providing the ovule.
    (a) green
    (b) white
    (c) white, variegated, or green (as illustrated below)
    (d) green

6. If the two are crossed as stated in the problem, then one would expect, in the diploid zygote, the *segregational* allele to be "covered" by normal alleles from the neutral strain. On the other hand, as the nuclear genes are again "exposed" in the haploid state of the ascospores, one would expect a 1:1 ratio of normals to petites. The petite phenoytpe is caused by the nuclear, *segregational* gene.

8. The fact that all of the offspring (F_1) showed a dextral coiling pattern indicates that one of the parents (maternal parent) contains the *D* allele. Taking these offspring and seeing that their progeny (call these F_2) occur in a 1:1 ratio indicates that half of the offspring (F_1) are *dd*. In order to have these results, one of the original parents must have been *Dd*, while the other must have been *dd*.

10. Since there is no evidence for segregation patterns typical of chromosomal genes and Mendelian traits, some form of extranuclear inheritance seems possible. If the *lethargic* gene is dominant, then a maternal effect may be involved. In that case, some of the F_2 progeny would be hyperactive because maternal effects are temporary, affecting only the immediate progeny. If the lethargic condition is caused by some infective agent, then perhaps injection experiments could be used. If caused by a mitochondrial defect, then the condition would persist in all offspring of lethargic mothers, through more than one generation.

12. The endosymbiotic theory states that mitochondria and chloroplasts arose independently around 2 billion years ago from free-living protobacteria. These bacteria brought the capacity for aerobic respiration and photosynthesis to primitive eukaryotic cells. Because such organelles have prokaryotic origins, a deeper understanding of extranuclear DNA is possible.

14. Developmental phenomena that occur early are more likely to be under maternal influence than those occurring late. Anterior/posterior and dorsal/ventral orientations are among the earliest to be established, and in organisms where their study is experimentally and/or genetically approachable, they often show considerable maternal influence. Maternal-effect genes yield products that are not carried over for more than one generation as is the case with organelle and infectious heredity. Crosses that illustrate the transient nature of a maternal effect could include the following.

    Female *Aa* × male *aa* -----> all offspring of the "A" phenotype. Take a female "A" phenotype from the above cross and conduct the following mating (note, only half of the offspring from the above cross will be *aa*): *aa* × male *Aa*.

    All offspring will be of the "a" phenotype because all of the offspring will reflect the *genotype* of the mother, not her *phenotype*. This cross illustrates that maternal effects last only one generation. In actual practice, the results of this cross may give a typical 1:1 ratio because the mother may be *Aa* or *aa*.

16. Because of sampling error due to relatively small numbers of offspring, this pedigree could represent a typical Mendelian dominant or recessive gene. However, because the thrust of this chapter is on extranuclear inheritance, the condition is probably a case of extranuclear inheritance. The phenotype of the offspring (of II-4 and II-5) will therefore follow the characteristics of the mother, in this case, all normal.

18. (a) Note, to provide a meaningful pedigree, several individuals are added to the pedigree that are not listed in the table. Refer to Chapter 3 for information regarding symbols used.

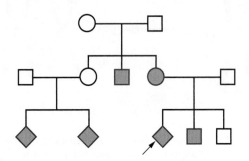

(b,c) If one looks solely at the above pedigree, one could argue that the pattern follows a typical Mendelian recessive. If that's the case, all individuals not showing the phenotype would be heterozygous. However, given that a mitochondrial DNA mutation was identified among affected family members, the pedigree is also consistent with organelle inheritance. Heteroplasmy explains variation in expression and transmission. While this may seem like an erroneous conclusion at first, consider the range of mutant mitochondria presented in the table. The maternal grandmother has 56 percent mutant mtDNA and could pass a mixed population of mitochondria to her offspring, two of which are symptomatic with percentages of mutant mtDNA above 90 percent. The two maternal cousins (of unknown sex) inherited, by chance, pools of mitochondria with relatively high frequencies (90 and 91 percent) of mutant mtDNA and are therefore symptomatic. It appears as if the threshold for phenotypic expression is 85 percent and above.

(d) Transmission of a trait by organelle heredity occurs primarily through the mother, whereas single-gene mutations (albinism) can be transmitted through either parent. In addition, because of heteroplasmy, phenotypic expression may vary and transmission patterns may complicate interpretations from pedigrees.

Chapter 10

ANSWERS TO NOW SOLVE THIS

10-1. In theory, the general design would be appropriate in that some substance, if labeled, would show up in the progeny of transformed bacteria. However, since the amount of transforming DNA is extremely small compared with the genomic DNA of the recipient bacterium and its progeny, it would be technically difficult to assay for the labeled nucleic acid. In addition, it would be necessary to know that the small stretch of DNA that caused the genetic transformation was actually labeled.

10-2. Guanine = 17.5%, Adenine and Thymine both = 32.5%.

10-3. Assuming the value of 1.13 is statistically different from 1.00, one can conclude that rubella is a single-stranded RNA virus.

SOLUTIONS TO PROBLEMS AND DISCUSSION QUESTIONS

2. *Replication* is the process that leads to the production of identical copies of the existing genetic information. Since daughter cells contain essentially exact copies (with some exceptions) of genetic information of the parent cell, and through the production and union of gametes, offspring contain copies (with variation) of parental genetic information, the genetic material must make copies of (replicate) itself. Replication is accomplished during the S phase of interphase. The genetic material is capable of *expression* through the production of a phenotype. Through transcription and translation, proteins are produced that contribute to the phenotype of the organism. The genetic material must be stable enough to maintain information in "*storage*" from one cell to the next and from one organism to the next. Because the genetic material is not "used up" in the processes of transcription and translation, genetic information can be stored and used constantly. Above, it was stated that the genetic material must be stable enough to store genetic information; however, variation through *mutation* provides the raw material for evolution. The genetic material is capable of a variety of changes, both at the chromosomal and nucleotide levels.

4. Griffith performed experiments with different strains of *Diplococcus pneumoniae* in which a heat-killed pathogen, when injected into a mouse with a live nonpathogenic strain, eventually led to the mouse's death. A summary of this experiment is provided in the text. Examination of the dead mouse revealed living pathogenic bacteria. Griffith suggested that the heat-killed virulent (pathogenic) bacteria transformed the avirulent (nonpathogenic) strain into a virulent strain. Avery and coworkers systematically searched for the transforming principle originating from the heat-killed pathogenic strain and determined it to be DNA. Taylor showed that transformed bacteria are capable of serving as donors of transforming DNA, indicating that the process of transformation involves a stable alteration in the genetic material (DNA).

6. Nucleic acids contain large amounts of phosphorus and no sulfur, whereas proteins contain sulfur and no phosphorus. Therefore, the radioisotopes ^{32}P and ^{35}S will selectively label nucleic acids and proteins, respectively. The Hershey and Chase experiment was based on the premise that the substance injected into the bacterium is the substance responsible for producing the progeny phage and therefore, must be the hereditary material. The experiment demonstrated that most of the ^{32}P-labeled material (DNA) was injected, while the phage ghosts (protein coats) remained outside the bacterium. Therefore, the nucleic acid must be the genetic material.

8. The early evidence would be considered indirect in that at no time was there an experiment, like transformation in bacteria, in which genetic information in one organism was transferred to another using DNA. Rather, by comparing DNA content in various cell types (sperm and somatic cells) and observing that the *action* and *absorption* spectra of ultraviolet light were correlated, DNA was considered to be the genetic material. This suggestion was supported by the fact that DNA was shown to be the genetic material in bacteria and some phages. Direct evidence for DNA being the genetic material comes from a variety of observations including gene transfer, which has been facilitated by recombinant DNA techniques.

10. The structure of deoxyadenylic acid is given below and in the text. Linkages among the three components require the removal of water (H_2O).

12.

Guanine:	2-amino-6-oxypurine
Cytosine:	2-oxy-4-aminopyrimidine
Thymine:	2,4-dioxy-5-methylpyrimidine
Uracil:	2,4-dioxypyrimidine

14. The following are characteristics of the Watson-Crick double-helix model for DNA:

The base composition is such that A = T, G = C and (A + G) = (C + T). Bases are stacked, 0.34 nm (3.4 Angstoms) apart, and in a plectonic, antiparallel manner. There is one complete turn for each 3.4 nm that constitutes 10 bases per turn. Hydrogen bonds hold the two polynucleotide chains together, each being formed by phosphodiester linkages between the five-carbon sugars and the phosphates. There are two hydrogen bonds forming the A to T pair and three forming the G to C pair. The double helix exists as a twisted structure, approximately 20 Angstroms in diameter, with a topography of major and minor grooves. The hydrophobic bases are located in the center of the molecule, while the hydrophilic phosphodiester backbone is on the outside.

16. Because in double-stranded DNA, A=T and G=C (within limits of experimental error), the data presented would have indicated a lack of pairing of these bases in favor of a single-stranded structure or some other nonhydrogen-bonded structure.

Alternatively, from the data it would appear that A=C and T=G, which would negate the chance for typical hydrogen bonding since opposite charge relationships do not exist. Therefore, it is quite unlikely that a tight helical structure would form at all. In conclusion, Watson and Crick might have concluded that hydrogen bonding is not a significant factor in maintaining a double-stranded structure.

18. Three main differences between RNA and DNA are the following:
   (1) uracil in RNA replaces thymine in DNA,
   (2) ribose in RNA replaces deoxyribose in DNA, and
   (3) RNA often occurs as both single- and partially double-stranded forms, whereas DNA most often occurs in a double-stranded form.

20. The nitrogenous bases of nucleic acids (nucleosides, nucleotides, and single- and double-stranded polynucleotides), absorb UV light maximally at wavelengths 254 to 260 nm. Using this phenomenon, one can often determine the presence and concentration of nucleic acids in a mixture. Since proteins absorb UV light maximally at 280 nm, this is a relatively simple way of dealing with mixtures of biologically important molecules. UV absorption is greater in single-stranded molecules (hyperchromic shift) as compared with double-stranded structures; therefore, one can easily determine, by applying denaturing conditions, whether a nucleic acid is in the single- or double-stranded form. In addition, A-T rich DNA denatures more readily than G-C rich DNA; therefore, one can estimate base content by denaturation kinetics.

22. Double-stranded DNA is changed to single-stranded DNA.

24. Because G-C base pairs are formed with three hydrogen bonds, while A-T base pairs by two such bonds, it takes more energy (higher temperature) to separate G-C pairs.

26. In one sentence of Watson and Crick's paper in *Nature*, they state,

> It has not escaped our notice that the specific pairing we have postulated immediately suggests a possible copying mechanism for the genetic material.

28. left side (a) = right, right side (b) = left

30. Under this condition, the hydrolyzed 5-methyl cytosine becomes thymine.

32. One of the basic principles of gel electrophoresis is that shorter molecules migrate at a faster rate through a given gel than longer ones. While a number of factors other than acrylamide concentration would be involved, depending on the length of the gel, it might be wise to use the 12 percent acrylamide recipe so that the short fragments will not reach the end of the gel, and enter the buffer, before the longer ones leave the wells.

34. Without knowing the exact bonding characteristics of hypoxanthine or xanthine, it may be difficult to predict the likelihood of each pairing type. It is likely that both are of the same class (purine or pyrimidine) because the names of the molecules indicate a similarity. In addition, the diameter of the structure is constant, which, under the model to follow, would be expected. In fact, hypoxanthine and xanthine are both purines.

Because there are equal amounts of A, T, and H, one could suggest that they are hydrogen bonded to each other; the same may be said for C, G, and X. Given the molar equivalence of erythrose and phosphate, an alternating sugar-phosphate-sugar backbone, as in "earth-type" DNA, would be acceptable. A model of a triple helix would be acceptable, since the diameter is constant. Given the chemical similarities to "earth-type" DNA, it is probable that the unique creature's DNA follows the same structural plan.

36. The way the question is stated suggests that DNA that is separated electrophoretically is of the same shape (long rod). In fact, DNA can exist in a variety of shapes as seen in supercoiled plasmids, relaxed (nicked) plasmids, and linear molecules. Size comparisons with DNA must be such that linear molecules are compared with linear molecules and supercoiled with supercoiled, *etc*. In comparing DNA migration with RNA, even though RNA molecules have the same charge-to-mass ratios, they also exist in a variety of shapes. Complementary intra-strand base pairing can make more compact structures compared with the more relaxed, open conformation. For electrophoretic-size comparisons, RNA molecules must be denatured to eliminate secondary structural variables.

38. In general, as the %GC pairs increase, the T_m increases. This is to be expected because there are three hydrogen bonds that hold GC pairs together instead of two that hold AT pairs. Therefore, it takes more energy to break GC pairs than AT pairs.

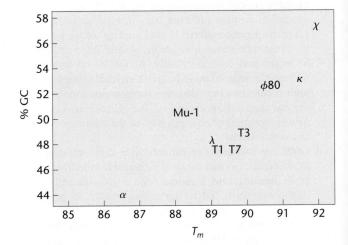

Chapter 11

ANSWERS TO NOW SOLVE THIS

11-1. After one round of replication in the ^{14}N medium, the conservative scheme can be ruled out. After one round of replication in ^{14}N under a dispersive model, the DNA is of intermediate density, just as it is in the semiconservative model. However, in the next round of replication in ^{14}N medium, the density of the DNA is between the intermediate and "light" densities, thus ruling out the dispersive model.

11-2. If the DNA contained parallel strands in the double helix and the polymerase would be able to accommodate such parallel strands, there would be continuous synthesis and no Okazaki fragments. Several other possibilities exist. If the DNA was replicated as single strands, the synthesis could begin at the free ends and there would be no need for Okazaki fragments.

SOLUTIONS TO PROBLEMS AND DISCUSSION QUESTIONS

2. *Conservative:* In the conservative scheme, the original double helix remains as a complete unit, and the new DNA double helix is produced as a single unit. The old DNA is completely conserved.

Semiconservative: Each daughter strand is composed of one old DNA strand and one new DNA strand. Separation of hydrogen bonds is required.

Dispersive: In the dispersive scheme, the original DNA strand is broken into pieces, and the new DNA in the daughter strand is interspersed among the old pieces. Separation of the individual covalent, phosphodiester bonds is required for this mode of replication.

4. (a) Under a conservative scheme, all of the newly labeled DNA will go to one sister chromatid, while the other sister chromatid will remain unlabeled. In contrast to a semiconservative scheme, the first replicative round would produce one sister chromatid, which has labels on both strands of the double helix.

(b) Under a dispersive scheme, all of the newly labeled DNA will be interspersed with unlabeled DNA. Because these preparations (metaphase chromosomes) are highly coiled and condensed structures derived from the "spread out" form at interphase (which includes the S phase). It is impossible to detect the areas where label is not found. Rather, both sister chromatids would appear as evenly labeled structures.

6. The *in vitro* replication requires a DNA template, a divalent cation (Mg^{++}), and all four of the deoxyribonucleoside triphosphates: dATP, dCTP, dTTP, and dGTP. The lowercase "d" refers to the deoxyribose sugar.

8. Because *base composition* can be similar without reflecting sequence similarity, the least stringent test was the comparison of base composition. By comparing *nearest neighbor frequencies*, Kornberg determined that there is a very high likelihood that the product is of the same base sequence as the template.

10. An exposed 3'-OH group is necessary for the attachment of the next nucleotide. The 3'-OH group is eventually removed in the form of water, and a covalent bond is formed to the 5'-phosphate of the added nucleotide.

12. All three enzymes share several common properties. First, none can *initiate* DNA synthesis on a template, but all can *elongate* an existing DNA strand assuming there is a template strand as shown in the figure below. Polymerization of nucleotides occurs in the 5' to 3' direction where each 5' phosphate is added to the 3' end of the growing polynucleotide.

    All three enzymes are large, complex proteins with a molecular weight in excess of 100,000 daltons, and each has 3' to 5' exonuclease activity. Refer to the text.

DNA polymerase I:
    5' – 3' exonuclease activity
    present in large amounts
    relatively stable
    removal of RNA primer

DNA polymerase II:
    possibly involved in repair function

DNA polymerase III:
    essential for replication
    complex molecule

14. Given a stretch of double-stranded DNA, one could initiate synthesis at a given point and either replicate strands in one direction only (unidirectional) or in both directions

(bidirectional) as shown below. Notice that in the text the synthesis of complementary strands occurs in a *continuous 5'>3'* mode on the leading strand in the direction of the replication fork, and in a *discontinuous 5'>3'* mode on the lagging strand opposite the direction of the replication fork. Such discontinuous replication forms Okazaki fragments.

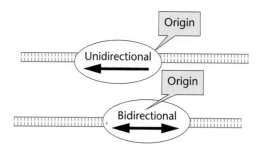

16. (a) *Okazaki fragments* are relatively short (1000 to 2000 bases in prokaryotes) DNA fragments that are synthesized in a discontinuous fashion on the lagging strand during DNA replication. Such fragments appear to be necessary because template DNA is not available for 5'>3' synthesis until some degree of continuous DNA synthesis occurs on the leading strand in the direction of the replication fork. The isolation of such fragments provides support for the scheme of replication shown in the text.

(b) DNA *ligase* is required to form phosphodiester linkages in gaps, which are generated when DNA polymerase I removes RNA primer and meets newly synthesized DNA ahead of it. The discontinuous DNA strands are ligated together into a single continuous strand.

(c) *Primer* RNA is formed by RNA primase to serve as an initiation point for the production of DNA strands on a DNA template. None of the DNA polymerases are capable of initiating synthesis without a free 3' hydroxyl group. The primer RNA provides that group and thus can be used by DNA polymerase III.

18. Eukaryotic DNA is replicated in a manner that is very similar to that of *E. coli*. Synthesis is bidirectional, continuous on one strand and discontinuous on the other, and the requirements of synthesis (four deoxyribonucleoside triphosphates, divalent cation, template, and primer) are the same. Okazaki fragments of eukaryotes are about one-tenth the size of those in bacteria.

    Because there is a much greater amount of DNA to be replicated and DNA replication is slower, there are multiple initiation sites for replication in eukaryotes (and increased DNA polymerase per cell) in contrast to the single replication origin in prokaryotes. Replication occurs at different sites during different intervals of the S phase.

20. (a) No repair from DNA polymerase I and/or DNA polymerase III (b) no DNA ligase activity (c) no primase activity (d) only DNA polymerase I activity (e) no DNA gyrase activity.

22. *Gene conversion* is likely to be a consequence of genetic recombination in which nonreciprocal recombination yields products in which it appears that one allele is "converted" to another. Gene conversion is now considered a result of heteroduplex formation, which is accompanied by mismatched bases. When these mismatches are corrected, the "conversion" occurs.

24. Telomerase activity is present in germ-line tissue to maintain telomere length from one generation to the next. In other words, telomeres cannot shorten indefinitely without eventually eroding genetic information.

26. If replication is conservative, the first autoradiographs (metaphase I) would have label distributed only on one side (chromatid) of the metaphase chromosome as shown below.

Conservative Replication

28. It is possible that the alien organism contains DNA that has parallel strands or only one DNA strand. In either case, the telomere problem would occur at one end of the chromosome. Since prokaryotic DNA is generally circular, therefore having no free ends, the alien organism is most likely a eukaryote.

30. Conservative replication can be eliminated.

32. In strain A, DNA synthesis is reduced at both 30 and 42°C, indicating that *strain A* is temperature sensitive. In addition, *strain A* is sensitive to novobiocin. Therefore *strain A* is *gyr*^{ts}. Strain B is resistant to novobiocin but temperature sensitive, and is therefore *gyr*^{ts,r}. Strain C is sensitive to novobiocin but not temperature sensitive, and is therefore *wild type*. Strain D is resistant to novobiocin and not temperature sensitive and is therefore *gyr*^{r}.

34.

Initial Labeled Base	Labeled Base after Spleen Phosphodiesterase Digestion	
	ANTIPARALLEL	PARALLEL
G	A,T	C,T
C	G,A,G	G,A
T	C,T,G	C,T,A,G
A	T,C,A,T	T,C,A,G

One can determine which model occurs in nature by comparing the pattern in which the labeled phosphate is shifted following spleen phosphodiesterase digestion. Focus your attention on the antiparallel model and notice that the frequency of which "C" (for example) is the 5′ neighbor of "G" is not necessarily the same as the frequency of which "G" is the 5′ neighbor of "C." However, in the parallel model (b), the frequency of which "C" is the 5′ neighbor of "G" is the same as the frequency of which "G" is the 5′ neighbor of "C." By examining such "digestion frequencies," it can be determined that DNA exists in the opposite polarity.

Chapter 12

ANSWERS TO NOW SOLVE THIS

12-1. By having a circular chromosome, no free ends present the problem of linear chromosomes, namely, complete replication of terminal sequences.

12-2. Since eukaryotic chromosomes are "multirepliconic" in that there are multiple replication forks along their lengths, one would expect to see multiple clusters of radioactivity if labeled for a short period of time.

12-3. Volume of the nucleus = 4/3 πr^3

$$= 4/3 \times 3.14 \times (5 \times 10^3 \text{ nm})^3$$

$$= 5.23 \times 10^{11} \text{ nm}^3$$

Volume of the chromosome = $\pi r^2 \times$ length

$$= 3.14 \times 5.5 \text{ nm} \times 5.5 \text{ nm} \times (2 \times 10^9 \text{ nm})$$

$$= 1.9 \times 10^{11} \text{ nm}^3$$

Therefore, the percentage of the volume of the nucleus occupied by the chromatin is

$$= (1.9 \times 10^{11} \text{ nm}^3 / 5.23 \times 10^{11} \text{ nm}^3) \times 100$$

$$= \text{about } 36.3\%$$

SOLUTIONS TO PROBLEMS AND DISCUSSION QUESTIONS

2. Bacteriophage λ has a linear, double-stranded DNA while in the phage coat and, upon infection, closes to form a circular chromosome. It contains about 50 kb. T2 phage also has a linear, double-stranded DNA chromosome; it is less than 200 kb. *E. coli* has a circular, double-stranded DNA chromosome of about 4.2×10^3 kb. Both intact phages are about 1/150 the size of *E. coli*. Since phages are obligate parasites of bacteria, they are dependent on their hosts for the manufacture of materials for their replication. Bacteria contain all genetic information for metabolism, replication, and *de novo* synthesis of numerous life-supporting materials. Phages, on the other hand, contain relatively few genes—namely, those needed to adsorb, inject, and produce progeny using primarily bacterial materials.

4. Most puffs represent active genes as evidenced by staining and uptake of labeled RNA precursors as assayed by autoradiography.

6. Lampbrush chromosomes are typically present in vertebrate oocytes. They are also found in spermatocytes of some insects. They are found as diplotene-stage structures and are active uncoiled versions of condensed meiotic chromosomes.

8. Digestion of chromatin with endonucleases, such as micrococcal nuclease, gives DNA fragments of approximately

200 base pairs or multiples of such segments. X-ray diffraction data indicate a regular spacing of DNA in chromatin. Regularly spaced bead-like structures (nucleosomes) were identified by electron microscopy.

10. As chromosome condensation occurs, a 300 Å fiber is formed. It appears to be composed of 5 or 6 nucleosomes coiled together. Such a structure is called a solenoid. These fibers form a series of loops that further condense into the chromatin fiber, which are then coiled into chromosome arms making up each chromatid.

12. (a) Since there are 200 base pairs per nucleosome (as defined in this problem) and 10^9 base pairs, there would be 5×10^6 nucleosomes. (b) Since there are 5×10^6 nucleosomes and nine histones (including H1) per nucleosome, there must be $9(5 \times 10^6)$ histone molecules: 4.5×10^7. (c) Since there are 10^9 base pairs present and each base pair is 3.4 Å, the overall length of the DNA is 3.4×10^9 Å. Dividing this value by the packing ratio (50) gives 6.8×10^7 Å.

14. One base pair occupies 0.34 nm, therefore the equation would be as follows:

$$52 \ \mu m/(0.34 \ nm/bp) \times 1000 \ nm/\mu m$$
$$= 152{,}941 \text{ base pairs}$$

16. (a) Since pseudogenes do not produce an observable phenotype, their identification by classical mutation analysis is not possible. In the absence of a product, the identification of pseudogenes is mainly dependent on extensive mining of DNA sequence databases to identify sections of DNA that resemble functional genes. In addition, the search for ORF and/or exon sequences is inhibited because of mutations that make up pseudogenes. (b) Because gene spacing, chromatin folding, and various silencer/enhancer domains of chromosomes clearly interact in normal genomic function, pseudogenes may play important roles by sponsoring or inhibiting such interactions. Recent evidence indicates that what can appear to be an inactive pseudogene in one setting (tissue or organism) may well have function in another. Some pseudogenes are transcribed and may serve in gene regulation.

18. First, the methylation patterns displayed by a particular disease must be stable and not change dramatically over time. Second, because DNA methylation patterns differ among individuals, specific patterns must be disease specific and not individual specific. Third, because disease-related changes in DNA methylation patterns often involve hundreds of genes, appropriate interpretation of pattern specifics requires expertise that is presently beyond our capabilities. However, some progress is being made.

20. Cancer cells undergo a variety of changes in histone acetylation compared to normal cells. Overacetylation (hyperacetylation) is often associated with increased gene activity. By introducing HDAC inhibitors, the attempt is to generally decrease gene activity. While the mode of action of HDAC inhibitors is obscure, they are thought to act by promoting cell-cycle arrest by reducing the output of a relatively small number of genes.

22. The intimate relationships among histones, nucleosomes, and DNA in chromatin clearly account for structural remodeling of chromosomes as the cell cycle proceeds from interphase to metaphase. That nucleosomes are associated with chromatin during periods of gene activity begs the question as to the possible roles they play in influencing not only chromosome structure but also gene function. The findings that natural chemical modification of nucleosomal components, as indicated in the question, increases gene activity suggests that changes in the binding of nucleosomes to DNA (in this case due to methylation) enables genes to be more accessible to factors that promote gene function. In addition, the finding that heterochromatin, containing fewer genes and more repressed genes, is undermethylated, further supports the suggestion that histone modification is functionally related to changes in gene activity.

24. Assuming a random distribution, dividing 3×10^9 base pairs by 10^6 gives approximately 3000 base pairs or 3 kbp between *Alu* sequences.

26. The distribution of microsatellites varies in a taxon-related manner. Microsatellites are more common within genes of yeast and fungi and quite infrequent in genes of mammals. There appears to be a general decrease in within-gene microsatellites in more recently evolved organisms.

28. In every case, the child includes at least one locus from the father and one from the mother, which is expected if the alleged father is indeed the real father. Without information as to the frequency of each marker in the general population, it is difficult to draw definitive conclusions.

Chapter 13

ANSWERS TO NOW SOLVE THIS

13-1. (a) The way to determine the fraction of each triplet that will occur with a random incorporation system is to determine the likelihood that each base will occur in each position of the codon (first, second, third), then multiply the individual probabilities (fractions) for a final probability (fraction).

$$GGG = 3/4 \times 3/4 \times 3/4 = 27/64$$
$$GGC = 3/4 \times 3/4 \times 1/4 = 9/64$$
$$GCG = 3/4 \times 1/4 \times 3/4 = 9/64$$
$$CGG = 1/4 \times 3/4 \times 3/4 = 9/64$$

$$CCG = 1/4 \times 1/4 \times 3/4 = 3/64$$
$$CGC = 1/4 \times 3/4 \times 1/4 = 3/64$$
$$GCC = 3/4 \times 1/4 \times 1/4 = 3/64$$
$$CCC = 1/4 \times 1/4 \times 1/4 = 1/64$$

(b) Glycine:
GGG and one G_2C (adds up to 36/64)
Alanine:
one G_2C and one C_2G (adds up to 12/64)
Arginine:
one G_2C and one C_2G (adds up to 12/64)
Proline:
one C_2G and CCC (adds up to 4/64)
(c) With the wobble hypothesis, variation can occur in the third position of each codon.
Glycine: GGG, GGC
Alanine: CGG, GCC, CGC, GCG
Arginine: GCG, GCC, CGC, CGG
Proline: CCC, CCG

13-2. Assume that you have introduced a copolymer (ACACA-CAC . . .) to a cell-free protein-synthesizing system. There are two possibilities for establishing the reading frames:

ACA, if one starts at the first base; and CAC, if one starts at the second base. These would code for two different amino acids (ACA = threonine; CAC = histidine) and would produce repeating polypeptides that would alternate *thr-his-thr-his . . .* or *his-thr-his-thr . . .*

Because of a triplet code, a trinucleotide sequence will, once initiated, remain in the same reading frame and produce the same code all along the sequence regardless of the initiation site.

Given the sequence CUACUACUACUA, notice the different reading frames producing three different sequences, each containing the same amino acid.

Codons:	CUA	CUA	CUA	CUA...
Amino Acids:	leu	leu	leu	leu...
	UAC	UAC	UAC	UAC...
	tyr	tyr	tyr	tyr...
	ACU	ACU	ACU	ACU...
	thr	thr	thr	thr...

If a tetranucleotide is used, such as ACGUACGUACGU...

Codons:	ACG	UAC	GUA	CGU	ACG
Amino Acids:	thr	tyr	val	arg	thr
	CGU	ACG	UAC	GUA	CGU
	arg	thr	tyr	val	arg
	GUA	CGU	ACG	UAC	GUA
	val	arg	thr	tyr	val
	UAC	GUA	CGU	ACG	UAC
	tyr	val	arg	thr	tyr

Notice that the sequences are the same except that the starting amino acid changes.

13-3. Apply complementary bases, substituting U for T:

(a)

Sequence 1:   3′-GAAAAAACGGUA-5′
Sequence 2:   3′-UGUAGUUAUUGA-5′
Sequence 3:   3′-AUGUUCCCAAGA-5′

(b)

Sequence 1:   *met-ala-lys-lys*
Sequence 2:   *ser-tyr-(termination)*
Sequence 3:   *arg-thr-leu-val*

(c) Apply complementary bases:

3′-GAAAAAACGGTA-5′

SOLUTIONS TO PROBLEMS AND DISCUSSION QUESTIONS

2. In eukaryotes, protein synthesis occurs primarily in the cytoplasm, far from the location of DNA and the encoded information. In addition, while some of the basic amino acids would be able to associate directly with DNA, the acidic amino acids would be unable to do so. Thus, some sort of "adaptor" system was needed for DNA to direct amino acid assembly.

4. The UUACUUACUUAC tetranucleotide sequence will produce the following triplets depending on the initiation point: UUA = leu; UAC = tyr; ACU = thr; CUU = leu. Notice that because of the degenerate code, two codons correspond to the amino acid leucine.

The UAUCUAUCUAUC tetranucleotide sequence will produce the following triplets depending on the initiation point: UAU = tyr; AUC = ileu; UCU = ser; CUA = leu. Notice that in this case, degeneracy is not revealed, and all the codons produce unique amino acids.

6. Given that AGG = arg, then information from the AG copolymer indicates that AGA also codes for arg and GAG must therefore code for glu. Coupling this information with that of the AAG copolymer, GAA must also code for glu, and AAG must code for lys.

8.

Original		Substitutions
threonine	----->	*alanine*
_AC(U,C,A, or G)		_GC(U,C,A, or G)
glycine	----->	*serine*
_GG(U or C)		_AG(U or C)
isoleucine	----->	*valine*
_AU(U, C or A)		_GU(U, C or A)

10. Polynucleotide phosphorylase generally functions in the degradation of RNA; however, in an *in vitro* environment, with high concentrations of the ribonucleoside diphosphates, the direction of the reaction can be forced toward polymerization. *In vivo*, the concentration of ribonucleoside diphosphates is low and the degradative process is favored.

12. Applying the coding dictionary, the following sequences are "decoded":

Sequence 1:   met-pro-asp-tyr-ser-(term)
Sequence 2:   met-pro-asp-(term)

The 12th base (a uracil) is deleted from Sequence #1, thereby causing a frameshift mutation, which introduced a terminating triplet UAA.

14.

G G A gly	G G U gly
U G A **term**	G G C gly
C G A **arg**	G G A gly
A G A **arg**	G G G gly
G U A **val**	U G U **cys**
G C A **ala**	C G U **arg**
G A A **glu**	A G U **ser**
G G U gly	G U U **val**
G G C gly	G C U **ala**
G G A gly	G A U **asp**

16. The number of codons for each particular amino acid (synonyms) is directly related to the frequency of amino acid incorporation stated in the problem.

18. First, DNA, the genetic material, is located in the nucleus of a eukaryotic cell, whereas protein synthesis occurs in the cytoplasm. DNA, therefore, does not directly participate in protein synthesis. Second, RNA, which is chemically similar to DNA, is synthesized in the nucleus of eukaryotic cells. Much of the RNA migrates to the cytoplasm, the site of protein synthesis. Third, there is generally a direct correlation between the amounts of RNA and protein in a cell. More direct support was derived from experiments showing that an RNA other than that found in ribosomes was involved in protein synthesis; shortly after phage infection, an RNA species is produced that is complementary to phage DNA.

20. Ribonucleoside triphosphates and a DNA template in the presence of RNA polymerase and a divalent cation (Mg^{++}) produce a ribonucleoside monophosphate polymer, DNA, and pyrophospate (diphosphate). Equimolar amounts of precursor ribonucleoside triphosphates, and product ribonucleoside monophosphates and pyrophosphates (disphosphates) are formed. In *E. coli*, transcription and translation can occur simultaneously. Ribosomes add to the 5′ end nascent mRNA and progress to the 3′ end during translation.

22. The immediate product of transcription of an RNA destined to become a mRNA often involves modification of the 5′ end to which a 7-methylguanosine cap is added. In addition, a stretch of as many as 250 adenylic acid residues is often added to the 3′ end after removal of a AAUAAA sequence. The vast majority of eukaryotic pre-mRNAs also contain intervening sequences that are removed, often in a variety of combinations, during the maturation process. In some organisms, RNA editing occurs in one of two ways: substitution editing where nucleotides are altered and insertion/deletion editing that changes the total number of bases.

Processing location	Example
5′-end	-addition of 7-mG
3′-end	-poly-A addition
internal	-removal of internal sequences
	-RNA editing
	-substitution
	-insertion/deletion

24. Substitution editing occurs when individual nucleotide bases are altered. It is very common in mitochondrial and chloroplast RNAs as well as some nuclear-derived eukaryotic RNAs. Apolipoprotein B occurs in a long and short form, even though a single gene encodes both forms. The initial transcript is edited, which generates a stop codon that terminates the polypeptide at about half its length. The other category is insertion/deletion editing. The parasite that causes African sleeping sickness uses insertion/deletion editing of its mitochondrial RNAs in forming the initiation codon that then places the remaining sequence in a proper reading frame.

26. (a) The first two bases in the triplet code are common to more codons than the last base. If fewer amino acids were used in earlier times, perhaps the first two bases were primarily involved.

(b) Interestingly, all the amino acids mentioned as primitive use guanine as the first base in each codon. We might therefore suppose that the GNN configuration was the starting point of the present-day code. Within the GNN format, the second base is used to distinguish among the amino acids mentioned as primitive.

(c) It is interesting to note that what are considered as the most primitive amino acids are GNN coded, while the later-arriving amino acids are U(U,A,G)(U,C) coded. It would seem that some phenomenon could explain the differences in codon structure between primitive and late-arriving amino acids. It is likely that the addition of new amino acids to the codon field was a slow and mutation-prone process. Did the addition of late-arriving amino acids displace earlier codon assignments, or were some of the late-arriving amino acids able to make use of codons that originally did not code for an amino acid? If this is the case, fewer and more restricted codon assignments would be available for late-arriving amino acids. Notice that each of the late-arriving amino acids has the same starting nucleotide as the present-day stop codons. Is it possible that what were once stop codons were used for late-arriving amino acids because this would be less disruptive to protein synthesis than displacing assignments at the time of their introduction? Much is left to be discovered regarding the structure of the genetic code.

28.
(a) #1: *nonsense mutation*
#2: *missense mutation*
#3: *frameshift mutation*
(b) #1: mutation in third position to A or G
#2: change U to C in third triplet
#3: removal of a G in the UGG triplet (trp)
(c) termination
(d) All of the amino acids can be assigned specific triplets, including the third base of each triplet. Compare the sequences for the wild type and mutant #2. After removal of a G in the UUG triplet of tryptophan, the frameshift mutation shifts the first base of the following triplet to the third (often ambiguous) base of the previous triplet. The only tricky solution is with serine, which has six triplet possibilities, but it can still be resolved.

AUG UGG UAU CGU GGU AGU CCA ACA

(e) The mutation may be in a promoter or enhancer, although many posttranscriptional alterations are possible. Depending on the gene and the organism, the mutation may be in an intron/exon splice site, etc.

30. Consider the following diagram representing the possibility described in the problem. If the promoter is not transcribed, as is the case of typical protein-coding genes, retrotransposition would produce a "daughter" *Alu* void of a promoter. Such an *Alu* would be a "dead-end" because it would not be capable of giving rise to its own *Alu* sequences. Perhaps an *Alu* sequence inserted 3′ to a promoter would produce daughters, but this would likely be rare.

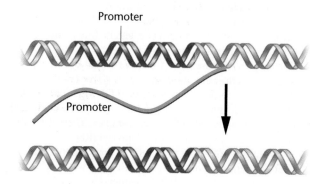

32. The advantage would be that if sequence homologies can be identified for a variety of HIV isolates, then perhaps a single or a few vaccines could be developed for the multitude of subtypes that infect various parts of the world. In other words, the wider the match of a vaccine to circulating infectives is, the more likely the efficacy. On the other hand,

the more finely aligned a vaccine is to the target, the more likely it is that new or previously undiscovered variants will escape vaccination attempts.

Chapter 14

ANSWERS TO NOW SOLVE THIS

14-1. One can conclude that the amino acid is not involved in recognition of the codon.

14-2.

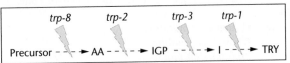

14-3. With the codes for valine being GUU, GUC, GUA, and GUG, single base changes from glutamic acid's GAA and GAG can cause the glu>>>val switch. The normal glutamic acid is a negatively charged amino acid, whereas valine carries no net charge and lysine is positively charged. Given these significant charge changes, one would predict some, if not considerable, influence on protein structure and function. Such changes could stem from internal changes in folding or interactions with other molecules in the RBC, especially other hemoglobin molecules.

SOLUTIONS TO PROBLEMS AND DISCUSSION QUESTIONS

2. A functional polyribosome will contain the following components: mRNA, charged tRNA, large and small ribosomal subunits, elongation and perhaps initiation factors, peptidyl transferase, GTP, Mg^{++}, nascent proteins, and possibly GTP-dependent release factors.

4. It was reasoned that there would not be sufficient affinity between amino acids and nucleic acids to account for protein synthesis. For example, acidic amino acids would not be attracted to nucleic acids. With an adaptor molecule, specific hydrogen bonding could occur between nucleic acids, and specific covalent bonding could occur between an amino acid and a nucleic acid tRNA.

6. Since there are three nucleotides that code for each amino acid, there would be 423 code letters (nucleotides), 426 including a termination codon. This assumes that other features, such as the poly-A tail, the 5′cap, and noncoding leader sequences are omitted.

8. An amino acid in the presence of ATP, Mg^{++}, and a specific aminoacyl synthetase produces an amino acid-AMP enzyme complex (+ PP_i). This complex interacts with a specific tRNA to produce the aminoacyl tRNA.

10. Isoaccepting tRNAs are those tRNAs that recognize and accept only one type of amino acid. In some way, each of the 20 different aminoacyl tRNA synthetases must be able to recognize either the base composition and/or tertiary structure of each of the isoaccepting tRNA species. Otherwise, the fidelity of translation would be severely compromised. The most direct solution to the problem would be to have each synthetase recognize each anticodon. Another reasonable consideration might involve the variable loop, which, in conjunction with the anticodon, might enable such specificity. In reality, several characteristics of each tRNA

are involved: one or more of the anticodon bases, portions of the acceptor arm, and a particular base that lies near the CCA terminus.

12. Both phenylalanine and tyrosine can be obtained from the diet. Most natural proteins contain these amino acids.

14. When an expectant mother returns to consumption of phenylalanine in her diet, she subjects her baby to higher than normal levels of phenylalanine throughout its development. Since increased phenylalanine is toxic, many (approximately 90 percent) newborns are severely and irreversibly retarded at birth. Expectant mothers (who are genetically phenylketonurics) should return to a low-phenylalanine intake during pregnancy.

16.

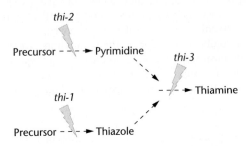

18. The electrophoretic mobility of a protein is based on a variety of factors, primarily the net charge of the protein and to some extent, the conformation in the electrophoretic environment. Both are based on the type and sequence (primary structure) of the component amino acids of a protein. The interactions (hydrogen bonds) of the components of the peptide bonds, hydrophobic, hydrophilic, and covalent interactions (as well as others) are all dependent on the original sequence of amino acids and take part in determining the final conformation of a protein. A change in the electrophoretic mobility of a protein would therefore indicate that the amino acid sequence had been changed.

20. Sickle-cell anemia is called a *molecular* disease because it is well understood at the molecular level; there is a base change in DNA, which leads to an amino acid change in the β chain of hemoglobin. It is a *genetic* disease in that it is inherited from one generation to the next. It is not contagious as might be the case of a disease caused by a microorganism. Diseases caused by microorganisms may not necessarily follow family blood lines, whereas genetic diseases do.

22. It is possible for an amino acid to change without changing the electrophoretic mobility of a protein under standard conditions. If the amino acid is substituted with an amino acid of like charge and similar structure, there is a chance that factors that influence electrophoretic mobility (primarily net charge) will not be altered. Other techniques such as chromatography of digested peptides may detect subtle amino acid differences.

24. All of the substitutions involve one base change.

26. *Colinearity* refers to the sequential arrangement of subunits, amino acids, and nitrogenous bases in proteins and DNA, respectively. Sequencing of genes and products in MS2 phages and studies on mutations in the A subunit of the *tryptophan synthetase* gene indicate a colinear relationship.

28. *Primary*: the linear arrangement or sequence of amino acids. This sequence determines the higher level structures.

   Secondary: α-helix and β-pleated sheet structures generated by hydrogen bonds between components of the peptide bond.

Tertiary: folding that occurs as a result of interactions between the amino acid side chains. These interactions include, but are not limited to, the following: covalent disulfide bonds between cysteine residues, interactions of hydrophilic side chains with water, and interactions of hydrophobic side chains with each other.

Quaternary: the association of two (dimer) or more polypeptide chains. Called *oligomeric*, such a protein is made up of more than one protein chain.

30. Enzymes function to regulate catabolic and anabolic activities of cells. They influence (lower) the *energy of activation*, thus allowing chemical reactions to occur under conditions that are compatible with living systems. Enzymes possess active sites and/or other domains that are sensitive to the environment. The active site is considered to be a crevice, or pit, which binds reactants, thus enhancing their interaction. The other domains mentioned above may influence the conformation and, therefore, the function of the active site.

32. Even though three gene pairs are involved, notice that because of the pattern of mutations, each cross may be treated as monohybrid (a) or dihybrid (b,c).

(a) F_1: AABbCC = speckled

   F_2: 3 AAB_CC = speckled
   1 AAbbCC = yellow

(b) F_1: AABbCc = speckled

   F_2: 9 AAB_C_ = speckled
   3 AAB_cc = green
   3 AAbbC_ = yellow ⎫
   1 AAbbcc = yellow ⎬4

(c) F_1: AaBBCc = speckled

   F_2: 9 A_BBC_ = speckled
   3 A_BBcc = green
   3 aaBBC_ = colorless ⎫
   1 aaBBcc = colorless ⎬4

34. A cross of the following nature would satisfy the data:

$$AABBCC \times aabbcc$$

Offspring in the F_2:

   27 A_B_C_ = purple
   9   A_B_cc = pink
   9   A_bbC_= rose
   9   aaB_C_ = orange
   3   A_bbcc = pink
   3   aaB_cc = pink
   3   aabbC_ = rose
   1   aabbcc = pink

c	*b*	*a*
pink ---> rose ---> orange ---> purple		

The above hypothesis could be tested by conducting a backcross as given below:

$$AaBbCc \times aabbcc$$

The cross should give a
4(pink):2(rose):1(orange):1(purple) ratio

36.

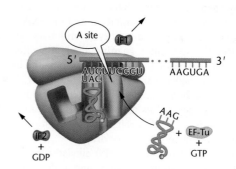

38. (a) It would appear that the gene is inherited as a dominant because to be recessive, all three individuals from outside the blood line would have to be heterozygous, an unlikely event. In addition, one would expect a mutation in a tRNA synthetase to have considerable negative impact on glycine-bearing protein synthesis.

   (b) On the other hand, because affected individuals in the pedigree are heterozygous, sufficient glycyl tRNA synthetase activity must be present to provide vital functions.

   (c) Mutational and physiological perturbations, especially when occurring early in development, often cause neuropathologies. This is because of the highly sensitive nature of nerve cell development, and perhaps because axons are typically quite long, reduction of synthesis of glycine-bearing proteins inhibits the required widespread distribution of such proteins within axons.

Chapter 15

ANSWERS TO NOW SOLVE THIS

15-1. The phenotypic influence of any base change is dependent on a number of factors, including its location in coding or noncoding regions, its potential in dominance or recessiveness, and its interaction with other base sequences in the genome. If a base change is located in a noncoding region, there may be no influence on the phenotype. However, some noncoding regions in a traditional sense may influence other genes and/or gene products. If a mutation that occurs in a coding region acts as a full recessive, there should be no influence on the phenotype. If mutant gene acts as a dominant, then there would be an influence on the phenotype. Some genes interact with other genes in a variety of ways that would be difficult to predict without additional information.

15-2. If a gene is incompletely penetrant, it may be present in a population and only express itself under certain conditions. It is unlikely that the gene for hemophilia behaved in this manner. If a gene's expression is suppressed by another mutation in an individual, it is possible that offspring may inherit a given gene and not inherit its suppressor. Such offspring would have hemophilia. It is possible that the mutation in Queen Victoria's family was new, arising in the father. Lastly, it is possible that the mother was heterozygous, and by chance, no other individuals in her family were unlucky enough to receive the mutant gene.

15-3. Any agent that inhibits DNA replication, either directly or indirectly, through mutation and/or DNA cross linking, will suppress the cell cycle and may be useful in cancer therapy. Since guanine alkylation often leads to mismatched bases, they can often be repaired by a variety of mismatched repair mechanisms. However, DNA cross linking can be repaired by recombinational mechanisms. Thus, for such agents to be successful in cancer therapy, suppressors of DNA repair systems are often used in conjunction with certain cancer drugs.

15-4. Ethylmethane sulfonate (EMS) alkylates the keto groups at the sixth position of guanine and at the fourth position of thymine. In each case, base-pairing affinities are altered and transition mutations result. Altered bases are not readily repaired, and once the transition to normal bases occurs through replication, such mutations avoid repair altogether.

SOLUTIONS TO PROBLEMS AND DISCUSSION QUESTIONS

2. Mutations are the "windows" through which geneticists look at the normal function of genes, cells, and organisms. When a mutation occurs, it allows the investigator to formulate questions as to the function of the normal allele of that mutation. At a different level, mutations provide "markers" with which biologists can study the genetics and dynamics of populations.

4. When conducting genetic screens, one assumes that all the cells of an organism are genetically identical. Therefore, the organism responds to the screen enabling detection of the mutation. Since a somatic mutation first appears in a single cell, it is highly unlikely that the organism will be sufficiently altered to respond to a screen because none of the other cells in that organism will have the same mutation.

6. A gene is likely to be the product of perhaps a billion or so years of evolution. Each gene and its product function in an environment that has also evolved, or coevolved. A coordinated output of each gene product is required for life. Deviations from the norm, caused by mutation, are likely to be disruptive because of the complex and interactive environment in which each gene product must function. However, on occasion a beneficial variation occurs.

8. A *conditional* mutation is one that produces a wild-type phenotype under one environmental condition and a mutant phenotype under a different condition. A conditional *lethal* is a gene that under one environmental condition leads to premature death of the organism.

10. All three of the agents are mutagenic because they cause base substitutions. Deaminating agents oxidatively deaminate bases such that cytosine is converted to uracil and adenine is converted to hypoxanthine. Uracil pairs with adenine, and hypoxanthine pairs with cytosine. Alkylating agents donate an alkyl group to the amino or keto groups of nucleotides, thus altering base-pairing affinities. 6-ethyl guanine acts like adenine, thereby pairing with thymine. Base analogs such as 5-bromouracil and 2-amino purine are incorporated as thymine and adenine, respectively, yet they base-pair with guanine and cytosine, respectively.

12. X rays are of higher energy and shorter wavelength than UV light. They have greater penetrating ability and can create more disruption of DNA.

14. *Photoreactivation* can lead to repair of UV-induced damage. An enzyme, the photoreactivation enzyme, will absorb a photon of light to cleave thymine dimers. *Excision repair* involves the products of several genes, DNA polymerase I and DNA ligase, to clip out the UV-induced dimer, fill in, and join the phosphodiester backbone in the resulting gap. The excision repair process can be activated by damage that distorts the DNA helix.

Recombinational repair is a system that responds to DNA that has escaped other repair mechanisms at the time of replication. If a gap is created on one of the newly synthesized strands, a "rescue operation or SOS response" allows the gap to be filled. Many different gene products are involved in this repair process: *rec*A and *lex*A. In SOS repair, the proofreading by DNA polymerase III is suppressed, and this therefore is called an "error-prone system."

16. There are numerous regions upstream from coding regions in a gene that are sensitive to mutation. Many mutations upset the regions that signal transcription factor and/or polymerase binding, thereby influencing transcription. Mutations within introns may affect intron splicing or other factors that determine mRNA stability or translation.

18. *Xeroderma pigmentosum* is a form of human skin cancer caused by perhaps several rare autosomal genes, which interfere with the repair of damaged DNA. Studies with heterokaryons provided evidence for complementation, indicating that there may be as many as seven different genes involved. The photoreactivation repair enzyme appears to be involved. Since cancer is caused by mutations in several types of genes, interfering with DNA repair can enhance the occurrence of these types of mutations.

20. It is possible that through the reduction of certain environmental agents that cause mutations, mutation rates might be reduced. On the other hand, certain industrial and medical activities actually concentrate mutagens (radioactive agents and hazardous chemicals). Unless human populations are protected from such agents, mutation rates might actually increase. If one asks about the accumulation of mutations (not rates) in human populations as a result of improved living conditions and medical care, then it is likely that as the environment becomes less harsh (through improvements), more mutations will be tolerated as selection pressure decreases. In addition, as individuals live longer and have children at a later age, some studies indicate that older males accumulate more gametic mutations.

22. Transposons cause changes in DNA in a variety of ways, including massive chromosomal alterations. In most cases, changes in DNA are harmful to organisms, while in rare cases, an evolutionary advantage occurs because the new genetic variation confers a selective advantage.

24.

XP1	XP4	XP5
XP2		XP6
XP3		XP7

The groupings (complementation groups) indicate that at least three "genes" form products necessary for unscheduled DNA synthesis. All of the cell lines that are in the same complementation group are defective in the same product.

26. Approximately 78 percent of humans' radiation exposure comes from natural sources. While diagnostic X rays do contribute about 10 percent of the exposure, other forms of man-made forms of radiation contribute only a relatively small amount. That's not to say that man-made radiation

exposure is not a factor in causing mutations; rather, it is not a major factor.

28. Since Betazoids have a 4-letter genetic code and the gene is 3,332 nucleotides long, the protein involved must be 833 amino acids in length.

(*mr-1*) Codon 829 specifies an amino that is very close to the end (carboxyl) of the gene. While a nonsense mutation would terminate translation prematurely, the protein would only be shortened by five amino acids. Thus, the protein's ability to fold and perform its cellular function must not be seriously altered. Because of the direction of translation (5′ to 3′ on the mRNA) the carboxyl terminal amino acids in a protein are the last to be included in folding priorities and are sometimes less significant in determining protein function.

(*mr-2*) Since the phenotype is mild, this amino acid change does not completely inactivate the protein, but it does change its activity to some extent. Perhaps the substitution causes the protein to fold in a slightly aberrant manner, allowing it to have some residual function but preventing it from functioning entirely normally. In addition, even if the protein folds similar to the wild-type protein, charge or structural differences in the protein's active site may be only mildly influenced.

(*mr-3*) This deletion contains a total of 68 nucleotides, which account for 17 amino acids. Since Betazoids' codons contain four nucleotides, the mRNA reading frames are maintained subsequent to the deletion. Protein function significantly depends on the relative positions of secondary levels of structure: α-helices and β-sheets. If the deleted section is a "benign" linker between more significant protein domains, then perhaps the protein can tolerate the loss of some amino acids in a part of the protein without completely losing its function.

(*mr-4*) Amino acid specified by codon 192 must be critical to the function of the protein. Altering this amino acid must disrupt a critical region of the protein, thus causing it to lose most or all of its activity. If the protein is an enzyme, this amino acid could be located in its active site and be critical for the ability of the enzyme to bind and/or influence its substrate. One might expect that the amino acid alteration is rather radical such as one sees in the generation of sickle-cell anemia. HbS is caused by the substitution of a valine (no net charge) for glutamic acid (negatively charged).

(*mr-5*) A deletion of 11 base pairs, a number that is not divisible by four, will shift the reading frames subsequent to its location. Even though this deletion is smaller than the deletion discussed above (83–150) and is located in the same region, it causes a reading frame shift. In addition, some or all of the amino acids that are added downstream from the mutation may be different from those in the normal protein. All will not likely change because of synonyms in the code. There is also the possibility that a nonsense triplet may be introduced in the "out-of-phase" region, thus causing premature chain termination. Because this mutation occurs early in the gene, most of the protein will be affected. This may well explain the severe insensitive phenotype.

30. It is probable that the IS occupied or interrupted the normal function of a controlling region related to the *galactose* genes, which are in an operon with one controlling upstream element.

32. Nonsense mutation in coding regions: shortened product somewhat less than 375 amino acids depending on where the chain termination occurred. Insertion in Exon 1, causing frameshift: a variety of amino acid substitutions and possible chain termination downstream. Insertion in Exon 7, causing frameshift: a variety of amino acid substitutions and possible chain termination downstream involving less of the protein than the insertion in Exon 1. Missense mutation: an amino acid substitution. Deletion in Exon 2, causing frameshift: depending on the size of the deletion, a few too many amino acids may be missing. The frameshift would cause additional amino acid changes and possible termination. Deletion in Exon 2, in frame: amino acids missing from Exon 2 without additional changes in the protein. Large deletion covering Exons 2 and 3: significant loss of amino acids toward the N-terminal side of the protein.

Chapter 16

ANSWERS TO NOW SOLVE THIS

16-1. (a) Due to the deletion of a base early in the *lac* Z gene there will be "frameshift" of all the reading frames downstream from the deletion thereby altering many amino acids. It is likely that either premature chain termination of translation will occur (from the introduction of a nonsense triplet in a reading frame) or the normal chain termination will be ignored. Regardless, a mutant condition for the *Z* gene will be likely. If such a cell is placed on a lactose medium, it will be incapable of growth because β-galactosidase is not available. (b) If the deletion occurs early in the *A* gene, one might expect impaired function of the *A* gene product, but it will not influence the use of lactose as a carbon source.

16-2. (a) With no lactose and no glucose, the operon is off because the *lac* repressor is bound to the operator and although CAP is bound to its binding site, it will not override the action of the repressor.

(b) With lactose added to the medium, the *lac* repressor is inactivated and the operon is transcribing the structural genes. With no glucose, the CAP is bound to its binding site, thus enhancing transcription.

(c) With no lactose present in the medium, the *lac* repressor is bound to the operator region, and since glucose inhibits adenyl cyclase, the CAP protein will not interact with its binding site. The operon is therefore "off."

(d) With lactose present, the *lac* repressor is inactivated; however, since glucose is also present, CAP will not interact with its binding site. Under this condition transcription is severely diminished and the operon can be considered to be "off."

16-3. (a) With tryptophan being abundant, the assumption is that charged tRNA^trp is also present. TRAP should be saturated with tryptophan and be actively bound to the 5′ end of the nascent mRNA. Therefore, the structural genes should not be expressed. The *AT* gene is not induced.

(b) If tryptophan is scarce, even though tRNA^trp is present it should not be charged. TRAP is present, but it is not saturated with tryptophan. The structural genes should be expressed. Uncharged tRNA^trp leads to the induction of the *AT* gene.

(c) The answer to this situation needs some qualification. If tryptophan is abundant, but tRNA^{trp} is scarce, the assumption is that charged tRNA^{trp} is also scarce. This could be due to a nonfunctional tryptophanyl-tRNA synthetase. In that case, the uncharged tRNA^{trp} can induce *AT*, which then binds to tryptophan-saturated TRAP. This prevents TRAP from binding to the leader RNA sequence, thus allowing expression of the tryptophan operon.

(d) With no TRAP, there can be no termination of transcription. Therefore under this condition, even with abundant tryptophan, there is expression of the operon. The *AT* gene is not induced because it is assumed that with tryptophan abundant, charged tRNA^{trp} would also be abundant.

SOLUTIONS TO PROBLEMS AND DISCUSSION QUESTIONS

2. The enzymes of the lactose operon are needed to break down and use lactose as an energy source. If lactose is the sole carbon source, the enzymes are synthesized to use that carbon source. With no lactose present, there is no "need" for the enzymes. The tryptophan operon contains structural genes for the *synthesis* of tryptophan. If there is little or no tryptophan in the medium, the tryptophan operon is "turned on" to manufacture tryptophan. If tryptophan is abundant in the medium, then there is no "need" for the operon to be manufacturing "tryptophan synthetases."

4. In an *inducible system*, the repressor that normally interacts with the operator to inhibit transcription is inactivated by an *inducer*, thus permitting transcription. In a *repressible system*, a normally inactive repressor is *activated* by a co-repressor, thus enabling it (the activated repressor) to bind to the operator to inhibit transcription. Because the interaction of the protein (repressor) has a negative influence on transcription, the systems described here are forms of *negative control*.

6. $I^+ \ O^+ \ Z^+$ = Because of the function of the active repressor from the I^+ gene, and no lactose to influence its function, there will be **No Enzyme Made**.

$I^+ \ O^c \ Z^+$ = There will be a **Functional Enzyme Made** because the constitutive operator is in *cis* with a *Z* gene. The lactose in the medium will have no influence because of the constitutive operator. The repressor cannot bind to the mutant operator.

$I^- \ O^+ \ Z^-$ = There will be a **Nonfunctional Enzyme Made** because with I^- the system is constitutive, but the *Z* gene is mutant. The absence of lactose in the medium will have no influence because of the nonfunctional repressor. The mutant repressor cannot bind to the operator.

$I^- \ O^+ \ Z^-$ = There will be a **Nonfunctional Enzyme Made** because with I^- the system is constitutive, but the *Z* gene is mutant. The lactose in the medium will have no influence because of the nonfunctional repressor. The mutant repressor cannot bind to the operator.

$I^- \ O^+ \ Z^+/F' \ I^+$ = There will be **No Enzyme Made** because in the absence of lactose, the repressor product of the I^+ gene will bind to the operator and inhibit transcription.

$I^+ \ O^c \ Z^+/F' \ O^+$ = Because there is a constitutive operator in *cis* with a normal *Z* gene, there will be a **Functional Enzyme Made**. The lactose in the medium will have no influence because of the mutant operator.

$I^+ \ O^+ \ Z^-/F' \ I^+ \ O^+ \ Z^+$ = Because there is lactose in the medium, the repressor protein will not bind to the operator and transcription will occur. The presence of a normal *Z* gene allows a **Functional and Nonfunctional Enzyme to be Made**. The repressor protein is diffusible, working in trans.

$I^- \ O^+ \ Z^-/F' \ I^+ \ O^+ \ Z^+$ = Because there is no lactose in the medium, the repressor protein (from I^+) will repress the operators and there will be **No Enzyme Made**.

$I^s \ O^+ \ Z^+/F' \ O^+$ = With the product of I^s there is binding of the repressor to the operator and therefore **No Enzyme Made**. The lack of lactose in the medium is of no consequence because the mutant repressor is insensitive to lactose.

$I^+ \ O^c \ Z^+/F' \ O^+ \ Z^+$ = The arrangement of the constitutive operator (O^c) with the *Z* gene will cause a **Functional Enzyme to be Made**.

8. A single *E. coli* cell contains very few molecules of the *lac* repressor. However, the *lac* I^q mutation causes a 10X increase in repressor protein production, thus facilitating its isolation. With the use of dialysis against a radioactive gratuitous inducer (IPTG), Gilbert and Muller-Hill were able to identify the repressor protein in certain extracts of *lac* I^q cells. The material that bound the labeled IPTG was purified and shown to be heat labile and have other characteristics of protein. Extracts of *lac* I^- cells did not bind the labeled IPTG.

10. (a) Because activated CAP is a component of the cooperative binding of RNA polymerase to the *lac* promoter, absence of a functional *crp* would compromise the positive control exhibited by CAP.

(b) Without a CAP binding site there would be a reduction in the inducibility of the *lac* operon.

12. Attenuation functions to reduce the synthesis of tryptophan when it is in full supply. It does so by reducing transcription of the *tryptophan* operon. The same phenomenon is observed when tryptophan activates the repressor to shut off transcription of the *tryptophan* operon.

14. Neelaredoxin appears to be a protein that defends anaerobic and perhaps aerobic organisms from oxidative stress brought on by the metabolism of oxygen. The generation of oxygen free radicals (creates the oxidative stress) is dependent on several molecular species including O_2 and H_2O_2. Apparently, relatively high levels of neelaredoxin are produced at all times (*constitutively expressed*) even when potential inducers of gene expression are not added to the system. Additional neelaredoxin gene expression is not responsive (*induced*) as a result of O_2 and H_2O_2 treatment.

16. When arabinose is present in the medium, the structural genes for the *arabinose* operon are transcribed. If the structural genes for the *lac* operon replaced the structural genes for the *ara* operon, then in the presence of arabinose, the *lac* structural genes would be transcribed and β-galactosidase would be produced at induced levels.

18. Because a repressor stops transcription, one would consider this system, as described here, as being under negative control.

20. The system is *inducible*. *C* codes for the **structural gene**. *B* must be the **promoter**. The *A* locus is the **operator**, and the *D* locus is the **repressor** gene.

22. The first two sentences in the problem indicate an inducible system where oil stimulates the production of a protein, which turns on (positive control) genes to metabolize oil. The different results in strains #2 and #4 suggest a *cis*-acting system. Because the operon by itself (when mutant as in strain #3) gives constitutive synthesis of the structural genes, *cis*-acting system is also supported. The *cis*-acting element is most likely part of the operon.

24. If one could develop an assay for the other gene products under SOS control, with a *lexA*⁻ strain the other gene products should be present at induced levels.

26. (a) The simplest model for the action of R and D in *Chlamydia* would be to have the repressor element (R) become ineffective in binding the *cis*-acting element (D) when heat-shocked. This could happen in two ways. Either the supercoiled DNA alters its conformation and becomes ineffective at binding R, or the D-binding efficiency of the R protein is altered by heat. In either case, the genes for infectivity are transcribed in the presence of the heat shock. (b) The most straightforward comparison between the heat shock R and D system in *Chlamydia* and the heat-shock sigma factor in *E. coli* would be where the R-D system is inactivated in *Chlamydia* and a sigma factor is activated by heat in *E. coli*.

Chapter 17

ANSWERS TO NOW SOLVE THIS

17-1. Cancer cells often originate under the influence of mutations in tumor-suppressor genes or proto-oncogenes. Should hypermethylation occur in one of many DNA repair genes, the frequency of mutation would increase because the DNA repair system is compromised. The resulting increase in mutations might occur in tumor-suppressor genes or proto-oncogenes.

17-2. General transcription factors associate with a promoter to stimulate transcription of a specific gene. Some *trans*-acting elements, when bound to enhancers, interact with coactivators to enhance transcription by forming an enhanceosome that stimulates transcription initiation. Transcription can be repressed when certain proteins bind to silencer DNA elements and generate repressive chromatin structures. The same molecule may bind to a different chromosomal regulatory site (enhancer or silencer), depending on the molecular environment of a given tissue type.

17-3. Mutations in the *tra* gene of *Drosophila* can dramatically alter development such that a normal female is produced if the TRA protein is present and a male develops when the TRA protein is absent. A null *tra* allele would produce males because of male-specific splicing while a constitutively active *tra* gene would produce females.

SOLUTIONS TO PROBLEMS AND DISCUSSION QUESTIONS

2. Eukaryotic cells contain greater amounts of DNA, and this DNA is associated with various proteins, including histones and nonhistone chromosomal proteins. *Chromatin* as such does not exist in prokaryotes. In addition, whereas there is usually only one chromosome in prokaryotes, eukaryotes

have more than one chromosome all enclosed in a membrane (nuclear membrane). This nuclear membrane separates, both temporally and spatially, the processes of transcription and translation, thus providing an opportunity for posttranscriptional, pretranslational regulation.

While prokaryotes respond genetically to changes in their external environment, cells of multicellular eukaryotes interact with each other as well as the external environment. The structural and functional diversity of cells of a multicellular eukaryote, coupled with the finding that all cells of an organism contain a complete complement of genes, suggests that in some cells certain genes are active that are not active in other cells.

It is often difficult to study eukaryotic gene regulation because of the complexities mentioned above, especially tissue specificity and the various levels at which regulation can occur. Obtaining a homogeneous group of cells from a multicellular organism often requires a significant alteration of the natural environment of the cell. Thus, results from studies on isolated cells must be interpreted with caution. In addition, because of the variety of intracellular components (nuclear and cytoplasmic), it is difficult to isolate, free of contamination, certain molecular species. Even if such isolation is accomplished, it is difficult to interpret the actual behavior of such molecules in an artificial environment.

4. Chromatin is remodeled when there are significant changes in chromatin organization. Such remodeling involves changes in DNA methylation and interaction of DNA with histones in nucleosomes. Nucleosome remodeling complexes alter nucleosome structure and position by a number of processes, including histone modification and the action of enhancers/silencers.

6. When DNA is transcriptionally active, it is in a less condensed state and as such, more open to DNase digestion.

8. In general, chromatin is remodeled when there are significant changes in chromatin organization. Such remodeling involves changes in DNA methylation and interaction of DNA with histones in nucleosomes. Nucleosome remodeling complexes alter nucleosome structure and position by a number of processes, including histone modification.

10. Transcription factors are proteins that are *necessary* for the initiation of transcription. However, they are not *sufficient* for the initiation of transcription. To be activated, RNA polymerase II requires a number of transcription factors. Transcription factors contain at least two functional domains: one binds to the DNA sequences of promoters and/or enhancers, while the other interacts with RNA polymerase or other transcription factors. Some transcription factors bind to other transcription factors without themselves binding to DNA.

12. Generally, one determines the influence of various regulatory elements by removing necessary elements or adding extra elements. In addition, examining the outcome of mutations within such elements often provides insight as to function. Assay systems determine the relative levels of gene expression after such alterations.

14. RNA interference begins with a double-stranded RNA being processed by a protein called Dicer that, in combination with RISC, generates short interfering RNA (siRNA). Unwinding of siRNA produces an antisense strand that combines with a protein to cleave mRNA complementary sequences. Short

RNAs called microRNAs pair with the 3'-untranslated regions of mRNAs and blocks their translation.

16. Inr, BRE, DPE, and MTE sequences are double-stranded DNA elements located in the transcription start site (Inr), immediately upstream or downstream from the TATA box (BRE), or downstream (DPE, MTE) from the transcription site (+18 to +27 and +28 to +33, respectively).

18. Given that DNA methylation plays a role in gene expression in mammals, any change in DNA methylation, plus or minus, can potentially have a negative impact on progeny development. In addition, since m5C >>> thymine, transitions are likely to cause mutations in coding regions of DNA, when methylation patterns change, new sites for mutation arise. Should mutations occur at a higher rate in previously unmethylated sites (genes), embryonic development is likely to be affected.

20. Since there are multiple routes that lead to cancer, one would expect complex regulatory systems to be involved. More specifically, while in some cases, downregulation of a gene, such as an oncogene, may be a reasonable cancer therapy, downregulation of a tumor-suppressor gene would be undesirable in therapy.

22. Below is a sketch of several RNA polymerase molecules (filled circles) in what might be a transcription factory. This diagram shows eight RNA pol II molecules being transcribed. Nascent transcripts are shown projecting from the RNA pol II molecules. For simplicity, only one promoter is shown and one structural gene is shown.

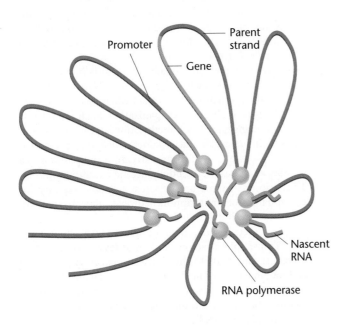

24. Since mRNA stability is directly related to the likelihood of translation, one could use a number of different constructs as shown below and test for luciferase activity in the assay system described in the problem.

5'-cap	3'-tail	mRNA (no cap, no tail)
−	−	+
+	−	+
−	+	+
+	−	−
−	+	−

26. If mRNA transport is achieved by diffusion, then regulation could be achieved by the positioning of different concentrations of ribosomes. The closer a ribosome pool is to the source of a particular mRNA source, the higher the likelihood that a given mRNA would be translated. In addition, it may be that certain territories in cells are less hostile to traveling mRNAs that would also favor the translation of certain mRNAs. If nuclear territories purge their RNAs at different locations around the nucleus, then even if diffusion is the mode of mRNA transport, different cellular domains would receive different types of mRNAs. If such domains have different translation efficiencies, then genetic regulation is achieved.

28. When splice specificity is lost, one might observe several classes of altered RNAs: (1) a variety of nonspecific variants producing RNA pools with many lengths and combinations of exons and introns, (2) incomplete splicing where introns and exons are erroneously included or excluded in the mRNA product, and (3) a variety of nonsense products, which result in premature RNA decay or truncated protein products. It is presently unknown as to whether cancer-specific splices initiate or result from tumorigenesis. Given the complexity of cancer induction and the maintenance of the transformed cellular state, gene products that are significant in regulating the cell cycle may certainly be influenced by alternative splicing and thus contribute to cancer.

30. The most direct way to determine whether the newly discovered GRE sequence in the human β-globin is necessary for transcription would be to assemble an assay system that carries the GRE sequence and compare it to one that lacks the sequence. Because GRE (glucocorticoid response element) is influenced by a glucocorticoid receptor protein only when bound to the glucocorticoid hormone, additional experimental approaches are available. Simply put, is the human β-globin gene responsive to the stimulatory action of the glucocorticoid hormone? If not, it is not likely to be necessary for accurate human β-globin expression. In the hMTIIA environment, gene transcription is stimulated when the cytoplasmic receptor binds to the hormone that allows it to enter the nucleus and bind to GRE. Addition of the glucocorticoid hormone to both the human β-globin and hMTIIA systems would indicate, in assays, whether each system responded similarly.

Chapter 18

ANSWERS TO NOW SOLVE THIS

18-1. It is possible that your screen was more inclusive; that is, it identified more subtle alterations than the screen of others. In addition, your screen may have included some zygotic effect mutations, which were dependent on the action of maternal-effect genes. You may have identified several different mutations (multiple alleles) in some of the same genes.

18-2. Because in *ftz/ftz* embryos, the engrailed product is absent and in *en/en* embryos *ftz* expression is normal, one can conclude that the *ftz* gene product regulates, either directly or indirectly, *en*. Because the *ftz* gene is expressed normally in *en/en* embryos, the product of the *engrailed* gene does not regulate expression of *ftz*.

18-3. Because signal-receptor interactions depend on membrane-bound structures, the pathway can only work with adjacent cells. The advantage of such a system is that only cells in a certain location will be influenced—those in contact. A disadvantage would occur if large groups of cells are to be induced into a particular developmental pathway or if cells not in contact need to be induced.

SOLUTIONS TO PROBLEMS AND DISCUSSION QUESTIONS

2. *Determination* refers to early developmental and regulatory events that set eventual patterns of gene activity. Determination is not the end result of the regulatory activity; rather, it is the process by which the developmental fate of a particular cell type is fixed. *Differentiation*, on the other hand, follows determination and is the manifestation, in terms of genetic, physiological, and morphological changes, of the determined state.

4. The syncytial blasterm is formed as nuclei migrate to the egg's outer margin or cortex, where additional divisions take place. Plasma membranes organize around each of the nuclei at the cortex, thus creating the cellular blastoderm.

6. (a,b) Zygotic genes are activated or repressed depending on their response to maternal-effect gene products. Three subsets of zygotic genes divide the embryo into segments. These segmentation genes are normally transcribed in the developing embryo, and their mutations have embryonic lethal phenotypes.

(c) The maternal genotype contains zygotic genes, and these are passed to the embryo as with any other gene.

8. Because the polar cytoplasm contains information to form germ cells, one would expect such a transplantation procedure to generate germ cells in the anterior region. Work done by Illmensee and Mahowald in 1974 verified this expectation.

10. First, one may determine whether levels of hnRNA are consistent among various cell types of interest. This is often accomplished by either direct isolation of the RNA and assessment by northern blotting or by use of *in situ* hybridization. If the hnRNA pools for a given gene are consistent in various cell types, then transcriptional control can be eliminated as a possibility. Support for translational control can be achieved directly by determining, in different cell types, the presence of a variety of mRNA species with common sequences. This can be accomplished only in cases where sufficient knowledge exists for specific mRNA trapping or labeling. Clues as to translational control *via* alternative splicing can sometimes be achieved by examining the amino acid sequence of proteins. Similarities in certain structural/functional motifs may indicate alternative RNA processing.

12. A dominant gain-of-function mutation is one that changes the specificity or expression pattern of a gene or gene product. The "gain-of-function" *Antp* mutation causes the wild-type *Antennapedia* gene to be expressed in the eye-antenna disc, and mutant flies have legs on the head in place of antenna.

14. Because of the regulatory nature of *homeotic* genes in fundamental cellular activities of determination and differentiation, it would be difficult to ignore their possible impact on oncogenesis. Homeotic genes encode DNA binding domains, which influence gene expression; any factor that influences gene expression may, under some circumstances, influence cell cycle control.

However attractive this model, there have been no homeotic transformations noted in mammary glands, so the typical expression of mutant homeotic genes in insects is not revealed in mammary tissue, according to Lewis (2000). A substantial number of experiments will be needed to establish a functional link between homeotic gene mutation and cancer induction. Mutagenesis and transgenesis experiments are likely to be most productive in establishing a cause-effect relationship.

16. Three classes of flower *homeotic* genes are known that are activated in an overlapping pattern to specify various floral organs. Class *A* genes give rise to sepals. Expression of *A* and *B* class genes specify petals, *B* and *C* genes control stamen formation, and expression of *C* genes gives rise to carpels.

18.

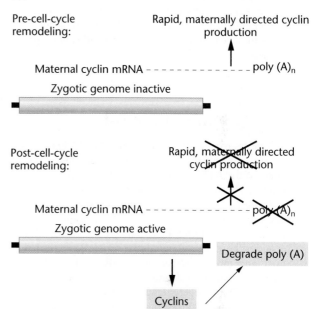

20. (a) The term *rescued* is often used when the introduction of genes from an outside source (within or among species) restores the wild-type phenotype from a mutant organism.

(b) Results such as these, and there are many like them, indicate the extreme conservation of protein structure and function across phylogenetically distant organisms. Such results attest to the conservation from a distant common ancestor of fundamental molecular species during development. Failure to adhere to a common developmental theme is rewarded by death.

22. Since *her-1⁻* mutations cause males to develop into hermaphrodites, and *tra-1⁻* mutations cause hermaphrodites to develop into males, one may hypothesize that the *her-1⁺* gene produces a product that suppresses hermaphrodite development, while the *tra-1⁺* gene product is needed for hermaphrodite development.

24. (a) The patterns somewhat follow expectations in that genes involved in photosynthesis appear to be most active in the leaf and flower. In addition, high levels of protein synthesis in root is expected. Since pollen and seeds would be called upon early in development, it would seem reasonable to have such tissues show strong expression of transcriptional regulators.

(b) Two factors may be driving the expression of photosynthetic gene expression in flowers and seeds. First, since development in plants is more continuous or sequential, genes may not necessarily be shut off when their products are not in demand. Second, perhaps various posttranscriptional modifications redirect the function of photosynthetic gene products. In other words, alternative uses for such gene products may have evolved in plants much like that observed in some animal tissues.

(c) Having a global view of gene expression in an organism and its various tissues and organs allows one to estimate the relative contribution a particular gene may have in development. Since developmental and transcriptional networks are common in development, it is helpful to know the context in which a particular gene may function. Lastly, the only way in which development will be understood at the molecular level is to have a broad view of classes and clusters of gene activity.

Chapter 19

ANSWERS TO NOW SOLVE THIS

19-1. Several approaches are used to combat CML. One includes the use of a tyrosine kinase inhibitor that binds competitively to the ATP binding site of ABL kinase, thereby inhibiting phosphorylation of BCR-ABL and preventing the activation of additional signaling pathways. In addition, real-time quantitative reverse transcription-polymerase chain reaction (Q-RT-PCR) allows one to monitor the drug responses of cell populations in patients so that less toxic and more effective treatments are possible. Being able to distinguish leukemic cells from healthy cells allows one to not only target therapy to specific cell populations, but also to quantify responses to therapy. Because such cells produce a hybrid protein, it may be possible to develop a therapy, perhaps an immunotherapy, based on the uniqueness of the BCR/ABL protein.

19-2. *p53* is a tumor-suppressor gene that protects cells from multiplying with damaged DNA. It is present in its mutant state in more than 50 percent of all tumors. Since the immediate control of a critical and universal cell cycle checkpoint is mediated by *p53*, mutation will influence a wide range of cell types. *p53*'s action is not limited to specific cell types.

19-3. Cancer is a complex alteration in normal cell-cycle controls. Even if a major "cancer-causing" gene is transmitted, other genes, often new mutations, are usually necessary in order to drive a cell toward tumor formation. Full expression of the cancer phenotype is likely to be the result of an interplay among a variety of genes and therefore show variable penetrance and expressivity.

19-4. Unfortunately, it is common to spend enormous amounts of money dealing with diseases after they occur rather than concentrating on disease prevention. Too often, pressure from special interest groups or lack of political will retard advances in education and prevention. Obviously, it is less expensive, both in terms of human suffering and money, to seek preventive measures for as many diseases as possible. However, having gained some understanding

of the mechanisms of disease, in this case cancer, it must also be stated that no matter what preventive measures are taken it will be impossible to completely eliminate disease from the human population. It is extremely important, however, that we increase efforts to educate and protect the human population from as many hazardous environmental agents as possible. A balanced, multipronged approach seems appropriate.

SOLUTIONS TO PROBLEMS AND DISCUSSION QUESTIONS

4. Kinases regulate other proteins by adding phosphate groups. Cyclins bind to the kinases, switching them on and off. CDK4 binds to cyclin D, moving cells from G1 to S. At the G2/mitosis border a CDK1 (cyclin-dependent kinase) combines with another cyclin (cyclin B). Phosphorylation occurs, bringing about a series of changes in the nuclear membrane *via* caldesmon, cytoskeleton, and histone H1.

6. To say that a particular trait is inherited conveys the assumption that when a particular genetic circumstance is present, it will be revealed in the phenotype. When one discusses an inherited predisposition, one usually refers to situations where a particular phenotype is expressed in families in some consistent pattern. However, the phenotype may not always be expressed or may manifest itself in different ways. In retinoblastoma, the gene is inherited as an autosomal dominant, and those that inherit the mutant *RB* allele are predisposed to develop eye tumors. However, approximately 10 percent of the people known to inherit the gene do not actually express it, and in some cases expression involves only one eye, rather than two.

8. Apoptosis, or programmed cell death, is a genetically controlled process that leads to death of a cell. It is a natural process involved in morphogenesis and a protective mechanism against cancer formation. During apoptosis, nuclear DNA becomes fragmented, cellular structures are disrupted, and the cells are dissolved. Caspases are involved in the initiation and progress of apoptosis.

10. The nonphosphorylated form of pRB binds to transcription factors such as E2F, causing inactivation and suppression of the cell cycle. Phosphorylation of pRB activates the cell cycle by releasing transcription factors (E2F) to advance the cell cycle. With the phosphorylation site inactivated in the PSM-RB form, phosphorylation cannot occur, thereby leaving the cell cycle in a suppressed state.

12. Various kinases can be activated by breaks in DNA. One kinase, called ATM and/or a kinase called Chk2, phosphorylates BRCA1 and p53. The activated p53 arrests replication during the S phase to facilitate DNA repair. The activated BRCA1 protein, in conjunction with BRCA2, mRAD51, and other nuclear proteins, is involved in repairing the DNA.

14. Mutations that produce oncogenes alter gene expression either directly or indirectly and act in a dominant capacity. Proto-oncogenes are those that normally function to promote or maintain cell division. In the mutant state (oncogenes), they induce or maintain uncontrolled cell division; that is, there is a gain-of-function. Generally this gain-of-function takes the form of increased or abnormally continuous gene output. On the other hand, loss-of-function is generally attributed to mutations in tumor-suppressor genes, which function to halt passage through the cell cycle. When such genes are mutant, they have lost their capacity to halt the cell cycle. Such mutations are generally recessive.

16. To encourage infected cells to undergo growth and division, viruses often encode genes that stimulate growth and division. Many viruses either inactivate tumor-suppressor genes of the host or bring in genes that stimulate cell growth and division. By inactivating tumor-suppressor genes, the normal breaking mechanism of the cell cycle is destroyed.

18. Normal cells are often capable of withstanding mutational assault because they have checkpoints and DNA repair mechanisms in place. When such mechanisms fail, cancer may be a result. Through mutation, such protective mechanisms are compromised in cancer cells, and as a result they show higher than normal rates of mutation, chromosomal abnormalities, and genomic instability.

20. Epigenetic effects can be caused by DNA methylation and/or histone modifications, including acetylation and/or phosphorylation. As such, they can silence or activate chromosomes (X chromosome, for example) or certain chromosomal regions and be responsible for parental imprinting or influencing gene activity in heterochromatin. Patterns of nucleotide demethylation and hypermethylation are often different when cancer cells are compared to normal cells.

22. No, she will still have the general population risk of about 10 percent. In addition, it is possible that genetic tests will not detect all breast cancer mutations.

24. Proteases in general and serine proteases, specifically, are considered tumor-promoting agents because they degrade proteins, especially those in the extracellular matrix. When such proteolysis occurs, cellular invasion and metastasis is encouraged. Consistent with this observation are numerous observations that metastatic tumor cells are associated with higher than normal amounts of protease expression. Inhibitors of serine proteases are often tested for their anticancer efficacy.

26. (a) Because one is working with somatic cells, the usual tests for heterozygosity through crosses are not available. Therefore, one must rely on chemical/physical approaches to answer the question. A genomic library could be constructed of both osteosarcoma cell DNA and noncancerous cells from the same organism. You could then screen the library using labeled probes from the clones carrying the *RB1* gene available to you as stated in the problem. At this point, some indications might emerge because if there is a significant alteration in mutant *RB1* genes, probes may not successfully hybridize to any clones in the cancerous cell DNA library. Assuming that control hybridization occurs in the noncancerous cells, lack of hybridization in the library derived from the osteosarcoma cell line might indicate deletions. However, assuming that hybridization does allow one to identify clones containing putative *RB1* alleles, subcloning into appropriate vectors would allow sequencing to reveal sequence changes in the *RB1* alleles when compared with nonmutant genes. A second approach combines an immunoassay described in part (b) of this problem. Assuming that one can successfully make antibodies to the normal RB1 gene product (pRB), lack of cross-reactivity of the pRB antibodies to proteins from the cancerous cell line would indicate that both *RB1* alleles are mutant.

(b) As indicated in the last portion of part (a) above, one can make antibodies to pRB from the noncancerous cells and test these antibodies for reactivity against proteins from the cancerous cell lines. A pRB-antibody reaction would indicate that the pRB protein is made.

(c) To determine whether addition of a normal *RB1* gene will change the cancer-causing potential of osteosarcoma cells, one could transfer the cloned normal *RB1* gene into the cells by transformation or transfection (often by electroporation or ultrasound). Transformed cells would then be introduced into the cancer-prone mice to determine whether their cancer-causing potential had been altered.

28. (a,b) Even though there are changes in the *BRCA1* gene, they do not always have physiological consequences. Such neutral polymorphisms make screening difficult in that one cannot always be certain that a mutation will cause problems for the patient.

(c) The polymorphism in *PM*2 is probably a silent mutation because the third base of the codon is involved.

(d) The polymorphism in *PM*3 is probably a neutral missense mutation because the first base is involved. However, because there is some first codon position degeneracy, it is possible for the mutation to be silent.

Chapter 20

ANSWERS TO NOW SOLVE THIS

20-1. (a) Bacteria that have been transformed with the recombinant plasmid will be resistant to tetracycline, and therefore tetracycline should be added to the medium.

(b) Colonies that grow on a tetracycline medium only should contain the insert. Those bacteria that do not grow on the ampicillin medium probably contain the *Drosophila* DNA insert.

(c) Resistance to both antibiotics by a transformed bacterium could be explained in several ways. First, if cleavage with the *Pst*I was incomplete, then no change in biological properties of the uncut plasmids would be expected. Also, it is possible that the cut ends of the plasmid were ligated together in the original form with no insert.

20-2. A filter is used to bind the DNA from the colonies containing recombinant plasmids. A labeled probe is constructed for the protein sequence of EF1a. Since it is highly conserved, it should show considerable complementation to the human EF-1a cDNA. It is used to detect, through hybridization, the DNA of interest. Cells with the desired clone are then picked from the original plate and the plasmid is isolated from the cells.

20-3. Because there are three hydrogen bonds involved in GC pairs, it takes more energy to separate GC-rich DNA than AT-rich DNA. In annealing a primer to the template, it is usually the practice to select the highest, yet most stable, annealing temperature. This practice minimizes the likelihood of imperfect annealing. Knowing the percentage of GC base pairs allows one to maximize precise annealing.

20-4. Often, the cloning vector itself contains known sequences that flank inserted DNA. Under that condition, it is easy to use standard primers for that particular vector site. Since much is known about mammalian genes in general, it is likely that genes of interest have already been studied in some detail in related species. If that is the case, sequences could be selected that flank or include the DNA of interest and successful PCR cloning might be achieved. In some cases of sequence uncertainty, a set of degenerate primers can be used. Since primers contain an array of similar sequences, one of the set might be efficient for priming a PCR.

SOLUTIONS TO PROBLEMS AND DISCUSSION QUESTIONS

2. Particular enzymes, called *restriction endonucleases*, cut DNA at specific sites and often yield "sticky" ends for additional interaction with DNA molecules cut with the same class of enzyme. Isolated from bacteria, restriction enzymes fall into several classes, each having peculiarities as to structure and interaction with DNA. A vector may be a plasmid, bacteriophage, or cosmid that receives, through ligation, a piece or pieces of foreign DNA. The recombinant vector can transform (or transfect) a host cell (bacterium, yeast cell, etc.) and be amplified in number.

4. In a DNA cloning experiment, DNA ligase is used to generate the covalent bonds of the phosphodiester backbone to yield an intact double-stranded DNA molecule. Restriction enzymes, on the other hand, break such bonds.

6. This segment contains the palindromic sequence of GGATCC, which is recognized by the restriction enzyme *Bam*HI. The double-stranded sequence is the following:

    CCTAGG
    GGATCC

8. Plasmids were the first to be used as cloning vectors, and they are still routinely used to clone relatively small fragments of DNA. Because of their small size, they are relatively easy to separate from the host bacterial chromosome, and they have relatively few restriction sites. They can be engineered fairly easily (i.e., polylinkers and reporter genes added). For cloning larger pieces of DNA such as entire eukaryotic genes, cosmids are often used. For instance, when modifications are made in the bacterial virus lambda (λ), relatively large inserts of about 20 kb can be cloned. This is an important advantage when one needs to clone a large gene or generate a genomic library from a eukaryote. In addition, some cosmids will only accept inserts of a limited size, which means that small, less meaningful perhaps, fragments will not be cloned unnecessarily. Both plasmids and cosmids suffer from the limitation that they can only use bacteria as hosts. BACs are artificial bacterial chromosomes that can be engineered for certain qualities.

    YACs (yeast artificial chromosomes) contain telomeres, an origin of replication, and a centromere and are extensively used to clone DNA in yeast. With selectable markers (TRP1 and URA3) and a cluster of restriction sites, DNA inserts ranging from 100 kb to 1000 kb can be cloned and inserted into yeast. Since yeast, being a eukaryote, undergoes many of the typical RNA and protein processing steps of other, more complex eukaryotes, the advantages are numerous when working with eukaryotic genes.

10. No. The tumor-inducing plasmid (Ti) that is used to produce genetically modified plants is specific for the bacterium *Agrobacterium tumifaciens*, which causes tumors in many plant species. There is no danger that this tumor-inducing plasmid will cause tumors in humans.

12. The total number of molecules after 15 cycles would be 16,384 or $(2)^{14}$.

14. Using the human nucleotide sequence, one can produce a probe to screen the library of the African okapi. Second, one can use the amino acid sequence and the genetic code to generate a complementary DNA probe for screening of the library. The probe is used, through hybridization, to identify the DNA that is complementary to the probe and allow one to identify the library clone containing the DNA of interest. Cells with the desired clone are then picked from the original plate and the plasmid is isolated from the cells.

16. The genomic clone probably contained a number of introns that were removed from the mRNA during processing. The cDNA was prepared from mRNA.

18. Several factors may contribute to the lack of representation of the 5′ end of the mRNA. One has to deal with the possibility that the reverse transcriptase may not completely synthesize the DNA from the RNA template. The other reason may be that the 3′ end of the copied DNA tends to fold back on itself, thus providing a primer for the DNA polymerase. Additional preparation of the cDNA requires some digestion at the folded region. Since this folded region corresponds to the 5′ end of the mRNA, some of the message is often lost.

20. Assuming that one has knowledge of the amino acid sequence of the protein product or the nucleotide sequence of the target nucleic acid, a degenerate set of DNA strands can be made that can be prepared for cloning into an appropriate vector or amplified by PCR. A variety of labeling techniques can then be used, through hybridization, to identify complementary base sequences contained in the genomic library. One must know at least a portion of the amino acid sequence of the protein product or its nucleic acid sequence in order for the procedure to be applied. Some problems can occur through degeneracy in the genetic code (not allowing construction of an appropriate probe), the possible existence of pseudogenes in the library (hybridizations with inappropriate related fragments in the library), and variability of DNA sequences in the library due to introns (causing poor or background hybridization). To overcome some of these problems, one can construct a variety of relatively small probes of different types that take into account the degeneracy in the code. By varying the conditions of hybridization (salt and temperature), one can reduce undesired hybridizations.

22. (a) Heating to 90–95°C denatures the double-stranded DNA so that it dissociates into single strands. It usually takes about five minutes, depending on the length and GC content of the DNA.

    (b) Lowering the temperature to 50–70°C allows the primers to bind to the denatured DNA.

    (c) Bringing the temperature to 70–75°C allows the heat-stable DNA polymerase an opportunity to extend the primers by adding nucleotides to the 3′ ends of each growing strand. Each PCR is designed with specific temperatures (not ranges) based on the characteristics of the DNAs (template and primers).

24. ddNTPs are analogs of the "normal" deoxyribonucleotide triphosphates (dNTPs), but they lack a 3′-hydroxyl group. As DNA synthesis occurs, the DNA polymerase occasionally inserts a ddNTP into a growing DNA strand. Since

there is no 3′-hydroxyl group, chain elongation cannot take place, and resulting fragments are formed, which can be separated by electrophoresis. Where the ddNTP was incorporated, the length of each strand, and therefore the position of the particular ddNTP, is established and used to eventually provide the base sequence of the DNA.

26. FISH involves the hybridization of a labeled probe to a complementary stretch of DNA in a chromosome. As such, it can be used to locate a specific DNA sequence (often a gene or gene fragment) in a chromosome. Spectral karyotyping uses FISH to detect individual chromosomes, a distinct advantage in identifying chromosomal abnormalities.

28. $T_m(°C) = 81.5 + 0.41(\%GC) - (675/N) = 81.5 + 0.41(33.3) - (675/21) =$ about 63°C. Subtracting 5°C gives us a good starting point of about 58°C for PCR with this primer. Notice that as the % of GC and length increase, the $T_m(°C)$ increases. GC pairs contain three hydrogen bonds rather than two as between AT pairs.

30. (a) Short tandem repeats of the Y chromosome (Y-STRs) vary considerably among individuals and populations. By amplifying Y-STRs by PCR and separating the amplified products by electrophoresis, one can genotypically type an individual as one does with a standard fingerprint. Because tissue samples are often left at the scene of a violent crime, DNA fingerprints are sometimes more available than standard fingerprints. Linking an individual with the time and place of a significant event has multiple forensic applications. Eliminating an individual as a suspect also has important forensic applications.

(b) The nonrecombining region of the Y is maintained strictly in the male population. Of special relevance in forensic applications would be the elimination of half the population (females) from a suspect group.

(c) Because different ethnic groups show different levels of Y-STR polymorphism, different final probabilities occur as products of individual probabilities. Since these probabilities are used to match individuals in forensics, ethnic variations must be taken under consideration.

(d) While there are many potential uses of DNA samples, generally a "match" is determined by multiplying the occurrence probabilites of each haplotype to arrive at the overall probability (product) of a genotype occurring in a population. If an individual's genotype matches that found in DNA at a crime scene, depending on the frequencies of the haplotypes, one might be able to say that the individual was at the crime scene. However, contamination, inappropriate genotyping, and laboratory expertise may give both false positive or negative results. Identical twins will have identical DNA fingerprints and may complicate forensic applications.

Chapter 21

ANSWERS TO NOW SOLVE THIS

21-1. Open reading frames are identified by computer programs based on identification of start (ATG) and stop (TAA, TGA, TAG) codons. Notice that the percentage of GC pairs compared with AT pairs is quite low in such

punctuation triplets. Therefore, when scanning DNA sequences for ORFs with high AT content, many short sequences are obtained that are clearly not likely to be involved in protein production. However, when DNA is GC rich, the likelihood of long ORFs similar to protein-coding size is increased. Therefore, the likelihood of falsely considering a sequence "protein-coding" increases with increasing GC content, as indicated in the figure.

21-2. (a) Assuming an average gene size of 5000 base pairs, there would be about 6.7×10^7 base pairs comprising genes. Subtracting this value from 116.8 Mb gives 49.8 Mb between genes. Dividing 49.8 Mb by 13,379 genes gives about 3700 bases between genes.

(b) 54,934/13,379 = 4.11 exons

(c) 48,257/13,379 = 3.61 introns

(d) There is a marked increase in the number of genes involved in alternative transcripts.

21-3. Since structural and chemical factors determine the function of a protein, it is likely to have several proteins share a considerable amino acid sequence identity, but not be functionally identical. Since the *in vivo* function of such a protein is determined by secondary and tertiary structures, as well as local surface chemistries in active or functional sites, the nonidentical sequences may have considerable influence on function. Note that the query matches to different site positions within the target proteins. A number of other factors suggesting different functions include: associations with other molecules (cytoplasmic, membrane, or extracellular), chemical nature and position of binding domains, posttranslational modification, signal sequences, and so on.

21-4. Because blood is relatively easy to obtain in a pure state, its components can be analyzed without fear of tissue-site contamination. Second, blood is intimately exposed to virtually all cells of the body and may therefore carry chemical markers to certain abnormal cells it represents. Theoretically, it is an ideal probe into the human body. However, when blood is removed from the body, its proteome changes, and those changes are dependent on a number of environmental factors. Thus, what might be a valid diagnostic for one condition might not be so for other conditions. In addition, the serum proteome is subject to change depending on the genetic, physiologic, and environmental state of the patient. Age and sex are additional variables that must be considered. Validation of a plasma proteome for a particular cancer would be strengthened by demonstrating that the stage of development of the cancer correlates with a commensurate change in the proteome in a relatively large, statistically significant pool of patients. Second, the types of changes in the proteome should be reproducible and, at least until complexities are clarified, involve tumorigenic proteins. It would be helpful to have comparisons with archived samples of each individual at a disease-free time.

SOLUTIONS TO PROBLEMS AND DISCUSSION QUESTIONS

2. Functional genomics seeks to understand functional components with the genome and similarities of genomes across phylogenetic and evolutionary distances. Comparative genomics analyzes the arrangement and organization of families of genes within and among genomes.

4. Genomes of both types of organisms are composed of double-stranded DNA (larger in eukaryotes) associated with proteins (more transient in prokaryotes). Both contain open reading frames, but those of prokaryotes are more densely packed. Both have some genes in clusters, but clustered genes are much more pronounced in prokaryotes (operons). There are a few repetitive sequences in prokaryotes, but this trend is much more common in eukaryotes. Both contain informational sequences, but those of eukaryotes are often interrupted (introns). Almost all genomes contain transposable elements

6. The main goals of the Human Genome Project are to establish, categorize, and analyze functions for human genes. As stated in the text:

> To analyze genetic variations between humans, including the identification of single-nucleotide polymorphisms (SNPs)
>
> To map and sequence the genomes of several model organisms used in experimental genetics, including *E. coli*, *S. cerevisiae*, *C. elegans*, *D. melanogaster*, and *M. musculus* (the mouse)
>
> To develop new sequencing technologies, such as high-throughput computer-automated sequencers, to facilitate genome analysis
>
> To disseminate genome information, both among scientists and the general public

8. One initial approach to annotating a sequence is to compare the newly sequenced genomic DNA to the known sequences already stored in various databases. The National Center for Biotechnology Information (NCBI) provides access to BLAST (Basic Local Alignment Search Tool) software that directs searches through databanks of DNA and protein sequences. A segment of DNA can be compared to sequences in major databases such as GenBank to identify matches that align in whole or in part. One might seek similarities of a sequence on chromosome 11 in a mouse and find that or similar sequences in a number of taxa. BLAST will compute a similarity score or identity value to indicate the degree to which two sequences are similar. BLAST is one of many sequence alignment algorithms (RNA-RNA, protein-protein, etc.) that may sacrifice sensitivity for speed.

10. The human genome is composed of over 3 billion nucleotides in which about 2 percent code for genes. Genes are unevenly distributed over chromosomes with clusters of gene-rich separated by gene-poor ones (deserts). Human genes tend to be larger and contain more and larger introns than in invertebrates such as *Drosophila*. It is estimated that at least half of the genes generate products by alternative splicing. Hundreds of genes have been transferred from bacteria into vertebrates. Duplicated regions are common, which may facilitate chromosomal rearrangement. The human genome appears to contain approximately 20,000 protein-coding genes. However, there is still uncertainty as to the total number.

12. Because many repetitive regions of the genome are not directly involved in production of a phenotype, they tend to be isolated from selection and show considerable variation in redundancy. Length variation in such repeats is unique among individuals (except for identical twins) and, with various detection methods, provides the basis for DNA fingerprinting. Single-nucleotide polymorphisms also occur frequently in the genome and can be used to distinguish individuals.

14. The Personal Genome Project (PGP) provides individual sequences of diploid genomes, and results of such projects indicate that the HGP may underestimate genome variation by as much as fivefold. Genome variation between individuals may be 0.5 percent rather than the 0.1 percent estimated from the HGP. Since the PGP provides sequence information on individuals, fundamental questions about human diversity and evolution may be more answerable.

16. A number of new subdisciplines of molecular biology will provide the infrastructure for major advances in our understanding of living systems. The following terms identify specific areas within that infrastructure:

> proteomics—proteins in a cell or tissue
> metabolomics—enzymatic pathways
> glycomics—carbohydrates of a cell or tissue
> toxicogenomics—toxic chemicals
> metagenomics—environmental issues
> pharmacogenomics—customized medicine
> transcriptomics—expressed genes
> Many other "-omics" are likely in the future.

18. Most microarrays, known also as gene chips, consist of a glass slide that is coated, using a robotic system, with single-stranded DNA molecules. Some microarrays are coated with single-stranded sequences of expressed sequenced tags or DNA sequences that are complementary to gene transcripts. A single microarray can have as many as 20,000 different spots of DNA, each containing a unique sequence. Researchers use microarrays to compare patterns of gene expression in tissues under different conditions or to compare gene-expression patterns in normal and diseased tissues. In addition, microarrays can be used to identify pathogens. Microarray databases allow investigators to compare any given pattern to others worldwide.

20. Increased protein production from approximately 20,000 genes is probably related to alternative splicing and various posttranslational processing schemes. In addition, a particular DNA segment may be read in a variety of ways and in two directions.

22. While the β-globin gene family is a relatively large (60 kb) sequence and restriction analyses show that it is composed of six genes, one is a pseudogene and therefore does not produce a product. The five functional genes each contain two similarly sized introns, which, when included with noncoding flanking regions (5′ and 3′) and spacer DNA between genes, account for the 95 percent mentioned in the question.

24. Any time a DNA sequence is conserved in other species, it is likely that that sequence has an influence on similar phenotypes. The higher the number of species that conserve the sequence, the higher the likelihood of determining its function. Coupled with mutation analysis and physical mapping, comparative genomics provides a powerful method for linking DNA sequences with complex human diseases.

26. The issue here is whether the organism under consideration is independent and self-reproducing. It appears that the minimum number of genes for a free-living organism is in the range of 250–350. Symbionts can have much smaller genomes and exist with fewer genes because of materials

supplied by the host cell. As long as one defines the life style (free-living or symbiont) of the organism in question, it is informative to consider how many genes are needed to accomplish the task of "living."

Chapter 22

ANSWERS TO NOW SOLVE THIS

22-1. Antigens are usually quite large molecules, and in the process of digestion, they are sometimes broken down into smaller molecules, thus becoming ineffective in stimulating the immune system. Some individuals are allergic to the food they eat, testifying to the fact that all antigens are not completely degraded or modified by digestion. In some cases, ingested antigens do indeed stimulate the immune system (oral polio vaccine) and provide a route for immunization. Localized (intestinal) immunity can sometimes be stimulated by oral introduction of antigens and in some cases this can offer immunity to ingested pathogens.

22-2. To generate glyphosate resistance in crop plants, a fusion gene was created that introduced a viral promoter to control the EPSP synthetase gene. The fusion product was placed into the Ti vector and transferred to *A. tumifaciens*, which was used to infect crop cells. Calluses were selected on the basis of their resistance to glyphosate. Resistant calluses were later developed into transgenic plants. There is a remote possibility that such an "accident" can occur as suggested in the question. However, in retracing the steps to generate the resistant plant in the first place, it seems more likely that the trait will not "escape" from the plant; rather, the engineered *A. tumifaciens* may escape, infect, and transfer glyphosate resistance to pest species.

22-3. The child in question is a carrier of the deletion in the β-globin gene, just as the parents are carriers. Its genotype is therefore $\beta^A \beta^o$.

22-4. It will hybridize by base complementation to the normal DNA sequence.

22-5. (a,b) One of the main problems with gene therapy is delivery of the desired virus to the target tissue in an effective manner. Several of the problems involving the use of retroviral vectors are the following: (1) Integration into the host must be cell specific so as not to damage nontarget cells. (2) Retroviral integration into host cell genomes only occurs if the host cell is replicating. (3) Insertion of the viral genome might influence nontarget, but essential, genes. (4) Retroviral genomes have a low cloning capacity and cannot carry large inserted sequences as are many human genes. (5) There is a possibility that recombination with host viruses will produce an infectious virus that may do harm.

(c) It would certainly be more efficient (although perhaps more difficult technically) to engineer germ tissue, for once it is done in a family, the disease would be eliminated. However, there are considerable ethical problems associated with germ-line therapy. It recalls previous attempts of the eugenics movements of past decades, which involved the use of selective breeding to purify the human stock. Some present-day biologists have said publically that germ-line gene therapy will not be conducted.

SOLUTIONS TO PROBLEMS AND DISCUSSION QUESTIONS

2. There are concerns about proper testing of GM crops and foods for allergenicity, environmental impact, and the possibility of cross-pollination leading to the contamination of native species. If certain crops become the standard and under the control of a few manufacturers, it is likely that the world's supply of genetic variability might be reduced. Concern would increase if such crops routinely contained antibiotic-resistant genetic markers and genes conferring toxicity to pests. A broader concern is that the design and patenting of crops might allow domination of the world food supply by a few companies.

4. One method is to use the amino acid sequence of the protein to produce the gene synthetically. Alternatively, since the introns are spliced out of the pre-mRNA during the maturation of mRNA, if mRNA can be obtained, it can be used to make DNA (cDNA) through the use of reverse transcriptase. The resulting cDNA can then be used (free of introns) to make the desired product.

6. In general, bacteria do not process eukaryotic proteins in the same manner as eukaryotes. Transgenic eukaryotes are more likely to correctly process eukaryotic proteins, thus increasing the likelihood of their normal biological activity.

8. (a) Both the saline and column extracts of Lkt50 appear to be capable of inducing at least 50 percent neutralization of toxicity when injected into rabbits.

(b) First, the immunogen must be stably incorporated into the host plant hereditary material, and the host must express only that immunogen. During feeding, the immunogen must be transported across the intestinal wall unaltered, or altered in such a way as to stimulate the desired immune response. There must be guarantees that potentially harmful by-products of transgenesis have not been produced. In other words, broad ecological and environmental issues must be addressed to prevent a transgenic plant from becoming an unintended vector for harm to the environment or any organisms feeding on the plant (directly or indirectly).

10. Even though you have developed a method for screening seven of the mutations described, it is possible that negative results can occur when the person carries the gene for CF. In other words, the specific probes (or allele-specific oligonucleotides) that have been developed will not necessarily be useful for screening all mutant alleles. In addition, the cost-effectiveness of such a screening proposal would need to be considered.

12. Since both mutations occur in the CF gene, children who possess both alleles will suffer from CF. With both parents heterozygous, each child born will have a 25 percent chance of developing CF.

14. Using restriction enzyme analysis to detect point mutations in humans is a tedious trial-and-error process. Given the size of the human genome in terms of base sequences and the relatively low number of unique restriction enzymes, the likelihood of matching a specific point mutation, separate from other normal sequence variations, to a desired gene is low.

16. As with all therapies, the cure must be less hazardous than the disease. In the case of viral-mediated gene therapy, the antigenicity of the virus must not interfere with the delivery system; such antigenicity can cause inflammation or more severe immunologic responses. Combating the host immune response may involve the use of immunosuppressive drugs or modification of the vector. The duration of desired gene expression at the diseased site is an issue. Short-period expression may require repeated exposure to the vehicle, which may present undesired responses. For some diseases, local gene therapy through inhalation or injection may produce fewer side effects than systemic exposure. Adenoviruses appear to be particularly useful for gene therapy because they can infect nondividing cells and they can accept relatively large amounts of additional DNA (30 kb or more).

18. While some success has been achieved using gene therapy, major questions remain. First, the vectors that deliver the desired gene must not trigger adverse reactions. Second, at present integration of some vectors is dependent on DNA replication in the host. All cells in the body are therefore not available to integration of some vectors. Third, precise target integration must be achieved to reduce the introduction of new mutations. Fourth, most human genes are large, while many vectors in use today can carry only small inserts. Large cargos will be needed to achieve a broad range of therapies. Finally, desired products of the vector should have long-lasting effects, and the vectors themselves must remain incapable of reverting to an infectious virus.

20. p53 and pRB are tumor-suppressor proteins and are required by the cell to effectively monitor the cell cycle. Reduction in their activity would diminish normal cell-cycle controls and most likely lead to cancer. It would be especially important if such viral-vectors are intended to treat cancer where it is likely that cell-cycle control is already compromised.

22. The two major problems described here are common concerns related to genetic engineering. The first is the localization of the introduced DNA into the target tissue and target location in the genome. Inappropriate targeting may have serious consequences. In addition, it is often difficult to control the output of introduced DNA. Genetic regulation is complicated and subject to a number of factors, including upstream and downstream signals as well as various posttranscriptional processing schemes. Artificial control of these factors will prove difficult.

24. Given the use to which genetic tests are put and their extreme personal nature, it would seem that FDA regulation is one way to decrease the distribution of misinformation that may be vital to individuals and families.

26. *Sample size:* Is the sample size sufficiently robust to minimize spurious correlations? Epigenetic factors: Since the genome is strongly influenced by epigenetic factors, how might this influence a study based on SNPs? *Environmental factors:* Have samples been compiled in such a way as to negate significant differences in environmental exposure? *Genetic background:* Since the entire genome is not compared, what suppressors and/or modifiers are present, but missed, that mask or enhance the expression of certain genes? What emergent properties are present in the entire allelic architecture? *Population stratification:* Might there be allelic frequency differences peculiar to the case population as compared with the control group? Such population frequency differences may result from differences in ancestry, depending on the composition of the cases and the controls.

28. There are several reasons as to why CF mice do not exhibit airway symptoms similar to humans with CF. First, the distribution of cell types, CFTR presence, and chloride/sodium handling differ in the upper and lower airways of mice and humans. Second, mice tend not to suffer from airway bacterial infections that severely influence human pathology. A complete review of this topic is available in Grubb and Boucher, 1999 *Physiological Reviews,* vol. 71, 5193–5214.

Chapter 23

ANSWERS TO NOW SOLVE THIS

23-1 (a) Since 1/256 of the F_2 plants are 20 cm and 1/256 are 40 cm, there must be four gene pairs involved in determining flower size.

(b) Since there are nine size classes, one can conduct the following backcross:

$$AaBbCcDd \times AABBCCDD$$

The frequency distribution in the backcross would be:

$$1/16 = 40 \text{ cm}$$
$$4/16 = 37.5 \text{ cm}$$
$$6/16 = 35 \text{ cm}$$
$$4/16 = 32.5 \text{ cm}$$
$$1/16 = 30 \text{ cm}$$

23-2. (a) Taking the sum of the values and dividing by the number in the sample gives the following means:

mean sheep fiber length = 7.7 cm
mean fleece weight = 6.4 kg

The variance for each is:

variance sheep fiber length = 6.097
variance fleece weight = 3.12

The standard deviation is the square root of the variance:

sheep fiber length = 2.469
fleece weight = 1.766

(b,c) The covariance for the two traits is 30.36/7, or 4.34, while the correlation coefficient is + 0.998.

(d) There is a very high correlation between fleece weight and fiber length, and it is likely that this correlation is not by chance. Even though correlation does not mean cause-and-effect, it would seem logical that as you increased fiber length, you would also increase fleece weight. It is probably safe to say that the increase in fleece weight is directly related to an increase in fiber length.

23-3. *Monozygotic twins* are derived from a single fertilized egg and are thus genetically identical to each other. They provide a method for determining the influence of genetics and environment on certain traits. *Dizygotic twins* arise from two eggs fertilized by two sperm cells. They have the same genetic relationship as siblings. The role of genetics and the role of the environment can be studied by comparing the expression of traits in monozygotic and dizygotic twins.

The higher concordance value for monozygotic twins as compared with the value for dizygotic twins indicates a significant genetic component for a given trait. Notice that for traits including blood type, eye color, and mental retardation, there is a fairly significant difference between MZ and DZ groups. However, for measles, the difference is not as significant, indicating a greater role of the environment. Hair color has a significant genetic component as do idiopathic epilepsy, schizophrenia, diabetes, allergies, cleft lip, and club foot. The genetic component of mammary cancer is present but minimal according to these data.

SOLUTIONS TO PROBLEMS AND DISCUSSION QUESTIONS

2. In *discontinuous* variation the influences of each gene pair are not additive, and more typical Mendelian ratios such as 9:3:3:1 and 3:1 result. In *continuous* variation, different gene pairs interact (usually additively) to produce a phenotype that is less "stepwise" in distribution. Inheritance involving polygenic systems follows a more continuous distribution.

4. (a) Because a dihybrid result has been identified, there are two loci involved in the production of color. There are two alleles at each locus for a total of four alleles.

(b,c) Because the description of red, medium red, and so on, gives us no indication of a *quantity* of color in any form of units, we would not be able to actually quantify a unit amount for each change in color. We can say that each gene (additive allele) provides an equal unit amount to the phenotype and the colors differ from each other in multiples of that unit amount. The number of additive alleles needed to produce each phenotype is as follows:

$$1/16 = \text{dark red} \qquad = AABB$$
$$4/16 = \text{medium-dark red} = 2AABb$$
$$\qquad\qquad\qquad\qquad\qquad 2AaBB$$
$$6/16 = \text{medium red} \qquad = AAbb$$
$$\qquad\qquad\qquad\qquad\qquad 4AaBb$$
$$\qquad\qquad\qquad\qquad\qquad aaBB$$
$$4/16 = \text{light red} \qquad = 2aaBb$$
$$\qquad\qquad\qquad\qquad\qquad 2Aabb$$
$$1/16 = \text{white} \qquad = aabb$$

(d)
$F_1 = $ all light red
$F_2 = 1/4$ medium red
2/4 light red
1/4 white

6. (a,b) See that where four gene pairs act additively, the proportion of one of the extreme phenotypes to the total number of offspring is 1/256. The same may be said for the other extreme type. The extreme types in this problem are the 12 cm and 36 cm plants. From this observation one would suggest that there are four gene pairs involved.

(c) If there are four gene pairs, there are nine $(2n+1)$ phenotypic categories and eight increments between these categories. Since there is a difference of 24 cm between the extremes, 24 cm/8 = 3 cm for each increment (each of the additive alleles).

(d) Because the parents are inbred, it is expected that they are fully homozygous. An example:

$$AABBccdd \times aabbCCDD$$

(e) Since the *aabbccdd* genotype gives a height of 12 cm and each uppercase allele adds 3 cm to the height, there are many possibilities for an 18 cm plant:

AAbbccdd,
AaBbccdd,
aaBbCcdd, etc.

Any plant with seven uppercase letters will be 33 cm tall:

AABBCCDd,
AABBCcDD,
AABbCCDD, for example.

8. For height, notice that average differences between MZ twins reared together (1.7 cm) and MZ twins reared apart (1.8 cm) are similar (meaning little environmental influence) and considerably less than differences of DZ twins (4.4 cm) or sibs (4.5) reared together. These data indicate that genetics plays a major role in determining height.

However, for weight, notice that MZ twins reared together have a much smaller (1.9 kg) difference than MZ twins reared apart, indicating that the environment has a considerable impact on weight. By comparing the weight differences of MZ twins reared apart with DZ twins and sibs reared together, one can conclude that the environment has almost as much an influence on weight as genetics.

For ridge count, the differences between MZ twins reared together and those reared apart are small. For the data in the table, it would appear that ridge count and height have the highest heritability values.

10. Many traits, especially those we view as quantitative, are likely to be determined by a polygenic mode with possible environmental influences. The following are some common examples: height, general body structure, skin color, and perhaps most common behavioral traits including intelligence.

12. (a) *For back fat:*

Broad-sense heritability $= H^2 = 12.2/30.6 = .398$
Narrow-sense heritability $= h^2 = 8.44/30.6 = .276$

For body length:

Broad-sense heritability $= H^2 = 26.4/52.4 = .504$
Narrow-sense heritability $= h^2 = 11.7/52.4 = .223$

(b) Selection for back fat would produce more response.

14. (a) For Vitamin A: $h_A^2 = 0.097$
For Cholesterol: $h_A^2 = 0.223$

(b) Cholesterol content should be influenced to a greater extent by selection.

16. $h^2 = (7.5 - 8.5/6.0 - 8.5) = 0.4$
(realized heritability)

18. $h^2 = 0.3 = (M_2 - 60/80 - 60)$
$M_2 = 66$ grams

20. (a, b) In many instances, a trait may be clustered in families, yet, traditional mapping procedures may not be applicable because the trait might be influenced by a number of genes. In general, researchers look for associations to particular DNA sequences (molecular markers). When the trait cosegregates with a particular marker and it statistically associates with that trait above chance, a likely QTL has been identified. Markers such as RFLPs, SNPs, and microsatellites are often used because they are highly variable, relatively easy to assess, and present in all individuals.

22. (a) There are two ways to answer this section, a hard way and an easy way. The hard way would to take a big sheet of paper, make the cross (*AaBbCcDdEeFf* × *AaBbCcDdEeFf*), collect the genotypes, and calculate the ratios.

This method would be very laborious and error-prone. The easy way would be to re-read the material on the binomial expansion and note the pattern preceding each expression. Notice that all numbers other than the 1's are equal to the sum of the two numbers directly above them. By enlarging the numbers to include six gene pairs, you can arrive at the 13 classes and their frequencies:

$$3'' = 1 \qquad 4'' = 12 \qquad 5'' = 66$$
$$6'' = 220 \quad 7'' = 495 \quad 8'' = 792$$
$$9'' = 924 \quad 10'' = 792 \quad 11'' = 495$$
$$12'' = 220 \quad 13'' = 66 \quad 14'' = 12$$
$$15'' = 1$$

To check your calculations, be certain that your frequencies total 4096. You will also notice an additional shortcut in that since the distribution is symmetrical, you need only calculate to the center and the remainder will be in the reverse order.

(b) To determine the outcome of a cross of the F_1 plants in the testcross, apply the formula that allows you to calculate any set of components: $n!/(s!t!)$ where n = total number of events (6), s = number of events of outcome a and t = number of events of outcome b. For example, to determine how many 6'' plants would be recovered from the cross *AaBbCcDdEeFf* × *aabbccddeeff*, we are really asking how many will have three additive alleles (uppercase) and three nonadditive alleles (lowercase).

$$6!/(3!3!) = 20$$

Applying this formula throughout gives the following frequencies:

$$3'' = 1 \quad 4'' = 6 \quad 5'' = 15$$
$$6'' = 20 \quad 7'' = 15 \quad 8'' = 6$$
$$9'' = 1$$

The total is 64. You can check your logic by considering that there should be only 1/64 with no additive alleles (3'') and 1/64 with all additive alleles (9'').

24. As with many traits that are caused by numerous loci acting additively, some genes have more influence on expression than others. In addition, environmental factors may play a role in the expression of some polygenic traits. In the case of brachydactyly, there are numerous modifier genes in the genome that can influence brachydactyly expression. Examination of OMIM (*Online Mendelian Inheritance of Man*) through http://www.ncbi.nlm.nih.gov/ will illustrate this point.

26. The greater the genetic variation in a species, the more likely and dramatic the response to selection. Therefore, one would expect a greater response to selection in the wild population.

28. Breeders attempt to "select" out this disorder by first maintaining complete and detailed breeding records of afflicted strains. Second, they avoid breeding dogs whose close relatives are afflicted. The molecular-developmental mechanism that causes the "month of birth" effect in canine hip dysplasia is unknown. However, with many, perhaps all quantitative traits, it is clear that there is a significant environmental influence on both the penetrance and/or expression of the phenotype. With many genes acting in various ways to influence a phenotype, there are opportunities for varied molecular and developmental intraorganismic microenvironments. Stated another way, the longer and more complex the molecular distance from the genome to the phenotype, the greater the likelihood for environmental factors to be involved in expression.

Chapter 24

ANSWERS TO NOW SOLVE THIS

24-1. If one can select inbred strains for a certain behavior, it might be possible to analyze molecular and physiological characteristics that differ from normal mice. Like mutation analysis, inbred strains provide an opportunity for comparative studies. Differences in alcohol tolerance, metabolism, and physiology can be examined. In addition, refined studies on known factors (receptors, etc.) related to addiction are open to investigation. Once major genes are located (by QTL analysis in this case), the genes, their regulation and products, can be studied. Such a molecular approach provides an opportunity to determine the exact differences between the normal state and the addictive state. Both approaches can provide insight into the nature of behavioral traits, and because of the complexity of most behaviors, both approaches are often necessary.

24-2. One of the easiest ways to determine whether a genetic basis exists for a given abnormality is to cross the abnormal fly to a normal fly. If the trait is determined by a dominant gene, that trait should appear in the offspring—probably half of them if the gene was in the heterozygous state. If the gene is recessive and homozygous, then one may not see expression in the offspring of the first cross. However, if one crosses the F_1, the trait might appear in approximately 1/4 of the offspring. Such ratios would not be expected if the trait is polygenic, which generates complexities in any study. Modifications of these patterns would be expected if the mode of inheritance were X-linked or showed other modifications of typical Mendelian ratios. One might hypothesize that the trait influences the nervous, cuticular, or muscular system.

SOLUTIONS TO PROBLEMS AND DISCUSSION QUESTIONS

2. One method, a behavior-first approach, attempts to correlate existing behavioral differences with general genetic differences. It is a comparative approach in which a particular behavior is examined among several (to many) closely related but genetically different strains of organisms. If the environment is held constant, yet the behavioral differences persist, there must be a genetic component to the behavior. A second approach involves selection for certain behaviors. When selection and interbreeding of a behavior yields a consistent phenotype, then the genetics of the behavior can be examined. A third approach takes advantage of the fact that some behaviors are strongly influenced by a

major locus. In a genetics-first approach, mutagens can be used to induce mutations from which those behaviors of interest can be studied. If an induced mutation produces a consistent behavioral change, it is often the most useful way to analyze a behavior. The latter two approaches can only be used if the organism of interest can be genetically manipulated, which is not always the case (primates for example). In addition, once selection and mutation approaches are applied, the natural expression of most behaviors is usually altered. On the positive side, selection and mutation analyses offer unique opportunities to examine genetic and physiological factors at the molecular levels.

4. One limitation of this approach is the intensity of effort required to establish genetically uniform strains reflecting a particular behavior. Another complexity is that genes often have multiple functions or may function somewhat differently in different strains and/or environments. In addition, since geotaxis is a complex trait, selection experiments offer little indication of the number and location of selected genes. New technologies, including microarrays, have been applied to this problem and enable the analysis of the expression patterns of many genes simultaneously. Strains having high and low responses to geotaxis selection showed reproducible differences in gene expression and revealed some of the molecular and physiological characteristics of geotaxis. However, it is important to state that association of a particular microarray pattern with a particular behaviour may not indicate a cause-effect relationship. Some genes having little or nothing to do with geotaxis may "hitch-hike" near global regulatory genes that just happen to influence geotaxis.

6. Because of the rigidity of development in *Caenorhabditis* and the extensive knowledge of cellular fates and connectivity, it represents an excellent experimental organism for the study of development. In addition, because of their fixed fate, cells can be altered, physically and/or genetically, and resulting development and behavior can be studied. However, the behavioral repertoire of *C. elegans* is somewhat narrow. In addition, *C. elegans* has more protein-coding genes than *Drosophila*, another organism whose behavioral genetics has been highly studied. Such an additional number of protein-coding genes coupled with a sparse behavioral repertoire creates a disadvantage when using *C. elegans*.

8. A number of issues limit the study of human behavior including the following:
   (a) With relatively small numbers of offspring per mating, standard genetic methods are difficult.
   (b) Records on family illnesses, especially behavioral illnesses, are difficult to obtain.
   (c) The long generation time makes longitudinal studies difficult.
   (d) The scientist cannot direct matings.
   (e) There are limits to the experimental treatments that can be applied.
   (f) Traits that are interesting to study are often complex and difficult to quantify.
   (g) Culture and family background may strongly influence behavior.

10. Recessive mutations are typically only observed at the phenotypic level when homozygous. Self-fertilization, the ultimate form of inbreeding, greatly enhances the likelihood that recessive genes will become homozygous. Homozygous strains are of considerable advantage when studying complex traits.

12. One might compare gene expression through microarray analysis between space reared flies with a control population on earth. By comparing behavioral (metabolic, mating, feeding, walking, flying) rituals between the two populations (space and earth), one might be able to correlate genetic expression profiles and determine which genes are most likely to respond to which behavioral alteration. While a number of parameters would be interesting to study in this way, one might concentrate the study on genes known to influence muscle and nerve function in flies and humans. A second study might focus on the immune system since it is known that there are surprisingly high genetic correlations between the immune response of flies and mammals (Hoffmann, 2003). Since it is known that microgravity negatively influences the human immune system (reduces mitogenic activity of T cells), a comparison of the gene-expression profiles of the immune-response system would also be worthy of study. The behavioral side of the experiment would focus on determining whether changes in gravitational stress alters behavior (inactivity, muscle and nerve use) to a point where the immune response is altered. Coupling the two studies, gene-expression profiles associated with behavioral and the immune-response, might provide insight into a number of significant problems that humans encounter in space flight, and thereby be justifiable and relatively inexpensive.

14. Genome-wide association (GWA) studies make use of millions of SNPs in an attempt to find associations to a given trait such as schizophrenia. Such studies indicate that no single gene or allele is a defining contributor to this disorder.

16. Data in the table indicate that females (*Gasterosteus aculeatus*) spend more time with males having the optimal MHC constitution than those that do not. As with other organisms, it appears that genetic diversity for the MHC is a selective advantage. Since immunological functions are associated with the MHC complex, mate selection mechanisms that support the maintenance of such diversity seems reasonable from an evolutionary viewpoint.

Chapter 25

ANSWERS TO NOW SOLVE THIS

25-1.

$$q = .544$$
$$p = .456$$

The frequencies of the genotypes are determined by applying the formula $p^2 + 2pq + q^2$ as follows:

Frequency of *AA* = .208 or 20.8%

Frequency of *Aa* = .496 or 49.6%

Frequency of *aa* = .296 or 29.6%

25-2. (a) q is 0.01
   (b) $p = 1 - q$ or .99
   (c) $2pq = 2(.01)(.99)$
      $= 0.0198$ (or about 1/50)
   (d) $2pq \times 2pq$
      $= 0.0198 \times 0.0198$
      $= 0.000392$ or about 1/255

25-3. The overall probability of the couple producing a CF child is $98/2500 \times 2/3 \times 1/4 = .00653$, or about 1/153.

SOLUTIONS TO PROBLEMS AND DISCUSSION QUESTIONS

2. Because of degeneracy in the code, there are some nucleotide substitutions, especially in the third base, that do not change amino acids. In addition, if there is no change in the overall charge of the protein, it is likely that electrophoresis will not separate the variants. If a positively charged amino acid is replaced by an amino acid of like charge, then the overall charge on the protein is unchanged. The same may be said for other negatively charged and neutral amino acid substitutions.

4. There are many sections of DNA in a eukaryotic genome that are not reflected in a protein product. Indeed, there are many sections of DNA that are not even transcribed and/or have no apparent physiological role. Such regions are more likely to tolerate nucleotide changes compared with those regions with a necessary physiological impact. Introns, for example, show sequence variation, which is not reflected in a protein product. Exons, on the other hand, code for products that are usually involved in production of a phenotype and, as such, are subject to selection.

6. For each of these values, one merely takes the square root to determine q, then one computes p, and finally one "plugs" the values into the $2pq$ expression.
   (a) $q = .08$; $2pq = 2(.92)(.08)$
       $= .1472$ or 14.72%
   (b) $q = .009$; $2pq = 2(.991)(.009)$
       $= .01784$ or 1.78%
   (c) $q = .3$; $2pq = 2(.7)(.3)$
       $= .42$ or 42%
   (d) $q = .1$; $2pq = 2(.9)(.1)$
       $= .18$ or 18%
   (e) $q = .316$; $2pq = 2(.684)(.316)$
       $= .4323$ or 43.23%
   (Depending on how one rounds off the decimals, slightly different answers will occur.)

8. Assuming that the population is in Hardy-Weinberg equilibrium, if one has the frequency of individuals with the dominant phenotype, the remainder have the recessive phenotype (q^2). With q^2 one can calculate q, and from this value one can arrive at p. Applying the expression $p^2 + 2pq + q^2$ will allow a solution to the question.

10. 5 percent

12. The general equation for responding to this question is

$$q_n = q_o /(1 + nq_o)$$

where $n =$ the number of generations, $q_o =$ the initial gene frequency, and $q_n =$ the new gene frequency.
   (a) **n = 1**
       $q_n = .33$ $p_n = .67$
   (b) **n = 5**
       $q_n = .143$ $p_n = .857$
   (c) **n = 10**
       $q_n = .083$ $p_n = .917$
   (d) **n = 25**
       $q_n = .037$ $p_n = .963$
   (e) **n = 100**
       $q_n = .0098$ $p_n = .9902$
   (f) **n = 1000**
       $q_n = .00099$ $p_n = .99901$

14. The frequency of heterozygosity is $2pq$ or approximately .02, as also stated in the problem. The probability for one of the grandparents to be heterozygous would therefore be 0.02 + 0.02 or 0.04 or 1/25. (*Note:* If one considers the probability of both parents being carriers, 0.02×0.02, the answer differs slightly.) If one of the grandparents is a carrier, then the probability of the offspring from a first-cousin mating being homozygous for the recessive gene is 1/16. Multiplying the two probabilities together gives $1/16 \times 1/25 = 1/400$.

   Following the same analysis for the second-cousin mating gives $1/64 \times 1/25 = 1/1600$. Notice that the population at large has a frequency of homozygotes of 1/10,000; therefore, one can easily see how inbreeding increases the likelihood of homozygosity.

16. 2/100,000   or 2×10^{-5}.

18. Reproductive isolating mechanisms are grouped into prezygotic and postzygotic and include the following:
   • geographic or ecological
   • seasonal or temporal
   • behavoral
   • mechanical
   • physiological
   • hybrid inviability or weakness
   • developmental hybrid sterility
   • segregational hybrid sterility
   • F_2 breakdown

20. The neutral theory of molecular evolution considers the possibility that some mutations are functionally equivalent to the alleles that they replace. In general, the frequency of neutral alleles in a population is determined by mutation rates and random genetic drift, and not by natural selection.

22.
$$F = (H_e - H_o)/H_e$$
$$F = (38.475 - 20)/38.475 = 0.48$$

24.
$$F = (H_e - H_o)/H_e$$
$$F = (.375 - .34)/.375 = 0.093$$

This problem could also be solved using the actual numbers of sheep in each category where there would be 18.75 heterozygotes expected $(2pq)(50)$:

$$F = (18.75 - 17)/18.75 = 0.093$$

26. (a) The gene is most likely recessive because all affected individuals have unaffected parents and the condition clearly runs in families. For the population, since $q^2 = .002$, then $q = .045$, $p = .955$, and $2(pq) = 0.086$. For the community, since $q^2 = .005$, then $q = .07$, $p = .93$, and $2(pq) = 0.13$.
   (b) The "founder effect" is probably operating here. Relatively small, local populations that are relatively isolated in a reproductive sense tend to show differences in gene frequencies when compared with larger populations. In such small populations, homozygosity is increased as a gene has a higher probability of "meeting itself."

28. When a population bottleneck occurs and the number of effective breeders is reduced in a population, two phenomena usually follow. First, because the population is small,

wide fluctuations in genotypic frequencies occur, thereby revealing deleterious alleles by chance. Second, inbreeding often occurs in small populations, thereby increasing the chance for homozygosity. With increased homozygosity comes an increased likelihood that recessive alleles will be expressed. Since many disease-producing genes are recessive, an increase in genetic diseases is a likely aftermath to a population bottleneck.

30. Approach this problem by writing the possible codons for all the amino acids (except Arg and Asp, which show no change) in the human cytochrome c chain. Then determine the minimum number of nucleotide substitutions required for each changed amino acid in the various organisms. Once listed, count up the numbers for each organism: horse, 3; pig, 2; dog, 3; chicken, 3; bullfrog, 2; fungus, 6.

32. (a) Since noncoding genomic regions are probably silent genetically, it is likely that they contribute little, if anything, to the phenotype. Selection acts on the phenotype; therefore, such noncoding regions are probably selectively neutral.

(b) These polymorphism data indicate that all the Lake Victoria area (lake and contributing rivers) cichlids are related by recent ancestry, whereas those from neighboring lakes are more distantly related. In addition, since Lake Victoria dried out about 14,000 years ago, it is likely that it was repopulated by a relatively small sample of cichlids.

34. In general, speciation involves the gradual accumulation of genetic changes to a point where reproductive isolation occurs. Depending on environmental or geographic conditions, genetic changes may occur slowly or rapidly. They can involve point mutations or chromosomal changes.

36. (a,b,c) The pattern of genetic distances through time indicates that from the present to about 25,000 years ago, modern humans and Cro-Magnons show an approximately constant number of differences. Conversely, there is an abrupt increase in genetic distance seen in comparing modern humans and Cro-Magnons with Neanderthals. The results indicate a clear discontinuity between modern humans, Cro-Magnons, and Neanderthals with respect to genetic variation in the mitochondrial DNAs sampled. Assuming that the sampling and analytical techniques used to generate the data are valid, it appears that Neanderthals made little, if any, genetic contributions to the Cro-Magnon or modern European gene pool. It could be argued that the absence of Neanderthal mtDNA lineages in living humans is a consequence of random drift or lineage extinction since the disappearance of Neanderthals. However, the examination of mtDNA in ancient Cro-Magnon mtDNA shows no evidence of a historical relationship and suggests that Neanderthals were not genetically related to the ancestors of modern humans.

Chapter 26

ANSWERS TO NOW SOLVE THIS

26-1. $N_e = 4(N_mN_f)/N_m + N_f = 6.9$
26-2. (a) $N_e = 25.21$
   (b) $H_t = .5073$
   (c) $F = 0.0776$

26-3. (a) 0.396
   (b) To determine N_e use the expression:

$$N_e/N = .42$$
$$N_e = .42 \times 50 = 21$$
$$H_t = (1-1/2N_e)^t H_o$$
$$= (1-1/42)^5 \times 0.0198$$
$$= 0.01755$$
$$H_t/H_o = .01755/.0198 = 0.886$$

Therefore there is a loss of approximately 11.4% heterozygosity after five generations.

SOLUTIONS TO PROBLEMS AND DISCUSSION QUESTIONS

2. (a,b) First, if detailed records are kept of the breeding partners of the captive birds, then knowledge of heterozygotes should be available. Breeding programs could be established to restrict matings between those carrying the lethal gene. Such "kinship management" is often used in captive populations. If kinship records are not available, it is often possible to establish kinship using genetic markers such as DNA microsatellite polymorphisms. Using such markers, one can often identify mating partners and link them to their offspring.

By coupling knowledge of mating partners with the likelihood of producing a lethal genetic combination, selective matings can often be used to minimize the influence of a deleterious allele. In addition, such markers can be used to establish matings that optimize genetic mixing, thus reducing inbreeding depression.

4. Both genetic drift and inbreeding tend to drive populations toward homozygosity. Genetic drift is more common when the effective breeding size of the population is low. When this condition prevails, inbreeding is also much more likely. They are different in that inbreeding can occur when certain population structures or behaviors favor matings between relatives, regardless of the effective size of the population.

6. Inbreeding depression, over time, reduces the level of heterozygosity, usually a selectively advantageous quality of a species. When homozygosity increases (through loss of heterozygosity), deleterious alleles are likely to become more of a load on a population. Outbreeding depression occurs when there is a reduction in fitness of progeny from genetically diverse individuals. It is usually attributed to offspring being less well-adapted to the local environmental conditions of the parents.

Even though forced outbreeding may be necessary to save a threatened species, where population numbers are low, it significantly and permanently changes the genetic make-up of the species.

8. Often, molecular assays of overall heterozygosity can indicate the degree of inbreeding and/or genetic drift. As inbreeding, and genetic drift for that matter, occur, the degree of heterozygosity decreases. An allele that has its frequency dictated by inbreeding will not be uniquely influenced. That is, other alleles would be characterized by decreased heterozygosity as well. So, if the genome in general has a relatively high degree of heterozygosity, the gene is probably influenced by selection rather than inbreeding and/or genetic drift.

10. Generally, threatened species are captured and bred in an artificial environment until sufficient population numbers are achieved to ensure species survival. Next, genetic management strategies are applied to breed individuals in such a way as to increase genetic heterozygosity as much as possible. If plants are involved, seed banks are often used to maintain and facilitate long-term survival.

12. Allozymes are variants of a particular protein often detected by electrophoresis. Such variation may or may not impact on the fitness of an individual. The greater the allozyme variation, the more genetically heterogeneous the individual. It is generally agreed that such genetic diversity is essential for long-term survival. All other factors being equal, allozyme variation is more likely to reflect physiological variation than RFLP variation because RFLP regions are not necessarily found in protein-coding regions of the genome. RFLP analysis allows one to detect very small amounts of genetic diversity in a population and is unlikely to encounter an organism that is not in some way variable in terms of RFLP with respect to other organisms (within and among species).

14. Data from *Antechinus* provide insight into the significance of genetic diversity to the survival of a species. Because such mechanisms (i.e., sperm mixing) are in place, there must be considerable evolutionary rewards. In this case, maintaining diversity must offset the cost (if any) of evolving such a mechanism.

16. (a) Since prairie dogs are the main food source for black-footed ferrets, a reduction in the population of prairie dogs would likely stress black-footed ferrets. Unless alternate food sources are available and are utilized by the ferrets, their numbers would decline.

(b) The fact that a population survives a population bottleneck does not mean that the population is in a healthy state. Usually bottlenecks reduce genetic variability upon which survival and adaptation depends. A second bottleneck, while perhaps not having immediate ramifications, would likely have a negative impact on the long-term survival of the species. One would expect additional reductions in genetic diversity.

(c) Because extinction of an organism is irreversible, one might consider the fate of the ferret as the highest priority. If the prairie dog population is reduced very slowly, the ferret population may succeed in finding alternative food sources, but this is doubtful. Since the ferret population is the most fragile of the three (ferret, prairie dog, cattle) and represents one of America's most endangered mammals, this case will test the strength of laws designed to protect such species. In some situations, compromise to the point of mutual agreement is not possible.

18. DNA profiles indicate the degree of heterogeneity in DNA sequences and therefore the degree of genetic variation. While noncoding DNA sequences represent the bulk of sequence diversity, such information can be helpful in determining gene flow, ancestry, and overall inter- and intrapopulation diversity. Since diversity per se appears to be essential for long-term species survival in natural environments, one would expect that the assessment of diversity by any tool will be a useful predictor.

20. A census of population size and range would be needed to establish levels of habitat exploitation and probable number of effective breeding pairs. From this information, an estimation of long-term habitat support can be provided along with the probability of genetic drift eroding genetic variability. Effective breeder estimates will also provide a method for estimating inbreeding depression. It would be helpful to conduct surveys on a season-to-season and year-to-year basis to decrease the possibility of sampling in an atypical season or year. It would be important to determine the age and stage-specific structure of the endangered population. Few young or juveniles might indicate that reproductive capacities are in decline. It would be important to determine the general-level biodiversity and carrying capacity of the habitat as well as the genetic diversity of the species in question. Nuclear, mitochondrial, and chloroplast DNA profiles can be used to assess intrapopulation and interpopulation variation as well as to aid in determining migration and breeding patterns.

22. While flagship species (often large mammals) may make it possible to gather considerable public support and funding, they may reduce support for species that may have a greater impact on a community of species. Primary producers (plants) are a necessary component of a diverse and supportive habitat. If one focuses on a flagship species within an area, it is possible that other areas will suffer more dramatically because foundational species are lost. Using umbrella species to protect a large geographic area in hopes of protecting other species in that area is a reasonable approach. However, the size of an area is not necessarily a primary factor in determining species success. Diversity and productivity of a habitat are major contributors to species success. Since land is at a premium, it may be wiser in the long run to select umbrella species in diverse and productive habitats rather than on the basis of land size. By selecting sets of species that show considerable biodiversity, one increases the likelihood of protecting a sufficiently rich habitat to support many species. Such habitats are often of considerable economic value, thereby making their availability limited.

Glossary

abortive transduction An event in which transducing DNA fails to be incorporated into the recipient chromosome. See *transduction*.

accession number An identifying number or code assigned to a nucleotide or amino acid sequence for entry and cataloging in a database.

acentric chromosome Chromosome or chromosome fragment with no centromere.

acridine dyes A class of organic compounds that bind to DNA and intercalate into the double-stranded structure, producing local disruptions of base pairing. These disruptions result in nucleotide additions or deletions in the next round of replication.

acrocentric chromosome Chromosome with the centromere located very close to one end. Human chromosomes 13, 14, 15, 21, and 22 are acrocentric.

active site The substrate-binding site of an enzyme; in other proteins, the portion whose structural integrity is required for function.

adaptation A heritable component of the phenotype that confers an advantage in survival and reproductive success. The process by which organisms adapt to current environmental conditions.

additive genes See *polygenic inheritance*.

additive variance Genetic variance attributed to the substitution of one allele for another at a given locus. This variance can be used to predict the rate of response to phenotypic selection in quantitative traits.

A-DNA An alternative form of right-handed, double-helical DNA. Its helix is more tightly coiled than the more common B-DNA, with 11 base pairs per full turn. In the A form, the bases in the helix are displaced laterally and tilted in relation to the longitudinal axis. It is not yet clear whether this form has biological significance. See *B-DNA*.

albinism A condition caused by the lack of melanin production in the iris, hair, and skin. In humans, it is most often inherited as an autosomal recessive trait.

alkaptonuria An autosomal recessive condition in humans caused by lack of the enzyme homogentisic acid oxidase. Urine of homozygous individuals turns dark upon standing because of oxidation of excreted homogentisic acid. The cartilage of homozygous adults blackens from deposition of a pigment derived from homogentisic acid. Affected individuals often develop arthritic conditions.

allele One of the possible alternative forms of a gene, often distinguished from other alleles by phenotypic effects.

allele-specific oligonucleotide (ASO) Synthetic nucleotides, usually 15–20 bp in length, that under carefully controlled conditions will hybridize only to a perfectly matching complementary sequence.

allelic exclusion In a plasma cell heterozygous for an immunoglobulin gene, the selective action of only one allele.

allelism test See *complementation test*.

allolactose A lactose derivative that acts as the inducer for the *lac* operon.

allopatric speciation Process of speciation associated with geographic isolation.

allopolyploid Polyploid condition formed by the union of two or more distinct chromosome sets with a subsequent doubling of chromosome number.

allosteric effect Conformational change in the active site of a protein brought about by interaction with an effector molecule.

allotetraploid An allopolyploid containing two genomes derived from different species.

allozyme An allelic form of a protein that can be distinguished from other forms by electrophoresis.

alternative splicing Generation of different protein molecules from the same pre-mRNA by incorporation of a different set and order of exons into the mRNA product.

***Alu* sequence** A DNA sequence of approximately 300 bp found interspersed within the genomes of primates that is cleaved by the restriction enzyme *Alu* I. In humans, 300,000–600,000 copies are dispersed throughout the genome and constitute some 3–6 percent of the genome. See *short interspersed elements*.

amber codon The codon UAG, which does not code for an amino acid but for chain termination. One of the stop codons.

Ames test A bacterial assay developed by Bruce Ames to detect mutagenic compounds; it assesses reversion to histidine independence in the bacterium *Salmonella typhimurium*.

amino acids Aminocarboxylic acids comprising the subunits that are covalently linked to form proteins.

aminoacyl tRNA A covalently linked combination of an amino acid and a tRNA molecule. Also referred to as a charged tRNA.

amniocentesis A procedure in which fluid and fetal cells are withdrawn from the amniotic layer surrounding the fetus; used for genetic testing of the fetus.

amphidiploid Same as *allotetraploid*.

anabolism The metabolic synthesis of complex molecules from less complex precursors.

analog A chemical compound that differs structurally from a similar compound but whose chemical behavior is the same. Used experimentally to provide improved detection during analysis. See also *base analog*.

anaphase Stage of mitosis or meiosis in which chromosomes begin moving to opposite poles of the cell.

anaphase I The stage in the first meiotic division during which members of homologous pairs of chromosomes separate from one another.

aneuploidy A condition in which the chromosome number is not an exact multiple of the haploid set.

annotation Analysis of genomic nucleotide sequence data to identify the protein-coding genes, the nonprotein-coding genes, and the regulatory sequences and function(s) of each gene.

antibody Protein (immunoglobulin) produced in response to an antigenic stimulus with the capacity to bind specifically to an antigen.

anticipation See *genetic anticipation*.

anticodon In a tRNA molecule, the nucleotide triplet that binds to its complementary codon triplet in an mRNA molecule.

antigen   A molecule, often a cell-surface protein, that is capable of eliciting the formation of antibodies.

antiparallel   A term describing molecules in parallel alignment but running in opposite directions. Most commonly used to describe the opposite orientations of the two strands of a DNA molecule.

antisense oligonucleotide   A short, single-stranded DNA or RNA molecule complementary to a specific sequence.

antisense RNA   An RNA molecule (synthesized *in vivo* or *in vitro*) with a ribonucleotide sequence that is complementary to part of an mRNA molecule.

apoptosis   A genetically controlled program of cell death, activated as part of normal development or as a result of cell damage.

artificial selection   See *selection*.

ascospore   A meiotic spore produced in certain fungi.

ascus   In fungi, the sac enclosing the four or eight ascospores.

asexual reproduction   Production of offspring in the absence of any sexual process.

assortative mating   Nonrandom mating between males and females of a species. Positive assortative mating selects mates with the same genotype; negative selects mates with opposite genotypes.

ATP   Adenosine triphosphate, a nucleotide that is the main energy source in cells.

attached-X chromosome   Two conjoined X chromosomes that share a single centromere and thus migrate together during cell division.

attenuator   A nucleotide sequence between the promoter and the structural gene of some bacterial operons that regulates the transit of RNA polymerase, reducing transcription of the related structural gene.

autogamy   A process of self-fertilization resulting in homozygosis.

autoimmune disease   The production of antibodies that results from an immune response to one's own molecules, cells, or tissues. Such a response results from the inability of the immune system to distinguish self from nonself. Diseases such as arthritis, scleroderma, systemic lupus erythematosus, and juvenile-onset diabetes are autoimmune diseases.

autonomously replicating sequences (ARS)   Origins of replication, about 100 nucleotides in length, found in yeast chromosomes. ARS elements are also present in organelle DNA.

autopolyploidy   Polyploid condition resulting from the duplication of one diploid set of chromosomes.

autoradiography   Production of a photographic image by radioactive decay. Used to localize radioactively labeled compounds within cells and tissues or to identify radioactive probes in various blotting techniques. See *Southern blotting*.

autosomes   Chromosomes other than the sex chromosomes. In humans, there are 22 pairs of autosomes.

autotetraploid   An autopolyploid condition composed of four copies of the same genome.

auxotroph   A mutant microorganism or cell line that requires a nutritional substance for growth that can be synthesized and is not required by the wild-type strain.

backcross   A cross between an F_1 heterozygote and one of the P_1 parents (or an organism with a genotype identical to one of the parents).

bacteriophage   A virus that infects bacteria, using it as the host for reproduction (also, *phage*).

balanced lethals   Recessive, nonallelic lethal genes, each carried on different homologous chromosomes. When organisms carrying balanced lethal genes are interbred, only organisms with genotypes identical to the parents (heterozygotes) survive.

balanced polymorphism   Genetic polymorphism maintained in a population by natural selection.

balanced translocation carrier   An individual with a chromosomal translocation in which there has been an exchange of genetic information with no associated extra or missing genetic material.

balancer chromosome   A chromosome containing one or more inversions that suppress crossing over with its homolog and which carries a dominant marker that is usually lethal when homozygous.

Barr body   Densely staining DNA-positive mass seen in the somatic nuclei of mammalian females. Discovered by Murray Barr, this body represents an inactivated X chromosome.

base analog   A purine or pyrimidine base that differs structurally from one normally used in biological systems but whose chemical behavior is the same. Used experimentally to provide improved detection during analysis, for example, 5-bromouracil, which "looks like" thymidine, substitutes for it, and can be detected because of its increased mass. See also *analog*.

base pair   See *nucleotide pair*.

base substitution   A single base change in a DNA molecule that produces a mutation. There are two types of substitutions: *transitions*, in which a purine is substituted for a purine, or a pyrimidine for a pyrimidine; and *transversions*, in which a purine is substituted for a pyrimidine or vice versa.

B-DNA   The conformation of DNA which is most often found in cells and serves as the basis of the Watson-Crick double-helical model. There are 10 base pairs per full turn of its right-handed helix, with the nucleotides stacked 0.34 nm apart. The helix has a diameter of 2.0 nm.

β-galactosidase   A bacterial enzyme, encoded by the *lacZ* gene, that converts lactose into galactose and glucose.

bidirectional replication   A mechanism of DNA replication in which two replication forks move in opposite directions from a common origin.

biodiversity   The genetic diversity present in populations and species of plants and animals.

bioinformatics   A field that focuses on the design and use of software and computational methods for the storage, analysis, and management of biological information such as nucleotide or amino acid sequences.

biometry   The application of statistics and statistical methods to biological problems.

biotechnology   Commercial and/or industrial processes that utilize biological organisms or products.

bivalents   Synapsed homologous chromosomes in the first prophase of meiosis.

BLAST (Basic Local Alignment Search Tool)   Any of a family of search engines designed to compare or query nucleotide or amino acid sequences against sequences in databases. BLAST also calculates the statistical significance of the matches.

Bombay phenotype   A rare variant of the ABO antigen system in which affected individuals do not have A or B antigens and thus appear to have blood type O, even though their genotype may carry unexpressed alleles for the A and/or B antigens.

bottleneck   See *population bottleneck*.

bovine spongiform encephalopathy (BSE)   A fatal, degenerative brain disease of cattle (transmissible to humans and other animals) caused by prion infection. Also known as mad cow disease.

BrdU (5-bromodeoxyuridine) A mutagenically active analog of thymidine in which the methyl group at the $5'$ position in thymine is replaced by bromine; also abbreviated BUdR.

broad heritability That proportion of total phenotypic variance in a population that can be attributed to genotypic variance.

buoyant density A property of particles (and molecules) that depends on their actual density, as determined by partial specific volume and degree of hydration. It provides the basis for density-gradient separation of molecules or particles.

CAAT box A highly conserved DNA sequence found in the untranslated promoter region of eukaryotic genes. This sequence is recognized by transcription factors.

cancer stem cells Tumor-forming cells in a cancer that can give rise to all the cell types in a particular form of cancer. These cells have the properties of normal stem cells: self-renewal and ability to differentiate into multiple cell types.

CAP Catabolite activator protein; a protein that binds cAMP and regulates the activation of inducible operons.

capillary electrophoresis A group of analytical methods that separates large and small charged molecules in a capillary tube by their size to charge ratio. Analysis of separated components takes place in the capillary usually by use of a UV detector.

carcinogen A physical or chemical agent that causes cancer.

carrier An individual heterozygous for a recessive trait.

catabolism A metabolic reaction in which complex molecules are broken down into simpler forms, often accompanied by the release of energy.

catabolite activator protein See *CAP*.

catabolite repression The selective inactivation of an operon by a metabolic product of the enzymes encoded by the operon.

***cdc* mutation** A class of cell division cycle (*cdc*) mutations in yeasts that affect the timing of and progression through the cell cycle.

cDNA (complementary DNA) DNA synthesized from an RNA template by the enzyme reverse transcriptase.

cDNA library A collection of cloned cDNA sequences.

cell cycle The sequence of growth phases of an individual cell; divided into G1 (gap 1), S (DNA synthesis), G2 (gap 2), and M (mitosis). A cell may temporarily or permanently be withdrawn from the cell cycle, in which case it is said to enter the G0 stage.

cell-free extract A preparation of the soluble fraction of cells, made by lysing cells and removing the particulate matter, such as nuclei, membranes, and organelles. Often used to carry out the synthesis of proteins by the addition of specific, exogenous mRNA molecules.

CEN The DNA region of centromeres critical to their function. In yeasts, fragments of chromosomal DNA, about 120 bp in length, that when inserted into plasmids confer the ability to segregate during mitosis.

centimorgan (cM) A unit of distance between genes on chromosomes representing 1 percent crossing over between two genes. Equivalent to 1 map unit (mu).

central dogma The concept that genetic information flow progresses from DNA to RNA to proteins. Although exceptions are known, this idea is central to an understanding of gene function.

centric fusion See *Robertsonian translocation.*

centriole A cytoplasmic organelle composed of nine groups of microtubules, generally arranged in triplets. Centrioles function in the generation of cilia and flagella and serve as foci for the spindles in cell division.

centromere The specialized heterochromatic chromosomal region at which sister chromatids remain attached after replication, and the site to which spindle fibers attach to the chromosome during cell division. Location of the centromere determines the shape of the chromosome during the anaphase portion of cell division. Also known as the primary constriction.

centrosome Region of the cytoplasm containing a pair of centrioles.

CFTR The protein (cystic fibrosis transmembrane conductance regulator) encoded by the *CFTR* gene. The protein regulates the movement of chloride ions across the plasma membrane of epithelial cells. Mutations in *CFTR* lead to cystic fibrosis.

chaperone A protein that regulates the folding of a polypeptide into a functional three-dimensional shape.

character An observable phenotypic attribute of an organism.

charon phages A group of genetically modified lambda phages designed to be used as vectors (carriers) for cloning foreign DNA. Named after the ferryman in Greek mythology who carried the souls of the dead across the River Styx.

chemotaxis Movement of a cell or organism in response to a chemical gradient.

chiasma (pl., chiasmata) The crossed strands of nonsister chromatids seen in diplotene of the first meiotic division. Regarded as the cytological evidence for exchange of chromosomal material, or crossing over.

ChIP-on-chip A technique that combines chromatin immunoprecipitation (ChIP) with microarrays (chips) to identify and localize DNA sites that bind DNA-binding proteins of interest. These sites may be enhancers, promoters, transcription initiation sites, or other functional DNA regions.

chi-square (χ^2) analysis Statistical test to determine whether or not an observed set of data is equivalent to a theoretical expectation.

chloroplast A self-replicating cytoplasmic organelle containing chlorophyll. The site of photosynthesis.

chorionic villus sampling (CVS) A technique of prenatal diagnosis in which chorionic fetal cells are retrieved intravaginally and used to detect cytogenetic and biochemical defects in the embryo.

chromatid One of the longitudinal subunits of a replicated chromosome.

chromatin The complex of DNA, RNA, histones, and nonhistone proteins that make up uncoiled chromosomes, characteristic of the eukaryotic interphase nucleus.

chromatin immunoprecipitation (ChIP) An analytical method used to identify DNA-binding proteins that bind to DNA sequences of interest. In ChIP, antibodies to specific proteins are used to isolate DNA sequences that bind these proteins. See *ChIP-on-chip.*

chromatin remodeling A process in which the structure of chromatin is altered by a protein complex, resulting in changes in the transcriptional state of genes in the altered region.

chromatography Technique for the separation of a mixture of solubilized molecules by their differential migration over a substrate.

chromocenter In polytene chromosomes, an aggregation of centromeres and heterochromatic elements where the chromosomes appear to be attached together.

chromomere A coiled, beadlike region of a chromosome, most easily visualized during cell division. The aligned chromomeres of polytene chromosomes are responsible for their distinctive banding pattern.

chromosomal aberration   Any duplication, deletion, or rearrangement of the otherwise diploid chromosomal content of an organism.

chromosomal mutation   The process resulting in the duplication, deletion, or rearrangements of the diploid chromosomal content of an organism. See also *chromosomal aberration*.

chromosomal polymorphism   Possession of alternative structures or arrangements of a chromosome among members of a population.

chromosome   In prokaryotes, a DNA molecule containing the organism's genome; in eukaryotes, a DNA molecule complexed with RNA and proteins to form a threadlike structure containing genetic information arranged in a linear sequence; a structure that is visible during mitosis and meiosis.

chromosome banding   Technique for the differential staining of mitotic or meiotic chromosomes to produce a characteristic banding pattern; or selective staining of certain chromosomal regions such as centromeres, the nucleolus organizer regions, and GC- or AT-rich regions. Not to be confused with the banding pattern present in polytene chromosomes, which is produced by the alignment of chromomeres.

chromosome map   A diagram showing the location of genes on chromosomes.

chromosome puff   A localized uncoiling and swelling in a polytene chromosome, usually regarded as a sign of active transcription.

chromosome theory of inheritance   The idea put forward independently by Walter Sutton and Theodore Boveri that chromosomes are the carriers of genes and the basis for the Mendelian mechanisms of segregation and independent assortment.

chromosome walking   A method for analyzing long stretches of DNA. The end of a cloned segment of DNA is subcloned and used as a probe to identify other clones that overlap the first clone.

cis-acting sequence   A DNA sequence that regulates the expression of a gene located on the same chromosome. This contrasts with a trans-acting element where regulation is under the control of a sequence on the homologous chromosome. See also *trans-acting element*.

cis configuration   The arrangement of two genes (or two mutant sites within a gene) on the same homolog, such as

$$\frac{a^1 \quad a^2}{+ \quad +}$$

This contrasts with a *trans* arrangement, where the mutant alleles are located on opposite homologs. See also *trans configuration*.

cis–trans test   A genetic test to determine whether two mutations are located within the same cistron (or gene).

cistron   That portion of a DNA molecule, often referring to a gene, coding for a single-polypeptide chain; defined by a genetic test as a region within which two mutations cannot complement each other.

cline   A gradient of genotype or phenotype distributed over a geographic range.

clonal selection theory   Proposed explanation in immunology that antibody diversity precedes exposure to the antigen and that the antigen functions to select the cells containing its specific antibody to undergo proliferation.

clone   Identical molecules, cells, or organisms derived from a single ancestor by asexual or parasexual methods; for example, a DNA segment that has been inserted into a plasmid or chromosome of a phage or a bacterium and replicated to produce many copies, or an organism with a genetic composition identical to that used in its production.

codominance   Condition in which the phenotypic effects of a gene's alleles are fully and simultaneously expressed in the heterozygote.

codon   A triplet of nucleotides that specifies a particular amino acid or a start or stop signal in the genetic code. Sixty-one codons specify the amino acids used in proteins, and three codons, called stop codons, signal termination of growth of the polypeptide chain. One codon acts as a start codon in addition to specifying an amino acid.

coefficient of coincidence   A ratio of the observed number of double crossovers divided by the expected number of such crossovers.

coefficient of inbreeding   The probability that two alleles present in a zygote are descended from a common ancestor.

coefficient of selection (s)   A measurement of the reproductive disadvantage of a given genotype in a population. For example, for genotype *aa* if only 99 of 100 individuals reproduce, then the selection coefficient is 0.01.

cohesin   A protein complex that holds sister chromatids together during mitosis and meiosis and facilitates attachments of spindle fibers to kinetochores.

colchicine   An alkaloid compound that inhibits spindle formation during cell division. In the preparation of karyotypes, it is used for collecting a large population of cells inhibited at the metaphase stage of mitosis.

colinearity   The linear relationship between the nucleotide sequence in a gene (or the RNA transcribed from it) and the order of amino acids in the polypeptide chain specified by the gene.

Combined DNA Index System (CODIS)   A standardized set of 13 short tandem repeat (STR) DNA sequences used by law enforcement and government agencies in preparing DNA profiles.

comparative genomic hybridization (CGH)   A microarray-based method for the analysis of copy number variations in genomic DNA or in specific cell types, such as tumor cells.

competence   In bacteria, the transient state or condition during which the cell can bind and internalize exogenous DNA molecules, making transformation possible.

complementarity   Chemical affinity between nitrogenous bases of nucleic acid strands as a result of hydrogen bonding. Responsible for the base pairing between the strands of the DNA double helix and between DNA and RNA strands during genetic expression in cells and during the use of molecular hybridization techniques.

complementation test   A genetic test to determine whether two mutations occur within the same gene (or cistron). If two mutations are introduced into a cell simultaneously and produce a wild-type phenotype (i.e., they complement each other), they are often nonallelic. If a mutant phenotype is produced, the mutations are noncomplementing and are often allelic.

complete linkage   A condition in which two genes are located so close to each other that no recombination occurs between them.

complexity (X)   The total number of nucleotides or nucleotide pairs in a population of nucleic acid molecules as determined by reassociation kinetics.

complex locus   A gene within which a set of functionally related pseudoalleles can be identified by recombinational analysis (e.g., the *bithorax* locus in *Drosophila*).

complex trait   A trait whose phenotype is determined by the interaction of multiple genes and environmental factors.

concatemer   A chain or linear series of subunits linked together. The process of forming a concatemer is called concatenation (e.g., multiple units of a phage genome produced during replication).

concordance   Pairs or groups of individuals with identical phenotypes. In twin studies, a condition in which both twins exhibit or fail to exhibit a trait under investigation.

conditional mutation   A mutation expressed only under a certain condition; that is, a wild-type phenotype is expressed under certain (permissive) conditions and a mutant phenotype under other (restrictive) conditions.

conjugation   Temporary fusion of two single-celled organisms for the sexual transfer of genetic material.

consanguineous   Related by a common ancestor within the previous few generations.

consensus sequence   The sequence of nucleotides in DNA or amino acids in proteins most often present in a particular gene or protein under study in a group of organisms.

conservation genetics   The branch of genetics concerned with the preservation and maintenance of genetic diversity and existing species of plants and animals in their natural environments.

contig   A continuous DNA sequence reconstructed from overlapping DNA sequences derived by cloning or sequence analysis.

continuous variation   Phenotype variation in which quantitative traits range from one phenotypic extreme to another in an overlapping or continuous fashion.

copy number variation (CNV)   DNA segments larger than 1 kb that are repeated a variable number of times in the genome.

cosmid   A vector designed to allow cloning of large segments of foreign DNA. Cosmids are composed of the *cos* sites of phage λ inserted into a plasmid. In cloning, the recombinant DNA molecules are packaged into phage protein coats, and after infection of bacterial cells, the recombinant molecule replicates and can be maintained as a plasmid.

coupling conformation   Same as *cis configuration.*

covalent bond   A nonionic chemical bond, formed by the sharing of electrons.

CpG island   A short region of regulatory DNA found upstream of genes that contain unmethylated stretches of sequence with a high frequency of C and G nucleotides.

Creutzfeldt–Jakob disease (CJD)   A progressive degenerative and fatal disease of the brain and nervous system caused by mutations that lead to aberrant forms of the encoded protein (prions). CJD is inherited as an autosomal dominant trait.

cri-du-chat syndrome   A clinical syndrome produced by a deletion of a portion of the short arm of chromosome 5 in humans. Afflicted infants have a distinctive cry that sounds like a cat.

crossing over   The exchange of chromosomal material (parts of chromosomal arms) between homologous chromosomes by breakage and reunion. The exchange of material between nonsister chromatids during meiosis is the basis of genetic recombination.

cross-reacting material (CRM)   A nonfunctional enzyme produced by a mutant gene, yet recognized by antibodies made against the normal enzyme.

C-terminal amino acid   In a peptide chain, the terminal amino acid that carries a free carboxyl group.

C terminus   In a polypeptide, the end that carries a free carboxyl group on the last amino acid. By convention, the structural formula of polypeptides is written with the C terminus at the right.

C **value**   The haploid amount of DNA present in a genome.

C **value paradox**   The apparent paradox that there is no relationship between the size of the genome and the evolutionary complexity of species. For example, the *C* value (haploid genome size) of amphibians varies by a factor of 100.

cyclic adenosine monophosphate (cAMP)   An important regulatory molecule in both prokaryotic and eukaryotic organisms.

cyclins   In eukaryotic cells, a class of proteins that are synthesized and degraded in synchrony with the cell cycle and regulate passage through stages of the cycle.

cytogenetics   A branch of biology in which the techniques of both cytology and genetics are used in genetic investigations.

cytokinesis   The division or separation of the cytoplasm during mitosis or meiosis.

cytoplasmic inheritance   Non-Mendelian form of inheritance in which genetic information is transmitted through the cytoplasm rather than the nucleus, usually by DNA-containing, self-replicating cytoplasmic organelles such as mitochondria and chloroplasts.

cytoskeleton   An internal array of microtubules, microfilaments, and intermediate filaments that confers shape and motility to a eukaryotic cell.

dalton (Da)   A unit of mass equal to that of the hydrogen atom, which is 1.67×10^{-24} gram. A unit used in designating molecular weights.

Darwinian fitness   Same as *fitness.*

Database of Genomic Variants (DGV)   A catalog of structural variants present in the human genome including copy number variations and inversions.

deficiency   A chromosomal mutation, also referred to as a deletion, involving the loss of chromosomal material.

degenerate code   The representation of a given amino acid by more than one codon.

deletion   A chromosomal mutation, also referred to as a deficiency, involving the loss of chromosomal material. See also *deficiency.*

deme   A local interbreeding population.

denatured DNA   DNA molecules that have separated into single strands.

de novo   Newly arising; synthesized from less complex precursors rather than being produced by modification of an existing molecule.

density gradient centrifugation   A method of separating macromolecular mixtures by the use of centrifugal force and solutions of varying density. In buoyant density gradient centrifugation using cesium chloride, the centrifugal field produces a density gradient in a cesium solution through which a mixture of macromolecules such as DNA will migrate until each component reaches a point equal to its own density. Sometimes referred to as equilibrium sedimentation centrifugation.

deoxyribonuclease (DNAse)   A class of enzymes that breaks down DNA into oligonucleotide fragments by introducing single-stranded or double-stranded breaks into the double helix.

deoxyribonucleic acid (DNA)   A macromolecule usually consisting of nucleotide polymers comprising antiparallel chains in which the sugar residues are deoxyribose and which are held together by hydrogen bonds. The primary carrier of genetic information.

deoxyribose   The five-carbon sugar associated with the deoxyribonucleotides in DNA.

dermatoglyphics   The study of the surface ridges of the skin, especially of the hands and feet.

determination Establishment of a specific pattern of gene activity and developmental fate for a given cell, usually prior to any manifestation of the cell's future phenotype.

diakinesis The final stage of meiotic prophase I, in which the chromosomes become tightly coiled and compacted and move toward the periphery of the nucleus.

dicentric chromosome A chromosome having two centromeres, which can be pulled in opposite directions during anaphase of cell division.

dicer An enzyme (a ribonuclease) that cleaves double-stranded RNA (dsRNA) and pre-micro RNA (miRNA) to form small interfering RNA (siRNA) molecules about 20–25 nucleotides long that serve as guide molecules for the degradation of mRNA molecules with sequences complementary to the siRNA.

dideoxynucleotide A nucleotide containing a deoxyribose sugar lacking a 3′ hydroxyl group. It stops further chain elongation when incorporated into a growing polynucleotide and is used in the Sanger method of DNA sequencing.

dideoxynucleotide chain-terminating sequencing See *Sanger sequencing*.

differentiation The complex process of change by which cells and tissues attain their adult structure and functional capacity.

dihybrid cross A genetic cross involving two characters in which the parents possess different forms of each character (e.g., yellow, round × green, wrinkled peas).

diploid (2*n*) A condition in which each chromosome exists in pairs; having two of each chromosome.

diplotene The stage of meiotic prophase I immediately following pachytene. In diplotene, the sister chromatids begin to separate, and chiasmata become visible. These cross-like overlaps move toward the ends of the chromatids (terminalization).

directed mutagenesis See *gene targeting*.

directional selection A selective force that changes the frequency of an allele in a given direction, either toward fixation or toward elimination.

discontinuous replication of DNA The synthesis of DNA in discontinuous fragments on the lagging strand during replication. The fragments, known as Okazaki fragments, are subsequently joined by DNA ligase to form a continuous strand.

discontinuous variation Pattern of variation for a trait whose phenotypes fall into two or more distinct classes.

discordance In twin studies, a situation where one twin expresses a trait but the other does not.

disjunction The separation of chromosomes during the anaphase stage of cell division.

disruptive selection Simultaneous selection for phenotypic extremes in a population, usually resulting in the production of two phenotypically discontinuous strains.

dizygotic twins Twins produced from separate fertilization events; two ova fertilized independently. Also known as fraternal twins.

DNA See *deoxyribonucleic acid*.

DNA cloning vectors See *vector*.

DNA fingerprinting A molecular method for identifying an individual member of a population or species. A unique pattern of DNA fragments is obtained by restriction enzyme digestion followed by Southern blot hybridization using minisatellite probes. See also *DNA profiling; STR sequences*.

DNA footprinting A technique for identifying a DNA sequence that binds to a particular protein, based on the idea that the phosphodiester bonds in the region covered by the protein are protected from digestion by deoxyribonucleases.

DNA gyrase One of a class of enzymes known as topoisomerases that converts closed circular DNA to a negatively supercoiled form prior to replication, transcription, or recombination. The enzyme acts during DNA replication to reduce molecular tension caused by supercoiling.

DNA helicase An enzyme that participates in DNA replication by unwinding the double helix near the replication fork.

DNA ligase An enzyme that forms a covalent bond between the 5′ end of one polynucleotide chain and the 3′ end of another polynucleotide chain. It is also called polynucleotide-joining enzyme.

DNA microarray An ordered arrangement of DNA sequences or oligonucleotides on a substrate (often glass). Microarrays are used in quantitative assays of DNA–DNA or DNA–RNA binding to measure profiles of gene expression (for example, during development or to compare the differences in gene expression between normal and cancer cells).

DNA polymerase An enzyme that catalyzes the synthesis of DNA from deoxyribonucleotides utilizing a template DNA molecule.

DNA profiling A method for identification of individuals that uses variations in the length of short tandem repeating DNA sequences (STRs) that are widely distributed in the genome.

DNase Deoxyribonucleosidase, an enzyme that degrades or breaks down DNA into fragments or constitutive nucleotides. See also *endonuclease; exonuclease; restriction endonuclease*.

DNase I A nuclease that preferentially cleaves DNA at phosphodiester bonds adjacent to pyrimidine nucleotides. DNase I can act on single- and double-stranded DNA and on DNA in chromatin.

dominance The expression of a trait in the heterozygous condition.

dominant negative mutation A mutation whose gene product acts in opposition to the normal gene product, usually by binding to it to form dimers.

dosage compensation A genetic mechanism that equalizes the levels of expression of genes at loci on the X chromosome. In mammals, this is accomplished by random inactivation of one X chromosome, leading to Barr body formation.

double crossover Two separate events of chromosome breakage and exchange occurring within the same tetrad.

double helix The model for DNA structure proposed by James Watson and Francis Crick, in which two antiparallel hydrogen-bonded polynucleotide chains are wound into a right-handed helical configuration 2 nm in diameter, with 10 base pairs per full turn.

driver mutation A mutation in a cancer cell that contributes to tumor progression.

Duchenne muscular dystrophy An X-linked recessive genetic disorder caused by a mutation in the gene for dystrophin, a protein found in muscle cells. Affected males show a progressive weakness and wasting of muscle tissue. Death ensues by about age 20 caused by respiratory infection or cardiac failure.

duplication A chromosomal aberration in which a segment of the chromosome is repeated.

dyad The products of tetrad separation or disjunction at meiotic prophase I. Each dyad consists of two sister chromatids joined at the centromere.

dystrophin A protein that attaches to the inside of the muscle cell plasma membrane and stabilizes the membrane during muscle contraction. Mutations in the gene-encoding dystrophin cause Duchenne and Becker muscular dystrophy. See *Duchenne muscular dystrophy*.

effective population size   In a population, the number of individuals with an equal probability of contributing gametes to the next generation.

effector molecule   A small biologically active molecule that regulates the activity of a protein by binding to a specific receptor site on the protein.

electrophoresis   A technique that separates a mixture of molecules by their differential migration through a stationary medium (such as a gel) under the influence of an electrical field.

electroporation   A technique that uses an electric pulse to move polar molecules across the plasma membrane into the cell.

ELSI (Ethical, Legal, Social Implications)   A program established by the National Human Genome Research Institute in 1990 as part of the Human Genome Project to sponsor research on the ethical, legal, and social implications of genomic research and its impact on individuals and social institutions.

embryonic stem cells   Cells derived from the inner cell mass of early blastocyst mammalian embryos. These cells are pluripotent, meaning they can differentiate into any of the embryonic or adult cell types characteristic of the organism.

endocytosis   The uptake by a cell of fluids, macromolecules, or particles by pinocytosis, phagocytosis, or receptor-mediated endocytosis.

endomitosis   An increased DNA content in multiples of the haploid amount occurring in the absence of nuclear or cytoplasmic division. Polytene chromosomes are formed by endomitosis.

endonuclease   An enzyme that hydrolyzes internal phosphodiester bonds in a single- or double-stranded polynucleotide chain.

endoplasmic reticulum (ER)   A membranous organelle system in the cytoplasm of eukaryotic cells. In rough ER, the outer surface of the membranes is ribosome-studded; in smooth ER, it is not.

endopolyploidy   The increase in chromosome sets within somatic nuclei that results from endomitotic replication.

endosymbiont theory   The proposal that self-replicating cellular organelles such as mitochondria and chloroplasts were originally free-living organisms that entered into a symbiotic relationship with nucleated cells.

enhancer   A DNA sequence that enhances transcription and the expression of structural genes. Enhancers can act over a distance of thousands of base pairs and can be located upstream, downstream, or internal to the gene they affect, differentiating them from promoters.

environment   The complex of geographic, climatic, and biotic factors within which an organism lives.

enzyme   A protein or complex of proteins that catalyzes a specific biochemical reaction by lowering the energy of activation that is normally required to initiate the reaction.

epidermal growth-factor receptor   A cell-surface receptor for proteins belonging to the family of epidermal growth factors. Binding of a growth factor to the receptor stimulates a signal transduction pathway. In humans, this protein is encoded by the *HER* gene, and mutant alleles are associated with cancer.

epigenesis   The idea that an organism or organ arises through the sequential appearance and development of new structures, in contrast to preformationism, which holds that development is the result of the assembly of structures already present in the egg.

epigenetics   The study of modifications in an organism's gene function or phenotypic expression that are not attributable to alterations in the nucleotide sequence (mutations) of the organism's DNA.

episome   In bacterial cells, a circular genetic element that can replicate independently of the bacterial chromosome or integrate and replicate as part of the chromosome.

epistasis   Nonreciprocal interaction between nonallelic genes such that one gene influences or interferes with the expression of another gene, leading to a specific phenotype.

epitope   That portion of a macromolecule or cell that acts to elicit an antibody response; an antigenic determinant. A complex molecule or cell can contain several such sites.

equational division   A division stage where the number of centromeres is not reduced by half but where each chromosome is split into longitudinal halves that are distributed into two daughter nuclei. Chromosome division in mitosis and the second meiotic division are examples of equational divisions. See also *reductional division*.

equatorial plate   See *metaphase plate*.

euchromatin   Chromatin or chromosomal regions that are lightly staining and are relatively uncoiled during the interphase portion of the cell cycle. Euchromatic regions contain most of the structural genes.

eugenics   A movement advocating the improvement of the human species by selective breeding. Positive eugenics refers to the promotion of breeding between people thought to possess favorable genes, and negative eugenics refers to the discouragement of breeding among those thought to have undesirable traits.

eukaryotes   Organisms having true nuclei and membranous organelles and whose cells divide by mitosis and meiosis.

euphenics   Medical or genetic intervention to reduce the impact of defective genotypes.

euploid   Polyploid with a chromosome number that is an exact multiple of a basic chromosome set.

evolution   Descent with modification. The emergence of new kinds of plants and animals from preexisting types.

excision repair   Removal of damaged DNA segments followed by repair. Excision can include the removal of individual bases (base repair) or of a stretch of damaged nucleotides (nucleotide repair).

exon   The DNA segments of a gene that contain the sequences that, through transcription and translation, are eventually represented in the final polypeptide product.

exonuclease   An enzyme that breaks down nucleic acid molecules by breaking the phosphodiester bonds at the 3'- or 5'-terminal nucleotides.

expressed sequence tag (EST)   All or part of the nucleotide sequence of a cDNA clone. ESTs are used as markers in the construction of genetic maps.

expression vector   Plasmids or phages carrying promoter regions designed to cause expression of inserted DNA sequences.

expressivity   The degree to which a phenotype for a given trait is expressed.

extranuclear inheritance   Transmission of traits by genetic information contained in cytoplasmic organelles such as mitochondria and chloroplasts.

5-methylcytosine   A methylated cytosine base that represents an epigenetic modification that alters gene expression.

F$^-$ cell   A bacterial cell that does not contain a fertility factor and that acts as a recipient in bacterial conjugation.

F$^+$ cell   A bacterial cell that contains a fertility factor and that acts as a donor in bacterial conjugation.

F factor   An episomal plasmid in bacterial cells that confers the ability to act as a donor in conjugation. See also *fertility factor*.

F' factor   A fertility factor that contains a portion of the bacterial chromosome.

F₁ generation First filial generation; the progeny resulting from the first cross in a series.

F₂ generation Second filial generation; the progeny resulting from a cross of the F₁ generation.

F pilus On bacterial cells possessing an F factor, a filament-like projection that plays a role in conjugation.

familial trait A trait transmitted through and expressed by members of a family. Usually used to describe a trait that runs in families, but whose precise mode of inheritance is not clear.

fate map A diagram of an embryo showing the location of cells whose developmental fate is known.

fertility factor See *F factor*.

fetal cell sorting A noninvasive method of prenatal diagnosis that recovers and tests fetal cells from the maternal circulation.

filial generations See *F₁, F₂ generations*.

fingerprint The unique pattern of ridges and whorls on the tip of a human finger. Also, the pattern obtained by enzymatically cleaving a protein or nucleic acid and subjecting the digest to two-dimensional chromatography or electrophoresis. See also *DNA fingerprinting*.

FISH See *fluorescence* in situ *hybridization*.

fitness A measure of the relative survival and reproductive success of a given individual or genotype.

fixation In population genetics, a condition in which all members of a population are homozygous for a given allele.

fluctuation test A statistical test developed by Salvadore Luria and Max Delbrück demonstrating that bacterial mutations arise spontaneously, in contrast to being induced by selective agents.

fluorescence *in situ* **hybridization (FISH)** A method of *in situ* hybridization that utilizes probes labeled with a fluorescent tag, causing the site of hybridization to fluoresce when viewed using ultraviolet light.

flush–crash cycle A period of rapid population growth followed by a drastic reduction in population size.

f-met See *formylmethionine*.

folded-fiber model A model of eukaryotic chromosome organization in which each sister chromatid consists of a single chromatin fiber composed of double-stranded DNA and proteins wound like a tightly coiled skein of yarn.

footprinting See *DNA footprinting*.

forensic DNA fingerprinting In law enforcement, the use of a standard set of short tandem repeats (STRs) to identify an individual. Also known as DNA typing. See also *DNA profiling*.

forensic science The use of laboratory scientific methods to obtain data used in criminal and civil law cases.

formylmethionine (f-met) A molecule derived from the amino acid methionine by attachment of a formyl group to its terminal amino group. This is the first monomer used in the synthesis of all bacterial polypeptides. Also known as *N*-formyl methionine.

forward genetics The classical approach used to identify a gene controlling a phenotypic trait in the absence of knowledge of the gene's location in the genome or its DNA sequence. Accomplished by isolating mutant alleles and mapping the gene's location, most traditionally using recombination analysis. Once mapped, the gene may be cloned and further studied at the molecular level. An approach contrasted with *reverse genetics*.

founder effect A form of genetic drift. The establishment of a population by a small number of individuals whose genotypes carry only a fraction of the different kinds of alleles in the parental population.

fragile site A heritable gap, or nonstaining region, of a chromosome that can be induced to generate chromosome breaks.

fragile X syndrome A human genetic disorder caused by the expansion of a CGG trinucleotide repeat and a fragile site at Xq27.3 within the *FMR-1* gene. Fragile X syndrome is the most common form of mental retardation in males. Affected males have distinctive facial features and are mentally retarded. Carrier females lack physical symptoms but as a group have higher rates of mental retardation than normal individuals.

frameshift mutation A mutational event leading to the insertion of one or more base pairs in a gene, shifting the codon reading frame in all codons that follow the mutational site.

fraternal twins Same as *dizygotic twins*.

G1 checkpoint A point in the G1 phase of the cell cycle when a cell becomes committed to initiating DNA synthesis and continuing the cycle or withdraws into the G0 resting stage.

G0 A nondividing but metabolically active state that cells may enter from the G1 phase of the cell cycle.

gain-of-function mutation A mutation that produces a phenotype different from that of the normal allele and from any loss of function alleles.

gamete A specialized reproductive cell with a haploid number of chromosomes.

gap genes Genes expressed in contiguous domains along the anterior–posterior axis of the *Drosophila* embryo that regulate the process of segmentation in each domain.

GC box In eukaryotes, a region in a promoter containing a 5′-GGGCGG-3′ sequence, which is a binding site for transcriptional regulatory proteins.

gene The fundamental physical unit of heredity, whose existence can be confirmed by allelic variants and which occupies a specific chromosomal locus. A DNA sequence coding for a single polypeptide.

gene amplification The process by which gene sequences are selected and differentially replicated either extrachromosomally or intrachromosomally.

gene chip See *DNA microarray*.

gene conversion The process of nonreciprocal recombination by which one allele in a heterozygote is converted into the corresponding allele.

gene duplication An event leading to the production of a tandem repeat of a gene sequence during replication.

gene family A number of closely related genes derived from a common ancestral gene by duplication and sequence divergence over evolutionary time.

gene flow The gradual exchange of genes between two populations; brought about by the dispersal of gametes or the migration of individuals.

gene frequency The percentage of alleles of a given type in a population.

gene interaction Production of novel phenotypes by the interaction of alleles of different genes.

gene knockout The introduction of a *null mutation* into a gene that is subsequently introduced into an organism using transgenic techniques, whereby the organism loses the function of the gene. Often used in mice. See also *gene targeting*.

gene mutation See *point mutation*.

gene pool The total of all alleles possessed by the reproductive members of a population.

gene redundancy The presence of several genes in an organism's genome that all have variations of the same function.

gene targeting A transgenic technique used to create and introduce a specifically altered gene into an organism. In mice, gene targeting often involves the induction of a specific mutation in a cloned gene that is then introduced into the genome of a gamete involved in fertilization. The organism produced is bred to produce adults homozygous for the mutation, for example, the creation of a *gene knockout.*

generalized transduction   The transfer of any gene from a bacterial host to a bacterial recipient in a process mediated by a bacteriophage. See also *specialized transduction.*

genetically modified organism (GMO)   A plant or animal whose genome carries a gene transferred from another species by recombinant DNA technology and expressed to produce a gene product.

genetic anticipation   The phenomenon in which the severity of symptoms in genetic disorders increases from generation to generation and the age of onset decreases from generation to generation. It is caused by the expansion of trinucleotide repeats within or near a gene and was first observed in myotonic dystrophy.

genetic background   The impact of the collective genome of an organism on the expression of a gene under investigation.

genetic code   The deoxynucleotide triplets that encode the 20 amino acids or specify termination of translation.

genetic counseling   Analysis of risk for genetic defects in a family and the presentation of options available to avoid or ameliorate possible risks.

genetic drift   Random variation in allele frequency from generation to generation, most often observed in small populations.

genetic engineering   The technique of altering the genetic constitution of cells or individuals by the selective removal, insertion, or modification of individual genes or gene sets.

genetic equilibrium   A condition in which allele frequencies in a population are neither increasing nor decreasing.

genetic erosion   The loss of genetic diversity from a population or a species.

genetic fine structure analysis   Intragenic recombinational analysis that provides intragenic mapping information at the level of individual nucleotides.

genetic load   Average number of recessive lethal genes carried in the heterozygous condition by an individual in a population.

genetic polymorphism   The stable coexistence of two or more distinct genotypes for a given trait in a population. When the frequencies of two alleles for such a trait are in equilibrium, the condition is called a balanced polymorphism.

Genetic Testing Registry (GTR)   A centralized, online database, sponsored by the U.S. government to provide information about the scientific basis and availability of genetic tests.

genetic variability   A measure of the tendency of the genotypes of individuals in a population to vary from one another. Genetic variability can be measured by determining the rate of mutation of specific genes.

genetics   The branch of biology concerned with study of inherited variation. More specifically, the study of the origin, transmission, and expression of genetic information.

genome   The set of hereditary information encoded in the DNA of an organism, including both the protein-coding and non–protein-coding sequences.

genome-wide association studies (GWAS)   Analysis of genetic variation across an entire genome, searching for linkage (associations) between variations in DNA sequences and a genome region encoding a specific phenotype.

genomic imprinting   The process by which the expression of an allele depends on whether it has been inherited from a male or a female parent. Also referred to as parental imprinting.

genomic library   A collection of clones that contains all the DNA sequences of an organism's genome.

genomics   A subdiscipline of the field of genetics generated by the union of classical and molecular biology with the goal of sequencing and understanding genes, gene interaction, genetic elements, as well as the structure and evolution of genomes.

genotype   The allelic or genetic constitution of an organism; often, the allelic composition of one or a limited number of genes under investigation.

germ line   An embryonic cell lineage that forms the reproductive cells (eggs and sperm).

germ plasm   Hereditary material transmitted from generation to generation.

GMO   See *genetically modified organism (GMO).*

Goldberg–Hogness box   A short nucleotide sequence 20–30 bp upstream from the initiation site of eukaryotic genes to which RNA polymerase II binds. The consensus sequence is TATAAAA. Also known as a TATA box.

gynandromorphy   An individual composed of cells with both male and female genotypes.

gyrase   See *DNA gyrase.*

haploid (*n*)   A cell or an organism having one member of each pair of homologous chromosomes. Also referred to as the gametic chromosome number.

haploinsufficiency   In a diploid organism, a condition in which an individual possesses only one functional copy of a gene with the other inactivated by mutation. The amount of protein produced by the single copy is insufficient to produce a normal phenotype, leading to an abnormal phenotype. In humans, this condition is present in many autosomal dominant disorders.

haplotype   A set of alleles from closely linked loci carried by an individual inherited as a unit.

HapMap Project   An international effort by geneticists to identify haplotypes (closely linked genetic markers on a single chromosome) shared by certain individuals as a way of facilitating efforts to identify, map, and isolate genes associated with disease or disease susceptibility.

Hardy–Weinberg law   The principle that genotype frequencies will remain in equilibrium in an infinitely large, randomly mating population in the absence of mutation, migration, and selection.

heat shock   A transient genetic response following exposure of cells or organisms to elevated temperatures. The response includes activation of a small number of loci, inactivation of some previously active loci, and selective translation of heat-shock mRNA.

helicase   See *DNA helicase.*

helix–turn–helix (HTH) motif   In DNA-binding proteins, the structure of a region in which a turn of four amino acids holds two α helices at right angles to each other.

hemizygous   Having a gene present in a single dose in an otherwise diploid cell. Usually applied to genes on the X chromosome in heterogametic males.

hemoglobin (Hb)   An iron-containing, oxygen-carrying multimeric protein occurring chiefly in the red blood cells of vertebrates.

hemophilia   An X-linked trait in humans that is associated with defective blood-clotting mechanisms.

heredity   Transmission of traits from one generation to another.

heritability   A relative measure of the degree to which observed phenotypic differences for a trait are genetic.

heterochromatin   The heavily staining, late-replicating regions of chromosomes that are prematurely condensed in interphase.

heteroduplex   A double-stranded nucleic acid molecule in which each polynucleotide chain has a different origin. It may be produced as an intermediate in a recombinational event or by the *in vitro* reannealing of single-stranded, complementary molecules.

heterogametic sex   The sex that produces gametes containing unlike sex chromosomes. In mammals, the male is the heterogametic sex.

heterogeneous nuclear RNA (hnRNA)   The collection of RNA transcripts in the nucleus, consisting of precursors to and processing intermediates for rRNA, mRNA, and tRNA. Also includes RNA transcripts that will not be transported to the cytoplasm, such as snRNA.

heterokaryon   A somatic cell containing nuclei from two different sources.

heterozygote   An individual with different alleles at one or more loci. Such individuals will produce unlike gametes and therefore will not breed true.

Hfr   Strains of bacteria exhibiting a high frequency of recombination. These strains have a chromosomally integrated F factor that is able to mobilize and transfer part of the chromosome to a recipient F⁻ cell.

high-throughput DNA sequencing   A collection of DNA sequencing methods that outperform the standard (Sanger) method of DNA sequencing by a factor of 100–1000 and reduce sequencing costs by more than 99 percent. Also called *next generation sequencing*.

histocompatibility antigens   See *HLA*.

histone code   Various chemical modifications applied to histone tails (the free ends of histone molecules, projecting from nucleosomes). These modifications influence DNA–histone interactions and promote or repress transcription.

histones   Positively charged proteins complexed with DNA in the nucleus. They are rich in the basic amino acids arginine and lysine, and function in coiling DNA to form nucleosomes.

HLA   Cell-surface proteins produced by histocompatibility loci and involved in the acceptance or rejection of tissue and organ grafts and transplants.

hnRNA   See *heterogeneous nuclear RNA*.

Holliday structure   In DNA recombination, an intermediate seen in transmission electron microscope images as an X-shaped structure showing four single-stranded DNA regions.

homeobox   A sequence of about 180 nucleotides that encodes a sequence of 60 amino acids called a *homeodomain*, which is part of a DNA-binding protein that acts as a transcription factor.

homeotic mutation   A mutation that causes a tissue normally determined to form a specific organ or body part to alter its pathway of differentiation and form another structure.

homogametic sex   The sex that produces gametes that do not differ with respect to sex-chromosome content; in mammals, the female is homogametic.

homologous chromosomes   Chromosomes that synapse or pair during meiosis and that are identical with respect to their genetic loci and centromere placement.

homozygote   An individual with identical alleles for a gene or genes of interest. These individuals will produce identical gametes (with respect to the gene or genes in question) and will therefore breed true.

horizontal gene transfer   The nonreproductive transfer of genetic information from an organism to another, across species and higher taxa (even domains). This mode is contrasted with vertical gene transfer, which is the transfer of genetic information from parent to offspring. In some species of bacteria and archaea, up to 5 percent of the genome may have originally been acquired through horizontal gene transfer.

H substance   The carbohydrate group present on the surface of red blood cells to which the A and/or B antigen may be added. When unmodified, it results in blood type O.

hot spots of mutation   Genome regions where mutations are observed with a high frequency. These include a predisposition toward single-nucleotide substitutions or unequal crossing over.

human immunodeficiency virus (HIV)   An RNA-containing human retrovirus associated with the onset and progression of acquired immunodeficiency syndrome (AIDS).

hybrid   An individual produced by crossing parents from two different genetic compositions or strains.

hybrid vigor   The general superiority of a hybrid over a purebred.

hydrogen bond   A weak electrostatic attraction between a hydrogen atom covalently bonded to an oxygen or a nitrogen atom and an atom that contains an unshared electron pair.

hypervariable regions   The regions of antibody molecules that attach to antigens. These regions have a high degree of diversity in amino acid content.

identical twins   See *monozygotic twins*.

Ig   See *immunoglobulin*.

immunoglobulin (Ig)   The class of serum proteins having the properties of antibodies.

imprinting   See *genomic imprinting*.

inborn error of metabolism   A genetically controlled biochemical disorder; usually an enzyme defect that produces a clinical syndrome.

inbreeding   Mating between closely related organisms.

inbreeding depression   A decrease in viability, vigor, or growth in progeny after several generations of inbreeding.

incomplete dominance   Expressing a heterozygous phenotype that is distinct from the phenotype of either homozygous parent. Also called *partial dominance*.

independent assortment   The independent behavior of each pair of homologous chromosomes during their segregation in meiosis I. The random distribution of maternal and paternal homologs into gametes.

inducer   An effector molecule that activates transcription.

inducible enzyme system   An enzyme system under the control of an inducer, a regulatory molecule that acts to block a repressor and allow transcription.

initiation codon   The nucleotide triplet AUG that in an mRNA molecule codes for incorporation of the amino acid methionine as the first amino acid in a polypeptide chain.

insertion sequence   See *IS element*.

***in situ* hybridization**   A cytological technique for pinpointing the chromosomal location of DNA sequences complementary to a given nucleic acid or polynucleotide.

interference (I)   A measure of the degree to which one crossover affects the incidence of another crossover in an adjacent region of the same chromatid. Negative interference increases the chance of another crossover; positive interference reduces the probability of a second crossover event.

interphase   In the cell cycle, the interval between divisions.

intervening sequence   See *intron*.

intron   Any segment of DNA that lies between coding regions in a gene. Introns are transcribed but are spliced out of the RNA product and are not represented in the polypeptide encoded by the gene.

inversion   A chromosomal aberration in which a chromosomal segment has been reversed.

inversion loop   The chromosomal configuration resulting from the synapsis of homologous chromosomes, one of which carries an inversion.

in vitro   Literally, *in glass*; outside the living organism; occurring in an artificial environment.

in vivo   Literally, *in the living*; occurring within the living body of an organism.

IS element   A mobile DNA segment that is transposable to any of a number of sites in the genome.

isoagglutinogen   An antigenic factor or substance present on the surface of cells that is capable of inducing the formation of an antibody (e.g., the A and B antigens on the surface of human red blood cells).

isochromosome   An aberrant chromosome with two identical arms and homologous loci.

isolating mechanism   Any barrier to the exchange of genes between different populations of a group of organisms. In general, isolation can be classified as spatial, environmental, or reproductive.

isotopes   Alternative forms of atoms with identical chemical properties that have the same atomic number but differ in the number of neutrons (and thus their mass) contained in the nucleus.

isozyme   Any of two or more distinct forms of an enzyme with identical or nearly identical chemical properties but differ in some property such as net electrical charge, pH optima, number and type of subunits, or substrate concentration.

κ particles   DNA-containing cytoplasmic particles found in certain strains of *Paramecium aurelia* capable of releasing a toxin, paramecin, that kills other, sensitive strains.

K-RAS   A plasma-membrane associated protein encoded by the *KRAS* gene. This protein is important in cellular signal transduction events. Mutant alleles of *KRAS* are oncogenes present in many forms of cancer.

karyokinesis   The process of nuclear division.

karyotype   The chromosome complement of a cell or an individual. Often used to refer to the arrangement of metaphase chromosomes in a sequence according to length and centromere position.

kilobase (kb)   A unit of length consisting of 1000 nucleotides.

kinetochore   A fibrous structure with a size of about 400 nm, located within the centromere. It is the site of microtubule attachment during cell division.

Klinefelter syndrome   A genetic disorder in human males caused by the presence of one or more extra X chromosomes. Klinefelter males are usually XXY instead of XY. This syndrome often includes enlarged breasts, small testes, sterility, and mild mental retardation.

knockout mice   Mice created by a process in which a normal gene is cloned, inactivated by the insertion of a marker (such as an antibiotic resistance gene), and transferred to embryonic stem cells, where the altered gene will replace the normal gene (in some cells). These cells are injected into a blastomere embryo, producing a mouse that is then bred to yield mice homozygous for the mutated gene. See also *gene targeting*.

Kozak sequence   A short nucleotide sequence adjacent to the initiation codon that is recognized as the translational start site in eukaryotic mRNA.

lac repressor protein   A protein that binds to the operator in the *lac* operon and blocks transcription.

lagging strand   During DNA replication, the strand synthesized in a discontinuous fashion, in the direction opposite of the replication fork. See also *Okazaki fragment*.

lampbrush chromosomes   Meiotic chromosomes characterized by extended lateral loops that reach maximum extension during diplotene. Although most intensively studied in amphibians, these structures occur in meiotic cells of organisms ranging from insects to humans.

lariat structure   A structure formed during pre-mRNA processing by formation of a 5′ to 3′ bond in an introns, leading to removal of that intron from an mRNA molecule.

leader sequence   That portion of an mRNA molecule from the 5′ end to the initiating codon, often containing regulatory or ribosome binding sites.

leading strand   During DNA replication, the strand synthesized continuously in the direction of the replication fork.

leptotene   The initial stage of meiotic prophase I, during which the replicated chromosomes become visible; they are often arranged with one or both ends gathered at one spot on the inner nuclear membrane (the so-called bouquet configuration).

lethal gene   A gene whose expression results in premature death of the organism at some stage of its life cycle.

leucine zipper   In DNA-binding proteins, a structural motif characterized by a stretch in which every seventh amino acid residue is leucine, with adjacent regions containing positively charged amino acids. Leucine zippers on two polypeptides may interact to form a dimer that binds to DNA.

linkage   The condition in which genes are present on the same chromosome, causing them to be inherited as a unit, provided that they are not separated by crossing over during meiosis.

linking number   The number of times that two strands of a closed, circular DNA duplex cross over each other.

locus (pl., loci)   The site or place on a chromosome where a particular gene is located.

lod score (LOD)   A statistical method used to determine whether two loci are linked or unlinked. A lod (log of the odds) score of 4 indicates that linkage is 10,000 times more likely than nonlinkage. By convention, lod scores of 3–4 are signs of linkage.

long interspersed elements (LINEs)   Long, repetitive sequences found interspersed in the genomes of higher organisms.

long terminal repeat (LTR)   A sequence of several hundred base pairs found at both ends of a retroviral DNA.

loss-of-function mutation   Mutations that produce alleles that encode proteins with reduced or no function.

Lyon hypothesis   The proposal describing the random inactivation of the maternal or paternal X chromosome in somatic cells of mammalian females early in development. All daughter cells will have the same X chromosome inactivated as in the cell they descended from, producing a mosaic pattern of expression of X chromosome genes.

lysis   The disintegration of a cell brought about by the rupture of its membrane.

lysogenic bacterium   A bacterial cell carrying the DNA of a temperate bacteriophage integrated into its chromosome.

lysogeny   The process by which the DNA of an infecting phage becomes repressed and integrated into the chromosome of the bacterial cell it infects.

lytic phase   The condition in which a bacteriophage invades, reproduces, and lyses the bacterial cell. Lysogenic bacteria may be induced to enter the lytic phase.

major histocompatibility (MHC) loci   Loci encoding antigenic determinants responsible for tissue specificity. In humans, the HLA complex; and in mice, the H2 complex are within the MHC complex. See also *HLA*.

mapping functions   A mathematical formula that relates map distances to recombination frequencies.

map unit   A measure of the genetic distance between two genes, corresponding to a recombination frequency of 1 percent. See also *centimorgan*.

maternal effect   Phenotypic effects in offspring attributable to genetic information transmitted through the oocyte derived from the maternal genome.

maternal influence   Same as *maternal effect*.

maternal inheritance   The transmission of traits strictly through the maternal parent, usually due to DNA found in the cytoplasmic organelles, the mitochondria, or chloroplasts.

mean ($\overline{X}$)   The arithmetic average.

median   In a set of data points, the point below and above which there are equal numbers of other data points.

meiosis   The process of cell division in gametogenesis or sporogenesis during which the diploid number of chromosomes is reduced to the haploid number.

melting profile (T_m)   The temperature at which a population of double-stranded nucleic acid molecules is half-dissociated into single strands.

merozygote   A partially diploid bacterial cell containing, in addition to its own chromosome, a chromosome fragment introduced into the cell by transformation, transduction, or conjugation.

messenger RNA (mRNA)   An RNA molecule transcribed from DNA and translated into the amino acid sequence of a polypeptide.

metabolism   The collective chemical processes by which living cells are created and maintained. In particular, the chemical basis of generating and utilizing energy in living organisms.

metacentric chromosome   A chromosome that has a centrally located centromere and therefore chromosome arms of equal lengths.

metafemale   In *Drosophila*, a poorly developed female of low viability with a ratio of X chromosomes to sets of autosomes that exceeds 1.0. Previously called a *superfemale*.

metagenomics   The study of DNA recovered from organisms collected from the environment as opposed to those grown as laboratory cultures. Often used for estimating the diversity of organisms in an environmental sample.

metamale   In *Drosophila*, a poorly developed male of low viability with a ratio of X chromosomes to sets of autosomes that is below 0.5. Previously called a *supermale*.

metaphase   The stage of cell division in which condensed chromosomes lie in a central plane between the two poles of the cell and during which the chromosomes become attached to the spindle fibers.

metaphase plate   The structure formed when mitotic or meiotic chromosomes collect at the equator of the cell during metaphase.

metastasis   The process by which cancer cells spread from the primary tumor and establish malignant tumors in other parts of the body.

methylation   Enzymatic transfer of methyl groups from S-adenosylmethionine to biological molecules, including phospholipids, proteins, RNA, and DNA. Methylation of DNA is associated with the regulation of gene expression and with epigenetic phenomena such as imprinting.

MHC   See *major histocompatibility loci*.

microarray   See *DNA microarray*.

microRNA   Single-stranded RNA molecules approximately 20–23 nucleotides in length that regulate gene expression by participating in the degradation of mRNA.

microsatellite   A short, highly polymorphic DNA sequence of 1–4 base pairs, widely distributed in the genome, that are used as molecular markers in a variety of methods. Also called *simple sequence repeats (SSRs)*.

migration coefficient   A measure of the proportion of migrant genes entering the population per generation.

minimal medium   A medium containing only the essential nutrients needed to support the growth and reproduction of wild-type strains of an organism. Usually comprised of inorganic components that include a carbon and nitrogen source.

minisatellite   Series of short tandem repeat sequences (STRs) 10–100 nucleotides in length that occur frequently throughout the genome of eukaryotes. Because the number of repeats at each locus is variable, the loci are known as variable number tandem repeats (VNTRs). Used in DNA fingerprinting and DNA profiles. See also *VNTRs* and *STR sequences*.

mismatch repair   A form of excision repair of DNA in which the repair mechanism is able to distinguish between the strand with the error and the strand that is correct.

missense mutation   A mutation that changes a codon to that of another amino acid and thus results in an amino acid substitution in the translated protein. Such changes can make the protein nonfunctional.

mitochondrial DNA (mtDNA)   Double-stranded, self-replicating circular DNA found in mitochondria that encodes mitochondrial ribosomal RNAs, transfer RNAs, and proteins used in oxidative respiratory functions of the organelle.

mitochondrion   The so-called power house of the cell—a self-reproducing, DNA-containing, cytoplasmic organelle in eukaryotes involved in generating the high-energy compound ATP.

mitogen   A substance that stimulates mitosis in nondividing cells.

mitosis   A form of cell division producing two progeny cells identical genetically to the progenitor cell—that is, the production of two cells from one, each having the same chromosome complement as the parent cell.

mode   In a set of data, the value occurring with the greatest frequency.

model genetic organism   An experimental organism conducive to efficiently conducted research whose genetics is intensively studied on the premise that the findings can be applied to other organisms; for example, the fruit fly (*Drosophila melanogaster*) and the mouse (*Mus musculus*) are model organisms used to study the causes and development of human genetic diseases.

molecular clock   In evolutionary studies, a method that counts the number of differences in DNA or protein sequences as a way of measuring the time elapsed since two species diverged from a common ancestor.

monohybrid cross   A genetic cross involving only one character (e.g., $AA \times aa$).

monophyletic group   A taxon (group of organisms) consisting of an ancestor and all its descendants.

monosomic   An aneuploid condition in which one member of a chromosome pair is missing; having a chromosome number of $2n - 1$.

monozygotic twins   Twins produced from a single fertilization event; the first division of the zygote produces two cells, each of which develops into an embryo. Also known as *identical twins*.

mRNA   See *messenger RNA*.

mtDNA   See *mitochondrial DNA*.

multigene family   A set of genes descended from a common ancestral gene usually by duplication and subsequent sequence divergence. The globin genes are an example of a multigene family.

multiple alleles   In a population of organisms, the presence of three or more alleles of the same gene.

multiple-factor inheritance   Same as *polygenic inheritance*.

mu (μ) bacteriophage   A phage group in which the genetic material behaves like an insertion sequence (i.e., capable of insertion, excision, transposition, inactivation of host genes, and induction of chromosomal rearrangements).

mutagen   Any agent that causes an increase in the spontaneous rate of mutation.

mutant   A cell or organism carrying an altered or mutant allele.

mutation   The process that produces an alteration in DNA or chromosome structure; in genes, the source of new alleles.

mutation rate   The frequency with which mutations take place at a given locus or in a population.

muton   In fine structure analysis of the gene, the smallest unit of mutation, corresponding to a single base change.

myotonic dystrophy   An inherited progressive neuromuscular disorder characterized by muscle weakness and myotonia (inability of muscles to relax after contraction).

natural selection   Differential reproduction among members of a species owing to variable fitness conferred by genotypic differences.

neutral mutation   A mutation with no immediate adaptive significance or phenotypic effect.

next-generation sequencing (NGS) technologies   See *high-throughput DNA sequencing.*

nonautonomous transposon   A transposable element that lacks a functional transposase gene.

noncrossover gamete   A gamete whose chromosomes have undergone no genetic recombination.

nondisjunction   A cell division error in which homologous chromosomes (in meiosis) or the sister chromatids (in mitosis) fail to separate and migrate to opposite poles; responsible for defects such as monosomy and trisomy.

noninvasive prenatal genetic diagnosis (NIPGD)   A noninvasive method of fetal genotyping that uses a maternal blood sample to analyze thousands of fetal loci using fetal DNA fragments present in the maternal blood.

nonsense codons   The nucleotide triplets (UGA, UAG, and UAA) in an mRNA molecule that signal the termination of translation.

nonsense mutation   A mutation that changes a codon specifying an amino acid into a termination codon, leading to premature termination during translation of mRNA.

NOR   See *nucleolar organizer region.*

normal distribution   A probability function that approximates the distribution of random variables. The normal curve, also known as a Gaussian or bell-shaped curve, is the graphic display of a normal distribution.

northern blot   An analytic technique in which RNA molecules are separated by electrophoresis and transferred by capillary action to a nylon or nitrocellulose membrane. Specific RNA molecules can then be identified by hybridization to a labeled nucleic acid probe.

N-terminal amino acid   In a peptide chain, the terminal amino acid that carries a free amino group.

N terminus   The free amino group of the first amino acid in a polypeptide. By convention, the structural formula of polypeptides is written with the N terminus at the left.

nuclease   An enzyme that breaks bonds in nucleic acid molecules.

nucleoid   The DNA-containing region within the cytoplasm in prokaryotic cells.

nucleolar organizer region (NOR)   A chromosomal region containing the genes for rRNA; most often found in physical association with the *nucleolus.*

nucleolus   The nuclear site of ribosome biosynthesis and assembly; usually associated with or formed in association with the DNA comprising the *nucleolar organizer region.*

nucleoside   In nucleic acid chemical nomenclature, a purine or pyrimidine base covalently linked to a ribose or deoxyribose sugar molecule.

nucleosome   In eukaryotes, a nuclear complex consisting of four pairs of histone molecules wrapped by two turns of a DNA molecule. The major structure associated with the organization of chromatin in the nucleus.

nucleotide   In nucleic acid chemical nomenclature, a nucleoside covalently linked to one or more phosphate groups. Nucleotides containing a single phosphate linked to the 5′ carbon of the ribose or deoxyribose are the building blocks of nucleic acids.

nucleotide pair   The base-paired nucleotides (A with T or G with C) on opposite strands of a DNA molecule that are hydrogen-bonded to each other.

nucleus   The membrane-bound cytoplasmic organelle of eukaryotic cells that contains the chromosomes and nucleolus.

null allele   A mutant allele that produces no functional gene product. Usually inherited as a recessive trait.

null hypothesis (H_0)   Used in statistical tests, the hypothesis that there is no real difference between the observed and expected datasets. Statistical methods such as chi-square analysis are used to test the probability associated with this hypothesis.

nullisomic   Term describing an individual with a chromosomal aberration in which both members of a chromosome pair are missing.

nutrigenomics   The study of how food and components of food affect gene expression.

Okazaki fragment   The short, discontinuous strands of DNA produced on the lagging strand during DNA synthesis.

oligonucleotide   A linear sequence of about 10–20 nucleotides connected by 5′-3′ phosphodiester bonds.

oncogene   A gene whose activity promotes uncontrolled proliferation in eukaryotic cells. Usually a mutant gene derived from a *proto-oncogene.*

Online Mendelian Inheritance in Man (OMIM)   A database listing all known genetic disorders and disorders with genetic components. It also contains a listing of all known human genes and links genes to genetic disorders.

open reading frame (ORF)   A nucleotide sequence organized as triplets that encodes the amino acid sequence of a polypeptide, including an initiation codon and a termination codon.

operator region   In bacterial DNA, a region that interacts with a specific repressor protein to regulate the expression of an adjacent gene or gene set.

operon   A genetic unit consisting of one or more structural genes encoding polypeptides, and an adjacent operator gene that regulates the transcriptional activity of the structural gene or genes.

origin of replication (*ori*)   Sites where DNA replication begins along the length of a chromosome.

orthologs   Genes with sequence similarity found in two or more related species that arose from a single gene in a common ancestor.

outbreeding depression   Reduction in fitness in the offspring produced by mating genetically diverse parents. It is thought to result from a lowered adaptation to local environmental conditions.

overdominance   The phenomenon in which heterozygotes have a phenotype that is more extreme than either homozygous genotype.

overlapping code   A hypothetical genetic code in which any given triplet is shared by more than one adjacent codon.

pachytene   The stage in meiotic prophase I when the synapsed homologous chromosomes split longitudinally (except at the centromere), producing a group of four chromatids called a tetrad.

pair-rule genes   Genes expressed as stripes around the blastoderm embryo during development of the *Drosophila* embryo.

palindrome   In genetics, a sequence of DNA base pairs that reads the same backward or forward. Because strands run antiparallel to one another in DNA, the base sequences on the two strands read the same backward and forward when read from the 5′ end. For example:

$$5'\text{-GAATTC-}3'$$
$$3'\text{-CTTAAG-}5'$$

Palindromic sequences are noteworthy as recognition and cleavage sites for restriction endonucleases.

paracentric inversion   A chromosomal inversion that does not include the region containing the centromere.

paralogs   Two or more genes in the same species derived by duplication and subsequent divergence from a single ancestral gene.

parental gamete   Same as *noncrossover gamete.*

parthenogenesis   Development of an egg without fertilization.

partial diploids   See *merozygote.*

partial dominance   See *incomplete dominance.*

patroclinous inheritance   A form of genetic transmission in which the offspring have the phenotype of the father.

pedigree   In human genetics, a diagram showing the ancestral relationships and transmission of genetic traits over several generations in a family.

P element   In *Drosophila,* a transposable DNA element responsible for hybrid dysgenesis.

penetrance   The frequency, expressed as a percentage, with which individuals of a given genotype manifest at least some degree of a specific mutant phenotype associated with a trait.

peptide bond   The covalent bond between the amino group of one amino acid and the carboxyl group of another amino acid.

pericentric inversion   A chromosomal inversion that involves both arms of the chromosome and thus the centromere.

Personal Genome Project   A project to enroll 100,000 individuals to share their genome sequence, personal information, and medical history with researchers and the general public to increase understanding of the contribution of genetic and environmental factors to genetic traits.

phage   See *bacteriophage.*

pharmacogenomics   The study of how genetic variation influences the action of pharmaceutical drugs in individuals.

PharmGKB database   A repository for genetic, molecular, cellular, and genomic information about individuals who have participated in clinical pharmacogenomic research studies. It is designed to assist researchers in studying how genetic variation among individuals contributes to differences in drug reactions among individuals in a population.

phenotype   The overt appearance of a genetically controlled trait.

phenylketonuria (PKU)   A hereditary condition in humans associated with the inability to metabolize the amino acid phenylalanine due to the loss of activity of the enzyme phenylalanine hydroxylase.

Philadelphia chromosome   The product of a reciprocal translocation in humans that contains the short arm of chromosome 9, carrying the *C-ABL* oncogene, and the long arm of chromosome 22, carrying the *BCR* gene.

phosphodiester bond   In nucleic acids, the system of covalent bonds by which a phosphate group links adjacent nucleotides, extending from the 5′ carbon of one pentose sugar (ribose or deoxyribose) to the 3′ carbon of the pentose sugar in the neighboring nucleotide. Phosphodiester bonds create the backbone of nucleic acid molecules.

photoreactivation enzyme (PRE)   An exonuclease that catalyzes the light-activated excision of ultraviolet-induced thymine dimers from DNA.

photoreactivation repair   Light-induced repair of damage caused by exposure to ultraviolet light. Associated with an intracellular enzyme system.

phyletic evolution   The gradual transformation of one species into another over time; so-called vertical evolution.

pilus   A filamentlike projection from the surface of a bacterial cell. Often associated with cells possessing F factors.

plaque   On an otherwise opaque bacterial lawn, a clear area caused by the growth and reproduction of a single bacteriophage.

plasmid   An extrachromosomal, circular DNA molecule that replicates independently of the host chromosome.

pleiotropy   Condition in which a single mutation causes multiple phenotypic effects.

ploidy   A term referring to the basic chromosome set or to multiples of that set.

pluripotent   See *totipotent.*

point mutation   A mutation that can be mapped to a single locus. At the molecular level, a mutation that results in the substitution of one nucleotide for another. Also called a *gene mutation.*

polar body   Produced in females at either the first or second meiotic division of gametogenesis, a discarded cell that contains one of the nuclei of the division process, but almost no cytoplasm as a result of an unequal cytokinesis.

polycistronic mRNA   A messenger RNA molecule that encodes the amino acid sequence of two or more polypeptide chains in adjacent structural genes.

polygenic inheritance   The transmission of a phenotypic trait whose expression depends on the additive effect of a number of genes.

polylinker   A segment of DNA that has been engineered to contain multiple sites for restriction enzyme digestion. Polylinkers are usually found in engineered vectors such as plasmids.

polymerase chain reaction (PCR)   A method for amplifying DNA segments that depends on repeated cycles of denaturation, primer annealing, and DNA polymerase–directed DNA synthesis.

polymerases   Enzymes that catalyze the formation of DNA and RNA from deoxynucleotides and ribonucleotides, respectively.

polymorphism   The existence of two or more discontinuous, segregating phenotypes in a population.

polynucleotide   A linear sequence of 20 or more nucleotides, joined by 5′-3′ phosphodiester bonds. See also *oligonucleotide.*

polypeptide   A molecule composed of amino acids linked together by covalent peptide bonds. This term is used to denote the amino acid chain before it assumes its functional three-dimensional configuration.

polyploid   A cell or individual having more than two haploid sets of chromosomes.

polyribosome   See *polysome.*

polysome   A structure composed of two or more ribosomes associated with an mRNA engaged in translation. Also called a *polyribosome.*

polytene chromosome   Literally, a many-stranded chromosome; one that has undergone numerous rounds of DNA replication without separation of the replicated strands, which remain in exact parallel register. The result is a giant chromosome with aligned chromomeres displaying a characteristic banding pattern, most often studied in *Drosophila* larval salivary gland cells.

population   A local group of actually or potentially interbreeding individuals belonging to the same species.

population bottleneck   A drastic reduction in population size and consequent loss of genetic diversity, followed by an increase in population size. The rebuilt population has a gene pool with reduced diversity caused by genetic drift.

positional cloning   The identification and subsequent cloning of a gene in the absence of knowledge of its polypeptide product or function. The process uses cosegregation of mutant phenotypes with DNA markers to identify the chromosome containing the gene; the position of the gene is identified establishing linkage with additional markers.

position effect   Change in expression of a gene associated with a change in the gene's location within the genome.

posttranslational modification   The processing or modification of the translated polypeptide chain by enzymatic cleavage, or the addition of phosphate groups, carbohydrate chains, or lipids.

posttranscriptional modification   Changes made to pre-mRNA molecules during conversion to mature mRNA. These include the addition of a methylated cap at the $5'$ end and a poly-A tail at the $3'$ end, excision of introns, and exon splicing.

postzygotic isolation mechanism   A factor that prevents or reduces inbreeding by acting after fertilization to produce nonviable, sterile hybrids or hybrids of lowered fitness.

preadaptive mutation   A mutational event that later becomes of adaptive significance.

preformationism   See *epigenesis*.

preimplantation genetic diagnosis (PGD)   The removal and genetic analysis of unfertilized oocytes, polar bodies, or single cells from an early embryo (3–5 days old).

prezygotic isolation mechanism   A factor that reduces inbreeding by preventing courtship, mating, or fertilization.

Pribnow box   In prokaryotic genes, a 6-bp sequence to which the sigma (σ) subunit of RNA polymerase binds, upstream from the beginning of transcription. The consensus sequence for this box is TATAAT.

primary protein structure   The sequence of amino acids in a polypeptide chain.

primary sex ratio   Ratio of males to females at fertilization, often expressed in decimal form (e.g., 1.06).

primer   In nucleic acids, a short length of RNA or single-stranded DNA required for initiating synthesis directed by polymerases.

prion   An infectious pathogenic agent devoid of nucleic acid and composed of a protein, PrP, with a molecular weight of 27,000–30,000 Da. Prions are known to cause scrapie, a degenerative disease in sheep; bovine spongiform encephalopathy (BSE, or mad cow disease) in cattle; and similar diseases in humans, including kuru and Creutzfeldt–Jakob disease.

probability   A way of expressing the mathematical likelihood that a particular event will occur or has occurred.

proband   An individual who is the focus of a genetic study leading to the construction of a pedigree tracking the inheritance of a genetically determined trait of interest. Formerly known as a *propositus*.

probe   A macromolecule such as DNA or RNA that has been labeled and can be detected by an assay such as autoradiography or fluorescence microscopy. Probes are used to identify target molecules, genes, or gene products.

product law   In statistics, the law holding that the probability of two independent events occurring simultaneously is equal to the product of their independent probabilities.

progeny   The offspring produced from a mating.

prokaryotes   Organisms lacking nuclear membranes and true chromosomes. Bacteria and blue–green algae are examples of prokaryotic organisms.

promoter element   An upstream regulatory region of a gene to which RNA polymerase binds prior to the initiation of transcription.

proofreading   A molecular mechanism for scanning and correcting errors in replication, transcription, or translation.

prophage   A bacteriophage genome integrated into a bacterial chromosome that is replicated along with the bacterial chromosome. Bacterial cells carrying prophages are said to be *lysogenic* and to be capable of entering the *lytic cycle*, whereby the phage is replicated.

propositus (female, proposita)   See *proband*.

protein   A molecule composed of one or more polypeptides, each composed of amino acids covalently linked together. Proteins demonstrate *primary, secondary, tertiary*, and often, *quaternary structure*.

protein domain   Amino acid sequences with specific conformations and functions that are structurally and functionally distinct from other regions on the same protein.

proteome   The entire set of proteins expressed by a cell, tissue, or organism at a given time.

proteomics   The study of the expressed proteins present in a cell at a given time.

proto-oncogene   A gene that functions to initiate, facilitate, or maintain cell growth and division. Proto-oncogenes can be converted to *oncogenes* by mutation.

protoplast   A bacterial or plant cell with the cell wall removed. Sometimes called a *spheroplast*.

prototroph   A strain (usually of a microorganism) that is capable of growth on a defined, minimal medium. Wild-type strains are usually regarded as prototrophs and contrasted with *auxotrophs*.

pseudoalleles   Genes that behave as alleles to one another by complementation but can be separated from one another by recombination.

pseudoautosomal region   A region on the human Y chromosome that is also represented on the X chromosome. Genes found in this region of the Y chromosome have a pattern of inheritance that is indistinguishable from genes on autosomes.

pseudodominance   The expression of a recessive allele on one homolog owing to the deletion of the dominant allele on the other homolog.

pseudogene   A nonfunctional gene with sequence homology to a known structural gene present elsewhere in the genome. It differs from the functional version by insertions or deletions and by the presence of flanking direct-repeat sequences of 10–20 nucleotides.

puff   Same as *chromosome puff*.

punctuated equilibrium   A pattern in the fossil record of long periods of species stability, punctuated with brief periods of species divergence.

pyrosequencing   A high-throughput method of DNA sequencing that determines the sequence of a single-stranded DNA molecule by synthesis of a complementary strand.

During synthesis, the sequence is determined by the chemilumines-cent detection of pyrophosphate release that accompanies nucleotide incorporation into a newly synthesized strand of DNA.

quantitative inheritance   Same as *polygenic inheritance*.

quantitative real-time PCR (qPCR)   A variation of PCR (polymerase chain reaction) that uses fluorescent probes to quantitate the amount of DNA or RNA product present after each round of amplification.

quantitative trait loci (QTLs)   Two or more genes that act on a single polygenic trait in a quantitative way.

quantum speciation   Formation of a new species within a single or a few generations by a combination of selection and drift.

quaternary protein structure   Types and modes of interaction between two or more polypeptide chains within a protein molecule.

race   A genotypically or geographically distinct subgroup within a species.

rad   A unit of absorbed dose of radiation with an energy equal to 100 ergs per gram of irradiated tissue.

radioactive isotope   An unstable isotope with an altered number of neutrons that emits ionizing radiation during decay as it is transformed to a stable atomic configuration. See also *isotope*.

random amplified polymorphic DNA (RAPD)   A PCR method that uses random primers about 10 nucleotides in length to amplify unknown DNA sequences.

random mating   Mating between individuals without regard to genotype.

reactive oxygen species (ROS)   Ions or molecules formed by incomplete reduction of oxygen. They include hydroxyl radicals, superoxides, and peroxides. ROS cause oxidative damage to DNA and other macromolecules.

reading frame   A linear sequence of codons in a nucleic acid.

reannealing   Formation of double-stranded DNA molecules from denatured single strands.

recessive   An allele whose potential genetic expression is overridden in the heterozygous condition by a dominant allele.

reciprocal cross   A pair of crosses in which the genotype of the female in one is present as the genotype of the male in the other, and vice versa.

reciprocal translocation   A chromosomal aberration in which nonhomologous chromosomes exchange parts.

recombinant DNA technology   A collection of methods used to create DNA molecules by *in vitro* ligation of DNA from two different organisms, and the replication and recovery of such recombinant DNA molecules.

recombinant gamete   A gamete containing a new combination of alleles produced by crossing over during meiosis.

recombination   The process that leads to the formation of new allele combinations on chromosomes.

recon   A term utilized in fine structure analysis studies to denote the smallest intragenic units between which recombination can occur.

reductional division   The chromosome division that halves the number of centromeres and thus reduces the chromosome number by half. The first division of meiosis is a reductional division. See also *equational division*.

redundant genes   Gene sequences present in more than one copy per haploid genome (e.g., ribosomal genes).

regulatory site   A DNA sequence that functions in the control of expression of other genes, usually by interaction with another molecule.

rem   Radiation equivalent in humans; the dosage of radiation that will cause the same biological effect as one roentgen of X rays.

renaturation   The process by which a denatured protein or nucleic acid returns to its normal three-dimensional structure.

repetitive DNA sequence   A DNA sequence present in many copies in the haploid genome.

replicating form (RF)   Double-stranded nucleic acid molecules present as an intermediate during the reproduction of certain RNA-containing viruses.

replication   The process whereby DNA is duplicated.

replication fork   The Y-shaped region of a chromosome associated with the site of DNA replication.

replicon   The unit of DNA replication, beginning with DNA sequences necessary for the initiation of DNA replication. In bacteria, the entire chromosome is a replicon.

replisome   The complex of proteins, including DNA polymerase, that assembles at the bacterial replication fork to synthesize DNA.

repressible enzyme system   An enzyme or group of enzymes whose synthesis is regulated by the intracellular concentration of certain metabolites.

repressor   A protein that binds to a regulatory sequence adjacent to a gene and blocks transcription of the gene.

reproductive isolation   Absence of interbreeding between populations, subspecies, or species. Reproductive isolation can be brought about by extrinsic factors, such as behavior, or intrinsic barriers, such as hybrid inviability.

resistance transfer factor (RTF)   A component of R plasmids that confers the ability to transfer the R plasmid between bacterial cells by conjugation.

restriction endonuclease   A bacterial nuclease that recognizes specific nucleotide sequences in a DNA molecule, often a *palindrome*, and cleaves or nicks the DNA at those sites. Provides bacteria with a defense against invading viral DNA and are widely used in the construction of *recombinant DNA molecules*.

restriction fragment length polymorphism (RFLP)   Variation in the length of DNA fragments generated by restriction endonucleases. These variations are caused by mutations that create or abolish cutting sites for restriction enzymes. RFLPs are inherited in a codominant fashion and are extremely useful as genetic markers.

restriction site   A DNA sequence, often palindromic, recognized by a restriction endonuclease. The enzyme binds to the restriction site and cleaves the DNA at the restriction site.

restrictive transduction   See *specialized transduction*.

retrotransposon   Mobile genetic elements that are major components of many eukaryotic genomes, that are copied by means of an RNA intermediate and inserted at a distant chromosomal site.

retrovirus   A type of virus that uses RNA as its genetic material and employs the enzyme reverse transcriptase during its life cycle.

reverse genetics   An experimental approach used to discover gene function after the gene has been identified, cloned, and sequenced. The cloned gene may be knocked out (e.g., by *gene targeting*) or have its expression altered (e.g., by *RNA interference* or *transgenic overexpression*) and the resulting phenotype studied. An approach contrasted with *forward genetics*.

reverse transcriptase   A polymerase that uses RNA as a template to transcribe a single-stranded DNA molecule as a product.

reversion   A mutation that restores the wild-type phenotype.

R factor (R plasmid)   A bacterial plasmid that carries antibiotic resistance genes. Most R plasmids have two components: an r-determinant which carries the antibiotic resistance genes, and the resistance transfer factor (RTF).

RFLP   See *restriction fragment length polymorphism*.

Rh factor An antigenic system first described in the rhesus monkey. Recessive r/r individuals produce no Rh antigens and are Rh negative, while R/R and R/r individuals have Rh antigens on the surface of their red blood cells and are classified as Rh positive. This is the genetic basis of the immunological incompatibility disease called erythroblastosis fetalis (hemolytic disease of the newborn).

ribonucleic acid (RNA) A nucleic acid similar to DNA but characterized by the pentose sugar ribose, the pyrimidine uracil, and the single-stranded nature of the polynucleotide chain. Several forms are recognized, including ribosomal RNA, messenger RNA, transfer RNA, and a variety of small regulatory RNA molecules.

ribose The five-carbon sugar associated with ribonucleosides and ribonucleotides associated with RNA.

ribosomal RNA (rRNA) The RNA molecules that are the structural components of the ribosomal subunits. In prokaryotes, these are the 16S, 23S, and 5S molecules; in eukaryotes, they are the 18S, 28S, and 5S molecules. See also *Svedberg coefficient (S)*.

ribosome A ribonucleoprotein organelle consisting of two subunits, each containing RNA and protein molecules. Ribosomes are the site of translation of mRNA codons into the amino acid sequence of a polypeptide chain.

RNA See *ribonucleic acid*.

RNA editing Alteration of the nucleotide sequence of an mRNA molecule after transcription and before translation. There are two main types of editing: substitution editing, which changes individual nucleotides, and insertion/deletion editing, in which individual nucleotides are added or deleted.

RNA interference (RNAi) Inhibition of gene expression in which a protein complex (RNA-induced silencing complex, or RISC), containing a partially complementary RNA strand binds to an mRNA, leading to degradation or reduced translation of the mRNA.

RNA polymerase An enzyme that catalyzes the formation of an RNA polynucleotide strand using the base sequence of a DNA molecule as a template.

RNase A class of enzymes that hydrolyzes RNA.

Robertsonian translocation A chromosomal aberration created by breaks in the short arms of two acrocentric chromosomes followed by fusion of the long arms of these chromosomes at the centromere. Also called *centric fusion*.

roentgen (R) A unit of measure of radiation emission, corresponding to the amount that generates 2.083×10^9 ion pairs in one cubic centimeter of air at 0°C and an atmospheric pressure of 760 mm of mercury.

rolling circle model A model of DNA replication in which the growing point, or replication fork, rolls around a circular template strand; in each pass around the circle, the newly synthesized strand displaces the strand from the previous replication, producing a series of contiguous copies of the template strand.

rRNA See *ribosomal RNA*.

RTF See *resistance transfer factor*.

S_1 nuclease A deoxyribonuclease that cuts and degrades single-stranded molecules of DNA.

Sanger sequencing DNA sequencing by synthesis of DNA chains that are randomly terminated by incorporation of a nucleotide analog (dideoxynucleotides) followed by sequence determination by analysis of resulting fragment lengths in each reaction.

satellite DNA DNA that forms a minor band when genomic DNA is centrifuged in a cesium salt gradient. This DNA usually consists of short sequences repeated many times in the genome.

SCE See *sister chromatid exchange*.

secondary protein structure The α-helical or β-pleated-sheet formations in a polypeptide, dependent on hydrogen bonding between certain amino acids.

secondary sex ratio The ratio of males to females at birth, usually expressed in decimal form (e.g., 1.05).

secretor An individual who secretes soluble forms of the blood group antigens A and/or B in saliva and other body fluids. This condition is caused by a dominant autosomal gene independent of the *ABO* locus.

sedimentation coefficient (S) See *Svedberg coefficient unit*.

segment polarity genes Genes that regulate the spatial pattern of differentiation within each segment of the developing *Drosophila* embryo.

segregation The separation of maternal and paternal homologs of each homologous chromosome pair into gametes during meiosis.

selection The changes that occur in the frequency of alleles and genotypes in populations as a result of differential reproduction.

selection coefficient (s) A quantitative measure of the relative fitness of one genotype compared with another. Same as *coefficient of selection*.

selfing In plant genetics, the fertilization of a plant's ovules by pollen produced by the same plant. Reproduction by self-fertilization.

semiconservative replication A mode of DNA replication in which a double-stranded molecule replicates in such a way that the daughter molecules are each composed of one parental (old) and one newly synthesized strand.

semisterility A condition in which a percentage of all zygotes are inviable.

sex chromatin body See *Barr body*.

sex chromosome A chromosome, such as the X or Y in humans, which is involved in sex determination.

sexduction Transmission of chromosomal genes from a donor bacterium to a recipient cell by means of the F factor.

sex-influenced inheritance Phenotypic expression conditioned by the sex of the individual. A heterozygote may express one phenotype in one sex and an alternate phenotype in the other sex (e.g., pattern baldness in humans).

sex-limited inheritance A trait that is expressed in only one sex even though the trait may not be X-linked.

sex ratio See *primary sex ratio* and *secondary sex ratio*.

sexual reproduction Reproduction through the fusion of gametes, which are the haploid products of meiosis.

Shine–Dalgarno sequence The nucleotides AGGAGG that serve as a ribosome-binding site in the leader sequence of prokaryotic genes. The 16S RNA of the small ribosomal subunit contains a complementary sequence to which the mRNA binds.

short interspersed elements (SINEs) Repetitive sequences found in the genomes of higher organisms. The 300-bp *Alu* sequence is a SINE element.

short tandem repeats See *STR sequences*.

shotgun cloning The cloning of random fragments of genomic DNA into a vector (a plasmid or phage), usually to produce a library from which clones of specific interest can be selected for use, as in sequencing.

shugoshins A class of proteins involved in maintaining cohesion of the centromeres of sister chromatids during mitosis and meiosis.

sibling species Species that are morphologically similar but reproductively isolated from one another.

sickle-cell anemia   A human genetic disorder caused by an autosomal recessive allele, often fatal in the homozygous condition if untreated. Caused by a nucleotide substitution that alters one amino acid in the β chain of globin.

sickle-cell trait   The phenotype exhibited by individuals heterozygous for the sickle-cell allele.

sigma (σ) factor   In RNA polymerase, a polypeptide subunit that recognizes the DNA binding site for the initiation of transcription.

signal transduction   An intercellular or intracellular molecular pathway by which an external signal is converted into a functional biological response.

single molecule sequencing   A high-throughput sequencing by synthesis involving direct sequencing of single DNA or RNA molecules without the need for amplification by PCR before sequencing.

single-nucleotide polymorphism (SNP)   A variation in a single nucleotide pair in DNA, as detected during genomic analysis. Present in at least 1 percent of a population, a SNP is useful as a genetic marker.

single-stranded binding proteins (SSBs)   In DNA replication, proteins that bind to and stabilize the single-stranded regions of DNA that result from the action of unwinding proteins.

sister chromatid exchange (SCE)   A crossing over event in meiotic or mitotic cells involving the reciprocal exchange of chromosomal material between sister chromatids joined by a common centromere. Such exchanges can be detected cytologically after BrdU incorporation into the replicating chromosomes.

site-directed mutagenesis   A process that uses a synthetic oligonucleotide containing a mutant base or sequence as a primer for inducing a mutation at a specific site in a cloned gene.

small nuclear RNA (snRNA)   Abundant species of small RNA molecules ranging in size from 90 to 400 nucleotides that in association with proteins form RNP particles known as snRNPs or *snurps*. Located in the nucleoplasm, snRNAs have been implicated in the processing of pre-mRNA and may have a range of cleavage and ligation functions.

SNP   See *single-nucleotide polymorphism.*

snurps   See *small nuclear RNA.*

solenoid structure   A feature of eukaryotic chromatin conformation generated by nucleosome supercoiling.

somatic cell genetics   The use of cultured somatic cells to investigate genetic phenomena by means of parasexual techniques involving the fusion of cells from different organisms.

somatic cells   All cells other than the germ cells or gametes in an organism.

somatic mutation   A nonheritable mutation occurring in a somatic cell.

somatic pairing   The pairing of homologous chromosomes in somatic cells.

SOS response   The induction of enzymes for repairing damaged DNA in *Escherichia coli.* The response involves activation of an enzyme that cleaves a repressor, activating a series of genes involved in DNA repair.

Southern blotting   Developed by Edwin Southern, a technique in which DNA fragments produced by restriction enzyme digestion are separated by electrophoresis and transferred by capillary action to a nylon or nitrocellulose membrane. Specific DNA fragments can be identified by hybridization to a complementary radioactively labeled nucleic acid probe using the technique of *autoradiography.*

spacer DNA   DNA sequences found between genes. Usually, these are repetitive DNA segments.

specialized transduction   Genetic transfer of specific host genes by transducing phages. See also *generalized transduction.*

speciation   The process by which new species of plants and animals arise.

species   A group of actually or potentially interbreeding individuals that is reproductively isolated from other such groups.

spectral karyotype   A display of all the chromosomes in an organism as a karyotype with each chromosome stained in a different color.

spheroplast   See *protoplast.*

spindle fibers   Cytoplasmic fibrils formed during cell division that attach to and are involved with separation of chromatids at the anaphase stage of mitosis and meiosis as well as their movement toward opposite poles in the cell.

spliceosome   The nuclear macromolecule complex within which splicing reactions occur to remove introns from pre-mRNAs.

spontaneous mutation   A mutation that is not induced by a mutagenic agent.

spore   A unicellular body or cell encased in a protective coat. Produced by some bacteria, plants, and invertebrates, spores are capable of surviving in unfavorable environmental conditions and give rise to a new individual upon germination. In plants, spores are the haploid products of meiosis.

SRY   The sex-determining region of the Y chromosome, found near the chromosome's pseudoautosomal boundary. Accumulated evidence indicates that this gene's product is the testis-determining factor (TDF).

stabilizing selection   Preferential reproduction of individuals with genotypes close to the mean for the population. A selective elimination of genotypes at both extremes.

standard deviation (SD)   A quantitative measure of the amount of variation in a sample of measurements from a population calculated as the square root of the variance.

standard error (SE)   A quantitative measure of the amount of variation around the mean in a sample of measurements from a population.

strain   A group of organisms with common ancestry with physiological or morphological characteristics of interest for genetic analysis or domestication.

STR sequences   Short tandem repeats 2–9 base pairs long that are found within minisatellites. These sequences are used to prepare DNA profiles in forensics, paternity identification, and other applications.

structural gene   A gene that encodes the amino acid sequence of a polypeptide chain.

sublethal gene   A mutation causing lowered viability, with death before maturity in less than 50 percent of the individuals carrying the gene.

submetacentric chromosome   A chromosome with the centromere placed so that one arm of the chromosome is slightly longer than the other.

subspecies   A morphologically or geographically distinct interbreeding population of a species.

sum law   The law that holds that the probability of one of two mutually exclusive outcomes occurring, where that outcome can be achieved by two or more events, is equal to the sum of their individual probabilities.

supercoiled DNA   A DNA configuration in which the helix is coiled upon itself. Supercoils can exist in stable forms only when the ends of the DNA are not free, as in a covalently closed circular DNA molecule.

superfemale See *metafemale*.

supermale See *metamale*.

suppressor mutation A mutation that acts to completely or partially restore the function lost by a mutation at another site.

Svedberg coefficient unit (*S*) A unit of measure for the rate at which particles (molecules) sediment in a centrifugal field. This rate is a function of several physicochemical properties, including size and shape. A rate of 1×10^{-13} sec is defined as one Svedberg coefficient unit.

symbiont An organism coexisting in a mutually beneficial relationship with another organism.

sympatric speciation Speciation occurring in populations that inhabit, at least in part, the same geographic range.

synapsis The pairing of homologous chromosomes at meiosis.

synaptonemal complex (SC) A sub-microscopic structure consisting of a tripartite nucleoprotein ribbon that forms between the paired homologous chromosomes in the pachytene stage of the first meiotic division.

syndrome A group of characteristics or symptoms associated with a disease or an abnormality. An affected individual may express a number of these characteristics but not necessarily all of them.

synkaryon The fusion of two gametic or somatic nuclei. Also, in somatic cell genetics, the product of nuclear fusion.

syntenic test In somatic cell genetics, a method for determining whether two genes are on the same chromosome.

synthetic biology A field of research that combines science and engineering to understand the complexity of living systems and to construct biological-based systems that do not exist in nature.

synthetic genome A genome assembled from chemically synthesized DNA fragments that is transferred to a host cell without a genome.

systems biology A field that identifies and analyzes gene and protein networks to gain an understanding of intracellular regulation of metabolism, intra- and intercellular communication, and complex interactions within, between, and among cells.

TATA box See *Goldberg–Hogness box*.

tautomeric shift A reversible isomerization in a molecule, brought about by a shift in the location of a hydrogen atom. In nucleic acids, tautomeric shifts in the bases of nucleotides can cause changes in other bases at replication and are a source of mutations.

TDF (testis-determining factor) The product of the *SRY* gene on the Y chromosome; it controls the developmental switch point for the development of the indifferent gonad into a testis.

telocentric chromosome A chromosome in which the centromere is located at its very end.

telomerase The enzyme that adds short, tandemly repeated DNA sequences to the ends of eukaryotic chromosomes.

telomere The heterochromatic terminal region of a chromosome.

telomere repeat-containing RNA (TERRA) Large noncoding RNA molecules transcribed from telomere repeats that are an integral part of telomeric heterochromatin.

telophase The stage of cell division in which the daughter chromosomes have reached the opposite poles of the cell and reverse the stages characteristic of prophase, re-forming the nuclear envelopes and uncoiling the chromosomes. Telophase ends with cytokinesis, which divides the cytoplasm and splits cell into two.

telophase I The stage in the first meiotic division when duplicated chromosomes reach the poles of the dividing cell.

temperate phage A bacteriophage that can become a prophage, integrating its DNA into the chromosome of the host bacterial cell and making the latter lysogenic.

temperature-sensitive mutation A conditional mutation that produces a mutant phenotype at one temperature range and a wild-type phenotype at another.

template The single-stranded DNA or RNA molecule that specifies the sequence of a complementary nucleotide strand synthesized by DNA or RNA polymerase.

terminalization The movement of chiasmata toward the ends of chromosomes during the diplotene stage of the first meiotic division.

tertiary protein structure The three-dimensional conformation of a polypeptide chain in space, specified by the polypeptide's *primary structure*. The tertiary structure achieves a state of maximum thermodynamic stability.

testcross A cross between an individual whose genotype at one or more loci may be unknown and an individual who is homozygous recessive for the gene or genes in question.

tetrad The four chromatids that make up paired homologs in the prophase of the first meiotic division. In eukaryotes with a predominant haploid stage (some algae and fungi), a tetrad denotes the four haploid cells produced by a single meiotic division.

tetrad analysis A method that analyzes gene linkage and recombination in organisms with a predominant haploid phase in their life cycle. See also *tetrad*.

tetranucleotide hypothesis An early theory of DNA structure proposing that the molecule was composed of repeating units, each consisting of the four nucleotides represented by adenine, thymine, cytosine, and guanine.

theta (*θ*) structure An intermediate configuration in the bidirectional replication of circular DNA molecules. At about midway through the cycle of replication, the intermediate resembles the Greek letter theta.

thymine dimer In a polynucleotide strand, a lesion consisting of two adjacent thymine bases that become joined by a covalent bond. Usually caused by exposure to ultraviolet light, this lesion inhibits DNA replication.

Ti plasmid A bacterial plasmid used as a vector to transfer foreign DNA to plant cells.

T_m See *melting profile*.

topoisomerase A class of enzymes that converts DNA from one topological form to another. During replication, a topoisomerase, *DNA gyrase*, facilitates DNA replication by reducing molecular tension caused by supercoiling upstream from the *replication fork*.

totipotent The capacity of a cell or an embryo part to differentiate into all cell types characteristic of an adult. This capacity is usually progressively restricted during development. Used interchangeably with *pluripotent*.

trait Any detectable phenotypic variation of a particular inherited character.

***trans*-acting element** A gene product (usually a diffusible protein or an RNA molecule) that acts to regulate the expression of a target gene.

***trans* configuration** An arrangement in which two mutant alleles are on opposite homologs, such as

$$\frac{a^1 \quad +}{+ \quad a^2}$$

in contrast to a *cis arrangement*, in which the alleles are located on the same homolog.

transcription   Transfer of genetic information from DNA by the synthesis of a complementary RNA molecule using a DNA template.

transcription associated factor (TAF)   A protein containing DNA-binding domains by which the protein attaches to specific DNA sequences adjacent to the genes the TAF regulates.

transcriptome   The set of mRNA molecules present in a cell at any given time.

transdetermination   Change in developmental fate of a cell or group of cells.

transduction   Virally mediated bacterial recombination. Also used to describe the transfer of eukaryotic genes mediated by a retrovirus.

transfer RNA (tRNA)   A small ribonucleic acid molecule with an essential role in *translation*. tRNAs contain: (1) a three-base segment (anticodon) that recognizes a codon in mRNA; (2) a binding site for the specific amino acid corresponding to the anticodon; and (3) recognition sites for interaction with ribosomes and with the enzyme that links the tRNA to its specific amino acid.

transformation   Heritable change in a cell or an organism brought about by exogenous DNA. Known to occur naturally and also used in *recombinant DNA* studies.

transgenic organism   An organism whose genome has been modified by the introduction of external DNA sequences into the germ line.

transition   A mutational event in which one purine is replaced by another or one pyrimidine is replaced by another.

translation   The derivation of the amino acid sequence of a polypeptide from the base sequence of an mRNA molecule in association with a ribosome and tRNAs.

translocation   A chromosomal mutation associated with the reciprocal or nonreciprocal transfer of a chromosomal segment from one chromosome to another. Also denotes the movement of mRNA through the ribosome during translation.

transmission genetics   The field of genetics concerned with heredity and the mechanisms by which genes are transferred from parent to offspring.

transposable element   A DNA segment that moves to other sites in the genome, essentially independent of sequence homology. Usually, such elements are flanked at each end by short inverted repeats of 20–40 base pairs. Insertion into a structural gene can produce a mutant phenotype. Insertion and excision of transposable elements depend on two enzymes, transposase and resolvase. Such elements have been identified in both prokaryotes and eukaryotes.

transversion   A mutational event in which a purine is replaced by a pyrimidine or a pyrimidine is replaced by a purine.

trinucleotide repeat   A tandemly repeated cluster of three nucleotides (such as CTG) within or near a gene. Certain diseases (myotonic dystrophy, Huntington disease) are caused by expansion in copy number of such repeats.

triploidy   The condition in which a cell or an organism possesses three haploid sets of chromosomes.

trisomy   The condition in which a cell or an organism possesses two copies of each chromosome except for one, which is present in three copies (designated $2n + 1$).

tRNA   See *transfer RNA*.

true single molecule sequencing   See *single molecule sequencing*.

tumor-suppressor gene   A gene that encodes a product that normally functions to suppress cell division. Mutations in tumor-suppressor genes result in the activation of cell division and tumor formation.

Turner syndrome   A genetic condition in human females caused by a 45,X genotype. Such individuals are phenotypically female but are sterile because of undeveloped ovaries.

unequal crossing over   A crossover between two improperly aligned homologs, producing one homolog with three copies of a region and the other with one copy of that region.

unique DNA   DNA sequences that are present only once per genome.

universal code   A genetic code used by all life forms. Some exceptions are found in mitochondria, ciliates, and mycoplasmas.

unwinding proteins   Nuclear proteins that act during DNA replication to destabilize and unwind the DNA helix ahead of the replicating fork.

variable number tandem repeats (VNTRs)   Short, repeated DNA sequences (of 2–20 nucleotides) present as tandem repeats between two restriction enzyme sites. Variation in the number of repeats creates DNA fragments of differing lengths following restriction enzyme digestion. Used in early versions of *DNA fingerprinting*.

variable region (V)   Portion of an immunoglobulin molecule that exhibits many amino acid sequence differences between antibodies of differing specificities.

variance (s^2)   A statistical measure of the variation of values from a central value, calculated as the square of the standard deviation.

variegation   Patches of differing phenotypes, such as color, in a tissue.

vector   In recombinant DNA, an agent such as a phage or plasmid into which a foreign DNA segment will be inserted and used to transform host cells.

vertical gene transfer   The transfer of genetic information from parents to offspring generation after generation.

viability   The ability of individuals in a given phenotypic class to survive as compared to another class (usually wild type).

virulent phage   A bacteriophage that infects, replicates within, and lyses bacterial cells, releasing new phage particles.

VNTRs   See *variable number tandem repeats*.

western blot   An analytical technique in which proteins are separated by gel electrophoresis and transferred by capillary action to a nylon membrane or nitrocellulose sheet. A specific protein can be identified through hybridization to a labeled antibody.

whole genome shotgun cloning   See *shotgun cloning*.

wild type   The most commonly observed phenotype or genotype, designated as the norm or standard.

wobble hypothesis   An idea proposed by Francis Crick, stating that the third base in an anticodon can align in several ways to allow it to recognize more than one base in the codons of mRNA.

W, Z chromosomes   The sex chromosomes in species where the female is the heterogametic sex (WZ).

X chromosome   The sex chromosome present in species where females are the homogametic sex (XX).

X chromosome inactivation   In mammalian females, the random cessation of transcriptional activity of the maternally or paternally derived X chromosome. This event, which occurs early in development, is a mechanism of dosage compensation, and all progeny cells inactivate the same X chromosome. See also *Barr body*, *Lyon hypothesis*, *XIST*.

XIST   A locus in the X-chromosome inactivation center that controls inactivation of the X chromosome in mammalian females.

X-linkage   The pattern of inheritance resulting from genes located on the X chromosome.

X-ray crystallography   A technique for determining the three-dimensional structure of molecules by analyzing X-ray diffraction patterns produced by bombarding crystals of the molecule under study with X-rays.

YAC   A cloning vector in the form of a yeast artificial chromosome, constructed using chromosomal components including telomeres (from a ciliate), and centromeres, origin of replication, and marker genes from yeast. YACs are used to clone long stretches of eukaryotic DNA.

Y chromosome   The sex chromosome in species where the male is heterogametic (XY).

Y-linkage   Mode of inheritance shown by genes located on the Y chromosome.

Z-DNA   An alternative "zig-zag" structure of DNA in which the two antiparallel polynucleotide chains form a left-handed double helix. Implicated in regulation of gene expression.

zein   Principal storage protein of maize endosperm, consisting of two major proteins with molecular weights of 19,000 and 21,000 Da.

zinc finger   A class of DNA-binding domains seen in proteins. They have a characteristic pattern of cysteine and histidine residues that complex with zinc ions, throwing intermediate amino acid residues into a series of loops or fingers.

zygote   The diploid cell produced by the fusion of haploid gametic nuclei.

zygotene   A stage of meiotic prophase I in which the homologous chromosomes synapse and pair along their entire length, forming bivalents. The synaptonemal complex forms at this stage.

Credits

Photo Credits

Chapter 1: Introduction to Genetics

Chapter opener-1: Sinclair Stammers/Photo Researchers. Chapter opener-2: Mark Smith/Photo Researchers. Chapter opener-3: CC-BY-SA photo: Alberto Salguero Quiles en Getafe, España. 1.1: National Library of Medicine. 1.2: Biophoto Associates/Photo Researchers. 1.4ab: From: Learning to Fly: Phenotypic Markers in *Drosophila*. A poster of common phenotypic markers used in *Drosophila* genetics. Jennifer Childress, Richard Behringer, and Georg Halder. 2005. Genesis 43(1). Cover illustration. 1.5: Biozentrum, University of Basel/Photo Researchers. 1.9: Oliver Meckes & Nicole Ottawa/Photo Researchers. 1.11: Roslin Institute. 1.12: SSPL/The Image Works. 1.14: CC-BY-SA photo: Guillaume Paumier. 1.15a: John Paul Endress, Pearson. 1.15b: Herman Eisenbeiss/Photo Researchers. 1.16a: Jeremy Burgess/Photo Researchers. 1.16b: David McCarthy/SPL/Photo Researchers.

Chapter 2: Mitosis and Meiosis

Chapter opener: Andrew S. Bajer, University of Oregon. 2.2: CNRI/SPL/Photo Researchers. 2.4: David C. Ward, Yale University. 2.7a–f: Andrew S. Bajer, University of Oregon. 2.14a: Biophoto Associates/Photo Researchers. 2.14b: Andrew Syred/SPL/Photo Researchers.

Chapter 3: Mendelian Genetics

Chapter opener: Images from the History of Medicine (NLM). 3.UN05.1, 3.UN05.2: Mike Dunton, The Victory Seed Company.

Chapter 4: Extensions of Mendelian Genetics

Chapter opener: Juniors Bildarchiv/Alamy. 4.1a: John Kaprielian/Photo Researchers. 4.4bc: The Jackson Laboratory. 4.9: Rosemary Buffoni/iStockphoto. 4.12bc: From: Learning to Fly: Phenotypic Markers in *Drosophila*. A poster of common phenotypic markers used in *Drosophila* genetics. Jennifer Childress, Richard Behringer, and Georg Halder. 2005. *Genesis* 43(1). Cover illustration. 4.14a: Ishihara, Test for Color Deficiency. Courtesy Kanehara & Co., Ltd. Offered Exclusively in The USA by Graham-Field Health Products, Atlanta, GA. 4.15: Hans Reinhard/Bruce Coleman/Photoshot. 4.16: Debra P. Hershkowitz/Bruce Coleman/Photoshot. 4.17a: Tanya Wolff, Washington University School of Medicine. 4.17b: Joel C. Eissenberg, Dept. of Biochemistry and Molecular Biology, St. Louis University Medical Center. 4.17c, 4.18a: Tanya Wolff, Washington University School of Medicine. 4.18b: Steven Henikoff, Howard Hughes Medical Institute, Fred Hutchinson Cancer Research Center, Seattle, Washington. 4.19a: Jane Burton/Bruce Coleman/Photoshot. 4.19b: William S. Klug. 4.UN.2: Ralph Somes. 4.UN.2: J. James Bitgood, Animal Sciences Dept., University of Wisconsin. 4.UN.2: Ralph Somes. 4.UN.2: J. James Bitgood, Animal Sciences Dept., University of Wisconsin. 4.UN.5: Shout It Out Design/Shutterstock. 4.UN.5: Zuzule/Shutterstock. 4.UN.5: Julia Remezova/Shutterstock.

Chapter 5: Chromosome Mapping in Eukaryotes

Chapter opener: James Kezer. 5.16: Sheldon Wolff. 5.17b: Biophoto Associates/Photo Researchers. 5.18b: Matthew L. Springer, University of California, San Francisco. 5.18c: Science Source/Photo Researchers. 5.19b: Namboori B. Raju, Stanford University. Fig. 3C pg 29 from Raju, N.B. 2008. Six decades of *Neurospora* ascus biology at Stanford. *Fung. Biol. Rev.* 22: 26–35.

Chapter 6: Genetic Analysis and Mapping in Bacteria and Bacteriophages

Chapter opener: Charles C. Brinton, Jr. 6.2: Pearson Science. 6.11a: Gopal Murti/SPL/Photo Researchers. 6.13a: M. Wurtz/Biozentrum, University of Basel/SPL/Photo Researchers. 6.15b: Christine Case. 6.18: William S. Klug.

Chapter 7: Sex Determination and Sex Chromosomes

Chapter opener: Wellcome Trust Medical Photographic Library. 7.1b: Biophoto Associates/Photo Researchers. 7.2b: Artem Solovev/Fotolia. 7.3a: Maria Gallegos, University of California, San Francisco. 7.5ab: Greenwood Genetic Center (GGC). 7.6ab: Catherine G. Palmer. 7.8ab: Michael Abbey/Photo Researchers. 7.10a: Sari O'Neal/Shutterstock. 7.10b: William S. Klug. 7.13: From: Dosage compensation: the beginning and end of generalization. *Nature Reviews/Genetics*, Vol 8, January 2007. Peter Becker, University of Munich. 7.UN02: Texas A & M University/AP Images.

Chapter 8: Chromosome Mutations: Variation in Number and Arrangement

Chapter opener: National Institutes of Health. 8.2: Arco Images GmbH/Alamy. 8.3a: Greenwood Genetic Center (GGC). 8.3b: Kristy-Anne Glubish/Design Pics/Alamy. 8.5: David D. Weaver, M.D., Indiana University. 8.9: National Cotton Council of America. 8.12a: Courtesy of University of Washington Medical Center Pathology. 8.12b: Ray Clarke, Cri Du Chat Syndrome Support Group, UK. 8.14abc: Mary Lilly/Carnegie Institution of Washington. 8.18b: Jorge Yunis. From Yunis and Chandler, 1979. 8.19: Christine J Harrison, LRF Cytogenetics, Northern Institute for Cancer Research, Newcastle University, UK.

Chapter 9: Extranuclear Inheritance

Chapter opener: From The shape of mitochondria and the number of mitochondrial nucleoids during the cell cycle of *Euglena gracilis*. Y. Hayashi and K. Ueda. *Journal of Cell Science*, 93: 565–570, Copyright © 1989 by Company of Biologists. 9.1b: Ryushi Abura. 9.2a: Dartmouth Electron Microscope Facility. 9.3: Matthew L. Springer, University of California, San Francisco. 9.4ab: Ronald A. Butow, Department of Molecular Biology and Oncology, University of Texas Southwestern Medical Center. 9.6a: Dana-Farber Cancer Institute. 9.7a: Don W. Fawcett/Kahri/Dawid/Science Source/Photo Researchers. 9.9ab: Alan Pestronk, Dept of Neurology, Washington University School of Medicine, St. Louis. 9.11a: Kuroda Laboratory. 9.12: Uwe Irion, Max-Planck-Institute for Developmental Biology.

Chapter 10: DNA Structure and Analysis

Chapter opener: Richard Megna/Fundamental Photographs. 10.2ab: Bruce Iverson, Photomicrography. 10.4a: Oliver Meckes/Max-Planck-Institut-Tubingen, Photo Researchers. 10.11: M. H. F. Wilkins. 10.15a: Ken Edward/Science Source/Photo Researchers. 10.19: Ventana Medical Systems. 10.20a: William S. Klug.

Chapter 11: DNA Replication and Recombination

Chapter opener: Gopal Murti/SPL/Photo Researchers. 11.5a: Walter H. Hodge/Peter Arnold/Photolibrary.com. 11.5ab: Figure from *Molecular Genetics*, Pt. 1 pp. 74–75, J. H. Taylor (ed). Copyright ©1963 and renewed 1991, reproduced with permission from Elsevier Science Ltd. 11.6a: Reprinted from *CELL*, Vol. 25, 1981, pp 659, Sundin and Varshavsky, (1 figure), with permission from Elsevier Science. Courtesy of A. Varshavsky. 11.14: Reproduced by permission from H. J. Kreigstein and D. S. Hogness, *Proceedings of the National Academy of Sciences* 71: 136 (1974), p. 137, Fig. 2. 11.15: Harold Weintraub, Howard Hughes Medical Institute, Fred Hutchinson Cancer Center/*Essential Molecular Biology* 2e, Freifelder & Malachinski, Jones & Bartlett, Fig. 7-24, pp. 141. 11.18a: David Dressler, Oxford University, England.

Chapter 12: DNA Organization in Chromosomes

Chapter opener: Don. W. Fawcett/Photo Researchers. 12.1a: M. Wurtz/Biozentrum, University of Basel/SPL/Photo Researchers. 12.1b: William S. Klug. 12.2: Science Source/Photo Researchers. 12.3: Gopal Murti/SPL/Photo Researchers. 12.5: From: Identification of the *Drosophila* interband-specific protein Z4 as a DNA-binding zinc-finger protein determining chromosomal structure. H. Eggert, A. Gortchakov, H. Saumweber. *J Cell Sci.* 2004 Aug 15;117(Pt 18): 4253–64. Epub 2004 Aug 3. 12.6b: John Ellison, Richardson Lab, Integrative Biology, The University of Texas at Austin. 12.7a: Omikron/Photo Researchers. 12.7b: William S. Klug. 12.8: Oscar L Miller, Jr. 12.11, 12.12: David Adler/University of Washington Department of Pathology. 12.16: William S. Klug.

Chapter 13: The Genetic Code and Transcription

Chapter opener: Oscar L. Miller/SPL/Photo Researchers. 13.11a: Bert W. O'Malley, Baylor College of Medicine. 13.15a: O.L. Miller, Jr. Barbara A. Hamkalo, C. A. Thomas, Jr. *Science* 169:392–395, 1970 by the American Association for the Advancement of Science. F:2.

Chapter 14: Translation and Proteins

Chapter opener: Reprinted from the front cover of *Science*, Vol. 292, May 4, 2001 Crystal structure of a *Thermus thermophilus* 70S ribosome containing three bound transfer RNAs(top) and exploded views showing its different molecular components (middle and bottom). Image provided by Dr. Albion Baucom (baucom@biology.ucsc.edu). Copyright American Association for the Advancement of Science. 14.9a: Cold Spring Harbor Laboratory Archives. 14.9b:

E. V. Kiseleva. 14.13: Sebastian Kaulitzki/Shutterstock. 14.19: Kenneth Eward/BioGrafx/Science Source/Photo Researchers.

Chapter 15: Gene Mutation, DNA Repair, and Transposition

Chapter opener: MaizeGDB/Courtesy M.G. Neuffer. 15.15ab: W. Clark Lambert, University of Medicine & Dentistry of New Jersey.

Chapter 16: Regulation of Gene Expression in Prokaryotes

Chapter opener: Johnson Research Foundation.

Chapter 17: Regulation of Gene Expression in Eukaryotes

Chapter opener: T. Cremer, I. Solovei, and F. Habermann, Biozentrum (LMU). 17.2b: Klaus Boller/Photo Researchers.

Chapter 18: Developmental Genetics

Chapter opener: Edward B. Lewis, California Institute of Technology. 18.1a: F. Rudolf Turner, Indiana University. 18.1b: From: Learning to Fly: Phenotypic Markers in *Drosophila*. A poster of common phenotypic markers used in *Drosophila* genetics. Jennifer Childress, Richard Behringer, and Georg Halder. 2005. *Genesis* 43(1). Cover illustration. 18.6: Stephen Paddock. 18.8ab: Peter A. Lawrence. 18.9: Jim Langeland, Stephen Paddock, and Sean Carroll, University of Wisconsin at Madison. 18.10: Reprinted by permission from Macmillan Publishers Ltd.: Fig. 1 on p244 from: *British Dental Journal* 195: 243–48 2003. Greenwood, M. and Meechan, J. G. General medicine and surgery for dental practitioners. Copyright © Macmillan Magazines Limited. 18.11ab: Reproduced by permission from Fig. 2 on p. 766 in *Cell* 89: 765–71, May 30, 1997. Copyright © by Elsevier Science Ltd. Image courtesy of Mike J. Owen. 18.12ab: Reproduced by permission from T. Kaufmann, et al., *Advanced Genetics* 27:309–362, 1990. Image courtesy of F. Rudolf Turner, Indiana University. 18.16: From Organizing Axes in Time and Space: 25 Years of Collinear Tinkering. M. Kmita and D. Duboule. *Science* 2003 v.301. p. 332. © American Association for the Advancement of Science. 18.17: P. Barber/Custom Medical Stock Photo. 18.18: Elliot M. Meyerowitz, California Institute of *Technolgy*, Division of Biology. 18.19ab: Max-Planck-Institut fur Entwicklungsbiologie. 18.21a–d: Elliot M. Meyerowitz, California Institute of *Technolgy*, Division of Biology.

Chapter 19: Cancer and Regulation of the Cell Cycle

Chapter opener: SPL/Photo Researchers. 19.1ab: National Institute of Health Genetics/Ried 8340062. 19.3ab: Manfred Schwab. 19.8a: Gopal Murti/Photo Researchers.

Special Topics in Modern Genetics, Genomics and Personalized Medicine ST2a, ST2.2b: From HER-2/neu testing in breast carcinoma: a combined immunohistochemical and fluorescence in situ hybridization approach. R. L. Ridolfi, M. R. Jamehdor, J. M. Arber. *Mod Pathol.* 2000 Aug;13(8):866-73. ST2.4a: dra_schwartz/iStockphoto. ST2.4b: Paphrag. ST2.5: From pharmgkb.org. Used with permission from PharmGKB and Stanford University. © PharmGKB. ST2.8ab: Copyright 2010. Helicos BioSciences Corporation. All rights reserved.

Special Topics in Modern Genetics, Epigenetics ST3.8: From: Maternal genistein alters coat color and protects Avy mouse off-spring from obesity by modifying the fetal epigenome. D. C. Dolinoy et al. *Environ Health Perspect*. 2006 Apr;114(4):567–72.

Special Topics in Modern Genetics, Stem Cells ST4.1a.Frank Geisler/medicalpicture/Alamy. ST4.1b: SPL/Photo Researchers. ST4.4: CNRI/SPL/Photo Researchers. ST4.7a: From Developmental biology. Neuron research leaps ahead. R. H. Brown Jr. *Science*. 2008 Aug 29;321(5893): 1169–70. ST4.9: Wake Forest University School of Medicine. ST4.11: Reproductive and Therapeutic Cloning-Photograph by Anne Bower and Manfred Baetscher, Transgenic core laboratory, Oregon Health and Science-University of Portland Oregon.

Chapter 20: Recombinant DNA Technology

Chapter opener: Pascal Goetgheluck/Photo Researchers. 20.3a: Gopal Murti/SPL/Photo Researchers. 20.5b: Custom Medical Stock Photo. 20.10: National Institutes of Health/Custom Medical Stock Photo. 20.12ab: From A cdc5+ homolog of a higher plant, *Arabidopsis thaliana*. T. Hirayama, K. Shinozaki. *Proc Natl Acad Sci U S A*. 1996 Nov 12;93(23):13371–6; Fig. 6. 20.13: The American Society for Cell Biology from *Molecular & Cellular Biology* (20: 8536–47). © the American Society for Cell Biology. Image courtesy of Dr. Gideon Dreyfuss. 20.14a–c: from Hostikka S. L., Capecchi M. R., The mouse Hoxc11 gene: genomic structure and expression pattern. *Mech Dev*. 1998 Jan;70(1–2): 133–45. 20.16: Autoradiograph of a dideoxy sequencing gel is Trade Marked to Pfizer Inc. Reproduced with permission.

Chapter 21: Genomics, Bioinformatics, and Proteomics

21.8a: Aurélien Boisson-Dernier, Institute of Plant Biology, University of Zurich, Switzerland. 21.15b: Eric Isselée/Shutterstock. 21.21b: dra_schwartz/iStockphoto. 21.21c: National Cancer Institute. 21.22a: From Hierarchical clusters for a microarray. G. Sherlock et al. (2001) The Stanford Microarray Database. *Nucleic Acids Research*, 29(1): 152–155. Fig. 2. 21.22b: Reproduced by permission from Claridge-Chang A, Wijnen H, Naef F, Boothroyd C, Rajewsky N, Young MW. Circadian regulation of gene expression systems in the *Drosophila* head. *Neuron*. 2001 Nov 20;32(4):657–71. Copyright by Elsevier Science Ltd. Image courtesy of Adam Claridge-Chang/The Rockefeller University. 21.23a: Tammy616/iStockphoto. 21.25: Swiss Institute of Bioinformatics.

Chapter 22: Applications and Ethics of Genetic Engineering and Biotechnology

Chapter opener: GFPgroup, The Roslin Institute. 22.1a: SIU BioMed/Custom Medical Stock Photo. 22.3: John Doebley/University of Wisconsin. 22.5a: Michael A. Palladino. 22.5b: C. Boerboom, North Dakota State University. 22.7a: Peter Beyer, University of Freiburg, Germany. 22.8a: University of California, Irvine Transgenic Mouse Facility collection. 22.8b: R.L. Brinster/Peter Arnold/Photolibrary.com. 22.9: Choy Hew, National University of Singapore, and Garth Fletcher, Memorial University of Newfoundland, St. John's, Newfoundland. 22.11: GloFish® (www.glofish.com). 22.12c: From Creation of a Bacterial Cell Controlled by a Chemically Synthesized. Genome. D. G. Gibson et al. Originally published in *Science Express* on 20 May 2010, doi: 10.1126/.science.1190719 *Science* 2 July 2010:vol. 329 no. 5987 52–56. 22.12c: From Creation of a Bacterial Cell Controlled by a Chemically Synthesized.

Genome. D. G. Gibson et al. Originally published in *Science Express* on 20 May 2010, doi: 10.1126/. science.1190719 *Science* 2 July 2010: vol. 329 no. 5987 52–56. 22.16a–d: From: Preimplantation Genetic Diagnosis for alpha thalassemia: experience in Siriraj Hospital. J. Prechapanich et al. *Siriraj Medical Journal*. Volume 60, Number 6, 2008. 22.18: Affymetrix. 22.20ab: From: Evaluation of the performance of a p53 sequencing microarray chip using 140 previously sequenced bladder tumor samples. Fredrik P. Wikman et al. *Clinical Chemistry* 46:10, 2000. 22.21c: National Cancer Institute. 22.22a: from Ash Alizadeh, *Nature Magazine*: 403, pgs. 503–511, fig.4. © 2000 Macmillian Magazines Limited. 22.24: From Application of High-Density Microarrays for Expression and Genotype Analysis in Infectious Disease. M. C. Lorence. *American Biotechnology Laboratory*, p. 10–12, January 2006. 22.27a: Van De Silva.

Chapter 23: Quantitative Genetics and Multifactorial Traits

Chapter opener: Ed Reschke/Peter Arnold/Photolibrary.com. 23.9ab: From: fw2.2: A Quantitative Trait Locus Key to the Evolution of Tomato Fruit Size. Frary, et al. *Science* 7 July 2000: 85–88. DOI:10.1126/science.289.5476.85. Courtesy of Steven D. Tanksley. 23.UN02: Nature's Images/Photo Researchers.

Chapter 24: Genetics of Behavior

Chapter opener: From: Eve. G. Stringham and E. Peter M. Candido, *Journal of Experimental Zoology* 266: 227–233 (1993). Publication of Wiley-Liss Inc. University of British Columbia. 24.1: Jonathan Flint. 24.8: From: Cellular bases of behavioral plasticity: establishing and modifying synaptic circuits in the *Drosophila* genetic system. J. Rohrbough, D. K. O'Dowd, R. A. Baines, and K. Broadie. *J Neurobiol*. 2003 Jan;54(1):254–71.

Chapter 25: Population and Evolutionary Genetics

Chapter opener: Edward S. Ross. 25.7: Michel Samson, Frederick Libert, et al., Resistance to HIV-1 infection in Caucasian individuals bearing mutant alleles of the CCR-5 chemokine receptor gene. Reprinted with permission from *Nature* [vol. 382, 22 August 1996, p. 725, Fig. 3]. Copyright 1996 Macmillan Magazines Limited. 25.15ab: Courtesy of Murray Brilliant, A 122.5 kilobase deletion of P gene underlies the high prevalence of oculocutaneous albinism type 2 in the Navajo population: From: *American Journal Human Genetics* 72: 62–72, fig. 2 p. 66. Published by University of Chicago Press. 25.16: Courtesy of Murray Brilliant, A 122.5 kilobase deletion of P gene underlies the high prevalence of oculocutaneous albinism type 2 in the Navajo population: From: *American Journal Human Genetics* 72: 62–72, fig. 2 p. 66. Published by University of Chicago Press. 25.18: Carl C. Hansen and Nancy Knowlton, Smithsonian Institution Photo Services. 25.20: Fig. 5 in *Molecular Biology Evolution* 15(4) pages 391–407, Kazuhiko Takahasi/Norihiro Okada. The Society for Molecular Biology and Evolution.

Chapter 26: Conservation Genetics

Chapter opener: Scott Bauer, USDA/ARS/Agricultural Research Service. 26.2: John Mitchell/Photo Researchers. 26.3: Thomas Pickard/iStockphoto. 26.4a: Ingvars Birznieks/Shutterstock. 26.4b: Louie Schoeman/Shutterstock. 26.5: Sarah Ward/Department of Soil and Crop Services/Colorado State University. 26.6: Alison Wilson/iStockphoto. 26.9: Art Wolfe/Stone Allstock/Getty Images. 26.9: Photo West/Shutterstock. 26.10: William Leaman/Alamy. 26.13: NCTC Image Libary/US Fish and Wildlife Service. 26.14: Larry Richardson, US Fish and Wildlife Service.

Text Credits

Chapter 5: Chromosome Mapping in Eukaryotes

p133 Reprinted by permission from Macmillan Publishers Ltd: *Nature*, "Why Didn't Gregor Mendel Find Linkage?" by Stig Blixt, *Nature*, Vol. 256, p. 206, Copyright © 1975. www.nature.com

Chapter 6: Genetic Analysis and Mapping in Bacteria and Bacteriophages

p172, #30 Lawley, T.D. et al. 2002. Functional and Mutational Analysis of Conjugative Transfer Region 1 (Tra1) from the IncHI1 Plasmid R27, *Journal of Bacteriology* 184: 2173–2180, fig. 1 and table 3. Copyright © 2002 American Society for Microbiology.

Chapter 7: Sex Determination and Sex Chromosomes

Fig 7.11 Lawley, T.D. et al. 2002. Functional and Mutational Analysis of Conjugative Transfer Region 1 (Tra1) from the IncHI1 Plasmid R27, *Journal of Bacteriology* 184: 2173–2180, fig. 1 and table 3. Copyright © 2002 American Society for Microbiology.

Chapter 9: Extranuclear Inheritance

Fig 9.06b E. Passarge, *Color Atlas of Genetics*, 3rd Edition, Thieme Medical Publishing, 2006. Reprinted by permission. Fig 9.07b E. Passarge, *Color Atlas of Genetics*, 3rd Edition, Thieme Medical Publishing, 2006. Reprinted by permission.

Chapter 10: DNA Structure and Analysis

p. 255 Reprinted by permission from Macmillan Publishers Ltd: *Nature*, "Molecular Structure of Nucleic Acids: A Structure for Deoxyribose Nucleic Acid" by F.H.C. Crick and J.D. Watson, *Nature*, Vol. 171, No. 4356, pp. 737–38. Copyright 1953. www.nature.com

Chapter 11: DNA Replication and Recombination

Fig 11.05ab J.H. Taylor, *Molecular Genetics*, Pt. 1, pp. 74–75, 1963. Copyright © 1963 and renewed 1991 by Elsevier. Used with permission. **F11.06.01** Reprinted from *Cell*, Volume 25, Issue 3, O. Sundin and A. Varshavsky, "Arrest of segregation leads to accumulation of highly intertwined catenated dimers: Dissection of the final stages of SV40 DNA replication," pp. 659–669, Copyright 1981, with permission from Elsevier. www.sciencedirect.com/science/journal/00928674

Chapter 17: Regulation of Gene Expression in Eukaryotes

Fig 17.14 Reprinted from *Cell*, vol. 103, D.L. Black, "Protein Diversity from Alternative Splicing: A Challenge for Bioinformatics and Post-Genome Biology," pp. 367–370, figure 1, p. 368. Copyright 2000, with permission from Elsevier. http://www.sciencedirect.com/science/journal/00928674

Chapter 18: Developmental Genetics

p472 Reprinted from *Cell*, vol. 103, D.L. Black, "Protein Diversity from Alternative Splicing: A Challenge for Bioinformatics and Post-Genome Biology," pp. 367–370, figure 1, p. 368. Copyright 2000, with permission from Elsevier. http://www.sciencedirect.com/science/journal/00928674

Chapter 19: Cancer and Regulation of the Cell Cycle

Fig 19.02 Reprinted from *Trends in Genetics*, vol. 9, Vogelstein, B. and Kinzler, K.W., "The multistep nature of cancer," pp. 138–141, Figure 1, Copyright 1993, with permission from Elsevier. http://www.sciencedirect.com/science/journal/01689525 **p492 #27** From Y. Miki et al., "A strong candidate for the breast and ovarian cancer susceptibility gene BRCA1 (Table 2)," *Science* 266: 66–71 (1994). Reprinted by permission from AAAS and author. www.sciencemag.org/content/266/5182/66.abstract

Special Topics in Modern Genetics, DNA Forensics

Fig ST1.02 Butler, J. M., from STR DNA Internet Database at http://www.cstl.nist.gov/div831/strbase/fbicore.htm. Used by permission. **Fig ST1.03** Butler, J. M., from STR DNA Internet Database at http://www.cstl.nist.gov/div 831/strbase/kits/Identifiler.htm. Used by permission. **Fig ST 1.05** Copyright © 2011 Life Technologies Corporation. Used under permission. (www.lifetechnologies.com)

Special Topics in Modern Genetics, Genomics and Personalized Medicine

Fig ST 2.01 Source: The Case for Personalized Medicine, Figure 1, page 5. Copyright © May 2009 Personalized Medicine Coalition. Used by permission. **Table ST 2.2** Mancinelli, L. et al., 2000. Pharmacogenomics: The Promise of Personalized Medicine. *AAPS PharmSci* 2(1): Article 4, Table 1. Used by permission. **Fig ST 2.03** Adapted by permission from Macmillan Publishers Ltd: *Nature Reviews Drug Discovery*, "Pharmacogenomics: Bench to Bedside," by R. Weinshilboum and L. Wang. *Nature Reviews Drug Discovery* 3: 739–748, fig. 3a. Copyright 2004, www.nature.com/nrd **Fig ST 2.05** Source: www.PharmGKB.org. Used by permission. **Fig ST 2.06** In: GeneTests: Medical Genetics Information Resource (database online). Educational Materials: Gene Tests: Growth of Laboratory Directory from GeneTests database (2009). Copyright University of Washington, Seattle. 1993–2011. Available at http://www.genetests.org. Accessed 2/3/11. **Fig ST 2.07** From "True Direct DNA Measurement," Fig. 2 by Helicos BioSciences (www.helicosbio.com). © Copyright 2010. Helicos BioSciences Corporation. All rights reserved. Used by permission of Helicos BioSciences Corporation.

Special Topics in Modern Genetics, Stem Cells

Fig ST 4.01 Reprinted with permission from *Stem Cells and the Future of Regenerative Medicine* (2002), p. 8, Table 1, by the National Academy of Sciences, Courtesy of the National Academies Press, Washington, D.C. **Fig ST 4.02** Reprinted with permission from *Understanding Stem Cells* (page 3) by the National Academy of Sciences, Courtesy of the National Academies Press, Washington, D.C. **Fig ST 4.05** Adapted from Fig. 7, p. 11 in HOGAN, KELLY A; PALLADINO, MICHAEL A., *STEM CELLS AND CLONING*, 2nd, ©2009. Printed and Electronically reproduced by permission of Pearson Education, Inc., Upper Saddle River, New Jersey. **Fig ST 4.06** Adapted by permission from Macmillan Publishers Ltd: *Nature*, "Nuclear reprogramming to a pluripotent state by three approaches," by S. Yamanaka and H.M. Blau. *Nature*,

465: 704–712, figs. 1 & 4. Copyright 2010, www.nature.com. **Fig ST 4.07**.2 From R. H. Brown, Jr., *Science* (2008): Neuron Research Leaps Ahead. Vol. 321 no. 5893 pp. 1169–1170. Reprinted with permission from AAAS. www.sciencemag.org/content/321/5893/1169.summary **Fig ST 4.08** Adapted from THIEMAN, WILLIAM J.; PALLADINO, MICHAEL A., *INTRODUCTION TO BIOTECHNOLOGY*, 2nd, ©2009. Printed and Electronically reproduced by permission of Pearson Education, Inc., Upper Saddle River, New Jersey. **Fig ST 4.10** Reprinted from *Cell*, Vol. 132, Issue 1, "Molecular Medicine Select," p. 5, Copyright 2008, with permission from Elsevier. www.sciencedirect.com/science/journal/00928674 **Fig ST 4.11** THIEMAN, WILLIAM J.; PALLADINO, MICHAEL A., *INTRODUCTION TO BIOTECHNOLOGY*, 2nd, ©2009. Printed and Electronically reproduced by permission of Pearson Education, Inc., Upper Saddle River, New Jersey.

Chapter 20: Recombinant DNA Technology

Fig 20.16 THIEMAN, WILLIAM J.; PALLADINO, MICHAEL A., *INTRODUCTION TO BIOTECHNOLOGY*, 2nd, ©2009. Printed and Electronically reproduced by permission of Pearson Education, Inc., Upper Saddle River, New Jersey. **Fig 20.18** 454 Sequencing © Roche Diagnostics **Fig 20.19** Adapted from "Personal genome sequencing: current approaches and challenges" by Michael Snyder, Jiang Du and Mark Gerstein, *Genes & Development* 24: 423–431, fig. 1, p. 424. © 2010 by Cold Spring Harbor Laboratory Press. Used by permission of Cold Spring Harbor Laboratory Press and author.

Chapter 21: Genomics, Bioinformatics, and Proteomics

Chapter opener BLAST search of the GenBank database: http://www.ncbi.nlm.nih.gov/BLAST **Fig 21.04** BLAST search of the GenBank database: http://www.ncbi.nlm.nih.gov/BLAST **Fig 21.08b** Adapted by permission from Macmillan Publishers Ltd: *Nature*, "Analysis of the genome sequence of the flowering plant *Arabidopsis Thaliana*," *Nature* 408: 796–815. Copyright 2000, www.nature.com. **Fig 21.10** Adapted by permission from Macmillan Publishers Ltd: *Nature Reviews Genetics*, "Emerging technologies in yeast genomics," by A. Kumar and M. Snyder. *Nature Reviews Genetics* 2(4): 302–312, fig. 3. Copyright 2001, www.nature.com/nrg. **Fig 21.13** Adapted by permission from Macmillan Publishers Ltd: *Nature*, "Sequencing the nuclear genome of the extinct woolly mammoth," by Miller et al. *Nature*, 456: 387–390, fig. 2, pg. 388. Copyright 2008, www.nature.com. **Fig 21.14** Adapted by permission from Macmillan Publishers Ltd: *Nature*, "Human Genome at Ten: The Sequence Explosion," *Nature*, 464, April 2010, pg. 671. Copyright 2010, www.nature.com. **Fig 21.19** Yooseph S, Sutton G, Rusch DB, Halpern AL, Williamson SJ, et al. (2007) The Sorcerer II Global Ocean Sampling Expedition: Expanding the Universe of Protein Families. *PLoS Biol* 5(3): 432–466, fig 1 p. 436. Used by permission. **Fig 21.20** Yooseph S, Sutton G, Rusch DB, Halpern AL, Williamson SJ, et al. (2007) The Sorcerer II Global Ocean Sampling Expedition: Expanding the Universe of Protein Families. *PLoS Biol* 5(3): 432–466, fig 1 p. 436. Used by permission. **Fig 21.22a** From G. Sherlock et al., "The Stanford Microarray Database," *Nucleic Acids Research*, 2001, 29 (1): 152–155, fig. 2. Used by permission of Oxford University Press. **Fig 21.23b** From M. P. Samanata et al., "The Transcitome of the sea urchin embryo", *Science* 314: 960–962, fig. 2, p. 961, (2006). Reprinted with permission from AAAS. www.sciencemag.org/content/314/5801/960.abstract **Fig 21.28** From J. M. Asara et al. (2007). Protein sequences from *Mastodon* and *Tyrannosaurus rex* revealed by mass spectrometry. *Science* 316: 280–285, Fig. 3, p. 283. Reprinted with permission from AAAS. www.sciencemag.org/content/316/5822/280.abstract **Fig 21.29** Adapted by permission from Macmillan Publishers Ltd: *Molecular Systems Biology*, "Human disease classification in the postgenomic era: a complex systems approach to human pathobiology," by Loscalzo, J., Kohane, I., and Barabasi, A-L. *Molecular Systems Biology* 3: 124, pp. 32–42, 2007. Figure 4, p. 39. Copyright 2007, www.nature.com/msb.

Chapter 22: Applications and Ethics of Genetic Engineering and Biotechnology

Fig 22.02 Reprinted with permission from The Biodesign Institute, Arizona State University, Box 5001, Tempe, AZ 85287–5001. **Fig 22.04a** Adapted by permission from Macmillan Publishers Ltd: *Nature Biotechnology*, "Agbio keeps on growing," by S. Lawrence. *Nature Biotechnology*, 23 (3), p. 281. Copyright 2005, www.nature.com/nbt. **Fig 22.04b** Adapted by permission from Macmillan Publishers Ltd: March 2, 2011, *Nature* 471, 10–11, Copyright 2011, www.nature.com. **Fig 22.07b** Adapted by permission from Macmillan Publishers Ltd: March 2, 2011, *Nature* 471, 10–11, Copyright 2011, www.nature.com. **Fig 22.10** Adapted by permission from Macmillan Publishers Ltd: *Nature Biotechnology*, "Tackling mastitis in dairy cows," by P. Rainard. *Nature Biotechnology* 23 (4): 430–432. Copyright 2005, www.nature.com/nbt. **Fig 22.12a** Adapted by permission from Macmillan Publishers Ltd: *Nature Biotechnology*, "A synthetic DNA transplant," by M. Itaya. *Nature Biotechnology*, 28(7):687–689, Panel A from figure 1, p. 688. Copyright 2010, www.nature.com/nbt. **Fig 22.12b** From Gibson, D. G. et al. Creation of a bacterial cell controlled by a chemically synthesized genome. *Science* 329: 52–56, Panel B from figure 1, p. 53. (2010) Reprinted with permission from AAAS. www.sciencemag.org/content/329/5987/52.abstract **Fig 22.25** Adapted by permission from Macmillan Publishers Ltd: *Nature Genetics*, " Twelve type 2 diabetes susceptibility loci identified through large-scale association analysis," by Voight et al. *Nature Genetics* vol. 42 (7) pp. 579–89. Copyright 2010, www.nature.com/ng. **p 656** Adapted by permission from Macmillan Publishers Ltd: Nature, "Genome-wide association study in alopecia areata implicates both innate and adaptive immunity," by Petukhova et al. *Nature* 466: 113–117, p. 113, figure 1. Copyright 2010, www.nature.com.

Chapter 23: Quantitative Genetics and Multifactorial Traits

Fig 23.08 Adapted by permission from Macmillan Publishers Ltd: Nature Reviews Genetics, "Quantitative trait loci in *Drosophila*," by Mackay, T. *Nature Reviews Genetics* 2: 11–20, Fig. 3, p. 13. Copyright 2001, www.nature.com/nrg. **Table 23.5** Used with permission of *Annual Reviews of Genetics*, from "Mapping Polygenes" by S.D. Tanksley, *Annual Review of Genetics*, Vol. 27: 205–233, Table 1, December 1993. Permission conveyed through Copyright Clearance Center, Inc.

Chapter 24: Genetics of Behavior

Fig 24.02 Adapted by permission from Macmillan Publishers Ltd: *European Journal of Human Genetics*, "The genetic basis of emotional behavior in mice," by S.A.G. Willis-Owen and J. Flint. *European Journal of Human Genetics* 14: 721–728, plot 1, p. 723. Copyright 2006, www.nature.com/ejhg. **Table 24.2** Adapted by permission from Macmillan Publishers Ltd: *Nature Reviews Genetics*, "Drosophila: Genetics Meets Behavior," by M.B. Sokolowski. *Nature Reviews Genetics* 2: 879–890, table 1, p. 882. Copyright 2001, www.nature.com/nrg. **Fig 24.10** Adapted by permission from Macmillan Publishers Ltd: *European Journal of Human Genetics*, "Molecular genetic studies of schizophrenia," by B. Riley and K.S. Kendler. *European Journal of Human Genetics*, 14: 669–680, fig. 1, p. 670. Copyright 2006, www.nature.com/ejhg. **Fig 24.11**

Adapted by permission from Macmillan Publishers Ltd: *Nature,* "Common variants on chromosome 6p22.1 are associated with schizophrenia," by J. Shi, et al. *Nature* 460: 753–757, fig. 1, p. 756. Copyright 2009, www.nature.com. **Table 24.3** From Crespi, B. et al., 2010, "Comparative genomics of autism and schizophrenia." *Proc. Nat. Acad. Sci.* 107: 1736–1741, table 1, p. 1737. Used by permission.

Chapter 25: Population and Evolutionary Genetics

Fig 25.02 Adapted by permission from Macmillan Publishers Ltd: *Nature,* "Nuceotide polymorphism at the alcohol dehydrogenase locus of the *Drosophila melanogaster,*" by M. Kreitman. *Nature* 304 (4): 412–417, table 1, p. 413. Copyright 1983, www.nature.com. **Fig 25.03** Reprinted from *Trends in Genetics,* Vol. 8, Issue 11, L. Tsui, "The spectrum of cystic fibrosis mutations," pp. 392–398, fig. 1, Copyright 1992, with permission from Elsevier. www.sciencedirect.com/science/journal/01689525 **Table 25.4** From *Processes of Organic Evolution,* 3rd Edition by G. Ledyard Stebbins, © 1977, p. 143. Reprinted by permission of Pearson Education, Inc. **Table 25.5, Fig 25.22** From W.M. Fitch and E. Margoliash, "Construction of phylogenetic trees," *Science* 155: 279–284, January 20, 1967.

Reprinted with permission from AAAS. www.sciencemag.org/content/155/3760/279.extract**Fig 25.22** From W.M. Fitch and E. Margoliash, "Construction of phylogenetic trees," *Science* 155: 279–284, Fig. 2, January 20, 1967. Reprinted with permission from AAAS. www.sciencemag.org/content/155/3760/279.extract **Fig 25.23** From W.M. Fitch et al., 1991. "Positive Darwinian evolution in human influenza A viruses," *Proc. Natl. Acad. Sci.* 88: 4270–4273, Fig. 1. Used by permission. **Fig 25.24a** Reprinted from *Cell,* Vol. 90, Issue 1, Krings, M., et al., "Neandertal DNA Sequences and the Origin of Modern Humans," pp. 19–30, Figure 7A, p. 26, Copyright 1997, with permission from Elsevier. www.sciencedirect.com/science/journal/00928674 **Fig 25.24b** Adapted by permission from Macmillan Publishers Ltd: *Nature,* "Molecular analysis of Neanderthal DNA from the northern Caucasus," by I.V. Ovchinnikov, et al. *Nature* 404: 490–493. Copyright 2000, www.nature.com. **Fig 25.25** From Noonan, J.P. et al., 2006. "Sequencing and analysis of Neanderthal genomic DNA." *Science* 314: 1113–1118, fig. 6, p. 1117. Reprinted with permission from AAAS. www.sciencemag.org/content/314/5802/1113.abstract **p724 #36** Reprinted from D. Caramelli et al., 2003. "Evidence for a genetic discontinuity between Neanderthals and 24,000-year-old anatomically modern Europeans," *Proc. Natl. Acad. Sci. USA* 100: 6593–6597, fig. 2, p. 6596. Copyright 2003 National Academy of Sciences, USA. Used by permission.

Index

Numbers

2-amino-purine (2-AP), 382
2-deoxyribose, 248
3′–5′ direction, 280
5′–3′ direction, 276
5-azacytidine, 431
5-methylcytosine, 431
5-bromouracil (5-BU), 382
5-methyl cytosine, 304
7-methylguanosine (7-mG)
     cap, 332, 354
9mers and 13mers, 278
14/21, D/G translocation, 214
45,X syndrome, 179–180, 181
45,X/46,XX karyotype, 180
45,X/46,XY karyotype, 180
47,XXX syndrome, 181
47,XXY syndrome, 179–180, 181
47,XYY condition, 181–182
48,XXXX karyotype, 181
48,XXXY karyotype, 180
48,XXYY karyotype, 180
49,XXXX karyotype, 181
49,XXXXY karyotype, 180
49,XXXYY karyotype, 180

A

A-DNA, 256
abdominal A and Abdominal B genes in
     Drosophila, 459
ABO blood groups, 75–77
   A and B antigens, 75–76
   and albinism in humans, 80
   Bombay phenotype, 77, 81
Abortions (spontaneous), in human
     conceptions, 203
Abortive transduction, 160
Absorption spectrum, 246
Absorption of ultraviolet light (UV),
     258, 259
Acceptor and donor sequences, 336
Accession number, 580
Acentric chromatids, 211, 212
Acetylation, 303, 520
Achondroplasia, 385, 399, 709
Acrocentric chromosomes, 21, 22
Action spectrum (UV light), 246
Actin, 21, 366
Activator (Ac)-dissociation (Ds) system in
     maize, 395
Activators, 436
   interaction with general transcription
     factors at promoter, 436–437
Active site, 367
Acute transforming retroviruses, 487
Adaptive enzymes, 404
Adaptation hypothesis, 144
Adaptive mutations, 376–377

ADAR (adenosine deaminase acting on RNA)
     enzymes, 337
Additive allele, 661
   basis for continuous variation, 662
Additive variance, 668
Adduct-forming agents, 383
Adelberg, Edward, 152
Adenine (A), 6, 248, 251–252
   adenine-thymine base pairs, 254
Adenine methylase, 387
Adenomatous polyposis (APC) gene, 486
Adenosine deaminase (ADA), 647
Adenosine triphosphate (ATP), 21, 249–250, 411
Adenosine-uracil rich element (ARE), 442
Adenovirus, 648
Adenyl cyclase, 411
Adjacent segregation-1, 213
Adult-derived stem cells, 532
   repairing damaged heart with, 539
Adverse drug reactions, 508–510
Aerobic respiration, mitochondrial mutations
     in *Neurospora*, 224
Aflatoxin, 488
Agarose gel, 263, 559, 562
Age of Genetics, 14–15
Age of onset, gene expression, 93–94
Aging
   age-related cancers, 475, 476
   human health, and mitochondria, 230
   telomeres and, 289
Agricultural biotechnology, 626
AIDS, 247, 708
Albinism, 62, 78
   ABO blood group and, 80
   in Native American population, 710–712
Alcohol dehydrogenase gene (ALD), in
     Drosophila, 698–700
Alignment, 576
Aliquot, 157
Alkaptonuria, 354
Alkylating agents, 382, 383
Allele-specific oligonucleotides, 636–638
Alleles, 45
   additive and nonadditive, 661
   alteration of phenotypes, 72–73
   calculating frequency of multiple alleles in
     population, 704–705
   calculating frequency in population,
     702–703, 704
   calculating frequency for X-linked traits,
     705–706
   codominance, 74–75
   defined, 5, 22
   frequency in a population, 700–702
   incomplete, or partial, dominance, 74
   migration and gene flow altering
     frequencies, 710
   lethal, 77–78
   multiple, 75–77

mutation creates new alleles in gene pool,
     709–710
natural selection driving allele frequency,
     706–709
symbols for, 73
Allopolyploidy, 203, 204–206
Allosteric, 406
Allosteric molecule, 413, 414
Allotetraploid, 205
Alloway, J. Lionel, 242
Alpha (α)
   α chains, 360
   α DNA polymerase, 283
   α-globin gene, 600
   α helix, 362, 364
   αI-antitrypsin protein, 624
   α subunit of DNA polymerase III
     holoenzyme, 277
ALK/EML4 fusion gene, 506
Alphoid family, 307–308
Alternative lengthening of telomeres
     (ALT), 289
Alternate segregation, 213
Alternative splicing, 190, 336, 439–442, 588
   and human diseases, 440–441
   of mRNA, 439–440
   sex determination in Drosophila, 441–442
Alu element, 397
Alu family, 309
Alzheimer disease, 366
Ames test, 392–393
Amino acids, 362
   analysis of, in hemoglobin from sickle-cell
     anemia studies, 360
   assignments to specific trinucleotides from
     triple-binding assay, 321, 322
   chemically similar, sharing middle base in
     triplet codons, 324
   coding dictionary, 323–325
   modifications in histones, 520
   results and interpretation of mixed
     copolymer experiment, 320
   twenty amino acids encoded by living
     organisms, 362, 363
Amino group, 362
   conversion to keto group, in
     deamination, 381
Aminoacyl tRNA synthetases, 228, 348
Aminoacyladenylic acid, 348
Amniocentesis, 202, 635
Amniotic fluid-derived stem cells, 532
Amphibian (*Xenopus laevis*), abundance of
     ribosomes, 209
AmpliChip® CYP450 assay, 509
Amplicon, 184
Amtzen, C. J., 167
Amphidiploid, 205
Amyotrophic lateral sclerosis (ALS, or Lou
     Gehrig's Disease), 536